Fundamentals of Engineering Thermodynamics

McGraw-Hill Series in Mechanical Engineering

Consulting Editors
Jack P. Holman, Southern Methodist University
John R. Lloyd, Michigan State University

Anderson: *Modern Compressible Flow: With Historical Perspective*
Arora: *Introduction to Optimum Design*
Bray and Stanley: *Nondestructive Evaluation: A Tool for Design, Manufacturing, and Service*
Culp: *Principles of Energy Conversion*
Dally: *Packaging of Electronic Systems: A Mechanical Engineering Approach*
Dieter: *Engineering Design: A Materials and Processing Approach*
Donaldson: *An Introduction to the Analysis of Aerospace and Other Vehicular Structures*
Eckert and Drake: *Analysis of Heat and Mass Transfer*
Edwards and McKee: *Fundamentals of Mechanical Component Design*
Heywood: *Internal Combustion Engine Fundamentals*
Hinze: *Turbulence*
Howell and Buckius: *Fundamentals of Engineering Thermodynamics*
Hutton: *Applied Mechanical Vibrations*
Juvinall: *Engineering Considerations of Stress, Strain, and Strength*
Kane and Levinson: *Dynamics: Theory and Applications*
Kays and Crawford: *Convective Heat and Mass Transfer*
Kimbrell: *Kinematics Analysis and Synthesis*
Martin: *Kinematics and Dynamics of Machines*
Norton: *Design of Machinery*
Phelan: *Fundamentals of Mechanical Design*
Raven: *Automatic Control Engineering*
Reddy: *Introduction to the Finite Element Method*
Rosenberg and Karnopp: *Introduction to Physics*
Schlichting: *Boundary-Layer Theory*
Shames: *Mechanics of Fluids*
Sherman: *Viscous Flow*
Shigley: *Kinematic Analysis of Mechanisms*
Shigley and Uicker: *Theory of Machines and Mechanisms*
Shigley and Mischke: *Mechanical Engineering Design*
Stiffler: *Design with Microprocessors for Mechanical Engineers*

Stoecker and Jones: *Refrigeration and Air Conditioning*
Vanderplaats: *Numerical Optimization: Techniques for Engineering Design, with Applications*
White: *Viscous Fluid Flow*
Zeid: *CAD/CAM Theory and Practice*

Also Available from McGraw-Hill

Schaum's Outline Series in Mechanical Engineering

Most outlines include basic theory, definitions, and hundreds of solved problems and supplementary problems with answers.

Titles on the Current List Include:

Acoustics
Basic Equations of Engineering
Continuum Mechanics
Engineering Economics
Engineering Mechanics, 4th edition
Fluid Dynamics, 2d editon
Fluid Mechanics & Hydraulics, 2d edition
Heat Transfer
Introduction to Engineering Calculations
Lagrangian Dynamics
Machine Design
Mathematical Handbook of Formulas & Tables
Mechanical Vibrations
Operations Research
Statics & Mechanics of Materials
Strength of Materials, 2d edition
Theoretical Mechanics
Thermodynamics, 2d edition

Schaum's Solved Problems Books

Each title in this series is a complete and expert source of solved problems containing thousands of problems with worked-out solutions.

Related Titles on the Current List Include:

3000 Solved Problems in Calculus
2500 Solved Problems in Differential Equations
2500 Solved Problems in Fluid Mechanics and Hydraulics
1000 Solved Problems in Heat Transfer
3000 Solved Problems in Linear Algebra
2000 Solved Problems in Mechanical Engineering Thermodynamics
2000 Solved Problems in Numerical Analysis
700 Solved Problems in Vector Mechanics for Engineers: Dynamics
800 Solved Problems in Vector Mechanics for Engineers: Statics

Fundamentals of Engineering Thermodynamics

Second Edition

John R. Howell
Department of Mechanical Engineering
University of Texas at Austin

Richard O. Buckius
Department of Mechanical and Industrial Engineering
University of Illinois at Urbana-Champaign

McGraw-Hill, Inc.
New York St. Louis San Francisco Auckland Bogotá
Caracas Lisbon London Madrid Mexico Milan
Montreal New Delhi Paris San Juan
Singapore Sydney Tokyo Toronto

FUNDAMENTALS OF ENGINEERING THERMODYNAMICS

2 3 4 5 6 7 8 9 0 DOH DOH 9 0 9 8 7 6 5 4 3 2

P/N 030751-2
PART OF
ISBN 0-07-909369-8

This book was set in Times Roman by Progressive Typographers, Inc.
The editors were John Corrigan and Jack Maisel;
the production supervisor was Richard A. Ausburn.
The cover was designed by Caliber/Phoenix Color Corporation.
New drawings were done by Fine Line Illustrations, Inc.
R. R. Donnelley & Sons Company was printer and binder.

Library of Congress Cataloging-in-Publication Data

Howell, John R.
Fundamentals of engineering thermodynamics / John R. Howell, Richard O. Buckius.—2nd ed.
p. cm.—(McGraw-Hill series in mechanical engineering)
Includes bibliographical references and index.
ISBN 0-07-909369-8 (set)
1. Thermodynamics. I. Buckius, Richard O. II. Title. III. Series.
TJ265.H726 1992 <MRCRR>
621.402'1—dc20 91-3840

About the Authors

John R. Howell received his B.S. (1958), M.S. (1960), and Ph.D. (1962) in chemical engineering from Case Institute of Technology (now Case Western Reserve University) in Cleveland, Ohio. He worked at the NASA Lewis Research Center in Cleveland on fundamental heat transfer research until 1968, when he joined the Department of mechanical Engineering at the University of Houston, Texas. He joined the Department of Mechanical Engineering at the University of Texas at Austin in 1978, where he is presently Baker-Hughes Professor and has served as Chairman and Director of the Center for Energy Studies.

Dr. Howell has published widely in various areas of heat transfer, including more than 130 technical papers and reports as well as text and reference books. He was awarded the Ralph Coats Roe award as outstanding teacher of mechanical engineering in 1987 by the American Society of Engineering Education, and he received the Thermophysics Award of the American Institute of Aeronautics and Astronautics in 1990. He is a Fellow of ASME.

Richard O. Buckius received his B.S (1972), M.S. (1973), and Ph.D. (1975) in mechanical engineering from the University of California, Berkeley. He then joined the Department of Mechanical and Industrial Engineering at the University of Illinois at Urbana-Champaign as an assistant professor. He has been the associate head of the department, program director with the National Science Foundation, and Associate Vice Chancellor for Research.

Dr. Buckius has published numerous technical articles in the areas of heat transfer and combustion. He has received various teaching awards, including the Campus Award for Excellence in Undergraduate Teaching from the University of Illinois and the Western Electric Fund Award from the American Society for Engineering Education. He is a Fellow of ASME.

Contents

CHAPTER 7 Vapor Cycle Analysis 496

CHAPTER 8 Analysis Using the Second Law of Thermodynamics 564

CHAPTER 9 General Property Relations and Equations of State 608

CHAPTER 11 Chemical Reactions and Combustion 714

CHAPTER 12 Phase and Chemical Equilibrium 772

Preface to the Second Edition

The attributes of the first edition of this work have been maintained and expanded, particularly the use of worked examples, two-color format to distinguish processes from properties on diagrams and to enhance the perception of availability on equipment diagrams, and the use of contemporary equipment and systems to illustrate the applications of this fascinating subject. As with any work of this kind, however, we have seen ways to improve the usefulness of the text, and we believe that this edition contains some significant improvements for both students and instructors.

A major change has been to convert to the traditional sign convention used in the engineering community that work produced by a system is positive. There are certainly strong arguments for the use of either the positive or the negative sign, or for a system of choice depending on analysis of a particular problem. We have chosen to go with the conventional system based on an overwhelming response by the community.

The continuing and probably never-ending debate on how best to introduce students to the second law of thermodynamics is reflected in our treatment in Chap. 5. We believe that the treatment based on a logical interpretation of commonly observed phenomena as presented in the first edition is an effective method, and we urge others to try it. We still present that method as the primary thrust in this edition, but we also provide a section on the traditional derivation of entropy based on the Carnot cycle for those who believe this to be a more effective presentation.

The number and scope of the homework problems have been enhanced in this edition by the addition of over 500 problems. We wish to acknowledge the United States Military Academy for providing many of these from their storehouse of auxiliary teaching materials.

To aid both instructor and student in having a choice of material, many new and expanded applications sections have been included. Particularly applications to multistage turbines and compressors, nozzles, and other components are treated. The chapter on cycles in the first edition was greatly expanded and split into two chapters that deal with gas cycles and vapor cycles, respectively. This allowed material on internal combustion engines to be expanded to include the concept of mean effective pressure, much more coverage of variable-property effects, and the addition of other gas cycles. In the vapor cycle chapter, additional material on the Kalina cycle and on cogeneration and combined cycle plants is included.

The chapter on second law analysis was revised to include discussions and definitions of the second law efficiency of devices and processes. In the coverage of combustion, a section on combustion chemistry was added. Updated and enhanced material on mixing, chemical reactions, and psychrometry was included.

In the tabular material, the steam tables were revised and all tables, problems, and the computerized tables in this edition are based on the NIST/NRC standard tables, which the authors believe to be the most accurate available. The computerized tables for steam were completely reprogrammed to provide a more useful tool than was available in the first edition.

Because many users of the text do not carry out a review of the mathematical concepts pertinent to thermodynamics, but wish to refer to these concepts as needed throughout the course, the material on mathematical concepts has been removed from Chap. 1 and placed in an appendix for easy reference. Those who wish to use this material as a review will find homework problems at the end of App. H.

Many additional changes, both substantive and cosmetic, have been made. Many of these, such as retaining running titles on multipage property tables, were made to aid in reducing errors or increasing understanding and will not be obvious to the casual user.

We again wish to thank our families for putting up with evening and weekend sessions used in preparing this edition. They make the work worthwhile.

McGraw-Hill and the authors would like to thank the following reviewers for their many helpful comments and suggestions: Nasser Ashgriz, State University of New York, Buffalo; Louis Bermeister, University of Kansas; Sushel Bhavanani, Auburn University; J.C. Cantrell, United States Military Academy, John Francis, University of Oklahoma; Sang Lu Han, Tennessee Technological University; Jai Krishna Kadambi, Case Western Reserve University; Thomas Litzinger, Pennsylvania State University; Jack Lloyd, Michigan State University; David Miller, Drexel University; G.P. Peterson, Texas A&M University; Harold Shock, Michigan State University; and Vijay Varadan, Pennsylvania State University.

John R. Howell
Richard O. Buckius

Preface to the First Edition

This text provides an introduction to engineering thermodynamics from the so-called classical approach. The organization follows a logical sequence which is considerably different from the historical development of thermodynamics. However, we proceed in a manner that allows the student to build an understanding of the fundamentals and applications, proceeding from simple but useful relations and applications from simple substances to the necessarily more complex relations for mixtures and materials with chemical reactions.

We have included many fully worked examples to illustrate the application of the theory presented in the text, and we have found these very helpful to the student. In these examples, we have followed the problem-solving methodology presented in Chaps. 1, 4, and 6. This methodology emphasizes the careful structuring of the problem, the use of property diagrams to visualize the processes involved, and the use of state tables for defining the process end states. These allow the student to see exactly what information is given and what must be generated through use of themodynamic relations. Such an approach can help the student to see to the heart of the problems posed and to develop orderly solution procedures.

Because of the availability of computerized property data for microcomputers, a wide range of problems can be treated by students that were not feasible for an earlier generation. We have provided a mix of problems; most can be solved by "hand" calculation, and others require so much interpolation from tabular data that only the use of computerized tables makes complete solution feasible. These problems have been flagged in the problems sections so that the instructor will not inadvertently assign them. We believe that the "hand" problems, making use of tabular data where necessary, are required to help the student gain the necessary insight into thermodynamics. The problems requiring computerized data generally show the behavior of a particular control mass or device under parametric variations of conditions; such problems also can provide great insight but have often been ignored in introductory texts because of the time formerly involved in solution. However, care must be taken to help the student develop the ability to critically examine what the computer is providing, so that erroneous computer results will not be blindly accepted and used.

The computerized tables available with this text cover the range of properties encompassed by the problems and are discussed in some detail in App. G.

At the beginning of each chapter are photographs, cutaways, and diagrams of the equipment analyzed in the text. These are included because many students entering the engineering curriculum have not seen such equipment and have little concept of its scale and complexity. These pictures give some idea of what lies within the blocks shown on the cycle diagrams in the body of the text.

Two versions of this text are available—this one, which contains both SI and U.S. conventional (USCS) systems of units, and one that is solely in SI.

Although the presentation in the first 11 chapters is generally from the classical viewpoint (with some microscopic interpretations where they appear helpful), Chap. 12 deals with the statistical interpretation of thermodynamics. It is organized so that a more detailed statistical viewpoint can be introduced along with the classical material if the instructor desires. Alternatively, Chap. 12 can be used to review the classical relations from another viewpoint following completion of the first 11 chapters. In any case, the statistical treatment is presented as an aid to understanding how properties can be computed from a fundamental understanding of structure; the interpretation of entropy in terms of uncertainty; the ideas of the increase of the entropy of the universe from the microscopic viewpoint; and a microscopic interpretation of the first and second laws of thermodynamics. No attempt is made to provide a complete treatment of statistical thermodynamics.

Finally, we have observed that thermodynamics is often the first course in which mathematical concepts from courses in partial differential equations are applied to engineering problems. We have tried to aid in the transition from abstract concepts to concrete applications with a section in Chap. 1 on the mathematics needed in this course. Some instructors may wish to omit this section, using it for reference as required.

We are indebted to our colleagues at the University of Texas at Austin and the University of Illinois at Urbana-Champaign. Their comments, criticisms, and suggestions have helped us improve this final product. We are also indebted to the academic reviewers listed on page ii of the first edition. We acknowledge the efforts of Kumbae Lee and Larry Lister for their detailed review of the problems and text. We are also grateful for the excellent typing of the entire manuscript and revisions by Angela Ehrsam.

We now understand the reason that authors invariably thank their families for their encouragement and support. Their contributions is a very real one. We are very grateful to Susan and Kathy, who have endured many periods of doubt and second thoughts about this project with great grace and understanding. And we are very grateful to our children. Reid, Keli, and David, who have passed to adulthood, and Sarah and Emily, who have passed through their preschool years, during this writing.

John R. Howell
Richard O. Buckius

Nomenclature

a	activity, specific Helmholtz function, acceleration
A	Helmholtz function, area
AFR	air-fuel ratio
c_P	specific heat at constant pressure
c_P^*	temperature mean specific heat at constant pressure
c_P^*, LM	temperature log mean specific heat at constant pressure
c_v	specific heat at constant volume
c_v^*	temperature mean specific heat at constant volume
$c_{v,LM}^*$	temperature log mean specific heat at constant volume
COP	coefficient of performance
d	distance
e	specific energy
E	energy, Young's modulus
E	electric potential
f	fugacity
f^*	ideal solution fugacity
$\mathbf{F}$	vector force
F	force
F	generalized force
FAR	fuel-air ratio
g	acceleration due to gravity, specific Gibbs function, degeneracy
g_c	constant that relates force, mass, length, and time in USCS units
G	Gibbs function
h	specific enthalpy, Planck's constant
H	enthalpy
i	electric current, specific irreversibility
I	irreversibility
k	specific heat ratio c_P/c_v, rate constant, Boltzmann constant
k_s	spring constant
K	equilibrium constant
KE	kinetic energy
L	length
m	mass
$\dot{m}$	mass rate of flow
M	molecular weight
n	number of moles, polytropic exponent

N	number
N_a	Avogadro's number
N_p	number of particles
P	pressure
P_i	partial pressure of component i
PE	potential energy
q	heat transfer per unit mass
Q	heat transfer
$\mathbf{Q}$	charge
$\dot{Q}$	rate of heat transfer
R	particular gas constant
$\overline{R}$	universal gas constant
$\mathbf{s}$	displacement
s	specific entropy
S	entropy
S_{gen}	entropy generation
t	time
T	temperature
u	specific internal energy
U	internal energy
v	specific volume
V	volume
$\mathbf{V}$	velocity
$\mathbf{V}$	speed
$\mathbf{V}_{\text{ref}}$	speed with reference to surroundings
w	work per unit mass
W	work
$\dot{W}$	rate of work, or power
W_{rev}	reversible work between two states
x	quality, mass fraction
$\mathbf{X}$	generalized displacement
y	mole fraction
z	compressibility factor
Z	elevation, partition function

Greek Letters

α	residual volume, extent of reaction
β	coefficient of thermal expansion
ε	strain, particle energy
η	efficiency
θ	angle between surface normal and direction of vector
κ	isothermal compressibility
μ	chemical potential, Joule-Thomson coefficient, degree of saturation
υ	stoichiometric coefficient
ρ	density

σ	stress, surface tension
τ	sheer stress
ϕ	nonflow availability per unit mass, relative humidity
Φ	nonflow availability
ψ	flow availability per unit mass, wave function
Ψ	flow availability
ω	humidity ratio, acentric factor

Subscripts

a	actual, air
comb	combustion
c	components, compressor
C	carnot
cr	critical point
CM	control mass
CV	control volume
d	dew point
f	formation, fuel
g	property of saturated vapor
H	high-temperature source
in	state of a substance entering a control volume
i	component
irr	irreversible
j	phase
L	low-temperature source
l	property of saturated liquid
lg	difference in property for saturated vapor and saturated liquid
out	state of a substance leaving a control volume
p	product
ph	phases
pr	properties
r	reduced property, reactant
ref	with surroundings as reference
rev	reversible
s	isentropic process
v	vapor
0	property of the surroundings, zero pressure

Superscripts

¯	bar over symbol denotes property on a molal basis, partial molal property
°	property at standard reference state
·	dot over symbol denotes rate

Fundamentals of Engineering Thermodynamics

1

Introduction

A theory is more impressive the greater is the simplicity of its premises, the more different are the things it relates, and the more extended its range of applicability. Therefore, the deep impression which classical thermodynamics made on me. It is the only physical theory of universal content which I am convinced, that within the framework of applicability of its basic concepts will never be overthrown.
Albert Einstein

A 12,600-hp gas turbine, with fifteen-stage air compressor, single annular combustor, two-stage gas producer turbine assembly, and two-stage power producer turbine assembly. (*Solar Turbines Incorporated.*)

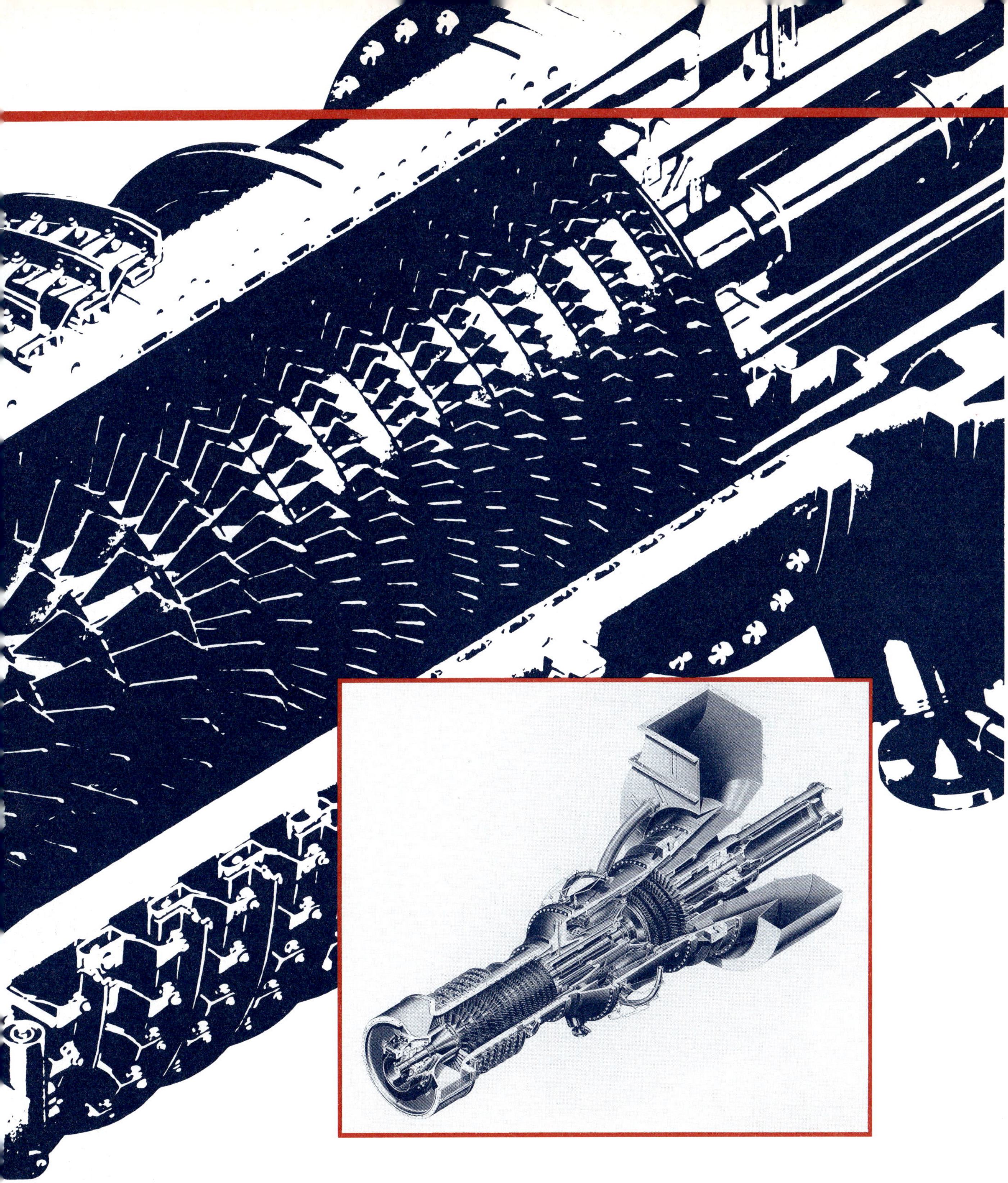

1.1 Energy and Society

Thermodynamics is defined as the study of energy, its forms and transformations, and the interactions of energy with matter. Before we begin this study, it is useful to think about the place and usefulness of this discipline, not only within the curriculum of the engineer and scientist but also within the framework of society itself.

1.1.1 Value of Energy

The availability of energy and people's ability to harness that energy in useful ways have transformed our society. A few hundred years ago, the greatest fraction of the population struggled to subsist by producing food for local consumption. Now, in many countries a small fraction of the total work force produces abundant food for the entire population, and much of the population is thus freed for other pursuits. We are able to travel great distances in short times by using a choice of conveyances (including trips to earth orbit as well as our nearest natural satellite for some); we can communicate instantaneously with persons anywhere on earth; and we control large amounts of energy at our personal whim in the form of automobiles, electric tools and appliances, and comfort conditioning in our dwellings.

How did these changes come about? They have resulted from a combination of inventiveness and ingenuity, coupled with a painstaking construction of theory by some of the great scientists and engineers throughout the years. The story of this development of the underlying science and engineering that so affects us now is one of interest and inspiration, but is too lengthy to reproduce here. A short history of the development of classical thermodynamics is included in App. A.

As a result of the development of the science and application of thermodynamics, our ability to obtain energy, transform it, and apply it to society's needs has brought about the change from agrarian to modern society. From our definition of thermodynamics, then, clearly the study of this discipline not only is useful to engineers in their professional lives but also has played, and will continue to play, a vital role in how society evolves.

1.1.2 Need to Understand Energy and Its Forms

Because of its generality, thermodynamics is the underlying science that forms the framework for the study of most other engineering subjects. The most obvious are *heat transfer,* the study of how energy is transferred from a material or location at a certain temperature to another material or location at a different temperature; *fluid mechanics,* which deals with the motion of fluids under externally applied forces and the transformation of energy between mechanical and thermal forms during this motion; many parts of *materials science,* such as those that consider the relative amounts of various structural forms of materials present in solids and how these relative amounts change under different conditions; and, in a sense, all subjects that treat energy in any of its forms.

Another way of seeing the pervasiveness of thermodynamics in studies of interest to engineers is to examine the many and diverse areas of application. These include power plants (fossil fuel, nuclear fission, nuclear fusion, solar, geothermal, etc.); engines (steam, gasoline, diesel, stationary and propulsion gas turbines, rockets, etc.); air-conditioning and refrigeration systems of all sorts; furnaces, heaters, and chemical process equipment; the design of electronic equipment (for example, to avoid overheating and failure of individual components, circuit boards, and larger assemblies and to understand the chemistry of semiconductor behavior); design of mechanical equipment (for example, in lubrication of bearings to predict the overheating and subsequent failure due to excessive applied loads and in brake design to predict lining wear rates due to frictional heating and erosion); and in manufacturing processes (where, for example, the wear of tool bits is often due to frictional heating of the cutting edge). Indeed, it is fairly easy to make a case that thermodynamics in its broadest sense is the underlying science in most fields of engineering. Even the fields of pure mechanics use energy conservation relations which are subsets of more general thermodynamic principles.

1.2 Energy Balance Approach — Applications in Engineering

The principle "energy can be neither created nor destroyed" is one of the conservation relations that is explored in detail in later chapters. It is one of the building blocks which, when carefully expanded and explored, forms the basis for a good deal of our study of thermodynamics. The conservation of energy principle can be made true in any situation by simply changing or redefining what we mean by energy, so that it is indeed conserved in all situations. This was, in fact, what happened in the historical development of thermodynamics.

The basic conservation principle has two important suppositions. The first is that energy is something that is "contained." A certain defined system "has" energy. The second supposition is that there is a well-specified system where the energy is contained. To apply the energy conservation principle, the user must define the space or material of interest that "has" the energy.

The accounting system that we set up to allow for the transformation of energy should be a useful one for the types of problems that we plan to solve. The various energy transfer mechanisms that have been found to be most useful for our purposes are explored in Sec. 1.3 as well as in greater detail in Chap. 2.

Without worrying at this point about how we classify energy, we can immediately invoke the conservation of energy principle to solve some problems in thermodynamics. For example, let us consider a power plant used for the production of electricity.

Example Problem 1.1

A power plant uses 1 unit of fuel energy to produce 0.4 energy units of electricity. What is the net energy transferred to the environment in the conversion of fuel to electricity?

System Diagram

Solution

For this type of problem, the energy inside the power plant boundary remains constant according to the conservation of energy principle (since it cannot be created or destroyed). We could write in this case that

$$\begin{pmatrix}\text{Energy}\\\text{transferred}\\\text{in}\end{pmatrix}+\begin{pmatrix}\text{Energy}\\\text{transferred}\\\text{out}\end{pmatrix}=0$$

where *in* and *out* refer to the direction in which the energy is crossing the plant boundary. Now we can expand these terms to include each of the energy transfer arrows on the diagram, or

$$\begin{pmatrix}\text{Energy}\\ \text{transferred}\\ \text{in}\end{pmatrix}_{\text{fuel}} + \begin{pmatrix}\text{Energy}\\ \text{transferred}\\ \text{out}\end{pmatrix}_{\text{electricity}} + \begin{pmatrix}\text{Energy}\\ \text{transferred}\\ \text{in}\end{pmatrix}_{\substack{\text{cooling}\\ \text{water}}} + \begin{pmatrix}\text{Energy}\\ \text{transferred}\\ \text{out}\end{pmatrix}_{\substack{\text{cooling}\\ \text{water}}} + \begin{pmatrix}\text{Energy}\\ \text{transferred}\\ \text{out}\end{pmatrix}_{\substack{\text{stack}\\ \text{gas}}} = 0$$

or

$$1\text{ unit} - 0.4\text{ unit} + \begin{pmatrix}\text{Energy}\\ \text{transferred}\\ \text{in}\end{pmatrix}_{\substack{\text{cooling}\\ \text{water}}} + \begin{pmatrix}\text{Energy}\\ \text{transferred}\\ \text{out}\end{pmatrix}_{\substack{\text{cooling}\\ \text{water}}} + \begin{pmatrix}\text{Energy}\\ \text{transferred}\\ \text{out}\end{pmatrix}_{\substack{\text{stack}\\ \text{gas}}} = 0$$

or finally

$$\begin{pmatrix}\text{Energy}\\ \text{transferred}\\ \text{out}\end{pmatrix}_{\text{environment}} = \begin{pmatrix}\text{Energy}\\ \text{transferred}\\ \text{out}\end{pmatrix}_{\substack{\text{cooling}\\ \text{water}}} + \begin{pmatrix}\text{Energy}\\ \text{transferred}\\ \text{in}\end{pmatrix}_{\substack{\text{cooling}\\ \text{water}}} + \begin{pmatrix}\text{Energy}\\ \text{transferred}\\ \text{out}\end{pmatrix}_{\substack{\text{stack}\\ \text{gas}}} = -0.6\text{ unit}$$

Comments

This very simple example illustrates a number of points about the energy conservation principle. First, when we apply the principle, we must carefully define the location to which we are going to apply it, in this case the power plant. Second, we must define a convention for how we assign the sign of the energy transfers. Here, we simply chose energy transfers into the plant as carrying a positive sign, so that energy transferred out is negative. Finally, we have tacitly kept the units (dimensions) of each quantity in the energy balance consistent.

The power of a simple energy balance is obvious from Example Problem 1.1; however, in more practical problems, the energy transfers in the various terms of the energy balance have different forms. For example, in the problem above, the energy transferred in might be chemical energy in a fossil fuel such as coal, oil, or natural gas; it might be the binding energy of the nuclei of atoms for a nuclear plant; it might be the energy transfer from the sun in a solar power plant or from energy stored in the earth in a geothermal plant. The electric energy is in the form of an electric current carried by the transmission lines leaving the plant. The energy transfer to the cooling water is usually in the form of thermal energy added to cooling water or the atmosphere, which then leaves the plant. Finally, the energy from the stack is transported in the hot flowing gas from the plant into the atmosphere. Thus, we need to classify the energy transfer across the plant boundary so that we can properly set up our energy accounting system.

One additional point is that we assumed (without stating the assumption) that the energy entering the plant was exactly balanced at each instant by the energy leaving. However, this is not always the case. Consider a new system boundary for a coal-fired plant that includes a coal stockpile, where coal is often delivered to the plant for later use at periods of high electrical demand. In such a case, our energy conservation equation must be extended to account for an energy storage term. Alternatively, the boundary could be redrawn to include only the powerhouse itself, excluding the coal pile, so that the original conservation equation could still be used. Thus, how we draw the boundary around our energy system defines the form of the equation to be used.

Let us now look at a useful way of classifying energy transfers for engineering problems.

1.3 Work and Heat Transfer

The conservation of energy is tied to a defined system. As indicated in Example Problem 1.1, the plant boundary surrounded the considered system, and there were energy transfers into and out of the plant. The electricity that is transported across the plant boundary can be viewed as a form of *work*. The transfer to the cooling water is through an energy transfer mechanism termed *heat transfer*.

Chapter 2 addresses these two energy transfers, *work* and *heat transfer*, in detail, but an important distinction between the two is made here. Work is viewed as an organized transfer mechanism. This transfer can be used to raise weights, move diaphragms, turn shafts, etc., which is often the desired product of a thermodynamic system. An energy transport in the form of a heat transfer is viewed as a disorganized transfer mechanism. Heat transfer cannot be directly employed to raise a weight, turn a shaft, etc. This distinction is important for classifying the energy terms that make up the conservation of energy principle.

A last point concerning these energy transfers is that they are *not* "contained" by a space or material. They should be thought of as transfers and are therefore necessarily coupled with system boundaries where these transfers take place. This point is developed in much more detail in Chap. 2.

1.4 Macroscopic versus Microscopic Viewpoint

A microscopic viewpoint is used to attempt to understand a process or system by considering the particle nature of matter. This viewpoint might focus on molecules, atoms, or even an electron and nucleus. A complete description would require an enormous effort with suitable approximations. A macroscopic consideration addresses the appropriate observable *averages* of the microscopic phenomena. For example, the microscopic momentum transfer between gas molecules and a surface is observed on the macroscopic level as the gas pressure on the surface. Clearly, the macroscopic viewpoint is of direct consequence to the engineer.

Classical thermodynamics is a macroscopic science. The fundamental statements, or laws, concern the macroscopic properties of matter. Any atomic or microscopic concept must be exhibited in the macroscopic behavior of the system. This does not imply that a microscopic viewpoint is inappropriate for thermodynamics. A clear understanding of macroscopic phenomena is often possible through microscopic concepts. Yet the overriding goal of engineering thermodynamics is to address macroscopic properties.

This text stresses the fundamental concepts from a macroscopic viewpoint. References are made to microscopic behavior where it is helpful to clearly present the material. An introduction to microscopic concepts is contained in Chap. 13.

1.5 Problem Solving

One of the main objectives of this text is to present a logical methodology to solve engineering problems. The subject of thermodynamics is composed of a few basic principles which can be applied to many different problems. Some problems are quite complex; yet, through a careful and logical approach, the solutions are generally straightforward.

We present an approach to solving problems which can be as important as the solution obtained. The student must learn the basic principles as well as the method to apply the principles and must not view the solution to problems as a routine substitution into an appropriate equation. Further, the student should seek the generality in the problem, whether it is an engineering problem being solved or an example in the text. Numerous examples are presented throughout the text to demonstrate these concepts and approaches.

There are many ways to subdivide the general approach to solving problems. The specific categories are not as important as the inclusion of all the basic steps. Grouping these basic steps into three categories yields the following elements:

1. ***Problem statement***
   Carefully evaluate the information presented. What are the unknowns? What is given or known? Determine the principal parts of the problems and those parts which are secondary. An essential element is a figure of the physical system with the considered boundaries and diagrams of the states (precisely defined later) indicating known and unknown information. Carrying out the details of a problem that is not clearly understood might yield a correct answer, but will not help in learning the fundamental principles or applying the approach to new problems.

   Large complex systems involving many complicated subcomponents require a systematic methodology to obtain the desired result. The given information for the component processes should be carefully presented so that the known and unknown information is clearly defined. The individual processes are then considered separately, and the entire system's behavior is built from the subcomponents.
2. ***Analysis***
   The plan of attack to obtain the unknown must be formulated and carried out. The plan is composed of a mix of physical laws or principles, material properties, and assumptions. The specific proportion depends on the problem and its complexity. This planning stage generally turns out to be an iterative procedure, particularly at the early stages of this course. Try to relate the given information to the basic principles or to a previously considered problem. Restating the problem differently might yield a possible direction for solution. Once the plan is formulated, the solution must be carried out correctly. Make sure that *each* step is correct. It is frustrating to abandon a correct approach because of an incorrect step. Recheck each step to ensure that it is necessary.

   Quite often it is useful to carry a solution as far as possible in algebraic form, because quantities may cancel or simplify. Early substitution of numerical values offers extra chances for numerical errors.
3. ***Review***
   This essential step is often neglected. It is important from the viewpoint of both obtaining the correct result and solving the problem. First, try an alternate approach to the problem to check the results. Second, does the result make sense physically? Does the result have the correct dependencies. This type of thinking will be valuable in our approach to new problems. Last, try to generalize the analysis and consolidate the knowledge. What are the key elements of the problem?

An instrument used in the first chapters of this book is a tabular presentation of the states and processes. The approach is not essential, but it has been found very useful by the first-time student. This table forces you to carefully understand the problem statement. When this table is combined with a diagram of the states and processes, a direction for the solution of the problem should be evident.

1.6 Units

Engineers and scientists need to communicate with their peers not only through carefully defined words but also through numeric descriptions of the magnitude of certain quantities. The magnitude of a quantity such as volume depends on the system of units used to make such a description; i.e., we can describe volume in terms of cubic centimeters, cubic feet, gallons, barrels, etc. So we must define our quantities carefully, but we must be equally careful to use a set of measurement units that is universally understood and accepted. Two unit systems are in general use among scientists and engineers: the U.S. Conventional System (USCS), or English system of units (sometimes called the customary engineering system), and SI (Système International d'Unités), or international system. The latter system is used almost universally outside the United States. Despite efforts to make SI accepted worldwide, in the United States much of engineering practice is still carried out according to the USCS.

It is generally agreed that SI has certain advantages over the USCS, notably that fewer conversion factors must be memorized and that the decimal base of the system leads to simple choices of the scale of units needed to descirbe a given quantity.

Any system of units can be subdivided into *base units* and *derived units.* The base units are prescribed, and then the derived units are obtained from the prescribed set. Table 1.1 shows the base units and some of the derived units for both USCS and SI. Conversion factors between the SI and USCS values are given in App. B. The key points to note are the two dashes in this table The USCS specifies force as a base unit while SI treats force as a derived unit. In the USCS, the original definition of certain units leads to the necessity of carrying a conversion factor in many equations.

This is made clear by an examination of Newton's second law for a constant mass subjected to a single force F and accelerated at a in the direction of F. In USCS, force and mass are unfortunately both given in units called pounds. To denote the funadmental difference between these quantities, the unit of force is always called the *pound-force* (lbf), and the unit of mass is called the *pound-mass* (lbm). To avoid confusion, both words should always be used in discussing these quantities, i.e., *pound-force* or *pound-mass,* never simply *pound.* Newton's second law in USCS would now be written as

$$F \text{ lbf} = m \text{ lbm} \times a \text{ ft/s}^2 \tag{1.1}$$

TABLE 1.1

Quantity	USCS	SI
	Base Units	
Length	foot, ft	meter, m
Mass	pound-mass, lbm	kilogram, kg
Time	second, s	second, s
Temperature	degree fahrenheit, °F	kelvin, K
Force	pound-force, lbf	—
	Derived Units	
Force	—	newton, N
Pressure	atmosphere, 1 atm = 14.696 lbf/in^2	pascal, Pa
Energy	Btu = 778.16 lbf·ft	joule, J
Power	Btu/s	watt, W
Specific heat	Btu/(lbm·°F)	J/(kg·K)
Area	ft^2	m^2
Volume	ft^3	m^3
Density	lbm/ft^3	kg/m^3
Velocity	ft/s	m/s

SI Multipliers

Number	Prefix
10^{12}	tera, T
10^{9}	giga, G
10^{6}	mega, M
10^{3}	kilo, k
10^{-2}	centi, c
10^{-3}	milli, m
10^{-6}	micro, μ
10^{-9}	nano, n
10^{-12}	pico, p

However, one difficulty remains. The pound-force is a *base* unit in the USCS. One pound-force is defined as the force that accelerates a mass of 1 lbm at a rate of 32.1740 ft/s^2. If these values are substituted directly into Eq. (1.1), the result is

$$1 \text{ lbf} = 1 \text{ lbm} \times 32.1740 \text{ ft/s}^2$$

which, because all terms are base units, is inconsistent in both defined units and magnitude. Thus, Eq. (1.1) must be modified by including the appropriate conversion factor, often explicitly denoted by the symbol g_c, so that Eq. (1.1) becomes

$$F = \frac{ma}{g_c} \tag{1.2}$$

where g_c has the value 32.1740 ft·lbm/(lbf·s^2). In SI, this conversion problem is avoided, because the unit of force, the newton, is a *derived* unit, derived from Newton's second law as the force that will accelerate 1 kg of mass at a rate of 1 m/s^2. Substituting these values into Eq. (1.2) will show that the value of g_c is then simply 1 kg·m/(N·s^2), and g_c is not considered in equations that use exclusively SI units.

The equations in this version of the text show the conversion factor g_c where it is needed. *Remember* that the SI does not need this conversion factor and the USCS does.

Example Problem 1.2

A 1-lbm mass is attached to a spring scale on the earth's surface. The scale and the mass are then moved to the surface of the moon by an enterprising astronaut. The gravitational acceleration on the surface of the moon is one-sixth that on the surface of the earth. What will the spring scale read in each of the two cases?

Solution

The mass of an object is unaffected by the local gravitational field. In both cases, the force exerted by the mass on the spring scale equals the force exerted on the mass by gravity (the weight), or

$$F = \frac{ma}{g_c} = \frac{mg}{g_c}$$

On the earth's surface, the scale will then read (using $g = 32.17$ ft/s^2)

$$F = \frac{(1\ \text{lbm})(32.17\ \text{ft/s}^2)}{32.17\ \text{ft}\cdot\text{lbm/(lbf}\cdot\text{s}^2)} = 1\ \text{lbf} = 4.45\ \text{N}$$

and on the moon

$$F = \frac{(1\ \text{lbm})(\frac{1}{6})(32.17\ \text{ft/s}^2)}{32.17\ \text{ft}\cdot\text{lbm/(lbf}\cdot\text{s}^2)} = \frac{1}{6}\ \text{lbf} = 0.74\ \text{N}$$

Comments

Note that the mass is not changed in either case and that the weight indicated by the spring scale F depends on the gravitational field in which the scale is used. Now think about using a chemical beam balance to determine the mass and weight of the object used in the above experiment.

1.7 Scope of Text

Thermodynamics deals with energy transformations. The presentation of the concepts requires an initial understanding of the basic definitions of systems, processes, and states. To understand the energy transformations of a system, energy transfer mechanisms in the form of work and heat transfer must be defined. Then the fundamental principles of thermodynamics can be presented. The portion of thermodynamics that includes the basic definitions, concepts, and laws forms the core of the subject.

Another important aspect of any introductory thermodynamics course is the set of required thermodynamic properties. This aspect includes the graphical, tabular, and equation forms of the quantities that describe the states of a substance. The use of thermodynamic properties is essential to solving problems, yet the detailed properties are not part of the foundations of the laws of thermodynamics.

The third portion of the study of thermodynamics is the application of the core concepts to particular problems of interest to the engineer.

These three elements are the basis of any presentation of engineering thermodynamics. Clearly the core element including the basic concepts must be presented first. A logical presentation might then discuss properties followed by applications. This text has a slightly different, yet equally logical, organization. The core concepts, properties, and applications are presented simultaneously, so that a working knowledge of the properties is obtained as the basic concepts are introduced. Practical applications of the fundamental principles can be carried out as the laws are presented.

Problems

1.1 One system of units takes the base units of length, time, and force as foot, second, and pound-force, respectively. The set of derived units includes mass, and the derived mass unit is called a *slug.* With Newton's second law $F = ma$, determine the relation between the slug and the length, time, and force units above.

1.2 Determine the SI values for the following quantities: 6 ft, 200 lbm, 70°F, 25 lbf/ft^2, and 1 atm.

1.3 Determine the SI values for the following energy-related quantities: 1 Btu/(ft^2·h). 13,000 Btu/lbm, and 50,000 Btu/h.

1.4 If an object weighs 170 lbf on earth, find its mass in pound-mass and kilograms and its weight in newtons.

1.5S On the surface of the earth a man weighs 750 N. What is his mass on earth? What is his mass on the moon?

1.5E On the surface of the earth a woman weighs 150 lbf. What is her mass on earth? What is her mass on the moon?

1.6S A man weighs 810 N on his home planet where the acceleration due to gravity is 10.2 m/s^2. How much would he weigh on a planet where the acceleration due to gravity was 7 m/s^2?

1.6E A woman weighs 180 lbf on her home planet where the acceleration due to gravity is 35 ft/s^2. How much would she weigh on a planet where the acceleration due to gravity was 25 ft/s^2?

1.7S An astronaut who weighs 900 N on earth flies to a planet where the acceleration due to gravity is 5 m/s^2. While on the planet she picks up 100 kg of rocks. While she is on the planet, it is noted that her weight drops to 450 N because of the physical activity. When the astronaut arrives back on earth, how much do she and the rocks weigh?

1.7E An astronaut who weighs 200 lbf on earth flies to a planet where the acceleration due to gravity is 20 ft/s^2. While on the planet he picks up 20 lbm of rocks. While he is on the planet, it is noted that his weight drops to 100 lbf because of the physical activity. When the astronaut arrives back on earth, how much do he and the rocks weigh?

1.8S A spacecraft weighing 273,000 N leaves earth with 900,000 N of fuel on board and flies to a planet where the acceleration due to gravity is 4 m/s^2. During the flight to the planet, two-thirds of the fuel is consumed. How much thrust does the rocket need to ensure lift-off from the planet?

1.8E A spacecraft weighing 50,000 lbf leaves earth with 180,000 lbf of fuel on board and flies to a planet where the acceleration due to gravity is 12 ft/s^2. During the flight to the planet, two-thirds of the fuel is consumed. How much thrust does the rocket need to ensure lift-off from the planet?

1.9S An astronaut weighs 700 N on earth where the acceleration due to gravity is equal to 9.5 m/s^2. She then goes to the moon ($g = 1.6$ m/s^2) and collects a bag of rocks weighing 50 N. Calculate the combined weight of the astronaut and rocks on Mars, where the local acceleration due to gravity is 6.5 m/s^2.

1.9E An astronaut weighs 160 lbf on earth where the acceleration due to gravity is equal to 32.2 ft/s^2. He then goes to the moon ($g = 5.4$ ft/s^2) and collects a bag of rocks weighing 10 lbf. Calculate the combined weight of the astronaut and the rocks on Mars, where the local acceleration due to gravity is 20 ft/s^2.

1.10S With what force is a mass of 50 kg attracted to the earth at a point where the gravitational acceleration is 9.75 m/s^2?

1.10E With what force is a mass of 100 lbm attracted to the earth at a point where the gravitational acceleration is 30.0 ft/s^2?

1.11 An experiment package, when tested during development on earth, required a force of 50 lbf for deployment from a lunar landing vehicle. When it is actually deployed on the moon, what force will be required to put it into position? ($g_{moon} = 1.6$ m/s^2)

1.12S Determine the force required to give an acceleration of 30 m/s^2 to a mass of 25 kg.

1.12E Determine the force required to give an acceleration of 100 ft/s^2 to a mass of 75 lbm.

1.13S An external force of 225 N is acting vertically upward on a body whose mass is 5 kg. In the absence of friction, is the body rising or falling, and what is its acceleration? The body is initially at rest, and the acceleration due to gravity is 9.8 m/s^2.

1.13E An external force of 50 lbf is acting vertically upward on a body whose mass is 12 lbm. In the absence of friction, is the body rising or falling, and what is its acceleration? The body is initially at rest, and the acceleration due to gravity is 32.2 ft/s^2.

1.14S A kilogram of mass is "weighed" with a beam balance at a point where $g = 9.7$ m/s^2. What reading is expected? If it is weighed with a spring scale that reads correctly for standard gravity, what scale reading is expected?

1.14E A pound of mass is "weighed" with a beam balance at a point where $g = 30$ ft/s^2. What reading is expected? If it is weighed with a spring scale that reads correctly for standard gravity, what scale reading is expected?

2

Energy and Energy Transfer

Normally screws are so cheap and small and simple you think of them as unimportant. But now, as your Quality awareness becomes stronger, you realize that this one, individual, particular screw is neither cheap nor small nor unimportant. Right now this screw is worth exactly the selling price of the whole motorcycle, because the motorcycle is actually valueless until you get the screw out. With this reevaluation of the screw comes a willingness to expand your knowledge of it.
Robert M. Pirsig, *Zen and the Art of Motorcycle Maintenance,* William Morrow and Co., Inc., New York, 1974, used with permission.

Bare bent-tube integral furnace boiler. For service to 975 psi, saturation temperature to 900°F. Gas flow is from right to left. (*Babcock and Wilcox.*)

2.1 Introduction

All specialized fields, such as law and economics, use terms with quite specific meanings, and these meanings may not be the same as those used in everyday conversation. To be sure of accurate communication in the classroom as well as with other engineers and scientists, the definitions used in thermodynamics must be held in common among us, and they must be as complete and accurate as possible.

The purpose of this chapter is to define the basic terms of thermodynamics. In later chapters, additional terms and concepts are introduced and defined; however, the terms defined here are sufficient to begin the study of thermodynamics.

2.2 Concepts and Definitions

To begin our study of thermodynamics, we must have in mind some definable quantity of matter which we wish to consider. Then the behavior of energy as it interacts with this matter can be examined.

2.2.1 System and Surroundings

We might want to determine, for example, the temperature of shaving cream as it leaves a pressurized container or to find the work output of a steam turbine in a power plant. What "definable quantity of matter" shall we examine in each of these cases?

Any such quantity of matter is obviously a subclass of all matter. Suppose we begin by defining the *universe* as the totality of matter that exists.

Now, to get down to the part of the universe that we want to study for a particular reason, we define a *system* as the part of the universe set apart for examination and analysis. In most interesting problems that we examine in thermodynamics, the system under study interacts with the rest of the universe through an exchange of energy and/or mass. Yet the great bulk of the universe can be considered to be unaffected when, for example, we allow shaving cream to expand from its container. It is useful, then, to look at a further subset of the universe, called the *surroundings,* that is made up of that portion of the universe which strongly interacts with the system under study. These definitions are shown schematically in Fig. 2.1.

Figure 2.1 The system, the surroundings, and the universe.

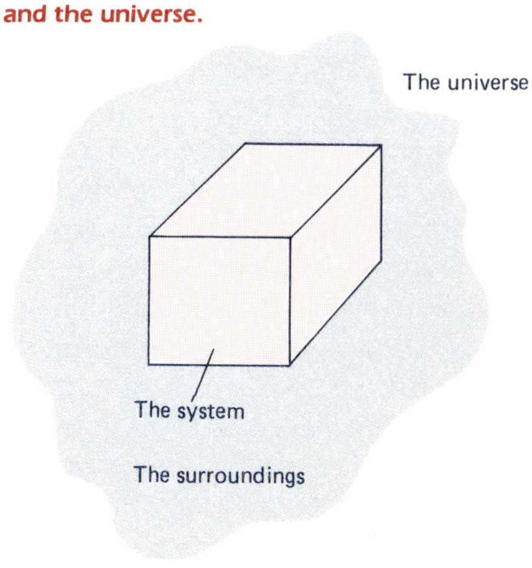

In summary, then, we perform a thermodynamic study of a system that interacts with its surroundings; the system and its surroundings form a part of the universe.

It is critical in all thermodynamic studies to carefully define the particular system under examination. It sounds as if this task would be obvious and easy, and in many cases it is. In other cases, judicious choice of the system to be studied can greatly reduce the problem-solving effort. For example, in the shaving cream example, what system should be studied? Is it the container? The container plus its contents? The contents only? If the last, how do we treat the fact that the separation between the system and its surroundings is moving in space as the contents expand out of the can? All these choices have benefits and deficiencies that will be addressed soon; for now, though, clearly we need some further ways to describe the system. In particular, how can the separation between the system and the surroundings be described?

Generally, we define the *system boundary* as the surface that separates the system from its surroundings. Thus the shaving cream as a system is separated from the air (the surroundings) as the cream leaves the container by an imaginary surface, or boundary. The boundary expands with the shaving cream, so the system always contains all the initial mass of the system (the shaving cream).

That approach seems straightforward enough. However, consider the other example mentioned earlier, the power plant turbine. What do we use as the system and its boundary? If we follow the shaving cream example, we might define the steam flowing through the turbine as the system. The system boundary, however, would then extend throughout all components of the power plant that contain the steam (and the water from which it is generated) before and after the steam passes through the turbine. Such a system definition is not useful for many problems. To study hardware characteristics, it is often more useful to look at a boundary that is roughly coincident with the hardware being studied (the turbine casing for example). In that case, however, mass is crossing the boundary!

Our definition of a system does not preclude mass crossing the system boundary. We must simply realize that the system under study has different mass within its boundary at each instant of time; alternatively, we can imagine the study to be of a series of constant-volume systems, each with a different inventory of mass.

We have seen that special circumstances will determine how to treat the system boundary. In the first case (shaving cream), no mass crosses the system boundary, and the boundary is allowed to expand in space so that the system mass is always contained within the boundary and is constant. In the second case, it is more convenient to fix the boundary in space and to allow the inventory of mass within the system to be different at each time. In such a case, the *amount* of mass within the system may or may not be constant, but it will be *different* mass at each time. We will see that there is no fundamental difference in the basic principles as we analyze these different cases. The former type of system (no mass crossing the bound-

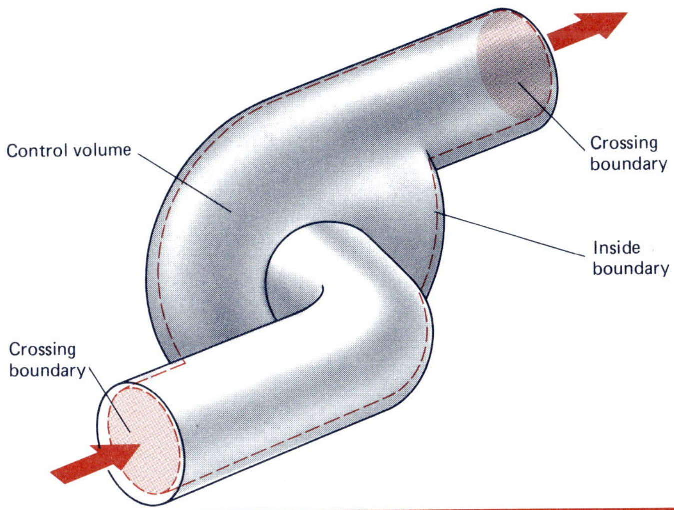

Figure 2.2 Control volume showing the system boundary composed of inside and crossing boundaries.

ary) is sometimes referred to as a *closed system,* or *control mass* (CM), while a system with mass crossing the boundary is called an *open system,* or *control volume* (CV).

In a system that has mass crossing its boundary, usually a part of the boundary allows no mass to cross, while another part of the boundary allows mass flow. Such a boundary configuration is typical of engineering hardware such as a turbine or compressor. In such equipment, the inside of the casing of the equipment is impervious to mass flow (the *inside boundary*) while the inlet and exit ports to the equipment encompass all the mass flow streams that enter or leave the system (the *crossing boundary*). A complete volume that is fixed in space and has a crossing boundary is called a control volume. Figure 2.2 shows the control volume and its boundaries for an idealized pump.

2.2.2 System Description

Now that we have defined the nomenclature that describes the matter under study, we need some means of describing the behavior of the system. In other words, how do we tell whether changes have occurred in the system? What will allow us to quantitatively describe these changes?

Let us examine for a moment a particular system that has no mass crossing its boundary and no energy crossing its boundary. Such a system is called *isolated.* Assume further that the system is *uniform*; that is, its properties are everywhere the same. From experience, we would expect such a system, after sufficient time had passed, to be unchanging. But how do we define *unchanging?* Our experience indicates that certain measurable quantities such as pressure, temperature, and volume will be constant for an isolated system. If we measure these or certain other quantities, a particular *set* of measured values will result. Let us agree that this set of values defines the *state* of the system. If we allow mass or energy to cross the system boundary, some of the measured values will change, and the system will be in a new state.

What quantities define the state of a system, and how many such quantities are necessary and sufficient to define the state? The answer is given later, but here we note some attributes of the quantities describing the state of a system. These attributes attempt to describe the *condition* of the system.

Let us define a *property* as any quantity that describes the state of a system and, conversely, as any quantity whose value depends solely on the state of the system under study. This is a quantity that is "possessed" by a system or that the system "has." A number of familiar quantities meet this definition. For example, the volume of the shaving cream before and after expansion from the container depends on only the states at the beginning and end of the expansion process; therefore, volume is a property. Pressure and temperature also meet the requirements of being properties. Other less familiar properties will be defined as they are found useful.

Properties have other characteristics that result from this definition. Since properties fix the state of a system, they are independent of how the system reaches a given state. Thus, all properties have the mathematical characteristics of exact differential quantities; i.e., the change in their values between two system states is independent of how the system state was changed. The mathematical properties presented in App. H, Eq. (H. 29) are valid for properties. These characteristics are used in later chapters.

Properties are conveniently divided into two categories. The first contains those properties with values that depend on the amount of mass contained in the system in a given state. From the list of properties mentioned so far, volume falls into this category. If the mass of the system is doubled while temperature and pressure are constant and uniform within the system, then the volume will also double. Properties that depend on the system's mass, or the "extent" of the system, are called *extensive* properties. Properties that are independent of the mass contained in the system boundaries are called *intensive* properties.

Both intensive and extensive properties can be denoted by capital letters. Thus P will stand for pressure (intensive), T for temperature (intensive), and V for volume (extensive). Extensive properties of a given system in a given state, however, can be denoted in an alternate intensive form. The volume of the shaving cream, for example, can be assigned the symbol V. However, for a system in a given state with properties that are uniform throughout, the volume is directly proportional to the system mass m. Therefore, the quantity V/m is also a property (since the ratio depends solely on the system state). The ratio V/m is much more convenient to tabulate than V itself, since one parameter, m, is eliminated from the tables. This is true of all extensive thermodynamic properties. Whenever any extensive property X is used or tabulated in the form X/m, the ratio is assigned the corresponding lowercase letter x, and the property is called a *specific* property. Thus, $v \equiv V/m$ is the *specific volume,* which is the inverse of the density $v = 1/\rho$. Specific extensive properties are intensive properties, since their value no longer depends on the mass of the system.

2.2.3 Equilibrium States and Quasi-Equilibrium Processes

The concept of equilibrium is important to thermodynamics and is closely tied to the definitions of properties and states. For a system the properties describing the state of an equilibrium system will be constant if the system is not allowed to interact with the surroundings or if the system is allowed to interact completely with unchanging surroundings. Such a state is termed an *equilibrium* state, and the properties are equilibrium properties. When a system is in equilibrium with its surroundings, it will not change unless the surroundings change.

Specific types of equilibrium refer to individual properties. When a single property is unchanging within the system, the equilibrium is specific to that property. Typical examples include *thermal equilibrium* (constant T), *mechanical equilibrium* (constant P), and so on. When all possible properties are unchanging, the system is in *thermodynamic equilibrium.*

Much of the study of classical thermodynamics deals with equilibrium states. The properties that we use (and develop in Sec. 2.3 and Chaps. 3 and 9) must be equilibrium properties. In fact, we refer to properties of particular materials with the understanding that there is a system that contains the material and that the system is in equilibrium. We then represent the locus of states for these materials as surfaces in a space with coordinates denoted by the properties.

Various systems that will be addressed may not meet all the conditions for equilibrium when the entire system is viewed. Such a system can be subdivided to yield local, smaller systems that can be treated as being in equilibrium. This will be important when we analyze more complex problems, particularly open systems or control volumes where changes occur throughout the volume.

The equilibrium state is described by equilibrium properties and the locus of possible states is viewed as a surface with coordinates of properties. A particular state is then a point on this surface. If the system is altered so that the state of the system moves along the surface from one equilibrium position to another equilibrium position, the process is termed a *quasi-equilibrium process.* Every position on this surface is in equilibrium, and so every step of the process is in equilibrium. Given the concept of equilibrium defined by unchanging properties in relation to the surroundings, the system must be allowed to reach equilibrium at every step or state. This is achieved by visualizing the process as occurring at an infinitely slow rate, so that there is only a slight difference in properties between the system and surroundings, and equilibrium is achieved at each state along the process. This is an idealizaton, but it is very useful in many problems.

2.3 Some Common Properties

We have discussed one property of a system that is quite familiar, the volume. We now discuss specific volume and some other familiar properties in more detail. All the properties discussed in this chapter have a misleading attribute — they are all quantities with which we have extensive experience before beginning the study of thermodynamics. In addition, the properties of volume, temperature, and pressure are all measurable quantities.† The student should not infer that properties are all directly measurable in the sense that P, V, and T are. Indeed, many of the properties useful in thermodynamics are *not* directly measurable. However, every property that is introduced in classical thermodynamics does share the attributes noted in our general definition.

2.3.1 Pressure P

Because many systems studied in thermodynamics involve gases or vapors, pressure is a useful property for describing the state of the system. *Pressure* is defined as the normal force exerted per unit area on a real or a fictitious surface within the system.

In classical thermodynamics, we do not consider effects that occur on a microscopic scale, so that we can treat only pressures that exist on areas which are large relative to intermolecular spacings. Thus the fluid is viewed as continuous, and this is the so-called continuum approximation. This approximation must be questioned in vacuum systems where molecular spacings become large.

With the restriction that the area over which the force is applied must not become smaller than some minimum value a (because of the continuum approximation), the mathematical definition of the local pressure is

$$P = \lim_{\lim \Delta A \rightarrow a} \frac{\Delta F}{\Delta A} \tag{2.1}$$

For engineering work, pressures are often measured with respect to atmospheric pressure rather than with respect to an absolute vacuum. The

† The concept of "measurable" is not as straightforward as implied here. Temperature, for example, is really not measured directly, but is found by measuring the length of a mercury column or some other quantity proportional to temperature.

former, or *gauge* pressure, is related to the atmospheric pressure by

$$P_{gauge} = P_{abs} - P_{atm} \tag{2.2}$$

Unless specifically noted otherwise, all pressures used in the text and tables of this book are absolute pressures.

Some pressure-measuring devices measure the height of a fluid column to evaluate the pressure at a particular point. The relation for the pressure change within a fluid is obtained from a force balance on a fluid element. Figure 2.3*a* presents a portion of fluid within the fluid located at an arbitrary position y above a reference plane. The pressure forces act on the top and bottom surfaces, and these forces balance the gravitational force on the fluid within the element. The force balance yields

$$P_yA - P_{y+dy}A - \frac{mg}{g_c} = 0 \tag{2.3}$$

The mass is expressed in terms of the fluid density as $m = \rho A\, dy$, where ρ is the fluid density. By expressing the pressure P_{y+dy} as $P_{y+dy} = P_y + (dP_y/dy)dy$, Eq. (2.3) becomes

$$P_yA - \left(P_y + \frac{dP_y}{dy}\,dy\right)A - \frac{\rho A\, dy\, g}{g_c} = 0 \tag{2.4}$$

or

$$\frac{dP_y}{dy} = \frac{-\rho g}{g_c} \tag{2.5}$$

This relates the change in pressure with elevation to the density of the fluid element and the gravitational acceleration. Integration between elevations y_1 and y_2 for a constant-density fluid yields

$$P_2 - P_1 = -\int_{y_1}^{y_2} \frac{\rho g}{g_c}\,dy = \frac{-\rho g}{g_c}(y_2 - y_1) \tag{2.6}$$

This expression gives the *difference* in pressure at two elevations in terms of density of the fluid and difference in elevation and is diagrammed in Fig. 2.3*b*.

In SI, force is given in newtons and area in square meters. The derived unit for pressure is the pascal (Pa), where 1 Pa is defined as 1 N/m^2. This turns out to be a very small unit for engineering purposes, so most pressures are tabulated in terms of kilopascals (1 kPa $= 1 \times 10^3$ Pa) or megapascals (1 MPa $= 1 \times 10^6$ Pa). One standard atmosphere (1 atm) is equal to 101.325 kPa. Another common unit for pressure is the bar, which is $10^5 \times$ Pa. This is not strictly an SI unit, but it is used in some applications. In the USCS, pressures are given in pounds-force per square foot or pounds-force per square inch (psi). It is also common in the USCS to indicate gauge pressure by pounds-force per square inch gauge (psig) and absolute pressure by pounds-force per square inch absolute (psia). One standard atmosphere in USCS is 14.696 psia, or 2116.2 lbf/ft^2.

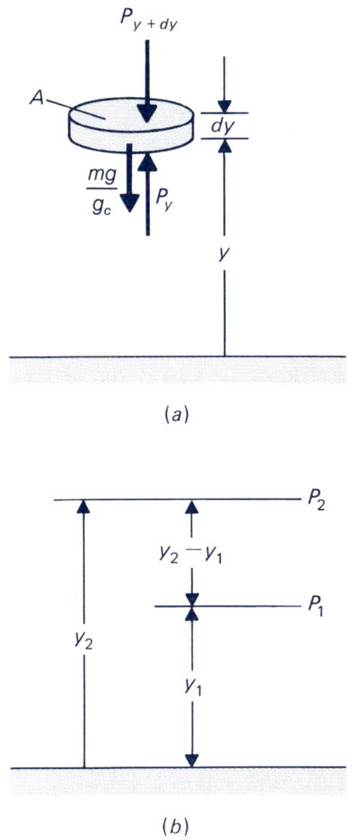

Figure 2.3 Pressure measurement. (*a*) Force balance in a fluid element. (*b*) Pressure difference between two elevations.

2.3.2 Specific Volume v

Specific volume must also be defined in classical thermodynamics in light of the continuum restriction. Thus, the mathematical definition of *specific volume* is

$$v = \lim_{\Delta m \to \mu} \frac{\Delta V}{\Delta m} \tag{2.7}$$

where μ is a minimum amount of mass that is large compared with the mass of the individual molecules of which it is composed. Again, such a restriction causes few difficulties unless gases at high-vacuum conditions or very small-volume systems are examined. The specific volume is the inverse of density, or $v = 1/\rho$.

2.3.3 Temperature T

Temperature may be the property which we think is most familiar, yet it is among the most difficult to define exactly. Our senses are not very reliable in determining temperature. For example, we might go swimming on what we would describe as a hot day. Upon emerging from the water, we find the air suddenly feels quite chilly, and we prefer to stay in the water that felt cold when we first entered. We doubt that either the air or water temperature has actually changed, even though our senses tell us that they seem to be different.

We might also observe that when we take a cold bottle of milk from the refrigerator and set it on the table next to a hot cup of coffee, both liquids will approach room temperature if we wait long enough. The coffee and milk are then said to be in *thermal equilibrium* with the room, and our senses tell us that thermal equilibrium is attained when all the materials are at the same temperature. We have observed here a general law based on this and other experiences: *Two bodies which are each in thermal equilibrium with a third body are in thermal equilibrium with each other.* This observation, because it is the basis for the measurement of temperature and precedes the first and second laws of thermodynamics, is often called the *zeroth law*.

The zeroth law now assures us that if we place a thermometer or other temperature sensor in thermal equilibrium with body (or system) A, and we similarly place the sensor in thermal equilibrium with system B, then systems A and B are at the same temperature if the sensor reads the same temperature in both cases.

We now need to define an adequate scale of temperature, so that all engineers can present their measurements on a common basis. For most such measurements, it is convenient to define a scale that is a linear function of some measurable quantity (such as the length of a mercury column), at least over some range of temperatures between fixed points. There are good reasons in various theoretical studies to use other "temperature" scales, such as one that is the inverse of the common scales.

The two scales most useful in thermodynamics are the so-called absolute scales. The absolute scale for SI is the *Kelvin scale,* named after William Thomson (1824–1907), who became Lord Kelvin. This scale is a 1-point scale based on the second law of thermodynamics. This is presented in Chap. 5. The single point is the triple point of water, where ice, liquid water, and water vapor coexist in a closed system in the absence of air. The Kelvin scale replaced the original scale which was based on a linear function between two selected points. Note that the kelvin unit does not use the degree symbol, but simply the symbol K. Temperatures on this scale are referred to, for example, as "36 kelvins," not "36 degrees kelvin." The other absolute scale is called the *Rankine scale* after W. J. M. Rankine (1820–1872). It is related to the Kelvin scale by

$$1.8°\text{R} = 1\ \text{K} \tag{2.8}$$

where a degree Rankine is denoted by °R.

Two other commonly used scales are the Fahrenheit scale, after Gabriel D. Fahrenheit (1686–1736), and the Celsius scale, after the Swedish astronomer Anders Celsius (1701–1744).

The Fahrenheit scale is also linear, and it was originally based on two defined points: 32° as the temperature at which a system of air-saturated water and ice coexist and 212° as the temperature of a system containing water and steam at a pressure of 1 atm. The symbol for the Fahrenheit degree is °F. This original definition has been replaced by the following relations to the Kelvin scale:

$$T°\text{C} = T\,\text{K} - 273.15 \tag{2.9}$$

$$T°\text{F} = (1.8T)°\text{C} + 32 \tag{2.10}$$

where a degree on the Celsius scale is denoted by °C.

The Celsius scale was originally defined by a single fixed point and a defined size for the degree. The fixed point was the triple point of water, and it was defined to have a temperature of 0.01 °C. The size of the degree chosen for the Celsius scale comes from the Kelvin scale and yields the temperature of the steam point to be 100.00°C at 1 atm. This original definition was replaced by the relation with the Kelvin scale given in Eq. (2.9).

Figure 2.4 compares the various scales. Note that the size of 1 K is the same as 1 degree on the Celsius scale; similarly, the size of the degrees on the Rankine and Fahrenheit scales are equal.

2.3.4 The Ideal Gas

Because we use the concept of the ideal gas to illustrate some applications of work in this chapter, we briefly review the ideal gas relation here. A more complete discussion is found in Chap. 3.

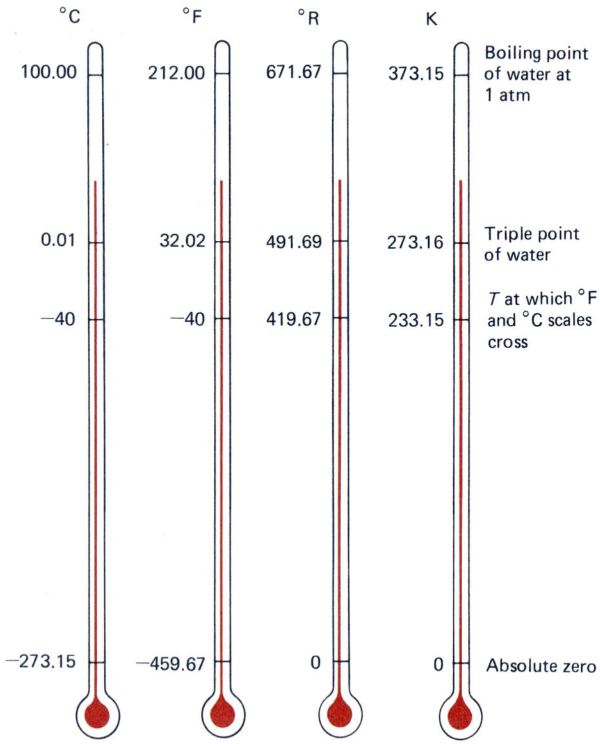

Figure 2.4 Comparison of temperature scales.

In almost every introductory science course, the concept of an ideal gas is introduced. An *ideal gas* is defined as a gas that obeys the relation

$$PV = n\bar{R}T \tag{2.11}$$

where n is the number of moles and $\bar{R}$ is the universal gas constant (Table B.2). Alternately using $v = V/m$ and noting that the number of moles is equal to the mass of gas m divided by the molecular weight of the gas M, we have

$$Pv = \frac{\bar{R}T}{M} = RT \tag{2.12}$$

where $R \equiv \bar{R}/M$ is a constant which is different for each gas. This difference in R occurs because we have chosen to write the ideal gas relation on a mass, rather than a molar, basis. The gas constant is given in Table C.2 for common gases.

We will see in the discussion in Chap. 3 that Eqs. (2.11) and (2.12) are restricted in their range of accurate application; however, the ideal gas relation does describe the relation among P, V (or v), and T over a useful range for some applications.

Example Problem 2.1

The mass of a system is a property which must be specified when the state of the system is being described. Because it is possible to divide certain equations by the system mass, thus converting the extensive properties in those equations to intensive specific properties, the system mass can sometimes be handled in such a way that it need not be specifically given or found. Demonstrate this for the ideal gas relation.

Solution

An ideal gas in a closed container is to be treated as a system. The gas has its properties related by

$$PV = mRT$$

where the symbols are as used earlier. Dividing the equation through by m results in

$$Pv = RT$$

Thus, the mass has disappeared as an explicit system property, and the resulting ideal gas equation provides a relation among the temperature, pressure, and specific volume of the system. Specifying any two of the system properties P, v, and T apparently fixes the other, without regard to the mass of the system.

2.4 Energy

Now that we have defined a system and can describe the state of that system by some set of its properties, we can discuss the quantity called *energy*. In particular, we discuss the various classifications of the forms of energy.

First, let us look again at the special case of an isolated system. Remember that such a system by definition has a fixed state, and no mass or energy is crossing its boundary. What forms of energy does such a system have? From studies of mechanics in introductory physics courses or later in engineering mechanics courses, we know that the mass of the system carries kinetic energy that depends on the overall system speed $\mathbf{V}_{ref}$ *relative to its surroundings* (which we take as the reference frame) of $m\mathbf{V}_{ref}^2/(2g_c)$. In addition, the position of the system relative to a reference plane in the surroundings imparts the system with potential energy that depends on the local gravitational acceleration g as well as the height of the system above the reference plane Z. The potential energy has magnitude mgZ/g_c. The potential energy is a property of the system, so the gravitational field is viewed as part of the system. This is convenient since many problems in engineering are considered within the gravitational field. Recall that the conversion factor g_c is 32.17 ft·lbm/(lbf·s^2) in the USCS, and it is replaced by unity in the SI.

Are other forms of energy possessed by the isolated system? Consider an isolated system composed of a compressed gas at high pressure. Certainly if we remove the restriction of the system being isolated, we could conceive of a number of ways of using this high pressure to operate some form of engine to perform useful work. This stored energy could be tapped without altering the system's velocity, position, or gravitational field and so does not fall in the categories of kinetic or potential energy. Nevertheless, it is apparently an energy that is possessed by the isolated system, and it seems to depend on the state of the system. We call this stored energy *internal* energy, and we give it the symbol U. *Internal energy incorporates the microscopic forms which result from the molecular motions* and is a thermodynamic property of the system. We examine the attributes and numerical values of U in later chapters.

The total energy contained within the system boundary is given the symbol E and is made up of the kinetic energy (KE), the potential energy (PE), and the internal energy U:

$$E = \mathrm{KE} + \mathrm{PE} + U = \frac{m\mathbf{V}^2_{\mathrm{ref}}}{2g_c} + \frac{mgZ}{g_c} + U \tag{2.13}$$

For systems with a mass that may undergo chemical or nuclear reactions, E must contain other terms as well. We will limit consideration to systems without such reactions until Chap. 11. The intensive form for the total energy, or the specific total energy, is

$$e = \frac{E}{m} = \frac{\mathrm{KE}}{m} + \frac{\mathrm{PE}}{m} + \frac{U}{m} = \frac{\mathbf{V}^2_{\mathrm{ref}}}{2g_c} + \frac{gZ}{g_c} + u \tag{2.14}$$

where u is the specific internal energy. (We are ignoring, for the moment, chemical, nuclear, magnetic, and some other forms of stored energy.)

2.5 Energy Transfer

To change the value of the system energy E, we must examine a system that is not isolated from its surroundings. How can energy be changed for the system?

Note that energy crossing the system boundary could be categorized in many ways. The categories historically chosen are used because they help in problem solving, and allow the engineer or scientist to easily tie the results of calculations to the performance of equipment such as turbines, compressors, and heat engines. Thus we use these categories here.

2.5.1 Work

It is useful to define *work* as an energy transfer across the system boundary in an organized form such that its sole use could be the raising of a weight. Such a definition encompasses the definitions used in mechanics, for example, a force acting through a distance. In mathematical terms, the amount of work δW done by a force $\mathbf{F}$ in moving through a differential distance $d\mathbf{s}$ directed outward from the system boundary is

$$\delta W = \mathbf{F} \cdot d\mathbf{s} \tag{2.15}$$

where the boldface center dot denotes the dot product. The work done over a finite path between points s_1 and s_2 is

$$_1W_2 = \int_1^2 \mathbf{F} \cdot d\mathbf{s} \tag{2.16}$$

where $\mathbf{F}$ is the external force exerted in the direction $\mathbf{s}$ along which motion occurs. Additionally, however, the thermodynamic definition of work includes other phenomena; electricity flowing across the system boundary could be used to turn an electric motor and thus lift a weight. Electric energy thus is classified as work when it crosses the system boundary.

Note that unless the path is specified from the initial to the final state of the system, we are not able to calculate the work done. In other words, the work done in going between the initial and final states can have *any* value, depending on the path chosen. This is not surprising in the light of mechanics. Consider the frictional work done in pushing a sandpaper block from point A to point B; certainly the amount of work will depend on whether the path taken is direct or roundabout. Work is therefore called a *path function* in thermodynamics and an *inexact differential* quantity in mathematics (as discussed in App. H). It is therefore denoted by δW. The value of work depends not only on the initial and final states of the system but also on the path followed. Work is not a property, since it cannot be specified from knowing only the state of the system.

Also note the sign convention for work relative to the system boundary. Any sign convention can be adopted for defining the direction of work relative to the system boundary. The sign convention adopted in this text for thermodynamic work is that work done *by* the system *on* the surroundings is positive, while work done *on* the system *by* the surroundings is negative. Thus, any energy crossing the system boundary from the system into the surroundings in the form of work has a positive sign.

Also the process used to compute the value of work is based on the force exerted on the surroundings and is equal to the force within the system for a quasi-equilibrium process. Recall that a quasi-equilibrium process is an idealized process which occurs very slowly. If the particular application is the motion of a piston in a cylinder, then the quasi-equilibrium process is motion due to a force which is counterbalanced at every step of the process.

Finally the units for work must be specified. Work is an energy transfer, so the units are the same as for energy. In SI, this is the joule (J); in USCS, it is a pound-force–foot (lbf·ft) or the British thermal unit (Btu), defined in Sec. 2.5.2.

P dV Work

Many problems of interest in thermodynamics are carried out in terms of *P*, *v*, and *T* rather than in terms of force and distance. So it is convenient to cast our mathematical relations for work in terms of variables *P*, *v*, and *T*. This is conveniently done by multiplying the numerator and denominator of the work equation by the cross-sectional area A_x normal to which the force is applied, to obtain

$$ {}_1W_2 = +\int_1^2 \mathbf{F}\cdot d\mathbf{s} = +\int_1^2 \frac{\mathbf{F}}{A_x}\cdot(d\mathbf{s}\,A_x) \tag{2.17} $$

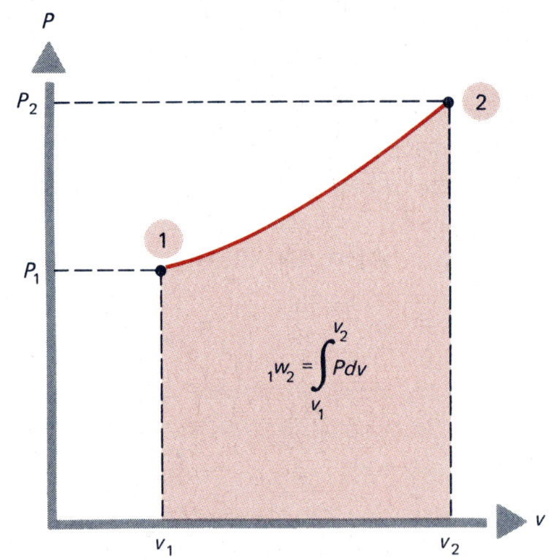

Figure 2.5 Pressure vs. specific volume.

With the definition of pressure as the normal force per unit area and with $d\mathbf{s}$ denoted by dx which is in the direction of **F**, the dot product in the integrand becomes $P\,dV$ (since volume increases as work is done *by* the system, so that the positive sign is required by our convention), and

$$ {}_1W_2 = \int_1^2 P\,dV \tag{2.18} $$

If Eq. (2.18) is divided by mass, then the work per unit mass ${}_1w_2$ is

$$ {}_1w_2 = \int_1^2 P\,dv \tag{2.19} $$

This pressure is the normal force of the system on the container, so a compression is negative work by the system.

Work is expressed in terms of the system pressure times the change in volume, integrated over the *P-v* curve between the initial and final states of the system. The work done in going between the initial volume v_1 and the final volume v_2 is a line integral that has been reduced to an ordinary integral and is visualized by using a diagram of pressure versus specific volume (or volume). Such a plot is shown in Fig. 2.5. The area under the curve of pressure versus specific volume is, according to Eq. (2.19), simply the work done per unit mass in going between the initial and final specific volumes.

Example Problem 2.2

A system has an initial volume of 10 ft³ and expands to a final volume of 20 ft³. The initial pressure is 14.7 psia. Find the work done if the system is at the initial temperature of 75°F and the pressure between the initial and the final states of the system is (*a*) constant, (*b*) inversely proportional to volume (*c*) given by the ideal gas relation (Eq. 2.12).

System Diagram

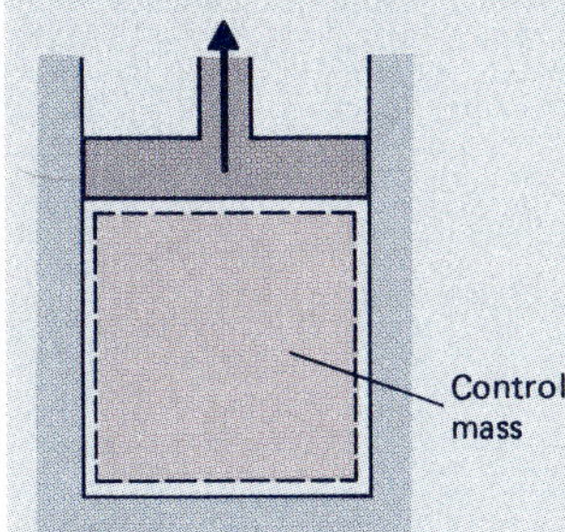

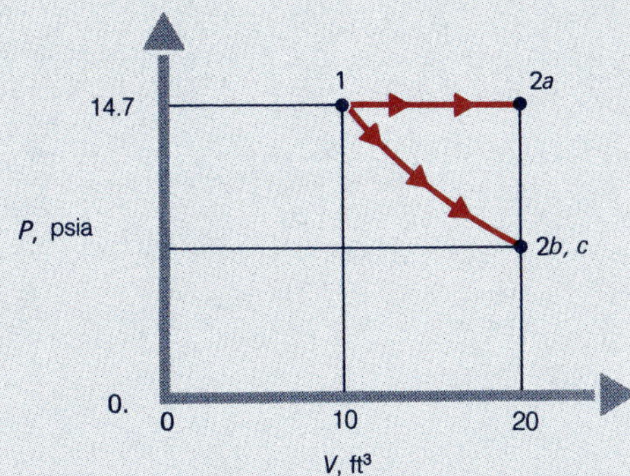

Solution

The state information is tabulated as follows:

State	*P*, psia	*V*, ft³	Process
1	14.7	10	
2*a*	—	20	$P = \text{constant}$
2*b*	—	20	$P = c/V$
2*c*	—	20	$P = mRT/V$

The processes from state 1 to the final state are denoted at the final state.

In each case, the work is given by Eq. (2.18). For (*a*), the pressure is constant at P_1, and we can simply write

$$
\begin{aligned}
{}_1W_2 &= \int_1^2 P\,dV = P_1(V_2 - V_1) \\
&= \left(14.7\,\frac{\text{lbf}}{\text{in}^2}\right)\left(144\,\frac{\text{in}^2}{\text{ft}^2}\right)(20. - 10.)(\text{ft}^3) \\
&= 21{,}200\ \text{ft}\cdot\text{lbf} = 28.7\ \text{kJ}
\end{aligned}
$$

For (*b*), the pressure varies so that P cannot be removed from the integral, and we must write that $P = c/V$, where c is a proportionality constant. The work is then

$$
{}_1W_2 = \int_1^2 P\,dV = \int_1^2 \frac{c}{V}\,dV = c\ln\frac{V_2}{V_1} = c\ln 2
$$

Because both the pressure and volume are given at state 1, the value of c must be $c = P_1 V_1 = 21{,}200$ ft·lbf, and the work done in (*b*) is then ${}_1W_2 = 14{,}700$ ft·lbf $= 19.90$ kJ.

For (*c*), the ideal gas relation may be written $P = mRT/V$, and the work equation becomes

$$ {}_1W_2 = \int_1^2 mRT\frac{dV}{V} = mRT\ln\frac{V_2}{V_1} = mRT\ln 2 $$

Now m can be determined because $m = P_1V_1/(RT)$ in state 1; substituting, we find

$$ {}_1W_2 = P_1V_1 \ln 2 $$

which is the same result as for (*b*).

Comments

The work done in (*a*) is different from that found in (*b*) and (*c*) because the path was different even though the initial and final volumes were the same in all cases. The results for (*b*) and (*c*) are the same because the pressure dependence on volume for a constant temperature ideal gas is the inverse proportionality assumed in (*b*). Note that the temperature of the ideal gas did not enter into the problem explicitly, nor did the mass of the system. Both were eliminated with the other given information. Also since the system volume increased in all cases, work was done by the system on the surroundings and was therefore positive.

Example Problem 2.3

An ideal gas [Eq. (12.12)] is contained within a cylinder with a movable piston as in Example Problem 2.2. Evaluate the work per unit mass by the gas for the path indicated in the system diagram on the next page. Take P and T as known at states 1 and 3.

Solution

This problem is evaluated by two different approaches.

Approach 1

The area under the path on a P-v diagram represents the work done by the system. The process on the P-T diagram is redrawn on a P-v diagram. The constant-P process from 1 to 2 is again a horizontal line, but

System Diagram

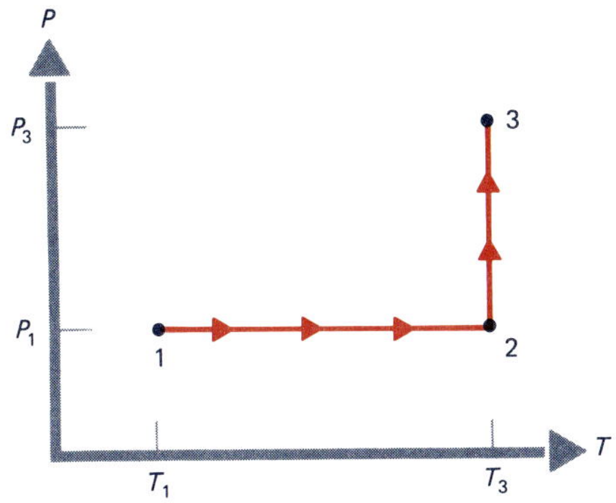

State	Properties			Process
	P	T	v	
1	P_1	T_1	v_1	P = constant = P_1
2	P_1	T_3		T = constant = T_3
3	P_3	T_3	v_3	

the constant-T process from 2 to 3 is given by $P = RT/v$. The work per unit mass is

$$
\begin{aligned}
{}_1w_3 &= \int_1^3 P\,dv = \int_1^2 P\,dv + \int_2^3 P\,dv \\
&= P_1(v_2 - v_1) + \int_2^3 \frac{RT}{v}\,dv \\
&= P_1(v_2 - v_1) + RT_3 \ln \frac{v_3}{v_2}
\end{aligned}
$$

This can be expressed as a function of the endpoints by noting that $v_2 = RT_2/P_2$; but $T_2 = T_3$ and $P_2 = P_1$, so $v_2 = RT_3/P_1$. Therefore, with $v = RT/P$,

$$
\begin{aligned}
{}_1w_3 &= P_1\left(\frac{RT_3}{P_1} - \frac{RT_1}{P_1}\right) + RT_3 \ln \frac{RT_3/P_3}{RT_3/P_1} \\
&= RT_3\left(1 - \frac{T_1}{T_3} + \ln \frac{P_1}{P_3}\right)
\end{aligned}
$$

System Diagram

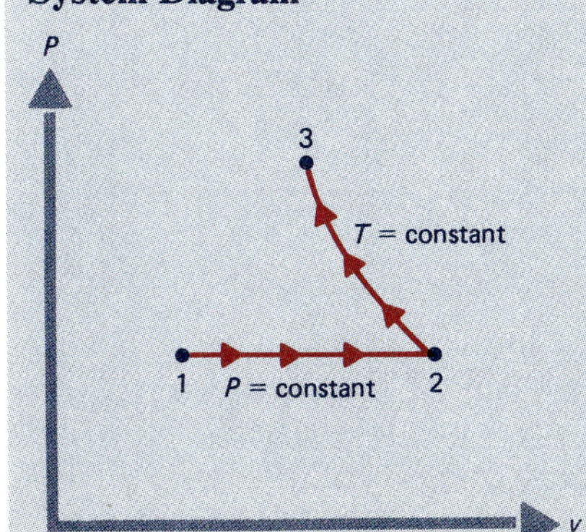

Approach 2

Integrate along the original path by expressing dv in terms of dT and dP. With $v = RT/P$,

$$dv = \left(\frac{\partial v}{\partial T}\right)_P dT + \left(\frac{\partial v}{\partial P}\right)_T dP$$

$$= \frac{R}{P}\,dT - \frac{RT}{P^2}\,dP$$

Then

$${}_1w_3 = \int_1^3 P\,dv = \int_1^3 \left(R\,dT - \frac{RT}{P}\,dP\right)$$

Separating the integral into its elements yields

$${}_1w_3 = \int_1^2 \left(R\,dT - \frac{RT}{P}\,dP\right) + \int_2^3 \left(R\,dT - \frac{RT}{P}\,dP\right)$$

In the process from 1 to 2, $P = P_1$, and for the process from 2 to 3, $T = T_3$; thus

$${}_1w_3 = R\int_1^2 dT - RT_3 \int_2^3 \frac{dP}{P}$$

$$= R(T_2 - T_1) - RT_3 \ln\frac{P_3}{P_2}$$

With $T_2 = T_3$ and $P_2 = P_1$,

$${}_1w_3 = RT_3\left(1 - \frac{T_1}{T_3} + \ln\frac{P_1}{P_3}\right)$$

Comments

The result is the same regardless of the approach. Work is a line integral, as indicated by the second approach; yet it can also be represented as an area integration on a P-v diagram. Also, for an ideal gas, constant-pressure work between states 1 and 2 results in $P_1(V_2 - V_1)$, and isothermal work between states 2 and 3 yields $RT_2 \ln(P_1/P_3)$. What is the value for constant-volume work? (It must be zero.)

Example Problem 2.4

A piston encloses a gas within a cylinder and is restrained by a linear spring as shown. The initial pressure and volume of the gas are 150 kPa and 0.001 m^3, respectively. The spring touches the piston but exerts no force at the initial position. The gas is heated until the volume is tripled, and the pressure is 1000 kPa.

(*a*) Draw the *P*-*V* diagram for the process.
(*b*) Calculate the work done by the gas.
(*c*) What is the work done against the piston and spring?

Diagrams

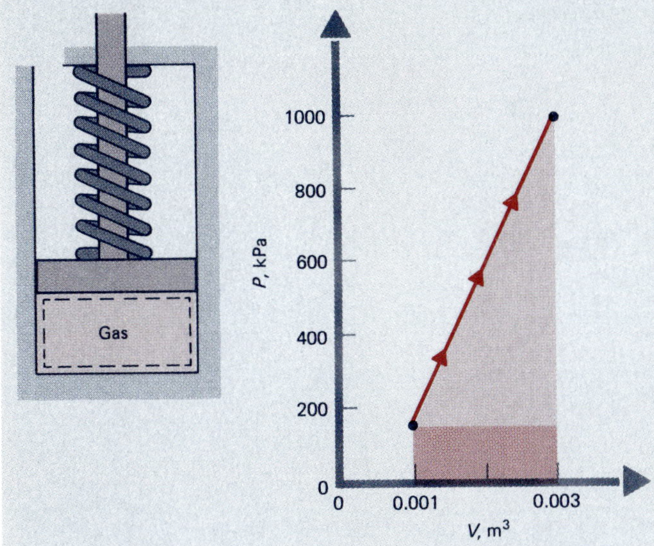

Solution

The state and process information is given in the problem statement. The states are specified by pressure and volume, and the process is governed by the linear spring. Before the state and process table is given, the process is considered. The force relation for a linear spring is given as

$$F_{\text{spring}} = k_s d$$

where k_s is the spring constant and d is the distance displaced from the position of zero force. The pressure exerted *on* the gas by the piston, atmosphere, and spring is the pressure exerted *by* the gas on the piston, atmosphere, and spring (since the process is in quasi-equilibrium). Thus, the pressure in the gas is

$$P = P_{\text{piston}} + P_{\text{spring}} + P_{\text{atm}}$$

where P_{piston} is the weight of the piston per unit area, which depends on the mass and area of the piston and the gravitational acceleration, i.e., $F_p/A = mg/(g_c A)$. Since neither the piston mass nor g changes during the process, P_{piston} is a constant. The piston area is given as A, so the gas pressure is

$$P = P_{\text{piston}} + \frac{F_{\text{spring}}}{A} + P_{\text{atm}} = P_{\text{piston}} + k_s \frac{d}{A} + P_{\text{atm}}$$

Expressing the distance displaced in terms of the volume of the gas yields

$$d = \frac{V - V_1}{A}$$

since the spring exerts no force on the gas initially, that is, $F = k_s d = (k_s/A)(V - V_1) = 0$ when $V = V_1$. Therefore, the pressure-volume relation for the process is

$$P = P_{\text{piston}} + P_{\text{atm}} + \frac{k_s}{A^2}(V - V_1)$$

The state and process chart is as follows:

State	P, kPa	V, m³	Process
1	150	0.001	$P = P_{\text{piston}} + P_{\text{atm}} + \frac{k_s}{A^2}(V - V_1)$
2	1000	0.003	

The three questions are now considered.

(*a*) This straight-line process on a *P-V* diagram is indicated in the diagram. The slope of this line is k_s/A^2.

(*b*) The values for the constants in the pressure-volume relation can be obtained by either writing the equation of a straight line directly from the *P-V* diagram or, equivalently, using the state chart to obtain the numerical values. The pressure expression *at* state 1 is

$$P_1 = P_{\text{piston}} + P_{\text{atm}} + \frac{k_s}{A^2}(0) = P_{\text{piston}} + P_{\text{atm}} = 150 \text{ kPa}$$

The expression at state 2 is

$$P_2 = P_{\text{piston}} + P_{\text{atm}} + \frac{k_s}{A^2}(V_2 - V_1)$$

$$= 150 + \frac{k_s}{A^2}(0.002 \text{ m}^3) = 1000 \text{ kPa}$$

Solving for k_s/A^2 yields

$$\frac{k_s}{A^2} = 425{,}000 \text{ kPa/m}^3 = 425{,}000 \text{ kN/m}^5$$

Thus, the pressure-volume relation is

$$P = 150 + 425{,}000\,(V - 0.001) \text{ kPa}$$

where V is in cubic meters.

The work is given by

$$
\begin{aligned}
{}_1W_2 &= \int_1^2 P\,dV \\
&= \int_1^2 \left[P_{\text{piston}} + P_{\text{atm}} + \frac{k_s}{A^2}(V - V_1) \right] dV
\end{aligned}
$$

$$
\begin{aligned}
{}_1W_2 &= (P_{\text{piston}} + P_{\text{atm}})(V_2 - V_1) + \frac{k_s}{2A^2}(V_2 - V_1)^2 \\
&= 150(0.002) + \frac{425{,}000}{2}(0.002)^2 \\
&= 0.3 + 0.85 \\
&= 1.15\ \text{kN}\cdot\text{m} = 1.15\ \text{kJ} = 848\ \text{ft}\cdot\text{lbf}
\end{aligned}
$$

This is simply the area under the process on the P-V diagram.

(*c*) The work required to raise the piston without the spring is

$$
\begin{aligned}
{}_1W_{2,\text{piston}} &= \int_1^2 (P_{\text{piston}} + P_{\text{atm}})\,dV = (P_{\text{piston}} + P_{\text{atm}})(V_2 - V_1) \\
&= 150(0.002) = 0.300\ \text{kJ} = 221\ \text{ft}\cdot\text{lbf}
\end{aligned}
$$

The work done against the spring is

$$
\begin{aligned}
{}_1W_{2,\text{spring}} &= \int_1^2 P_{\text{spring}}\,dV = \int_1^2 \frac{k_s}{A^2}(V - V_1)\,dV \\
&= \frac{k_s}{A^2}\frac{(V_2 - V_1)^2}{2} = 0.850\ \text{kJ} = 627\ \text{ft}\cdot\text{lbf}
\end{aligned}
$$

These contributions sum to the total work done by the gas (1.15 kJ = 848 ft·lbf). The spring contribution represents 73.9 percent of the total work.

Comments

The process is given by the linear spring equation which implies a linear relation between gas pressure and system volume. Note that no force is exerted by the spring initially. This fact specifies the location for the distance displaced d. If the spring had exerted a force initially, then $d = 0$ would be not at $V = V_1$ but at a system volume $V < V_1$. The pressure-volume relation is still linear, but the constants in the P-V relation would be different.

The work for the process is given as the area under the process on the P-V diagram. The spring work is represented by the triangular area above $P = P_1 = 150$ kPa. The rectangular area below $P = 150$ kPa represents the work done to displace the piston.

Polytropic Process

Many practical applications involve processes described by a specific relationship between pressure and volume. The piston-cylinder type of problem within an internal combustion engine can often be described by an experimentally determined pressure-volume relation. A process that is particularly important is the *polytropic process,* which is described by

$$PV^n = \text{constant} \tag{2.20}$$

where n is a specified constant. Between state 1 and 2, the polytropic process is also represented as

$$\frac{P_1}{P_2} = \left(\frac{V_2}{V_1}\right)^n \tag{2.21}$$

The $P\,dV$ work for a polytropic process is

$$_1W_2 = \int_1^2 P\,dV = \begin{cases} \dfrac{P_1V_1 - P_2V_2}{n-1} & \text{for } n \neq 1 \\ P_1V_1 \ln \dfrac{V_2}{V_1} & \text{for } n = 1 \end{cases} \tag{2.22}$$

Generally, the value of n is known from other process information or is obtained empirically from data.

The process described by $n = 0$ is a constant-pressure one, and this process is a horizontal line on a P-V diagram. A constant-volume process is given by $n = \pm\infty$ and is a vertical line on a P-V diagram. Intermediate values of n, both positive and negative, are possible. Negative values are only encountered when the P-V processes are approximated that involve internal energy generation by chemical reaction or when significant heat transfer to the control mass or control volume is present. In such cases, the pressure may increase as the system volume increases. Example Problem 2.4 shows such a process.

Example Problem 2.5

The combustion products in an internal combustion engine undergo an expansion process from an initial volume of 0.007 ft^3 and initial pressure of 2000 psia. The final volume is 0.045 ft^3. The experimental data for the polytropic process indicate that $n = 1.45$. Evaluate the work performed by the expanding gas.

Diagrams

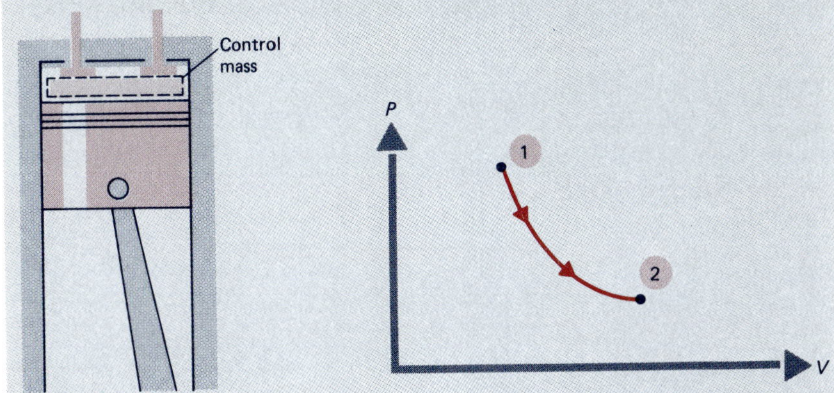

Solution

The state and process information is given:

State	P, psia	V, ft³	Process
1	2000	0.007	$PV^{1.45} =$ constant
2	—	0.045	

The work for the process is given in Eq. (2.22) as

$$ {}_1W_2 = \frac{P_1V_1 - P_2V_2}{n-1} $$

The value of P_2 is obtained from the fact that the process is polytropic, and Eq. (2.21) yields

$$ P_2 = P_1\left(\frac{V_1}{V_2}\right)^n = 2000\left(\frac{0.007}{0.045}\right)^{1.45} = 135 \text{ psia} = 0.93 \text{ MPa} $$

The work is

$$ {}_1W_2 = \frac{2000(0.007) - 135(0.045)}{0.45}\,144 = 2540 \text{ ft}\cdot\text{lbf} = 3.44 \text{ kJ} $$

The work is performed by the gas, so the value is positive.

Comments

First note that the system is a control mass, so the problem could be considered on a specific volume basis, that is, $Pv^n =$ constant and ${}_1w_2 = (P_1v_1 - P_2v_2)/(n-1)$. This is a convenient form for some problems. Also the value of n is given in this problem, but an alternative problem statement could give P_1, V_1, P_2, and V_2 for the polytropic process. Then n could be evaluated from the P-V data.

Example Problem 2.6

During the expansion of the gas in a piston-cylinder system, these data were taken:

Pressure, kPa	Volume, m^3
10	1
4.06	2
2.39	3
1.65	4
1.23	5
0.974	6
0.797	7
0.669	8
0.575	9
0.501	10

Find the work done during this process.

Solution

In this case, the relation between pressure and volume is known through the data provided, but no analytical expression relating P and V is given. Thus, the value of the work must be obtained from a numerical integration of the work integral. Numerical integration is simply an approximation of the area under the P-V curve obtained by summing small area increments. Many forms of numerical integration can be used, such as *Simpson's rule,* the *trapezoidal rule,* or *gaussian quadrature* [Ref. 1]. For the data presented here, the ΔV increments are evenly spaced, and the trapezoidal rule is easily applied. The trapezoidal-rule approximation to the integral is defined by

$$_1W_2 = \int_{V_1}^{V_2} P\,dV = \Delta V\left(\frac{P_0}{2} + \sum_{i=1}^{N-1} P_i + \frac{P_N}{2}\right)$$

Here N is the number of ΔV increments used to divide the range of integration, P_0 is the pressure at the left-hand end of the range of integration, P_N is the pressure at the right-hand end, and the P_i are the pressures at all points between. The data are plotted below to show the meaning of the terms.

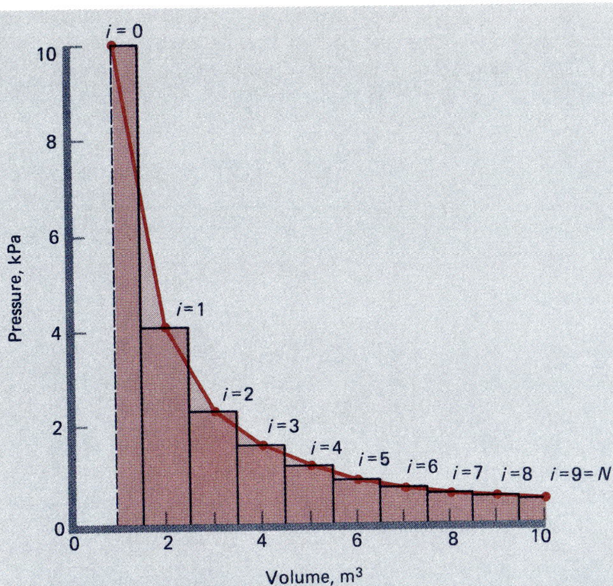

For the given data,

$$ {}_1W_2 = (1\ \text{m}^3)\left[\left(\frac{10}{2} + 4.06 + 2.39 + 1.65 + 1.23 + 0.974 + 0.797 + 0.669 + 0.575 + \frac{0.501}{2}\right)\text{kPa}\right] $$

$$ = 1 \times 17.596\ \text{kPa}\cdot\text{m}^3 = 17.60\ \text{kJ} $$

From the figure, note that the trapezoidal rule is simply an orderly process for approximating the sums of the areas under successive portions of the P-V curve (each portion having the same ΔV increment except for the two end increments, which have increments of $\Delta V/2$) and adding these areas. The greater the value of N, the more exact the method will become. This method can be applied whenever equal ΔV increments are chosen, and it is easily programmed for a calculator or computer.

Comments

Numerical integration is often needed in thermodynamic calculations. Many of the approximations traditionally used in thermodynamics were originally invoked to avoid numerical integrations; however, the use of the computer makes these approximations unnecessary, for accurate numerical integration is easily done. We will still use some approximate solutions in many cases simply because they aid in understanding.

The data in the table above were generated by using the function $P = 10/V^{1.3}$, so it is possible to obtain an analytical solution for the work, which is, from Eq. (2.22),

$$_1W_2 = 10\int_{V_1}^{V_2} \frac{1}{V^{1.3}}\,dV = \frac{P_0V_0 - P_NV_N}{1.3-1}$$

$$= \frac{10\times 1 - 0.501\times 10}{0.3} = 16.63\ \text{kJ}$$

so the error in the numerical integration is

$$\text{Percentage of error} = \frac{|16.63 - 17.60|}{16.63}\times 100 = 5.8\ \text{percent}$$

As noted, breaking the range of integration into more ΔV increments will improve the accuracy of the numerical integration.

Work on Solids

A force applied to a solid can alter the length of the solid. The normal stress acting to extend the solid is denoted by σ_z in the direction z. The work *by* the solid is

$$\delta W = -\mathbf{F}\cdot d\mathbf{s} = -\sigma_z A_0\,dz \tag{2.23}$$

where A_0 is the initial cross-sectional area. The minus sign is required since the system (solid) restrains the motion; and if dz is positive (extension), work is done on the system. Alternatively this can be expressed in terms of the strain. The strain ε_z is the deformation per unit length $d\varepsilon_z = dz/L_0$, so

$$\delta W = -\sigma_z A_0 L_0\,d\varepsilon_z = -\sigma_z V_0\,d\varepsilon_z \tag{2.24}$$

where V_0 is the initial volume. This is integrated between two points to yield

$$_1W_2 = -\int_1^2 \sigma_z V_0\,d\varepsilon_z \tag{2.25}$$

Another form of work that can be done on a solid is torsional work, i.e., the work done to twist one end of a rod through an angular displacement of $d\theta$ while holding the other end fixed (Fig. 2.6). This is done by applying a force F normal to the rod at some distance r from the rod centerline. The product of the force and the distance is called the *torque* τ, and the resulting work is given by

$$\delta W_{\text{torque}} = -Fr\,d\theta = -\tau\,d\theta \tag{2.26}$$

The work done in torsion can be used in the analysis of some automotive suspension systems.

Figure 2.6 Work done by applying torsion to a rod.

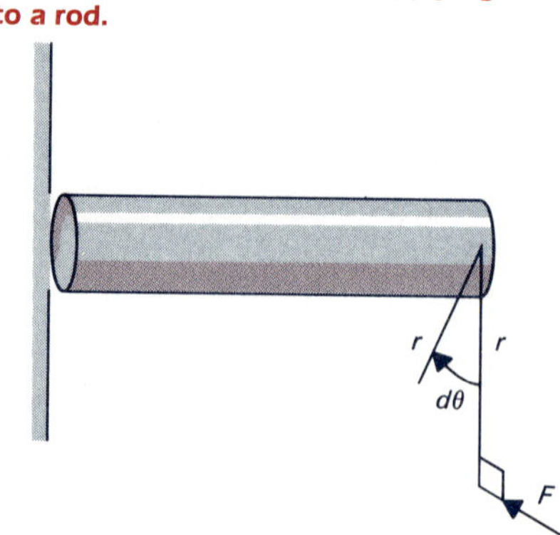

Example Problem 2.7

Calculate the work required to stretch an unstressed steel wire from 1 to 1.001 m. Young's modulus is $E = \sigma_z / \varepsilon_z = 2 \times 10^{11}$ N/m^2, and the cross-sectional area is $A_0 = 3 \times 10^{-6}$ m^2.

Solution

Equation (2.24) gives

$$\delta W = -\sigma_z A_0 L_0 \, d\varepsilon_z$$

Note that Young's modulus yields $\sigma_z = E\varepsilon_z$. Then

$$\delta W = -E A_0 L_0 \varepsilon_z \, d\varepsilon_z$$

Integrating from the initial unstressed length to the final length gives

$$_1W_2 = -\int_{\varepsilon_{z1}=0}^{\varepsilon_{z2}} E A_0 L_0 \varepsilon_z \, d\varepsilon_z$$

Since A_0 and L_0 represent the initial values and E is a constant,

$$_1W_2 = -\frac{E A_0 L_0}{2} \varepsilon_{z2}^2$$

Substituting the numerical values gives

$$_1W_2 = -\left[\frac{1}{2}(2 \times 10^{11}) \frac{\text{N}}{\text{m}^2}\right](3 \times 10^{-6}\ \text{m}^2)(1\ \text{m}) \left(\frac{0.001}{1}\right)^2$$

$$= -0.3\ \text{J} = -0.22\ \text{ft} \cdot \text{lbf}$$

Work to Stretch a Film

This type of work arises in stretching a two-dimensional sheet such as a soap film on a wire frame. The force is the surface tension times the length of the line where the force is acting, L. The surface tension is denoted by σ_t, N/m or lbf/ft, so that the force is $\sigma_t L$. Work performed *by* the system composed of the film is

$$\delta W = -\sigma_t L \, dz \tag{2.27}$$

The sign results from exactly the same ideas as in the case of extending a solid. The change in the surface area of the film as it is being stretched is $dA = L \, dz$, so

$$\delta W = -\sigma \, dA \tag{2.28}$$

and the work *by* the film for a finite change is

$$_1W_2 = -\int_1^2 \sigma \, dA \tag{2.29}$$

Work by Electric Energy

We have indicated that a possible work mode is in the form of electric energy flowing across the system boundary. The potential difference **E** results in a transfer of charge $d\mathbf{Q}$. The process yields work on the system as

$$\delta W = -\mathbf{E}\, d\mathbf{Q} \tag{2.30}$$

for this quasi-equilibrium process. With the current i represented as $i = d\mathbf{Q}/dt$, where t is time,

$$\delta W = -\mathbf{E} i\, dt \tag{2.31}$$

so the work done is

$$_1W_2 = -\int_1^2 \mathbf{E} i\, dt \tag{2.32}$$

Example Problem 2.8

An automotive battery is charged with a battery charger. The charger operates for 1 h at 12 V and a current of 25 A. Calculate the work done on the battery.

Solution

The work is given by Eq. (2.32) as

$$W = -\int_1^2 \mathbf{E} i\, dt$$

The potential difference and current are taken to be constants for the charging period, so

$$_1W_2 = -\mathbf{E} i(t_2 - t_1)$$

Noting that 1 V = 1 N·m/C = 0.7376 ft·lbf/C and 1 A = 1 C/s, then

$$\begin{aligned} _1W_2 &= -(12)(0.7376 \text{ ft}\cdot\text{lbf/C})(25 \text{ C/s})(1 \text{ h})(3600 \text{ s/h}) \\ &= -796{,}600 \text{ lbf}\cdot\text{ft} = -1080 \text{ kJ} \end{aligned}$$

Comment

The work is negative indicating that the charger does work on the battery. Note that the *discharge* of the battery will occur at a somewhat lower measured potential than required for charging, because of the presence of internal resistances in the battery.

Other Work Forms

Other forms of work that can be done by a system include *work* done by the system by a change in the magnetic field

$$\delta W_{\text{mag}} = -\mu_0 V\mathbf{H} \cdot d\,\mathbf{M} \tag{2.33}$$

where μ_0 is the magnetic permeability of a vacuum, V is the system volume, $\mathbf{H}$ is the magnetic field strength, and $\mathbf{M}$ is the magnetization of the system per unit volume.

Work can be done by an electric field to polarize a dielectric material. This *polarization work* is given by

$$\delta W_{\text{pol}} = -V\mathbf{E} \cdot d\mathbf{P} \tag{2.34}$$

where V is the system volume, $\mathbf{E}$ is the electric field strength, and $\mathbf{P}$ is the polarization of the system per unit volume, or the electric dipole moment of the system.

General Work Expressions

A number of mechanisms for energy transfer in the form of work have been discussed. This list is not complete, and other work modes exist for other quasi-equilibrium processes. The common feature of all the work modes is that work is the product of a generalized force, which is an intensive property, and a generalized displacement, which is an extensive or specific intensive property. This can be represented mathematically as

$$\delta W_i = \mathbf{F}_i\, d\mathbf{X}_i \tag{2.35}$$

where δW_i = work mode i

$\mathbf{F}_i$ = generalized force i

$\mathbf{X}_i$ = generalized displacement i

A general system could have many possible work modes, so a general work expression is

$$\begin{aligned} \delta W_{\text{gen}} &= \sum_i \delta W_i = \sum_i \mathbf{F}_i\, d\mathbf{X}_i \\ &= P\,dV - \sigma_x V\,d\varepsilon_x - \sigma\,dA - \mathbf{E}\,d\mathbf{Q} + \cdots \end{aligned} \tag{2.36}$$

2.5.2 Heat Transfer

Energy that may cross the system boundary because of a temperature difference falls into a separate category. When a can of frozen orange juice is placed in a warm room, experience shows us that the orange juice (the system) will increase in temperature until it approaches the temperature of the room. The state of the system is changing because energy from the surroundings is crossing the system boundary owing to the temperature difference between the room and the juice. This energy transfer is not work, since we cannot envision a way in which its sole effect is the raising of a weight. We thus define this disorganized form of energy that crosses the system boundary because of a temperature difference between the system and its surroundings as *heat transfer.* Heat transfer is given the symbol Q, and the heat transfer per unit mass is $q = Q/m$. The SI unit for heat transfer is, as for work, the joule; in USCS, it is the pound-force–foot (lbf·ft). An alternative energy unit in the USCS is the British thermal unit, or Btu. The original definition of the Btu was the necessary amount of heat transfer to raise 1 lbm of water through a temperature change of 1 °F, starting at some given temperature. This definition has been abandoned in favor of the relation 1 Btu ≡ 778.16 ft·lbf = 1.0550 kJ. This definition is preferred because both work and heat transfer must have the same energy units.

In this text, values for heat transfer in various problems either are specified in the problem statement or are an unknown to be calculated from the thermodynamics. The calculation of the amount of energy entering or leaving a system as it depends on various mechanisms and the temperature difference between the system and its surroundings is outside the realm of classical thermodynamics and is the subject of courses in heat transfer or transport processes.

An important distinction between heat transfer and work is the "disorganization" on the molecular level. Heat transfer is viewed as a disorganized energy transfer on a molecular level. This energy transfer results from molecular activity which is not directly useful to raise a weight. Work, however, is viewed as an organized energy transfer to raise a weight. This difference is discussed in more detail in Chap. 13.

As noted in the discussion on work, any consistent convention can be chosen for assigning a sign to the direction of energy transfer. In this text, the sign convention for heat transfer differs from that for work; i.e., a heat transfer is positive *into* the system and negative *out of* the system. This can be remembered since a heat loss, being undesirable in many problems, is considered negative.

An impotant yet limiting process for heat transfer is the *adiabatic* process. The process permits no heat transfer. It occurs only in an idealized case, but is approached in very well-insulated systems and in cases where heat transfer is small compared with other energy transfers or changes.

2.5.3 Power

To design practical devices, not only is it important to know how much energy is needed to drive a device or how much energy can be obtained from the device, but also we need to know the rate at which energy can be obtained. It may be possible to run a very small gasoline engine to move freight, but a much larger engine would allow the movement to be done more quickly. The rate at which energy is transferred is called the *power*. We denote the rate of various energy forms by placing a dot over the symbol; i.e., the rate of doing work is

$$\dot{W} = \lim_{\Delta t \to 0} \frac{\delta W}{\Delta t} \tag{2.37}$$

and the rate of heat transfer across a boundary is

$$\dot{Q} = \lim_{\Delta t \to 0} \frac{\delta Q}{\Delta t} \tag{2.38}$$

The overdot implies the rate of transfer, not the time derivative of a system property. In SI, the unit of power is the watt (W), which is equal to an energy rate of 1 J/s. The common units of power in the USCS are the Btu/h, the lbf·ft/s, or the horsepower (550 lbf·ft/s).

2.6 Energy—What Is It?

To this point, we have used the term *energy* without giving it a careful definition. Energy can have many forms, and we are so used to working with the concept of energy that we have an intuitive feel for its meaning.

In mechanics, energy is often defined as the capacity for doing work. As we will see during our study of the second law of thermodynamics, such a definition is not adequate in thermodynamics.

What attributes have we discovered in our definitions of the particular energy forms of internal energy, kinetic energy, and potential energy and the energy transfers of work and heat transfer that might help with a more general definition? One thing that the energy *transfers* have in common is that each has the capacity to change the state of the system for which it is defined. If we do work on a system or a heat transfer to a system occurs, then the state of the system is changed (assuming that we do not add, subtract, or change more than one of the energy forms at a time so that their effects cancel).

Thus we can define *energy* as a system quantity that describes the thermodynamic state of the system and which is changed by transfers to and from the system. Energy has many forms, and it can be used in any of these forms. In the remaining chapters of this text, we see how the properties of materials in a system are changed by the energy transfers; how energy can be transformed from one form to another; what the limitations are on these transformations; and how we can make use of the ability to transform energy to help us in designing useful devices.

Problems

2.1S A force of 3000 N is exerted uniformly over a plate of area 3 cm^2 at an angle of 30° to the normal. What is the pressure, in pascals, exerted on the plate?

2.1E A force of 700 lbf is exerted uniformly over a plate of area 0.5 in^2 at an angle of 30° to the normal. What is the pressure, in psia, exerted on the plate?

2.2S The pressure of a gas in a tank is desired. A manometer is attached to the tank, and the level of the fluid in the manometer is measured. When the manometer is filled with mercury ($\rho = 13{,}550$ kg/m^3), the level reading is $h = 2$ m. Evaluate the absolute and gauge pressures in the tank. What would be the level reading if the fluid were water ($\rho = 1000$ kg/m^3)? Take the gravitational acceleration as 9.8 m/s^2.

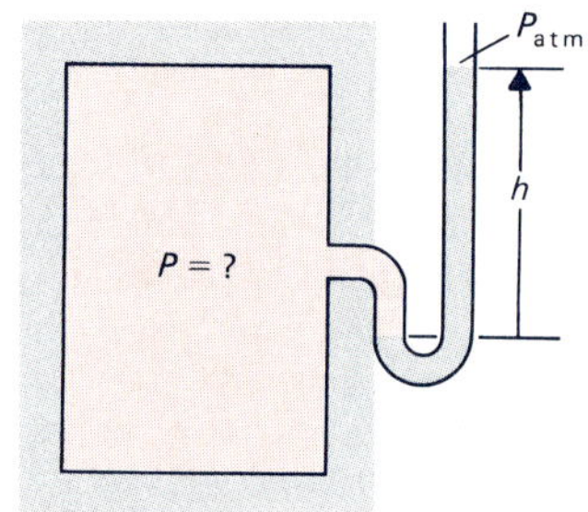

Figure P2.2

2.2E The pressure of a gas in a tank is desired. A manometer is attached to the tank, and the level of the fluid in the manometer is measured. When the manometer is filled with mercury ($\rho = 833$ lbm/ft^3), the level reading is $h = 6$ ft. Evaluate the absolute and gauge pressures in the tank. What would be the level reading if the fluid were water ($\rho = 62.4$ lbm/ft^3)? Take the gravitational acceleration as 32.2 ft/s^2.

2.3S The gauge pressure of H_2O is measured to be 5 lbf/in^2 (vacuum); i.e., 5 psi below zero gauge pressure. If atmospheric pressure is 0.100 MPa, calculate the absolute pressure of the H_2O in psia.

2.3E The gauge pressure of H_2O is measured to be 30 kPa (vacuum); i.e., 30 kPa below zero gauge pressure. If atmospheric pressure is 14.7 psia, calculate the absolute pressure of the H_2O in kilopascals.

2.4 The exit pressure from a steam turbine is measured to be 12 in Hg vacuum. If $P_{atm} = 101.3$ kPa, calculate the absolute pressure of the H_2O.

2.5 Attached to the containers shown in Fig. P2.5 are two pressure gauges. Determine the absolute pressure in compartment 2.

Figure P2.5

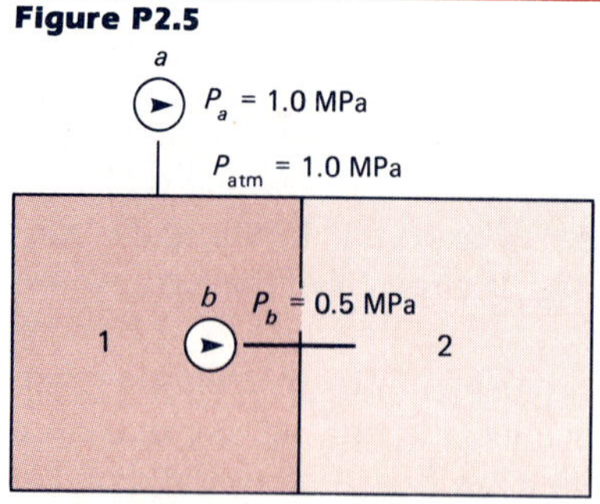

2.6S Given the piston-cylinder device shown in Fig. P2.6S, determine the absolute pressure inside the device.

2.6E Given the piston-cylinder device shown in Fig. P2.6E, determine the absolute pressure inside the device.

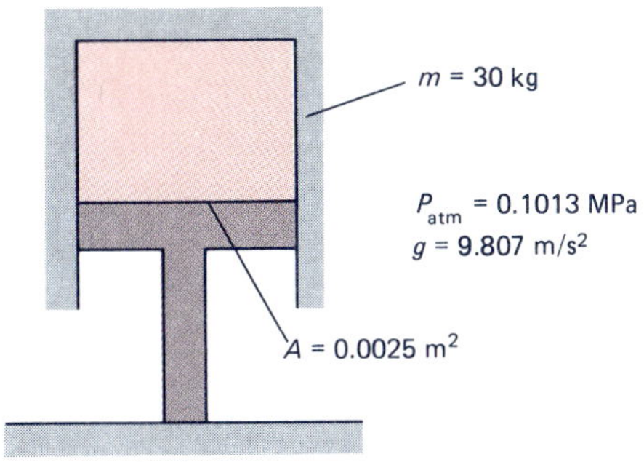

Figure P2.6S

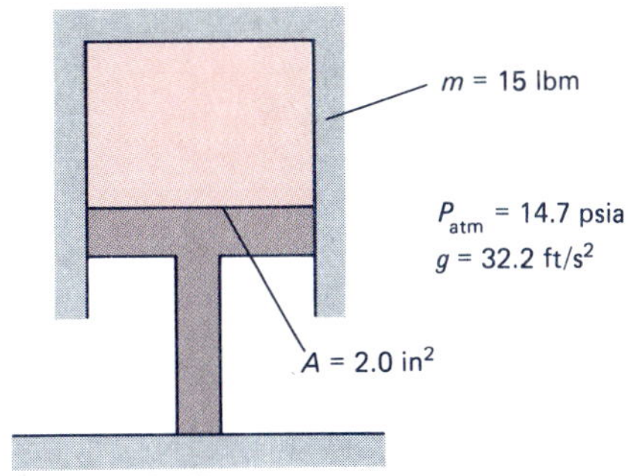

Figure P2.6E

2.7S A cylinder which weighs 1000 N on earth is transported to another planet. The new planet has no atmosphere but has an acceleration due to gravity of 4.5 m/s^2. Determine the absolute pressure of the air enclosed by the cylinder and piston shown.

2.7E A cylinder which weighs 2000 lbf on earth is transported to another planet. The new planet has no atmosphere but has an acceleration due to gravity of 15 ft/s^2. Determine the absolute pressure of the air enclosed by the cylinder and piston shown.

Figure P2.7S

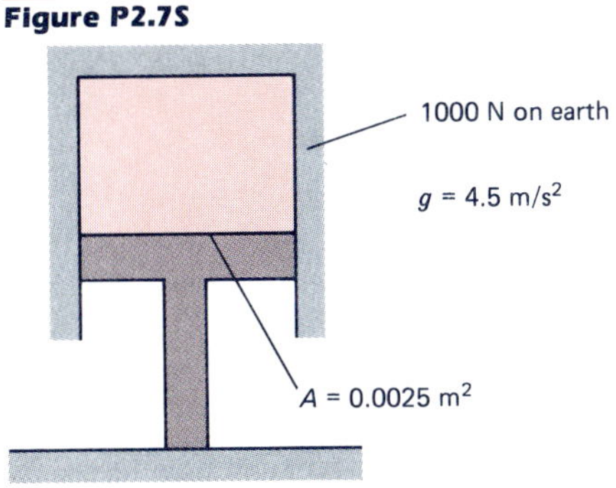

Figure P2.7E

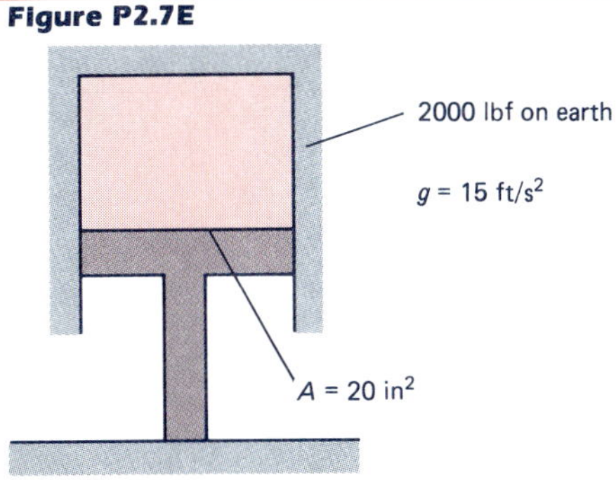

2.8S If the absolute pressure inside of the piston-cylinder device shown in Fig. P2.8S is 2 MPa, determine the mass of the piston.

2.8E If the absolute pressure inside of the piston-cylinder device shown in Fig. P2.8E is 30 psia, determine the mass of the piston.

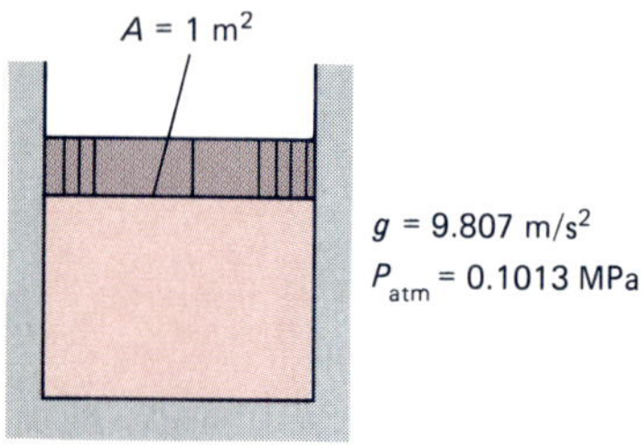

Figure P2.8S

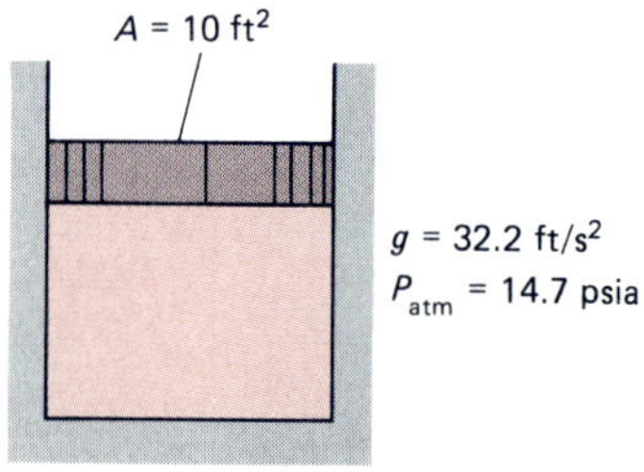

Figure P2.8E

2.9S In the piston-cylinder device shown in Fig. 2.9S, determine the force necessary to produce an absolute pressure of 5 MPa within the device.

2.9E In the piston-cylinder device shown in Fig. 2.9E, determine the force necessary to produce an absolute pressure of 750 psia within the device.

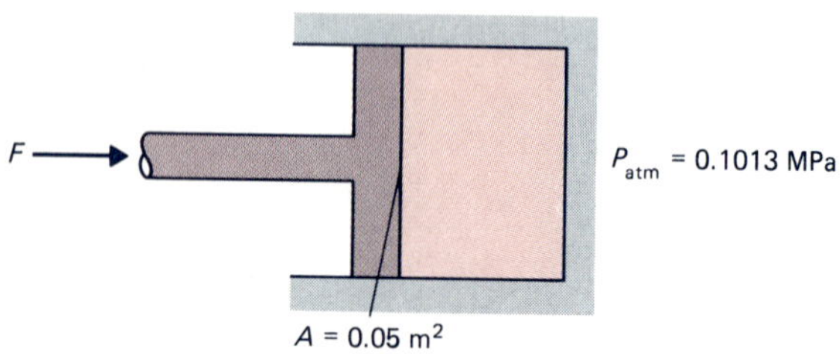

Figure P2.9S

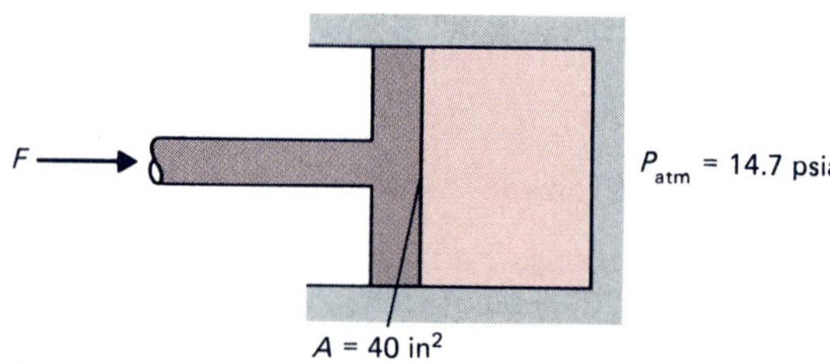

Figure P2.9E

2.10S A person starts an outing at sea level where the pressure is 0.101 MPa. Calculate the final pressure for each of the following excursions.

(*a*) The person climbs to the top of Mt. Everest where the elevation is 8848 m. Take the average air density between sea level and the mountaintop to be 0.754 kg/m³.

(*b*) The person dives to the bottom of a sea where the depth is 395 m. Take the water density to be 1000 kg/m³.

(*c*) The person travels to the bottom of Death Valley where the elevation is 86 m below mean sea level. Take the air density to be 1.30 kg/m³.

Neglect the fluid density variations and local gravity variations for each excursion.

2.10E A person starts an outing at sea level where the pressure is 14.7 psia. Calculate the final pressure for each of the following excursions.

(*a*) The person climbs to the top of Mt. Everest where the elevation is 29,028 ft. Take the average air density between sea level and the mountaintop to be 0.050 lbm/ft³.

(*b*) The person dives to the bottom of a sea where the depth is 1296 ft. Take the water density to be 62.4 lbm/ft³.

(*c*) The person travels to the bottom of Death Valley where the elevation is 282 ft below mean sea level. Take the air density to be 0.0772 lbm/ft³.

Neglect the fluid density variations and local gravity variations for each excursion.

2.11S A cylinder encloses a gas with a piston as shown. The area of the piston is 0.01 m². Take the atmospheric pressure to be 0.101 MPa and the local gravitational acceleration as 9.8 m/s². If the piston supports a mass of 50 kg (including the mass of the piston), what is the gas pressure? Will the gas pressure change if the gas volume beneath the piston is doubled?

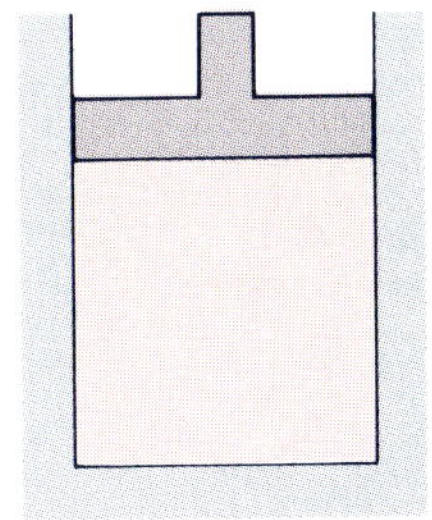

Figure P2.11

2.11E A cylinder encloses a gas with a piston as shown. The area of the piston is 0.1 ft². Take the atmospheric pressure to be 14.7 psia and the local gravitational acceleration as 32.2 ft/s². If the piston supports a mass of 100 lbm (including the mass of the piston), what is the gas pressure? Will the gas pressure change if the gas volume beneath the piston is doubled?

2.12S If the density of H_2O in a container is 600 kg/m³, calculate the mass of the H_2O if the container has a total volume of 70 cm³.

2.12E If the density of H_2O in a container is 62.4 lbm/ft³, calculate the mass of the H_2O if the container has a total volume of 30 in³.

2.13S The 50-kg piston shown in Fig. P2.13S is free to move in the vertical cylinder and has a diameter of 12 cm. If atmospheric pressure is equal to 0.1 MPa and the acceleration due to gravity is 9.81 m/s², determine the absolute pressure of the H_2O.

2.13E The 100-lbm piston shown in Fig. P2.13E is free to move in the vertical cylinder and has a diameter of 5 in.

Figure P2.13S

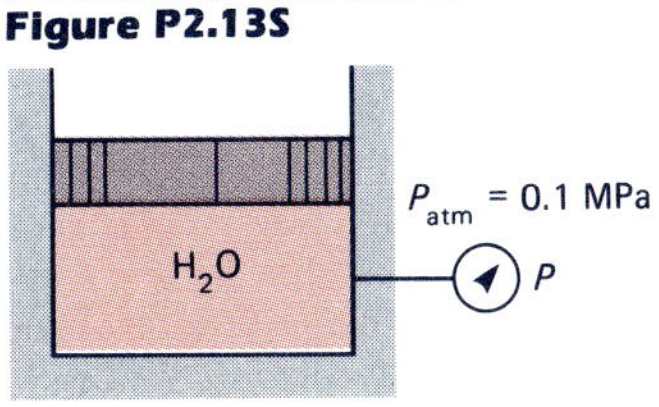

Figure P2.13E

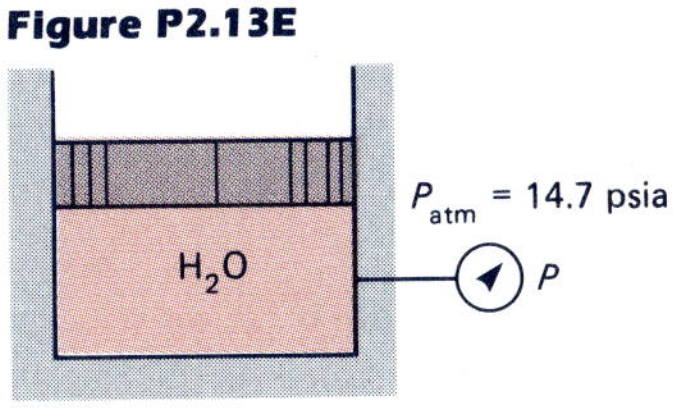

If atmospheric pressure is equal to 14.7 psia and the acceleration due to gravity is 32.2 ft/s², determine the absolute pressure of the H_2O.

2.14S What is the weight of the H_2O occupying a volume of 0.85 m³ if $v = 0.021$ m³/kg? The acceleration due to gravity is 9.807 m/s².

2.14E What is the weight of the H_2O occupying a volume of 24 ft³ if $v = 0.016$ ft³/lbm? The acceleration due to gravity is 32.2 ft/s².

2.15S A cylindrical tank 1.25 m long and 0.40 m in diameter contains 750 g of H_2O. Calculate the specific volume and density of the H_2O.

2.15E A cylindrical tank 10 ft long and 1.5 ft in diameter contains 2.0 lbm of H_2O. Calculate the specific volume and density of the H_2O.

2.16S What is the total volume occupied by 660 g of H_2O when its specific volume is 0.225 m³/kg?

2.16E What is the total volume occupied by 1.5 lbm of H_2O when its specific volume is 2.7 ft³/lbm?

2.17S A frictionless piston-cylinder device has a cross-sectional area of 100 cm². Find the piston mass such that the absolute pressure in the cylinder is 1.0 MPa. Atmospheric pressure is 0.1 MPa, and the local acceleration of gravity is 9.807 m/s².

2.17E A frictionless piston-cylinder device has a cross-sectional area of 15 in². Find the piston mass such that the absolute pressure in the cylinder is 150 psia. Atmospheric pressure is 14.7 psia, and the local acceleration of gravity is 32.2 ft/s².

2.18S Given the piston-cylinder shown in Fig. 2.18S, determine the absolute pressure inside the device.

2.18E Given the piston-cylinder shown in Fig. 2.18E, determine the absolute pressure inside the device.

Figure P2.18S

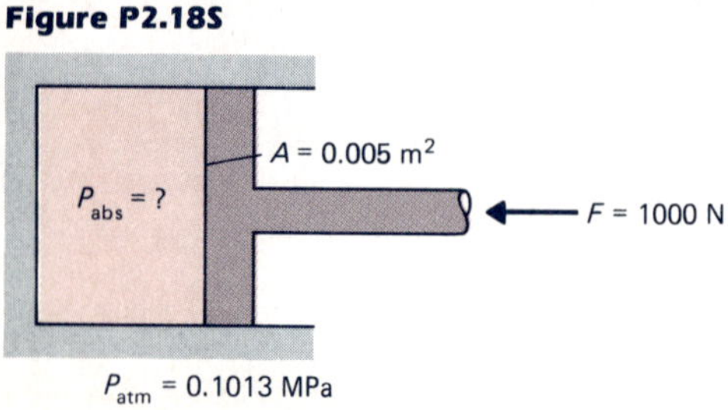

Figure P2.18E

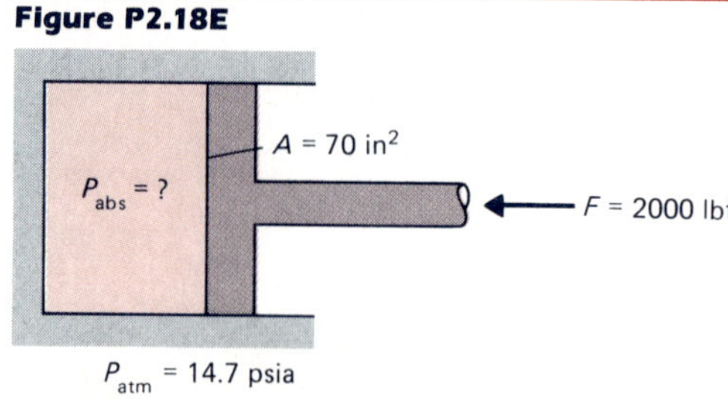

2.19S Steam in a container is separated by a membrane; the specific volume of the steam in *B* is 0.5951 m^3/kg. The membrane breaks, and the resulting specific volume is 0.7964 m^3/kg. Find the original specific volume in *A*.

2.19E Steam in a container is separated by a membrane; the specific volume of the steam in *B* is 17.352 ft^3/lbm. The membrane breaks, and the resulting specific volume is 34.741 ft^3/lbm. Find the original specific volume in *A*.

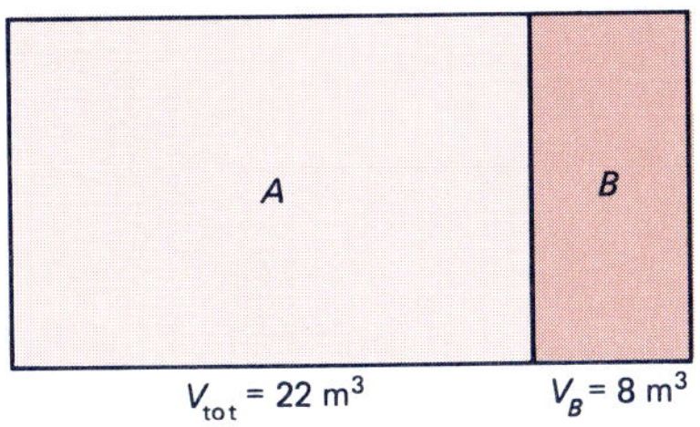

Figure P2.19S

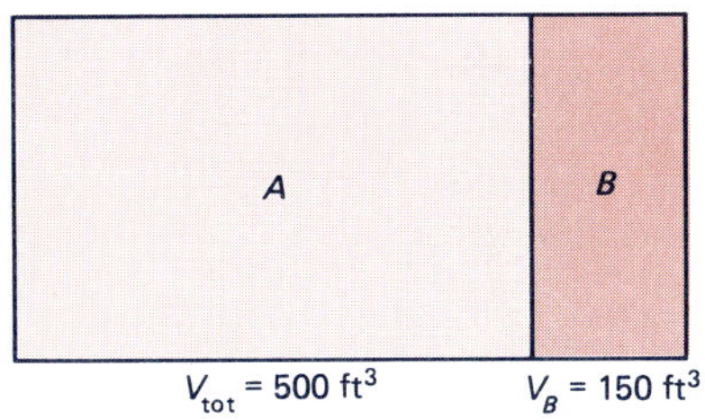

Figure P2.19E

2.20S Steam in compartments *A* and *B* is separated by a membrane; $v_A = 17.196\ m^3/kg$, and $v_B = 3.418\ m^3/kg$. If the membrane is broken, find the resulting specific volume.

2.20E Steam in compartments *A* and *B* is separated by a membrane; $v_A = 150\ ft^3/lbm$, and $v_B = 35\ ft^3/lbm$. If the membrane is broken, find the resulting specific volume.

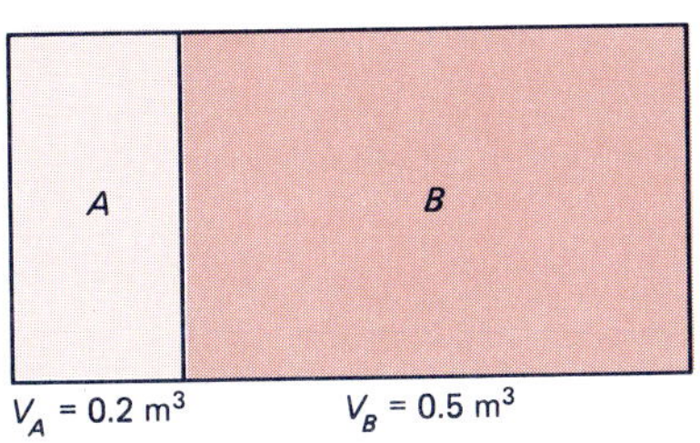

Figure P2.20S

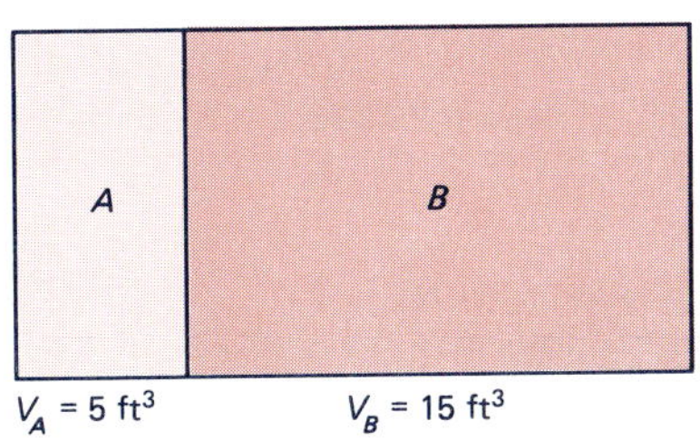

Figure P2.20E

2.21S Some 4 kg of H_2O is contained in the piston-cylinder device shown in Fig. 2.21S. The initial volume enclosed is 2500 cm^3. If the piston rises, causing the total volume to double, calculate the final specific volume of H_2O.

2.21E Some 10 lbm of H_2O is contained in the piston-cylinder device shown in Fig. 2.21E. The initial volume enclosed is 150 in^3. If the piston rises, causing the total volume to double, calculate the final specific volume of H_2O.

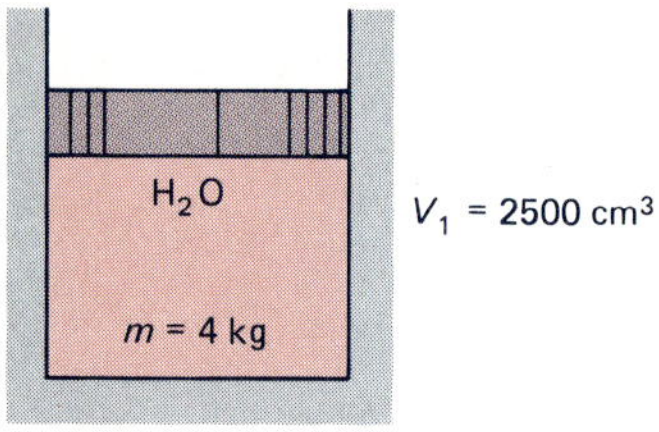

Figure P2.21S

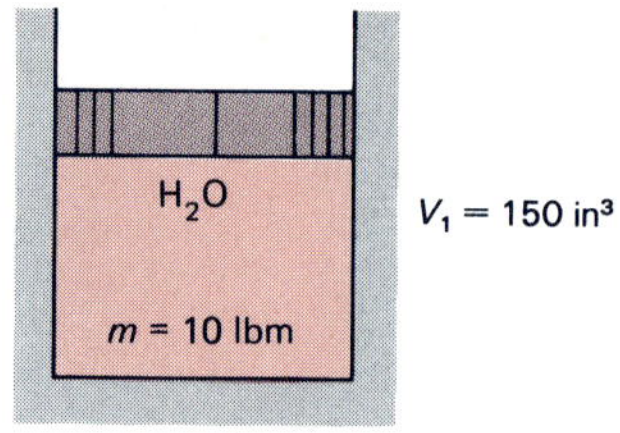

Figure P2.21E

2.22S The device shown in Fig. 2.22S has a piston with holes positioned between the two chambers. If the piston is moved so that x is one-fourth of the entire length, determine the final mass of air in chambers A and B.

2.22E The device shown in Fig. 2.22E has a piston with holes positioned between the two chambers. If the piston is moved so that x is one-fourth of the entire length, determine the final mass of air in chambers A and B.

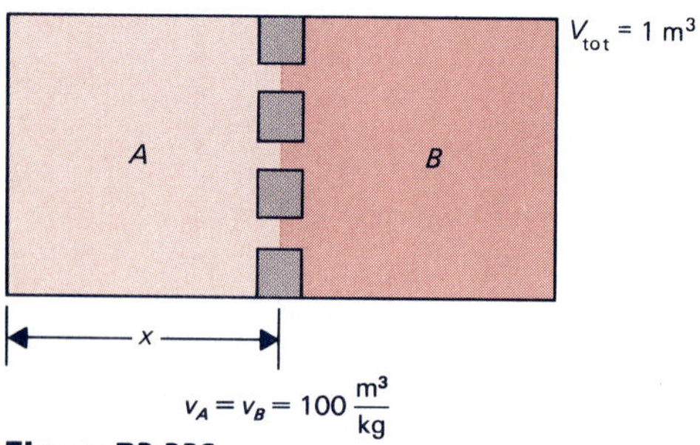

Figure P2.22S

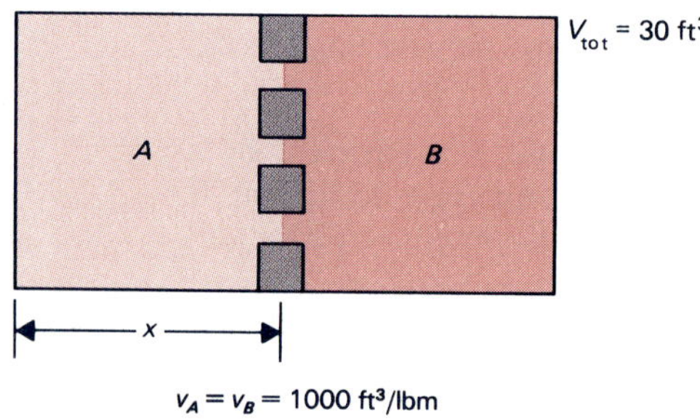

Figure P2.22E

2.23S The device shown in Fig. P2.23S has a free-moving piston between the two chambers. The initial total volumes of A and B are equal with $v_A = 100\ \text{m}^3/\text{kg}$ and $v_B = 50\ \text{m}^3/\text{kg}$. If the piston is moved so that x is one-fourth of the entire length, determine the final specific volumes of chambers A and B.

2.23E The device shown in Fig. P2.23E has a free-moving piston between the two chambers. The initial total volumes of A and B are equal with $v_A = 1000\ \text{ft}^3/\text{lbm}$ and $v_B = 500\ \text{ft}^3/\text{lbm}$. If the piston is moved so that x is one-fourth of the entire length, determine the final specific volumes of chambers A and B.

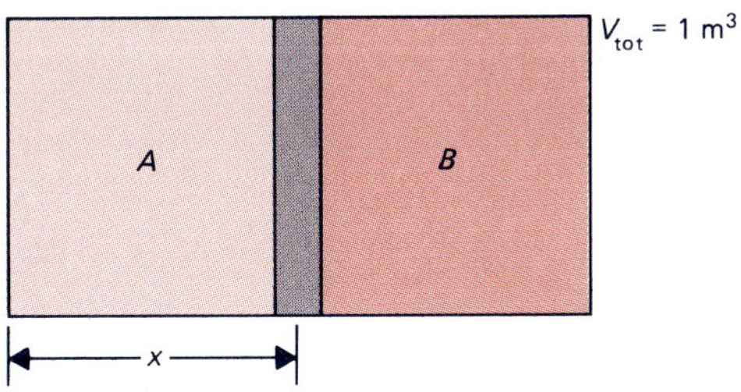

Figure P2.23S

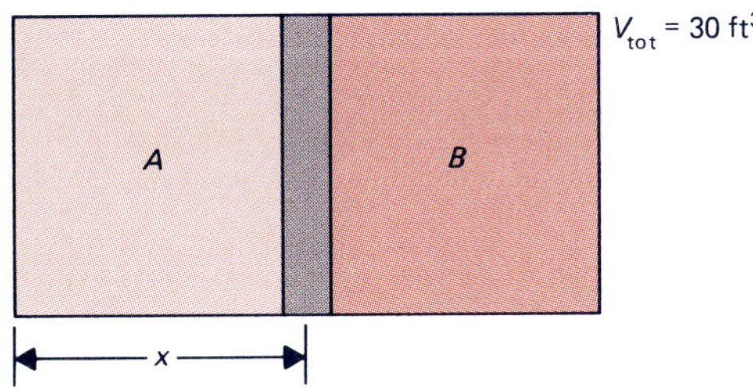

Figure P2.23E

2.24S What is the density of the unknown fluid? The piston is weightless. The piston diameter is 10 cm.

2.24E What is the density of the unknown fluid? The piston is weightless. The piston diameter is 4 in.

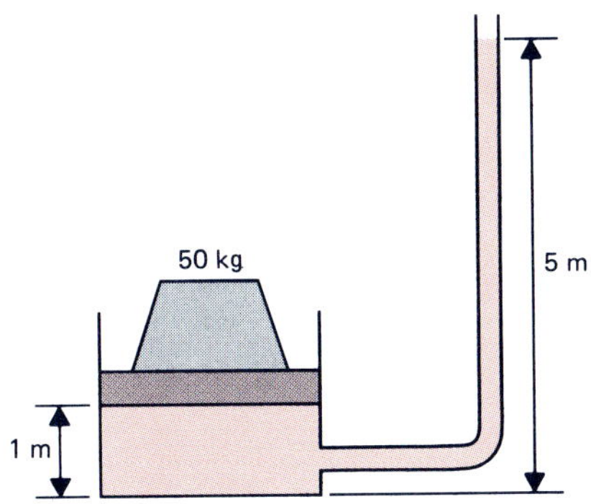

Figure P2.24S

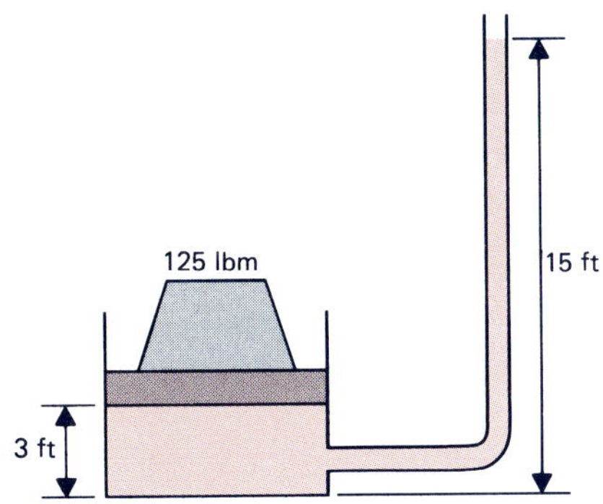

Figure P2.24E

2.25S What is the difference in the pressure readings between gauges A and B?

Figure P2.25S

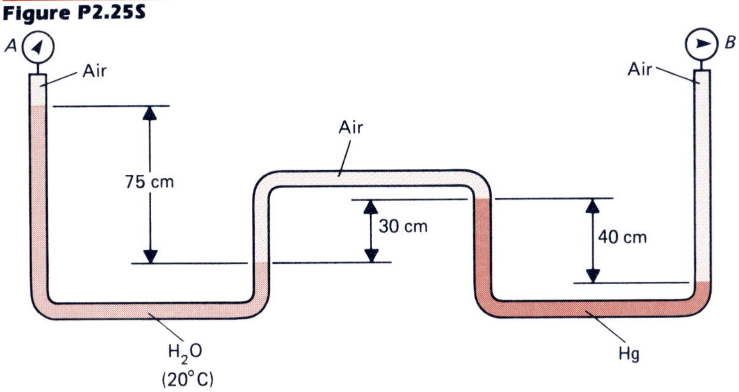

2.25E What is the difference in the pressure readings between gauges A and B?

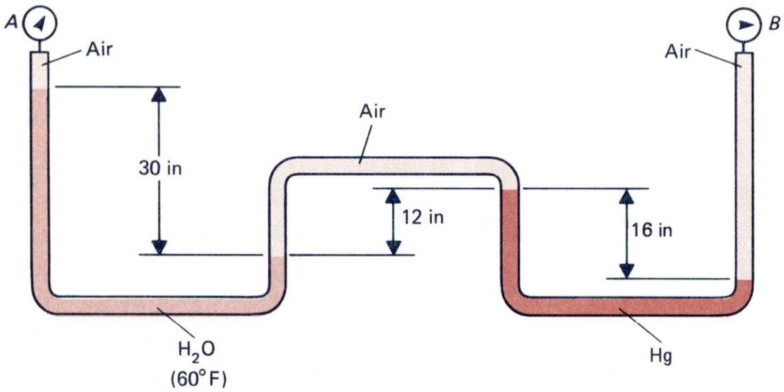

Figure P2.25E

2.26S Kerosene at 20°C flowing in a pipe has a pressure drop of 55.16 kPa across an orifice. What will be the height of the mercury deflection in the differential manometer? The specific volume of kerosene is 1.67×10^{-3} m³/kg.

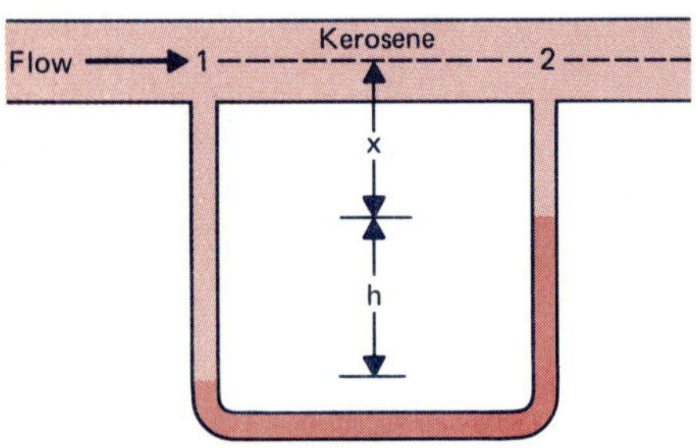

Figure P2.26

2.26E Kerosene at 60°F flowing in a pipe has a pressure drop of 8 psi across an orifice. What will be the height of the mercury deflection in the differential manometer? The specific volume of kerosene is 0.027 ft³/lbm.

2.27S Container x, holding 0.5 m³ of light oil, is held in equilibrium by container y, holding benzene as shown in Fig. P2.27S. If the area of piston A is 2.8 m² and the area of piston B is 1.4 m², how many cubic meters of benzene are in container y? Neglect the weight of the container and the pistons.

2.27E Container x, holding 15 ft³ of light oil, is held in equilibrium by container y, holding benzene as shown in Fig. P2.27E. If the area of piston A is 26 ft² and the area of piston B is 13 ft², how many cubic feet of benzene are in container y? Neglect the weight of the container and the pistons.

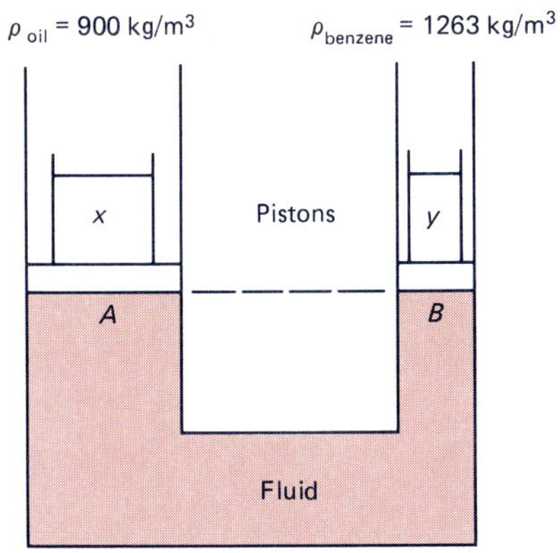

Figure P2.27S

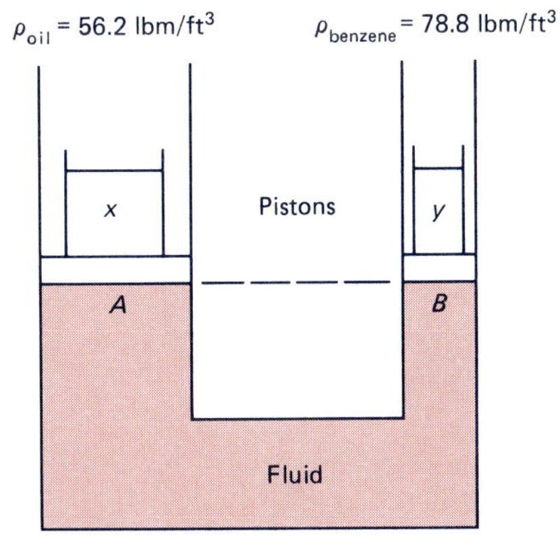

Figure P2.27E

2.28S The Empire State Building in New York City is 380 m high above the street. The static pressure in a waterline at the top of the building must be 170 kPa. What is the required pressure in the water main 10 m below the street?

2.28E The Empire State Building in New York City is 1170 ft high above the street. The static pressure in a waterline at the top of the building must be 25 psi. What is the required pressure in the water main 30 ft below the street?

2.29S The sketch shows a sectional view through a submarine. Calculate the depth of submergence y. Assume the density of seawater is 1030 kg/m³.

Figure P2.29S

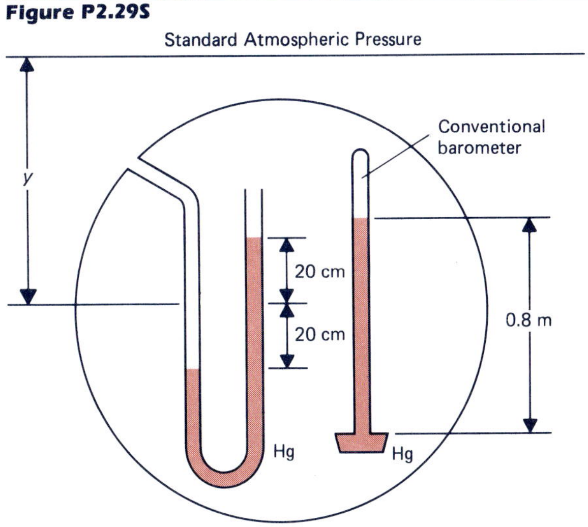

2.29E The sketch shows a sectional view through a submarine. Calculate the depth of submergence y. Assume the density of seawater is 64.5 lbm/ft³.

2.30S What is the difference in pressure between the surface of a lake and a point 20 m below the surface?

2.30E What is the difference in pressure between the surface of a lake and a point 60 ft below the surface?

2.31 A mercury barometer reads 29.90 in. A pressure gauge on a steam turbine indicates a pressure of 380 kPa. What is the absolute pressure of the steam?

2.32S A mercury manometer is to be used to measure the pressure drop between two points along a horizontal pipe in which water flows. The maximum pressure drop to be measured is expected to be 140 kPa. What is the corresponding manometer deflection? Assume $\rho =$ 13,500 kg/m³ for mercury.

2.32E A mercury manometer is to be used to measure the pressure drop between two points along a horizontal pipe in which water flows. The maximum pressure drop to be measured is expected to be 25 psia. What is the corresponding manometer deflection? Assume $\rho = 842$ lbm/ft³ for mercury.

2.33S Find the three gauge readings L_1, P_2, and L_3 if the gauge pressure at A is 200 kPa.

2.33E Find the three gauge readings L_1, P_2, and L_3 if the gauge pressure at A is 30 psia.

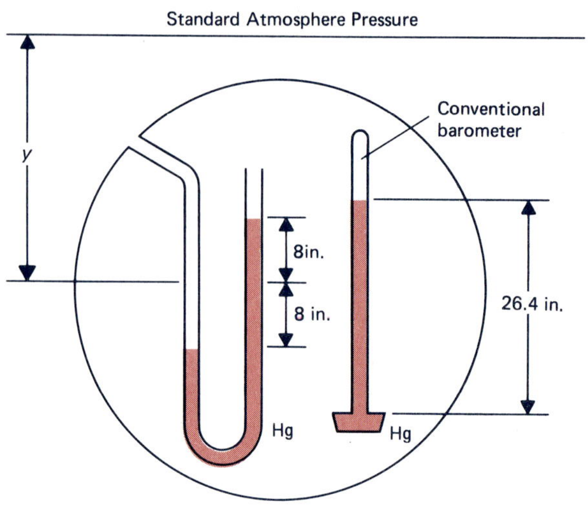

Figure P2.29E

Figure P2.33S

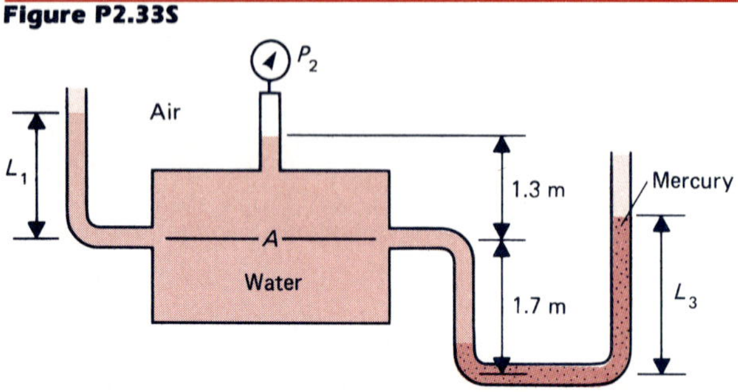

Figure P2.33E

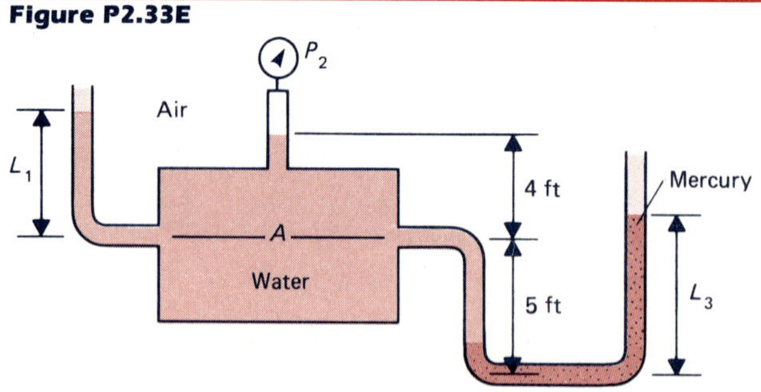

2.34 What is the density of fluid A. Elevations are in meters.

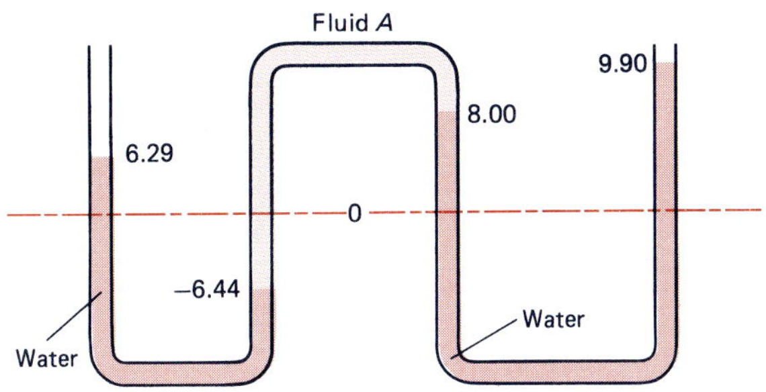

Figure P2.34

2.35 Calculate the absolute pressure in the tank above the water surface. Barometric pressure is 30 in Hg. The temperature of the water and mercury is 20°C.

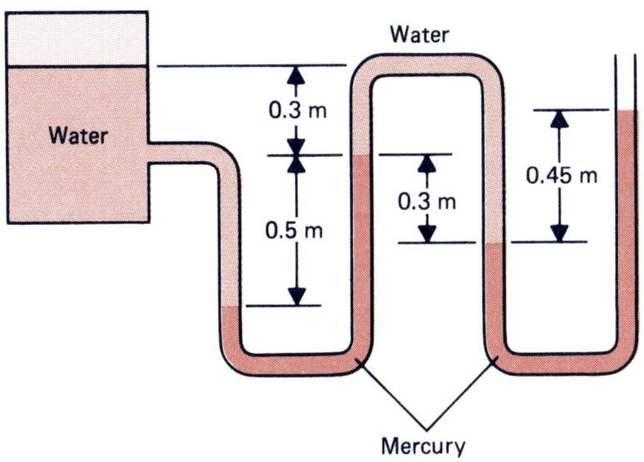

Figure P2.35

2.36 Calculate the pressure gauge reading. The density of the oil is 850 kg/m^3. The barometric pressure is 29.75 in Hg.

Figure P2.36

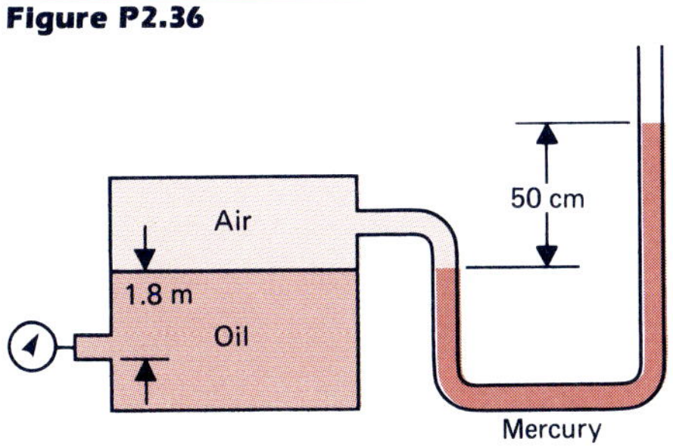

2.37S Calculate the elevation of point *A*. (Elevation is given in meters.)

2.37E Calculate the elevation of point *A*. (Elevation is given in feet.)

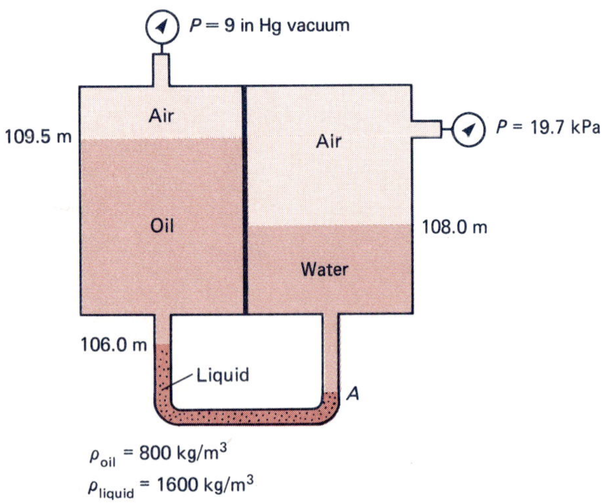

Figure P2.37S

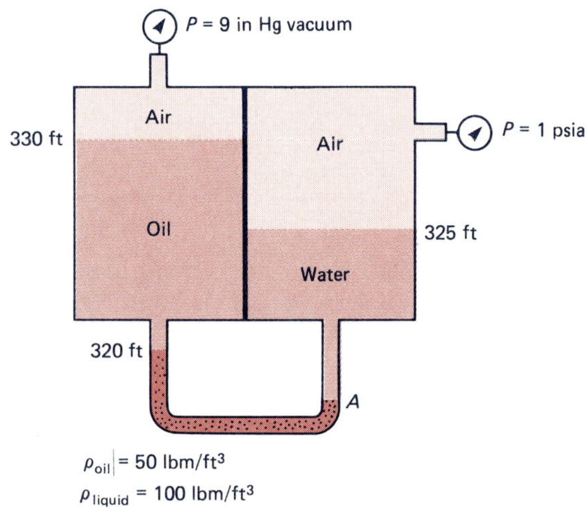

Figure P2.37E

2.38 Determine an expression for the unknown mass density.

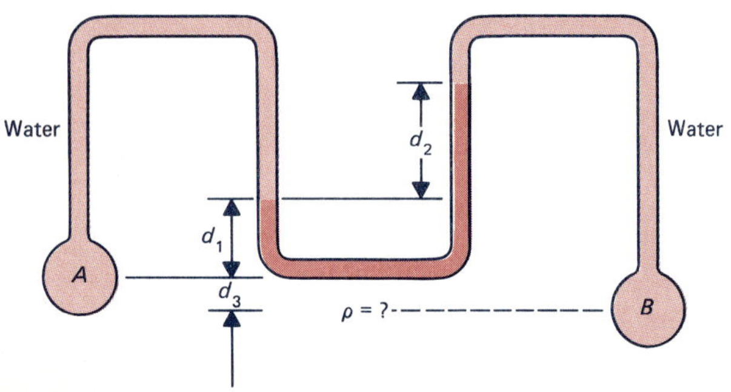

Figure P2.38

2.39 Determine the pressure difference between tanks *A* and *B*, $(P_A - P_B)$.

2.40 Water flows down an 8-cm-diameter pipe as shown in Fig. P2.40S. Attached over the top of the pipe is a mercury manometer. Knowing that the pressure in the main pipe drops 60 kPa in going from point 1 to point 2, determine the height of mercury in the manometer *h*.

Figure P2.39

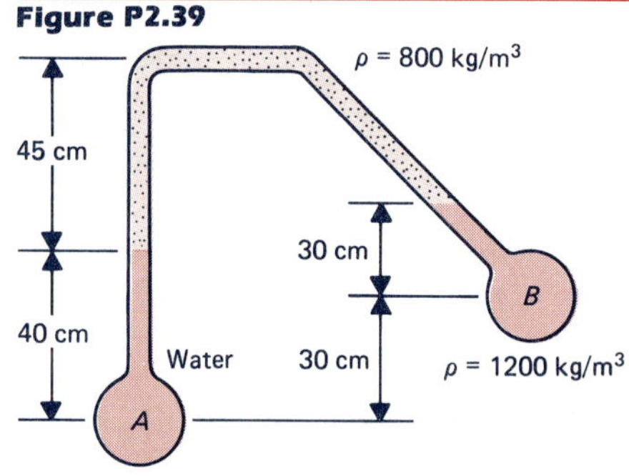

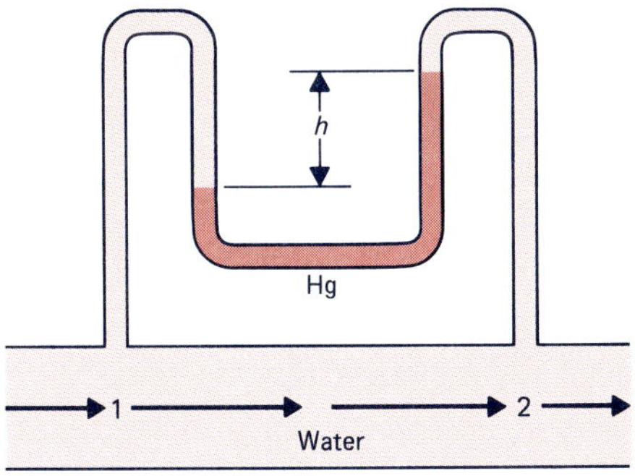

Figure P2.40

2.41 What is the pressure difference between points A and B in terms of h's and ρ's, where ρ is the density.

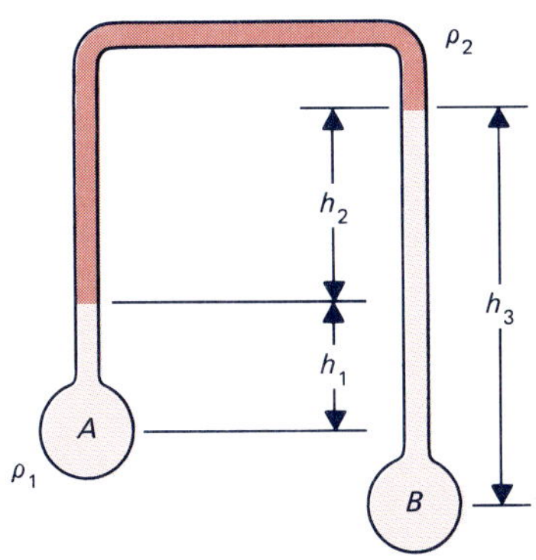

Figure P2.41

2.42 Calculate the distance A.

Figure P2.42

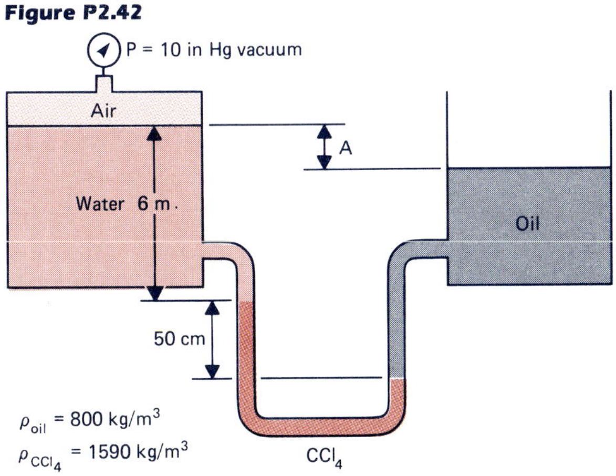

2.43S Calculate the pressure gauge reading.

2.43E Calculate the pressure gauge reading.

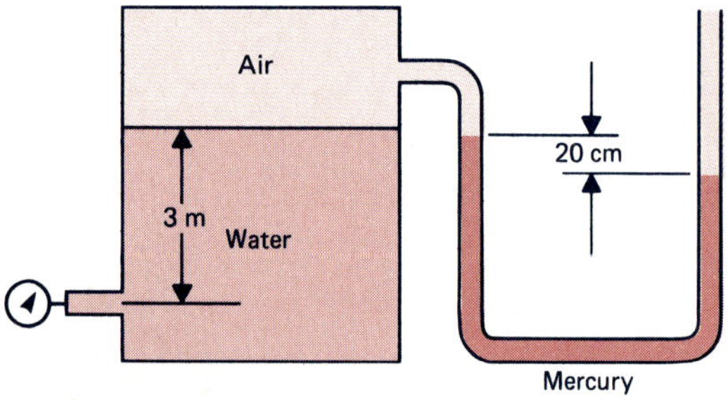

Figure P2.43S

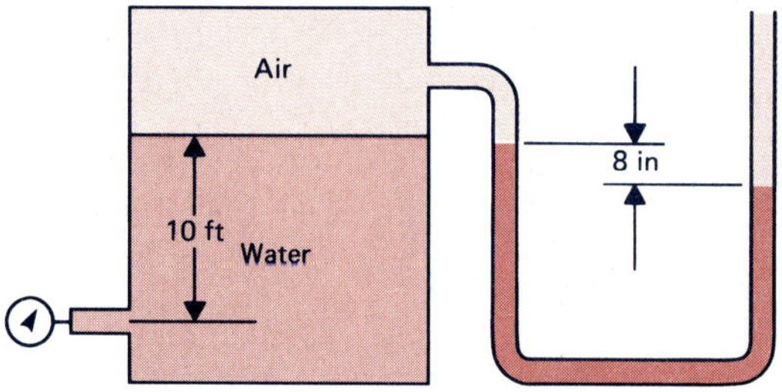

Figure P2.43E

2.44S What is the density of the unknown fluid?

Figure P2.44S

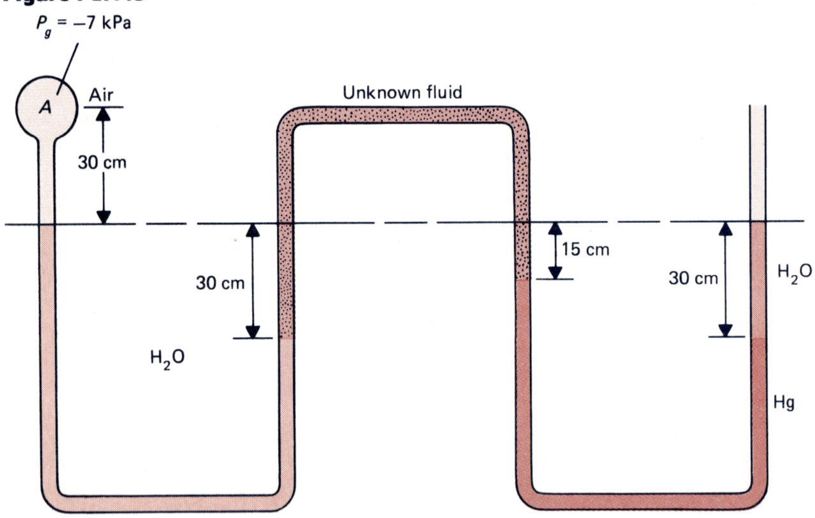

2.44E What is the density of the unknown fluid?

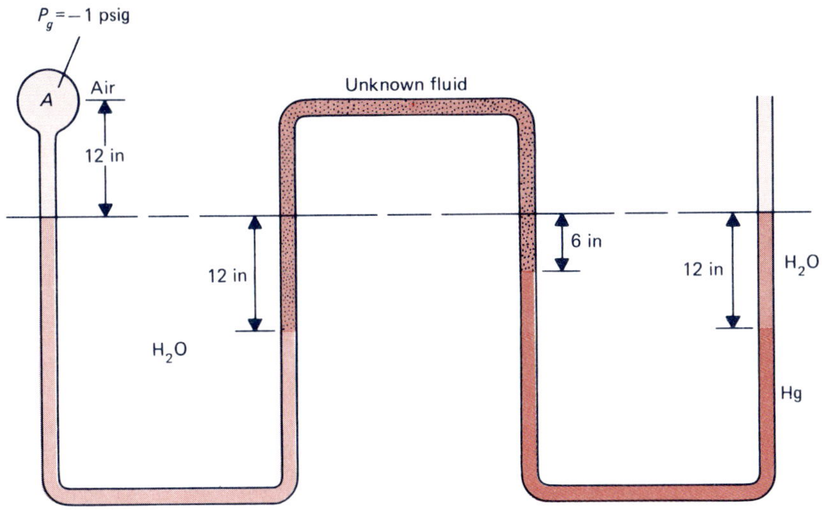

Figure P2.44E

2.45 From the manometer reading R and the depth L, compute the depth of water in the diving bell.

Figure P2.45

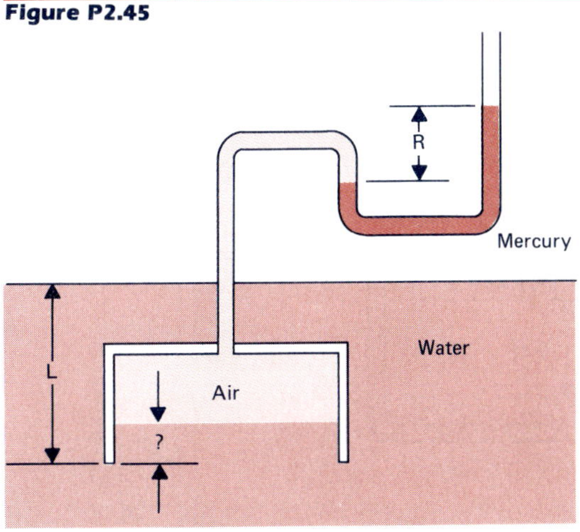

2.46S Determine the pressure gauge reading at A.

2.46E Determine the pressure gauge reading at A.

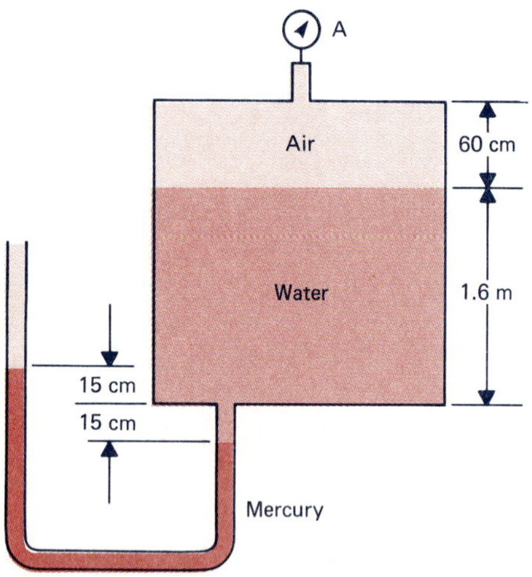

Figure P2.46S

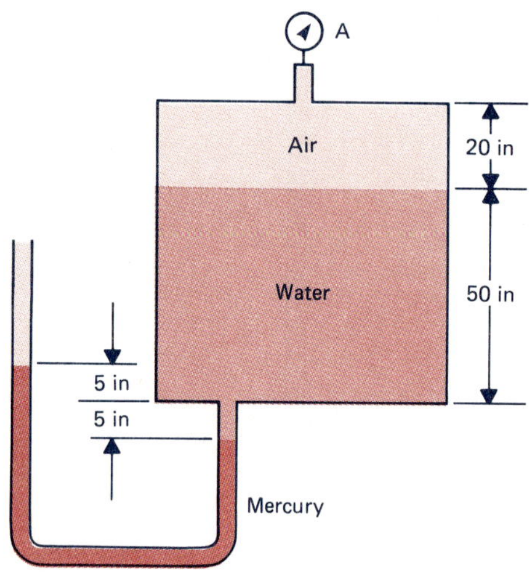

Figure P2.46E

2.47S What is the difference in pressure between points A and B?

2.47E What is the difference in pressure between points A and B?

Figure P2.47S

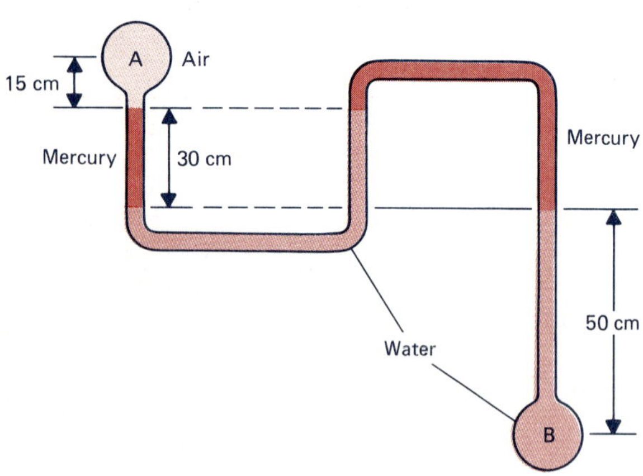

Figure P2.47E

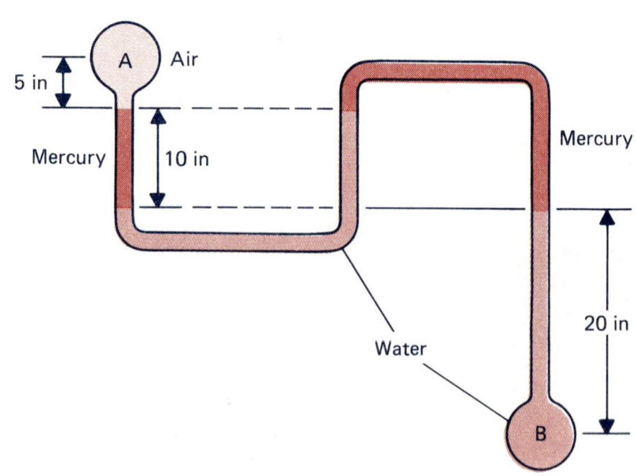

2.48 The device is located on the moon where g is one-sixth that on earth. What is the height h of the mercury column?

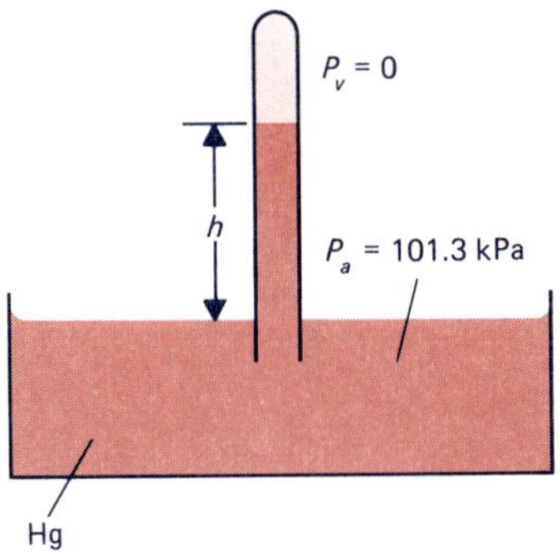

Figure P2.48

2.49S What is the reading on the pressure gauge?

2.49E What is the reading on the pressure gauge?

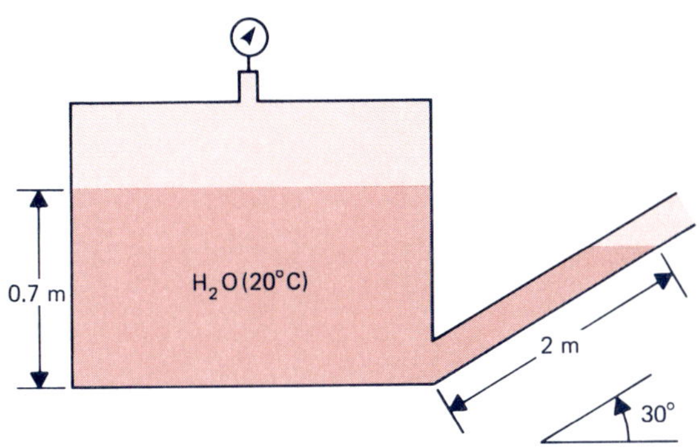

Figure P2.49S

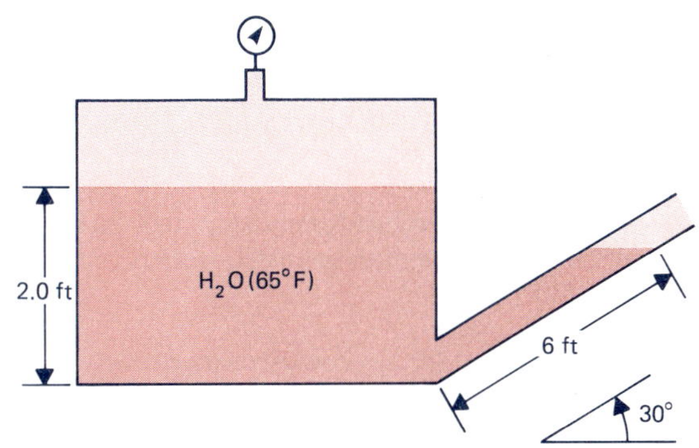

Figure P2.49E

2.50S The two pistons are weightless and have diameters of 0.15 m. What is the distance l?

Figure P2.50S

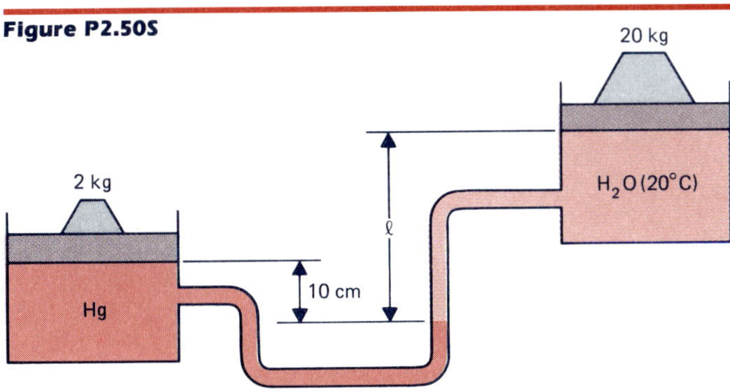

2.50E The two pistons are weightless and have diameters of 6 in. What is the distance l?

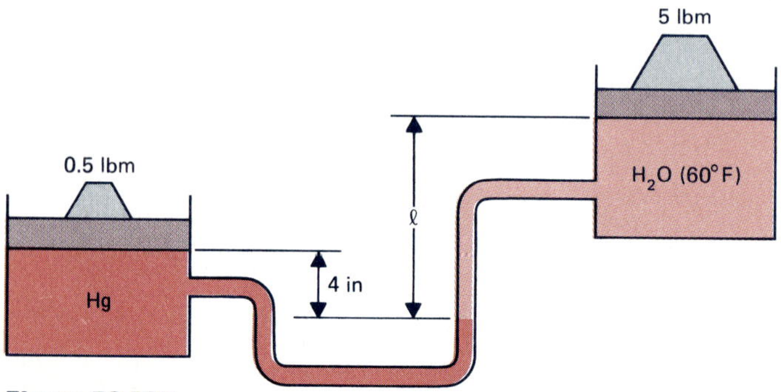

Figure P2.50E

2.51 How would the height of a water barometer on a warm summer day at 40°C compare with that on a cold day at 10°C, assuming the same atmospheric pressure exists on both days? Explain.

2.52S In a given process, the absolute temperature of a fluid doubles. What is the final temperature if the initial temperature was 50°C?

2.52E In a given process, the absolute temperature of a fluid doubles. What is the final temperature if the initial temperature was 110°F?

2.53S A colleague reports that she has developed a new material which will withstand an absolute temperature 3 times as great as her previous best material, which could endure a temperature of 1200 K. What is this new upper limit in kelvins and degrees Celsius?

2.53E A colleague reports that he has developed a new material which will withstand an absolute temperature 3 times as great as his previous best material, which could endure a temperature of 2000°R. What is this new upper limit in degrees Rankine and Fahrenheit?

2.54 At the beginning of a process, the thermometer reads 15°F. You want to stop the experiment when the absolute temperature has doubled. What is the final temperature in degrees Celsius?

2.55S A mass of 2 kg is moving at a speed of 3 m/s at a height 10 m above a reference plane. What are the values of the kinetic and potential energy of the mass, and what is the weight of the mass?

2.55E A mass of 5 lbm is moving at a speed of 10 ft/s at a height 30 ft above a reference plane. What are the values of the kinetic and potential energy of the mass, and what is the weight of the mass?

2.56S The measured pressure within a piston-cylinder arrangement similar to that diagrammed in Prob. 2.8 is 0.5 MPa, and the piston area is 0.08 m². Determine the mass supported by the gas.

2.56E The measured pressure within a piston-cylinder arrangement similar to that diagrammed in Prob. 2.8 is 75 psia, and the piston area is 2.5 ft². Determine the mass supported by the gas.

2.57S The 10 g of water in the piston-cylinder device shown undergoes a constant-pressure process. The initial pressure is 1.0 MPa, and the initial volume is 2000 cm³. If the final volume is 10 cm³, find the work done. Is the work done on or by the system?

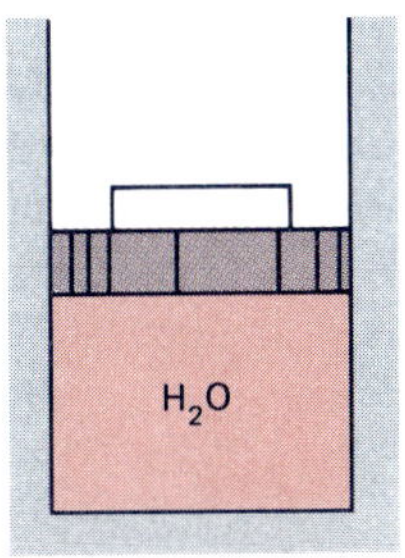

Figure P2.57

2.57E The 0.2 lbm of water in the piston-cylinder device shown undergoes a constant-pressure process. The initial pressure is 150 psia, and the initial volume is 0.5 ft³. If the final volume is 0.002 ft³, find the work done. Is the work done on or by the system?

2.58 One kilogram of water in a closed, rigid container undergoes a process such that its temperature increases from 10 to 50°C. The volume of the container is 25,000 cm³. Find the amount of work done during the process.

2.59S Air in a system undergoes the process shown on the *P-v* diagram in Fig. P2.59S. Find the work done. Is it work done on or by the system? What would the work be if the medium were changed to H_2O?

2.59E Air in a system undergoes the process shown on the *P-v* diagram in Fig. P2.59E. Find the work done. Is it work done on or by the system? What would the work be if the medium were changed to H_2O?

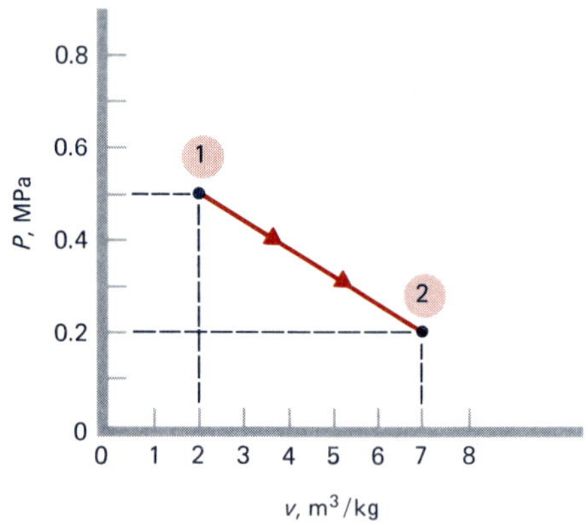

Figure P2.59S

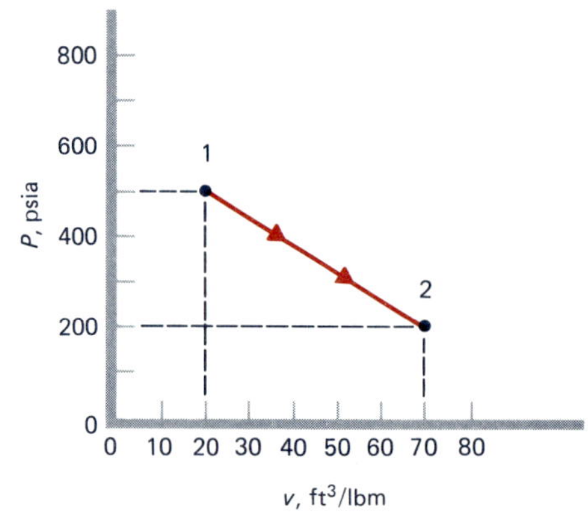

Figure P2.59E

2.60 An ideal gas ($PV = mRT$) is contained within a piston-cylinder system as shown. The piston is weightless and frictionless. The atmospheric pressure outside the cylinder is 1 atm (101.3 kPa). Initially, a spring inside the cylinder is held in a compressed state by the piston to a length of 5 m. The spring is then allowed to expand, slowly pushing the piston outward against a restraining force $F(y)$ until the spring reaches a length of 10 m, where it is fully extended and just barely touches the piston. Throughout this process, the gas in the cylinder is held at constant temperature. The spring constant is 1×10^3 kN/m, and the spring has negligible volume. [Take R for the gas as 0.2870 J/(kg·K) (air).]

(*a*) Sketch the *P-V* diagram for the gas for the process described, showing numerical values for *P* and *V* at each end of the process.

(*b*) Calculate the work done by the piston on the spring-gas system.

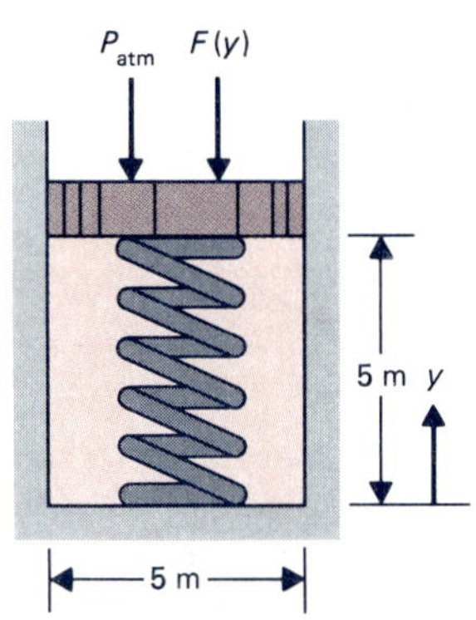

Figure P2.60

2.61 Air with mass of 1 kg and initially at $P = 101.3$ kPa and $T = 300$ K is contained within a cylinder as shown in Fig. P2.61. The cup on top of the piston is then filled at a constant rate until it contains 50 kg of water, thus compressing the gas. The compression proceeds slowly, so that heat transfer to the surroundings maintains the temperature of the air at 300 K. Assume air is an ideal gas.

(*a*) Draw a *P-V* diagram of the process. Label the initial state *A* and the final state *B*.

(*b*) Calculate the work done on the gas during the process.

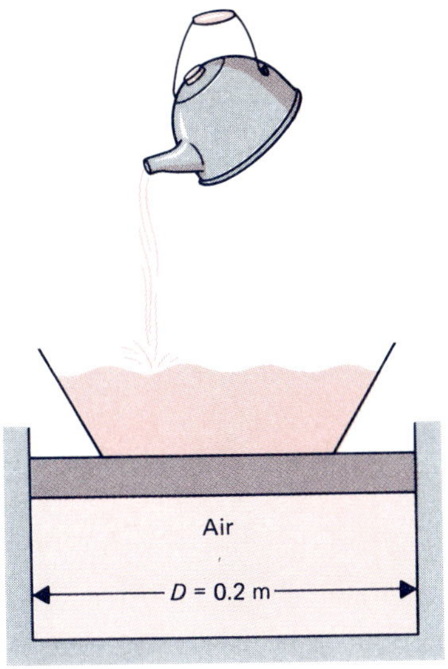

Figure P2.61

2.62 A gas is contained within a piston-cylinder arrangement. The initial pressure is 100 kPa, and the initial specific volume is 1 m^3/kg. The gas undergoes a process that obeys the relation

$$Pv^{1.4} = \text{const}$$

The final specific volume is 0.1 m^3/kg.

(*a*) What is the value of the final pressure, in kilopascals?

(*b*) Sketch the process on the *P-v* diagram.

(*c*) Calculate the work done per unit mass by the gas during the process in kilojoules per kilogram.

2.63 An ideal gas is contained within a piston-cylinder arrangement as shown. A spring exerts a linearly varying force on the piston. The work done on the piston in moving the piston from position $x = 0$, where the spring just touches the piston, to $x = 1$ m, where the spring is compressed, is 600 kJ. The gas in the cylinder is held at constant pressure of 100 kPa by varying the temperature during the compression process. The diameter of the piston face is 2 m.

(*a*) Find the value of the spring constant, in kilonewtons per meter.

(*b*) Sketch a diagram of the force exerted to move the piston versus displacement. Crosshatch the area under the resulting curve that is due to compression of gas.

Figure P2.63

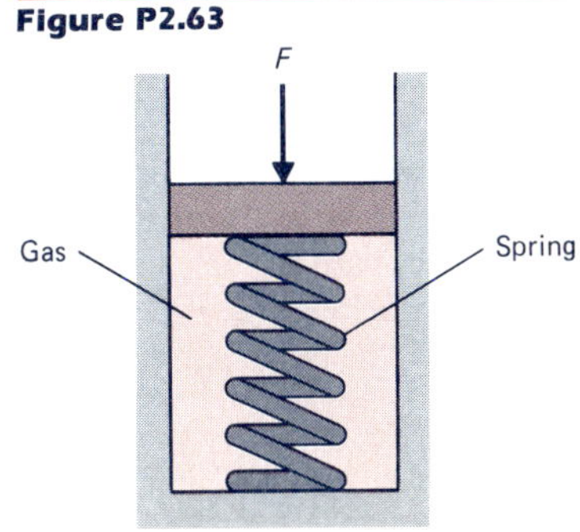

2.64 A gas initially at 100 kPa is contained within a piston-cylinder arrangement. At state 1, the gas has volume of 2 m^3. The gas is expanded isobarically to state 2, where the volume is 4 m^3. The gas pressure is then changed by heat transfer at constant volume to state 3. The gas is then compressed by a process that follows the relation

$$Pv^{1.4} = \text{const}$$

until it returns to state 1.
(*a*) Sketch the process on the *P-V* diagram below.
(*b*) Find the values of *P* and *V* at state 3.
(*c*) Find the net work done in kilojoules in completing the cycle of processes 1-2-3-1 outlined in the problem statement.

2.65 A beryllium carbide rode is 5 cm in diameter and 20 cm long. The rod is stretched by 0.1 percent. The amount of work done is 49.1 J. What is the modulus of elasticity of the rod (in kilonewtons per square meter)?

2.66S Find the work done in kilojoules, by an ideal gas in going from state *A* to state *C* along the path shown in the *P-V* diagram.

2.66E Find the work done, in Btu, by an ideal gas in going from state *A* to state *C* along the path shown in the *P-V* diagram.

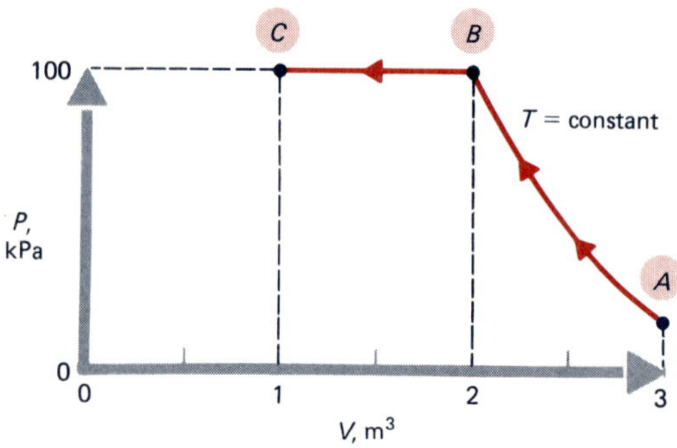

Figure P2.66S

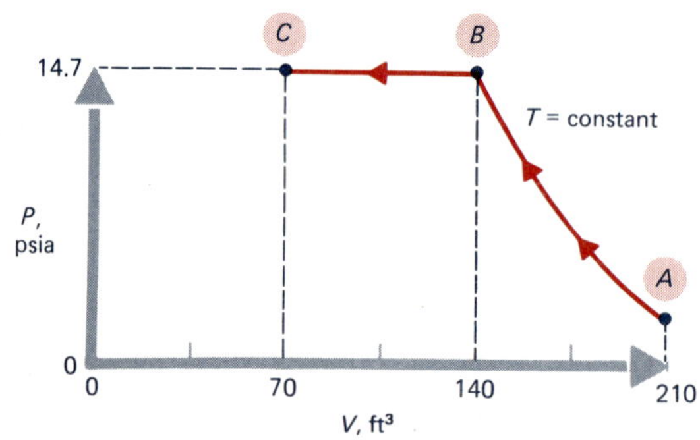

Figure P2.66E

2.67S A 25-kg piston is required to travel a specific distance within the piston-cylinder arrangement shown in Fig. P2.67S. The piston initially rests on the bottom stops. The gas within the cylinder is heated, and the piston moves until it touches the upper stops, as shown. The initial pressure and temperature are $P_1 = P_{atm} = 0.101$ MPa and $T_1 = 20°C$ (neglect piston volume).
(*a*) Draw the *P-V* diagram for the process.
(*b*) Calculate the work done by the gas.
(*c*) If the gas is an ideal gas ($PV = mRT$), determine the final temperature of the gas.

2.67E A 50-lbm piston is required to travel a specific distance within the piston-cylinder arrangement shown in Fig. P2.67E. The piston initially rests on the bottom stops. The gas within the cylinder is heated, and the piston moves until it touches the upper stops, as shown. The initial pressure and temperature are $P_1 = P_{atm} = 14.7$ psia and $T_1 = 68°F$ (neglect piston volume).

(*a*) Draw the *P-V* diagram for the process.
(*b*) Calculate the work done by the gas.
(*c*) If the gas is an ideal gas ($PV = mRT$), determine the final temperature of the gas.

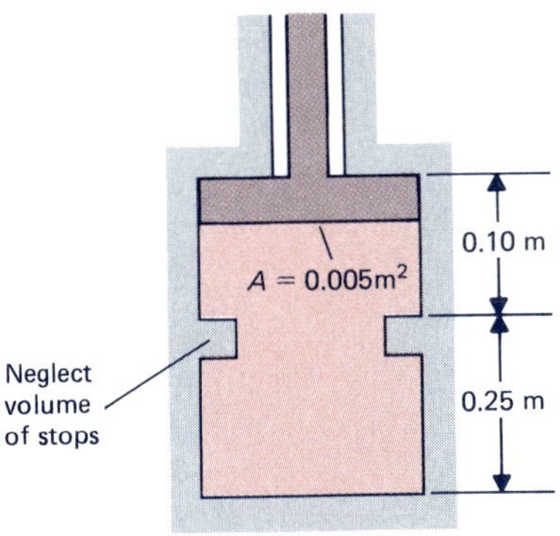

Figure P2.67S

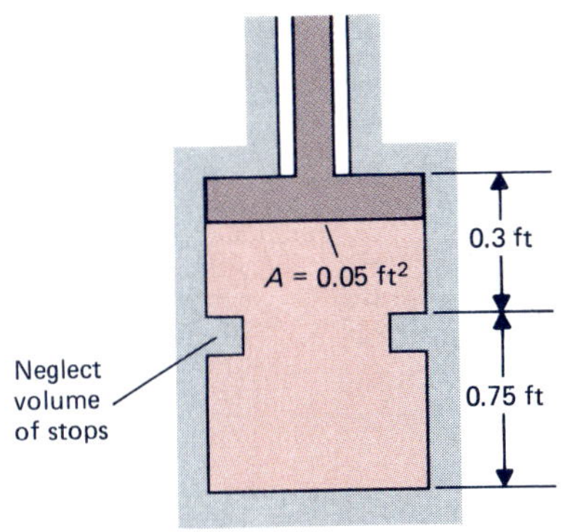

Figure P2.67E

2.68S A piston travels between two stops and is restrained by a linear spring, as shown. The cross-sectional area of the piston is 0.05 m² and the initial volume is 0.01 m³. At the top stops, the volume is 0.03 m³. The spring constant is 10 kN/m. The gas within the chamber is initially at 0.5 MPa, and the pressure required to just raise the piston from the stops is 1.0 MPa. The final pressure in the chamber is 6.5 MPa.

(*a*) Draw an accurate *P-V* diagram.
(*b*) Evaluate the work done by the gas.

Figure P2.68

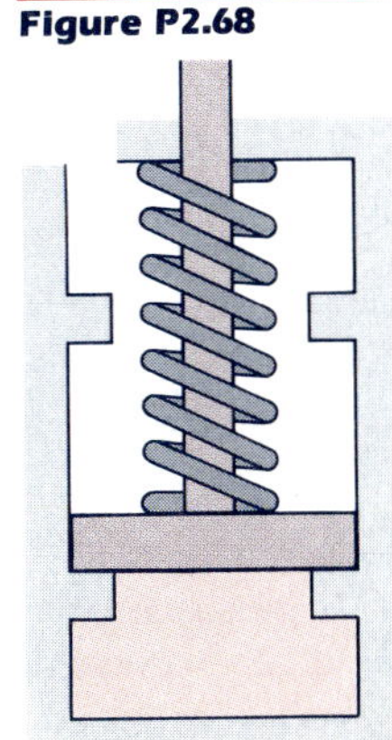

2.68E A piston travels between two stops and is restrained by a linear spring, as shown. The cross-sectional area of the piston is 0.5 ft², and the initial volume is 0.3 ft³. At the top stops, the volume is 0.9 ft³. The spring constant is 750 lbf/ft. The gas within the chamber is initially at 75 psia, and the pressure required to just raise the piston from the stops is 150 psia. The final pressure in the chamber is 975 psia.

(*a*) Draw an accurate *P-V* diagram.
(*b*) Evaluate the work done by the gas.

2.69S A weightless and frictionless piston moves 10 cm against an ideal gas (i.e., a gas with properties that obey $Pv = RT$) contained in a cylinder that has an inside diameter of 15 cm. A spring is also within the cylinder, positioned as shown. The spring exerts no force on the piston in the initial position, but is touching the piston. The spring has a spring constant of 2×10^4 N/m. During the piston motion, the gas in the cylinder is maintained at 1 atm.

(*a*) How much work was done by the piston?

(*b*) What fraction of the work is done in compressing the gas?

(*c*) What fraction of the total work done by the piston is done in the first 5 cm of piston travel?

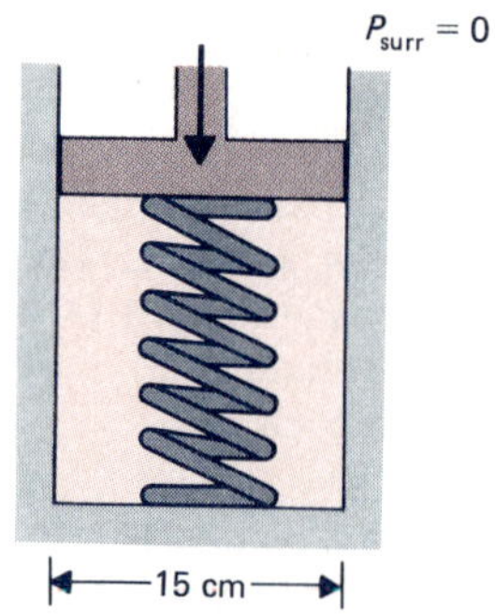

Figure P2.69S

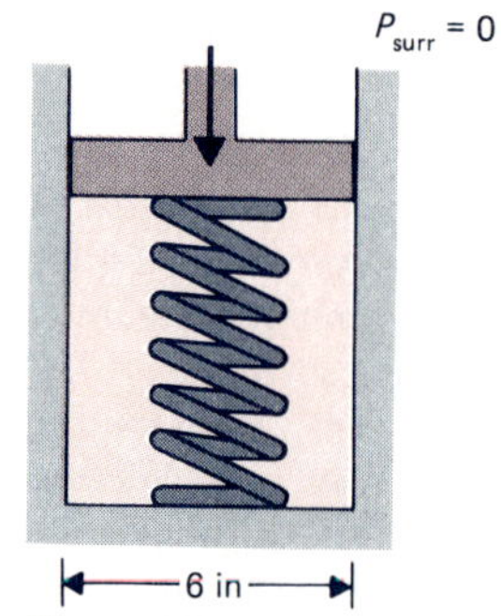

Figure P2.69E

2.69E A weightless and frictionless piston moves 4 in against an ideal gas (i.e., a gas with properties that obey $Pv = RT$) contained in a cylinder that has an inside diameter of 6 in. A spring is also within the cylinder, positioned as shown. The spring exerts no force on the piston in the initial position, but is touching the piston. The spring has a spring constant of 1500 lbf/ft. During the piston motion, the gas in the cylinder is maintained at 1 atm.

(*a*) How much work was done by the piston?

(*b*) What fraction of the work is done in compressing the gas?

(*c*) What fraction of the total work done by the piston is done in the first 2 in of piston travel?

2.70 The gas within a cylinder-piston arrangement undergoes a process with the gas pressure given by a polytropic process, that is $PV^n = \text{constant}$, where n is a specified constant. Evaluate the work done per unit mass between points A and B. [This problem seeks the derivation of Eq. (2.22).]

2.71S An aluminum rod 2 cm in diameter and 10 cm long is compressed under a load. What is the work done on the rod if the strain is 1 percent ? The modulus of elasticity is 7×10^{10} N/m^2.

2.71E An aluminum rod 1 in in diameter and 4 in long is compressed under a load. What is the work done on the rod if the strain is 1 percent? The modulus of elasticity is 1.75×10^9 lbf/ft^2.

2.72S An ideal electric motor (100 percent conversion of electric to shaft work) operates through a lossless gear train to stretch a solid aluminum rod. The motor is operated at a power input of 1 kW for 15 s, and in that interval the rod is stretched so that its length changes by 2×10^{-1} m. The rod is 1 cm^2 in cross section and 9 m in length. What is the value of Young's modulus for the rod?

2.72E An ideal electric motor (100 percent conversion of electric to shaft work) operates through a lossless gear train to stretch a solid aluminum rod. The motor is operated at a power input of 1.3 hp for 15 s, and in that interval the rod is stretched so that its length changes by 8 in. The rod is 0.16 in^2 in cross section and 30 ft in length. What is the value of Young's modulus for the rod?

2.73S An alien fungus in the form of a thin sheet is found stretched across the jaws of a shop vise in the engineering shop. The fungus is holding itself in place by the surface tension force exerted by its contact with the vise jaws. A student finds that a force of 10 N must be applied to the end of the vise handle (with radius 0.3 m—see the figure) to move the frictionless vise. She then turns the handle 24 times, causing the jaws to move 10 cm farther apart than they were initially. The vise jaws are 0.2 m in length. What is the surface tension of the alien fungus?

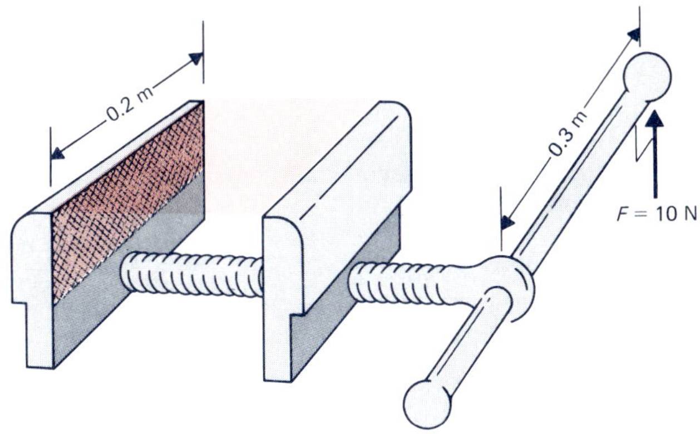

Figure P2.73S

2.73E An alien fungus in the form of a thin sheet is found stretched across the jaws of a shop vise in the engineering shop. The fungus is holding itself in place by the surface tension force exerted by its contact with the vise jaws. A student finds that a force of 2.2 lbf must be applied to the end of the vise handle (with radius 1 ft—see the figure) to move the frictionless vise. She then turns the handle 24 times, causing the jaws to move 0.3 ft farther apart than they were initially. The vise jaws are 0.7 ft in length. What is the surface tension of the alien fungus?

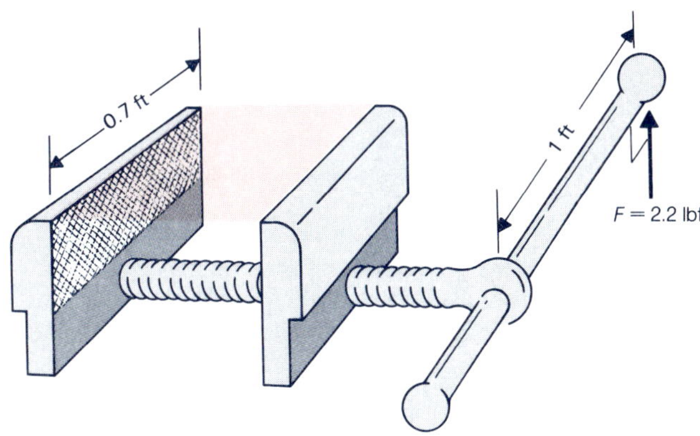

Figure P2.73E

2.74 An electric wet storage battery is used for powering the lights of a sailboat during a night race. The lights consist of three 10-W bulbs for the running lights plus a 20-W mast light. The race will last 6 h (if the present winds do not change). A battery trickle charger is available that provides a constant charge rate of 2 A to the 12-V battery while the boat is tied up at the dock. Assume that the charge-discharge cycle of the battery is 100 percent efficient (i.e., all energy stored can be recovered).

(*a*) For how long should the trickle charger be used to ensure lights for the entire race?

(*b*) How much will the energy stored in the battery change during the charging process?

2.75S An elevator has a mass of 1000 kg. It is at the top of a 60-m-high shaft when the cable snaps. The air in the elevator shaft is at 27°C and remains at that temperature as the elevator car falls. None of the air leaks past the car during its descent. The shaft is 2.5 m^2 in cross section, and the air in the shaft is initially at 1 atm. See the figure.

(*a*) Sketch the *P-v* diagram for the air in the shaft, labeling the initial and final states with numerical values.

(*b*) What will be the equilibrium position of the elevator car?

(*c*) How much work is done by the elevator car in reaching the equilibrium position?

(*d*) Are any assumptions necessary to arrive at your answer other than those given?

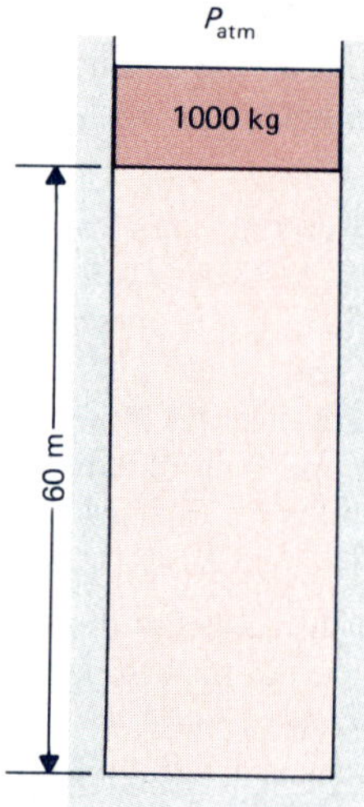

Figure P2.75S

2.75E An elevator has a mass of 2200 lbm. It is at the top of a 200-ft-high shaft when the cable snaps. The air in the elevator shaft is at 80°F and remains at that temperature as the elevator car falls. None of the air leaks past the car during its descent. The shaft is 25 ft^2 in cross section, and the air in the shaft is initially at 1 atm. See the figure.

(*a*) Sketch the *P-v* diagram for the air in the shaft, labeling the initial and final states with numerical values.

(*b*) What will be the equilibrium position of the elevator car?

(*c*) How much work is done by the elevator car in reaching the equilibrum position?

(*d*) Are any assumptions necessary to arrive at your answer other than those given?

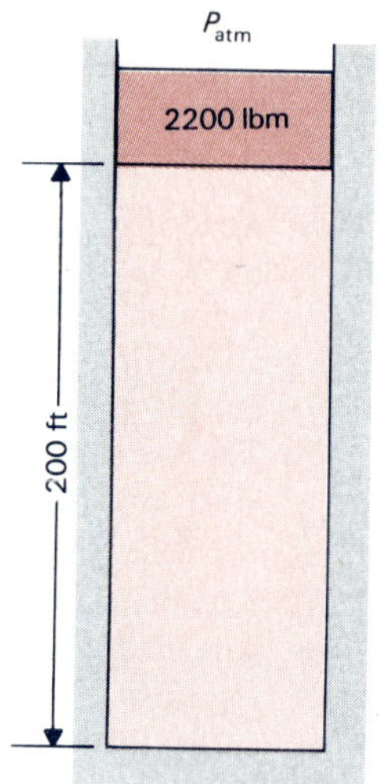

Figure P2.75E

2.76S An ideal frictionless isothermal compressor takes air at a pressure of 1 atm and compresses it to 10 atm at a temperature of 30°C. Each revolution of the compressor handles 1 kg of air, and the compressor operates at 200 rpm. What minimum size electric motor should be ordered to run the compressor?

2.76E An ideal frictionless isothermal compressor takes air at a pressure of 1 atm and compresses it to 10 atm at a temperature of 86°F. Each revolution of the compressor handles 2 lbm of air, and the compressor operates at 200 rpm. What minimum size electric motor should be ordered to run the compressor?

2.77S A 12-V storage battery for an aircraft is completely discharged. It is then placed on a quick-charger, and 10 A is fed to the battery for a period of 1 h. The battery, which has a mass of 10 kg, is then loaded aboard an aircraft for delivery. The aircraft flies at 700 km/h at an altitude of 2000 m above the rom nd.

(*a*) What is the change in the ,otal energy E of the battery from its discharged state on the ground to its state aboard the aircraft during flight?

(*b*) What percentage of the energy change is due to the various energy forms (kinetic, potential, and stored)?

(*c*) How much work was done during the charging process?

2.77E A 12-V storage battery for an aircraft is completely discharged. It is then placed on a quick-charger, and 10 A is fed to the battery for a period of 1 h. The battery, which has a mass of 22 lbm, is then loaded aboard an aircraft for delivery. The aircraft flies at 400 mi/h at an altitude of 6500 ft above the ground.

(*a*) What is the change in the total energy E of the battery from its discharged state on the ground to its state aboard the aircraft during fight?

(*b*) What percentage of the energy change is due to the various energy forms (kinetic, potential, and stored)?

(*c*) How much work was done during the charging process?

2.78 The total differential of $P\,dV$ work could be written as $\delta W = P\,dV + 0\,dP$. Show that work is an inexact differential.

2.79S A 1-kg mass is moved from the floor to a position on a staircase 10 m above the floor. The new position is reached by the following paths:

(*a*) The mass is lifted directly upward by a frictionless pulley in a block-and-tackle system.

(*b*) The mass is moved on a frictionless dolly to an elevator, which is also frictionless. The elevator moves to the height necessary, and then the dolly is used to move the mass back to the final position.

In these two cases find the work done on the mass in moving it to the final position. Discuss whether the work, which is a path function, depended upon the path in these two cases, and if not, what characteristics were present that made work independent of the path.

2.79E A 2-lbm mass is moved from the floor to a position on a staircase 30 ft above the floor. The new position is reached by the following paths:

(*a*) The mass is lifted directly upward by a frictionless pulley in a block-and-tackle system.

(*b*) The mass is moved on a frictionless dolly to an elevator, which is also frictionless. The elevator moves to the height necessary, and then the dolly is used to move the mass back to the final position.

In these two cases, find the work done on the mass in moving it to the final position. Discuss whether the work, which is a path function, depended upon the path in these two cases, and if not, what characteristics were present that made work independent of the path.

Reference

1. Steven C. Chapra and Raymond P. Canale, *Numerical Methods for Engineers,* 2nd Ed., McGraw-Hill Book Co., New York, 1988

3

Properties of Common Substances

When you can measure what you are speaking about, and express it in numbers, you know something about it; but when you cannot measure it, when you cannot express it in numbers, your knowledge is of a meager and unsatisfactory kind: It may be the beginning of knowledge, but you have scarcely, in your thoughts, advanced to the stage of science.
Lord Kelvin

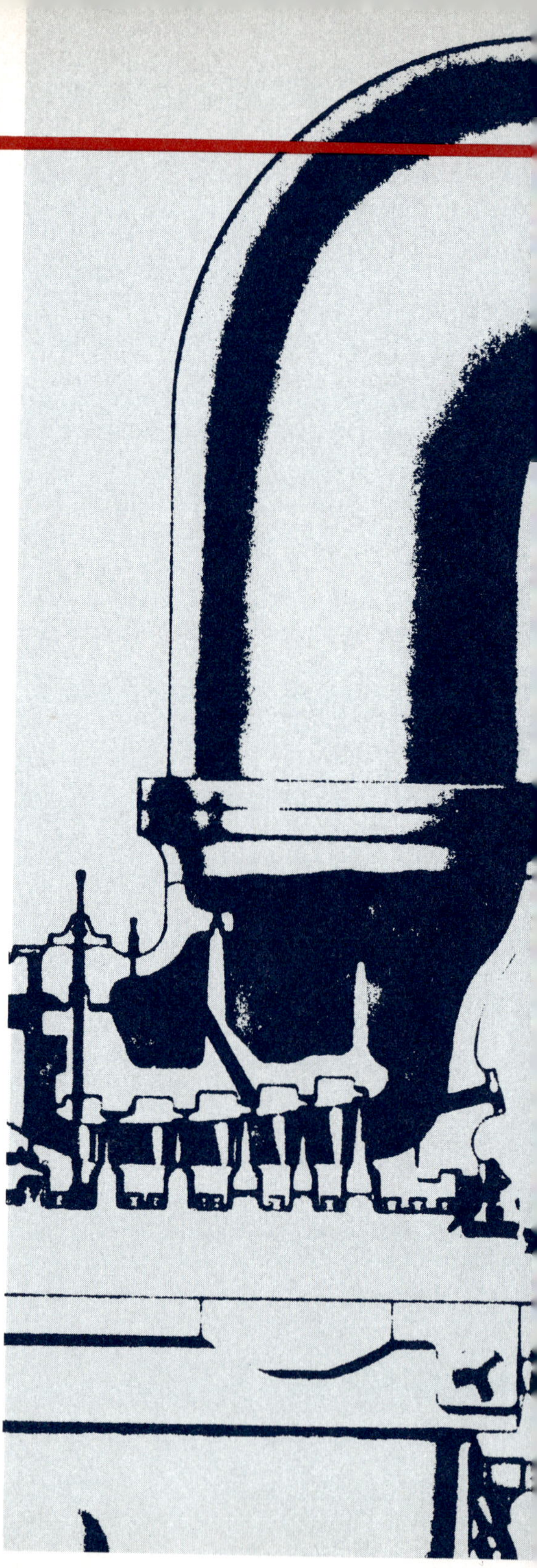

Cross section of 3600-rpm tandem-compound two-flow reheat steam turbine; 300,000- to 350,000-kW rating range. (*General Electric.*)

3.1 Introduction

In Chap. 2, certain common properties were defined so that the concepts of state, process, and energy could be discussed. However, other properties of a system are useful in thermodynamics in addition to those defined in Chap. 2. In this chapter, some new properties are introduced, and the behavior of various properties is discussed for a particular class of substances. Still other properties are introduced in later chapters as the need arises; however, the concepts discussed here apply to all other properties.

3.2 State Postulate — Applications to Property Relations

In Chap. 2, we noted that properties could be classified as either *intensive* or *extensive*. Further, properties are defined only for a system that is in thermodynamic equilibrium. Study of a large number of substances and their behaviors during different processes reveals that the more substances present in the system and the more ways that energy can be exchanged between the system and its surroundings, the more properties are necessary to describe the stable equilibrium state of the system.

We limit discussion in this chapter to *pure simple compressible substances*. The words *simple* and *compressible* imply that the only work mode considered is the $P\,dV$ work form. *Pure* implies substances composed of a single chemical species (such as H_2O, ammonia, and nitrogen), and so we exclude all mixtures of two or more substances (such as salt water, whiskey, and humid air). The exception is air, which is a nonreacting mixture of N_2, O_2, and other molecules and can be treated in many practical cases as a pure substance. In addition, the substances are taken to be of fixed molecular composition throughout our study; i.e., no chemical reactions are taking place. In later chapters, we study mixtures and systems in which the species can undergo chemical transformations.

For a pure simple compressible substance, repeated observations and experiments show that two independent properties are necessary and sufficient to establish the stable equilibrium state of a system. This observation is not derivable from more basic theorems using classical approaches to thermodynamics; but no violations have been found, and it can be accorded the status of the other thermodynamic laws studied later.

The observed behavior of a pure simple compressible substance is summarized in the *state postulate:*

The values of any two independent thermodynamic properties are sufficient to establish the stable thermodynamic state of a control mass composed of a pure simple compressible substance.

Thermodynamic properties are properties such as pressure, temperature, and volume, not geometric properties such as shape or elevation and not velocity. *Stable state* means an equilibrium state, so that the state postulate gives no information concerning nonequilibrium states. The above statement is for a control mass or a specified amount of matter. The state postulate can also be interpreted on a per unit mass basis or on an intensive property basis. Recalling that extensive properties can be expressed in an alternative intensive form, we see that the state postulate can be stated in a very useful form:

The values of any two independent intensive thermodynamic properties are sufficient to establish the stable thermodynamic state of a pure simple compressible substance.

There are a number of important generalizations to these forms of the state postulate. These statements apply only to simple compressible substances, so the only work mode possible is $P\,dV$. If there are other quasi-equilibrium work modes, then the number of independent thermodynamic properties needed to determine the thermodynamic state is 1 plus the number of work modes. Also for a control mass that contains more than one substance or that has chemically reacting components, we expect it is necessary to specify more than two independent properties to establish the state. This is explored more fully in Chaps. 10 and 11.

The state postulate indicates the information needed to specify the state of the substance in a particular system. The properties considered thus far are P, v, T, and u, and more are presented in later sections. The state postulate says that if two of these properties are independent and known, then all others are uniquely specified. If T and v are known for a pure compressible substance, then P and u have unique known values. Mathematically this is given as

$$P = P(T, v) \tag{3.1}$$

and

$$u = u(T, v) \tag{3.2}$$

The choice of the independent variables in the expressions is arbitrary as long as they are independent, so that equivalent alternative statements are

$$P = P(u, T) \quad \text{or} \quad P = P(u, v) \tag{3.3}$$

and

$$u = u(P, T) \quad \text{or} \quad u = u(P, v) \tag{3.4}$$

Since thermodynamics deals with energy and its transformation, Eqs. (3.2) and (3.4) are very important for thermodynamic analysis.

3.3 Simple Compressible Substances

Simple compressible substances are used in many engineering systems including power plants, many refrigeration systems, and thermal distribution systems using water or steam to transport energy. In addition, internal and external combustion engines can be usefully studied by assuming that they operate with simple compressible substances as the working fluid, even though in practice this is not the case. Finally, some nonreacting mixtures of pure substances, for example, dry air, can be treated as pure substances with little error, which permits considerable practical extension and application of the property relations that we develop for pure substances.

3.3.1 Liquid Phases

Suppose that, having set the values of the two independent variables, we wish, for a particular pure simple substance, to determine the *value* of a dependent variable. How could this be done, and how could the results be best presented?

As an illustration, we take a very familiar substance, cold water, and place it in a cylinder. Then we remove all air from the water (remember, we are limited to a pure substance), and we place a piston in contact with the water surface (Fig. 3.1). The water is considered to be the control mass, and we can measure the pressure in the control mass with a pressure gauge and use a thermometer to measure the control mass temperature. We have thus set the two independent properties—temperature and specific volume (the latter is found by measuring the total volume contained by the cylinder-piston arrangement and dividing by the mass of the water placed in the system). The dependent property (pressure) that results at the set values of specific volume and temperature is found from the pressure gauge. All three properties, P, v, and T, are thus measurable in this state.

Suppose now that we want to change the state of the control mass. We could perform many types of processes. For example, let us submerge the piston-cylinder arrangement within a large bath of fluid at a fixed temperature $T_1 = 100°C = 212°F$ that is equal to the control mass temperature. The water and surroundings will be in thermal equilibrium at the temperature of these surroundings. We impose an initial force on the piston sufficient to give the desired specific volume of v_A. Now we increase the volume of the control mass by moving the piston upward, and we measure the resulting pressure by reading the gauge attached to the cylinder wall. If we take care to increase the volume slowly, the water will stay at constant temperature. If the control mass temperature should change infinitesimally during the process, there will be a heat transfer from the control mass to or from the surrounding bath, and the control mass will come back to thermal equilibrium with the bath at the bath temperature.

Figure 3.1 Experimental apparatus for property measurement.

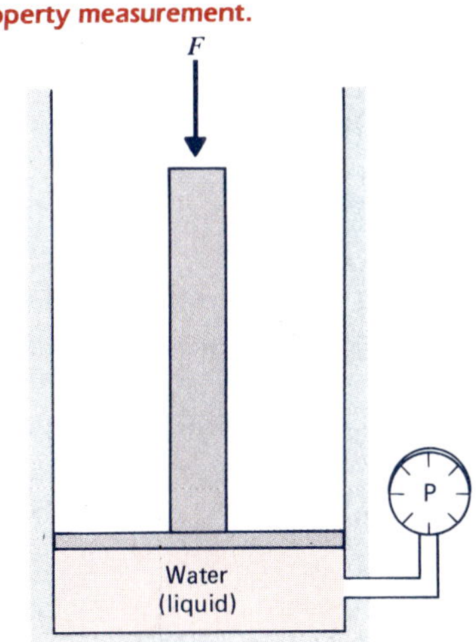

The independent variables during this process are T and v, and we are measuring the value of the dependent variable P that exists at the given T and v. Thus, we are determining the states given by $P = P(T, v)$. If we plot the path of this process on a graph of pressure versus volume, it will appear as shown in Fig. 3.2, where A denotes the initial state and B the final state, and the process between the states proceeds along an *isotherm,* or constant-temperature line, which passes through a series of states between A and B that are each in thermodynamic equilibrium. Such a process that passes through a series of equilibrium states has been defined as a *quasi-equilibrium* process.

Note in Fig. 3.2 the very great change in water pressure for a very small change in specific volume. The explanation is that liquid water is very nearly *incompressible,* and it is often assumed for convenience that the value of v is constant for liquid water (and most other liquids) as the pressure is changed at constant temperature.

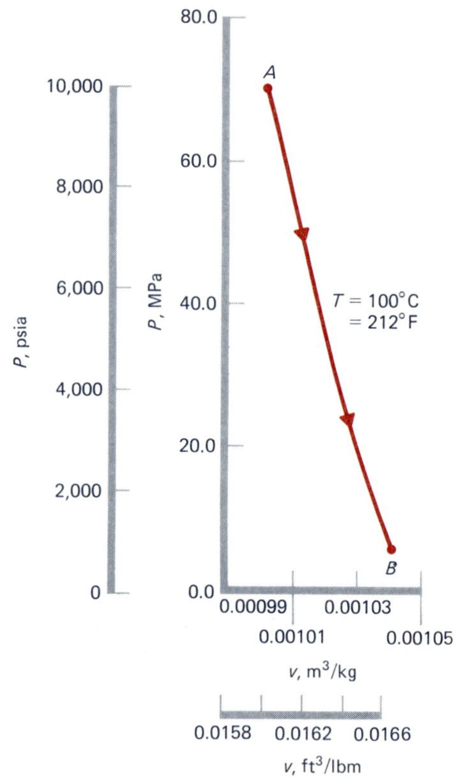

Figure 3.2 Constant-temperature expansion of water.

3.3.2 Saturation and Phases

As the volume of the system is increased further, we find that after the control mass reaches a particular volume corresponding to state C (Fig. 3.3), *the pressure of the control mass remains constant.* We can change the volume by moving the piston, but the piston can be stopped at a position corresponding to any volume along line CD, and the measured pressure remains at the same value found at point C; i.e., the pressure in the control mass is the same at each state along CD. Observation of the water in the control mass at some point between C and D shows that some of the liquid present at point C has changed into vapor, thus filling the increased volume. As the volume is further increased, more of the liquid is converted to vapor, until point D is reached, where all the liquid has just disappeared and all the water is in the form of vapor.

Note that, for water, the specific volume increases dramatically during the vaporization process, and we are now forced to plot the specific volume on a logarithmic scale to keep sufficient detail at small specific volumes. The isothermal expansion of the liquid, path A-B-C, is seen to be very nearly vertical on this scale.

The physical state or form of the control mass at point C is referred to as being a *saturated liquid;* and the water temperature at this state is the *boiling point,* or *saturation temperature,* at the control mass pressure. The water in states along line CD is a mixture of two physically distinct forms, liquid and vapor. These physically distinct forms for a component are called *phases,* and states along line CD are referred to as *two-phase* states. Because the control mass has given P, v, and T values at each state along line CD, each intensive property (P and T) is the same for each phase. However, the specific intensive property v, though known for the entire two-phase mixture (the volume over the mass), is *different* for each of the

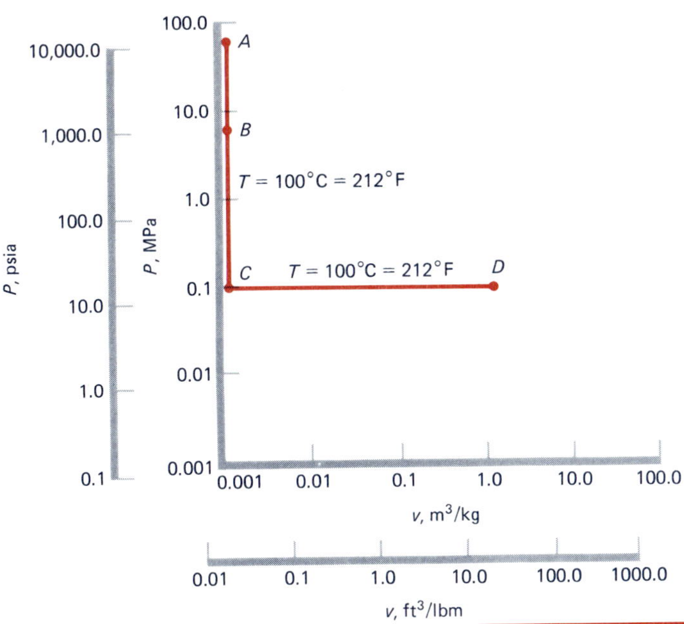

Figure 3.3 Isothermal expansion in the two-phase region.

two phases. The specific volume of the liquid phase can be found by measurement. We can remove a small sample of the liquid phase at some state between C and D and measure the volume and mass of the sample while holding its temperature at the boiling point for that pressure; a similar measurement can be made on the vapor phase.

Let us think about these observations in light of the state postulate that two independent properties must be specified in order to define the state of the system. Along line CD, if the pressure is specified, the temperature is automatically known; i.e., the boiling point is dependent on only the pressure of the control mass and cannot be varied independently. Thus, *along the two-phase line CD, T and P are not independent properties*. From the set of P, v, and T, then, the only independent property sets that can be used to specify the state of a two-phase system are the sets P and v or T and v.

3.3.3 Quality

Each of the two phases is at the boiling point at the existing control mass pressure at every state along line CD. Thus, if we measure the specific volume of the liquid phase at *any* state along line CD, we find the same value: the specific volume of saturated water at the control mass temperature, $v_l(T)$, is equal to v_C. Similarly, the specific volume of the vapor at any state along line CD is found to be the specific volume of the saturated vapor, $v_g(T) = v_D$. The subscripts l and g denote the liquid and vapor (gas) phases, respectively. [Some references use the subscript f (fluid) for the liquid state. Because the word *fluid* is commonly used for both liquid and vapor states, we use the subscript l for the liquid state throughout this text.]

It is relatively simple to tabulate the specific volume of saturated liquid and saturated vapor as a function of temperature. It would thus be useful to define the specific volume v of the total mass defined by the control mass at any point along the two-phase line CD in terms of the specific volumes of the individual phases v_l and v_g. To do this, we introduce a property called the *quality* of the two-phase mixture. The quality is defined as the mass of the vapor present in the two-phase mixture divided by the total mass, and it is given the symbol x. Thus

$$x = \frac{m_g}{m} \tag{3.5}$$

For a two-phase control mass, then, the quality ranges from 0 for a control mass composed of only saturated liquid to 1.0 for a control mass composed of only saturated vapor. Quality is also often expressed as a percentage. Note that quality is undefined except for a two-phase mixture of vapor and liquid.

The system volume along the two-phase line CD is

$$V = V_l + V_g \tag{3.6}$$

Dividing by the total mass gives

$$\frac{V}{m} = v = \frac{V_l}{m} + \frac{V_g}{m}$$

$$= \frac{m_l}{m}\frac{V_l}{m_l} + \frac{m_g}{m}\frac{V_g}{m_g}$$

or

$$v = (1 - x)v_l + xv_g \tag{3.7a}$$

and the specific volume of the control mass is specified in terms of tabulated properties and the quality. An alternative way of writing Eq. (3.7*a*) is

$$v = v_l + x(v_g - v_l) = v_l + xv_{lg} \tag{3.7b}$$

where v_{lg} is defined as the difference between the specific volumes of the saturated vapor and the saturated liquid at the control mass temperature, or $v_g - v_l$.

Because the specific volume and, as we shall see, other specific properties of the control mass are determined once x is specified at a given P or T, it is common practice to use x and one other independent property to specify the state along the two-phase line.

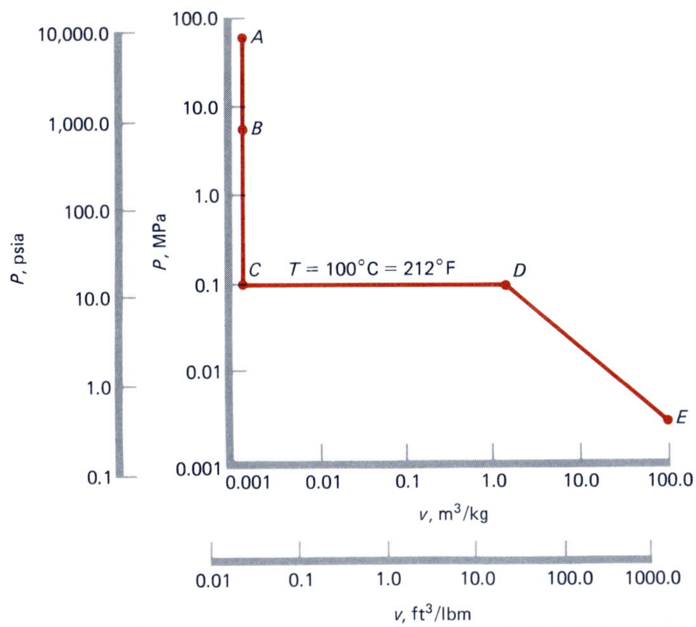

Figure 3.4 Expansion into the superheat region.

3.3.4 Superheated Vapor

Let us return to Fig. 3.3. Here the process of constant temperature expansion was halted at state *D*, where the control mass had just been completely changed to the vapor phase. Suppose that we now continue the expansion by further withdrawing the piston so that the volume of the control mass is increased while the temperature is held constant by the surrounding bath. As we do so, we find that the pressure in the system decreases, and the *P-v* trace continues along path *DE* (Fig. 3.4). In the *DE* series of states, the vapor is referred to as being *superheated,* because the temperature is above the boiling point at any pressure that exists along the path. The volume can be continuously increased past v_E, and the pressure will continue to fall. The process can be continued until the restrictions on our definition of pressure or specific volume [see Eqs. (2.1) and (2.7)] are no longer valid.

In these states note that *P* and *T* are independent. Thus, unlike the saturation states, *P* and *T* can be used as two independent properties to specify the state of a superheated vapor.

3.3.5 *P-v* Diagram

Suppose that we now return the piston of the experimental apparatus of Fig. 3.1 to its original position, and we change the temperature of the

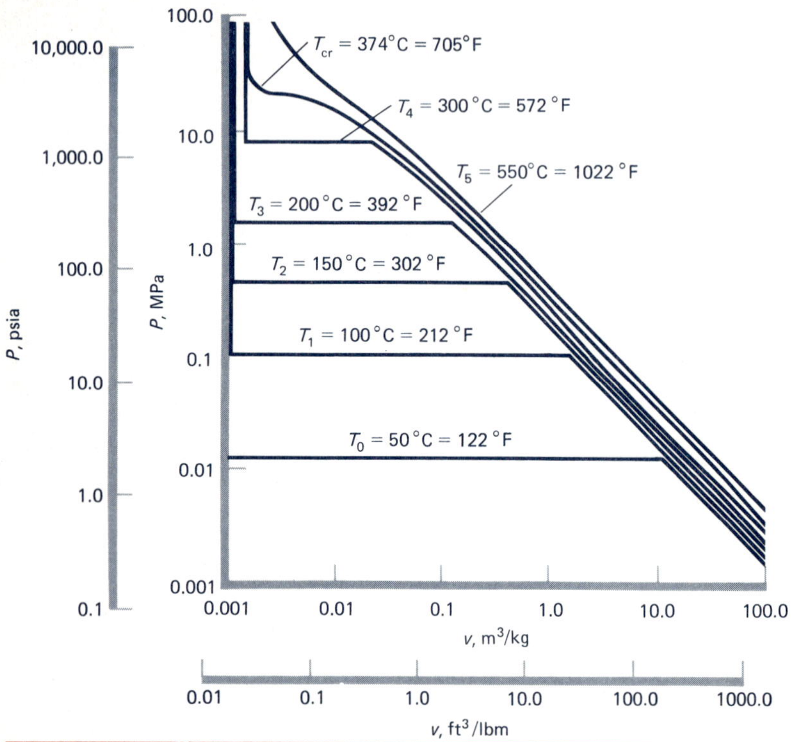

Figure 3.5 Isotherm for expansion at various temperatures.

surrounding bath to some higher value $T_2 > T_1$. By moving the piston slowly to increase the volume of the contol mass while holding the control mass temperature constant at T_2, a new isotherm is found on the P-v diagram. This procedure can be repeated at higher and higher temperatures, and the isotherms for each temperature can be drawn, as in Fig. 3.5.

Critical Temperature

At higher temperatures, the length of the horizontal two-phase line decreases. Finally, for water, the isotherm in the two-phase region corresponding to a temperature $T = 373.976°C = 705.17°F$ has zero length. That is, no discernible two-phase region exists, and as the volume is increased, the water undergoes a transition from liquid to vapor without the appearance of a distinct two-phase region. Experimentally, if we watch the original liquid held at or above this temperature while we expand its volume, we cannot tell when the liquid becomes a vapor, because no interface is formed between the phases—no liquid surface can be seen. This phenomenon occurs at such a high pressure and temperature for water that it is outside our everyday experience. At the high pressures present in the region of the P-v graph where this occurs, as the specific volume of the liquid is increased along an isotherm, the pressure of the control mass decreases continuously. The liquid specific volume increases, and the liquid transitions smoothly into a gas of low specific

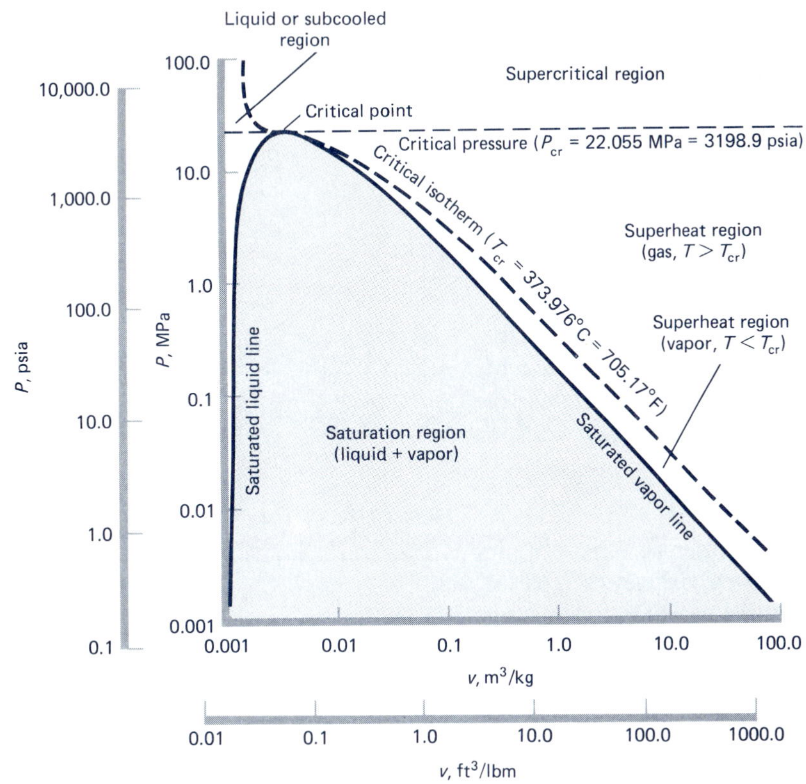

Figure 3.6 The *P-v* liquid-vapor phase diagram for water.

volume, with no separation of the phases being visible, which we can envision because both "phases" have the same specific volume (and thus density) and are thus indistinguishable from each other. The same phenomenon occurs for all isotherms above the critical temperature.

The lowest temperature for which the isotherm shows no phase transition is called the *critical temperature* T_{cr}. Values of the critical temperature for various substances are given in Table C.4 of App. C.

It is common to call a vapor that is above the critical temperature a *gas* and to retain the word *vapor* for the phase that could exist in equilibrium with its liquid phase at the vapor temperature if the pressure were increased sufficiently.

P-v Liquid-Vapor Phase Diagram

If all the points on the various isotherms that show the onset of the two-phase region are joined and all the points corresponding to the appearance of 100 percent saturated vapor are joined, then the resulting diagram is shown on logarithmic coordinates in Fig. 3.6. The line joining saturated liquid points joins smoothly to the saturated vapor line at the point where the critical isotherm touches both lines.

Within the dome-shaped region are all states that contain a two-phase mixture of saturated liquid and saturated vapor. This part of the diagram is referred to as the *two-phase*, or *saturation, region.* Only within this region can the property of quality be defined and used.

To the left of the dome are all states in which the control mass exists in only the liquid phase. In this latter region, the liquid is at a temperature below its boiling point at any *P-v* state; states in this region are referred to as *subcooled,* and the liquid is referred to as a *subcooled* or *compressed* liquid. (Such a definition has no meaning above the critical isotherm, since the phenomenon of boiling cannot be observed for supercritical temperatures.)

The region to the right of the dome is called the *superheat region.*

The *critical point* is the pont at which the critical isotherm is tangent to the saturation curve. The *critical pressure* is the pressure at this tangent point (values are found in Table C.4). The region above the critical isotherm and critical pressure line (critical isobar) is called the *supercritical region.*

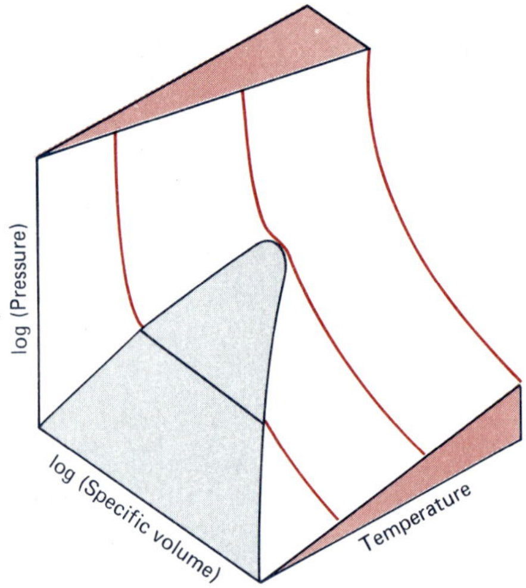

Figure 3.7 *P-v-T* surface for the liquid-vapor region for water.

P-v-T Surface

It is useful to plot the data developed for water or other simple substances on *P, v, T* coordinates rather than on the *P-v* diagram. If this is done, a *surface* of equilibrium *P-v-T* states results. For water, the portion of the surface that we have described so far appears as in Fig. 3.7. The projection of the surface onto the *P-v* plane results in the *P-v* diagram shown in Fig. 3.6.

Projection onto the *P-T* and *T-v* Surfaces

The *P-v-T* surface can be projected onto the other coordinate planes as well as onto the *P-v* plane. The phase diagram on these coordinates

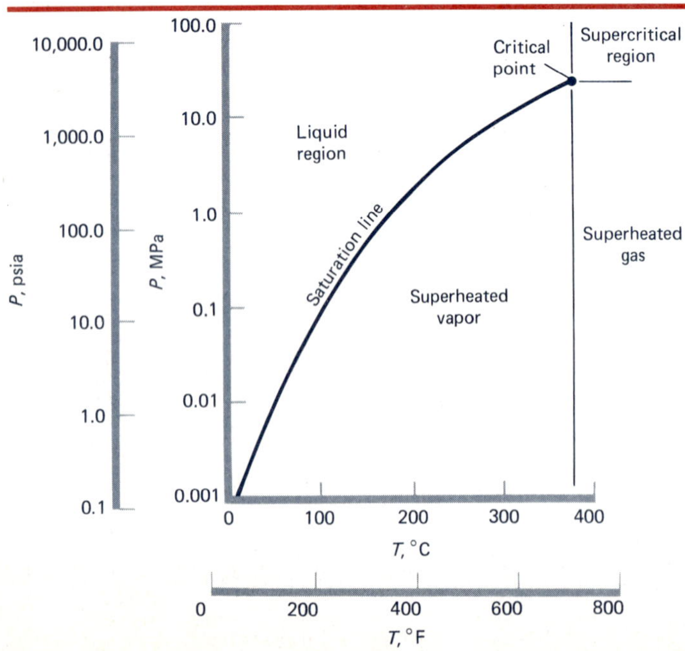

Figure 3.8 *P-T* diagram for vapor-liquid equilibrium of water.

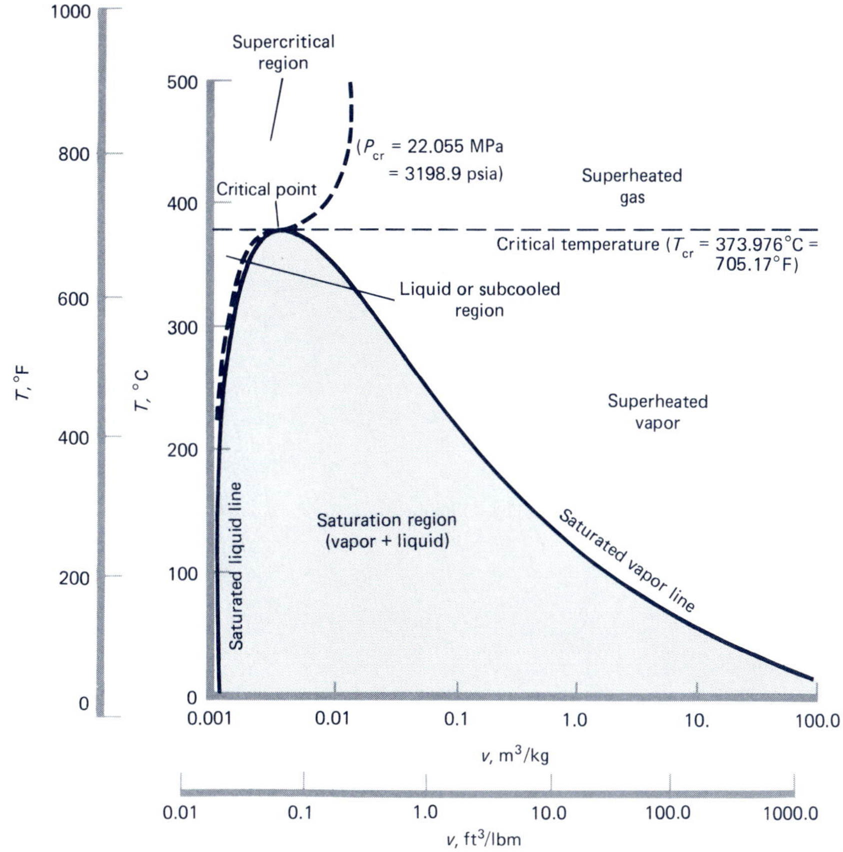

Figure 3.9 *T-v* **diagram for vapor-liquid equilibrium of water.**

often provides useful insight. Examination of Fig. 3.7 shows, for example, that the region of the *P-v-T* surface examined so far would project onto the *P-T* surface as shown in Fig. 3.8, and the *T-v* surface would appear as shown in Fig. 3.9. Both diagrams are useful in presenting experimental data, as we see in the next section. Note that on all these diagrams, *T* is plotted on a linear scale, while *P* and *v* are plotted logarithmically.

Figure 3.10 Apparatus for constant-pressure experiment.

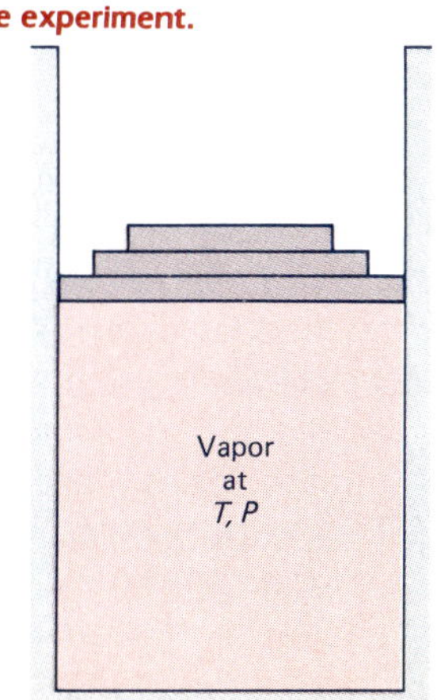

Other Regions on the *P-v-T* Diagram

We have directed attention so far to states in which only liquid, gas, vapor, or a combination of liquid and vapor phases exists. Such a treatment, of course, ignores the existence of the other common phase of pure simple substances—the solid phase. Every simple substance will solidify if the temperature is lowered sufficiently. [However, some substances have solidification temperatures that are extremely low; helium, for example, freezes at temperatures below 1 K (1.8°R) unless the control mass is at a pressure above 2 MPa (290 psia).]

Let us look at the *T-v* diagram of Fig. 3.9. If we choose a new experiment as shown in Fig. 3.10, in which a fixed control mass is held at

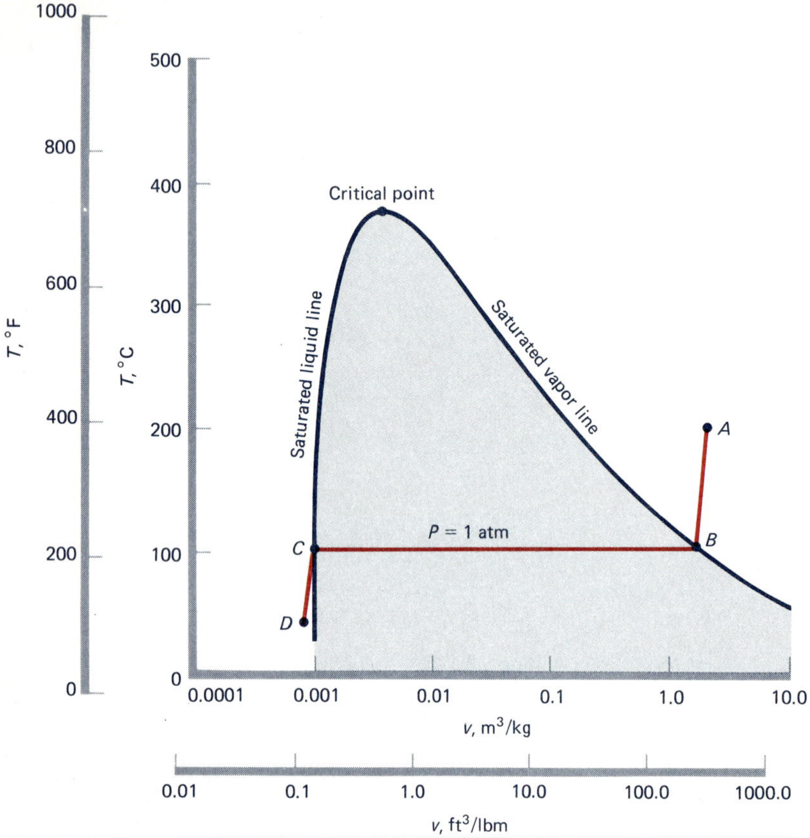

Figure 3.11 Constant-pressure process on the *T-v* diagram.

constant pressure by a weighted piston that is allowed to move, we could vary the control mass temperature by placing the experimental cylinder into a series of constant-temperature baths. Then the resulting control mass volume could be measured at each P and T by noting the resulting piston displacement. We are thus carrying out a *constant-pressure,* or *isobaric,* process in which the independent variables are pressure and temperature and the dependent variable is specific volume. Thus, we are determining $v = v(T, P)$. Such a process is plotted as a horizontal line on the *P-v* diagram of Fig. 3.6. Let us choose a pressure and temperature combination that corresponds to a state in the superheated vapor region, state A of Fig. 3.11 ($P = 1$ atm and $T = 200°\text{C} = 392°\text{F}$). If we continuously lower the control mass temperature by placing the piston-cylinder arrangement containing it in a series of baths with each successive bath having a slightly lower temperature than the previous one, then path *A-B-C-D* is traced. Point B occurs at the saturated-vapor specific volume corresponding to our constant pressure of 1 atm (14.7 psia); point C corresponds to the saturated-liquid specific volume; and point D corresponds to a point somewhere in the subcooled liquid region.

Suppose we continue to lower the temperature of the control mass (Fig. 3.12, where the scales have been greatly expanded). At some temperature (point E), the system undergoes a transition to the solid phase (i.e., it freezes). This occurs at a particular temperature dependent on the pressure chosen for the experiment. Note a curious phenomenon: The freezing process occurs at constant temperature, but as the control mass freezes, the specific volume *increases,* until the control mass is completely solid at point F. This is not the case for all materials; it is a consequence of having chosen water for the experimental substance. (If this volume did not increase during freezing, ice would not float, since its specific volume would be less than that of the water in which it is floating!) Most substances experience a *decrease* in specific volume during the freezing process, and water is a curious exception.

If the control mass temperature is lowered still further, the specific volume changes only slightly (the specific volume of most solids is a very weak function of both pressure and temperature).

If we repeat the experiment with various initial pressures, a complete T-v diagram can be constructed in a manner similar to that used for the P-v diagram. The regions are shown and labeled in Fig. 3.13.

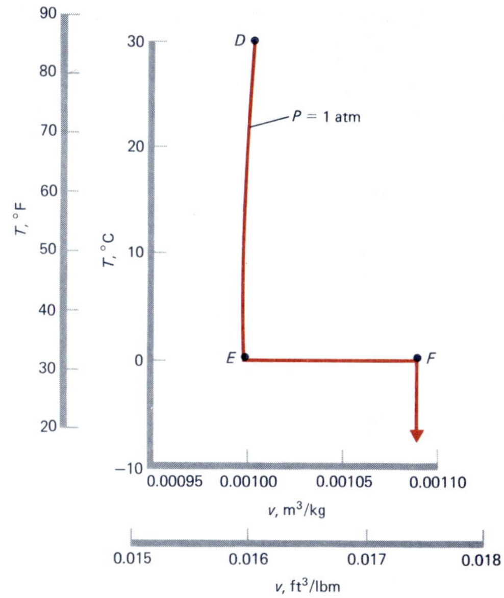

Figure 3.12 *T-v* **diagram for the freezing of water.**

Solid-Vapor Region

In Fig. 3.13, a large region of states exists in which ice and vapor coexist. This is generally the case below 0.01°C = 32.02°F. If large

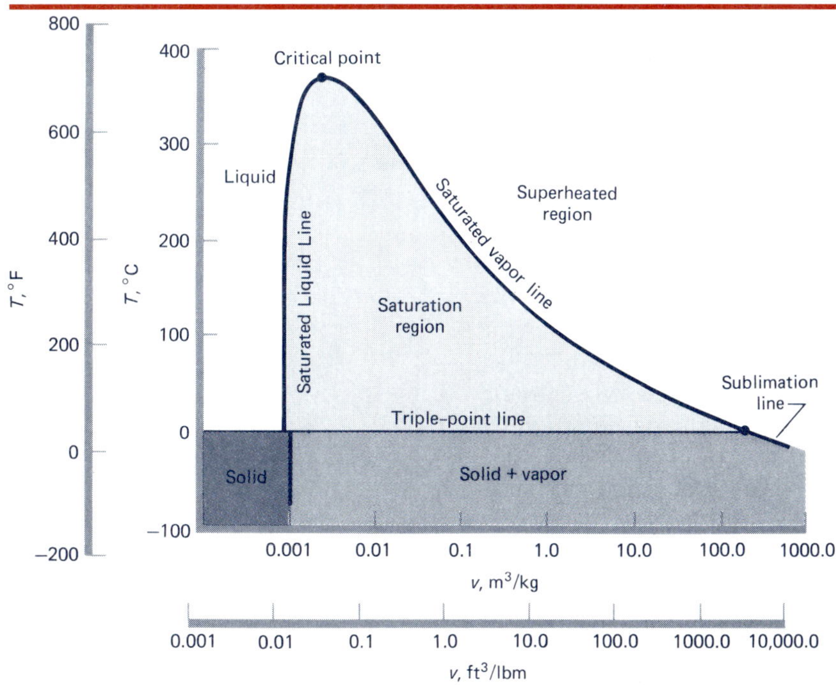

Figure 3.13 *T-v* **phase diagram for water.**

enough pressures are applied so that the specific volume of the control mass is reduced to 0.001091 $m^3/kg = 0.0175\ ft^3/lbm$ or below, then only the solid phase exists.

Sublimation Line

At extremely large specific volumes (which correspond to very low pressures) and low temperatures, it is possible during an expansion for a control mass to transform from a solid to a vapor without the appearance of the liquid phase. This process is called *sublimation.* The *sublimation line* marked on Fig. 3.13 indicates the boundary between the solid-vapor and superheated vapor states where this transition can occur. For example, a constant-volume heating process (a vertical line on the T-v diagram) at $v = 600\ m^3/kg = 9610\ ft^3/lbm$ causes the control mass to undergo sublimation. The reverse of this process is responsible for the appearance of frost on cold surfaces as water vapor in the air (at very low effective water vapor pressures) deposits as solid ice on an appropriate cold surface without passing through the liquid phase. Another commonly observed example is the vaporization (sublimation) of solid carbon dioxide (dry ice), which passes directly from the solid to the vapor state as a result of heat transfer from the environment at commonly observed conditions of room pressure.

Triple Point

Note that a horizontal line (constant temperature) joins the saturation region and the solid-gas region in Fig. 3.13. This indicates that only at this particular temperature can the liquid, solid, and vapor phases coexist. This temperature is called the *triple-point temperature,* and it is 0.01°C = 32.02°F for water. It is a convenient calibration point for thermometers. Note that this discussion is limited to a pure simple substance. However, a laboratory ice bath used for temperature calibration is usually open to the atmosphere, and the liquid-solid mixture will be saturated with dissolved air, as well as exposed to humid air rather than the vapor phase of pure water. Such an air-saturated system will not be at the triple point. The *ice point,* or temperature of an *air-saturated* ice-liquid bath in equilibrium with saturated air at standard conditions, is 0.0°C = 32.00°F. The conditions needed to exactly maintain the ice point are more difficult to achieve in the laboratory than for a carefully constructed triple-point cell; however, for less acccurate calibration, an ice bath is much more convenient to produce.

The values of the triple-point properties for various pure substances are given in Table C.1.

Complete P-v-T Surface

If the entire state surface, including solid, liquid, and vapor phases, is plotted on P, v, T coordinates for a substance such as water which expands upon freezing, the diagram appears as in Fig. 3.14.

Figure 3.14 ***P-v-T* phase diagram for a substance that expands upon freezing.**

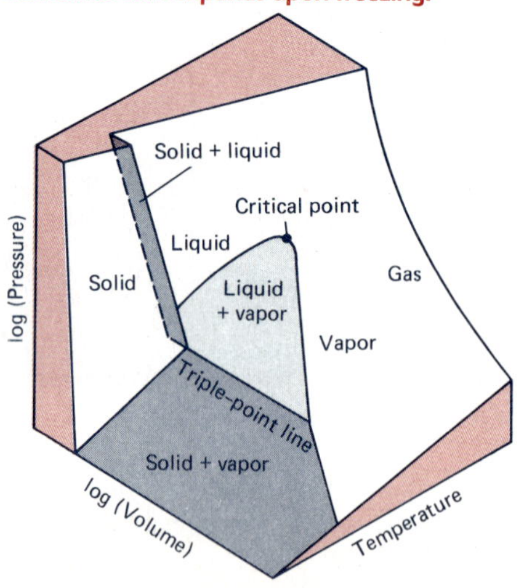

The solid phase of water can actually exist in various physically distinct forms, definable by their differing crystal structures. These structures also fall under our definition of phase, and the regions of their existence can be shown in the various diagrams.

It should not be inferred that most substances expand upon freezing. In fact, the opposite is true. For substances that contract upon freezing, the *P-v-T* phase diagram appears as shown in Fig. 3.15.

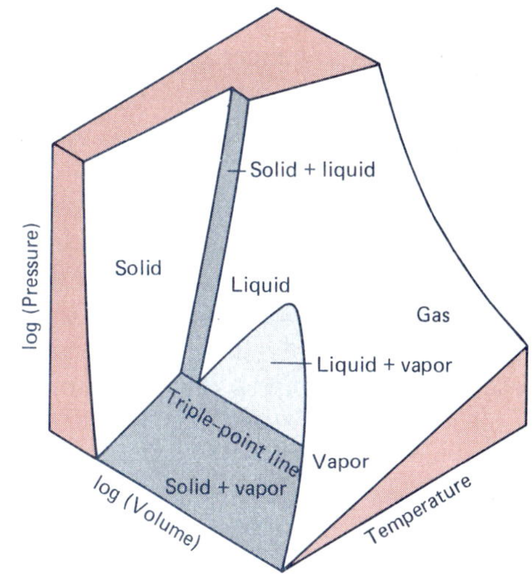

Figure 3.15 ***P-v-T* diagram for a substance that contracts during the freezing process.**

3.4 Other Thermodynamic Properties

3.4.1 Internal Energy and Enthalpy

Temperature, pressure, and specific volume are, of course, not the only properties of interest in the study of thermodynamics. For example, the property of internal energy, given the symbol U, was introduced in Chap. 2. The specific internal energy $u = U/m$ can be found for any simple compressible substance as a function of any two independent variables according to the state postulate. We could thus construct, for example, a *u-P-T* surface by the methods used to construct the *P-v-T* surface, assuming that we can invent a suitable method of measuring or determining u as a function of P and T. Then phase diagrams based on this surface could be constructed.

In many problems in thermodynamics, the group of properties $u + Pv$ occurs repeatedly. To evaluate this group at a given state, values of u and v must be found from some data source. For convenience, we can define a *new* property, called *enthalpy* and given the symbol h, as

$$h \equiv u + Pv \tag{3.8}$$

Because all the properties on the right-hand side of Eq. (3.8) can be found as a function of two independent variables, it follows that h can be tabulated as a function of the same two variables. Thus, we can evaluate and tabulate h directly at any given state, saving a great deal of effort in retrieving property data for the problems where $u + Pv$ appears.

Any specific extensive property or intensive property such as h and u can be evaluated for a state in the saturation region by relating it to the saturation properties through the quality, as was done for specific volume in Eq. (3.7*b*). Thus, for *any* intensive property z, we can write

$$z = z_l + xz_{lg} \tag{3.9}$$

Note that properties such as density, ρ, do not follow Eq. (3.9). This is because ρ, the inverse of specific volume v, is on a *per unit volume*, not per unit mass, basis.

The properties u and h are most often used in thermodynamic relations in terms of the *difference* of u or h values between two states, i.e., in terms of $u_2 - u_1$ or $h_2 - h_1$. Thus, it is immaterial whether we add a constant value to all tabulated values of u or h, since the *difference* in properties between two states will be the same because the constants will cancel when differences are taken. For the same reason, any state we wish can be assigned an arbitrary value of $u = u_{ref}$ or $h = h_{ref}$. Such a state is called a *reference state.* Often u_{ref} and h_{ref} are set equal to zero at the reference state. Because such states are arbitrary and have no effect on our thermodynamic calculations, reference states are often chosen for convenience in constructing property tables. This can lead to some difficulties when the properties of mixtures are computed from the properties of the individual components, or when the components of a system change because of the presence of chemical reactions, unless the properties of each substance are based on the same reference state.

3.4.2 Specific Heats

Suppose we wish to find the change in internal energy for a substance undergoing a process. By using Eq. (3.2), the change in internal energy is expressed [see Eq. (H.13) of App. H] as

$$du = \left(\frac{\partial u}{\partial T}\right)_v dT + \left(\frac{\partial u}{\partial v}\right)_T dv \tag{3.10}$$

Complete specification of the state is assumed by the state postulate if T and v are independent. Any two independent thermodynamic properties could have been used in Eq. (3.10), yet T and v are two very useful measurable properties. The partial derivatives in Eq. (3.10) are needed and will be important to thermodynamic analysis. The first partial derivative is particularly important and is considered here. The change in internal energy per degree of temperature change for a constant-volume process is this first partial derivative and is called the *specific heat at constant volume* (or sometimes the *specific heat capacity* at constant volume). It is given the symbol c_v and, from the above discussion, is defined mathematically as

$$c_v \equiv \left(\frac{\partial u}{\partial T}\right)_v \tag{3.11}$$

Values of c_v range from about 0.31 kJ/(kg·K) [0.07 Btu/(lbm·°R)] for argon gas to about 4.18 kJ/(kg·K) [1 Btu/(lbm·°R)] for liquid water. The values of c_v depend on the temperature of the control mass. Because the specific heat is a specific property, the property on a mass basis can be written as

$$C_v = mc_v \tag{3.12}$$

If the change in enthalpy is required, then this is considered in the same manner as the internal energy. Take the two independent variables to be P and T (this change will be quite useful), so that

$$h = h(T, P) \tag{3.13}$$

and the change in the enthalpy is

$$dh = \left(\frac{\partial h}{\partial T}\right)_P dT + \left(\frac{\partial h}{\partial P}\right)_T dP \tag{3.14}$$

For a constant-pressure process, the first partial derivative is significant. This property is the *specific heat at constant pressure,* defined as

$$c_P = \frac{C_P}{m} \equiv \left(\frac{\partial h}{\partial T}\right)_P \tag{3.15}$$

The two specific heats defined in Eqs. (3.11) and (3.15) will prove quite useful when we study the energy flows for various thermodynamic processes. For the moment, however, note that if u or h data as a function of T at given v or P are known, then c_v and c_P can be readily calculated from the slopes of the data curves, and no additional data are needed.

3.5 Development of Property Data

Each pure substance has its distinctive P, v, T, u, h, c_v, and c_P data, and usually these data must be determined experimentally. It is possible to generate the values of all properties from some minimum set of experimental measurements, and the methods of doing this are addressed in Chap. 9. Also the properties of some substances that have simple microscopic structure can be determined for some regions of the phase diagram without the use of direct experiment if some auxiliary spectroscopic data are available. This can be achieved by application of the methods of statistical thermodynamics (Chap. 13). Nevertheless, for most substances, careful and detailed experimental measurements are necessary to obtain the values of the minimum set of dependent properties as a function of a chosen set of independent properties.

Presentation of the experimental data relating all the properties in all regions of interest for each substance is obviously a difficult task. We need to compress as many data as possible into a small space. Many methods have been tried, and some of the most useful are described here. Others are introduced as properties are needed.

3.5.1 Graphical Data Presentation

It is very useful to be able to visualize thermodynamic processes on coordinates of various pairs of thermodynamic properties. We have done so on the P-v diagram in developing the background and definitions in this chapter. Graphical representation of processes on the phase diagrams plotted in terms of other property pairs enables visualization and analysis of many practical systems such as engines.

One very useful phase diagram can be plotted on coordinates of pressure versus enthalpy. An example appears in Fig. 3.16. On the P-h

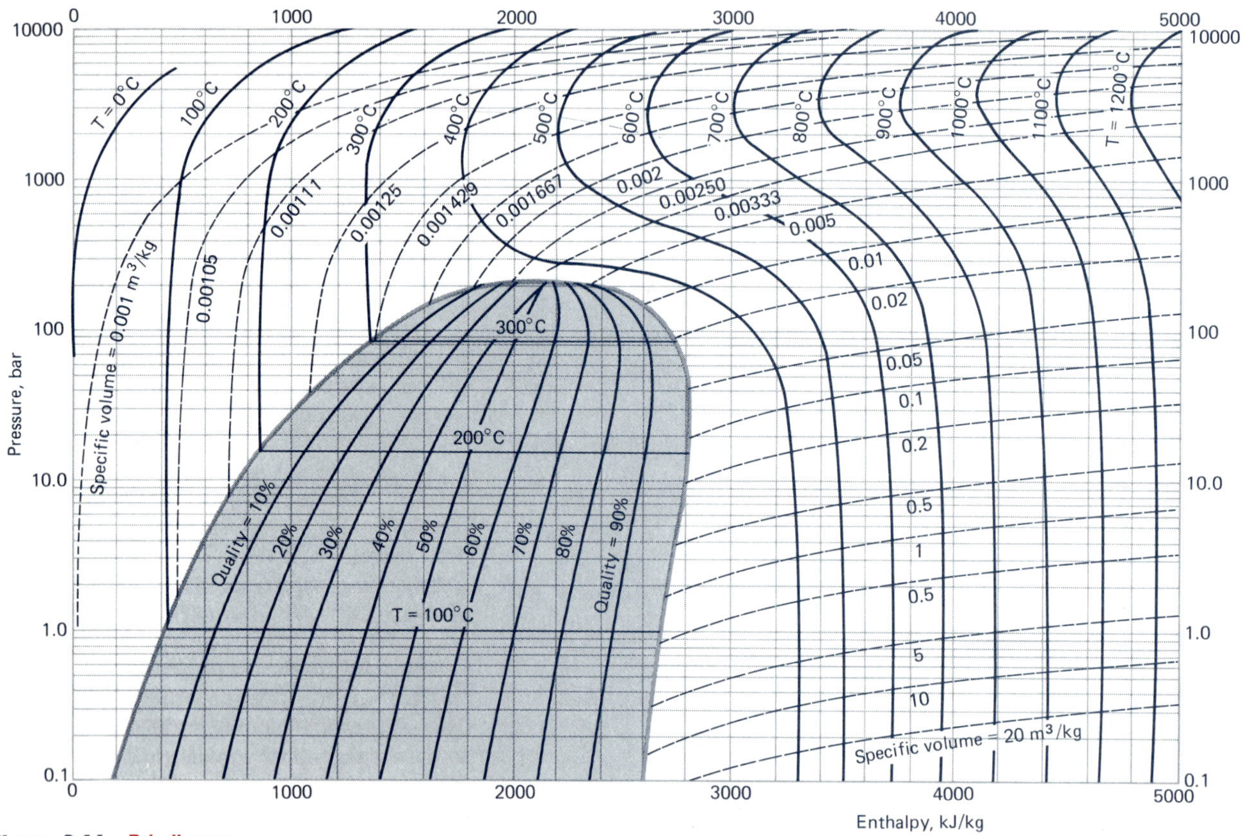

Figure 3.16 *P-h diagram.*

diagram, we can immediately see the change in enthalpy that occurs during any constant-pressure process, since the change in h then appears as the length of a horizontal line. Such insights are very helpful when we begin the study of thermodynamic processes and cycles. Diagrams on other property coordinates also prove useful and are discussed as other properties are introduced. (Note that in Fig. 3.16 the reference state for h is taken to be $h = 0$ for saturated liquid water at $T = 0°C$).

3.5.2 Equation of State

It is possible to develop mathematical relations among P-v-T data or other property sets for a particular substance. Such relations can be simple curve fits to experimental points or sophisticated relations based on knowledge developed from theory of the expected variation of a dependent property as a function of the independent properties. Such relations are called *equations of state.* When they are particularly simple and have good accuracy, they provide a convenient way to present P-v-T or other property data.

The *steam tables,* compiled by the American Society of Mechanical Engineers [1] and others [2, 3], are based on empirical curve fits to voluminous experimental data. The curve fits were used to generate table entries at discrete values of temperature and pressure. Computerized steam tables are generally based on the use of these equations rather than on the raw data themselves.

Simplified equations of state can often be used to describe property behavior over some limited region of states. However, the student *must* be aware of the region of validity and not use the simplified relations outside that region.

Ideal Gas Relation

Probably the most famous and the most useful equation of state is the *ideal gas relation,* which relates P, V, and T. This relation was introduced in Sec. 2.3.4.

$$PV = n\overline{R}T \tag{3.16}$$

where n is the number of moles of the ideal gas and $\overline{R}$ is the universal gas constant [8.31441 kJ/(kmol·K) or 1.98586 Btu/(lbmol·°R) from Table B.2]. This form is often modified for engineering practice, where we usually deal in mass rather than in moles, as is common in physics and chemistry. To convert to the engineering form, note that the number of moles of gas n is equal to the mass of gas m divided by the molecular weight of the gas M. Thus, Eq. (3.16) becomes

$$Pv = \frac{\overline{R}T}{M} = \frac{\overline{R}}{M}T = RT \tag{3.17}$$

where $R \equiv \overline{R}/M$ is the particular gas constant for the particular gas under study. Thus R is simply the universal gas constant divided by the gas molecular weight (see Tables C.2 and C.4). Figure 3.17 shows the surface described by the ideal gas relation. Also shown are the projections onto the two-dimensional plots.

The microscopic description of an ideal gas must be consistent with the above macroscopic relation. This requires a number of idealizations of actual molecular phenomena. The first is that the gas is composed of a large number of very small particles. Thus, the fraction of the total volume occupied by the particles is small. The second idealization is that the particles are independent of one another and are in continuous random motion. The time spent between collisions is much greater than the collision time. The final idealization is that no energy is dissipated. These concepts are helpful in interpreting the ideal gas relation and, in fact, are used in Chap. 13 to derive the ideal gas relation from a model of a gas composed of elastic hard spheres.

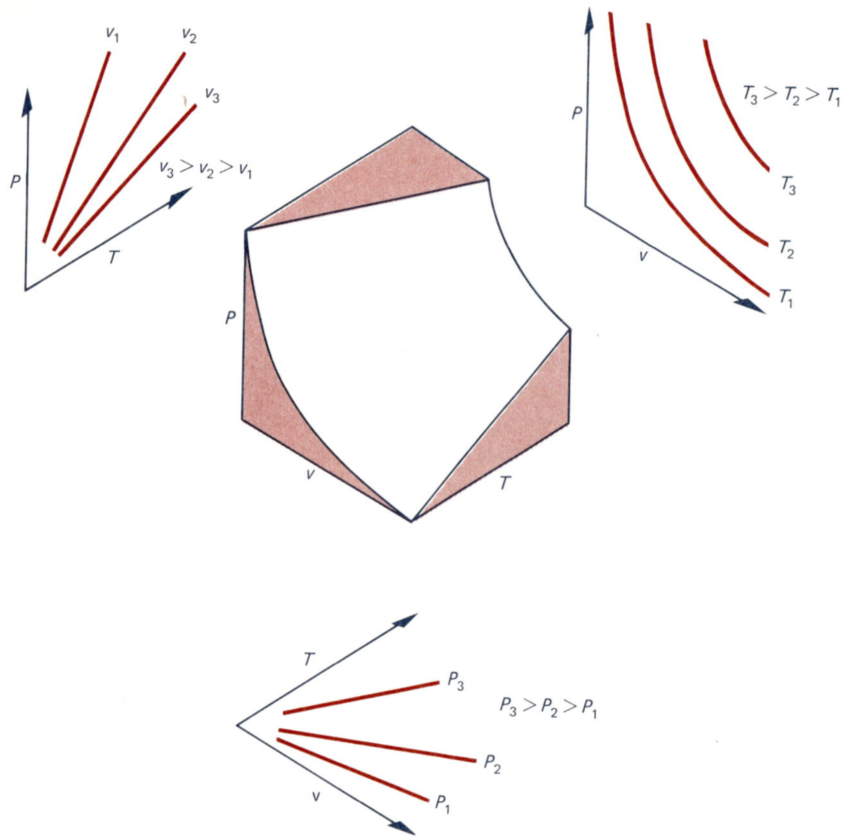

Figure 3.17 ***P-v-T* surface for an ideal gas.**

Under what conditions does this simple equation apply? The simplified microscopic description indicates that the equation does *not* apply to any substance that is in a solid, liquid, or saturation region state, or is in a state near the critical point. In the superheated and supercritical regions away from the critical point, the ideal gas relation, Eq. (3.17), is sufficiently accurate to be used for many gases that are well above their critical temperature at room conditions (CO_2, N_2, O_2, the noble gases, H_2, CO, and others). Specifically, if $P < P_{cr}$ and $T > 1.4T_{cr}$ for the particular gas, then Eq. (3.17) will be within 10 percent of the experimental values. Additionally if $P < 10P_{cr}$ and $T > 10T_{cr}$, then the 10 percent criterion is met. The higher the temperature and the lower the pressure within the prescribed range, the more accurate will be the predictions based on ideal gas relations.

Other Properties of the Ideal Gas

Although the ideal gas can be defined as a substance that follows the ideal gas relation, we can infer some other property behavior for the ideal gas. Although the material in Chap. 5 is necessary for the proof, the ideal gas internal energy can be shown to be *a function of temperature only;* i.e.,

$$u_{\text{ideal gas}} = u(T) \tag{3.18}$$

If u for the ideal gas depends on *only* temperature, then Eqs. (3.10) and (3.11) yield

$$du = c_v\, dT \tag{3.19}$$

or

$$c_v = \left(\frac{\partial u}{\partial T}\right)_v = \frac{du}{dT} \tag{3.20}$$

From the definition of enthalpy and the use of the ideal gas relation,

$$h = u + Pv = u(T) + RT \tag{3.21}$$

and it follows that, for an ideal gas, h is also a function of temperature only. Then Eqs. (3.13), (3.14), and (3.15) yield

$$dh = c_P\, dT \tag{3.22}$$

and

$$c_P = \left(\frac{\partial h}{\partial T}\right)_P = \frac{dh}{dT} \tag{3.23}$$

Now using Eq. (3.21) in Eq. (3.23) yields

$$\frac{dh}{dT} = c_P = \frac{du}{dT} + R = c_v + R \tag{3.24}$$

or, for an ideal gas,

$$c_P - c_v = R \tag{3.25}$$

For any ideal gas, then, c_P must always be greater than c_v; and if either specific heat is known, the other is easily found from Eq. (3.25). The values for specific heats are given for many gases in Table C.2. The superscript zero implies the evaluation is at 300 K or 540°R; that is, c_P^0 and c_v^0 are the values at 300 K or 540°R. Also, representations for the temperature dependence of the specific heat at constant presssure are given by polynomials in Table D.1 (SI) and in Table E.1 (USCS). (Note that the values in Tables D.1 and E.1 are for $\bar{c}_P$; that is, they are per mole rather than per unit mass.) For ideal gases, it is possible to predict the values of c_v [and thus c_P, by using Eq. (3.25)] from the use of the kinetic theory of gases and, more exactly, from statistical thermodynamcis. The basis of these predictions is given in Chap. 13, and it was an early triumph of the microscopic approach to thermodynamics by construction of simple models of molecules.

The predictions are quite accurate in some cases. For example, for monatomic gases such as argon, neon, krypton, etc., kinetic theory predicts that $c_v = 3/2R$ (and therefore, using $c_P - c_v = R$, that $c_P = 5/2R$). This value is predicted to be temperature-independent, and it holds until the gas begins to ionize, which occurs at quite high temperatures for these gases. Comparison with the tables for argon, for example (which are based on a more complete statistical theory and which agree very well with experimental data), will show that the predicted value of c_v is 0.3124 kJ/(kg·K) [0.0746 Btu/(lbm·°R)]. The value of c_v from kinetic theory is $3/2R = 3/2 \times 0.20813 = 0.31220$ kJ/(kg·K) [0.07465 Btu/(lbm·°R)].

For symmetric diatomic molecules such as N_2, O_2, or H_2, kinetic theory predicts that c_v will lie between the values of $5/2R$ and $7/2R$. Data for nitrogen are shown in Fig. 13.5.

For more complex molecules, the theory provides some guidance, but there are no simple rules that work as well as those for the simpler molecules. When no data are available, however, the simple kinetic theory relations can be used for reliable estimation of the properties of monatomic and symmetric diatomic gases. More accurate statistical thermodynamic relations that account for energy stored in electronic energy states give essentially exact agreement between predictions from theory and experiment for simple gases.

Example Problem 3.1

Find the change in enthalpy and the change in internal energy for nitrogen from 30 to 200°C. Take the pressure at atmospheric conditions.

Solution

The ideal gas expressions are valid since $P = 0.1$ MPa is less than $P_{cr} = 3.39$ MPa and the minimum temperature of $T = 303$ K is far greater than $T_{cr} = 126.2$ K (critical values from Table C.4). Thus

$$dh = c_P\,dT$$

and

$$h_2 - h_1 = \int_1^2 dh = \int_1^2 c_P\,dT$$

Also

$$du = c_v\,dT$$

With $c_v = c_P - R$ in Eq. (3.25),

$$du = (c_P - R)\,dT$$

Integration yields

$$u_2 - u_1 = \int_1^2 du = \int_1^2 (c_P - R)\, dT$$

$$= \int_1^2 c_P\, dT - R(T_2 - T_1)$$

This last expression can also be obtained from Eq. (3.21). The changes in h and u require the specific heat at constant pressure. One method is to assume c_P to be a constant and use this single value; the other method is to integrate the temperature dependence of the specific heat.

For a constant specific heat at constant pressure,

$$h_2 - h_1 = c_P(T_2 - T_1)$$

and

$$u_2 - u_1 = c_P(T_2 - T_1) - R(T_2 - T_1)$$

From Table C.2, $c_P = 1.0404$ kJ/(kg·K), so

$$h_2 - h_1 = 1.0404(200 - 30) = 176.9 \text{ kJ/kg} = 76.05 \text{ Btu/lbm}$$

$$u_2 - u_1 = 1.0404(200 - 30) - 0.29680(200 - 30) = 126.4 \text{ kJ/kg}$$

$$= 54.34 \text{ Btu/lbm}$$

The temperature dependence of the specific heat is given in Table D.1 on a mole basis as

$$\bar{c}_P = 28.90 - (0.1571 \times 10^{-2})(T) + (0.8081 \times 10^{-5})(T^2) - (2.873 \times 10^{-9})(T^3)$$

with $\bar{c}_P$ kJ/(kmol·K) and T K. The change in enthalpy on a mole basis (the overbar indicates a unit mole basis) is

$$\bar{h}_2 - \bar{h}_1 = 28.90 \int_1^2 dT - (0.1571 \times 10^{-2}) \int_1^2 T\, dT + (0.8081 \times 10^{-5}) \int_1^2 T^2\, dT - (2.873 \times 10^{-9}) \int_1^2 T^3\, dT$$

Integration between $T_1 = 273.15 + 30 = 303.15$ K and $T_2 = 273.15 + 200 = 473.15$ K yields

$$\bar{h}_2 - \bar{h}_1 = 4990 \text{ kJ/kmol}$$

With the molecular weight of 28.016, the change in enthalpy is $h_2 - h_1 = 178.1$ kJ/kg $= 76.57$ Btu/lbm. The change in u is evaluated directly from the above integration by combining the above equations:

$$u_2 - u_1 = \int_1^2 c_P\,dT - R(T_2 - T_1)$$

But the integral is given above, so

$$u_2 - u_1 = (h_2 - h_1) - R(T_2 - T_1)$$

$$= 178.1 - 0.29680(170) = 127.6 \text{ kJ/kg} = 54.86 \text{ Btu/lbm}$$

Comments

In the region of the properties of this problem, the ideal gas relation is a very accurate approximation. The temperature variation of the specific heat is relatively unimportant over the temperature range of the problem and yields a value less than 1 percent higher than the constant specific heat solution. The constant specific heat solution is given for a c_P evaluated at 300 K (80.6°F). A more accurate solution is obtained if the value of c_P is taken at the average of the temperatures within the needed range.

Many attempts have been made to provide an accurate *P-v-T* relation that covers a broader range of conditions than the ideal gas relation. Such relations generally gain in accuracy at the expense of increased complexity of form and the need for more data to describe the particular gas. Some of the relations extend *P-v-T* behavior to the liquid region. We examine three such representative equations of state.

van der Waals' Equation of State

The van der Waals equation is an attempt to modify some obvious theoretical difficulties in the ideal gas relation. In 1873, J. D. van der Waals argued that the volume of a gas should not go to zero when the pressure gets very large, as implied by the ideal gas relation, because the atoms of the gas should still take up some specific volume b. The ideal gas relation then becomes

$$v = \frac{RT}{P} + b$$

and the gas volume approaches b as P becomes very large.

In addition, the pressure measured at a surface is not the true gas pressure and is actually too small. Away from surfaces, gas molecules are exposed to short-range attractive forces from nearby gas molecules on all sides (Fig. 3.18*a*); near a surface, gas molecules are exposed to these forces from molecules on only one side, thus reducing the effective pressure exerted on the wall (Fig. 3.18*b*). Thus the gas pressure should be *increased*

Figure 3.18 Attractive forces between molecules. (*a*) Forces far from boundary. (*b*) Forces near boundary.

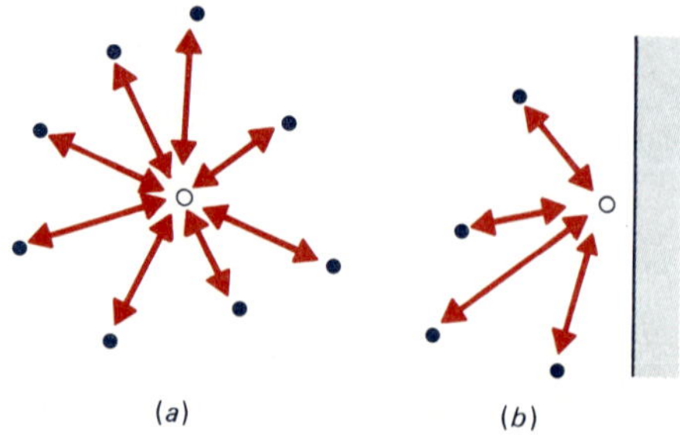

to a value above the measured pressure P by a certain amount. The increase in P should depend on the specific volume, since less mass and thus fewer molecules are present in a given volume to provide the attractive forces as the specific volume increases. van der Waals proposed that the ideal gas pressure P be replaced by $P + a/v^2$ to account for these forces. The van der Waals equation then becomes

$$P = \frac{RT}{v - b} - \frac{a}{v^2} \tag{3.26}$$

The constants a and b must be determined for each gas. We extend our interpretation of Eq. (3.26) when we discuss other property relations in Chap. 9. There it is shown that a and b are given by

$$a = \frac{27R^2T_{cr}^2}{64P_{cr}} \tag{3.27}$$

$$b = \frac{RT_{cr}}{8P_{cr}} \tag{3.28}$$

Equation (3.26) is obviously based on a physical model of the behavior of gases that is somewhat more detailed than the ideal gas model. However, van der Waals' equation does not provide a precise relation between gas properties but does offer insight into the reasons for deviation from ideal gas predictions.

Paynter [4] has provided an extension of van der Waals' equation that has the form

$$P_r = \left[1 + \frac{(1/v_r) - 1)^4}{1 - b_P/v_r}\right] T_r - \left(\frac{a_P}{v_r^2}\right)(1 - T_r) \tag{3.29}$$

where P_r, T_r, and v_r are dimensionless ("reduced") pressure, temperature, and volume, defined by $P_r = P/P_{cr}$, $T_r = T/T_{cr}$ and $v_r = v/v_{cr}$. Paynter suggests taking $a_P = 7$ and $b_P = 0.25$. In this form, Eq. (3.29) requires knowledge of the specific volume at the critical point, which is generally the least accurately known of the critical constants.

The Redlich-Kwong Equation of State

A simple but accurate equation of state is the Redlich-Kwong equation

$$P = \frac{RT}{v - b_{RK}} - \frac{a_{RK}}{v(v + b_{RK})\,T^{1/2}} \tag{3.30}$$

where $a_{RK} = 0.4275R^2T_{cr}^{2.5}/P_{cr}$ and $b_{RK} = 0.0867RT_{cr}/P_{cr}$. This equation is quite accurate at temperatures above the critical temperature, but deviates from experimental data at lower temperatures.

Benedict-Webb-Rubin Equation of State

An example of a more accurate but more complex equation that has some basis in the physical behavior of the properties of gases but is largely based on empiricism is the Benedict-Webb-Rubin equation:

$$P = \frac{RT}{v} + \left(B_0RT - A_0 - \frac{C_0}{T^2}\right)\left(\frac{1}{v^2}\right) + (bRT - a)\left(\frac{1}{v^3}\right) + \frac{\alpha a}{v^6} + c\frac{1 + \gamma/v^2}{v^3T^2}\exp\left(\frac{-\gamma}{v^2}\right) \quad (3.31)$$

Here A_0, B_0, C_0, a, b, c, α, and γ are constants specific to the particular gas. With eight constants available for describing the P-v-T behavior, the range of validity of this equation of state is obviously much greater than that for the ideal gas relation. It is chiefly applied to hydrocarbons, and values of the constants are given in Table C.5.

Example Problem 3.2

Predict the specific volume of methane gas at 100 atm and 0°F by using the ideal gas relation, the van der Waals equation [Eq. (3.26)], and the Benedict-Webb-Rubin (B-W-R) equation.

Solution

The ideal gas relation predicts

$$v = \frac{RT}{P} = \frac{[96.35\ \text{ft}\cdot\text{lbf/(lbm}\cdot{}^\circ\text{R)}](459.67^\circ\text{R})}{100(2116.2\ \text{lbf/ft}^2)} = 0.2093\ \text{ft}^3\text{/lbm}$$

$$= 0.01307\ \text{m}^3\text{/kg}$$

where R for methane is taken from Table C.2. For the van der Waals equation, the constants for methane are [from Eqs. (3.27) and (3.28)] $a = 4787.4\ \text{ft}^4\cdot\text{lbf/lbm}^2$ and $b = 0.04296\ \text{ft}^3\text{/lbm}$. If the van der Waals equation is rewritten in terms of v,

$$v^3 - v^2\left(\frac{RT}{P} + b\right) + v\left(\frac{a}{P}\right) - \frac{ab}{P} = 0$$

Solving for the positive real root gives

$$v = 0.1404\ \text{ft}^3\text{/lbm} = 0.00877\ \text{m}^3\text{/kg}$$

The B-W-R relation constants for methane are given in Table C.5. The B-W-R equation is implicit in v, as was the van der Waals equation, and must be solved by iteration. Assuming various values of v and substituting into the right-hand side of Eq. (3.31) until the right-hand side is equal to the prescribed value of P yield $v = 0.1565\ \text{ft}^3\text{/lbm} = 0.00977\ \text{m}^3\text{/kg}$.

Comments

The B-W-R result, which we would expect to be most accurate, predicts specific volume that is well below the ideal gas prediction but is 10.6 percent above the van der Waals prediction. The variations among the methods change with the particular gas and conditions chosen, but the B-W-R result for this example agrees within 1 percent with experimental data.

The increasing accuracy of the predicted values for specific volume comes at the expense of increasing difficulty in use of the equations of state. The equation to be used must be chosen based on the required accuracy for the problem at hand.

Other Equations of State

Many equations of state have been proposed, some based on the empirical behavior of P-v-T data, others based on the behavior predicted from the statistical mechanical approach. One relation based on the latter approach is the *virial equation*

$$P = \frac{RT}{v} + \frac{RTB}{v^2} + \frac{RTC}{v^3} + \cdots \tag{3.32}$$

where B, C, . . . are functions of T. Note that for $B = C = \cdots = 0$, this reduces to the ideal gas relation. Other equations of state can also be placed in this general form.

Other equations of state are discussed in Chap. 9. For a comprehensive discussion of various equations, their range of accuracy, and their limitations, see Reid, Prausnitz, and Poling [5].

Principle of Corresponding States

van der Waals noted that every pure substance has a critical state, and he proposed that the P-v-T data for all substances be presented in terms of the critical properties. The most easily measured of the critical properties are the critical temperature T_{cr} and the critical pressure P_{cr}. The *principle of corresponding states* declares that all substances obey the same equation of state expressed in terms of the critical properties.

If the principle is true, then we should be able to express the behavior of any simple single-phase pure substance by

$$\frac{Pv}{RT} = z(P_r, T_r) \tag{3.33}$$

where z, called the *compressibility factor,* is a function of only the reduced properties $P_r = P/P_{cr}$ and $T_r = T/T_{cr}$ of the particular substance. The value of z is a measure of the deviaton of the behavior of the P-v-T data of the substance from that of an ideal gas.

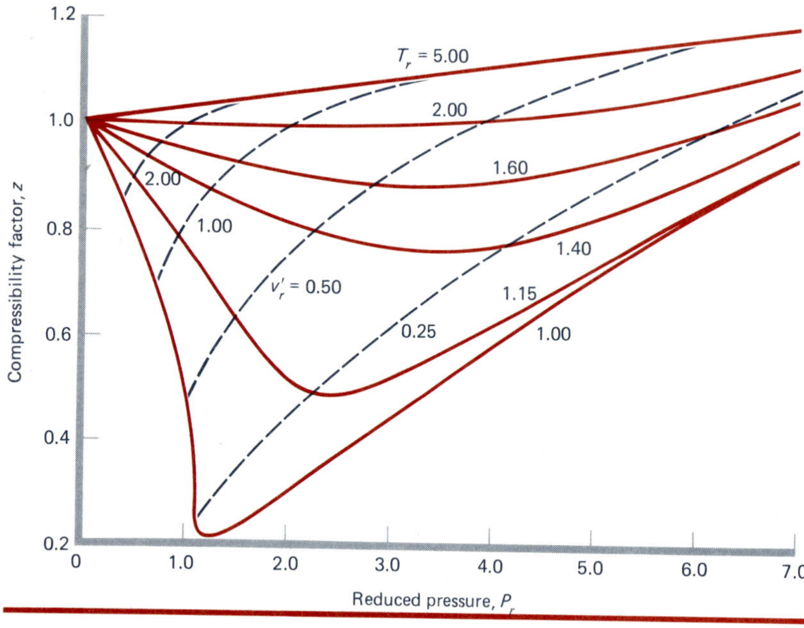

Figure 3.19 Compressibility chart.

If the principle of corresponding states is correct, then carefully measured *P-v-T* data for *any* gas could be used to obtain a plot of z as a function of P_r and T_r; that same plot should then apply to all other gases. Unfortunately, the values for z as computed by Eq. (3.33) for different substances may vary by up to 6 percent, with even greater variations near the critical point. Table C.4 shows values of z at the critical point as well as the critical properties of many substances. The form of the z chart is shown in Fig. 3.19. Detailed charts of $z(P_r, T_r)$ have been prepared based on data for 30 gases and are presented in Fig. C.1.

In these charts are cross-plotted values of the "pseudo-reduced specific volume" v_r', defined by

$$v_r' \equiv \frac{v}{RT_{cr}/P_{cr}} = z\frac{T_r}{P_r} \tag{3.34}$$

This is *not* the true reduced specific volume $v_r = v/v_{cr}$. The critical volume is difficult to measure accurately; in addition, v_r' can be computed from only v, T_r, and P_r, and the charts remain a function of only two reduced quantities.

The principle of corresponding states has wider application than to only *P-v-T* data; other properties should also follow the principle. We follow up with applications to other properties in Chap. 9.

Example Problem 3.3

Find the value of the specific volume of methane at 250 K and 10 MPa by using the compressibility factor charts.

Solution

First, the critical temperature and pressure for methane are found from Table C.4 as $T_{cr} = 190.4$ K and $P_{cr} = 4.60$ MPa. The reduced properties are then $T_r = 250/190.4 = 1.31$ and $P_r = 10/4.60 = 2.17$. Figure C.1 indicates $z = 0.69$. Using Eq. (3.33) results in

$$v = \frac{zRT}{P} = \frac{0.69(0.51835)(250)}{10{,}000} = 0.0089\ \text{m}^3/\text{kg} = 0.1425\ \text{ft}^3/\text{lbm}$$

Comments

The compressibility factor allows relatively accurate correction for nonideal gas effects over a very wide range of states. Only the critical properties need to be known to apply the factor in determining one dependent property from any two independent properties in the collection of P, v, and T for any gas. This factor is very useful in many engineering applications where the larger amount of data necessary for the more detailed equations of state (such as that of Benedict, Webb, and Rubin) is not available.

Observe that the specific volume calculated in this example is much smaller than that predicted by the ideal gas relation, but is only 6.3 percent larger than that predicted by the van der Waals equation in Example Problem 3.2. Agreement with the B-W-R relation is nearly exact and with experimental data is within 3 percent.

Example Problem 3.4

Use the compressibility factor to determine whether the following states can be treated as an ideal gas:

(*a*) Water at 1800°F, 20 atm
(*b*) Water at 900°F, 3000 atm
(*c*) Water at 600°F, 2 atm
(*d*) Oxygen at 900°F, 1000 atm
(*e*) Oxygen at 100°F, 15 atm
(*f*) Methane at 50°F, 100 atm
(*g*) Carbon dioxide at 650°F, 10 atm

Solution

The critical-point data given in Table C.4 are as follows:

Substance	T_{cr}, °R	P_{cr}, atm
Water	1165	217.7
Oxygen	278.3	49.7
Methane	342.7	45.4
Carbon dioxide	547.4	72.8

To use the compressibility charts in Fig. C.1, the reduced temperature and pressure, $T_r = T/T_{cr}$ and $P_r = P/P_{cr}$, must be found. The problem statement can be expressed in terms of reduced properties as follows:

Problem Part	T_r	P_r	z
(*a*)	1.940	0.09	0.995
(*b*)	1.167	13.78	1.53
(*c*)	0.910	0.009	0.996
(*d*)	4.886	19.96	1.38
(*e*)	2.011	0.30	0.99
(*f*)	1.487	2.20	0.82
(*g*)	2.027	0.14	0.993

The values of z have been obtained from the charts in Fig. C.1.

The compressibility factor is given as

$$z = \frac{Pv}{RT}$$

so that $z \simeq 1.0$ is an ideal gas. Thus, the states given in the above table imply that (*a*), (*c*), (*e*), and (*g*) can be treated as an ideal gas. Also (*b*), (*d*), and (*f*) should not be considered an ideal gas.

Comments

The compressibility factor chart can be used to estimate when a state or process is accurately approximated as an ideal gas. The previous recommendation of $P < P_{cr}$ and $T > 1.4T_{cr}$ can be extended to all locations within $0.9 \leq z \leq 1.1$ if the desired accuracy is within $\pm$ 10 percent.

Incompressible Liquids and Solids

Many liquids and solids exhibit a very small change in density or specific volume if the other properties are altered. These substances are termed *incompressible* and are represented by

$$v = \text{constant} \tag{3.35}$$

The other properties for incompressible liquids and solids are also needed. We show in Chap. 9 that the constant-volume and constant-pressure specific heats are equal for a given incompressible liquid or solid. Therefore,

$$c_P = c_v = c \tag{3.36}$$

The change in the internal energy is given by Eq. (3.10). Equation (3.35) indicates that $dv = 0$, so Eq. (3.10) yields

$$du = c_v\, dT = c\, dT \tag{3.37a}$$

and

$$u_2 - u_1 = \int_1^2 c\, dT \tag{3.37b}$$

The enthalpy is $h = u + Pv$, so

$$dh = du + P\, dv + v\, dP \tag{3.38}$$

Again, $dv = 0$ for an incompressible liquid or solid, so with Eqs. (3.37*a*) and (3.37*b*)

$$dh = c\, dT + v\, dP \tag{3.39a}$$

and

$$h_2 - h_1 = \int_1^2 c\, dT + v(P_2 - P_1) = u_2 - u_1 + v(P_2 - P_1) \tag{3.39b}$$

These equations are useful in determining the property changes for incompressible liquids and solids.

Example Problem 3.5

The following data are adapted from data similar to that in Table D.10 for the subcooled liquid properties of water. Determine the error in the incompressible liquid approximation for the temperatures and pressures given.

Subcooled Liquid Properties for H_2O

P, MPa	T, °C	v, m³/kg	u, kJ/kg	h, kJ/kg
5	100	0.0010409	417.54	422.75
	260	0.0012751	1127.95	1134.33
10	100	0.0010384	416.13	426.52
	260	0.0012650	1121.14	1133.69
20	100	0.0010336	413.40	434.07
	260	0.0012469	1108.5	1133.4
50	100	0.0010201	405.84	456.84
	260	0.0012032	1077.7	1137.9

Solution

The incompressible fluid approximation takes $v = \text{constant}$, so that $du = c\,dT$ and $dh = c\,dT + v\,dP$. The state is specified in the subcooled liquid region by the temperature T and pressure P. If a constant-temperature line is considered from the subcooled liquid state to the saturated liquid line, then

$$u - u_l = \int_{T_l}^{T} c\,dT = 0 \qquad \text{for an isotherm}$$

and

$$u(T, P) - u_l(T) = \int_{T_l}^{T} c\,dT = 0 \qquad \text{for an isotherm}$$

Thus,

$$u(T, P) = u_l(T)$$

Also

$$h(T, P) - h_l(T) = u(T, P) - u_l(T) + v_l[P - P_l(T)]$$

or

$$h(T, P) - h_l(T) = 0 + v_l[P - P_{sat}(T)]$$

$$h(T, P) = h_l(T) + v_l[P - P_{sat}(T)] \qquad \text{for an isotherm}$$

where v_l is evaluated at the state temperature $v = v_l(T)$.

To approximate the first line of the above table, the above equations and Table D.10 (discussed in Sec. 3.5.3) are used. At 100°C, $P_{sat}(100°\text{C}) = 101.32$ kPa, $v_l(100°\text{C}) = 0.001043$ m^3/kg, $u_l(100°\text{C}) =$ 418.96 kJ/kg, and $h_l(100°\text{C}) = 419.06$ kJ/kg. Thus, at the first state of $P = 5$ MPa and $T = 100°$C, the approximated values are

$$v = v_l(100°\text{C}) = 0.001043 \text{ m}^3/\text{kg}$$

$$u = u_l(100°\text{C}) = 418.96 \text{ kJ/kg}$$

$$h = h_l(100°\text{C}) + v_l(100°\text{C})[P - P_{sat}(100°\text{C})]$$

$$= 419.06 + 0.001043(5000 - 101.32)$$

$$= 424.2 \text{ kJ/kg}$$

For the second state at $P = 5$ MPa and $T = 260°$C, the approximated values are $v = v_l(260°\text{C}) = 0.001276$ m^3/kg, $u = u_l(100°\text{C}) = 1128.4$ kJ/kg, and $h = 1140.7$ kJ/kg. All the values required in the above table are at two temperatures. So the approximated values of u and v are the same, and only the enthalpy must be calculated for each P and T. These calculations are carried out for all the values in the table. With the error calculated as

$$\text{Error} = \frac{\text{approximate property} - \text{tabulated property}}{\text{tabulated property}}$$

the following table is generated:

P, MPa	T, °C	v Error, %	u Error, %	h Error, %
5	100	0.202	0.340	0.336
	260	0.071	0.040	0.557
10	100	0.443	0.680	0.672
	260	0.870	0.657	1.18
20	100	0.909	1.34	1.32
	260	2.33	1.80	2.33
50	100	2.24	3.23	3.12
	260	5.96	4.70	5.29

Comments

The table indicates the errors in the incompressible approximation. Below the critical pressure, this approximation is accurate to within a few percent. The error in the specific volume is usually the largest of the property errors. Therefore, the incompressible approximation can be used in many applications with a high degree of accuracy.

3.5.3 Tabular Data

When detailed property data are to be presented for a wide range of states, so that a single equation of state is not sufficient, it is customary to present the data in tabular form. Because the commonly measured properties are temperature and pressure, usually data are presented with T and P used as the independent properties and the other properties treated as dependent. The intervals between table entries in complete tabulations are usually chosen so that linear interpolation will provide accurate values for states that fall between entries.

Use of Steam Tables

Because of the many uses of steam in industrial practice, extensive tables are available for steam properties [1–3]. Abbreviated data computed from relations in Ref. 3 are included in Tables D.10 through D.12 (SI) and Tables E.10 through E.12 (USCS). The reference state for these tables is saturated liquid at 0°C (32°F).

Example Problem 3.6

Find the value of the enthalpy of water or steam at the following states:

(*a*) $P = 0.2$ MPa, $T = 120.24$°C
(*b*) $P = 0.2$ MPa, $x = 0.59$
(*c*) $T = 250$°F, $v = 10$ ft³/lbm
(*d*) $P = 350$ psia, $T = 700$°F
(*e*) $T = 700$°F, $v = 2.2267$ ft³/lbm
(*f*) $T = 750$°F, $v = 2.1$ ft³/lbm
(*g*) $P = 2.0$ MPa, $u = 2900$ kJ/kg
(*h*) $P = 2.5$ MPa, $T = 100$°C

Solution

To begin, the region of the phase diagram in which the state lies must be determined. The tables are organized so that the saturation properties are given separately from the properties of superheated steam and subcooled liquid. In addition, the saturation properties are presented in two ways—with the saturation pressure given at even intervals for easy interpolation and with the saturation temperature given at even intervals for the same reason.

The region in which the prescribed state lies usually is found easily by referring to one of the two tables for saturation properties. If, at the given T or P, the given specific intensive property lies outside the range of properties that can exist for saturated liquid, saturated vapor, or their mixtures, then the state must lie in either the superheated or the subcooled region.

(*a*) For the state prescribed, the pressure is given at an even increment, so we first refer to the saturation property table that is presented as a function of pressure (Table D.11). An excerpt of that table is given here:

Pressure Table for Saturated Water

		Specific Volume, m³/kg			Enthalpy, kJ/kg		
P_{sat}, kPa	T_{sat}, °C	v_l	v_{lg}	v_g	h_l	h_{lg}	h_g
101.32	100.00	0.001043	1.6727	1.6737	419.06	2256.6	2675.7
125.00	105.99	0.001048	1.3742	1.3752	444.38	2240.7	2685.1
150.00	111.38	0.001053	1.1584	1.1595	467.18	2226.2	2693.4
175.00	116.07	0.001057	1.0027	1.0038	487.08	2213.3	2700.4
200.00	120.24	0.001060	0.8848	0.8859	504.80	2201.7	2706.5
225.00	124.01	0.001064	0.7923	0.7934	520.83	2191.2	2712.0

For each saturation pressure, the corresponding saturation temperature (boiling point) is given along with the values of specific volume and enthalpy for both saturated liquid and saturated vapor. The enthalpy value in the h_{lg} column is the difference in enthalpy between the saturated vapor and saturated liquid states.

For the particular state given in (*a*), $P = 0.2$ MPa and $T = 120.24°C$, what is the value of enthalpy? It so happens that the temperature given is exactly the saturation temperature. If the temperature were greater at this pressure, then the state would be in the superheated vapor region; if T were lower, the state would be in the subcooled liquid (or possibly solid) region. All we can say with the data given, however, is that the state is in the two-phase region. The T and P given are not independent in this case, so the enthalpy could have any value from 504.80 to 2706.5 kJ/kg, depending on the relative amounts of saturated liquid and vapor present.

(*b*) For this case, P and x are given. Such a state must lie in the saturation region unless the state has been misspecified, since x is defined only in the saturation region. (The only misspecification possible for x given in the range $0 < x < 1$ would be if P were greater than P_{cr}.)

To find h in this state, we use Eq. (3.9):

$$h = h_l + xh_{lg}$$

and introducing the data from the abbreviated table at 0.2 MPa, we find

$$h = 504.80 + 0.59(2201.7) = 1803.8 \text{ kJ/kg}$$

(*c*) For this case, we again first refer to the tables of saturation properties, to determine which region contains the specified state. In this case, however, the temperature is given, so we refer for convenience to Table E.10, a part of which is excerpted here:

Temperature Table for Saturated Water

		Specific Volume, ft³/lbm			Enthalpy, Btu/lbm		
T_{sat}, °F	P_{sat}, psia	v_l	v_{lg}	v_g	h_l	h_{lg}	h_g
250	29.814	0.01700	13.813	13.830	218.62	945.5	1164.1
260	35.411	0.01708	11.755	11.772	228.79	938.8	1167.6
270	41.835	0.01717	10.052	10.069	239.00	931.9	1170.9
280	49.173	0.01726	8.6350	8.6523	249.23	924.9	1174.1
290	57.516	0.01735	7.4505	7.4679	259.49	917.7	1177.2

Because the specified v lies between the saturated liquid and saturated vapor values, the state does lie in the saturation region. Equation (3.7*b*) can be used with the table values to find the quality from

$$x = \frac{v - v_l}{v_{lg}}$$

to give

$$x = \frac{10.0 - 0.017}{13.813} = 0.723$$

Now the enthalpy can be found by using the known quality:

$$h = h_l + xh_{lg} = 218.62 + (0.723)(945.5) = 900 \text{ Btu/lbm}$$

(*d*) For $P = 350$ psia, $T = 700°$F, we check either of the saturation tables and find that for $P = 350$ psia, the temperature is above the saturation temperature of 431.81°F or for $T = 700°$F, the pressure is below the saturation pressure of 3090.8 psia. In either case, the indication is that the state is in the superheat region. We then proceed to the superheated steam tables (Table E.12), a portion of which is shown now:

Properties for Superheated Steam

T, °F	v, ft³/lbm	h, Btu/lbm
	$P = 300$ psia ($T_{sat} = 417.43$°F)	
500	1.7656	1257.1
600	2.0037	1314.2
700	2.2267	1368.1
800	2.4419	1421.0
900	2.6525	1473.6
	$P = 350$ psia ($T_{sat} = 431.81$°F)	
500	1.4908	1251.1
600	1.7021	1310.3
700	1.8973	1365.3
800	2.0844	1418.8
900	2.2667	1471.9

For this case, h is read directly as 1365.3 Btu/lbm.

(*e*) Again, we check the saturation tables at the given temperature, and we find that the specific volume is greater than that for a saturated vapor at that temperature. The state is thus in the superheat region, and from the abbreviated table above the corresponding enthalpy is 1368.1 Btu/lbm.

(*f*) This state is also found to be in the superheat region and its values to lie within the abbreviated table given above. However, neither the temperature nor the specific volume lies on a table entry. We are thus required to perform a double interpolation to find the corresponding enthalpy. This is done by first finding the values of v and h that occur at $T = 750°F$ at the pressures that bracket known values of v. Thus, at $P = 300$ psia and $T = 750°F$,

$$v = 2.2267 + \frac{750 - 700}{800 - 700}(2.4419 - 2.2267) = 2.3343 \text{ ft}^3/\text{lbm}$$

and

$$h = 1368.1 + \frac{750 - 700}{800 - 700}(1421.0 - 1368.1) = 1394.6 \text{ Btu/lbm}$$

A similar interpolation at $P = 350$ psia for $T = 750°F$ gives $v = 1.9909$ ft³/lbm and $h = 1392.1$ Btu/lbm. Now, since we have assumed that linear interpolation is acceptable, we also assume that h lies between the known values at the two pressures in the same proportion that v lies between the two values, or

$$\frac{v - v(300 \text{ psia})}{v(350 \text{ psia}) - v(300 \text{ psia})} = \frac{h - h(300 \text{ psia})}{h(350 \text{ psia}) - h(300 \text{ psia})}$$

or

$$\frac{2.1 - 2.3343}{1.9909 - 2.3343} = \frac{h - 1394.6}{1392.1 - 1394.6}$$

which gives $h = 1392.9$ Btu/lbm.

(*g*) Some tables provide internal energy as a tabulated property, but many do not. If u is not tabulated, then it must be computed from h, and this can lead to considerable tail chasing. Note first by examination of any portion of the superheat tables that both u and h are weak functions of pressure and strong functions of temperature. Thus, if P and h or P and u are given to define the state, usually it is fairly easy to find the corresponding temperature and then the other state properties (the saturation tables are first examined to determine the state region). If T and u or h are given, it is sometimes tedious to converge on the corresponding properties to good accuracy.

In the present case, we assume various temperatures and find the corresponding h and v values at $P = 2.0$ MPa. The corresponding u values are then computed from $u = h - Pv$. This effectively adds another column to the table given above, which for 2.0 MPa becomes

Properties of Superheated Steam

T, °C	v, m³/kg	u, kJ/kg	h, kJ/kg
	$P = 2000$ kPa ($T_{sat} = 212.42$°C)		
250	0.1114	2678.8	2901.6
300	0.1254	2771.8	3022.7
350	0.1386	2859.4	3136.6
400	0.1512	2945.1	3247.5

For the given $u = 2900$ kJ/kg, the temperature must lie between 350 and 400°C. Again, assuming that linear interpolation is adequate, we can determine h at the given state from

$$\frac{h - h(350°\text{C})}{h(400°\text{C}) - h(350°\text{C})} = \frac{u - u(350°\text{C})}{u(400°\text{C}) - u(350°\text{C})}$$

or

$$\frac{h - 3136.6}{3247.5 - 3136.6} = \frac{2900 - 2859.4}{2945.1 - 2859.4}$$

which gives $h = 3189.1$ kJ/kg.

(*h*) From the saturation states given inTable D.10 the saturation pressure corresponding to 100°C is 0.10132 MPa. Since the desired pressure is 2.5 MPa, the water must be in the subcooled liquid state. This state may be estimated from the saturation tables or from the subcooled liquid tables if available.

The saturated liquid state represents the liquid in the saturated region, and pressure and temperature are dependent. The subcooled state's relation to the saturation state is interpreted by using Fig. 3.11. Constant-pressure lines on this T-v diagram are very close to the saturated liquid line for all pressures. Thus large pressure variations do not significantly alter the state. The temperature, however, fairly accurately represents the state. Thus, the state in the subcooled region is approximated by the saturated liquid value at the corresponding temperature.

The enthalpy is evaluated by assuming the liquid is an incompressible fluid, as in Example Problem 3.5. Equations (3.37) and (3.39) are valid. Considering a constant-temperature line from the saturated liquid line to the subcooled liquid state, we can evaluate the difference in enthalpy from Eq. (3.39). Integrating this expression yields

$$h - h_l = 0 + v_l(P - P_{sat})$$

$$h = h_l + v_l(P - P_{sat})$$

$$= 419.06 + 0.001043(2500 - 101.32)$$
$$= 419.06 + 2.502 = 421.56 \text{ kJ/kg}$$

Using the more accurate compressed liquid properties for water (Table D.10) and interpolating first on T at 100°C at the given pressures of 2.0 and 3.0 MPa and then interpolating on pressure give

$$h = \frac{1}{2}\left(\frac{392.39 + 475.87}{2} + \frac{400.16 + 483.24}{2}\right) = 437.92 \text{ kJ/kg}$$

and the error in using the compressed liquid approximation in this case is

$$\text{Error} = \frac{(421.56 - 437.92) \times 100}{437.92} = -3.74 \text{ percent}$$

Comments

We find in later chapters that it is often necessary to find the change in certain specific intensive properties between two states. To do so requires that the property values in each state be found to four or five significant figures, so that the differences will retain acceptable accuracy. Use of the tables can provide such accurate property values if tabular values are available at small increments. Complete tables [1–3, 6] should therefore be used for accurate work. The abbreviated tables presented in Appendixes D and E can introduce errors in interpolation and are presented chiefly as an aid to solving homework problems.

The major difficulty in use of the tables lies in the inevitable interpolations required for accurate solution of real problems. The use of computerized tables offers great savings in time and reduces numerical errors.

Table for Other Substances

The exact data for the thermodynamic properties of each substance are different. Tables or other data presentations of the properties must be available for each substance if careful thermodynamic analysis is to be done. Abridged tables are shown in SI in Tables D.9 through D.15 and in Tables E.9 through E.15 in the USCS for the properties of water and refrigerant 12. These tables cover the saturated liquid, two-phase, and superheat regions. The ammonia tables are Tables D.16 through D.18 in SI and Tables E.16 through E.18 in USCS. Table C.13 presents data for the heat of fusion at standard pressure (1 atm) for selected substances.

Gas Tables

As noted in Sec. 3.5.2, the internal energy and enthalpy of an ideal gas are independent of pressure and so can be presented as a function of temperature only. This makes data presentation much simpler than for substances that cannot be treated as ideal gases, such as steam. Tables of the thermodynamic properties of ideal gases based on the statistical thermodynamics approach (Chap. 13) can be prepared, and they are very accurate in comparison with experimental measurement of the properties. Such calculated properties depend on a limited amount of data that can be accurately obtained by analysis of the spectra emitted or absorbed by the gases. For monatomic gases (argon, neon, krypton, etc.) and diatomic gases (carbon monoxide, nitrogen, oxygen, hydrogen, etc.), the calculation of the properties is fairly straightforward. For more complex molecules (CO_2, H_2O in the ranges of ideality, CH_4, etc.), the calculation of the properties is more difficult. Such tables are available in Ref. 7.

In Tables D.2 through D.8 (SI) and Tables E.2 through E.8 (USCS), the properties of air, argon, nitrogen, oxygen, superheated water vapor, carbon dioxide, and carbon monoxide are presented under the assumption that these substances behave as ideal gases. The data are computed from Ref. 8. The values for air are computed from values for nitrogen, oxygen, and argon by assuming that air acts as an ideal mixture of ideal gases. These tables are in essentially exact agreement with the tables of Ref. 7, since both references use the fundamental statistical thermodynamics equations for their calculated properties.

We have noted that the state to which we assign the value of zero for u or h is arbitrary insofar as classical thermodynamics is concerned, because only *changes* in u or h are of interest. For the gas tables, it is common practice to assign zero values to u and h at an absolute temperature of $T = 0$ K or 0°R so that u and h always have positive values. These tables are denoted as being for "low pressure" because they do not apply for states where u and h become pressure-dependent, i.e., when the behavior of the gas becomes nonideal. The gases presented in the tables follow ideal gas behavior for temperatures above about 30°C (85°F) and pressures below about 14 atm with negligible error.

3.5.4 Computerized Property Data Retrieval

Values for the thermodynamic properties of many substances are available in the form of computer programs that can provide the properties at any given state. This avoids the need for interpolation and/or auxiliary computation, such as when only h is given in a table and u must be calculated from $u = h - Pv$. Such programs are included with this text for the properties of air, nitrogen, and argon (as ideal gases); for steam; for refrigerant 12; and for various hydrocarbons. Except for the steam tables, the programs are abridged forms of more general programs that cover many more substances and a broader range of variables. These more general programs [8, 9] have other capabilities as well, such as the ability to carry out some combustion and cycle analyses. A complete discussion of these programs is found in App. G.

The student is urged to use the property programs available with this text in evaluating the properties necessary for carrying out the homework problems, some of which can be very tedious if only tables are available for solution. The programs should help in broadening the scope of problems that can be usefully approached. The programs provide data of accuracy equivalent to the tables of properties in the appendixes.

In this text, we have avoided using programs that carry out cycle analysis, combustion analysis, etc. The ability to understand the basis of such calculations and to determine the assumptions and limitations commonly made to facilitate them is part of the value of an engineering education. After these skills are mastered, *then* fully computerized solution becomes a valuable tool.

3.6 Remarks

All the properties defined in classical thermodynamics are limited to a control mass in thermodynamic equilibrium. That is, the control mass is in mechanical, chemical, and thermal equilibrium and has no gradients across it, so that the properties are the same in each portion of the control mass that we examine. The control mass, in other words, is at a *uniform* state. However, most problems of interest in engineering applications are not in a uniform state in thermodynamic equilibrium. How do we justify the use of temperature, for example, for a control mass where there is heat transfer occurring, since heat transfer, by definition, requires the presence of a temperature difference between the control mass and the surroundings, and the control mass is therefore not in thermal equilibrium?

Classical thermodynamics does not really answer this question in a satisfactory way, except to note that the use of the properties that we have defined (and will define) leads to the results that we expect from application of the thermodynamic relations *even in cases that are quite far from thermodynamic equilibrium.* We must take another viewpoint, that of statistical thermodynamics as introduced in Chap. 13, to show *why* we can extend the use of properties defined only for the case of thermodynamic equilibrium to cases of nonequilibrium.

There are indeed cases in which the idea of assigning a single temperature to a control mass that is far from thermal equilibrium is simply not valid. In the plasma necessary to sustain a fusion reaction, for example, the conditions are so far from thermodynamic equilibrium, even in the smallest control mass taken from the plasma, that the equilibrium properties do not apply, and a more detailed microscopic analysis must be involved. However, for the vast majority of engineering problems, we can assume that application of the property information developed from the classical approach is valid.

Problems

Problems marked with a star are lengthy; thus, computerized steam tables or an accurate Mollier chart should be available if these problems are to be completely worked.

3.1S Find P_1/P_2 for the states $v_1 = 50$ m^3/kg, $T_1 = 500$°C and $v_2 = 10$ m^3/kg, $T_2 = 1200$°C for ethane, using (*a*) the ideal gas relation, (*b*) the van der Waals equation, (*c*) the Benedict-Webb-Rubin equation, (*d*) the compressibility factor charts.

3.1E Find P_1/P_2 for the states $v_1 = 600$ ft^3/lbm, $T_1 = 950$°F and $v_2 = 120$ ft^3/lbm, $T_2 = 2400$°F for ethane, using (*a*) the ideal gas relation, (*b*) the van der Waals equation, (*c*) the Benedict-Webb-Rubin equation, (*d*) the compressibility factor charts.

3.2S Can the following fluids in the specified states be treated as ideal gases?
(*a*) Air at 0.1 MPa and 20°C
(*b*) Air at 13 MPa and 900°C
(*c*) Methane at 2 MPa and 1000°C
(*d*) Water at 0.1 MPa and 20°C
(*e*) Water at 0.01 MPa and 30°C
(*f*) Refrigerant 12 at 1 MPa and 50°C

3.2E Can the following fluids in the specified states be treated as ideal gases?
(*a*) Air at 14.7 psia and 68°F
(*b*) Air at 2000 psia and 1650°F
(*c*) Methane at 300 psia and 1850°F
(*d*) Water at 14.7 psia and 68°F
(*e*) Water at 1.5 psia and 86°F
(*f*) Refrigerant 12 at 150 psia and 120°F

3.3S Determine whether the following processes for the specified gas can be treated as ideal gas processes. If the process is an ideal gas process, then evaluate $u_2 - u_1$ and $h_2 - h_1$, assuming constant specific heats.
(*a*) Isothermal process (T = constant) for water from 1200 K and 2 MPa to 10 MPa
(*b*) Isometric process (v = constant) for refrigerant 12 from 405 K and 4.0 MPa to 470 K
(*c*) Isobaric process (P = constant) for nitrogen from 250 K and 17.0 MPa to 280 K

3.3E Determine whether the following processes for the specified gas can be treated as ideal gas processes. If the process is an ideal gas process, then evaluate $u_2 - u_1$ and $h_2 - h_1$, assuming constant specific heats.
(*a*) Isothermal process (T = constant) for water from 1700°F and 300 psia to 1500 psia

(*b*) Isometric process (v = constant) for refrigerant 12 from 270°F and 590 psia to 390°F

(*c*) Isobaric process (P = constant) for nitrogen from −10°F and 1500 psia to 45°F

3.4S Find the specific volume of sulfur dioxide at $T = 480$ K and $P = 15$ MPa.

3.4E Find the specific volume of sulfur dioxide at $T = 850°$R and $P = 2100$ psia.

3.5S Nitrogen gas is originally at $P = 200$ atm and $T = 252.4$ K. It is cooled at constant volume to $T = 189.3$ K. What is the pressure at the lower temperature?

3.5E Nitrogen gas is originally at $P = 200$ atm and $T = 455°$R. It is cooled at constant volume to $T = 340°$R. What is the pressure at the lower temperature?

3.6 Plot the $T_r = 1.30$ isotherm for n-butane on a plot of compressibility factor versus P_r, using the Benedict-Webb-Rubin equation to find z.

3.7 A virial equation of state has the form

$$\frac{P}{T} = \frac{R}{v}\left[1 + \frac{B(T)}{v} + \frac{C(T)}{v^2} + \cdots\right]$$

where $B(T)$, $C(T)$, etc., are functions of temperature only and are called the first virial coefficient, the second virial coefficient, and so on. Find expressions for the virial coefficients for gases that obey (*a*) the ideal gas relation, (*b*) the van der Waals equation, and (*c*) the Benedict-Webb-Rubin equation.

3.8 The numerical values of properties x, y, and z are given. Linearly interpolate and obtain the specified property.
(*a*) Given:

x	y
110	1.2074
115	1.0351

Find y for $x = 112.3$.
(*b*) Given:

$x = 10$		$x = 50$	
y	z	y	z
150	19.513	150	3.8894
200	21.820	200	4.3561

Find z for $x = 27$ and $y = 150$.

(*c*) Given: Same data as in (*b*). Find z for $x = 32$ and $y = 181$.

3.9S Indicate whether the following states for water are in the liquid, saturation, or superheated region. Specify the quality of the states in the saturation region.

State	P, kPa	T, °C	v, m^3/kg
1	1700	200	—
2	1200	—	0.0010
3	—	75	3.0
4	500	202	—
5	350	—	0.005

3.9E Indicate whether the following states for water are in the liquid, saturation, or superheated region. Specify the quality of the states in the saturation region.

State	P, psia	T, °F	v, ft^3/lbm
1	250	390	—
2	175	—	0.012
3	—	166	36.0
4	75	400	—
5	50	—	0.060

3.10S An ideal gas (molecular weight of 29) occupies a rigid container at the given conditions. Determine (*a*) the molal specific volume in cubic meters per kilomole, (*b*) the number of moles in kilomoles, and (*c*) the gas constant R.

$T = 21°C \qquad P = 0.1\ MPa \qquad V = 57\ m^3$

3.10E An ideal gas (molecular weight of 29) occupies a rigid container at the given conditions. Determine (*a*) the molal specific volume in cubic feet per pound-mole, (*b*) the number of moles in pound-moles, and (*c*) the gas constant R.

$T = 72°F \qquad P = 14.7\ psia \qquad V = 1650\ ft^3$

3.11S Helium occupies a rigid container at the given conditions. Determine the specific volume in cubic meters per kilogram and the mass in kilograms.

$T = 20°C \qquad P = 0.1\ MPa \qquad V = 60\ m^3$

3.11E Helium occupies a rigid container at the given conditions. Determine the specific volume in cubic feet per pound-mass and the mass in pound-mass.

$T = 68°F \qquad P = 14.7\ psia \qquad V = 1810\ ft^3$

3.12S Air occupies a rigid container at the conditions shown below. Determine the temperature in kelvins and the total volume of the container.

$P = 0.1$ MPa $\quad \rho = 1.00$ kg/m^3 $\quad m = 453.6$ kg

3.12E Air occupies a rigid container at the conditions shown below. Determine the temperature in degrees Rankine and the total volume of the container.

$P = 14.7$ psia $\quad \rho = 0.0200$ lbm/ft^3
$m = 1000$ lbm

3.13S Given an ideal gas with the following data, determine the molecular weight of the gas.

$P = 2.1$ MPa $\quad T = 40°$C $\quad v = 0.0946$ m^3/kg

3.13E Given an ideal gas with the following data, determine the molecular weight of the gas.

$P = 300$ psia $\quad T = 100°$F $\quad v = 1.16$ ft^3/lbm

3.14S In a rigid tank 1.0 m^3 of O_2 is heated from an initial pressure of 0.25 MPa and temperature of 50°C to a final temperature of 325°C. Determine the mass of the gas and its final pressure.

3.14E In a rigid tank 30 ft^3 of O_2 is heated from an initial pressure of 40 psia and temperature of 120°F to a final temperature of 620°F. Determine the mass of the gas and its final pressure.

3.15S The Liter-than-Air Balloon Company has a helium balloon that will be used in an attempted crossing of the Atlantic Ocean. The helium will initially be at 20°C and 500 kPa. The balloon, a rigid sphere, has a radius of 25 m. Determine the mass of the helium in kilograms.

3.15E The Liter-than-Air Balloon Company has a helium balloon that will be used in an attempted crossing of the Atlantic Ocean. The helium will initially be at 70°F and 75 psia. The balloon, a rigid sphere, has a radius of 75 ft. Determine the mass of the helium in pound-mass.

3.16S As the balloon in Prob. 3.15S rises, the helium is heated by solar radiation to a temperature of 300°C. If the safe pressure limit is 1 MPa, are the balloonists in danger?

3.16E As the balloon in Prob. 3.15E rises, the helium is heated by solar radiation to a temperature of 600°F. If the safe pressure limit is 150 psia, are the balloonists in danger?

3.17S If a newer model of the balloon in Prob. 3.15S were to have its allowable mass cut by 26,690 kg, what gas could be used in place of helium?

3.17E If a newer model of the balloon in Prob. 3.15E were to have its allowable mass cut by 46,300 lbm, what gas could be used in place of helium?

NOTE: For Probs. 3.18 through 3.35, assume the substance is water and find the required properties, using the necessary tables. Compare the results with values from the computerized tables.

3.18S Given water at $P = 10.0$ MPa and $u = 418.96$ kJ/kg. Find T and h.

3.18E Given water at $P = 2000$ psia and $u = 180.14$ Btu/lbm. Find T and h.

3.19S Given water at $T = 200°C$ and $h = 852.38$ kJ/kg. Find P and v.

3.19E Given water at $T = 400°F$ and $h = 375.08$ kJ/kg. Find P and v.

3.20S Given water at $P = 0.08$ MPa and $h = 1000$ kJ/kg. Find T and v.

3.20E Given water at $P = 9$ psia and $h = 600$ Btu/lbm. Find T and v.

3.21S Given water at $P = 0.3$ MPa and $h = 334.90$ kJ/kg. Find v.

3.21E Given water at $P = 75$ psia and $h = 225.36$ Btu/lbm. Find v.

3.22S Given water at $P = 15.0$ MPa and $u = 3507.8$ kJ/kg. Find h and T.

3.22E Given water at $P = 2500$ psia and $u = 1512.5$ Btu/lbm. Find h and T.

3.23S Given water at $T = 170°C$ and $x = 0.45$. Find P and v.

3.23E Given water at $T = 350°F$ and $x = 0.45$. Find P and v.

3.24S Given water at $T = 300°C$ and $u = 2804.4$ kJ/kg. Find P and h.

3.24E Given water at $T = 500°F$ and $u = 1167.6$ Btu/lbm. Find P and h.

3.25S Given water at $T = 200°C$ and $P = 10.0$ MPa. Find u.

3.25E Given water at $T = 400°F$ and $P = 1500$ psia. Find u.

3.26S Given water at $P = 0.2$ MPa and $x = 0.75$. Find T and u.

3.26E Given water at $P = 30$ psia and $x = 0.75$. Find T and u.

3.27S Given water at $T = 120°C$ and $P = 0.5$ MPa. Find h.

3.27E Given water at $T = 250°F$ and $P = 75$ psia. Find h.

3.28S Given water at $T = 400°C$ and $P = 2.0$ MPa. Find v.

3.28E Given water at $T = 750°F$ and $P = 300$ psia. Find v.

3.29S Given water at $T = 200°C$ and $h = 1800$ kJ/kg. Find P and v.

3.29E Given water at $T = 500°F$ and $h = 800$ Btu/lbm. Find P and v.

3.30S Given water at $P = 8$ MPa and $v = 0.006952$ m^3/kg. Find x and h.

3.30E Given water at $P = 1200$ psia and $v = 0.06857$ ft^3/lbm. Find x and h.

3.31S Given water at $T = 210°C$ and $P = 3.0$ MPa. Find u and h.

3.31E Given water at $T = 400°F$ and $P = 2000$ psia. Find u and h.

3.32S Given water at $T = 90°C$ and $v = 0.001036$ m^3/kg. Find h and P.

3.32E Given water at $T = 200°F$ and $v = 0.01663$ ft^3/lbm. Find h and P.

3.33S Given water at $P = 15$ MPa and $v = 0.001006$ m^3/kg. Find T and h.

3.33E Given water at $P = 2000$ psia and $v = 0.01844$ ft^3/kg. Find T and h.

3.34S Given water at $P = 16$ MPa and $x = 1.0$. Find T and u.

3.34E Given water at $P = 2500$ psia and $x = 1.0$. Find T and u.

3.35S Given water at $T = 400°C$ and $v = 0.5137$ m^3/kg. Find x, P, and h.

3.35E Given water at $T = 800°F$ and $v = 7.4451$ ft^3/lbm. Find x, P, and h.

3.36S Find the value of internal energy for the substances in the states indicated.
(*a*) Water at 0.4 MPa and 725°C
(*b*) Water at 3.0 MPa and 0.01 m^3/kg
(*c*) Refrigerant 12 at 130°C and 125 kPa
(*d*) Water at 1.0 MPa and 100°C

3.36E Find the value of internal energy for the substances in the states indicated.
(*a*) Water at 60 psia and 1350°F
(*b*) Water at 450 psia and 0.12 ft^3/lbm
(*c*) Refrigerant 12 at 265°F and 18 psia
(*d*) Water at 150 psia and 212°F

3.37E Indicate whether the following states for refrigerant 12 are in the subcooled, saturation, or superheated regions. The reference values for u and h are those used for Tables D.13 through D.15.

State	P, kPa	T, °C	v, m³/kg	u, kJ/kg	h, kJ/kg
1	2000	—	—	—	220
2	—	100	0.04	—	—
3	—	—	0.02	325	—
4	2000	—	—	—	300
5	600	200	—	—	—
6	—	—	0.30	—	365

3.37S Indicate whether the following states for refrigerant 12 are in the subcooled, saturation, or superheated regions. The reference values for u and h are those used for Tables E.13 through E.15.

State	P, psia	T, °F	v, ft³/lbm	u, Btu/lbm	h, Btu/lbm
1	300	—	—	—	20
2	—	212	0.64	—	—
3	—	—	0.3	140	—
4	300	—	—	—	130
5	90	400	—	—	—
6	—	—	5.0	—	160

3.38S Find the following properties for steam from the steam tables. Show all interpolation calculations. Compare the results with those from computerized tables.
(*a*) $u(T = 400°\text{C}, P = 1500 \text{ kPa})$
(*b*) $h(T = 200°\text{C}, P = 2000 \text{ kPa})$
(*c*) $T(h = 2100 \text{ kJ/kg}, P = 6900 \text{ kPa})$
(*d*) $x(P = 500 \text{ kPa}, h = 2000 \text{ kJ/kg})$
(*e*) $h(u = 2000 \text{ kJ/kg}, x = 0.65)$
(*f*) $h(u = 3100 \text{ kJ/kg}, v = 0.1 \text{ m}^3\text{/kg})$

3.38E Find the following properties for steam from the steam tables. Show all interpolation calculations. Compare the results with those from computerized tables.
(*a*) $u(T = 750°\text{F}, P = 200 \text{ psia})$
(*b*) $h(T = 400°\text{F}, P = 300 \text{ psia})$
(*c*) $T(h = 900 \text{ Btu/lbm}, P = 1010 \text{ psia})$
(*d*) $x(P = 75 \text{ psia}, h = 860 \text{ Btu/lbm})$
(*e*) $h(u = 860 \text{ Btu/lbm}, x = 0.65)$
(*f*) $h(u = 1350 \text{ Btu/lbm}, v = 1.6 \text{ ft}^3\text{/lbm})$

3.39S Find the change in value of enthalpy h for water (steam) between an initial state where $P = 100$ kPa and

$v = 1.6500\ m^3/kg$ and a final state where $P = 3000$ kPa and $T = 525°C$.

3.39E Find the change in value of enthalpy h for water (steam) between an initial state where $P = 14.7$ psia and $v = 26$ ft^3/lbm and a final state where $P = 450$ psia and $T = 980°F$.

3.40S In certain ranges of pressure and temperature, steam does act as an ideal gas. Using the steam tables from either App. D or computerized tables, plot an accurate graph of pressure versus specific volume for pressure in the range of 0.01 to 0.3 times the critical pressure (Table C.4) for 200, 300, and 400°C. On the same graph, plot P versus v for the same temperatures, assuming that P and T are independent variables and v is calculated by using the ideal gas relation. Discuss your results. How do the results compare if T and v are chosen as the independent variables and P is calculated from the ideal gas relation? Use multicycle log-log paper for your plots.

3.40E In certain ranges of pressure and temperature, steam does act as an ideal gas. Using the steam tables from either App. E or computerized tables, plot an accurate graph of pressure versus specific volume for pressure in the range of 0.01 to 0.3 times the critical pressure (Table C.4) for 400, 600, and 800°F. On the same graph, plot P versus v for the same temperatures, assuming that P and T are independent variables and v is calculated by using the ideal gas relation. Discuss your results. How do the results compare if T and v are chosen as the independent variables and P is calcualted from the ideal gas relation? Use multicycle log-log paper for your plots.

3.41S Calculate the value of c_P for air at 100 and 1500°C, using only the enthalpy data from Table D.2. Compare with the results from the polynomial expressions for c_P from Table D.1.

3.41E Calculate the value of c_P for air at 200 and 2800°F, using only the enthalpy data from Table E.2. Compare with the results from the polynomial expressions for c_P from Table E.1.

3.42S Calculate c_P and c_v for steam at $T = 400°C$ and $P = 200$ kPa, using data for h and u from the steam tables of App. D or computerized steam tables. Check whether $c_P - c_v = R$, which would be an indication that steam acts as an ideal gas in this state.

3.42E Calculate c_P and c_v for steam at $T = 800°F$ and $P = 40$ psia, using data for h and u from the steam tables of App. E or computerized steam tables. Check whether

$c_p - c_v = R$, which would be an indication that steam acts as an ideal gas in this state.

3.43 (*a*) From the data in the tables, find the value of the specific heat at constant pressure c_p of ammonia at 40 psia and 100°F.
(*b*) Find the value of the quality of saturated ammonia when the temperature is 40°F and the enthalpy is 500 Btu/lbm.
(*c*) Find the enthalpy of superheated ammonia at 25 psia and 105°F.

3.44 Find the value of the compressibility factor z for saturated ammonia vapor at 60 psia. The molecular weight of ammonia is 17.031, and the universal gas constant is 10.73 psia·ft³/(lbmol·°R).

3.45 Find the required properties in the given states:
(*a*) $v(T = 568.8\text{ K}, P = 12.3\text{ atm})$ for octane
(*b*) P or x at $T = 20°\text{C}$, $h = 250$ kJ/kg for refrigerant 12, whichever is appropriate to define the state
(*c*) $u(P = 550\text{ kPa}, T = 95°\text{C})$ for ammonia

3.46 A particular hydrocarbon has van der Waals' constants with values $a = 9.315$ bar·(m³/kmol)² and $b = 0.0900$ m³/kmol. What is the value of the compressibility factor z for this gas at a temperature of 415 K and a volume per mole of 0.2 m³/kmol? (Do not use the graphs of compressibility factor to determine your answer.) *Note:* The universal gas constant is 8.31441 kPa·m³/(kmol·K), and 1 bar = 100 kPa.

3.47S Calculate the change in enthalpy for nitrogen between $T_1 = 20°\text{C}$ and $T_2 = 1500°\text{C}$, using the data in Table D.4. Compare the result to that found by using the polynomial expressions for c_p from Table D.1 and to that found by using a constant value of c_p based on the mean temperature between T_1 and T_2.

3.47E Calculate the change in enthalpy for nitrogen between $T_1 = 70°\text{F}$ and $T_2 = 2800°\text{F}$, using the data in Table E.4. Compare the result to that found by using the polynomial expressions for c_p from Table E.1 and to that found by using a constant value of c_p based on the mean temperature between T_1 and T_2.

***3.48S** Using the data from Prob. 3.40S and actual P-v-T data from the steam tables or the computerized tables, compute the compressibility factor z for steam in the same ranges of P, v, and T as for Prob. 3.40S. Plot the result as z versus P with T as a parameter. On the same graph, plot z from Fig. C.1 for the same ranges of the variables. Discuss the comparison between the results. What type of graph paper did you choose for plotting the results? Why?

***3.48E** Using the data the Prob. 3.40E and actual P-v-T data from the steam tables or the computerized tables, compute the compressibility factor z for steam in the same ranges of P, v, and T as for Prob. 3.40E. Plot the result as z versus P with T as a parameter. On the same graph, plot z from Fig. C.1 for the same ranges of the variables. Discuss the comparison between the results. What type of graph paper did you choose for plotting the results? Why?

3.49S Find the values of u and h for water at $T = 100°\text{C}$ and $P = 300$ kPa.

3.49E Find the values of u and h for water at $T = 212°\text{F}$ and $P = 45$ psia.

3.50S Steam is at $T = 600°\text{C}$ and $P = 2000$ kPa. Find the change in the enthalpy of the steam when (*a*) T is held constant and P is changed to 100 MPa and (*b*) P is held constant and T is changed to 800°C. Discuss the result in terms of the relative sensitivity of h to P or T. What would you conclude about the relative magnitudes of the partial derivatives of the internal energy, $(\partial u/\partial T)_P$ and $(\partial u/\partial P)_T$? Examine Fig. D.2, and discuss whether the results for the state chosen in this problem are representative of the behavior of steam in all regions of the P-h diagram.

3.50E Steam is at $T = 1100°\text{F}$ and $P = 300$ psia. Find the change in the enthalpy of the steam when (*a*) T is held constant and P is changed to 1500 psia and (*b*) P is held constant and T is changed to 1500°F. Discuss the result in terms of the relative sensitivity of h to P or T. What would you conclude about the relative magnitudes of the partial derivatives of the internal energy, $(\partial u/\partial T)_P$ and $(\partial u/\partial P)_T$? Examine Fig. E.2, and discuss whether the results for the state chosen in this problem are representative of the behavior of steam in ali regions of the P-h diagram.

3.51S A sealed volume of 2.0 cm^3 contains saturated water. The container is heated from an initial pressure of 0.10 MPa until the vessel contains a single phase. Determine the final state of the water if the vessel contains (*a*) 0.10 g and (*b*) 1.0 g.

3.51E A sealed volume of 0.15 in^3 contains saturated water. The container is heated from an initial pressure of 14.7

psia until the vessel contains a single phase. Determine the final state of the water if the vessel contains (*a*) 2.3×10^{-4} lbm and (*b*) 2.3×10^{-3} lbm.

3.52S Nitrogen is contained in a bottle. Initially the nitrogen is at 42 atm and 130 K. The bottle has a volume of 0.02 m^3. The bottle is then heated until the nitrogen reaches 202 K.

(*a*) Can the nitrogen be treated as an ideal gas? Why or why not?
(*b*) What is the pressure of the nitrogen after heating?
(*c*) What is the mass of the nitrogen in the bottle?

3.52E Nitrogen is contained in a bottle. Initially the nitrogen is at 42 atm and 234°R. The bottle has a volume of 0.7 ft^3. The bottle is then heated until the nitrogen reaches 450°R.

(*a*) Can the nitrogen be treated as an ideal gas? Why or why not?
(*b*) What is the pressure of the nitrogen after heating?
(*c*) What is the mass of the nitrogen in the bottle?

3.53S Water (0.5 kg) contained within a constant-volume container ($V = 0.01$ m^3) is heated. The initial pressure is 100 kPa, and the water is heated until it exists in the saturated vapor state. Plot the quality and volume occupied by the liquid versus pressure for this process (that is, x and V_l versus P).

3.53E Water (1 lbm) contained within a constant-volume container ($V = 0.3$ ft^3) is heated. The initial pressure is 14.7 psia, and the water is heated until it exists in the saturated vapor state. Plot the quality and volume occupied by the liquid versus pressure for this process (that is, x and V_l versus P).

3.54S The piston-cylinder device shown contains 1 kg of water at a temperature of 200°C and pressure of 0.8 MPa. The water undergoes an isobaric process until it becomes a saturated liquid. Find the total work done in the process. Is it work done on or by the system? Show the process on a T-v diagram.

3.54E The piston-cylinder device shown contains 2 lbm of water at a temperature of 400°F and a pressure of 120 psia. The water undergoes an isobaric process until it becomes a saturated liquid. Find the total work done in the process. Is it work done on or by the system? Show the process on a T-v diagram.

Figure P3.54

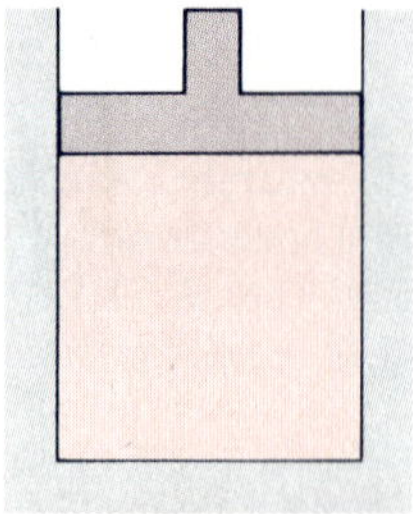

3.55S A piston encloses 2 kg of water in an initial volume of 0.02 m^3 (see Fig. P3.55). The initial temperature is 50°C. The piston leaves the stops at a pressure of 1.0 MPa. The water is heated from its initial state to a final temperature of 200°C.

(*a*) Plot the process on a *P-v* diagram.
(*b*) Evaluate the work done by the H_2O.
(*c*) What volume is occupied by the liquid at the initial state and when the piston leaves the stops?

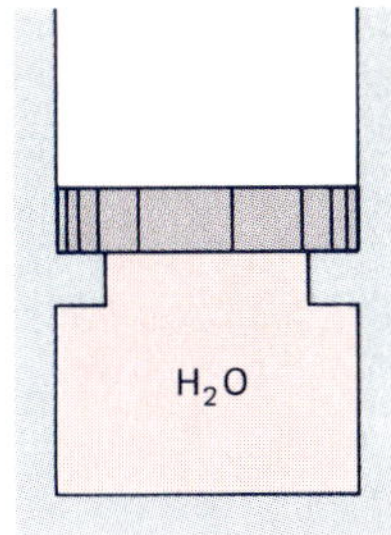

Figure P3.55

3.55E A piston encloses 1 lbm of water in an initial volume of 0.3 ft^3 (see Fig. P3.55). The initial temperature is 120°F. The piston leaves the stops at a pressure of 150 psia. The water is heated from its initial state to a final temperature of 400°F.

(*a*) Plot the process on a *P-v* diagram.
(*b*) Evaluate the work done by the H_2O.
(*c*) What volume is occupied by the liquid at the initial state and when the piston leaves the stops?

3.56S Refrigerant 12 is heated isothermally from an initial state of 200 kPa and 200°C to a final pressure of 800 kPa. Evaluate the work done per unit mass. Also evaluate the work done per unit mass, assuming refrigerant 12 to be an ideal gas.

3.56E Refrigerant 12 is heated isothermally from an initial state of 30 psia and 390°F to a final pressure of 120 psia. Evaluate the work done per unit mass. Also evaluate the work done per unit mass, assuming refrigerant 12 to be an ideal gas.

Figure P3.57

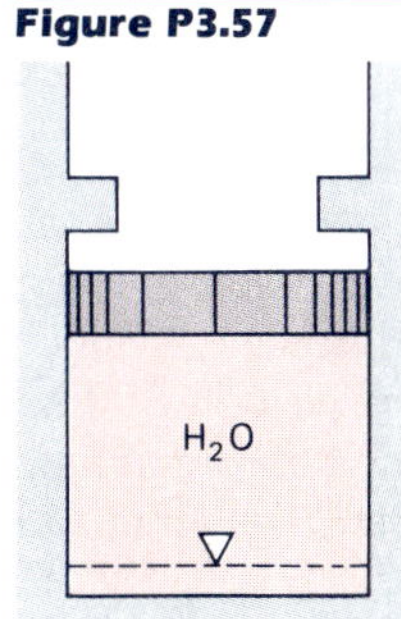

3.57S A piston encloses 1 kg of saturated water which is initially 10.0 percent liquid on a mass basis. The initial pressure is 1.0 MPa. The water is heated in the system shown in Fig. P3.57 until the water is at the saturated vapor state. The volume at the top stops is 0.19 m^3. Determine the work done by the water and the final temperature and pressure.

3.57E A piston encloses 0.5 lbm of saturated water which is initially 10.0 percent liquid on a mass basis. The initial pressure is 150 psia. The water is heated in the system shown in Fig. P3.57 until the water is at the saturated vapor state. The volume at the top stops is 1.5 ft^3. Determine the work done by the water and the final temperature and pressure.

3.58S A piston of mass equal to 900 kg and area equal to 9.8×10^{-3} m^2 is supported against stops (see Fig. P3.58) by water at the saturated vapor state and 20 MPa (state 1). The water is cooled until half the container is filled with liquid and half is vapor by volume (state 3). Determine the work per unit mass for the process ${}_1w_3$. Draw the process on the *P-v* diagram.

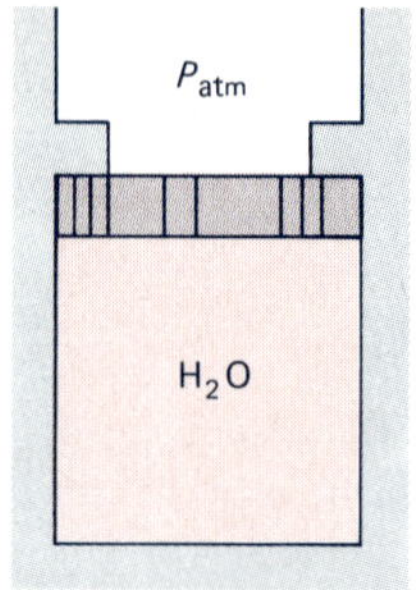

Figure P3.58

3.58E A piston of mass equal to 2000 lbm and area equal to 0.1 ft^2 is supported against stops (see Fig. P3.58) by water at the saturated vapor state and 3000 psia (state 1). The water is cooled until half the container is filled with liquid and half is vapor by volume (state 3). Determine the work per unit mass for the process ${}_1w_3$. Draw the process on the *P-v* diagram.

3.59S Heat transfer takes place isothermally to ammonia between the saturated vapor state at $P = 500$ kPa and a final pressure of 100 kPa. Draw a careful *P-v* diagram of the process, and evaluate the work done per unit mass. Numerical or graphical integration will be required.

3.59E Heat transfer takes place isothermally to ammonia between the saturated vapor state at $P = 75$ psia and a final pressure of 15 psia. Draw a careful *P-v* diagram of the process, and evaluate the work done per unit mass. Numerical or graphical integration will be required.

References

1. *Steam Tables,* American Society of Mechanical Engineers, New York, 1967.
2. J. H. Keenan, F. G. Keyes, P. G. Hill, and J. G. Moore, *Steam Tables, English Version,* Wiley, New York, 1969.
3. L. Haar, J. S. Gallagher, and G. S. Kell, *NBS/NRC Steam Tables,* Hemisphere, New York, 1984.
4. Henry M. Paynter, "Simple Veridical State Equations for Fluid Simulaton: Generalization and Improvements upon van der Waals," *ASME J. Dyn. Sys., Meas., and Cont.,* vol. 107, pp. 233–234, December 1985.
5. Robert C. Reid, John M. Prausnitz, and Bruce E. Poling, *The Properties of Liquids and Gases,* 4th ed., McGraw-Hill, New York, 1987.
6. Richard B. Stewart and Victor J. Johnson, *A Compendium of the Properties of Materials at Low Temperatures (Phase II),* National Bureau of Standards Cryogenic Engineering Lab. Rep. WADD G0-56, Part IV, U.S. Air Force Systems Command, Wright-Patterson Air Force Base, Ohio, December 1961.
7. J. H. Keenan, J. Chao, and J. Kaye, *Gas Tables (English Units),* 2d ed., Wiley, New York, 1980.
8. *GASPROPS, A Computer Program for the Calculation of the Thermodynamic Properties of Ideal Gases and Combustion Products,* Wiley, New York, 1984.
9. *REFRIG, A Computer Program for the Thermodynamic Properties of Common Refrigerants,* Wiley, New York, 1984.

4

First Law of Thermodynamics

Energy is an Eternal Delight.
William Blake

Homopolar generators convert kinetic energy in a spinning flywheel into large current pulses by magnetic braking. Shown is a bank of six individual generators providing 60 MJ, at up to 9 MA. (*Center for Electromechanics, University of Texas at Austin.*)

4.1 Introduction

Conservation relations form the foundation of thermodynamics. The concept of a conserved quantity is important in many areas of science and is based on common experiences. The conservation of energy is discussed briefly in Chap. 1 and is formally presented as the first law of thermodynamics in this chapter. Conservation of energy and the first law of thermodynamics are different names for the same principle. Other conservation relations include the conservation of momentum and the conservation of mass.

Momentum conservation was probably first encountered in an elementary physics course. We will revisit it briefly when we dicuss the analysis of rocket motors in Chap. 5. Mass conservation was probably also presented at the same time, and it is one conservation principle that we all take for granted. The water entering one end of a garden hose is expected to exit from the other end. This type of physical expectation or "physical feeling" for a phenomenon can be expressed in conservation statements or laws. This chapter focuses on the first law of thermodynamics, or conservation of energy principle. The relationship to other conserved quantities is also discussed.

A clear picture of the relationship between the states of the system and energy transfer is required. For a simple compressible substance, two independent intensive properties specify the state. The system changes from one state to another by an energy transfer in the form of work and/or heat transfer. This energy transfer must cross a system boundary. The amount of energy transfer depends on the path; i.e., work depends on the specific path followed in performing the work. The state of a substance depends not on how the state is reached, but on its properties only.

The first law of thermodynamics as well as the principle of conservation of mass are presented. The fundamental principles are first presented for a control mass and are subsequently transformed for use in a control volume.

4.2 Conservation Principles and the First Law of Thermodynamics

All the conservation statements in this section are presented for a control mass. Thus, we are considering the mass within a boundary surface which can undergo translation, rotation, and deformation, but no mass crosses the boundary. The substance contained within a cylinder with a sliding piston is a typical example. Also a fluid within a balloon is a control mass which can be used to visualize the following statements.

4.2.1 Conservation of Mass

We have chosen to make the fundamental conservation statements about a control mass so that no mass crosses the system boundary. Therefore, this conservation principle is as follows:

> The mass of a control mass never changes.

This is expressed mathematically as

$$dm = 0 \tag{4.1}$$

or

$$m = \text{constant} \tag{4.2}$$

The change in the mass of the control mass between state 1 and state 2 is

$$m_2 - m_1 = 0 \tag{4.3}$$

With the specific volume and density related as $\rho = 1/v$, a uniform distribution of mass within the control volume can also be represented as $m = \rho V = V/v$, where V is the volume occupied by the control mass.

The conservation of mass is also expressed on a rate basis by considering the time interval dt. This principle is then

$$\frac{dm}{dt} = 0 \tag{4.4}$$

All the above equations express the conservation of mass, and the differences are in the point of view and the notation used. This statement of conservation of mass is only a reiteration of our definition of a control mass. The usefulness of conservation of mass becomes apparent in Sec. 4.3.

The conservation of mass statement has been employed previously. The problem which considers a control mass within a cylinder with a sliding piston has always made use of a known quantity of mass. If water as a liquid is contained within a piston-cylinder arrangement and the volume is increased, the mass will be constant as the liquid changes to a vapor.

4.2.2 First Law of Thermodynamics

This law is alternatively termed the *conservation of energy*. This energy balance is stated as follows:

> A change of the total energy (kinetic, potential, and internal) of a control mass is equal to the heat transfer to the control mass minus work done by the control mass.

This is expressed mathematically as

$$dE = \delta Q - \delta W \tag{4.5}$$

The sign convention is that all heat transfer *into* the system is positive, while work done *by* the system *on* the surroundings is positive (Fig. 4.1). There is no reference to a particular path. Thus, the change in energy of the control mass from state 1 to state 2 is equal to the heat transfer to the control mass for any path minus the work done by the control mass, again, along any path. Between these two states, Eq. (4.5) is integrated to obtain

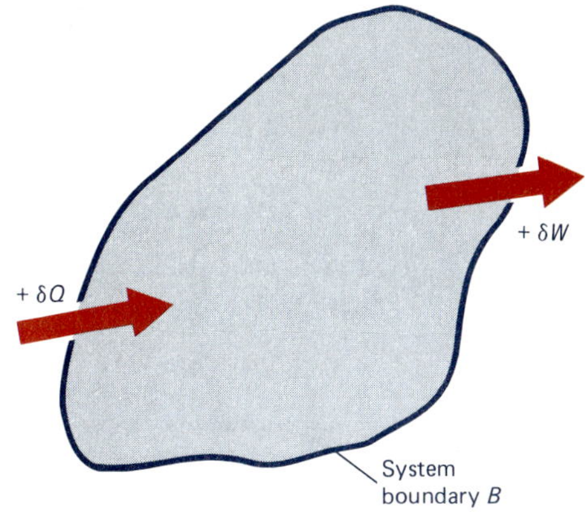

Figure 4.1 Sign convention.

$$E_2 - E_1 = {}_1Q_2 - {}_1W_2 \tag{4.6}$$

The total energy in all the above equations is the sum of the internal, kinetic, and potential energy. These are included explicitly to yield

$$dU + d(\mathrm{KE}) + d(\mathrm{PE}) = \delta Q - \delta W \tag{4.7}$$

Integrating between states 1 and 2 results in

$$(U + \mathrm{KE} + \mathrm{PE})_2 - (U + \mathrm{KE} + \mathrm{PE})_1 = {}_1Q_2 - {}_1W_2 \tag{4.8}$$

Substituting in the expressions for kinetic and potential energy [see Eq. (2.13)] gives the conservation of energy as

$$\left(U + m\frac{\mathbf{V}^2_{\mathrm{ref}}}{2g_c} + \frac{mgZ}{g_c}\right)_2 - \left(U + m\frac{\mathbf{V}^2_{\mathrm{ref}}}{2g_c} + \frac{mgZ}{g_c}\right)_1 = {}_1Q_2 - {}_1W_2 \tag{4.9}$$

If the mass within the volume of the control mass is uniform in space, then these equations are alternatively expressed on a specific basis or unit-mass basis as

$$d(me) = \delta(mq) - \delta(mw) \tag{4.10}$$

But the mass is constant in a control mass, so

$$de = \delta q - \delta w \tag{4.11}$$

For a process between states 1 and 2, the conservation of energy relation becomes

$$e_2 - e_1 = {}_1q_2 - {}_1w_2 \tag{4.12}$$

With the internal, kinetic, and potential energy expressions this is

$$\left(u + \frac{1}{2}\frac{\mathbf{V}^2_{\mathrm{ref}}}{g_c} + \frac{gZ}{g_c}\right)_2 - \left(u + \frac{1}{2}\frac{\mathbf{V}^2_{\mathrm{ref}}}{g_c} + \frac{gZ}{g_c}\right)_1 = {}_1q_2 - {}_1w_2 \tag{4.13}$$

As in Eq. (4.9), the left-hand side of Eq. (4.13) is an integration of an exact differential which represents the change in properties of the control mass. This change in total energy is independent of the path between the states. The right-hand sides of these equations are integrals of inexact differentials and generally depend on the particular path. Note again that $\mathbf{V}_{ref}$ is the velocity of the control mass relative to the surroundings, which are taken as the reference.

Example Problem 4.1

A frictionless piston is raised by slowly heating the gas contained in the cylinder. The initial pressure is 2 atm, and the initial volume is 35 ft³. The heat transfer to the tank is 2000 Btu, and the final volume is 70 ft³. Determine the change in internal energy.

Diagrams

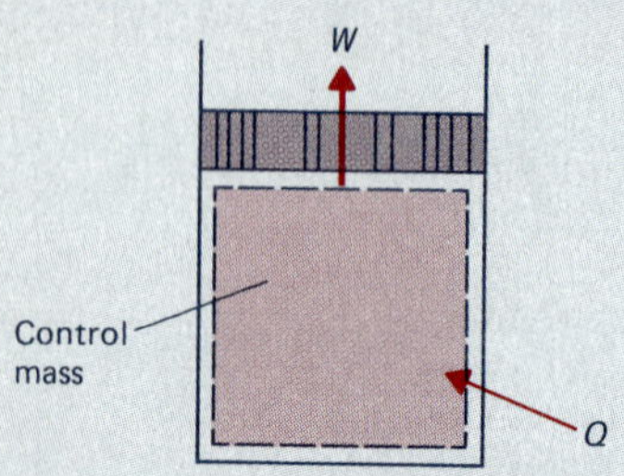

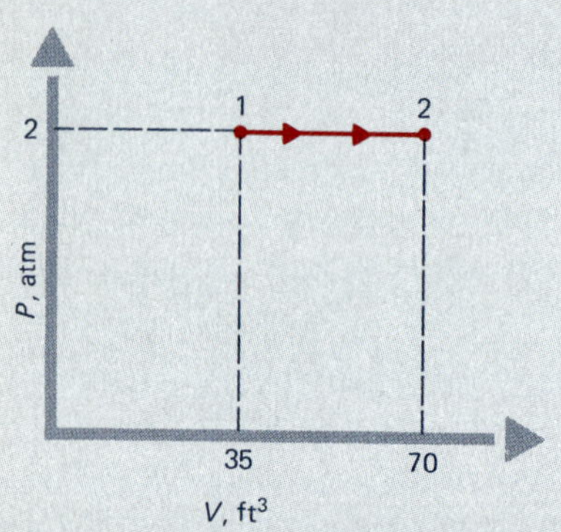

Solution

The gas is taken as a control mass, as indicated. This problem does not require property evaluation for a particular substance since all states are given. The work is calculated for the constant-pressure process, and the heat transfer is specified. The change in total energy is evaluated from the first law of thermodynamics, and the internal energy is obtained directly, since the changes in kinetic and potential energy are negligible.

The states are indicated in this table:

State	P, atm	V, ft³	U, Btu	Process
1	2.0	35.0	—	$P =$ constant
2	2.0	70.0	—	

The work done by the system is

$$
{}_1W_2 = \int_1^2 P\,dV = P_1(V_2 - V_1)
$$

$$
= (4232.4 \text{ lbf/ft}^2)[(70.0 - 35.0) \text{ ft}^3]
$$

$$
= 148{,}134 \text{ ft}\cdot\text{lbf} = 190.4 \text{ Btu} = 200.8 \text{ kJ}
$$

The first law yields

$$
U_2 - U_1 = {}_1Q_2 - {}_1W_2 = 2000 \text{ Btu} - 190.4 \text{ Btu} = +1810 \text{ Btu}
$$

Comments

The internal energy of state 2 is 1810 Btu greater than the internal energy of state 1.

The kinetic energy is related to the velocity of the system, and the velocities at states 1 and 2 are zero, so the kinetic energy contribution is also zero. The change in potential energy has been neglected, and from the stated information, the change in potential energy can only be estimated. The change in height of the gas center of mass depends on the specific geometric configuration, but it is probably no more than a few feet. The densities of various gases can be found from the appendixes to be of the order of a few hundredths of a pound-mass per cubic foot. With the gravitational acceleration as 32.174 ft/s², the change in potential energy ($mg\,\Delta Z/g_c$) is of the order of 0.1 Btu. This is negligible in the above calculation.

Example Problem 4.2

Evaluate the heat transfer necessary to evaporate water and to warm the vapor in the piston-cylinder experiment shown. The process is a quasi-equilibrium process. The initial state 1 is a saturated liquid at 1.0 MPa, and the evaporation process ends at state 2 on the saturated vapor line. Subsequent heat transfer takes the water to 600°C at state 3. Find ${}_1q_2$, ${}_2q_3$, and ${}_1q_3$.

Diagrams

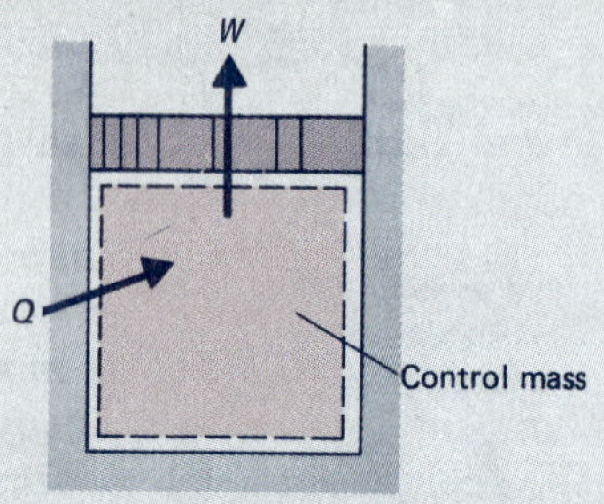

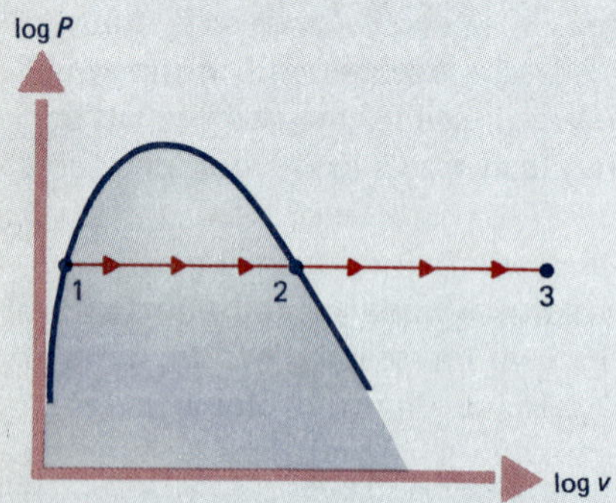

Solution

The control mass undergoes a constant-pressure process, as indicated on the system diagram. The following table gives the information. States 1 and 2 are known, since both P and the quality are given; v and u are found in Tables D.11 and D.12. State 3 is specified by P_3 and T_3.

State	P, MPa	T, °C	v, m³/kg	u, kJ/kg	x	Process
1	1.0†	—	(0.001127)	(761.75)	0.0	
2	1.0	—	(0.1944)	(2583.3)	1.0	P = constant
3	1.0	600	(0.4011)	(3297.0)	—	P = constant

† Values not in parentheses determine the state.

States 1 and 2 are obtained from the saturated steam tables, and state 3 is obtained from the superheated vapor tables.

The kinetic and potential energy changes are neglected, so the first law yields

$$e_2 - e_1 = u_2 - u_1 = {}_1q_2 - {}_1w_2$$

Thus

$${}_1q_2 = u_2 - u_1 + {}_1w_2$$

The work done for a constant-pressure process is

$$ {}_1w_2 = \int_1^2 P\,dv = P_1(v_2 - v_1) $$

Note that this is a special case of the polytropic process ($n = 0$) given in Eq. (2.22). For the process from 1 to 2,

$$ \begin{aligned} {}_1w_2 &= 1.0(0.1944 - 0.001127)\ \text{MJ/kg} = 193.3\ \text{kJ/kg} \\ &= 83.10\ \text{Btu/lbm} \end{aligned} $$

Thus the heat transfer is

$$ {}_1q_2 = 2583.3 - 761.75 + 193.3 = 2014.9\ \text{kJ/kg} = 866.2\ \text{Btu/lbm} $$

Repeating for the process between 2 and 3 yields

$$ {}_2w_3 = 206.7\ \text{kJ/kg} = 88.9\ \text{Btu/lbm} $$

and

$$ {}_2q_3 = 920.4\ \text{kJ/kg} = 395.7\ \text{Btu/lbm} $$

Thus the total heat transfer is

$$ {}_1q_3 = {}_1q_2 + {}_2q_3 = 2935.3\ \text{kJ/kg} = 1261.9\ \text{Btu/lbm} $$

Because ${}_1q_3$ is positive, this heat transfer is *from* the surroundings *to* the system. Note that work was done *by* the system *on* the surroundings as indicated by the positive sign on ${}_1w_2$ and ${}_2w_3$.

Comments

The work performed by the control mass in each segment of this constant-pressure process is nearly equal, yet the heat transfer for evaporation (state 1 to state 2) is 68.6 percent of the total heat transfer.

The above analysis could also be expressed in terms of the enthalpy. With constant-pressure work given as

$$ {}_1w_2 = P_1(v_2 - v_1) = P_2v_2 - P_1v_1 $$

and the first law given as

$$ u_2 - u_1 = {}_1q_2 - {}_1w_2 $$

we have

$$ \begin{aligned} {}_1q_2 &= u_2 - u_1 + {}_1w_2 = u_2 - u_1 + P_2v_2 - P_1v_1 \\ &= (u_2 + P_2v_2) - (u_1 + P_1v_1) \end{aligned} $$

The enthalpy is defined as $h = u + Pv$, so

$$ {}_1q_2 = h_2 - h_1 $$

In a control mass, this expression is valid *only* for a *constant-pressure* work process. Reading the enthalpy from Tables D.11 and D.12 yields

$$_1q_2 = h_g - h_l = h_{lg} = 2014.8 \text{ kJ/kg}$$

and from state 2 to state 3

$$_2q_3 = 3698.1 - 2777.7 = 920.4 \text{ kJ/kg}$$

These are the values obtained above to within the accuracy of the tables.

The conservation of energy expression has been given for a process between states 1 and 2. If the control mass undergoes a complete cycle that takes the control mass from state 1 back to state 1, then Eq. (4.5) yields

$$\oint dE = \oint \delta Q - \oint \delta W \tag{4.14}$$

But total energy is a property, and according to the mathematical attributes of a property (App. H), $\oint dE = 0$, so

$$\oint \delta Q = \oint \delta W \tag{4.15}$$

Thus, the net heat transferred to a control mass undergoing a cycle is equal to the net work done. This is the starting point for some presentations of the first law of thermodynamics.

The conservation of energy can also be presented on a rate basis. For a time interval Δt, the expression is

$$\frac{dE}{dt} = \frac{\delta Q}{\Delta t} - \frac{\delta W}{\Delta t} \tag{4.16}$$

With the definitions presented in Sec. 2.5.3, this can be written as

$$\frac{dE}{dt} = \dot{Q} - \dot{W} \tag{4.17}$$

Example Problem 4.3

A flywheel is used for the storage of energy in the form of kinetic energy. The flywheel is constructed of a Kevlar-epoxy composite with a density of 1400 kg/m^3, allowing the flywheel rim to achieve a safe tangential rim speed of $\mathbf{V}_t = 1200$ m/s at 38,000 rpm. Effectively all the mass of the flywheel, 100 kg, is concentrated at the rim. If energy is removed by connecting the flywheel to an ideal electric generator, how much electricity can be generated in bringing the flywheel to rest? What is the specific energy that can be stored in the flywheel, in kilojoules per kilogram, at 38,000 rpm? What is the radius of the flywheel?

Solution

The kinetic energy stored in the flywheel when it is operating at maximum safe speed is

$$(\text{KE})_1 = \tfrac{1}{2}m\mathbf{V}_t^2 = \left(\frac{1}{2}\right)(100\text{ kg})(1200\text{ m/s})^2 \times 1\,\frac{\text{MJ}}{10^6\text{ J}}$$

$$= 72\text{ MJ}$$

From Eq. (4.13), assuming that there is no change in the internal energy of the flywheel (i.e., its temperature is constant) and that there is no heat transfer during the deceleration,

$$(\text{KE})_2 - (\text{KE})_1 = -{}_1W_2$$

or since $(\text{KE})_2 = 0$ if the flywheel is brought to rest, the total work produced in bringing the flywheel to rest is

$${}_1W_2 = (\text{KE})_1 = 72\text{ MJ}$$

For an ideal electric generator, this will also be the electric energy produced from the system. The specific energy that can be stored in the flywheel, assuming that the mass of the flywheel peripherals and the mass of the electric generator are small compared with the mass of the flywheel itself, is

$$w = \frac{(\text{KE})_1}{m} = \frac{72\text{ MJ}}{100\text{ kg}} = 720\text{ kJ/kg}$$

The radius of this particular flywheel is found from

$$\mathbf{V}_t = r\omega = \frac{2\pi r \times \text{rpm}}{60\text{ s/min}}$$

where $\omega = 2\pi \times \text{rpm}/(60\text{ s/min})$ is the angular velocity in radians per second, or, by solving for the radius,

$$r = \frac{30}{\pi}\,\frac{1200\text{ m/s}}{38{,}000\text{ rpm}} = 0.302\text{ m}$$

Comments

The use of kinetic energy storage using high-speed flywheels is an active research area. The system described above would require a flywheel design capable of maintaining a tangential (hoop) stress at a safe velocity of approximately

$$\frac{F_t}{A_x} = \rho(r\omega)^2 = \left(1400\text{ kg/m}^3)(0.302\text{ m} \times 2\pi \times \frac{38{,}000}{60}\right)^2 \times \frac{1\text{ MPa}}{10^6\text{ Pa}}$$

$$= 2020\text{ MPa}$$

where A_x is the cross-sectional area of the flywheel. This value is attainable with available high-strength filament-wound composite materials; values of about 5000 MPa are expected in the near future.

Example Problem 4.4

A control mass undergoes a three-process power cycle with air as the working fluid. The fluid undergoes an isothermal compression, a constant-pressure heating, and a polytropic process given by $Pv^k =$ constant, where $k = c_p/c_v$. Determine q and w for each individual process. Assume the specific heats are constant and $P_1 = 1$ atm, $T_1 = 70°F$, and $P_2 =$ 6 atm.

Diagram

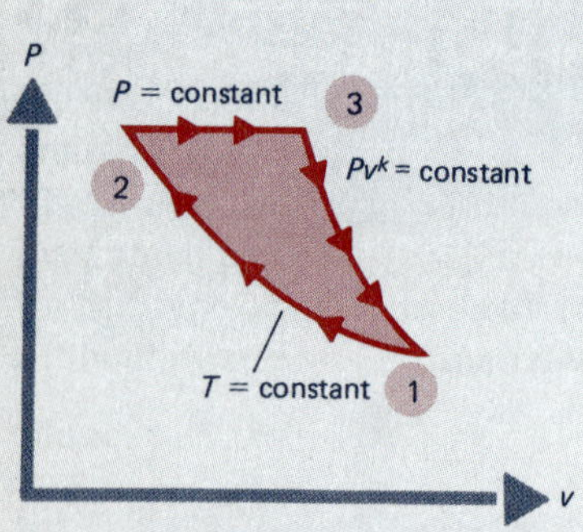

Solution

First, the ideal gas approximation must be checked for air for the conditions of this process. Examination of Fig. C.1, part 3, shows that when $P_r > 10$, the most critical check occurs at the lowest temperature and *largest* pressure. When $1 < P_r < 10$, the most critical check is at the lowest temperature and the *lowest* pressure. For $P_r < 1$, again the lowest temperature and highest pressure give the greatest deviation from ideal behavior. With the critical properties given from Table C.4 as $T_{cr} =$ 238.5°R and $P_{cr} = 37.2$ atm, then $T_r = 529.67/238.5 = 2.22$ and $P_{r,1} = 1.0/37.2 = 0.027$ and $P_{r,2} = 6.0/37.2 = 0.161$. The compressibility factor from Fig. C.1 yields for both states 1 and 2,

$$z \simeq 1.0$$

Thus the ideal gas approximation is very accurate. The state information is given:

State	P, atm	T, °F	v, ft³/lbm	u, Btu/lbm	h, Btu/lbm	Process
1	1.0	70	—	—	—	
						$T=$ constant
2	6.0	70	—	—	—	
						$P=$ constant
3	6.0	—	—	—	—	
						$Pv^k=$ constant
1	1.0	70	—	—	—	

For air, $R = 53.34$ ft·lbf/(lbm·°R) and $k = 1.400$ (Table C.2), thus

$$v_1 = \frac{RT_1}{P_1} = \frac{(53.34)(529.67)}{2116.2} = 13.351 \text{ ft}^3/\text{lbm}$$

Also

$$v_2 = \frac{RT_2}{P_2} = 2.2251 \text{ ft}^3/\text{lbm}$$

The state table indicates that both state 1 and state 2 are completely specified, but state 3 requires one more property. This is evaluated from the process information between states 1 and 3. Thus

$$P_1 v_1^k = P_3 v_3^k$$

or

$$v_3 = v_1 \left(\frac{P_1}{P_3}\right)^{1/k}$$

Thus

$$v_3 = (13.351)(\tfrac{1}{6})^{1/1.4} = 3.7127 \text{ ft}^3/\text{lbm}$$

and state 3 is then specified. Also $T_3 = P_3 v_3/R = 883.76°\text{R} = 424.09°\text{F}$.

The work is evaluated from Eq. (2.19) as

$${}_1w_2 = \int_1^2 P\,dv$$

For the process from 1 to 2, $P = RT/v$, so for this isothermal process [see Eq. (2.22)]

$${}_1w_2 = \int_1^2 \frac{RT}{v}\,dv = RT_2 \ln\frac{v_2}{v_1} = -50{,}620 \text{ ft·lbf/lbm}$$

$$= -65.05 \text{ Btu/lbm} = -151.3 \text{ kJ/kg}$$

For the process from 2 to 3, $P=$ constant and

$$ {}_2w_3 = P_2(v_3 - v_2) = 18{,}890 \text{ ft}\cdot\text{lbf/lbm} = 24.27 \text{ Btu/lbm} $$
$$ = 56.45 \text{ kJ/kg} $$

For the process from 3 to 1, $Pv^k = \text{constant} = P_1 v_1^k = P_3 v_3^k$, so

$$ {}_3w_1 = \int_3^1 P\,dv = \text{const} \int_3^1 \frac{dv}{v^k} $$
$$ = \text{const}\left(\frac{v_1^{1-k} - v_3^{1-k}}{1-k}\right) $$

Using $P_1 v_1^k = P_3 v_3^k = \text{constant}$ yields [see Eq. (2.22)]

$$ {}_3w_1 = \frac{P_3 v_3 - P_1 v_1}{k-1} = 47{,}220 \text{ ft}\cdot\text{lbf/lbm} = 60.68 \text{ Btu/lbm} $$
$$ = 141.14 \text{ kJ/kg} $$

The heat transfer for each process is evaluated from the first law with the work expressions given above. Conservation of energy for no kinetic or potential energy changes [Eq. (4.13)] states that

$$ u_2 - u_1 = {}_1q_2 - {}_1w_2 $$

so

$$ {}_1q_2 = u_2 - u_1 + {}_1w_2 $$

The change in internal energy is evaluated from Eq. (3.19) as

$$ u_2 - u_1 = \int_1^2 c_v\,dT = c_v(T_2 - T_1) $$

Thus

$$ {}_1q_2 = c_v(T_2 - T_1) + {}_1w_2 $$

For the isothermal process from state 1 to state 2

$$ {}_1q_2 = 0 + {}_1w_2 = 65.05 \text{ Btu/lbm} = 151.3 \text{ kJ/kg} $$

With $c_v = 0.1716$ Btu/(lbm·°R), the constant-pressure process from state 2 to state 3 yields

$$ {}_2q_3 = 0.1716(883.76 - 529.67) + 24.27 \text{ Btu/lbm} = 85.04 \text{ Btu/lbm} $$
$$ = 197.8 \text{ kJ/kg} $$

The process from state 3 to state 1 yields

$$ {}_3q_1 = 0.1716(529.67 - 883.78) + 60.68 \text{ Btu/lbm} $$
$$ = -0.09 \text{ Btu/lbm} \approx 0 \text{ Btu/lbm} $$

The process described by $Pv^k =$ constant is shown in Chap. 5 to be an adiabatic process, so the finite value is a result of property uncertainty.

In summary,

State	w, Btu/lbm	q, Btu/lbm
1 to 2	−65.05	−65.05
2 to 3	24.27	85.04
3 to 1	60.68	−0.09
Total	19.90	+19.90

This satisfies Eq. (4.15), or the heat transfer is equal to the work done for a cyclic process.

Comments

The constant-pressure process can also be considered in terms of enthalpy. The heat transfer from state 2 to state 3 is

$$\begin{aligned} {}_2q_3 &= u_3 - u_2 + {}_2w_3 \\ &= u_3 - u_2 + P_2(v_3 - v_2) \\ &= (u_3 + P_3 v_3) - (u_2 + P_2 v_2) \\ &= h_3 - h_2 \end{aligned}$$

With Eq. (3.22),

$${}_2q_3 = c_P(T_3 - T_2) = 0.2401(883.76 - 529.67) = 85.02 \text{ Btu/lbm}$$

This is the value obtained above to within the accuracy of the numbers used.

4.2.3 Generalized Control Mass Forms for Conservation of Mass and the First Law

The conservation relations discussed in the preceding two subsections are presented without specific consideration of the distribution of mass and energy within the control volume. This subsection generalizes the expressions by including each element within the control volume and summing all the elements to obtain the total contribution.

The mass within the boundary of the control volume is the sum of all elemental contributions, so the conservation of mass in Eq. (4.1) is also given as

$$dm = d\left(\int_m dm\right) = 0 \tag{4.18}$$

With the specific volume and density related as $\rho = 1/v$, Eq. (4.1) can also be represented as a volume integral

$$d\left(\int_V \rho\, dV\right) = d\left(\int_V \frac{dV}{v}\right) = 0 \tag{4.19}$$

The rate form of the conservation of mass in Eq. (4.4) is also expressed in terms of the integral over the mass elements and volume as

$$\frac{dm}{dt} = \frac{d}{dt}\left(\int_m dm\right) = \frac{d}{dt}\left(\int_V \rho\, dV\right) = 0 \tag{4.20}$$

The expressions for mass in Eqs. (4.18) and (4.19) have been used to obtain the various forms in Eq. (4.20).

Conservation of energy can also be rewritten to include the contributions of the elements within the control volume. The total energy contained by the control mass is the sum of the contribution of all the elements, so the total energy is represented as

$$E = \int_m e\, dm \tag{4.21}$$

With the density and specific volume, this is given as a volume integral:

$$E = \int_V e\rho\, dV = \int_V e \frac{dV}{v} \tag{4.22}$$

Thus the final representation is

$$d\left(\int_V e\rho\, dV\right) = \delta Q - \delta W \tag{4.23}$$

Also the rate form of the conservation of energy in Eq. (4.17) is

$$\frac{d}{dt}\left(\int_V e\rho\, dV\right) = \dot{Q} - \dot{W} \tag{4.24}$$

where Eq. (4.22) has been used to obtain these last equations.

4.2.4 Other Conservation Relations

Two conservation statements can also be made for momentum: linear momentum and the moment of momentum. The conservation of linear momentum has been used previously in obtaining relations for pressure measurement. These principles form the foundation for solid mechanics and fluid mechanics. They are not covered in this text except for some rocket propulsion problems. They are mentioned here because the form of these two conservation relations is very similar to that of the statements of conservation of mass and conservation of energy. The conservation of linear momentum is outside the realm of classical thermodynamics and is studied in a course on fluid mechanics or transport phenomena. Yet this equation is necessary to evaluate the detailed velocity distributions within flowing systems. This text focuses on thermodynamics, and any needed velocity distributions are specified; thus, the solution to the linear momentum equation is not required.

4.3 Control Volume Formulation

Many systems are most conveniently analyzed as a system fixed in space which permits mass to cross the boundaries. Such a system is defined to be a *control volume* (CV). The boundary surface B can move, and mass passes through the boundary at the crossing boundaries. Most flow systems including pumps, turbines, heat exchangers, and valves are analyzed with control volumes.

The conservation statements for mass and energy are presented for a control mass (CM) in Sec. 4.2. The mass under consideration is always the same, and the conservation statement specifies what is necessary to change the property in question. An alternative, yet equivalent, approach considers a control volume and specifies what is necessary to change the property in the control volume. The difference is that the changes in the properties resulting from flows at the inlets and outlets must be included.

This section restates the conservation principles presented in Sec. 4.2 for a control volume. The conservation principles are exactly the same; yet since mass crosses the boundary, extra contributions must be included. The mathematical justification for the extra terms is presented in App. F.

4.3.1 Conservation of Mass

The mass flowing into a specified control volume must equal the mass leaving the control volume; otherwise, the mass within the control volume must change. When a garden hose is initially turned on, there is no water leaving the hose since water is filling the volume of the hose. This is stated as follows:

> The change of mass within a control volume is equal to the mass entering minus the mass leaving.

This is simply a "bookkeeping" of the mass within the control volume at any time.

Let the symbol $\dot{m}$ stand for the mass flow rate. All flow streams entering the control volume across the system boundary are added to get the total rate at which mass enters the system:

$$\dot{m}_{\text{in}} = \sum_{\text{in}} \dot{m} \tag{4.25}$$

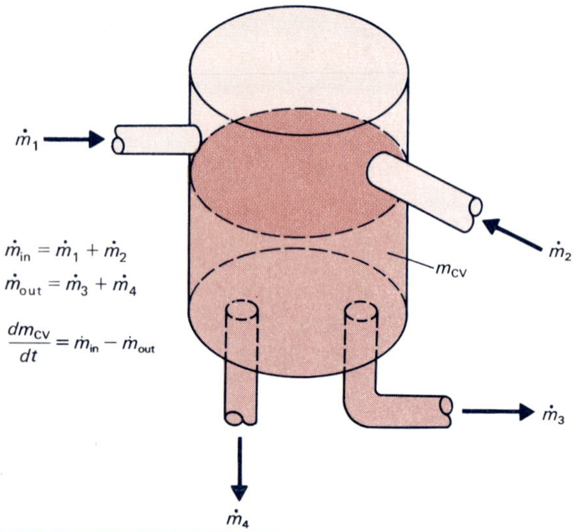

Figure 4.2 Conservation of mass for a control volume.

Here the subscript *in* denotes mass flow at the inlets made up of contributions from the multiple inlet streams, each with individual mass flow rate $\dot{m}$. Similarly for the outlet streams, the mass flow rate at the outlets is

$$\dot{m}_{out} = \sum_{out} \dot{m} \tag{4.26}$$

where the subscript *out* denotes outlets.

Now the mass balance on the control volume (CV) becomes simply

$$\frac{dm_{CV}}{dt} = \dot{m}_{in} - \dot{m}_{out} = \sum_{in} \dot{m} - \sum_{out} \dot{m} \tag{4.27}$$

This states that the rate of change of mass within the control volume is equal to the mass that enters the control volume per unit time minus the mass that leaves the control volume per unit time. In short, what stays within is equal to what goes in minus what comes out. This is shown schematically in Fig. 4.2.

The inlet and outlet mass flow rates at the boundaries can be expressed in terms of the density, velocity, and area. The volume swept out for a time increment Δt as the mass moves through a port (Fig. 4.3) is

Figure 4.3 Mass flow at a port.

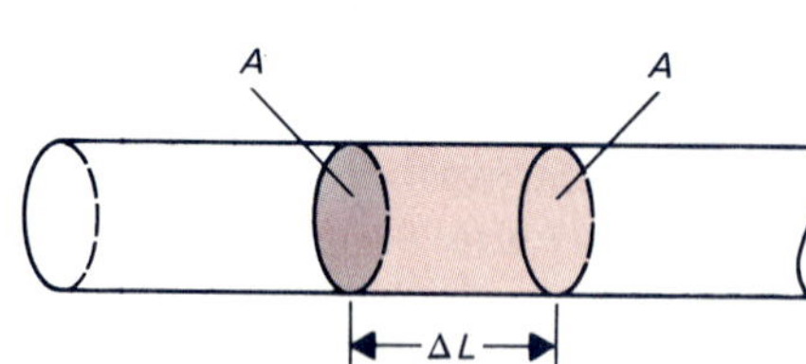

$$\dot{V} = A\frac{\Delta L}{\Delta t} \tag{4.28}$$

As Δt goes to zero, $\Delta L/\Delta t$ is the magnitude of the velocity, denoted as $\mathbf{V}$. This is the velocity of the fluid *relative to the control volume;* the velocity used in determining the kinetic energy of the control volume itself, $\mathbf{V}_{ref}$, is relative to the surroundings.

$$\dot{V} = A\mathbf{V} \tag{4.29}$$

and the mass flow rate for a port is given by the volume swept out times the density or

$$\dot{m} = \rho A \mathbf{V} = \rho \dot{V} \tag{4.30}$$

The area of the crossing boundary at the inlet is denoted by A_{in} and at the outlet by A_{out}. The mass flow rates are

$$\dot{m}_{\text{in}} = (\rho A \mathbf{V})_{\text{in}} \tag{4.31}$$

and

$$\dot{m}_{\text{out}} = (\rho A \mathbf{V})_{\text{out}} \tag{4.32}$$

The conservation of mass statement is then

$$\frac{dm_{\text{CV}}}{dt} = \sum_{\text{in}} (\rho A \mathbf{V})_{\text{in}} - \sum_{\text{out}} (\rho A \mathbf{V})_{\text{out}} \tag{4.33}$$

There are other important forms for these equations as well as a number of useful idealizations. The particular expression employed depends on the problem considered. These are addressed in detail in Sec. 4.4. One special case considers no mass entering or leaving the control volume. The conservation of mass reduces to

$$\frac{dm_{\text{CV}}}{dt} = 0 \tag{4.34}$$

or, by integrating with respect to time,

$$m_{\text{CV}} = \text{constant} \tag{4.35}$$

which we have assumed all along for such a system. This is exactly the expression for a control mass [Eq. (4.2)].

4.3.2 Conservation of Energy

The mass flowing into the control volume carries energy with it, and the same is true for the outlets from the control volume. The net gain in energy from the inlets and outlets must equal the change in energy within the control volume plus the heat transfer minus the work done. The work term requires special attention since the fluid moving into and out of the control volume is doing work. This is the conservation statement:

> The change in energy within the control volume is equal to the net energy transported to the control volume plus the heat transferred minus the work done by the control volume.

This is shown schematically in Fig. 4.4.

Let $\dot{E}$, termed the *energy rate,* denote the energy transported with the fluid at the inlets and outlets. The overdot indicates the rate of flow of

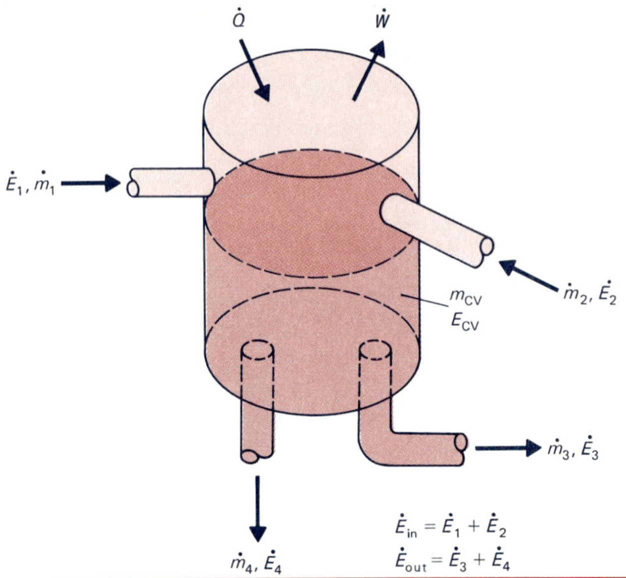

Figure 4.4 Conservation of energy for a control volume.

energy as in the mass flow rate. The energy within the control volume is denoted by E_{CV}, so the conservation of energy is

$$\frac{dE_{CV}}{dt} = \dot{E}_{net} + \dot{Q}_{CV} - \dot{W} \tag{4.36}$$

or, in terms of inlets and outlets,

$$\frac{dE_{CV}}{dt} = \dot{E}_{in} - \dot{E}_{out} + \dot{Q}_{CV} - \dot{W} \tag{4.37}$$

Each term requires detailed consideration.

The energy change within the control volume is given by dE_{CV}/dt. This represents the change of total energy within the control volume and is often called the *energy storage* term, since any accumulation of energy within the control volume is included in this expression.

The mass flow transports energy into and out of the control volume. These contributions are denoted by $\dot{E}_{in}$ and $\dot{E}_{out}$, respectively. The development of the mass flow rate term for the conservation of mass is extended to include the energy transport. Equations (4.31) and (4.32) represent the mass flow rate per unit-area element of the boundary. The fluid carries energy across each element, so the energy transported with the fluid element is given by Eqs. (4.31) and (4.32), multiplied by the specific total energy, or

$$\dot{E}_{in} = (e\dot{m})_{in} \tag{4.38}$$

and

$$\dot{E}_{out} = (e\dot{m})_{out} \tag{4.39}$$

Therefore, the net energy transported into the control volume by fluid motion is

$$\dot{E}_{net} = (e\dot{m})_{in} - (e\dot{m})_{out} \tag{4.40}$$

Equation (4.36) is given with Eq. (4.40) as

$$\frac{dE_{CV}}{dt} = (e\dot{m})_{in} - (e\dot{m})_{out} + \dot{Q}_{CV} - \dot{W}_{CV} \tag{4.41}$$

For multiple inlets and multiple outlets this is

$$\frac{dE_{CV}}{dt} = \sum_{in} (e\dot{m})_{in} - \sum_{out} (e\dot{m})_{out} + \dot{Q}_{CV} - \dot{W}_{CV} \tag{4.42}$$

Heat Transfer Contribution

The heat transfer to the control volume is the energy transport across any portion of the boundary because of a temperature difference between the control volume and the surroundings. This energy transfer is the same as that considered for the control mass. It generally occurs across the inside boundary rather than the crossing boundary of the control volume.

Work Contribution

The work term in the conservation of energy includes the general formulation contained in Sec. 2.5.1. The contributions of $P\,dV$ work, electrical work, and other work forms which transport energy across the boundary are incorporated into the energy balance. Three additional forms of work were not included in Sec. 2.5.1 but must be included in the work contribution when fluid motion is permitted. The general definition for work is

$$\delta W = \mathbf{F} \cdot d\mathbf{s}$$

The rate of doing work is

$$\dot{W} = \lim_{\Delta t \to 0} \frac{\delta W}{\Delta t} = \lim_{\Delta t \to 0} \frac{\mathbf{F} \cdot d\mathbf{s}}{\Delta t} \tag{4.43}$$

With the velocity defined as

$$\mathbf{V} = \lim_{\Delta t \to 0} \frac{d\mathbf{s}}{\Delta t} \tag{4.44}$$

the rate of doing work is

$$\dot{W} = \mathbf{F} \cdot \mathbf{V} \tag{4.45}$$

Shear work results from fluid flowing over the boundary where a force is exerted on it by the boundary element dA. This force is expressed in terms of the shear stress as $\tau\, dA$. Thus, the work done by the fluid is the shear work and is

$$\delta \dot{W}_{\text{shear}} = -\tau\, dA \cdot \mathbf{V} \tag{4.46}$$

The shear work done on the entire boundary B is the integral of Eq. (4.46) over the entire boundary, or

$$\dot{W}_{\text{shear}} = \int_B \delta \dot{W}_{\text{shear}} = -\int_B \tau \cdot \mathbf{V}\, dA \tag{4.47}$$

The particular choice of the boundary prescribes the contribution of this term. At the inside boundaries which coincide with a solid surface, the fluid "sticks" to the walls, so the velocity is zero. Therefore, the shear work contribution for an inside boundary is zero. At inlets and outlets to the control volume, the flow is generally normal to the area element. Since the normal component of the shear stress is zero, the shear work is zero for this case. If a control volume includes a shaft that crosses the boundary and interacts with the fluid through a shearing motion, this contribution is included under a different label, *shaft work* (see the next paragraph) and is not included here. Other contributions to the shear work term are possible but are rarely significant, so they are not included.

Shaft work is the contribution resulting from the rotation of a paddle wheel or turbine rotor within the control volume. The shaft crosses the boundary and transfers energy between the control volume and the surroundings. This shaft work term is truly a result of shear forces acting on the paddle wheel, yet the measurable response from the control volume is in the rotation of the shaft. This work contribution is termed *shaft work* $\dot{W}_{\text{shaft}}$. This work is usually the desired result in a work calculation because it is directly usable.

Flow work is related to the normal force at the boundary. The normal force includes the contribution from the pressure and viscous effects, yet the pressure force is the dominant contribution in most applications. At the inside boundary, the net work is zero since the wall counterbalances the fluid force. At the crossing boundary, the pressure force is $-P\, dA$, and it acts on an area element which is perpendicular to the normal velocity component $\mathbf{V} \cdot \mathbf{n}$. Thus, the flow work for the crossing boundaries is

$$\dot{W}_{\text{flow}} = \int_B P\mathbf{V} \cdot \mathbf{n}\, dA \tag{4.48}$$

With the mass flow rate expression in Eqs. (4.31) and (4.32) and multiplying by $\rho v = 1$, we find

$$\dot{W}_{\text{flow}} = \int_B Pv\rho\mathbf{V} \cdot \mathbf{n}\, dA = (\dot{m}Pv)_{\text{out}} - (\dot{m}Pv)_{\text{in}} \tag{4.49}$$

In summary, the rate of work done in the control volume formulation is the sum of the shear, shaft, and flow work plus the general forms discussed in Sec. 2.5.1. Since the shear work is typically zero for the reasons just discussed, the control volume work is

$$\begin{aligned}\dot{W} &= \dot{W}_{\text{shear}}^{\nearrow 0} + \dot{W}_{\text{shaft}} + \dot{W}_{\text{gen}} + \dot{W}_{\text{flow}} \\ &= \dot{W}_{\text{shaft}} + \dot{W}_{\text{gen}} + (\dot{m}Pv)_{\text{out}} - (\dot{m}Pv)_{\text{in}} \end{aligned} \tag{4.50}$$

The $\dot{W}_{\text{gen}}$ term includes the $P\,dV$ work, electrical work, etc. from Eq. (2.36), and it is combined with $\dot{W}_{\text{shaft}}$ and expressed in the following equations as $\dot{W}_{\text{CV}}$. Thus,

$$\begin{aligned}\dot{W} &= \dot{W}_{\text{shaft}} + \dot{W}_{\text{gen}} + (\dot{m}Pv)_{\text{out}} - (\dot{m}Pv)_{\text{in}} \\ &= \dot{W}_{\text{CV}} + (\dot{m}Pv)_{\text{out}} - (\dot{m}Pv)_{\text{in}} \end{aligned} \tag{4.51}$$

Conservation of Energy Expression

The conservation of energy for a control volume is obtained from Eq. (4.41) with Eq. (4.51). These substitutions yield

$$\frac{dE_{\text{CV}}}{dt} = [(e + Pv)\dot{m}]_{\text{in}} - [(e + Pv)\dot{m}]_{\text{out}} + \dot{Q}_{\text{CV}} - \dot{W}_{\text{CV}} \tag{4.52}$$

This conservation statement says that the change of energy within the control volume is equal to the entering energy contributions minus the leaving energy contributions. The energy balance is also expressed in terms of multiple inlet streams and multiple outlet streams as

$$\frac{dE_{\text{CV}}}{dt} = \sum_{\text{in}} [(e + Pv)\dot{m}]_{\text{in}} - \sum_{\text{out}} [(e + Pv)\dot{m}]_{\text{out}} + \dot{Q}_{\text{CV}} - \dot{W}_{\text{CV}} \tag{4.53}$$

The results for conservation of mass in Eq. (4.33) compare with the conservation of energy in Eq. (4.53). Both describe the change in the property within the control volume—the property is mass in Eq. (4.33) and energy in Eq. (4.53). These changes are equal to the incoming contributions minus the outgoing contributions. In the case of the conservation of mass, the contributions are mass flow rates. For conservation of energy with the sign convention indicated in Fig. 4.1, the incoming contribution is heat transfer, energy flux, and flow work, and the outgoing contribution is the energy flux, control volume work, and flow work.

A very important term appears in the flux terms. It is the sum of the specific total energy and the product Pv. The Pv term results from the flow work performed as fluid enters and leaves the control volume. The total energy is the sum of the internal, kinetic, and potential energy, so

$$e + Pv = \frac{E}{m} + Pv = u + \frac{1}{2}\frac{\mathbf{V}^2}{g_c} + \frac{gZ}{g_c} + Pv \tag{4.54}$$

Enthalpy is defined as $h = u + Pv$, so

$$e + Pv = h + \frac{1}{2}\frac{\mathbf{V}^2}{g_c} + \frac{gZ}{g_c} \tag{4.55}$$

This indicates another important application for enthalpy. Equation (4.53) is written as

$$\frac{dE_{\text{CV}}}{dt} = \sum_{\text{in}}\left[\left(h + \frac{\mathbf{V}^2}{2g_c} + \frac{gZ}{g_c}\right)\dot{m}\right]_{\text{in}} - \sum_{\text{out}}\left[\left(h + \frac{\mathbf{V}^2}{2g_c} + \frac{gZ}{g_c}\right)\dot{m}\right]_{\text{out}} + \dot{Q}_{\text{CV}} - \dot{W}_{\text{CV}} \tag{4.56}$$

In short, what stays within is equal to what goes in minus what goes out.

As in the case of conservation of mass, a number of important idealizations are of practical significance. The first consideration is at the inlet and outlet streams. The others involve the control volume. These are addressed in more detail in Sec. 4.4. For the special case of no mass entering or leaving the control volume, all $\dot{m}$'s are zero, and Eq. (4.56) reduces to the control mass form:

$$\frac{dE_{\text{CV}}}{dt} = \dot{Q}_{\text{CV}} - \dot{W}_{\text{CV}} \tag{4.57}$$

Since $\dot{W}_{\text{gen}}$ is included in the work term, the same expression results as previously given for a control mass [Eq. (4.17)]. If there is no energy transfer, the result is

$$\frac{dE_{\text{CV}}}{dt} = 0 \tag{4.58}$$

which is the isolated system result.

4.3.3 Generalized Control Volume Forms for Conservation of Mass and the First Law

This section presents a more detailed view of the mass and energy that are within the control volume and that cross the control surface. The conservation relations discussed in the preceding two subsections are generalized to include each element within the control volume and on the control surface, summing all the elements to obtain the total contribution.

The mass within the control volume is expressed as

$$m_{\text{CV}} = \int_{m_{\text{CV}}} dm = \int_{V_{\text{CV}}} \rho\, dV = \int_{V_{\text{CV}}} \frac{dV}{v} \tag{4.59}$$

This is the same term as in the control mass formulation given in Eq. (4.19).

The boundary terms are expressed in terms of the local velocity and area elements. Consider an individual area element dA with a mass flow rate per unit area across this area element of $d\dot{m}/dA$. Then

$$\dot{m} = \int_A \frac{d\dot{m}}{dA}\, dA$$

The mass flow rate per unit area is also expressed in terms of the velocity. The volume swept out as the fluid particle moves (Fig. 4.5) is

$$\Delta L \cos\theta\, dA \tag{4.60}$$

For a time increment Δt, the volume flow rate across the boundary area element is

$$\frac{\Delta L \cos\theta\, dA}{\Delta t} \tag{4.61}$$

but the ratio of ΔL to Δt as the time increment goes to zero is the magnitude of the velocity:

$$\lim_{\Delta t \to 0} \frac{\Delta L \cos\theta}{\Delta t} = |\mathbf{V} \cdot \mathbf{n}| \tag{4.62}$$

Velocity is a vector, so the dot product with the outward normal from the crossing boundary gives the normal velocity as $\mathbf{V}\cos\theta$. Thus, the volume flow rate is

$$\lim_{\Delta t \to 0} \frac{\Delta L}{\Delta t} \cos\theta\, dA = |\mathbf{V} \cdot \mathbf{n}|\, dA \tag{4.63}$$

The mass flow rate per unit area of the crossing boundary is

$$d\dot{m} = \rho |\mathbf{V} \cdot \mathbf{n}|\, dA \tag{4.64}$$

For the boundary B with the crossing boundary we have

$$\dot{m}_{\text{in}} = \int_{A_{\text{in}}} \rho |\mathbf{V} \cdot \mathbf{n}|\, dA = \int_{A_{\text{in}}} d\dot{m} \tag{4.65}$$

and

$$\dot{m}_{\text{out}} = \int_{A_{\text{out}}} \rho |\mathbf{V} \cdot \mathbf{n}|\, dA = \int_{A_{\text{out}}} d\dot{m} \tag{4.66}$$

The conservation of mass statement in Eq. (4.33) is

$$\frac{d}{dt}\left(\int_{V_{\text{CV}}} \rho\, dV\right) = \int_{A_{\text{in}}} \rho |\mathbf{V} \cdot \mathbf{n}| dA - \int_{A_{\text{out}}} \rho |\mathbf{V} \cdot \mathbf{n}|\, dA \tag{4.67}$$

For multiple inlet and outlet ports, this becomes

$$\frac{d}{dt}\left(\int_{V_{\text{CV}}} \rho\, dV\right) = \sum_{\text{in}} \int_{A_{\text{in}}} d\dot{m} - \sum_{\text{out}} \int_{A_{\text{out}}} d\dot{m} \tag{4.68}$$

This is the generalized form that incorporates the contributions of non-uniform and unsteady properties.

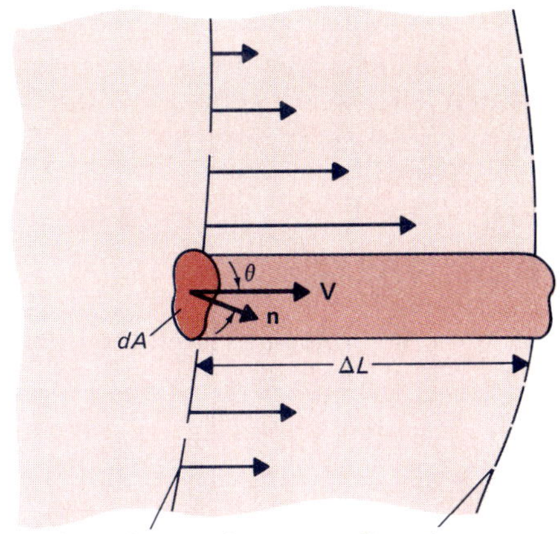

Figure 4.5 Flow at the outlet.

Conservation of energy also can be rewritten to include the contributions of the elements within the control volume and crossing the boundary. The energy change within the control volume on the left-hand side of Eq. (4.41) is given by

$$\frac{dE_{CV}}{dt} = \frac{d}{dt}\left(\int_{V_{CV}} \rho e \, dV\right) \tag{4.69}$$

The sum, or integral, over all the inlet and outlet boundaries is

$$\dot{E}_{in} = \int_{A_{in}} e\rho |\mathbf{V} \cdot \mathbf{n}| \, dA = \int_{A_{in}} e \, d\dot{m} \tag{4.70}$$

and

$$\dot{E}_{out} = \int_{A_{out}} e\rho |\mathbf{V} \cdot \mathbf{n}| \, dA = \int_{A_{out}} e \, d\dot{m} \tag{4.71}$$

Equation (4.41) is given with Eqs. (4.70) and (4.71) as

$$\frac{d}{dt}\left(\int_{V_{CV}} \rho e \, dV\right) = \int_{A_{in}} e \, d\dot{m} - \int_{A_{out}} e \, d\dot{m} + \dot{Q}_{CV} - \dot{W} \tag{4.72}$$

For multiple inlets and multiple outlets this is

$$\frac{d}{dt}\left(\int_{V_{CV}} \rho e \, dV\right) = \sum_{in} \int_{A_{in}} e \, d\dot{m} - \sum_{out} \int_{A_{out}} e \, d\dot{m} + \dot{Q}_{CV} - \dot{W} \tag{4.73}$$

The work term is also composed of contributions from each crossing area element. The generalized form follows from Eqs. (4.49) through (4.51) as

$$\dot{W} = \dot{W}_{CV} + \int_{A_{out}} Pv \, d\dot{m} - \int_{A_{in}} Pv \, d\dot{m} \tag{4.74}$$

Combining this last equation with Eqs. (4.72) and (4.73) yields

$$\frac{d}{dt}\left(\int_{V_{CV}} \rho e \, dV\right) = \int_{A_{in}} (e + Pv) \, d\dot{m} - \int_{A_{out}} (e + Pv) \, d\dot{m} + \dot{Q}_{CV} - \dot{W}_{CV} \tag{4.75}$$

and for multiple inlet streams and multiple outlet streams,

$$\frac{d}{dt}\left(\int_{V_{CV}} \rho e \, dV\right) = \sum_{in} \int_{A_{in}} (e + Pv) \, d\dot{m} - \sum_{out} \int_{A_{out}} (e + Pv) \, d\dot{m} + \dot{Q}_{CV} - \dot{W}_{CV} \tag{4.76}$$

4.4 Control Volume Analysis

The analysis of a system requires detailed consideration of the inlets and outlets to the control volume as well as the state within the control volume. The spatial considerations are addressed at both the inlets and outlets and the volume itself. This section considers various idealizations used in control volume analysis.

Uniform properties are spatially constant but may vary with time. At inlets and outlets this means that the state across the crossing boundary is described at any time by single values of the properties. Uniform properties within the control volume indicate that all elements within the control volume have the same value for the properties at any time. *Nonuniform properties* vary spatially. The time variation must also be addressed at the inlets and outlets as well as within the control volume. *Steady state* flow conditions at the inlets and outlets mean that there is no variation of the properties with time. The density, velocity, specific internal energy, etc. have constant values for all times. Steady state properties within the control volume indicate that there is no accumulation within the control volume that causes the properties within the control volume to vary with time. *Unsteady state* conditions imply time variation either at the inlets and outlets or within the control volume. In summary, uniform and nonuniform describe the spatial variation while steady state and unsteady state describe the time variation.

4.4.1 Inlet and Outlet Considerations

The general form for the mass flow rate for an inlet or outlet is given by [Eq. (4.65) or (4.66)]

$$\dot{m} = \int_A d\dot{m} = \int_A \rho |\mathbf{V} \cdot \mathbf{n}| \, dA \tag{4.77}$$

where all the quantities in the integrand can vary with time and space. Thus, this is the general nonuniform unsteady form for an inlet or outlet. Applying appropriate idealizations reduces this term to a form useful for the particular problem. A few common idealizations are addressed here.

If the *velocity and density (or specific volume) are uniform across the inlets and outlets and the velocity is normal to the ports,* then $|\mathbf{V} \cdot \mathbf{n}| = \mathbf{V}$ and the mass flow rates in Eq. (4.77) reduce to

$$\dot{m}_{\text{in}} = (\rho \mathbf{V} A)_{\text{in}} = \left(\frac{\mathbf{V} A}{v}\right)_{\text{in}} \tag{4.78}$$

$$\dot{m}_{\text{out}} = (\rho \mathbf{V} A)_{\text{out}} = \left(\frac{\mathbf{V} A}{v}\right)_{\text{out}} \tag{4.79}$$

for a single inlet and outlet. These expressions can be used if the stated information is interpreted as average values. Although these equations for averaged values are given in Eqs. (4.31) and (4.32), they are discussed here to clearly show the assumptions made. Many problems specify only the inlet and outlet pipe sizes and velocities, so Eqs. (4.78) and (4.79) are directly applicable.

Instrumentation for flow measurements is often calibrated to read *volumetric flow rate* $\dot{V}$. Because conservation of mass in its simplest form is written in terms of mass flow rate $\dot{m}$, it is convenient to have relations between $\dot{m}$ and $\dot{V}$. The volumetric flow rate is easily converted to mass flow rate if the density or specific volume is uniform at the inlets and outlets. Dividing by the specific volume of the flowing material at the port yields

$$\dot{m} = \frac{\dot{V}}{v} \tag{4.80}$$

Unsteady considerations at inlets and outlets result from the time variation of the mass flow rate, usually through a time variation of the fluid velocity. These conditions generally require the summation of all time increments or an integration of the inlet and outlet mass flow rates in time. The mass transported into or out of the control volume from t_1 to t_2 is

$$\int_{t_1}^{t_2} \dot{m}\, dt = m \tag{4.81}$$

The unsubscripted mass without an overdot on the right-hand side represents the total mass crossing the control volume for the time interval $t_2 - t_1$.

The inlet and outlet terms for conservation of energy [see Eq. (4.76)] are

$$\int_A (e + Pv)\, d\dot{m} = \int_A (e + Pv)\rho|\mathbf{V} \cdot \mathbf{n}|\, dA \tag{4.82}$$

where $e = u + \frac{1}{2}\mathbf{V}^2/g_c + gZ/g_c$. This general form incorporates nonuniform properties and unsteady time variations. Again, simplifying idealizations are useful as in the case of the mass flow rate.

If the *properties (h and v) and velocity are uniform* at the inlets and outlets, then the crossing boundary terms are

$$\left(h + \frac{1}{2}\frac{\mathbf{V}^2}{g_c} + \frac{gZ}{g_c}\right)\dot{m} \tag{4.83}$$

which can also be combined with Eqs. (4.78) and (4.79) for $\dot{m}$. This yields

$$\left(h + \frac{1}{2}\frac{\mathbf{V}^2}{g_c} + \frac{gZ}{g_c}\right)_{\text{in}} (\rho \mathbf{V} A)_{\text{in}} \tag{4.84}$$

TABLE 4.1 Mass and Energy Terms for a Control Volume

	Nonuniform and Unsteady	Uniform	Unsteady Time Integration
Inlets and outlets†	$\dot{m} = \int_A d\dot{m} = \int_A \rho \lvert \mathbf{V} \cdot \mathbf{n} \rvert \, dA$	$\dot{m} = \rho \mathbf{V} A$	$\int_{t_1}^{t_2} \dot{m} \, dt = m$
	$\int_A (e + Pv) \, d\dot{m}$ $= \int_A (e + Pv)\rho \lvert \mathbf{V} \cdot \mathbf{n} \rvert \, dA$	$(e + Pv)\dot{m}$	$\int_{t_1}^{t_2} \int_A (e + Pv) \, d\dot{m} \, dt$ $= (e + Pv)m$ with uniform and steady state properties
Within the control volume‡	$m_{CV} = \int_{V_{CV}} \rho \, dV$	$m_{CV} = (\rho V)_{CV}$	$\int_{t_1}^{t_2} \frac{dm_{CV}}{dt} \, dt = m_2 - m_1$
	$E_{CV} = \int_{V_{CV}} \rho e \, dV$	$E_{CV} = (\rho e V)_{CV} = (me)_{CV}$	$\int_{t_1}^{t_2} \frac{dE_{CV}}{dt} \, dt = E_2 - E_1$ $= (em)_2 - (em)_1$ with uniform properties

† Here $e = u + \frac{1}{2}\mathbf{V}^2/g_c + gZ/g_c$.
‡ Here $e = u + \frac{1}{2}\mathbf{V}_{\text{ref}}^2/g_c + gZ/g_c$.

and

$$\left(h + \frac{1}{2}\frac{\mathbf{V}^2}{g_c} + \frac{gZ}{g_c}\right)_{\text{out}} (\rho \mathbf{V} A)_{\text{out}} \tag{4.85}$$

When the *time variation* must be included, time integration of Eq. (4.82) is often required. This contribution takes the form

$$\int_{t_1}^{t_2} \int_A (e + Pv) \, d\dot{m} \, dt \tag{4.86}$$

which represents all the contributions of $(e + Pv) \, d\dot{m}$ for $t_2 - t_1$. If the inlet or outlet has *uniform, steady state* properties ($e + Pv$ = constant), then this term is

$$(e + Pv)m = \left(h + \frac{1}{2}\frac{\mathbf{V}^2}{g_c} + \frac{gZ}{g_c}\right) m \tag{4.87}$$

Equation (4.87) requires $e + Pv$ to be uniform and steady but for negligible kinetic energy contributions, $d\dot{m}$ could be nonuniform as well as unsteady.

All the above forms along with the required idealization are shown in Table 4.1.

4.4.2 Considerations within the Control Volume

The idealizations addressed at the inlets and outlets also must be considered for the terms describing the properties within the control volume.

The general form for the mass within the control volume [Eq. (4.59)] is

$$m_{CV} = \int_{V_{CV}} \rho \, dV \tag{4.88}$$

This includes nonuniform property variations.

If the *density (or specific volume) is uniform* throughout the control volume, then

$$m_{CV} = (\rho V)_{CV} \tag{4.89}$$

Also the properties within the control volume vary with time in some applications. This requires the integration of the conservation of mass over the time interval considered. The variation of mass with time or *unsteady* contribution yields

$$\int_{t_1}^{t_2} \frac{dm_{CV}}{dt} \, dt = m_2 - m_1 \tag{4.90}$$

The right-hand side of Eq. (4.90) represents the mass within the control volume at time t_2 minus the mass within the control volume at t_1.

The energy within the control volume is given by

$$E_{CV} = \int_{V_{CV}} \rho e \, dV \tag{4.91}$$

where $e = u + \frac{1}{2}\mathbf{V}_{\text{ref}}^2/g_c + gZ/g_c$. This term represents the energy contained within the control volume in its most general form, including possible nonuniform and unsteady variations.

The total energy and density within the control volume might be idealized as *uniform* in space. This yields

$$E_{CV} = (\rho e V)_{CV} = (me)_{CV} \tag{4.92}$$

where Eq. (4.89) has been used. This is an important expression since only one value describes the complete control volume energy.

Start-up and filling problems describe the accumulation of mass and energy within the control volume. This *unsteady* consideration results in an energy storage term. The time integration is generally required, and the needed expression is

$$\int_{t_1}^{t_2} \frac{dE_{CV}}{dt} \, dt = E_2 - E_1 \tag{4.93}$$

where E_2 and E_1 are the energies within the control volume at t_2 and t_1, respectively. If it is also assumed that the properties are *uniform* throughout the volume at any instant, then

$$\int_{t_1}^{t_2} \frac{dE_{CV}}{dt} \, dt = (em)_2 - (em)_1 \tag{4.94}$$

All the above expressions with the appropriate idealizations are given in Table 4.1.

The inlets and outlets are considered separately from the properties within the control volume since the terms have very different meanings. They also result in different forms for the various idealizations. Table 4.1 summarizes the above assumptions, but not all possible conditions have been addressed. Two types of problems that occur very often in engineering analysis combine the idealizations given above and are considered next. Both assume uniform properties.

4.4.3 Steady State Analysis

The analysis of the long-term operation of a particular component or complete system eliminates the need for describing the initial start-up period. This permits consideration of the steady state operation where no mass or energy accumulates within the control volume. Thus

$$\frac{dm_{\text{CV}}}{dt} = 0 \tag{4.95}$$

$$\frac{dE_{\text{CV}}}{dt} = 0 \tag{4.96}$$

The conservation of mass and of energy in Eqs. (4.27) and (4.52) are further reduced by assuming steady state conditions and uniform properties at each inlet and outlet to yield (see Table 4.1)

$$\dot{m}_{\text{in}} = \dot{m}_{\text{out}} \tag{4.97}$$

and

$$\left(h + \frac{1}{2}\frac{\mathbf{V}^2}{g_c} + \frac{gZ}{g_c}\right)_{\text{in}} \dot{m}_{\text{in}} + \dot{Q}_{\text{CV}} = \left(h + \frac{1}{2}\frac{\mathbf{V}^2}{g_c} + \frac{gZ}{g_c}\right)_{\text{out}} \dot{m}_{\text{out}} + \dot{W}_{\text{CV}} \tag{4.98}$$

These expressions are the basis of much of the cycle component analysis presented in the following chapters. The subscripts *in* and *out* represent all the inlets and outlets. In short, both these expressions state, "What goes in equals what comes out."

4.4.4 Unsteady State Analysis

Time variation occurs in filling or discharging problems as well as in start-up and shutdown considerations. Unsteady state analysis is needed when there is accumulation or storage within the control volume and/or time variation at the inlets and outlets. Integrating the conservation statements in Eqs. (4.27) and (4.52) from t_1 to t_2 and using Table 4.1 yield

$$\int_{t_1}^{t_2} \frac{dm_{\mathrm{CV}}}{dt}\, dt = \int_{t_1}^{t_2} \dot{m}_{\mathrm{in}}\, dt - \int_{t_1}^{t_2} \dot{m}_{\mathrm{out}}\, dt \tag{4.99}$$

or

$$m_2 - m_1 = m_{\mathrm{in}} - m_{\mathrm{out}} \tag{4.100}$$

and

$$\int_{t_1}^{t_2} \frac{dE_{\mathrm{CV}}}{dt}\, dt = \int_{t_1}^{t_2} (e\dot{m})_{\mathrm{in}}\, dt - \int_{t_1}^{t_2} (e\dot{m})_{\mathrm{out}}\, dt + \int_{t_1}^{t_2} \dot{Q}_{\mathrm{CV}}\, dt - \int_{t_1}^{t_2} \dot{W}_{\mathrm{CV}}\, dt \tag{4.101}$$

If the properties are considered to be uniform (at the inlets and outlets and within the control volume), the inlets and outlets are steady, and the kinetic energy of the inlet and outlet streams is negligible in comparison with the enthalpy and potential energy of these streams, then Table 4.1 yields the simplified expressions. The heat transfer and work terms are

$$\int_{t_1}^{t_2} \dot{Q}\, dt = {}_1Q_2 \tag{4.102}$$

and

$$\int_{t_1}^{t_2} \dot{W}\, dt = {}_1W_2 \tag{4.103}$$

where the right-hand sides represent the heat transferred and work done for $t_2 - t_1$.

This results in the conservation of energy as

$$E_{\mathrm{CV},2} - E_{\mathrm{CV},1} = (em)_{\mathrm{in}} - (em)_{\mathrm{out}} + {}_1Q_2 - {}_1W_2 \tag{4.104}$$

or

$$\left(u + \frac{1}{2}\frac{\mathbf{V}_{\mathrm{ref}}^2}{g_c} + \frac{gZ}{g_c}\right)_2 m_2 - \left(u + \frac{1}{2}\frac{\mathbf{V}_{\mathrm{ref}}^2}{g_c} + \frac{gZ}{g_c}\right)_1 m_1 = \left(h + \frac{1}{2}\frac{\mathbf{V}^2}{g_c} + \frac{gZ}{g_c}\right)_{\mathrm{in}} m_{\mathrm{in}} + {}_1Q_2 - {}_1W_2 - \left(h + \frac{1}{2}\frac{\mathbf{V}^2}{g_c} + \frac{gZ}{g_c}\right)_{\mathrm{out}} m_{\mathrm{out}} \tag{4.105}$$

This equation includes the unsteady storage term within the control volume and all inlet and outlet flow terms (terms with subscripts *in* and *out*). In short, the conservation of energy states that the change of energy within equals energy in minus energy out.

These two idealizations for conservation of mass and energy are summarized in Table 4.2.

4.5 Control Volume Applications

There are three major subdivisions to thermodynamic analysis.

1. *Control volume location*
   The control volume must be indicated on the system drawing. Carefully distinguish between the crossing boundary and inside boundary. Denote the energy transfers. The boundary must be drawn so that the transfer process of interest (mass flow, work, heat transfer) crosses the boundary if the solution for that quantity is sought. The conservation statements connect the desired result to other quantities. These other quantities must be known if the desired quantity is to be evaluated.
2. *Conservation statements*
   The conservation of mass and that of energy are applied to the control volume. The first point to consider is the time variation of the problem. Is unsteady or steady state analysis appropriate? The second point is the spatial variation of the inlets and outlets as well as within the control volume. Are the properties uniform within the control volume or at the inlets and outlets? All idealizations must be carefully considered.
3. *Properties*
   Two independent properties for a simple compressible substance are required to determine each state. Tabular information or equations of state are needed to obtain the results.

These three steps are many times done simultaneously or iteratively depending on the problem.

The results presented in Secs. 4.3 and 4.4 are applied to typical engineering components in this section. They are subdivided into four subsections depending on whether the component requires unsteady state or steady state analysis and whether there is a work interaction. The starting equations for all applications are given in Secs. 4.3 and 4.4, with important applications given in Table 4.2.

TABLE 4.2 Special Cases for Control Volume Analysis

Steady State with Uniform Properties

Mass: $$\sum_{\text{in}} \dot{m}_{\text{in}} = \sum_{\text{out}} \dot{m}_{\text{out}}$$

Energy: $$\sum_{\text{in}} \left(h + \frac{1}{2}\frac{\mathbf{V}^2}{g_c} + \frac{gZ}{g_c}\right)\dot{m} + \dot{Q}_{\text{CV}} = \sum_{\text{out}} \left(h + \frac{1}{2}\frac{\mathbf{V}^2}{g_c} + \frac{gZ}{g_c}\right)\dot{m} + \dot{W}_{\text{CV}}$$

Unsteady State with Uniform Properties and Steady State Port Properties

Mass: $$m_2 - m_1 = m_{\text{in}} - m_{\text{out}}$$

Energy: $$\left(u + \frac{1}{2}\frac{\mathbf{V}_{\text{ref}}^2}{g_c} + \frac{gZ}{g_c}\right)_2 m_2 - \left(u + \frac{1}{2}\frac{\mathbf{V}_{\text{ref}}^2}{g_c} + \frac{gZ}{g_c}\right)_1 m_1 = \sum_{\text{in}} \left(h + \frac{1}{2}\frac{\mathbf{V}^2}{g_c} + \frac{rgZ}{g_c}\right) m + {}_1Q_2 - {}_1W_2 - \sum_{\text{out}} \left(h + \frac{1}{2}\frac{\mathbf{V}^2}{g_c} + \frac{gZ}{g_c}\right) m$$

4.5.1 Steady State Work Applications

Turbines, compressors, pumps, and fans are typical steady state devices that have a work interaction with the surroundings. These devices generally have a single inlet and a single outlet. Also the change in potential energy from inlet to outlet is usually very small. Thus, Table 4.2 yields

Mass:

$$\dot{m}_{\text{in}} = \dot{m}_{\text{out}} \tag{4.106a}$$

Energy:

$$\left(h + \frac{1}{2}\frac{\mathbf{V}^2}{g_c}\right)_{\text{in}} \dot{m}_{\text{in}} + \dot{Q}_{\text{CV}} = \left(h + \frac{1}{2}\frac{\mathbf{V}^2}{g_c}\right)_{\text{out}} \dot{m}_{\text{out}} + \dot{W}_{\text{CV}}$$

or

$$\left(h + \frac{1}{2}\frac{\mathbf{V}^2}{g_c}\right)_{\text{in}} + q_{\text{CV}} = \left(h + \frac{1}{2}\frac{\mathbf{V}^2}{g_c}\right)_{\text{out}} + w_{\text{CV}} \tag{4.106b}$$

A turbine is a power-producing component which does work through the rotation of a shaft. The high-pressure fluid expands to a low pressure, doing work against the turbine blades. The kinetic and potential energy changes are usually, but not always, small. The heat transfer to the surroundings is also generally small compared to the work produced. A schematic of a turbine is shown in Fig. 4.6*a*.

Compressors require a work input to produce a high pressure at the outlet from the low pressure at the inlet. They are usually used for a gaseous working fluid. Fans, on the other hand, are employed to move fluid and alter velocities rather than produce large pressure changes. Pumps also require work input and are generally associated with liquids. All three devices require work input and typically have small heat transfers. Schematic representations of these steady state work applications are shown in Fig. 4.6 *b*, *c*, and *d*.

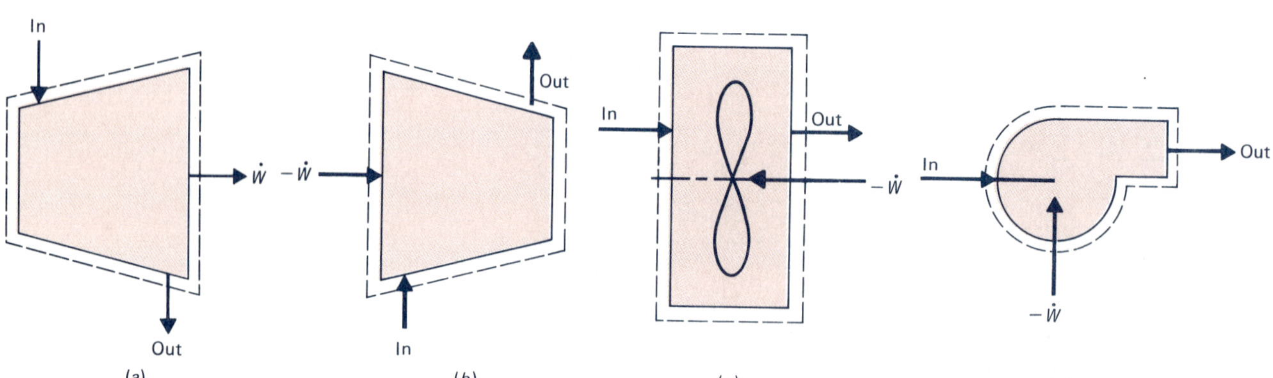

Figure 4.6 Steady state work applications. (*a*) Turbine; (*b*) compressor; (*c*) fan; (*d*) pump.

Example Problem 4.5

A steam turbine is designed to have a power output of 9 MW for a mass flow rate of 17 kg/s. The inlet state is 3 MPa, 450°C, and 200 m/s, and the outlet state is 0.5 MPa, saturated vapor, and 80 m/s. What is the heat transfer for this turbine? See the diagrams.

Diagrams

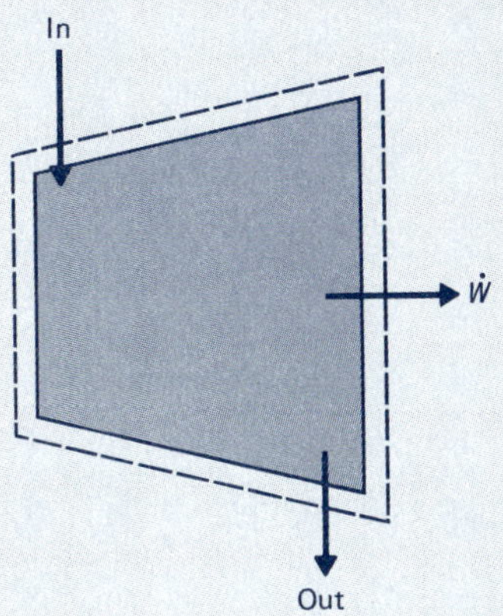

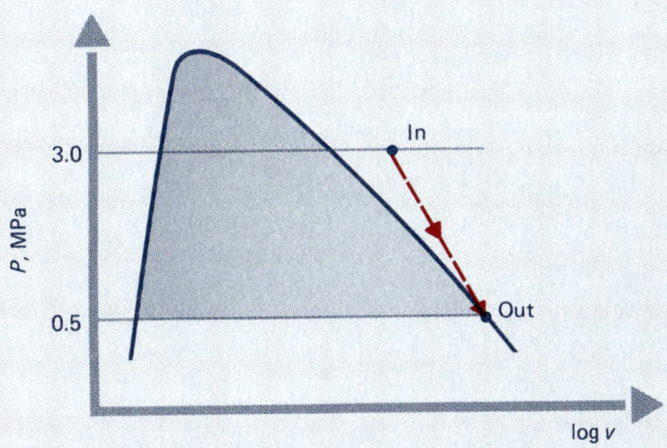

Solution

The conservation statements for a single inlet and a single outlet are

$$\text{Mass:}\ \left(\frac{dm}{dt}\right)_{CV} = \dot{m}_{in} - \dot{m}_{out}$$

$$\text{Energy:}\ \left(\frac{dE}{dt}\right)_{CV} = \dot{m}_{in}\left(h + \frac{1}{2}\frac{\mathbf{V}^2}{g_c} + \frac{gZ}{g_c}\right)_{in}$$

$$+ \dot{Q}_{CV} - \dot{W}_{CV} - \dot{m}_{out}\left(h + \frac{1}{2}\frac{\mathbf{V}^2}{g_c} + \frac{gZ}{g_c}\right)_{out}$$

All statements in the problem indicate steady operation, so the storage terms (within the control volume terms) are zero. Also there is no time variation at the ports, so the analysis is steady state. Only single values of properties are given for the inlet and outlet, so the ports are assumed to be uniform and the terms are given by Eq. (4.83). Thus the governing equations are

Mass:
$\dot{m}_{\text{in}} = \dot{m}_{\text{out}}$

Energy:

$$\left(h + \frac{1}{2}\frac{\mathbf{V}^2}{g_c} + \frac{gZ}{g_c}\right)_{\text{in}} \dot{m}_{\text{in}} + \dot{Q}_{\text{CV}} = \left(h + \frac{1}{2}\frac{\mathbf{V}^2}{g_c} + \frac{gZ}{g_c}\right)_{\text{out}} \dot{m}_{\text{out}} + \dot{W}_{\text{CV}}$$

They are the same as those given in Table 4.2 for a single inlet and a single outlet. Assuming the potential energy change is small compared with the other terms, we find

$$\dot{Q}_{\text{CV}} = \dot{W}_{\text{CV}} + \dot{m}(h_{\text{out}} - h_{\text{in}}) + \frac{\dot{m}}{2g_c}(\mathbf{V}^2_{\text{out}} - \mathbf{V}^2_{\text{in}})$$

The property information is given from Tables D.11 and D.12:

State	P, MPa	T, °C	x	h, kJ/kg
In	3.0	450	—	(3344.1)
Out	0.5	—	1.0	(2748.6)

Substituting numerical values yields

$$\begin{aligned}\dot{Q}_{\text{CV}} &= 9000 + 17(2748.6 - 3344.1) + \tfrac{17}{2}(80^2 - 200^2)\tfrac{1}{1000} \\ &= 9000 - 10{,}124 - 286 \text{ kW} \\ &= -1410 \text{ kW} = -4.81 \text{ MBtu/h}\end{aligned}$$

Note that the change in the kinetic energy term requires the conversion from watts to kilowatts when the mass flow rate is in kilograms per second and the velocity is in meters per second.

Comments

First, the rate of heat transfer is from the turbine to the surroundings. The temperature of the steam is greater than that of the surroundings, so this is expected. The heat transfer is 1410/9000 = 15.6 percent of the work from the turbine. Also the kinetic energy contribution is only 286/10,124 = 2.8 percent of the enthalpy change and is often neglected.

Example Problem 4.6

A compressor is required to provide 5 atm and 400°F air for a stationary power plant. The air intake is at 1 atm and 70°F. The outlet velocity is 80 ft/s. Evaluate the work per unit mass needed for an adiabatic compressor.

Diagram

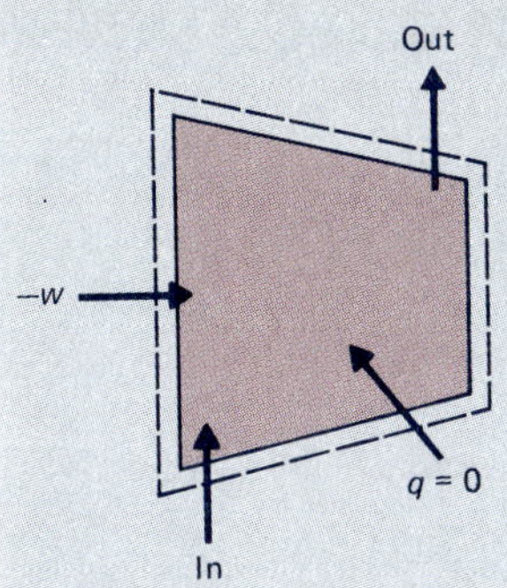

Solution

The state information is as follows:

State	P, atm	T, °F	h, Btu/lbm
In	1	70	
Out	5	400	

Thus, two independent properties are known for each state, and the other information can be obtained.

The conservation statements are given in Eq. (4.106) as

Mass:

$$\dot{m}_{\text{in}} = \dot{m}_{\text{out}}$$

Energy:

$$\left(h + \frac{1}{2}\frac{\mathbf{V}^2}{g_c}\right)_{\text{in}} + q_{\text{CV}} = \left(h + \frac{1}{2}\frac{\mathbf{V}^2}{g_c}\right)_{\text{out}} + w_{\text{CV}}$$

The inlet area to a compressor is often very large, so the velocity at the inlet is approximately zero for a fixed $\dot{m} = \rho A \mathbf{V}$. The control surface could be placed at another location in the compressor inlet with a smaller area, but then the inlet velocity would be needed. Also the compressor is adiabatic, so $q = 0$. Thus,

$$w = (h_{\text{in}} - h_{\text{out}}) - \frac{1}{2}\frac{\mathbf{V}_{\text{out}}^2}{g_c}$$

The change in enthalpy for air is determined by treating air as an ideal gas (this is accurate, which can be determined from the generalized compressibility chart in Fig. C.1). For an ideal gas [Eq. (3.22)]

$$dh = c_p\, dT$$

So if c_P is constant, Table C.2 yields

$$w = c_P(T_{\text{in}} - T_{\text{out}}) - \frac{1}{2}\frac{\mathbf{V}^2_{\text{out}}}{g_c}$$

$$= (0.2401)(70 - 400) - \frac{1}{2}\frac{(80)^2}{32.174}\frac{1}{778.16}$$

$$= -79.23 \text{ Btu/lbm} - 0.13 \text{ Btu/lbm} = -79.36 \text{ Btu/lbm}$$
$$= -184.3 \text{ kJ/kg} - 0.3 \text{ kJ/kg} = -184.6 \text{ kJ/kg}$$

Comments

The outlet kinetic energy contribution to the work is 0.2 percent for this example. A more accurate approach includes the temperature dependence of the specific heat. Table E.1 gives

$$\bar{c}^{o}_{P} = 6.713 + (0.02609 \times 10^{-2})(T) + (0.03540 \times 10^{-5})(T^2) - (0.08052 \times 10^{-9})(T^3) \qquad \text{Btu/(lbmol}\cdot{}^{\circ}\text{R)}$$

Then

$$w = -\int_{\text{in}}^{\text{out}} c_P\, dT - \frac{1}{2}\frac{\mathbf{V}^2_{\text{out}}}{g_c}$$

yields

$$w = -80.19 \text{ Btu/lbm} - 0.13 \text{ Btu/lbm} = -80.32 \text{ Btu/lbm}$$

The difference between the constant c_P solution and variable c_P solution is very small since the temperature dependence of c_P near 20°C indicates that c_P is almost constant in this temperature range.

4.5.2 Magnitude of Various Terms in the First Law

In the steady state form of the first law [Eq (4.98)], various terms can often be neglected in specific applications. Care must be taken, however, because there are exceptions to almost all the rules of thumb commonly used to decide which terms are negligible.

Kinetic Energy Change of the Working Fluid

For example, examine the change in the kinetic energy terms relative to the change in enthalpy. The general criterion for omitting the kinetic energy term relative to the enthalpy is

$$\frac{\Delta(\mathbf{V}^2)}{2g_c} \ll \Delta h \tag{4.107}$$

Or, more quantitatively, if we wish the kinetic energy change to be less than 1 percent of the enthalpy change so that it may be neglected,

$$\frac{\Delta \mathbf{V}^2}{2g_c} < 0.01\ \Delta h \tag{4.108}$$

Suppose that one of the velocities being considered, either $\mathbf{V}_{in}$ or $\mathbf{V}_{out}$, has a magnitude of 0. Then, in order to neglect the kinetic energy change, the magnitude of the remaining mean flow stream velocity must be less than

$$\mathbf{V} = (0.02 g_c\ \Delta h)^{1/2} \tag{4.109}$$

Example Problem 4.7

Find the maximum inlet velocity that may be neglected when steam is used as the working fluid in a turbine if the exit velocity is negligible.

Solution

For steam turbines, the enthalpy change across the turbine is often on the order of a few hundred kilojoules per kilogram. Taking even a relatively low value of $\Delta h = 100$ kJ/kg, we see that the velocity change (relative to zero velocity) required to make the kinetic energy change equal to 1 percent of the enthalpy change is

$$\mathbf{V} = [0.02 \times 100\ \text{kJ/kg} \times 1000\ \text{m}^2/(\text{s}^2 \cdot \text{kJ} \cdot \text{kg})]^{1/2} = 45\ \text{m/s}\ (145\ \text{ft/s})$$

This is a relatively high velocity, yet the associated kinetic energy change of the steam is negligible when compared with a relatively small change in enthalpy.

Potential Energy Change of the Working Fluid

The height change required to make the potential energy change equal to 1 percent of the enthalpy change is

$$\Delta Z = \left(\frac{0.01 g_c \Delta h}{g}\right)^{1/2} \tag{4.110}$$

For a change in enthalpy of, say, $\Delta h = 100$ kJ/kg, the required height change of the flow stream across the control volume (on the earth's surface) for the potential energy change to be significant is

$$\Delta Z = \left[0.01 \times 100\ \text{kJ/kg} \times \frac{1000\ \text{m}^2/(\text{s}^2 \cdot \text{kJ} \cdot \text{kg})}{9.8\ \text{m/s}^2}\right]^{1/2} = 10\ \text{m} = 32\ \text{ft}$$

Thus the potential energy change of the flow stream can often be neglected when we are analyzing typical equipment such as turbines, pumps, and compressors where significant enthalpy changes occur because of the presence of large shaft work. An exception is the analysis of hydroelectric turbines, where, depending on the choice of control volume, the work output can be computed chiefly from the change in potential energy of the working fluid.

Heat Transfer to the Control Volume

If a large temperature difference is present between the control volume and the surroundings, then heat transfer may be significant unless thermal insulation is provided. From thermodynamics, the amount of heat transfer can be computed from the first law only if all other terms in the equation are known; this implies that we must usually have actual performance data on a device to determine whether heat transfer is important. The prediction of the amount of heat transfer based on the temperature difference between the control volume and the surroundings and the mechanisms of heat transfer (conduction, convection, and radiation) is the aim of courses in heat transfer, and it is beyond the scope of this text.

For most equipment, the heat transfer between the equipment and the surroundings is small compared with the work done on or by the equipment. If this were not the case in a particular application, then the designer would normally take steps to make it the case. A significant amount of heat transfer usually shows up in a decrease of work output from a turbine, or an increase in the work input to a pump or compressor in order to achieve given outlet conditions. Thus, the designer will insulate these devices if heat transfer is significant. Neglect of heat transfer is thus usually justified in analyzing these devices, unless specific information is provided about the heat transfer.

4.5.3 Steady State Flow Applications

Heat exchangers, condensers, steam generators, diffusers and nozzles, throttling valves, and pipes are devices that operate at steady state and do not produce or consume work. Most experience very small changes in potential energy, so

Mass:

$$\dot{m}_{\text{in}} = \dot{m}_{\text{out}} \tag{4.111a}$$

Energy:

$$\left(h + \frac{1}{2}\frac{\mathbf{V}^2}{g_c}\right)_{\text{in}} \dot{m}_{\text{in}} + \dot{Q}_{\text{CV}} = \left(h + \frac{1}{2}\frac{\mathbf{V}^2}{g_c}\right)_{\text{out}} \dot{m}_{\text{out}} \tag{4.111b}$$

The number of inlets and outlets depends on the device as well as on the location of the control volume.

A heat exchanger is used to transfer energy from one fluid to another. There is no work interaction with the surroundings, and the changes in potential and kinetic energies are typically small. The heat

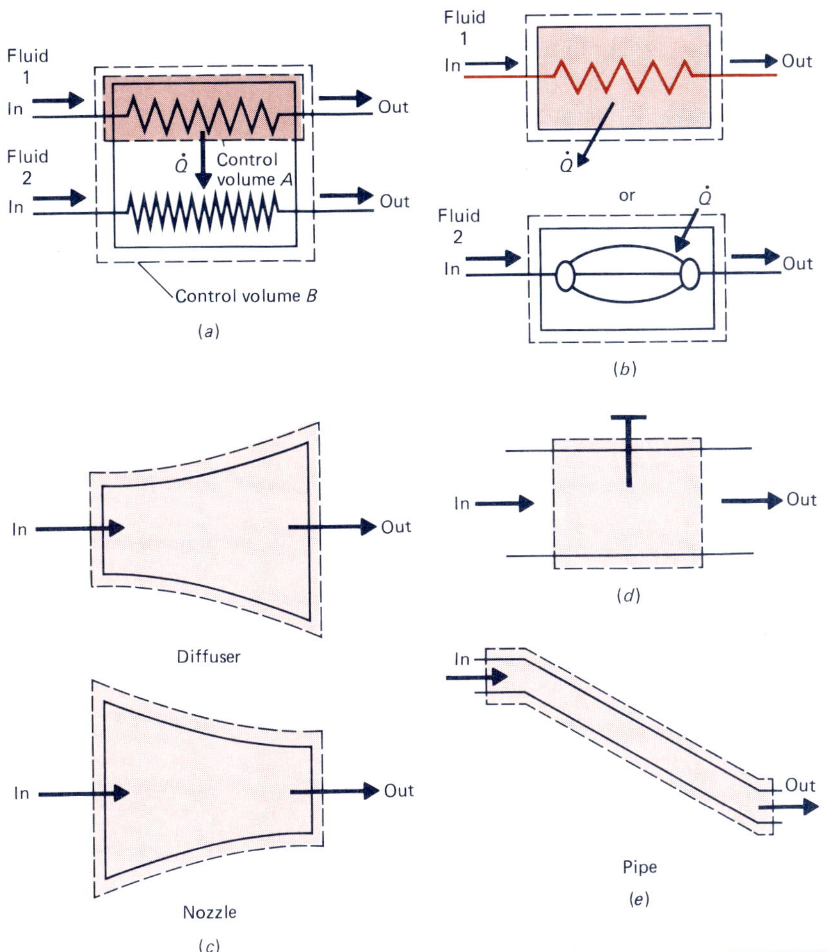

Figure 4.7 Steady state flow applications. (*a*) Heat exchanger; (*b*) steam generator; (*c*) nozzle and diffusor; (*d*) throttling valve; (*e*) pipe.

transferred is computed from the enthalpy difference, and the analysis is very dependent on the choice of the control volume. Figure 4.7*a* indicates two possible control volumes, and the specific one used depends on the desired and known information. For control volume A, the first law is,

$$q = h_{1,\text{out}} - h_{1,\text{in}} \tag{4.112}$$

and for control volume B,

$$\dot{m}_1 h_{1,\text{in}} + \dot{m}_2 h_{2,\text{in}} = \dot{m}_1 h_{1,\text{out}} + \dot{m}_2 h_{2,\text{out}} \tag{4.113}$$

A steam generator and a condenser are types of heat exchangers that perform a specific task. A steam generator uses an energy source such as the burning of pulverized fuel or a nuclear reactor to create a high-temperature fluid. There is heat transfer from this fluid to water which passes through tube banks to form a superheated vapor. This is diagrammed in Fig. 4.7*b*. A condenser circulates cold water to condense a high-quality mixture to the liquid state. For the steam generator and condenser, the first law reduces to

$$q = h_{\text{out}} - h_{\text{in}} \tag{4.114}$$

For a condenser, it is useful in some situations to have the control volume as shown in Fig. 4.7*a* (control volume *B*).

Diffusers and nozzles are used to control the velocity of the fluid. A diffuser decreases the velocity while increasing the pressure, and the opposite is true for a nozzle. These devices are simply area-changing devices, as shown in Fig. 4.7*c*. Since kinetic energy is typically important and there is typically a single inlet and outlet, Eq. (4.111) reduces to

$$q = (h_{\text{out}} - h_{\text{in}}) + \frac{1}{2g_c}(\mathbf{V}_{\text{out}}^2 - \mathbf{V}_{\text{in}}^2) \tag{4.115}$$

where q might also be negligible.

A throttling valve is simply a flow constriction, as shown in Fig. 4.6*d*. This reduces the pressure but performs no work. Generally, the heat transfer is small. If a control volume sufficiently far from the actual constriction is chosen, then the change in kinetic energy is small. Equation (4.111) yields

$$h_{\text{in}} = h_{\text{out}} \tag{4.116}$$

This is a very important result which is applied to any valve.

One final steady state flow device is a simple pipe. Assuming that potential energy changes are important and that there is no work or heat transfer, we see from Table 4.2 that

$$\left(h + \frac{1}{2}\frac{\mathbf{V}^2}{g_c} + \frac{gZ}{g_c}\right)_{\text{in}} = \left(h + \frac{1}{2}\frac{\mathbf{V}^2}{g_c} + \frac{gZ}{g_c}\right)_{\text{out}}$$

or

$$h_{\text{in}} - h_{\text{out}} + \frac{1}{2g_c}(\mathbf{V}_{\text{in}}^2 - \mathbf{V}_{\text{out}}^2) + \frac{g}{g_c}(Z_{\text{in}} - Z_{\text{out}}) = 0 \tag{4.117}$$

If the fluid is incompressible and isothermal, then Eq. (3.39*a*) yields

$$dh = c\,dT + v\,dP = v\,dP$$

so

$$\frac{P_{\text{in}} - P_{\text{out}}}{\rho} + \frac{1}{2g_c}(\mathbf{V}_{\text{in}}^2 - \mathbf{V}_{\text{out}}^2) + \frac{g}{g_c}(Z_{\text{in}} - Z_{\text{out}}) = 0 \tag{4.118}$$

This is termed *Bernoulli's equation.* This derivation of Bernoulli's equation is *not* general since Bernoulli's equation is generally derived from

Newton's second law and does not require the isothermal assumption. It is presented here to show the connection with fluid mechanics.

Example Problem 4.8

A high-velocity water jet at $T = 30°C$ is proposed for use as a cutting device with possible applications in metal, concrete, and plastic. A sapphire nozzle with dimensions shown in the diagram uses liquid water at a mass flow rate of 0.0065 kg/s. The water discharges to 0.1 MPa. Assume the flow is isothermal, and determine the pressure of the water source needed to operate this device.

Diagram

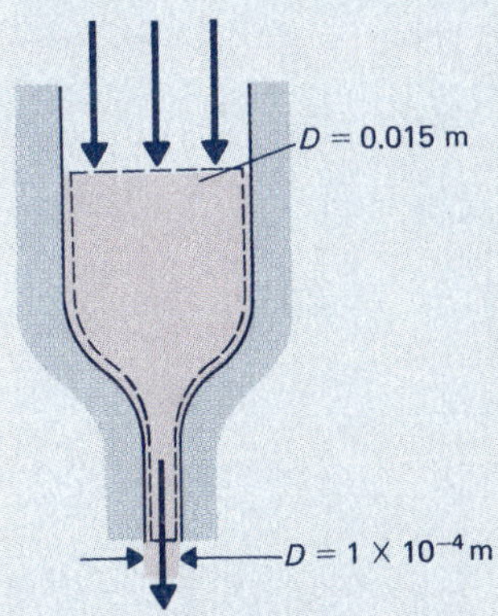

Solution

The properties are evaluated by assuming the liquid is incompressible. This process is idealized as adiabatic, isothermal, and steady state. The inlet and outlet are assumed to have uniform properties; thus conservation of mass yields

$$\dot{m} = \rho A \mathbf{V} \qquad \text{or} \qquad \mathbf{V} = \frac{\dot{m}}{\rho A} = \frac{v\dot{m}}{A}$$

Equation (4.118), neglecting changes in potential energy, yields

$$0 = v(P_{out} - P_{in}) + \frac{\dot{m}^2 v^2}{2g_c}\left[\left(\frac{1}{A_{out}}\right)^2 - \left(\frac{1}{A_{in}}\right)^2\right]$$

With $A = \pi D^2/4$,

$$P_{in} = P_{out} + \frac{8\dot{m}^2 v}{\pi^2 g_c}\left(\frac{1}{D_{out}^4} - \frac{1}{D_{in}^4}\right)$$

With numerical values, taking $v = 0.001004\ \mathrm{m^3/kg}$ at $T = 30°\mathrm{C}$, we have

$$P_{in} = 100 + \frac{3(0.0065)^2(0.001004)}{\pi^2(1000)(1)}\left[\frac{1}{(1 \times 10^{-4})^4} - \frac{1}{(0.015)^4}\right]$$

$$= 100 + 343{,}800\ \mathrm{kPa} = 343.9\ \mathrm{MPa} = 3394.0\ \mathrm{atm}$$

Comments

The required pressure is very high so the incompressible assumptions might introduce some error. Also the dependence of the inlet pressure on the nozzle diameter controls the water outlet speed. For a nozzle diameter of 1×10^{-4} m, the outlet speed is

$$\mathbf{V} = \frac{v\dot{m}}{A} = \frac{(0.001004)(0.0065)}{\pi(10^{-8})/4} = 831\ \mathrm{m/s} = 2727\ \mathrm{ft/s} = 1859\ \mathrm{mi/h}$$

This very large speed is required for this application.

4.5.4 Unsteady State Work Applications

Short-operation turbines and the start-up of systems require an unsteady analysis. Another type of unsteady work problem involves the increase of fluid within a tank required to move a piston. This type of problem is diagramed in Fig. 4.8*a*. Conservation of mass and energy from Table 4.2 for this type of problem yields (for negligible kinetic and potential energy in the flow streams compared with the enthalpy)

Mass:

$$m_2 - m_1 = m_{in} - m_{out} \tag{4.119a}$$

Energy:

$$u_2 m_2 - u_1 m_1 = h_{in} m_{in} + {}_1Q_2 - {}_1W_2 - h_{out} m_{out} \tag{4.119b}$$

Remember that the state within the control volume has been assumed to be uniform at any instant.

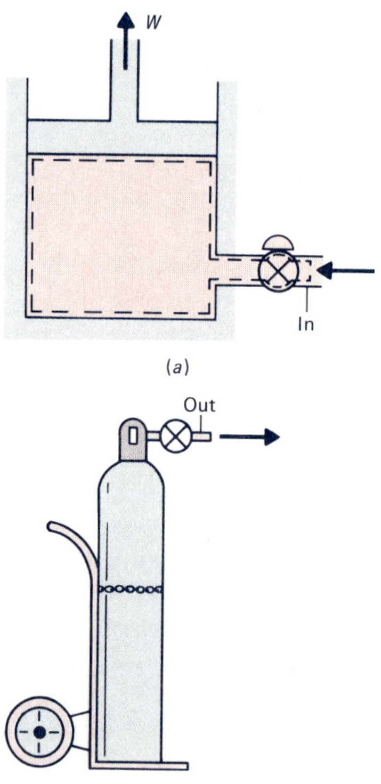

Figure 4.8 Unsteady state applications. (*a*) Unsteady state work applications; (*b*) unsteady state flow application.

Example Problem 4.9

An insulated 10-ft³ cylinder is used to displace a load through the motion of a piston, as shown in the diagram. The piston has negligible volume and is initially at the bottom of the cylinder. The valve to a steam line (200 psia and 700°F) is opened to raise the load, and it is closed when the piston reaches the top of the cylinder. The total work done is 50 Btu. What is the final pressure in the cylinder if the final mass is 0.5 lbm?

Diagram

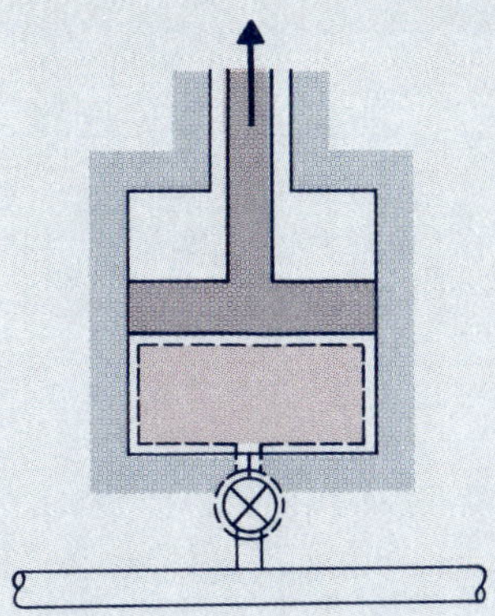

Solution

The states are indicated in the table:

State	P, psia	T, °F	v, ft³/lbm	u, Btu/lbm	h, Btu/lbm
1	—	—	0	—	—
2	—	—	20		
In	200	700	—		

The control volume is indicated in the diagram, and it includes the valve. The initial state within the control volume is not required, since there is no fluid within the control volume. The inlet is completely specified, but the final state needs one more property. The control volume is placed around the steam under the piston and enlarges with time to move the piston. The cylinder is insulated, so there is no heat transfer during filling. All properties are taken as uniform. Table 4.2 yields

$$m_2 - m_1 = m_{in} - m_{out}$$

but $m_1 = 0$ and $m_{out} = 0$, so

$$m_2 = m_{in}$$

Thus from Table 4.2, conservation of energy states that

$$u_2 m_2 = h_{in} m_{in} - {}_1W_2$$

Eliminating m_{in} with conservation of mass yields

$$u_2 m_2 = h_{in} m_2 - {}_2W_2$$

or, by dividing through by m_2,

$$u_2 = h_{in} - {}_1w_2$$
$$= h_{in} - \frac{50}{0.5}\frac{\text{Btu}}{\text{lbm}}$$
$$= h_{in} - 100 \qquad \text{Btu/lbm}$$

Table E.12 yields $h_{in} = 1373.7$ Btu/lbm, so $u_2 = 1273.7$ Btu/lbm. The final state is specified by u and v. Interpolation yields $u = 1139.8$ Btu/lbm for $P = 25$ psia and $v = 20$ ft^3/lbm and $u = 1465.2$ Btu/lbm for $P = 50$ psia and $v = 20$ ft^3/lbm. Thus linearly interpolating between the u values yields

$$P_2 = 35.29 \text{ psia} = 243.3 \text{ kPa}$$

Also $T_2 = 756°\text{F} = 402°\text{C}$.

Comments

If the control volume had been chosen without including the valve, the enthalpy on the tank side of the valve would be needed in the energy balance. However, the enthalpy on the pipe side is known. To relate these two enthalpies, a control volume around the valve is needed, as shown in the margin.

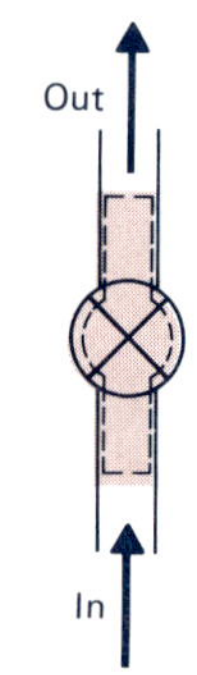

This is an unsteady flow problem without mass and energy accumulation. Since no work or heat transfer crosses the control volume boundary, Table 4.2 reduces to

Mass:

$$m_{in} = m_{out}$$

Energy:

$$0 = h_{in}m_{in} - h_{out}m_{out}$$

The energy equation yields

$$h_{in} = h_{out}$$

even if the inlet flow is unsteady. This is the connection between enthalpies necessary to complete the problem if the control volume is chosen around the tank.

Observe that the final steam temperature in the tank is *greater* than that in the steam line. Why should this be so?

4.5.5 Unsteady State Flow Applications

Charging and discharging of tanks without work are important applications. The purpose is to add or withdraw fluid from a tank. The discharge process is diagrammed in Fig. 4.8*b*. The kinetic and potential energy changes are usually small, so conservation of mass and conservation of energy for the diagramed problem are

Mass:

$$m_2 - m_1 = -m_{out} \tag{4.120a}$$

Energy:

$$u_2 m_2 - u_1 m_1 = {}_1Q_2 - h_{out} m_{out} \tag{4.120b}$$

The energy equation [Eq. (4.120*b*)] contains the assumption that h_{out} is steady. If h_{out} varies with time, then the final term in Eq. (4.120*b*) must be replaced by a term of the form in Eq. (4.86).

Example Problem 4.10

Water flows through a pipe with a 5-cm inside diameter (ID) at a velocity of 0.5 m/s into an initially empty 0.2-m³ drum (approximately a 55-gal drum). How long will it take to fill the drum? What is the volumetric flow rate of the water through the pipe?

Solution

The water is assumed to have a uniform velocity profile in the pipe, so the velocity given is used as the fluid velocity. An alternative interpretation is to assume that the velocity stated is the average velocity in the pipe. The specific volume of water at atmospheric conditions is $v = 0.001$ m³/kg. The mass flow rate is

$$\dot{m} = \frac{\mathbf{V}A}{v} = \frac{0.5\pi(0.025)^2}{0.001} = 0.982 \text{ kg/s}$$

The volumetric flow rate in the pipe is

$$\dot{V} = \dot{m}v = (0.982)(0.001) = 0.982 \times 10^{-3} \text{ m}^3/\text{s} = 0.03468 \text{ ft}^3/\text{s}$$

The conservation of mass from Eq. (4.68) yields

$$\frac{d}{dt}\left(\int_{CV} \rho \, dV\right) = \int_{A_{in}} d\dot{m} - \int_{A_{out}} d\dot{m}$$

This is an unsteady filling problem which requires an integration over time. The inlet and outlet terms are given by Eq. (4.81), and the control volume term is given by Eq. (4.90). The integrated conservation of mass expression is $m_2 - m_1 = m_{in} - m_{out}$, which is the same as that given in Table 4.2. Since $m_{out} = 0$,

$$m_2 - m_1 = m_{in}$$

But the flow is steady, so $m_{in} = \dot{m}_{in}(t_2 - t_1)$ and

$$m_2 - m_1 = \dot{m}_{in}(t_2 - t_1)$$

If the drum is initially empty at $t_1 = 0$, then $m_1 = 0$. At t_2 the drum is full, so $m_2 = V/v = 0.2/0.001 = 200$ kg. Then the conservation of mass yields

$$200 \text{ kg} - 0 \text{ kg} = (0.982 \text{ kg/s})(t_2 \text{ s})$$

Solving for t_2 yields

$$t_2 = \frac{200}{0.982} = 204 \text{ s} = 3.39 \text{ min}$$

Example Problem 4.11

A tank of volume V is to be filled with an ideal gas. Initially the tank is at P_1 and T_1. The port is regulated with a valve, and the port properties are constant at T_{in}. The process is adiabatic. Determine the final temperature T_2 to obtain P_2.

Diagram

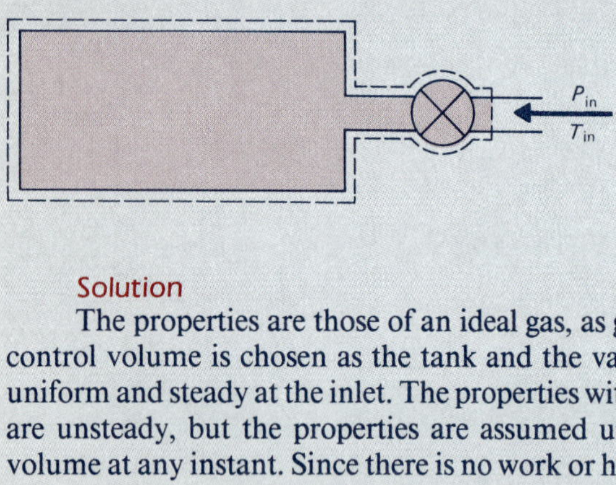

Solution

The properties are those of an ideal gas, as given in Sec. 3.5.2. The control volume is chosen as the tank and the valve. The properties are uniform and steady at the inlet. The properties within the control volume are unsteady, but the properties are assumed uniform throughout the volume at any instant. Since there is no work or heat transfer, the conservation statements are (Table 4.2)

Mass:

$$m_2 - m_1 = m_{in}$$

Energy:

$$m_2 u_2 - m_1 u_1 = h_{in} m_{in}$$

The inlet mass flow is not specified, so it is eliminated from the above equations, and

$$m_2 u_2 - m_1 u_1 = h_{\text{in}}(m_2 - m_1)$$

or

$$m_2(u_2 - h_{\text{in}}) = m_1(u_1 - h_{\text{in}})$$

This is the expression needed to evaluate the final temperature through $u_2 m_2$.

The properties are determined for the ideal gas, by assuming constant specific heats:

$$PV = mRT \qquad \text{or} \qquad m = \frac{PV}{RT}$$

and

$$du = c_v\, dT$$

With $h = u + Pv = u + RT$ and $c_P - c_v = R$, we have

$$u_2 - h_{\text{in}} = u_2 - u_{\text{in}} - RT_{\text{in}} = c_v(T_2 - T_{\text{in}}) - RT_{\text{in}} = c_v T_2 - c_P T_{\text{in}}$$

Substituting this into the above expression and performing the same manipulations for $u_1 - h_{\text{in}}$ yield

$$\frac{P_2 V}{RT_2}(c_v T_2 - c_P T_{\text{in}}) = \frac{P_1 V}{RT_1}(c_v T_1 - c_P T_{\text{in}})$$

With V and R as constants and $k = c_P/c_v$,

$$P_2 - P_2 k \frac{T_{\text{in}}}{T_2} = P_1 - P_1 k \frac{T_{\text{in}}}{T_1}$$

so

$$T_2 = \frac{1}{(P_2 - P_1)/(kP_2 T_{\text{in}}) + P_1/(P_2 T_1)}$$

Comments

As P_2 approaches P_1, then T_2 approaches T_1. But this is expected, since there is not much filling when P_2 is close to P_1. If P_2 is much larger than P_1, then the second term in the denominator is negligible, and P_1 is negligible compared with P_2. In this case,

$$T_2 \rightarrow kT_{\text{in}}$$

Because k is always greater than unity $[k = c_P/c_v = (R + c_v)/c_v = 1 + R/c_v]$, the final temperature is always above the inlet value.

4.6 Other Statements of the First Law

This chapter discusses conservation of mass and energy for both a control mass and a control volume. The conservation principles are the same, but the resulting equations have different terms.

The first law of thermodynamics has many different forms, and all are equivalent. We have chosen to develop the first law from a general control mass viewpoint, as given in Eq. (4.5). We stated the first law as "energy can be neither created nor destroyed" for an isolated system in Chap. 1. We could alternatively state that the energy is constant for an isolated system. Another statement of the first law is that the work done by a system in a cyclic process is equal to the heat transferred to the system. A different approach is to start with the idea that a perpetual-motion machine of the first kind is impossible. A perpetual-motion machine of the first kind is a continously operating device that produces a continuous supply of energy without receiving energy input. The equivalency of these statments is shown here.

The control mass form of conservation of energy [Eq. (4.5)] is

$$dE = \delta Q - \delta W$$

This expresses the change in energy for a process. This form reduces to the alternative statement that the energy is constant for an isolated system. An isolated system is one with no energy transfers with the surroundings. Thus, if $\delta Q = \delta W = 0$, then Eq. (4.5) yields

$$dE = 0$$

as stated.

Another form states that an extensive property E exists whose increment is equal to the work received by the system while surrounded by an adiabatic wall. For an adiabtic wall, $\delta Q = 0$, and the work received is negative by our sign convention, so

$$dE = -\delta W$$

Therefore, the work is equal to the change in total energy. The total energy is an extensive property. Note that for this particular process (adiabatic), dE is as usual an exact differential; therefore, δW for this process also must be an exact differential.

The perpetual-motion machine principle states that a continuously operating device which produces a continuous supply of energy without receiving energy input (a perpetual-motion machine of the first kind) is impossible. If a device has no energy transfers, then $\delta Q = 0$ and $\delta W = 0$. Thus it is impossible to have dE equal anything but zero. Another form based on cycles states that the work done by a system in a cyclic process is proportional to the heat transfer. This equality was demonstrated in Eq. (4.15). All these statements of the first law are equivalent, and any could be used as the starting point for the development in this chapter.

Problems

4.1S An ideal gas (CO_2) with molecular weight 44.01 is expanded slowly and isobarically. The gas has a mass of 0.1 kg, and the expansion occurs at 1 atm. If the expansion occurs from an initial temperature of 50°C to a final temperature of 150°C, find the average value of c_v for the gas during the process.

4.1E An ideal gas (CO_2) with molecular weight 44.01 is expanded slowly and isobarically. The gas has a mass of 0.2 lbm, and the expansion occurs at 1 atm. If the expansion occurs from an initial temperature of 100°F to a final temperature of 300°F, find the average value of c_v for the gas during the process.

4.2S Water is heated in a piston-cylinder arrangement from the saturated liquid state to the saturated vapor state (Fig. P4.2). Evaluate the heat transfer per unit mass for (*a*) 0.05 MPa, (*b*) 0.10 MPa, (*c*) 0.5 MPa, and (*d*) 20 MPa.

4.2E Water is heated in a piston-cylinder arrangement from the saturated liquid state to the saturated vapor state (Fig. P4.2). Evaluate the heat transfer per unit mass for each of the following pressures: (*a*) 7 psia, (*b*) 14.7 psia, (*c*) 70 psia, and (*d*) 3000 psia.

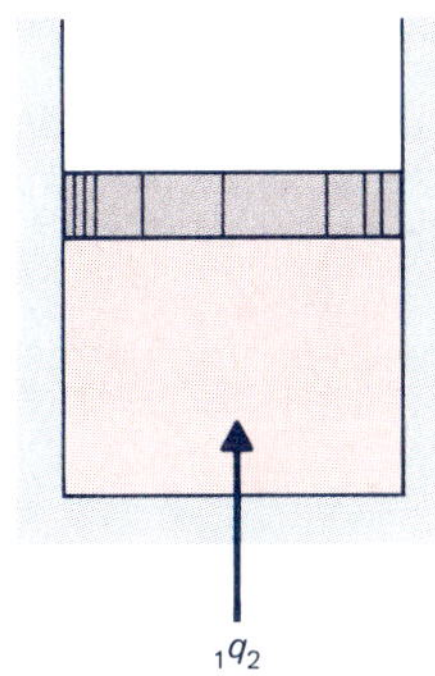

Figure P4.2

4.3S A piston-cylinder arrangement initially contains water at $P = 2$ MPa and $T = 300$°C. The piston is moved to a final position, where the volume is twice the initial volume. During the piston movement, there is heat transfer to the water in such a way that the pressure in the cylinder remains constant.

(*a*) Sketch the process on a P-v diagram, indicating the relative position of the saturation curve.
(*b*) Find the work done per kilogram of water during the expansion.
(*c*) Find the heat transfer per kilogram of water during the expansion.
(*d*) Find the value of the enthalpy of the water after expansion.

4.3E A piston-cylinder arrangement initially contains water at $P = 300$ psia and $T = 500$°F. The piston is moved to a final position where the volume is twice the initial volume. During the piston movement, there is a heat transfer to the water in such a way that the pressure in the cylinder remains constant.

(*a*) Sketch the process on a P-v diagram, indicating the relative position of the saturation curve.

(*b*) Find the work done per pound of water during the expansion.
(*c*) Find the heat tranfer per pound of water during the expansion.
(*d*) Find the value of the enthalpy of the water after expansion.

4.4S A cylinder-piston system contains 2 kg of H_2O at 150 kPa and has a volume of 0.35 m³. The piston is moved so that the final volume of the cylinder is 2.319 m³. During the piston motion from the initial to the final state, there is heat transfer to the cylinder in such a way as to hold the temperature contant.
(*a*) What is the final pressure in the cylinder?
(*b*) How much work was done by the steam?
(*c*) Evaluate the heat transfer during the process.
(*d*) Sketch the process on P-v and P-h diagrams, labeling the initial state 1 and the final state 2.

4.4E A cylinder-piston system contains 5 lbm of H_2O at 20 psia and has a volume of 10 ft³. The piston is moved so that the final volume of the cylinder is 100.47 ft³. During the piston motion from the initial to the final state, there is a heat transfer to the cylinder in such a way as to hold the temperature constant.
(*a*) What is the final pressure in the cylinder?
(*b*) How much work was done by the steam?
(*c*) Evaluate the heat transfer during the process.
(*d*) Sketch the process on P-v and P-h diagrams, labeling the initial state as 1 and the final state as 2.

4.5S One kilogram of nitrogen fills a piston-cylinder system. The piston is weightless and frictionless and separates the nitrogen from the surroundings, which are at 1 atm. The initial volume in the cylinder is 1 m³ at 1 atm. See Fig. P4.5. Heat transfer to the nitrogen occurs until the volume is doubled. Assuming the specific heat of nitrogen can be taken as constant during the process, evaluate the heat transfer to the nitrogen during the process.

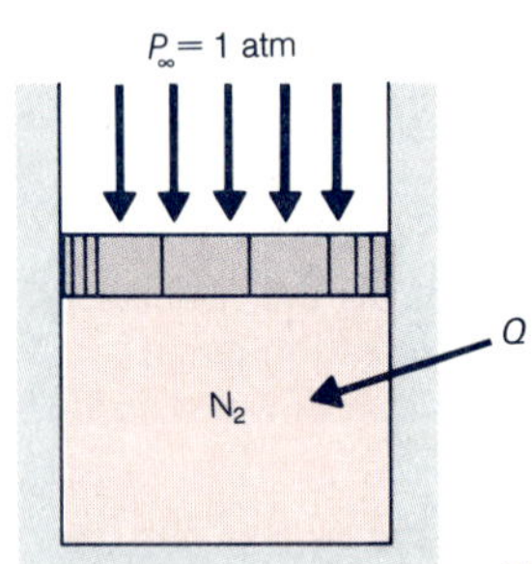

Figure P4.5

4.5E Two pounds mass of nitrogen fills a piston-cylinder system. The piston is weightless and frictionless, and separates the nitrogen from the surroundings, which are at 1 atm. The initial volume in the cylinder is 35 ft³ at 1 atm. See Fig. P4.5. Heat transfer to the nitrogen occurs until the volume is doubled. Assuming the specific heat of nitrogen can be taken as constant during the process, evaluate the heat transfer to the nitrogen during the process.

4.6S One kilogram of water is contained in a sealed kettle with a volume of 3 m³ at $T = 20$°C. An electric immersion heater inside the kettle is turned on and remains on until the kettle is filled with saturated vapor. Electricity costs $0.15 per kilowatthour.

(*a*) Draw a sketch defining the system you choose for analysis. Also sketch a *P-v* diagram of the process undergone by that system. (Show the saturation dome on your sketch.) Note any necessary assumptions.

(*b*) Find the final pressure in the kettle.

(*c*) Find the cost of the electricity necessary to carry out the process.

4.6E Two pounds-mass of water is contained in a sealed kettle with a volume of 100 ft^3, at $T = 70°F$. An electric immersion heater inside the kettle is turned on and remains on until the kettle is filled with saturated vapor. Electricity costs $0.15 per kilowatthour.

(*a*) Draw a sketch defining the system you choose for analysis. Also sketch a *P-v* diagram of the process undergone by that system. (Show the saturation dome on your sketch.) Note any necessary assumptions.

(*b*) Find the final pressure in the kettle.

(*c*) Find the cost of the electricity necessary to carry out the process.

4.7 Develop an equation for the heat transfer to an ideal gas during a constant-pressure expansion in terms of the initial pressure P_1, the initial and final volumes V_1 and V_2, and c_P (assumed temperature-independent) and R for the gas.

4.8 Develop an equation for the change in internal energy U of an ideal gas during a constant-pressure expansion in terms of the initial pressure P_1, the initial and final volumes V_1 and V_2, and c_v and R for the gas.

4.9 In Prob. 2.75, an elevator is allowed to fall to an equilibrium position. For that problem, determine the heat transfer from the air in the elevator shaft in order for the assumption of an isothermal process to be justified.

4.10S Two blocks A and B are initially at 100 and 500°C, respectively (Fig. P4.10). They are brought together and isolated from the surroundings. Determine the final equilibrium temperature of the blocks. Block A is aluminum [$c_P = 0.900$ kJ/(kg·K)] with $m_A = 0.5$ kg, and block B is copper [$c_P = 0.386$ kJ/(kg·K)] with $m_B = 1.0$ kg.

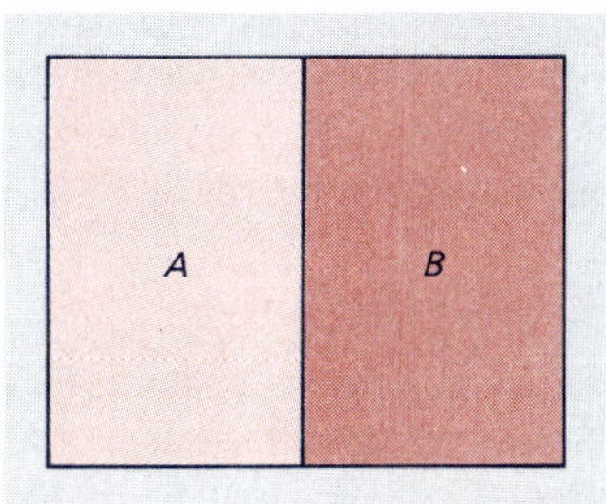

Figure P4.10

4.10E Two blocks A and B are initially at 212 and 930°F, respectively (Fig. P4.10). They are brought together and isolated from the surroundings. Determine the final equilibrium temperature of the blocks. Block A is aluminum [$c_P = 0.210$ Btu/(lbm·°F)] with $m_A = 1$ lbm, and block B is copper [$c_P = 0.0939$ Btu/(lbm·°F)] with $m_B = 2.0$ lbm.

4.11S Refrigerant 12 is initially contained within the cylinder shown at 20°C and 0.7 MPa (Fig. P4.11). There is a heat transfer to the refrigerant 12 until the volume is 21 times the initial value. The piston is then locked in place and the fluid is cooled to 20°C. Determine the heat transfer per unit mass for these processes.

4.11E Refrigerant 12 is initially contained within the cylinder shown at 67°F and 100 psia (Fig. P4.11). There is a heat transfer to the refrigerant 12 until the volume is 21 times the initial value. The piston is then locked in place, and the fluid is cooled to 67°F. Determine the heat transfer per unit mass for these processes.

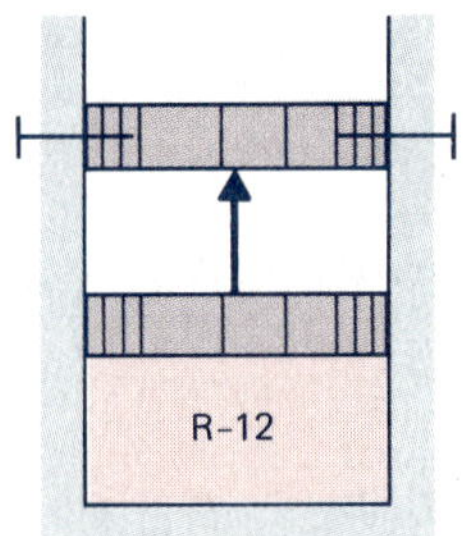

Figure P4.11

4.12S The cylinder shown in Fig. P4.12 contains 0.1 kg of water at 0.150 MPa. The volume initially enclosed is 0.005 m³. The spring touches the piston but exerts no force at this initial state. The piston rises owing to a heat transfer to the water until the water is in the saturated vapor state. The spring constant is 150 kN/m, and the piston area is 0.02 m². Determine the final temperature and pressure in the water, and evaluate the heat transfer.

4.12E The cylinder shown in Fig. P4.12 contains 0.2 lbm of water at 20 psia. The volume initially enclosed is 0.15 ft³. The spring touches the piston but exerts no force at this initial state. The piston rises owing to a heat transfer to the water until the water is in the saturated vapor state. The spring constant is 12×10^3 lbf/ft, and the piston area is 0.2 ft². Determine the final temperature and pressure in the water, and evaluate the heat transfer.

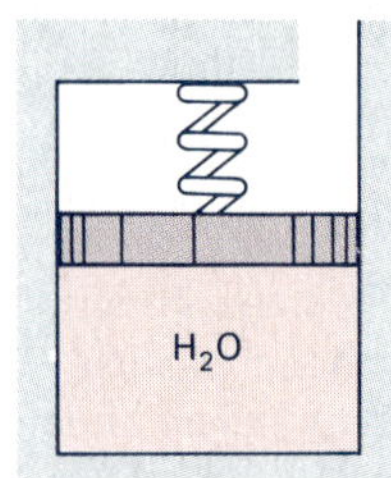

Figure P4.12

4.13S Water (1 kg) at 0.2 MPa is initially enclosed within a volume of 0.10 m³, and the piston rests on the stops (Fig. P4.13). The piston will move when the pressure is 1.0 MPa. A total heat transfer of 2500 kJ is added to the water. Determine the work done and the final state of the water.

Figure P4.13

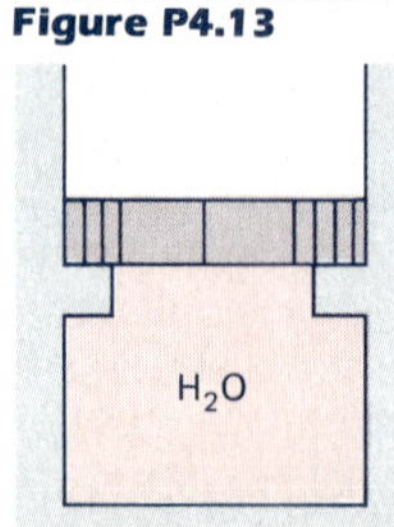

4.13E Water (2 lbm) at 30 psia is initially enclosed within a volume of 3 ft³, and the piston rests on the stops (Fig. P4.13). The piston will move when the pressure is 150 psia. A total heat transfer of 2400 Btu is added to the water. Determine the work done and the final state of the water.

4.14 Heat transfer (20 kJ) occurs to 1 kg of nitrogen gas contained in a piston-cylinder arrangement shown in Fig. P.4.14. The nitrogen is initially at $T = 100°C$ and $P = 1$ atm. Find the change in temperature of the gas if the heat transfer process occurs with (*a*) the piston held in its original position during the heat transfer process and (*b*) the weightless piston allowed to move freely during the heat transfer process. Use data from the ideal gas tables to find your results.

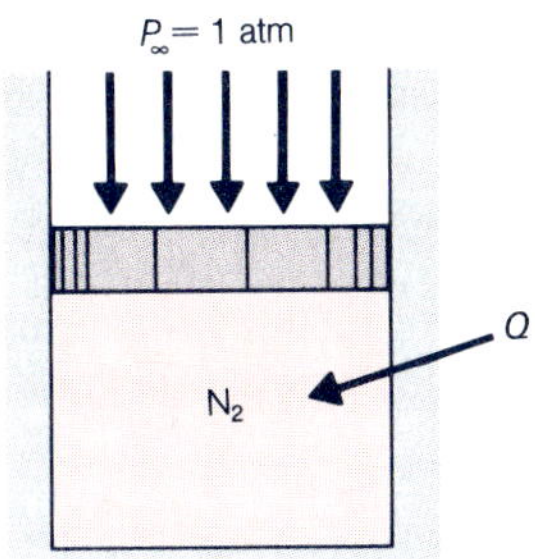

Figure P4.14

4.15 A storage tank contains air at the same temperature as the surroundings (35°C). Suddenly it begins to rain, and the surrounding temperature drops to 20°C. The tank was initially at a pressure of 200 kPa, and the tank volume is 10 m^3.

(*a*) What is the final pressure in kilopascals in the tank after the air cools in the tank to the temperature of the surroundings?

(*b*) How much heat transfer in kilojoules occurred to or from the air during the cooling?

4.16 Ten kilograms of refrigerant 12 is contained in a cylinder by a piston. Initially the cylinder volume is 0.01 m^3, and the refrigerant 12 temperature is 30°C. Heat transfer occurs to the cylinder until the refrigerant 12 volume is doubled.

(*a*) Sketch the *P-v* diagram for the process.

(*b*) How much heat transfer in kilojoules occurred during the process?

4.17 Saturated refrigerant 12 vapor (1 kg) is contained in a rigid canister of volume of 0.01 m^3. Heat transfer occurs until the quality of the refrigerant 12 is 0.5.

(*a*) Sketch the process on a *P-v* diagram.

(*b*) Find the initial temperature and pressure in the canister.

(*c*) Find the final specific enthalpy h of the refrigerant 12.

4.18 A block of gold has a mass of 2 kg and an initial temperature of 600 K. It is brought into contact with a block of nickel inside an insulated container. The nickel block had an initial temperature of 300 K. The blocks reach thermal equilibrium with each other at 400 K. What is the mass of the nickel block?

Figure P4.19

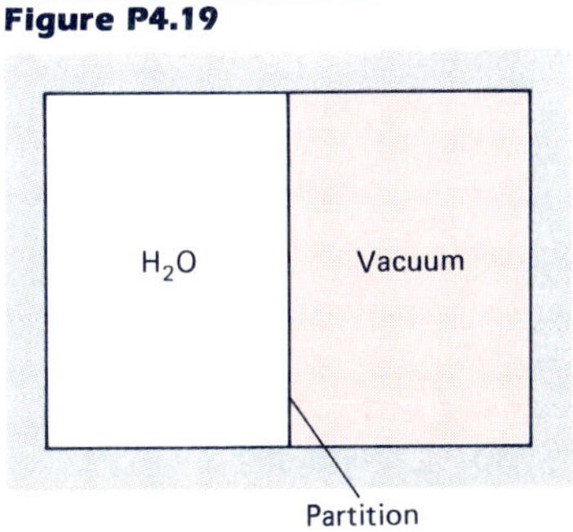

4.19S Steam occupies one-half of the partitioned insulated chamber (Fig. P4.19). The initial pressure and temperature of the steam are 0.3 MPa and 350°C. The other half of the chamber contains a vacuum. The partition is removed, and the steam fills the entire volume at the end of the process. Determine the final temperature of the steam.

4.19E Steam occupies one-half of the partitioned insulated chamber shown (Fig. P4.19). The initial pressure and temperature of the steam are 50 psia and 500°F. The other half of the chamber contains a vacuum. The partition is removed, and the steam fills the entire volume at the end of the process. Determine the final temperature of the steam.

4.20S A sealed container [$c = 1.0$ kJ/(kg·K), $\rho = 2500$ kg/m^3] containing water at 0.10 MPa and $x = 0.10$ is unknowingly put into a high-temperature furnace. The volume occupied by the water is 0.03 m^3, and the volume of the container material is 0.005 m^3. Calculate the heat transfer (to the container and water together) necessary for the water to reach 3.0 MPa.

4.20E A sealed container [$c = 0.24$ Btu/(lbm·°R), $\rho = 160$ lbm/ft^3] containing water at 14.7 psia and $x = 0.10$ is unknowingly put into a high-temperature furnace. The volume occupied by the water is 1 ft^3, and the volume of the container material is 0.15 ft^3. Calculate the heat transfer (to the container and water together) necessary for the water to reach 450 psia.

4.21S Water (2 kg) is contained within a piston-cylinder arrangement. See Fig. P4.21. The initial temperature is 105°C, and there are equal masses of liquid and vapor initially. The system is heated to a position where the piston is locked, and then the system is cooled to the saturated vapor state at $T = 50$°C. Draw the process on a P-v diagram, and evaluate the work done during the process.

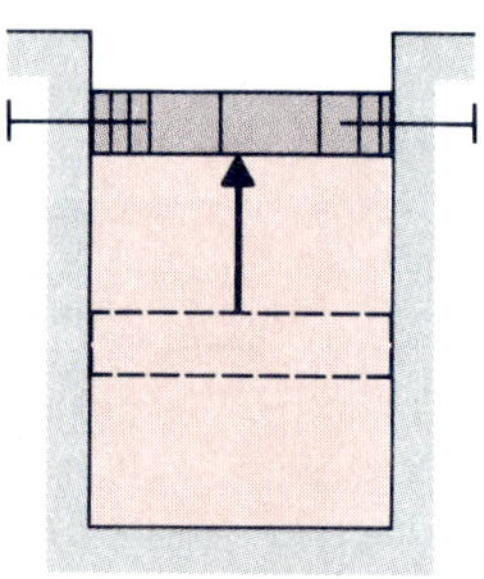

Figure P4.21

4.21E Water (5 lbm) is contained within a piston-cylinder arrangement. See Fig. P4.21. The initial temperature is 220°F, and there are equal masses of liquid and vapor initially. The system is heated to a position where the piston is locked, and then the system is cooled to the saturated vapor state at $T = 120$°F. Draw the process on a P-v diagram, and evaluate the work done during the process.

Figure P4.22S

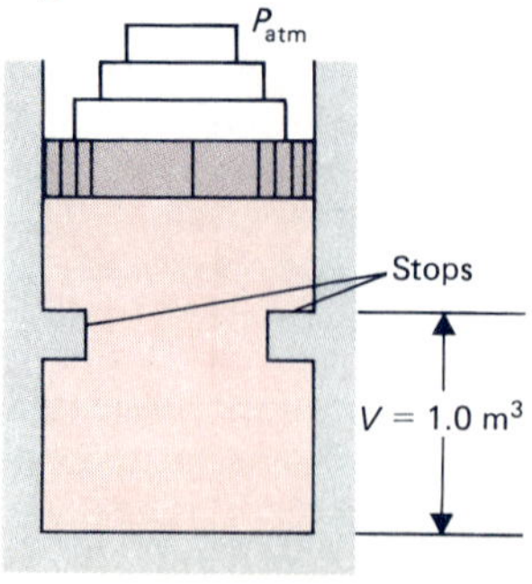

4.22S A piston and weights of total mass equal to 299 kg and area of 0.0098 m^2 contain 5.0 kg of H_2O at 500°C. See Fig. P4.22S. The water is cooled until the volume is 1.0 m^3, where the piston rests on stops. Then the H_2O is cooled while the piston remains on the stops to a state that could support a piston and weights of mass equal to 199 kg. At this point weights are removed to yield

199 kg, and the stops are removed. The system is then cooled to the saturated liquid state. Draw the process on a *P-v* diagram, and evaluate the work done.

4.22E A piston and weights of total mass equal to 1000 lbm and area of 0.300 ft² contain 10 lbm of H_2O at 900°F. See Fig. P4.22E. The water is cooled until the volume is 30 ft³, where the piston rests on stops. Then the H_2O is cooled while the piston remains on the stops to a state that could support a piston and weights of mass equal to 500 lbm. At this point, weights are removed to yield 500 lbm, and then the stops are removed. The system is then cooled to the saturated liquid state. Draw the process on a *P-v* diagram, and evaluate the work done.

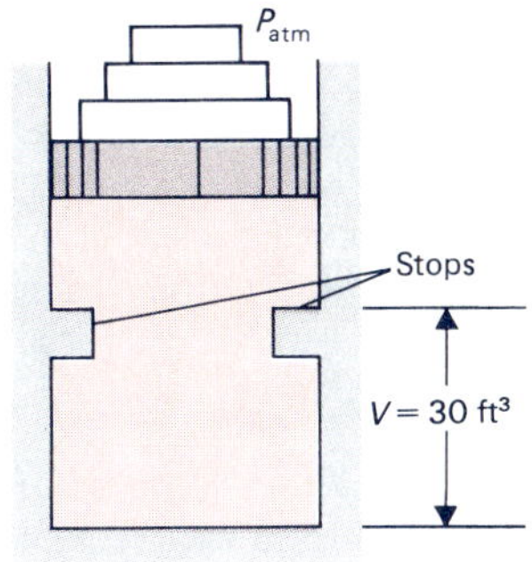

Figure P4.22E

4.23S Water (1.0 kg) is enclosed in a cylinder with a movable piston (cross-sectional area of 1.0 m²). See Fig. P4.23. Initially, the piston rests on stops, and the water is at 0.1 MPa and $x = 0.6374$. The water is heated, and the piston rises after a heat transfer of 905.4 kJ to the water. There is an additional heat transfer until $T =$ 750°C. Find the total heat transfer and work done. Also plot the process on the *P-v* diagram.

4.23E Water (2 lbm) is enclosed in a cylinder with a movable piston (cross-sectional area of 10 ft²). See Fig. P4.23. Initially the piston rests on stops, and the water is at 14.7 psia and $x = 0.8483$. The water is heated, and the piston rises after a heat transfer of 479 Btu to the water. There is an additional heat transfer until $T = 1400$°F. Find the total heat transfer and work done. Also plot the process on the *P-v* diagram.

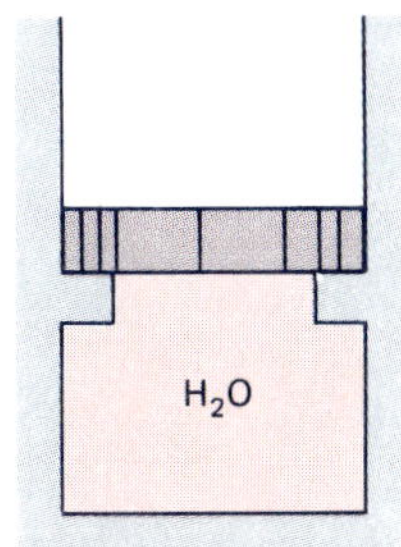

Figure P4.23

4.24S Consider the system diagrammed in Fig. P4.24 which is composed of air (subscript *a*) and H_2 (subscript *H*) separated by a movable adiabatic piston. The initial data are $V_a = 1$ m³, $T_a = 25$°C, $P_a = 0.15$ MPa, $P_H =$ 0.15 MPa, and $m_H = 0.10$ kg. There is a total heat transfer of 80 kJ to the air resulting in a final air temperature of $T_a = 45$°C. Find the temperature *increase* in the H_2 and the work done by the H_2. Assume both gases to be ideal.

4.24E Consider the system diagrammed in Fig. P4.24 which is composed of air (subscript *a*) and H_2 (subscript *H*) separated by a movable adiabatic piston. The initial data are $V_a = 30$ ft³, $T_a = 77$°F, $P_a = 25$ psia, $P_H = 25$ psia, and $m_H = 0.20$ lbm. There is a total heat transfer of 75 Btu to the air resulting in a final air temperature of $T_a = 110$°F. Find the temperature *increase* in the H_2 and the work done by the H_2. Assume both gases to be ideal.

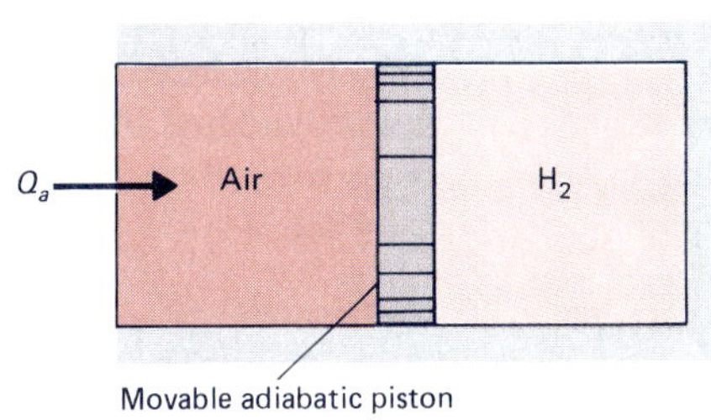

Figure P4.24

4.25S Consider a system of carbon dioxide enclosed in an expandable vessel. The carbon dioxide undergoes a complete cycle in the following manner: For the expansion, the final value minus the initial value of the internal energy is found to be −150 kJ. For the compression process, there is 35 kJ of heat transfer from the system. Find the amount of work done by the carbon dioxide in the compression process.

4.25E Consider a system of carbon dioxide enclosed in an expandable vessel. The carbon dioxide undergoes a complete cycle in the following manner: For the expansion, the final value minus the initial value of the internal energy is found to be −140 Btu. For the compression process, there is 33 Btu of heat transfer from the system. Find the amount of work done by the carbon dioxide in the compression process.

4.26S A piston performs work on air within the system shown in Fig. P4.26. There is a heat transfer between the chambers, but there is no heat transfer with the surroundings. Initially, the air is at 500 kPa and 200°C, and the N_2 is initially at 1500 kPa. The initial volumes are $V_{N_2} = 0.01$ m^3 and $V_{air} = 0.01$ m^3. Determine the work performed, final temperature, and heat transfer between chambers if the process ends when the N_2 pressure is 1580 kPa.

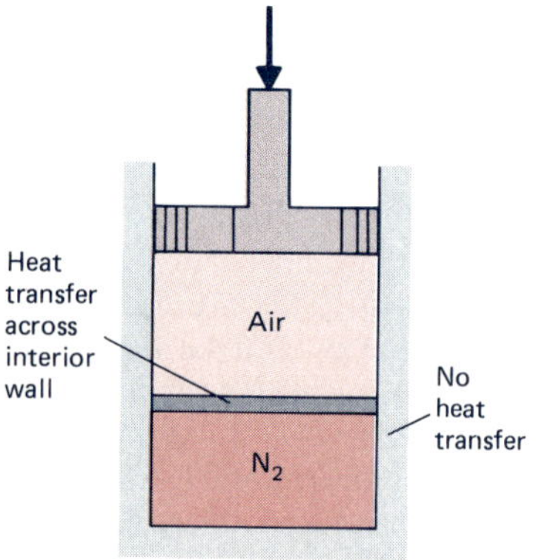

Figure P4.26

4.26E A piston performs work on air within the system shown in Fig. P4.26. There is a heat transfer between the chambers, but there is no heat transfer with the surroundings. Initially, the air is at 75 psia and 400°F, and the N_2 is initially at 220 psia. The initial volumes are $V_{N_2} = 0.3$ ft^3 and $V_{air} = 0.3$ ft^3. Determine the work performed, final temperature, and heat transfer between chambers if the process ends when the N_2 pressure is 370 psia.

4.27S Saturated water vapor at 200°C is contained in a cylinder fitted with a piston. The initial volume of the steam is 0.05 m^3. The H_2O is then compressed in an isothermal process until the quality is 20 percent. Determine the total work done in kilojoules on the system during this process.

4.27E Saturated water vapor at 400°F is contained in a cylinder fitted with a piston. The initial volume of the steam is 1.5 ft^3. The H_2O is then compressed in an isothermal process until the quality is 20 percent. Determine the total work done in Btu on the system during this process.

4.28S Water (3.10 kg) in the piston-cylinder device shown has an initial pressure of 0.2 MPa and temperature of 70°C. Energy is added until the piston just reaches the upper stops where the total volume is 4.015 m^3. It takes a pressure of 0.4 MPa to move the piston. Calculate the work done in kilojoules per kilogram.

4.28E Water (6 lbm), contained in the piston-cylinder device shown, is initially at a pressure of 15 psia and a quality of 0.50. Energy is added until the piston just reaches the upper stops where the total volume is 120 ft^3. A pressure of 30 psia is required to lift the frictionless piston from the stops. Determine the total work done.

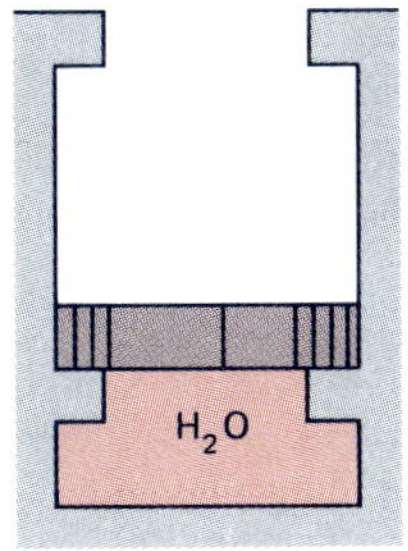

Figure P4.28

4.29S Water is cooled in a closed rigid vessel from a temperature of 600°C and pressure of 4.0 MPa to a temperature of 100°C.
(*a*) How much work is done?
(*b*) What is the change in pressure?

4.29E Water is cooled in a closed rigid vessel from a temperature of 1000°F and pressure of 600 psia to a temperature of 212°F.
(*a*) How much work is done?
(*b*) What is the change in pressure?

4.30S Water (15 kg) in the piston-cylinder device shown has an initial total pressure of 1.0 MPa and specific enthalpy of 3698.1 kJ/kg. The H_2O is cooled until the temperature is 80°C. It takes a pressure of 0.6 MPa to support the piston. How much work is done?

4.30E Water (35 lbm) in the piston-cylinder device shown has an initial total pressure of 150 psia and specific enthalpy of 1636.8 Btu/lbm. The H_2O is cooled until the temperature is 150°F. It takes a pressure of 100 psia to support the piston. How much work is done?

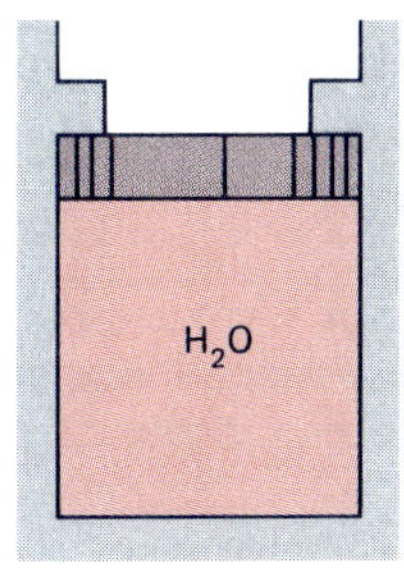

Figure P4.30

4.31S Water is contained in a piston-cylinder device, and the piston is free to move. Initially the H_2O is at a pressure of 1.0 MPa and a temperature of 600°C. The H_2O is cooled until the specific volume is 0.205 m^3. Determine the work done on the system in kilojoules per kilogram and the mass of H_2O if the total work done on the system is 30 kJ.

4.31E Water is contained in a piston-cylinder device, and the piston is free to move. Initially the H_2O is at a pressure of 150 psia and a temperature of 900°F. The H_2O is cooled until the specific volume is 3.20 ft^3/lbm. Determine the work done on the system (pound force-feet per pound mass) and the mass of H_2O if the total work done on the system is 21,000 lbf/ft.

4.32 A radiator with a volume of 1 m^3 contains H_2O as a saturated vapor at 5.0 MPa. The valves are closed, and as a result of heat transfer to the atmosphere, the pressure drops to 0.5 MPa. Determine the total heat transfer, and show the process on a T-v diagram.

4.33 A rigid aluminum container with a volume of 0.755 m^3 contains 30 kg of water at an initial pressure and temperature of 25.033 kPa and 65°C. If the manufacturer's specifications state that the container will rupture at a pressure of 20.0 MPa, what is the total heat transfer necessary to rupture the container? Show the process on a T-v diagram.

4.34S A pressure cooker has a volume of 0.1 m^3. When the valve on the cooker begins to release steam, it maintains the pressure in the cooker at 150 kPa. Assume 10 kg of water is initially in the cooker.

(*a*) What is the quality of the saturated mixture in the cooker when steam begins to escape?

(*b*) What is the quality of the saturated mixture in the cooker when 7 kg of steam has escaped through the valve?

(*c*) If the cooker is placed on a stove with an initial charge of 10 kg of water (liquid plus vapor) at 20°C, evaluate the heat transfer to the water between the time the stove is turned on and the pressure valve begins to release steam.

4.34E A pressure cooker has a volume of 4 ft^3. When the valve on the cooker begins to release steam, it maintains the pressure in the cooker at 20 psia. Assume 20 lbm of water is initially in the cooker.

(*a*) What is the quality of the saturated mixture in the cooker when steam begins to escape?

(*b*) What is the quality of the saturated mixture in the cooker when 14 lbm of steam has escaped through the valve?

(*c*) If the cooker is placed on the stove with an initial charge of 20 lbm of water (liquid plus vapor) at 70°F, evaluate the heat transfer to the water between the time the stove is turned on and the pressure valve begins to release steam.

4.35 A vertical frictionless piston-cylinder device containing water has an initial temperature and pressure of 10°C and 40 N/cm^2. There is a heat transfer to the cylinder until the final temperature is 550°C. What is the final specific volume, and how much heat transfer is added? Show the process on a T-v diagram.

Figure P4.35

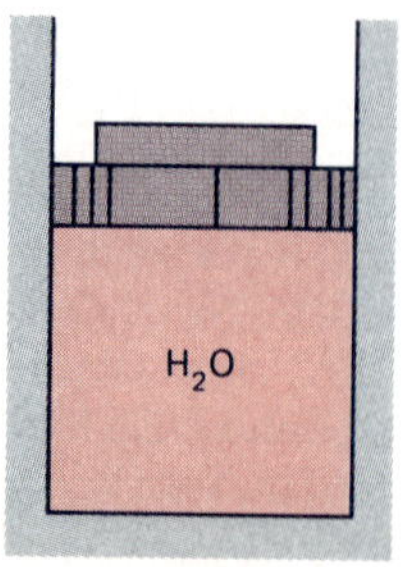

4.36 Water (2 kg) is contained in a vertical frictionless piston-cylinder device. The mass of the piston is such that the H_2O exists as a saturated liquid at 10.0 MPa. There is a heat transfer to the cylinder until the piston reaches some stops, at which point the total volume is 0.02 m³. There is an additional heat transfer to the water until the H_2O exists as a saturated vapor. Determine the total heat transfer and the total work, and show the process on a *T-v* diagram.

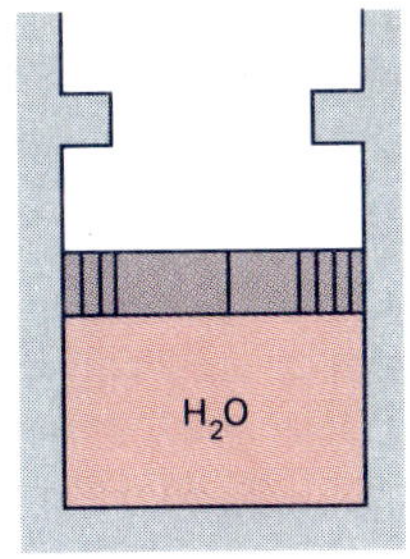

Figure P4.36

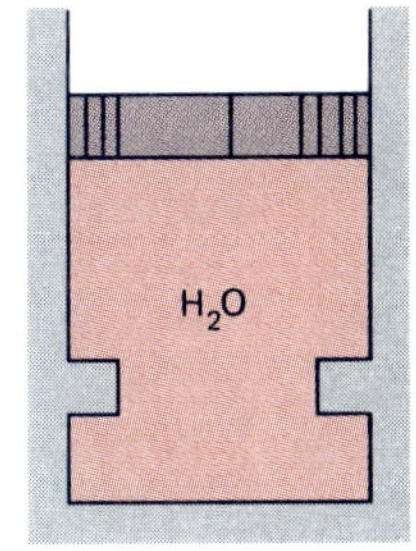

Figure P4.37

4.37 Water (1 kg) is contained in a vertical frictionless piston-cylinder device. The mass of the piston is such that the H_2O exists at a pressure of 20.0 MPa and a temperature of 850°C. There is a heat transfer from the device until the piston just rests on stops, at which time the volume inside the cylinder is 1.560×10^{-3} m³. Determine the total work and heat transferred for this process, and show the process on a *T-v* diagram.

4.38 Water (14 kg) is contained in a piston-cylinder device. The H_2O is heated from an initial temperature of 150°C to a final temperature and pressure of 450°C and 0.6 MPa.

(*a*) Plot and label the process (*T-v* plot).
(*b*) What is the total work done in kilojoules?
(*c*) What is the change in internal energy?

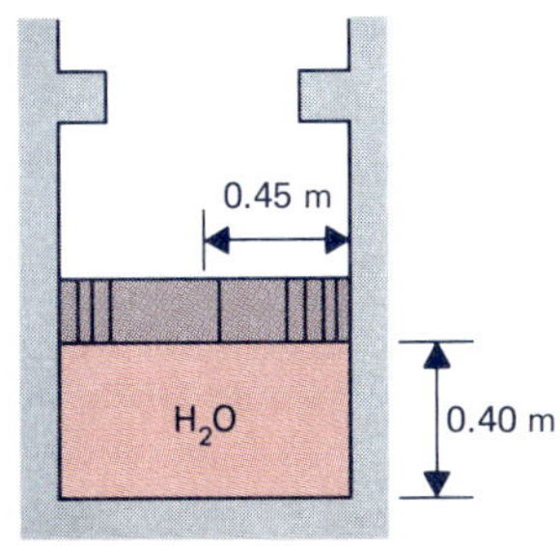

Figure P4.38

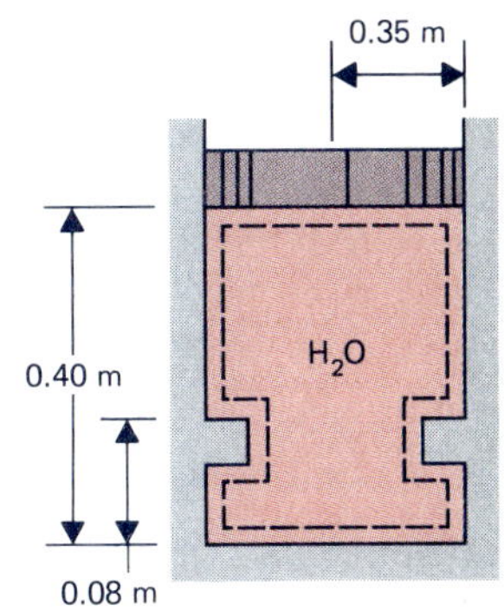

Figure P4.39

4.39 Water is contained in the frictionless piston-cylinder device shown. The H_2O is cooled from an initial temperature of 300°C and pressure of 0.8 MPa to a final pressure of 0.06 MPa. Plot and label the processes on a *T-v* diagram. Determine (*a*) the mass of the H_2O, and (*b*) the total work done in joules.

4.40 A sealed rigid container has a volume of 7.854×10^{-4} m³. It contains H_2O as a saturated vapor at 350°C. It is then cooled to 200°C. Determine (*a*) the mass of the H_2O, (*b*) the work done during the process (in joules per gram), and (*c*) the change in internal energy (in joules per gram).

4.41 A piston-cylinder device with one set of stops has dimensions as shown. The piston is free to move. The H_2O inside is initially a saturated vapor at 270°C. It is cooled until half the mass is a liquid. The piston requires a pressure of 5.4987 MPa to support it. Draw and label the processes on a *T-v* diagram. Determine (*a*) the initial and final mass of the H_2O and (*b*) the total work done, in kilojoules.

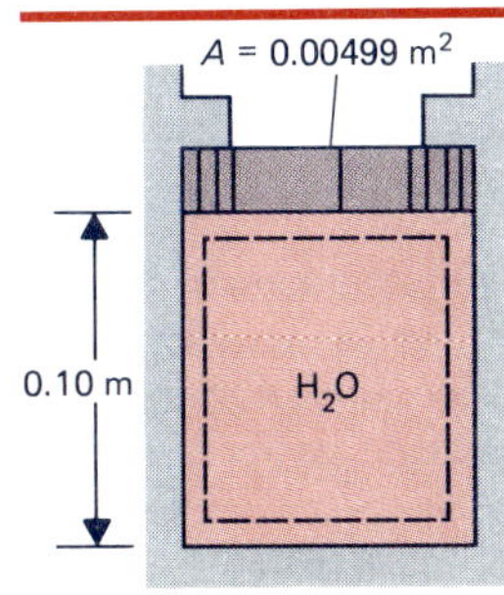

Figure P4.41

4.42 A rigid spherical container has a volume of 10^{-3} m³. It is filled with 46.14 g of H_2O initially at 260°C. This medium is then heated to a saturated vapor. Plot the process(es) on a T-v diagram and then determine (a) the initial and final quality (b) the work done (in kilojoules per kilogram), and (c) the final temperature and pressure.

4.43 Water in the piston-cylinder device shown is initially 40 percent liquid and 60 percent vapor by mass and is at 0.2 MPa. A pressure of 0.4 MPa is required to raise the piston. The medium is heated to a final temperature of 250°C. Plot and label the process(es) on a T-v diagram. Determine (a) the mass of the H_2O in the device and (b) the total work done in kilojoules.

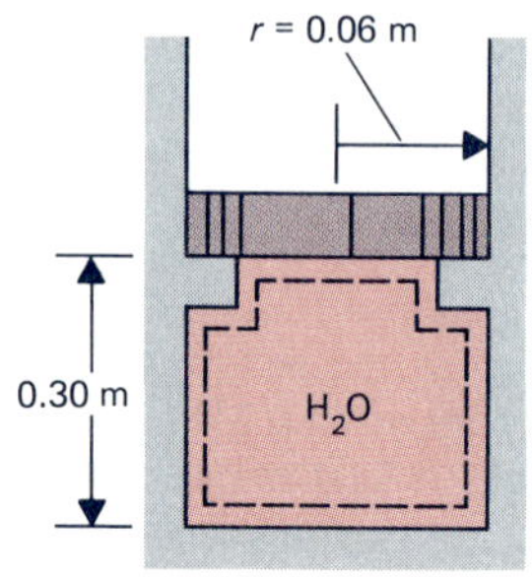

Figure P4.43

4.44 A closed rigid container has 6 kg of H_2O at an initial pressure of 101.3 kPa and volume of 26.58 m^3. The H_2O undergoes a process to a final pressure of 20 kPa. In what form (heat transfer or work) did the energy transfer take place? How much total energy was transferred, in kilojoules? Was it into or out of the system?

4.45 A frictionless piston-cylinder device containing H_2O is shown. The initial conditions are $P_1 = 80$ kPa and $T_1 = 90$°C. A pressure of 101.3 kPa is required to move the piston. If the H_2O is heated to a final temperature of 300°C, calculate the heat transfer and work (in kilojoules per kilogram). Also draw and label the processes on a T-v plot.

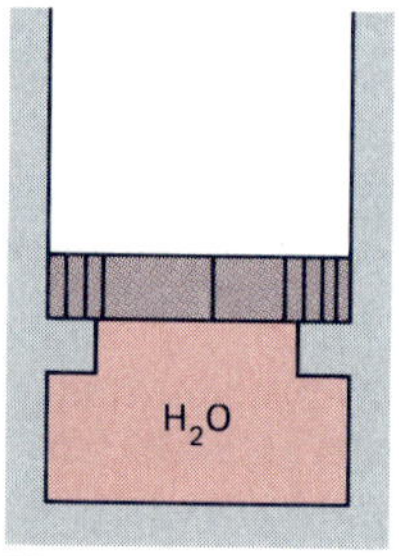

Figure P4.45

4.46 A frictionless piston-cylinder device containing H_2O is shown. The initial conditions are $P = 200$ kPa, $T = 100$°C, and $V = 0.01$ m^3. There is a heat transfer until $T = 900$°C. When the piston strikes the upper stops, $V = 8.603$ m^3. Draw and label the process on a T-v plot. Calculate the total heat transfer and work transferred (in kilojoules).

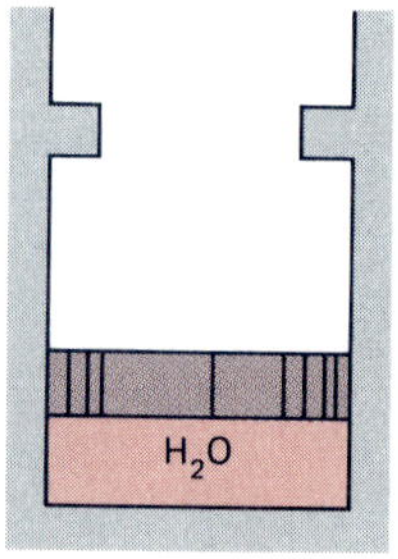

Figure P4.46

4.47 A frictionless piston-cylinder device containing H_2O is shown. The initial conditions are $P = 400$ kPa and $T = 800$°C. A pressure of 0.1 MPa is required to support the piston. There is a heat transfer from the device until $T = 80$°C. Draw and label the process on a T-v plot, and calculate the heat transfer and work (kJ/kg) transferred.

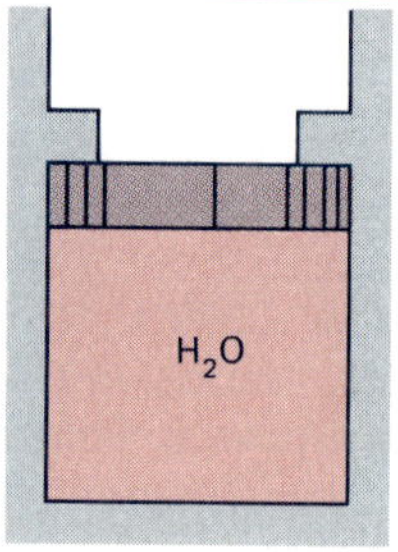

Figure P4.47

4.48S A piston-cylinder device undergoes a quasi-static isothermal process from an initial pressure of 50 kPa and specific volume of 0.00103 m^3/kg to a final specific volume of 2.7 m^3/kg. Draw and label the process on a P-v plot, and calculate the heat transfer and work (kJ/kg). The medium is H_2O.

4.48E A piston-cylinder device undergoes a quasi-static isothermal process from an initial pressure of 9 psia and specific volume of 0.01656 ft³/lbm to a final specific volume of 40 ft³/lbm. Draw and label the process on a *P-v* plot, and calculate the heat transfer and work (Btu/lbm). The medium is H_2O.

4.49S A control mass undergoes a cycle; 10 kJ of work is done by the control mass, and there is a heat transfer of 50 kJ to the control mass for the first portion of the cycle. The cycle is completed with no work being done during the last portion. What is the heat transfer on this last portion of the cycle?

4.49E A control mass undergoes a cycle; 7400 ft · lbf of work is done by the control mass, and there is a heat transfer of 200 Btu to the control mass for the first portion of the cycle. The cycle is completed with no work being done during the last portion. What is the heat transfer on this last portion of the cycle?

4.50S A control mass undergoes the process from state 1 to state 2 shown in Fig. P4.50S. The heat transfer is ${}_1Q_2 = 200$ kJ. The control mass returns adiabatically from state 2 to state 1 by another process. What is the work on the return process, that is, ${}_2W_1$?

4.50E A control mass undergoes the process from state 1 to state 2 shown in Fig. P4.50E. The heat transfer is ${}_1Q_2 = 200$ Btu. The control mass returns adiabatically from state 2 to state 1 by another process. What is the work on the return process, that is, ${}_2W_1$?

Figure P4.50S

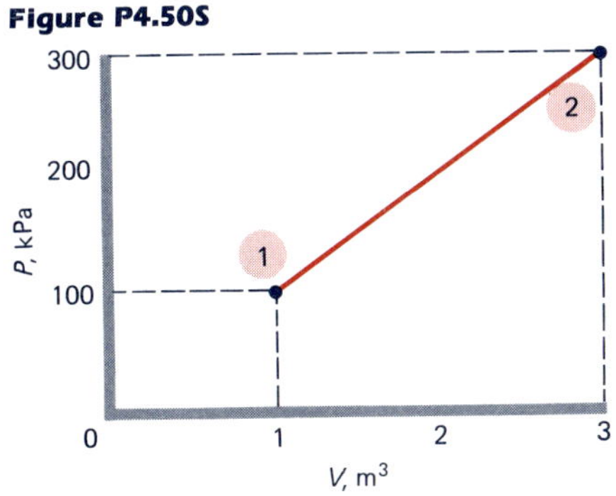

Figure P4.50E

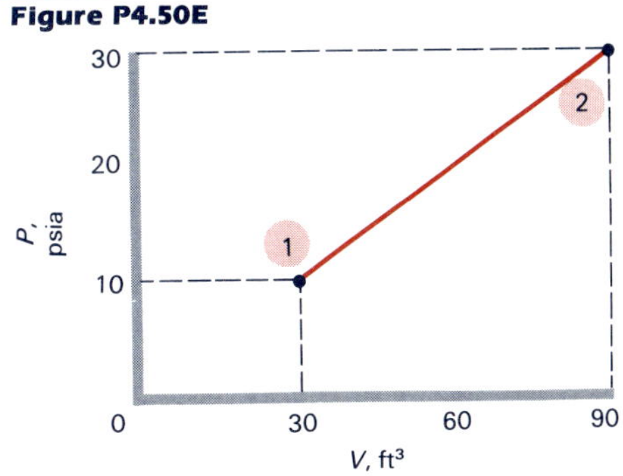

4.51 With a heat transfer of 750 kJ/kg from the H_2O-filled frictionless piston-cylinder device shown, determine the shaft work (kJ/kg). The initial and final states of the H_2O are $T_1 = 300°C$, $P_1 = 0.6$ MPa, and $T_2 = 450°C$, respectively.

4.52 When 500 kJ/kg of shaft work is transferred to the H_2O-filled frictionless piston-cylinder device shown, determine the heat transfer. The initial and final states of the H_2O are $T_1 = 200°C$, $P_1 = 0.4$ MPa, and $T_2 = 250°C$, respectively.

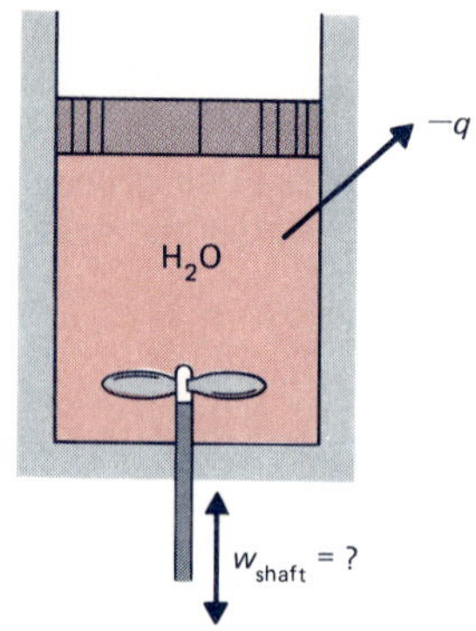

Figure P4.51

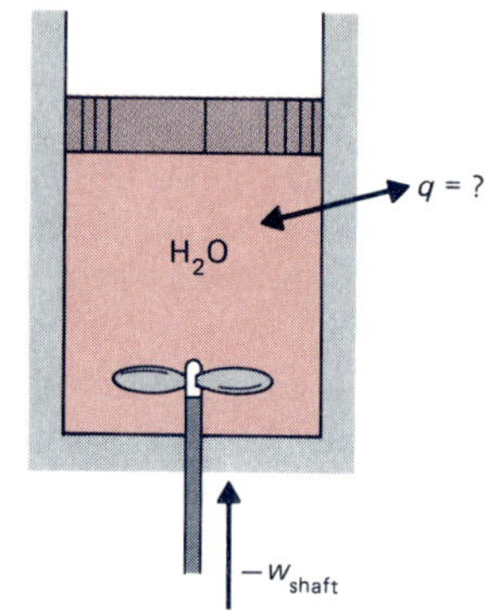

Figure P4.52

4.53 The free-moving piston-cylinder device shown contains 2 kg of H_2O at an initial temperature of 350°C. When there is a heat transfer of 500 kJ from the medium, the temperature rises to 600°C. Determine the amount of shaft work (kJ) necessary for the process to occur.

4.54 The free-moving piston-cylinder device shown contains 5 kg of H_2O at a temperature of 400°C. When 3000 kJ of shaft work is transferred to the medium, the temperature rises to 650°C. Determine the heat transferred (kJ) during the process.

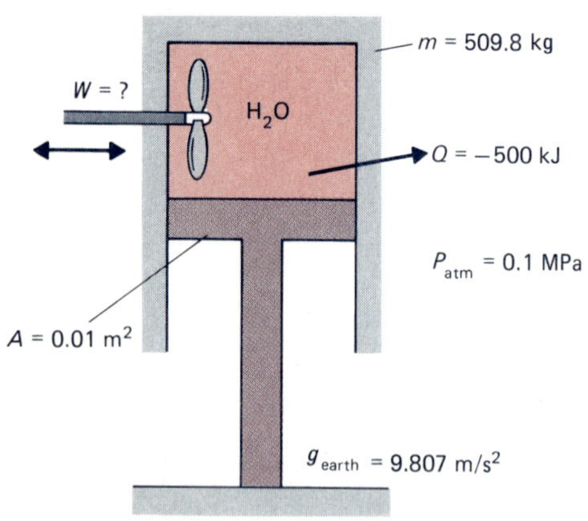

Figure P4.53

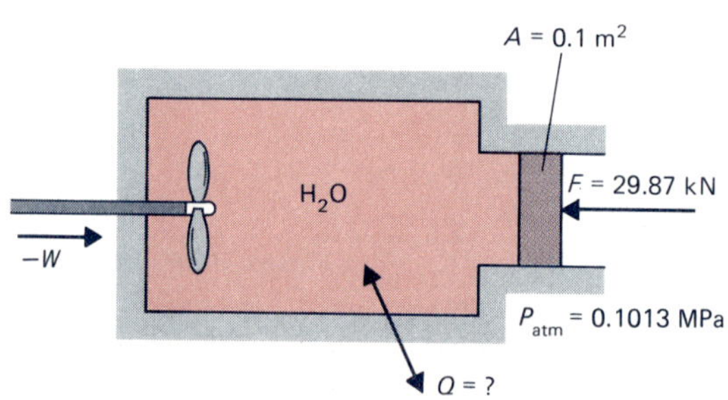

Figure P4.54

4.55S A "fan in a box" system (see Fig. P4.55), using H_2O as the medium, is operated for a period during which 209 kJ/kg of shaft work is added to the system. Initially the H_2O is at $P_1 = 0.1013$ MPa and $T_1 = 20°C$. The final pressure is 150 kPa. Determine the final temperature.

4.55E A fan-in-a-box system (see Fig. P4.55), using H_2O as the medium, is operated for a period during which 120 Btu/lbm of shaft work is added to the system. Initially the H_2O is at $P_1 = 14.7$ psia and $T_1 = 70°F$. The final pressure is 25 psia. Determine the final temperature.

4.56S A fan-in-a-box system (see Fig. P4.56) with water as the fluid is operated from an initial condition of $P_1 = 0.1013$ MPa and $T_1 = 30°C$ to a final temperature of 50°C. If the shaft work added is 2000 kJ/kg, determine the heat transferred (kJ/kg).

4.56E A fan-in-a-box system (see Fig. P4.56) with water as the fluid is operated from an initial condition of $P_1 = 14.7$ psia and $T_1 = 90°F$ to a final temperature of 120°F. If the shaft work added is 1000 Btu/lbm, determine the heat transferred (Btu/lbm).

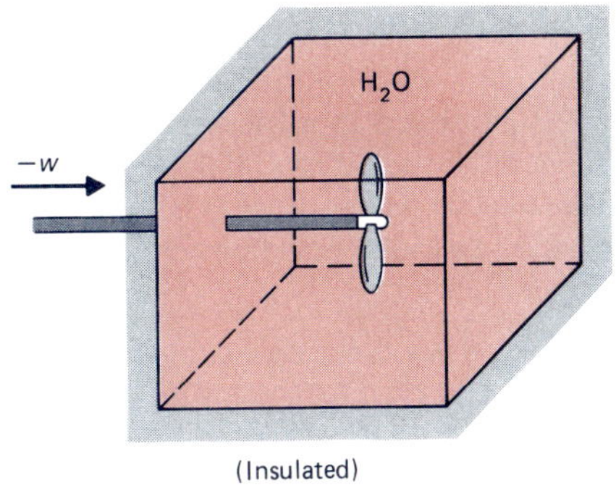

Figure P4.55

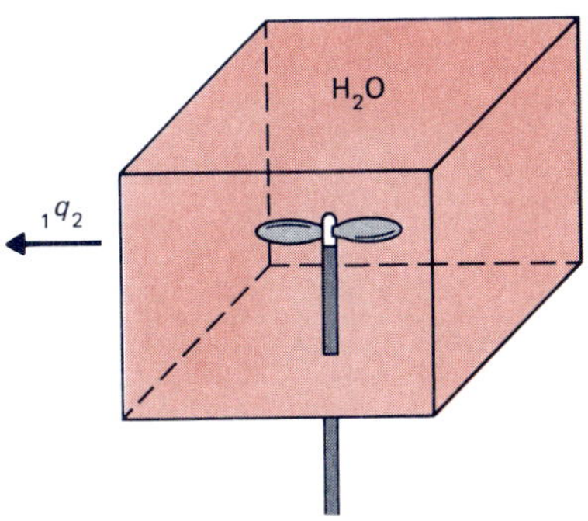

Figure P4.56

4.57S Water in a rigid vessel initially at the critical point is cooled until the pressure is 20 MPa. Determine the work and heat transfer (kJ/kg) for the process.

4.57E Water in a rigid vessel initially at the critical point is cooled until the pressure is 3000 psia. Determine the work and heat transfer (Btu/lbm) for the process.

4.58S Water is contained in a frictionless piston and cylinder initially at 1 MPa and 600°C. The H_2O is cooled until the specific volume is 0.206 m³/kg. What is the mass of H_2O if the total heat transferred is 1000 kJ?

4.58E Water is contained in a frictionless piston and cylinder initially at 150 psia and 1000°F. The H_2O is cooled until the specific volume is 3.5 ft³/lbm. What is the mass of H_2O if the total heat transferred is 1000 Btu?

4.59S Water (2.5 kg) is contained in a frictionless piston-cylinder device at 30°C and 250 kPa. If there is a heat transfer until the total volume of the H_2O is 1.2 m³, find the work (kJ/kg) and total heat transferred in the process.

4.59E Water (5 lbm) is contained in a frictionless piston-cylinder device at 80°F and 30 psia. If there is a heat transfer until the total volume of the H_2O is 30 ft^3, find the work (Btu/lbm) and total heat transferred in the process.

4.60S Water (15 kg) is heated in a closed rigid container from 60 to 400°C and 2 MPa. Find the total work and heat transfer.

4.60E Water (35 lbm) is heated in a closed rigid container from 150 to 750°F and 300 psia. Find the total work and heat transfer.

4.61S Water (3 kg) in the frictionless piston-cylinder device shown is initially at a pressure of 1 MPa and a quality of 75 percent. There is a heat transfer until the temperature is 500°C. It takes a pressure of 2 MPa to lift the piston. What is the total heat transferred in the process?

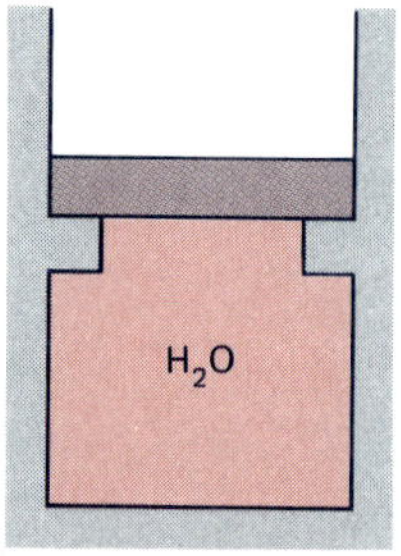

Figure P4.61

4.61E Water (8 lbm) in the frictionless piston-cylinder device shown is initially at a pressure of 100 psia and a quality of 75 percent. There is a heat transfer until the temperature is 950°F. It takes a pressure of 150 psia to lift the piston. What is the total heat transferred in the process?

4.62S Water existing as a saturated liquid at 200°C is expanded isothermally in a frictionless piston-cylinder device until it is saturated vapor. Determine the heat transferred in this process (kJ/kg).

4.62E Water existing as a saturated liquid at 350°F is expanded isothermally in a frictionless piston-cylinder device until it is saturated vapor. Determine the heat transferred in this process (Btu/lbm).

4.63 A cocktail party is attended by 25 people. The party is in a room that is perfectly insulated and 20 × 20 × 8 ft^3 and each person has a volume of 4.5 ft^3. The metabolic energy added to the room air by heat transfer is 300 Btu/h per person. The room air is initially at 70°F. What temperature will the room air reach after 15 min?

4.64 Oxygen (8 kmol) is cooled isometrically from 200°C and 1.25 MPa to 10°C. Determine the total heat transferred (kJ), if the process occurs in a system, and the total volume (m^3).

4.65 Air in a nonflow device is heated isobarically from 10°C and 200 kPa to 400°C. Determine the work (kJ/kg) and heat transferred (kJ/kg).

4.66 Hydrogen is contained in a 0.5 m^3 tank at an initial temperature of 20°C and pressure of 100 kPa. The rigid tank is heated to 200°C. Determine the *total* heat transferred (kJ) and the final pressure.

4.67 Air is contained in a vertical cylinder fitted with a frictionless piston and a set of stops as shown. The cross-sectional area of the piston is 0.05 m^2. The air, whose initial state is given below, is cooled as a result of heat transfer to the surroundings. Find the temperature of the air inside the cylinder when the piston reaches the stops and the pressure inside the cylinder if cooling continues until the temperature is 70°C.

$P_1 = 0.21$ MPa $\quad T_1 = 430°C$

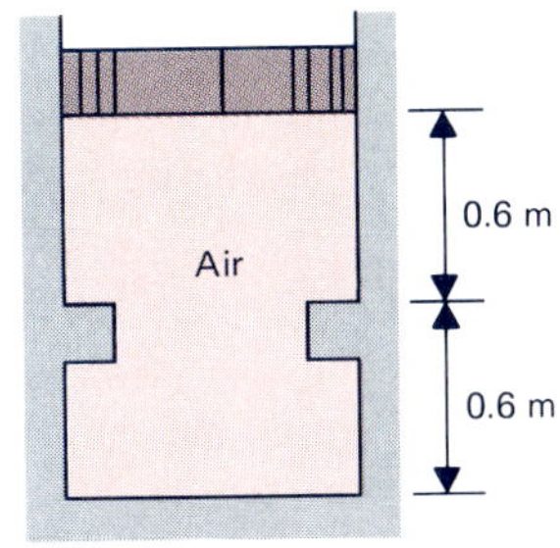

Figure P4.67

4.68 Nitrogen is heated in the piston-cylinder device shown in a quasi-equillibrium process from state 1 to state 2. If the total pressure at state 1 is 200 kPa, determine the mass of the piston. Also determine the final temperature and the mass of the air.

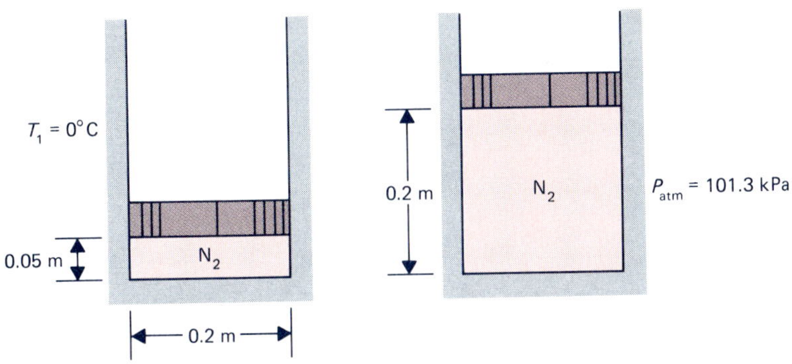

Figure P4.68

4.69 Oxygen (0.1 kg) is contained in a cylinder fitted with a piston. The initial conditions are 150 kPa and 20°C. Weights are then added to the piston, and the O_2 is slowly compressed isothermally until the final pressure is 500 kPa. Determine the change in internal energy (kJ/kg) and the final volume (ft^3).

4.70 Hydrogen (10 g) in the piston-cylinder device shown undergoes a constant-pressure process. The initial pressure is 1.0 MPa, and the initial volume is 2000 cm^3. If the final volume is 10 cm^3, find the work done. Is the work done on or by the system?

Figure P4.70

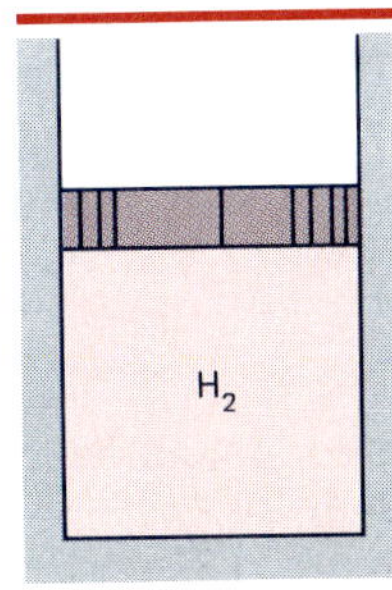

4.71 One kilogram of air in a closed, rigid container undergoes a process such that its temperature increases from 10 to 50°C. The volume of the container is 25,000 cm^3. Find the amount of work done during the process.

4.72 The piston-cylinder device shown contains 1 kg of helium at an initial pressure and volume of 0.8 MPa and 1.228 m^3. The helium undergoes an isobaric process until it reaches a final temperature of 10°C. Find the total work done in the process. Is it work done on or by the system? Show the process on a *P-v* diagram.

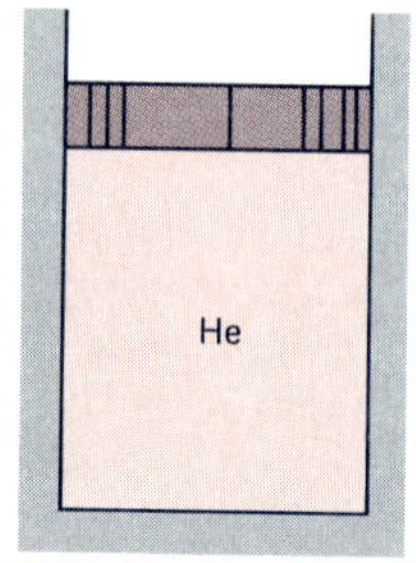

Figure P4.72

4.73 Oxygen in a system undergoes the process shown on the *P-v* diagram. Find the work done. Is it work done on or by the system?

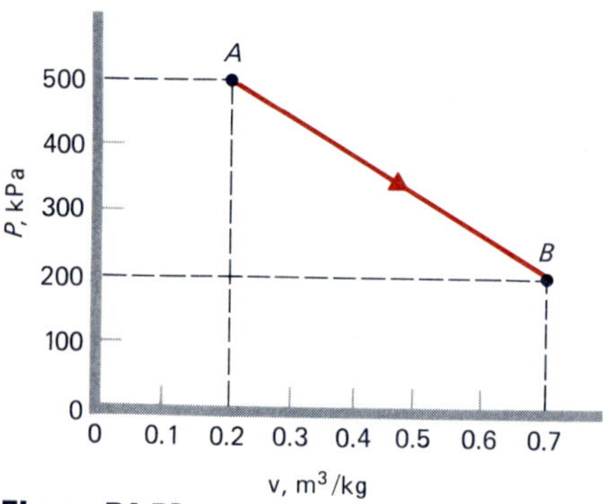

Figure P4.73

4.74 Carbon dioxide in a system undergoes the processes from point 1 to point 2 as shown.
(*a*) In which path is the most work done?
(*b*) In which path is the least work done?
(*c*) If the process were reversed, which path would require the least amount of work?

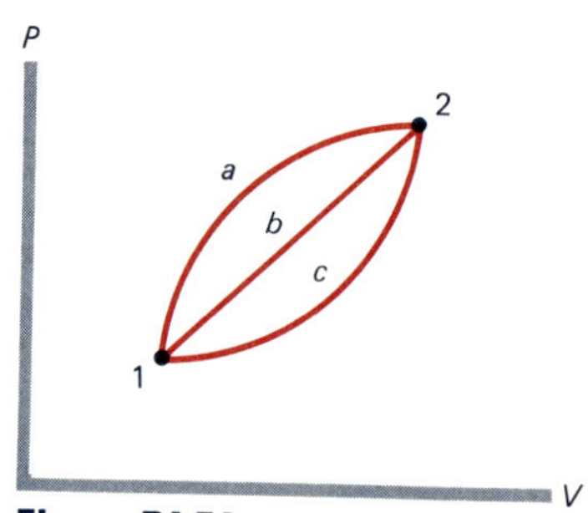

Figure P4.74

4.75 Argon (100 g) is in the piston-cylinder device shown. The initial pressure is 5.0 MPa and temperature is 200°C. There is a heat transfer to the argon, causing the piston to rise until it hits the stops. There is an additional heat transfer until the final pressure is 8.0 MPa and temperature is 800°C.
(*a*) Draw the process on a *P-v* diagram.
(*b*) Find the total work done in the process.
(*c*) Is it work done on or by the system?

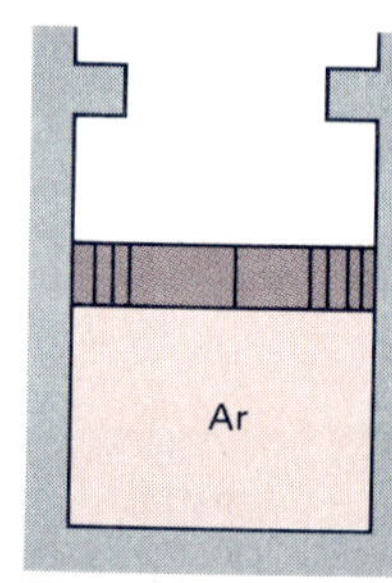

Figure P4.75

4.76 Air (0.1105 kg) at 200°C is contained in a cylinder fitted with a piston. The initial volume of the air is 0.05 m^3. The air is then compressed isobarically until the volume is 0.01 m³. Determine the work (kJ/kg) done on the system during this process.

4.77 Nitrogen (3.10 kg) in the piston-cylinder device shown has an initial pressure of 0.2 MPa and temperature of 70°C. Energy is added until the piston just reaches the upper stops where the total volume is 4.015 m³. It takes a pressure of 0.4 MPa to move the piston. Calculate the work (kJ/kg) done.

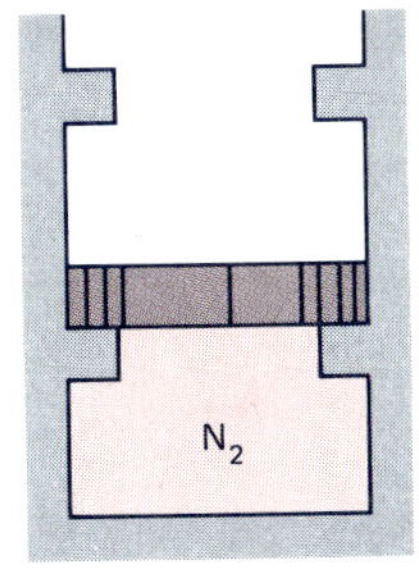

Figure P4.77

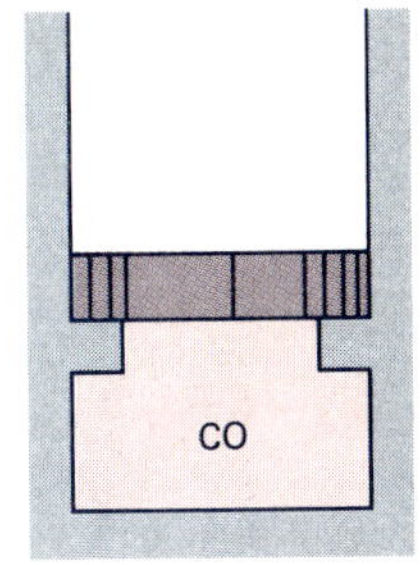

Figure P4.78

4.78 Carbon monoxide (2 kg), contained in the piston-cylinder device shown, is initially at a pressure of 1.0 MPa and a temperature of 50°C. Energy is added until the final temperature is 500°C and the pressure is 2.0 MPa. A pressure of 2.0 MPa is required to lift the frictionless piston from the stops. Determine the total work done.

4.79 Air is cooled in a closed rigid vessel from a temperature of 600°C and pressure of 4.0 MPa to a temperature of 100°C.
(*a*) How much work is done?
(*b*) What is the final pressure?

4.80 Oxygen (15 kg) in the piston-cylinder device shown has an initial temperature and pressure of 496.8°C and 1 MPa. The oxygen is cooled until the temperature is 73.4°C. It takes a pressure of 0.6 MPa to support the piston. How much work is done (kJ/kg)?

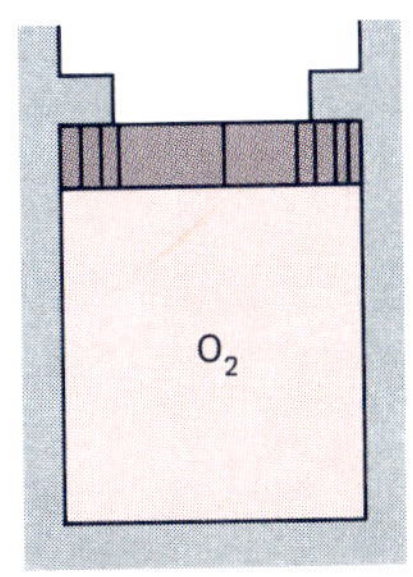

Figure P4.80

4.81 Oxygen is contained in a piston-cylinder device, and the piston is free to move. Initially the oxygen is at a pressure of 0.6 MPa and a temperature of 188.9°C. The oxygen is cooled unil the specific volume is 0.15 m³/kg. Determine the work done on the system (kJ/kg) and the mass of the oxygen, if the total work done on the system is 30 kJ.

4.82 A heat exchanger with a volume of 1.0 m³ contains hydrogen at an initial pressure of 5.0 MPa and temperature of 2000°C. The valves are closed, and as a result of heat transfer to the atmosphere, the pressure drops to 0.9 MPa. Determine the total heat transfer, and show the process on a *P-v* diagram.

4.83 A rigid aluminum container with a volume of 0.755 m³ contains 25 kg of carbon dioxide at an initial pressure of 2.0 MPa. If the manufacturer's specifications state that the container will rupture at 20.09 MPa pressure, what is the total heat transfer required to rupture the container? Show the process on a *P-v* diagram.

4.84 A vertical frictionless piston-cylinder device shown contains hydrogen at an initial temperature and pressure of 10°C and 40 N/cm. There is a heat transfer to the cylinder until the final temperature is 550°C. What is the final specific volume and heat transfer? Show the process on a *P-v* diagram.

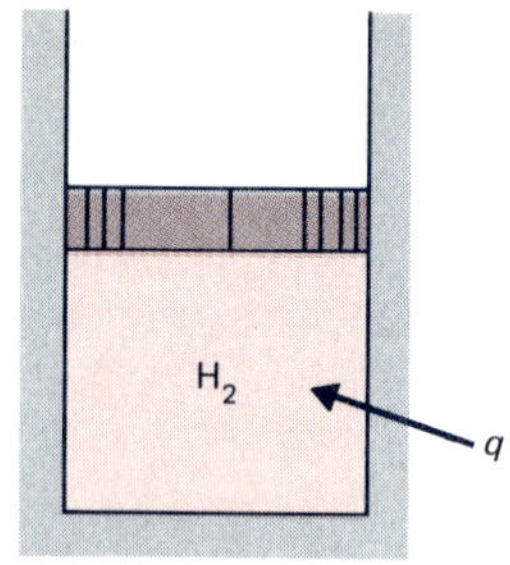

Figure P4.84

4.85 Air (2 kg) is contained in a vertical frictionless piston-cylinder device shown. The mass of the piston is such that the air has a pressure and temperature of 10.0 MPa and 75.5°C. There is a heat transfer to the cylinder until the piston reaches some stops, at which point the total volume is 0.04 m^3. There is an additional heat transfer to the air until the pressure is 15.0 MPa. Determine the total heat transfer and the total work, and show the process on a *P-v* diagram.

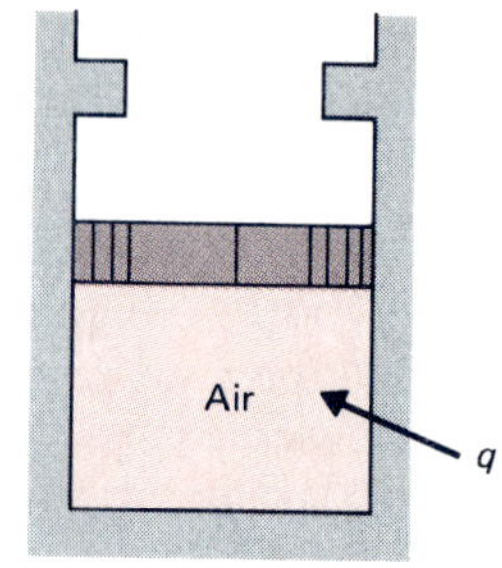

Figure P4.85

4.86 Argon (1 kg) is contained in a vertical frictionless piston-cylinder device shown. The mass of the piston is such that the argon exists at a pressure of 20 MPa and a temperature of 1150°C. There is a heat transfer from the device until the piston just rests on stops, at which time the volume inside the cylinder is 1.5607×10^{-3} m^3. Determine the total work and heat transferred for this process, and show the process on a *P-v* diagram.

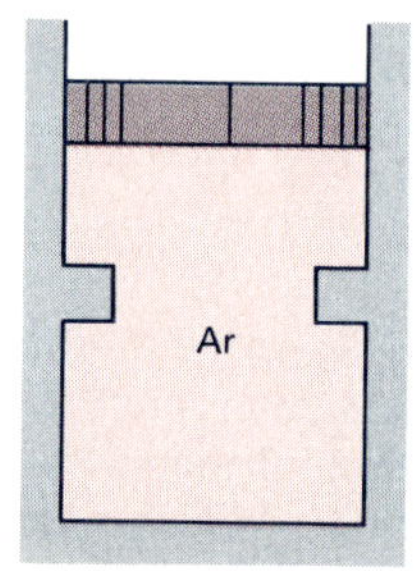

Figure P4.86

4.87 Oxygen (14 kg) is contained in a piston-cylinder device shown. The oxygen is heated from an initial temperature of 146.8°C to a final temperature and pressure of 1476°C and 10.0 MPa.

(*a*) Show the processes on a *P-v* diagram.
(*b*) What is the total work done?
(*c*) What is the total heat transfer?

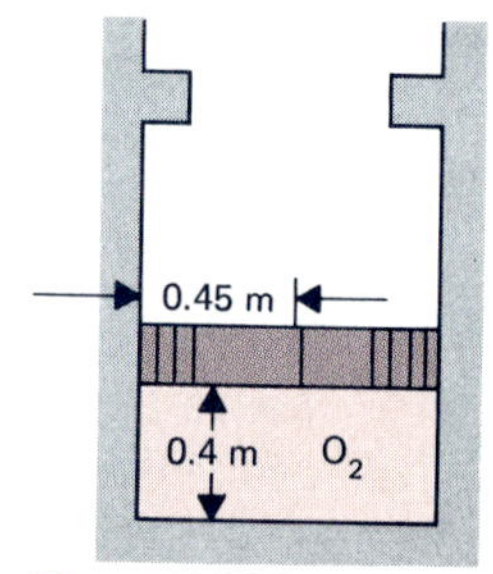

Figure P4.87

4.88 Nitrogen is contained in the frictionless piston-cylinder device shown. The nitrogen is cooled from an initial temperature and pressure of 2450°C and 6 MPa to a final pressure of 2 MPa. Plot and label the processes on a *P-v* diagram. Determine (*a*) the mass of the nitrogen, (*b*) the total work done, and (*c*) the total heat transfer.

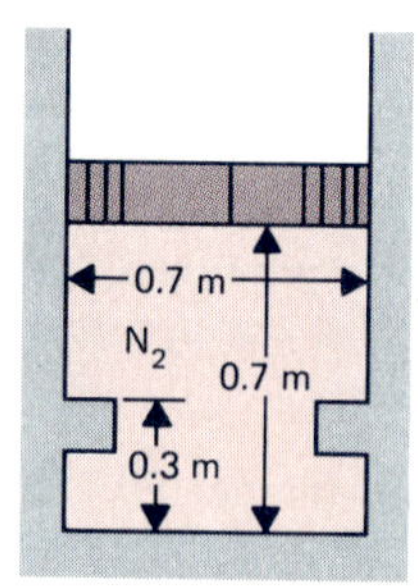

Figure P4.88

4.89 A sealed rigid container has a volume of 8.03×10^{-2} m^3. It contains carbon monoxide at an initial temperature and pressure of 500°C and 20 MPa. It is then cooled to 200°C. Determine (*a*) the mass of the carbon monoxide, (*b*) the work done during the process, and *c*) the total heat transfer.

4.90 A frictionless piston-cylinder device with one set of stops has the dimensions shown. The hydrogen inside is initially at 10 MPa and 2152°C. It is cooled until the temperature is 182°C. The piston requires a pressure of 6 MPa to support it. Draw and label the processes on a *P-v* diagram. Determine, (*a*) the initial and final mass of the hydrogen, (*b*) the total work done, and (*c*) the total heat transfer.

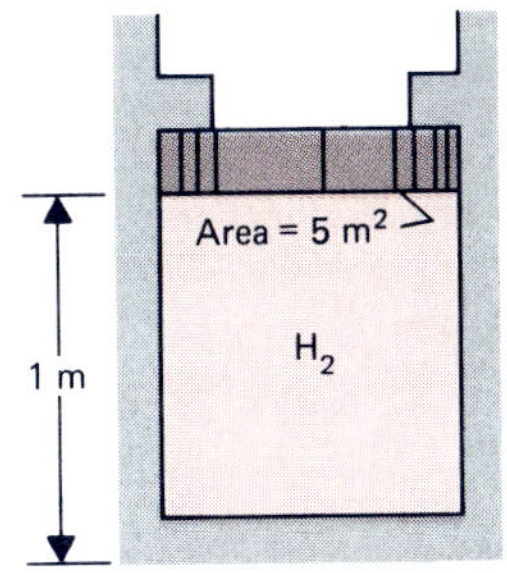

Figure P4.90

4.91 A rigid spherical container has a volume of 1 m^3. It is filled with 46.14 g of hydrogen initially at 260°C. The medium is heated to a temperature of 300°C. Plot the process on a *P-v* diagram and determine (*a*) the work done (kJ/kg) and (*b*) the change in pressure (kPa).

4.92 Nitrogen in the piston-cylinder device shown is initially at 200 kPa and 250°C. A pressure of 0.3 MPa is required to raise the piston. The medium is heated to a final temperature of 600°C. Determine (*a*) the mass of the nitrogen in the device and (*b*) the total work done (kJ). (*c*) Plot these processes on a *P-v* diagram.

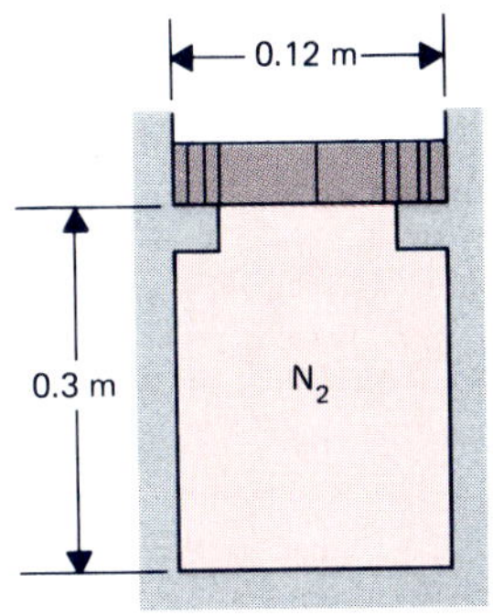

Figure P4.92

4.93 A closed rigid container has 6 kg of air at an initial pressure of 100 kPa and volume of 25 m^3. The air undergoes a process to a final pressure of 20 kPa. In what form (heat transfer or work) did the energy transfer take place? How much energy was transferred? Was it into or out of the system?

4.94 A frictionless piston-cylinder device containing neon is shown. The initial conditions are $P_1 = 100$ kPa and $T_1 = 20$°C. A pressure of 150 kPa is required to move the piston. If the neon is heated to a final temperature of 300°C, calculate the heat transfer and work (kJ/kg). Also draw and label the processes on a *P-v* plot.

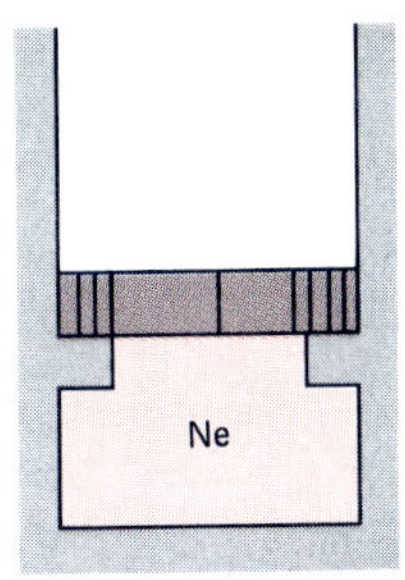

Figure P4.94

4.95 A frictionless piston-cylinder device (initially 3.0 m^3) containing argon is shown. The initial conditions are $P = 300$ kPa and $T = 100$°C. There is a heat transfer until $T = 1000$°C. When the piston strikes the upper stops, $V = 6.30$ m^3. Draw and label the processes on a *P-v* plot, and calculate the heat and work transferred (kJ).

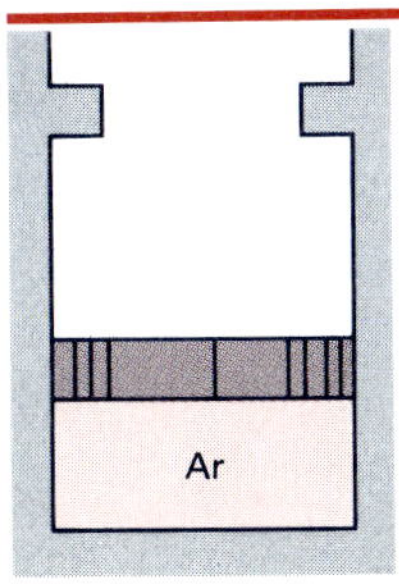

Figure P4.95

4.96 A frictionless piston-cylinder device containing helium is shown. The initial conditions are $P = 300$ kPa and $T_1 = 800°C$. A pressure of 200 kPa is required to support the piston. There is a heat transfer from the device until $T = 80°C$. Draw and label the processes on a P-v plot, and calculate the heat and work (kJ/kg) transferred.

4.97 A fan-in-the-box system (see Fig. P4.97) using air as the medium is operated for a time during which 190 kJ/kg of shaft work is added. Initially the air is at $P_1 = 101.3$ kPa and $T_1 = 20°C$. Assuming the process is isometric and the box is insulated, determine the final temperature T_2.

4.98 A fan-in-the-box system (see Fig. P4.98) is operated from an initial condition of $P_1 = 101.3$ kPa and $T_1 = 30°C$ to a final temperature of 450°C. If the shaft work added is 200 kJ/kg, determine the heat transfer (kJ/kg). The medium is carbon dioxide.

4.99 Helium in a rigid vessel initially at $T_1 = 20°C$ and $P_1 = 150$ kPa is cooled until the pressure is 100 kPa. Determine the work and heat transfer (kJ/kg) for the process.

4.100 Air is contained in a frictionless piston-cylinder device initially at 1 MPa and 500°C. The air is cooled until the specific volume is 0.1 m³/kg. What is the mass of air if the total heat transferred is 1000 kJ? (Assume no shaft work.)

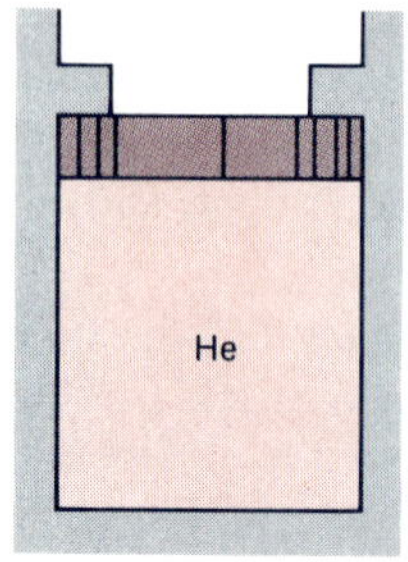

Figure P4.96

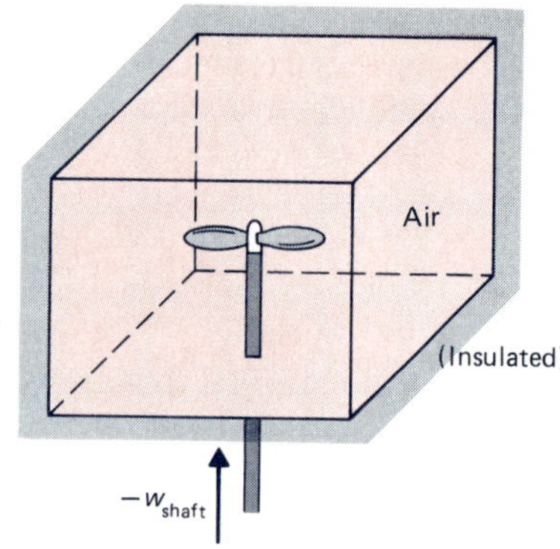

Figure P4.97

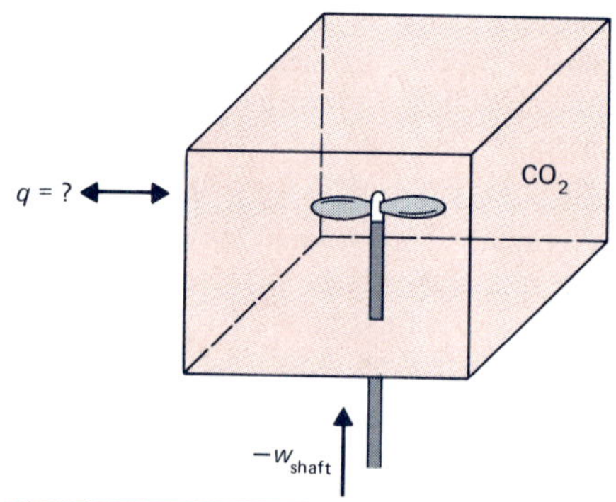

Figure P4.98

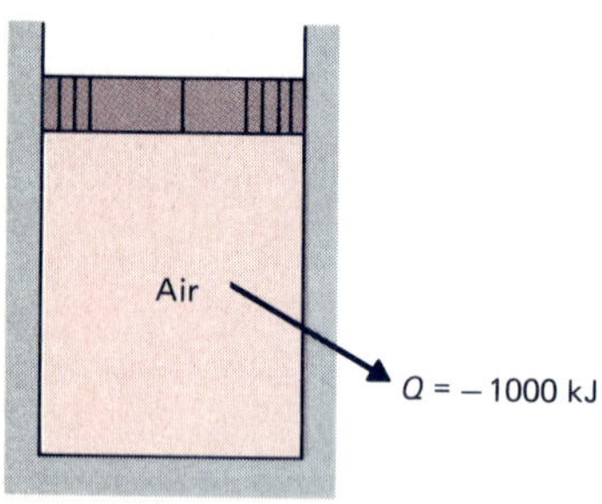

Figure P4.100

4.101 Nitrogen (2.5 kg) is contained in a frictionless piston-cylinder device at 30°C and 250 kPa. If there is a heat transfer until the total volume of N_2 is 1.2 m³, find the work (kJ/kg) and total heat transfer in the process.

4.102 With a heat transfer of 750 kJ/kg from the nitrogen-filled frictionless piston-cylinder device shown, determine the shaft work (kJ/kg). The initial and final states of the N_2 are $T_1 = 300°C$. $P_1 = 600$ kPa and $T_2 = 450°C$, $P_2 = 600$ kPa, respectively.

Figure P4.101

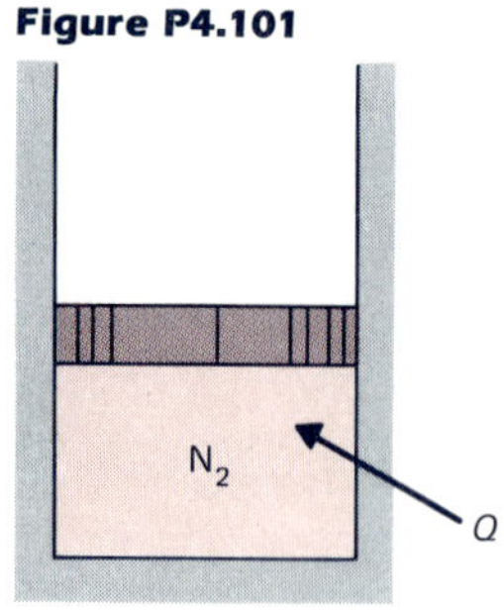

Figure P4.102

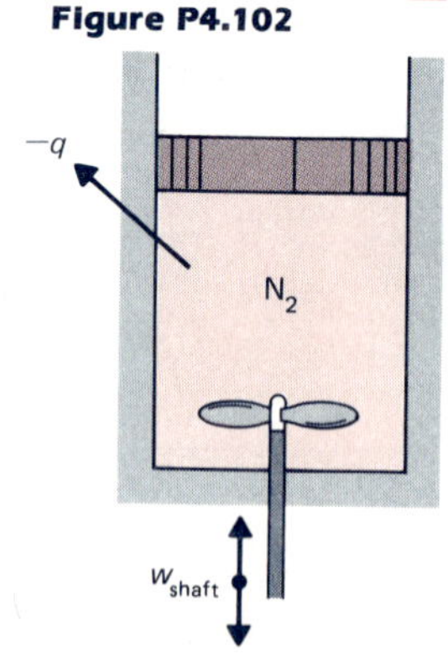

4.103 Shaft work (500 kJ/kg) is transferred to the air-filled frictionless piston-cylinder device shown. Determine the heat transfer (kJ/kg). The initial and final states of the air are $T_1 = 200°C$, $P_1 = 400$ kPa and $T_2 = 250°C$, respectively.

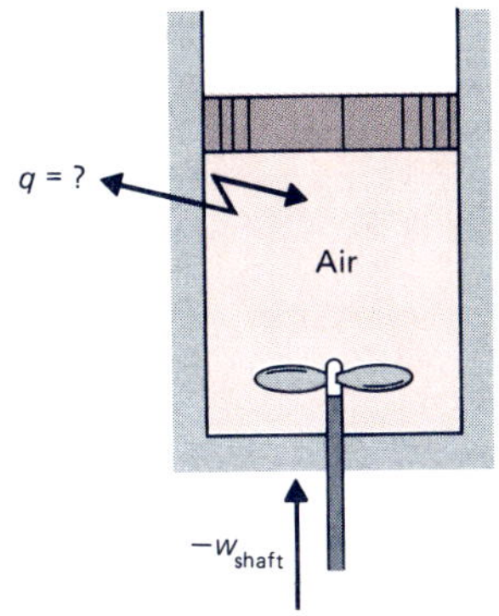

Figure P4.103

4.104 The free-moving piston-cylinder device shown contains 2 kg of carbon monoxide at an initial temperature of 350°C. With a heat transfer of 500 kJ from the medium, the temperature rises to 600°C. Determine the amount of shaft work (kJ) necessary for the process to occur.

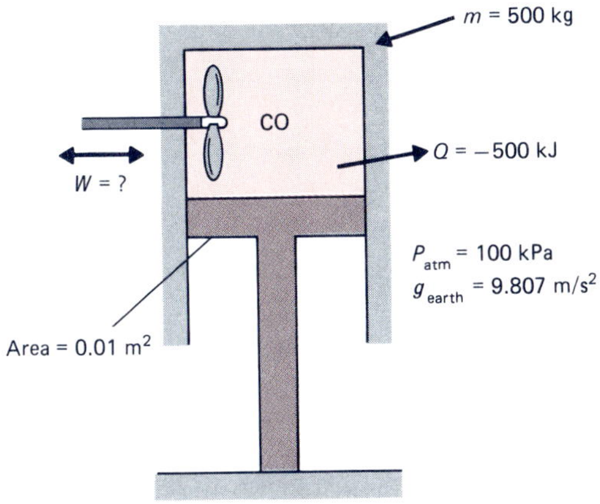

Figure P4.104

4.105 The free-moving piston-cylinder device shown contains 5 kg of hydrogen at a temperature of 400°C. When 4000 kJ of shaft work is transferred to the medium, the temperature rises to 450°C. Determine the amount of heat transferred (kJ) during the process.

Figure P4.105

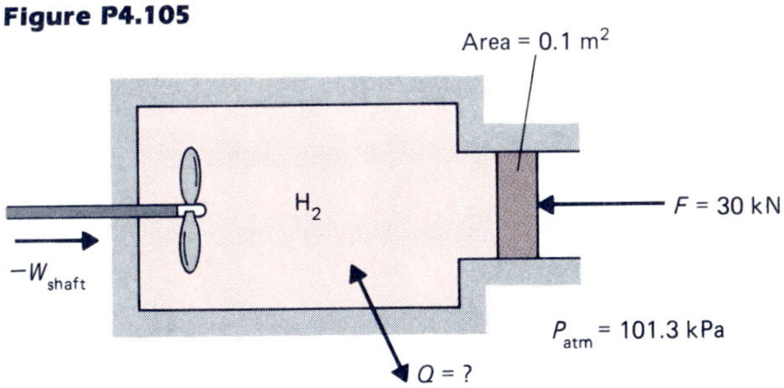

4.106 Helium (15 kg) is heated in a closed rigid container from 60 to 400°C and 2 MPa. Find the total work and heat transfer (kJ).

4.107 Air (3 kg) in the frictionless piston-cylinder device shown is initially at a pressure of 1 MPa and a temperature of 100°C. There is a heat transfer to the air until the temperature is 500°C. It takes a pressure of 2 MPa to lift the piston. What is the total heat transferred in the process?

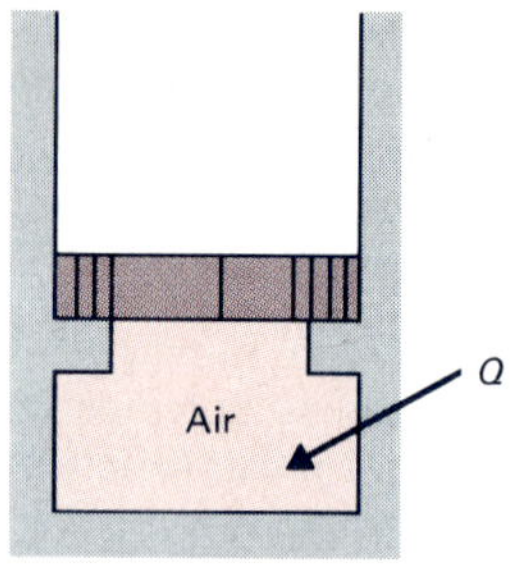

Figure P4.107

4.108 Initially, air at $T_1 = 200°C$ and $P = 101.3$ kPa is expanded isothermally in a frictionless piston-cylinder device to a temperature of 400°C. Determine the heat transferred in this process (kJ/kg).

4.109 Air enters an adiabatic turbine at a pressure and temperature of 2.5 MPa and 500°C, respectively. It leaves the turbine at a pressure and temperature of 0.15 MPa and 200°C, and the mass flow rate is 72 kg/s. Determine the turbine power output.

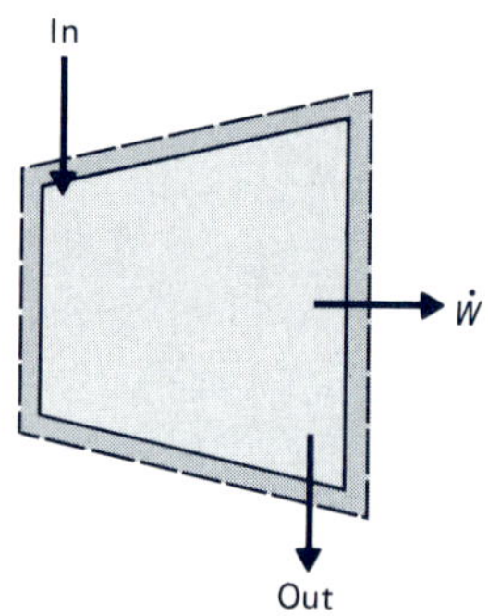

Figure P4.109

4.110 Air in a piston–cylinder device is cooled from an initial temperature and pressure of 8.0 MPa and 700°C to a final temperature of 300°C. If the process is isobaric, find the heat transfer.

4.111S Water flows in a 0.05-m-diameter pipe. At the inlet to a tank, the water is at 20°C and atmospheric pressure. See Fig. P4.111. Evaluate the mass flow rate $\dot{m}$ and $\int_A h \, d\dot{m}$ at the inlet for the following two velocity profiles:

(*a*) $\mathbf{V}_z(r) = \mathbf{V}_{z,\max}\left[1 - \left(\frac{r}{r_0}\right)^2\right]$

where

$\mathbf{V}_{z,\max} = 1$ m/s and $r_0 = \frac{D_{\text{pipe}}}{2}$

(*b*) $\mathbf{V}_z(r) = \mathbf{V}_{z\max}$

Note that in the pipe cross-section, $dA = 2\,\pi r\,dr$.

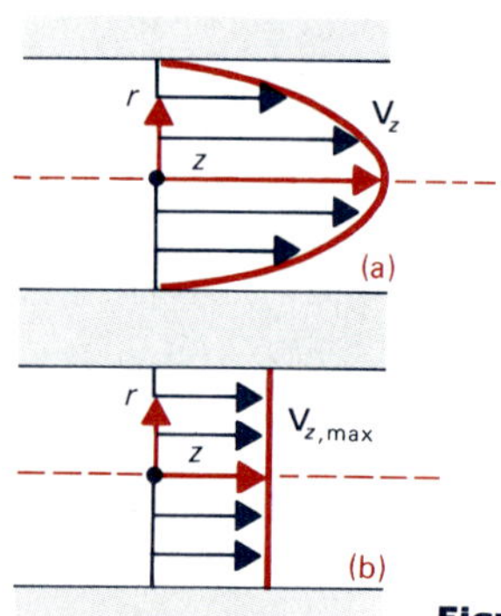

Figure P4.111

4.111E Water flows in a 2-in-ID pipe. See Fig. P4.111. At the inlet to a tank, the water is at 67°F and atmospheric pressure. Evaluate the mass flow rate $\dot{m}$ and $\int_A h \, d\dot{m}$ at the inlet for the following two velocity profiles:

(*a*) $\mathbf{V}_z(r) = \mathbf{V}_{z,\max}\left[1 - \left(\frac{r}{r_0}\right)^2\right]$

where

$\mathbf{V}_{z,\max} = 3$ ft/s and $r_0 = \frac{D_{\text{pipe}}}{2}$

(*b*) $\mathbf{V}_z(r) = \mathbf{V}_{z,\max}$

Note that in the pipe cross-section, $dA = 2\pi r\,dr$.

4.112 A storage tank is shown in Fig. P4.112. The tank has an inlet pipe carrying water at a mass flow rate of 100 kg/min. An outlet pipe at the bottom of the tank is 3 cm in inside diameter and carries water to a pump at a velocity of 2 m/s. The tank takes 20 min to change level by 2 m. What is the tank diameter?

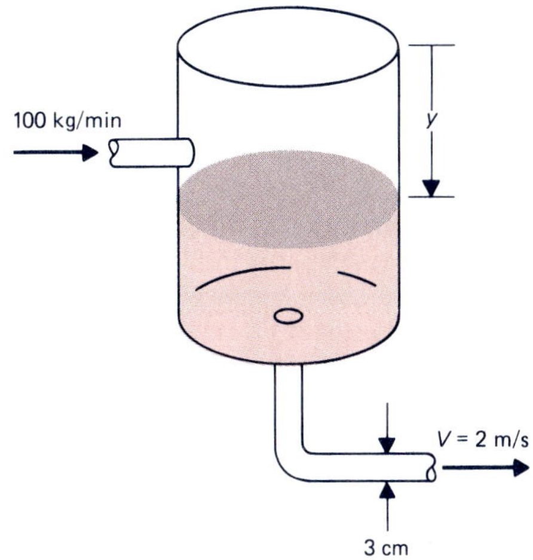

Figure P4.112

4.113S A crude-oil storage tank is in the shape of a right circular cylinder. The tank is 20 m high and holds 2000 m^3 when full. Crude oil has a specific volume of 0.0015 m^3/kg. At 10 a.m. on June 16, the tank contains 1000 m^3, and crude oil is being pumped into the tank through one pipe at a rate of 2 m^3/min. Oil is being pumped out of the tank at a velocity of 1.5 m/s in another pipe of 0.15-m ID. What will the height of the oil in the tank be at noon on June 17? What volume of oil will be in the tank at that time?

4.113E A crude-oil storage tank is in the shape of a right circular cylinder. The tank is 60 ft high and holds 70,000 ft^3 when full. Crude oil has a specific volume of 0.024 ft^3/lbm. At 10 a.m. on June 16, the tank contains 35,000 ft^3, and crude oil is being pumped into the tank through one pipe at a rate of 70 ft^3/min. Oil is being pumped out of the tank at a velocity of 5 ft/s in another pipe of 6-in ID. What will the height of the oil in the tank be at noon on June 17? What volume of oil will be in the tank at that time?

4.114S A conical storage tank is oriented with its axis vertical and its apex toward the ground. It has a base 1 m in diameter and is 2 m high. Water enters the cone at a rate of 5.0 kg/s. How fast is the water rising in the cone when it reaches a height of 1 m? Take the specific volume of water as 0.001003 m^3/kg.

4.114E A conical storage tank is oriented with its axis vertical and its apex toward the ground. It has a base 3 ft in diameter and is 6 ft high. Water enters the cone at a rate of 10 lbm/s. How fast is the water rising in the cone when it reaches a height of 3 ft? Take the specific volume of water as 0.01607 ft^3/lbm.

4.115S A steam turbine is designed to operate at a mass flow rate of 1.5 kg/s. The inlet conditions are $P_1 = 2$ MPa, $T_1 = 400°C$, and $\mathbf{V}_1 = 60$ m/s; the outlet conditions are $P_2 = 0.1$ MPa, $x_2 = 0.98$, and $\mathbf{V}_2 = 150$ m/s. The change in elevation from inlet to outlet is a drop of 1 m, and the heat loss is 50 kW.

(*a*) Evaluate the power output of the turbine.

(*b*) Evaluate the power output if the changes in kinetic and potential energy are neglected.

(*c*) What are the diameters of the inlet pipe and exhaust pipe?

4.115E A steam turbine is designed to operate at a mass flow rate of 3 lbm/s. The inlet conditions are $P_1 = 300$ psia, $T_1 = 800°F$, and $\mathbf{V}_1 = 180$ ft/s; the outlet conditions are $P_2 = 14.7$ psia, $x_2 = 0.98$, and $\mathbf{V}_2 = 450$ ft/s. The change in elevation from inlet to outlet is a drop of 3 ft, and the heat loss is 1.7×10^5 Btu/h.

(*a*) Evaluate the power output of the turbine.

(*b*) Evaluate the power output if the changes in kinetic and potential energy are neglected.

(*c*) What are the diameters of the inlet pipe and exhaust pipe?

4.116 A steam turbine operates with inlet steam conditions of 500°C and 2 MPa and outlet conditions of 10 kPa and $x = 0.95$. What rate of heat transfer from the turbine (kJ/kg) can be allowed if the required turbine work per kilogram of steam flow is 900 kJ/kg?

4.117 Steam enters an adiabatic turbine as saturated vapor at a pressure of 2.5 MPa. It exists at $x_4 = 0.98$ and a pressure of 0.15 MPa. If $\dot{m}$ is 72 kg/s, what is the turbine power output?

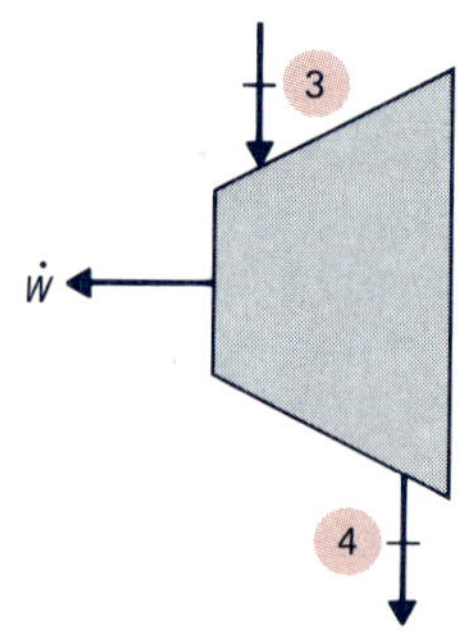

Figure P4.117

4.118 Two components of a steam engine are shown below. The turbine is adiabatic.

(*a*) Determine the heat transferred in the boiler (kJ/kg).

(*b*) Determine the power output of the turbine (kW).

Figure P4.118

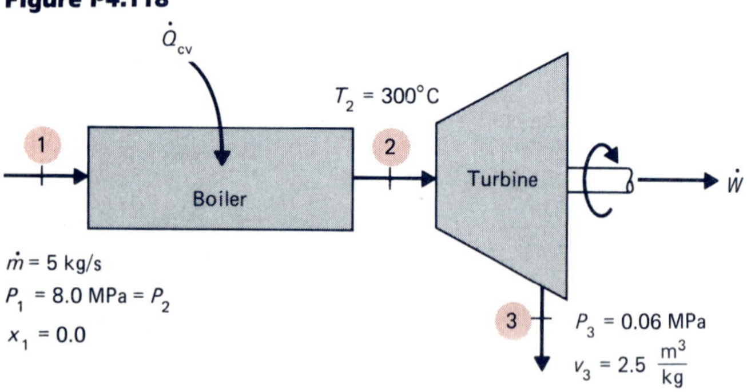

4.119 Steam enters a turbine at a temperature of 300°C and a pressure of 0.4 MPa. If it leaves as a saturated vapor at 0.03 MPa, what is the change in enthalpy between the two state points?

4.120 Steam flows through a turbine as shown, producing shaft work. The walls of the turbine are not insulated, so that there is a heat transfer through the walls as shown. Calculate the power output of the turbine in horsepower.

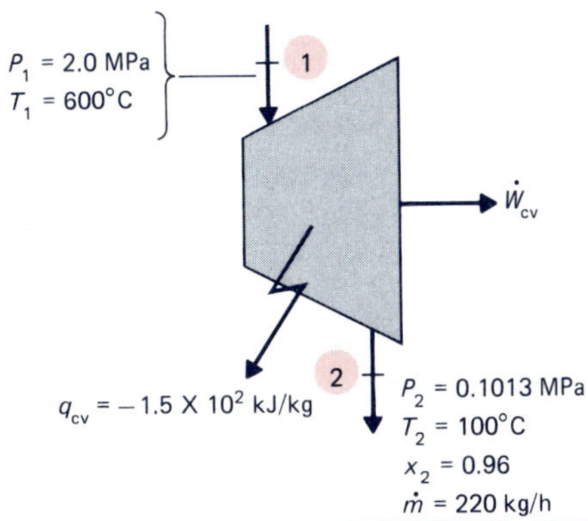

Figure P4.120

4.121 A nonadiabatic steam turbine operates as shown. Determine the rate of heat transfer (kW).

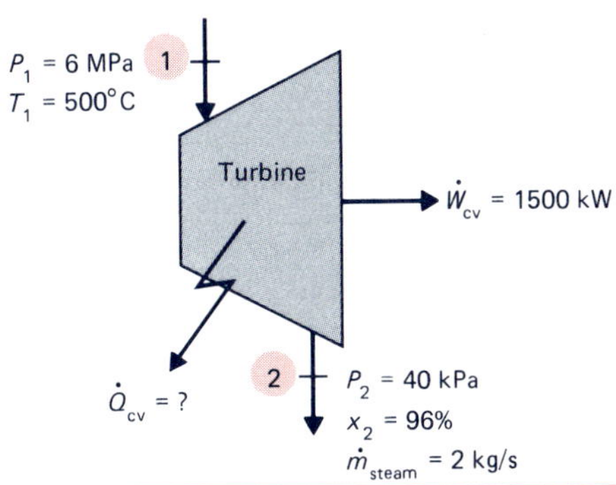

Figure P4.121

4.122 Determine the power (kW) output of the steam turbine operating as shown.

Figure P4.122

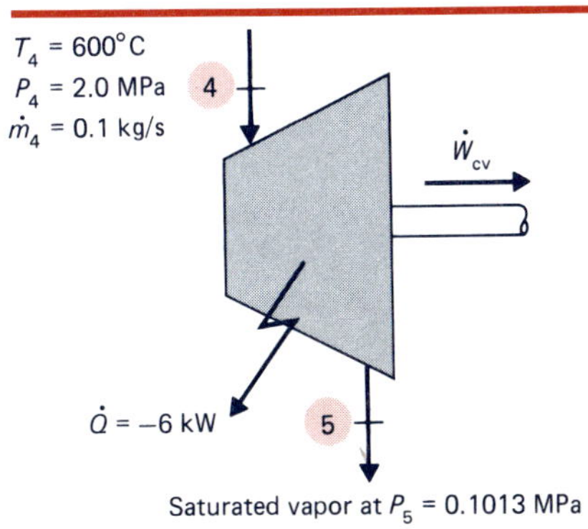

4.123 Determine the work (kJ/kg) output of the adiabatic steam turbine operating as shown.

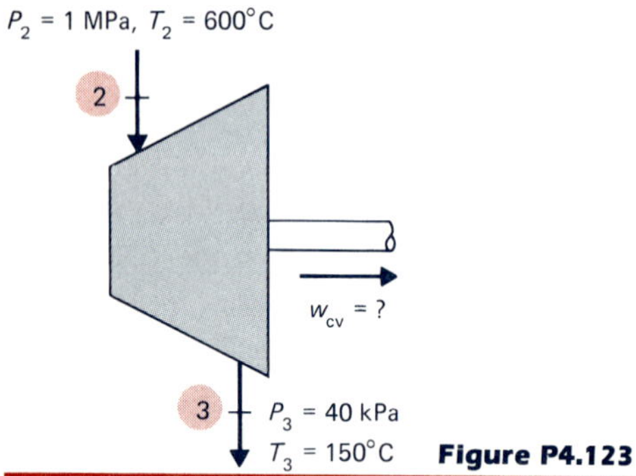

Figure P4.123

4.124 A nonadiabatic helium turbine operates as shown below. Calculate the power output in horsepower.

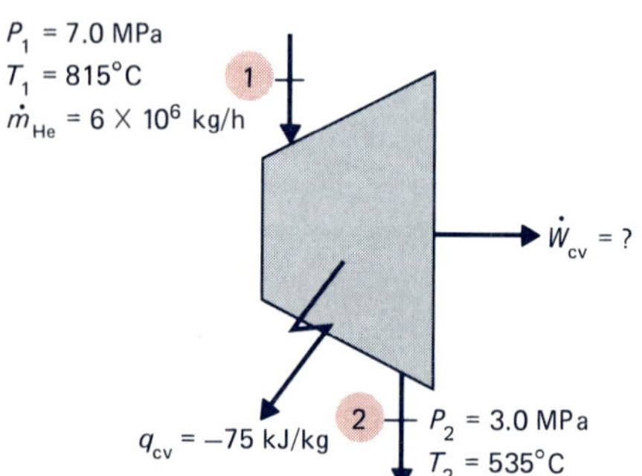

Figure P4.124

4.125 Determine the power output (kW) of the air turbine operating as shown.

Figure P4.125

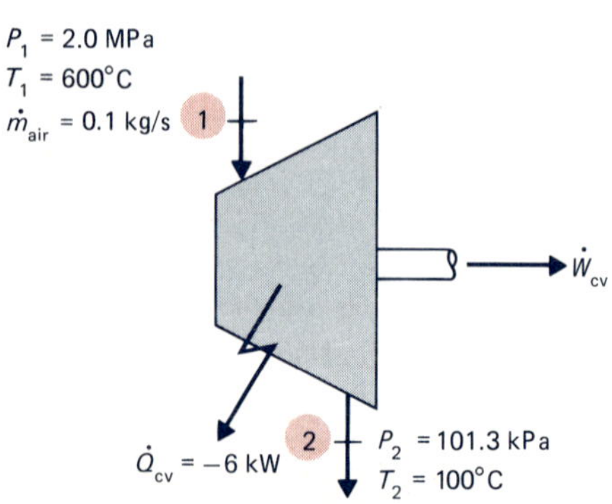

4.126 Determine the work (kJ/kg) done by the adiabatic helium turbine operating as shown.

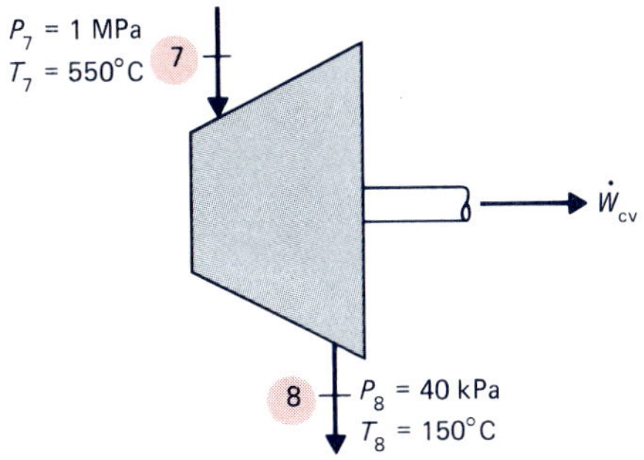

Figure P4.126

4.127 Determine the work (kJ/kg) required to operate the adiabatic helium compressor operating as shown.

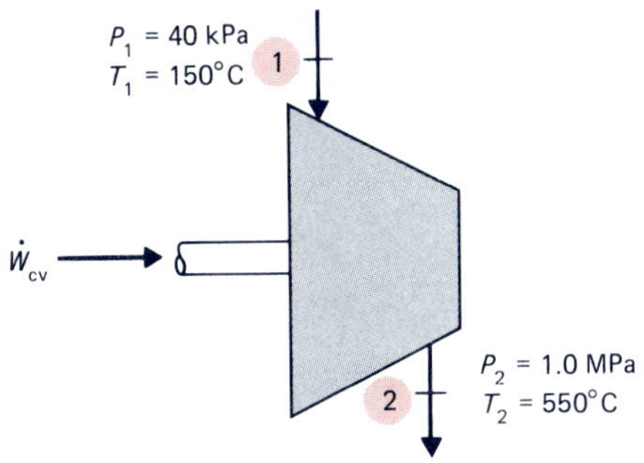

Figure P4.127

4.128S What flow rate of steam is required to produce 500 kW from an adiabatic turbine with inlet conditions of 800 kPa and 400°C and exit conditions of 6 kPa and $x =$ 95 percent?

4.128E What flow rate of steam is required to produce 700 hp from an adiabatic turbine with inlet conditions of 120 psia and 800°F and exit conditions of 5 psia and $x =$ 95 percent?

4.129S A compressor is to be purchased for the new mechanical engineering building. It must compress air from atmospheric pressure and 25°C to 10 atm and 600°C; and the outlet velocity from the compressor must not exceed 10 m/s. Assume the compressor to be adiabatic and frictionless.

(*a*) How much work is required for each kilogram of air that is compressed?

(*b*) What power is required to drive the compressor if 2 kg/s of air is to be compressed?

4.129E A compressor is to be purchased for the new mechanical engineering building. It must compress air from atmospheric pressure and 77°F to 10 atm and 1100°F; and the outlet velocity from the compressor must not exceed 30 ft/s. Assume the compressor to be adiabatic and frictionless.

(*a*) How much work is required for each pound-mass of air that is compressed?

(*b*) What power is required to drive the compressor if 4.5 lbm/s of air is to be compressed?

4.130S An adiabatic compressor requires 10 kW to compress 0.05 kg/s of oxygen from 1 atm and 50°C to a final pressure of 5 atm. Using the ideal gas tables for all properties, find the temperature of the oxygen as it leaves the compressor.

4.130E An adiabatic compressor requires 14 hp to compress 0.1 lbm/s of oxygen from 1 atm and 100°F to a final pressure of 5 atm. Using the ideal gas tables for all properties, find the temperature of the oxygen as it leaves the compressor.

4.131 Steam enters an isothermal compressor at 400°C and 0.1013 MPa. If the exit pressure is 10 MPa, determine the change in enthalpy.

4.132 Determine the work (kJ/kg) necessary to operate the adiabatic steam compressor shown.

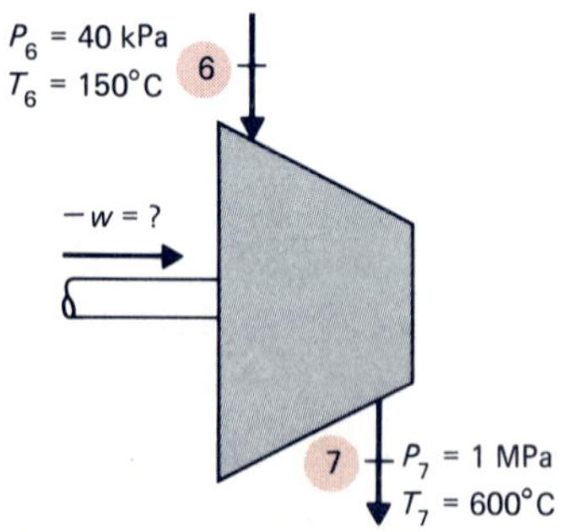

Figure P4.132

Figure P4.133

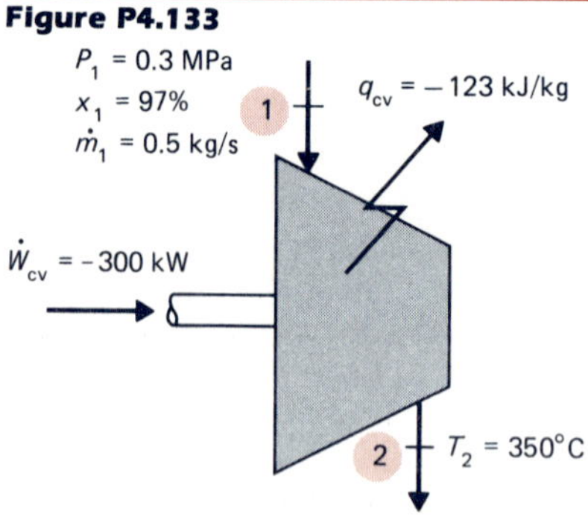

4.133 A nonadiabatic steam compressor operates as shown. Determine the pressure at the compressor exit.

4.134 The nonadiabatic air turbine operates as shown. Calculate the rate of heat transfer (kW).

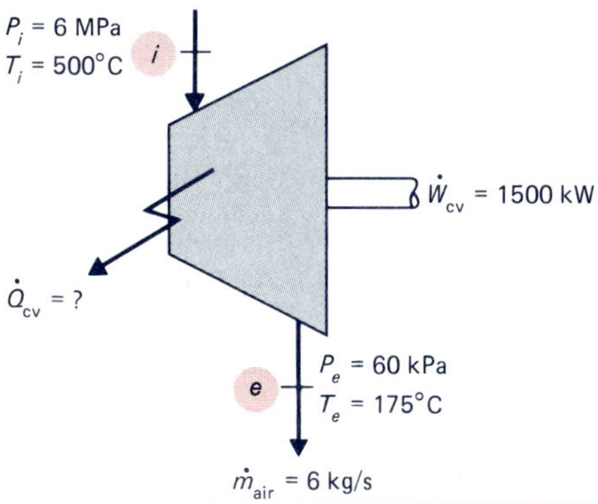

Figure P4.134

4.135 A nonadiabatic air compressor operates as shown. What is the exit temperature?

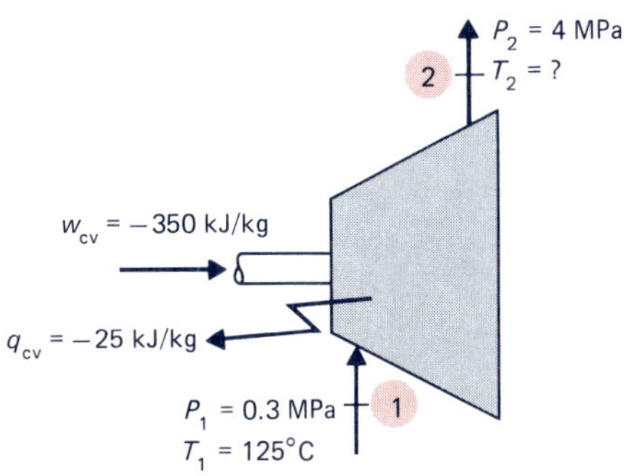

Figure P4.135

4.136 A helium compressor operates as shown. Determine the power required.

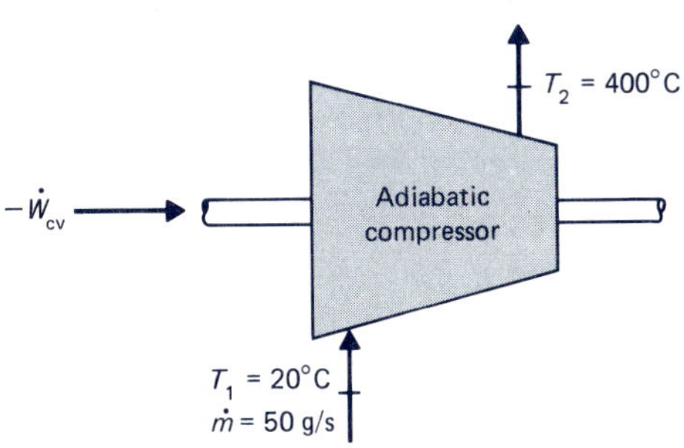

Figure P4.136

4.137 (*a*) An air compressor requires a power input of 100 kW to take air at 25°C and 1 atm and compress it adiabatically to a final pressure of 20 atm and a final temperature of 400°C. What mass flow rate of air is being compressed (kg/s)?

(*b*) Suppose that the outlet pressure of the compressor is changed to 15 atm, but the inlet conditions, mass flow rate of air, and the outlet temperature remain the same as in (*a*). What power input to the compressor is now required (kW)? Give a *short* explanation of why your answer is reasonable.

4.138 Air (2 kg/s) enters an adiabatic compressor through a large-diameter feed pipe. The air is compressed from atmospheric pressure and 25°C to a final pressure of 20 atm and leaves the compressor at a speed of 200 m/s. The compressor has a power input of 1400 kW.

(*a*) What is the temperature of the air leaving the compressor (K)?

(*b*) What is the inside diameter of the outlet pipe from the compressor (m)?

4.139 A fan is designed to take air from a large room at a volumetric flow rate (based on the inlet state to the fan) of 200 m^3/min. The fan must raise the inlet pressure by 1.0 kPa. The air enters the fan at 25°C and increases in temperature by a negligible amount in passing through the fan. The air leaves the fan in a duct of 1.00-m^2 cross section.

What power (kW) is required to drive the fan (neglecting frictional effects)? *Note any assumptions that you make in obtaining your answer.*

4.140S A large pump is used to take water at 25°C from a nearby lake at a rate of 1 m^3/s and raise its pressure from 120 to 700 kPa so that it can be fed into a fire-safety main. If the pump is adiabatic and frictionless, how much power is necessary to drive the pump?

4.140E A large pump is used to take water at 66°F from a nearby lake at a rate of 35 ft^3/s and raise its pressure from 17 to 100 psia so that it can be fed into a fire-safety main. If the pump is adiabatic and frictionless, how much power is necessary to drive the pump?

4.141 An adiabatic and frictionless pump is rated at 600 kW when providing a pressure change of 600 kPa to water flowing through the pump. What volumetric flow rate (m^3/s) and what mass flow rate (kg/s) of water can be pumped at these conditions?

4.142 A pump is used to pump 5 kg/min of liquid refrigerant 12 from inlet conditions of 1 atm and 20°C to outlet conditions of 5 atm and 20°C. How much power (kW) is required to drive the pump if it is frictionless?

4.143 Steam enters an isobaric device with a quality of 100 percent and a pressure of 1.5 MPa. If the exit temperature is 350°C, determine the change in enthalpy.

4.144 An inventor has proposed the insulated device shown which operates with H_2O as the medium. Determine T_3.

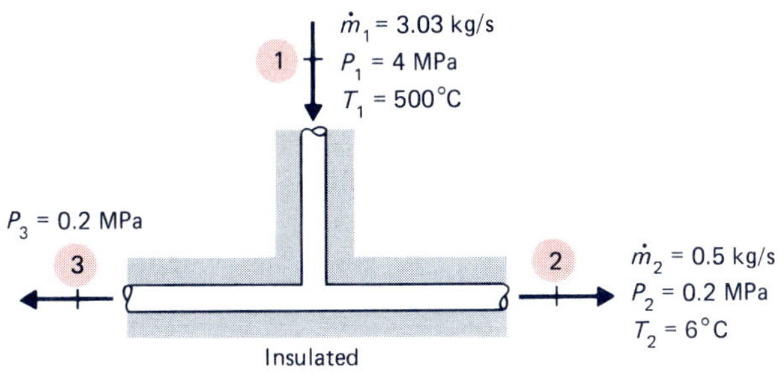

Figure P4.144

4.145 An inventor has proposed the insulated device shown with air as the medium. Determine T_3.

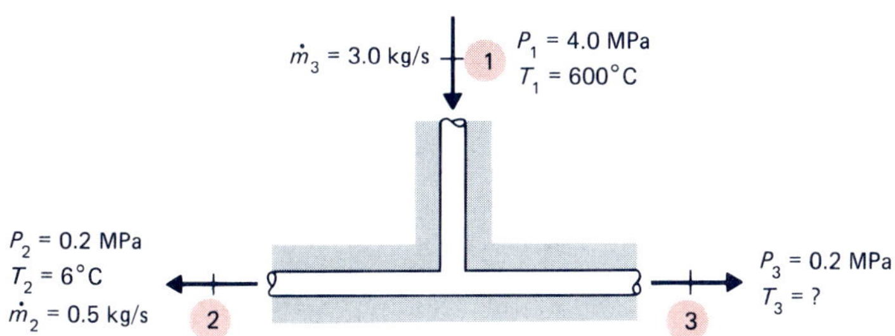

Figure P4.145

4.146 For the device shown below, determine the flow rate (kg/s) of state 2 necessary to ensure a rate of change of enthalpy (inlets to exit) equal to 2000 kW.

Figure P4.146

4.147 Determine the total rate of change of enthalpy from the inlets to the exit of the isobaric device shown.

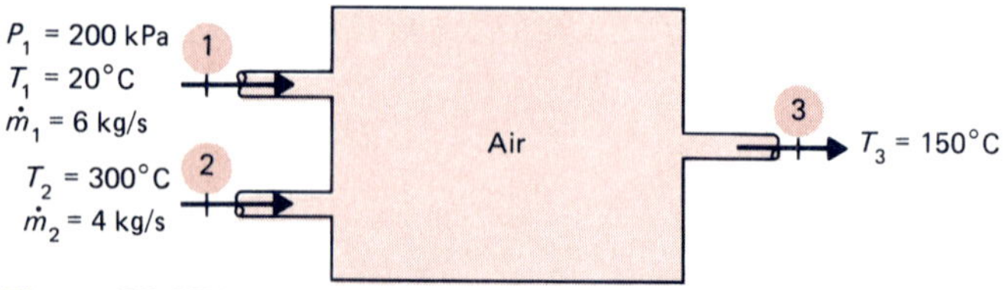

Figure P4.147

4.148 Determine the total rate of change of enthalpy from the inlets to the exits of the device shown.

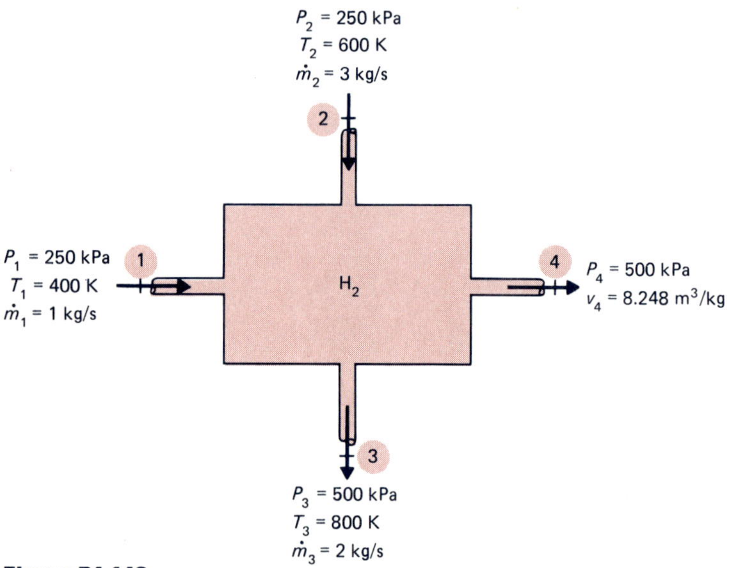

Figure P4.148

4.149 Find the rate of change of enthalpy from the inlet to the exits of the CO_2 device shown.

Figure P4.149

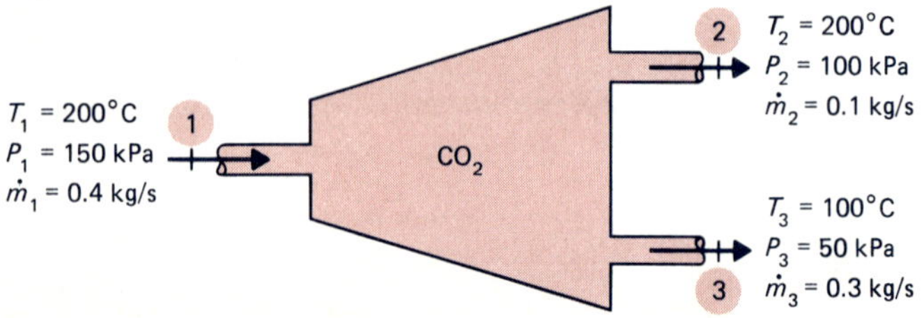

4.150 In the device shown below, the total rate of change of enthalpy from the inlets to the exit is 1000 kW. Determine T_2.

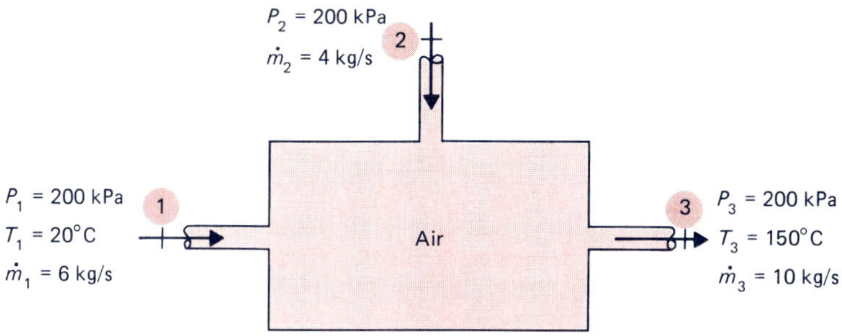

Figure P4.150

4.151 Determine the mass flow rate and temperature of the air at the exit of the adiabatic device shown.

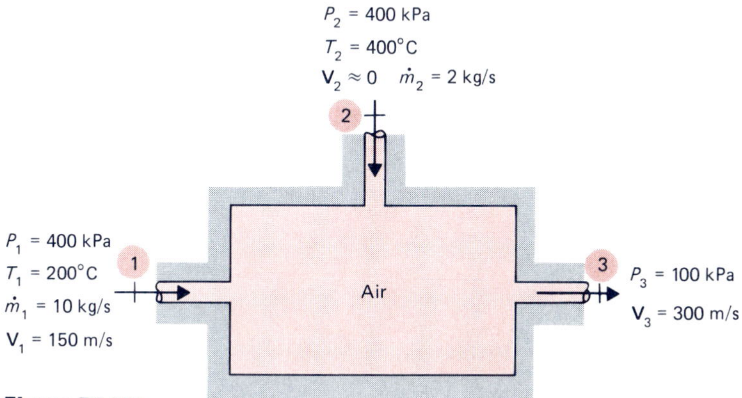

Figure P4.151

4.152 Determine the exit temperature T_3 of the vortex-tube refrigeration device shown. The device is insulated, and the medium is air.

Figure P4.152

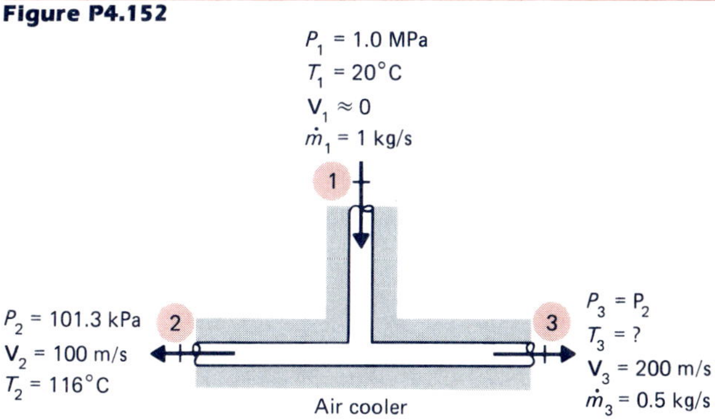

4.153 Find the heat transfer to or from the following device. (*Hint:* The inlet velocity is *not* zero.) The medium is air.

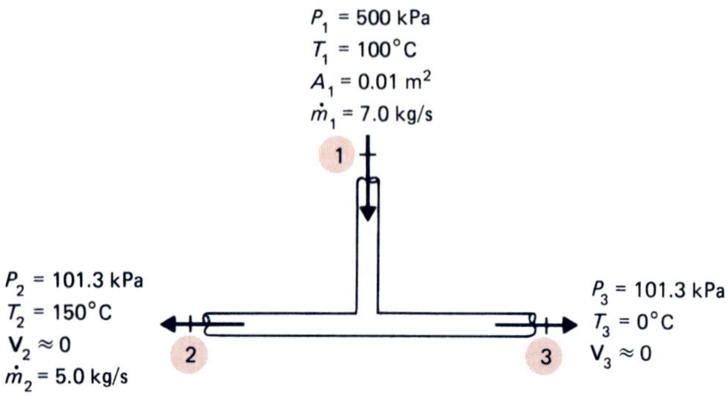

Figure P4.153

4.154 Determine the amount of power produced or consumed by the device below. The medium is air.

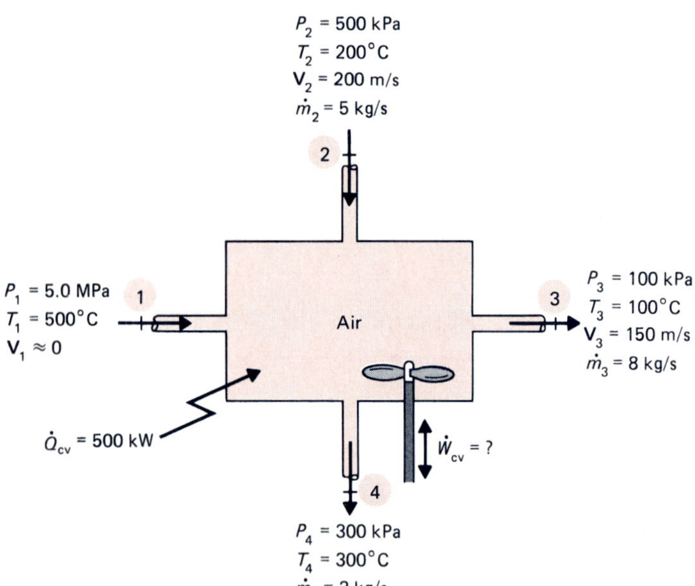

Figure P4.154

4.155 Consider the steady-state steady flow device shown in Fig. P4.155. The medium is steam. Calculate the shaft power.

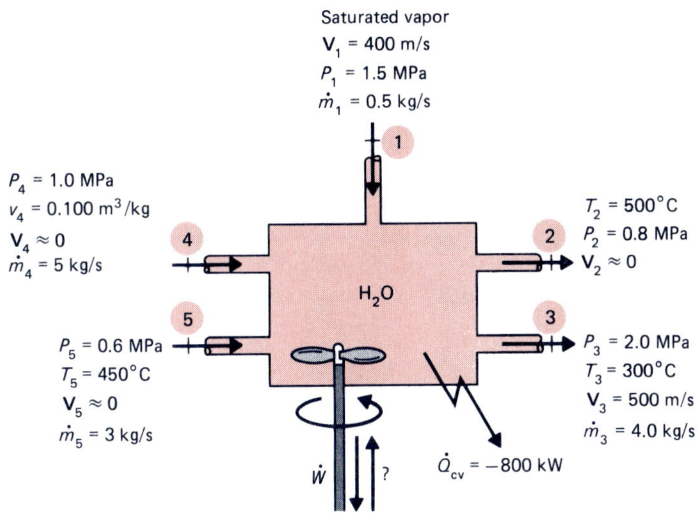

Figure P4.155

4.156 A nonadiabatic steady-state steady flow device operates as shown. If the medium is H_2O, determine $\dot{W}_{CV}$ (kW).

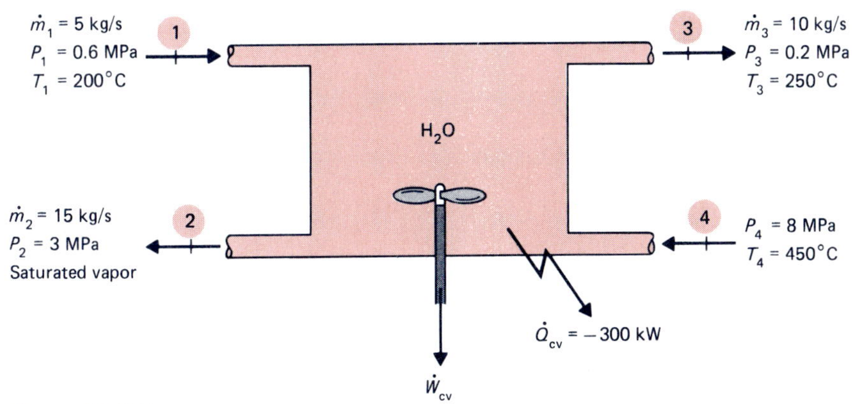

Figure P4.156

4.157 Helium flows through the steady-state steady flow device shown in Fig. P4.157. Determine the rate of heat transfer (kW).

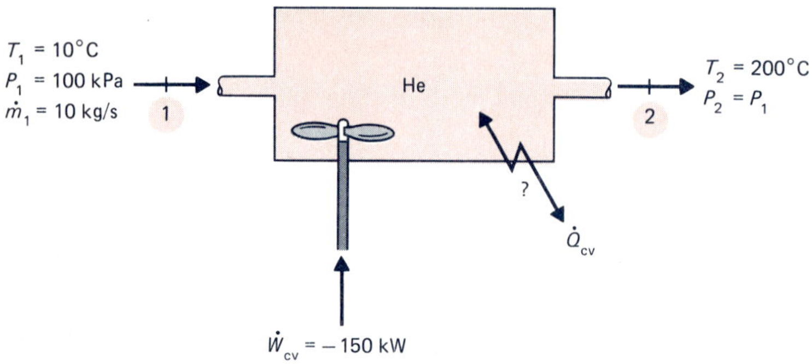

Figure P4.157

4.158 A nonadiabatic steady-state steady flow device operates as shown. If the medium is air, determine $\dot{W}_{CV}$ (kW).

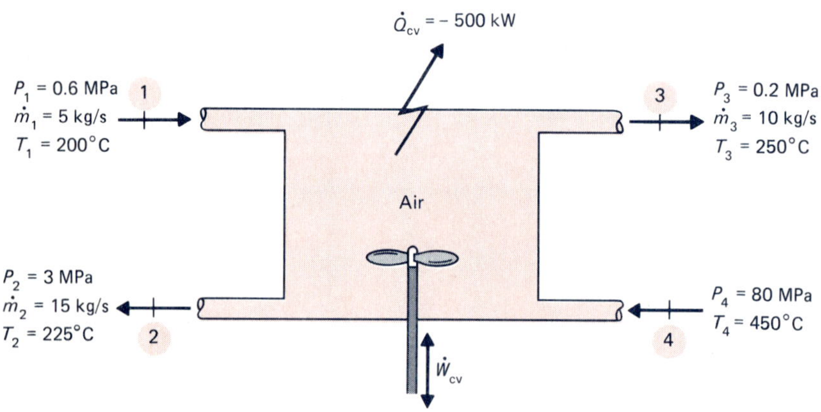

Figure P4.158

4.159 Steam (H_2O) enters and exits the nozzle as shown. Find the exit velocity in meters per second.

Figure P4.159

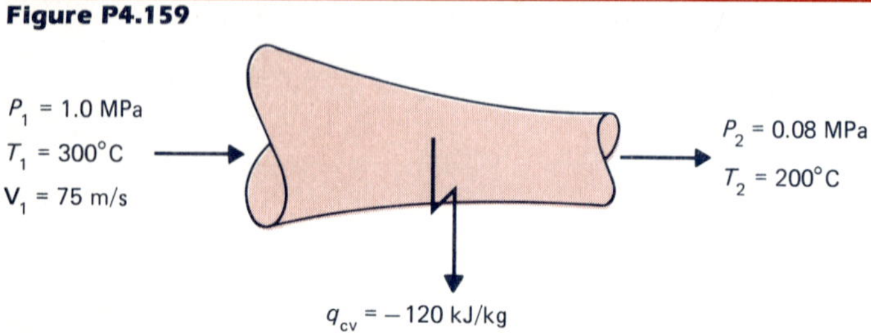

4.160 Determine the exit velocity and exit mass flow rate of steam for the nozzle operating as shown.

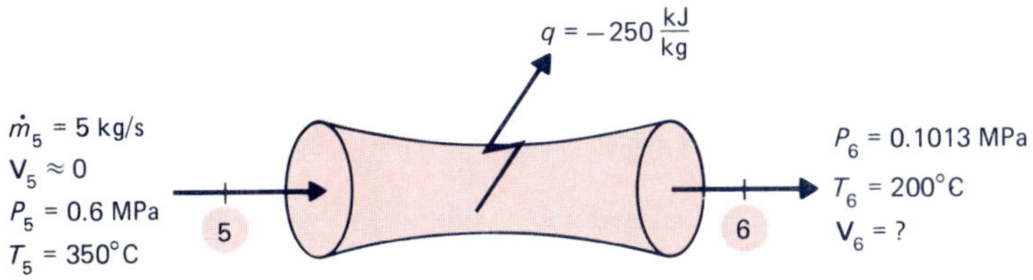

Figure P4.160

4.161 Steam expands through the adiabatic steady-state steady flow nozzle shown. For the given conditions, find the exit velocity V_2.

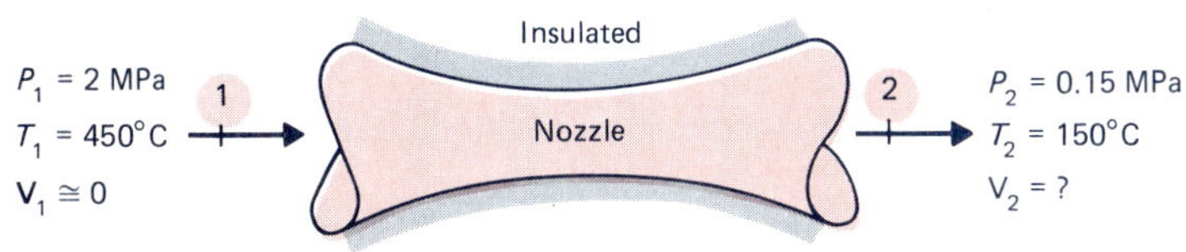

Figure P4.161

4.162 Steam flows through the nonadiabatic nozzle shown. Determine the exit velocity (m/s).

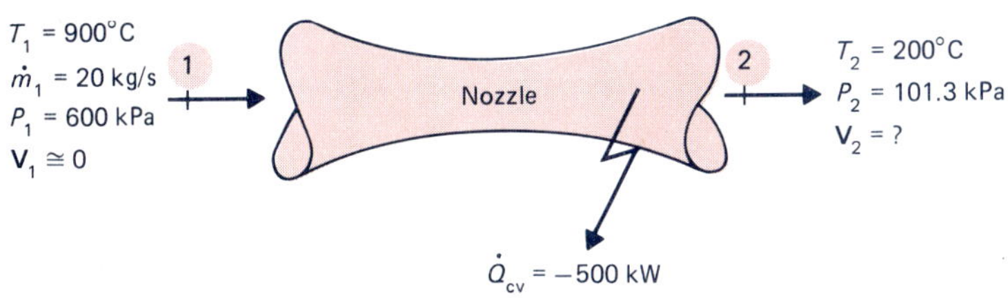

Figure P4.162

4.163 An air nozzle operates as shown below. Find the exit velocity (m/s).

Figure P4.163

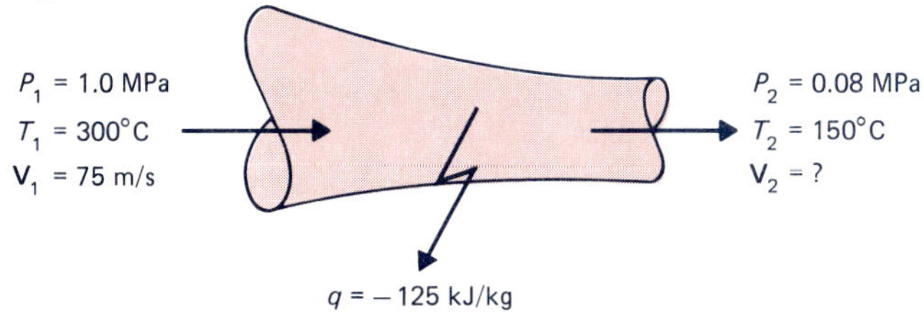

4.164 Air expands through the adiabatic steady-state steady flow nozzle shown. Calculate the exit velocity (m/s).

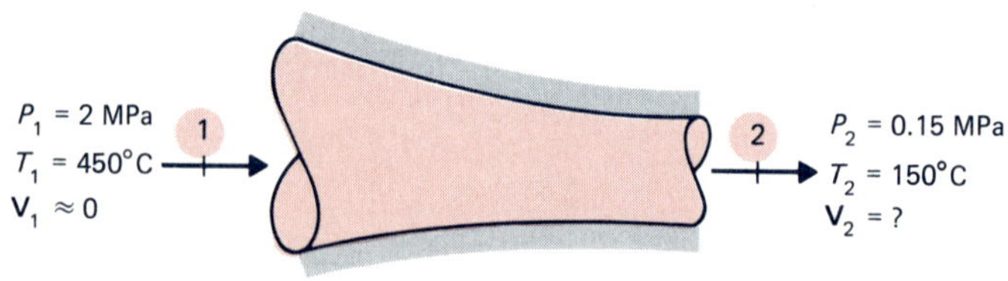

Figure P4.164

4.165 Determine the exit velocity and exit area of the carbon dioxide nozzle shown below.

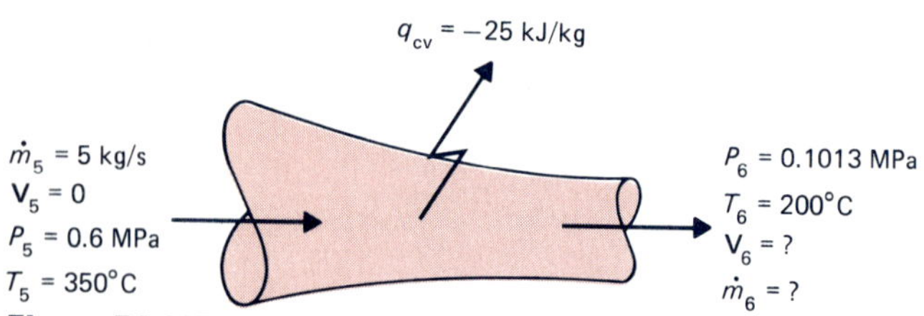

Figure P4.165

4.166 Air flows through a nonadiabatic nozzle as shown. Determine the exit velocity (m/s).

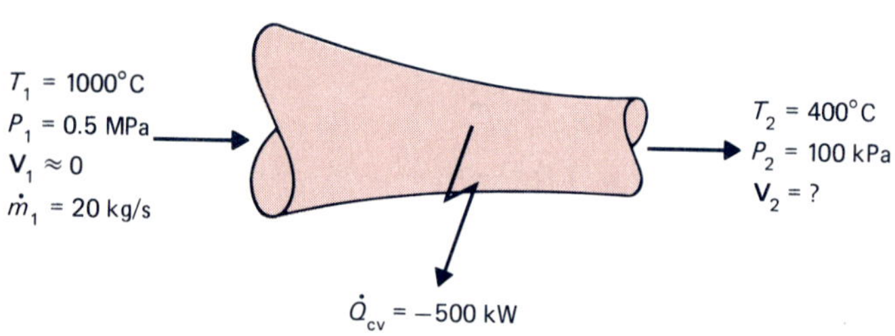

Figure P4.166

4.167 In an air conditioning system, saturated liquid refrigerant 12 at $T = 40°C$ flows through an expansion valve and reaches a final pressure of 300 kPa. What is the quality of the refrigerant 12 after the expansion?

4.168 Ammonia enters the piping system shown in Fig. P4.168 (all pipes are of circular cross section) at $T_{in} = 200°F$, $P_{in} = 100$ psia, and $\mathbf{V}_{in} = 350$ ft/s. The ammonia exits at a pressure of 20 psia and a velocity of $\mathbf{V}_{out} = 25$ ft/s. What is the exit temperature (°F)? (Assume the flow is subsonic at all points.) What is the diameter of the outlet?

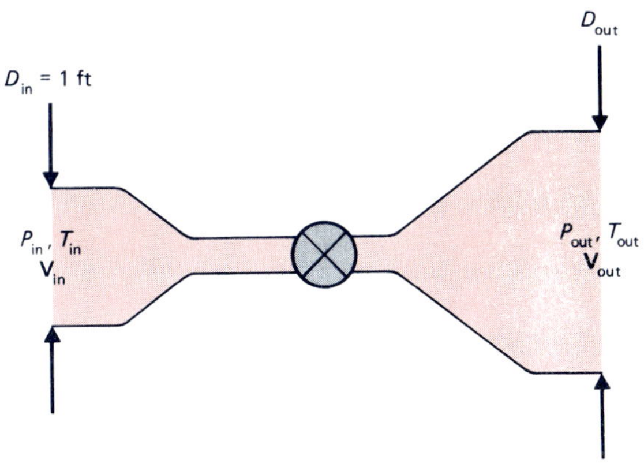

Figure P4.168

4.169S Ammonia flows at 15 kg/min through a pipe 0.5 cm in diameter and encounters a diffuser. Upstream of the diffuser the pressure P_A is 0.9 MPa, and the temperature T_A is 20°C. Downstream the pressure P_B is 0.3689 MPa, and the speed $\mathbf{V}_B$ is 100 m/s. See Fig. P4.169S.

(*a*) Find the downstream temperature or quality. *Note:* Only one of these will appropriately fix the downstream state.

(*b*) Find the downstream diffuser diameter.

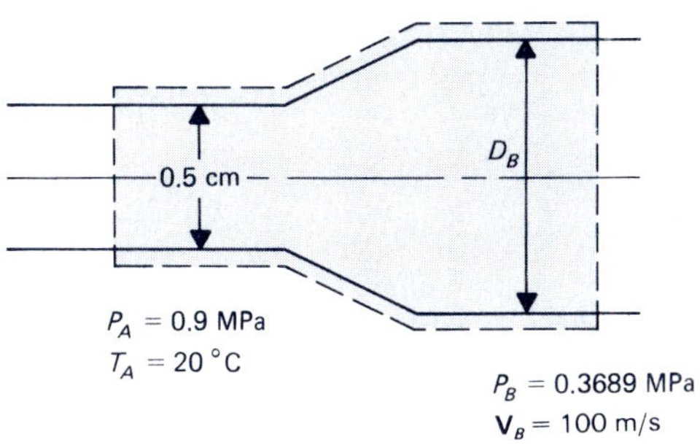

Figure P4.169S

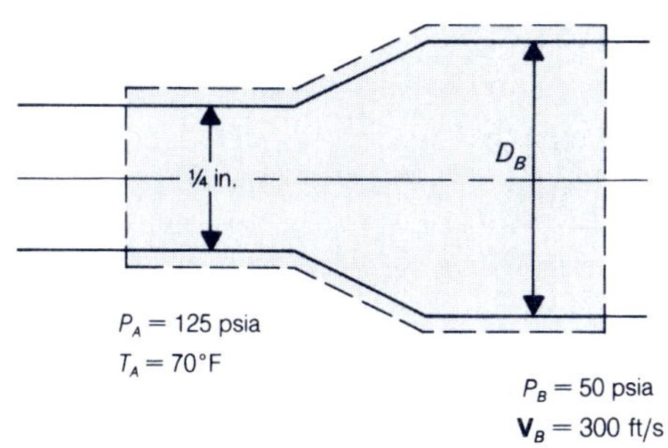

Figure P4.169E

4.169E Ammonia flows at 30 lbm/min through a pipe which is $\frac{1}{4}$ in in diameter and encounters a diffuser. Upstream of the diffuser, the pressure P_A is 125 psia, and the temperature T_A is 70°F. Downstream the pressure P_B is 50 psia, and the speed $\mathbf{V}_B$ is 300 ft/s. See Fig. P4.169E.

(*a*) Find the downstream temperature or quality. *Note:* Only one of these will appropriately fix the downstream state.

(*b*) Find the downstream diffuser diameter.

4.170S Steam at atmospheric pressure is used to heat air in the heat exchanger shown in Fig. P4.170. Steam enters as a saturated vapor and leaves as a saturated liquid. The air enters the heat exchanger at atmospheric pressure and 20°C and undergoes a constant-pressure heating. If the mass flow rate of air is 1 kg/s and the outlet temperature of the air is 80°C, what is the required water mass flow rate?

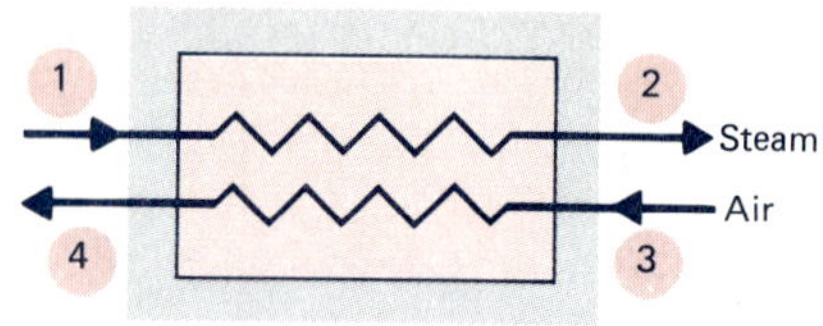

Figure P4.170

4.170E Steam at atmospheric pressure is used to heat air in the heat exchanger shown in Fig. P4.170. The steam enters as a saturated vapor and leaves as a saturated liquid. The air enters the heat exchanger at atmospheric pressure and 70°F and undergoes a constant-pressure heating. If the mass flow rate of air is 2 lbm/s and the outlet temperature of the air is 180°F, what is the required water mass flow rate?

4.171S A fluid flows through a valve. The inlet conditions are 1.0 MPa and 400°C, and the exit pressure is 0.3 MPa. See Fig. P4.171. Evaluate the exit temperature for (*a*) steam and (*b*) air as the fluid.

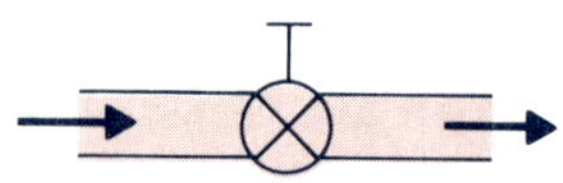

Figure P4.171

4.171E A fluid flows through a valve. The inlet conditions are 150 psia and 800°F, and the exit pressure is 50 psia. See Fig. P4.171. Evaluate the exit temperature for (*a*) steam and (*b*) air as the fluid.

4.172S An adiabatic mixing chamber is designed to obtain refrigerant 12 in the saturated vapor state at 0.1 MPa. See Fig. P4.172. Gaseous refrigerant 12 at 1.0 MPa and 150°C passes through a valve and into the chamber. Liquid refrigerant 12 enters the chamber at 0.1 MPa and −40°C. If the mass flow rate of refrigerant 12 gas is 0.080 kg/s, determine the mass flow rate of liquid.

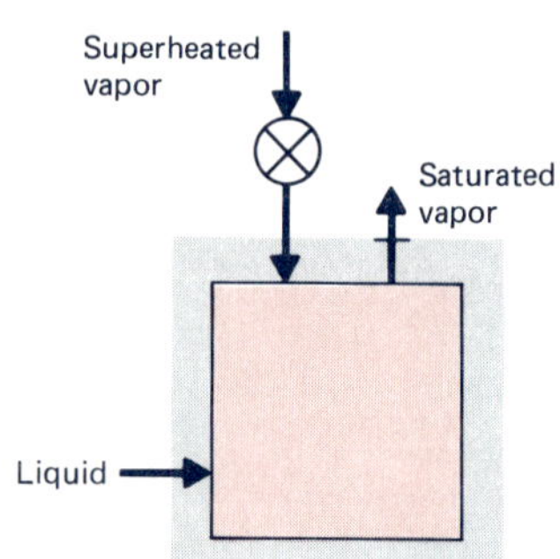

Figure P4.172

4.172E An adiabatic mixing chamber is designed to obtain refrigerant 12 in the saturated vapor state at 14.7 psia. See Fig. P4.172. Gaseous refrigerant 12 at 150 psia and 300°F passes through a valve and into the chamber. Liquid refrigerant 12 enters the chamber at 14.7 psia and −40°F. If the mass flow rate of refrigerant 12 gas is 0.18 lbm/s, determine the mass flow rate of liquid.

4.173S A heat exchanger transfers energy by means of heat transfer from a hotter to a colder fluid without allowing the fluids to physically mix. Such an exchanger is used to cool liquid ethyl alcohol (ethanol) (Table C.3) from 40 to 30°C. The alcohol flows at a rate of 10 kg/s. Cooling water is available at 20°C at a flow rate of $\dot{m}_{H_2O}$. See Fig. P4.173S.

(*a*) Consider the control volume(s) that you wish to specify in order to determine the outlet water temperature. Write the first law for all such control volumes.

(*b*) Prepare a plot of cooling water outlet temperature versus mass flow rate of the water $\dot{m}_{H_2O}$. Do not plot for any flow rates that predict that the water outlet temperature can exceed 40°C.

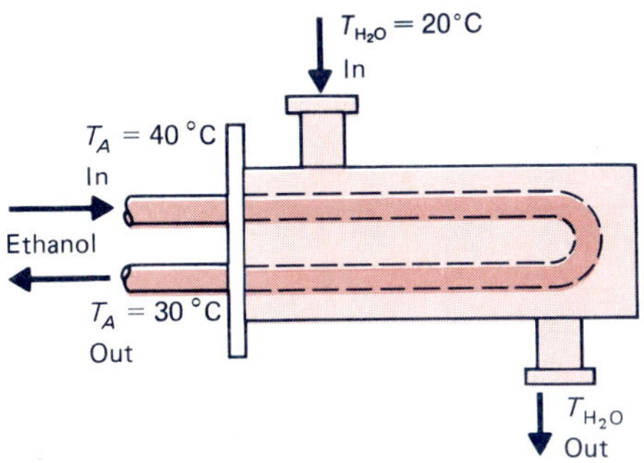

Figure P4.173S

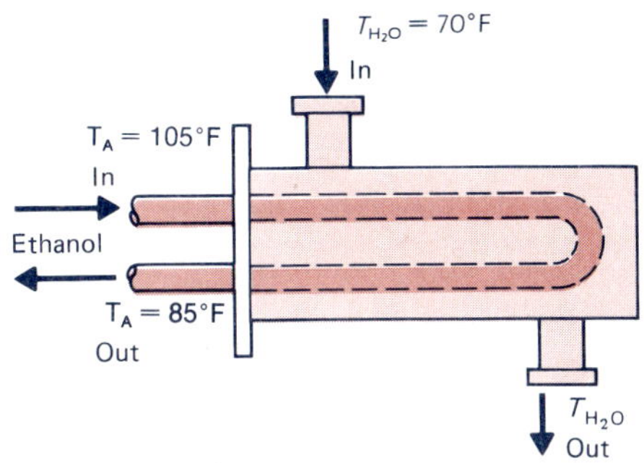

Figure P4.173E

4.173E A heat exchanger transfers energy by means of heat transfer from a hotter to a colder fluid without allowing the fluids to physically mix. Such an exchanger is used to cool liquid ethyl alcohol (ethanol) (Table C.3) from 105 to 85°F. The alcohol flows at a rate of 20 lbm/s. Cooling water is available at 70°F at a flow rate of $\dot{m}_{H_2O}$. See Fig. P4.173E.

(*a*) Consider the control volume(s) that you wish to specify in order to determine the outlet water temperature. Write the first law for all such control volumes.

(*b*) Prepare a plot of cooling water outlet temperature versus mass flow rate of the water, $\dot{m}_{H_2O}$. Do not plot for any flow rates that predict that the water outlet temperature can exceed 105°F.

4.174 A condenser is a counterflow heat exchanger. The pressure in the device is constant. The device shown is insulated. Determine the final water temperature T_4.

Figure P4.174

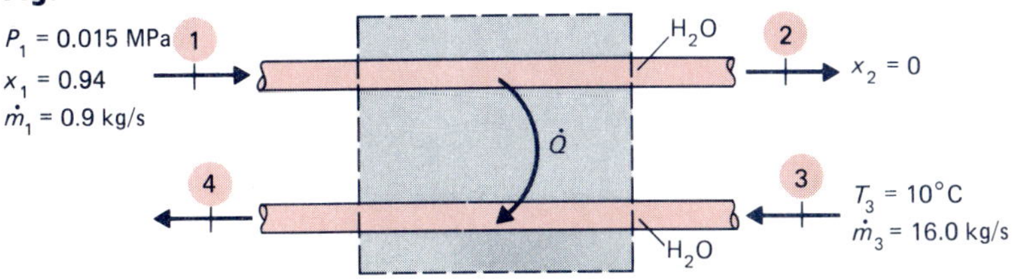

4.175 An adiabatic condenser operates as shown. If the process is steady-state steady flow, determine the mass flow rate of cooling water required (kg/s)

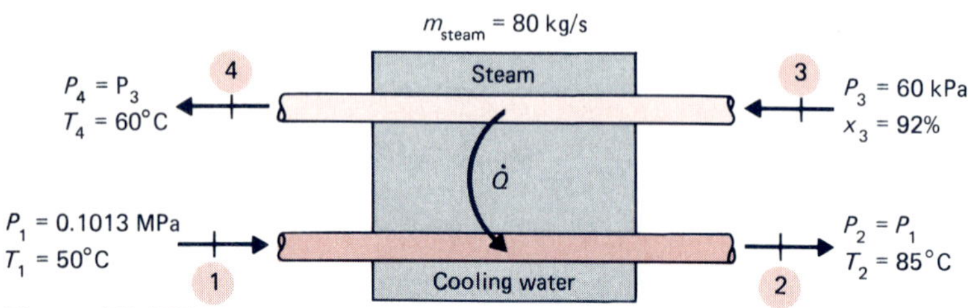

Figure P4.175

4.176 A regenerator is often a counterflow heat exchanger. The device operates isobarically and is insulated. Determine the temperature T_4 of the exiting air.

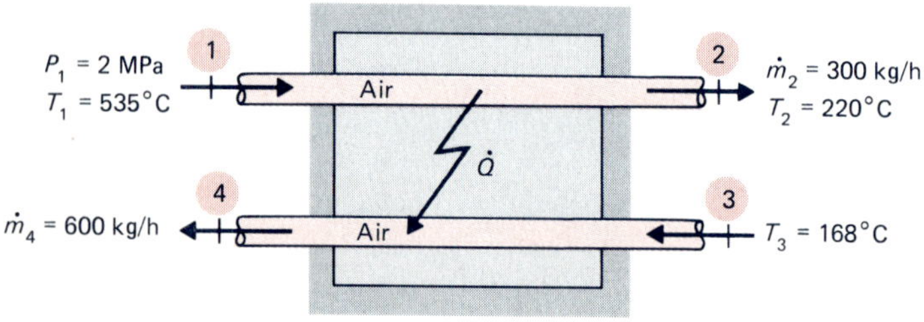

Figure P4.176

4.177 An adiabatic gas-turbine regenerator operates as shown. Determine the mass flow rate of the cold air (kg/s). Both fluids are air.

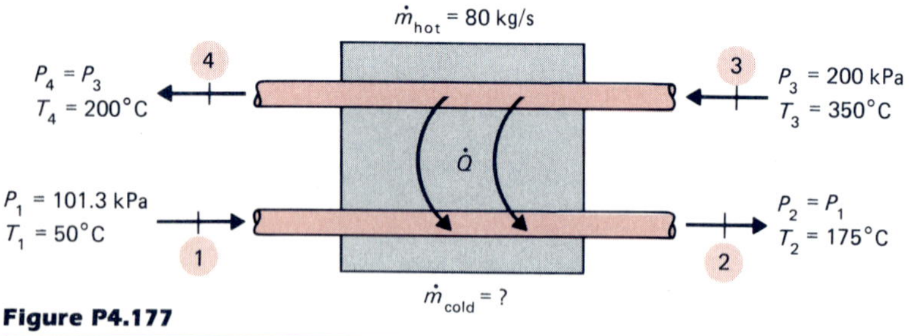

Figure P4.177

4.178 Determine the exit temperature of the helium of the helium–CO_2 heat exchanger shown in Fig. P4.178. Assume the device is adiabatic.

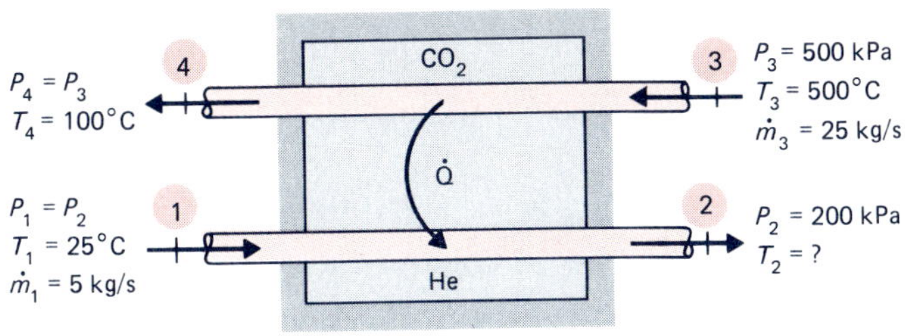

Figure P4.178

4.179S A heat exchanger is a device for transferring energy from a hotter fluid to a colder fluid. The conditions present in one such device are shown in Fig. P4.179S. The hot stream is air; the cold stream is water. How many kilograms per second of cooling water are required to cool 2 kg/s of air from 150 to 100°C if the temperature change of the water is 10°C? Assume constant specific heats.

4.179E A heat exchanger is a device for transferring energy from a hotter fluid to a colder fluid. The conditions present in one such device are shown in Fig. P4.179E. The hot stream is air; the cold stream is water. How many pounds-mass per hour of cooling water are required to cool 2 lbm/h of air from 300 to 212°F if the temperature change of the water is 50°F? Assume constant specific heats.

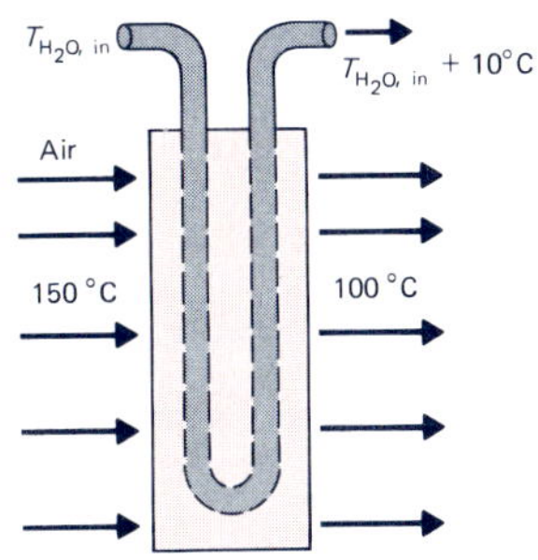

Figure P4.179S

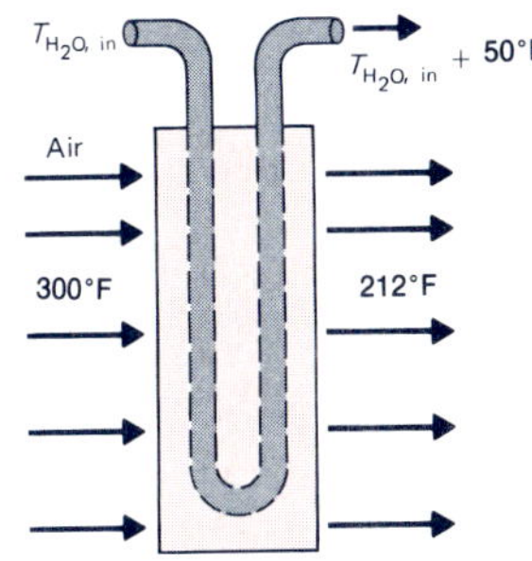

Figure P4.179E

4.180S Nitrogen flows through a pipe in which a porous plug has been inserted. The pressure drop across the plug is large. See Fig. P4.180. If the pressure upstream of the plug is 5 atm, derive an expression for the change in temperature of the nitrogen across the plug. Note any assumptions that you make.

4.180E Nitrogen flows through the pipe in which a porous plug has been inserted. The pressure drop across the plug is large. See Fig. P4.180. If the pressure upstream of the plug is 5 atm, derive an expression for the change in temperature of the nitrogen across the plug. Note any assumptions that you make.

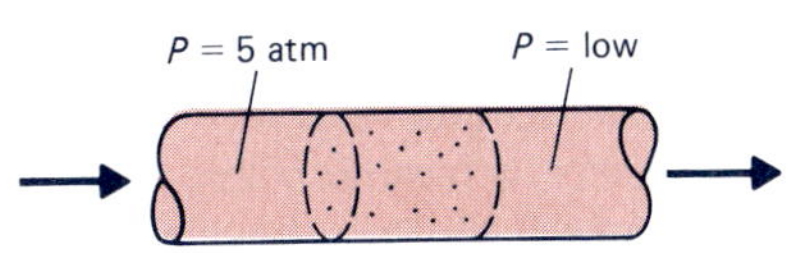

Figure P4.180

4.181S Steam is continuously removed through an orifice from a steam line that contains saturated steam at 1 MPa. The bleed steam enters a chamber kept at constant pressure by an exhaust line. The temperature and pressure of the bleed steam in the exhaust chamber are measured as 200 kPa and 150°C. See Fig. P4.181.

(*a*) What is the quality of the steam in the steam line? (The device described is called a *throttling calo-*

rimeter and is used for the purpose outlined, i.e., finding the quality of steam in a supply line.)

(*b*) Can you think of possible limitations on the range of quality in the line that can be measured?

4.181E Steam is continuously removed through an orifice from a steam line that contains saturated steam at 150 psia. The bleed steam enters a chamber kept at constant pressure by an exhaust line. The temperature and pressure of the bleed steam in the exhaust chamber are measured as 25 psia and 300°F. See Fig. P4.181.

(*a*) What is the quality of the steam in the steam line? (The device described above is called a *throttling calorimeter,* used for the purpose outlined, i.e. finding the quality of steam in a supply line.)

(*b*) Can you think of possible limitations on the range of quality in the line that can be measured?

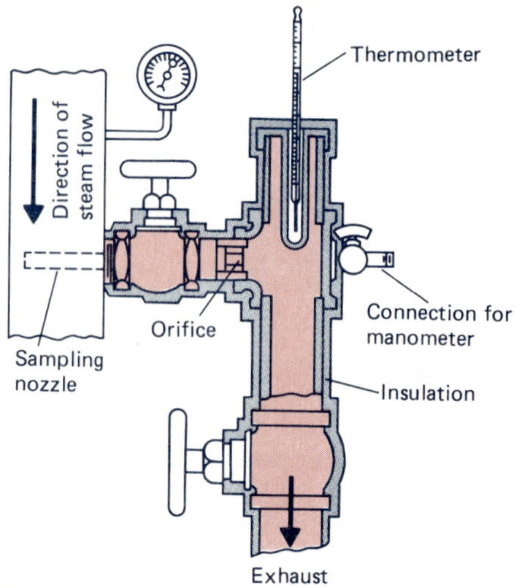

Figure P4.181

4.182S Steam at $T_A = 400°C$ and 1 MPa enters a subsonic nozzle at a speed of $\mathbf{V}_A = 30$ m/s and leaves the nozzle at a speed of $\mathbf{V}_B = 100$ m/s. After leaving the nozzle at state *B*, the steam enters a turbine, where it is expanded to state *C*. At state *C*, the steam has a pressure of 8 kPa, a quality of $x_C = 0.98$, and a speed $\mathbf{V}_C = 10$ m/s. If the mass flow rate entering the nozzle is 2 kg/s, what is the power output of the turbine?

4.182E Steam at $T_A = 800°F$ and 160 psia enters a subsonic nozzle at a speed of $\mathbf{V}_A = 100$ ft/s and leaves the nozzle at a speed of $\mathbf{V}_B = 300$ ft/s. After leaving the nozzle at state *B*, the steam enters a turbine, where it is expanded to state *C*. At state *C*, the steam has a pressure of 1 psia, a quality of $x_C = 0.98$, and a speed $\mathbf{V}_C = 30$ ft/s. If the mass flow rate entering the nozzle is 4.5 lbm/s, what is the power output of the turbine?

4.183S A continuous copper sheet passes through a furnace at a velocity of 5 m/s. The sheet is 2 m × 0.05 m in cross section. It is heated from a uniform temperature at the inlet of 27°C to a uniform temperature of 977°C at the furnace outlet. The furnace is heated by electric resistance heaters, and heat transfer from the furnace to the surroundings is negligible. How large a power supply must be provided to feed the heaters?

4.183E A continuous copper sheet passes through a furnace at a velocity of 15 ft/s. The sheet is 6 ft × 2 in in cross section. It is heated from a uniform temperature at the inlet of 80°F to a uniform temperature of 2060°F at the furnace outlet. The furnace is heated by electric resistance heaters, and heat transfer from the furnace to the surroundings is negligible. How large a power supply must be provided to feed the heaters?

4.184S Steam is used in a lift as shown in Fig. P4.184. The steam source is at 1.0 MPa and 300°C. The initial state is 0.5 MPa, saturated vapor, and 0.002 m^3. The steam slowly fills the volume until the volume is 0.040 m^3, and the mass added is 0.15 kg. Determine the final state and the heat transfer.

4.184E Steam is used in a lift as shown in Fig. P4.184. The steam source is at 150 psia and 600°F. The initial state is 75 psia, saturated vapor, and 0.06 ft^3. The steam slowly expands the volume until the volume is 1.2 ft^3 and the mass added is 0.30 lbm. Determine the final state and the heat transfer.

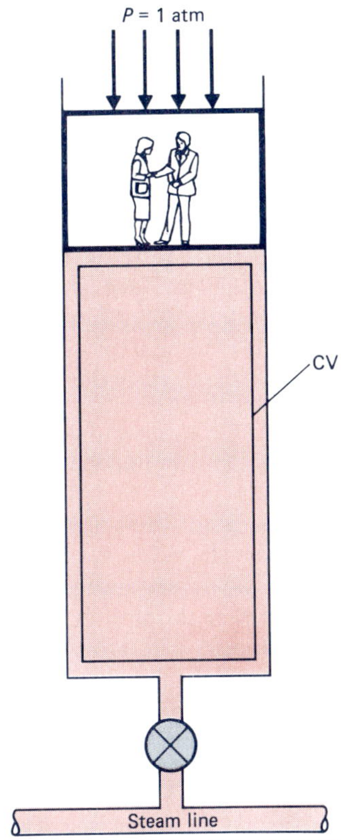

Figure P4.184

4.185S A balloon is to be filled from a helium tank. The tank is large, and the tank pressure remains effectively constant while the balloon is filled. A valve between the tank and the balloon controls the helium flow to a constant rate of 0.05 kg/s. The pressure in the balloon varies during the filling process. The temperature of the helium entering the balloon is 20°C. See Fig. P4.185.

(*a*) How much work is done in filling the balloon to a final pressure of 500 kPa, where the final radius is 2 m if the balloon is adiabatic? (It takes 8 min to fill the balloon.)

(*b*) What is the final temperature in the balloon?

4.185E A balloon is to be filled from a helium tank. The tank is large, and the tank pressure remains effectively constant while the balloon is filled. A valve between the tank and the balloon controls the helium flow to a constant rate of 0.1 lbm/s. The pressure in the balloon varies during the filling process. The temperature of the helium entering the balloon is 70°F. See Fig. P4.185.

(*a*) How much work is done in filling the balloon to a final pressure of 60 psia where the final radius is 5 ft if the balloon is adiabatic? (It takes 8 min to fill the balloon.)

(*b*) What is the final temperature in the balloon?

Figure P4.185

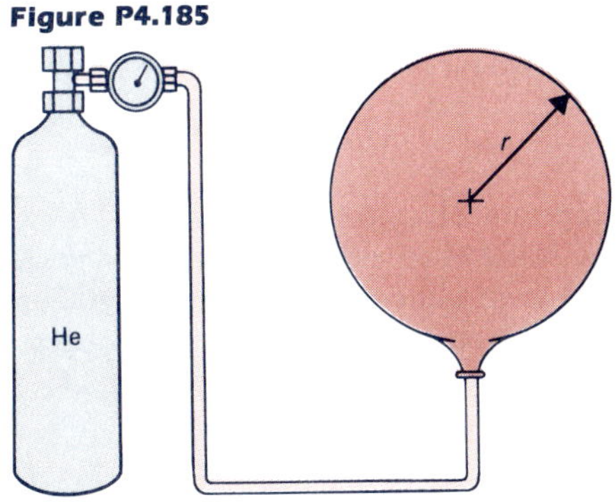

4.186S A steam line with very large capacity contains saturated vapor at $P = 1$ MPa. The line is connected through a valve to an evacuated and insulated tank with a volume of 5 m^3. At $t = 0$, the valve is cracked open, and steam flows into the tank until the pressure in the tank equalizes with that in the steam line. The valve is immediately shut. At the moment the valve is shut, what is the temperature of the steam in the tank?

4.186E A steam line with very large capacity contains saturated vapor at $P = 150$ psia. The line is connected through a valve to an evacuated and insulated tank with a volume of 175 ft^3. At $t = 0$, the valve is cracked open, and steam flows into the tank until the pressure in the tank equalizes with that in the steam line. The valve is immediately shut. At the moment the valve is shut, what is the temperature of the steam in the tank?

4.187S Air at 1 atm and 20°C initially fills a bottle of 0.1-m^3 volume. The bottle is attached to an air line that provides air at 20°C and 50 atm, and the bottle is "charged" to a pressure of 50 atm. There is a heat transfer from the bottle, so the air within is held at 20°C throughout the process. Identify an appropriate control volume. What is the heat transfer for your control volume during this process?

4.187E Air at 1 atm and 60°F initially fills a bottle of 3-ft^3 volume. The bottle is attached to an air line that provides air at 60°F and 50 atm and the bottle is "charged" to a pressure of 50 atm. There is a heat transfer from the bottle so that the air within is held at 60°F throughout the process. Identify an appropriate control volume. What is the heat transfer for your control volume during this process?

4.188S Find the temperature increase ΔT that results during the adiabatic filling of an evacuated tank. The gases enter the tank from a line at 1.0 MPa and 20°C. The process stops when the pressure in the tank is equal to the line pressure. Consider (*a*) air, (*b*) ethane, and (*c*) argon.

4.188E Find the temperature increase ΔT that results during the adiabatic filling of an evacuated tank. The gases enter the tank from a line at 150 psia and 60°F. The process stops when the pressure in the tank is equal to the line pressure. Consider (*a*) air, (*b*) ethane, and (*c*) argon.

4.189S A tank initially contains H_2O at 0.05 MPa and 100°C and is filled from a steam line at 0.60 MPa and 200°C. The tank is filled to a level of 90 percent liquid on a volume basis. The 1.0-m^3 tank is maintained at 100°C during the process. Evaluate the required heat transfer to keep the tank isothermal.

4.189E A tank initially contains H_2O at 5 psia and 200°F and is filled from a steam line at 100 psia and 400°F. The tank is filled to a level of 90 percent liquid on a volume basis. The 30-ft³ tank is maintained at 200°F during the process. Evaluate the required heat transfer to keep the tank isothermal.

4.190S A steam pipeline carries saturated vapor at pressure P. An evacuated and insulated 3-m³ tank is connected to the pipeline through a valve. At $t = 0$ the valve is opened, and steam flows into the tank. When the tank pressure just reaches the pipeline pressure P, the temperature of the steam in the tank is measured to be 250°C. What is the pressure in the pipeline? (Give your answer within a 100-kPa span, for example, $1000 < P < 1100$ kPa.)

4.190E A steam pipeline carries saturated vapor at pressure P. An evacuated and insulated tank of volume 10 ft³ is connected to the pipeline through a valve. At $t = 0$, the valve is opened and steam flows into the tank. When the tank pressure just reaches the pipeline pressure P, the temperature of the steam in the tank is measured to be 500°F. What is the pressure in the pipeline? (Give your answer as within a 25-psia span, for example, $150 < P < 175$ psia.)

4.191 A nitrogen supply line provides nitrogen at any required rate at a temperature of 300°C and a pressure of 10 atm. At a given time t_1, a valve is opened from the nitrogen line, and nitrogen flows through a gas turbine into an initially evacuated tank. When the

Figure P4.191

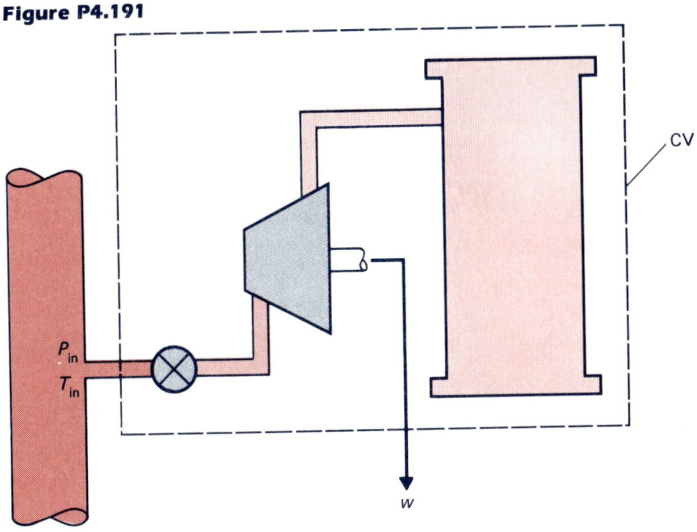

pressure in the tank just reaches the supply line pressure, the temperature of the nitrogen in the tank is measured to be 250°C. The valve is then closed. The work output of the gas turbine during this process is measured to be 45.0 kJ.

What is the heat transfer per kilogram of nitrogen entering the control volume during this process (kJ/kg) for the control volume shown?

4.192S A power cycle uses refrigerant 12 as the working fluid. It operates between pressures of 60 and 600 kPa, and the states at the process endpoints are shown on the *P-h* diagram. The quality at state 3 is $x_3 = 0.98$. See Fig. P4.192S.

Construct a state table for this heat engine, and find the net work output divided by the heat transfer to the working fluid, that is, $|w_{net}|/{}_1q_2$. (Use Tables D.13 to D.15 or computerized tables for all properties.)

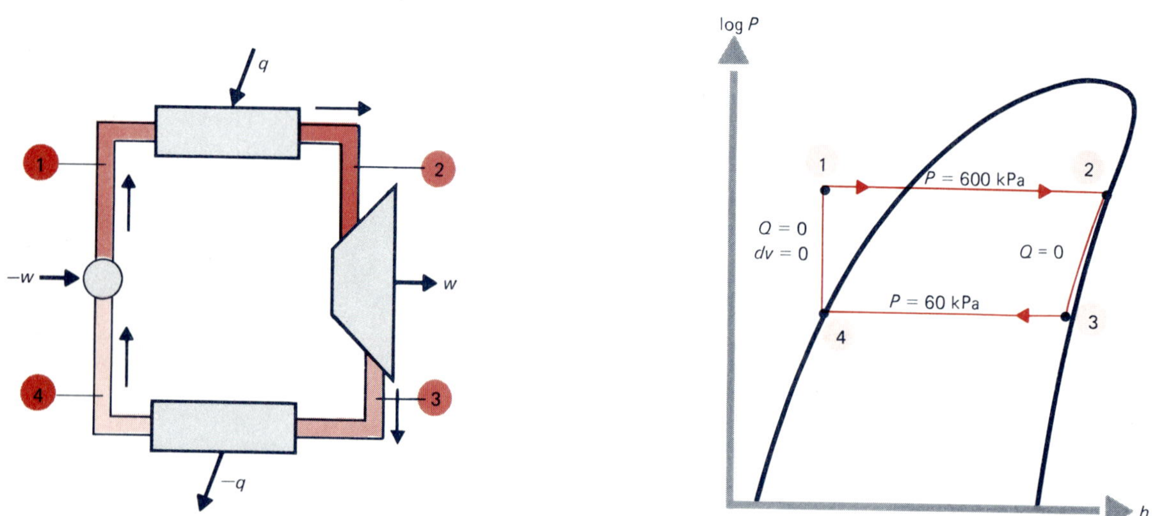

Figure P4.192S

4.192E A power cycle uses refrigerant 12 as the working fluid. It operates between pressures of 10 and 90 psia, and the states at the process endpoints are shown on the *P-h* diagram. The quality at state 3 is $x_3 = 0.98$. See Fig. P4.192E.

Construct a state table for this heat engine, and find the net work output divided by the heat transfer to the working fluid, that is, $|w_{net}|/{}_1q_2$. (Use Tables E.13 to E.15 or computerized tables for all properties.)

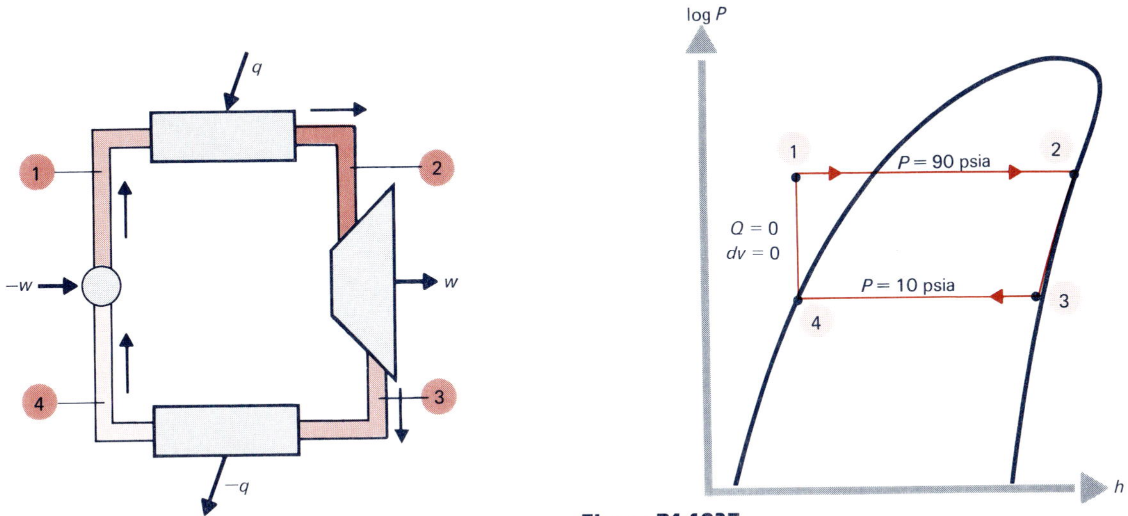

Figure P4.192E

4.193S A refrigeration cycle operates between the same two pressures as the heat engine of Prob. 4.192. The *P-h* diagram is shown in Fig. P4.193S. The quality $x_3 = 0.8$. Construct a state table for this refrigeration cycle.

(*a*) Find the value of ${}_4q_3/{}_3w_2$.

(*b*) Find the rate at which refrigerant 12 must circulate in kilograms per second to provide 3 tons of cooling capacity. (*Note:* 1 ton = 12,000 Btu/h. See App. B for conversion factors.)

Figure P4.193S

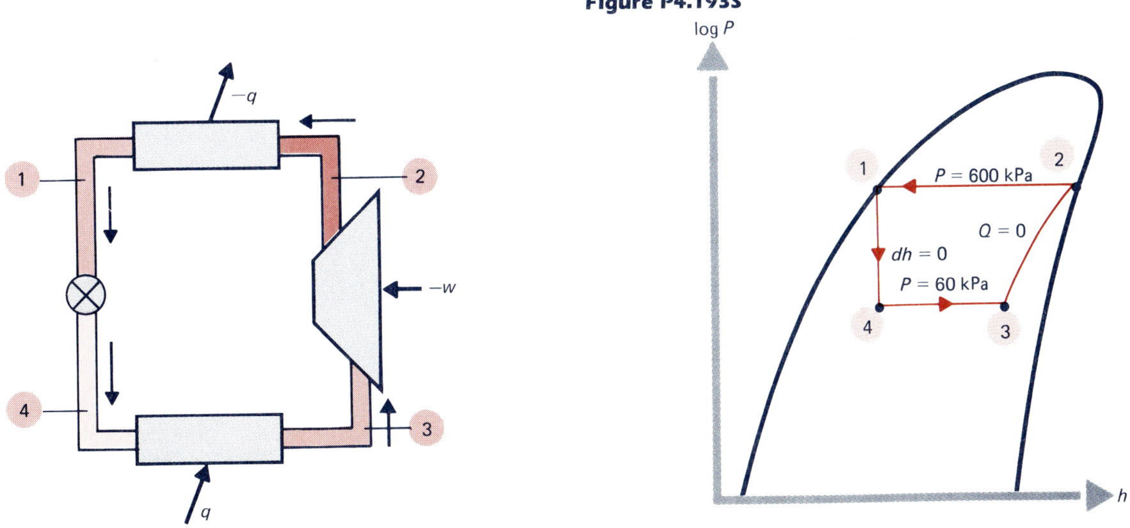

4.193E A refrigeration cycle operates between the same two pressures as the heat engine of Prob. 4.192. The *P-h* diagram is shown in Fig. P4.193E. The quality $x_3 = 0.8$. Construct a state table for this refrigeration cycle.

(*a*) Find the value of ${}_4q_3/{}_3w_2$.

(*b*) Find the rate at which refrigerant 12 must circulate in pounds-mass per second to provide 3 tons of cooling capacity. (1 ton = 12,000 Btu/h. See App. B for conversion factors.)

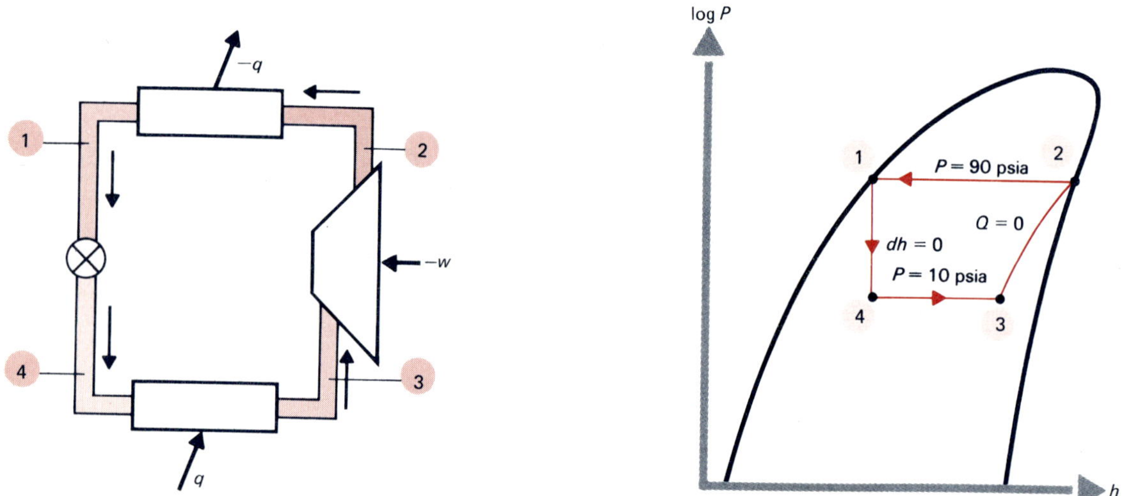

Figure P4.193E

4.194 A series of processes is carried out as follows: Saturated liquid water at 300°F (state *A*) is expanded at constant pressure to a quality of 90 percent (state *B*). The resulting saturated mixture is then cooled at constant volume until the final pressure is 14.70 psia (state *C*). The mixture at state *C* is then cooled at constant pressure until the specific volume is the same as at the initial state *A* (state *D*). Finally, the mixture at state *D* is heated at constant volume until the temperature of the mixture reaches 300°F.

(*a*) Show this cycle on the *P-v* diagram.

(*b*) Calculate the net work per pound-mass of steam done in this cycle (lbf·ft/lbm).

4.195 Air is flowing in the circular duct shown. At point 1 the pressure is 150 kPa, the temperature is 25°C, and the velocity is 25 m/s. Assuming incompressible flow and ideal gas, find the pressure, temperature, and velocity at point 2.

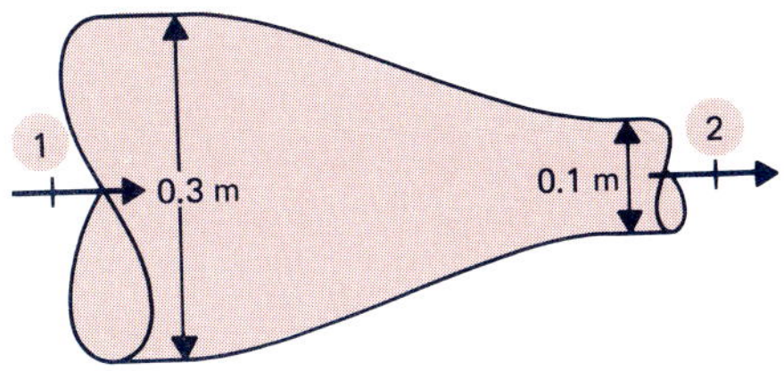

Figure P4.195

4.196 A pitot tube is used to measure the velocity of water flowing in a pipeline. The deflection of a mercury-water manometer is as shown. What is the velocity in the pipeline? (Neglect the height difference between the static and the stagnation port.)

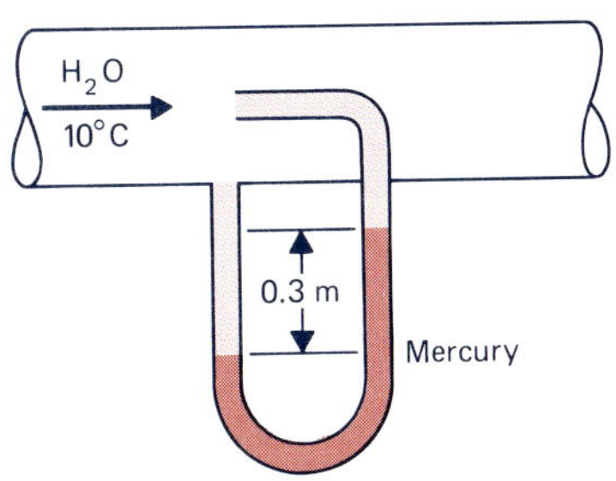

Figure P4.196

4.197 Assuming a constant level for the reservoir, calculate the flow rate in the pipe and the pressure at *A* and *B*.

Figure P4.197

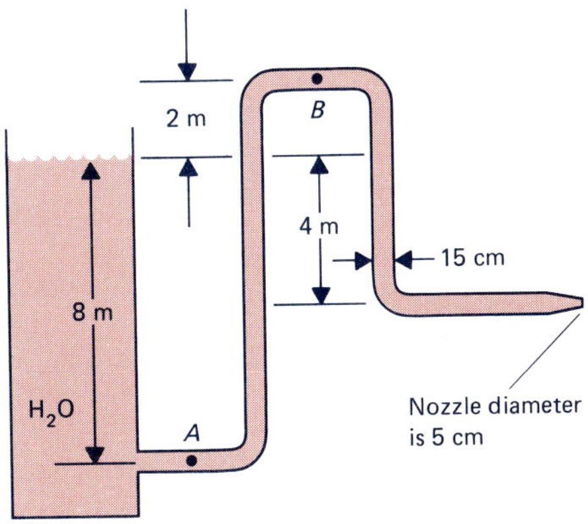

4.198 This device is designed to measure the flow rate by recording a pressure difference. Calculate the flow rate (m^3/s) if ethyl alcohol is flowing. (Neglect the height difference between 1 and 2.)

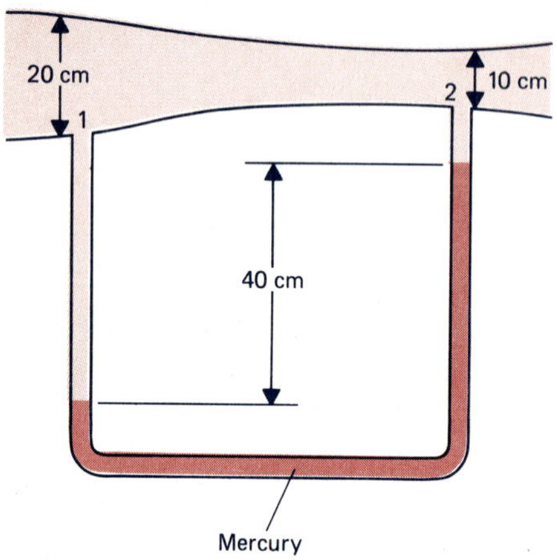

Figure P4.198

4.199 Calculate the flow rate (m^3/s) of H_2O through the pipe shown.

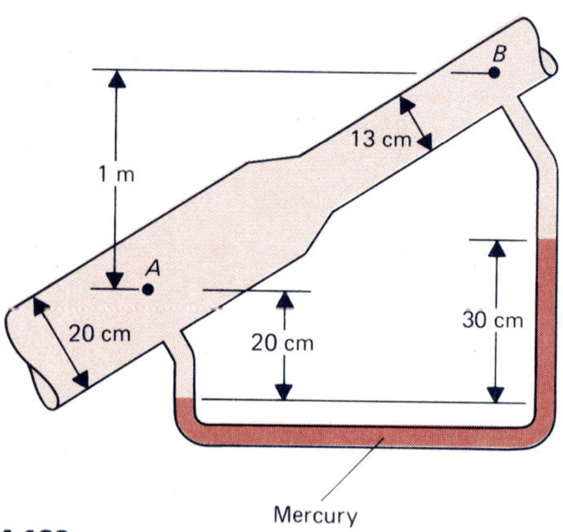

Figure P4.199

5

Entropy and the Second Law of Thermodynamics

While the earth remaineth, seedtime and harvest, and cold and heat, and summer and winter, and day and night shall not cease.
Gen. 8:22.

Cutaway of a hot-water to air heat exchanger for a commercial-scale heating system. (*the Trane Company.*)

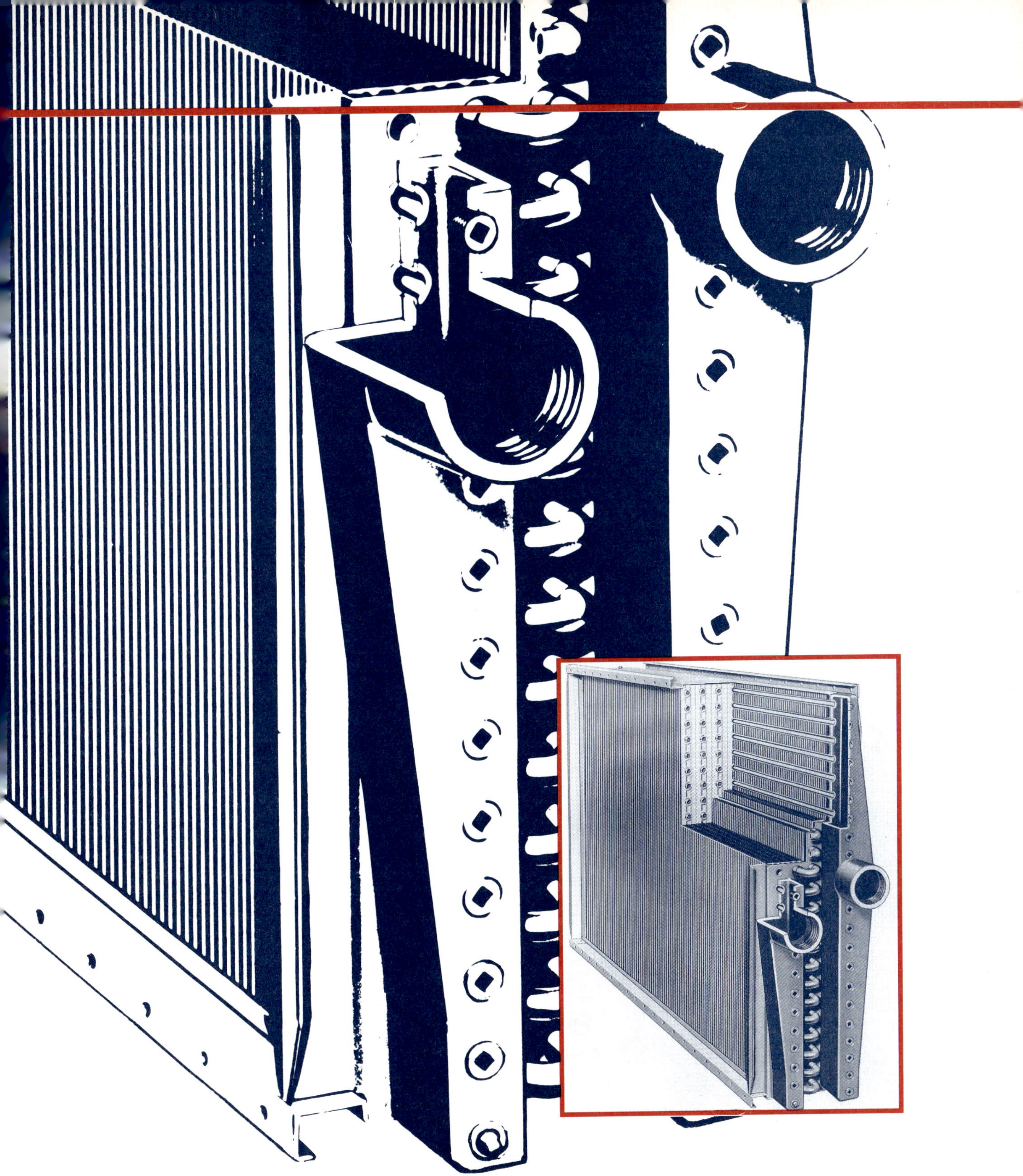

5.1 Introduction

The first law of thermodynamics, like the conservation of mass principle, describes the conservation of a particular property. The first law expresses the manner in which a system's energy is altered by energy transfer across the system boundary and mass transport into and out of the system. The first law is a strict accounting procedure that describes the changes in the system's energy. The conservation of mass quantifies the changes in the system's mass. Both conservation statements directly relate the change in a system's property to transfers at a boundary. The second law of thermodynamics also relates a system property to energy transfer at a boundary, but the relation simply specifies the direction of change. Like all other physical laws used in classical thermodynamics, the second law cannot be proved but is a statement of observed phenomena.

5.1.1 Physical Observations

Two examples are presented to demonstrate the inability of the first law of thermodynamics and the conservation of mass statement to completely describe systems. Both examples consider isolated systems, so the heat transfers and work across the system boundary are zero. Figure 5.1*a* presents an isolated chamber with steam on one side of a partition and a vacuum on the other side. This is the initial state of the system and is indicated as state 1. The partition is suddenly removed, and the system reaches equilibrium in the final state, where the steam fills the entire chamber. This final state is denoted as state 2. The conservation of mass and the first law are presented in the figure, indicating that the transformation is a constant-mass and constant-internal-energy process. Clearly these conservation statements are satisfied, and the process is physically reasonable. In Fig. 5.1*b*, the steam completely fills the chamber in its initial state, state 1. The system transforms itself to two subsystems with the steam residing in one-half of the chamber and a perfect vacuum in the other half. A partition is inserted, and this final state is state 2. This case is also described as a constant-mass and constant-internal-energy process, as indicated in Fig. 5.1. Again the principles of conservation of mass and energy are satisfied; yet, from our experience, the process in Fig. 5.1*b* is impossible. Regardless of the time allowed for this process, it will never occur in an isolated system. The inability of the first law and the conservation of mass to explain this impossibility indicates the need for another fundamental law.

Figure 5.2 shows an isolated system containing two blocks of material, labeled A and B. In Fig. 5.2*a* block A is at a high temperature T_H, and

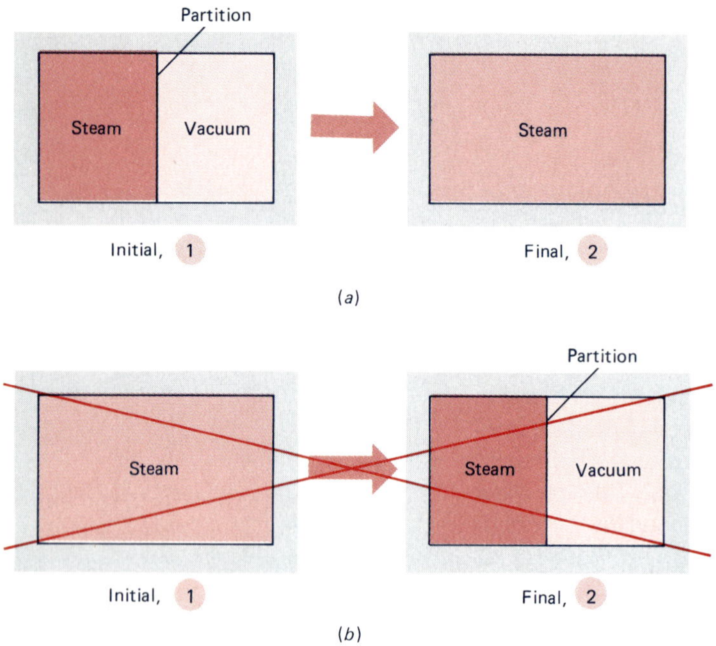

Mass: $m_1 = m_2$

Energy: $U_2 - U_1 = {}_1Q_2 - {}_1W_2 = 0$

$U_2 = U_1$

Figure 5.1 Expansion into a vacuum for an isolated system.

block B is at a low temperature T_L. They are initially brought together and isolated from all surroundings, as indicated by state 1. There is a subsequent heat transfer between block A and block B until they reach a common intermediate equilibrium temperature T. This final state is denoted as state 2. The statements of conservation of mass and energy are presented in Fig. 5.2, indicating that this is a constant-mass and constant-internal-energy process. The temperature of the hotter block is expected to decrease, and the temperature of the colder block is expected to increase to the equilibrium temperature. The process satisfies the conservation statements as well as our physical expectations. In Fig. 5.2*b*, the two blocks are initially brought together at the same temperature, and the final state results when block A is at the high temperature T_H and block B is at the low temperature T_L. Again the conservation statements given in Fig. 5.2 are satisfied, yet our experience leads us to expect that this is an impossible process.

These are only a few examples that demonstrate that there is a fundamental principle, the second law of thermodynamics, which is not described by the principles of conservation of mass and energy. These examples present the two key features of the second law of thermodynamics. First, the second law prescribes the *direction* for the processes; second, some thermodynamic characteristic of the system, a system *prop-*

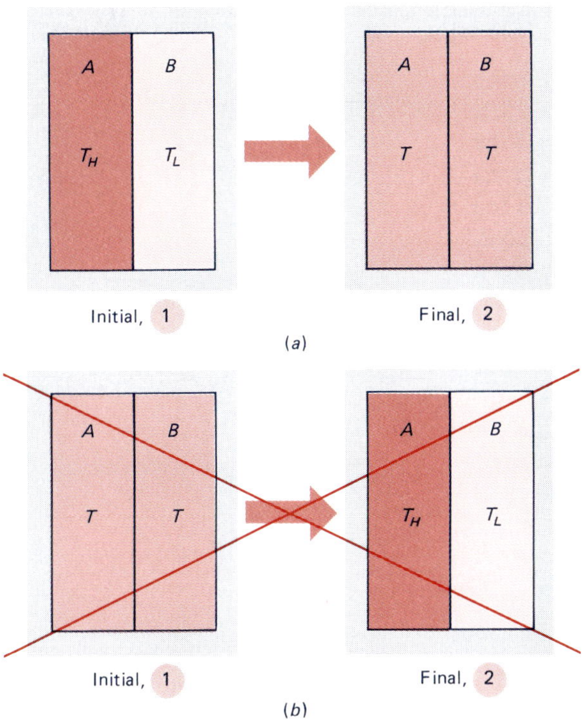

Mass: $$m_{A1} + m_{B1} = m_{A2} + m_{B2}$$
$$m_1 = m_2$$

Energy: $$(U_{A2} + U_{B2}) - (U_{A1} + U_{B1}) = {}_1Q_2 - {}_1W_2 = 0$$
$$U_{A2} + U_{B2} = U_{A1} + U_{B1}$$
$$U_2 = U_1$$

Figure 5.2 Heat transfer between two blocks within an isolated system.

erty, exists and is always changing in the specified direction. These features are discussed below.

The example in Fig. 5.1 indicates that the system (Fig. 5.1*a*) tends to proceed to a more random state and that we expect the reverse process (Fig. 5.1*b*), which proceeds to a more organized state (i.e., a state with more individual control masses, each with different intensive properties), to be impossible. Thus, there is a specific direction to the process. The direction appears to be related to the disorder, randomness, or uncertainty of the microscopic scale of the system. There is an inability of the system to self-organize. The example in Fig. 5.2 also demonstrates that the organized, or structured, state of a separate high-temperature material and low-temperature material in state 1 tends toward a more disorganized, or less structured, state of a single homogeneous temperature in state 2. Again the reverse process is not possible. Thus, the second law needs to describe the direction for processes, and the direction is described in terms of a system property that characterizes the system's randomness, disorganization, or uncertainty.

Observe that the randomness, disorganization, or uncertainty in the two systems of Figs. 5.1 and 5.2 was changed in different ways. In Fig. 5.1, the state of the steam was changed by rearranging the positions of the molecules of the steam. Because more positions are available for a given molecule after the expansion than before it, our uncertainty as to the position of a given molecule is greater following the expansion. In Fig. 5.2, the state of the system was changed by allowing heat transfer between subsystems, with no spatial rearrangement of the system. By statistical mechanics, the microscopic disorder of a material depends on the temperature. The heat transfer process between the subsystems increases our uncertainty about the molecular disorder in the two subsystems compared with the disorder before the process. (Proving this is outside the scope of classical thermodynamics, but this discussion is meant to show where the idea of uncertainty comes from in these processes. These ideas are discussed quantitatively and in greater detail in Chap. 13.)

Thus, the disorder of the system can be changed in at least two ways: geometric rearrangement and heat transfer.

5.1.2 Increasing Disorder by Heat Transfer

The above examples focus on isolated systems, and this is the starting point for the second law presentation; yet, energy transfers are fundamental to thermodynamics and require special attention. The example in Fig. 5.2 indicates that Fig. 5.2*a* is the possible process and that the isolated system tends to a less structured, or more random, state. If the system is redefined to be either block A or block B, then a heat transfer across the system boundary exists between the blocks. This is diagramed in Fig. 5.3. For control mass A, the process starts from a high-temperature initial state and ends at a lower-temperature final state as a result of heat transfer from

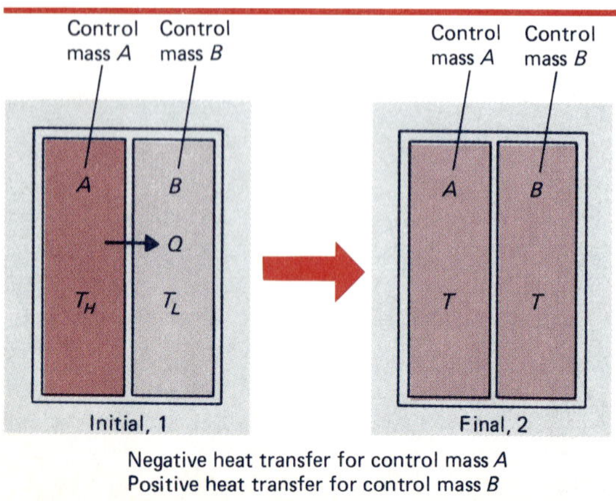

Figure 5.3 Heat transfer between control mass A and control mass B.

this control mass to block *B*. Physically, block *A* proceeds from a high-temperature, or less organized, state (from a molecular viewpoint) to a lower-temperature, or more organized, state (on the molecular scale). This process, which tends to organize or reduce the random molecular nature of block *A*, is a result of heat loss from this control mass, or a negative energy transfer. In Sec. 1.3, heat transfer is associated with a disorganized energy transfer, so a heat loss from the control mass is consistent with a decrease in the randomness of block *A*. However, control mass *B* receives the heat transfer from block *A*, and this is interpreted as positive disorganized energy transfer. Thus control mass *B* increases from an initial low-temperature organized state to a more random high-temperature state as a result of heat transfer. Remember that the possible process is one where the complete isolated system tends to a more random (or uncertain) state, but the subsystems can individually experience an increase or a decrease in disorder.

This last discussion links heat transfer or disorganized energy transfer to the system property that characterizes the randomness, uncertainty, or disorder of the state. This connection between heat transfer and an uncertainty property which shows the directionality of a process is the second law of thermodynamics. The property that describes the randomness or uncertainty is called *entropy*. The direction of the process or change in state as specified by the change in entropy is tied to the direction of the heat transfer. The subsequent sections of this chapter present the macroscopic viewpoint of the second law of thermodynamics. The microscopic calculation and interpretation is presented in Chap. 13.

5.2 Entropy and the Second Law for an Isolated System

The microscopic disorder of a system is described by a system property called *entropy*. The previous examples indicate that an isolated system, after removal of internal constraints, undergoes processes that are not quantified by the first law of thermodynamics. These facts are combined into the statement of the second law of thermodynamics:

> The entropy *S*, an extensive equilibrium property, must always increase or remain constant for an isolated system.

This is expressed mathematically as

$$dS_{\text{isolated}} \geq 0$$

or (5.1)

$$(S_{\text{final}} - S_{\text{initial}})_{\text{isolated}} \geq 0$$

Entropy, like our other thermodynamic properties, is defined only at equilibrium states or for quasi-equilibrium processes. Equation (5.1) shows that the entropy of the final state is never less than that of the initial state for any process which an isolated system undergoes. In the examples presented in Sec. 5.1, the final state of the isolated system results after the removal of an internal constraint. The final state has the largest value of entropy for the complete isolated system. Thus, an isolated system composed of subsystems tends toward a state which increases the total entropy of the composite or isolated system. This is the maximum possible entropy subject to any constraints that are imposed.

Expressions for changes in the property S as a function of other system properties must be developed before we can apply the second law. However, some characteristics of entropy can be examined first. Entropy is an extensive property of the system, so any system composed of subsystems yields

$$S = m_A s_A + m_B s_B + \cdots = \sum_i m_i s_i \tag{5.2}$$

where s is the specific entropy or entropy per unit mass and the subscript denotes the specific subsystem. The second law, expressed as an integral over all the elements of an isolated system, is

$$dS_{\text{isolated}} = d\left(\int_m s\, dm\right)_{\text{isolated}} = d\left(\int_V \rho s\, dV\right)_{\text{isolated}} \geq 0 \tag{5.3}$$

This property is specified by the state of the system, or it is used to specify the state of the system. As prescribed by the state postulate for a simple compressible substance in Chap. 3, the entropy is completely specified by two independent properties of the system; or it can be used as an independent property, combined with another property, to specify all other properties. Entropy is a measure of the molecular disorder of the substance. Larger values of entropy imply greater disorder or uncertainty, and lower values imply more microscopically organized states. The units of entropy S are kilojoules per kelvin or Btu per degree Rankine.

A key difference between the second law and the first law is the inequality sign. The first law specifies a direct relationship between energy and energy transfer, while the second law indicates only the direction for the change in molecular disorder or uncertainty. An alternative, yet equivalent, presentation which is quite useful is to define the term *entropy production* or *entropy generation* S_{gen} and to eliminate the inequality sign:

$$(dS - \delta S_{\text{gen}})_{\text{isolated}} = 0 \tag{5.4}$$

or

$$(S_{\text{final}} - S_{\text{gen}} - S_{\text{initial}})_{\text{isolated}} = 0 \tag{5.5}$$

Here δS_{gen} is the entropy generated during a change in system state and is always positive or zero. Equation (5.4) is exactly the same as Eq. (5.1), where δS_{gen} represents the "amount of inequality" as prescribed by the second law. The negative sign preceding the entropy generation is necessary to "reduce" the change in entropy to zero. As is demonstrated in subsequent sections, δS_{gen} is a path-dependent function or inexact differential unlike dS, which represents the change in a property and is independent of path or an exact differential. [The relationship between an exact differential and inexact differential in Eq. (5.4) should not be surprising since the first law connects the exact differential dU and the inexact differentials δQ and δW. From Sec. 4.6, for an adiabatic process, $dU = -\delta W$.]

The second law is also expressed on a rate basis by considering the time interval dt. This law is then

$$\left.\frac{dS}{dt}\right|_{\text{isolated}} = \frac{d}{dt}\left(\int_V \rho s\, dV\right)_{\text{isolated}} \geq 0 \tag{5.6}$$

or

$$\left.\frac{dS}{dt}\right|_{\text{isolated}} - \dot{S}_{\text{gen}} = 0 \tag{5.7}$$

where the overdot on $\dot{S}_{\text{gen}}$ indicates the rate of entropy generation.

The above expressions are the starting point for further considerations of the second law. The changes in entropy of a control mass and control volume are developed from the isolated system. Before these expressions are presented, the special case of an equality versus an inequality in Eq. (5.1), or zero entropy generation versus finite entropy generation in Eq. (5.4), must be discussed. Also the thermodynamic definitions of temperature and pressure in terms of entropy are required.

5.3 Reversible and Irreversible Processes

The concepts of reversible and irreversible processes play a fundamental role in thermodynamics, and the reduction of irreversibilities is a goal of many analyses. To better understand these concepts, consider the isolated system diagramed in Fig. 5.4. The isolated system is composed of a number of subsystems, four in this example, labeled A, B, C, and D. The subsystems might communicate thermally or mechanically, or they might even transfer mass across their boundaries (except for the outside boundary of the isolated system). Therefore, the total system is isolated when all four subsystems are considered. But if the system is taken as one subsystem, then the other three are viewed as surroundings to the chosen subsystem. If subsystem A has energy transfers in the form of heat transfer and work with the other subsystems, then subsystem A is a control mass and subsystem B, C, and D are the surroundings to control mass A. Reversible and irreversible processes are strictly applicable for only isolated systems and must be interpreted for subsystems (such as subsystem A) with appropriate terminology. This leads to a set of subcategories of reversible and irreversible processes.

Figure 5.4 An isolated system composed of four subsystems.

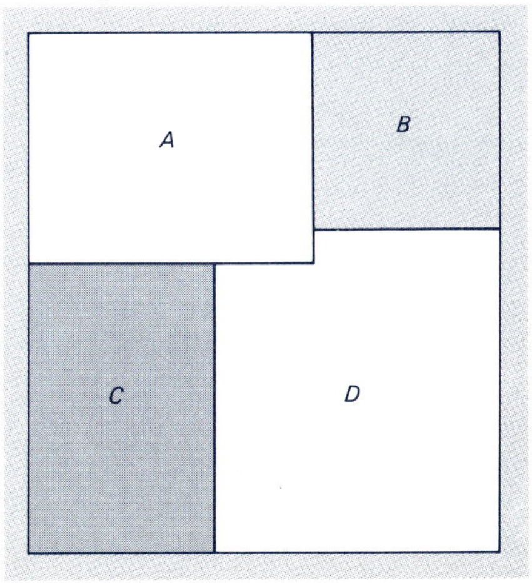

The second law puts a direction on all isolated system processes:

$S_{final} \geq S_{initial}$. Thus there is a forward and backward direction to all processes. When $S_{final} > S_{initial}$, the forward direction is defined and the backward direction is impossible, as stated by the second law. Such a process is termed *irreversible.* When $S_{final} = S_{initial}$, the forward and backward processes are equally possible, and the process is termed *reversible.* The inequality sign is the key distinguishing feature of reversible and irreversible processes. Alternatively, a reversible process produces no entropy generation, so S_{gen} is zero; and an irreversible process has a net entropy generation, so S_{gen} is finite and positive.

The previous discussion pertains to an isolated system which could be composed of subsystems. Reversible and irreversible processes are defined in terms of the *composite* entropy of all the subsystems as required by the second law. These processes can also be referred to as *totally reversible* or *totally irreversible.*

If a particular subsystem is considered as the system, say, subsystem A in Fig. 5.4, then the concepts of internally reversible or irreversible and externally reversible or irreversible are needed. An *internally reversible process* considers the system (not surroundings), and the process can be reversed at any point, and the system then proceeds in reverse through the same states. Note that this requires that all transfers with the surroundings be reversible, but the surroundings do not necessarily undergo a reversible process. An *internally irreversible* process has irreversibilities within the system boundaries. An *externally reversible process* (or *externally irreversible process*) pertains to the surroundings of the system. In our example, the surroundings are subsystems B, C, and D, so an externally reversible process (or externally irreversible process) specifies the particular type of process for these subsystems.

Physically, a totally reversible process can be reversed at any point, and the system and surroundings (making up the isolated system) proceed backward along the same process. The totally reversible or simply reversible process must include both the system *and* the surroundings. This differs significantly from an internally reversible process, which views only the system as reversible and excludes the surroundings from consideration. Clearly the totally reversible process is more difficult to achieve.

The reversible process is an idealization, and all actual processes are irreversible and approach reversible processes only in special cases. Figure 5.5 shows some examples of irreversible and corresponding reversible processes. The first example considers the expansion of a substance against a piston and inelastic spring. As the substance expands, energy is dissipated by the frictional work during the interaction of the piston and wall. The disorder within the piston and wall has increased, so the entropy is increased and the process is irreversible. Friction is a very common contributor to irreversible processes. The inelastic spring also dissipates energy and increases the disorder. This dissipation contributes to the irreversible process. If the ideal case of frictionless motion and an elastic

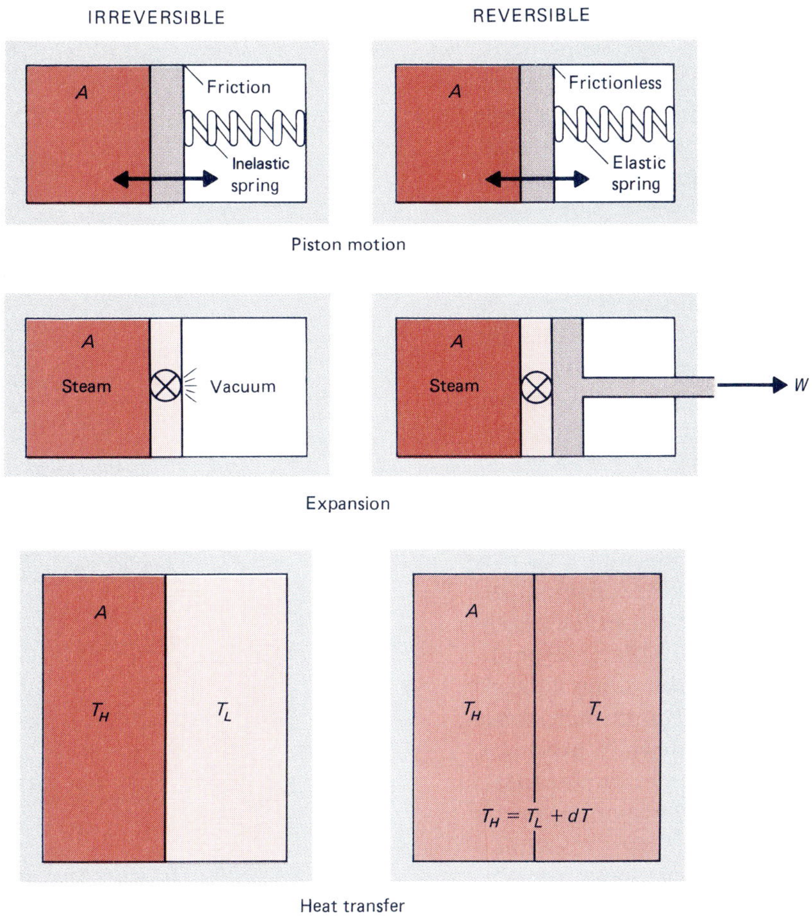

Figure 5.5 Irreversible and reversible processes.

spring are considered, then the process can be reversed at any point in the process and restored to its original state. Thus this process is reversible. The unrestrained expansion of steam into a vacuum increases the disorder and is clearly irreversible, whereas a controlled, slow expansion does work against the surroundings, which can be reversed to achieve a reversible process. The heat transfer across a finite temperature difference is irreversible since the disorder of the entire system increases. Only for the very idealized case of heat transfer across an infinitesimal temperature difference can a reversible heat transfer be obtained. This case is considered in greater detail in succeeding sections, but it should be viewed as an ideal heat transfer similar to the ideal work of a constant-pressure piston motion.

In each of the above cases, the viewpoint can be a subsystem denoted by subsystem A in each case. It is possible for subsystem A to be internally reversible while the irreversibilities occur exterior to the system boundary. Specifically the steam in subsystem A undergoing unrestrained gas expansion (Fig. 5.5) could be proceeding through an internally reversible process, but the process for the complete isolated system is irreversible. This case indicates that this is an externally irreversible process. Similar subsets of reversible and irreversible processes are possible for the other examples.

The main features of irreversible processes are that there are dissipative effects which are not directly useful and that the processes can proceed through nonequilibrium states. These both tend to increase the molecular disorder, or entropy, of the systems. Both contributions make processes irreversible. However, the reversible process progresses through equilibrium states, and there are no dissipative effects. Therefore, reversible processes are idealizations of real situations.

A clear understanding of the relationship of reversible and irreversible processes with the quasi-equilibrium process presented in Sec. 2.2.3 is important. The previous definition of a quasi-equilibrium process pertained to only the system and indicated nothing about the surroundings. *A totally reversible process must specify reversibility for both the system and the surroundings.* Therefore, a reversible process is always a quasi-equilibrium process since both the system and the surroundings undergo reversible processes and thus pass through a series of equilibrium states. Also all internally reversible processes are quasi-equilibrium processes. Conversely, a quasi-equilibrium process is an internally reversible process since these both pertain to the system only. A quasi-equilibrium process is not necessarily a reversible process (totally reversible), since irreversibilities can occur outside the system boundaries.

5.4 Temperature and Pressure Definitions

Temperature and pressure are discussed in Sec. 2.3, yet formal definitions were lacking. The development to this point has relied on our intuitive concepts of these properties. Now specific definitions are possible. The entropy property requires the definition of both temperature and pressure. In this section we state the definitions for temperature and pressure and show that these definitions are consistent with our concepts of these properties [1, 2].

The definitions of temperature and pressure are developed from the differential form of the system energy. As indicated by the state postulate, the energy of a pure simple compressible substance is specified by any two independent properties. Thus, $E = E(S, V) = U(S, V) + KE + PE$, or if we exclude the potential and kinetic energies, $E = U = U(S, V)$. The differential of this property, an exact differential, is

$$dU = \left(\frac{\partial U}{\partial S}\right)_V dS + \left(\frac{\partial U}{\partial V}\right)_S dV \tag{5.8}$$

The two partial derivatives in this expression are intensive parameters since the ratio of two extensive parameters is an intensive parameter.

5.4.1 Temperature

Temperature is defined in classical thermodynamics as the first partial derivative in Eq. (5.8)

$$T \equiv \left(\frac{\partial U}{\partial S}\right)_V \tag{5.9}$$

This is a rather abstract definition for temperature and is appropriate only if it satisfies our intuitive ideas of temperature. The two intuitive concepts that must be demonstrated from the above definition are thermal equilibrium as stated by the zeroth law (Sec. 2.3.3) and our perceptions of the direction of heat transfer. Both are demonstrated with the two-block examples shown in Fig. 5.2.

Figure 5.2 diagrams an isolated system of high-temperature block A and low-temperature block B which are brought into contact and then isolated from the surroundings (state 1). The final state 2 results after the blocks are in thermal equilibrium at T. The first law states

$$dU = d(U_A + U_B) = 0 \qquad \text{or} \qquad dU_A = -dU_B \tag{5.10}$$

The second law says

$$dS = d(S_A + S_B) \geq 0 \tag{5.11}$$

The entropy change is positive as the system undergoes the irreversible heat transfer process until the final equilibrium state is reached. Then further entropy change is equal to zero, when infinitesimal deviations from the final equilibrium state are considered. For thermal equilibrium in an isolated system, the change in entropy is

$$dS = d(S_A + S_B) = 0 \tag{5.12}$$

The volumes are taken as constants without loss of generality. Thus $S = S(U, V)$ reduces to $S = S(U)$, and

$$dS = dS_A + dS_B = \left(\frac{\partial S_A}{\partial U_A}\right)_{V_A} dU_A + \left(\frac{\partial S_B}{\partial U_B}\right)_{V_B} dU_B = 0 \tag{5.13}$$

Applying the first law [Eq. (5.10)], we obtain

$$dS = \left[\left(\frac{\partial S_A}{\partial U_A}\right)_{V_A} - \left(\frac{\partial S_B}{\partial U_B}\right)_{V_B}\right] dU_A = 0 \tag{5.14}$$

But the partial derivatives are defined as temperature in Eq. (5.9), so Eq. (5.14) is

$$dS = \left(\frac{1}{T_A} - \frac{1}{T_B}\right) dU_A = 0 \tag{5.15}$$

Thermal equilibrium results when $dS = 0$, so this requires that

$$T_A = T_B \tag{5.16}$$

which is the expected requirement for thermal equilibrium. Thus the temperature definition in Eq. (5.9) together with the second law satisfies the concept of thermal equilibrium.

As the two blocks proceed toward the final thermal equilibrium state in Fig. 5.2, the second law and the temperature definition combine to describe the change in entropy, similar to Eq. (5.15), as

$$dS = \left(\frac{1}{T_A} - \frac{1}{T_B}\right) dU_A \geq 0 \tag{5.17}$$

From Fig. 5.3, the first law for control mass A yields

$$dU_A = \delta Q_A - \delta W_A \tag{5.18}$$

but the block does not do any work, so

$$dU_A = \delta Q_A \tag{5.19}$$

Substituting Eq. (5.19) into Eq. (5.17) yields

$$dS = \left(\frac{1}{T_A} - \frac{1}{T_B}\right) \delta Q_A \geq 0 \tag{5.20}$$

This example was assumed to initially have

$$T_H = T_A > T_B = T_L \tag{5.21}$$

So with the second law requiring $dS \geq 0$, the heat transfer is

$$\delta Q_A \leq 0 \tag{5.22}$$

Thus, the heat transfer is from the higher-temperature block $T_A = T_H$ to the lower-temperature block $T_B = T_L$. This is exactly as expected and as presented in Sec. 5.1.

The temperature definition in Eq. (5.9) is therefore consistent with our intuitive concepts of temperature and heat transfer. The actual measurement of temperature and the construction of the temperature scale are deferred to Sec. 5.13.

5.4.2 Pressure

The other partial derivative in Eq. (5.8) is used to define pressure. The formal definition is

$$P \equiv -\left(\frac{\partial U}{\partial V}\right)_S \tag{5.23}$$

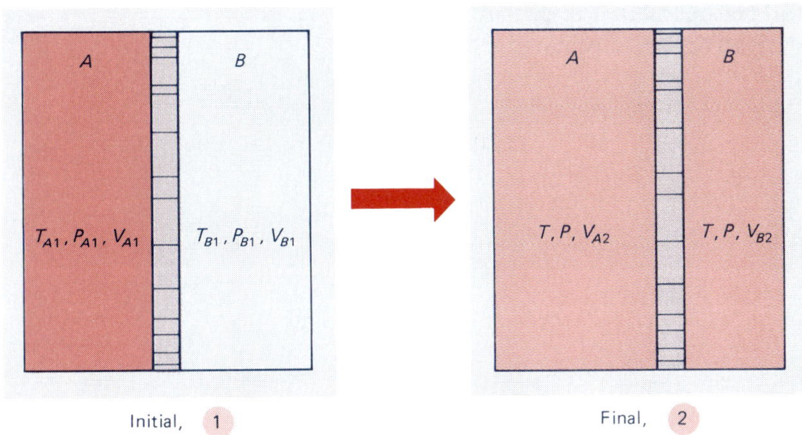

Figure 5.6 Two subsystems that communicate thermally and mechanically.

Like the previous definition for temperature, this rather abstract definition for pressure must satisfy our previous concepts. The two intuitive concepts of pressure that must be demonstrated are mechanical equilibrium at equal pressures and the equivalence of the defined pressure with the previous force-related pressure [Eq. (2.1)].

The system considered is an isolated system composed of two subsystems which are separated by a movable piston. The two subsystems also communicate thermally. Figure 5.6 indicates the initial state, denoted 1, and the final state, denoted 2. The first law is given as

$$dU = d(U_A + U_B) = 0 \qquad \text{or} \qquad dU_A = -dU_B \tag{5.24}$$

and the total volume of the chamber is constant, so

$$dV = d(V_A + V_B) = 0 \qquad \text{or} \qquad dV_A = -dV_B \tag{5.25}$$

As in the thermal equilibrium example above, the final equilibrium state is prescribed from the second law as

$$dS = d(S_A + S_B) = 0 \tag{5.26}$$

The state postulate yields $S = S(U, V)$, so

$$dS = \left(\frac{\partial S_A}{\partial U_A}\right)_{V_A} dU_A + \left(\frac{\partial S_A}{\partial V_A}\right)_{U_A} dV_A + \left(\frac{\partial S_B}{\partial U_B}\right)_{V_B} dU_B + \left(\frac{\partial S_B}{\partial V_B}\right)_{U_B} dV_B \tag{5.27}$$

With Eqs. (5.24) and Eq. (5.25),

$$dS = \left[\left(\frac{\partial S_A}{\partial U_A}\right)_{V_A} - \left(\frac{\partial S_B}{\partial U_B}\right)_{V_B}\right] dU_A + \left[\left(\frac{\partial S_A}{\partial V_A}\right)_{U_A} - \left(\frac{\partial S_B}{\partial V_B}\right)_{U_B}\right] dV_A \tag{5.28}$$

The first two partial derivatives in Eq. (5.28) are temperatures, as defined in Eq. (5.9). The third and fourth partial derivatives are eliminated by noting [Eq. (H.14)]

$$\left(\frac{\partial S}{\partial V}\right)_U \left(\frac{\partial V}{\partial U}\right)_S \left(\frac{\partial U}{\partial S}\right)_V = -1 \tag{5.29}$$

Thus

$$\left(\frac{\partial S}{\partial V}\right)_U = \frac{-1}{(\partial V/\partial U)_S(\partial U/\partial S)_V} \tag{5.30}$$

or [Eq. (H.15)]

$$\left(\frac{\partial S}{\partial V}\right)_U = -\frac{(\partial U/\partial V)_S}{(\partial U/\partial S)_V} \tag{5.31}$$

Thus the definitions for temperature and pressure give

$$\left(\frac{\partial S}{\partial V}\right)_U = \frac{P}{T} \tag{5.32}$$

and Eq. (5.28) is

$$dS = \left(\frac{1}{T_A} - \frac{1}{T_B}\right) dU_A + \left(\frac{P_A}{T_A} - \frac{P_B}{T_B}\right) dV_A \tag{5.33}$$

Since $dS = 0$ at thermodynamic equilibrium and dU_A and dV_A are independent, each of the parenthetical expressions must be zero, or

$$\frac{1}{T_A} - \frac{1}{T_B} = 0 \qquad \text{or} \qquad T_A = T_B \tag{5.34}$$

$$\frac{P_A}{T_A} - \frac{P_B}{T_B} = 0 \qquad \text{or} \qquad P_A = P_B \tag{5.35}$$

This last expression is the required equality for mechanical equilibrium.

The definitions for temperature and pressure given in Eqs. (5.9) and (5.23), respectively, are substituted into Eq. (5.8) to yield

$$dU = T\,dS - P\,dV \tag{5.36}$$

This is an extremely important relation in thermodynamics and is used extensively. It is known as the *Gibbs equation* and is considered in Sec. 5.5. It applies generally to a control mass. The first law for a control mass is

$$dU = \delta Q - \delta W \tag{5.37}$$

For an adiabatic reversible process, Eq. (5.36) yields

$$dU = -P\,dV \tag{5.38}$$

and Eq. (5.37) yields

$$dU = -\delta W \tag{5.39}$$

Comparing these last expressions indicates that the previous force definition of pressure in Eq. (2.1) which is used in the work relation in Eq. (2.18) is in agreement with the pressure definition in Eq. (5.23).

Thus the definitions for temperature and pressure in terms of partial derivatives of internal energy [Eqs. (5.9) and (5.23)] are consistent with the expected behavior of these properties. The next task is to present values and relations for entropy determination.

5.5 Entropy—The Property

Entropy is a property that is specified for every equilibrium state of a substance. Entropy represents the disorder, or uncertainty, of the microscopic scale, yet macroscopically it is used as all other properties. Thus, the relations presented in Chap. 2 for other properties are interpreted here for this property. The state postulate for a pure simple compressible substance states that two independent intensive properties specify the state. Therefore, entropy is an additional property that can be used to specify a state. As with all previous properties, data for entropy can be found in tabular, graphical, equation, and computer retrieval forms. The steam tables provide tabular information for the entropy of steam, and the interpolation methods described in Chap. 3 for obtaining v, u, and h also apply for determining s. Within the saturation region, the entropy is evaluated with the quality as

$$s = s_l + xs_{lg} \tag{5.40}$$

The tables of thermodynamic properties present numerical values for entropy in Apps. D and E for various substances.

The graphical presentation of property information is used extensively in thermodynamic analysis. The analysis to this point in the text focuses on the *P-v* diagram. The *P-v* diagram is very important to simple reversible compressible work calculation since the graphical area under the process representation is the work done. Much of the thermodynamic analysis performed in the subsequent sections uses the *T-s* and *h-s* diagrams. Typical *T-s* and *h-s* diagrams are shown in Fig. 5.7 (specific substances are presented in Apps. D and E). The lines of constant volume, pressure, and enthalpy are shown on the *T-s* diagram. The *h-s* diagram is called the *Mollier chart,* and the constant-volume, constant-pressure, and constant-temperature lines are indicated. For ideal gases with temperature-independent c_P, the *T-s* and *h-s* diagrams are similar, since h is directly proportional to T.

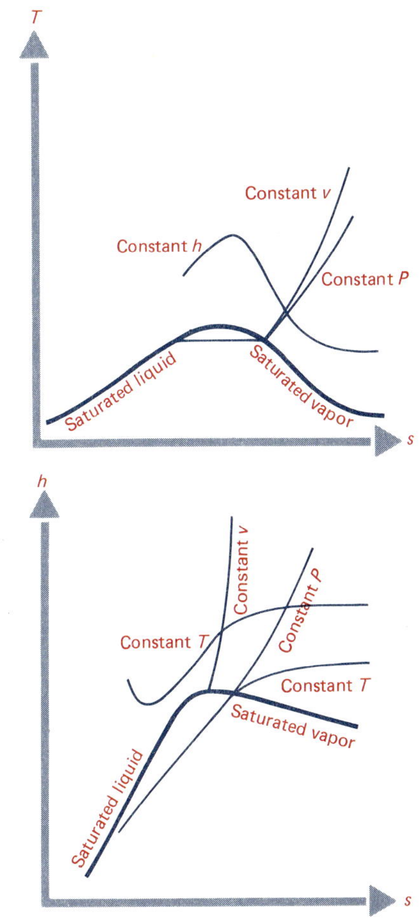

Figure 5.7 The *T-s* and *h-s* diagrams.

Example Problem 5.1

Evaluate the entropy (per unit mass) for water and refrigerant 12 at the following states with $P = 100$ kPa: saturated liquid, saturated vapor,

$T = 150°C$, and $T = 200°C$. Use Tables D.11 and D.12 for water and Tables D.14 and D.15 for refrigerant 12.

Solution

All the states are specified, and the properties are given in the indicated tables. Thus, in kilojoules per kilogram-kelvin,

	s_l	s_g	s_{150}	s_{200}
Water	1.3027	7.3589	7.6129	7.8335
Refrigerant 12	0.89454	1.5749	1.9273	2.0075

Comments

Comparisons between the absolute values of s for different substances are not possible since the reference state for each is chosen to be different in the various reference tables. The entropy increases from left to right in the above table, which satisfies the physical concepts of increasing molecular disorder. The saturated liquid is more ordered than the saturated vapor state. The superheated states are more molecularly disordered than the saturation states.

Example Problem 5.2

Steam occupies one-half of the partitioned insulated chamber shown. The initial pressure and initial temperature of the steam are 50 psia and 700°F. The other half of the chamber is a vacuum. The partition is removed, and the steam fills the entire volume at the end of the process. Determine the final temperature of the steam and the increase in entropy (entropy generation) of the process.

Diagrams

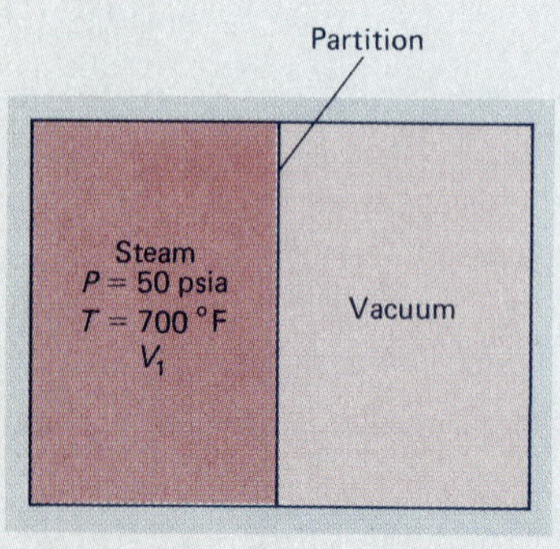

Initial, 1

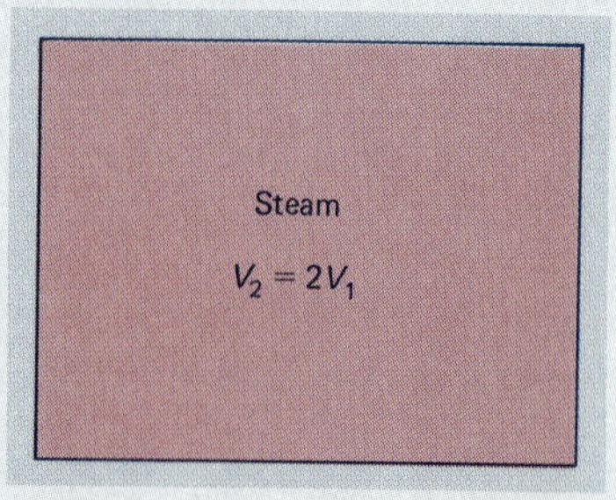

Final, 2

Solution

The initial state is completely specified, and the final volume is known:

State	P, psia	T, °F	V, ft³	u, Btu/lbm	s, Btu/(lbm·°R)
1	50	700	V_1		
2			$2V_1$		

The control mass is chosen to be the steam in the entire chamber, so the mass is constant and

First law: $U_1 = U_2$

Second law: $S_2 - S_1 \geq 0$ or $S_{gen} = S_2 - S_2 \geq 0$

or, on a unit-mass basis,

First law: $u_1 = u_2$

Second law: $s_2 - s_1 \geq 0$ or $s_{gen} = s_2 - s_1 \geq 0$

Thus the first law specifies the second needed property, u_2, for state 2. Table E.12 yields $v_1 = 13.742$ ft³/lbm, $u_1 = 1254.7$ Btu/lbm, and $s_1 = 1.8813$ Btu/(lbm·°R). The final state is specified by $v_2 = 2v_1 = 27.484$ ft³/lbm (since the mass is constant) and $u_2 = u_1 = 1254.2$ Btu/lbm. Interpolating first on u, then on v from Table E.12 yields

$$T_2 = 697.6°\text{F} \qquad P_2 = 25.13 \text{ psia} \qquad s_2 = 1.9570 \text{ Btu/(lbm·°R)}$$

Therefore, the entropy generation is

$$s_{gen} = s_2 - s_1 = 0.0757 \text{ Btu/(lbm·°R)} = 0.3169 \text{ kJ/(kg·K)}$$

and the process is clearly irreversible.

Comments

The process is almost isothermal since T_2 is only a fraction of 1 percent lower than T_1. Checking the compressibility chart reveals that the states are well approximated by the ideal gas equation of state. A constant-internal-energy process is also isothermal since

$$du = c_v \, dT$$

Therefore, $T_1 \approx T_2$ as evaluated in this example is expected. This fact helps in determining the final state from the property tables.

5.5.1 Entropy Relations

In Sec. 5.4 we define the equilibrium properties of temperature and pressure. The state postulate yields $U = U(S, V)$, so the change in internal energy combined with the temperature and pressure definitions yields

$$dU = T\,dS - P\,dV \tag{5.36}$$

This is the Gibbs equation which relates equilibrium thermodynamic properties. Thus this expression depends not on the path between any two states but on the substance only. A companion expression is obtained from enthalpy as

$$dH = d(U + PV) = dU + P\,dV + V\,dP \tag{5.41}$$

Substituting Eq. (5.36) for dU yields

$$dH = T\,dS + V\,dP \tag{5.42}$$

These expressions are also represented on an intensive basis as

$$du = T\,ds - P\,dv \tag{5.43}$$

$$dh = T\,ds + v\,dP \tag{5.44}$$

The changes in entropy are obtained directly from these equations as

$$dS = \frac{dU}{T} + \frac{P}{T}\,dV \tag{5.45}$$

$$dS = \frac{dH}{T} - \frac{V}{T}\,dP \tag{5.46}$$

or

$$ds = \frac{du}{T} + \frac{P}{T}\,dv \tag{5.47}$$

$$ds = \frac{dh}{T} - \frac{v}{T}\,dP \tag{5.48}$$

Example Problem 5.3

Consider the vaporization of water at 100 kPa. Using Eqs. (5.47) and (5.48), evaluate the change in entropy per unit mass (i.e., the change in entropy from the saturated liquid to the saturated vapor state s_{lg}) from the tabular values of u, h, P, T, and v.

Solution

The constant-pressure evaporation of water is also at constant temperature. Equation (5.47) is integrated from the saturated liquid state to the saturated vapor state. Taking P and T as constants, we have

$$\int_{s_l}^{s_g} ds = \int_{u_l}^{u_g} \frac{du}{T} + \int_{v_l}^{v_g} \frac{P}{T}\,dv$$

$$s_g - s_l = \frac{1}{T_{\text{sat}}}(u_g - u_l) + \frac{P_{\text{sat}}}{T_{\text{sat}}}(v_g - v_l)$$

$$s_{lg} = \frac{u_{lg}}{T_{sat}} + \frac{P_{sat}}{T_{sat}} v_{lg}$$

$$= \frac{2088.3}{372.78} + \frac{100}{372.78}(1.6933) = 6.0562 \text{ kJ/(kg·K)}$$

$$= 1.4465 \text{ Btu/(lbm·°R)}$$

Since the evaporation is at constant pressure, Eq. (5.48) yields

$$\int_{s_l}^{s_g} ds = \int_{h_l}^{h_g} \frac{dh}{T} - \int_{P_{sat}}^{P_{sat}} \frac{v}{T} dP = \int_{h_l}^{h_g} \frac{dh}{T}$$

$$s_g - s_l = \frac{h_g - h_l}{T_{sat}} = \frac{h_{lg}}{T_{sat}} = \frac{2257.6}{372.78} = 6.0561 \text{ kJ/(kg·K)}$$

$$= 1.4465 \text{ Btu/(lbm·°R)}$$

These results are to be compared with the tabulated value of 6.0562 kJ/(kg·K) = 1.4465 Btu/(lbm·°R).

Comments

Equations (5.47) and (5.48) are viewed as required relations between thermodynamic properties. They describe the property surfaces and depend on only the state.

Equations of state are presented in Chap. 3 to mathematically relate the properties of a particular substance. They are shown in Sec. 3.5.2 to be very useful for particular substances. The changes in entropy for the ideal gas and incompressible fluid idealizations are considered here.

5.5.2 Ideal Gas Relations

The equation of state for an ideal gas is $Pv = RT$, and as stated in Sec. 3.5.2, the change in internal energy is a function of temperature only. In mathematical terms, we wish to show that $(\partial u/\partial v)_T = 0$ for an ideal gas, because then

$$du = \left(\frac{\partial u}{\partial v}\right)_T dv + \left(\frac{\partial u}{\partial T}\right)_v dT = \left(\frac{\partial u}{\partial T}\right)_v dT = \left(\frac{du}{dT}\right) dT$$

which is only a function of temperature even if the volume changes. This can be done using Eq. (H.21), which may be written for the variables of interest as

$$\left(\frac{\partial u}{\partial v}\right)_T = \left(\frac{\partial u}{\partial v}\right)_s + \left(\frac{\partial u}{\partial s}\right)_v \left(\frac{\partial s}{\partial v}\right)_T = -P + T\left(\frac{\partial s}{\partial v}\right)_T \tag{5.49}$$

where the definitions of pressure and temperature from Eqs. (5.9) and (5.23) have been substituted. This is a general relation between properties at this point. To evaluate the remaining partial derivative on the right-hand-side of Eq. (5.49), define a new combination of properties, $a = u - Ts$. (The new property a is called the Helmholtz function, and will be used extensively in later chapters. For the moment, consider it as simply equal to the property relation given.) The total derivative of a is then

$$da = du - T\,ds - s\,dT \tag{5.50a}$$

Substituting the Gibbs relation, Eq. (5.36), this becomes

$$da = (T\,ds - P\,dv) - T\,ds - s\,dT = -P\,dv - s\,dT \tag{5.50b}$$

Because this is an exact differential, it follows from Eq. (H.29*a*) that $(\partial P/\partial T)_v = (\partial s/\partial v)_T$. Substituting this result into Eq. (5.49) results in

$$\left(\frac{\partial u}{\partial v}\right)_T = -P + T\left(\frac{\partial P}{\partial T}\right)_v \tag{5.51a}$$

For an ideal gas, $(\partial P/\partial T)_v = R/v$, and Eq. (5.51*a*) reduces to

$$\left(\frac{\partial u}{\partial v}\right)_T = -P + T\left(\frac{R}{v}\right) = -P + P = 0 \tag{5.51b}$$

This is what we set out to prove: The internal energy of an ideal gas does not vary with volume, and is therefore a function of temperature only.

Thus, Eqs. (3.19) and (3.22) yield

$$du = c_v\,dT \qquad \text{and} \qquad dh = c_P\,dT$$

These relations and the equation of state are combined with the Gibbs equation in Eqs. (5.47) and (5.48) to give

$$ds = c_v \frac{dT}{T} + R\frac{dv}{v} \tag{5.52}$$

$$ds = c_P \frac{dT}{T} - R\frac{dP}{P} \tag{5.53}$$

Finite changes in entropy are evaluated by integrating these last equations to yield

$$s_2 - s_1 = \int_{T_1}^{T_2} c_v \frac{dT}{T} + R \ln\frac{v_2}{v_1} \tag{5.54}$$

$$s_2 - s_1 = \int_{T_1}^{T_2} c_P \frac{dT}{T} - R \ln\frac{P_2}{P_1} \tag{5.55}$$

Either of the previous expressions for the change in entropy is used, depending on what information is known. Unlike the internal energy and enthalpy for an ideal gas which require only temperature for their evaluation, entropy requires two properties [see Eq. (5.52) or (5.53)]. Completion of the evaluation of the change in entropy requires the functional form of the specific heat variation with temperature.

The specific heats are often accurately approximated by a single average value. Thus, for an ideal gas with *constant specific heats,*

$$s_2 - s_1 = c_v \ln \frac{T_2}{T_1} + R \ln \frac{v_2}{v_1} \tag{5.56}$$

$$s_2 - s_1 = c_P \ln \frac{T_2}{T_1} - R \ln \frac{P_2}{P_1} \tag{5.57}$$

Table C.2 presents specific heats at 300 K for various gases.

More accurate values for changes in entropy are obtained if the temperature variation of the specific heats is used in the integration in Eqs. (5.54) and (5.55). The temperature dependence of the specific heat at constant pressure is given in Tables D.1 and E.1 for various gases. These expressions are substituted into the integrals, and the values for the change are obtained. Note that $c_P - c_v = R$, so only the c_P integration needs to be performed. When the temperature dependence of c_P is included, the results are termed *variable specific heat* solutions.

A few differences occur in the determination of s from the gas tables, Tables D.2 through D.8 or Tables E.2 through E.8, as compared with the properties discussed in Chap. 3. Because these tables are based on an ideal gas relation, the values for u and h are functions of T only. The tables are thus presented as a function of the *single* property T, unlike the steam tables (Tables D.9 through D.12 and Tables E.9 through E.12), which require *two* independent properties to define a state. However, for the entropy of an ideal gas, Eqs. (5.54) and (5.55) show that entropy is not a function of T only, but that a second independent property (usually v or P) must be specified. However, by choosing P as the second independent variable, the value of s is obtained by writing Eq. (5.55) as

$$s - s_{\text{ref}} = \int_{T_{\text{ref}}}^{T} c_P \frac{dT}{T} - R \ln \frac{P}{P_{\text{ref}}} = s_0(T) - R \ln \frac{P}{P_{\text{ref}}} \tag{5.58}$$

where

$$s_0(T) \equiv \int_{T_{\text{ref}}}^{T} c_P \frac{dT}{T}$$

In this form, s is determined from summing $s_0(T)$, which is a function of T only, with the second term, which is a function of P only. The subscript 0 in $s_0(T)$ denotes that this is the temperature-dependent portion of the entropy of an ideal gas and so is *not* a reference value. In the gas tables, $s_0(T)$, often given the symbol ϕ, is tabulated, and the value of $s - s_{\text{ref}}$ is simply obtained by subtracting the pressure term $R \ln (P/P_{\text{ref}})$ from $s_0(T)$.

The difference in entropy for two states is

$$s_2 - s_1 = s_0(T_2) - s_0(T_1) - R \ln \frac{P_2}{P_{\text{ref}}} + R \ln \frac{P_1}{P_{\text{ref}}} \tag{5.59}$$

or

$$s_2 - s_1 = s_0(T_2) - s_0(T_1) - R \ln \frac{P_2}{P_1} \tag{5.60}$$

The reference state is chosen at absolute zero temperature and atmospheric pressure.

5.5.3 Incompressible Fluid or Solid Relations

The equation of state for an incompressible fluid or solid, as stated in Eq. (3.35), is $v =$ constant. Equation (3.37*a*) states that

$$du = c\,dT$$

This expression is substituted into Eq. (5.47) to give the change in entropy

$$ds = c\frac{dT}{T} + \frac{P}{T}dv \tag{5.61}$$

but $v =$ constant, so

$$ds = c\frac{dT}{T} \tag{5.62}$$

Finite changes in entropy are

$$s_2 - s_1 = \int_{T_1}^{T_2} c\frac{dT}{T} \tag{5.63}$$

If the specific heat is approximated by a constant, then

$$s_2 - s_1 = c \ln \frac{T_2}{T_1} \tag{5.64}$$

otherwise the temperature integration must be performed. With $dh = du + P\,dv + v\,dP$ and Eq. (5.48), Eq. (5.64) again results.

Example Problem 5.4

Two blocks A and B which are initially at 200 and 1000°F, respectively, are brought into contact and isolated from the surroundings. They are allowed to reach a final state of thermal equilibrium. Determine the entropy change of each block and of the isolated system. Block A is aluminum [$c_P = 0.215$ Btu/(lbm·°R)] with $m_A = 1$ lbm, and block B is copper [$c_P = 0.092$ Btu/(lbm·°R)] with $m_B = 2$ lbm.

Solution

The diagram of the system is as presented in Fig. 5.2. The blocks are incompressible, so only temperature is required to specify the states. The initial temperatures are specified, and the final temperature is unknown. The first and second laws for the isolated system are

First law: $dU = \delta Q - \delta W = 0$

$$U_2 - U_1 = 0$$

$$U_{2A} + U_{2B} = U_{1A} + U_{1B}$$

Second law: $dS \geq 0 \quad \text{or} \quad dS - \delta S_{\text{gen}} = 0$

$$S_{\text{gen}} = S_2 - S_1 = (S_{2A} + S_{2B}) - (S_{1A} + S_{1B})$$

or

$$S_{\text{gen}} = (S_{2A} - S_{1A}) + (S_{2B} - S_{1B})$$

The properties are obtained from Eqs. (3.37*a*) and (5.64) as

$$U_2 - U_1 = mc(T_2 - T_1)$$

$$S_2 - S_1 = mc \ln \frac{T_2}{T_1}$$

The first law is used to determine the final state. Rearranging the first law expression above yields

$$U_{2A} - U_{1A} = U_{1B} - U_{2B}$$

$$m_A c_A (T_{2A} - T_{1A}) = m_B c_B (T_{1B} - T_{2B})$$

With $T_{2A} = T_{2B} = T_2$, then,

$$T_2 = \frac{m_A c_A T_{1A} + m_B c_B T_{1B}}{m_A c_A + m_B c_B} = 568.9°\text{F} = 298.3°\text{C}$$

The entropy generated is given above as

$$S_{\text{gen}} = m_A c_A \ln \frac{T_2}{T_{1A}} + m_B c_B \ln \frac{T_2}{T_{1B}}$$

$$= +0.0955 \text{ Btu/°R} - 0.0644 \text{ Btu/°R} = +0.0311 \text{ Btu/°R}$$

$$= +0.0591 \text{ kJ/K}$$

Thus there is an increase in the entropy of block A of $+0.0955$ Btu/°R and a decrease in the entropy of block B of -0.0644 Btu/°R. The entropy generation of the isolated system is greater than zero, as stated by the second law. Therefore, this is an irreversible process.

Comments

The final temperature is a weighted average of the initial values as specified by the first law. The final temperature, with the material properties, determines the entropy generated within each block. The entropy generated must be zero or positive for the isolated system, but each block undergoes different directional changes. Block *A* is increasing in temperature; so it is increasing in molecular disorder, and the entropy increases. Block *B* is decreasing in temperature, or it is becoming more molecularly ordered, and the entropy decreases. Since a heat transfer from block *B* to block *A* is associated with a transfer of disorganization, these increases and decreases are expected.

Also if the initial temperature of block *B* is closer to that of block *A*, say, $T_{1B} = 250°\text{F}$, then the entropy generation is ($T_2 = 223.1°\text{F}$)

$$S_{gen} = +0.00740 \text{ Btu/°R} - 0.00711 \text{ Btu/°R} = +0.00029 \text{ Btu/°R}$$

Thus, the entropy changes for each block are much smaller than for the case of large temperature difference. The process is irreversible; but since the initial block temperatures are closer, it is approaching a reversible process.

5.6 Control Mass Formulation

The second law is presented for an isolated system, and as indicated by irreversible and reversible processes, the isolated system processes are directional. This directionality of the entropy change of an isolated system is a fundamental and important aspect of the second law. But the isolated system is somewhat abstract, and the primary interest of much of thermodynamic analysis lies in control mass and control volume systems. Therefore, the second law must be transformed to these viewpoints. This section focuses on the control mass formulation, and Sec. 5.8 considers the control volume formulation.

Many of the previous examples in this chapter consider an isolated system as a set of subsystems. One subsystem is considered the control mass, and the others comprise the surroundings. From this viewpoint, a general control mass undergoes a process with arbitrary energy transfers with the surroundings, and the control mass and the surroundings form the isolated system. The change in entropy for the control mass is desired. Thus the surroundings are idealized and specified, so the only unspecified entropy changes are those associated with the control mass.

A control mass exchanges energy with the surroundings by work and heat transfer. Therefore, two types of idealized subsystems of the surroundings are conceived: a *reversible work reservoir* and a *reversible heat transfer reservoir.* Both reservoirs are reversible, indicating that they proceed slowly through equilibrium states and can be restored to the original state at any point in the process. Both are control masses and are termed *reservoir* to indicate a very large system that is unchanging during work or heat transfer.

The reversible heat transfer reservoir participates in heat transfer only (not in work transfers). Therefore, this reservoir is a constant-volume control mass. All the heat transfers to the reversible heat transfer reservoir transform this system through equilibrium states which are specified by a uniform temperature. The heat transfers can be to or from this reservoir. From the Gibbs equation [Eq. (5.36)] for constant-volume processes,

$$dS = \frac{dU}{T} \tag{5.65}$$

and the first law for a control mass that does no work is

$$dU = \delta Q \tag{5.66}$$

Thus, for the reversible heat transfer reservoir, the entropy change is

$$dS = \frac{\delta Q}{T} \tag{5.67}$$

where δQ is positive *into* the reversible heat transfer reservoir. The temperature is uniform throughout this reservoir.

The reversible work reservoir participates in only work interactions (no heat transfers), and for a compressible system, the work is a result of boundary expansion. The work, either positive or negative, is a quasi-equilibrium process. The Gibbs equation for the reversible work reservoir is given in Eq. (5.36) as

$$dS = \frac{dU}{T} + \frac{P}{T}\,dV$$

but the first law with only $P\,dV$ work and no heat transfer is

$$dU = \cancelto{0}{\delta Q} - \delta W = -P\,dV \tag{5.68}$$

Thus, substituting Eq. (5.68) into Eq. (5.36) yields

$$dS = -\frac{P\,dV}{T} + \frac{P}{T}\,dV = 0 \tag{5.69}$$

so the reversible work reservoir has no entropy change.

In summary, the control mass form of the second law is obtained by considering a general control mass that undergoes both work interactions with reversible work reservoirs and heat transfers with reversible heat transfer reservoirs. The composite system of control mass, reversible work reservoirs, and reversible heat transfer reservoirs forms an isolated system. The work reservoirs have no entropy change, and the heat transfer reservoirs' entropy change is $dS_i = dQ_i/T_i$. A general control mass is shown in Fig. 5.8, where there could be an arbitrary number of work and heat transfer reservoirs.

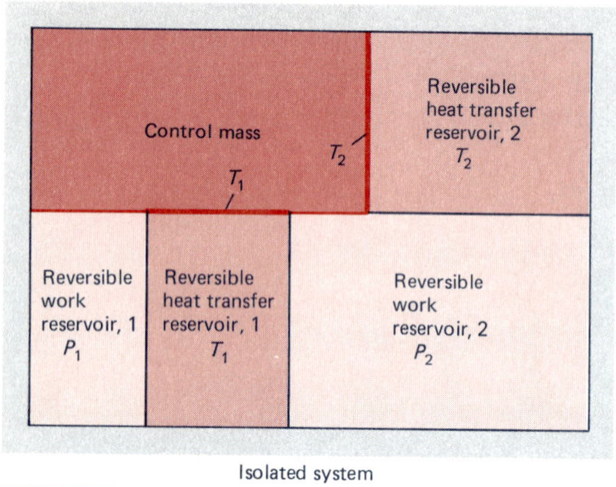

Figure 5.8 Control mass, reversible heat transfer reservoirs, and reversible work reservoirs within an isolated system.

The second law for the isolated system [control mass (CM) plus the reservoirs] is

$$dS_{\text{isolated}} \geq 0 \tag{5.70}$$

but

$$dS_{\text{isolated}} = dS_{\text{CM}} + \sum_i \overset{0}{\cancel{dS}}_{\substack{\text{reversible}\\ \text{work}\\ \text{reservoirs}}} + \sum_i dS_{\substack{\text{reversible}\\ \text{heat transfer}\\ \text{reservoirs}}} \tag{5.71}$$

The entropy change for each reversible heat transfer reservoir is $dS_i = \delta Q_i/T_i$, where δQ_i is positive into the reservoir and T_i is the reservoir temperature and so is the temperature of the boundary between the reversible heat transfer reservoir and the control mass. Thus, in terms of the control mass, the heat transfer to the control mass from each reservoir is

$$\delta Q_{\text{CM},\,i} = -\delta Q_{\substack{\text{reversible}\\ \text{heat transfer}\\ \text{reservoir},\,i}} \tag{5.72}$$

Thus, substituting Eqs. (5.71) and (5.72) into (5.70) yields

$$dS_{\text{isolated}} = dS_{\text{CM}} - \sum_i \left(\frac{\delta Q_i}{T_i}\right)_{\text{CM}} \geq 0 \tag{5.73}$$

or

$$dS_{\text{CM}} \geq \sum_i \left(\frac{\delta Q_i}{T_i}\right)_{\text{CM}} \tag{5.74}$$

since the change in entropy of all reversible work reservoirs is zero. Note especially that the temperature in the second law is the boundary temperature interfacing the control mass and heat transfer reservoirs. Thus this is the control mass form of the second law:

> The change in entropy S, an extensive property of the control mass, is greater than or equal to the sum of the heat transfer divided by the corresponding absolute temperatures of the reservoirs at the boundary.

Note that dS_{CM} can be positive or negative since the heat transfer can be positive or negative. Recall that heat transfer is a microscopically disorganized energy transfer, so positive heat transfer increases microscopic disorder and increases entropy while negative heat transfer decreases microscopic disorder and decreases entropy. Also there is no contribution to entropy change by work interactions. Since work is an organized energy transfer mechanism and entropy describes microscopic disorder, this is appropriate.

Equation (5.74) is the control mass form for the second law. The inequality applies to irreversible processes. Since the development is in terms of reversible surroundings, the irreversibilities are within the control mass; so this type of process is termed *internally irreversible.* The equality applies to an internally reversible process. The inequality is eliminated if the concept of entropy generation is used. The development is as above, but Eq. (5.4) is the starting point. Thus

$$(dS - \delta S_{gen})_{isolated} = 0 \tag{5.75}$$

and with Eqs. (5.67), (5.71), and (5.72),

$$dS_{CM} - \sum_i \left(\frac{\delta Q_i}{T_i}\right)_{CM} - \delta S_{gen} = 0 \tag{5.76}$$

or

$$\delta S_{gen} = dS_{CM} - \sum_i \left(\frac{\delta Q_i}{T_i}\right)_{CM} \geq 0 \tag{5.77}$$

This is the entropy generation by the process. Equations (5.74) and (5.77) are the forms of the second law that are the starting point for subsequent analysis. The importance of these expressions cannot be overemphasized.

Alternative forms can be obtained with simplifying assumptions. If the control mass has a uniform distribution of mass, then

$$d(ms)_{CM} \geq \sum_i \left[\frac{\delta(mq)_i}{T_i}\right]_{CM} \tag{5.78}$$

or

$$ds_{CM} \geq \sum_i \left(\frac{\delta q_i}{T_i}\right)_{CM} \tag{5.79}$$

In terms of entropy generation, this is

$$\delta s_{gen} = ds_{CM} - \sum_i \left(\frac{\delta q_i}{T_i}\right)_{CM} \geq 0 \tag{5.80}$$

The heat transfers are relative to the control mass, and T_i corresponds to the temperature at boundary i of the control mass where the heat transfer occurs. The sum includes all possible heat transfers with the surroundings. All real processes are irreversible to some extent; so the inequality sign is applicable, and the entropy generation is greater than zero. Only for the idealization of a reversible process is the equals sign appropriate, and $\delta s_{\text{gen}} = 0$. The control mass forms of the second law are expressed on a rate basis as

$$\frac{dS_{\text{CM}}}{dt} \geq \sum_i \left(\frac{\dot{Q}_i}{T_i}\right)_{\text{CM}} \tag{5.81}$$

or

$$\dot{S}_{\text{gen}} = \frac{dS_{\text{CM}}}{dt} - \sum_i \left(\frac{\dot{Q}_i}{T_i}\right)_{\text{CM}} \geq 0 \tag{5.82}$$

The second law connects the system property, entropy, to the heat transfer. If the *process is reversible* (rev), the entropy generation is zero, and

$$dS_{\text{CM, rev}} = \sum_i \left(\frac{\delta Q_i}{T_i}\right)_{\text{CM, rev}} \tag{5.83}$$

For a single heat transfer

$$dS_{\text{CM, rev}} = \left(\frac{\delta Q}{T}\right)_{\text{CM, rev}} \tag{5.84}$$

or, since the reversible heat transfer occurs across an infinitesimal temperature difference,

$$\delta Q_{\text{rev}} = T\,dS \tag{5.85}$$

This shows the importance of the T-s diagram to thermodynamic analysis. The heat transfer for a specified reversible process between states 1 and 2 is

$${}_1Q_{2,\text{ rev}} = \int_1^2 T\,dS \tag{5.86}$$

or

$${}_1q_{2,\text{ rev}} = \int_1^2 T\,ds \tag{5.87}$$

Thus the area under the process curve on a T-s diagram is the reversible heat transfer per unit mass. This is similar to the work representation on a P-v diagram. This is shown schematically in Fig. 5.9.

As previously presented for the conservation of mass and the first law, the forms of the second law are generalized to include each element

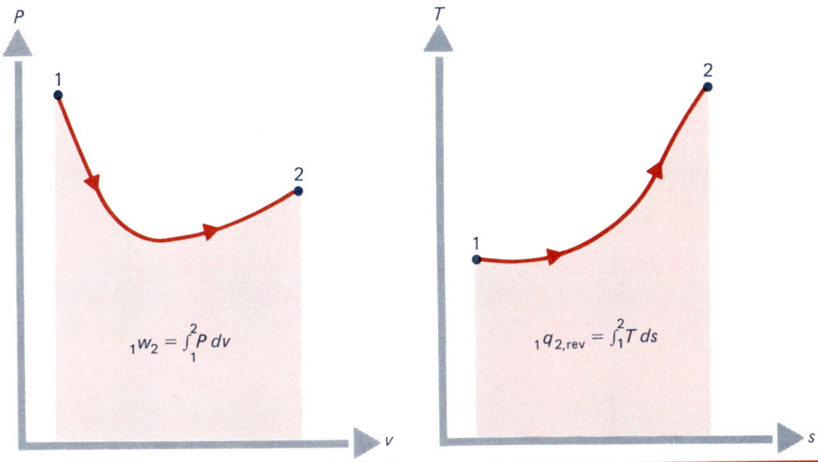

Figure 5.9 Heat transfer and work represented as areas on *T-s* and *P-v* diagrams, respectively.

within the control mass. The total entropy within the control mass is the sum of the elemental contributions, so the total entropy is represented as

$$S_{\text{CM}} = \int_V \rho s\, dV \tag{5.88}$$

Equation (5.74) can be expressed also as

$$dS_{\text{CM}} = d\left(\int_V \rho s\, dV\right) \geq \sum_i \left(\frac{\delta Q_i}{T_i}\right)_{\text{CM}} \tag{5.89}$$

In terms of entropy generation as given in Eq. (5.77), this is

$$\delta S_{\text{gen}} = d\left(\int_V \rho s\, dV\right) - \sum_i \left(\frac{\delta Q_i}{T_i}\right)_{\text{CM}} \geq 0 \tag{5.90}$$

The same expressions on a rate basis take the forms of

$$\frac{dS_{\text{CM}}}{dt} = \frac{d}{dt}\left(\int_V \rho s\, dV\right)_{\text{CM}} \geq \sum_i \left(\frac{\dot{Q}_i}{T_i}\right)_{\text{CM}} \tag{5.91}$$

and

$$\dot{S}_{\text{gen}} = \frac{d}{dt}\left(\int_V \rho s\, dV\right)_{\text{CM}} - \sum_i \left(\frac{\dot{Q}_i}{T_i}\right)_{\text{CM}} \geq 0 \tag{5.92}$$

Example Problem 5.5

Calculate the heat transfers required to evaporate water at 0.1 and 1.0 MPa. Assume the processes to be reversible.

Solution

All the states are known, and Table D.11 yields

State	P, MPa	T, °C	x	s, kJ/(kg·K)
1a	0.1†	(99.632)	0	(1.3027)
1b	1.0	(179.92)	0	(2.1388)
2a	0.1	(99.632)	1.0	(7.3589)
2b	1.0	(179.92)	1.0	(6.5859)

† Entries *not* in parentheses determine the state.

Diagram

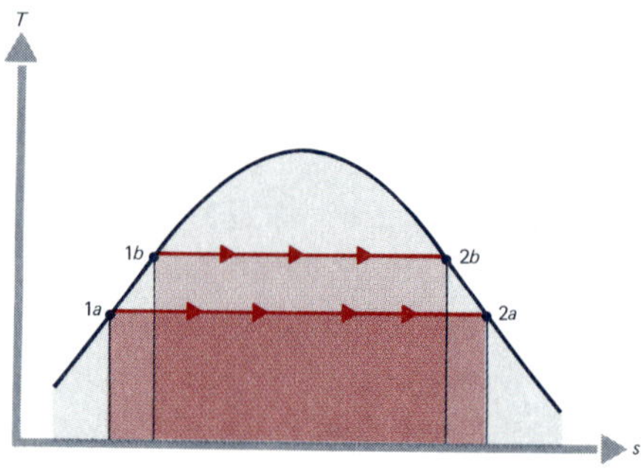

The system is chosen as the water within a container which is isothermal for the constant-pressure evaporation. The second law for a reversible process is

$$ds = \frac{\delta q_{\text{rev}}}{T} \quad \text{or} \quad \delta q_{\text{rev}} = T\,ds$$

Integration for an isothermal process yields

$${}_1q_{2,\,\text{rev}} = T_{\text{sat}}(s_2 - s_1) = T_{\text{sat}} s_{lg}$$

Using absolute temperature results in

$${}_1q_{2,\,\text{rev}} = \begin{cases} 2257.6 \text{ kJ/kg for } P = 0.1 \text{ MPa} & = 970.6 \text{ Btu/lbm for } P = 14.5 \text{ psia} \\ 2014.8 \text{ kJ/kg for } P = 1.0 \text{ MPa} & = 866.2 \text{ Btu/lbm for } P = 145.0 \text{ psia} \end{cases}$$

Comments

The area on the T-s diagram is shown. The s_{lg} is very different in each case, but since the temperatures are also different, the heat transfers are comparable.

An alternative approach to this problem starts with the first law for the control mass as

$$du = \delta q - \delta w$$

with $\delta w = P\,dv$,

$$\delta q = du + P\,dv$$

which for a constant-pressure process is

$$\delta q = dh$$

so

$${}_1q_2 = h_2 - h_1 = h_{lg}$$

From Table D.11,

$$ {}_1q_2 = \begin{cases} 2257.6 \text{ kJ/kg} & \text{for } P = 0.1 \text{ MPa} \\ 2014.8 \text{ kJ/kg} & \text{for } P = 1.0 \text{ MPa} \end{cases} $$

which agrees with the above analysis.

Example Problem 5.6

Water is transformed from the saturated vapor state to the saturated liquid state at $P = 14.7$ psia. This process occurs by a heat transfer across the container wall to the environment (heat transfer reservoir) at T_0. Calculate the entropy generation for the container per unit mass of water for the heat transfer with $T_0 = 70°\text{F}$ and $T_0 = 210°\text{F}$.

Diagram

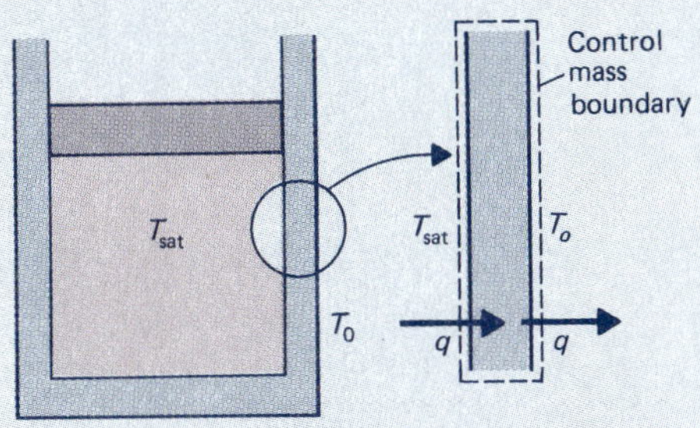

Solution

The heat transfer required to condense the water at 14.7 psia is obtained from the same analysis used to evaluate evaporation in Example Problem 5.5. The water was considered as the control mass, and the second law was applied. The process was assumed to be internally reversible in that example. The required heat transfer per unit mass is calculated as

$$ q = h_{lg} = 970.22 \text{ Btu/lbm} $$

and the water is isothermal at $T_{sat} = 212.02°\text{F}$. This is the heat transfer across the container wall in the present example.

The entropy generation for the heat transfer with the environment focuses on the wall as the system with T_{sat} at the inside surface and T_0 at the outside surface. The process is assumed to be at steady state, so q at the inside surface is the same as q at the outside surface. The second law is applied as given in Eq. (5.82),

$$ \dot{S}_{gen} = \frac{dS_{CM}}{dt} - \sum_i \left(\frac{\dot{Q}_i}{T_i}\right)_{CM} \geq 0 $$

The entropy change with time within the control mass (the wall) is zero. The temperature varies spatially across the wall, so the local value of the entropy varies spatially. But the values do not change in time, since steady state is assumed; so the change in entropy within the control mass is zero.

Integrating the second law with respect to time yields

$$S_{gen} = -\sum_{i=1}^{2}\left(\frac{Q_i}{T_i}\right)_{CM} \geq 0$$

The heat transfer into the wall Q must be equal to the heat transfer from the water $m_w q$, where m_w is the mass of the water. The heat transfer into the wall is positive at T_{sat}, and the heat transfer to the environment is negative at T_0, so

$$S_{gen} = -\frac{m_w q}{T_{sat}} + \frac{m_w q}{T_0} = m_w q\left(\frac{1}{T_0} - \frac{1}{T_{sat}}\right) \geq 0$$

Thus, with $T_0 = 70°F$

$$S_{gen} = m_w(970.22)\left(\frac{1}{529.67} - \frac{1}{671.69}\right) = m_w(0.3873)\ \text{Btu/°R}$$
$$= m_w(0.7355)\ \text{kJ/K}$$

and with $T_0 = 210°F$

$$S_{gen} = m_w(970.22)\left(\frac{1}{669.67} - \frac{1}{671.69}\right) = m_w(0.0044)\ \text{Btu/°R}$$
$$= m_w(0.0084)\ \text{kJ/K}$$

Comments

The entropy generation is greatly reduced as T_0 is increased from 70 to 210°F. Thus the heat transfer is approaching the reversible heat transfer idealization. If the environment is at $T_0 = T_{sat} - dT$, then

$$\frac{S_{gen}}{m_w} = q\left(\frac{1}{T_{sat} - dT} - \frac{1}{T_{sat}}\right)$$
$$= q\frac{T_{sat} - T_{sat} + dT}{(T_{sat} - dT)(T_{sat})}$$
$$= q\frac{dT}{(T_{sat} - dT)(T_{sat})}$$

Thus, as the infinitesimal temperature difference is reduced to zero, the heat transfer becomes reversible, or

$$S_{gen} \to 0 \quad \text{as} \quad dT \to 0$$

5.7 Classical Approach to the Second Law of Thermodynamics†

After a physical explanation of the general phenomena, the second law of thermodynamics was stated in Sec. 5.2 in terms of an equilibrium property, entropy, and a direction for the change of this property for an isolated system. Reversible and irreversible processes were then differentiated by the direction and magnitude of the change in entropy. Temperature is defined and justified in Sec. 5.4. The second law of thermodynamics for a control mass or closed system is then developed in Sec. 5.6. An equivalent approach to the development of the second law of thermodynamics that is more historically oriented and closely tied with the original development of the second law of thermodynamics is presented in this section.

Note that this presentation is independent of the material in Secs. 5.2 through 5.6, and it need not be considered for a complete understanding of the second law. Or this section can be used in lieu of those sections (students should return to Sec. 5.5 after completing this section if the previous material has not been covered). After completion of this section and/or Secs. 5.2, 5.3, 5.4, and 5.6, students should be at the same level of understanding of thermodynamics. Due to the nature of this presentation as compared to the presentation in the previous sections, the material contained here is somewhat repetitive. But if there is sufficient time for students to cover both perspectives of the second law, a deeper understanding will result.

As in any fundamental law, the second law of thermodynamics is an expression of our observations and experiences. The original statements of the second law were closely aligned with thermodynamic system performance where the systems of interest were typically operating in cycles. Thus the presentation in this section starts with a discussion of cycles before we state the second law of thermodynamics. The system statements of the second law and the conclusions drawn from these statements lead to a definition of temperature. The final step in obtaining the change in entropy for a control mass is to develop the relation between heat transfer and temperature for a cyclic process.

5.7.1 Cycles

A cyclic process is a series of individual processes that start at some initial state and return to this same state. The individual components of a general cycle vary from application to application but share the general characteristic that the working fluid returns to the initial state. In an idealized power cycle or engine, the working fluid at high-temperature high-pressure state is expanded through a turbine to produce work. This

† This section contains an alternative presentation of the second law of thermodynamics.

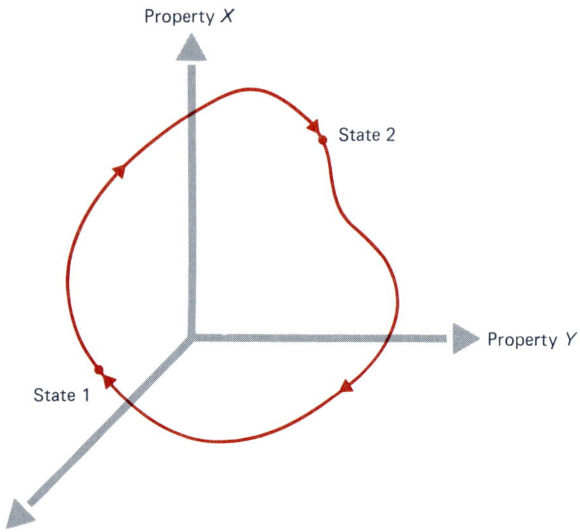

Figure 5.10 State diagram representation of a cyclic process.

same working fluid then undergoes a heat transfer to a cold reservoir, which produces a low-temperature working fluid. This fluid is then pumped or compressed to a high pressure and receives a heat transfer from a high-temperature source. The resultant fluid is at the original high-temperature high-pressure state that enters the turbine, and the cycle is complete. A general state diagram for a cycle is shown in Fig. 5.10.

The discussion of the power cycle above is also represented schematically in Fig. 5.11 as a two-temperature engine. The high-temperature reservoir at T_H is the energy source for the power cycle, and the low-temperature reservoir at T_L receives the heat transfer from the fluid. The net work produced by the engine W is the work output of the turbine minus the work input to the pump or compressor.

Figure 5.11 Two-temperature power cycle or engine.

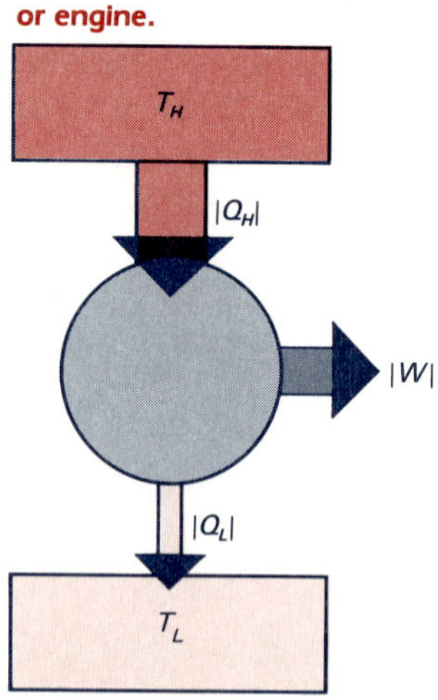

This presentation of a cyclic process has also introduced the concept of the thermal reservoir. This is a control mass system that is extremely large, so a heat transfer to or from such a system does not change the state of the system. System boundaries in contact with thermal reservoirs will always be at the reservoir temperature. In Fig. 5.11, the thermal reservoirs labeled T_H and T_L are closed systems that extend infinitely in one direction. Note that the two-temperature cycles discussed above are special cases of more general power cycles because heat transfers to practical cycles occur from or to thermal reservoirs at various temperatures, rather than simply two.

The efficiency of a cycle is an important measure of the attributes of the system under consideration. The *cycle efficiency* is defined as

$$\eta = \frac{\text{desired output}}{\text{required input}} \tag{5.93}$$

This definition compares the cycle's net desired energy output to the required energy input. In the discussion of the idealized power cycle or engine above, the desired output is the net work produced by the cycle (turbine output minus pump or compressor input), and the required input is the heat transfer from the high-temperature source. The power cycle efficiency is then

$$\eta = \frac{W}{Q_H} \tag{5.94}$$

The first law for a cycle yields

$$W = |Q_H| - |Q_L| \tag{5.95}$$

so that the power cycle efficiency is

$$\eta = 1 - \left|\frac{Q_L}{Q_H}\right| \tag{5.96}$$

The entire cycle can also be run in the opposite direction, so that work is the input to the cycle and there is a heat transfer from the low-temperature to the high-temperature reservoir. This type of cycle is referred to as a *refrigerator.* The cycle efficiency in this case has the same basic definition as given in Eq. (5.93) but is termed the *coefficient of performance* (COP). The definition is

$$\text{COP} = \frac{\text{desired output}}{\text{required input}} = \frac{Q_L}{W} \tag{5.97}$$

where the desired output of the refrigerator is the heat transfer from the cold space and the required input is the work necessary to produce such a heat transfer. The first law for this refrigerator cycle is

$$\text{COP} = \frac{|Q_L|}{|Q_H| - |Q_L|} = \frac{1}{|Q_H/Q_L| - 1} \tag{5.98}$$

The schematic of this refrigerator cycle is shown in Fig. 5.12.

Figure 5.12 Two-temperature refrigerator cycle.

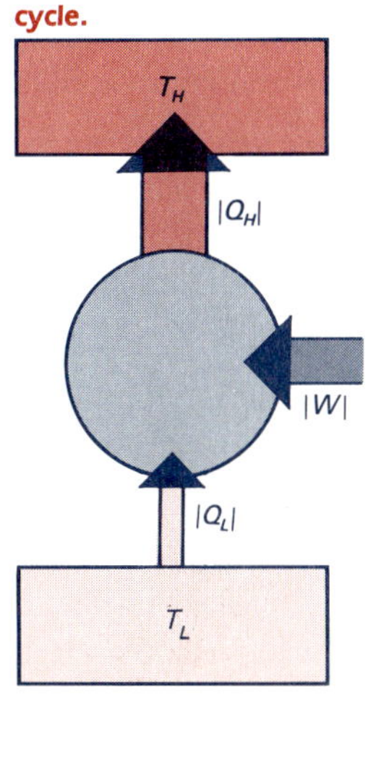

5.7.2 Reversible and Irreversible Processes

A reversible process is a highly idealized process which can be reversed at any point and returned through the same states to the original starting point. All actual processes are irreversible and will only approximate a reversible process in special cases. Referring to Fig. 5.5, we see that the unrestrained expansion of a gas into vacuum is a highly irreversible process, since at any instant the gas expansion cannot be reversed through the same states as occur in the expansion. If the gas expansion is restrained by the piston and the piston proceeds at an extremely slow rate, the piston motion can be reversed at any point and the gas will proceed in reverse through the same states as in the expansion. This latter case is a reversible process. Expansions against an inelastic spring or against friction forces are typically irreversible processes while expansions against elastic springs and frictionless surfaces are potentially reversible. Irreversible processes generally include dissipative effects and the possibility of the process proceeding through nonequilibrium states.

To proceed with a clear understanding of a reversible and irreversible process, it is important to clearly separate the system behavior from the characteristics of the surroundings. Strictly, a reversible process is one where both the system and the surroundings undergo processes that can be reversed at any point and proceed in reverse through the same states as in the initial direction. Such a reversible process is often termed a *totally reversible* process. A *totally irreversible* process (or irreversible process) requires that both the system and the surroundings undergo irreversible processes. These reversible and irreversible processes are differentiated from *internally reversible* and *internally irreversible* processes where only the system, and not the surroundings, undergoes the appropriate process. For an internally reversible process the system undergoes a reversible process and all the transfers with the surroundings are reversible, but the surroundings need not undergo a reversible process. Likewise, externally reversible applies to the surroundings. The internally irreversible and externally irreversible processes follow from the above definitions.

5.7.3 Statements of the Second Law of Thermodynamics and Resultant Conclusions

Various statements of the second law are appropriate starting points for the development of the useful expressions of the second law. The Kelvin-Planck statement [3] deals with the transformation of a single heat transfer to work. This second law statement is

> It is impossible to construct an engine that, operating continuously, will produce no effect other than the extraction of heat from a single reservoir and the performance of an equivalent amount of work.

There are a number of key elements in this statement. An engine operating continuously implies a cyclic process in which the system returns to its original state. This is particularly important since it places first law restrictions on the magnitude and direction of the heat transfer. The next key element is the heat transfer from only one reservoir.

The Clausius version [3] of the second the law states

> It is impossible to construct a device that, operating continuously, will produce no effect other than the transfer of heat from a cooler to a hotter body.

Again operating continuously implies a cyclic process with a heat transfer from the low-temperature reservoir to the device and a heat transfer to the

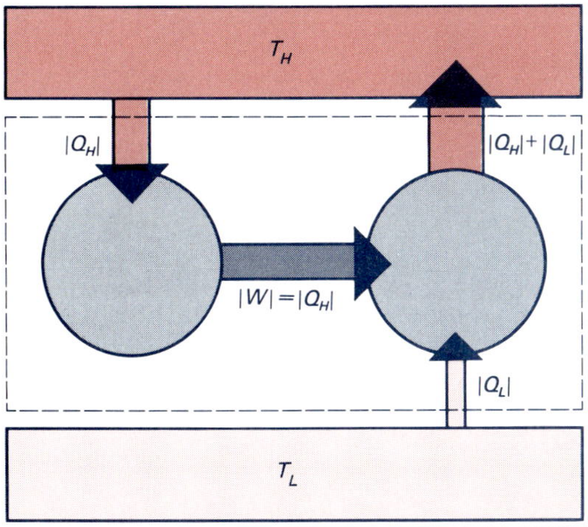

Figure 5.13 Composite cycle to demonstrate that a violation of the Kelvin-Planck statement leads to a violation of the Clausius statement.

high-temperature reservoir. There is no work transferred to perform this impossible cycle.

Both of the above statements of the second law are equivalent, i.e., a violation of the Kelvin-Planck statement implies a violation of the Clausius statement, and a violation of the Clausius statement leads to a violation of the Kelvin-Planck statement. These violations are demonstrated by a two-temperature cycle which is a composite of an ordinary two-temperature engine or refrigerator and either a Kelvin-Planck or Clausius violation device. For example, consider the composite cycle within the dotted box in Fig. 5.13. Within this composite system, we have a cyclic engine that receives a heat transfer from T_H and produces work $W = Q_H$ and a refrigerator that receives this work input, withdraws a heat transfer from T_L, and provides a heat transfer to T_H. The engine portion of this composite cycle violates the Kelvin-Planck statement of the second law. Focusing on the composite cycle within the dotted box, we see that the composite cycle withdraws a heat transfer from the low-temperature reservoir and transfers a heat transfer to the high-temperature reservoir. The net heat transfer from T_L is Q_L, and the net heat transfer to T_H is also Q_L. There is no work supplied to the composite cycle. This violates the Clausius statement of the second law. Alternately, the demonstration that a violation of the Clausius statement yields a violation of the Kelvin-Planck statement is left to the student.

From the Kelvin-Planck or Clausius statement of the second law, two important conclusions are drawn. These conclusions or principles are needed to define the temperature scale and to subsequently define entropy. One conclusion is that

> The cycle efficiency of an irreversible power cycle is always less than the cycle efficiency of a reversible power cycle when each operates between the same two reservoirs.

Mathematically,

$$\eta_{\text{rev}} > \eta_{\text{irrev}} \tag{5.99}$$

or, for the same heat transfers *to* the reversible and irreversible cycle Q_H,

$$W_{\text{rev}} > W_{\text{irrev}} \tag{5.100}$$

The first law yields

$$(Q_H - |Q_L|)_{\text{rev}} > (Q_H - |Q_L|)_{\text{irrev}} \tag{5.101}$$

Yet the input heat transfers Q_H have been taken to be the same, so that

$$|Q_{L,\text{ rev}}| < |Q_{L,\text{ irrev}}| \tag{5.102}$$

These facts will be needed below in the development of the Clausius inequality.

The second conclusion from the second law is

> All reversible power cycles operating between the same two thermal reservoirs have the same cycle efficiency.

This rather simple statement implies that the specific states, working fluid, or specific processes of the reversible power cycle are unimportant to the cycle efficiency. Reversibility and reservoir temperatures prescribe the cycle efficiency. These facts will be important in defining the thermodynamic temperature in the next subsection.

These last two conclusions follow directly from the Kelvin-Planck statement or the Clausius statement of the second law. To demonstrate the first conclusion indicated in Eq. (5.99), consider two separate engines (each diagrammed in Fig. 5.11) operating between the same high-temperature reservoir and the same low-temperature reservoir and each receiving a heat transfer of Q_H. One engine is reversible and produces W_{rev}, while the other engine is irreversible and produces W_{irrev}. Assume that the cycle efficiency of the irreversible engine is greater than that of the reversible engine, i.e.,

$$\eta_{\text{irrev}} > \eta_{\text{rev}} \tag{5.103}$$

or, because both engines receive a heat transfer of Q_H,

$$W_{\text{irrev}} > W_{\text{rev}} \tag{5.104}$$

If the reversible engine is reversed and operated as a refrigerator (see Fig. 5.12), the work input W_{rev} to the refrigerator produces a heat transfer Q_H to the high-temperature reservoir. By viewing the irreversible engine and the reversible refrigerator as a composite system, the net work *output* by the composite system is

$$W_{\text{net}} = W_{\text{irrev}} - W_{\text{rev}} > 0 \tag{5.105}$$

since W_{irrev} is assumed to be greater than W_{rev}. But the net heat transfer to the high-temperature reservoir is zero because the irreversible engine withdraws Q_H and the reversible refrigerator supplies Q_H. Thus the composite system produces a net work output from a heat transfer with a single heat transfer reservoir. This is a violation of the Kelvin-Planck statement of the second law. Therefore the assumption that $\eta_{rev} < \eta_{irrev}$ is false, and the first conclusion—that no engine operating between the same two thermal reservoirs has a greater cycle efficiency than a reversible engine—is valid.

The second conclusion can be demonstrated by a similar approach using a composite system of two engines.

5.7.4 Thermodynamic Temperature Scale

Because the cycle efficiency of a reversible power cycle operating between two thermal reservoirs is dependent on only the heat transfers with the thermal reservoirs, the temperatures of the thermal reservoirs are the only information needed to specify the efficiency. Thus it is possible to relate the temperature to the heat transfers. The cycle efficiency for a reversible power cycle is given in Eq. (5.96) as

$$\eta = f(T_H, T_L) = 1 - \left(\frac{|Q_L|}{Q_H}\right)_{\text{rev}} \tag{5.106}$$

where the subscript *rev* is added to indicate that a reversible power cycle is being considered.

The function $f(T_H, T_L)$ is unspecified at this point and is subject to the particular definition chosen for the temperature. If we use the standard Kelvin scale and define

$$f(T_H, T_L) = 1 - \frac{T_L}{T_H} \tag{5.107}$$

then

$$\left(\frac{|Q_L|}{Q_H}\right)_{\text{rev}} = \frac{T_L}{T_H} \tag{5.108}$$

Although the specification of the function may seem arbitrary, this choice satisfies all the intuitive constraints on the temperature.

The complete specification of the temperature scale requires the assignment of a reference point. With such an assignment the temperature scale is completed. In the case of the Kelvin scale, the choice of the reference point is 273.16 K for the triple point of water.

5.7.5 The Clausius Inequality

The Clausius inequality provides a cyclic relationship between heat transfer and temperature. From the definition of the thermodynamic temperature scale, the relation between heat transfer and the temperature for a two-temperature reversible cycle in Eq. (5.108) is rearranged to yield

$$\frac{|Q_{L,\,\text{rev}}|}{T_L} = \frac{|Q_{H,\,\text{rev}}|}{T_H} \tag{5.109}$$

For a reversible power cycle with constant-temperature thermal reservoirs, the cyclic integral of the quantity $\delta Q/T$ is

$$\oint \left(\frac{\delta Q}{T}\right)_{\text{rev}} = \frac{|Q_{H,\,\text{rev}}|}{T_H} - \frac{|Q_{L,\,\text{rev}}|}{T_L} = 0 \tag{5.110}$$

where the definition of the thermodynamic temperature scale given in Eq. (5.109) is used to give the final value of zero.

The above statements are valid for a reversible cycle. An irreversible power cycle gives $|Q_{L,\,\text{rev}}| < |Q_{L,\,\text{irrev}}|$ from Eq. (5.102). The cyclic integral of the quantity $\delta Q/T$ is given from Eq. (5.110) as

$$\oint \left(\frac{\delta Q}{T}\right) = \frac{|Q_H|}{T_H} - \frac{|Q_{L,\,\text{irrev}}|}{T_L} < 0 \tag{5.111}$$

Thus we can draw the important conclusion that

$$\oint \frac{\delta Q}{T} \le 0 \tag{5.112}$$

which is the *Clausius inequality.* Although it has not been demonstrated here, the same conclusion is obtained for a refrigeration cycle, so Eq. (5.112) is a general result.

5.7.6 Entropy

The Clausius inequality is a conclusion drawn from the second law of thermodynamics and expresses an important relation between heat transfer and absolute temperature. If we consider a reversible cycle, then the Clausius inequality yields

$$\oint \left(\frac{\delta Q}{T}\right)_{\text{rev}} = 0 \tag{5.113}$$

The quantity $(\delta Q/T)_{\text{rev}}$ has the mathematical properties of an exact differential, which indicates that the relation $(\delta Q/T)_{\text{rev}}$ is independent of path and therefore is a system property. The general discussion of the properties of exact differentials is given in App. H.

Figure 5.14 diagrams a cycle from state 1 to state 2 by process A and a return to state 1 via process B. The Clausius inequality for this process yields

$$\int_1^2 \left(\frac{\delta Q}{T}\right)_{\text{CM, rev, }A} + \int_2^1 \left(\frac{\delta Q}{T}\right)_{\text{CM, rev, }B} = 0 \tag{5.114}$$

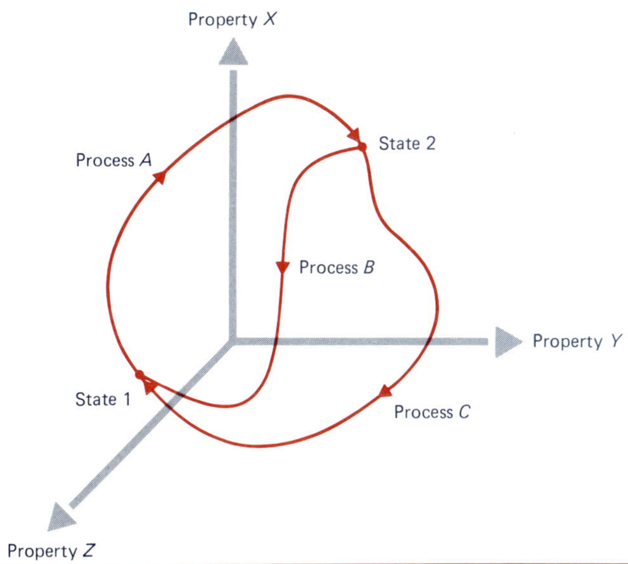

Figure 5.14 Reversible processes between states 1 and 2.

There is also a complete cycle from state 1 to state 2 to state 1 by process A and then process C. The Clausius inequality for this cycle is

$$\int_1^2 \left(\frac{\delta Q}{T}\right)_{\text{CM, rev, } A} + \int_2^1 \left(\frac{\delta Q}{T}\right)_{\text{CM, rev, } C} = 0 \tag{5.115}$$

Eliminating the integral for process A from Eqs. (5.114) and (5.115) yields

$$\int_1^2 \left(\frac{\delta Q}{T}\right)_{\text{CM, rev, } B} = \int_2^1 \left(\frac{\delta Q}{T}\right)_{\text{CM, rev, } C} \tag{5.116}$$

Since the processes are entirely arbitrary, this equality shows that the quantity $(\delta Q/T)_{\text{rev}}$ is independent of path and depends on only the initial and final states. Thus this quantity is a property of the system, and this new property is defined to be *entropy* S. Entropy is valued only for thermodynamic equilibrium, and the change in entropy between two states is given by

$$S_2 - S_1 = \int_1^2 \left(\frac{\delta Q}{T}\right)_{\text{int rev}} \tag{5.117}$$

The subscript indicating internally reversible rather than reversible is appropriate because entropy is a system property and not dependent on the surroundings. For a differential process the entropy change is

$$\delta S = \left(\frac{\delta Q}{T}\right)_{\text{int rev}} \tag{5.118}$$

The property, entropy, has the characteristics of all other properties so that values of entropy are determined by two independent properties for a simple compressible system. For more complex systems, the state postulate determines the number of independent properties that determine the state. Entropy in its extensive form has units of joules per kelvin or Btu per degree Rankine. Section 5.4 presents methods to determine the values for entropy as well as many useful relations involving entropy.

5.8 Control Volume Formulation and Analysis

Control volume analysis permits consideration of systems in which mass crosses the boundary. This mass transfer carries the property entropy into and out of the volume. Therefore, the change in entropy within the control volume is modified by the mass transport. The control mass form of the second law states that the change in entropy within the boundary is related to the quantity $\delta Q/T$ at the boundary. Thus, this is the control volume statement of the second law:

> The change in entropy within the control volume minus the net entropy transported into the control volume is greater than or equal to the sum of the heat transfers divided by the corresponding boundary absolute temperatures.

Let $\dot{S}$ denote the rate at which entropy is transported with the fluid at the inlets and outlets, where the overdot indicates the rate of flow of entropy. The entropy within the control volume is denote by S_{CV}, so the second law is

$$\frac{dS_{CV}}{dt} - \dot{S}_{net} \geq \sum_i \left(\frac{\dot{Q}_i}{T_i}\right)_{CV} \tag{5.119}$$

Or, with the net rate of flow of entropy equal to the inlet minus the outlet contributions,

$$\frac{dS_{CV}}{dt} - \dot{S}_{in} + \dot{S}_{out} \geq \sum_i \left(\frac{\dot{Q}_i}{T_i}\right)_{CV} \tag{5.120}$$

The terms on the left-hand side of the inequality require detailed consideration. The mass transport contributions into and out of the control volume, denoted $\dot{S}_{in}$ and $\dot{S}_{out}$, respectively, follow the development presented in the control volume formulation of conservation of mass and energy. The fluid element carries entropy across each boundary element, so the transported entropy is the mass flow rate in Eqs. (4.31) and (4.32) multiplied by the entropy per unit mass, or

$$\dot{S}_{\text{in}} = \int_{A_{\text{in}}} s\rho\,|\mathbf{V}\cdot\mathbf{n}|\,dA = \int_{A_{\text{in}}} s\,d\dot{m} = (s\dot{m})_{\text{in}} \tag{5.121}$$

and

$$\dot{S}_{\text{out}} = \int_{A_{\text{out}}} s\rho\,|\mathbf{V}\cdot\mathbf{n}|\,dA = \int_{A_{\text{out}}} s\,d\dot{m} = (s\dot{m})_{\text{out}} \tag{5.122}$$

Thus the net transport is

$$\dot{S}_{\text{net}} = (s\dot{m})_{\text{in}} - (s\dot{m})_{\text{out}} \tag{5.123}$$

The control volume form of the second law is given as

$$\frac{dS_{\text{CV}}}{dt} - (s\dot{m})_{\text{in}} + (s\dot{m})_{\text{out}} \geq \sum_i \left(\frac{Q_i}{T_i}\right)_{\text{CV}} \tag{5.124}$$

For multiple inlets and outlets, this is

$$\frac{dS_{\text{CV}}}{dt} - \sum_{\text{in}} (s\dot{m})_{\text{in}} + \sum_{\text{out}} (s\dot{m})_{\text{out}} \geq \sum_i \left(\frac{Q_i}{T_i}\right)_{\text{CV}} \tag{5.125}$$

The left-hand side of this expression should be compared with Eqs. (F.16) and (F.17); the Reynold's transport theorem could have been used to develop these expressions. Equation (5.124) is the general control volume form of the second law. The first term is the entropy storage term within the control volume. The next two terms are the entropy transported by mass transport into and out of the control volume. The right-hand side represents the entropy transport via heat transfer. The equivalent expression corresponding to Eq. (5.124) from using the entropy generation concept is

$$\dot{S}_{\text{gen}} = \frac{dS_{\text{CV}}}{dt} - (s\dot{m})_{\text{in}} + (s\dot{m})_{\text{out}} - \sum_i \left(\frac{Q_i}{T_i}\right)_{\text{CV}} \geq 0 \tag{5.126}$$

Before we proceed to the idealizations at the crossing boundaries and within the control volume, note that the above expressions [Eqs. (5.124) and (5.126)] reduce to the special cases considered in previous sections. For the case of no mass transport into or out of the control volume, these expressions reduce to

$$\frac{dS_{\text{CV}}}{dt} \geq \sum_i \left(\frac{\dot{Q}_i}{T_i}\right)_{\text{CV}}$$

$$\dot{S}_{\text{gen}} = \frac{dS_{\text{CV}}}{dt} - \sum_i \left(\frac{Q_i}{T_i}\right) \geq 0$$

These are the same expressions given in Eqs. (5.81) and (5.82). If the system is isolated so that there are no interactions with the surroundings, the heat transfers are zero, and

$$\frac{dS_{CV}}{dt} \geq 0$$

$$\dot{S}_{gen} \geq 0$$

These expressions are equivalent to Eqs. (5.6) and (5.7).

A more detailed view of the entropy within the control volume and that crosses the control surface is now presented. The change in entropy within the control volume is given as

$$\frac{dS_{CV}}{dt} = \frac{d}{dt}\left(\int_{V_{CV}} \rho s \, dV\right) \tag{5.127}$$

The entropy carried across the boundary element by the fluid is obtained by multiplying the entropy per unit mass and the mass flow rate per unit area given in Eq. (4.64) to yield

$$d\dot{S} = s\rho \,|\mathbf{V} \cdot \mathbf{n}|\, dA = s \, d\dot{m} \tag{5.128}$$

The net entropy transport into and out of the control volume is

$$\begin{aligned} \dot{S}_{net} &= \int_{A_{in}} s\rho \,|\mathbf{V} \cdot \mathbf{n}|\, dA - \int_{A_{out}} s\rho \,|\mathbf{V} \cdot \mathbf{n}|\, dA \\ &= \int_{A_{in}} s \, d\dot{m} - \int_{A_{out}} s \, d\dot{m} \end{aligned} \tag{5.129}$$

Substituting Eqs. (5.127) through (5.129) into the control volume forms of the second law in Eqs. (5.125) and (5.126) gives

$$\frac{d}{dt}\left(\int_{V_{CV}} \rho s \, dV\right) - \int_{A_{in}} s \, d\dot{m} + \int_{A_{out}} s \, d\dot{m} \geq \sum_i \left(\frac{\dot{Q}_i}{T_i}\right)_{CV} \tag{5.130}$$

and

$$\frac{d}{dt}\left(\int_{V_{CV}} \rho s \, dV\right) - \sum_{in} \int_{A_{in}} s \, d\dot{m} + \sum_{out} \int_{A_{out}} s \, d\dot{m} \geq \sum_i \left(\frac{\dot{Q}_i}{T_i}\right)_{CV} \tag{5.131}$$

The second law expression using the entropy generation concept is

$$\dot{S}_{gen} = \frac{d}{dt}\left(\int_{V_{CV}} \rho s \, dV\right) - \int_{A_{in}} s \, d\dot{m} + \int_{A_{out}} s \, d\dot{m} - \sum_i \left(\frac{\dot{Q}_i}{T_i}\right)_{CV} \geq 0 \tag{5.132}$$

which is obtained from Eq. (5.126).

A summary of the governing equations for the control mass and control volume formulations is given in Table 5.1. For convenience, the corresponding conservation of mass and energy relations are also given.

TABLE 5.1 General Conservation Relations

	Control Mass	Control Volume
Mass	$\frac{dm_{CM}}{dt} = 0$	$\frac{dm_{CV}}{dt} = \sum_{in} \dot{m}_{in} - \sum_{out} \dot{m}_{out}$
First law	$\frac{dE_{CM}}{dt} = \dot{Q} - \dot{W}$	$\frac{dE_{CV}}{dt} = \sum_{in} [(e + Pv)\dot{m}]_{in} - \sum_{out} [(e + Pv)\dot{m}]_{out} + \dot{Q}_{CV} - \dot{W}_{CV}$
Second law	$\frac{dS_{CM}}{dt} \geq \sum_i \frac{\dot{Q}_i}{T_i}$ or $\dot{S}_{gen} = \frac{dS_{CM}}{dt} - \sum_i \frac{\dot{Q}_i}{T_i} \geq 0$	$\frac{dS_{CV}}{dt} + \sum_{out} (s\dot{m})_{out} - \sum_{in} (s\dot{m})_{in} \geq \sum_i \left(\frac{\dot{Q}_i}{T_i}\right)_{CV}$ or $\dot{S}_{gen} = \frac{dS_{CV}}{dt} + \sum_{out} (s\dot{m})_{out} - \sum_{in} (s\dot{m})_{in} - \sum_i \left(\frac{\dot{Q}_i}{T_i}\right)_{CV} \geq 0$

Here $e = u + \frac{1}{2}\mathbf{V}^2/g_c + gZ/g_c$.

5.8.1 Spatial and Time Variation Idealizations

The analysis of a system requires detailed consideration of the crossing boundaries and the state within the control volume. *Uniform state* and *nonuniform state* describe spatial variation, while *steady state* and *unsteady state* describe time variation. These terms describe both the crossing boundaries and the state within the control volume. The idealizations parallel the development of the energy equations. Table 5.2 presents the

TABLE 5.2 Control Volume Relations

	Nonuniform and Unsteady	Uniform	Unsteady Time Integration
Inlets and outlets	$\dot{m} = \int_A d\dot{m} = \int_A \rho\lvert\mathbf{V}\cdot\mathbf{n}\rvert\, dA$	$\dot{m} = \rho\mathbf{V}A$	$\int_{t_1}^{t_2} \dot{m}\, dt = m$
	$\int_A (e + Pv)\, d\dot{m} = \int_A (e + Pv)\rho\lvert\mathbf{V}\cdot\mathbf{n}\rvert\, dA$	$(e + Pv)\dot{m}$	$\int_{t_1}^{t_2}\int_A (e + Pv)\, d\dot{m}\, dt = (e + Pv)m$ with uniform and steady state properties
	$\int_A s\, d\dot{m} = \int_A s\rho\lvert\mathbf{V}\cdot\mathbf{n}\rvert\, dA$	$s\dot{m}$	$\int_{t_1}^{t_2}\int_A s\, d\dot{m}\, dt = sm$ with uniform and steady state properties
Within the control volume	$m_{CV} = \int_{V_{CV}} \rho\, dV$	$m_{CV} = (\rho V)_{CV}$	$\int_{t_1}^{t_2} \frac{dm_{CV}}{dt}\, dt = m_2 - m_1$
	$E_{CV} = \int_{V_{CV}} \rho e\, dV$	$E_{CV} = (\rho eV)_{CV} = (me)_{CV}$	$\int_{t_1}^{t_2} \frac{dE_{CV}}{dt}\, dt = E_2 - E_1 = (em)_2 - (em)_1$ with uniform properties
	$S_{CV} = \int_{V_{CV}} \rho s\, dV$	$S_{CV} = (\rho sV)_{CV} = (ms)_{CV}$	$\int_{t_1}^{t_2} \frac{dS_{CV}}{dt}\, dt = S_2 - S_1 = (sm)_2 - (sm)_1$ with uniform properties

idealizations for the first and second laws as well as the conservation of mass (this is an expanded version of Table 4.1). Recall that the uniform state idealization eliminates the spatial integration and that the unsteady state analysis generally requires a time integration.

The two classifications of problems that often occur in engineering analysis are the steady state and unsteady state idealizations with uniform states at the crossing boundaries and within the control volume. For *steady state analysis* there is no time variation, so Eq. (5.131) reduces to

$$\sum_{\text{out}} \int_{A_{\text{out}}} s\, d\dot{m} - \sum_{\text{in}} \int_{A_{\text{in}}} s\, d\dot{m} \geq \sum_{i} \left(\frac{\dot{Q}_i}{T_i}\right)_{\text{CV}} \tag{5.133}$$

With uniform properties at the inlets and outlets, this results in

$$\sum_{\text{out}} s\dot{m} - \sum_{\text{in}} s\dot{m} \geq \sum_{i} \left(\frac{\dot{Q}_i}{T_i}\right)_{\text{CV}} \tag{5.134}$$

The entropy generation form for steady analysis is

$$\dot{S}_{\text{gen}} = \sum_{\text{out}} s\dot{m} - \sum_{\text{in}} s\dot{m} - \sum_{i} \left(\frac{\dot{Q}_i}{T_i}\right)_{\text{CV}} \geq 0 \tag{5.135}$$

The *unsteady state analysis* requires the time integration of Eq. (5.125). With uniform properties, this integration yields

$$s_2 m_2 - s_1 m_1 - \sum_{\text{in}} sm + \sum_{\text{out}} sm \geq \int_{t_1}^{t_2} \sum_{i} \left(\frac{\dot{Q}_i}{T_i}\right)_{\text{CV}} dt \tag{5.136}$$

The time integration of Eq. (5.126) with Table 5.2 yields

$${}_1S_{\text{gen},\,2} = s_2 m_2 - s_1 m_1 + \sum_{\text{out}} sm - \sum_{\text{in}} sm - \int_{t_1}^{t_2} \sum_{i} \left(\frac{\dot{Q}_i}{T_i}\right)_{\text{CV}} dt \geq 0 \tag{5.137}$$

These equations [Eqs. (5.134) through (5.137)] are summarized, along with the corresponding first law and conservation of mass, in Table 5.3.

5.8.2 Applications

The application of the second law to specific problems requires the same general approach as the first law analysis. A review of the general statements in Sec. 4.5 is appropriate at this point. A careful denotation of the system and boundary transfers is important. Now the second law is used as another governing relation. Properties available include the new property of entropy. A steady state application is presented in Example Problem 5.7, and unsteady applications are seen in Example Problems 5.8 and 5.9.

TABLE 5.3 Relations for Special Cases

Steady State with Uniform Properties

Mass: $$\sum_{\text{in}} \dot{m} = \sum_{\text{out}} \dot{m}$$

First law: $$\sum_{\text{in}} \left(h + \frac{1}{2}\frac{\mathbf{V}^2}{g_c} + \frac{gZ}{g_c}\right)\dot{m} + \dot{Q}_{\text{CV}} = \sum_{\text{out}} \left(h + \frac{1}{2}\frac{\mathbf{V}^2}{g_c} + \frac{gZ}{g_c}\right)\dot{m} + \dot{W}_{\text{CV}}$$

Second law: $$\sum_{\text{out}} s\dot{m} - \sum_{\text{in}} s\dot{m} \geq \sum_{i} \left(\frac{\dot{Q}_i}{T_i}\right)_{\text{CV}}$$

or

$$\dot{S}_{\text{gen}} = \sum_{\text{out}} s\dot{m} - \sum_{\text{in}} s\dot{m} - \sum_{i} \left(\frac{\dot{Q}_i}{T_i}\right)_{\text{CV}} \geq 0$$

Unsteady State with Uniform Properties and Steady State Port Properties

Mass: $$m_2 - m_1 = \sum_{\text{in}} m - \sum_{\text{out}} m$$

First law: $$\left(u + \frac{1}{2}\frac{\mathbf{V}_{\text{ref}}^2}{g_c} + \frac{gZ}{g_c}\right)_2 m_2 - \left(u + \frac{1}{2}\frac{\mathbf{V}_{\text{ref}}^2}{g_c} + \frac{gZ}{g_c}\right)_1 m_1$$
$$= \sum_{\text{in}} \left(h + \frac{1}{2}\frac{\mathbf{V}^2}{g_c} + \frac{gZ}{g_c}\right) m + {}_1Q_2 - {}_1W_2 - \sum_{\text{out}} \left(h + \frac{1}{2}\frac{\mathbf{V}^2}{g_c} + \frac{gZ}{g_c}\right) m$$

Second law: $$s_2 m_2 - s_1 m_1 + \sum_{\text{out}} sm - \sum_{\text{in}} sm \geq \int_{t_1}^{t_2} \sum_{i} \left(\frac{\dot{Q}_i}{T_i}\right)_{\text{CV}} dt$$

or

$${}_1S_{\text{gen, 2}} = s_2 m_2 - s_1 m_1 + \sum_{\text{out}} sm - \sum_{\text{in}} sm - \int_{t_1}^{t_2} \sum_{i} \left(\frac{\dot{Q}_i}{T_i}\right)_{\text{CV}} dt \geq 0$$

Example Problem 5.7

A steam turbine inlet state is given as 6 MPa and 500°C, and the outlet pressure is 10 kPa. Determine the work output per unit mass if the process is reversible and adiabatic. Evaluate the work output per unit mass if the turbine is adiabatic and the output quality is 0.90 and 1.0. Compare these outputs to the reversible adiabatic case, and calculate the entropy generation.

Diagram

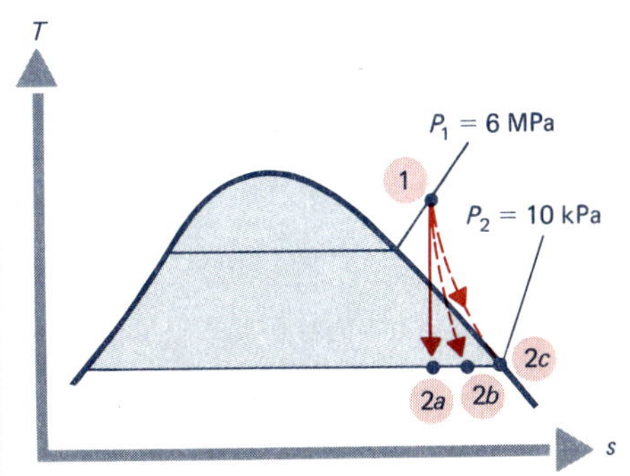

Solution

The state information given above and Tables D.11 and D.12 yield

State	P, kPa	T, °C	h, kJ/kg	x	s, kJ/(kg·K)
1	6000	500	(3422.3)	—	($s_1 = 6.8805$)
2a	10		(2179.3)	(0.8309)	s_1
2b	10		(2344.6)	0.9	(7.3983)
2c	10		(2583.8)	1.0	(8.1482)

The first law for an adiabatic turbine is

$$w = h_{in} - h_{out} = h_1 - h_2$$

The second law is

$$s_{gen} = s_{out} - s_{in} - \left(\frac{q}{T}\right)_{CV} \geq 0$$

but the turbine is adiabatic, so

$$s_{gen} = s_2 - s_1 \geq 0$$

The states are all specified, so the enthalpies are obtained. The results for w and s_{gen} are as follows:

Process	w, kJ/kg	w, Btu/lbm	$\frac{w}{_1w_{2a}}$, %	s_{gen}, kJ/(kg·K)	s_{gen}, Btu/(lbm·°R)
1-2a	1243.0	534.4	100.0	0.0	0.0
1-2b	1077.7	463.3	86.7	0.5178	0.1237
1-2c	838.4	360.4	67.4	1.2677	0.3028

Comments

The turbine is adiabatic, so the loss in work and the increase in the entropy generation are a result of the irreversibilities in the process. The percentage drop in the work output from the reversible case increases as the output moves toward the saturated vapor line.

The outlet state of the turbine is slightly "wet." Some power plants provide saturated steam to the turbine, and the turbines must be designed to remove even greater amounts of condensed liquids. This liquid must be removed before it passes through the turbine blades. This is done by including centrifugal liquid removal stages in the turbine that spin the liquid to the outer casing.

Example Problem 5.8

An initially evacuated tank of volume V is filled with an ideal gas. The inlet is controlled with the valve, and the inlet properties are constants at T_{in} and P_{in}. Determine the final tank temperature T_2 when the tank pressure is P_2. Evaluate the entropy generation for the filling process. The process is adiabatic.

Diagram

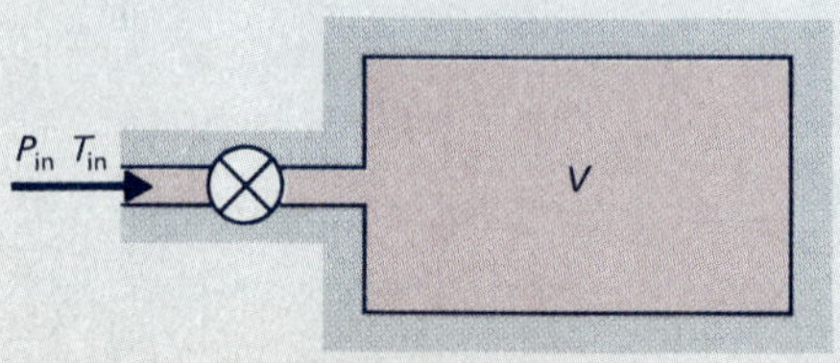

Solution

The state information is represented as follows:

State	P	T	m
In	P_{in}	T_{in}	
1	—	—	0.0
2	P_2		

The initial state is eliminated because nothing exists in the tank. The governing equations are used to specify the final state. The governing equations are

Mass:

$$m_2 - \cancelto{0}{m_1} = m_{in}$$

First law:

$$m_2 u_2 = h_{in} m_{in}$$

Second law:

$${}_1S_{gen,\,2} = m_2 s_2 - m_{in} s_{in} \geq 0$$

By eliminating m_{in}, the first and second laws are

$$m_2 u_2 = h_{in} m_2$$

and

$${}_1S_{gen,\,2} = m_2 s_2 - m_2 s_{in} = m_2(s_2 - s_{in}) \geq 0$$

Thus

$$u_2 = h_{in}$$

and

$${}_1S_{gen,\,2} = s_2 - s_{in} \geq 0$$

The properties for an ideal gas give $h = u + Pv = u + RT$, so the first law is

$$u_2 = u_{in} + RT_{in}$$

Then

$$du = c_v\,dT$$

and with $c_P = c_v + R$,

$$c_v T_2 = c_v T_{in} + RT_{in} = c_P T_{in}$$

Thus with $k = c_P/c_v$,

$$T_2 = \frac{c_P}{c_v} T_{in} = kT_{in}$$

The change in entropy is given in Eq. (5.53) as

$$ds = c_P \frac{dT}{T} - \frac{R}{P}\,dP$$

Therefore

$${}_1S_{\text{gen},\,2} = c_P \ln \frac{T_2}{T_{\text{in}}} - R \ln \frac{P_2}{P_{\text{in}}}$$

or

$${}_1S_{\text{gen},\,2} = c_P \ln k - R \ln \frac{P_2}{P_{\text{in}}}$$

Comments

The final temperature is always greater than the inlet temperature since $k > 1$. The entropy generation is always positive. This is seen from the last equation. Rearranging yields

$${}_1S_{\text{gen},\,2} = c_P \ln k + R \ln \frac{P_{\text{in}}}{P_2}$$

With $k > 1$ and $P_{\text{in}} \geq P_2$ for a filling process, the entropy generation is greater than zero. The filling process with an ideal gas is always irreversible.

Example Problem 5.9

A 0.10-ft^3 chamber contains refrigerant 12 at 100°F. Initially, one-tenth of the total volume is liquid, and the remainder is vapor. The chamber is filled by a refrigerant 12 source which is at $P_{\text{in}} = 150$ psia and 150°F. The tank is maintained at 100°F by a heat transfer with the surroundings at 100°F. The process ends when the tank is filled entirely with liquid. Determine the heat transfer and entropy generation during this process.

Diagram

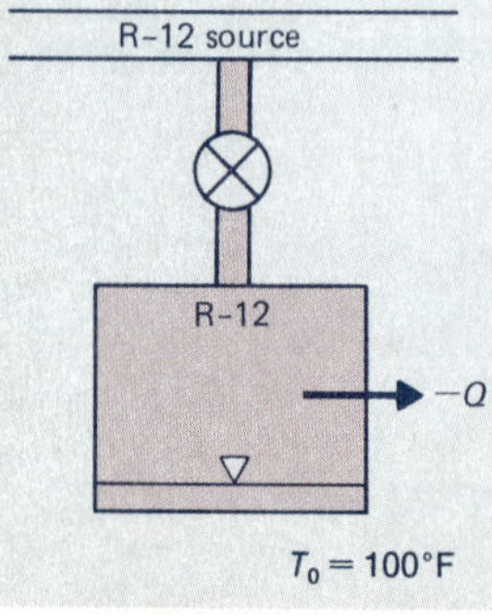

Solution

The necessary information is given to determine the states. The initial liquid volume and vapor volume give the quality of that state, since

$$x = \frac{m_g}{m} = \frac{m_g}{m_g + m_l} = \frac{1}{1 + m_l/m_g} = \frac{1}{1 + V_l v_g/(V_g v_l)}$$

The state information is tabulated as follows:

State	P, psia	T, °F	x	u, Btu/lbm	h, Btu/lbm	s, Btu/(lbm·°R)
In	150	150	—	—	(92.25)	(0.1767)
1	—	100	0.2705	(44.041)	—	(0.09096)
2	—	100	0	(30.859)	—	(0.06362)

The conservation statements for this unsteady filling problem are as follows:

Mass:

$$m_2 - m_1 = m_{in}$$

First law:

$$u_2 m_2 - u_1 m_1 = h_{in} m_{in} + {}_1Q_2$$

Second law:

$${}_1S_{gen,\,2} = s_2 m_2 - s_1 m_1 - s_{in} m_{in} - \int_{t_1}^{t_2} \frac{\dot{Q}}{T}\, dt \geq 0$$

The volume is given by $V = 0.1$ ft^3, so the initial mass and final mass are obtained with the specific volumes. Therefore, $m_1 = 1.0800$ lbm and $m_2 = 7.8790$ lbm, and conservation of mass yields $m_{in} = 6.7990$ lbm. The heat transfer is evaluated from the first law, which is rearranged as

$$\begin{aligned} {}_1Q_2 &= u_2 m_2 - u_1 m_1 - h_{in} m_{in} \\ &= 30.859(7.8790) - 44.041(1.0800) - 95.25(6.7990) \\ &= -452.0 \text{ Btu} \end{aligned}$$

This is the heat loss *to* the surroundings.

The entropy generation is determined from the second law. The temperature T_0 at the boundary is constant at 100°F, so the second law becomes

$${}_1S_{gen,\,2} = s_2 m_2 - s_1 m_1 - s_{in} m_{in} - \frac{{}_1Q_2}{T_0}$$

Substituting the values yields

$$ {}_1S_{\text{gen},\,2} = 0.06362(7.8790) - 0.09096(1.0800) - 0.1767(6.7990) - \left(\frac{-452.0}{559.67}\right) $$

$$ = +0.00926 \text{ Btu/°R} = 0.0176 \text{ kJ/K} $$

Comments

The refrigerant 12 and the surroundings are at the same temperature. Therefore, the chamber wall that contains refrigerant 12 is at the same constant temperature. The above analysis can be interpreted as including or not including the chamber wall (since $m_{\text{wall}}\, du_{\text{wall}}$ and $m_{\text{wall}}\, ds_{\text{wall}}$ are zero). If there is a temperature drop across the wall so that refrigerant 12 is at 100°F and the surroundings are at 90°F, then the entropy generation expression which includes the entire system of refrigerant 12 and the chamber wall requires an idealization. The heat transfer is idealized as being steady, so the change in internal energy and the entropy of the wall are zero (see Example Problem 5.6 for a similar approximation). Then the entropy generation for the refrigerant 12 and the chamber is the same as the above expression with the temperature of the surroundings replaced by (459.67 + 90)°R. The entropy generation is slightly greater for this case.

How would the above analysis be altered if the filling process were initially adiabatic to the point where the valve was closed and then the chamber was cooled as a control mass to the same final state? The states are all the same, and the governing equations are the same. Thus, the answers are the same.

5.9 Isentropic Process

An *isentropic* process is a constant-entropy process. If a control mass undergoes a process which is both reversible and adiabatic, then the second law [Eq. (5.74)] specifies the entropy change to be zero. A steady state reversible flow through an adiabatic control volume also has no entropy change from inlet to outlet [Eq. (5.126)]. Both examples are isentropic processes. Although an isentropic process might be an idealization of an actual process, it serves as a limiting case, for particular applications.

The comparison between the isentropic processes and the adiabatic but irreversible processes is shown in Fig. 5.15. The processes denoted by solid lines are the reversible processes, and the dashed lines indicate the irreversible processes. The irreversible processes proceed from the initial state of the process to larger values of entropy. The process for state 1 to

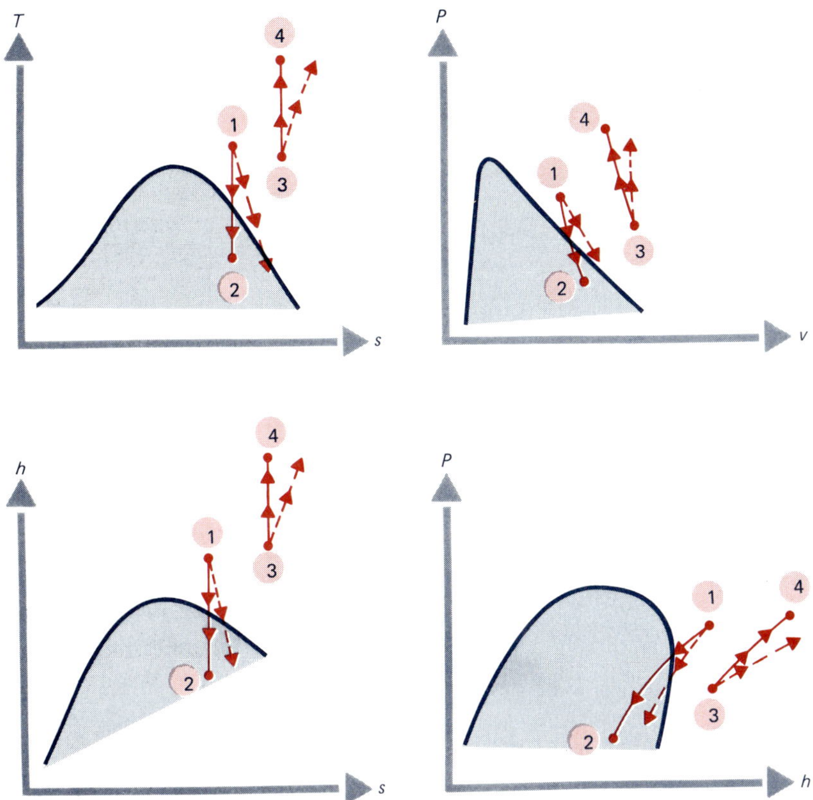

Figure 5.15 Process comparisons (1 to 2 and 3 to 4 are isentropic).

state 2 is an expansion process while the process from state 3 to state 4 is a compression process.

5.9.1 Isentropic Process for an Ideal Gas

The entropy change for an ideal gas is presented in Eqs. (5.54) and (5.55). These expressions are property relations that are independent of the particular system studied. Useful relations are obtained for the case when there is no entropy change. For such an isentropic process, $s_2 - s_1 = 0$, so these expressions become

$$0 = \int_{T_1}^{T_2} c_v \frac{dT}{T} + R \ln \frac{v_2}{v_1} \tag{5.138}$$

$$0 = \int_{T_1}^{T_2} c_P \frac{dT}{T} - R \ln \frac{P_2}{P_1} \tag{5.139}$$

The final form used depends on the approximation made for the temperature dependence of the specific heats.

Assuming that the specific heats are accurately approximated by constant values eliminates the integrals in the above equations. The *constant-specific-heat* equations are

$$0 = c_v \ln \frac{T_2}{T_1} + R \ln \frac{v_2}{v_1} \tag{5.140}$$

$$0 = c_P \ln \frac{T_2}{T_1} - R \ln \frac{P_2}{P_1} \tag{5.141}$$

Dividing by the specific heats and taking the exponential of each expression yield

$$\left(\frac{T_2}{T_1}\right)_s = \left(\frac{v_2}{v_1}\right)_s^{-R/c_v} = \left(\frac{v_1}{v_2}\right)_s^{R/c_v} \tag{5.142}$$

$$\left(\frac{T_2}{T_1}\right)_s = \left(\frac{P_2}{P_1}\right)_s^{R/c_P} \tag{5.143}$$

where the subscript indicates that the process occurs at constant entropy. The power on each expression is rewritten in terms of $k = c_P/c_v$ by noting that $c_P - c_v = R$, so $R/c_v = k - 1$ and $R/c_P = (k-1)/k$. Thus, Eqs. (5.142) and (5.143) become

$$\left(\frac{T_2}{T_1}\right)_s = \left(\frac{v_1}{v_2}\right)_s^{k-1} \tag{5.144}$$

$$\left(\frac{T_2}{T_1}\right)_s = \left(\frac{P_2}{P_1}\right)_s^{(k-1)/k} \tag{5.145}$$

A relation between pressure and volume is obtained by eliminating the ratio of temperatures in the above expressions. This yields

$$\left(\frac{P_2}{P_1}\right)_s = \left(\frac{v_1}{v_2}\right)_s^{k} \tag{5.146}$$

These last three equations are specific relations that are used for an ideal gas undergoing an isentropic process if the specific heats are taken as constants. These expressions are summarized in Table 5.4. They apply for either a control mass or a mass passing through a control volume and undergoing an isentropic process.

If the specific heats cannot be approximated as constants, then the temperature dependence of the specific heats must be included. The *variable-specific-heat* solution for an ideal gas undergoing an isentropic process is obtained from Eq. (5.60). For an isentropic process, this expression yields

$$s_2 - s_1 = 0 = s_0(T_2) - s_0(T_1) - R \ln \frac{P_2}{P_1} \tag{5.147}$$

TABLE 5.4 Ideal Gas Relations

$$Pv = RT$$
$$c_P - c_v = R$$

Variable Specific Heats	Constant Specific Heats
$u_2 - u_1 = \int_{T_1}^{T_2} c_v\,dT$	$u_2 - u_1 = c_v(T_2 - T_1)$
$h_2 - h_1 = \int_{T_1}^{T_2} c_P\,dT$	$h_2 - h_1 = c_P(T_2 - T_1)$
$s_2 - s_1 = \int_{T_1}^{T_2} c_v \frac{dT}{T} + R \ln \frac{v_2}{v_1}$	$s_2 - s_1 = c_v \ln \frac{T_2}{T_1} + R \ln \frac{v_2}{v_1}$
$s_2 - s_1 = \int_{T_1}^{T_2} c_P \frac{dT}{T} - R \ln \frac{P_2}{P_1}$	$s_2 - s_1 = c_P \ln \frac{T_2}{T_1} - R \ln \frac{P_2}{P_1}$
$s_2 - s_1 = s_0(T_2) - s_0(T_1) - R \ln \frac{P_2}{P_1}$	
For isentropic processes, use	For isentropic processes:
$\left(\frac{P_{0,1}}{P_{0,2}}\right)_s = \left(\frac{P_1}{P_2}\right)_s$	$\left(\frac{T_2}{T_1}\right)_s = \left(\frac{v_1}{v_2}\right)_s^{(k-1)}$
$\left(\frac{v_{0,1}}{v_{0,2}}\right)_s = \left(\frac{v_1}{v_2}\right)_s$	$\left(\frac{T_2}{T_1}\right)_s = \left(\frac{P_2}{P_1}\right)_s^{(k-1)/k}$
in Tables D.2 to D.8 and Tables E.2 to E.8.	$\left(\frac{P_2}{P_1}\right)_s = \left(\frac{v_1}{v_2}\right)_s^{k}$
	where $k = c_P/c_v$

where the first terms are a function of T only and are obtained from Tables D.2 to D.8 and Tables E.2 to E.8. Since $s_0(T)$ is available only in tabular form, the direct solution for the constant-entropy process using the above expression could require a lengthy iterative procedure. To avoid this difficulty, Tables D.2 through D.8 and Tables E.2 through E.8 include two additional entries at each state, denoted P_0 and v_0. These entries are useful for isentropic process.

Equation (5.147) is rearranged to yield

$$\ln \frac{P_2}{P_1} = \frac{s_0(T_2) - s_0(T_1)}{R} \tag{5.148}$$

The pressure ratio between two states connected by an isentropic process for an ideal gas depends on only the $s_0(T)$ values of the two states. Therefore, this pressure ratio is a *function of absolute temperature* only. The values of $s_0(T)$ are integrals of the specific heat at constant pressure from the reference state to the state in question. The difference in entropy between two states is evaluated by taking the difference in $s_0(T)$ values and including the pressure term [Eq. (5.147)]. A similar procedure is used in the evaluation of the pressure ratio in Eq. (5.148). The value of the pressure ratio that occurs if an isentropic process is carried out between the reference state and the desired state is denoted $P_0 = P/P_{ref}$. From the definition of $s_0(T)$ below Eq. (5.58), $s_0(T_{ref}) = 0$, so Eq. (5.148) yields

$$\ln P_0 = \ln \frac{P}{P_{ref}} = \frac{s_0(T) - 0}{r} = \frac{s_0(T)}{R} \tag{5.149}$$

Thus, this ratio is a function of temperature only, and it can be tabulated. The specific volume is useful in some applications, and the corresponding ratio is obtained from the ideal gas relation as

$$v_0 = \frac{v}{v_{ref}} = \frac{RT}{PR}\frac{P_{ref}}{T_{ref}} = \frac{T}{T_{ref}}\frac{P_{ref}}{P} = \frac{T}{T_{ref}}\frac{1}{P_0} \tag{5.150}$$

This is a function of temperature only and is tabulated also.

The specific values for $s_0(T)$, P_0, and v_0 are given in Tables D.2 to D.8 and Tables E.2 to E.8, and they all depend on temperature only. The use of $s_0(T)$ is discussed in Sec. 5.5. Note that differences in $s_0(T)$ are needed in Eq. (5.60) so that the specific choice of the reference value is eliminated for pure compressible substances; however, the tables are based on $s_0(T = 0\text{ K})$ as the reference value for use with mixtures or when chemical reactions occur (Chaps. 10 through 13). And P_0 and v_0 are used as ratios only, so the reference state is eliminated. For an isentropic process between a state 1 and another state 2, the ratio is

$$\left(\frac{P_1}{P_2}\right)_s = \left(\frac{P_1}{P_{ref}}\frac{P_{ref}}{P_2}\right)_s = \left(\frac{P_{0,1}}{P_{0,2}}\right)_s \tag{5.151}$$

Thus, for an ideal gas undergoing an isentropic process from a specified state to a desired pressure, the final pressure is evaluated by the pressure ratio in Eq. (5.151). Similarly, the specific volume ratio is

$$\left(\frac{v_1}{v_2}\right)_s = \left(\frac{v_{0,1}}{v_{0,2}}\right)_s \tag{5.152}$$

Note that entries in Tables D.2 to D.8 and Tables E.2 to E.8 for P_0 and v_0 are used for an isentropic process only. The following examples demonstrate the use of these tables.

The tabulated values of P_0 and v_0 in Apps. D and E are used in ratio form as given in Eqs. (5.151) and (5.152). Therefore, their absolute values are unimportant. The values presented in the tables have been adjusted by a constant to obtain reasonable magnitudes for presentation. This does not affect the use of P_0 and v_0.

Example Problem 5.10

Air is to be compressed isentropically from 1 atm and 20°C to a final pressure of 20 atm. What is the temperature of the air leaving the compressor?

Solution

If air is assumed to have constant specific heat, then Table 5.4 can be used directly to give

$$\left(\frac{T_2}{T_1}\right)_s = \left(\frac{P_2}{P_1}\right)_s^{(k-1)/k} = (20)^{0.4/1.4} = 2.354$$

or

$$T_2 = 2.354(20 + 273.15)\text{ K} = 690.1\text{ K} = 416.9°\text{C} = 782.5°\text{F}$$

Using the gas tables for air (Table D.2), which do not assume constant specific heat, we go to the entry for 20°C, which by interpolation gives $P_0 = 1.2991$. Since the ratios P_2/P_1 and $P_{0,2}/P_{0,1}$ must be equal [see Eq. 5.151)], we note that in state 2

$$P_{0,2} = 1.2991 \times 20 = 25.982$$

Because P_0 depends on T only, we search the table for the corresponding T; again, using linear interpolation, we find $T_2 = 409.9°\text{C} = 760.8°\text{F}$.

Comments

We find a difference of 12°C or 21.7°F between the constant-specific-heat result and the more accurate result obtained by using the tables. Note that the tables themselves are generated from statistical thermodynamic relations based on models of ideal gases with a few constants obtained from spectroscopic measurements. Comparison with experimental data for real gases shows the gas tables to be accurate within a few percent over the temperature range presented, unless the pressures are extremely high. However, the values for v_0 and P_0 are highly nonlinear with T, and linear interpolation can lead to significant error. The computerized tables or more complete tables are recommended for high accuracy.

5.9.2 Isentropic Process for an Incompressible Fluid or Solid

The entropy change for an incompressible fluid or solid is given in Eq. (5.62) as

$$ds = c\frac{dT}{T} \tag{5.153}$$

For an isentropic process, $ds = 0$, so $dT = 0$. Thus, an isentropic process is an isothermal process for an incompressible fluid or solid. Also the internal energy is given [Eq. (3.37a)] as

$$du = c\,dT \tag{5.154}$$

TABLE 5.5 Incompressible Fluid or Solid Relations

$v =$ constant

$c_P = c_v = c$

Variable Specific Heats	Constant Specific Heats
$u_2 - u_1 = \int_{T_1}^{T_2} c\,dT$	$u_2 - u_1 = c(T_2 - T_1)$
$h_2 - h_1 = \int_{T_1}^{T_2} c\,dT + v(P_2 - P_1)$	$h_2 - h_1 = c(T_2 - T_1) + v(P_2 - P_1)$
$s_2 - s_1 = \int_{T_1}^{T_2} c\frac{dT}{T}$	$s_2 - s_1 = c\ln\frac{T_2}{T_1}$
For isentropic processes:	For isentropic processes:
$T_2 = T_1$	$T_2 = T_1$
$u_2 = u_1$	$u_2 = u_1$
$h_2 - h_1 = v(P_2 - P_1)$	$h_2 - h_1 = v(P_2 - P_1)$

So $du = 0$ for an isentropic process. The change in enthalpy is

$$dh = du + P\,dv + v\,dP \tag{5.155}$$

But $du = 0$ for this process and $dv = 0$ for an incompressible fluid or solid, so

$$dh = v\,dP \tag{5.156}$$

This last expression is integrated to yield

$$h_2 - h_1 = v(P_2 - P_1) \tag{5.157}$$

since $v =$ constant. This last expression is particularly useful in adiabatic work considerations of liquid pumps.

Applicable relations for incompressible fluids or solids are summarized in Table 5.5.

5.10 Special Considerations

A few applications require special consideration. These specific applications are quite important and are discussed here. The first is the work done in a steady state flow problem, and the second is the unsteady discharging problem. The starting points are the differential forms of conservation of mass, first law and second law. The expressions in Table 5.1 are rewritten with Table 5.2 by considering a differential time element dt and differential area element dA. All terms are multiplied by dt, and the terms with an overdot become differentials. Thus,

Mass:

$$dm_{\text{CV}} = dm_{\text{in}} - dm_{\text{out}} \tag{5.158}$$

First law:

$$dE_{CV} = (e + Pv)_{in}\, dm_{in} + \delta Q_{CV} - \delta W_{CV} - (e + Pv)_{out}\, dm_{out} \tag{5.159}$$

Second law:

$$dS_{CV} + s_{out}\, dm_{out} - s_{in}\, dm_{in} \geq \left(\frac{\delta Q}{T}\right)_{CV}$$

or

$$\delta S_{gen} = dS_{CV} + s_{out}\, dm_{out} - s_{in}\, dm_{in} - \left(\frac{\delta Q}{T}\right)_{CV} \geq 0 \tag{5.160}$$

The summation signs have not been included for the inlets, outlets, or heat transfers.

The *steady state work application* involves a single inlet and outlet. The changes in properties within the control volume are zero, because the application is steady state. The process is also considered to be reversible, and the changes in potential and kinetic energy are neglected. Thus, the governing expressions are

Mass:

$$dm_{in} = dm_{out} \tag{5.161}$$

First law:

$$h_{in}\, dm_{in} + \delta Q_{CV,\,rev} = h_{out}\, dm_{out} + \delta W_{CV} \tag{5.162}$$

Second law:

$$s_{out}\, dm_{out} - s_{in}\, dm_{in} = \left(\frac{\delta Q}{T}\right)_{CV,\,rev} \tag{5.163}$$

Dividing the first law and second law by the conservation of mass yields

$$h_{in} + \delta q_{CV,\,rev} = h_{out} + \delta w_{CV} \tag{5.164}$$

and

$$s_{out} - s_{in} = \left(\frac{\delta q}{T}\right)_{CV,\,rev} \tag{5.165}$$

Expressing the changes in enthalpy and entropy as differentials for a control volume yields

$$\delta q_{CV,\,rev} = dh + \delta w_{CV} \tag{5.166}$$

$$ds = \left(\frac{\delta q}{T}\right)_{CV,\,rev} \tag{5.167}$$

Eliminating the heat transfer between Eqs. (5.166) and (5.167) yields

$$T\, ds = dh + \delta w_{CV} \tag{5.168}$$

TABLE 5.6 Work Relations for Polytropic Processes

$Pv^n = \text{constant}$
where $n = 0$, constant pressure
$n = 1$, constant temperature for an ideal gas
$n = k$, constant entropy for an ideal gas
$n \to \infty$, constant volume

$-v\,dP$	$P\,dv$
${}_{\text{in}}w_{\text{CV, out}} = -\int_{\text{in}}^{\text{out}} v\,dP$	${}_1w_2 = \int_1^2 P\,dv$
${}_{\text{in}}w_{\text{CV, out}} = \begin{cases} \dfrac{n}{n-1}[(Pv)_{\text{in}} - (Pv)_{\text{out}}] & n \neq 1 \\ (Pv)_{\text{out}} \ln \dfrac{P_{\text{in}}}{P_{\text{out}}} & n = 1 \end{cases}$	${}_1w_2 = \begin{cases} \dfrac{1}{1-n}[(Pv)_2 - (Pv)_1] & n \neq 1 \\ (Pv)_1 \ln \dfrac{v_2}{v_1} & n = 1 \end{cases}$

This expression is compared to Eq. (5.44), which is rearranged as

$$T\,ds = dh - v\,dP \tag{5.169}$$

Thus

$$\delta w_{\text{CV}} = -v\,dP \tag{5.170}$$

or

$$w_{\text{CV}} = -\int_1^2 v\,dP \tag{5.171}$$

This expression gives the shaft work within a control volume analysis as an integral of $-v\,dP$ and this integral is thus the area under a v-P (rather than a P-v) curve. The only assumption is that the device is *reversible* and therefore is *not* restricted to isothermal or adiabatic processes.

It is convenient to summarize the work expressions for reversible shaft work in steady state control volume analysis, $\delta w_{\text{CV}} = -v\,dP$, and the work resulting from boundary movement, $\delta w = P\,dv$. The shaft work requires the reversible idealization as well as steady state conditions. It is applicable for control volume analysis only. The boundary movement work is valid for both a control volume and a control mass, but the boundary of the control surface must move. The evaluation of the work between two states requires the integral of these expressions and therefore the dependence of v on P or P on v. Table 5.6 gives the results for both $-v\,dP$ and $P\,dv$ work for a polytropic process.

Example Problem 5.11

Evaluate the work per unit mass to (*a*) pump liquid water from 14.7 psia and 70°F to 200 psia and (*b*) compress steam isothermally from 14.7 psia and 600°F to 200 psia. Both processes are reversible and operating under steady state conditions.

Solution

(*a*) Liquid water is assumed to be incompressible, so Table 5.6 for $n \rightarrow \infty$ yields

$$w = v(P_1 - P_2)$$

The specific volume at 70°F is the saturated liquid value of $v = 0.01605$ ft^3/lbm, so

$$w = (0.01605)(14.7 - 200)(144) = -428.3 \text{ ft}\cdot\text{lbf/lbm}$$
$$= -0.550 \text{ Btu/lbm} = -1.28 \text{ kJ/kg}$$

(*b*) Steam at a reduced temperature of $T_r = 1059.67/1164.9 = 0.9097$ and $P_r = 13.6/218.0 = 0.062$ may be approximated as an ideal gas (Fig. C.1). At lower pressures, the approximation is more accurate. Table 5.6 gives the work relation with $n = 1 (Pv = RT = \text{constant}$ for an isothermal process). Thus

$$w = P_1 v_1 \ln \frac{P_1}{P_2} = RT \ln \frac{P_1}{P_2} = (85.76)(1059.67) \ln \frac{14.7}{200}$$
$$= -237{,}200 \text{ ft}\cdot\text{lbf/lbm} = -304.9 \text{ Btu/lbm} = -709.1 \text{ kJ/kg}$$

Comments

The general relations in Table 5.6 apply to all reversible processes. The ideal gas approximation and the polytropic processes are useful for various systems. These numerical examples present the incompressible and isothermal cases, but other processes (different *n*'s) are also possible.

A second important application is the discharge of a constant-volume vessel which is an *unsteady state application.* The different aspect of this problem is that the exit state, which is often assumed to be the same as the state within the control volume (an internally reversible process), changes with time and so cannot be assumed constant. If there is only one exit and there are no inlets, then the discharge is [see Eqs. (5.158) through (5.160)]

Mass:

$$dm_{CV} = -dm_{out} \tag{5.172}$$

First law:

$$dE_{CV} = \delta Q_{CV} - (e + Pv)_{out}\, dm_{out} \tag{5.173}$$

Second law:

$$dS_{CV} + s_{out}\, dm_{out} \geq \left(\frac{\delta Q}{T}\right)_{CV} \tag{5.174}$$

The changes in kinetic and potential energy are neglected, and the changes in control volume properties are expressed in their intensive form. Thus the first and second laws are

$$d(um)_{CV} = m_{CV}\, du_{CV} + u_{CV}\, dm_{CV} = \delta Q_{CV} + h_{out}\, dm_{CV} \tag{5.175}$$

and

$$d(sm)_{CV} - s_{out}\, dm_{CV} = s_{CV}\, dm_{CV} + m_{CV}\, ds_{CV} - s_{out}\, dm_{CV} \geq \left(\frac{\delta Q}{T}\right)_{CV} \tag{5.176}$$

where the conservation of mass has been used. The properties at the outlet and within the control volume vary with time, but if they are equal at every instant, then $u_{CV} = u_{out}$ and $s_{CV} = s_{out}$,

$$m_{CV}\, du_{CV} = \delta Q_{CV} + (Pv)_{CV}\, dm_{CV} \tag{5.177}$$

$$m_{CV}\, ds_{CV} \geq \left(\frac{\delta Q}{T}\right)_{CV} \tag{5.178}$$

These expressions govern the *discharge of a single-phase substance* from a vessel including the heat transfer contribution. For a two-phase substance, the enthalpy of the outgoing material is not generally equal to the enthalpy of the control volume, and the analysis must start from the first law (see the Comments in Example Problem 5.12).

Consider the discharge from a constant-volume adiabatic tank which is described by Eq. (5.177) and the Gibbs equation [Eq. (5.43)]. Equation (5.177), for an adiabatic process, is

$$m_{CV}\, du_{CV} = (Pv)_{CV}\, dm_{CV} \tag{5.179}$$

Since the volume is constant, $d(m_{CV}v_{CV}) = 0$ or $v_{CV}\, dm_{CV} = -m_{CV}\, dv_{CV}$. Thus

$$m_{CV}\, du_{CV} = -m_{CV}\, P_{CV}\, dv_{CV} \tag{5.180}$$

or

$$du_{CV} = -P_{CV}\, dv_{CV} \tag{5.181}$$

This is the Gibbs equation given in Eq. (5.43) when

$$ds_{CV} = 0 \tag{5.182}$$

Thus the constant-volume discharge of a single-phase fluid from an adiabatic tank is a constant-entropy process. However, we have assumed that the outlet entropy is equal to the entropy of the uniform mass within the control volume; this has just been shown to result in an internally reversible process with respect to the control volume. This is true in all cases for a single-phase fluid and does *not* require the assumption of a totally reversible process. This is important in discharge calculations since a property is known at every point for the process.

Example Problem 5.12

An insulated 1.0-m^3 tank contains steam at 2.0 MPa and 500°C. Vapor is withdrawn from the top of the tank for use in a related experiment. What are the final temperature and the mass withdrawn if the final pressue is 0.2 MPa?

Diagrams

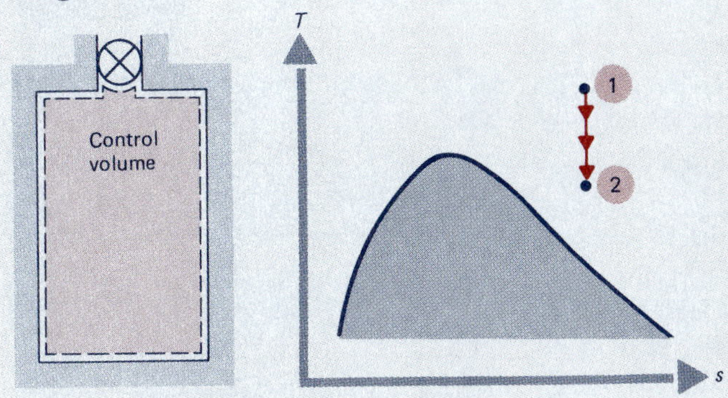

Solution

The initial state is specified by pressure and temperature. The final state is specified by constant entropy and pressure. Thus

State	P, MPa	T, °C	v, m^3/kg	s, kJ/(kg·K)
1	2.0	500	(0.1757)	(7.4318)
2	0.2	(183.6)	(1.0409)	$s_2 = s_1$

The states and process are specified so that the remaining properties are obtained from Table D.12. These properties are entered into the above table.

The final temperature is given by interpolation as 183.6°C = 362.5°F. The mass withdrawn is found through the change in specific volume as

$$m_{out} = m_1 - m_2 = \left(\frac{V}{v}\right)_1 - \left(\frac{V}{v}\right)_2$$

$$= V\left(\frac{1}{v_1} - \frac{1}{v_2}\right) = 1\left(\frac{1}{0.1757} - \frac{1}{1.0409}\right) = 4.73 \text{ kg} = 10.4 \text{ lbm}$$

Comments

The solution to this process is straightforward if $s_2 = s_1$ is used. An alternative approach to the solution, which is required in the analysis of the discharge of two-phase substances, starts with the first law. Equation (5.181) is

$$du = -P\,dv$$

for an adiabatic tank. This expression would have to be incremented through the process by using Table D.11. Specifically, for two intermediate states a and b,

$$u_b - u_a = P_{sat}(v_a - v_b)$$

where v_a and u_a are known from the previous step and u_b and v_b are evaluated through the quality, that is, $u_b = u_{lb} + x_b u_{lgb}$ and $v_b = v_{lb} + x_b v_{lgb}$. Thus the quality is obtained, and the solution is completed.

5.11 Component Efficiencies

In many components the adiabatic performance is the desired situation. A steam turbine seeks to produce work given a high-temperature and high-pressure inlet state and an outlet pressure. Any energy transfer from the turbine as heat transfer is a loss and reduces the work output. A compressor that takes a fluid from the inlet state to the desired outlet pressure requires a work input. A heat loss from a desired adiabatic compressor requires a larger work input. (Some compressors operate almost isothermally, so there is a heat transfer. These compressors require special consideration.) The inefficiencies in these components results from irreversible processes.

The component efficiencies are used to compare the actual process to the adiabatic and reversible processes. The reduction of the irreversibilities of a component process is desired to increase the efficiency of the process. Therefore, the component efficiency is usually defined as the ratio of the actual to the isentropic result, with the limiting value being unity. The isentropic device is taken as the standard for comparison of real adiabatic operation.

The thermodynamic component efficiency is thus a measure of the actual performance of a given device relative to an ideal standard. When the efficiency of a real device is measured or known, the engineer can use this information along with a calculation of the ideal performance to predict the actual performance.

5.11.1 Turbine Efficiency

The efficiency of a turbine is a comparison of the actual work produced with the work produced for an isentropic process. The inlet to the turbine

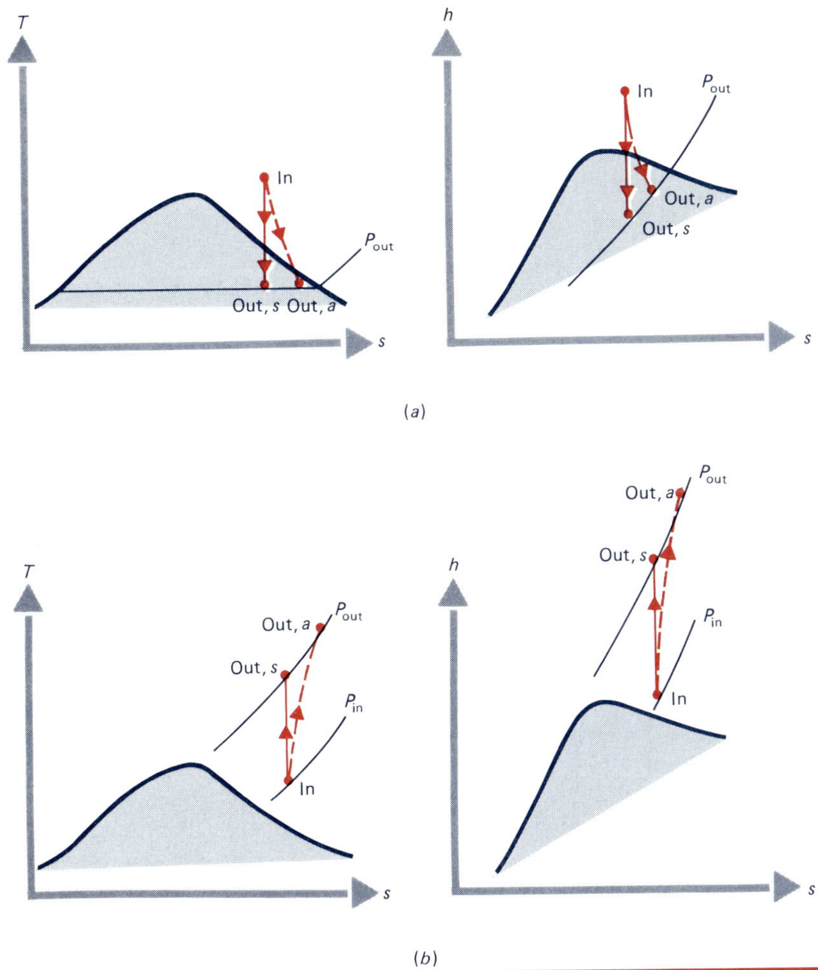

Figure 5.16 Component efficiency comparisons. (*a*) Turbine; (*b*) compressor.

is at a specified state, and the exit is at a specified pressure. The exhaust pressure is the outlet condition, since factors such as the available cooling water temperature to the condenser usually set the outlet pressure for a given application. The two processes compared in turbine efficiency are shown in Fig. 5.16*a*. The actual process and the isentropic process start at the same state. The exit state is different in each case, but the exit states for the actual and isentropic processes are at the same pressure. Denoting the actual work by w_a and the isentropic work by w_s, we define the turbine efficiency as

$$\eta_t \equiv \frac{w_a}{w_s} \tag{5.183}$$

The reduction of irreversibilities moves the actual process exit state toward the isentropic exit state. Applying the first law for steady state adiabatic flow with uniform properties at the inlet and exit, we find for the turbine efficiency

$$\eta_t = \frac{(h_{\text{in}} - h_{\text{out}})_a}{(h_{\text{in}} - h_{\text{out}})_s} \tag{5.184}$$

when the changes in kinetic and potential energy are negligible. Figure 5.16*a* shows the process for a steam turbine, but the definition is valid for a gas turbine also.

Example Problem 5.13

A steam turbine of efficiency η_t has an inlet state of 700°F and 550 psia. The mixture that emerges from the turbine exit is condensed in the condenser at 14.7 psia. The process is adiabatic, and the mass flow rate is 1 lbm/s. Determine the percentage of the initial mass that is liquid at the turbine exhaust and the turbine power as the turbine efficiency degrades from $\eta_t = 95$ to 80 to 60 percent.

Diagrams

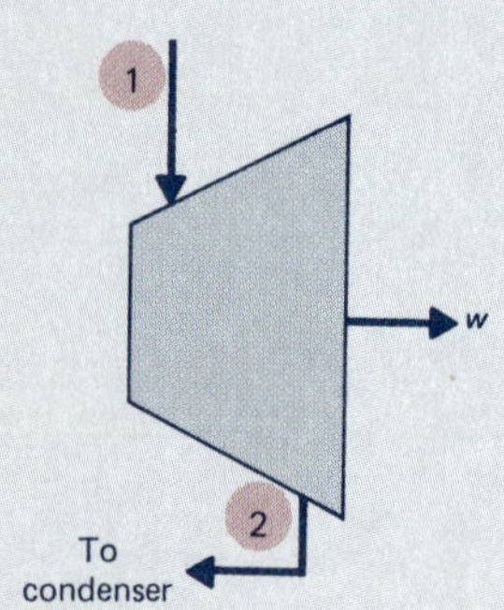

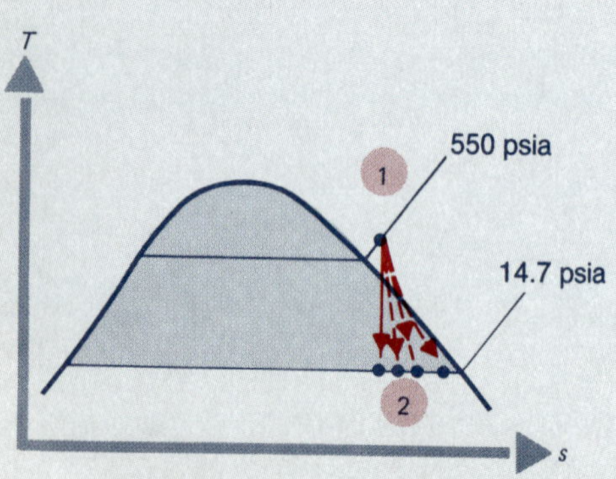

Solution

The initial state is specified. The outlet state is determined from the turbine efficiency and the outlet pressure. The turbine efficiency compares the isentropic process to the actual process, so that η_t and the isentropic path specify the actual outlet state. For an adiabatic turbine, Eq. (5.184) yields

$$\eta_t = \frac{(h_1 - h_2)_a}{(h_1 - h_2)_s}$$

or

$$h_{2a} = h_1 - \eta_t(h_1 - h_2)_s$$

Thus, with state 2*s* specified, the above expression gives the actual outlet state (with $P_2 = 14.7$ psia).

The state information is as follows:

State	P, psia	T, °F	x	h, Btu/lbm	s, Btu/(lbm·°R)
1	550	700	—	(1353.5)	(1.5986)
2_s	14.7	—	(0.8906)	(1044.3)	$s_{2s} = s_1$
2_{95}	14.7	—	(0.9066)	1059.8	—
2_{90}	14.7	—	(0.9225)	1075.2	—
2_{80}	14.7	—	(0.9546)	1106.1	—
2_{60}	14.7	—	—	1168.0	—

The percentage of the initial mass that is liquid at state 2 is ($\dot{m}_1 = \dot{m}_2 = \dot{m}$)

$$\frac{\dot{m}_{f2}}{\dot{m}} = \frac{\dot{m}_2 - \dot{m}_{g2}}{\dot{m}} = 1 - x$$

and the quality is given in the above table. The power output for this adiabatic turbine is

$$\dot{W} = \dot{m}(h_1 - h_2)_a = \dot{m}\eta_t(h_1 - h_2)_s$$

The table above is used to evaluate the percentage of the mass that is liquid and the power output. Thus

Turbine Efficiency, %	Mass Liquid, %	Power, MBtu/h	Power, kW
100	10.92	1.115	326.2
95	9.34	1.059	310.1
90	7.75	1.002	293.7
80	4.54	0.892	260.8
60	0.00	0.669	195.8

Comments

The reversible process is given by $\eta_t = 1.0$ and indicates the maximum work that is possible from the stated inlet to the outlet pressure. All other efficiencies must produce less. The percentage of liquid in the turbine exhaust is very dependent on the turbine efficiency. If the turbine is designed to operate at $\eta_t = 0.95$, then the condenser must handle 9.34 percent liquid.

If the water leaving the condenser is saturated liquid, an energy balance for the condenser requires a heat transfer of $\dot{Q} = \dot{m}(h_{out} - h_{in}) = \dot{m}(h_l - h_{2a})$. The transfer from the water changes from -3.11 MBtu/h for $\eta_t = 1.0$ to -3.56 MBtu/h for $\eta_t = 0.6$. Therefore, a degradation of the turbine performance requires changes in the other cycle component performance.

Multi-stage Turbine Analysis

Commercial turbines often have two or more *stages*, i.e., sets of blades through which the working fluid passes, and each stage thus handles a smaller pressure drop than is necessary in a single-stage turbine. Such a design allows the blades in each stage to be designed for a higher efficiency than can be obtained in a single-stage design for a large pressure drop, and the forces on the turbine shaft are distributed more evenly along the shaft length.

Consider the two-stage turbine shown in Fig. 5.17. Here the inlet working fluid is at T_1, P_1, and the exhaust pressure is fixed at $P_{3,4}$. If the *overall* efficiency of the turbine is specified, then analysis of the turbine is handled exactly as for a single-stage turbine by taking the control volume around the entire two-stage turbine. Because inlet conditions and the outlet pressure and overall turbine efficiency are known, the actual outlet conditions are found as for the single-stage device. The details of the pressure and temperature changes across the stages are not needed in this case.

Now consider a turbine in which the efficiencies of *each* stage η_1 and η_2 are specified, and it is necessary to find the outlet conditions from the first stage. (This is often the case when the outlet working fluid is to be partially removed, or the outlet fluid is to be reheated before addition to the second stage.) In addition, we wish to know the total work output expected from the turbine.

If the outlet pressure of the first (high-pressure) stage is assumed to be the inlet pressure to the second (low-pressure) stage, and if the outlet fluid from the first stage is fed directly into the second stage without intermediate heating, then the *h-s* diagram will appear as in Fig. 5.18. In this case, the actual outlet enthalpy from the first stage $h_{2,a}$ is found by treating the first stage with a separate control volume and using the identical approach applied for a single-stage turbine. This will be the actual inlet enthalpy to the second stage.

Figure 5.17 Two-stage turbine.

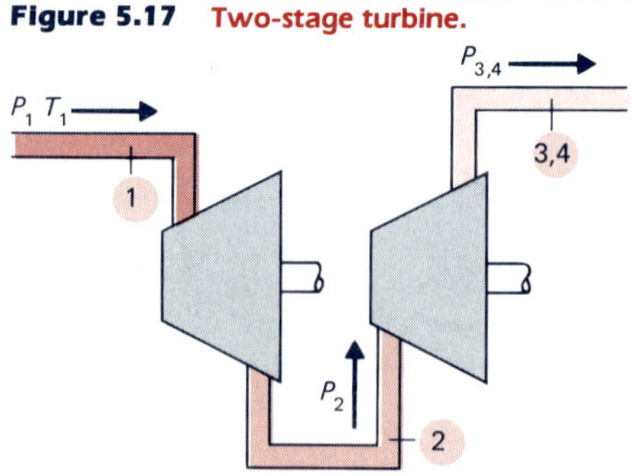

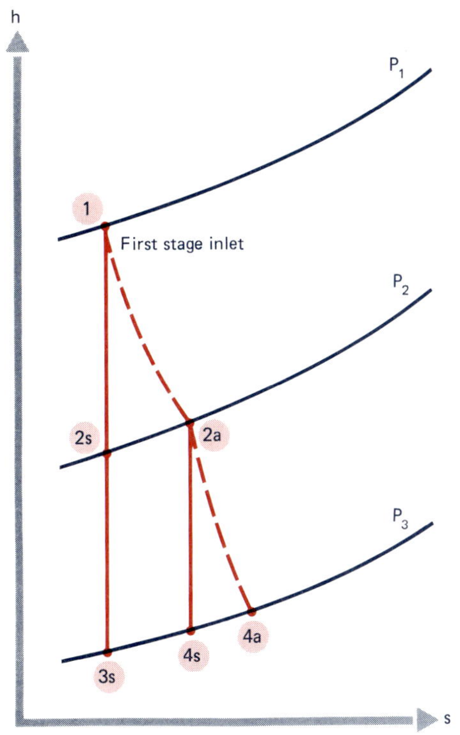

Figure 5.18 States for two-stage turbine on *h-s* diagram.

The second stage must now be analyzed for work output. The exit pressure from the second stage $P_{3,4}$ is usually known, and we have assumed that the second-stage efficiency is given. The only question remaining is how to correctly specify the inlet conditions to the second stage.

The work from the second stage is calculated by specifying the inlet condition to the second stage as the *pressure at the outlet of the first stage and the actual outlet temperature from the first stage.* This is state $2a$ in Fig. 5.18. The actual work output of the second stage is given by $w_{2,a} = \eta_{t,2}(h_{2,a} - h_{4,s}) = h_{2,a} - h_{4,a}$, as illustrated in Fig. 5.18. Note that the work output from a turbine with each stage of equal efficiency may differ from the work output of a two-stage turbine with the same overall efficiency! We will see why this occurs in Example Problem 5.14.

The outlet condition from the second stage is calculated by drawing a control volume around the second stage. The inlet conditions to the second stage are taken to be the actual outlet conditions from the first stage, state $2a$ in Fig. 5.18, *not* the pressure and isentropic outlet conditions from the first stage (state $2s$). The second stage is then treated as for a single-stage turbine, using the efficiency of the second stage to find the actual outlet conditions based on the actual inlet conditions.

Example Problem 5.14

A two-stage adiabatic steam turbine operates with the conditions shown in the diagram and a mass flow rate of 1 lbm/s. The efficiencies of the stages are $\eta_1 = 0.88$ and $\eta_2 = 0.92$. Find the outlet temperature from the first stage, the total work output of the turbine, the outlet state from the second stage, and the overall efficiency of the turbine. Repeat the solution for the case when both stages have the same efficiency of 0.90.

Diagram

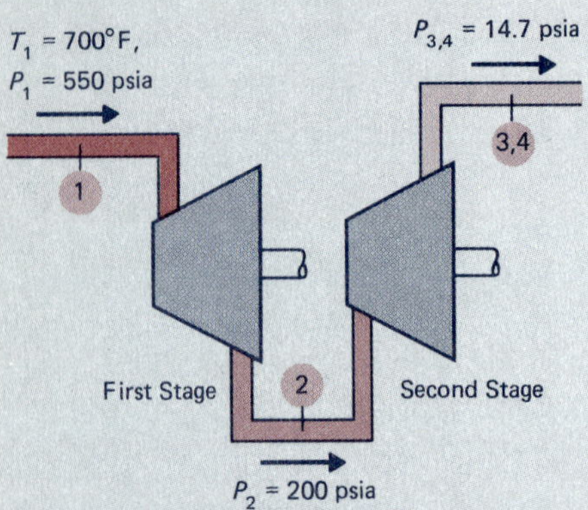

Solution

The state table for the first case is

State	P, psia	T, °F	x	h, Btu/lbm	s, Btu/(lbm · °R)
1	550	700	—	(1353.5)	(1.5986)
2s	200	(460.9)	—	(1245.8)	$s_{2,s} = s_1$
2a	200	(483.2)	—	1258.7	(1.6128)
3s	14.7	(212)	(0.8906)	(1044.3)	$s_{3,s} = s_{2,s}$
4s	14.7	(212)	(0.9004)	(1053.8)	$s_{4,s} = s_{2,a}$
4a	14.7	(212)	(0.9173)	1070.2	—

The total work output is then

$$w_{tot} = -[(h_{2,a} - h_1) + (h_{4,a} - h_{2,a})] = -[(1258.7 - 1353.5) + (1070.2 - 1258.7)]$$

$$= -(-94.8 - 188.5) = 283.3 \text{ Btu/lbm}$$

and the power output is then 1.020×10^6 Btu/h = 298.9 kW, and we see that the outlet condition from the first stage is $P_2 = 200$ psia, $T_{2a} =$

483.2°F. The overall turbine efficiency is

$$\eta_t = \frac{w_{\text{tot}}}{|h_{3,s} - h_1|}$$

$$= \frac{283.3}{|1044.3 - 1353.5|} = \frac{283.3}{309.2} = 0.916 \text{ or } 91.6\%$$

The actual outlet condition from the second stage is $P_{3,4} = 14.7$ psia, $x_{4,a} = 0.9173$.

For the second part of the problem, with both stages having an efficiency of 0.90, the state table becomes

State	P, psia	T, °F	x	h, Btu/lbm	s, Btu/(lbm·°R)
1	550	700	—	(1353.5)	(1.5986)
2s	200	(460.9)	—	(1245.8)	$s_{2,s} = s_1$
2a	200	(479.6)	—	1256.6	(1.6105)
3s	14.7	(212)	(0.8906)	(1044.3)	$s_{3,s} = s_{2,s}$
4s	14.7	(212)	(0.8989)	(1052.3)	$s_{4,s} = s_{2,a}$
4a	14.7	(212)	(0.9199)	1072.7	—

and the work output is

$$w_{\text{tot}} = -[(h_{2,a} - h_1) + (h_{4,a} - h_{2,a})] = -[(1256.6 - 1353.5) + (1072.7 - 1256.6)]$$

$$= -(-96.9 - 183.9) = 280.8 \text{ Btu/lbm}$$

and the power output is then 1.011×10^6 Btu/h = 296.3 kW, and we see that the outlet condition from the first stage is $P_2 = 200$ psia, $T_{2,a} =$ 479.6°F, and for the second stage is $P_{3,4} = 14.7$ psia, $x_{4,a} = 0.9199$.

The overall turbine efficiency for this case is

$$\eta_t = \frac{w_{\text{tot}}}{|h_{3,s} - h_1|}$$

$$= \frac{280.8}{|1044.3 - 1353.5|} = \frac{280.8}{309.2} = 0.908 \text{ or } 90.8\%$$

Comments

Using an equal efficiency of 0.90 for both stages gives a different answer than for the case of two efficiencies of 0.88 and 0.92 (which average 0.90). This is because the two stages do not have the same enthalpy difference for isentropic operation, so that the isentropic work output from each stage is different. Thus, the differing stage efficiencies cause a different total work output (and a different overall turbine efficiency) from that calculated by using a simple average of the two-stage efficiencies. In this particular case, the stage with the higher efficiency has a greater isentropic work output, so the total work output of the two-stage turbine exceeds that from a two-stage turbine in which each stage has the average efficiency.

Note the more surprising result that *for the two-stage turbine with equal stage efficiencies of 0.90, more work output, a higher overall efficiency, and a different outlet state are found than for the single-stage turbine with the same inlet conditions and outlet pressure but overall efficiency of 90 percent analyzed in Example Problem 5.13.* This occurs because the calculated work output from the second stage is based on an inlet condition of state 2*a*, which has a higher enthalpy remaining after the work extraction in the first stage than would occur in an isentropic turbine (see Fig. 5.18). Thus, more work can be extracted in the second stage if its efficiency is 90 percent.

The *overall* turbine efficiency is based on total actual work output divided by overall isentropic work output; the stage efficiencies are defined in terms of the stage actual work output over the isentropic work output of that stage. Thus, the efficiencies of the stages are not simply related to the overall turbine efficiency, as evidenced by the results of this example.

Turbine with Reheating

For a multistage turbine, it is possible to increase the work output by increasing the enthalpy of the working fluid entering the second stage. This can be done by passing the working fluid through a heat exchanger where heat transfer from a higher-temperature source is used to *reheat* the working fluid, as shown in Fig. 5.19.

Figure 5.19 Two-stage turbine with reheat.

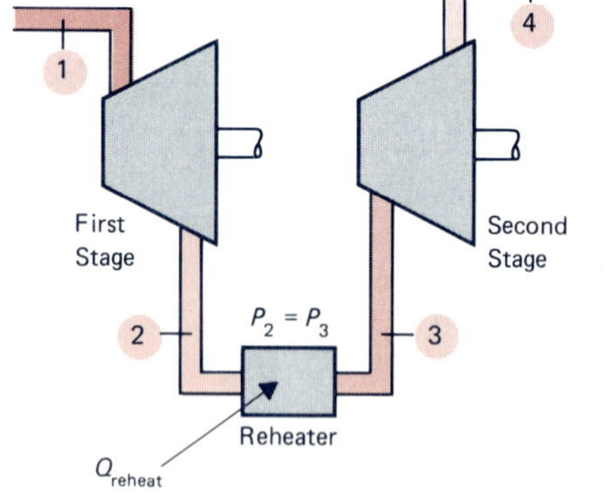

Such an arrangement is a practical approach to the ideal of a constant-temperature or isothermal turbine. Consider a two-stage turbine with exhaust pressure P_4. As shown in the temperature-entropy diagram of Fig. 5.20, an *isentropic two-stage turbine* would exit at pressure P_4 and temperature T_5.

An *isothermal two-stage turbine* with reheating occurring at constant pressure P_3 would have an exhaust temperature of T_6 and would require complex design so that heat transfer would exactly equal the work extraction at each point within the turbine.

The *T-s* diagram of a *two-stage turbine with reheat at $P_3 = P_2$, in which each turbine stage is isentropic,* will follow the process 1-2*s*-3-4*s* and will have an exit temperature of $T_{4,s}$. The required heat transfer in the reheat heat exchanger for this case will be

$$q_{\text{reheat}} = h_3 - h_{2,s} \tag{5.185}$$

For a *two-stage turbine with reheat at $P_3 = P_2$ with each stage having a known efficiency,* the process will follow the path 1-2*a*-3-4*a*, with an exhaust temperature of $T_{4,a}$. The required heat transfer to the heat exchanger is

$$q_{\text{reheat}} = h_3 - h_{2,a} \quad (5.186)$$

The required heat transfer will be less for the real turbine than for the ideal turbine with isentropic stages, because more of the inlet enthalpy is converted to work in the ideal turbine, making $h_{2,s} < h_{2,a}$. Note especially that the temperature within the reheat turbine is much closer to that in the isothermal turbine T_1 than is the case for the isentropic turbine. If the turbine were constructed with a large number of stages and with reheat to T_1 between each stage, then isothermal operation could be approached.

It can be shown that for a turbine operating with an ideal gas with temperature-independent specific heats as the working fluid, the reheat process should occur at

$$P_3 = P_2 = (P_1 P_4)^{1/2} \quad (5.187)$$

to maximize the turbine work output. For other working fluids such as steam, a trial and error process must be used to determine the optimum reheat pressure, using Eq. (5.187) as an initial guess.

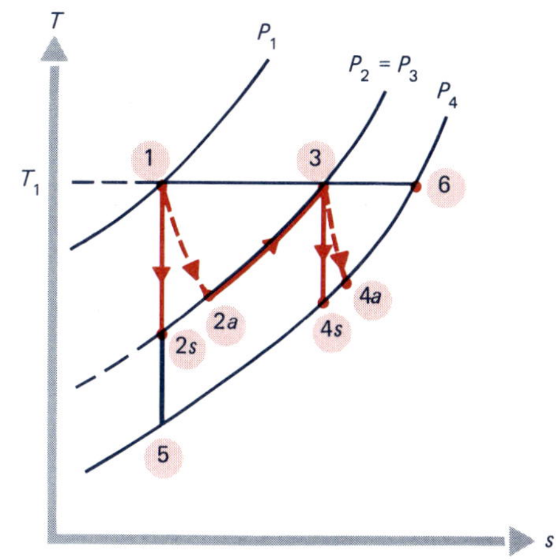

Figure 5.20 Temperature-entropy diagram for two-stage turbine with reheat.

Example Problem 5.15

A two-stage adiabatic steam turbine with reheat to the inlet temperature operates with the conditions shown in the diagram and a mass flow rate of 1 lbm/s. The efficiencies of the stages are $\eta_1 = 0.88$ and $\eta_2 = 0.92$. Find the total power of the turbine, the heat transfer required in the reheater, and the overall efficiency of the turbine.

Diagram

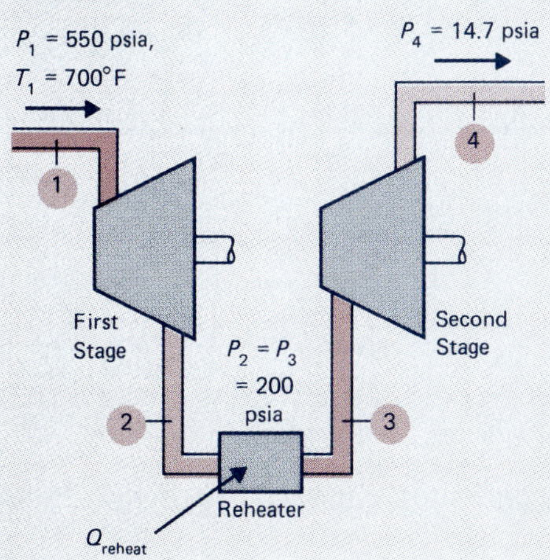

Solution

The state table is

State	P, psia	T, °F	x	h, Btu/lbm	s, Btu/(lbm·°R)
1	550	700	—	(1353.5)	(1.5986)
2s	200	(460.9)	—	(1245.8)	$s_{2,s} = s_1$
2a	200	(483.2)	—	1258.7	—
3	200	700	—	(1373.7)	(1.7233)
4s	14.7	(212)	(0.9769)	(1128.0)	$s_{4,s} = s_3$
4a	14.7	(212)	(0.9972)	1147.7	—

The total work output is then

$$\begin{aligned} w_{tot} &= -[(h_{2,a} - h_1) + (h_{4,a} - h_3)] \\ &= -[(1258.7 - 1353.5) + (1147.7 - 1373.7)] \\ &= -[(-94.8 - 226.0)] = 320.8 \text{ Btu/lbm} \end{aligned}$$

and the power output is then 1.155×10^6 Btu/h = 338.4 kW.

The heat transfer to the reheater, from Eq. (5.186), is

$$\begin{aligned} q_{reheat} &= h_3 - h_{2,a} \\ &= 1373.7 - 1258.7 = 115.0 \text{ Btu/lbm} \end{aligned}$$

or, for the given mass flow rate, 0.414×10^6 Btu/h.

Comments

The power output of the turbine has increased because of the addition of reheat (compare with Example Problem 5.13 which is identical except for reheat and gives a power output of 1.011×10^6 Btu/h). Of course, this increase is not free; it is paid for with the additional energy in the form of heat transfer required by the reheater. Of additional interest is that reheating has increased the exit quality to 0.9972 from 0.9173 for the identical turbine with no reheat analyzed in Example Problem 5.13.

Many nuclear reactor systems of the pressurized water type produce dry saturated steam [usually in the range of about 3 to 7 MPa (400 to 1000 psig)] rather than superheated steam at the turbine inlet. The quality of the steam may be reduced to unacceptable levels during its passage through the turbine, as we will see in Chap 7. The quality at the turbine outlet can decline to 0.75 if moisture is not removed, leading to turbine blade erosion, decline in turbine efficiency, and possible failure. The turbines are therefore designed to remove condensed moisture at one or more places.

Mechanically, the moisture removal is usually done in one of two ways: either through moisture removal between stages or by designing for moisture removal within the stages as an integral part of the turbine design.

In a two-stage turbine, the moisture can be removed at the outlet of the high-pressure stage. This is usually done with a centrifugal separator. At the Shippingport power station, a stationary centrifugal separator provided an entering steam quality to the second stage of 0.99. No reheat is used in this particular system. Newer systems that incorporate reheat between stages incorporate chevron vanes in the entry to the reheat heat exchanger to eliminate moisture.

To remove moisture within a given stage, particularly within the low-pressure turbine, moisture slung to the outer blade edges by centrifugal force is removed from the turbine. If regenerative heating (Chap. 7) is used in the power plant, steam is often removed from the turbine at intermediate pressures between the inlet and outlet. In this case, the moisture can be carried off with these extraction streams.

In analyzing these turbines, the mass removed in the form of liquid must be accounted for, and the power output from each succeeding portion of the turbine is correspondingly reduced because of the decreased mass flow. Otherwise, analysis proceeds as for the two-stage turbine, with or without reheat as required.

In practice, turbine efficiencies are on the order of 90 percent. Gas turbines for aircraft typically have two or three stages, while stationary gas turbine and vapor turbines in power plants can have a large number of stages.

5.11.2 Compressors and Pumps

Compressors

In a compressor, the comparison is again made between the actual and the isentropic case. The desired performance of the compressor is to produce the exit pressure with minimum work. Assuming that the actual process is adiabatic, we see that the irreversibilities require a larger work input. The compressor efficiency is

$$\eta_c \equiv \frac{w_s}{w_a} \tag{5.188}$$

The inlet state is the same for each process, and both processes exhaust to the same outlet pressure. The first law for adiabatic steady state flow yields

$$\eta_c = \frac{(h_{\text{in}} - h_{\text{out}})_s}{(h_{\text{in}} - h_{\text{out}})_a} \tag{5.189}$$

when the changes in kinetic and potential energies are not significant. The isentropic work is less than the actual work input, and the limit is a compressor efficiency of unity. The comparison of these two processes is shown in Fig. 5.16*b*.

Adiabatic single-stage compressors usually operate with ratios of outlet to inlet pressure of less than about 7 : 1. Centrifugal or radial flow compressors may achieve efficiencies of about 90 percent at pressure ratios of 3 to 4, with decreasing efficiency at greater pressure ratios. Axial flow compressors achieve somewhat higher efficiencies at comparable pressure ratios. Higher pressure ratios are achieved by placing multiple compressor stages in series. Aircraft compressors may have from 4 to as many as 16 compressor stages, achieving overall pressure ratios greater than 30.

Because compressors operate efficiently if the volume of gas being compressed is maintained at a minimum, it is often desirable to provide cooling within the compressor to minimize the expansion of the gas owing to temperature increases. For cooled compressors, a different definition for the efficiency is used where the isothermal reversible rather than the isentropic compressor is used for comparison.

For a cooled compressor, the efficiency is given by

$$\eta_T = \frac{w_T}{w_a} \tag{5.190}$$

where w_T is the work input to a reversible isothermal compressor and w_a is the actual work input to the compressor.

For the case of an ideal gas, the enthalpy does not change for an isothermal compressor, and the work required to operate the compressor is given from the first law as $w_T = q_T$ (neglecting changes in potential and kinetic energy in the flow stream). Equation (5.190) becomes, *for an ideal gas,*

$$\eta_T = \frac{q_T}{w_a} \tag{5.191}$$

where q_T is the heat transfer per unit mass flowing through the compressor that is required to maintain isothermal compression.

Suppose that the working fluid enters the compressor at state 1 of Fig. 5.21. Two paths are shown for compression to a final pressure P_2, one for an isentropic compression to state $2s$ and one for an isothermal reversible compression to state $2T$. Note that the work under the curves is given by Eq. (5.171), so the work for the isothermal curve is smaller than for the isentropic case by the amount of the shaded area between the curves.

The work required for isentropic compression from a given initial state 1 to a given final pressure P_2 is *always* greater than the isothermal work from the same initial state to the same final pressure. This ratio is given for an ideal gas with temperature-independent specific heats by using the relations in Table 5.6 to obtain

$$\frac{w_s}{w_T} = \frac{k}{k-1}\,\frac{P_2V_{2,s} - P_1V_1}{P_1V_1 \ln (P_2/P_1)} \tag{5.192}$$

Figure 5.21 Comparison of isentropic and isothermal compression.

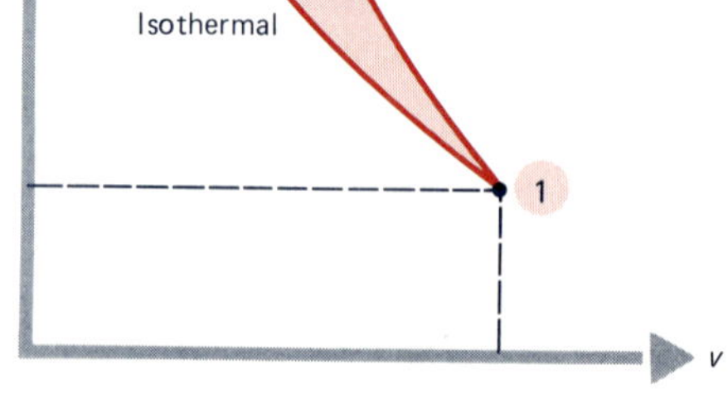

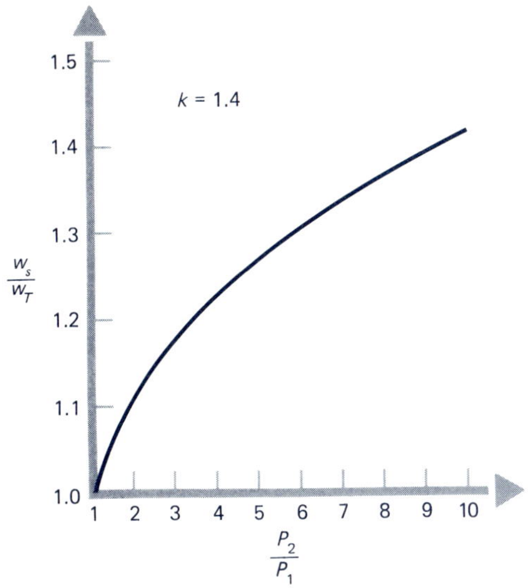

Figure 5.22 Ratio of isentropic to isothermal work in compression of an ideal gas.

Using the isentropic relations to eliminate $V_{2,s}$ gives

$$\frac{w_s}{w_T} = \frac{k}{k-1} \frac{(P_2/P_1)^{(k-1)/k} - 1}{\ln (P_2/P_1)} \tag{5.193}$$

For the case of $k = 1.4$, the ratio of w_s/w_T is plotted against P_2/P_1 in Fig. 5.22. The ratio of isentropic to isothermal reversible work is always greater than unity.

Compressor Intercooling

In practice, the fluid passing through a real compressor cannot be maintained at constant temperature, and the compression process is more nearly adiabatic than isothermal, so that the increased efficiency promised by isothermal compression cannot be reached. However, some of the benefits of an isothermal compressor can be obtained by cooling the working fluid between compressor *stages;* i.e., the working fluid leaving a given set of compressor blades in a turbocompressor, or from the low-pressure pistons in a reciprocating compressor, is passed through a heat exchanger and is thus cooled before entering the next stage of compression.

Examine the three cases shown in Fig. 5.23*a*, *b*, and *c*. In each case, it is assumed that the compression in each stage (processes 1-2 and 3-4) takes place isentropically and that intercooling takes place at constant pressure between the stages in process 2-3. During intercooling, it is also assumed that the working fluid temperature is returned to the entering temperature T_1, although this is not possible in practice. In Fig. 5.23*a*, the intercooling takes place after a first stage with a small compressor ratio. Little compressor work is saved over the isentropic case, as shown by the

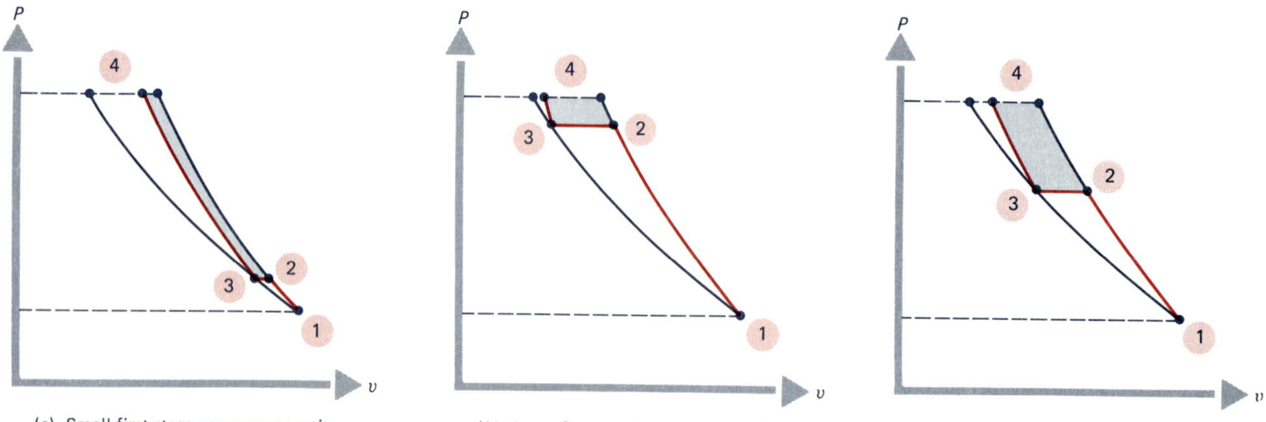

Figure 5.23 Effect of intercooler pressure on compressor work savings.

small shaded area. Figure 5.23*b* shows intercooling following a first stage with a very high compression ratio; again, the relative work savings is small. For the intermediate case, shown in Fig. 5.23*c*, more work is saved than for either of the other cases examined. Apparently there is some pressure ratio at which intercooling will provide the greatest benefit.

Example Problem 5.16

Find the pressure at which intercooling should be done to provide the greatest reduction in work input to a two-stage compressor for the case of an ideal gas with temperature-independent specific heat. Assume that each stage of compression is isentropic and that intercooling at constant pressure returns the working fluid to the inlet temperature before entering the second stage.

Solution

The net work required by the compressor (following the P-v diagram in Fig. 5.23) is

$$w_{\text{comp}} = c_P(T_4 - T_3) + c_P(T_2 - T_1)$$

Using the isentropic relations for each stage, we can write this as

$$w_{\text{comp}} = c_P\left\{T_3\left[\left(\frac{P_4}{P_3}\right)^{(k-1)/k} - 1\right] + T_1\left[\left(\frac{P_2}{P_1}\right)^{(k-1)/k} - 1\right]\right\}$$

By noting that $T_3 = T_1$ and $P_3 = P_2$, this can be recast in the form

$$w_{\text{comp}} = c_P T_1\left[\left(\frac{P_4}{P_2}\right)^{(k-1)/k} - 1 + \left(\frac{P_2}{P_1}\right)^{(k-1)/k} - 1\right]$$

To determine the pressure P_2 at which intercooling should occur, the derivative of w_{comp} is taken with respect to P_2 and the result set equal to zero to determine the maximum. The result is

$$P_2 = (P_1 P_4)^{1/2}$$

Comments

This result is valid for choosing the best intercooling pressure between any two stages of a multistage compressor [and is the same result noted for the use of reheat in a gas turbine in Eq. (5.187)]. In practice, intercooling is seldom used except in two-stage compressors. Some other methods of providing the effect of intercooling are discussed in Chap. 6 with respect to turbine engines.

Pumps

Pumps provide an interesting special case. For most situations involving pumps, the fluid flowing through the pump can be considered to be *incompressible*; that is $v =$ constant. The work for a pump is conveniently expressed from Eq. (5.171) as

$$w_s = -\int_1^2 v\, dP = v(P_1 - P_2) \tag{5.194}$$

which is valid for a *reversible* pump or other device through which an *incompressible* fluid is flowing. [This form of *Bernoulli's equation,* which is of fundamental importance in fluid mechanics, is extended from Eq. (4.118) to account for a pump in the line. It is also valid for an isothermal pump.] Therefore, the pump efficiency is

$$\eta_p = \frac{w_s}{w_a} = \frac{v(P_1 - P_2)}{w_a} \tag{5.195}$$

5.11.3 Nozzles

Another component efficiency is defined for a nozzle. A nozzle increases the velocity as a result of a pressure decrease. This component is generally adiabatic. The efficiency compares the actual exit kinetic energy to the kinetic energy at the exit for an isentropic process. The inlet state for each process and the exhaust pressure are the same for the compared processes. The nozzle efficiency is

$$\eta_N \equiv \frac{(\mathrm{KE})_a}{(\mathrm{KE})_s} = \frac{\mathbf{V}_a^2/(2g_c)}{\mathbf{V}_s^2/(2g_c)} \tag{5.196}$$

The component efficiencies compare two processes that have the same inlet state. The exit state is at the same pressure for the actual and ideal processes.

As the exit diameter of a converging nozzle is reduced, or the downstream pressure is reduced, or the upstream pressure is increased, the exit velocity of the fluid flowing through a converging nozzle is increased. If the fluid is a gas, however, the exit velocity cannot be increased by these means without limit; and the limit is reached when the exit velocity of the nozzle becomes equal to the *acoustic* or *sonic velocity* $\mathbf{V}_{ac}$, which is the velocity of the speed of sound in the fluid at the exit. The acoustic velocity is given by

$$\mathbf{V}_{ac} = (g_c kRT_t)^{1/2} \tag{5.197}$$

where T_t is the absolute fluid temperature at the nozzle exit. Further reductions in downstream pressure from that required to achieve $\mathbf{V}_{ac}$ at the nozzle exit will not increase the exit velocity from a converging nozzle, which in this case is also the minimum diameter or throat, and a condition known as *choked flow* will result. This occurs because the fluid is then moving at the same velocity as the velocity of sound, which is the velocity at which a pressure wave can be transmitted in the fluid. Thus, a pressure wave at these conditions cannot move upstream, and changes in downstream conditions can no longer be transmitted upstream. A reduction in downstream pressure cannot then affect the flow at the nozzle exit, which remains at the acoustic velocity.

A useful parameter in such problems is the *Mach number*, denoted Ma, of the flow, which is the ratio of the local fluid velocity to the local acoustic velocity, or

$$\mathrm{Ma} = \mathbf{V}/\mathbf{V}_{ac} \tag{5.198}$$

The mass flow rate through the nozzle is limited by the achievable velocity at the throat $\mathbf{V}_{ac}$ and can thus be determined from Eq. (4.31) as

$$\dot{m}_{\text{choked}} = \mathbf{V}_{ac} A_t / v_t \tag{5.199}$$

where v_t is the specific volume of the gas at the nozzle throat and A_t is the cross-sectional area at the throat. However, v_t is difficult to determine unless the pressure and temperature of the gas at the throat are known. If the flow of the ideal gas is assumed to be isentropic, then the throat pressure and temperature can be found by equating the nozzle exit velocity found from the first law [Eq. (4.118)] to the acoustic velocity [Eq. (5.197)] and solving for the temperature at the throat in terms of the nozzle entrance temperature. The isentropic ideal gas relation Eq. (5.146) is then used to relate the inlet and throat pressures to the known throat-to-inlet temperature ratio. The result is

$$T_t = T_{\text{in}}\left(\frac{2}{k+1}\right) \tag{5.200a}$$

$$P_t = P_{\text{in}}\left(\frac{2}{k+1}\right)^{k/(k-1)} \tag{5.200b}$$

and these relations can be substituted into Eq. (5.199) to give the mass flow rate at the throat in terms of the nozzle inlet conditions as

$$\dot{m}_{\text{choked}} = \Gamma A_t P_{\text{in}} \left(\frac{1}{RT_{\text{in}}}\right)^{1/2} \tag{5.201}$$

where $\Gamma = k^{1/2}[2/(k+1)]^{(k+1)/[2(k-1)]}$ and P_{in} and T_{in} are the nozzle entrance pressure and temperature (and the inlet gas is assumed to be at zero velocity). Note that Eq. (5.201) indicates that the mass flow rate through the nozzle is independent of the downstream pressure for this condition; i.e., we cannot increase the mass flow rate above that for choked flow by further lowering the downstream pressure.

Example Problem 5.17

A converging nozzle exhausts high-pressure air to the atmosphere with the conditions shown below. Find the mass flow rate through the nozzle and the air temperature and pressure at the nozzle throat. Assume isentropic flow.

Diagram

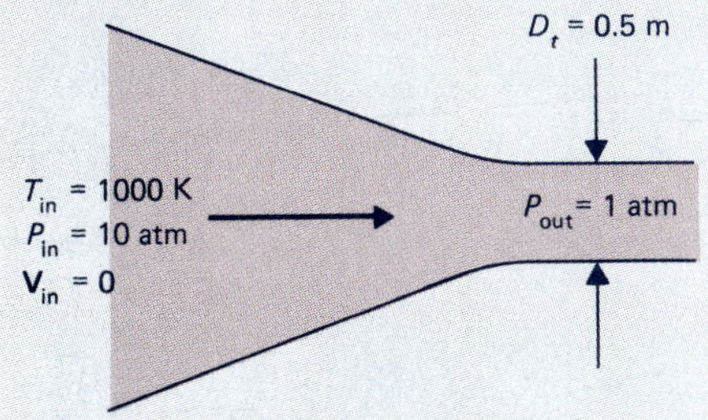

Solution

First, check to see if the nozzle is choked. If the nozzle is *not* in choked flow, then the exit temperature is found from the isentropic relation Eq. (5.145) as

$$T_t = T_{\text{in}}\left(\frac{P_t}{P_{\text{in}}}\right)^{(k-1)/k} = T_{\text{in}}\left(\frac{1}{10}\right)^{0.4/1.4} = 518\ \text{K}$$

and this can be substituted into the first law relation for an adiabatic nozzle Eq. (4.115) to give the exit velocity as

$$\begin{aligned}\mathbf{V}_{\text{out}} &= [2g_c c_P(T_{\text{in}} - T_t)]^{1/2}\\ &= \{[2 \times 1.0052\ \text{kJ/(kg}\cdot\text{K)}][(1000 - 518)\ \text{K}](1000\ \text{J/kJ})\}^{1/2}\\ &= 984\ \text{m/s}\end{aligned}$$

This can be checked against the acoustic velocity at the throat Eq. (5.197) (based on the calculated T_t for the unchoked case):

$$\begin{aligned}\mathbf{V}_{ac} &= (g_c kRT_t)^{1/2}\\ &= \{[1.4 \times 0.2870\ \text{kJ/(kg}\cdot\text{K)}](518\ \text{K})(1000\ \text{J/kJ})\}^{1/2}\\ &= 456\ \text{m/s}\end{aligned}$$

Because the predicted acoustic velocity at the nozzle throat is lower than the computed value for an unchoked nozzle, and because the actual velocity at the throat cannot exceed the acoustic velocity at the throat temperature, this nozzle must be operating in choked flow. We now calculate the mass flow rate through the nozzle for the choked flow case from Eq. (5.201):

$$\begin{aligned}\Gamma &= k^{1/2}\left(\frac{2}{k+1}\right)^{(k+1)/[2(k-1)]}\\ &= 1.4^{1/2}\left(\frac{2}{1.4+1}\right)^{(1.4+1)/[2(1.4-1)]} = 0.6847\end{aligned}$$

$$\begin{aligned}\dot{m}_{\text{choked}} &= \Gamma A_t P_{\text{in}}\left(\frac{1}{RT_{\text{in}}}\right)^{1/2}\\ &= 0.6847\,\frac{\pi D_t^2}{4}\,(\text{m}^2)(10\ \text{atm})(101.3\times 10^3\ \text{Pa/atm})\\ &\quad\times\left\{\frac{1}{[0.2870\ \text{kJ/(kg}\cdot\text{K)}](1000\ \text{K})(1000\ \text{J/kJ})}\right\}^{1/2}\\ &= \frac{0.6847\pi}{16}\times 10\times 101.3\times 10^3\times\left(\frac{1}{2.87\times 10^5}\right)^{1/2} = 254\ \text{kg/s}\end{aligned}$$

which is the desired result. The temperature at the throat for choked flow Eq. (5.200*a*)

$$T_t = T_{\text{in}}\left(\frac{2}{k+1}\right) = 1000\left(\frac{2}{2.4}\right) = 833\ \text{K}$$

which is quite different from the value for unchoked flow. This is another check for choked flow; if the predicted throat temperature is higher for the choked flow relations than is predicted by the unchoked relations, then choked flow exists and the choked flow predictions should be used.

The pressure at the throat is given for choked flow by Eq. (5.200*b*) as

$$P_t = P_{\text{in}}\left(\frac{2}{k+1}\right)^{k/(k-1)} = 10\left(\frac{2}{2.4}\right)^{1.4/0.4} = 5.28\ \text{atm}$$

which is well above the pressure of 1 atm of the surroundings just outside the nozzle throat. Checking the nozzle throat pressure for choked flow and comparing with the exit pressure are still another way of determining whether choked flow is present; if $P_{t,\,choked} > P_{out}$, then the flow is choked. Otherwise, the flow is subsonic (or if $P_{t,\,choked} = P_{out}$, the flow at the throat has just reached exactly Ma = 1).

To increase the nozzle exit velocity above $\mathbf{V}_{ac}$ (that is, above Ma = 1), the diameter of the nozzle past the minimum nozzle diameter must be *increased,* resulting in a converging-diverging nozzle shape. If the discharge pressure from a converging-diverging nozzle is low enough, then the flow in the diverging section will be increasingly supersonic (that is, Ma > 1) along the diverging section, as shown in Fig. 5.24. For some range of exit pressures, the flow may be choked at the nozzle throat and supersonic through an initial portion of the diverging section. And then a *shock layer* will form, and the flow between the shock and the nozzle exit

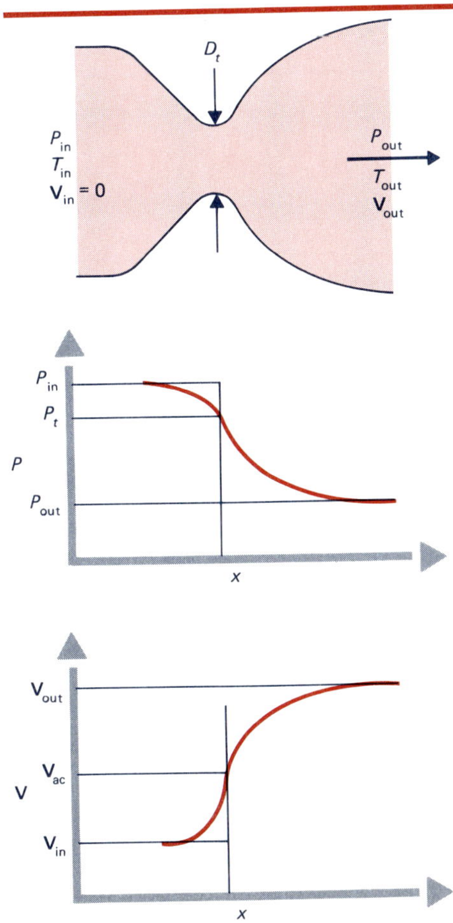

Figure 5.24 Pressure and velocity profiles of one-dimensional flow (without shock) through a converging-diverging nozzle.

will be subsonic. A shock layer is a thin layer of gas across which a step change in temperature, pressure, and velocity occurs. We do not examine this case.

Note that the relations derived above for nozzle throat conditions for choked flow are based on an adiabatic reversible (isentropic) flow. Most nozzles do operate at close to adiabatic conditions, because even if the nozzle walls are cooled, as for many rocket nozzles, the heat transfer at the nozzle surface is much smaller than the enthalpy and velocity-change terms in the first law. The heat transfer can thus be neglected in most cases and the flow treated as adiabatic. However, the flow is often quite irreversible because of fluid friction in the converging section and because of both fluid friction and flow separation in the diverging sectional. (Flow separation is a fluid-mechanical phenomenon in which a smoothly flowing fluid suddenly develops chaotic flow characteristics, resulting in a large dissipation of energy.) The irreversible nature of the flow can be treated through the use of the nozzle efficiency in the same way as described in Eq. (5.196).

Example Problem 5.18

A converging-diverging nozzle exhausts high-pressure air to the atmosphere with the conditions shown below. Find the mass flow rate through the nozzle, the air temperature and pressure at the nozzle throat, the exit velocity, and the nozzle efficiency. Assume isentropic flow in the converging section, but not in the diverging section.

Diagram

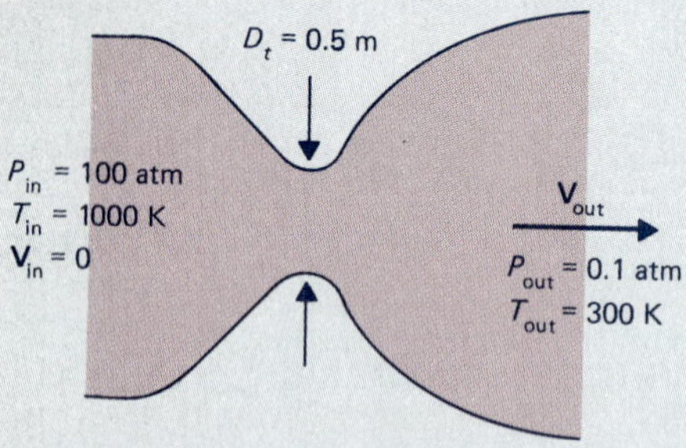

Solution

The throat conditions for temperature, pressure, and mass flow rate will be the same as in Example Problem 5.17 because the inlet conditions and the throat diameter for this problem are the same. Thus, $T_t = 833$ K, $P_t = 5.28$ atm, and $\dot{m} = 254$ kg/s. The actual outlet velocity may be found directly from the first law

$$\mathbf{V}_{\text{out},\,a} = [2g_c c_P(T_{\text{in}} - T_{\text{out},\,a})]^{1/2}$$
$$= \{[2 \times 1.0052\ \text{kJ/(kg}\cdot\text{K)}][(1000 - 300)(\text{K})](1000\ \text{J/kJ})\}^{1/2}$$
$$= 1186\ \text{m/s}$$

If the nozzle were isentropic throughout the diverging section, the outlet temperature at the prescribed outlet pressure would be

$$T_{\text{out},\,s} = T_t\left(\frac{P_{\text{out}}}{P_t}\right)^{(k-1)/k} = 833\left(\frac{0.1}{5.28}\right)^{0.4/1.4} = 268\ \text{K}$$

and the nozzle outlet velocity would then be

$$\mathbf{V}_{\text{out},\,s} = [2g_c c_P(T_{\text{in}} - T_{\text{out},\,s})]^{1/2}$$
$$= \{[2 \times 1.0052\ \text{kJ/(kg}\cdot\text{K)}][(1000 - 268)\ (\text{K})](1000\ \text{J/kJ})\}^{1/2}$$
$$= 1213\ \text{m/s}$$

and the nozzle efficiency is then

$$\eta_{\text{nozzle}} = \frac{\mathbf{V}^2_{\text{out},\,a}}{\mathbf{V}^2_{\text{out},\,s}} = \left(\frac{1186}{1213}\right)^2 = 0.956 = 95.6\%$$

The outlet diameter for this nozzle can be found from the continuity equation and the known mass flow rate to be

$$D_{\text{out}} = \left(\frac{4\dot{m}v_{\text{out}}}{\pi\mathbf{V}_{\text{out},\,a}}\right)^{1/2} = \left(\frac{4\dot{m}RT_{\text{out},\,a}}{\pi P_{\text{out}}\mathbf{V}_{\text{out},\,a}}\right)^{1/2}$$
$$= \left\{\frac{(4 \times 254\ \text{kg/s})[0.2870\ \text{kJ/(kg}\cdot\text{K)}](300\ \text{K})}{\pi(0.1)(101.3\ \text{kPa})(1186\ \text{m/s})}\right\}^{1/2}$$
$$= 1.52\ \text{m}$$

Because vapor turbines such as steam turbines operate by converting the kinetic energy of the vapor flow to work through momentum transfer, the kinetic energy of the flow stream entering the turbine itself should be as great as possible. Therefore, turbines are always preceded by a nozzle or more generally multiple nozzles, each carrying a portion of the total vapor flow. The nozzles serve to convert the enthalpy of the inlet vapor stream to kinetic energy at the entrance to the first row of turbine blades.

Nozzles for this purpose are limited by two contraints. First, if the pressure ratio across the nozzle will cause sonic flow at the nozzle throat, then the nozzle must be converging-diverging; otherwise, it can be simply a converging nozzle. Second, assuming the vapor at the entrance to the nozzle is superheated, the outlet pressure must be great enough that the outlet state of the vapor remains superheated or nearly so. Otherwise, if the outlet state is in the saturation region, liquid droplets will appear in the high-velocity flow, and destruction of the turbine blades can result.

The pressure at the entrance to the nozzle is controlled by a valve, called the *throttle*, which simply introduces a pressure drop in the vapor suppply line (with no change in enthalpy to the nozzle, according to the first law). The throttle also serves to change the cross-sectional area of the supply line and thus changes the mass flow rate to the nozzle and turbine. This controls the power output of the turbine.

In a gas turbine,the effect of the nozzles is incorporated by placing alternate rows of stationary and rotating turbine blades. The stationary blades serve to both change the direction of flow, to provide the best angle of flow with respect to the following turbine blade set, and to increase the flow kinetic energy, and thus they act as nozzles to the following turbine blades. This arrangement allows continual expansion of the gases passing through the turbine, with the reduction of gas enthalpy because of work done on any turbine blade set, followed by an increase in the gas kinetic energy in passing through the succeeding set of stationary blades acting as nozzles. The nozzle–turbine blade sequence is designed so that the maximum work can be extracted from the enthalpy of the gases entering the turbine.

For gas-turbine propulsion systems, the enthalpy remaining in the gases leaving the turbine can be converted to kinetic energy by passing these gases through a propulsion nozzle; the change in kinetic energy and therefore the change in momentum of the gas between the engine inlet and nozzle outlet provide the propulsive thrust in a turbojet engine. This application of nozzles is discussed more fully in Chap. 6.

5.11.4 Heat Exchangers

There are many types of heat exchangers, and their ability to transfer energy from one flow stream to another can be defined in various ways. A useful form for the efficiency of exchangers that involve two fluids flowing through the exchanger, neither of which changes phase, is the ratio of the actual enthalpy change of the fluid being heated to the change that would occur if the temperature of the heated fluid leaving the exchanger became equal to the inlet temperature of the fluid being cooled, or

$$\eta_{\text{he}} = \frac{h_{H,\text{in}} - h_{H,\text{out}}}{h_{H,\text{in}} - h_{C,\text{in}}} \tag{5.202}$$

The second law prevents the hot outlet fluid from ever having a lower temperature or enthalpy than the cold inlet fluid, so this efficiency is always positive and less than 100 percent. If the manufacturer provides the efficiency of a given exchanger for given flow rates and inlet conditions, then Eq. (5.202) along with application of the first law to the exchanger allows calculation of the temperatures of both outlet streams.

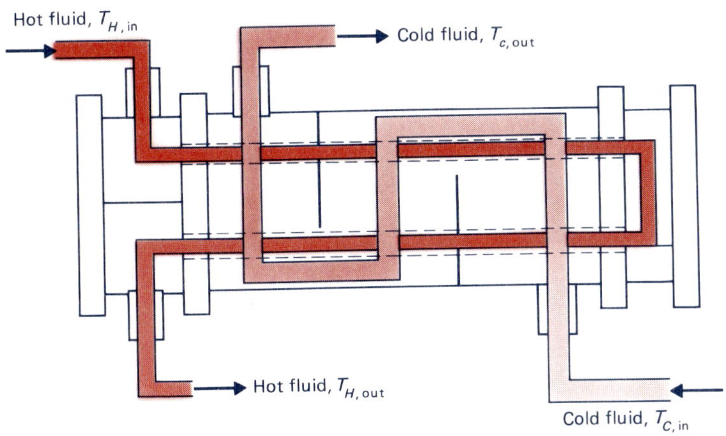

Figure 5.25 Shell-and-tube heat exchanger (showing single tubes from large tube bundle).

There are many types of heat exchangers. Those that use flow channels with solid surfaces to separate one fluid from the other are called *recuperators, preheaters, feedwater heaters, condensers, reboilers,* or simply *heat exchangers;* the name depends on the particular application. Common examples are the radiators used in automotive cooling systems (an example of a liquid-to-air or *finned-tube* heat exchanger).

When the two fluids are liquids at different pressures and both must be pumped because of process requirements (such as in refineries or chemical process plants), the typical heat exchanger is of the *shell-and-tube type*, as shown in Fig. 5.25. Here the higher-pressure fluid is pumped through many parallel tubes (only one is shown in the figure), and the lower-pressure fluid is passed over the tubes inside the shell, or pressure vessel. Heat transfer between the fluids occurs through the tube walls. Power plant condensers, used to condense the low-pressure steam leaving a steam turbine, are a modification of the shell-and-tube exchanger in which cooling water passes through the tubes and saturated vapor condenses on the tube surfaces inside the shell.

Some heat exchangers use a mechanically rotated barrel that contains a matrix of material which is heated while passing through the hot fluid stream and continually rotates to transfer the energy it has gained to the cold stream (Fig. 5.26). When used in air conditioning systems to transfer energy from an exhaust stream to an incoming airstream, such a device is often called an *enthalpy wheel;* when employed with a heat engine such as the Brayton cycle (Chap. 6), it is called a *regenerator.*

Figure 5.26 Rotating heat exchanger; enthalpy wheel or regenerator.

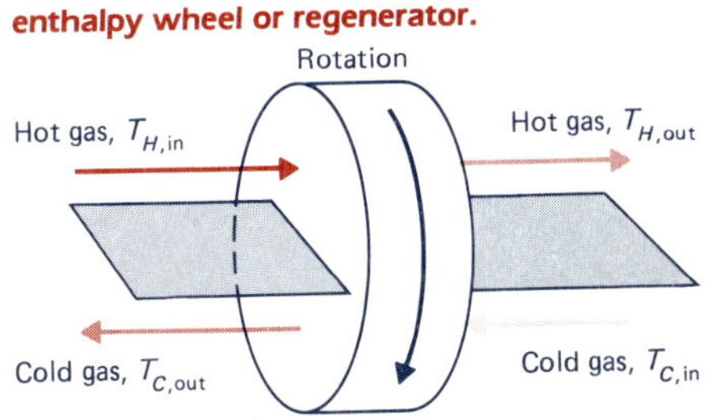

5.11.5 Control Mass Efficiency

The expansion or work stroke in an internal combustion engine requires a control mass analysis. An efficiency for this process or other piston-cylinder process compares the actual work output to the isentropic work to the same final pressure. This efficiency is expressed in terms of internal energy as

$$\eta_{CM} = \frac{w_a}{w_s} = \frac{(u_2 - u_1)_a}{(u_2 - u_1)_s} \tag{5.203}$$

The initial state is the same in this comparison, and kinetic and potential energy changes are neglected.

5.12 Cyclic Processes and the Carnot Cycle

Processes that return to their initial state are called *cyclic processes.* The individual processes that make up the elements of the cyclic process vary, and they depend on the particular application. An ideal steam power cycle is composed of constant-pressure heat transfer processes (to the working fluid in the steam generator and from the working fluid in the condenser) and adiabatic work processes (work addition by the pump and work output by the turbine). The idealized spark-ignition engine is composed of adiabatic and constant-volume processes. The fuel and air are compressed adiabatically, and the subsequent combustion is idealized as a constant-volume heating. The hot gases expand adiabatically, doing work. Then the exhaust gases are cooled at constant volume. In these idealized examples, the processes are usually assumed to be reversible. These examples (and there are many others) indicate that a cyclic process is composed of many different individual processes, and the combination depends on the application.

These example cycles have a common feature—they operate between two limiting temperatures. The high temperature results from the combustion process in the steam generator or within the cylinder. The low temperature results from the cooling process. The characteristics of these two-temperature cycles are shown from a general viewpoint as a high-temperature heat transfer reservoir or source at T_H and a low-temperature heat transfer reservoir at T_L. The cycle operating between these two temperatures is arbitrary. The general power cycle or engine is shown schematically in Fig. 5.27. The characteristics of such cycles are discussed in Sec. 5.7. A more complete presentation is given here.

Figure 5.27 Engine representation.

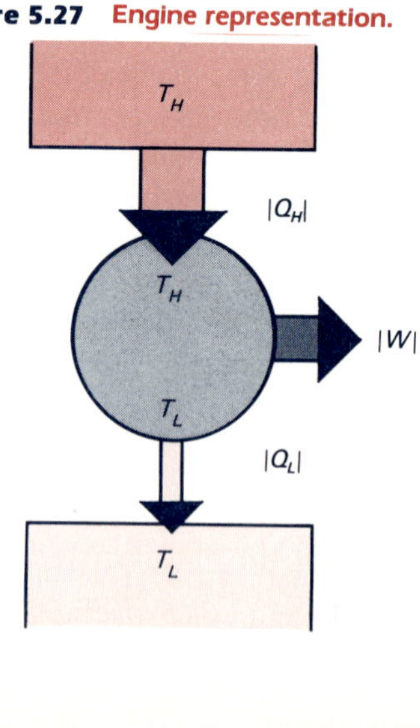

The first law for an arbitrary cycle states

$$\oint \delta W = \oint \delta Q \tag{5.204}$$

which is valid for an arbitrary collection of processes and for both reversible and irreversible cycles. For the two heat transfer reservoir cycle shown in Fig. 5.27,

$$W = \oint \delta W = Q_H - Q_L \tag{5.205}$$

Note that *for this discusson,* we follow the convention for signs on Q and W shown in Fig. 5.27, and the sign is explicitly indicated in the equations to show the direction of heat transfer. The second law for a cycle [Eq. (5.74)] states

$$\oint dS_{\mathrm{CM}} = 0 \geq \oint \sum_i \left(\frac{\delta Q_i}{T_i}\right)_{\mathrm{CM}} \tag{5.206}$$

where the zero results from the consideration of a cycle. Equations (5.204) and (5.206) are general equations for cycles. These expressions lead to a very important statement about cycles operating between two heat transfer reservoirs. For reversible heat transfers with the two heat transfer reservoirs, the second law is

$$0 \geq \frac{Q_H}{T_H} - \frac{Q_L}{T_L} \tag{5.207a}$$

or

$$\frac{Q_L}{Q_H} \geq \frac{T_L}{T_H} \tag{5.207b}$$

This last expression could also be obtained from the entropy generation expression in Eq. (5.77).

The *efficiency* of a cycle η is defined as

$$\eta \equiv \frac{\text{desired output}}{\text{required input}} \tag{5.208}$$

This efficiency should not be confused with the component efficiency defined in Sec. 5.11. The cycle efficiency compares the complete cycle's desired output to required input, whereas the component efficiency considers a process (not a cycle) and compares actual to reversible paths. A power cycle or engine as diagramed in Fig. 5.27 has a work output W and an input from the high-temperature reservoir Q_H. Thus, this efficiency is

$$\eta = \frac{W}{Q_H} \tag{5.209}$$

Equation (5.205) yields

$$\eta = 1 - \frac{Q_L}{Q_H} \tag{5.210}$$

The ratio of heat transfers is eliminated with Eq. (5.207*b*) to yield

$$\eta \leq 1 - \frac{T_L}{T_H} \tag{5.211}$$

where the equality applies to a reversible cycle and the inequality applies to an irreversible cycle. Thus

$$\eta_{\text{irr}} < \eta_{\text{rev}} = 1 - \frac{T_L}{T_H} \tag{5.212}$$

or the maximum efficiency of a heat engine operating between two heat transfer reservoirs occurs for a reversible cycle. This statement applies regardless of the specific processes of the cycle, but they must be reversible.

A very important reversible cycle that uses two heat transfer reservoirs is shown in Fig. 5.28*a* and *b*. The two specific examples undergo the same processes but operate with different components. The cycle includes isothermal energy exchange with the two heat transfer reservoirs, and the other two processes take place adiabatically. All processes are reversible. The heat transfers occur across infinitesimal temperature differences, so the temperature within the system is the same as the source or sink. The other two processes are reversible and adiabatic, so they are isentropic. The state diagram in Fig. 5.28*c* shows the following four processes:

1 to 2	Isothermal heat transfer at T_H
2 to 3	Isentropic expansion process
3 to 4	Isothermal heat transfer at T_L
4 to 1	Isentropic compression process

This cycle is called a *Carnot cycle.*

A control mass undergoing a Carnot cycle is shown in Fig. 5.28*a*. There is a heat transfer to the control mass from state 1 to state 2 at T_H, and some expansion work is obtained. The process from state 2 to state 3 is an adiabatic expansion to temperature T_L. Then there is a heat transfer from the control mass from state 3 to state 4 which occurs reversibly at T_L. The final process takes the control mass adiabatically from state 4 back to state 1. An alternate situation is a power plant which includes four control volume elements with the working fluid undergoing the processes as it moves through the elements (Fig. 5.28*b*). There is an isothermal heat transfer to the fluid at the high temperature in the boiler from state 1 to state 2. The fluid is expanded adiabatically through the turbine from state 2 to 3 to obtain a work output. The condenser is used for the isothermal heat transfer from state 3 to 4 at T_L. The adiabatic compression of the fluid to state 4 is achieved by the pump. Both types are represented by the general power cycle diagram shown in Fig. 5.27.

The above analysis considers a power cycle or engine used to produce a net output of work. A refrigerator is a cyclic device that seeks a heat transfer from a low-temperature region and requires work input. This

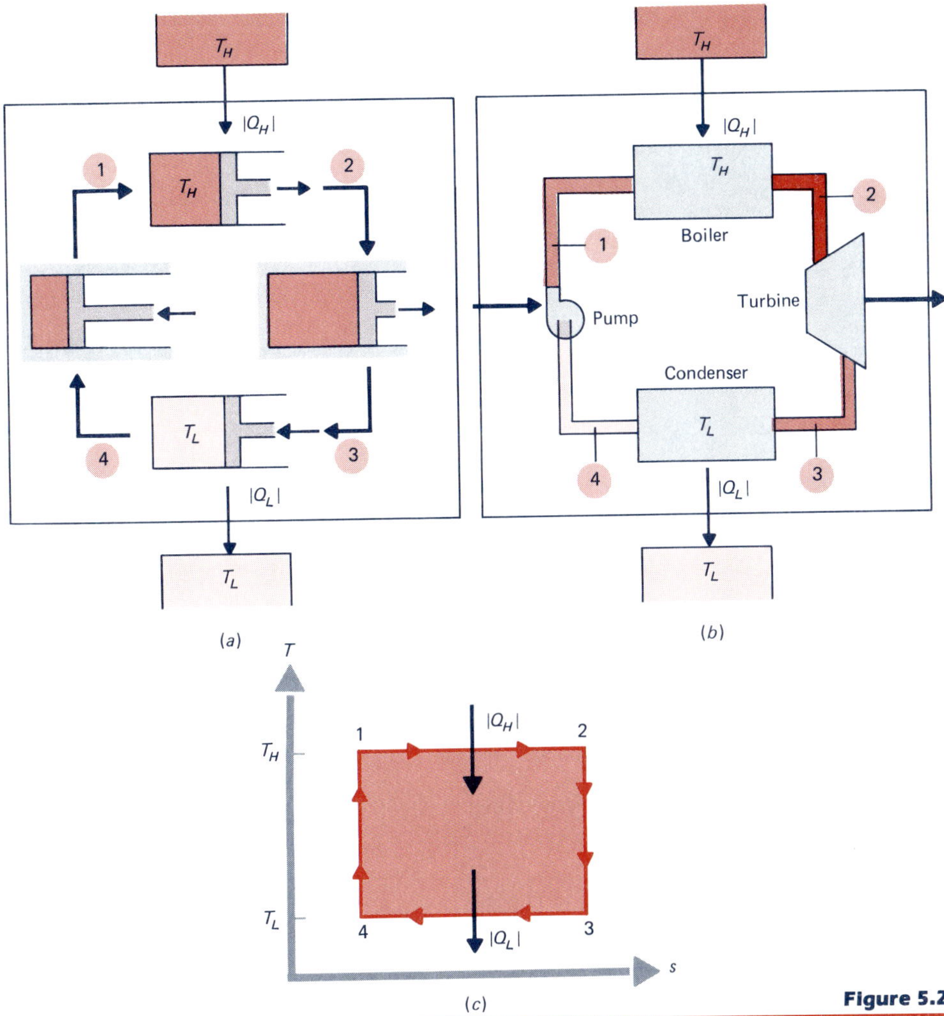

Figure 5.28 Carnot cycle.

device is shown schematically in Fig. 5.29. The efficiency for a refrigerator is termed the *coefficient of performance* and is given as

$$\text{COP} = \frac{Q_L}{W} \tag{5.213}$$

The first law states that

$$-W = Q_L - Q_H \tag{5.214}$$

or

$$\text{COP} = \frac{Q_L}{Q_H - Q_L} = \frac{1}{Q_H/Q_L - 1} \tag{5.215}$$

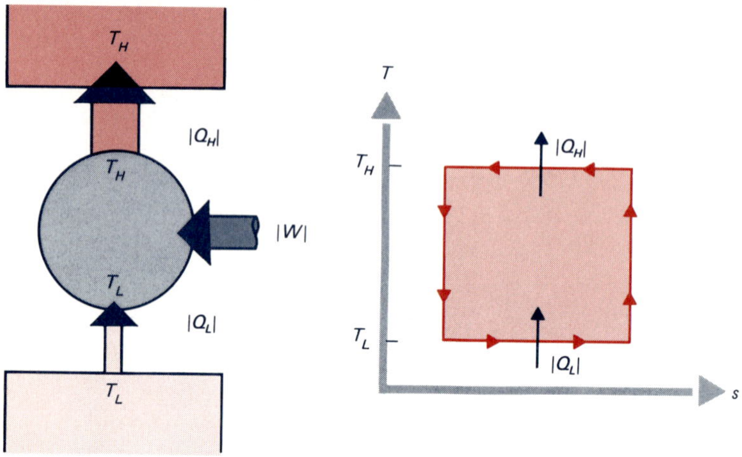

Figure 5.29 Refrigerator representation.

The second law yields

$$0 \geq \frac{Q_L}{T_L} - \frac{Q_H}{T_H} \qquad \text{or} \qquad \frac{Q_H}{Q_L} \geq \frac{T_H}{T_L} \tag{5.216}$$

Thus

$$\text{COP} \leq \frac{1}{T_H/T_L - 1} \tag{5.217}$$

The equality applies to the reversible refrigerator and the inequality to the irreversible refrigerator. Thus

$$\text{COP}_{\text{irr}} < \text{COP}_{\text{rev}} = \frac{1}{T_H/T_L - 1} \tag{5.218}$$

Note that the COP theoretically ranges from 0 to large values > 1. Further consideration is given to these cycles in Chaps. 6 and 7.

A heat pump (HP) operates as indicated in Fig. 5.29, but it seeks a heat transfer to a high-temperature region so the desired output is Q_H. The above analysis yields

$$\text{COP}_{\text{HP}} = \frac{Q_H}{W} = \frac{Q_H}{Q_H - Q_L} = \frac{1}{1 - Q_L/Q_H} \leq \frac{1}{1 - T_L/T_H} \tag{5.219}$$

where the equality sign applies to the reversible cycle and the inequality applies to the irreversible cycle. Therefore,

$$\text{COP}_{\text{HP, irr}} < \text{COP}_{\text{HP, rev}} = \frac{1}{1 - T_L/T_H} \tag{5.220}$$

Note that

$$\text{COP}_{\text{HP}} - \text{COP} = \frac{Q_H}{Q_H - Q_L} - \frac{Q_L}{Q_H - Q_L} = 1 \tag{5.221}$$

5.13 Temperature Measurement

The thermodynamic temperature is defined in Eq. (5.9) as a derivative incorporating internal energy and entropy. This definition satisfied our expectations of thermal equilibrium and our perceptions of hot and cold. This definition of thermodynamic temperature has led, through some analysis, to Eq. (5.207). For a reversible engine operating between two heat transfer reservoirs, this expression is

$$\frac{Q_L}{Q_H} = \frac{T_L}{T_H} \tag{5.222}$$

This expression is valid for an arbitrary reversible engine and is independent of the working fluid used in the engine. This expression, in principle, supplies a method to measure thermodynamic temperature.

The approach is to operate a reversible engine between a specified temperature reservoir and another temperature reservoir where the temperature is desired. Since the heat transfers are measurable, in principle, the unknown temperature is obtained. The Kelvin scale is based on the assigned value of the triple point of water as 273.16 K. The measurement of the ratio of the heat transfers for a cycle operating between this temperature and the desired temperature yields the value for the desired temperature. Therefore, any temperature is assigned by this reversible engine.

The practical approach to temperature measurement is to use an ideal gas thermometer. This thermometer is based on the ideal gas equation of state

$$Pv = RT \tag{5.223}$$

If a specified amount of gas occupies a specified volume, then

$$\frac{P}{T} = \text{constant} \tag{5.224}$$

Therefore, the measurement of pressure yields the temperature for a fixed-volume thermometer containing an ideal gas. Again, assigning 273.16 K to the triple point of water and measuring the pressure in this ideal gas thermometer yield the value for the constant in Eq. (5.224). Subsequent pressure measurements with this ideal gas thermometer yield the desired temperatures. Note that this method is based on the validity of the ideal gas equation of state. The verification of a specific gas obeying the ideal gas equation of state requires the measurement of temperature, which is possible with the reversible engine described above.

5.14 Other Statements of the Second Law

The original development of classical thermodynamics took into consideration engines and cyclic machines (as noted in Sec. 5.7), because the development of thermodynamics, like other disciplines, is based on physical observations. A short history of this development is presented in App. A. In this section we present a few of the other forms for the second law (including the forms presented in Sec. 5.7) to show the correspondence with the present approach. An understanding of the other forms is useful to a clear interpretation of the present approach. Also the other statements of the second law serve as good examples of thermodynamic analysis.

Statements of the second law are probably as numerous as the number of other authors. No attempt is made to present all possible statements. Also complete proofs of other statements of the second law from the present approach are not developed. The comparison of the various approaches is presented and discussed.

A number of variations of the second law deal with the impossibility of transforming a single heat transfer to work. This is the form referred to as the *Kelvin-Planck statement* [3]:

> It is impossible to construct an engine that, operating continuously, will produce no effect other than the extraction of heat from a single reservoir and the performance of an equivalent amount of work.

This statement is demonstrated by considering the cycle expression ("operating continuously" implies a cycle) in Sec. 5.12. The first law for a cycle with a single high-temperature reservoir is

$$W = Q_H$$

where the work is positive for the engine output and the heat transfer is positive into the engine. The second law for these same conditions is

$$0 \geq \frac{Q_H}{T_H}$$

This expression indicates that the heat transfer must be negative, or out of the engine. Therefore, there is a contradiction, and this leads to the impossibility. Note that it *is* possible to produce work from a single heat transfer if the system does not operate in a cycle.

Another statement that is similar to the Kelvin-Planck form is

> It is impossible to construct a machine which would work periodically and continuously and perform work at the expense of heat extracted from only one source.

This was formulated by Planck and is rephrased above from Kestin [4]. Another similar form is

> A perpetual-motion machine of the second kind (PMM2) is impossible.

where a PMM2 is a system which operates cyclically, delivering work while exchanging energy as a heat transfer with a single reservoir of uniform temperature. This statement is from Keenan [5].

An alternative approach considers the impossibility of a heat transfer from a low-temperature reservoir to a high-temperature reservoir. This second law statement is referred to as the *Clausius statement* [3]:

> It is impossible to construct a device that, operating continuously, will produce no effect other than the transfer of heat from a cooler to a hotter body.

The first law for a cycle that does not perform work, that accepts a heat transfer from a low-temperature body Q_L, and that has a heat transfer to a high-temperature body Q_H is

$$0 = Q_L - Q_H$$

or

$$Q_H = Q_L$$

The second law for a cycle is

$$0 \geq \frac{Q_L}{T_L} - \frac{Q_H}{T_H}$$

or

$$\frac{Q_L}{T_L} \leq \frac{Q_H}{T_H}$$

But the first law requires that $Q_L = Q_H$, so the second law requires

$$T_L \geq T_H$$

This contradicts the statement that T_L is lower than T_H and leads to the impossibility of such a device.

These examples are second law forms that result from the historical approach to classical thermodynamics. There are many other examples that are not presented here.

5.15 Summary

The basic elements of classical thermodynamics are now complete. The first and second laws are presented for both the control mass and control volume. These laws when combined with the state postulate and conservation of mass form the basis of thermodynamic analysis. The governing principles are summarized in Table 5.1. Some of the more common idealizations are presented in Tables 5.2 and 5.3. These expressions are applied to a clearly defined system indicating the boundary transfers. The properties required for the analysis have also been presented. The properties are contained in the appendixes. The ideal gas relations and incompressible fluid or solid relations are summarized in Tables 5.4 and 5.5, respectively. Also the Gibbs equation, Eq. (5.43), and Eq. (5.44) are important property relations. Be sure to separate the thermodynamic analysis from the property evaluation.

In the following chapters we apply the thermodynamic principles and properties to various systems and processes. New definitions are presented, but they are based on the elements delineated here.

Problems

5.1 Indicate whether the entropy of the control mass has increased, decreased, or remained constant when the following processes are complete. Explain your answer.

(*a*) A frictionless piston slowly compresses a gas in an adiabatic (perfectly insulated) cylinder.

(*b*) A frictionless piston slowly compresses an isothermal gas in a diabatic (permits heat transfer) cylinder.

(*c*) A piston that fits tightly in a cylinder slowly compresses an isothermal gas in a diabatic (allows heat transfer) cylinder.

(*d*) A cake of ice floating in water that is very near freezing increases in mass until all the water is frozen. (Take the control mass as the ice plus the water.)

(*e*) One kilogram of putty is dropped onto a floor tile and sticks without bouncing.

(*f*) A bead with a hole through it slides down a slack wire hung between two hooks that are at equal height, and it comes to rest at the lowest point on the wire. Take the bead as the control mass.

5.2 For the following processes, consider the total entropy generated in the control mass and in the surroundings. Describe the processes that change the entropy and the sign change of each, and show that $S_{gen} \geq 0$ in each case.

(*a*) A frictionless piston slowly compresses a gas in an adiabatic (perfectly insulated) cylinder.

(*b*) A frictionless piston slowly compresses an isothermal gas in a diabatic (permits heat transfer) cylinder.

(*c*) A piston that fits tightly in a cylinder slowly compresses an isothermal gas in a diabatic cylinder.

(*d*) A cake of ice floating in water that is very near the freezing temperature slowly increases in mass until all of the water is frozen. (Take the control mass as the ice plus the water.)

(*e*) One kilogram of putty is dropped onto a floor tile and sticks without bouncing.

(*f*) A bead with a hole through it slides down a slack wire hung between two hooks that are at equal height, and it comes to rest at the lowest point on the wire. Take the bead as the control mass.

5.3 Two reservoirs A and B contained within an isolated system are at temperatures T_A and T_B, respectively, and communicate thermally with each other. Prove that the entropy of the isolated system increases for this process regardless of whether $T_A > T_B$ or $T_B > T_A$.

5.4 Using the Mollier chart, determine the required properties for water in the given state.

Given: $x = 0.94$, $T = 100°\text{C}$
Find: h, s, P

5.5 Using the Mollier chart, determine the required properties for water in the given state.

Given: $s = 7.3$ kJ/(kg·K), $P = 6.0$ MPa
Find: T, h, x

5.6 Using the Mollier chart, determine the required properties for water in the given state.

Given: $P = 0.3$ MPa, $T = 400°\text{C}$
Find: s, h, x

5.7 Using the Mollier chart, determine the required properties for water in the given state.

Given: $x = 0.96$, $P = 0.02$ MPa
Find: h, s, T

5.8 Using the Mollier chart, determine the required properties for water in the given state.

Given: $x = 92$ percent, $T = 250°\text{C}$
Find: h, s, P

5.9 Using the Mollier chart, determine the required properties for water in the given state.

Given: $P = 1.0$ MPa, $h = 3480$ kJ/kg
Find: s, T

5.10 Using the Mollier chart, determine the required properties for water in the given state.

Given: $T = 700°\text{C}$, $s = 8.0$ kJ/(kg·K)
Find: P, h

5.11 Using the refrigerant 12 chart, determine the required properties for the given state.

Given: $P = 1.0$ MPa, $T = 60°\text{C}$
Find: h, s

5.12 Using the refrigerant 12 chart, determine the required properties for the given state.

Given: $h = 213$ kJ/kg, $x = 20$ percent
Find: T, P

5.13 Using the refrigerant 12 chart, determine the required properties for the given state.

Given: $s = 1.58$ kJ/(kg·K), $T = 40°C$
Find: h, P

5.14 Using the refrigerant 12 chart, determine the required properties for the given state.

Given: $x = 0$, $P = 1.5$ MPa
Find: T, h

5.15 A control mass in initial state 1 undergoes a process that results in final state 2. Because entropy is a property, we know that $(\Delta s)_{CM} = S_2 - S_1$ is the same regardless of the process that results in the change of state. Describe the entropy change of the surroundings $(\Delta S)_{surr}$ in terms of $(\Delta S)_{CM}$, i.e., greater than, less than, or equal to, if the process is:
(*a*) Heat transfer from the control mass to the surroundings
(*b*) Heat transfer from the surroundings to the control mass
(*c*) Adiabatic work done on the control mass with friction
(*d*) Adiabatic work done by the control mass without friction

5.16S At $T = 200°C$ and $P = 500$ kPa, find the specific entropy of:
(*a*) Nitrogen, using the gas tables (Table D.4)
(*b*) Steam, using the steam tables
(*c*) Steam, using the T-s diagram (Fig. D.1)
(*d*) Steam, using the P-h diagram (Fig. D.2)
(*e*) Steam, using the Mollier diagram (Fig. D.3)

5.16E At $T = 400°F$ and $P = 80$ psia, find the specific entropy of:
(*a*) Nitrogen, using the gas tables (Table E.4)
(*b*) Steam, using the steam tables
(*c*) Steam, using the T-s diagram (Fig. E.1)
(*d*) Steam, using the Mollier diagram (Fig. E.3)

5.17S Refrigerant 12 expands isothermally from an initial state at $P = 300$ kPa and $T = 40°C$ to a final state at $P = 150$ kPa. Find the change in specific entropy of the refrigerant 12 that occurs in this process, using the tables.

5.17E Refrigerant 12 expands isothermally from an initial state at $P = 45$ psia and $T = 100°F$ to a final state at $P = 20$ psia. Find the change in specific entropy of the refrigerant 12 that occurs in this process, using the tables.

5.18S Calculate the change in entropy for air between the states $P_1 = 1$ atm, $T_1 = 50°C$ and $P_2 = 10$ atm, $T_2 = 1000°C$, using Eq. (5.55) and the expression for variable specific heat from Table D.1. Compare your result with that from using Table D.2.

5.18E Calculate the change in entropy for air between the states $P_1 = 1$ atm, $T_1 = 120°F$ and $P_2 = 10$ atm, $T_2 = 1800°F$, using Eq. (5.55) and the expression for variable specific heat from Table E.1. Compare your result with that from using Table E.2.

5.19S A massless container of fixed volume encloses 1 kg of nitrogen at 1000 K. A second similar container encloses 3 kg of argon at $T_a < 1000$ K. Both containers are isolated from the surroundings. The containers are brought together, and there is heat transfer until the containers are in thermal equilibrium. Plot a curve of S_{gen}, the entropy generation during this process, versus T_a. (Here T_a is in the range $100 \le T_a \le 1000$ K.) Assume constant specific heats for argon and nitrogen.

5.19E A massless container of fixed volume encloses 2.2 lbm of nitrogen at 1800°R. A second similar container encloses 6.6 lbm of argon at $T_a < 1800°R$. Both containers are isolated from the surroundings. The containers are brought together, and there is heat transfer until the containers are in thermal equilibrium. Plot a curve of S_{gen}, the entropy generation during this process, versus T_a. (Here T_a is in the range $200 \le T_a \le 1800°R$.) Assume constant specific heats for argon and nitrogen.

5.20S A reversible adiabatic process occurs for steam between an intial state of $P_1 = 15$ kPa and $v_1 = 10.0$ m^3/kg and a final state with $P_2 = 100$ kPa. What is the temperature change for this process?

5.20E A reversible adiabatic process occurs for steam between an initial state of $P_1 = 1$ psia and $v_1 = 320$ ft^3/lbm and a final state with $P_2 = 14.7$ psia. What is the temperature change for this process?

5.21 A piston compresses an ideal gas in a cylinder from a state P_1, v_1 to a final pressure P_2. The gas has constant c_P and c_v.

(*a*) Derive an expression for the ratio of work done if the process is isentropic, w_s, to the work done if the process is isothermal, w_t, in terms of P, v, and k only.

(b) Sketch the two processes on a P-v diagram and a T-s diagram. Label all states.

5.22 An ideal gas in a cylinder undergoes a process from a state P_1, v_1 to a final temperature T_2.

(a) Derive an expression for the ratio of the work done if the process is isentropic, w_s, to the work done if the process is isobaric (constant pressure), w_p, in terms of k only.

(b) Sketch the two processes on P-v and T-s diagrams. Label all states.

5.23S A closed copper vessel has a volume of 2 m³. In the vessel is 4 kg of water (liquid plus vapor) at a pressure of 200 kPa. The surroundings are at 160°C. There is a heat transfer to the vessel until the water is all just evaporated.

(a) What is the final pressure in the vessel?

(b) What is the entropy change of the water in the vessel during the heat transfer process?

(c) What is the entropy change of the surroundings due to the heat transfer process?

(d) What is the entropy generation for the water due to the heat transfer process?

5.23E A closed copper vessel has a volume of 70 ft³. In the vessel is 9 lbm of water (liquid plus vapor) at 30 psia. The surroundings are at 320°F. There is a heat transfer to the vessel until the water is all just evaporated.

(a) What is the final pressure in the vessel?

(b) What is the entropy change of the water in the vessel during the heat transfer process?

(c) What is the entropy change of the surroundings due to the heat transfer process?

(d) What is the entropy generation for the water due to the heat transfer process?

5.24S A glass of ice water initially contains 50 percent ice and 50 percent water by mass. The ice water undergoes heat transfer with the surroundings, which are at T_0, until the water in the glass reaches very nearly a temperature of T_0.

(a) If $T_0 = 27°C$ and the entropy generation is $S_{gen} = 3$ kJ/K, what is the total mass of water in the glass during the process?

(b) At what value of T_0 would $S_{gen} = 0$ for the process described? (Prove your result.) The enthalpy of melting (*latent heat of fusion*) of ice is 335 kJ/kg.

5.24E A glass of ice water initially contains 50 percent ice and 50 percent water by mass. The ice water undergoes heat transfer with the surroundings, which are at T_0, until the water in the glass reaches very nearly a temperature of T_0.

(*a*) If $T_0 = 80°F$ and the entropy generation is $S_{gen} = 1.5$ Btu/°R, what is the total mass of water in the glass during the process?

(*b*) At what value of T_0 would $S_{gen} = 0$ for the process described? (Prove your result.) The enthalpy of melting (*latent heat of fusion*) of ice is 144 Btu/lbm.

5.25S A reversible isothermal process occurs in which a fixed amount of refrigerant 12 is evaporated and heated from an initial state with $T_1 = 40°C$ and $h_1 = 300$ kJ/kg to a final state at $P_2 = 200$ kPa. How much heat transfer occurs during this process?

5.25E A reversible isothermal process occurs in which a fixed amount of refrigerant 12 is evaporated and heated from an initial state with $T_1 = 100°F$ and $h_1 = 58$ Btu/lbm to a final state at $P_2 = 30$ psia. How much heat transfer occurs during this process?

5.26 Water (4 kg) is heated reversibly and isobarically in a nonflow process at a pressure of 1.0 MPa from a saturated liquid to a saturated vapor. Calculate the total heat transfer for the process.

5.27 An inventor proposes to build an adiabatic piston-cylinder device which will compress H_2O from a pressure of 0.2 MPa and 200°C to a final pressure of 1.0 MPa using 220.8 kJ/kg of work . Is this device reversible, irreversible, or impossible?

5.28 Water expands in a piston-cylinder device from an initial pressure and temperature of 10.0 MPa and 350°C to a final pressure of 2.0 MPa. If the process is reversible and adiabatic, calculate the work (kJ/kg) for the process.

5.29 Water is compressed in a reversible isothermal nonflow process from a pressure and temperature of 4.0 MPa and 400°C, respectively, to a final pressure of 8.0 MPa. Calculate the heat transfer (kJ/kg) and work (kJ/kg) for the process.

5.30 Calculate the total heat transfer during a nonflow process where 3.5 kg of H_2O is cooled reversibly and isobarically from a saturated vapor to a saturated liquid at 1400 kPa.

5.31 Water (1.5 m³) originally at a pressure of 0.2 MPa is compressed reversibly and adiabatically to a pressure and temperature of 0.6 MPa and 700°C. If the process takes place in a piston-cylinder device, calculate the total work required.

5.32 A piston-cylinder device is said to require 154.3 kJ/kg of work to compress H_2O from an initial pressure and temperature of 400 kPa and 300°C to a final pressure of 4 MPa. If a heat transfer of 233.8 kJ/kg is released during the process, is the process impossible, reversible, or irreversible?

5.33 Water in a piston-cylinder device is cooled from an initial temperature and pressure of 8.0 MPa and 700°C to a final temperature of 300°C. If the process is isobaric, find the heat transferred.

5.34 Water in a piston-cylinder device is compressed reversibly and isothermally from 1.0 MPa and 600°C to a final pressure of 7.0 MPa. If the initial volume is 1.750 m^3, find the total work done.

5.35 Water (2 kg) expands in a piston-cylinder device from a pressure of 0.4 MPa to a pressure and temperature of 0.2 MPa and 300°C. If the process is reversible and adiabatic, calculate the total work for the process.

5.36 Water in a piston-cylinder device undergoes a reversible adiabatic expansion from a pressure of 1 MPa and temperature of 250°C to a temperature of 100°C. Find the work (kJ/kg) done in the process.

5.37 Air (2 kg) in a system is expanded in a process from a pressure of 2.0 MPa and a temperature of 600 K to a pressure of 0.2 MPa and a volume of 0.7175 m^3. Determine the final temperature and the total change in entropy, assuming variable specific heats.

5.38 Air in a nonflow device undergoes the reversible process shown. Determine the heat transfer (kJ/kg) and work (kJ/kg) for each process.

Figure P5.38

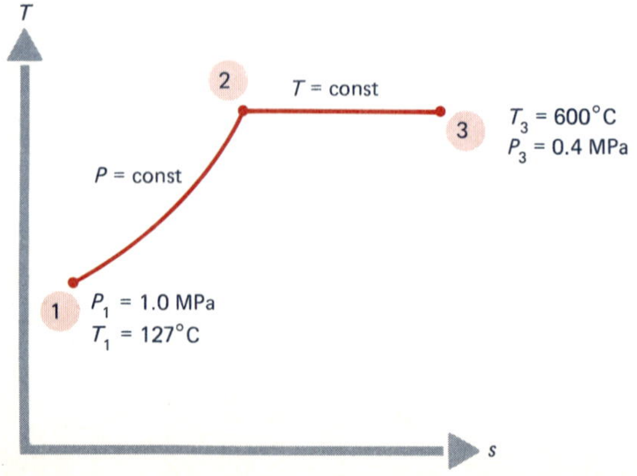

5.39 Helium in a nonflow device undergoes the reversible process shown. Determine the heat transfer (kJ/kg) and work (kJ/kg) for each process.

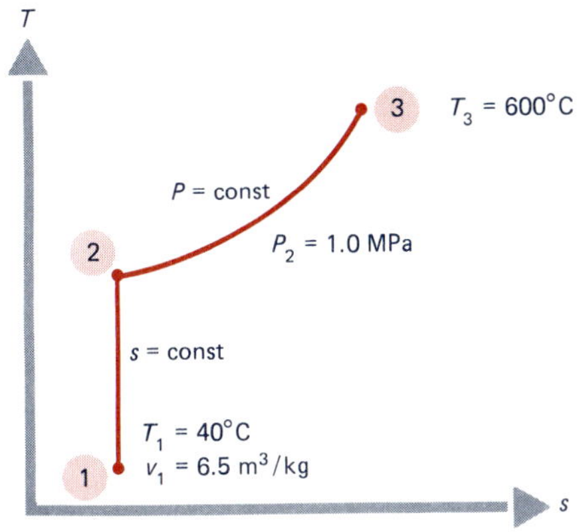

Figure P5.39

5.40 A nonflow device has been proposed to compress air adiabatically from 30°C and 100 kPa to 450 kPa with 120 kJ/kg of work. Is the device possible?

5.41 Air (0.5 g) in a system is allowed to expand adiabatically from a pressure of 5.0 MPa and volume of 25 cm^3 to a pressure of 0.15 MPa and temperature of 77°C. Determine the total work done and the total change in entropy, assuming *variable* specific heats.

5.42 Air in a piston-cylinder is cooled isobarically from a temperature of 800 K and pressure of 1.0 MPa to a temperature of 310 K. Determine the heat transfer per kg and the work done per kg during the process, assuming *variable* specific heats. Assume a reversible process.

5.43 Ammonia expands isobarically ($P =$ constant) from saturated liquid to saturated vapor at $P = 100$ kPa. The expansion is caused by heat transfer from the surroundings to the ammonia.

(*a*) What is the entropy change of the ammonia during the expansion process?

(*b*) What is the maximum entropy change of the surroundings that can occur during the process?

5.44 Ammonia undergoes an isentropic process from an initial pressure of 75 kPa to a final state of 150 kPa and 50°C. What is the specific enthalpy change of the ammonia for this process?

5.45 Refrigerant 12 undergoes an isentropic process from an initial pressure of 80 kPa to a final state of 600 kPa and 150°C. What is the change in specific volume of the refrigerant 12 for this process?

5.46 Air at the conditions given below is compressed isothermally to the final conditions (state 2). Find the change in entropy [kJ/(kg·K)] for the process.

$T_1 = 300$ K $\quad P_1 = 150$ kPa $\quad P_2 = 300$ kPa

5.47 Air at the conditions given below is compressed adiabatically to state 2. Calculate the change in entropy [kJ/(kg·K)] and w_{CV}. The process is steady-state–steady-flow.

$T_1 = 45°C$; $\quad P_1 = 100$ kPa;
$T_2 = 90°C$; $\quad P_2 = 400$ kPa

5.48 Air undergoes the following nonflow reversible processes (Fig. P5.48):

$T_1 = 300$ K; $\quad P_1 = 2.0$ MPa; $\quad$ 1 to 2: isobaric
$T_2 = 400$ K; $\quad T_3 = 500$ K; $\quad$ 2 to 3: isentropic

Find (*a*) $s_3 - s_1$ and (*b*) P_3.

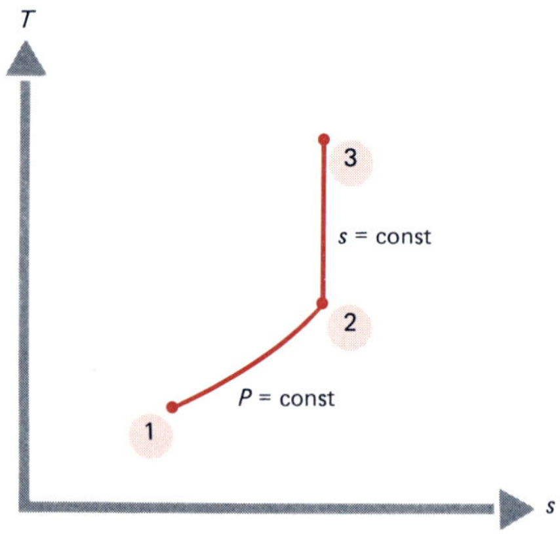

Figure P5.48

5.49 A piston-cylinder device undergoes a reversible isothermal process from an initial pressure of 50 kPa and specific volume of 10 m³/kg to a final specific volume of 40 m³/kg. The medium is air. Draw and label the process on a T-v plot, and calculate the heat transfer and work (kJ/kg).

5.50 A frictionless piston-cylinder device undergoes a reversible adiabatic process from an initial condition of 200 kPa and 100°C to a final state of 200 kPa and 250°C. The medium is helium. In what form did the energy transfer take place (heat transfer or work)? How much energy was transferred (kJ/kg)? How much boundary work was there? How much shaft work?

5.51 An inventor proposes to build a nonflow device that will compress air isothermally at a temperature of 400°C from a pressure of 4.0 MPa to a final pressure of 15.0 MPa. If the device requires 150 kJ/kg of work input, is it reversible, irreversible, or impossible?

5.52 Air (2 kg) expands in a piston-cylinder device from a pressure of 0.4 MPa to a pressure and temperature of 0.2 MPa and 300°C. If the process is reversible and adiabatic, calculate the total work for the process.

5.53 Air (4 kg) is heated reversibly and isothermally in a nonflow process at a pressure of 1.0 MPa to a pressure of 4 MPa. Calculate the total heat transfer for the process if the temperature is 25°C.

5.54 An inventor proposes to build an adiabatic piston-cylinder device which will compress air from pressure and temperature of 0.2 MPa and 200°C to a final pressure of 1.0 MPa using 240 kJ/kg of work. Is this device reversible, irreversible, or impossible?

5.55 Air expands in a piston-cylinder device from an initial pressure and temperature of 10 MPa and 350°C to a final pressure of 2.0 MPa. If the process is reversible and adiabatic, calculate the work (kJ/kg) for the process.

5.56 Air is compressed in a reversible isothermal nonflow process from a pressure and temperature of 4.0 MPa and 400°C to a pressure of 8.0 MPa. Calculate the heat transfer (kJ/kg) and work (kJ/kg) for the process.

5.57 Calculate the total heat transfer during a nonflow process where 3.5 kg of air is heated reversibly and isobarically from 400 to 800°C.

5.58 Air (1.5 m^3) originally at a pressure of 0.2 MPa is compressed reversibly and adiabatically to a pressure and temperature of 0.6 MPa and 700°C. If the process takes place in a piston-cylinder device, calculate the total work required.

5.59 Air in a piston-cylinder device is compressed reversibly and isothermally from 1.6 MPa and 600°C to a final pressure of 7.0 MPa. If the initial volume is 1.750 m^3, find the total work done.

5.60 Air is compressed in a reversible adiabatic nonflow process from an initial pressure and temperature of 0.80 MPa and 250°C to a final pressure of 4.5 MPa. Calculate the work for the process.

5.61S A 1-m^3 tank is one-half filled with refrigerant 12 liquid, the remainder with refrigerant 12 vapor. The initial pressure is 600 kPa. There is a heat transfer from the surroundings until one-half of the original liquid by mass is evaporated. An automatic valve allows saturated vapor to escape at constant tank pressure. The surroundings are at $T_0 = 27$°C. Determine the required heat transfer and S_{gen}.

5.61E A 30-ft^3 tank is one-half filled with refrigerant 12 liquid, the remainder with refrigerant 12 vapor. The initial pressure is 80 psia. There is a heat transfer from the surroundings until one-half of the original liquid by mass is evaporated. An automatic valve allows saturated vapor to escape at constant tank pressure. The surroundings are at $T_0 = 77$°F. Determine the required heat transfer and S_{gen}.

5.62S With a reversible isothermal heat transfer, refrigerant 12 goes from an initial state of 200 kPa and 200°C to a final pressure of 800 kPa in a piston-cylinder arrangement. Evaluate the work done per unit mass.

5.62E With a reversible isothermal heat transfer, refrigerant 12 goes from an initial state of 30 psia and 400°F to a final pressure of 120 psia in a piston-cylinder arrangement. Evaluate the work done per unit mass.

5.63S Saturated refrigerant 12 (liquid plus vapor) (5 kg) is in a 0.491-m^3 tank. A valve is opened, and the tank is heated so that refrigerant 12 vapor escapes at constant pressure until 1 kg of saturated vapor just fills the tank.
(*a*) Sketch the *T-s* diagram for this process, and label the states.
(*b*) What is the initial pressure in the tank?
(*c*) What is the entropy change of the control volume $(\Delta S)_{CV}$ if the control volume is taken as the tank interior?

5.63E Saturated refrigerant 12 (liquid plus vapor) (10 lbm) is in a 10-ft^3 tank. A valve is opened, and the tank is heated so the refrigerant 12 vapor escapes at constant pressure until 2 lbm of saturated vapor just fills the tank.
(*a*) Sketch the *T-s* diagram for this process, and label the states.
(*b*) What is the initial pressure in the tank?
(*c*) What is the entropy change of the control volume $(\Delta S)_{CV}$ if the control volume is taken as the tank interior?

5.64S Saturated water (liquid plus vapor) (3 kg) is in a tank of volume 6.1996 m^3. A valve is opened, and there is heat transfer to the tank in such a way that water vapor escapes at constant temperature until 1 kg of saturated vapor just fills the tank.
(*a*) Sketch the *T-s* diagram for the material in the tank during this process, and label the states.
(*b*) What is the initial quality of the water vapor in the tank?
(*c*) What is the entropy change of the control volume $(\Delta S)_{CV}$ if the control volume is taken as the tank interior?

5.64E Saturated water (liquid plus vapor) (6 lbm) is in a tank of volume 12.945 ft^3. A valve is opened, and there is heat transfer to the tank in such a way that water vapor escapes at constant temperature until 2 lbm of saturated vapor just fills the tank.

(*a*) Sketch the *T*-*s* diagram for the material in the tank during this process, and label the states.

(*b*) What is the initial quality of the water vapor in the tank?

(*c*) What is the entropy change of the control volume $(\Delta S)_{CV}$ if the control volume is taken as the tank interior?

5.65S A bladder initially contains 7 kg of refrigerant 12 (liquid plus vapor) at 100 kPa, and the volume is 1 m^3. The bladder (which expands at constant internal pressure) is heated until the refrigerant 12 is all saturated vapor. What is the change in the entropy of the refrigerant 12?

5.65E A bladder initially contains 15 lbm of refrigerant 12 (liquid plus vapor) at 14.7 psia, and the volume is 30 ft^3. The bladder (which expands at constant internal pressure) is heated until the refrigerant 12 is all saturated vapor. What is the change in the entropy of the refrigerant 12?

5.66S Neon is in an insulated container at an initial state defined by P_1 and T_1. Work done until the system comes to a final state at P_2 and T_2. One of the end states of the process has $P = 1$ atm and $T = 25°C$. The other has $P = 2$ atm and $T = 139.7°C$. Which state is the initial state? (Prove your result.)

5.66E Neon is in an insulated container at an initial state defined by P_1 and T_1. Work is done until the system comes to a final state at P_2 and T_2. One of the end states of the process has $P = 1$ atm and $T = 70°F$. The other has $P = 2$ atm and $T = 283°F$. Which state is the initial state? (Prove your result.)

5.67S Consider two possible expansion processes for steam. (See Fig. P5.67.) Steam is initially at 0.60 MPa and 400°C. The stream is partitioned into a chamber with the other side of the partition being a vacuum. The partition is removed, and the stream fills the complete chamber. Determine the temperature and pressure of the final state and the entropy generated if (*a*) $V_s = 0.1 V_v$ and (*b*) $V_s = 10 V_v$.

5.67E Consider two possible expansion processes for steam. Steam is initially at 90 psia and 750°F. The steam is partitioned into a chamber with the other side of the partition being a vacuum. The partition is removed, and the steam fills the complete chamber. Determine the temperature and pressure of the final state and the entropy generated if (*a*) $V_s = 0.1 V_v$ and (*b*) $V_s = 10 V_v$.

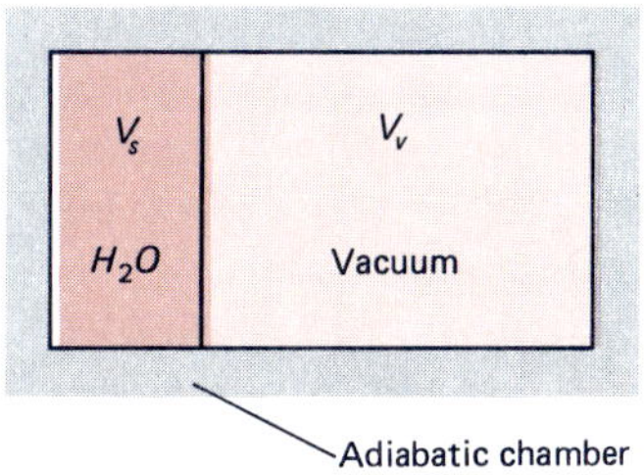

Figure P5.67

5.68S A reversible process for a piston-cylinder control mass is shown on the *P-v* and *T-s* diagrams. Evaluate the change in energy per unit mass for the process.

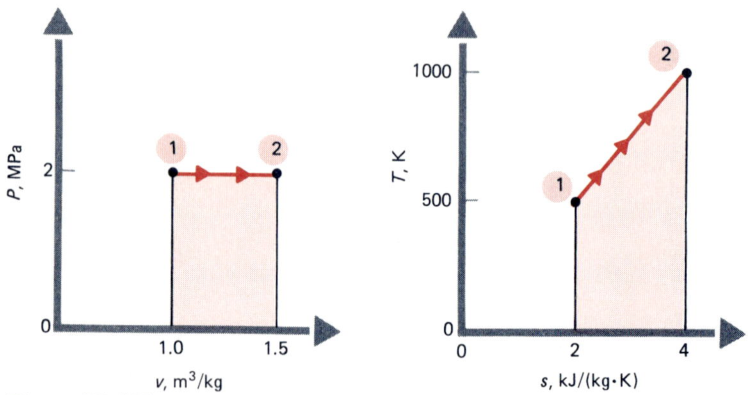

Figure P5.68S

5.68E A reversible process for a piston-cylinder control mass is shown on the *P-v* and *T-s* diagrams. Evaluate the change in energy per unit mass for the process.

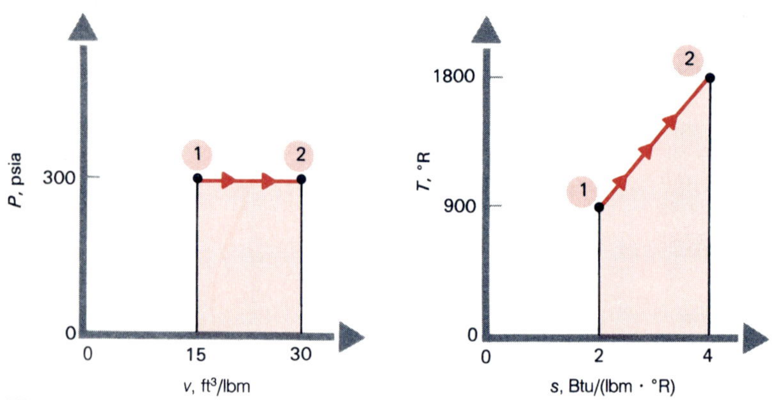

Figure P5.68E

5.69 Water is compressed in a reversible adiabatic nonflow process from an initial pressure and temperature of 0.8 MPa and 250°C to a final pressure of 4.0 MPa. Calculate the work for the process.

5.70 Air (0.2 kg) is allowed to expand in a piston-cylinder device from 5.0 MPa and 800 K to a pressure of 0.25 MPa. The process is reversible and isothermal. Determine the total heat transfer and total work done, assuming *constant* specific heats.

5.71 Air (0.2 kg) is allowed to expand in a piston-cylinder device from 5.0 MPa and 800 K to a pressure of 0.1 MPa. The process is reversible and isothermal. Determine the total heat transfer and the total work done, assuming *variable* specific heats.

5.72 Air in a piston-cylinder device undergoes a reversible adiabatic expansion from a pressure of 1 MPa and temperature of 250°C to a temperature of 100°C. Find the work done in the process (kJ/kg). What is the final pressure?

5.73 A turbine operates with ammonia as the working fluid. The inlet conditions are 1000 kPa and 40°C. At the turbine outlet, the ammonia is at 500 kPa and has a measured quality of 0.985. What is the efficiency of the turbine?

5.74 An adiabatic steam turbine operates with inlet steam conditions of 500°C and 2 MPa and a measured outlet condition of 10 kPa and $x = 0.95$.

(*a*) What would be the quality of the steam at the turbine outlet if the turbine operated isentropically? (Comment on whether this makes sense in terms of the quality given for the measured performance, and why.)

(*b*) What is the efficiency of the actual turbine at these operating conditions?

5.75 A turbine has an efficiency of 65 percent when operated with ammonia and when the inlet conditions to the turbine are 1000 kPa and 40°C and the outlet pressure is 500 kPa. What power (kW) is developed by the turbine when the ammonia mass flow rate through the turbine is 1.5 kg/s?

5.76 Determine if the adiabatic air turbine can operate as shown.

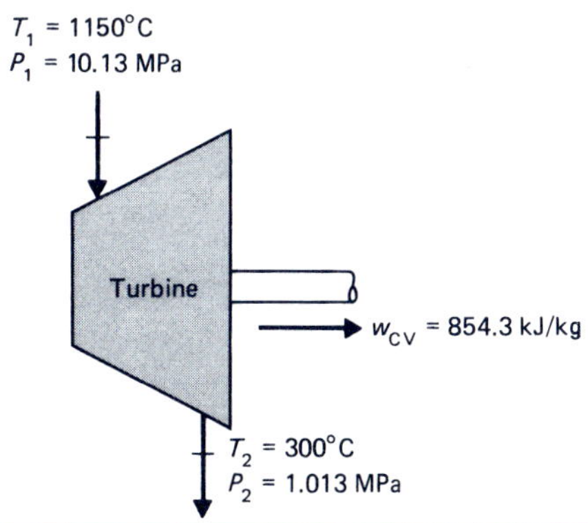

Figure P5.76

5.77 An air turbine with an inlet temperature and pressure of 800°C and 8 MPa has an efficiency of 70 percent when exhausting to a pressure of 1.1 MPa with a mass flow rate of 0.9 kg/s. It is proposed to use this turbine to drive a compressor which requires 350 hp. Will this turbine satisfy the compressor requirement?

5.78 For the adiabatic air turbine shown, determine the mass flow rate required to produce 9 kW.

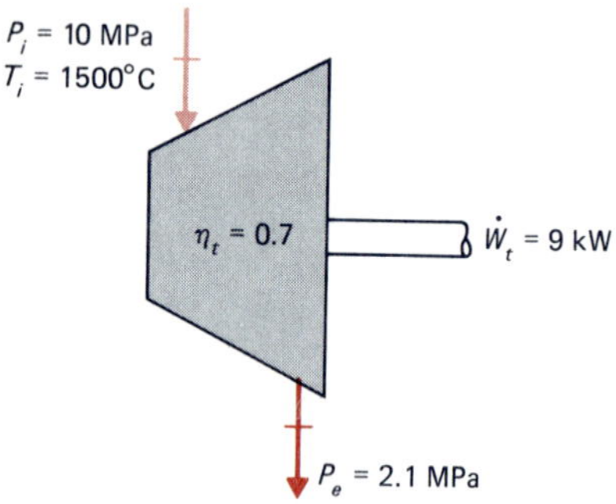

Figure P5.78

5.79 A reversible adiabatic turbine operates between the conditions shown. Find the turbine output in horsepower if the medium is air.

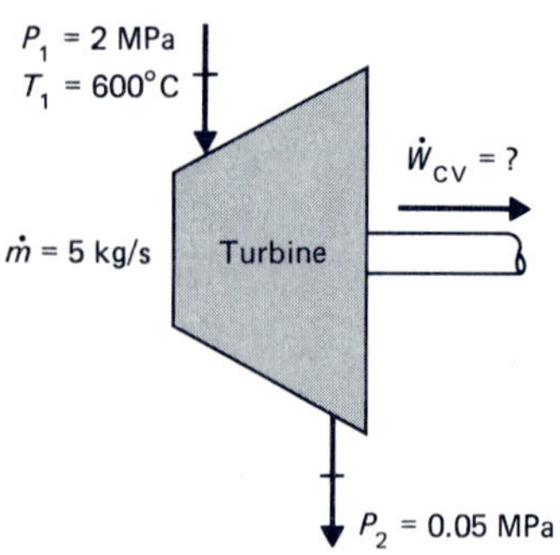

Figure P5.79

5.80 Air exits an adiabatic turbine at a pressure of 1 atm. The inlet condition are $T_1 = 1100°C$ and $P_1 = 5$ atm. The actual work output is 381.8 kJ/kg. Determine the isentropic efficiency.

5.81 Air enters an adiabatic turbine at 7.091 MPa and 1000°C and exits at a pressure of 2.0 MPa. If the turbine efficiency is 80 percent, what is the actual work output?

5.82 Nitrogen enters an adiabatic auxiliary turbine at 35 MPa and 20°C and exits at a pressure of 0.9 MPa and a temperature of −100°C. If the mass flow rate is 0.1 kg/s, determine the turbine efficiency and the power output.

5.83 An emergency turbine is driven by CO_2 from high-pressure tanks at 20°C and 10 atm. The manufacturer claims that a mass flow rate of 0.3 kg/s will produce 15 hp when exiting at 2 atm. Is his claim valid?

5.84 Helium is expanded isothermally through a reversible turbine. Inlet conditions are 2.5 MPa of pressure and a temperature of 200°C. The exit pressure is 0.2 MPa. Find the turbine work (kJ/kg).

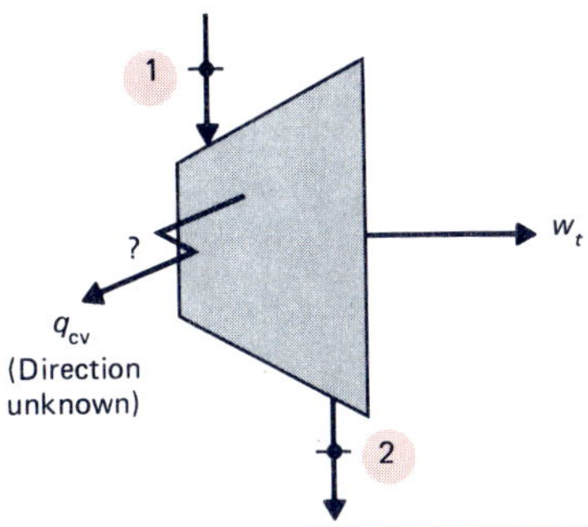

Figure P5.84

5.85 Air is expanded through a turbine from a pressure of 3.0 MPa and a temperature of 600°C to an exit temperature of 120°C. The process if found to be adiabatic and to produce 120 kW of power. If the turbine is 85 percent efficient, find the exit pressure and the volume flow rate (m^3/s) at the exit.

5.86 A gas turbine operates with air as the medium. Determine the actual exit temperature T_2 for the given conditions.

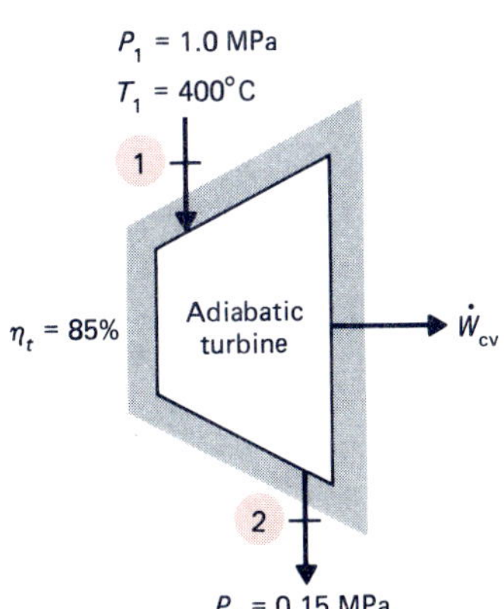

Figure P5.86

5.87 Determine if the adiabatic turbine can operate on steam as shown. Show work.

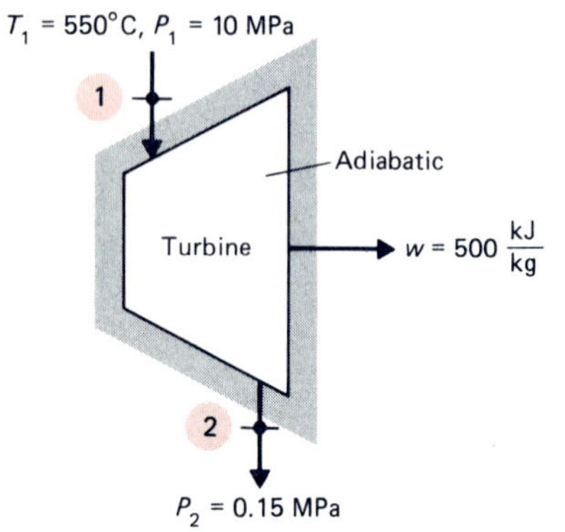

Figure P5.87

5.88 Steam exits an adiabatic turbine at 0.2 MPa and a quality of 98 percent. The inlet pressure is 6.0 MPa, and the actual work output is 520 kJ/kg. Determine the isentropic efficiency.

5.89 Steam enters an adiabatic turbine at 10 MPa and 550°C. Exit conditions are 0.06 MPa and a quality of 96 percent. Determine the isentropic efficiency and actual work output (hp) for a mass flow rate of 10 kg/s.

5.90 Steam enters an adiabatic turbine at 6.0 MPa and 500°C. The turbine efficiency is 90 percent and the exit pressure is 0.2 MPa. Determine the actual work output (kJ/kg) and the actual exit quality.

5.91 Manufacturer's specifications for two steam turbines are shown below. Which would you recommend? (Which is more efficient?) Use a Mollier Chart in your solution.

Figure P5.91

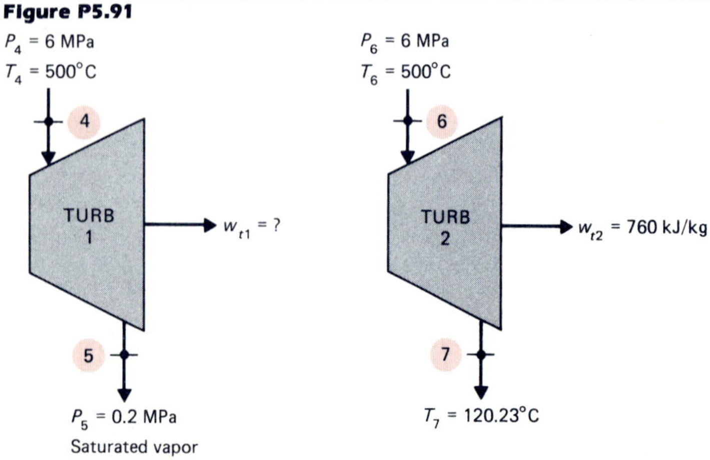

5.92 A steam turbine operates between 0.4 MPa and 10 kPa. The inlet temperature is 350°C. The adiabatic turbine has an efficiency of 95 percent. Excessive wear may occur if the moisture content exceeds 2 percent. Does it, in this case?

5.93 An adiabatic steam turbine has an efficiency of 70 percent and is shown. What mass flow rate is required for a turbine output of 1750 kW?

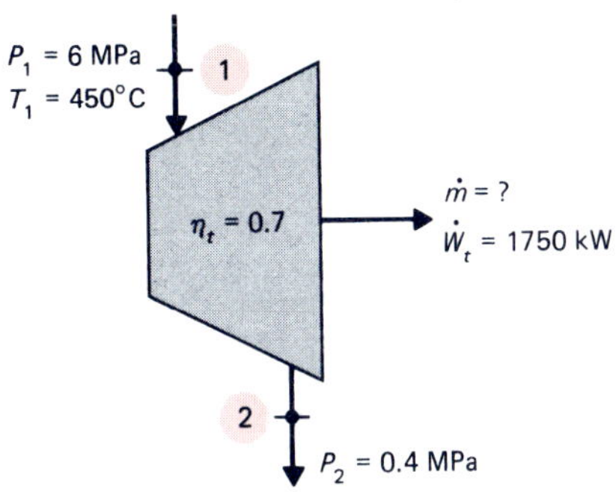

Figure P5.93

5.94 Given the isothermal air turbine shown, determine if the device is possible, assuming variable specific heats.

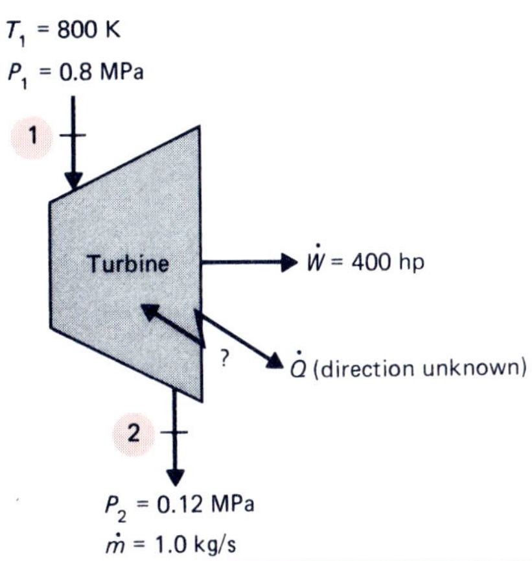

Figure P5.94

5.95 Determine the maximum power output for the steady-state–steady-flow air turbine shown, assuming variable specific heats.

5.96 Air expands through an insulated turbine as shown. If the turbine produces 112 kJ/kg of work, find the temperature of the air as it enters the turbine.

Figure P5.95

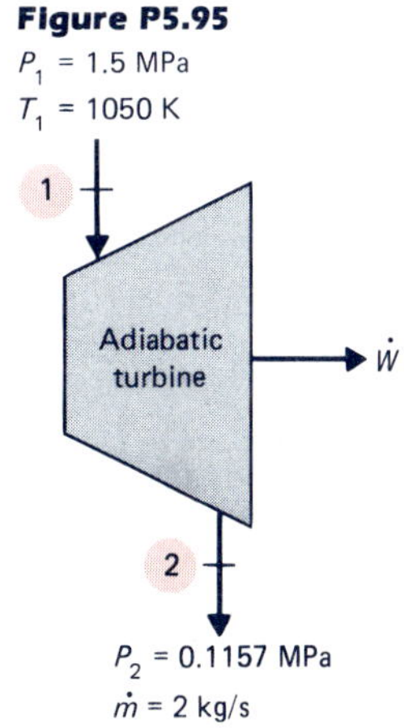

Figure P5.96

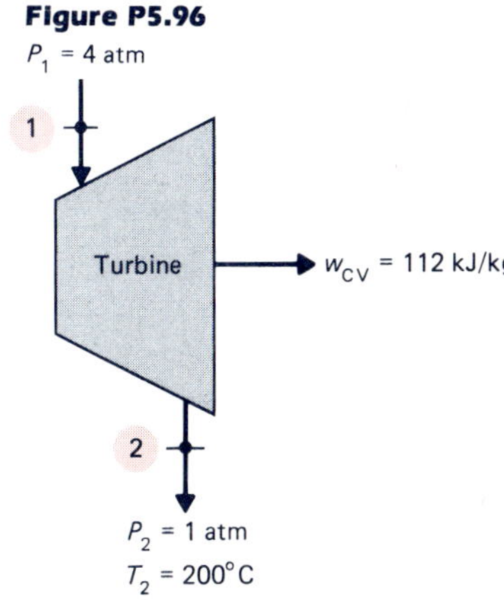

5.97 A reversible adiabatic steam turbine operates between the conditions shown. Find the turbine output in horsepower.

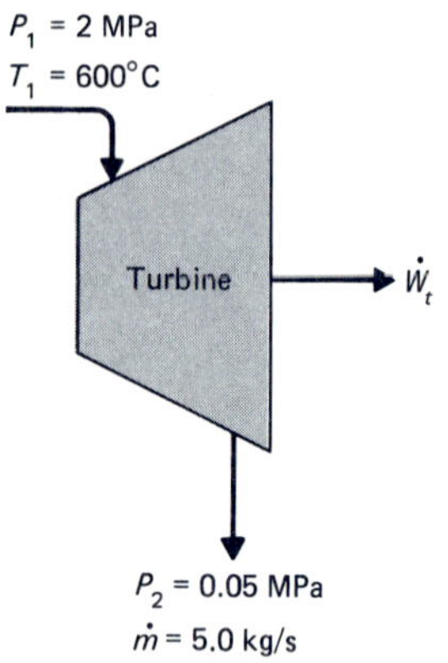

Figure P5.97

5.98 Consider the device shown. Steam flows in a line at 0.6 MPa and 350°C. From this steam line, steam flows through a turbine. The steam exhausts into a large chamber having a volume of 30 m³. Initially this chamber is evacuated. The turbine can operate until the pressure in the chamber is 0.6 MPa. At this point, the steam temperature is 300°C in the chamber. Assume that the entire process is adiabatic, and disregard any volume within the turbine. Determine the work done by the turbine during this process (kJ).

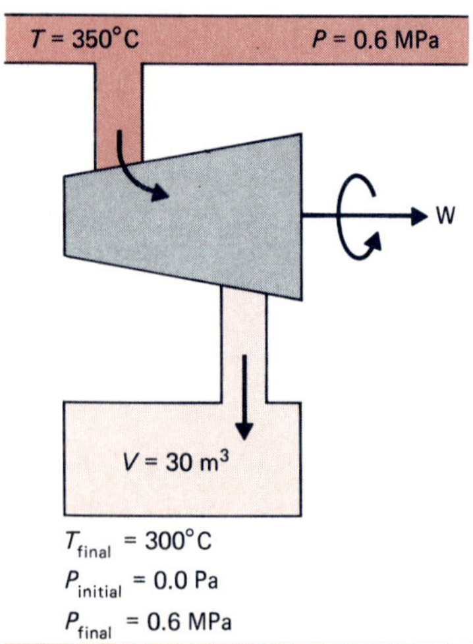

Figure P5.98

5.99 A steam turbine has an efficiency of 86 percent when it operates with inlet steam conditions of 500°C and 2 MPa and a measured outlet pressure of 10 kPa.
(*a*) What is the actual quality of the steam at the turbine outlet?
(*b*) What is the specific work (kJ/kg) of the actual turbine at these operating conditions?

5.100 Steam enters a turbine at 500°C and 10 MPa. The steam expands isentropically to a pressure of 0.2 MPa. Determine the change in enthalpy.

5.101S The inlet conditions to a steam turbine are $P = 6000$ kPa and $T = 400$°C. The outlet conditions are $P = 8$ kPa and $x = 0.904$. What is the turbine efficiency?

5.101E The inlet conditions to a steam turbine are $P = 900$ psia and $T = 750$°F. The outlet conditions are $P = 1$ psia and $x = 0.904$. What is the turbine efficiency?

5.102S A space solar power plant uses large solar mirrors to concentrate solar energy onto a boiler, which produces steam at $P = 600$ kPa and $T = 400$°C. The steam enters a reversible adiabatic turbine at a flow rate of 4 kg/s. The turbine exit pressure is 8 kPa.
(*a*) What is the quality of steam leaving the turbine?
(*b*) What is the turbine power output?

5.102E A space solar power plant uses large solar mirrors to concentrate solar energy onto a boiler, which produces steam at $P = 80$ psia and $T = 800$°F. The steam enters a reversible adiabatic turbine at a flow rate of 9 lbm/s. The turbine exit pressure is 1 psia.
(*a*) What is the quality of steam leaving the turbine?
(*b*) What is the turbine power output?

5.103S The inlet state to an adiabatic steam turbine is 2.0 MPa and 450°C, and the outlet pressure is 0.030 MPa. The turbine is rated so that the actual work output divided by the work output for reversible operation (between inlet state and given outlet pressure) is 0.85. Specify the outlet state (either x_2 or T_2 depending on the state location), and evaluate the actual work output.

5.103E The inlet state to an adiabatic steam turbine is 350 psia and 850°F, and the outlet pressure is 4.0 psia. The turbine is rated so that the actual work output divided by the work output for reversible operation (between inlet state and given outlet pressure) is 0.85. Specify the outlet state (either x_2 or T_2 depending upon the state location), and evaluate the actual work output.

5.104S Hydrogen enters a turbine at 1000 kPa and 400°C and exhausts at 200 kPa. Neglecting kinetic and potential energy changes and assuming an isentropic turbine, find (*a*) the exhaust temperature and (*b*) the work output.

5.104E Hydrogen enters a turbine at 150 psia and 750°F and exhausts at 30 psia. Neglecting kinetic and potential energy changes and assuming an isentropic turbine, find (*a*) the exhaust temperature and (*b*) the work output.

5.105S Tests are run on a newly designed steam turbine. At an inlet pressure of 60 bar and an inlet temperature of 430°C, the turbine exhausts to a pressure of 0.10 bar.
(*a*) What is the enthalpy of the steam leaving the turbine?
(*b*) What is the value of the turbine work?
(*c*) What is the turbine efficiency?

5.105E Tests are run on a newly designed steam turbine. At an inlet pressure of 60 atm and an inlet temperature of 800°F, the turbine exhausts to a pressure of 0.10 atm. The exit quality from the turbine is measured to be 0.90.
(*a*) What is the enthalpy of the steam leaving the turbine?
(*b*) What is the value of the turbine work?
(*c*) What is the turbine efficiency?

5.106 A turbine with efficiency of 80 percent operates at the same inlet conditions and outlet pressure as the one described in Prob. 5.105.
(*a*) What is the quality of steam leaving the turbine?
(*b*) What is the turbine work?
(*c*) What is the entropy generation for this turbine?
(*d*) What is the percentage change in quality and turbine work in comparison with an isentropic turbine operating at the same conditions?

5.107S Steam enters a turbine at an inlet temperature of 900°C and a pressure of 3.0 MPa. The outlet pressure is 7.5 kPa. The mass flow rate of the steam is 2 kg/s. The inlet and outlet velocities are small.
(*a*) If the turbine is isentropic, what is its power output?
(*b*) If the turbine efficiency is 85 percent, what is the power output?
(*c*) What is the outlet enthalpy of the steam from the real turbine?
(*d*) Explain why the enthalpy of the steam leaving the real turbine differs from the isentropic turbine (if it does).

5.107E Steam enters a turbine at an inlet temperature of 1600°F and a pressure of 400 psia. The outlet pressure is 1 psia. The mass flow rate of the steam is 5 lbm/s. The inlet and outlet velocities are small.

(*a*) If the turbine is isentropic, what is its power output?

(*b*) If the turbine efficiency is 85 percent, what is the power output?

(*c*) What is the outlet enthalpy of the steam from the real turbine?

(*d*) Explain why the enthalpy of the steam leaving the real turbine differs from the isentropic turbine (if it does).

5.108S A steam turbine provides a measured work output of 4.0 MW when a mass flow of 5 kg/s flows through it. (See the figure.) The outlet steam is at 25°C and has a quality of 90 percent and a speed of 100 m/s. The inlet steam pressure to the turbine is 2 MPa. What are (*a*) the actual inlet steam temperature and (*b*) the turbine efficiency? State any necessary assumptions.

5.108E A steam turbine provides a measured work output of 5400 hp when a mass flow of 10 lbm/s flows through it. (See the figure.) The outlet steam is at 70°F and has a quality of 90 percent and a speed of 300 ft/s. The inlet steam pressure to the turbine is 300 psia. What are (*a*) the actual inlet steam temperature and (*b*) the turbine efficiency? State any necessary assumptions.

Figure P5.108S

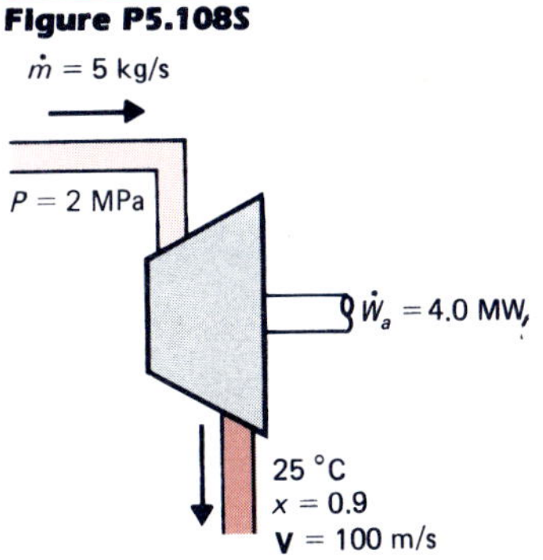

Figure P5.108E

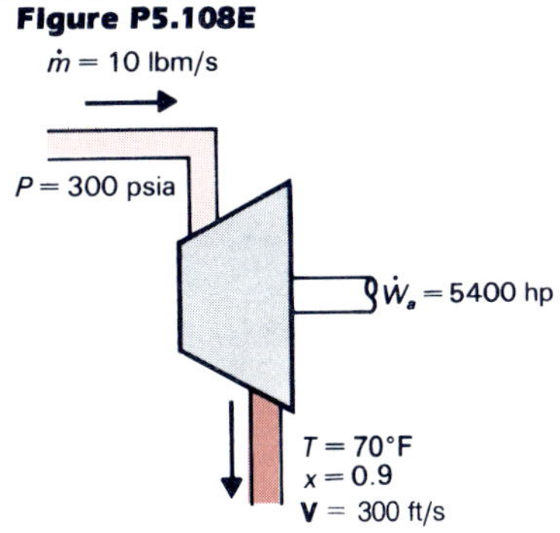

5.109S A compressor is used to compress steam in a steady flow process from 150 kPa and 120°C to 1.5 MPa and 200°C. The work input is measured to be 485.4 kJ/kg. KE and PE changes are negligible. The environment is at 27°C.

(*a*) Find the magnitude and direction of any heat transfer.

(*b*) Find the entropy change of the fluid flowing through the compressor.
(*c*) Find the total entropy change for the overall process.

5.109E A compressor is used to compress steam in a steady flow process from 25 psia and 250°F to 200 psia and 400°F. The work input is measured to be 1.6×10^5 ft·lbf/lbm. KE and PE changes are negligible. The environment is at 80°F.
(*a*) Find the magnitude and direction of any heat transfer.
(*b*) Find the entropy change of the fluid flowing through the compressor.
(*c*) Find the total entropy change for the overall process.

5.110S Air enters an adiabatic compressor at 20°C and 0.1 MPa. The outlet pressure is 10 times the inlet pressure. If the steady state compressor efficiency is 95 percent, determine the outlet temperature and the actual work per unit mass required, assuming (*a*) constant specific heats and (*b*) variable specific heats.

5.110E Air enters an adiabatic compressor at 70°F and 14.7 psia. The outlet pressure is 10 times the inlet pressure. If the steady state compressor efficiency is 95 percent, determine the outlet temperature and the actual work per unit mass required, assuming (*a*) constant specific heats and (*b*) variable specific heats.

5.111S An adiabatic air compressor has an efficiency of 75 percent and is required to compress 1 kg/s of ambient air ($P_1 = 1$ atm, $T_1 = 27$°C) to a final pressure of 5 atm. How much power input is required?

5.111E An adiabatic air compressor has an efficiency of 75 percent and is required to compress 1 lbm/s of ambient air ($P_1 = 1$ atm, $T_1 = 80$°F) to a final pressure of 5 atm. How much power input is required?

5.112S An air compressor has an isothermal efficiency of 75 percent and is required to compress 1 kg/s of ambient air ($P_1 = 1$ atm, $T_1 = 27$°C) to a final pressure of 5 atm. How much power input is required?

5.112E An air compressor has an isothermal efficiency of 75 percent and is required to compress 2.2 lbm/s of ambient air ($P_1 = 1$ atm, $T_1 = 80$°F) to a final pressure of 5 atm. How much power input is required?

5.113S An adiabatic air compressor has an efficiency of 75 percent. Air enters the compressor at 27°C and 1 atm and exits at 3 atm. What is the temperature of the air at the compressor exit?

5.113E An adiabatic air compressor has an efficiency of 75 percent. Air enters the compressor at 80°F and 1 atm and exits at 3 atm. What is the temperature of the air at the compressor exit?

5.114S Nitrogen is compressed adiabatically at a rate of 40 m^3/min (inlet conditions) in an axial flow compressor from 1 to 5 MPa. The temperature of the nitrogen entering the compressor is 40°C. The speed of nitrogen entering the compressor is 100 m/s and leaving it is 10 m/s. The compressor has an efficiency of 75 percent. Calculate the shaft power required to drive the compressor.

5.114E Nitrogen is compressed adiabatically at a rate of 1200 ft^3/min (inlet conditions) in an axial flow compressor from 150 to 750 psia. The temperature entering the compressor is 100°F. The speed of nitrogen entering the compressor is 300 ft/s and leaving it is 30 ft/s. The compressor has an efficiency of 75 percent. Calculate the shaft power required to drive the compressor.

5.115 Some 180 kJ/kg of work is required to compress air isothermally in a nonflow process from 27°C and 0.012 MPa to 1.2 MPa. Assuming variable specific heats, determine if the process is possible.

5.116 Air is compressed in an adiabatic compressor from a pressure of 0.05 MPa and a temperature of 80°C to a pressure of 1.7 MPa and a temperature of 780°C. Find the compressor efficiency η_c.

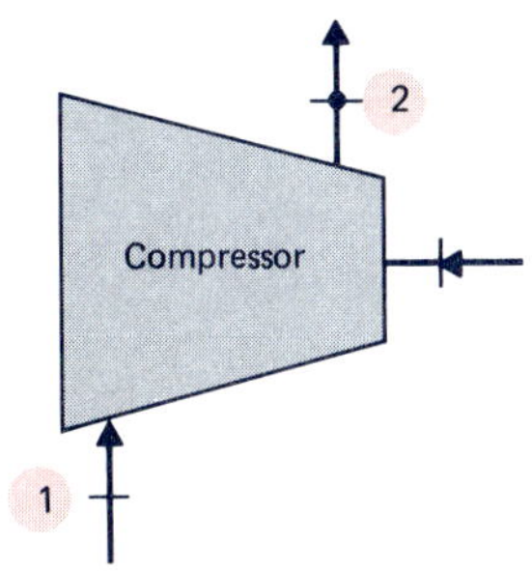

Figure P5.116

5.117 A steady-state–steady-flow compressor isothermally compresses air from the initial to the final conditions shown. Determine the total heat transfer rate and the power required, assuming *variable* specific heats.

Figure P5.117

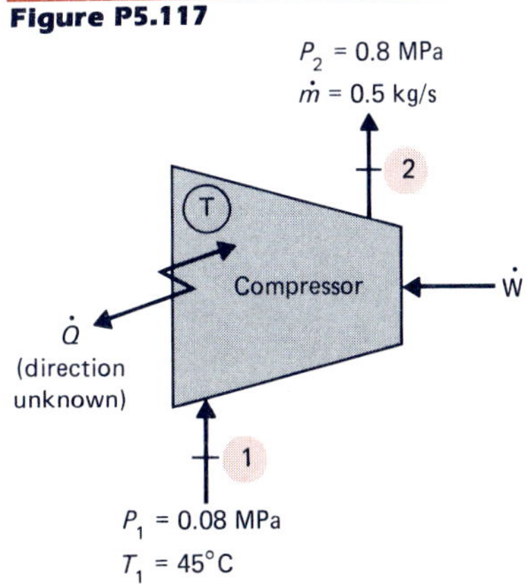

5.118 Air is compressed in a nonflow process from 0.15 MPa and 42°C to 1.02 MPa. Assuming a reversible, adiabatic process and *variable* specific heats, determine the final temperature and the work required (kJ/kg).

5.119 Air is compressed in a reversible adiabatic steady-state–steady-flow process from 0.2 MPa and 67°C to 1.43 MPa. Assuming *variable* specific heats, determine the work required (kJ/kg).

5.120 A 10-hp air compressor has an inlet air temperature of 15°C and raises the pressure from 1 to 5 atm. The mass flow rate is 0.02 kg/s. Assume the compressor is adiabatic.

(*a*) What is the compressor efficiency?

(*b*) What is the exit temperature?

(*c*) If the compressor were reversible, what horsepower would it require?

5.121 If an isentropic air compressor operates between the conditions shown, find the exit temperature of the air.

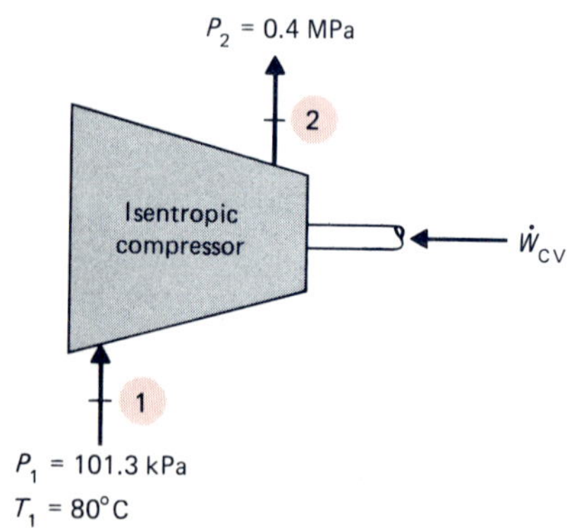

Figure P5.121

5.122 Air enters a compressor at 1 atm and 15°C. The exit pressure is 5 atm. if the compressor efficiency is 60 percent and the mass flow rate is 1 kg/s, determine the actual power input.

5.123 Helium enters an adiabatic compressor with a pressure of 1 MPa and a temperature of 20°C. The helium exits at 12 MPa and 600°C. Determine the isentropic efficiency.

5.124 Given the reversible air compressor shown, determine the mass flow rate.

5.125 For the reversible isentropic air compressor shown in Fig. P5.125, determine the rate of heat transfer and power required.

Figure P5.124

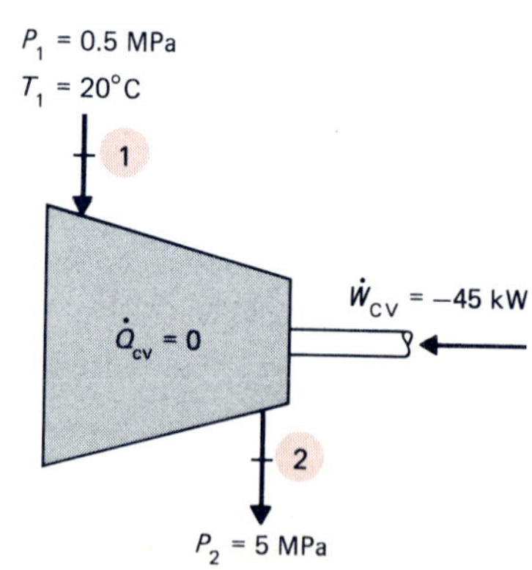

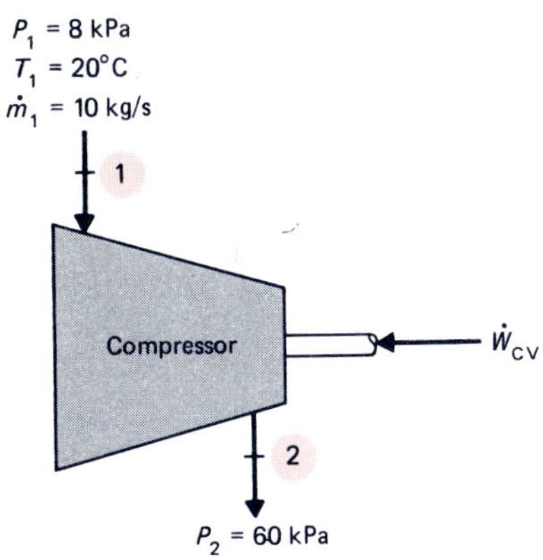

Figure P5.125

5.126 A hydrogen compressor operates as shown. Determine the compressor efficiency.

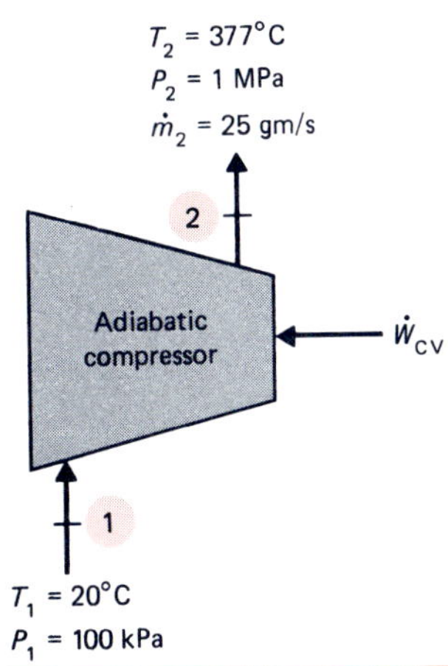

Figure P5.126

5.127 A reversible adiabatic hydrogen compressor operates as shown. Calculate the required compressor work input.

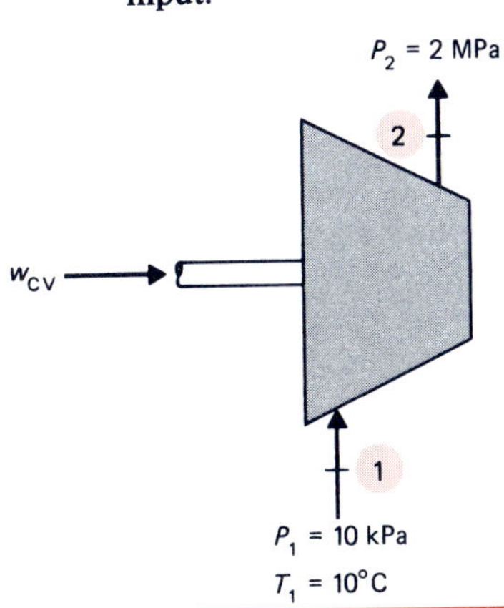

Figure P5.127

5.128 The helium compressor is designed to handle 3 kg/s; and $P = P_1 = P_2$, $T_a = T_1 = T_2$.

(a) What is the heat transfer rate $\dot{Q}_{CV}$ (kW).

(b) What mass flow rate is provided to the high-pressure (1-MPa) line?

(c) The design review says this is an impossible device. Is this correct?

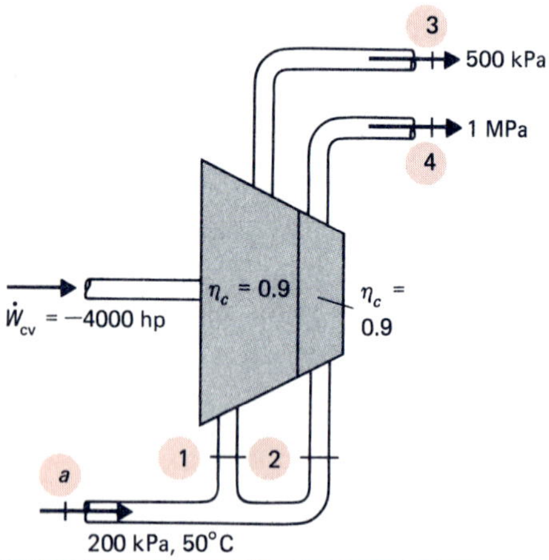

Figure P5.128

5.129 The Adiabatic Air Compressor Company has a 500-hp unit for sale. You need it to raise nitrogen from −20°C and 1 atm to 15 MPa at a mass flow rate of 366.7 g/s. How efficient must the unit be to meet your requirements?

5.130 Astronaut Base reports finding a gas turbine operating as shown. Local planet gas has a molecular weight of 35 and a $c_v = 2.0$ kJ/(kg·K). Find the turbine power output in kilowatts.

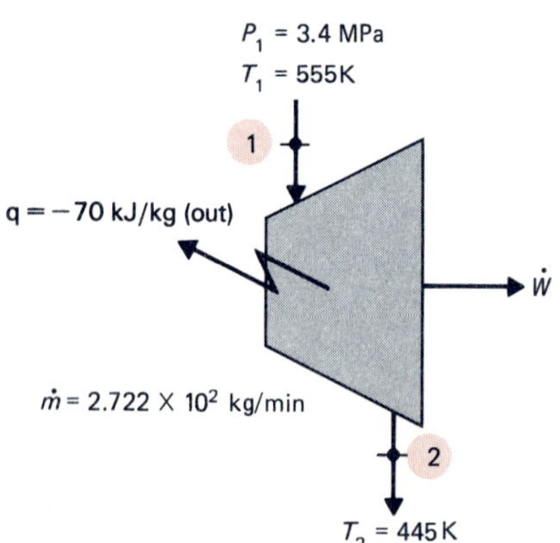

Figure P5.130

5.131 (*a*) If the steam compressor shown is operated reversibly and isothermally, determine the rate of heat transfer (kW) and power required (kW). (*b*) If the same steam compressor could be operated isentropically between the same conditions, determine the rate of the heat transfer (kW) and power required (kW).

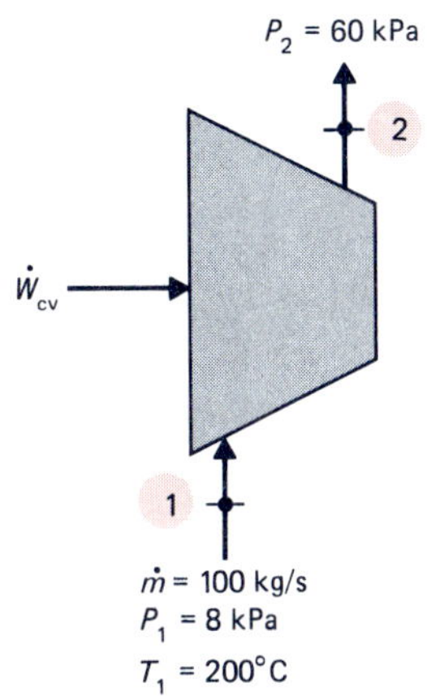

Figure P5.131

5.132 Steam enters an adiabatic compressor with a quality of 100 percent and a pressure of 2.0 MPa. The steam exits at 15.0 MPa and 550°C. Determine the isentropic efficiency.

5.133 The Adiabatic Steam Compressor Company offers a 100-hp unit for sale. You need it to raise saturated vapor at 100°C to 0.6 MPa. Also $\dot{m}_{steam} = 106.5$ g/s. How efficient must this unit be for your operating conditions?

5.134 An ideal compressor using H_2O as the thermodynamic medium undergoes what is known to be an isentropic process between 0.2 and 10.0 MPa. What is ΔT for the process?

5.135 A compressor is ideal (i.e., reversible and adiabatic) and is to be used to compress nitrogen from 101.32 kPa and 25°C to some final pressure P_a. The compressor has a work input per unit mass of nitrogen of 620.23 kJ/kg. What value of P_a (kPa) can be attained by the compressor? (Use the nitrogen tables.)

5.136S A small swimming pool pump is specified by the manufacturer to have an efficiency of 83 percent. The flow rate through the pump is 20 kg/min. The pump must increase the pressure of the inlet water by 200 kPa. What is the required size for the electric motor to drive the pump for the following cases?

(*a*) The inlet and outlet pipes to the pump each have 5-cm ID.
(*b*) The inlet pipe has a 2.5-cm ID, the outlet has a 5-cm ID.
(*c*) The inlet pipe has a 5-cm ID, the outlet has a 2.5-cm ID.

5.136E A small swimming pool pump is specified by the manufacturer to have efficiency of 83 percent. The flow rate through the pump is 50 lbm/min. The pump must increase the pressure of the inlet water by 30 psia. What is the required size for the electric motor to drive the pump for the following cases?
(*a*) The inlet and outlet pipes to the pump each have a 2-in ID.
(*b*) The inlet pipe has a 1-in ID, the outlet a 2-in ID.
(*c*) The inlet pipe has a 2-in ID, the outlet a 1-in ID.

5.137 A 10-hp pump has an inlet water temperature of 20°C and raises the pressure from 1 atm to 5 MPa. The mass flow rate is 0.89 kg/s. The pump is adiabatic.
(*a*) What is the pump efficiency?
(*b*) What is the exit temperature?
(*c*) If the pump were reversible, what horsepower would it require?

5.138 Water enters a pump at 0.6 MPa and 50°C. The exit pressure is 1 MPa, and the pump efficiency is 60 percent. If the mass flow rate is 3 kg/s, determine the actual power input (hp).

5.139 The feedwater pump in a steam power plant receives 5.0 kg/min of saturated liquid from the condenser at a pressure of 5 kPa. Find the power (hp) required to pump the water isentropically to a pressure of 3 MPa.

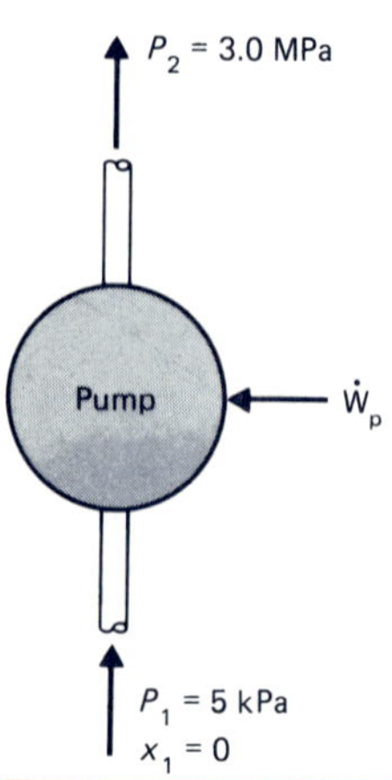

Figure P5.139

5.140 If an isentropic pump operates between the conditions shown, find the exit enthalpy of the water.

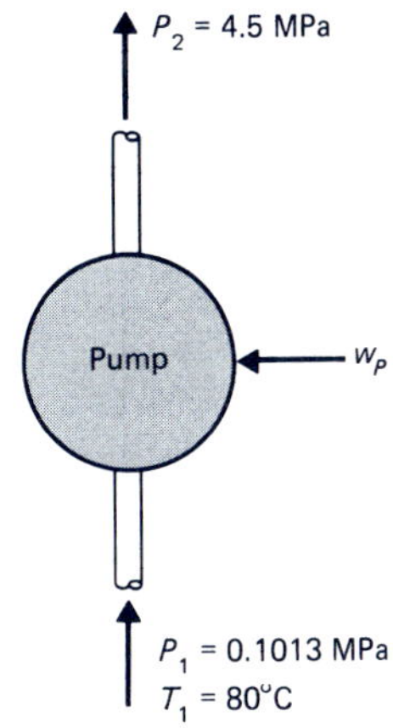

Figure P5.140

5.141S Water is flowing through a pump. It enters the pump at a pressure of 500 kPa, and the actual work done on the water passing through the pump is 6 kJ/kg. The water leaves the pump at 5 MPa and 30°C. The pump efficiency is 78 percent. What is the temperature of the water entering the pump?

5.141E Water is flowing through a pump. It enters the pump at a pressure of 70 psia, and the actual work done on the water passing through the pump is 20,000 ft·lbf/lbm. The water leaves the pump at 700 psia and 90°F. The pump efficiency is 78 percent. What is the temperature of the water entering the pump?

5.142 An isentropic pump is used to pump ammonia liquid at −40°C from an initial pressure of 100 kPa to a final pressure of 500 kPa. The inlet pipe to the pump is of large diameter, and the outlet line is of diameter D_0 = 3 cm. The velocity of the ammonia at the outlet is measured to be 50 m/s. What size motor (kW) is required to drive the pump?

5.143 The reversible adiabatic water pump operates as shown. Calculate the required pump work input (kJ/kg).

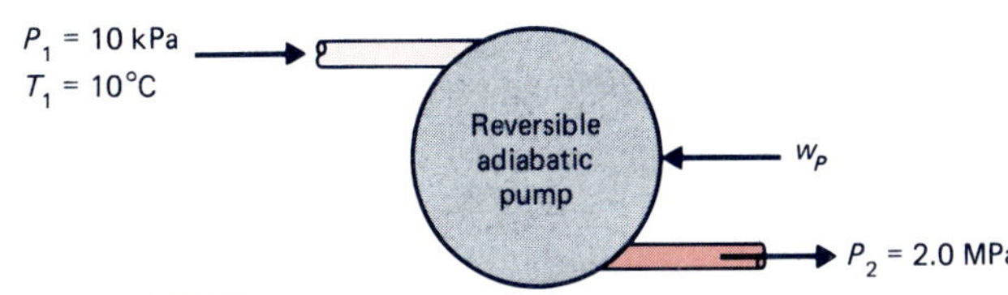

Figure P5.143

5.144 Given the reversible pump shown, determine the mass flow rate and volumetric flow rate of the water.

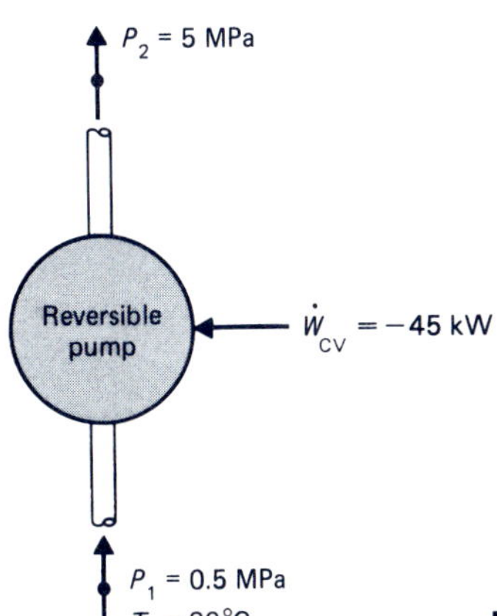

Figure P5.144

5.145S An air turbine drives a water pump with the conditions shown in Fig. P5.145S. The air undergoes an isentropic process ($k = 1.4$) in going through the turbine. The inlet water temperature is 25°C. The pump has an efficiency of 70 percent. What is the inlet temperature T_{in} of the air?

5.145E An air turbine drives a water pump with the conditions shown in Fig. P5.145E. The air undergoes an isentropic process ($k = 1.4$) in going through the turbine. The inlet water temperature is 70°F. The pump has an efficiency of 70 percent. What is the inlet temperature T_{in} of the air?

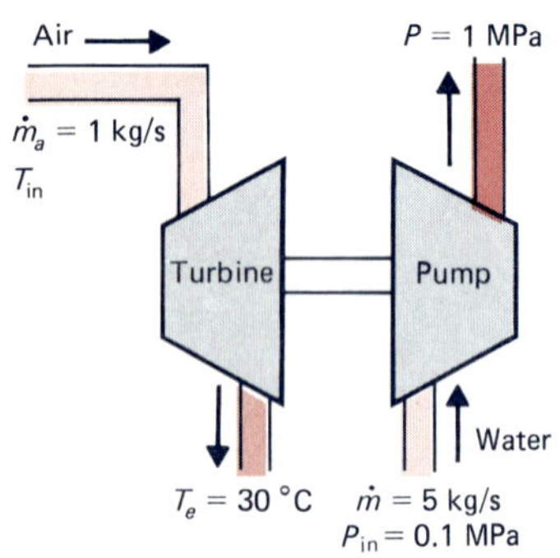

Figure P5.145S

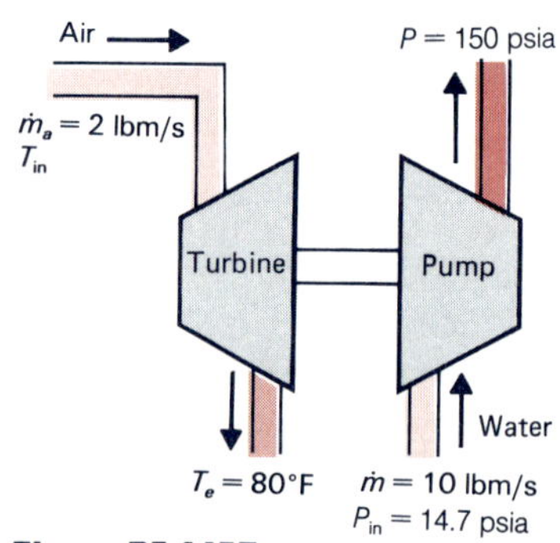

Figure P5.145E

5.146S A steam turbine with 70 percent efficiency drives an air compressor with an isentropic efficiency of 78 percent, as shown in Fig. P5.146S. What mass flow rate $\dot{m}$ of steam, in kilograms per hour, must be supplied to the turbine?

5.146E A steam turbine with 70 percent efficiency drives an air compressor with an isentropic efficiency of 78 percent, as shown in Fig. P5.146E What mass flow rate $\dot{m}$

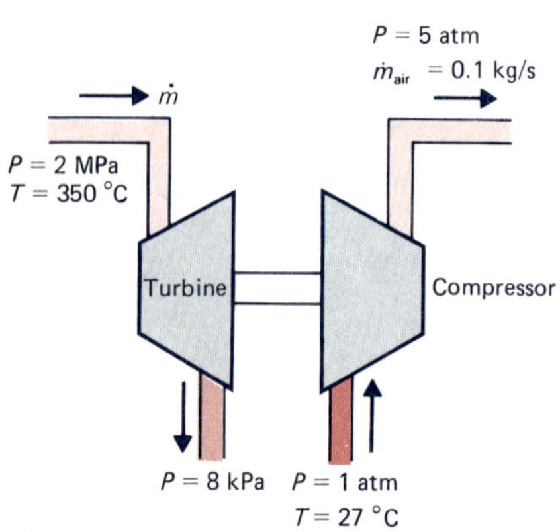

Figure P5.146S

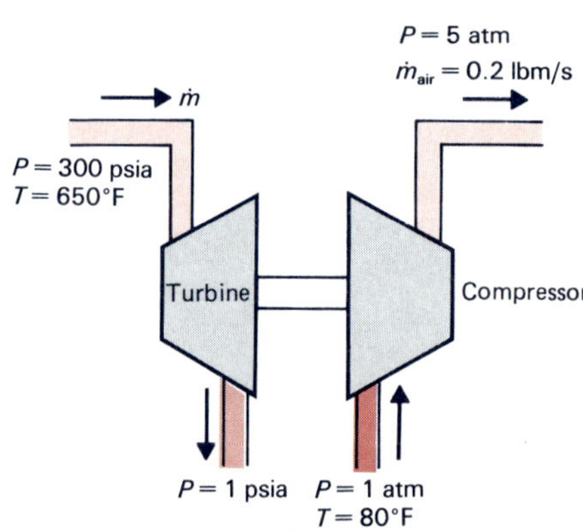

Figure P5.146E

of steam, in pounds-mass per hour, must be supplied to the turbine?

5.147S A steam turbine (efficiency 72 percent) drives an air compressor (efficiency 80 percent) as shown in Fig. P5.147S. What mass flow rate $\dot{m}$ of air, in kilograms per hour, can be compressed?

5.147E A steam turbine (efficiency 72 percent) drives an air compressor (efficiency 80 percent) as shown in Fig. P5.147E. What mass flow rate $\dot{m}$ of air, in pounds-mass per hour, can be compressed?

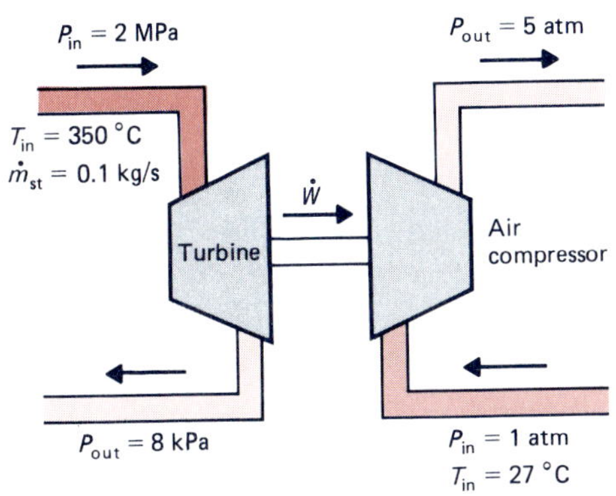

Figure P5.147S

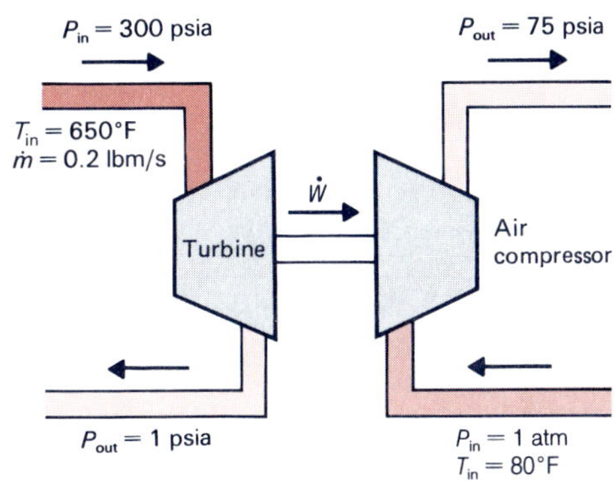

Figure P5.147E

5.148S A two-stage air compressor is compressing 2 kg/s at the conditions shown in Fig. P5.148S. The first stage is operated as an adiabatic compressor; the second stage is operated isothermally. The actual power required to drive the compressor is 650 kW.

(*a*) What would be T_{out} if each stage of the compressor were reversible?

(*b*) What power input would be required to drive the compressor if each stage were reversible?

(*c*) If T_{out} for the actual compressor were 250°C, how much power would be required to drive the second stage?

5.148E A two-stage air compressor is compressing 1 lbm/s at the conditions shown in Fig. P5.148E. The first stage is operated as an adiabatic compressor; the second stage is operated isothermally. The actual power required to drive the compressor is 250 hp.

(*a*) What would T_{out} be if each stage of the compressor were reversible?

(*b*) What power input would be required to drive the compressor if each stage were reversible?

(*c*) If T_{out} for the actual compressor were 500°F, how much power would be required to drive the second stage?

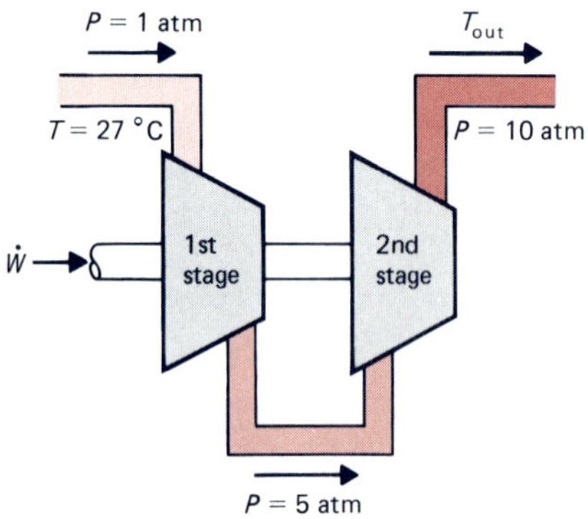

Figure P5.148S

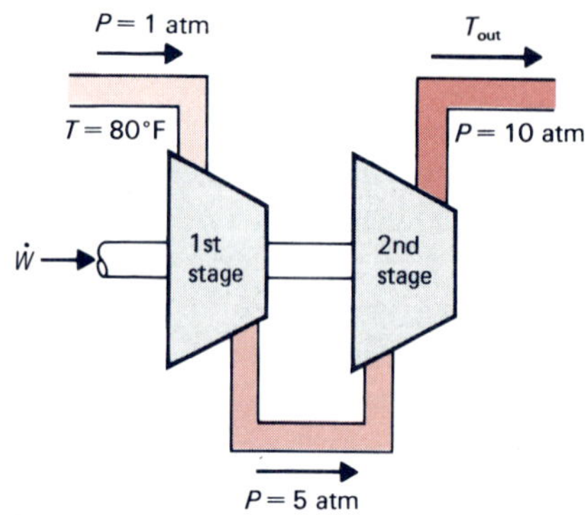

Figure P5.148E

5.149S Two adiabatic compressors are used to take inlet air at $P_1 = 0.1$ MPa and $T_1 = 300$ K to an outlet state of $P_4 = 0.7$ MPa and $T_4 = 700$ K. An adiabatic valve connects the compressors. Both compressor efficiencies have the same value of $\eta_c = 0.70$. See Fig. P5.149. If $P_2 = 0.3$ MPa, evaluate the work per unit mass for each compressor and the entropy generation per unit mass for this process. Determine T_3 and P_3. Draw the process on the T-s diagram. Assume air is an ideal gas.

5.149E Two adiabatic compressors are used to take inlet air at $P_1 = 14.7$ psia and $T_1 = 540°$R to an outlet state of $P_4 = 100$ psia and $T_4 = 1200°$R. An adiabatic valve connects the compressors. Both compressor efficiencies have the same value of $\eta_c = 0.70$. See Fig. P5.149. If $P_2 = 50$ psia, evaluate the work per unit mass for each compressor and the entropy generation per unit mass for this process. Determine T_3 and P_3. Draw the process on the T-s diagram. Assume air is an ideal gas.

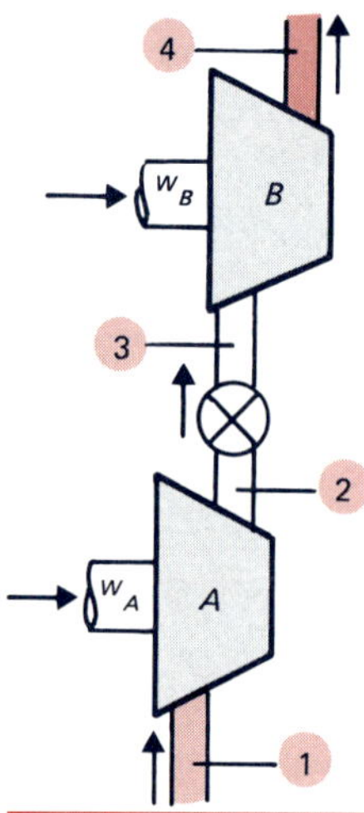

Figure P5.149

5.150S Two valves are used to control an adiabatic steady state turbine (see Fig. P5.150). The air (treat as an ideal gas) enters the first valve at $P_1 = 1000$ kPa and $T_1 = 800$ K. The turbine exhausts to a pressure of $P_3 = 140$ kPa. The outlet of the second valve is at $P_4 = 100$ kPa and $T_4 = 500$ K. The turbine efficiency is $\eta_c = 0.90$. Determine work per unit mass of the turbine, the entropy generation per unit mass for the process from 1 to 4, and the entropy generation for the turbine alone. Draw the process on a T-s diagram.

5.150E Two valves are used to control an adiabatic steady state turbine (see Fig. P5.150). The air (treat as an ideal gas) enters the first valve at $P_1 = 150$ psia and $T_1 = 1450°$R. The turbine exhausts to a pressure of $P_3 = 20$ psia. The outlet of the second valve is at $P_4 = 14.7$ psia and $T_4 = 900°$R. The turbine efficiency is $\eta_c = 0.90$.

Figure P5.150

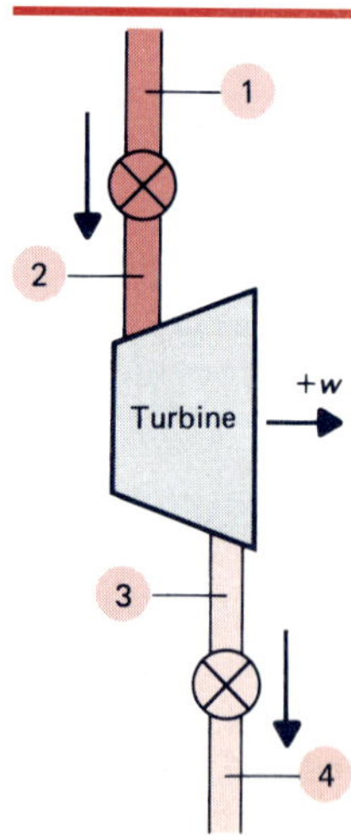

Determine work per unit mass of the turbine, the entropy generation per unit mass for the process from 1 to 4, and the entropy generation for the turbine alone. Draw the process on a *T-s* diagram.

5.151S The system shown in Fig. P5.151 is used to separate liquid water and water vapor while recovering some work from the steam. Vapor first enters a turbine at 350°C and 4.0 MPa. The mixture emerges from the turbine and enters a separator where saturated liquid and saturated vapor at 0.10 MPa are obtained as separated streams. The flow rate at state 1 is 0.5 kg/s in this adiabatic process.

(*a*) Find the percentage of initial mass flow converted to liquid water if the turbine produces 300 kW.

(*b*) Explain what happens if the turbine does not produce any work.

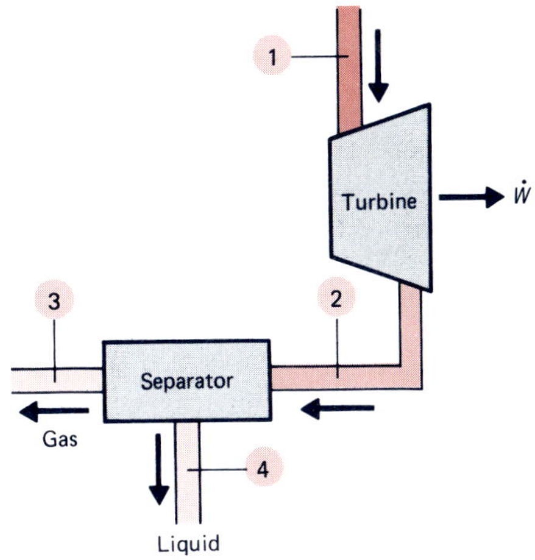

Figure P5.151

5.151E The system shown in Fig. P5.151 is used to separate liquid water and water vapor while recovering some work from the steam. Vapor first enters a turbine at 650°F and 600 psia. The mixture emerges from the turbine and enters a separator where saturated liquid and saturated vapor at 14.7 psia are obtained as separated streams. The flow rate at state 1 is 1 lbm/s in this adiabatic process.

(*a*) Find the percentage of initial mass flow converted to liquid water if the turbine produces 400 hp.

(*b*) Explain what happens if the turbine does not produce any work.

5.152 An adiabatic feedwater heater is also isobaric. Find the power input, and determine if the device is possible.

Figure P5.152

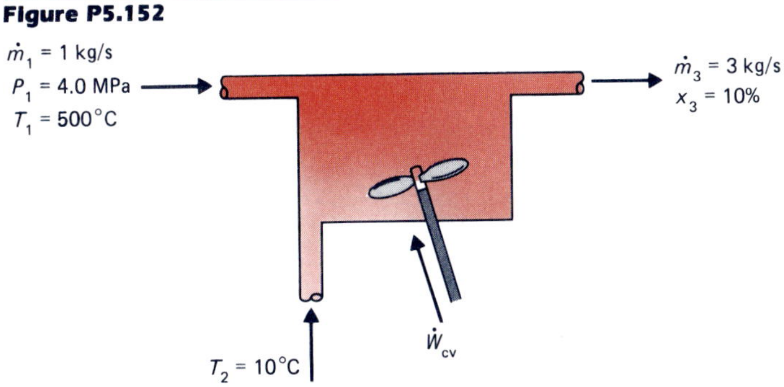

5.153 For the fully insulated control volume shown in Fig. P5.153, determine the *minimum* possible exit temperature of the air at state point 2, assuming a steady-state–steady-flow process and variable specific heats.

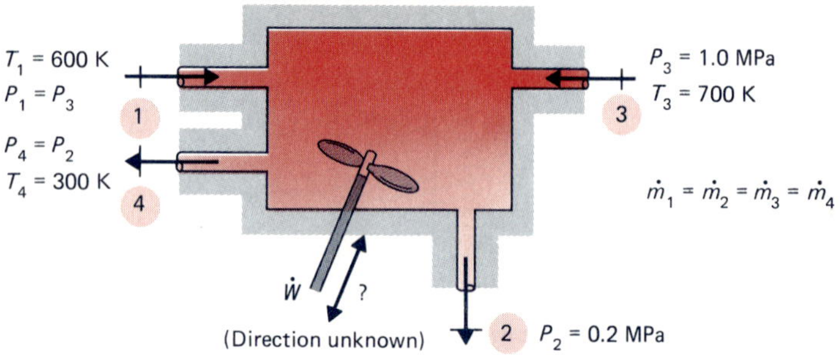

Figure P5.153

5.154 An adiabatic steady-state–steady flow device operates as shown below. There is work, but its direction is unknown. Determine the direction of *flow* through the device.

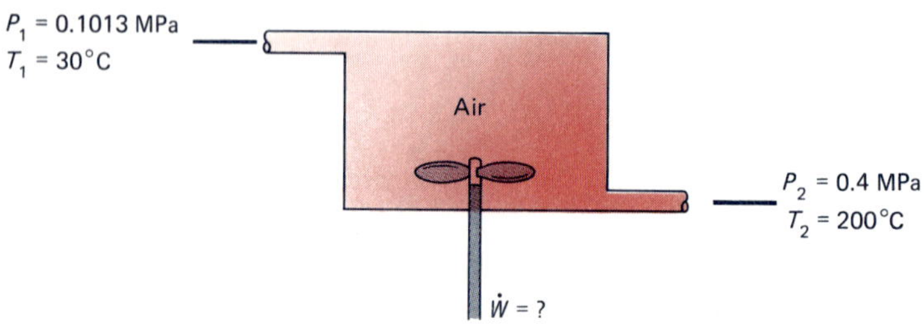

Figure P5.154

5.155 The adiabatic steady-state–steady-flow device below operates as shown, but the direction of flow and work are unknown. Find the magnitude and direction of the work (kJ/kg).

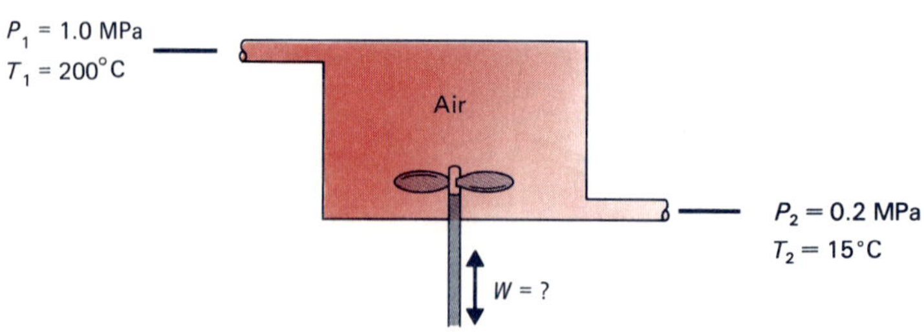

Figure P5.155

5.156 For the adiabatic air device shown in Fig. P5.156, (*a*) determine T_3. (*b*) Does the device violate the second law?

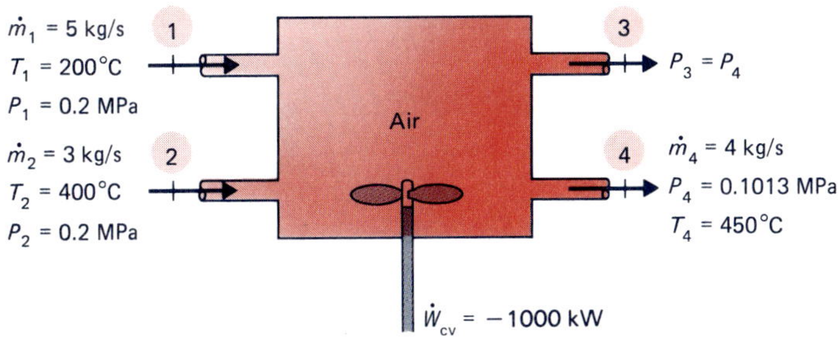

Figure P5.156

5.157 Is the following steady-state–steady-flow adiabatic device possible?

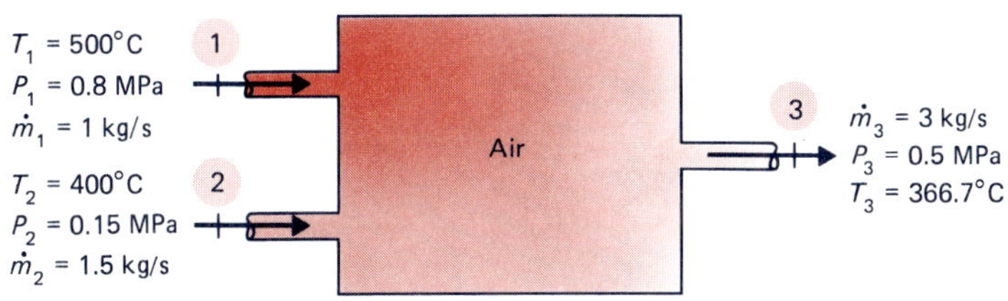

Figure P5.157

5.158 Is the following steady-state–steady-flow adiabatic device possible?

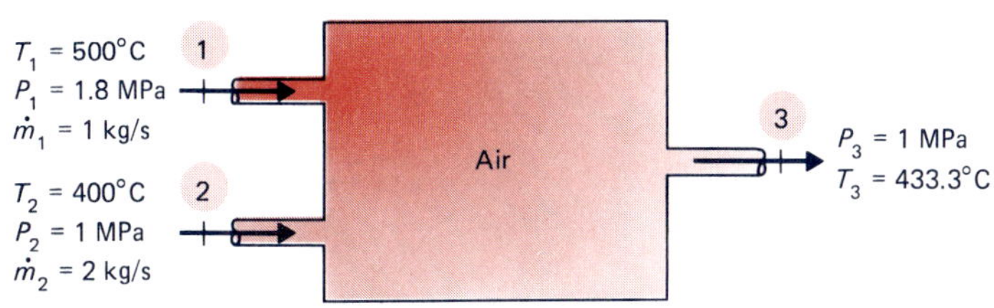

Figure P5.158

5.159 An *adiabatic* air heater is also isobaric. Find the shaft power input, and determine if the device is possible.

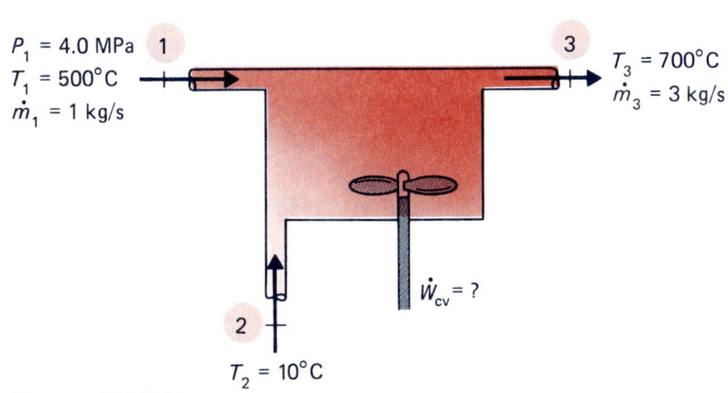

Figure P5.159

5.160 Given the adiabatic steady-state–steady-flow device shown, determine the direction of flow of the air, assuming variable specific heats.

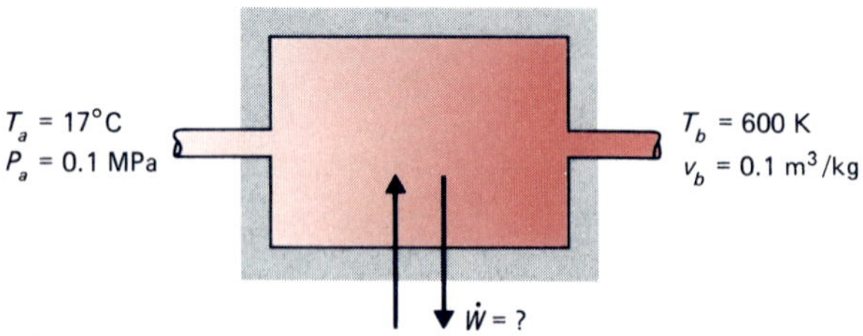

Figure P5.160

5.161 The adiabatic device shown operates with H_2O as the medium.

(*a*) Determine T_3.

(*b*) Does the device violate the second law of thermodynamics?

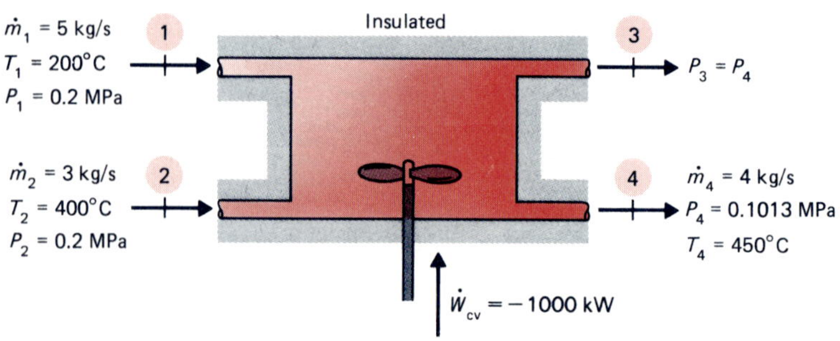

Figure P5.161

5.162 Is the following adiabatic device possible?

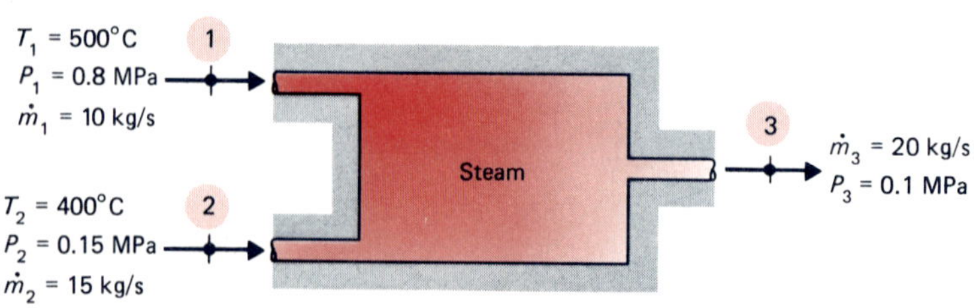

Figure P5.162

5.163 Is the adiabatic device shown in Fig. P5.163 possible? Show your work.

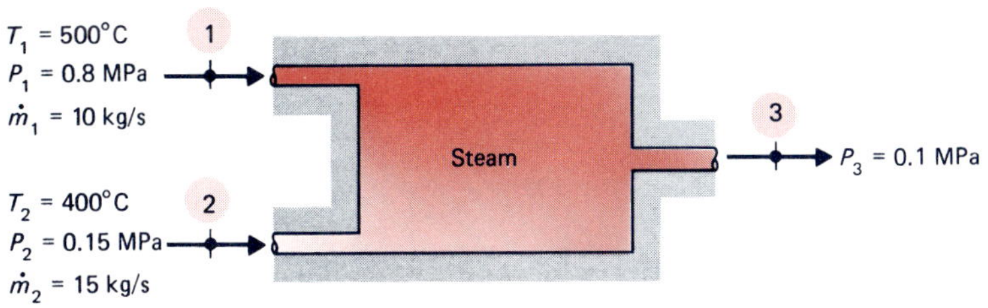

Figure P5.163

5.164 In the adiabatic steady-state–steady-flow device shown below, there is work, but its direction is unknown. Determine the direction of *flow* through the device.

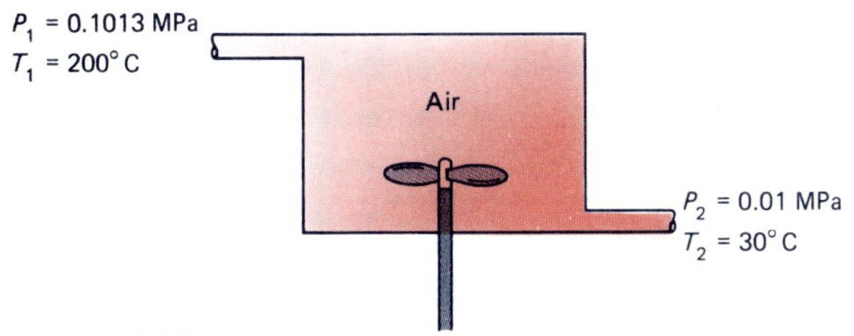

Figure P5.164

5.165 What are the magnitude and direction of the work in the isothermal steady-state–steady-flow device shown below if the medium is air? *Note*: flow direction is unknown.

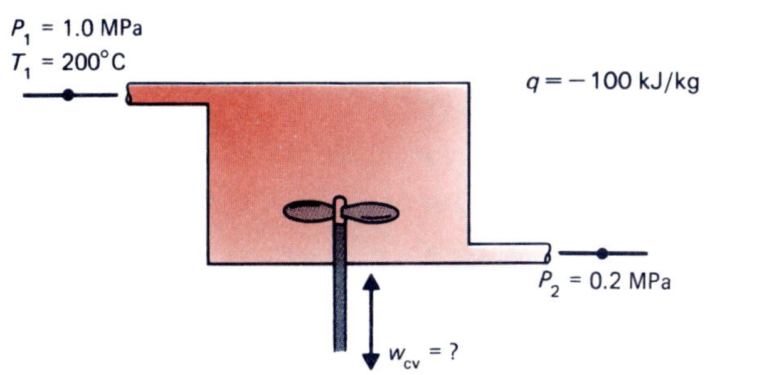

Figure P5.165

5.166S Air at 1 atm and 27°C is flowing in a pipe. An adiabatic porous plug in the pipe causes the air pressure to drop to 0.1 atm. What is the change in specific entropy of the air across the plug?

5.166E Air at 1 atm and 80°F is flowing in a pipe. An adiabatic porous plug in the pipe causes the air pressure to drop to 0.1 atm. What is the change in specific entropy of the air across the plug?

5.167S Water flows reversibly through the piping system shown in Fig. P5.167S. How large an electric motor (efficiency = 90 percent) should be selected to drive the pump (efficiency = 70 percent)?

5.167E Water flows reversibly through the piping system shown in Fig. P5.167E. How large an electric motor (efficiency = 90 percent) should be selected to drive the pump (efficiency = 70 percent)?

5.168S Steam flows reversibly into an adiabatic nozzle at 15 MPa and 500°C. The steam leaves at 400°C. If the inlet speed is 100 m/s, calculate the (*a*) work done, (*b*) heat transferred, (*c*) entropy change of the steam, (*d*) exit pressure, and (*e*) exit speed. Indicate any assumptions.

5.168E Steam flows reversibly into an adiabatic nozzle at 2200 psia and 900°F. The steam leaves at 750°F. If the inlet speed is 300 ft/s, calculate the (*a*) work done, (*b*) heat transferred, (*c*) entropy change of the steam, (*d*) exit pressure, and (*e*) exit speed. Indicate any assumptions.

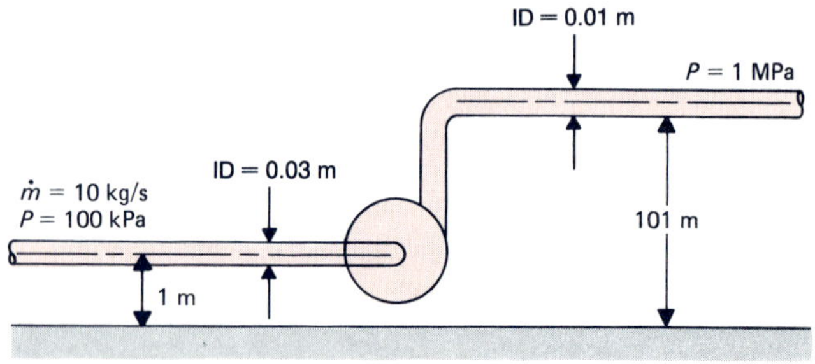

Figure P5.167S

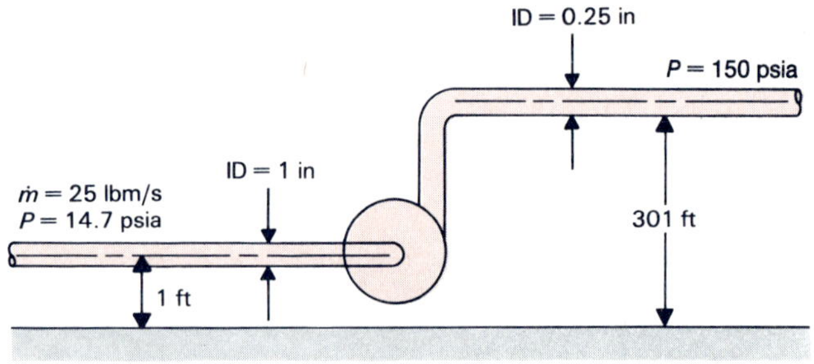

Figure P5.167E

5.169 An incompressible fluid flows through a throttling valve with $P_1 > P_2$. See Fig. P5.169. Develop the explicit relationship for the entropy generation for this process in terms of P_1, T_1, P_2, v, and c.

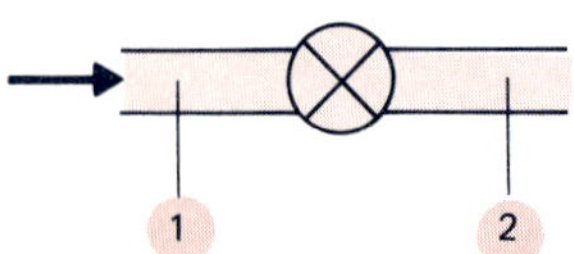

Figure P5.169

5.170 A reversible adiabatic air nozzle operates as shown, Determine the exit temperature and velocity.

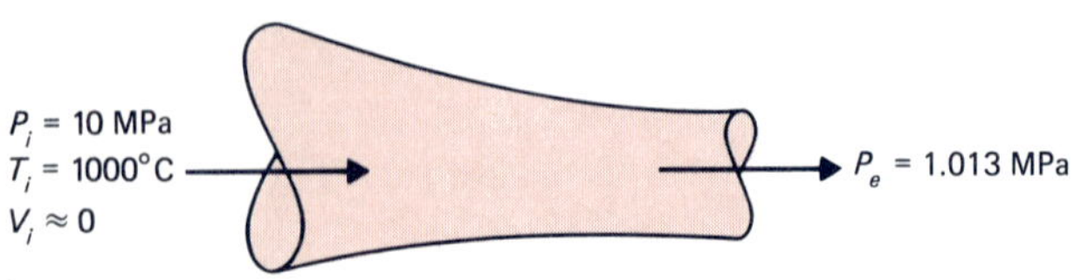

Figure P5.170

5.171 Air is expanded through the isothermal nozzle/diffuser shown in Fig. P5.171. Determine the exit velocity (m/s).

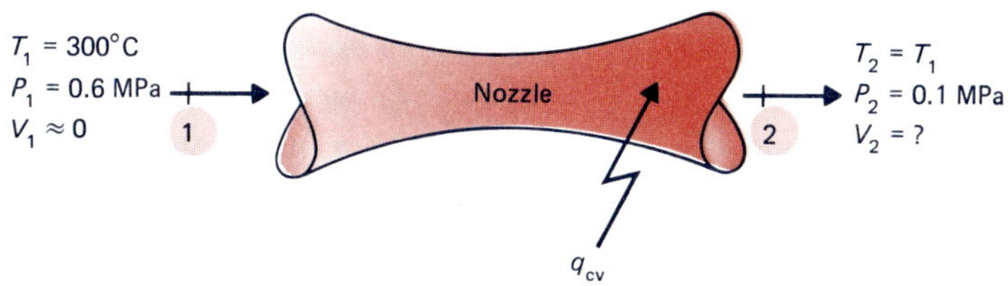

Figure P5.171

5.172 Air in a converging nozzle has the following stagnation ($\mathbf{V} = 0$) conditions: $P_0 = 160$ kPa and $T_0 = 320$ K. If the outlet pressure is $P_B = 80$ kPa, find M_E, P_E, $\mathbf{V}_E$, and $\dot{m}$. Also $A_E = 0.002$ m^2.

5.173 Air in a converging nozzle has the following stagnation ($\mathbf{V} = 0$) conditions: $P_0 = 200$ kPa and $T_0 = 370$ K. If the outlet pressure is $P_B = 101.3$ kPa, find M_E, P_E, and $\mathbf{V}_E$.

5.174 A converging nozzle discharges into atmospheric air at $P = 101.3$ kPa. Between what range of stagnation pressures will the exit Mach number be sonic?

5.175 Air in a converging nozzle has the following stagnation conditions: $P_0 = 230$ kPa and $T_0 = 400$ K. If the back pressure $P_B = 130$ kPa, find M_E, P_E, $\mathbf{V}_E$, and $\dot{m}$. Also $A_E = 0.03$ m^2.

5.176 Air flows in a converging nozzle. $A_E = 0.016$ m^2, $P_0 = 400$ kPa, the back-pressure $P_B = 101.3$ kPa, and $T_0 = 250$ K. Find M_E, P_E, $\mathbf{V}_E$, T_E, and $\dot{m}$.

5.177 Air flows in a converging nozzle. $A_E = 0.016$ m^2, $P_0 = 400$ kPa, and $T_0 = 250$ K. If $P_B = 220$ kPa, find M_E, P_E, $\mathbf{V}_E$, T_E, and $\dot{m}$.

5.178 Air flows in a converging nozzle. $A_E = 0.009$ m^2, $P_0 = 150$ kPa, and $T_0 = 450$ K. If the nozzle discharges into atmospheric air, $P = 101.3$ kPa, is the nozzle choked? Find M_E, P_E, T_E, and $\mathbf{V}_E$.

5.179 Air is expanded in an adiabatic nozzle/diffuser from the given inlet conditions. Calculate the exit velocity, assuming that the air has constant specific heats and that the inlet velocity is negligible.

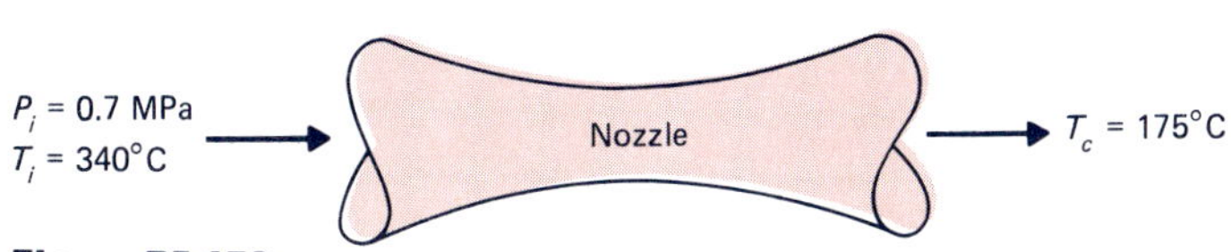

Figure P5.179

5.180 A reversible adiabatic steam nozzle operates as shown. Determine the exit temperature and velocity.

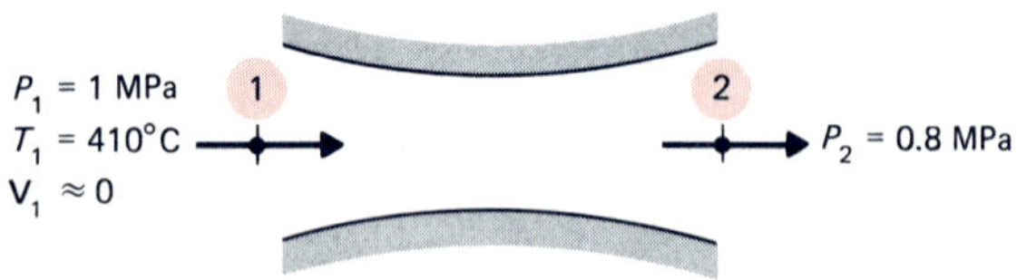

Figure P5.180

5.181 Steam is provided from two boilers for an adiabatic nozzle as shown. (*a*) If thrust equals $\dot{m}\mathbf{V}_e$, determine the nozzle thrust. (*b*) Is this device possible? Show your work.

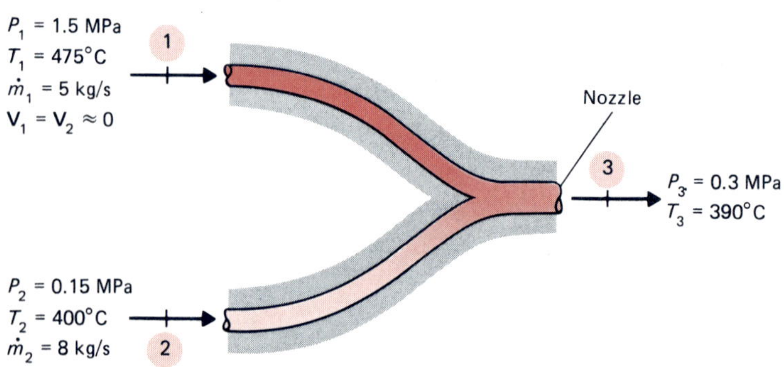

Figure P5.181

5.182 Air is provided from two combustors for an adiabatic nozzle as shown. (*a*) If thrust equals $\dot{m}\mathbf{V}_e$, determine the nozzle thrust. (*b*) Is this device possible? Show your work.

Figure P5.182

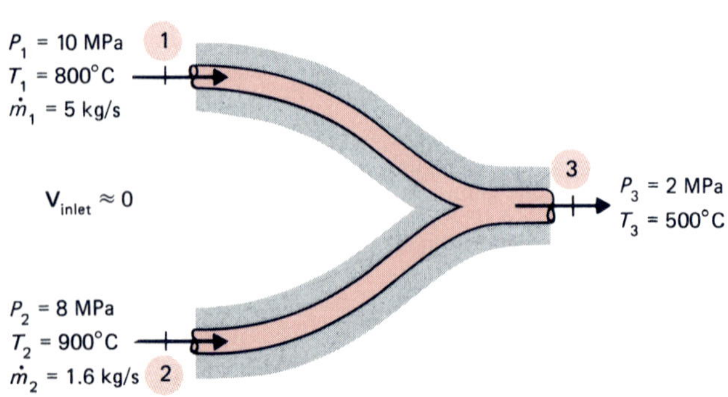

5.183 Air flows in a duct with $\mathbf{V} = 390$ m/s, $P = 113$ kPa, $T = 27°\text{C}$, and $A = 0.06\ \text{m}^2$. Determine the Mach number, P_0, T_0, and the area for the flow to become sonic, $M = 1$.

5.184 Air flows through a nozzle with inlet conditions of 100°F, 1.50 atm, $\mathbf{V}_{in} = 0$ (approximately). At the nozzle exit, the measured temperature is 77°F at the exit pressure of 1 atm. (Assume air acts as an ideal gas.) What is the nozzle efficiency η_N (defined as the square of the actual exit velocity over the square of the exit velocity for an isentropic nozzle with the same inlet conditions exhausting to the same pressure, $\eta_N = \mathbf{V}^2_{a,\,out}/\mathbf{V}^2_{s,\,out}$)?

5.185 Air flows through a porous plug at a constant flow rate with an inlet temperature and pressure of $T_{in} = 30°C$ and $P_{in} = 5$ atm. At the outlet of the plug, the pressure is $P_{out} = 1$ atm. What is the entropy change of the air across the plug $s_{out} - s_{in}$ [kJ/(kg·K)]?

5.186 An air-cooled water radiator operates as shown. Determine $\dot{Q}_{CV}$.

Figure P5.186

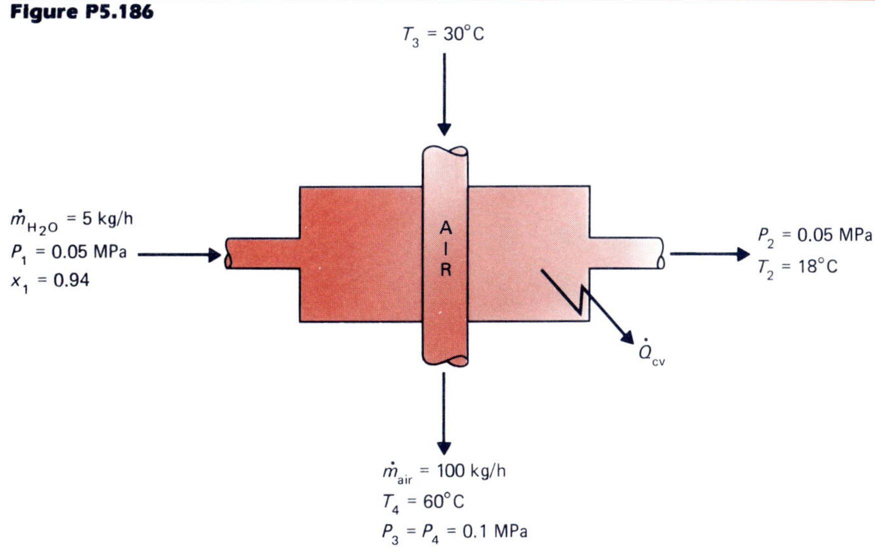

5.187 There is a heat transfer from the condenser of a refrigerator using refrigerant 12 by a flow of cooling water as shown in Fig. P5.187. The refrigerant and the water do not mix, and the condenser is adiabatic. (*a*) Find the required flow rate of cooling water (kg/min). (*b*) Is this device possible? Justify your answer with the second law of thermodynamics.

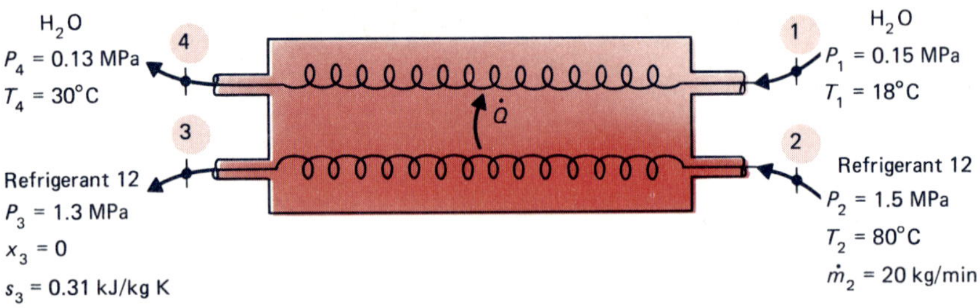

Figure P5.187

5.188S Argon gas is available from a large-capacity supply line in a chemical plant at $P = 500$ kPa, $T = 27°$C. See Fig. P5.188S. An evacuated 2-m³ tank is to be filled adiabatically from the supply line. (*a*) What is the temperature of argon in the tank when it reaches the supply line pressure? (*b*) How much entropy was generated during the process?

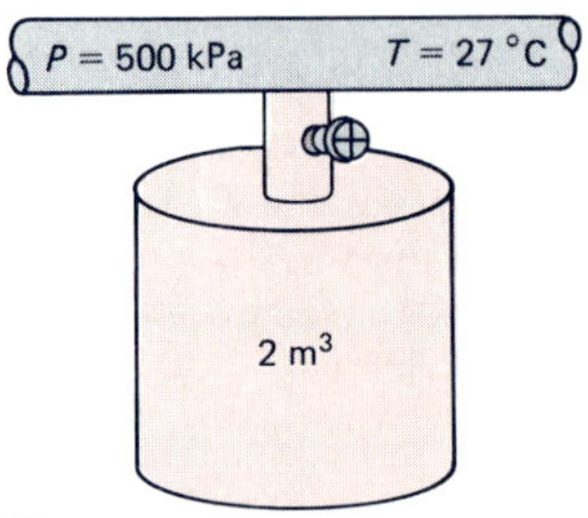

Figure P5.188S

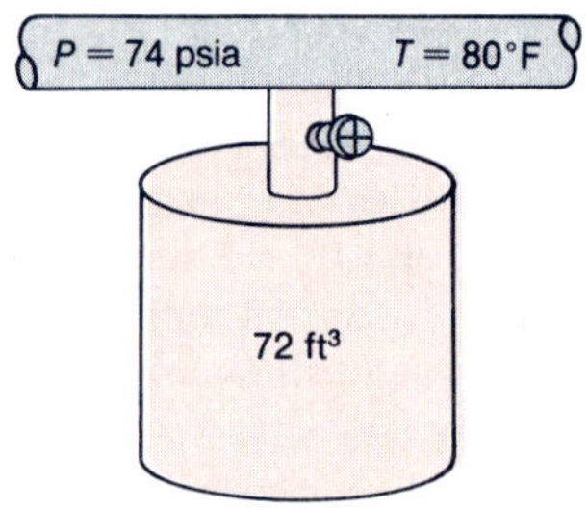

Figure P5.188E

5.188E Argon gas is available from a large-capacity supply line in a chemical plant at $P = 74$ psia, $T = 80°$F. See Fig. P5.188E. An evacuated 72-ft³ tank is to be filled adiabatically from the supply line. (*a*) What is the temperature of argon in the tank when it reaches the supply line pressure? (*b*) How much entropy was generated during the process?

5.189S The supply line described in Prob. 5.188S is connected to an identical tank, which initially contains argon at 100 kPa and 27°C. (*a*) What is the temperature of the argon in the tank when it reaches the supply line pressure? (*b*) How much entropy was generated during the process?

5.189E The supply line described in Prob. 5.188E is connected to an identical tank, which initially contains argon at 14.7 psia and 80°F. (*a*) What is the temperature of the argon in the tank when it reaches the supply line pressure? (*b*) How much entropy was generated during the process?

5.190 A supply line carries carbon dioxide gas at 200°C and 20 atm. An evacuated tank of volume 2 m³ is attached to the supply line, and a valve is opened between the supply line and the tank until the tank pressure just reaches the supply line pressure. What is the total entropy S (kJ/K) of the CO_2 in the tank at the final condition?

5.191 Air is contained in a 10-m³ tank at 3.5 MPa and 32°C. A valve at the top of the tank is opened, allowing the air to be withdrawn. During the process the tank is heated to maintain constant pressure. Assuming the air exits the tank at a temperature of 32°C during the process, determine the total heat transfer after half the air has been withdrawn.

5.192 A nitrogen tank is well insulated. At some initial time, the pressure gauge on the tank reads 1185 psig, and the tank valve is opened. After 5 min the tank valve is closed, and at that instant the gauge reads 285 psig and the temperature of the nitrogen remaining in the tank is 600°F. What was the initial temperature of the nitrogen in the tank?

5.193 A tank containing argon is well insulated. At some initial time, the pressure gauge on the tank reads 1485 psig, and the tank valve is opened. After 5 min the tank valve is closed, and at that instant the gauge reads 285 psig and the temperature of the nitrogen remaining in the tank is 600°F. What was the specific enthalpy change of the argon (Btu/lbm) in the tank during the discharge process?

5.194 A compressor is used for filling oxygen bottles. The compressor has an efficiency of 70 percent when taking oxygen from 25°C and 1 atm and compressing it to a final pressure of 10 atm. The outlet oxygen is passed through a pressure control valve and into a bottle for filling. The bottles are initially evacuated. The valve is closed after the pressure in the bottle reaches 10 atm. What is the temperature of the oxygen in the tank when the valve is closed (°C)?

Figure P5.194

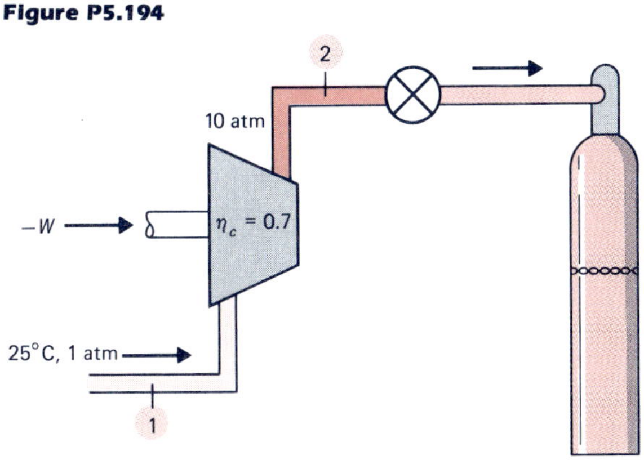

5.195 An argon storage tank is initially at 25°C and pressure P_1. The tank is well insulated. A valve is opened, and argon is discharged from the tank. After a short time the valve is closed, and the tank temperature is found to be 0°C and the pressure 1 atm. What was the initial tank pressure P_1 (atm)?

5.196 Air is available in a large supply line at $P_{in} = 5$ atm and $T_{in} = 27°C$. At time $t = 0$, valve 1 is opened and air is allowed to enter the evacuated and insulated tank. Valve 2 is initially closed. When the tank pressure reaches 5 atm, valve 2 is opened. What is the temperature of the air leaving valve 2 when it is first opened?

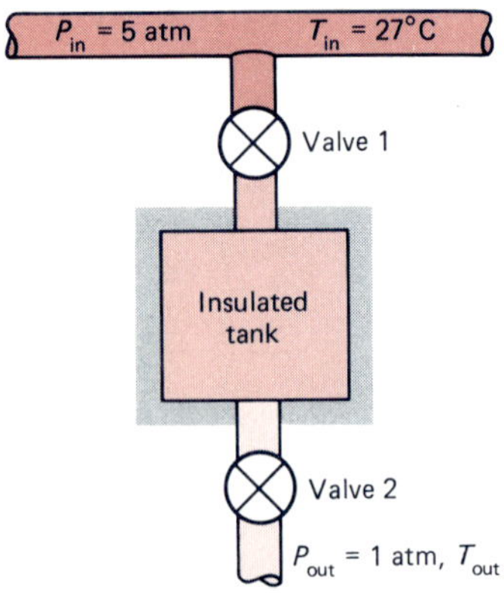

Figure P5.196

5.197S An evacuated tank has a volume of 0.5 m³. It is connected through a large valve to an air line in which air is available at 70°C and 5 MPa. The valve is opened, allowing air to flow adiabatically into the tank. The valve is closed when the tank pressure reaches 700 kPa. The tank is then allowed to sit until it comes to thermal equilibrium at room temperature ($T_{room} =$ 20°C). What is the final pressure inside the tank?

5.197E An evacuated tank has a volume of 20 ft³. It is connected through a large valve to an air line in which air is available at 160°F and 750 psia. The valve is opened, allowing air to flow adiabatically into the tank. The valve is closed when the tank pressure reaches 100 psia. The tank is then allowed to sit until it comes to thermal equilibrium at room temperature ($T_{room} = 70°F$). What is the final pressure inside the tank?

5.198S A tank initially contains very little water (assume $m_1 = u_1 = s_1 = 0$). The tank is attached to a steam line which is at 0.8 MPa and 400°C. The valve connecting the tank and steam line is opened until the tank pressure reaches 0.8 MPa, and then the valve is closed. During the process there is a heat transfer of 1000 kJ from the tank to the surroundings (surroundings are at 300 K) for each kilogram in the tank at the end of the process ($_1Q_2/m_2$). Determine the final state in the tank (give P and T, or P and x as appropriate) and the entropy generation per unit mass S_{gen}/m_2.

5.198E A tank initially contains very little water (assume $m_1 = u_1 = s_1 = 0$). The tank is attached to a steam line which is at 125 psia and 750°F. The valve connecting the tank and steam line is opened until the tank pressure reaches 125 psia, and then the valve is closed. During the process there is a heat transfer of 1000 Btu from the tank to the surroundings (surroundings are at 80°F) for each pound-mass in the tank at the end of the process ($_1Q_2/m_2$). Determine the final state in the tank (give P and T, or P and x as appropriate) and the entropy generation per unit mass S_{gen}/m_2.

5.199S The tank shown in Fig. P5.199 initially contains 0.3 kg of an ideal gas, N_2, at 530 K and an adiabatic, volumeless, weightless, and frictionless piston. The 0.25-m³ tank is connected to an infinite source of ideal gas, air, at 530 K and 0.7 MPa. Air enters the tank until the temperature of the air is 560 K and the final mass of air is 0.15 kg. Calculate the entropy generation of the process, assuming there is 8.6 kJ of heat transfer from the N_2 to the surroundings at 25°C.

5.199E The tank shown in Fig. P5.199 initially contains 0.7 lbm of an ideal gas, N_2, at 950°R and an adiabatic, volumeless, weightless, and frictionless piston. The 7-ft³ tank is connected to an infinite source of ideal gas, air, at 950°R and 100 psia. Air enters the tank until the temperature of the air is 1000°R and the final mass of air is 0.35 lbm. Calculate the entropy generation of the process, assuming there is 8 Btu of heat transfer from the N_2 to the surroundings at 80°F.

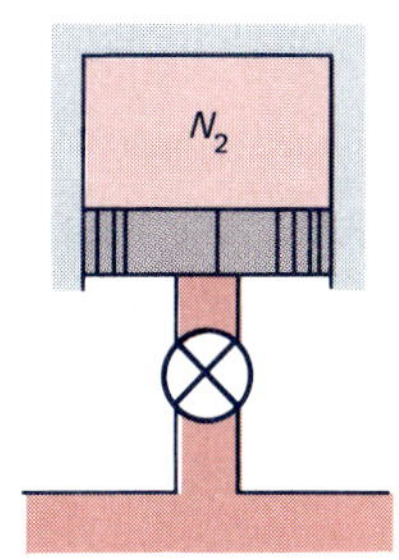

Figure P5.199

5.200S A chamber of volume 0.001 m³ is attached to a steam line as shown in Fig. P5.200. The steam line state is constant at $P_{in} = 0.600$ MPa and $x_{in} = 0.97$. A small portion of the fluid from the line is diverted into a valve, through the chamber, and out a second valve. The chamber is insulated, so the process is adiabatic.

(*a*) The mass flow rate into the chamber is equal to the mass flow rate out, and the chamber pressure is 0.050 MPa. Determine the fluid temperature within the chamber and the entropy generation if the outlet valve exhausts to 10 kPa.

(*b*) Then the outlet valve is quickly closed, and the filling process terminates when the chamber pressure is equal to the steam line pressure. The mass addition to the chamber is 0.002 kg. Determine the fluid temperature within the chamber and the entropy generation.

Figure P5.200

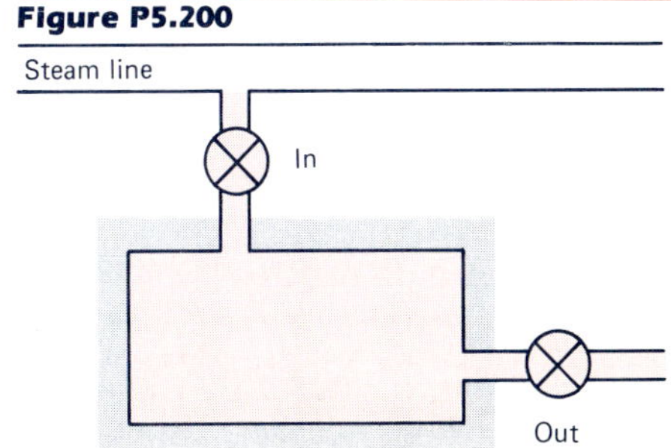

5.200E A chamber of volume 0.03 ft^3 is attached to a steam line as shown in Fig. P5.200. The steam line state is constant at $P_{in} = 90$ psia and $x_{in} = 0.97$. A small portion of the fluid from the line is diverted into a valve, through the chamber, and out a second valve. The chamber is insulated so the process is adiabatic.

(*a*) The mass flow rate into the chamber is equal to the mass flow rate out, and the chamber pressure is 8 psia. Determine the fluid temperature within the chamber and the entropy generation if the outlet valve exhausts to 1 psia.

(*b*) Then the outlet valve is quickly closed, and the filling process terminates when the chamber pressure is equal to the steam line pressure. The mass addition to the chamber is 0.005 lbm. Determine the fluid temperature within the chamber and the entropy generation.

5.201S A bottle of pressurized air is sitting in the laboratory at a temperature of 27°C and an initial pressure of 500 kPa and contains an initial mass of air of 1 kg. The valve on the tank is left open, and air escapes into the laboratory until the tank pressure reaches 100 kPa.

(*a*) Define the system or control volume for this problem, and write the first and second laws for it. Note any assumptions used to simplify the equations.

(*b*) Find the entropy generation for the process.

5.201E A bottle of pressurized air is sitting in the laboratory at a temperature of 80°F and an initial pressure of 70 psia and contains an initial mass of air of 2.2 lbm. The valve on the tank is left open, and air escapes into the laboratory until the tank pressure reaches 14.7 psia.

(*a*) Define the system or control volume for this problem, and write the first and second laws for it. Note any assumptions used to simplify the equations.

(*b*) Find the entropy generation for the process.

5.202S A tank with 10-m^3 volume contains a gas with a temperature-independent c_p of 0.3 kJ/(kg·K) and a gas constant of $R = 0.0857$ kJ/(kg·K). The tank is initially at $T = 327$°C and $P = 10$ atm, and $s = 6$ kJ/(kg·K) at this condition. The tank is discharged to a final pressure of 1 atm. Treat the gas as ideal.

(*a*) What is the final temperature of the gas in the tank?

(*b*) What is the change in total entropy of the gas in the tank $(\Delta S)_{CV}$?

5.202E A tank with 300-ft^3 volume contains a gas with a temperature-dependent c_p of 0.07 Btu/(lbm·°R) and a gas constant of $R = 0.02048$ Btu/(lbm · °R). The tank is initially at $T = 620$°F and $P = 10$ atm, and $s = 1.43$ Btu/(lbm · °R) at this condition. The tank is discharged to a final pressure of 1 atm. Treat the gas as ideal.

(*a*) What is the final temperature of the gas in the tank?

(*b*) What is the change in total entropy of the gas in the tank $(\Delta S)_{CV}$?

5.203S A stainless steel tank has a volume of 0.07 m^3. It initially contains pure nitrogen at 300°C and 600 kPa. At a particular time, a low-flying aircraft knocks the valve from the outlet of the tank, and nitrogen discharges adiabatically from the tank until the tank pressure just reaches 100 kPa.

(*a*) What is the temperature of the nitrogen remaining in the tank just as the tank pressure reaches 100 kPa?

(*b*) How much nitrogen was lost from the tank during the transient discharge?

(*c*) What is the change in total entropy $(\Delta S)_{CV}$ inside the tank during the discharge process?

(*d*) How much entropy is generated during the process?

5.203E A stainless-steel tank has a volume of 2.5 ft^3. It initially contains pure nitrogen at 600°F and 90 psia. At a particular time, a low-flying aircraft knocks the valve from the outlet of the tank, and nitrogen discharges adiabatically from the tank until the tank pressure just reaches 14.7 psia.

(*a*) What is the temperature of the nitrogen remaining in the tank just as the tank pressure reaches 14.7 psia?

(*b*) How much nitrogen was lost from the tank during the transient discharge?

(*c*) What is the change in total entropy, $(\Delta S)_{CV}$ inside the tank during the discharge process?

(*d*) How much entropy is generated during the process?

5.204S An insulated air storage tank initially is at 1.5 MPa and 30°C. The valve on the 0.4-m^3 tank is opened, and the air exhausts until the tank pressure is 0.5 MPa. What is the tank temperature at this state? How much mass is left in the tank? (Evaluate these quantities with both constant and variable specific heats.)

5.204E An insulated air storage tank initially is at 250 psia and 85°F. The valve on the 12-ft^3 tank is opened, and the air exhausts until the tank pressure is 75 psia. What is the tank temperature at this state? How much mass is left in the tank? (Evaluate these quantities with both constant and variable specific heats.)

5.205 A steam boiler has a volume of 0.5 m^3. After the boiler is heated to $P = 0.5$ MPa and $x = 10$ percent, a valve opens, allowing the saturated *vapor* to run to a steam line. There is an additional heat transfer to the boiler to maintain constant pressure. Determine the total mass of the mixture remaining in the boiler when 73 percent of the initial H_2O has run to the steam line.

5.206 Given the steam boiler in the preceding problem, determine the *total* heat transfer to the boiler when 75 percent of the H_2O in the boiler has run to the steam line. Assume $m_1 = 13$ kg and $x_2 = 40$ percent.

5.207S An inventor makes the following claims. Determine whether the claims are valid and explain why or why not.

(*a*) A flame at 1500 K is to be used as a heat source, and the low-temperature reservoir is at 300 K. The inventor indicates that 69 percent of the heat transfer to the cyclic device from the flame is returned as work.

(*b*) A building receives a heat transfer 50,000 kJ/h from a heat pump. The inside temperature is maintaned at 21°C, and the surroundings are at −1°C. The inventor claims a work input of 7000 kJ/h is required.

(*c*) An engine operates between 1000 and 400 K with a heat transfer into the engine of 550 kW. The inventor states that the heat transfer to the low-temperature reservoir is 250 kW and the work output is 250 kW.

5.207E An inventor makes the following claims. Determine whether the claims are valid and explain why or why not.

(*a*) A flame at 2700°R is to be used as a heat source, and the low-temperature reservoir is at 540°R. The inventor indicates that 69 percent of the heat transfer to the cyclic device from the flame is returned as work.

(*b*) A building receives a heat transfer of 48,000 Btu/h from a heat pump. The inside temperature is maintained at 76°F, and the surroundings are at 39°F. The inventor claims a work input of 2.4 hp is required.

(*c*) An engine operates between 1800 and 720°R with a heat transfer into the engine of 1,800,000 Btu. The inventor states that the heat transfer to the low-temperature reservoir is 820,000 Btu and the work output is 332 hp.

5.208 A cyclic device has a single work input of $|W|$ and a single heat transfer $|Q_H|$ to a high-temperature reservoir at T_H. Is the device possible? Why or why not?

5.209 A reversible heat engine operates between a high-temperature thermal reservoir at 400°C and low-temperature thermal reservoir at 40°C. Calculate the ratio of the absolute temperatures of the two reservoirs.

5.210 A proposed external combustion engine for a new line of motorized skateboards operates in a thermodynamic cycle and requires the combustion of 5 g/min of fuel. In burning this fuel, thermal energy amounting to 15 kJ is released for each gram consumed. This fuel is burned at 1200°C, and there is a heat transfer to the atmosphere at 20°C. The engine develops 0.5 hp. What is the cycle efficiency of this engine and what is the heat transfer?

5.211 A Carnot refrigerator rejects a heat transfer of 2.5×10^3 kJ at 80°C. If W_{net} for this refrigerator is 1.1×10^3 kJ, find the heat transferred, the low temperature in the cycle, and the coefficient of performance.

5.212 A Carnot heat engine device receives a heat transfer of 100 kJ from a source at 100°C. If there is a heat transfer to 20°C, calculate the work produced and the thermal efficiency of the Carnot engine.

5.213 A Carnot engine rejects a heat transfer of 230 kJ at 24°C. The net work for the cycle is 370 kJ. Find the thermal efficiency and T_H.

5.214 A Carnot refrigerator operates between the temperature extremes of −5 and 30°C. The compressor requires 4 kW of outside power to run. If a heat transfer of 30 kJ/kg is absorbed by the Carnot refrigerator, calculate (*a*) the coefficient of performance and (*b*) the mass flow rate (kg/h) of the working medium. (*c*) If the refrigerator is run as a heat pump, find the coefficient of performance.

5.215 An inventor claims she has a steam engine that operates between the temperature extremes of 700 and 10°C and produces 200 kJ of work while rejecting a heat transfer of 100 kJ. Is her heat engine possible?

5.216 Super Air Conditioner Company claims to have developed a new air conditioner. This device maintains a room at 22°C and rejects a heat transfer of 105 kJ at a maximum temperature of 42°C. The compressor requires 10 kJ or work. Is this device believable?

5.217 A Carnot cycle refrigerator operates in a room which has a temperature of 21°C. It must reject 75 kW to the room in order to maintain the refrigerated space at 10°C. What size motor (hp) is required?

5.218 An aspiring inventor has offered to sell the Student Union a food freezer unit which would maintain an inside temperature of −37°C and would have a COP of 4.0. Should the Union consider the purchase of this unit? (Assume the units would be located in a room having an average temperature of 20°C.)

5.219 A Carnot heat pump is designed to heat an apartment complex in an area where the expected winter temperature may be as low as −35°C. The expected COP_{HP} is 1.25. To what temperature could this unit conceivably heat the complex? Is this adequate?

5.220 A Carnot heat engine produces power at a rate of 2.5 kW. It rejects energy in the form of heat transfer to a river that is flowing at a mass flow rate of 1 kg/s, causing the river to increase in temperature by 2°C. The river is at an average temperature of 20°C.

(*a*) What is the rate of heat transfer required to supply the heat engine (kW)?
(*b*) What is the efficiency of the heat engine (%)?
(*c*) What is the temperature at which heat transfer occurs to the engine (°C)?

5.221 An engine operates on the Carnot cycle. The engine efficiency is 60 percent when 20 kJ/kg per engine cycle of thermal energy is rejected to a low-temperature reservoir at $T_L = 27$°C, and has a specific work output of 10 kJ/kg per cycle.

(*a*) How much energy is transferred to the engine from the high-temperature reservoir per cycle (kJ/kg)?
(*b*) At what temperature T_H is the high-temperature reservoir?

5.222 An engine operates on the Carnot cycle. The engine takes 50 kJ/kg per cycle of energy from a high-temperature reservoir with a heat transfer to a low-temperature reservoir at $T_L = 27$°C, and has a specific work output of 10 kJ/kg per cycle.

(*a*) How much energy is transferred from the engine to the low-temperature reservoir per cycle (kJ/kg)?
(*b*) At what temperature T_H is the high-temperature reservoir?
(*c*) What is the heat engine efficiency?

5.223 A heat engine operates between a higher temperature of T_H and a lower temperature of $T_L = 60°C$. The efficiency of the engine is said to be 45 percent.

(*a*) What is the minimum value of T_H (°C) for which the proposed efficiency is possible?

(*b*) If this engine operates on a Carnot cycle, how much heat transfer *to* the engine must occur for each kilojoule that is rejected at T_L?

(*c*) If the engine does not operate on a reversible cycle, would the heat transfer *to* the engine be greater or smaller than that found in part (*b*) for each kilojoule rejected at T_L? Explain your answer.

5.224 Two Carnot heat engines are connected in series as shown so that the waste thermal energy at constant temperature T_M from engine 1 is the thermal energy source for engine 2. The high-temperature reservoir for engine 1 is at T_H. The low-temperature reservoir for engine 2 is at $T_L = 27°C$, and $T_L \le T_M \le T_H$. The total work output $w_1 + w_2$ from this combined engine is 1.167 kJ/kg when the energy rejected from engine 1 (and used as input to engine 2) is $q_M = 1$ kJ/kg. This occurs when $T_M = 327°C$.

(*a*) What is the value of w_1 (kJ/kg)?

(*b*) What is the value of T_H (°C)?

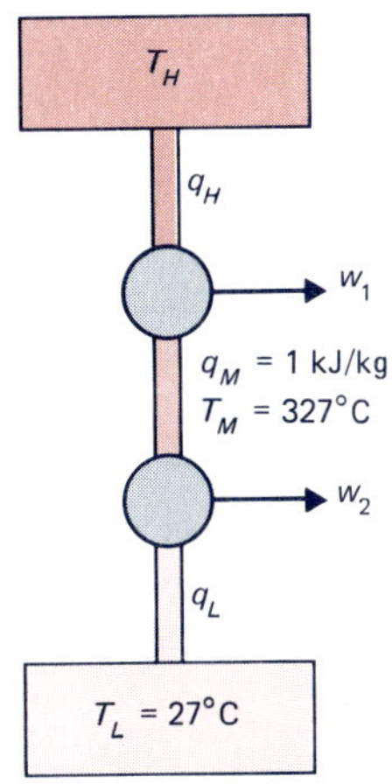

Figure P5.224

5.225 Two Carnot engines are hooked in series as shown, so that the waste thermal energy at constant temperature T_M from engine 1 is used as the thermal energy source for engine 2. The high-temperature reservoir for engine 1 is at $T_H = 327°C$. The low-temperature reservoir for engine 2 is at $T_L = 27°C$, and $T_L \le T_M \le T_H$. What value of T_M (°C) will provide the greatest total work output $w_1 + w_2$ for this combined engine? (Prove your answer.)

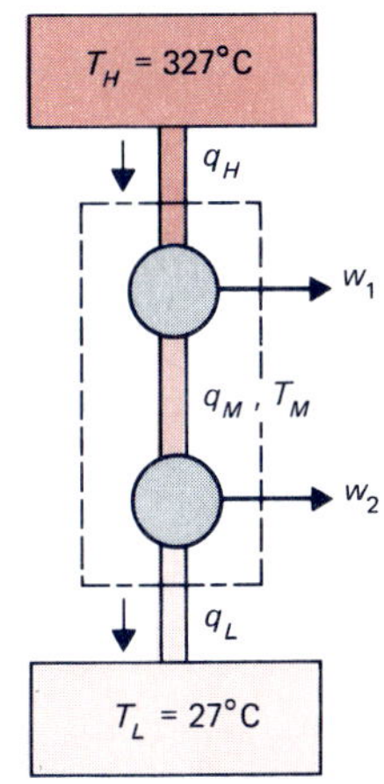

Figure P5.225

5.226 In a "selective energy system," a power plant produces electric power in sufficient quantity that the waste energy matches the thermal demands of the user. Any excess electric power is purchased from another source. Your power plant has η_{th} of 40 percent, and the village being supplied needs 20×10^3 kJ/s of *thermal* energy. If the village needs 20 MW of electricity, how much do you have to buy?

5.227 The thermal reservoirs to a Carnot engine are constant-pressure steam lines at $T_H = 300°C$ and $T_L = 25°C$, respectively. As heat transfer occurs from the high-temperature steam line, saturated vapor passing through the line at a mass flow rate of 3 kg/s is condensed to an exit quality of 95 percent.

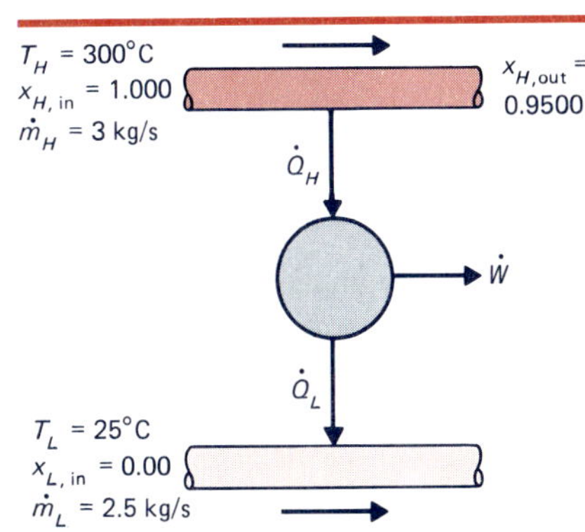

Figure P5.227

(*a*) If the steam line entering the low-temperature reservoir carries saturated liquid at a mass flow rate of 2.5 kg/s, what is the exit quality of the steam?
(*b*) What is the power output of the Carnot engine (kW)?

5.228 In a counterflow heat exchanger of a large heat engine using H_2O as the working medium, a heat transfer of 200 kJ/s is rejected to a second system employing an exotic fluid which has a very low boiling point compared to H_2O. Upon exiting from the heat exchanger, the exotic fluid passes through a turbine where shaft power is produced. Ultimately a heat transfer of 135 kJ/s is discharged to the surroundings from the second engine. What are the net work (kW) and thermal efficiency η_{th} of the second engine?

5.229 An electric utility company has a plant with a capacity of 1000 MW of electricity for sale to its customers. The overall thermal efficiency of the plant is 25 percent, and the oil (the fuel) costs 25¢ per gallon. The heating value of the oil is 4.65×10^6 kJ/kg, and there is 3.4 kg of oil per gallon What is the cost of running at maximum capacity for one week?

5.230 The mass flow rate of the coolant in a small boiling water reactor is 0.5 kg/s, and a heat transfer of 300 kJ/kg is rejected from the cycle in the nearby cooling pond. If 100 kW is produced by the turbine, how much energy is transferred to the coolant from the fission reaction? What is the thermal efficiency?

5.231 A Carnot heat pump delivers a heat transfer of 3×10^4 kJ at 50°C. The evaporator is at 15°C. Calculate W_{net} and the coefficient of performance for this Carnot heat pump.

5.232 A heat engine is designed to have a thermal efficiency of 27 percent. It will be used to supply power to a remote farm. The engine will be operated near a large pond, which will act as a low-temperature reservoir. However, due to environmental considerations, only a heat transfer of 50 kJ/min may be rejected to the pond. If the farm requires 1.8 kW of power will this engine provide sufficient power?

5.233 A steam power plant is constructed such that it produces 75 kW of power while rejecting 190.0 kJ/s to a low-temperature reservoir. (*a*) Find $\dot{Q}_H$. (*b*) Find η_{th}.

5.234 A refrigeration unit is built which will provide 100 kJ/s of cooling $\dot{Q}_L$. And 50 kW of compressor power is required to run the refrigerator. (*a*) Find $\dot{Q}_H$. (*b*) What is the coefficient of performance for the refrigerator? (c) If this same unit were run as a heat pump, what would be its coefficient of performance?

5.235 A retired engineer living in Arizona is attempting to heat his home by using a heat pump. He has determined the minimum heat transfer rate needed to heat his home to be 50.8 kJ/h. He has the chance to buy a heat pump which has a coefficient of performance of 5 and which requires 2.75 W of electric power to run. Assuming the power requirement does not present a problem, should the heat pump be bought?

5.236 A large municipal power plant is cooled with river water as shown in Fig. P5.236. Find the net rise in river temperature $T_2 - T_1$ that results from the thermal load of the power plant.

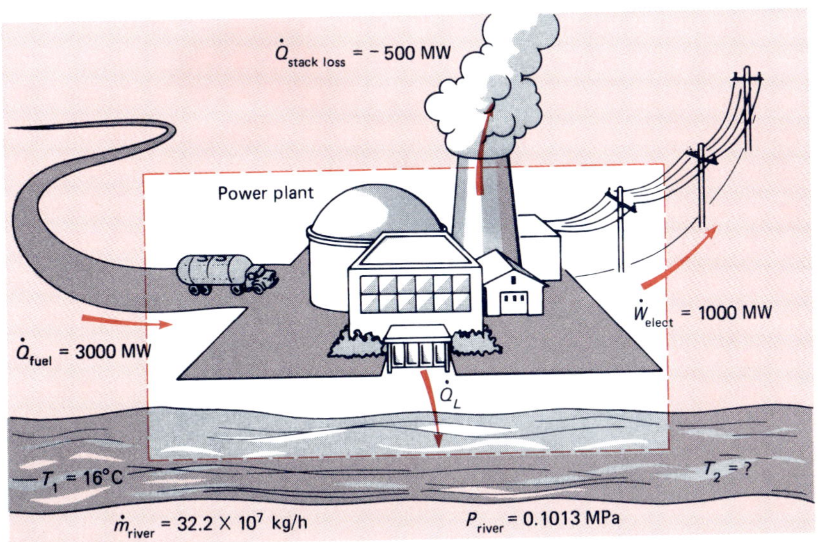

Figure P5.236

5.237 A large municipal power plant is cooled with river water as shown in Fig. P5.237. Find the required mass flow rate of the cooling water.

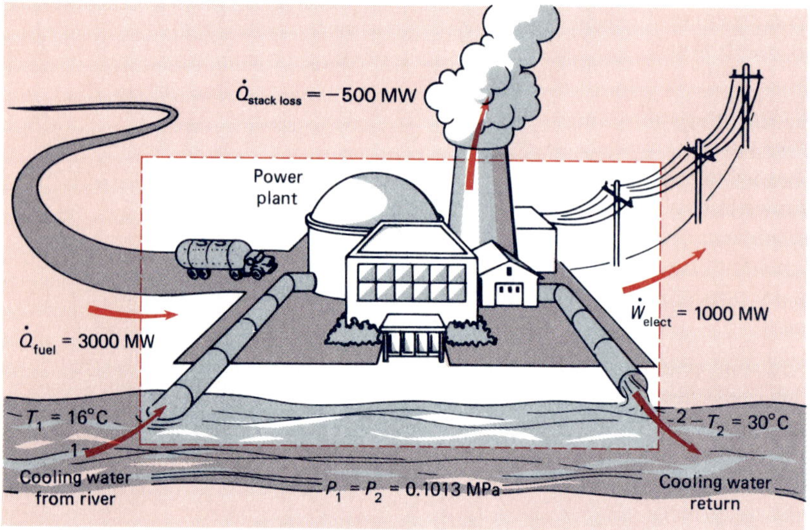

Figure P5.237

5.238 A large municipal power plant uses a cooling tower to precool the cooling water before dumping it back into the river (see Fig. P5.238). Federal regulations direct that the net rise in river temperature $T_2 - T_1$ not exceed 2°C. Find the required heat transfer in the cooling tower to *just meet* the regulations.

Figure P5.238

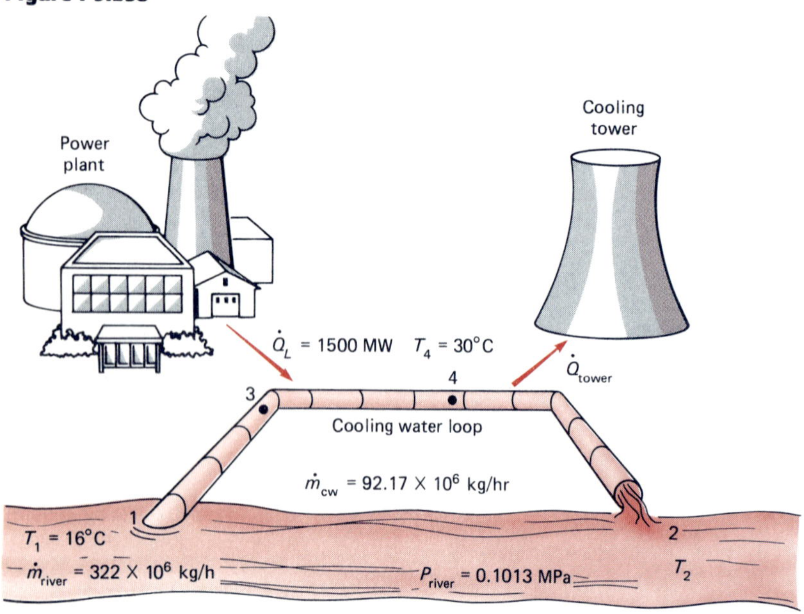

5.239 Hot-water wells in the Southwest provide a potential source of energy. If the temperature of the hot water in these wells is 200°C and the local ambient temperature is 25°C, what is the maximum thermal efficiency that a heat engine could provide?

5.240 Ocean thermal gradients have been proposed as one future source of energy to meet our needs. If a cyclic heat engine could operate based on an ocean surface temperature of 18°C and a temperature of 3°C at a depth of 300 m, what would be the maximum thermal efficiency of this engine?

5.241 A dormitory is to be maintained at 22°C by means of a heat pump pumping from the atmosphere. Heat transfer through the walls and doors are estimated at 5 kJ/s per degree Celsius temperature difference between the atmosphere and the building.

(*a*) If the outside temperature is −5°C, what is the minimum power required to drive the pump?

(*b*) If the same "heat pump" is used to cool the dormitory in the summer to the same temperature, with the same heat transfer rate through the walls and the same power to the pump, what is the maximum permissible atmospheric temperature?

5.242 A Carnot heat engine has a thermal efficiency of 40 percent. It is then reversed to operate as a refrigerator and provide 20 kW of cooling capacity. (*a*) What is the required net power (kW)? (*b*) What is the COP of the refrigerator?

5.243 Assume that a Carnot refrigerator is used to maintain the ice at a skating rink at −10°C. If the average ambient temperature in the winter is 10°C, what is the coefficient of performance for this device?

5.244 An inventor proposes to build a nonflow device which will compress H_2O isothermally at a temperature of 400°C from a pressure of 4.0 MPa to a final pressure of 15.0 MPa. If the device uses 150 kJ/kg of work input, is it reversible, irreversible, or impossible?

5.245S Solar thermal power plants use solar energy to increase the temperature of some working fluid to T_H. This fluid can be used to power a heat engine. Low-cost flat-plate solar collectors can provide working fluids at temperatures in the range of about $30 \leq T_H \leq 100$°C. (Concentrating collectors can provide much higher temperatures.) If cooling water is available at 27°C, plot the efficiency of a reversible heat engine powered by flat-plate collectors versus T_H, for T_H in the range given for flat-plate collectors.

5.245E Solar thermal power plants use solar energy to increase the temperature of some working fluid to T_H. This fluid can be used to power a heat engine. Low-costs flat-plate solar collectors can provide working fluids at temperatures in the range of about $85 \leq T_H \leq 212°F$. (Concentrating collectors can provide much higher temperatures.) If cooling water is available at 80°F, plot the efficiency of a reversible heat engine powered by flat-plate collectors versus T_H, for T_H in the range given for flat-plate collectors.

5.246S An inventor brings you a device. She claims that it will collect solar energy at 120°C and that each 1 m² of the collecting surface will provide power from the attached heat engine at a rate of 0.5 kW/m². You know that the solar energy on a clear day near noon in your area rarely exceeds 1 kW/m². You are asked to decide the following:

(*a*) Does the proposed device violate the first law?
(*b*) Does it violate the second law?
(*c*) Could the device operate as advertised if a reservoir were available at a low enough temperature?

5.246E An inventor brings you a device. He claims that it will collect solar energy at 150°F and that each square foot of the collecting surface will provide power from the attached heat engine at a rate of 32 ft·lbf/(s·ft²). You know that the solar energy on a clear day near noon in your area rarely exceeds 300 Btu/(h·ft²). You are asked to decide the following:

(*a*) Does the proposed device violate the first law?
(*b*) Does it violate the second law?
(*c*) Could the device operate as advertised if a reservoir were available at a low enough temperature?

5.247S A Carnot engine receives a heat transfer from a boiler at $T_H = 1000$ K, and it rejects a heat transfer of 20 kJ to a low-temperature reservoir at T_L. What is the value of the product $Q_H T_L$ for this engine?

5.247E A Carnot engine receives a heat transfer from a boiler at $T_H = 1800°R$ and rejects a heat transfer of 20 Btu to a low-temperature reservoir at T_L. What is the value of the product $Q_H T_L$ for this engine?

5.248S A Carnot engine is used in a nuclear power plant. It receives 1500 MW of power as a heat transfer from a source at 327°C, and it rejects thermal waste to a nearby river at 27°C. If the river temperature rises by 3°C because of this power rejection by the plant, (*a*) what is the mass flow rate of the river, (*b*) what is the efficiency of the power plant, and (*c*) what is the power output of the plant?

5.248E A Carnot engine is used in a nuclear power plant. It receives 5×10^9 Btu/h of power as a heat transfer from a source at 620°F, and it rejects thermal waste to a nearby river at 80°F. If the river temperature rises by 5°F because of this power rejection by the plant, (*a*) what is the mass flow rate of the river, (*b*) what is the efficiency of the power, and (*c*) what is the power output of the plant?

5.249S A Carnot engine provides power to a plant that stamps out engineering students. The engine receives a heat transfer from a steam line that contains saturated vapor at $P = 250$ kPa and rejects a heat transfer to saturated refrigerant 12 at $P = 10$ kPa. See Fig. P5.249S. If the engine receives 10 kW from the steam line and it requires 5 kJ to stamp out one student, how many students can be stamped out per second?

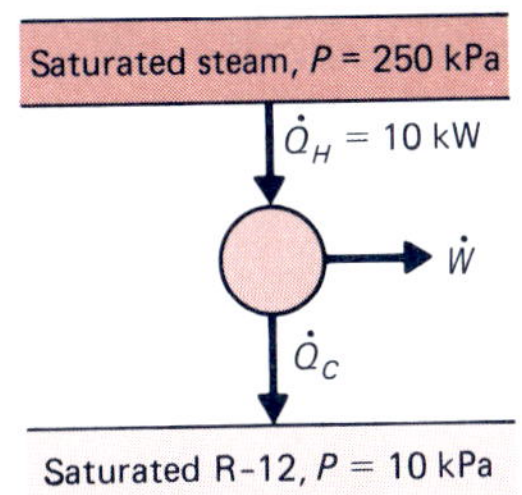

Figure P5.249S

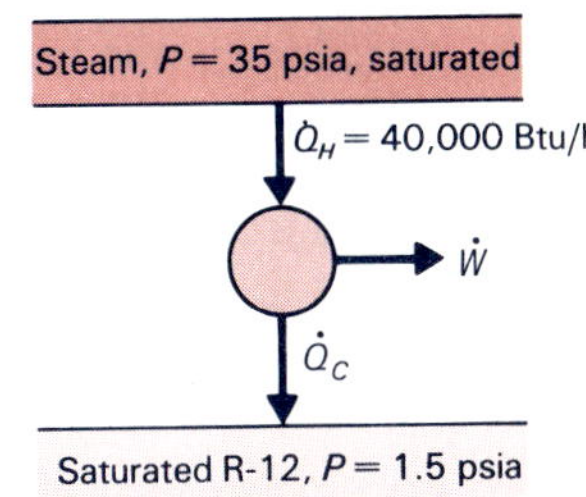

Figure P5.249E

5.249E A Carnot engine provides power to a plant that stamps out engineering students. The engine receives a heat transfer from a steam line that contains saturated vapor at $P = 35$ psia and rejects a heat transfer to saturated refrigerant 12 at $P = 1.5$ psia. See Fig. P5.249E. If the engine receives 40,000 Btu/h from the steam line and it requires 3500 ft·lbf to stamp out one student, how many students can be stamped out per second?

5.250S A reversible heat engine has a power output of 500 MW when receiving a heat transfer of 1000 MW from a heat transfer reservoir. The engine rejects a heat transfer at a constant temperature 27°C to a flowing river. The average river temperature increases by 2.0°C.

(*a*) What is the mass flow rate of the river?
(*b*) What is the temperature of the heat transfer reservoir providing energy to the heat engine?
(*c*) What is the efficiency of the heat engine?
(*d*) If the heat engine were irreversible, would the temperature change of the river be greater or smaller? Explain.

5.250E A reversible heat engine has a power output of 500 MW when receiving a heat transfer of 3000×10^6 Btu/h from a heat transfer reservoir. The engine rejects a heat transfer at a constant temperature of 80°F to a flowing river. If the average river temperature increases by 4°F,

(*a*) What is the mass flow rate of the river?
(*b*) What is the temperature of the heat transfer reservoir providing energy to the heat engine?

(*c*) What is the efficiency of the heat engine?
(*d*) If the heat engine were irreversible, would the temperature change of the river be greater or smaller? Explain.

5.251S A reversible heat engine receives energy from a solar collector at a temperature of 80°C, and rejects a heat transfer to the surroundings at 25°C (see Fig. 5.251S). The solar collector converts 50 percent of the incident solar energy into usable thermal energy. If 1 kW of solar energy strikes each 1 m² of the collectors, what collector area is necessary to provide 5 kW of power output from the heat engine?

5.251E A reversible heat engine receives energy from a solar collector at a temperature of 175°F, and rejects a heat transfer to the surroundings at 80°F (see Fig. 5.251E). The solar collector converts 50 percent of the incident solar energy to usable thermal energy. If 300 Btu/h of solar energy strikes each square foot of the collectors, what collector area is necessary to provide 7 hp of power output from the heat engine?

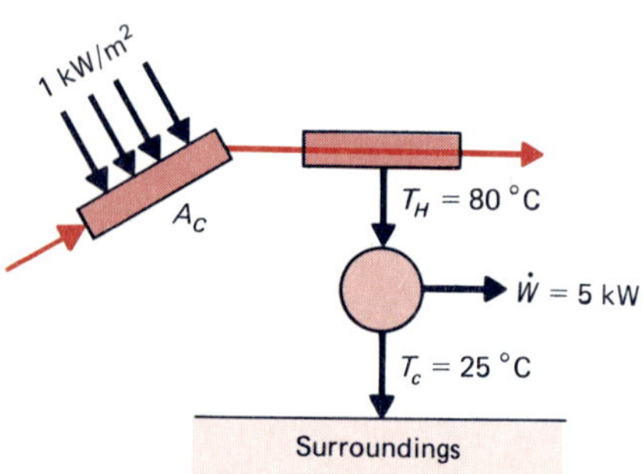

Figure P5.251S

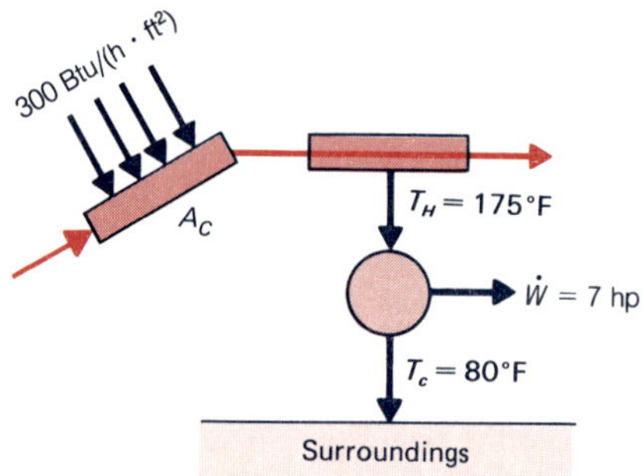

Figure P5.251E

5.252S A farmer buys a small heat engine from Carnot Products Co., Ottumwa, Iowa, to drive an irrigation pump. The engine takes a heat transfer from a nearby solar pond that is kept at $T_H = 73$°C. The pump is 70 percent efficient and pumps 1 kg/s of water at 27°C from a well. The water undergoes a pressure change of 2 MPa in the pump. The water leaving the pump is used as the reservoir for the heat transfer from the so-called Carnot engine.
(*a*) Sketch the system.
(*b*) Calculate the rate of heat transfer from the solar pond.
(*c*) State your assumptions.

5.252E A farmer buys a small heat engine from Carnot Products Co., Ottumwa, Iowa, to drive an irrigation pump. The engine takes a heat transfer from a nearby solar pond that is kept at $T_H = 160°F$. The pump is 70 percent efficient and pumps 2 lbm/s of water at 75°F from a well. The water undergoes a pressure change of 300 psia in the pump. The water leaving the pump is used as the reservoir for the heat transfer from the Carnot engine.
(*a*) Sketch the system.
(*b*) Calculate the rate of heat transfer from the solar pond.
(*c*) State your assumptions.

References

1. H. B. Callen, *Thermodynamics and an Introduction to Thermostatics,* 2d ed., Wiley, New York, 1985.
2. W. C. Reynolds and H. C. Perkins, *Engineering Thermodynamics,* McGraw-Hill, New York, 1977.
3. S. L. Soo, *Thermodynamics of Engineering Science,* Prentice-Hall, Englewood Cliffs, N.J., 1958.
4. J. Kestin, *A Course in Thermodynamics,* vol. 1, Blaisdell, New York, 1966.
5. J. H. Keenan, *Thermodynamics,* Wiley, New York, 1941.

6

Thermodynamic Cycle Analysis and Applications to Gas Cycles

When the Waters were dried an' the earth did appear, . . . The Lord He created the Engineer
Rudyard Kipling

PW4000-type high-bypass fanjet engine for commercial aircraft. Includes full digital control for thrust management to reduce fuel consumption. In modified versions, it can produce up to 100,000 lbf of thrust. (*Pratt and Whitney*.)

In this chapter, we examine in detail some of the common thermodynamic cycles used for converting heat transfer to work (heat engines) and some of the cycles used for converting work input to heat transfer (air conditioners, chillers, and heat pumps), and we discuss the hardware that can be used to implement these cycles. We present the limitation imposed on the ideal thermodynamic cycles by real hardware.

Attention is focused in this chapter on cycles that use gases as the working fluid. Cycles that involve phase changes of the working fluid are discussed in Chap. 7. The introductory material in this chapter, however, applies to all such cycles regardless of the type of working fluid.

6.1 Introduction to Heat Engine Cycles

Devices for converting heat transfer to work are called *heat engines*. One standard measure of performance for heat engines is the *efficiency* η, defined as the ratio of the net work output per cycle to the heat transfer *into* the engine per cycle, or $\eta = |w_{\text{net}}|/q_H$. In Chap. 5, the Carnot cycle was used to demonstrate the limitations that the second law places on the efficiency of all heat engine cycles which operate between the same heat addition and rejection temperatures. Although the Carnot cycle is useful in determining such ideal behavior, it is not a practical cycle to use in the design of a heat engine. It is not possible to design hardware that will allow a heat transfer to a working fluid at a constant temperature in a reversible process over a finite time. To do so would require that the temperature difference between the system and its surroundings be infinitesimally small; this has the unfortunate side effect of making the energy transfer require an extremely long time. To operate a Carnot engine at a reasonable power level, many cycles per unit time must be carried out; and if this is done, the engine operation is very far from the reversible ideal. The same difficulty arises in the heat rejection process, which for the true Carnot cycle must also be isothermal and reversible.

Because of these difficulties, other cycles have been developed which have *theoretical* efficiencies the same as or lower than that of the Carnot cycle. These cycles usually operate with nonisothermal heat transfers but have higher *actual* thermodynamic efficiencies when implemented in practical engines than are attainable from engines operating with near-isothermal heat transfer.

There are other reasons for developing heat engines that do not operate on the Carnot cycle. These include the characteristics of the energy source available to drive the heat engine, the characteristics of the working fluid chosen for the cycle, material limitations in the hardware, requirements for steady or varying work output, and other practical considerations. For these reasons and because of the difficulty of constructing an efficient heat engine that actually operates on an approximation to the Carnot cycle, a number of other heat engine cycles have been proposed and engines built that operate on these cycles.

In analyzing cycles, the states at the beginning and end of each process making up the cycle must be known, as must the path of the process connecting the states. As we have seen, definition of the states requires that two independent properties be known if the working fluid is a pure substance; then all other properties can be determined at that state. Thus, if the state at a cycle point is an unsaturated fluid, then *any* two properties will define the state of the system. However, if the fluid is saturated, then two properties are again required. But at most one of these can be an intensive property from the set of P and T, since these intensive properties are not independent for a saturated system. In other words, for a saturated system, either one extensive or specific intensive and one intensive property from the set P and T must be specified, or two extensive (or specific intensive) properties must be specified to define the state. Cycle analysis then depends on applying our knowledge of some of the end states along with the first and second laws to the processes connecting the end states; next the state of the system throughout the cycle can be calculated. The work done and heat transferred during the cycle can be found, and the cycle efficiency determined from these values. This approach is illustrated for various cycles.

Heat engine cycles can be categorized in a number of ways. A convenient division is into cycles that use a single-phase working fluid and those, such as the steam engine, in which the working fluid undergoes a change of phase during the cycle. These two types of engines are discussed in this chapter and Chap. 7, respectively.

We will see also that some practical heat engines partially convert heat transfer to the engine from an external source (the burning of fossil fuel in a boiler, nuclear energy, solar energy, geothermal energy, etc.) into useful work; other heat engines, usually classified as *internal combustion engines*, use the ignition of a combustible fuel within the engine itself as the source of energy which is partially converted to work. Although this division between external energy source and internal combustion engines is a convenient practical subdivision of heat engines, we will see that from the standpoint of an introductory treatment by thermodynamics, both the external energy source and internal combustion engines can be treated by the same methods.

6.1.1 Cycle Analysis Methodology

For each cycle to be analyzed for efficiency, a certain minimum amount of information must be given, so that the state at each point in the cycle can be determined. Then the efficiency and other information about the cycle behavior can be found. To carefully lay out the given information, it is useful to tabulate the properties at convenient state points in the cycle. It then becomes obvious which points have the necessary minimum of two properties known and which points require further analysis to determine the state. Unknown properties at a cycle point usually may be found from the first or second law relationships that describe the process connecting an unknown state point to an adjacent point with known properties. Steady state operation is always assumed for cycle analysis.

Example Problem 6.1

A system operates on the cycle shown with air as the working fluid. Assume that air acts as an ideal gas with temperature-independent specific heats. The processes making up the cycle are reversible and are noted on the figure. Find the cycle efficiency.

Cycle Diagram

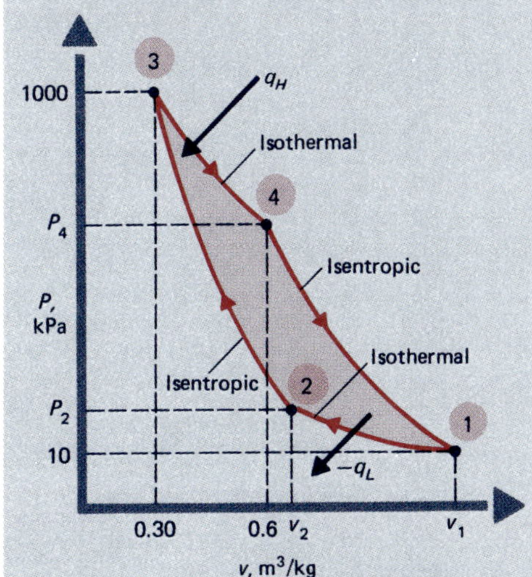

Solution

First, a general table is made of the known properties at points 1 through 4:

State	P, kPa	T, K	v, m³/kg	u, kJ/kg	s, kJ/(kg·K)
1	10	—	—	—	s_1
2	—	$T_2 = T_1$	—	—	s_2
3	1000	—	0.30	—	$(s_3 = s_2)$
4	—	$T_4 = T_3$	0.6	—	$(s_4 = s_1)$

These are all the direct entries that can be made in the table. Only state 3 is completely defined, and all properties in row 3 are fixed (since two properties are known). Since air at moderate conditions is the working fluid, the ideal gas law may be used to find T_3 immediately. So

$$T_3 = \frac{(Pv)_3}{R} = \frac{1000 \text{ kPa} \times 0.30 \text{ m}^3/\text{kg}}{0.2870} = 1045 \text{ K}$$

Because the process between states 3 and 4 is isothermal, $T_4 = T_3$, and these entries may be made in the table. Two properties, T_4 and v_4, are now known at state point 4, so all properties at state 4 can be found. From the ideal gas relation, for example,

$$P_4 = \frac{RT_4}{v_4} = 500 \text{ kPa}$$

The process connecting points 4 and 1 is adiabatic and reversible. Process 4-1 is therefore isentropic, and $s_1 = s_4$. For an ideal gas with temperature-independent properties,

$$s_1 - s_4 = c_P \ln \frac{T_1}{T_4} - R \ln \frac{P_1}{P_4} = 0$$

or (see Table 5.4)

$$T_1 = T_4 \left(\frac{P_1}{P_4}\right)^{(k-1)/k} = 1045 \left(\frac{10}{500}\right)^{(1.4-1)/1.4} = 342 \text{ K}$$

Now T_1 and P_1 are known, and all other properties can be found for state 1.

Because the cycle we are examining is exactly the Carnot cycle, the cycle efficiency can now be found as

$$\eta_c = 1 - \frac{T_1}{T_4} = 1 - \frac{342}{1045} = 0.673$$

It is left to the reader to determine P, v, and T at point 2.

Comments

A table of properties and knowledge of the processes that connect the states in the cycle have allowed us to find information about the cycle behavior in this case. The same method is used to follow through the behavior of other cycles.

6.1.2 Mean Effective Pressure

One way of comparing cycles and of gaining insight into strategies for improving performance of a particular cycle is through the concept of *mean effective pressure* P_{ME}. This is the constant pressure which, if exerted

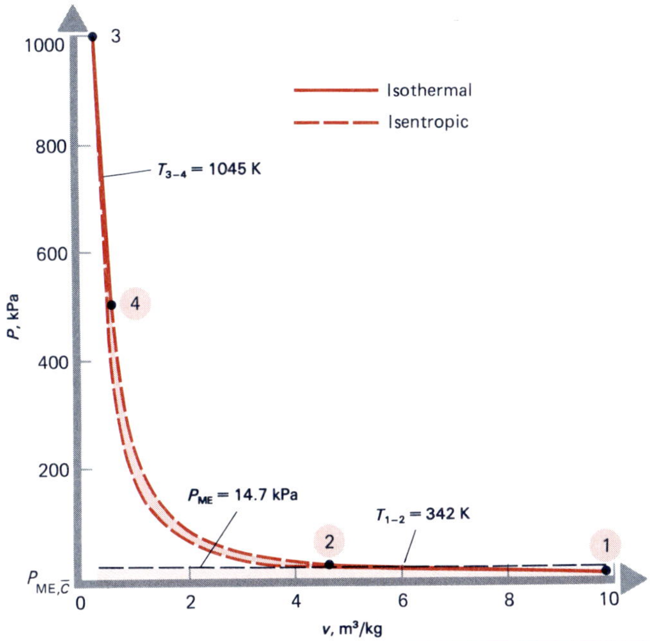

Figure 6.1 Mean effective pressure of Carnot cycle compared with actual *P-v* diagram.

through the work-producing expansion stroke of an engine between the largest specific volume v_f and the smallest volume v_i gives the same *net* specific work for the entire cycle as produced by the actual engine. Mathematically,

$$|w_{net}| = P_{ME}(v_f - v_i) = P_{ME}v_i\left(\frac{v_f}{v_i} - 1\right) = P_{ME}v_i(r_v - 1) \tag{6.1}$$

where $r_v = v_f/v_i$ is the volumetric expansion ratio for the engine. The net work output from a heat engine is thus proportional to the mean effective pressure, the specific volume at the beginning of the expansion process, and the volumetric expansion ratio. The higher the mean effective pressure of a given cycle, the larger will be its work output.

Thus, on a *P-v* diagram for a cycle, say for a Carnot cycle, we can superimpose the work done by the constant-pressure engine operating at P_{ME} during all work-producing expansion strokes as shown in Fig. 6.1. Note that the areas under the cycle curves for the real cycle (in this case, the Carnot cycle) and the constant-pressure expansion at P_{ME} must be the same.

In general, Eq. (6.1) can be written to find P_{ME} as

$$P_{ME} = \frac{|w_{net}|}{v_i(r_v - 1)} = \frac{q_{in}\eta}{v_i(r_v - 1)} \tag{6.2}$$

For any gas cycle, the mean effective pressure is increased by decreasing the volumetric expansion ratio r_v or by increasing the specific heat transfer q_{in} into the engine per cycle.

Example Problem 6.2

Find the mean effective pressure for the Carnot engine analyzed in Example Problem 6.1.

Solution

The state table is shown below:

State	P, kPa	T, K	v, m³/kg	u, kJ/kg	s, kJ/(kg·K)
1	10	342	—	—	s_1
2	—	342	—	—	s_2
3	1000	1045	0.30	—	$s_3 = s_2$
4	500	1045	0.60	—	$s_4 = s_1$

From the ideal gas law,

$$v_1 = v_4 \frac{P_4 T_1}{T_4 P_1} = 0.60 \times \frac{500 \times 342}{1045 \times 10} = 9.82 \text{ m}^3/\text{kg}$$

The cycle work output is given in terms of its relation to the mean effective pressure P_{ME} and the cycle efficiency η_C as

$$\begin{aligned} |w_{\text{net}}| &= q_H \eta_C = q_H \left(1 - \frac{T_1}{T_3}\right) \\ &= T_3(s_4 - s_3)\left(1 - \frac{T_1}{T_3}\right) \\ &= P_{\text{ME}}(v_1 - v_3) \end{aligned}$$

or, because $s_4 - s_3 = c_P \ln (T_4/T_3) - R \ln (P_4/P_3)$ and $T_4 = T_3$,

$$|w_{\text{net}}| = P_{\text{ME}}(v_1 - v_3) = -R(T_3 - T_1) \ln \frac{P_4}{P_3}$$

Substituting the ideal gas law to eliminate T results in

$$P_{\text{ME}} = \frac{|w_{\text{net}}|}{v_1 - v_3} = -\frac{P_3 v_3 - P_1 v_1}{v_1 - v_3} \ln \frac{P_4}{P_3}$$

Substituting the pressure and volume values for this problem from the state table gives

$$P_{ME} = -\frac{1000 \times 0.3 - 10 \times 9.82}{9.82 - 0.3} \ln \frac{500}{1000} = 14.7 \text{ kPa}$$

The *P-v* diagrams for the Carnot cycle and the equivalent cycle using P_{ME} are shown in Fig. 6.1.

6.1.3 Efficiency and Practicality

In Example Problem 6.2, note how low the value of $P_{ME} = 14.7$ kPa is in comparison with the actual pressure range of the problem. This can be seen easily from the *P-v* diagram. This is an indication that the Carnot cycle may not be a practical one, because it requires extremely large pressure changes over the cycle in order to produce the same net work output as the process working at P_{ME} over the same volume displacement. The work produced over the constant-temperature heat transfer process and part of the isentropic expansion occurs at high pressure in the ideal air-standard Carnot cycle. Careful mechanical design will be required that is sufficient to handle these high pressures. Also a large fraction of the work produced during the expansion processes is used during the low-temperature heat transfer and compression processes in order to return to state 1.

Although the ideal air-standard Carnot cycle is thermodynamically *efficient*, it has a relatively low work output per cycle $|W_{net}| = P_{ME}(V_f - V_i)$, as indicated by the low value of P_{ME}. The implication is that a large engine (i.e., a large displacement) will be required to obtain a large work output. If another cycle can gain the same $|W_{net}|$ through a combination of smaller displacement but larger P_{ME}, even though this new cycle has a lower efficiency than the Carnot cycle, it might be chosen. Such a cycle might be embodied in an actual engine that is more compact or less expensive than the Carnot engine. Thus, efficiency is not the only measure used in the choice of a heat engine.

Another way to look at engine output is to determine the engine *power*. If the engine operates at a given number of thermodynamic cycles per second (CPS), then the power output is

$$\dot{W} = |W_{net}| \times \text{CPS} = P_{ME}(V_f - V_i) \times \text{CPS} \tag{6.3}$$

Thus, a large power output can be obtained by choosing a cycle with a large P_{ME}, or a large displacement $V_f - V_i$, or by operating the engine at high speed (i.e., a large number of cycles per second). Note that thermodynamic CPS is not necessarily the same as engine revolutions per second, as we will see for the Otto and diesel cycles in particular.

In conclusion, thermodynamic efficiency is only one measure of performance to be considered in the choice of a particular cycle or in decisions about whether to add features to a cycle that might add efficiency at the expense of other desirable attributes. The engineer must be concerned with not only efficiency, but also cost, size, reliability, ease of maintenance, safety, type of fuel available, and characteristics of the application.

6.2 Gas Cycle Analysis

The arc of the balance wheel
flows like a curved rush of swallows, come over a hill . . .
Things lost come again in sudden new beauty
Look long on an engine. It is sweet to the eyes.
MacKnight Black

In this section, we apply tools from Chaps. 4 and 5, that is, the first and second law expressions for processes involving ideal gases, to the analysis of cycles that use ideal gases as the working fluid. A number of practical heat engines do indeed use gases as the working fluid at conditions where they can be treated as ideal, and other cycles can be treated as if they used such fluids in order to gain insight into their behavior.

6.2.1 Air-Standard Cycles

It is useful to study the efficiency of certain cycles by making simplifying assumptions about the behavior of the processes that make up the cycle. This approach allows us to separate the general behavior of the cycle from the properties of the particular working fluid in the cycle and from the irreversibilities caused by particular hardware. In addition, idealized heat transfer processes can be substituted for the combustion processes that occur in internal combustion engines. Thus, internal combustion engines can be treated as if they were heat engines driven by heat transfer from an external energy source. The general behavior of individual cycles and the comparison between cycles can then be demonstrated on a common basis. These idealized cycles, called *air-standard cycles,* are usually analyzed as in Example Problem 6.1, by assuming that the working fluid behaves as an ideal gas. In addition, the mechanical devices and heat transfer processes used in the cycle are often taken to be reversible. These restrictions can be relaxed if an actual device is to be modeled more carefully.

When we make these assumptions, the cycle efficiencies based on these analyses are greater than can be realized in practice. The particular causes of deviation from the air-standard efficiencies are pointed out for each cycle.

Note that it is not difficult to analyze the air-standard cycles by allowing for the temperature dependence of the properties of the working fluid. And this should always be done in analyzing a real cycle. The gas tables (App. D or E) can be used. The constant-specific-heat assumption is invoked in the analysis of air-standard cycles simply to allow comparison of different cycles.

6.2.2 Effect of the Assumption of Temperature-Independent Specific Heats

By assuming that c_P and c_v do not depend on temperature, the form of the equations derived for cycle efficiency and other measures of comparison between cycles become quite simple. Easy comparison among cycles is then possible. However, this assumption can lead to some wrong conclusions when the working fluid being used has strongly temperature-dependent properties.

When the monatomic gases (helium, argon, neon, etc.) are used as working fluids, the specific heats are indeed independent of temperature (you can see this for argon by looking at Tables D.3 or E.3, for example). This is true until the temperature is so large that the monatomic gases begin to ionize. Such temperatures are usually outside the practical limits set by the materials used in engines. Thus, for monatomic gases, the assumption of constant c_P and c_v is quite good.

For more complex gases and their mixtures (diatomic gases such as hydrogen, nitrogen, oxygen, etc., or the polyatomic gases such as carbon dioxide, ammonia, etc.), the temperature dependence of c_P and c_v becomes more pronounced. The ratio of specific heats $k = c_P/c_v$ tends to be less temperature-dependent than either c_P or c_v independently, because both c_P and c_v generally follow the same general trend with temperature. Thus, for some cycles where the ratio of specific heats appears, it may be useful to treat k as constant (i.e., independent of temperature), even though c_P and c_v exhibit fairly strong temperature dependence. However, *care must be taken even in this case, because the values of c_P and c_v usually appear from analysis of different processes within the cycle; they may thus come from processes that occur over quite different ranges of temperature.* The treatment of this case is outlined in some of the examples that follow.

Where temperature-dependent properties are present, it is always possible to do a detailed analysis without resorting to the simplifying assumption of temperature-independent properties, and this should be done.

Example Problem 6.3

A system operates on the cycle shown with air as the working fluid. The processes making up the cycle are reversible and are noted on the figure.

Find the cycle efficiency, using temperature-dependent properties for the air.

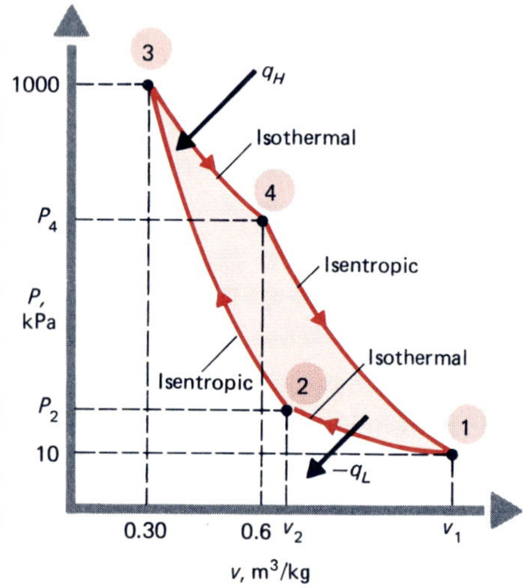

Solution

Following Example Problem 6.1, the temperature at state 1 and the pressure at state 2 can be found by application of the ideal gas law. The state table is thus

State	P, kPa	T, K	v, m³/kg	u, kJ/kg	s, kJ/(kg·K)
1	10	—	—	—	$s_1 = s_4$
2	—	$T_2 = T_1$	—	—	s_2
3	1000	(1045)	0.30	—	$(s_3 = s_2)$
4	(500)	1045	0.60	—	$(s_4 = 7.5646)$

We can use the air tables or the computer program to find s_4. Then since $s_1 = s_4$ and P_1 is known, we can use either the air tables or the computerized tables to find T_1. Using the computer tables gives $T_1(s_1 = s_4 = 7.5646$ kJ/(kg·K), $P_1 = 10$ kPa$) = 90°\text{C} = 363$ K. Substituting into the relation for Carnot efficiency gives

$$\eta_C = 1 - \frac{363}{1045} = 0.653$$

Comments

The constant-c_p solution of Example Problem 6.1 for the ideal air-standard Carnot cycle gave an efficiency of 0.673, a difference of 3 percent from the more accurate solution gained from using temperature-dependent properties. Note that the more accurate solution predicts a lower efficiency than was found for the idealized solution. Also note that the reason that the efficiency of the Carnot cycle in this problem differs numerically from that found in Example Problem 6.1 is that the heat transfer temperature T_1 is found to be higher in this case. If temperatures T_3 and T_1 were *set* by the problem statement, then consideration of temperature-dependent specific heats would not affect the computed cycle efficiency, because the efficiency of any reversible heat engine operating between the same two isothermal heat transfer reservoirs is the same!

6.2.3 The Air-Standard Carnot Cycle

The Carnot cycle is analyzed for a particular case in Example Problems 6.1 and 6.2 and in Chap. 5. However, a few remarks may deepen our understanding of the analysis of other cycles.

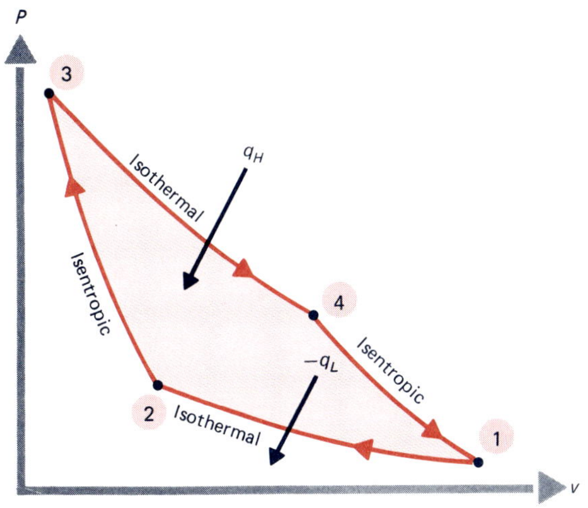

Figure 6.2 Pressure-volume diagram for Carnot cycle.

The *P-v* diagram for the ideal air-standard Carnot cycle is illustrated in Fig. 6.2. Let us examine the slopes of the lines on this diagram for consistency. For the isothermal processes, an ideal gas follows the relation

$$P = \frac{RT}{v} = \frac{C_1}{v} \tag{6.4}$$

Or, by taking the derivative,

$$\left(\frac{\partial P}{\partial v}\right)_T = \frac{-C_1}{v^2} \tag{6.5}$$

However, the constant $C_1 = RT$ is different for each isotherm but always positive. So the slope of the *P-v* curve will always be negative, and the slope will be greater for the higher-temperature isotherm at any given specific volume. For the isentropic processes, $Pv^k = C_2$, so the slopes of these curves are

$$\left(\frac{\partial P}{\partial v}\right)_s = \frac{-kC_2}{v^{1+k}} \tag{6.6}$$

which are again always negative, with the slope decreasing more rapidly with v (since $k > 1$ for all ideal gases) than for the isotherms. At point 4, the slope of the isentropic curve is greater than the slope of the isotherm by the ratio

$$\frac{(\partial P/\partial v)_s}{(\partial P/\partial v)_T} = \frac{kC_2/v^{1+k}}{C_1/v^2} \tag{6.7}$$

Evaluating the ratio at P_4 and v_4 and noting that $C_1 = RT_4 = P_4v_4$ and $C_2 = (Pv^k)_4$, we find

$$\frac{(\partial P/\partial v)_s}{(\partial P/\partial v)_T} = k \tag{6.8}$$

or the slope of the isentropic line at point 4 is greater than the slope of the isotherm at that point by a factor k. Similar analysis at other points leads to the shape of the Carnot cycle diagram shown in Fig. 6.2.

It is often useful to plot cycles in other coordinates than P and v. For example, the T-s diagram for the Carnot cycle plots as a rectangle (Fig. 6.3). This makes it very easy to determine the efficiency of the cycle, since the ratio q_L/q_H is available from the relations $q_H = T_3(s_4 - s_3)$ and $q_L = T_2(s_2 - s_1)$. This gives the cycle efficiency directly [Eq. (5.210)] as

$$\eta_C = 1 - \frac{|q_L|}{|q_H|} = 1 - \frac{T_L}{T_H} \tag{6.9}$$

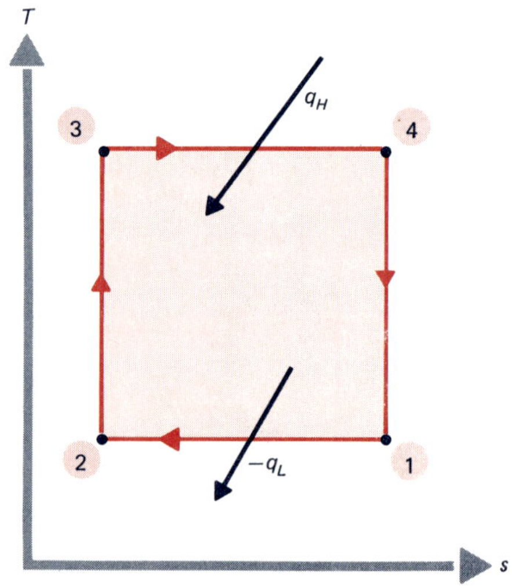

Figure 6.3 Temperature-entropy diagram for Carnot cycle.

Another useful plot is the P-h diagram, which allows the inlet and exit conditions to such equipment as compressors, turbines, and throttling valves to be displayed, aiding in the design process. The P-h diagram for the Carnot cycle is shown in Fig. 6.4. The isotherms are vertical because the enthalpy of an ideal gas is a function of temperature only. In general, the isentrope shape can be inferred from the definition

$$h = u + Pv \tag{6.10}$$

Taking the total derivative of $h(s, P)$ gives

$$dh = \left(\frac{\partial h}{\partial s}\right)_P ds + \left(\frac{\partial h}{\partial P}\right)_s dP \tag{6.11}$$

Equation (5.44) shows that

$$dh = T\,ds + v\,dP \tag{6.12}$$

For an isentropic process, comparison of Eqs. (6.11) and (6.12) shows that

$$\left(\frac{\partial h}{\partial P}\right)_s = v \qquad \text{or} \qquad \left(\frac{\partial P}{\partial h}\right)_s = \frac{1}{v} \tag{6.13}$$

Thus, the slope of the isentropes is always positive, and the slope is greater at higher pressure. As the system moves along the isentropic compression line from 2 to 3 on the P-h diagram, v decreases and the slope $(\partial P/\partial h)_s$ increases.

The h-s, or Mollier, diagram is also convenient for the description of cycles, particularly those using a vapor that undergoes a phase change for the working fluid. The change in enthalpy across an ideal compressor, turbine, or pump appears as a vertical line, while energy addition to heating or cooling devices is conveniently found as a change in vertical distance. The h-s diagram is explained in Sec. 7.2 on the Rankine engine.

Figure 6.4 Pressure-enthalpy diagram for Carnot cycle.

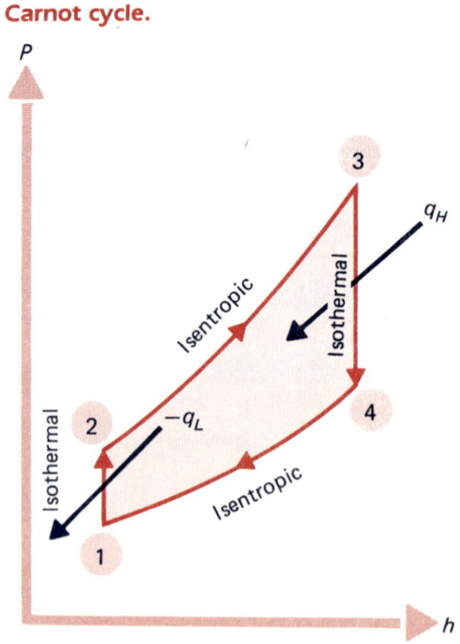

For air-standard cycles, the h-s diagram is similar to the T-s diagram because of the assumption of an ideal gas (often with temperature-independent specific heat) as the working fluid, which makes h directly proportional to T and independent of P.

Similar analyses can be performed on other cycles to predict the general shape of the diagrams in the various coordinates.

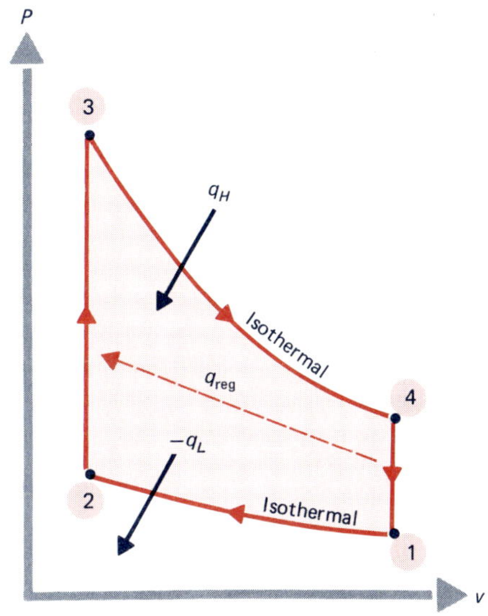

Figure 6.5 Pressure-volume diagram for Stirling cycle.

6.3 Gas Cycle Heat Engines

6.3.1 Stirling Cycle

This is the first of the practical heat engine cycles that we study. The Stirling cycle uses a compressible fluid as the working fluid, and the cycle is made up of a constant-temperature energy addition process followed by a constant-volume process, a constant-temperature energy rejection, and finally a return to the initial state by another constant-volume process, as shown in Fig. 6.5.

To accomplish this cycle requires some ingenious hardware. The cycle of operation is shown in Fig. 6.6. Two pistons A and B enclose the working gas, which is forced back and forth through a *regenerator* C. The regenerator is typically a chamber containing wire or ceramic mesh or fine metallic gauze and is used for the temporary storage of energy. Because the regenerator is within the system boundary, the energy stored and released by the regenerator during the cycle is treated not as a heat transfer, but simply as a change in the internal energy. Because the internal energy changes of the regenerator and the working fluid are always equal and opposite in sign, the changes do not affect the first law as written for the Stirling engine (i.e., the working fluid as the control mass) as a whole. The idea of regeneration is common to most external heat transfer engines.

Figure 6.6 Operation of Stirling cycle.

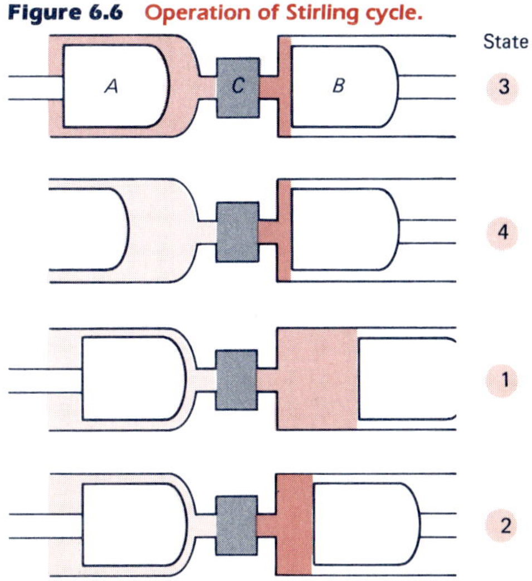

To carry through the cycle, there is a heat transfer to the working fluid initially at high pressure at state 3, in the cylinder containing piston A while piston A is moved to the left at a rate just sufficient to hold the temperature of the gas constant. This process of constant-temperature energy addition is continued until state 4 is reached (Fig. 6.5). When state 4 is reached, both pistons are moved to the right at the same rate. Thus the gas volume is held constant, and gas is forced through the gauze, which is at a lower temperature because of the previous step in the cycle. The gas is thus cooled by transferring some of its energy to the regenerator. When state 1 is reached, the cool gas is in the cylinder containing piston B. Piston B is then moved to the left while piston A is held in position. At the same time, there is a heat transfer from the gas at such a rate as to hold the gas temperature constant during compression. When state 2 is reached (at the same volume as state 3), both pistons are moved to the left at the same rate, again holding the gas volume constant while forcing the gas through the regenerator. The cold gas regains the energy it had previously given to the regenerator, thus cooling the regenerator material. The regenerator is then ready to receive energy in the next cycle, and the system is thus returned to state 3.

One interesting attribute of the Stirling engine is that the working fluid is a closed system, in that it is always contained within the engine, so no contamination need enter the lubricant. In addition, the working fluid can be chosen for its thermal properties since the inventory of working fluid is small enough that cost is not important. Furthermore, in the ideal cycle just described, since all heat transfer to the cycle takes place at constant temperature and all heat loss is also at constant temperature, the ideal Stirling cycle efficiency must equal the Carnot efficiency! This presupposes that all the processes in the cycle can be carried out reversibly, which, we noted in the beginning of the chapter, is not possible in real systems. It is, of course, possible to include compression and expansion efficiencies in the cycle if they are known.

To carry out the interrelated motion of the pistons or other devices needed to provide the processes making up the Stirling cycle, many ingenious methods have been tried. Stirling's original patent drawing is shown in Fig. 6.7, and present Stirling engine designs are no less complex.

Figures 6.8 and 6.9 show the *T-s* and *P-h* diagrams for the air-standard Stirling cycle.

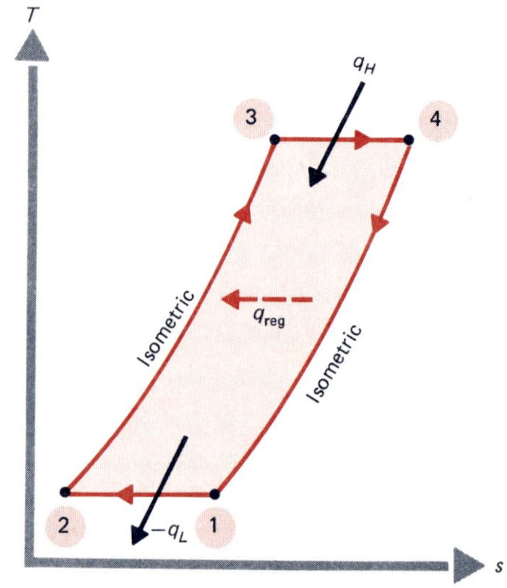

Figure 6.8 *T-s* diagram for Stirling cycle.

Figure 6.7 Original patent drawing of Stirling engine.

Figure 6.9 *P-h* diagram for Stirling cycle.

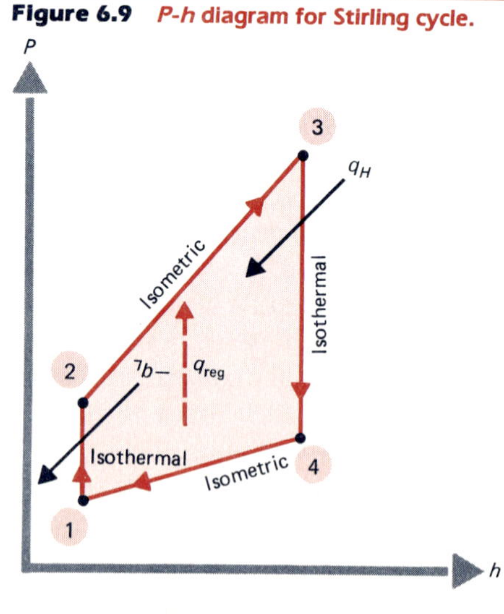

Example Problem 6.4

Determine general relations for the cycle efficiency and mean effective pressure for the ideal air-standard Stirling cycle by using cycle analysis.

Solution

The cycle efficiency can be determined by computing the cycle work (Fig. 6.5) as

$$w = \oint P\,dv = {}_1w_2 + {}_3w_4$$

$$= \int_{v_1}^{v_2} P\,dv + \int_{v_3}^{v_4} P\,dv$$

$$= RT_1 \ln \frac{v_2}{v_1} + RT_3 \ln \frac{v_4}{v_3}$$

However, $v_2 = v_3$ and $v_1 = v_4$, so

$$w = R(T_3 - T_1) \ln \frac{v_4}{v_3}$$

The heat transfer to the cycle is

$$q_H = T_3(s_4 - s_3) = RT_3 \ln \frac{v_4}{v_3}$$

and the cycle efficiency is then

$$\eta_s = \frac{|w|}{|q_H|} = 1 - \frac{T_1}{T_3}$$

which is exactly the Carnot efficiency, as expected.

To determine the mean effective pressure of the ideal air-standard Stirling cycle $P_{ME,\,S}$, observe that

$$P_{ME,\,S}(v_4 - v_3) = |w_{net}| = \eta_S q_H$$

$$= q_H\left(1 - \frac{T_1}{T_3}\right) = T_3(s_4 - s_3)\left(1 - \frac{T_1}{T_3}\right)$$

Because the energy addition process takes place at constant temperature, the entropy change is [from Eq. (5.56)]

$$s_4 - s_3 = R \ln \frac{v_4}{v_3}$$

so that

$$|w_{net}| = P_{ME,\,S}(v_4 - v_3) = R(T_3 - T_1) \ln \frac{v_4}{v_3}$$

Noting that $T_2 = T_1$ and using the ideal gas law, we have

$$|w_{net}| = P_{ME,\,S}(v_4 - v_3) = (P_3v_3 - P_2v_2)\ln\frac{v_4}{v_3}$$

Because $v_3 = v_2$ and by defining the expansion (or, in this case, also the compression ratio) as $r_v = v_4/v_3$, we can express this as

$$|w_{net}| = P_{ME,\,S}(v_4 - v_3) = (P_3 - P_2)v_3 \ln r_v$$

and the expression for mean effective pressure can be written as

$$P_{ME,\,S} = \frac{P_3 - P_2}{r_v - 1}\ln r_v$$

Comments

The greater the pressure change in process 2-3 due to heat transfer in the regenerator during the constant-volume process, the greater will be the value of $|w_{net}|$. Further, the larger the compression ratio and the larger the specific volume after compression v_3, the larger will be $|w_{net}|$.

For the ideal air-standard Stirling cycle, we can summarize the characteristics (based on the notation in Fig. 6.4) as

$$\eta_S = 1 - \frac{T_1}{T_3} \tag{6.14}$$

$$P_{ME,\,S} = \left[\frac{P_3 - P_2}{r_v - 1}\right]\ln r_v \tag{6.15}$$

$$|w_{net}| = P_{ME,\,S}(v_4 - v_3) = (P_3 - P_2)v_3 \ln r_v \tag{6.16}$$

6.3.2 Ericsson Cycle

The Ericsson cycle was originally developed, as was the Stirling cycle, in an effort to find a practical external heat transfer engine that could substitute air for steam as the working fluid. The cycle is composed of two constant-pressure processes connected by constant-temperature energy addition and rejection processes (Fig. 6.10). Again, in the ideal reversible (air-standard) cycle, the efficiency must approach that of the Carnot cycle, since the Ericsson cycle operates between two constant temperatures at which energy is added and rejected. The *T-s* and *P-h* diagrams for the Ericsson cycle are shown in Figs. 6.11 and 6.12.

Because of the low mean effective pressure in comparison with steam engines, the Ericsson cycle was originally used in very large marine piston engines—so large that they were impractical in competition with steam engines of similar power. However, it is possible to envision a

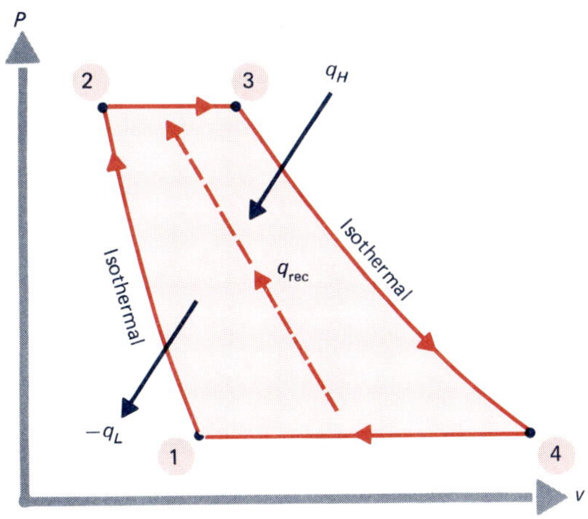

Figure 6.10 *P-v* diagram for Ericsson cycle.

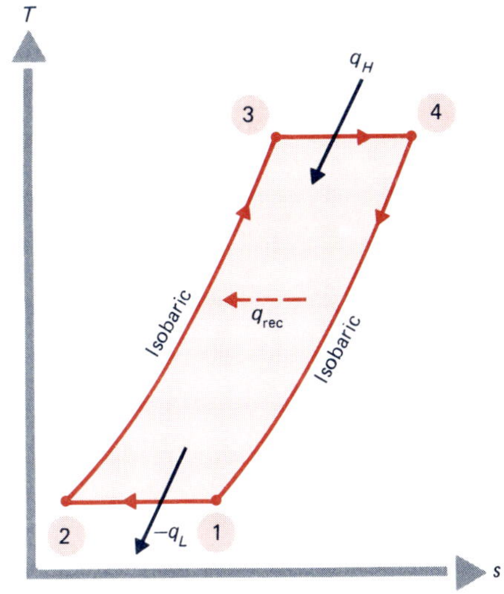

Figure 6.11 *T-s* diagram for Ericsson cycle.

Figure 6.12 *P-h* diagram for Ericsson cycle.

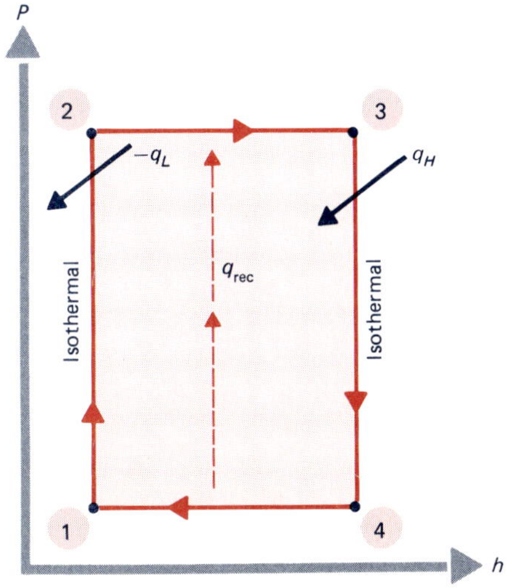

system using a turbine-compressor system employing the Ericsson cycle, as shown in Fig. 6.13. Here the working fluid enters the compressor at state 1. The compressor is designed so that the work done at each point is balanced by the heat transfer at each point. Thus, the working fluid is maintained at constant temperature and leaves the compressor at state 2. The fluid then enters the recuperator (heat exchanger), and energy is added to the fluid from the hotter fluid that has left the turbine. The fluid leaves the recuperator at state 3 and enters the turbine. The turbine is designed so that as the working fluid does work in the turbine, energy is added at exactly the same rate at which work is done at each point in the turbine. Thus the working fluid is kept at constant temperature. The fluid leaves the turbine at state 4 and enters the recuperator, where it transfers energy to the working fluid en route to the turbine. This energy exchange process is assumed to take place at constant pressure. The fluid exits the recuperator at state 1 and enters the compressor, thus completing the cycle.

In practice, it is difficult to design turbines and compressors that have the internal cooling profiles necessary to achieve isothermal conditions. In addition, of course, real components are not reversible; and, for example, fluid friction in the recuperator makes the pressure vary through the regeneration process. Also the recuperator cannot be designed in practice to bring the outlet temperature of the cold fluid up to the inlet temperature of the hot fluid, as the cycle diagram implies. This would require an infinite heat transfer surface area in the recuperator or zero resistance to heat transfer across the boundary between the fluids. Neither condition can be attained in practice. The actual efficiency will be below the Carnot efficiency for these reasons and others. However, it is possible to build components that approximate the Ericsson processes, as we will see.

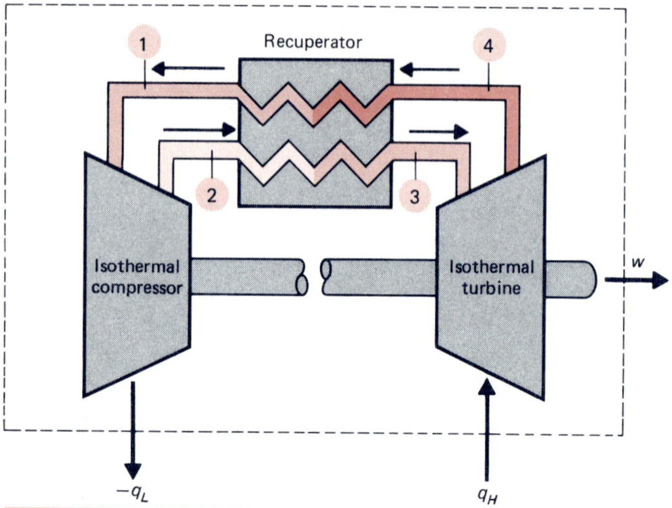

Figure 6.13 Turbocompressor version of Ericsson cycle.

Ericsson Cycle Efficiency and Mean Effective Pressure

The ideal air-standard Ericsson cycle with recuperation is a reversible cycle with heat transfer processes that occur at constant temperature, and so the air-standard cycle efficiency is equal to the Carnot efficiency $(1 - T_1/T_3)$.

The mean effective pressure for the Ericsson cycle $P_{\mathrm{ME},\,E}$ can be found from

$$|w_{\mathrm{net}}| = P_{\mathrm{ME},\,E}(v_4 - v_2) = \eta_E q_H = \left(1 - \frac{T_2}{T_4}\right) T_4(s_4 - s_3)$$

$$= -(T_4 - T_2)R \ln \frac{P_4}{P_3} = R(T_3 - T_1) \ln r_P \tag{6.17}$$

where $r_P = P_3/P_4$ and we have used $T_1 = T_2$ and $T_3 = T_4$. Then in summary for the Ericsson cycle, based on the notation of Fig. 6.12,

$$\eta_E = 1 - \frac{T_2}{T_4} \tag{6.18}$$

and, using some further substitutions,

$$P_{\mathrm{ME},\,E} = \frac{P_3[(r_v/r_P) - 1]}{r_v - 1} \ln r_P \tag{6.19}$$

and

$$|w_{\mathrm{net}}| = RT_3[(r_v/r_P) - 1] \ln r_P = RT_3\left(\frac{T_3}{T_4} - 1\right) \tag{6.20}$$

where $r_v = v_4/v_2$, the overall volumetric expansion ratio.

The greatest net work output per cycle is obtained with the smallest r_P, the pressure ratio across the turbine (and therefore compressor), the greatest volumetric expansion (or compression) ratio r_v, and when operating at the highest possible turbine inlet temperature T_3.

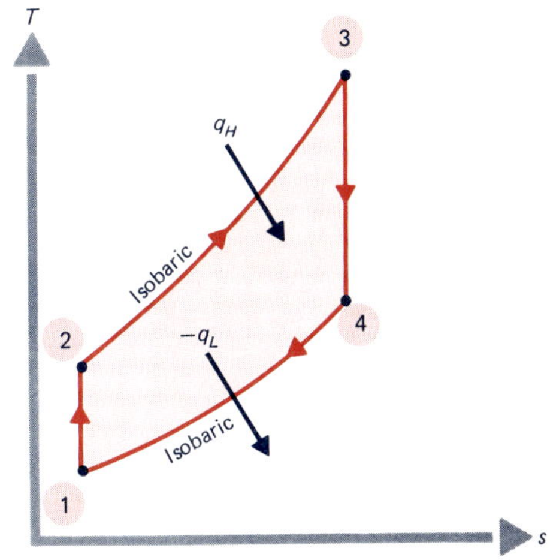

Figure 6.15 *T-s* diagram for closed Brayton cycle.

6.3.3 Brayton Cycle (External Heat Transfer)

The Brayton cycle, also called the *Joule cycle,* was developed originally for use in a piston engine with fuel injection, but its most common embodiment at present is in open- and closed-cycle turbine engines. The open-cycle engine can be used with either internal combustion or an external heat transfer, while the closed-cycle engine has an external energy source. First, we examine the closed-cycle air-standard Brayton engine.

The air-standard simple Brayton cycle is composed of constant-pressure heat transfer processes separated by isentropic expansion and compression processes, as shown in Fig. 6.14. The corresponding *T-s* and *P-h* diagrams are shown in Figs. 6.15 and 6.16.

Note that the Brayton cycle is the first heat engine cycle that we have examined in which heat transfer does not take place in an isothermal process. We might therefore expect the effect of temperature-dependent properties to be important.

The turbine compressor system that uses such a cycle is shown in Fig. 6.17. The working fluid enters the compressor in state 1. After compression it enters the high-temperature heat exchanger in state 2, and there is an energy addition in a constant-pressure process until the working fluid attains the higher temperature of state 3. The fluid then enters the turbine, where an isentropic expansion occurs, producing power. The

Figure 6.14 *P-v* diagram for Brayton (Joule) cycle.

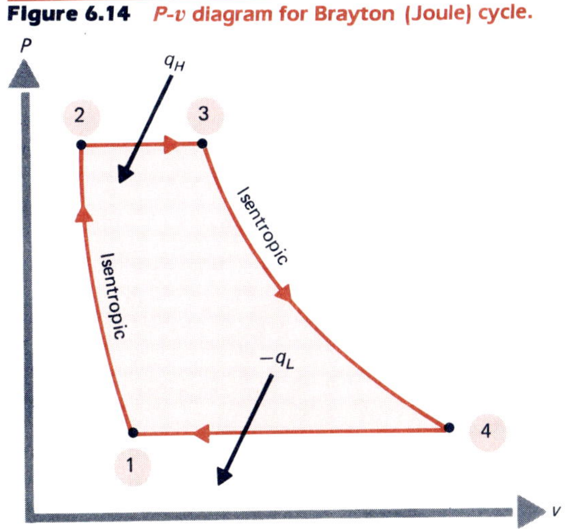

Figure 6.16 *P-h* diagram for Brayton cycle.

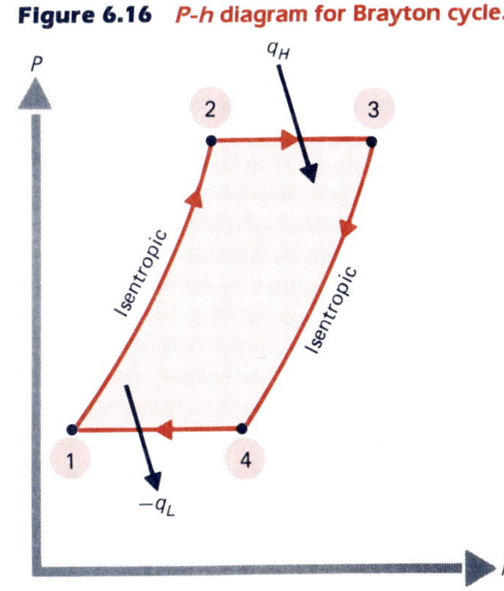

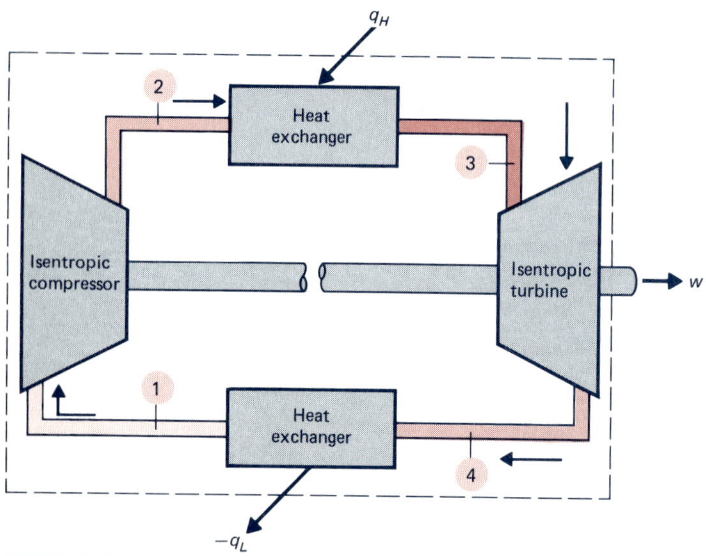

Figure 6.17 Diagram of simple Brayton cycle hardware.

fluid leaves the turbine at state 4, is further cooled at constant pressure in the low-temperature heat exchanger, and then enters the compressor at state 1. A considerable fraction of the turbine work is used to power the compressor.

The turbine and compressor are assumed to be isentropic in the ideal air-standard Brayton cycle, rather than isothermal as in the Ericsson cycle. A turbine can be built which is nearly adiabatic, although reversibility is much more difficult to approach. Thus the Brayton cycle is easier to implement in a practical device than is the Ericsson cycle. The Brayton cycle, however, provides heat transfer to the working fluid over a range of temperatures, as can be seen from the *T-s* diagram. The Brayton cycle does not have the capability in its simplest form, therefore, of approaching the Carnot efficiency when it operates between the same two reservoir temperatures, unlike the Ericsson cycle.

Example Problem 6.5

Determine the efficiency of an ideal air-standard closed-cycle Brayton engine in terms of (*a*) the compressor inlet and outlet temperatures, (*b*) the turbine inlet and outlet temperatures, and (*c*) the turbine (or compressor) inlet/outlet pressure ratio.

Solution

The cycle efficiency (Fig. 6.15) is given by

$$\eta_B = \frac{|w_{net}|}{|q_H|} = 1 - \frac{|q_L|}{|q_H|} = 1 - \frac{h_4 - h_1}{h_3 - h_2} = 1 - \frac{c_P(T_4 - T_1)}{c_P(T_3 - T_2)}$$

$$= 1 - \frac{T_1}{T_2}\frac{T_4/T_1 - 1}{T_3/T_2 - 1} \tag{6.21}$$

For simplicity, c_P has been assumed to be independent of temperature. For the isentropic turbine and compressor, Table 5.4 yields

$$\frac{P_3}{P_4} = \left(\frac{T_3}{T_4}\right)^{k/(k-1)}$$

$$\frac{P_2}{P_1} = \left(\frac{T_2}{T_1}\right)^{k/(k-1)}$$

However, $P_3/P_4 = P_2/P_1$, so it follows that

$$\frac{T_3}{T_4} = \frac{T_2}{T_1}$$

or

$$\frac{T_3}{T_2} = \frac{T_4}{T_1}$$

Substituting the latter two ratios into the efficiency equation yields

$$\eta_B = 1 - \frac{T_1}{T_2} = 1 - \frac{T_4}{T_3} \tag{6.22}$$

which provides the answers to (*a*) and (*b*). For (*c*), substitute for the temperature ratio in terms of the pressure ratio, to obtain

$$\eta_B = 1 - \frac{1}{(P_2/P_1)^{(k-1)/k}} = 1 - \frac{1}{r_P^{(k-1)/k}} \tag{6.23}$$

where r_P is the pressure ratio across the isentropic turbine or compressor P_2/P_1.

Comments

The efficiency of the closed-cycle Brayton engine is found to depend on the isentropic pressure ratio only. If we increase the inlet pressure to the turbine, the inlet temperature to the turbine is also increased. The inlet turbine temperature is usually limited by the material properties of the turbine blades, which places a practical upper limit on the cycle efficiency.

The closed-cycle (external heat addition) Brayton engine has received considerable attention for its use with nuclear and, more recently, high-temperature solar energy systems.

Note that if the ideal air-standard Brayton engine is operating at idle, i.e., there is no net cycle work output, then Eq. (6.21) implies that $T_4 - T_1 = T_3 - T_2$ (because the efficiency must be zero). Equations (6.22) and (6.23) also imply that $T_1 = T_2$, $T_3 = T_4$, and $P_1 = P_2$. Thus, if the engine is at idle, there is no work output from the turbine, and no work input is required for the compressor. The isentropic turbine and compressor simply spin while exchanging no work, and any energy added to the engine is rejected at the low-temperature heat exchanger.

Mean Effective Pressure and Work Ratio for Ideal Air-Standard Brayton Cycle

Using the general equation for P_{ME} [Eq. (6.2)], we can put the mean effective pressure for the ideal air-standard Brayton cycle in the form

$$\begin{aligned} P_{ME,\,B} &= \frac{|w_{net}|}{v_4 - v_2} = \frac{\eta_B q_H}{v_4 - v_2} \\ &= \frac{\eta_B c_P (T_3 - T_2)}{v_4 - v_2} \\ &= \eta_B \left(\frac{c_P}{R}\right) \frac{P_3(v_3/v_2 - 1)}{r_v - 1} \\ &= \eta_B \left(\frac{k}{k-1}\right) \frac{P_3(T_3/T_2 - 1)}{r_v - 1} \end{aligned} \tag{6.24}$$

Thus, higher inlet pressure and temperature to the turbine as well as a larger temperature ratio, resulting from greater heat transfer per cycle, will increase the mean effective pressure of the cycle. Because the compressor outlet and turbine inlet pressures are assumed equal, good design of the compressor will allow large pressure ratios and thus better $P_{ME,\,B}$. Generally, temperature limitations of the turbine blade materials set the upper limit on usable turbine inlet temperatures to about 1540°C (2800°F) in engines with cooled blades. Turbine inlet temperatures of up to 1700°C (3100°F) are projected for the near future in advanced turbine engines.

For the ideal air-standard Brayton cycle, a measure of the usefulness of the engine for producing net shaft work is the ratio of the compressor work input to the turbine work output, called the *work ratio*. For the simple ideal air-standard Brayton cycle with turbine and compressor efficiencies of 100 percent,

$$\text{Work ratio} = \frac{|w_c|}{|w_t|} = \frac{h_2 - h_1}{h_3 - h_4} \overset{\text{constant } c_P(T)}{=\!=\!=} \left(\frac{T_2}{T_3}\right) \tag{6.25}$$

The work ratio should be considerably less than 1 in a practical engine.

Effect of Temperature-Dependent Specific Heats

To analyze the effect of temperature-dependent properties, it is convenient to define two types of averaged specific heat. The first type, the *arithmetic mean specific heat*, is averaged over a finite temperature range and is defined for the constant-pressure and constant-volume specific heats by

$$c^*_{P,\, i-f} = \int_{T_i}^{T_f} \frac{c_P(T)\, dT}{T_f - T_i} \tag{6.26}$$

and

$$c^*_{v,\, i-f} = \int_{T_i}^{T_f} \frac{c_v(T)\, dT}{T_f - T_i} \tag{6.27}$$

These specific heats can be calculated for a particular temperature interval from the polynomial expressions for specific heats given in Tables D.1 or E.1. The enthalpy change and internal energy change for an ideal gas with temperature-dependent properties are now given by

$$h_f - h_i = \int_{T_i}^{T_f} c_P(T)\, dT = c^*_{P,\, i-f}(T_f - T_i) \tag{6.28}$$

$$u_f - u_i = \int_{T_i}^{T_f} c_v(T)\, dT = c^*_{v,\, i-f}(T_f - T_i) \tag{6.29}$$

Values of $c^*_{P,\, i-f}$ and $c^*_{v,\, i-f}$ can be plotted by using Eqs. (6.28) and (6.29) and finding the necessary values of $u(T)$ or $h(T)$ from the tables of properties for ideal gases. Table C.14 gives values of mean specific heat for air, nitrogen, carbon dioxide, and water vapor.

For isentropic processes, a second type of mean specific heat, the *log-mean* (LM) *specific heat*, may be defined as

$$c^*_{P,\, \mathrm{LM},\, i-f} = \int_{T_i}^{T_f} \frac{[c_P(T)/T]\, dT}{\ln (T_f/T_i)} \tag{6.30}$$

$$c^*_{v,\, \mathrm{LM},\, i-f} = \int_{T_i}^{T_f} \frac{[c_v(T)/T]\, dT}{\ln (T_f/T_i)} \tag{6.31}$$

Values of log-mean specific heat are also easily calculable. The entropy change of a process can now be expressed in terms of the log-mean specific heats by substitution into Eqs. (5.54) and (5.55) as

$$\begin{aligned} s_f - s_i &= \int_{T_i}^{T_f} \frac{c_P(T)}{T}\, dT - R \ln \frac{P_f}{P_i} \\ &= c^*_{P,\, \mathrm{LM},\, i-f} \ln \frac{T_f}{T_i} - R \ln \frac{P_f}{P_i} \end{aligned} \tag{6.32}$$

or

$$s_f - s_i = \int_{T_i}^{T_f} \frac{c_v(T)}{T}\,dT + R\ln\frac{v_f}{v_i}$$

$$= c^*_{v,\,\mathrm{LM},\,i-f}\ln\frac{T_f}{T_i} + R\ln\frac{v_f}{v_i} \tag{6.33}$$

It is easy to show [by substitution of $c_P(T) - c_v(T) = R$ into Eq. (6.32) or (6.33)] that $c^*_{P,\,\mathrm{LM},\,i-f} - c^*_{v,\,\mathrm{LM},\,i-f} = R$. If we now define $k^*_{i-f} = c^*_{P,\,\mathrm{LM},\,i-f}/c^*_{v,\,\mathrm{LM},\,i-f}$, then for isentropic processes we find

$$\frac{P_i}{P_f} = \left(\frac{T_i}{T_f}\right)^{k^*_{i-f}/(k^*_{i-f}-1)} \tag{6.34}$$

$$\frac{v_i}{v_f} = \left(\frac{T_f}{T_i}\right)^{1/(k^*_{i-f}-1)} \tag{6.35}$$

In using Eqs. (6.26) through (6.35), the values of $c^*_{P,\,i-f}$, $c^*_{v,\,i-f}$, $c^*_{P,\,\mathrm{LM},\,i-f}$, $c^*_{v,\,\mathrm{LM},\,i-f}$, and k^*_{i-f} must be evaluated based on the end-state temperatures T_i and T_f, and they vary in value depending on the choice of working fluid and the end-state temperatures. Note that

$$c^*_{v,\,\mathrm{LM},\,i-f}\ln\frac{T_f}{T_i} = c^*_{v,\,\mathrm{LM},\,\mathrm{ref}-f}\ln\frac{T_f}{T_{\mathrm{ref}}} - c^*_{v,\,\mathrm{LM},\,\mathrm{ref}-i}\ln\frac{T_i}{T_{\mathrm{ref}}}$$

$$= s_{0,f} - s_{0,i}$$

or

$$s_0(T_f) = c^*_{v,\,\mathrm{LM},\,\mathrm{ref}-f}\ln\frac{T_f}{T_{\mathrm{ref}}} \tag{6.36}$$

The ideal gas tables may now be used to find $s_0(T_f)$, and thus a plot of values of $c^*_{v,\,\mathrm{LM},\,\mathrm{ref}-f}$ versus T_f may be obtained for a given problem.

Application of Temperature-Dependent Specific Heats to the Brayton Cycle

For the Brayton cycle, the efficiency derived in Example Problem 6.5 can be expressed as

$$\eta_B = 1 - \frac{h_4 - h_1}{h_3 - h_2}$$

$$= 1 - \frac{c^*_{P,\,1-4}(T_4 - T_1)}{c^*_{P,\,2-3}(T_3 - T_2)} \tag{6.37}$$

Care must be taken in further substitutions. From Eq. (6.34),

$$\frac{P_3}{P_4} = \left(\frac{T_3}{T_4}\right)^{k^*_{3-4}/(k^*_{3-4}-1)} \tag{6.38}$$

$$\frac{P_2}{P_1} = \left(\frac{T_2}{T_1}\right)^{k^*_{1-2}/(k^*_{1-2}-1)} \tag{6.39}$$

or because $P_3/P_4 = P_2/P_1$,

$$\frac{T_3}{T_4} = \left(\frac{T_2}{T_1}\right)^{[k^*_{1-2}(k^*_{3-4}-1)]/[k^*_{3-4}(k^*_{1-2}-1)]} \tag{6.40}$$

Only in the case when $k^*_{1-2} = k^*_{3-4}$ will this expression reduce to that found for the ideal air-standard cycle.

For many gases over broad temperature ranges, k^*_{i-f} *is* fairly independent of temperature, much more so than $c^*_{P,\,\mathrm{LM},\,i-f}$ or $c^*_{v,\,\mathrm{LM},\,i-f}$. This is because the temperature dependence of the log-mean specific heats $c^*_{P,\,\mathrm{LM},\,i-f}$ and $c^*_{v,\,\mathrm{LM},\,i-f}$ tends to be about the same, and the ratio of the two as expressed in k^*_{i-f} is thus more nearly constant. This can be easily checked for a specific case.

If k^*_{i-f} is taken to be temperature-independent, then Eq. (6.40) reduces to

$$\frac{T_3}{T_4} = \frac{T_2}{T_1} \quad \text{or} \quad \frac{T_3}{T_2} = \frac{T_4}{T_1}$$

The Brayton cycle efficiency can then be put in the relatively simple form

$$\begin{aligned}\eta_B &= 1 - \frac{c^*_{P,\,4-1}T_1}{c^*_{P,\,2-3}T_2} \\ &= 1 - \frac{c^*_{P,\,1-4}T_4}{c^*_{P,\,2-3}T_3} \\ &= 1 - \frac{c^*_{P,\,1-4}}{c^*_{P,\,2-3}\,r_P^{(k-1)/k}}\end{aligned} \tag{6.41}$$

where $r_P = P_2/P_1$, the pressure ratio across the turbine or compressor.

For the case where k^*_{i-f} is temperature-*dependent,* it is perhaps easiest to use Eq. (6.37) directly in terms of enthalpy and to find any required unknown temperatures by using the gas tables or the auxiliary expressions (6.38) and (6.39). Note that use of the equations will require an iterative solution, because it is necessary to know the final temperature T_f in order to evaluate k^*_{i-f}.

Effect of Actual Turbine and Compressor Efficiencies

Real turbines and compressors are, of course, not isentropic. For air-standard cycles, the efficiency of each component of the cycle, as defined in Chap. 5, is easily included in the analysis. The real compressor and turbine are taken to exit at the same pressure as the isentropic device (Brayton turbine and compressor efficiencies are usually given with respect to isentropic, not isothermal, devices). The T-s diagram for the Brayton cycle (with no regeneration) as modified to include a nonisentropic compressor and turbine is shown in Fig. 6.18.

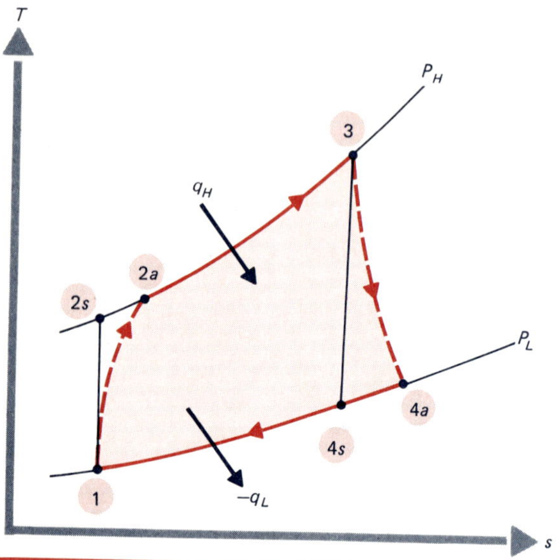

Figure 6.18 Brayton cycle with real compressor and turbine.

Example Problem 6.6

A simple ideal air-standard Brayton cycle operates between 1 and 5 atm. The inlet air to the compressor is at 80°F, and the turbine inlet is at 2040°F. The compressor is 65 percent efficient, and the turbine efficiency is 75 percent. What are the efficiency and work ratio of this Brayton cycle?

Solution

Referring to Fig. 6.18, the outlet compressor temperature is found from the compressor efficiency [taking $c_P \neq f(T)$] by

$$\eta_c = 0.65 = \frac{h_{2,s} - h_1}{h_{2,a} - h_1} = \frac{c_P(T_{2,s} - T_1)}{c_P(T_{2,a} - T_1)}$$

The value of $T_{2,s}$ is found from the isentropic relation

$$T_{2,s} = T_1\left(\frac{P_2}{P_1}\right)^{(k-1)/k} = (460 + 80)(5)^{(1.4-1)/1.4} = 855\text{°R}$$

which results in

$$T_{2,a} = T_1 + \frac{T_{2,s} - T_1}{0.65} = 540 + \frac{855 - 540}{0.65} = 1025\text{°R}$$

Similarly, for the turbine,

$$T_{4,s} = 2500(\tfrac{1}{5})^{0.4/1.4} = 1578\text{°R}$$

and

$$T_{4,a} = 2500 + 0.75(1578 - 2500) = 1809°\text{R}$$

The cycle efficiency is then

$$\eta_B = 1 - \frac{|Q_L|}{|Q_H|} = 1 - \frac{c_P(T_{4,a} - T_1)}{c_P(T_3 - T_{2,a})}$$

$$= 1 - \frac{1809 - 540}{2500 - 1025} = 0.140$$

This compares with the ideal air-standard cycle efficiency of

$$\eta_{B,\text{ ideal}} = 1 - \frac{1}{(P_1/P_4)^{(k-1)/k}} = 1 - \frac{1}{5^{0.4/1.4}} = 0.369$$

The work ratio for this Brayton engine is

$$\text{Work ratio} = \frac{|w_c|}{|w_t|} = \frac{T_{2,a} - T_1}{T_3 - T_{4,a}}$$

$$= \frac{1025 - 540}{2500 - 1809} = 0.702$$

For the ideal air-standard Brayton engine with reversible turbine and compressor, Eq. (6.25) gives a predicted work ratio of

$$(\text{Work ratio})_{\text{rev}} = \left(\frac{T_{2,s}}{T_3}\right) = \frac{855}{2500} = 0.342$$

Comments

The work ratio increases as the turbine and/or compressor efficiency decreases, and more of the turbine work goes to drive the compressor. As the work ratio approaches unity, the net work output of the Brayton engine approaches zero.

Obviously, if the turbine and/or compressor efficiency becomes low enough, the efficiency of the cycle can become zero (or even negative) and the work ratio can reach or exceed unity. So no net cycle work can be obtained, or work input to the cycle may be required for operation. This is not a desirable way to operate a power cycle!

Improving the Brayton Cycle Efficiency

The idea of recuperation can be incorporated into the Brayton cycle by modifying the hardware as shown in Fig. 6.19. For the simple closed Brayton cycle (Fig. 6.17), the turbine exhaust gases at state 4 may be at a higher temperature than the gases at the outlet of the compressor (state 2). It is then possible to use a portion of the rejected energy q_L to preheat the gases leaving the compressor and thereby reduce the heat transfer required to the cycle.

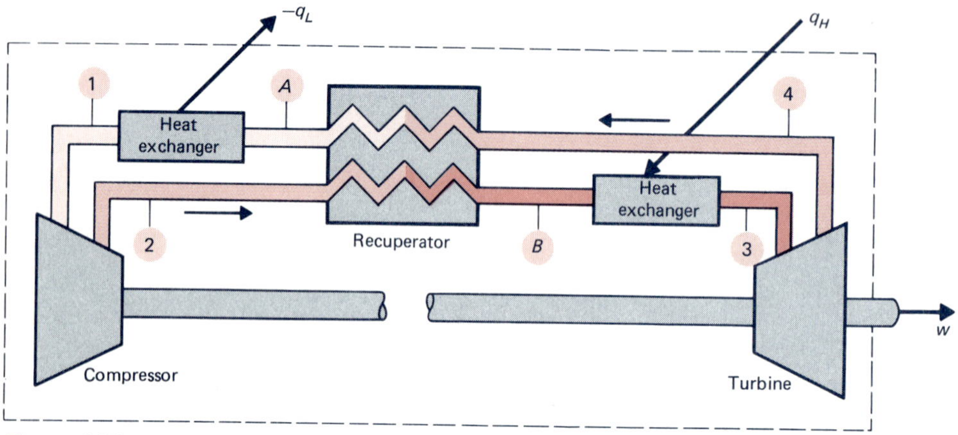

Figure 6.19 Brayton cycle with ideal recuperation.

The T-s diagram for an ideal air-standard Brayton cycle with recuperation is shown in Fig. 6.20. Note that the gases leaving the compressor could, at best, be heated by the recuperator to $T_B = T_4$, the turbine exhaust temperature. The turbine exhaust gas, in that case, will be cooled to $T_A = T_2$. For such an ideal case, the energy added to the cycle becomes ${}_Bq_3$, and the work output remains unchanged. The ideal air-standard Brayton cycle efficiency with ideal recuperator then is given by

$$\begin{aligned}\eta_{B,\,\text{rec, ideal}} &= 1 - \frac{|Q_L|}{|Q_H|} = \frac{|Q_H| - |Q_L|}{|Q_H|} \\ &= \frac{c_P(T_3 - T_B) - c_P(T_A - T_1)}{c_P(T_3 - T_B)} \\ &= \frac{(T_3 - T_4) - (T_2 - T_1)}{T_3 - T_4} \\ &= 1 - \frac{T_2 - T_1}{T_3 - T_4}\end{aligned} \tag{6.42}$$

If for a particular cycle $T_4 < T_2$, then recuperation by using turbine exhaust is not possible. It is also not possible to build a recuperator that will actually make $T_B = T_4$. For the real case, the compressor outlet temperature will be raised to some lower temperature $T_{B'}$ (Fig. 6.20). A *recuperator efficiency* η_{rec} can be defined as the fraction of the maximum enthalpy that is recovered by the real recuperator, or

$$\eta_{\text{rec}} = \frac{h_{B'} - h_2}{h_B - h_2} = \frac{h_4 - h_{A'}}{h_4 - h_A} \tag{6.43}$$

Figure 6.20 *T-s* diagram for a Brayton cycle with ideal regeneration.

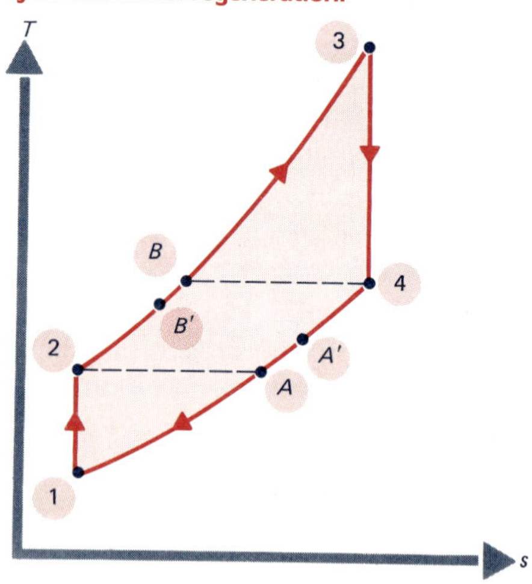

Or, for the constant-property air-standard cycle,

$$\eta_{\text{rec}} = \frac{T_{B'} - T_2}{T_4 - T_2} = \frac{T_4 - T_{A'}}{T_4 - T_2} \tag{6.44}$$

The cycle efficiency with the real regenerator is now

$$\eta_{B,\text{ rec}} = 1 - \frac{|Q_L|}{|Q_H|} = 1 - \frac{h_{A'} - h_1}{h_3 - h_{B'}}$$
$$\stackrel{c_p \neq F(T)}{=\!=\!=} 1 - \frac{c_P(T_{A'} - T_1)}{c_P(T_3 - T_{B'})} = 1 - \frac{T_{A'} - T_1}{T_3 - T_{B'}} \tag{6.45}$$

The values of $T_{A'}$ and $T_{B'}$ can be determined from known values of the recuperator efficiency.

Example Problem 6.7

Find the ideal air-standard cycle efficiency of a simple closed Brayton cycle, a Brayton cycle with an ideal recuperator, and with a recuperator having an efficiency of 70 percent. Compare the results. The cycle operates between 70 and 1800°F at a pressure ratio of 8.

Solution

For the simple Brayton cycle, Eq. (6.23) gives

$$\eta_B = 1 - \frac{1}{(P_2/P_1)^{(k-1)/k}} = 1 - \frac{1}{(8)^{(1.4-1)/1.4}} = 0.448$$

By adding an ideal recuperator, Eq. (6.42) gives

$$\eta_{B,\text{ rec, ideal}} = 1 - \frac{T_2 - T_1}{T_3 - T_4}$$

The values of T_3 and T_1 are given. The isentropic expansion relations can be used to find the other temperatures:

$$T_2 = T_1 \left(\frac{P_2}{P_1}\right)^{(k-1)/k}$$
$$= (70 + 460)(8)^{0.4/1.4}$$
$$= 960°\text{R} = 500°\text{F}$$

and

$$T_4 = \frac{1800 + 460}{8^{0.286}} = 1248°\text{R} = 788°\text{F}$$

The cycle efficiency is then

$$\eta_{B,\text{ rec, ideal}} = 1 - \frac{500 - 70}{1800 - 788} = 0.575$$

For a real recuperator, $T_{A'}$ and $T_{B'}$ are, from Eq. (6.44),

$$T_{A'} = T_4 - \eta_{\text{rec}}(T_4 - T_2) = 788 - 0.7(788 - 500) = 586°\text{F}$$

and

$$T_{B'} = T_2 + \eta_{\text{rec}}(T_4 - T_2) = 500 + 0.7(788 - 500) = 702°\text{F}$$

The cycle efficiency is then, from Eq. (6.45),

$$\eta_{B,\text{ rec}} = 1 - \frac{586 - 70}{1800 - 702} = 0.530$$

So the simple cycle has an efficiency of 44.8 percent, and adding an ideal recuperator to the cycle increases the efficiency to 57.5 percent. However, considering a real recuperator drops the predicted efficiency back to 53.0 percent.

Comments

Although the recuperator will improve cycle efficiency when the temperature constraints are met that allow its use, there are other real effects that we have not considered. Usually a significant pressure drop is incurred in the recuperator, and this must be overcome at the expense of useful work output from the engine.

The Brayton cycle efficiency can be improved in other ways. For example, if a multistage compressor is used, the working fluid can be cooled between each stage of compression. If enough stages of such *intercooling* are added, the compressor approaches isothermal operation. Another option is to use a multistage turbine and heat the working fluid between each stage. Again, isothermal heat transfer can be approached. Thus, the cycle diagram will approach the case of two isothermal heat transfer processes separated by two adiabatic (recuperative) constant-pressure processes, which is the Ericsson cycle. The Ericsson cycle has the same efficiency as the Carnot efficiency when operating between the same two temperatures. The Brayton cycle can thus be made to approach the Carnot efficiency, but at the expense of increasing complexity. Of course, real turbine and compressor efficiencies cause even a complex real cycle to depart considerably from the Carnot efficiency.

6.4 Internal Combustion Cycles

We now consider air-standard versions of cycles that gain their energy in real engines from the combustion of fuel within the engine. Such engines are treated in their air-standard version as if the energy were introduced from an external source. Then the process of energy addition is approximated as one of the standard types treated previously, i.e., via constant temperature, constant volume, etc.

6.4.1 Brayton Cycle (Internal Combustion)

If energy is added to the working fluid of the Brayton cycle by the combustion of a fuel rather than by heat transfer in a heat exchanger, then the air-standard cycle analysis is effectively unchanged from that shown in Figs. 6.14 through 6.16. However, the cycle must now be treated as an open cycle. The working fluid is replaced during each cycle, so the exhaust gases are purged, and new oxidant is introduced for further combustion to occur. In addition, the hardware used to carry through the cycle will be much different than for the closed cycle. An example of a possible hardware arrangement is shown in Fig. 6.21. Note that the exhaust state A is not necessarily the same as the inlet state 1 to the compressor.

The open Brayton cycle with internal combustion is the cycle used to analyze gas-turbine engines. In most stationary Brayton engines, the turbine work output is designed to be a maximum and thus exceed the required compressor work input. The excess shaft work is then used to drive, for example, an electric generator or pipeline compressor. The total

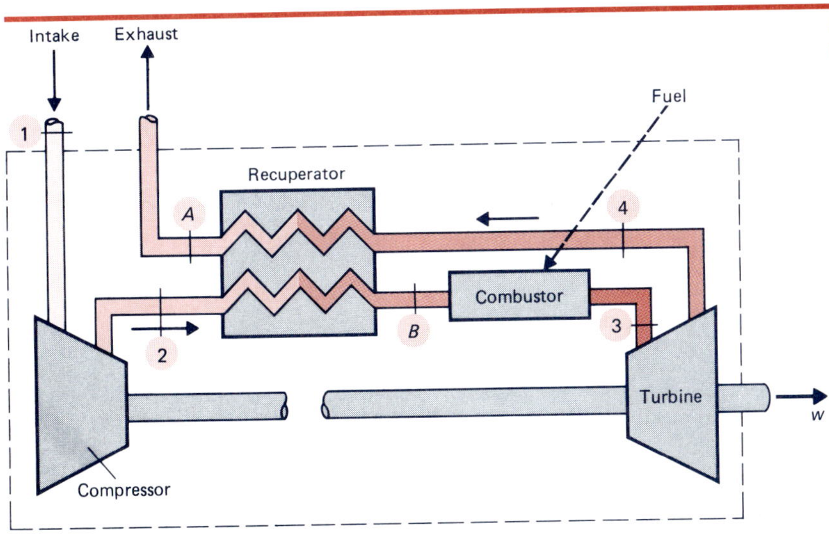

Figure 6.21 Internal combustion version of Brayton cycle.

energy of the gases leaving the turbine is minimized so that the turbine work output is maximized.

Another possible design is to *maximize* the energy of the exhaust gases, taking only enough energy from the turbine so that the turbine work is just sufficient to drive the compressor. The remaining energy of the exhaust gases is then used for propulsion, usually by ejecting the gases through a reaction nozzle, as in the turbojet engine. A compromise between the design for the jet engine and the stationary plant turbine is the turboprop engine. This engine uses much of the energy carried by the hot gases from the combustor for turbine work, to turn a propeller as well as to power the compressor. The remaining energy of the exhaust gases is used to provide additional thrust. These engines are discussed more fully in Sec. 6.4.2.

6.4.2 Applications of the Brayton Cycle

In Sec. 6.4.1, the ideal air-standard Brayton cycle was discussed, along with the effects on ideal behavior of the temperature dependence of the properties of the working fluid, the presence of nonidealities in the turbine and compressor, and the addition of a recuperator to the cycle. In this section, we examine further modifications and characteristics of Brayton cycles in certain common applications.

Propulsion Systems

The Brayton cycle is applied to propulsion of vehicles because of certain practical characteristics. Included is the capability of using low-grade refined crude oil or even pulverized coal as a fuel as well as the presence of rotating rather than reciprocating components, which can lead to long service life and ease of maintenance in comparison with reciprocating engines. We will examine some of the configurations of Brayton engines in these applications.

Automotive Turbine Engines

The Brayton cycle gas-turbine engine has been adapted to automotive propulsion systems, although its use is certainly less widespread than that of the Otto or diesel cycle for this purpose. The automotive gas-turbine engine has the general advantages of being able to burn fuels that require little refining and of having few moving parts. However, they tend to produce high levels of pollutants, and generally they have poorer efficiency than piston engines. To provide competitive thermal efficiencies, automotive gas-turbine engines usually incorporate recuperation, and many proposed engines also provide for intercooling and/or reheating.

One proposed cycle is shown in Fig. 6.22. Here a two-shaft gas generator (compressor-burner-turbine) system is used to provide high-enthalpy gas, which is then expanded through a separate power turbine that is used to provide propulsion. Intercooling is used between the compres-

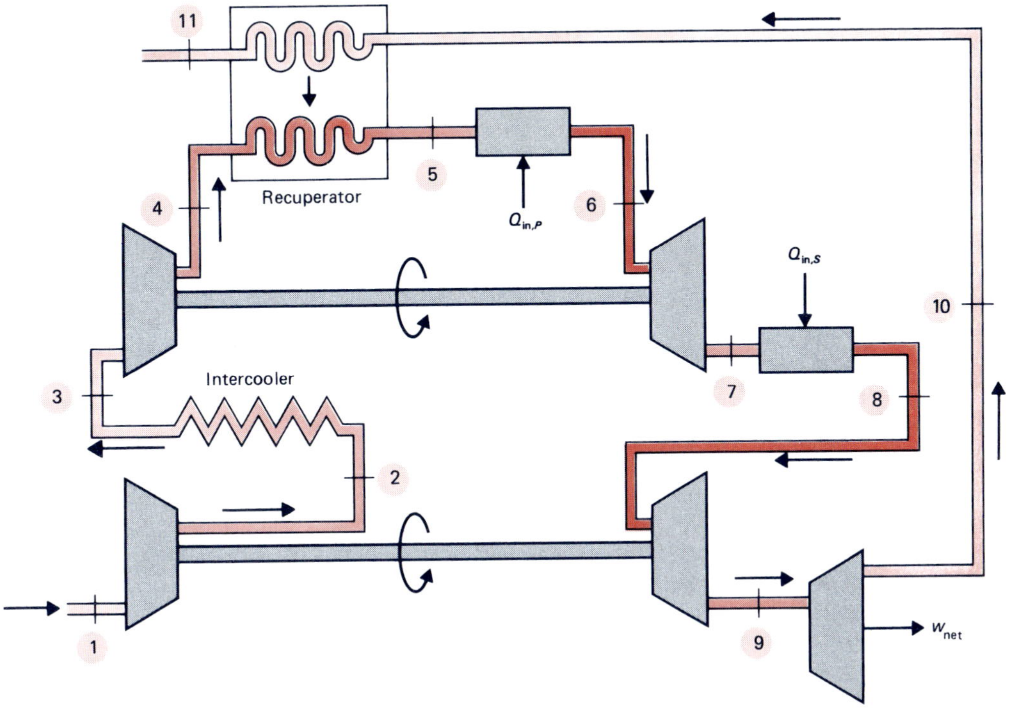

Figure 6.22 Automotive turbine engine configuration using intercooling, reheat, and recuperation.

sors, and reheat is used between one gas generator turbine and the power turbine. A recuperator is used to transfer some of the enthalpy carried by the exit gases from the final gas generator turbine stage to the air entering the combustion chamber. The T-s diagram for this engine is shown in Fig. 6.23.

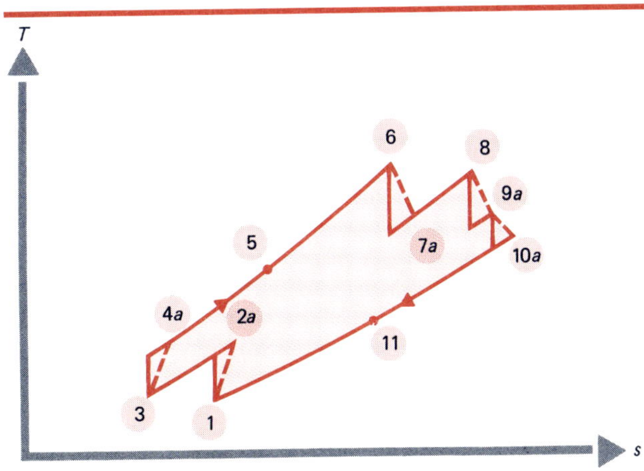

Figure 6.23 Temperature-entropy diagram of an intercooled-reheat-regenerative gas turbine cycle as shown in Fig. 6.22.

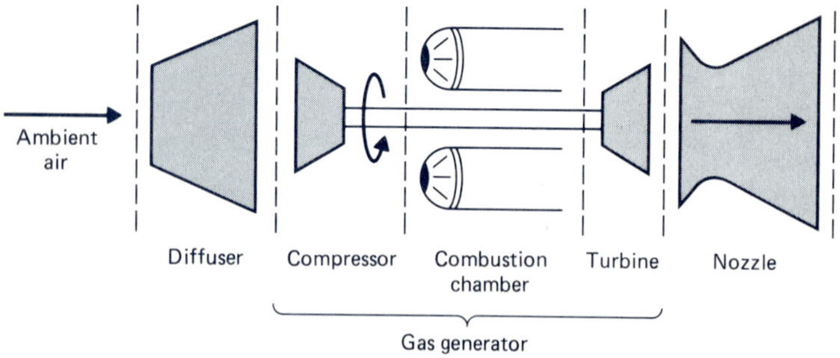

Figure 6.24 Schematic diagram for the turbojet engine.

One successful application of the gas turbine in a vehicle is in the M-1 and M-1A1 tanks. The turbine engines in these vehicles have a multistage compressor and turbine on a single shaft, followed by a power turbine which drives the gear box/power train. The engines generate about 1.2 MW (1500 hp). Exhaust gases pass through a recuperator, where heat transfer occurs to the incoming air from the compressor. The recuperator serves the dual purpose of increasing cycle efficiency while reducing the temperature of the exhaust gases in order to lower detectibility by infrared heat sensors.

Aircraft Turbine Engines

The Turbojet Engine. Aircraft Brayton cycle engines, commonly called turbojet (or simply jet) engines, operate by changing the momentum of the air traveling between the engine inlet and the exhaust. The thrust produced by this momentum change is used to propel the aircraft. Figure 6.24 shows the major components of the common axial flow jet engine. Often a diffuser is present at the engine inlet to slow the velocity of the inlet air relative to the compressor while increasing the air pressure; this enables more efficient compressor blade design. If the air velocity is too large relative to the speed and angle of the compressor blades, the compressor efficiency may be poor, or flow through the compressor may become "choked," i.e., may approach supersonic velocity at some stage in the compressor, causing the mass flow through the compressor to reach a maximum. However, if the inlet air velocity to the compressor is too small relative to the blade speed and angle, the blades can undergo *stall*, and the airflow will separate from the blades, again causing very poor compressor performance.

After the air enters the compressor, it is compressed to a higher pressure; fuel is then added to the air, and the mixture is ignited in the burner cans. This high-enthalpy mixture of air and combustion products then travels through the turbine, where sufficient work is produced to drive the compressor. The gases that exit the turbine retain a relatively high enthalpy, because no net shaft work is produced by the turbine above that required by the compressor. Thus, their velocity relative to the aircraft can be further increased by passing them through an exhaust nozzle. According to Eq. (4.117), the velocity of the exit gases can be increased by trading turbine exit enthalpy for kinetic energy.

In the Comments to Example Problem 6.7, we noted that the Brayton cycle efficiency can be increased in two ways (in addition to the possibility of recuperation). These methods were *intercooling* (heat transfer from the working fluid between each stage of compression so that the compressor approaches isothermal operation) and *reheating* (heat transfer to the working fluid between each set of turbine stages so that the turbine approaches isothermal operation). Adding actual heat exchange equipment to aircraft engines usually imposes too great a weight penalty to be practical; however, it is possible to use some lightweight but effective methods for approximating the effects of intercooling and reheating.

Intercooling is considered practical when the compressor ratio exceeds about 5. Intercooling can be approximated by injecting liquid water between compressor stages, and the resulting vaporization then cools the air. Because the necessary weight of water is large for long-term use, such *water injection* is used only for short periods of high-power operation, such as at takeoff or for combat operations. In practice, injection may actually occur at a point either before or after the air enters the compressor. A second benefit of water injection is that it lowers the peak inlet temperature to the turbine, which is usually a limiting factor in engine efficiency.

Reheating is approximated by injecting fuel into the airstream after it leaves the turbine but before entering the nozzle, in the area called the *afterburner*. Normally there is much more oxygen in the airstream than is needed to burn the fuel injected in the combustion cans, so that this extra fuel is easily burned. The increased enthalpy of the gas entering the nozzle because of this increased fuel addition can increase engine thrust by as much as 40 percent, with a major increase in fuel consumption.

The Fan-jet Engine. To improve the performance of the turbojet engine, the airflow through the engine may be split into two parts. The first undergoes the same history as in the jet engine. A second flow stream is accelerated by a set of large-diameter blades on the turbine compressor shaft. This second stream passes through a duct that normally surrounds the turbojet section of the engine, and then either it is passed through a separate nozzle (Fig. 6.25*a*) or it is first mixed with the turbojet turbine exhaust and then passed through the nozzle (Fig. 6.25*b*). An afterburner can be incorporated in either version; water injection is used only with the stream passing through the gas generator section of the engine.

The fan-jet is similar to the turboprop engine discussed in the next section, except that the fan operates at the high speed of the turbine compressor shaft, and the airflow in the bypass stream passes through a duct of controlled geometry and thus can be remixed with the gas generator exhaust.

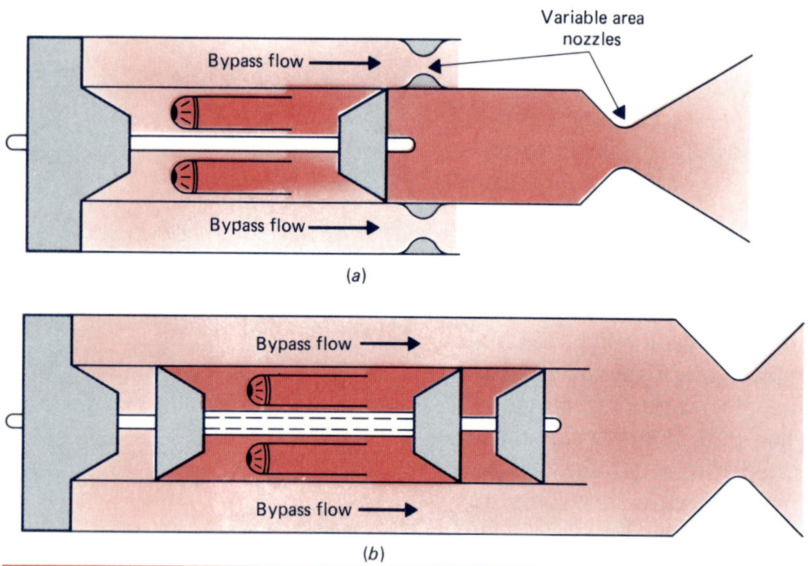

Figure 6.25 Fan-jet engine with (*a*) separate nozzles and (*b*) mixed flow, single nozzle.

Turboprop Engines. The turboprop or prop-jet engine extracts work from either the turbine that also drives the compressor or a separate power turbine, and it uses this work to drive the propeller. The propeller will normally have a much larger diameter than the fan in a fan-jet engine, so that much more air can undergo a momentum change with a prop-jet. A schematic is shown in Fig. 6.26. The propeller normally must be geared to the power turbine shaft to reduce the propeller rpm. Speed reductions of about 9 : 1 are commonly used. High propeller rpm would cause the blade tips to be inefficient, as they will easily be at supersonic speeds relative to the air without substantial gear reduction of rotational speed. Thus, although much more air is contacted by the lower-speed propeller of the prop-jet than by the blades of the turbofan or turbojet engine, the turboprop imparts a smaller velocity change to the air. Some additional thrust is provided to the engine by the exhaust gases, which are allowed to exit through a nozzle for additional thrust. For contemporary engines, the propeller provides about 90 percent of the total engine thrust at sea-level conditions.

The advantage of the turboprop is that it has a considerably lower fuel consumption when operated at lower aircraft speeds than the turbofan or turbojet. Development of efficient high-speed propeller designs may extend the competitive speed range of the turboprop to the high subsonic range in the near future.

Power Generation Systems

The Brayton engine is quite commonly used in stationary power plants. Its ability to operate with reasonable efficiency when running at

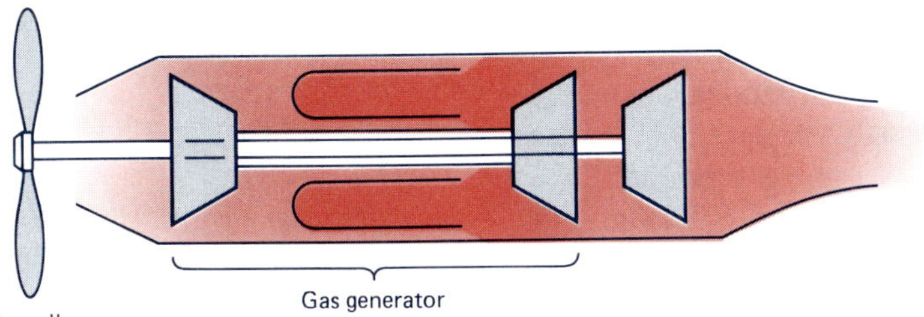

Figure 6.26 Schematic diagram of a turboprop gas turbine engine.

constant speed and load make it suitable for running electric generators. It can run on a variety of fuels from fuel oil to natural gas, and it can be put on-line in a very short time compared with coal-fired, gas-fired, or nuclear-powered steam generating plants. It is also low in capital cost for a given power output, and it can be installed quickly. Most often it is used for providing power during short periods when peak electrical loads occur.

In such stationary applications, it is usually cost-effective to include recuperation or regeneration, intercooling and reheat to improve overall plant thermal efficiency.

Combined-cycle plants, in which the remaining enthalpy in the gas-turbine exhaust is used to generate steam for a steam turbogenerator, can be quiet efficient. They are discussed more fully in Chap. 7.

It is possible to use recuperation, reheat, and intercooling on stationary internal combustion Brayton engines, as discussed in Sec. 6.3.3. These efficiency-improving techniques can also be employed in propulsion turbines, but the increase in complexity and weight must be carefully analyzed to assess the net benefit.

6.4.3 Air-Standard Otto Cycle

The Otto cycle is an idealization of the cycle used in spark-ignition gasoline engines, although its original use was in engines using natural gas or other gaseous fuels. The air-standard Otto cycle deviates substantially from the cycle realized in practice, but it is still instructive to examine the idealized cycle.

In the so-called two-stroke Otto cycle used in many small gasoline engines, the cycle is as shown in Fig. 6.27 for the *P-v* diagram and in Figs. 6.28 and 6.29 for the *T-s* and *P-h* diagrams. In these diagrams, at state point 1, the piston begins the compression stroke, assumed to be an isentropic compression process, and compression is continued until the piston reaches the top of its stroke at point 2. The combustion process is then assumed to occur instantaneously and is modeled as raising the pressure by energy addition at constant volume to state 3. The piston begins the power stroke, again isentropically, and the expansion con-

Figure 6.27 *P-v* diagram for two-stroke Otto cycle.

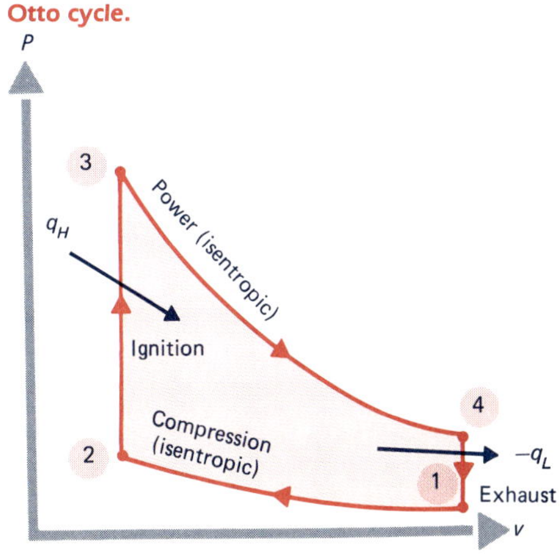

tinues to point 4, where the piston reaches the limit of its travel. The exhaust valve opens, reducing the pressure instantaneously to state 1, while the cylinder volume remains constant. This process is modeled as a constant-volume energy rejection. The fuel-air charge for the next cycle is then assumed to enter the cylinder at state 1, and the cycle is repeated.

In the four-stroke air-standard cycle, illustrated in Fig. 6.30, the piston compresses the working fluid from state 1 until it reaches top dead center, state 2, and then combustion occurs instantaneously. The pressure thus rises at constant volume to point 3. The piston then descends in an isentropic expansion on the power stroke to point 4, as in the two-stroke engine. Now the exhaust valve opens, and the pressure drops at constant volume to state. 1. The piston begins its upward travel, purging the cylinder of combustion products. This exhaust, or "scavenging," stroke occurs at essentially constant pressure between states 1 and 1′. At top dead center, the exhaust valve closes, and the intake valve opens. The piston now descends, again at effectively constant pressure, and a new charge of unburned fuel and air is drawn into the cylinder on the intake stroke between states 1′ and 1. Note that, as indicated on the P-v diagram, no net work is done on the combination of the exhaust and intake strokes. The intake valve closes at state 1, and the piston begins the upward compression stroke to state 2, continuing the cycle.

The efficiency of the ideal air-standard Otto cycle, either two- or four-stroke, is easily found by

$$\eta_O = 1 - \frac{|Q_L|}{|Q_H|} = 1 - \frac{mc_v(T_4 - T_1)}{mc_v(T_3 - T_2)}$$

For the isentropic processes,

$$\frac{T_4}{T_3} = \left(\frac{v_3}{v_4}\right)^{k-1} = \left(\frac{v_2}{v_1}\right)^{k-1} = \frac{T_1}{T_2}$$

and substituting the resulting ratio of temperatures into the efficiency relation gives

$$\eta_O = 1 - \frac{T_4}{T_3} = 1 - \left(\frac{v_3}{v_4}\right)^{k-1} = 1 - \frac{1}{r_v^{k-1}} \tag{6.46}$$

where r_v is the compression ratio v_4/v_3.

The efficiency of the ideal air-standard Otto cycle (either two- or four-stroke, since the additional two strokes in the four-stroke cycle do not change the thermodynamics of the ideal air-standard cycle) is thus dependent on only the compression ratio of the engine.

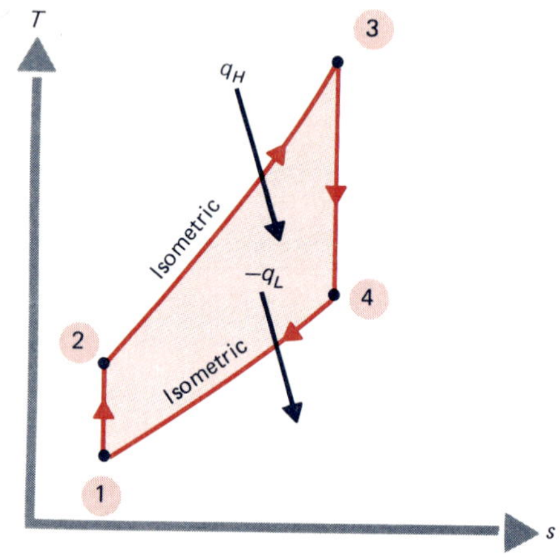

Figure 6.28 *T-s* diagram for two-stroke Otto cycle.

Figure 6.29 *P-h* diagram for two-stroke Otto cycle.

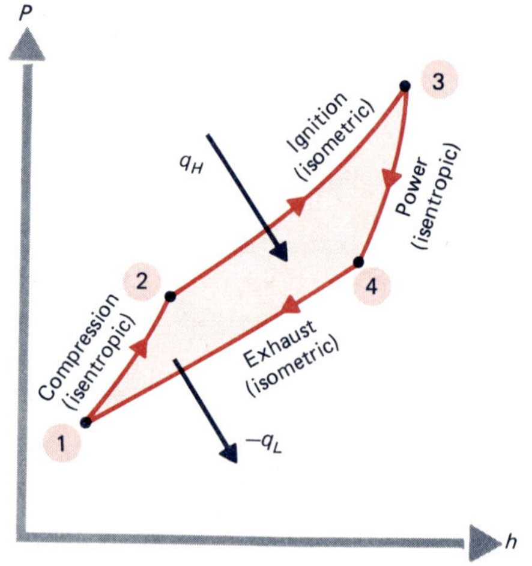

Mean Effective Pressure of the Air-Standard Otto Cycle

The mean effective pressure of the ideal air-standard Otto cycle is

$$\begin{aligned}P_{\text{ME},\,O} &= \frac{|w_{\text{net}}|}{v_4 - v_2} = \frac{\eta_O q_H}{v_4 - v_2} \\ &= \frac{\eta_O c_v (T_3 - T_2)}{v_4 - v_2} \\ &= \eta_O \frac{c_v}{R} \frac{P_3(1 - P_2/P_3)}{r_v - 1} \\ &= \eta_O \left(\frac{1}{k-1}\right) \frac{P_3(1 - T_2/T_3)}{r_v - 1}\end{aligned} \tag{6.47}$$

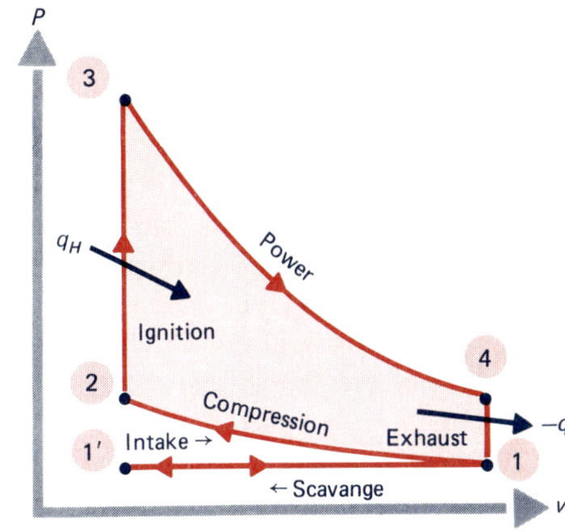

Figure 6.30 *P-v* diagram for four-stroke Otto cycle.

Improving the Performance of Otto Cycle Engines

Improving the *efficiency* of an air-standard Otto cycle engine depends chiefly on increasing the compression ratio according to Eq. (6.46).

Changes in the compression ratio can be examined in terms of the piston diameter D_p, the length of the piston stroke L, and the clearance volume V_c remaining in the cylinder when the piston is at top dead center (Fig. 6.31). The compression ratio is given in terms of these variables by

$$r_v = \frac{V_1}{V_2} = \frac{\pi D_p^2 L/4 + V_c}{V_c} = 1 + \frac{\pi D_p^2 L}{4V_c} \tag{6.48}$$

so we see that increasing the piston diameter or piston stroke, or decreasing the clearance volume, will increase the compression ratio.

The mean effective pressure relation, Eq. (6.47), can be rewritten in the form

$$w_{\text{net}} = P_{\text{ME},\,O} v_2 (r_v - 1) \tag{6.49}$$

Now the work output per cycle (and thus, for a given engine rpm, the power) also depends strongly on the compression ratio as well as the overall displacement of the engine through v_2.

For an existing engine, how can the compression ratio be changed to improve power and efficiency? Hot-rod modifications include increasing the displacement and thus power by drilling out the cylinders to a larger diameter and replacing the existing pistons with pistons of greater diameter ("boring"); replacing the existing crankshaft with a crankshaft with a greater "throw" and replacing the existing connecting rods with shorter ones, so that the value of L is increased, while not allowing the pistons to exceed their allowable travel at the top of the cylinder ("stroking"); and decreasing the clearance volume by removing material from the engine head at the interface between the head and the engine block ("milling the head"). Of course, the original engine designer presumably had optimized the mechanical design for the mechanical and thermal stresses expected

Figure 6.31 Compression ratio quantities.

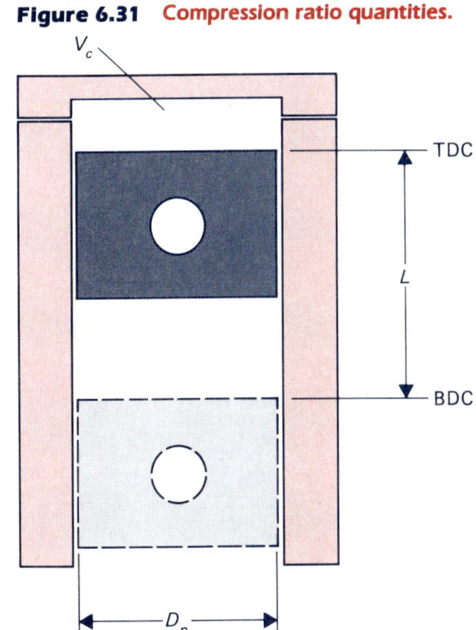

with the original compression ratio, and all these modifications reduce some of the design safety factors built into the original engine.

As with the previous cycles, the mean effective pressure can be increased by increasing the heat transfer into the engine per cycle (which will decrease the ratio T_2/T_3) or by increasing the "compression ratio" r_v, which will increase P_3 for any given initial value of P_1. If the initial pressure in the cycle P_1, which is usually near ambient pressure, could be increased, then P_3 would increase by the same amount. Operating at this higher pressure would then increase $P_{ME,\,O}$, according to Eq. (6.47). This pressure increase can be achieved in practice by *supercharging* the engine.

Supercharging

The Otto cycle can be modified by increasing the inlet pressure to a value above that of the exhaust pressure, so that the *P-v* diagram for the four-stroke cycle is as shown in Fig. 6.32. This can be done in a number of ways, such as by using a compressor on the inlet to the engine which is driven by power taken from a power takeoff from the engine itself (*supercharging*). Another option common on piston aircraft engines and some automotive engines is called *turbocharging*. In this method, a turbine driven by the exhaust gases is used to provide power to the compressor or blower at the inlet.

By following the notation on Fig. 6.32, it is easy to see that the net work output per unit mass of working fluid (air) from the Otto cycle has been increased by the shaded area. The apparent efficiency of the cycle is

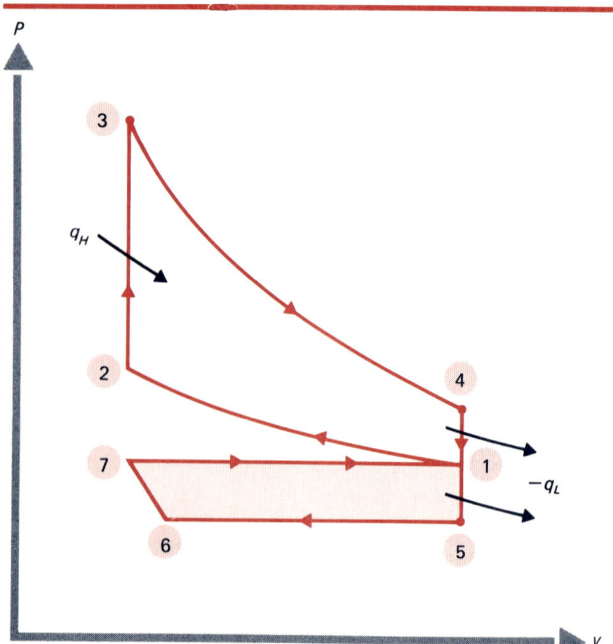

Figure 6.32 *P-v* diagram for supercharged ideal air-standard Otto cycle.

increased, because the work output of the conventional part of cycle 1-2-3-4 is the same as for the unsupercharged cycle [Eq. (6.46)], while more work output has been obtained for the same heat transfer to the cycle q_H. However, this apparent increase occurs only if the work to drive the supercharger is "free," such as from waste enthalpy from the exhaust products. A power take-off driven supercharger will use work from area 1-2-3-4 to drive the supercharger for producing the work in area 1-5-6-7, and no net gain in efficiency can be expected.

The major advantage of supercharging is not its increased efficiency, but rather the increased power and net work output available from an engine of given displacement. By increasing the mean effective pressure of the cycle [Eq. (6.47)], more net work and power can be obtained from an engine of given displacement. However, for an engine of given displacement, the specific volume of the working fluid is decreased at higher pressure, resulting in less net work output per cycle, as shown by Eq. (6.1).

The practical disadvantage of supercharging is that pressure P_3 will be greater in the supercharged engine, resulting in "knocking" or preignition in a carbureted engine. Most supercharged or turbocharged engines use *fuel injection* of gasoline directly into the combustion chamber for that reason, so that ignition of the fuel-air mixture from the carburetor cannot be triggered by the temperature increase during compression process 1-2. The increase in P_3 also results in more stringent mechanical design requirements for a super- or turbocharged engine.

One highly developed aircraft piston engine, the Pratt and Whitney R-2800, is a 28-cylinder engine with a bore of 14.61 cm (5.75 in) and a stroke of 15.24 cm (6 in). At cruise, the engine operates at 1800 rpm, with a mean effective pressure of 875 kPa (127 psia). When it was operating in this way, the energy of the fuel was measured to go to shaft work to drive the propeller (29 percent), exhaust energy loss (44 percent), losses to cooling by the airstream (18 percent), and losses to the oil cooler (9 percent).

Effect of Variable Properties on the Otto Cycle

To include the effect of temperature-dependent properties on the calculation of the efficiency of the air-standard Otto cycle, note that the heat transfer to and from the engine takes place at different temperatures. The efficiency can be written as

$$\eta_O = 1 - \left|\frac{Q_L}{Q_H}\right| = 1 - \frac{c^*_{v,\,1-4}(T_4 - T_1)}{c^*_{v,\,2-3}(T_3 - T_2)} \tag{6.50}$$

If it can be assumed that k^* is constant for air over a broad range of temperatures, then, as for the ideal air-standard cycle, $T_4/T_3 = T_1/T_2$, and the cycle efficiency can be approximated by

$$\eta_O = 1 - \left(\frac{c^*_{v,\,1-4}}{c^*_{v,\,2-3}}\right)\left(\frac{1}{r_v^{k^*-1}}\right) \tag{6.51}$$

6.4.4 Air-Standard Diesel Cycle

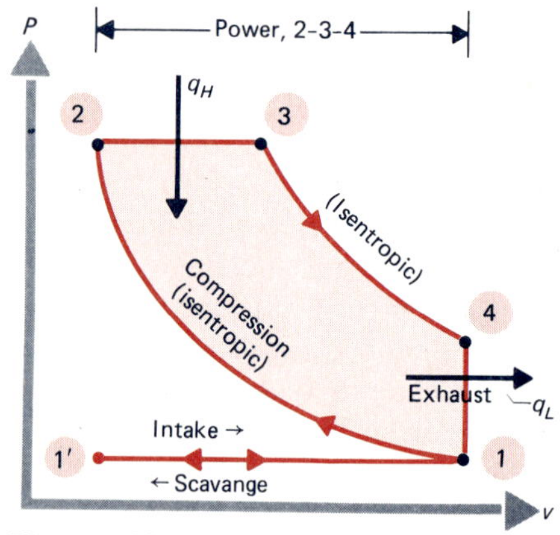

Figure 6.33 *P-v diagram for diesel cycle.*

The diesel cycle is an attempt to approach the Carnot efficiency as closely as possible in an internal combustion engine. Again, the actual cycle differs considerably from the behavior predicted by the air-standard version.

The *P-v* diagram for the ideal air-standard diesel cycle is shown in Fig. 6.33. At state 1, the piston begins an isentropic compression of the combustion air and continues to state 2. At state 2, fuel is injected into the air. If a sufficiently high compression ratio is used, the temperature of the compressed air is high enough that the fuel will ignite without an external ignition source. The ignition process is assumed to occur at such a rate that the energy addition to the cylinder is just sufficient to hold the pressure in the cylinder constant while the piston moves, until state 3 is reached. The combustion process is assumed to end at this point, and the expansion continues as an isentropic process to point 4. The exhaust valve then opens, and the pressure drops to the initial condition at state 1. If the four-stroke cycle is used, then scavenge or intake strokes occur between states 1 and 1′, at essentially constant pressure.

The diesel cycle efficiency is

$$\eta_D = 1 - \frac{|Q_L|}{|Q_H|} = 1 - \frac{c_v(T_4 - T_1)}{c_P(T_3 - T_2)} \tag{6.52}$$

For the ideal air-standard diesel cycle assuming c_P and c_v are not temperature-dependent, we can take $k = c_P/c_v$, and rearranging, we find

$$\eta_D = 1 - \frac{T_1}{kT_2}\frac{T_4/T_1 - 1}{T_3/T_2 - 1} \tag{6.53a}$$

If the compression ratio v_1/v_2 is denoted r_1 and the expansion ratio v_4/v_3 is denoted r_2, then the diesel ideal air-standard efficiency can be written as

$$\eta_D = 1 - \left(\frac{1}{r_1}\right)^{k-1}\frac{1}{k}\frac{(r_1/r_2)^k - 1}{r_1/r_2 - 1} = 1 - \left(\frac{1}{r_1}\right)^{k-1}\frac{1}{k}\frac{r_c^k - 1}{r_c - 1} \tag{6.53b}$$

where $r_c = r_1/r_2 = v_3/v_2$ is called the *cutoff ratio.*

The compression ratio of the diesel cycle is always greater than the expansion ratio (see Fig. 6.33). So for a given compression ratio r_1, the diesel engine always has lower efficiency than an Otto engine operating at the same compression ratio. The closer r_2 is made to r_1, the closer the diesel efficiency will approach that of the Otto engine.

Mean Effective Pressure of the Diesel Cycle

The mean effective pressure of the ideal air-standard diesel cycle can be written

$$\begin{aligned} P_{ME,\,D} &= \frac{|w_{\text{net}}|}{v_4 - v_2} = \frac{\eta_D q_H}{v_4 - v_2} \\ &= \eta_D c_P \frac{T_3 - T_2}{v_4 - v_2} \\ &= \eta_D \left(\frac{c_P}{R}\right) \frac{P_3[(v_3/v_2) - 1]}{r_1 - 1} \\ &= \eta_D \left(\frac{k}{k-1}\right) \frac{P_3[(T_3/T_2) - 1]}{r_1 - 1} \end{aligned} \tag{6.54}$$

Again, the mean effective pressure can be increased by operating at high compression or by increasing the heat transfer into the engine per cycle.

Effect of Temperature-Dependent Properties on the Diesel Cycle Efficiency

If temperature-dependent property effects are included, the diesel cycle efficiency can be approximated by

$$\eta_D = 1 - \frac{c^*_{v,\,1-4}}{c^*_{P,\,2-3}} \left(\frac{1}{r_1^{k-1}}\right) \frac{r_c^k - 1}{r_c - 1} \tag{6.55}$$

6.4.5 Dual Cycle

The nonideal behavior found in real engines causes significant changes from the ideal air-standard cycles discussed for the Otto and diesel cycles. To better show the real behavior of these cycles, we can construct a better approximation, called the *ideal air-standard dual* or *mixed cycle.*

In both the Otto and diesel air-standard cycles, a major assumption is that all heat transfer occurs in either a constant-volume process (the Otto cycle) or a constant-pressure process (the diesel cycle). In actual engines, this is not the case.

In the Otto cycle, the combustion process does not take place instantaneously when the piston is at top dead center (minimum volume), but rather continues during the power (expansion) stroke. Thus, a better approximation is to consider heat transfer as occurring in both the constant-volume process and a portion of the expansion process where the pressure is nearly constant.

In the air-standard diesel cycle, the assumption is that all heat transfer takes place at constant pressure during a portion of the expansion process. In actual engines, part of the combustion occurs during the constant-volume process of fuel injection and combustion at top dead center, and it continues at constant pressure throughout a portion of the expansion process.

Both the Otto and diesel cycles thus can be better approximated with the dual cycle, in which heat transfer occurs in part in both the constant-volume and the constant-pressure processes. The P-v and T-s diagrams

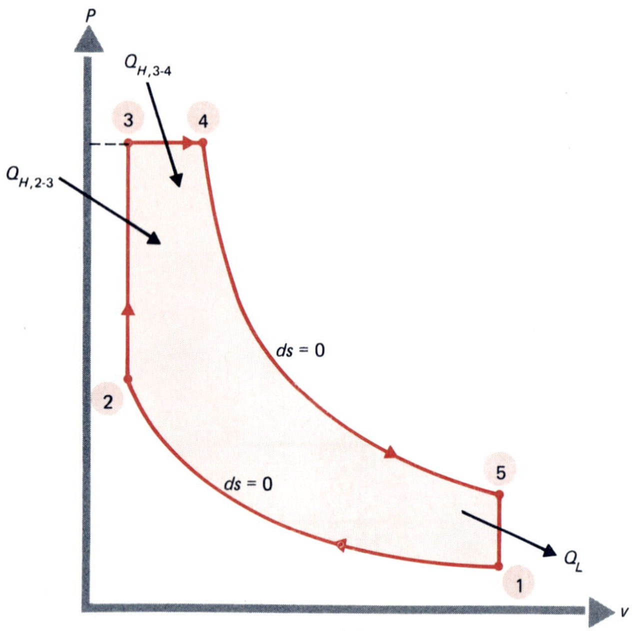

Figure 6.34a *P-v* diagram for the dual cycle.

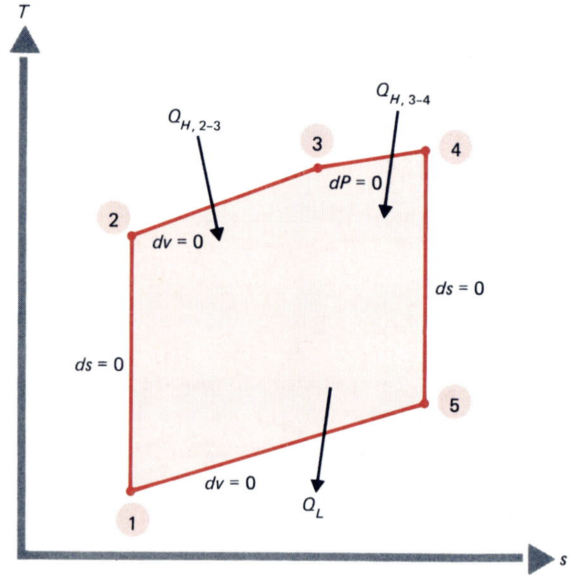

Figure 6.34b *T-s* diagram for the dual cycle.

for the ideal air-standard dual cycle are shown in Fig. 6.34*a* and *b*. Following the methods used in analysis of the ideal air-standard Otto and diesel cycles, we find that the efficiency of the dual cycle is

$$\eta_{\text{dual}} = 1 - \frac{[1/(r_1)^{k-1}][(P_3/P_2)(v_4/v_3)^k - 1]}{P_3/P_2 - 1 + k(P_3/P_2)(v_4/v_3 - 1)}$$

$$= 1 - \frac{[1/(r_1)^{k-1}][(P_3/P_2)(r_1/r_2)^k - 1]}{P_3/P_2 - 1 + k(P_3/P_2)(r_1/r_2 - 1)} \qquad (6.56)$$

In this expression, r_1 is the volume compression ratio v_1/v_2 and r_2 is the volume expansion ratio v_5/v_4. When $P_3 = P_2$, Eq. (6.55) reduces to the efficiency of the diesel cycle, and when $v_4 = v_3$ (that is, $r_1 = r_2$), it reduces to the Otto cycle efficiency.

Mean Effective Pressure for the Dual Cycle

From Eq. (6.2),

$$P_{\text{ME, dual}} = \frac{\eta_{\text{dual}} q_{\text{in}}}{v_5 - v_3} = \frac{\eta_{\text{dual}} q_{\text{in}}}{v_2(r_1 - 1)}$$

Noting that $q_{in} = c_v(T_3 - T_2) + c_P(T_4 - T_3)$ and $P_2 = P_3T_2/T_3$, also $T_4/T_3 = v_4/v_3 = r_1/r_2$, we have

$$P_{ME,\,dual} = \eta_{dual} \frac{P_3\{(1 - P_2/P_3)[1/(k-1)] + [k/(k-1)](r_1/r_2 - 1)\}}{r_1 - 1} \tag{6.57}$$

When $P_2 = P_3$, this reduces to the mean effective pressure for the diesel cycle, and when $r_1 = r_2$, it reduces to the mean effective pressure of the Otto cycle.

Example Problem 6.8

Compare the efficiencies and mean effective pressures of ideal air-standard Otto, diesel, and dual heat engines, each operating at a compression ratio of 15. The state at the beginning of the compression stroke in each cycle is 20°C and 1 atm. Each cycle has a maximum temperature of 1500 K. The dual-cycle pressure reaches a maximum that is midway between states 2 and 3 of the ignition process of the Otto cycle.

Solution

It is instructive to draw the T-s diagrams for the three cycles. Note first that, from Eqs. (5.52) and (5.53), the slopes of the isobars and isometric lines on the T-s diagram are given by

$$\left(\frac{\partial T}{\partial s}\right)_P = \frac{T}{c_P} \qquad \left(\frac{\partial T}{\partial s}\right)_v = \frac{T}{c_v}$$

so that the isobars always have the smaller slope for an ideal gas (because $c_P > c_v$), and the slopes increase with increasing temperature. The three cycles thus appear on the T-s diagram as shown.

The three cycles are coincident for compression process 1-2, and all return to the same final state 1 after completion of the exhaust process. At state 2 for all cycles,

$$P_2 = P_1(r_1)^k = 1\,(15)^{1.4} = 44.3 \text{ atm}$$

$$T_2 = T_1\left(\frac{P_2}{P_1}\right)^{(k-1)/k} = 293 \times (44.3)^{0.4/1.4} = 866 \text{ K}$$

For the Otto cycle, sufficient information is available for solution once the value of $P_{3,O}$ is determined. This is easily done by using the ideal gas law for a constant-volume process as

$$P_{3,O} = P_2\left(\frac{T_3}{T_2}\right) = (44.3 \text{ atm})\left(\frac{1500}{866}\right) = 76.7 \text{ atm}$$

Diagram

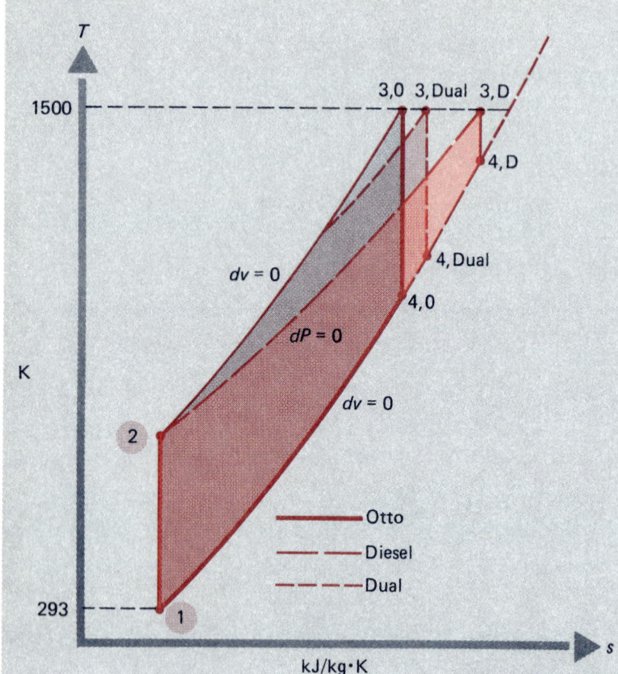

The cycle efficiency and mean effective pressure are given by Eqs. (6.46) and (6.47) as

$$\eta_O = 1 - \left(\frac{1}{r_1}\right)^{k-1} = 1 - \frac{1}{15^{0.4}} = 0.661 = 66.1\%$$

and

$$\begin{aligned} P_{\text{ME},\,O} &= \eta_O\left(\frac{1}{k-1}\right)\frac{P_3(1 - T_2/T_3)}{r_1 - 1} \\ &= 0.661\left(\frac{1}{0.4}\right)(76.7\ \text{atm})\frac{(1 - 877/1500)}{15 - 1} \\ &= 3.83\ \text{atm} \end{aligned}$$

For the diesel cycle, the expansion ratio $r_2 = v_4/v_3$ must be determined. It is easy to see that since $P_3 = P_2$,

$$v_{3,\,D} = \frac{RT_3}{P_3} = \frac{RT_3}{P_2} \qquad v_4 = v_1 = \frac{RT_1}{P_1}$$

so $r_2 = v_4/v_3 = P_2T_1/P_1T_3 = 44.3 \times 293/(1 \times 1500) = 8.65$. Then η_D and

$P_{ME,\,D}$ are found from Eqs. (6.53) and (6.54) as

$$\eta_D = 1 - \left(\frac{1}{r_1}\right)^{k-1}\left(\frac{1}{k}\right)\frac{(r_1/r_2)^k - 1}{r_1/r_2 - 1}$$
$$= 1 - \left(\frac{1}{15}\right)^{0.4}\left(\frac{1}{1.4}\right)\frac{(15/8.65)^{1.4} - 1}{15/8.64 - 1}$$
$$= 0.618 = 61.8\%$$

and

$$P_{ME,\,D} = \eta_D \frac{k}{k-1}\,\frac{P_3(T_3/T_2 - 1)}{r_v - 1}$$
$$= 0.618\left(\frac{1.4}{0.4}\right)(44.3\text{ atm})\,\frac{1500/866 - 1}{15 - 1}$$
$$= 5.01\text{ atm}$$

For the dual cycle, the problem statement defines $P_{3,\,\text{dual}}$ as

$$P_{3,\,\text{dual}} = \frac{P_2 + P_{3,\,O}}{2} = \frac{44.3 + 76.7}{2} = 60.5\text{ atm}$$

The expansion ratio r_2 for the dual cycle must be found, and it is

$$r_2 = \frac{v_5}{v_4} = \frac{v_1}{v_4} = \frac{P_4 T_1}{P_1 T_4} = \frac{P_3 T_1}{P_1 T_4}$$
$$= \frac{60.5 \times 293}{1 \times 1500} = 11.82$$

The values of η_{dual} and $P_{ME,\,\text{dual}}$ are found from Eqs. (6.56) and (6.57) as

$$\eta_{\text{dual}} = 1 - \frac{[1/(r_1)^{k-1}][(P_3/P_2)(r_1/r_2)^k - 1]}{P_3/P_2 - 1 + k(P_3/P_2)(r_1/r_2 - 1)}$$
$$= 1 - \frac{(1/15)^{0.4}[(60.5/44.3)(15/11.82)^{1.4} - 1]}{60.5/44.3 - 1 + 1.4(60.5/44.3)(15/11.82 - 1)}$$
$$= 0.651 = 65.1\%$$

and

$$P_{ME,\,\text{dual}} = \frac{\eta_{\text{dual}} P_3\{(1 - P_2/P_3)[1/(k-1)] + [k/(k-1)](r_1/r_2 - 1)\}}{r_1 - 1}$$
$$= \frac{0.651(60.5\text{ atm})[(1/0.4)(1 - 44.3/60.5) + (1.4/0.4)(15/11.82 - 1)]}{15 - 1}$$
$$= 4.53\text{ atm}$$

In summary, the comparison of the three cycles gives:

Cycle	Efficiency, %	Mean Effective Pressure, atm
Otto	66.1	3.83
Diesel	61.8	5.01
Dual	65.1	4.53

Comments

Note that the efficiency of the Otto cycle is the largest of the three, while the mean effective pressure of the diesel cycle is largest. This implies that, for the given operating conditions, a diesel engine could be made smaller than an Otto cycle engine of the same power output (if both operate at the same number of cycles per second), but the Otto engine would have lower fuel consumption. The performance of the dual cycle lies between the Otto and diesel performances.

6.4.6 Other Gas Cycles

Although the cycles covered so far are those commonly used for heat engines, other cycles have been proposed and used, and some hold potential interest. We discuss some of these briefly in this section.

The Atkinson Cycle

The Otto cycle has the potential for greater work output per cycle than shown in its common form, because at the end of the isentropic expansion process (power stroke) the gas in the cylinder has a relatively high enthalpy, which could be used to obtain work in a further expansion. This expansion could be continued until the gas pressure reached the pressure at the beginning of the compression process, and the cycle could be completed with a constant-pressure exhaust process. Such a cycle is called the *Atkinson cycle*, and the P-v and T-s diagrams for the cycle are shown in Fig. 6.35. The ideal air-standard Atkinson cycle thermodynamic efficiency is given by

$$\eta_A = 1 - \frac{k}{(r_v)^{k-1}} \frac{(r_P)^{1/k} - 1}{r_P - 1} \tag{6.58}$$

The mechanical means for attaining the Atkinson cycle usually involve a free piston reciprocating engine compressor, with a turbine for extracting the work from the exhaust gases of the piston engine.

The Humphrey Liquid Piston Pump

A very interesting internal combustion engine cycle that describes an engine with few moving mechanical parts is the *Humphrey pump cycle*.

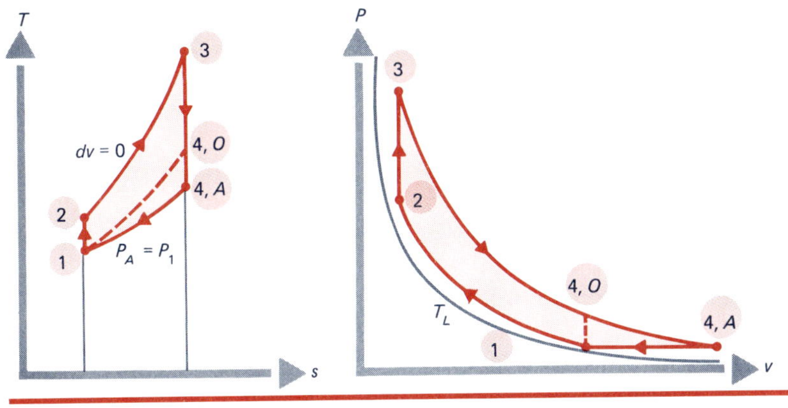

Figure 6.35 *T-s* and *P-v* diagrams for the Atkinson cycle.

The Humphrey pump operates as shown in Fig. 6.36. The only mechanical parts are the intake and exhaust valves for the gases and the intake valves to admit water to the J-pipe.

The engine is difficult to design for optimum performance because of the coupling of the fluid mechanical forces with the gas combustion. The gases have a tendency to penetrate the liquid-gas interface unless the system is carefully designed. Nevertheless, very large-scale Humphrey pumps have operated successfully in commercial service for many years.

The air-standard Humphrey pump cycle is approximated by an isentropic compression, a constant-volume heat transfer input, an isentropic expansion, and a constant-pressure exhaust stroke, as shown in Fig. 6.37. The resulting cycle efficiency is

$$\eta_H = 1 - k \frac{r_{v,2} - r_{v,1}}{r_{v,2}^k - r_{v,1}^k} \tag{6.59}$$

where $r_{v,2}$ is the expansion ratio $v_4/v_3 = v_4/v_2$ and $r_{v,1}$ is the compression ratio v_1/v_2. Note that $r_{v,2} > r_{v,1}$, which is the opposite of the diesel cycle, which combines the same processes as the Humphrey pump cycle but in a different order.

Rocket Engines

Rocket engines can be treated as control volumes in which a source of internal energy is converted to the kinetic energy of expelled propellant. The energy source is often chemical, but engines have also been developed and tested that use nuclear sources, and still others use electrical energy to ionize a gas which is accelerated by electrical or magnetic forces.

Conservation of momentum can be used to find the behavior of a vehicle powered by a rocket engine. We have applied conservation of mass and energy before; this is the first opportunity to approach a problem from the viewpoint of the conservation of momentum. Consider a spacecraft and its contained fuel moving at velocity **V** outside the earth's

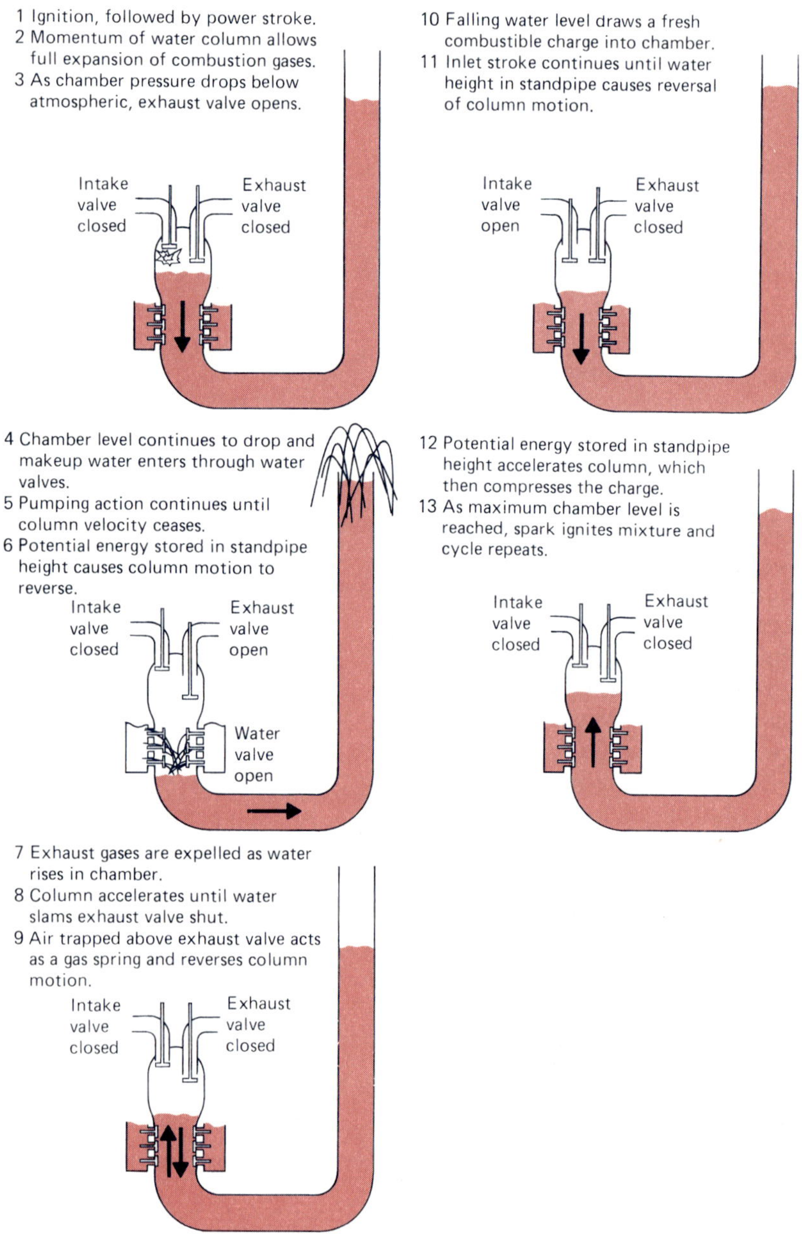

Figure 6.36 Humphrey pump operation.

atmosphere and in some region of space where it is not strongly affected by a gravitational field. Remember that velocity is a vector, so we are writing the conservation of momentum along the axis of motion of the spacecraft, and we assume that the rocket exhaust is aligned with this axis. In this case, since no external forces act on the spacecraft, the change in

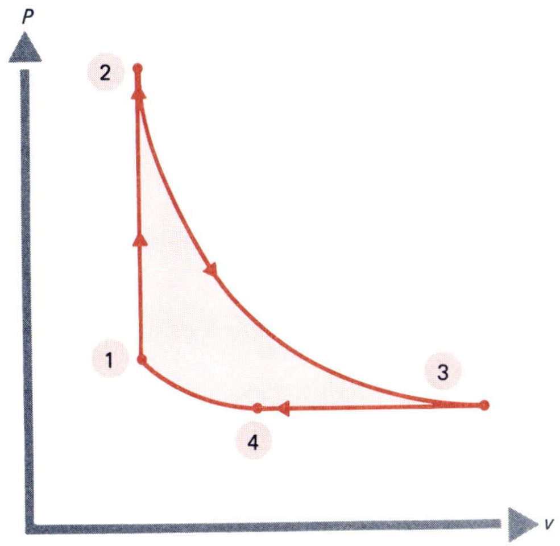

Figure 6.37 Air-standard Humphrey pump cycle.

momentum of the spacecraft plus its contained fuel is zero. At time $t = 0$, the rocket engine is ignited. Now the momentum of the expelled rocket fuel over a short time interval Δt plus the momentum of the spacecraft (less expelled fuel) must still sum to zero, because there are still no net *external* forces acting on the system. Let $\mathbf{V}_e$ be the velocity of the expelled fuel relative to the rocket. Then over the time interval Δt we can write

$$\text{Momentum change of rocket plus fuel} = \Delta \mathbf{M} = \mathbf{M}_2 - \mathbf{M}_1 = 0 \qquad (6.60)$$

The initial momentum of the spacecraft (sc) plus fuel is

$$\mathbf{M}_1 = \frac{(m_{sc} + m_{fuel})\mathbf{V}}{g_c} \qquad (6.61a)$$

and after time interval Δt, the momentum of the spacecraft plus the remaining fuel and of the expelled fuel is

$$\mathbf{M}_2 = \frac{(m_{sc} + m_{fuel} + dm_{fuel})(\mathbf{V} + d\mathbf{V})}{g_c} - \frac{dm_{fuel}\,(\mathbf{V} - \mathbf{V}_e)}{g_c} \qquad (6.61b)$$

Here we have followed the usual convention in calculus of taking all changes as positive; clearly, the numerical value of dm_{fuel} must be negative. Substituting Eqs. (6.61a) and (6.61b) into Eq (6.60), we find (after dropping second-order terms)

$$(m_{sc} + m_{fuel})\, d\mathbf{V} = -\mathbf{V}_e\, dm_{fuel} \qquad (6.62)$$

By taking the velocity of the propellant relative to the rocket $\mathbf{V}_e$ as constant and assuming that the amount of fuel at velocity $\mathbf{V}_0$ is $m_{fuel,\,0}$, Eq. (6.62) can be integrated to give

$$\frac{\mathbf{V} - \mathbf{V}_0}{\mathbf{V}_e} = \ln \frac{m_{sc} + m_{fuel,\,0}}{m_{sc} + m_{fuel}} \qquad (6.63)$$

Thus as the fuel remaining in the spacecraft is expended, thus reducing the denominator in the log term, the velocity $\mathbf{V}$ of the spacecraft increases. If the mass of the spacecraft is very small relative to the mass of fuel originally carried, then the final velocity of the spacecraft approaches $\mathbf{V} = \mathbf{V}_0 + \mathbf{V}_e \ln$ (argument$\rightarrow\infty$). Thus, according to this relation, if enough fuel can be carried, there is essentially no upper limit to the attainable velocity. (We neglect relativistic effects, which limit the attainable velocity to below the speed of light, and practical considerations, which keep the attainable velocities to considerably below that!)

Now we divide Eq. (6.62) by Δt and take the limit as $\Delta t \rightarrow 0$, and we obtain

$$(m_{sc} + m_{fuel}) \frac{d\mathbf{V}}{g_c\, dt} = -\mathbf{V}_e \left(\frac{dm_{fuel}}{g_c\, dt} \right) \tag{6.64}$$

The left-hand side is the instantaneous rate of change of momentum of the spacecraft and remaining fuel, or, according to Newton's second law, the *force* exerted on the spacecraft and remaining fuel. This is seen by Eq. (6.64) to be equal to a term that depends on the exit velocity of the propellant and the rate of propellant use; this term is called the *thrust* of the rocket engine, and obviously it has units of force.

Thus, the thrust F, or the force exerted *by* the rocket engine in which the exhaust gases are expanded through a nozzle until they reach the ambient pressure, is given by

$$F = \frac{\dot{m}_p \mathbf{V}_e}{g_c} \tag{6.65}$$

where $\dot{m}_p$ is the mass flow rate of propellant leaving the nozzle (the negative of the depletion rate of the fuel dm_{fuel}/dt) and $\mathbf{V}_e$ is the exit velocity of the propellant. An important measure of performance for a rocket engine is the *specific impulse I*, or the ratio of the thrust to the rate of propellant mass flow rate:

$$I = \frac{F}{\dot{m}_p} = \frac{\mathbf{V}_e}{g_c} \tag{6.66}$$

The units of I in SI will be N/(kg/s), or N·s/kg, while in the USCS the units are lbf/(lbm/s) or lbf·s/lbm. Obviously the higher the specific impulse, the more thrust per unit of fuel expenditure will be produced by the rocket engine. (In the English system, the units for specific impulse are often erroneously given as simply *seconds*, reflecting the confusion between pound-force and pound-mass.)

Nozzle Exit Velocity

Consider a rocket engine that uses a single-component propellant. This is the case, for example, for certain tested and proposed nuclear-powered rockets that use hydrogen as the propellant. A schematic is shown in

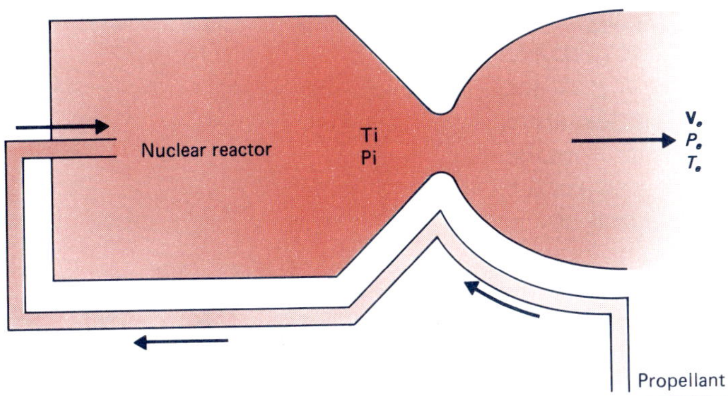

Figure 6.38 Nuclear-powered rocket motor using regenerative cooling by propellant.

Fig. 6.38. In such a device, cold propellant is used to regeneratively cool the rocket nozzle by passing it along the exterior surface of the nozzle. The propellant then enters a chamber where heat transfer from some high-temperature source (in this case, the nuclear reactor) occurs. The velocity of the propellant entering and leaving this chamber can be assumed to be small; the propellant is then exhausted through a nozzle to low pressure. Generally, the heat transfer to the cold propellant on the outside of the nozzle from the hot propellant inside the nozzle is small compared with the enthalpy of the hot propellant, and the propellant expansion through the nozzle can be approximately treated as an adiabatic process.

For a control volume around the rocket nozzle, the first law for a steady adiabatic process, assuming ideal gas behavior for the propellant so that h is independent of pressure, from Eq. (4.117) is

$$h(T_i) = h(T_e) + \frac{\mathbf{V}_e^2}{2g_c} \tag{6.67}$$

or, solving for the exit velocity,

$$\mathbf{V}_e = \{2g_c[h(T_i) - h(T_e)]\}^{1/2} \tag{6.68}$$

The expansion process can be considered to be adiabatic and internally reversible, as was the case for problems involving the discharge of a single-phase fluid from a tank. Thus, for the propellant $s_i = s_e$. If we assume temperature-independent specific heats for the propellant, then for the isentropic expansion

$$T_e = T_i\left(\frac{P_e}{P_i}\right)^{(k-1)/k} \tag{6.69}$$

and substituting this into Eq. (6.68) results in

$$\mathbf{V}_e = \left\{2g_c c_P T_i\left[1 - \left(\frac{P_e}{P_i}\right)^{(k-1)/k}\right]\right\}^{1/2} \tag{6.70}$$

Thus, the exit velocity is increased if the inlet temperature to the nozzle is made as large as possible and the ratio of exit pressure to nozzle inlet pressure is made as small as possible. Also a propellant with the largest possible c_P value should be chosen. Increasing the temperature requires careful design of the heat source and good heat transfer from the source to the propellant. The pressure ratio is affected by the design of the supersonic nozzle. And hydrogen gas has the largest specific heat and is the propellant of choice for this type of rocket motor.

If the specific heat of the propellant varies considerably during the expansion process through the nozzle, then the gas tables may be used to account for the temperature dependence.

Example Problem 6.9

A nuclear rocket motor uses hydrogen as the propellant. The propellant is heated in the reactor to 1000 K at a pressure of 30 atm and is then expelled from the nozzle at a pressure of 1 atm. The hydrogen flow rate is 50 kg/s. Find the exit velocity of the hydrogen, the specific impulse of the motor, and the thrust produced by the motor.

Solution

Assuming that the properties of hydrogen can be taken as temperature-independent, the exit velocity is, from Eq. (6.70),

$$\mathbf{V}_e = \left\{2g_c c_P T_i\left[1 - \left(\frac{P_e}{P_i}\right)^{(k-1)/k}\right]\right\}^{1/2}$$

$$= \{2(1)[14.32\ \text{kJ/(kg}\cdot\text{K)}](1000\ \text{K})[1 - (\tfrac{1}{30})^{0.4/1.4}]\}^{1/2}$$

$$= 4219\ \text{m/s}\ (13{,}840\ \text{ft/s})$$

The specific impulse is then, from Eq. (6.66), simply $I = \mathbf{V}_e/g_c =$ 4219 N·s/kg, or 430 lbf·s/lbm.

The thrust is, from Eq. (6.65),

$$F = \frac{\dot{m}_p \mathbf{V}_e}{g_c} = \frac{50(\text{kg/s})4219(\text{m/s})}{1\ (\text{kg}\cdot\text{m/N}\cdot\text{s}^2)}$$

$$= (50\ \text{kg/s})(4219\ \text{N}\cdot\text{s/kg}) = 21.09\ \text{kN}\ (4741\ \text{lbf})$$

If a chemical rocket motor is to be analyzed for which the combustion products are a mixture rather than the single-component propellant considered here, then we must consider the characteristics of the combustion process and the properties of the mixture of combustion products. These are considered in Chaps. 10 through 12.

6.5 Refrigeration, Air Conditioning, and Heat Pump Cycles

It is possible to have a heat transfer from a lower- to a higher-temperature reservoir by doing work on a system. Common domestic air conditioners and heat pumps work by implementing this statement of the second law in a practical device using a *vapor-compression cycle* (treated in detail in Chap. 7, because the working fluid undergoes a phase change). By combining a heat engine with a vapor compression cycle, a heat transfer can be used as the driving energy to the heat engine, which in turn drives the vapor compression cycle. Thus, heat transfer rather than work can be used to provide the energy to drive the cycle. Other methods of using energy to provide cooling are also possible. A few are addressed in this section.

Systems that cool and dehumidify air for commercial or domestic structures are called *air conditioners*; those that provide cold water or other liquid are usually referred to as *chillers*. Systems that provide space heating by a heat transfer from a low-temperature reservoir and using a cycle that provides a heat transfer at the desired temperature are called *heat pumps*.

6.5.1 Coefficient of Performance for Air Conditioners and Chillers

The measure of performance for air conditioning and chilling systems is the *coefficient of performance* (COP). COP is defined in Sec. 5.12 as the ratio of energy removed from the environment in the form of a heat transfer (the cooling provided) to the energy input. The energy input E_{in} to the system may be in the form of work or heat transfer and should include the energy necessary to drive all auxiliary equipment such as blowers, pumps, backup heaters, etc. The definition is then

$$\mathrm{COP} = \frac{|q_L|}{|E_{\mathrm{in}}|} \tag{6.71}$$

In the United States, a related figure of merit is the *energy efficiency rating* (EER). This is simply COP expressed in mixed units of Btu per hour of cooling per watt of energy input, resulting in the relation EER = 3.41 × COP. The higher the COP or EER of a given system, the better its efficiency.

The cooling capacity of air conditioning systems in the United States is often given in *tons*. One ton is the cooling rate necessary to freeze 1 ton (2000 lbm) of water in 1 day, which is 12,000 Btu/h or 3.517 kW.

For a reversible refrigeration or air conditioning cycle that removes energy from a low-temperature reservoir at T_L and rejects the energy to a reservoir at a higher temperature T_H,

$$(\mathrm{COP})_{\mathrm{rev}} = \frac{|q_L|}{|E_{\mathrm{in}}|} = \frac{|q_L|}{|q_H| - |q_L|} = \frac{1}{T_H/T_L - 1} \tag{6.72}$$

which is the ideal COP for a reverse Carnot cycle. All real systems operating between constant-temperature reservoirs must have a lower COP than predicted by Eq. (6.72).

6.5.2 Coefficient of Performance for Heat Pumps

Another use of the cycles described above is to take energy from a low-temperature source and make it available at a higher temperature by providing work to the cycle. In many cases, the work input is a small fraction of the energy made available at the high temperatures; the remaining energy is drawn from the low-temperature reservoir. The figure of merit for such a system, called a *heat pump* (HP), is again called $\mathrm{COP_{HP}}$. However, the definition is now changed to be the ratio of energy rejected in the form of a heat transfer *at the high temperature* to the energy input to the cycle in the form of work, or

$$\mathrm{COP_{HP}} = \frac{|q_H|}{|w|} \tag{6.73a}$$

For a reversible heat pump cycle as in Sec. 5.12,

$$\mathrm{COP_{HP,\,rev}} = \frac{|q_H|}{|q_H| - |q_L|} = \frac{1}{1 - T_L/T_H} \tag{6.73b}$$

which is the upper limit for real heat pump systems.

6.5.3 Gas Cooling Cycles Driven by Work Input

All the external energy source gas cycle heat engines discussed earlier in this chapter can be envisioned as operating with each process in the cycle proceeding in the reverse direction. The cycle diagrams will not be identical to the parent heat engine cycles when real effects such as turbine and compressor efficiencies are considered, since irreversibilities cause entropy increases that change the shapes of various process lines in the cooling cycles from those found for the heat engines.

Brayton Cooling Cycle

Perhaps the most widely used of the reverse cycles is the Brayton cooling cycle. It is used to provide cooling for aircraft cabins and avionics and serves as a good example of the use of the reverse cycles.

The cycle operates by using work input to the turbine compressor shaft, using air as the working fluid on an open cycle. The hardware arrangement for a general Brayton cooling cycle is shown in Fig. 6.39, and the corresponding cycle diagrams are in Fig. 6.40.

Figure 6.39 Open Brayton cooling system.

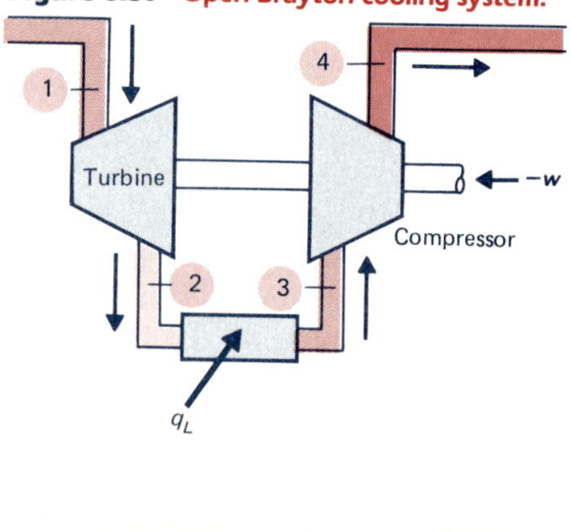

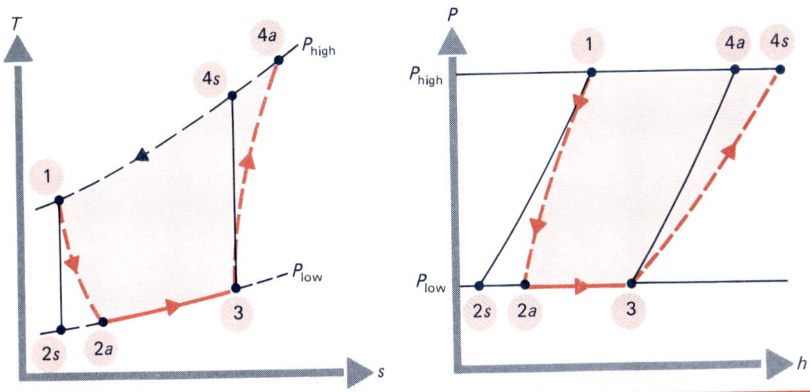

Figure 6.40 *T-s* and *P-h* diagrams for open Brayton cooling cycle.

The COP of the ideal cooling cycle (isentropic compressor and turbine) is given by

$$\begin{aligned} COP_B &= \frac{|q_L|}{|w|} = \frac{|q_L|}{|q_H| - |q_L|} \\ &= \frac{1}{(T_{4,s} - T_1)/(T_3 - T_{2,s}) - 1} \\ &= \frac{1}{(T_{4,s}/T_3)[(1 - T_1/T_{4,s})/(1 - T_{2,s}/T_3)] - 1} \end{aligned} \tag{6.74}$$

For an isentropic compressor and turbine,

$$\frac{T_1}{T_{2,s}} = \frac{T_{4,s}}{T_3} = \left(\frac{P_{high}}{P_{low}}\right)^{(k-1)/k} \tag{6.75}$$

or

$$\frac{T_1}{T_{4,s}} = \frac{T_{2,s}}{T_3} \tag{6.76}$$

Substituting into Eq. (6.75) gives

$$\begin{aligned} COP_B &= \frac{1}{T_{4,s}/T_3 - 1} \\ &= \frac{1}{(P_{high}/P_{low})^{(k-1)/k} - 1} \end{aligned} \tag{6.77}$$

For aircraft cabin cooling, the inlet air to the system is usually bleed air from the main engine compressor [or, when the aicraft is on the ramp, from an auxiliary compressor run by an auxiliary power unit (APU)]. This compressed air stream is cooled in a heat exchanger by an ambient air stream, and is then expanded through a turbine to produce a low-pressure cooled air stream. The turbine work output is low, and may be used

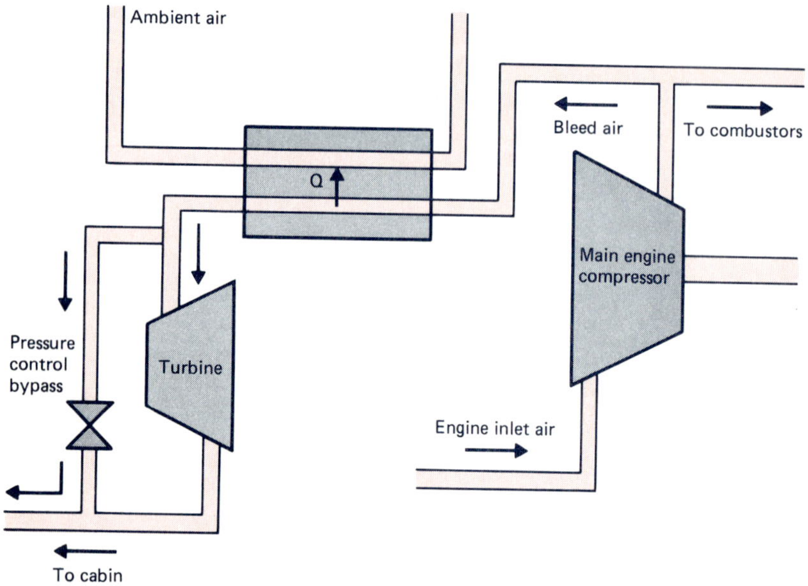

Figure 6.41 Open Brayton cooling cycle for aircraft.

to drive circulation fans in the system. The turbine outlet pressure can be adjusted to whatever level is required for cabin pressurization. The cooled cabin air supply can be provided directly in a once-through system, and this is the usual practice. The system diagram is shown in Fig. 6.41.

6.5.4 Cooling Cycles Driven by Heat Transfer

Heat-Engine-Driven Cycles

In some cases, heat transfer must be used to drive the cooling cycle. A convenient way to carry out the conversion of a high-temperature heat transfer into a cooling effect is to couple a heat engine with a work-driven refrigeration cycle. To determine the maximum performance of such a combined device, we can consider the case of a Carnot engine operating between the heat addition temperature of T_H and the environment temperature T_E. The Carnot engine drives a reverse Carnot cooling cycle that operates by a heat transfer addition at T_L and a subsequent rejection at the higher temperature T_E (Fig. 6.42).

For the heat engine, the work output is

$$|w| = |q_H|\left(1 - \frac{T_E}{T_H}\right) \tag{6.78}$$

The work input to drive the cooling machine must be [Eq. (6.72)]

$$|w| = |q_L|(T_E/T_L - 1) \tag{6.79}$$

Figure 6.42 Carnot cycle cooling machine.

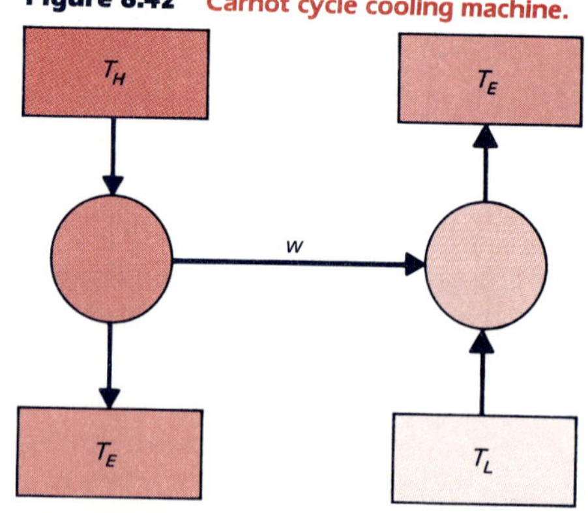

Now since the work output from the Carnot heat engine is equal to the work used to drive the Carnot cooling machine, $|w|$ from Eq. (6.78) may be substituted into Eq. (6.79) to give

$$\text{COP} = \frac{|q_L|}{|q_H|} = \frac{1 - T_E/T_H}{T_E/T_L - 1}$$
$$= \frac{\text{efficiency of Carnot engine}}{\text{COP of reversible cooling machine}} = \frac{\eta_C}{(\text{COP})_{\text{rev}}} \qquad (6.80)$$

where, as before, the COP for the overall cooling cycle is defined as the cooling effect over the energy input to drive the cycle. Note that the COP depends not only on the heat transfer temperature T_H and the low temperature at which energy is removed T_L, but also on the environment temperature T_E to which energy is rejected by both the heat engine and the cooling engine. Equation (6.80) predicts the upper limit of the COP for a cooling device driven by heat transfer rather than work.

Example Problem 6.10

A combined cycle Carnot refrigeration system is receiving heat transfer from a solar collector at $T_H = 400°\text{F}$. The environment is at 70°F, and the system is being used to cool a residence. The cooling coil is to operate at $T_L = 40°\text{F}$. What is the maximum COP of the system?

Solution

Using Eq. (6.80), we find

$$\text{COP} = \frac{500}{860}\frac{860 - 530}{530 - 500} = 6.40$$

Comments

This result indicates the maximum possible COP of a heat-engine-driven cooling machine, because both machines operate on reversible cycles between fixed heat addition and rejection temperatures.

It is useful to compare other proposed heat-transfer-driven cycles with the COP of the combined Carnot cycle, since the Carnot device provides the upper limit of performance. The COP of the Carnot combined-cycle cooling device is shown in Fig. 6.43. Note that

$$\frac{T_E}{T_H} > \frac{T_L}{T_H}$$

for any real system.

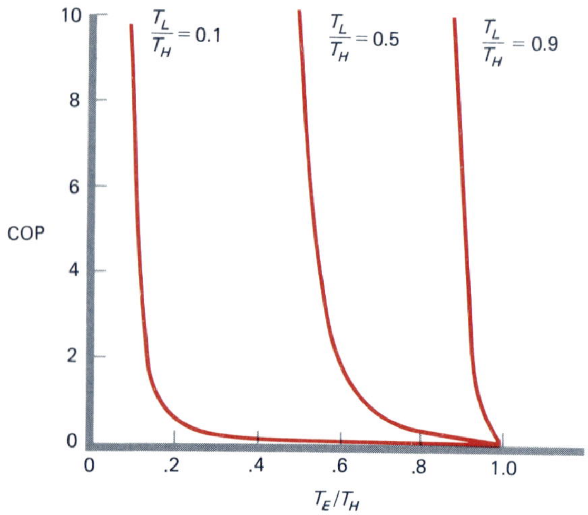

Figure 6.43 Coefficient of performance of a combined Carnot cycle cooling machine.

Other Cycles Driven by Heat Transfer

For some applications, it is useful to have a cycle that uses energy in the form of thermal energy rather than work to drive the cooling cycle, and to not go through a conventional heat engine in the process. For example, if thermal energy from a solar collector, a geothermal source, or a source of waste heat is available, it may be quite inefficient to first drive a heat engine with the waste heat and then use the work output from the heat engine to drive the cooling cycle, as indicated by the results shown in Fig. 6.43 for low values of T_H. A cycle that efficiently uses the heat transfer directly to provide a cooling effect would be more practical. Some cycles of that type are discussed in Chap. 7.

6.6 Concluding Remarks

In this chapter, heat engines that use a gas as the working fluid have been analyzed and described. In addition, cooling cycles based on gaseous working fluids have been introduced. Each of the cycles described here has been covered by using the first and second laws, but the treatment has been idealized through the use of the concept of ideal air-standard processes. More complete engine analysis, which is necessary for careful design and comparison, must include many of the factors neglected here. These include the effects of chemical reaction rates, fluid mechanical effects, mixing within the combustion chambers of internal combustion engines, fluid and mechanical friction, heat transfer among engine parts and working fluids, and many others. However, the material that we have covered provides a good introduction and insight into the important effects that govern the thermodynamics of gas cycle heat engines.

In the next chapter, we examine heat engines and cooling cycles in which the working fluid undergoes a phase change.

Problems

6.1 Find P, v, and T at point 2 in Example Problem 6.1.

6.2S Plot the cycle diagram for the air-standard Carnot cycle operating with 1 kg of working fluid between 20 and 200°C if the maximum volume reached in the cycle is 0.1 m^3 and the maximum pressure in the cycle is 14 MPa. Plot the P-v, T-s, P-h, and h-s diagrams. Assume that the specific heat of air can be taken as temperature-independent.

6.2E Plot the cycle diagram for the air-standard Carnot cycle operating with 1 lbm of working fluid between 60 and 400°F if the maximum volume reached in the cycle is 1 ft^3 and the maximum pressure in the cycle is 2000 psia. Plot the P-v, T-s, P-h, and h-s diagrams. Assume that the specific heat of air can be taken as temperature-independent.

6.3 Find the power output of the Carnot engine of Prob. 6.2 if it operates at 10 cycles/min.

6.4S Assuming temperature-independent specific heats:

(*a*) Draw on the same graph the P-v diagrams for Carnot cycles using working fluids of air and argon. The cycles both have these state points (corresponding to Fig. 6.2):

State	P, kPa	v, m^3/kg
1	3500	0.02
2	—	0.1
3	—	0.2
4	—	—

(*b*) Find the efficiency of each cycle.

6.4E Assuming temperature-independent specific heats:

(*a*) Draw on the same graph the P-v diagrams for Carnot cycles using working fluids of air and argon.

The cycles both have these state points (corresponding to Fig. 6.2):

State	P, psia	v, ft³/lbm
1	500	1
2	—	5
3	—	10
4	—	—

(*b*) Find the efficiency of each cycle.

6.5 An air-standard thermodynamic cycle is outlined below. The specific heats of air can be assumed constant throughout the cycle. The four processes that make up the cycle are:

Process 1-2: Air at standard conditions (state 1) (1 atm, 298 K) is heated at constant volume to state 2, where $P_2 = 10P_1$.

Process 2-3: The air at state 2 is expanded isobarically to state 3, where $v_3 = 3v_2$.

Process 3-4: The air is expanded isotropically through a turbine until it reaches P_1 at state 4.

Process 4-1: The air is cooled isobarically to return it to state 1.

(*a*) Sketch the P-v and T-s diagrams of this cycle. Pay special attention to the relative slopes of the process lines.

(*b*) Find the *numerical value* of the thermodynamic efficiency of this heat engine cycle.

6.6 An ideal air-standard engine is operating on a Carnot cycle as shown in Fig. 6.2 of the text. Beginning with the general definition of heat engine cycle efficiency

$$\eta = \frac{|W_{\text{net}}|}{Q_H}$$

derive the efficiency of the air-standard Carnot engine in terms of the temperatures of the high- and low-temperature heat transfer reservoirs. Use first law expressions for both the net work output of the cycle and the heat transfer input to begin the derivation (i.e., do *not* use $Q_{\text{rev}} = T\,\Delta S$ in your derivation). Equations (5.142) through (5.146) may be used if needed.

6.7 The Stirling cycle engine, with hydrogen as the working medium, can be analyzed as a system with four processes, two of which are shown in Fig. P6.7. Determine the change in internal energy in processes 1-2, 2-3, and 1-3.

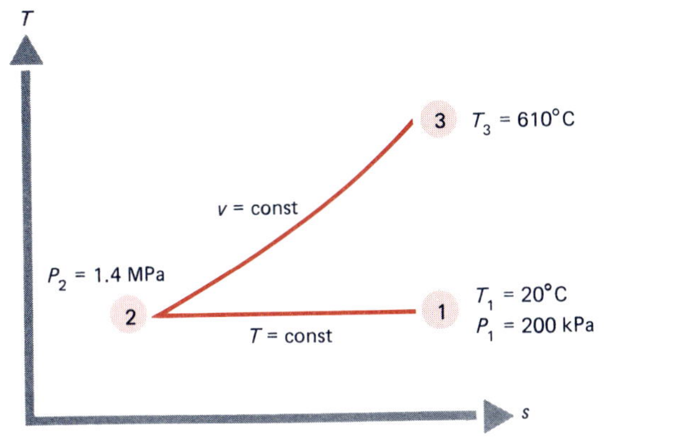

Figure P6.7

6.8 An engine operates on an air-standard Stirling cycle with regeneration. At the beginning of the isothermal compression, the pressure of the air is 0.15 MPa. At the end of the isothermal compression, the pressure is 0.3 MPa. The peak pressure and peak temperature of the cycle are 1.5 MPa and 850°C, respectively. (*a*) Determine q_H. (*b*) Determine q_L. (*c*) Determine η_s.

6.9 An engine operates on an air-standard Stirling cycle with regeneration. If the pressure and temperature at the beginning of the isothermal compression are 3.5 MPa and 150°C, respectively, find the heat transferred to the regenerator during the isometric heat rejection process if the thermal efficiency of the engine is 40 percent.

6.10 An engine operates on an air-standard Stirling cycle with regeneration. If the pressure and temperature at the beginning of the isothermal compression are 0.8 MPa and 100°C, respectively, calculate the net work of the cycle. The pressures at the beginning and end of the isometric heat transfer process are 1.6 and 3.2 MPa, respectively.

6.11 An engine operates on an air-standard Stirling cycle with regeneration. The pressure and temperature at the beginning of the isothermal compression are 0.7 MPa and 100°C, respectively. If the engine has a compression ratio of 3:1 and a mean effective pressure of 1.0 MPa, calculate the heat transferred to the regenerator during the constant-volume heat transfer process.

6.12S A Stirling engine operates between 25 and 450°C. The maximum pressure reached in the engine is limited to 50 atm, and the minimum operating pressure is 1 atm.

Plot the T-s diagram for the cycle, and calculate the cycle efficiency and the work output per revolution (*a*) when the working fluid is air and (*b*) when the working fluid is hydrogen. Assume in each case that the specific heat of the working fluid is temperature-independent and can be evaluated at the arithmetic average of the cycle temperature limits.

6.12E A Stirling engine operates between 80 and 800°F. The maximum pressure reached in the engine is limited to 50 atm, and the minimum operating pressure is 1 atm. Plot the T-s diagram for the cycle, and calculate the cycle efficiency and the work output per revolution (*a*) when the working fluid is air and (*b*) when the working fluid is hydrogen. Assume in each case that the specific heat of the working fluid is temperature-independent and can be evaluated at the arithmetic average of the cycle temperature limits.

6.13S A Stirling engine operates between 25 and 450°C and has a maximum operating pressure of 50 atm and a minimum operating pressure of 1 atm. The engine uses air as the working fluid. Determine the work output per cycle of the engine for two cases:

(*a*) The specific heat of air is temperature-independent.

(*b*) The properties of air are taken to be temperature-dependent.

6.13E A Stirling engine operates between 80 and 800°F and has a maximum operating pressure of 50 atm and a minimum operating pressure of 1 atm. The engine uses air as the working fluid. Determine the work output per cycle of the engine for two cases:

(*a*) The specific heat of air is temperature-independent.

(*b*) The properties of air are taken to be temperature-dependent.

6.14S An ideal air-standard gas-turbine cycle operates on a closed cycle between the pressures of 1 atm and P_2. Intake air to the compressor is at 27°C. The turbine is limited to a maximum temperature of 1227°C by materials limitations. Heat transfer occurs *to* the cycle at a value of $q_H = 200$ kJ/kg.

(*a*) What is the value of the allowable inlet pressure to the gas turbine?

(*b*) What is the cycle efficiency?

(*c*) What is the outlet temperature from the turbine?

6.14E An ideal air-standard gas-turbine cycle operates on a closed cycle between the pressures of 1 atm and P_2.

Intake air to the compressor is at 60°F. The turbine is limited to a maximum temperature of 2250°F by materials limitations. Heat transfer occurs *to* the cycle at a value of $q_H = 100$ Btu/lbm.

(*a*) What is the value of the allowable inlet pressure to the gas turbine?

(*b*) What is the cycle efficiency?

(*c*) What is the outlet temperature from the turbine?

6.15S An ideal air-standard closed Brayton cycle operates between the pressures of 1 atm and P_2. Intake air to the compressor is at 27°C and 1 atm. The turbine is limited to a maximum temperature of 1227°C. Heat transfer from the cycle is $q_L = 100$ kJ/kg.

(*a*) What is the value of the inlet pressure to the turbine?

(*b*) What is the cycle efficiency?

(*c*) What is the compressor outlet temperature?

6.15E An ideal air-standard closed Brayton cycle operates between the pressures of 1 atm and P_2. Intake air to the compressor is at 60°F and 1 atm. The turbine is limited to a maximum temperature of 2250°F. Heat transfer from the cycle is $q_L = 50$ Btu/lbm.

(*a*) What is the value of the inlet pressure to the turbine?

(*b*) What is the cycle efficiency?

(*c*) What is the compressor outlet temperature?

6.16S Points 2 and 4 on air-standard Stirling, Brayton, and Carnot cycle *P-v* diagrams all lie at the same values: $P_2 = 10$ atm, $v_2 = 0.05$ m^3/kg; $P_4 = 1$ atm, $v_4 = 0.40$ m^3/kg. What is the fraction of the ideal Carnot efficiency at which the other cycles will operate? Assume that specific heat values are temperature-independent.

6.16E Points 2 and 4 on air-standard Stirling, Brayton, and Carnot cycle *P-v* diagrams all lie at the same values: $P_2 = 10$ atm, $v_2 = 2$ ft^3/lbm; $P_4 = 1$ atm, $v_4 = 16.4$ ft^3/lbm. What is the fraction of the ideal Carnot efficiency at which the other cycles will operate? Assume that specific heat values are temperature-independent.

6.17S Air-standard Stirling, Ericsson, and simple Brayton external combustion engines all operate between minimum and maximum temperatures of 20 and 450°C. If each cycle has a maximum pressure of 4 MPa and a maximum specific volume of 0.06 m^3/kg, plot the *P-v*, *T-s*, and *P-h* diagrams for the three cycles. (Plot the cycles for the three engines on the same *P-v*, *T-s*, and

P-h graphs for comparison.) Assume that specific heat values are temperature-independent.

6.17E Air-standard Stirling, Ericsson, and simple Brayton external combustion engines all operate between minimum and maximum temperatures of 60 and 800°F. If each cycle has a maximum pressure of 600 psia and a maximum specific volume of 1 ft^3/lbm, plot the *P-v*, *T-s*, and *P-h* diagrams for the three cycles. (Plot the cycles for the three engines on the same *P-v*, *T-s*, and *P-h* graphs for comparison.) Assume that specific heat values are temperature-independent.

6.18 Air enters the compressor of an ideal gas turbine (Brayton cycle) at 25°C and 0.1 MPa. The pressure ratio is 6:1, and the maximum temperature in the cycle is 700°C. Determine the net work (kJ/kg) and the cycle thermal efficiency.

6.19 An ideal gas turbine (Brayton cycle) has a net power output of 100 kW. The working medium is air with $T_1 = 30$°C, $T_3 = 750$°C, and $T_4 = 300$°C. Determine (*a*) the compressor pressure ratio, (*b*) the compressor work (kJ/kg), (*c*) the mass flow rate of air (kg/s), and (*d*) the thermal efficiency.

6.20S A closed Brayton cycle has a two-stage turbine. The high-pressure (first) stage has inlet $T = 1100$ K at 10 atm and an efficiency of 90 percent. Argon is the working fluid. After leaving the first stage at 5 atm, the argon is reheated to 1100 K and enters the second stage, which also has an efficiency of 90 percent. The exhaust from the turbine is at 1 atm. The compressor in the cycle has efficiency of 95 percent, and the compressor inlet temperature is 27°C.

(*a*) Draw a *T-s* diagram for the cycle.
(*b*) Find the cycle efficiency (percentage).
(*c*) Find the cycle efficiency without reheat (percentage).

6.20E A closed Brayton cycle has a two-stage turbine. The high-pressure (first) stage has inlet $T = 2000$°F at 10 atm and an efficiency of 90 percent. Argon is the working fluid. After leaving the first stage at 5 atm, the argon is reheated to 2000°F and enters the second stage, which also has an efficiency of 90 percent. The exhaust from the turbine is at 1 atm. The compressor in the cycle has efficiency of 95 percent, and the compressor inlet temperature is 60°F.

(*a*) Draw a *T-s* diagram for the cycle.
(*b*) Find the cycle efficiency (percentage).
(*c*) Find the cycle efficiency without reheat (percentage).

6.21S A closed Brayton cycle has a two-stage compressor, with each stage having an efficiency (compared to an isentropic compressor) of 95 percent. The working fluid is argon. The compressor inlet is at 1 atm and 27°C. The compressor outlet is at 10 atm. Between the compressor stages, the argon is cooled at $P = 5$ atm to 100°C and then enters the second compressor stage. The single-stage turbine has an efficiency of 90 percent and an inlet temperature of 1100 K.

(*a*) Draw a *T-s* diagram for the cycle.
(*b*) Find the cycle efficiency (percentage).
(*c*) Find the cycle efficiency without intercooling (percentage).

6.21E A closed Brayton cycle has a two-stage compressor, with each stage having efficiency (compared to an isentropic compressor) of 95 percent. The working fluid is argon. The compressor inlet is at 1 atm and 60°F. The compressor outlet is at 10 atm. Between the compressor stages, the argon is cooled at $P = 5$ atm to 200°F and then enters the second compressor stage. The single-stage turbine has an efficiency of 90 percent and an inlet temperature of 2000°F.

(*a*) Draw a *T-s* diagram for the cycle.
(*b*) Find the cycle efficiency (percentage).
(*c*) Find the cycle efficiency without intercooling (percentage).

6.22S A closed Brayton cycle uses argon as the working fluid. The engine has a two-stage compressor with intercooling, and each stage has an isentropic efficiency of 95 percent. The turbine is a two-stage type with reheat, and each stage has an efficiency of 90 percent. Reheat and intercooling occur at 5 atm, and the cycle operates between 1 and 10 atm. The turbine inlet temperature is 1100 K, and the compressor inlet is at 27°C. Reheating brings the second-stage inlet temperature to 1100 K, and intercooling brings the second-stage compressor inlet temperature to 100°C.

(*a*) Draw the *T-s* diagram for the cycle.
(*b*) Find the cycle efficiency (percentage).
(*c*) Find the cycle efficiency without intercooling (percentage).
(*d*) Find the cycle efficiency without reheat (percentage).
(*e*) Find the cycle efficiency with *neither* reheat *nor* intercooling (percentage).

6.22E A closed Brayton cycle uses argon as the working fluid. The engine has a two-stage compressor with intercooling, and each stage has isentropic efficiency of 95 percent. The turbine is a two-stage type with reheat, and each stage has an efficiency of 90 percent. Reheat and intercooling occur at 5 atm, and the cycle operates between 1 and 10 atm. The turbine inlet temperature is 2000°F, and the compressor inlet is at 60°F. Reheating brings the second-stage inlet temperature to 2000°F, and intercooling brings the second-stage compressor inlet temperature to 200°F.

(*a*) Draw the *T-s* diagram for the cycle.
(*b*) Find the cycle efficiency (percentage).
(*c*) Find the cycle efficiency without intercooling (percentage).
(*d*) Find the cycle efficiency without reheat (percentage).
(*e*) Find the cycle efficiency with *neither* reheat *nor* intercooling (percentage).

6.23S An air-standard closed-cycle Brayton engine takes in air at 27°C and 1 atm, and the compressor pressure ratio is 10. The compressor efficiency is 80 percent. The inlet temperature to the turbine is 900°C. What turbine efficiency is necessary to provide zero work output from the engine?

6.23E An air-standard closed-cycle Brayton engine takes in air at 60°F and 1 atm, and the compressor pressure ratio is 10. The compressor efficiency is 80 percent. The inlet temperature to the turbine is 1650°F. What turbine efficiency is necessary to provide zero work output from the engine?

6.24 For the three cases of Example Problem 6.7, determine the fraction of the turbine work output and the fraction of the cycle work that go to drive the compressor in each case. At what value of the turbine efficiency will the compressor work equal the entire turbine work output (assume an ideal compressor)? For an ideal turbine, at what value of the compressor efficiency will the required compressor work equal the entire turbine work output?

6.25S Plot the cycle efficiency versus recuperator effectiveness for isentropic pressure ratios of 6, 8, and 10 for an air-standard Brayton cycle with ideal recuperation operating between $T_2 = 900$ and $T_4 = 300$ K. Discuss how you would choose an optimum isentropic pressure ratio for a given recuperator.

6.25E Plot the cycle efficiency versus recuperator effectiveness for isentropic pressure ratios of 6, 8, and 10 for an air-standard Brayton cycle with ideal recuperation operating between $T_2 = 1500$ and $T_4 = 500$°R. Discuss how you would choose an optimum isentropic pressure ratio for a given recuperator.

6.26S A Carnot engine is connected to a bank of solar collectors in such a way that the outlet fluid at temperature T_H from the solar collectors is used to provide energy to the Carnot engine. Energy is rejected from the Carnot engine at the environment temperature T_L.

The efficiency of a solar collector (sc) is defined by $\eta_{sc} = Q_u/Q_s$, where Q_u is the useful energy used to raise the enthalpy of the fluid passing through the collector and Q_s is the amount of solar energy incident on the collector. For the collector used here, the efficiency is defined at a given time by

$$\eta_{sc} = 0.8 - 0.003(T_H - T_L)$$

If T_L is 300 K, what should be the outlet temperature from the collectors T_H to maximize the ratio of work output from the engine to available solar energy? (Note that the efficiency of the Carnot engine increases with T_H, while the solar collector efficiency decreases.)

6.26E A Carnot engine is connected to a bank of solar collectors in such a way that the outlet fluid at temperature T_H from the solar collectors is used to provide energy to the Carnot engine. Energy is rejected from the Carnot engine at the environment temperature T_L.

The efficiency of a solar collector (sc) is defined by $\eta_{sc} = Q_u/Q_s$, where Q_u is the useful energy used to raise the enthalpy of the fluid passing through the collector and Q_s is the amount of solar energy incident on the collector. For the collector used here, the efficiency is defined at a given time by

$$\eta_{sc} = 0.8 - 0.005(T_H - T_L)$$

If T_L is 520°R, what should be the outlet temperature from the collectors T_H to maximize the ratio of work output from the engine to available solar energy? (Note that the efficiency of the Carnot engine increases with T_H, while the solar collector efficiency decreases.)

6.27 The schematic diagram of an actual gas turbine is shown in Fig. P6.27. Determine (*a*) w_c, (*b*) w_{net}, (*c*) η_{th}, and (*d*) η_t.

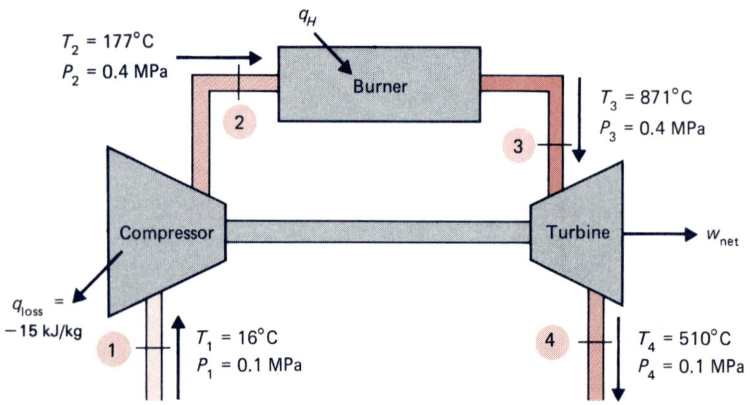

Figure P6.27

6.28 In a gas turbine engine, a 100 percent efficient regenerator transfers 350 kJ/kg from the exhaust stream to the air entering the combustion chamber. If the exhaust gas temperature is 200°C at the regenerator exit, find the inlet temperature to the combustion chamber. Assume that regenerator efficiency is

$$\eta_{reg} \equiv \frac{T_x - T_2}{T_5 - T_2}$$

6.29 The schematic diagram of an actual gas turbine is shown in Fig. P6.29. Determine (*a*) the compressor efficiency, (*b*) net work (kJ/kg), and (*c*) thermal efficiency.

Figure P6.29

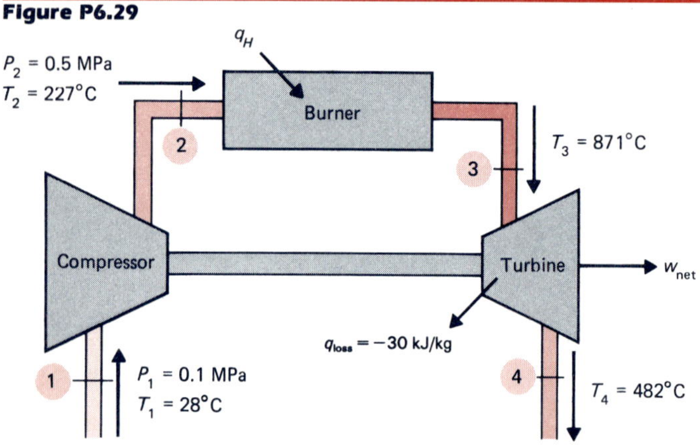

6.30 The schematic diagram of an actual gas turbine is shown in Fig. P6.30. The turbine and compressor are adiabatic. Determine (*a*) w_c, (*b*) T_2, and (*c*) η_{th}.

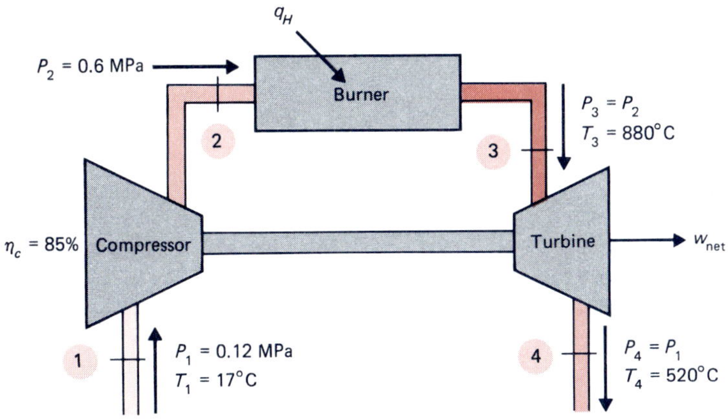

Figure P6.30

6.31 The schematic of an actual gas turbine is shown in Fig. P6.31. The turbine and compressor are adiabatic. Determine (*a*) w_c, (*b*) T_4, and (*c*) η_{th}.

Figure P6.31

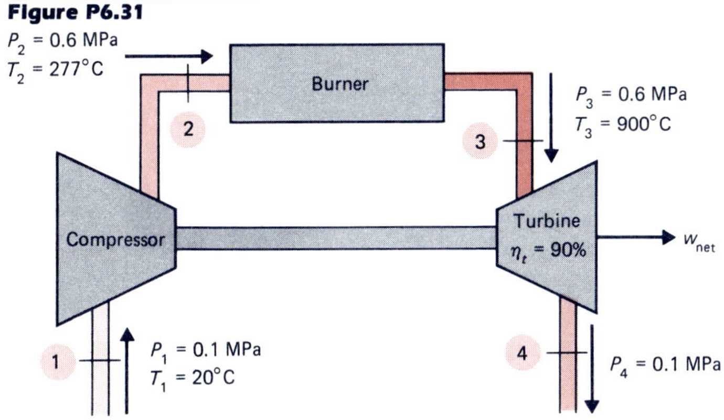

6.32 The compressor for an actual gas turbine requires 300 kJ/kg of work from the turbine to quadruple the inlet pressure. The inlet air temperature is 100°C. What are the compressor exit temperature and the efficiency?

6.33 An ideal gas turbine operates with an 8 : 1 pressure ratio over a range of temperatures of 20 to 1000°C. The combustor provides 0.2 MJ/s. (*a*) What is the airflow rate (kg/s)? (*b*) What percentage of the turbine work is used by the compressor?

6.34 Air at 20°C and 1 atm enters an ideal gas turbine which operates with a pressure ratio of 9. The airflow rate is 0.1 kg/s. Maximum turbine $T = 1400$°C.
(*a*) What rate of heat addition is required?
(*b*) The fuel provides 26 MJ/kg. What $\dot{m}_{\text{fuel}}$ is required?
(*c*) What is the air-fuel ratio for this gas turbine?

6.35 An ideal gas turbine takes 0.2 kg/s of air at 30°C and 50 kPa and raises the pressure 10-fold. The fuel is burned at a rate which provides 0.3 MJ/s in the combustor.
(*a*) How efficient is this turbine?
(*b*) If the fuel is burned twice as quickly, does this double w_{net}? What is the new efficiency?
(*c*) Could this result have been predicted? Why?

6.36 An actual gas turbine with a regenerator is shown in Fig P6.36. (*a*) What are the power output (hp) and the cycle efficiency? (*b*) What would the efficiency be without the regenerator? Assume T_2 and T_3 are unchanged.

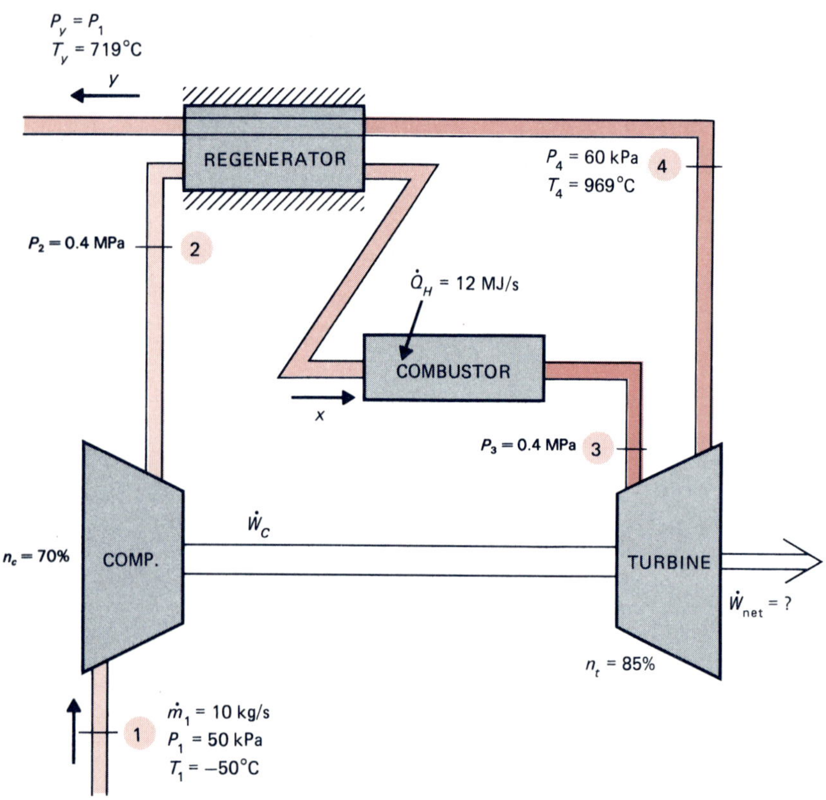

Figure P6.36

6.37 Consider the actual jet propulsion cycle shown in Fig. P6.37. The compressor is adiabatic. Find (*a*) η_c, (*b*) w_c (kJ/kg), and (*c*) $\mathbf{V}_5$ (m/s).

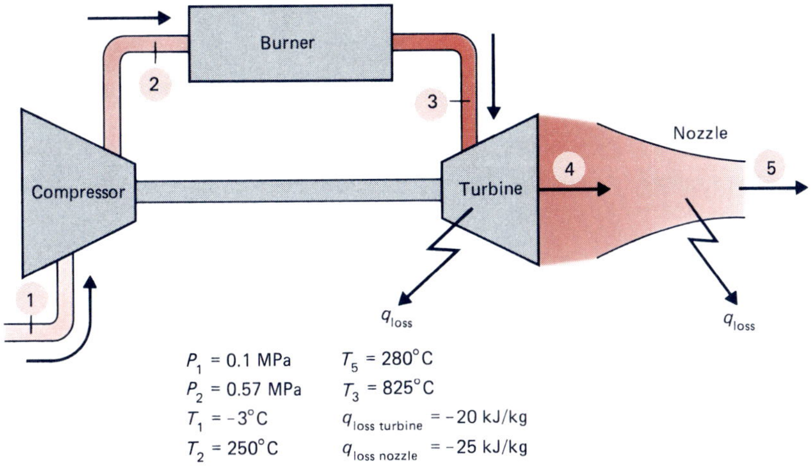

Figure P6.37

6.38 Air flows through the components of a gas turbine shown in Fig. P6.38. Determine (*a*) the mass flow rate of air required and (*b*) the rate of heat transfer in the combustor.

Figure P6.38

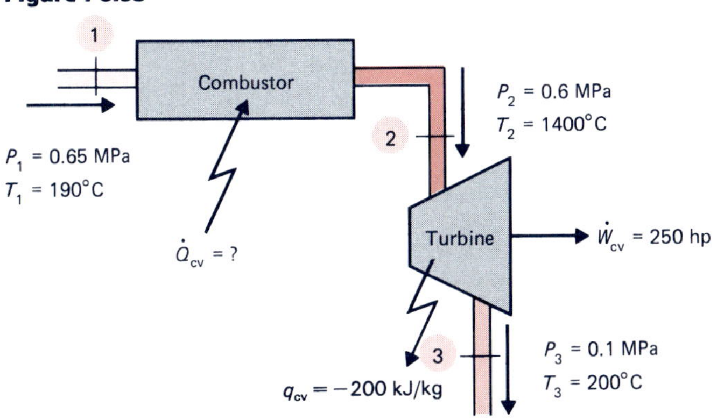

6.39 Match the letter of the specific type of jet engine listed in column A to the appropriate description listed in column B.

Column A		Column B
A Turbojet	________	1. Bypass ratio
B Turboprop	________	2. Typical commercial jet engine

C Turbofan

_______ 3. Compressor work = turbine work

_______ 4. Loses efficiency when **V** > 300 km/h

_______ 5. Afterburner

_______ 6. Ducted fan precedes compressor

_______ 7. Total thrust provided by nozzle

6.40 An air-standard closed-cycle Brayton engine has inlet conditions to the compressor of 25°C and 1 atm, and the compressor pressure ratio is 10. The turbine inlet temperature is 900°C. If the turbine efficiency is 0.6, what compressor efficiency is necessary to obtain zero work output from the engine? (See Fig. P6.40.)

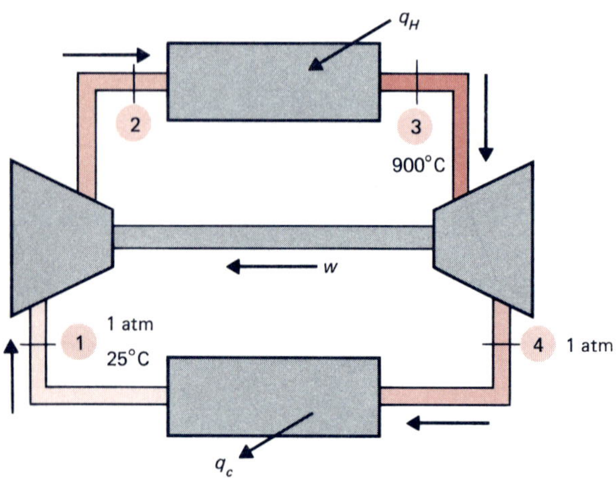

Figure P6.40

6.41 A Brayton engine is running at idle (that is, $w_{net} = 0$). The compressor efficiency is 65 percent at these conditions. Treating the engine as operating on an air-standard cycle with temperature-independent specific heats, find the turbine efficiency for the conditions shown in Fig. P6.41.

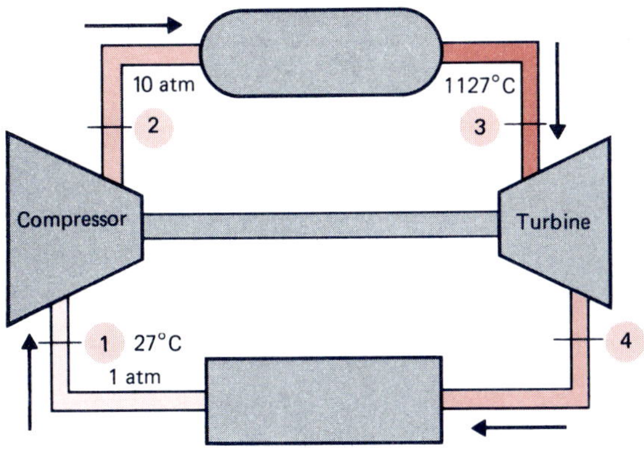

Figure P6.41

6.42 A Brayton engine is running with a net work output of $w_{net} = 110$ kJ/kg. The compressor efficiency is 65 percent at these conditions. Treating the engine as operating on an air-standard cycle with temperature-independent specific heats, find the turbine efficiency for the conditions shown in Fig. P6.42.

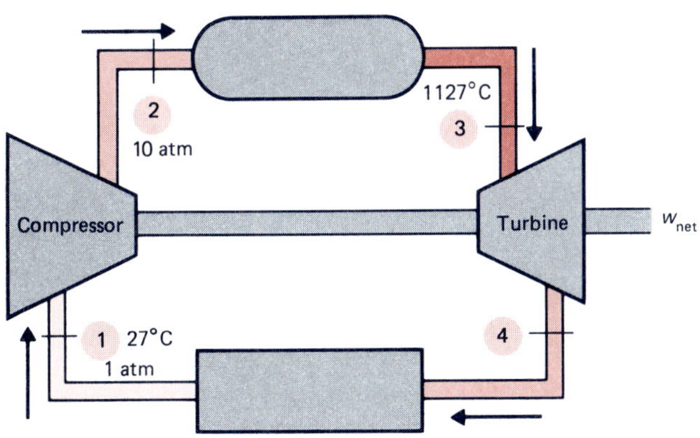

Figure P6.42

6.43 A gas-turbine engine operating on a simple air-standard Brayton closed cycle has an inlet condition to the compressor of 27°C and 1 atm. The *compressor* has an efficiency of 70 percent at a pressure ratio of 6, and the *turbine* has an efficiency of 80 percent. What is the specific net work output from the engine (kJ/kg) if the turbine inlet temperature is limited to 1227°C? Assume constant specific heats.

6.44 A gas-turbine engine operating on a simple air-standard Brayton closed cycle has an inlet condition to the compressor of 27°C and 1 atm. The *turbine* has an efficiency of 65 percent at a pressure ratio of 4. What is the minimum *compressor* efficiency that will allow net work output from the engine if the turbine inlet temperature is limited to 1227°C? Assume constant specific heats.

6.45 The ideal air-standard Brayton closed cycle shown in Fig. P6.45 has a *cycle* efficiency of 25 percent and a compressor efficiency of 60 percent. What is the turbine efficiency?

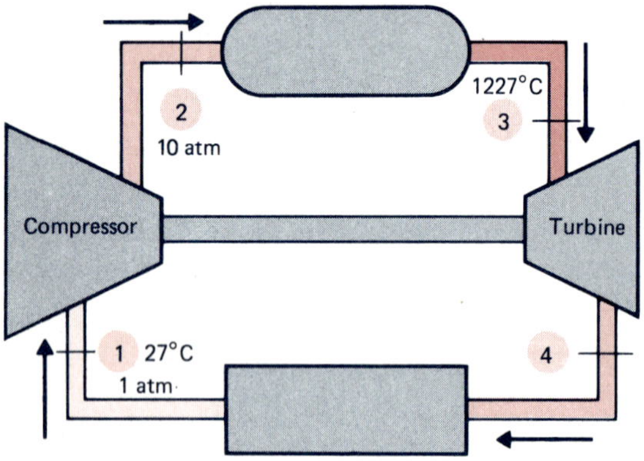

Figure P6.45

6.46 Two components of a gas turbine are shown in Fig. P6.46. Determine the heat transferred to the combustion chamber (kJ/kg) and the power output of the turbine (kW).

Figure P6.46

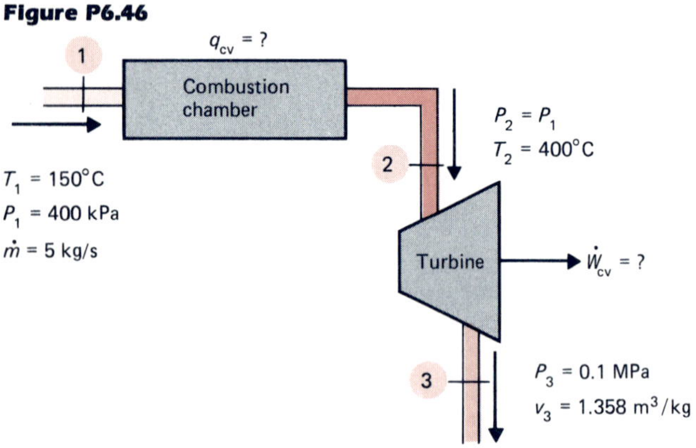

6.47 An ideal turbojet with afterburner operates at altitudes where the ambient pressure is 10 kPa. The inlet air temperature is −20°C, and the pressure ratio is 7.5 : 1. The maximum turbine temperature allowable is 1500°C. The airflow rate is 8.2 kg/s.

(*a*) What exit velocity is obtained without the afterburner?
(*b*) What is the rate of heat transfer required (MJ/s)?
(*c*) Fuel is consumed twice as quickly when the afterburner is used. What is the new exit velocity?

6.48 An ideal turbofan takes in air at 0.1 MPa and 10°C. The pressure ratio is 8 : 1, and the maximum temperature is 1200°C. The fan takes 201 kJ/kg to run. What is the nozzle exit velocity?

6.49 A turbofan takes in air at 100 m/s. As a turbojet, without extracting power to run the fan, an exit velocity of 915 m/s is possible. Now the hot exhaust gas velocity is 660 m/s, and the cold air exits at 140 m/s. For the turbofan to have the same thrust as the turbojet, what must the bypass ratio (BPR) be? Assume the nozzles are perfectly expanding and $\dot{m}_H$ is the same for both engines.

6.50 The schematic for an actual gas turbine is as shown in Fig. P6.50. If the compressor and turbine are adiabatic, determine (*a*) compressor work (kJ/kg), (*b*) net work (kJ/kg), (*c*) thermal efficiency, and (*d*) regenerator efficiency.

Figure P6.50

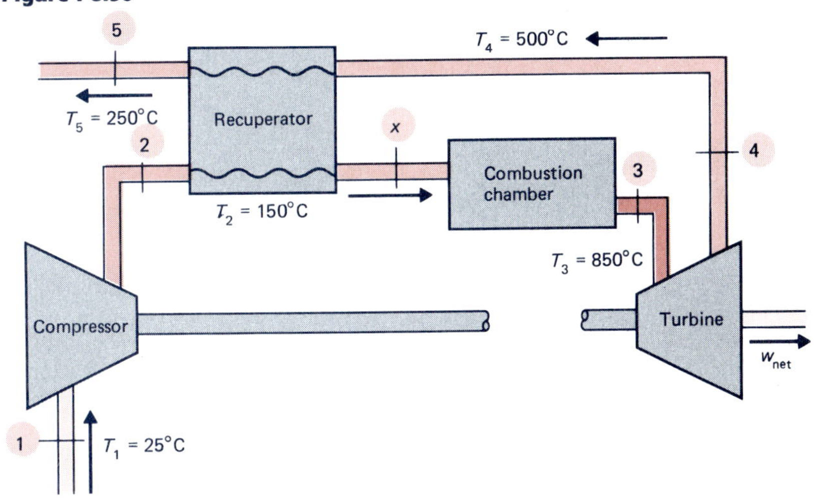

6.51 The schematic for a new gas turbine is as shown in Fig. P6.51. The turbine is adiabatic. Determine the following if $w_{turb} = 250$ kJ/kg: (*a*) the highest pressure to which the recuperator is subjected, (*b*) thermal efficiency η_{th}, and (*c*) turbine efficiency η_t. Is the device possible?

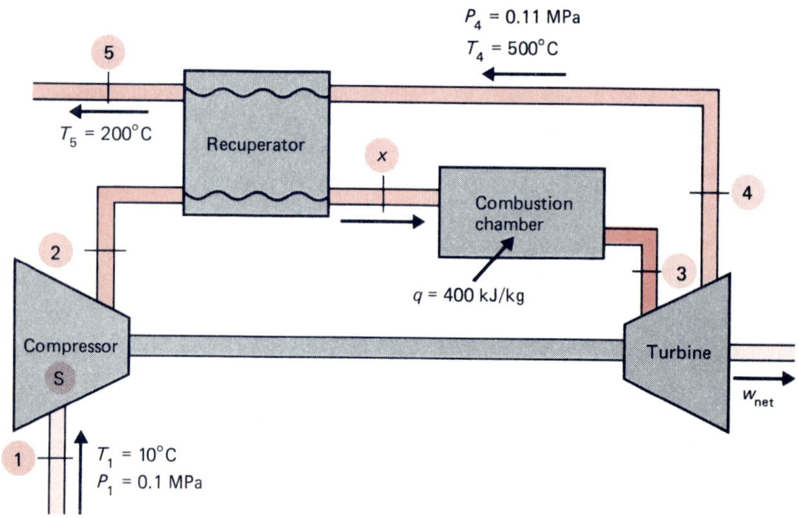

Figure P6.51

6.52 The VW microbus operates as shown in Fig. P6.52. Determine T_2, ${}_1w_2$, T_4, T_5, T_y, and η_{th}. The compressor and turbines are adiabatic.

Figure P6.52

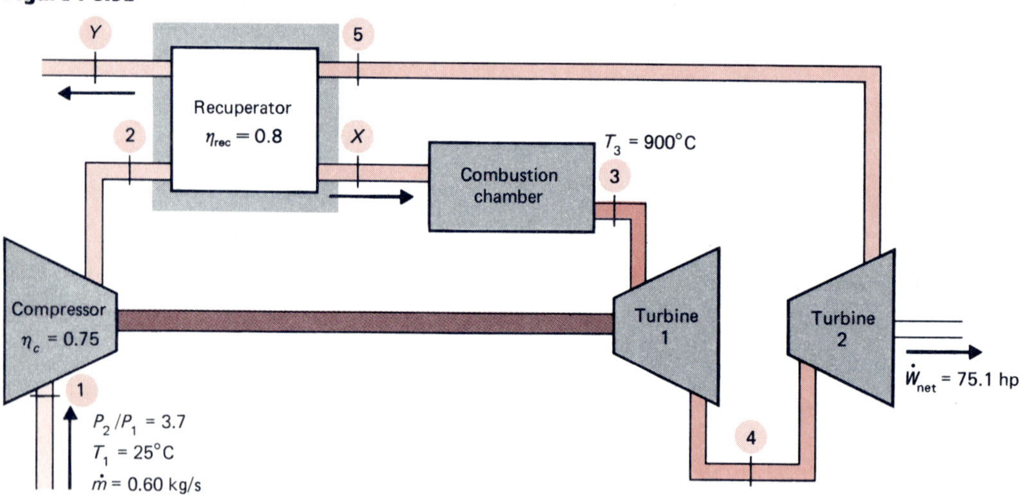

6.53 The sketch of an *actual* jet propulsion cycle is shown in Fig. P6.53. The compressor and nozzle are adiabatic. Determine (*a*) the compressor work, (*b*) T_2 (K), and (*c*) $\mathbf{V}_5$ (m/s).

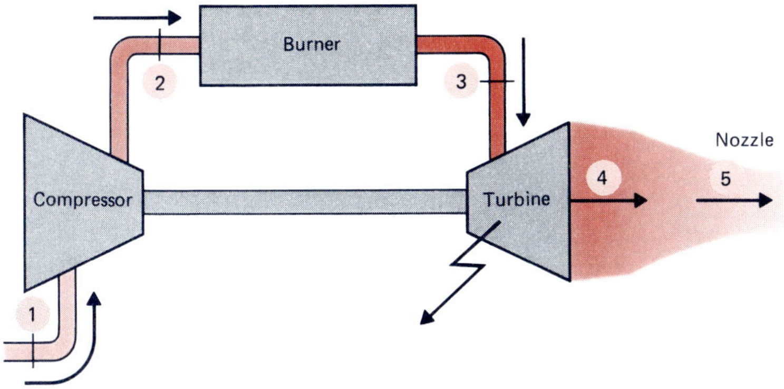

Figure P6.53

6.54 An *ideal* turbojet operates under the following conditions:

Compressor inlet temperature = 18°C
Turbine inlet temperature = 870°C
Nozzle exit temperature = 412°C
Compressor inlet pressure = 0.1 MPa
Compressor exit pressure = 0.6 MPa
Ambient pressure = 0.1 MPa

What is the exit velocity of the air from the nozzle?

6.55 A Boeing 747 engine has a bypass ratio of 6:1. (What kind of jet engine is this?) If the mass flow rate of the gas exiting the nozzle is 100 kg/s, what is the mass flow rate of the air entering the engine?

6.56S An internal combustion Brayton turbine engine is being designed for jet propulsion. The engine has no regeneration. The turbine and compressor have efficiencies of 80 percent, the maximum allowable temperature on the turbine blades is 750°C, and air entering the compressor may be assumed to be at 1.0 atm and 0°C. What is the kinetic energy (per kilogram) of the gases leaving the engine nozzle if they exit at 70°C? Assume that constant air specific heat and a constant mass flow rate can be used for the gases passing through the engine and that the gases leave the turbine at 1 atm.

6.56E An internal combustion Brayton turbine engine is being designed for jet propulsion. The engine has no regeneration. The turbine and compressor have efficiencies of 80 percent, the maximum allowable temperature on the turbine blades is 1400°F, and air entering the compressor may be assumed to be at 1.0 atm and 30°F. What is the kinetic energy (per pound-mass) of the gases leaving the engine nozzle if they exit at 150°F? Assume that constant air specific heat and a constant mass flow rate can be used for the gases passing through the engine and that the gases leave the turbine at 1 atm.

6.57 Repeat Prob. 6.56, using temperature-dependent properties for air.

6.58S A two-stroke motorcycle engine displaces 250 cm^3 and operates at 4000 rpm. For a compression ratio of 8:1, what is the air-standard cycle efficiency, and how many kilowatts will the air-standard engine develop? (Assume that at the end of the power stroke, the cylinder pressure has dropped to 350 kPa and the compression stroke begins at 1 atm.)

6.58E A two-stroke motorcycle engine displaces 250 cm^3 and operates at 4000 rpm. For a compression ratio of 8:1, what is the air-standard cycle efficiency, and how many horsepower will the air-standard engine develop? (Assume that at the end of the power stroke, the cylinder pressure has dropped to 50 psia and the compression stroke begins at 1 atm.)

6.59S A hot-rod mechanic has an engine from a 1966 British sports car. The engine displaces 4.2 liters, has an original compression ratio of 10.2:1, and has six cylinders. Assume that the pressure at the end of the power stroke is 300 kPa just before the exhaust valve opens. The pistons in the original engine are 12 cm in diameter.

(*a*) If the mechanic bores the cylinders and replaces the pistons with new pistons that are 0.5 cm larger in diameter than the originals, by what percentage will the engine efficiency change (assuming the 300-kPa pressure is unchanged)?

(*b*) If the mechanic replaces the engine crankshaft and piston connecting rods, so that the original pistons have their stroke increased by 0.5 cm, by what percentage will the engine efficiency change?

(*c*) If the cylinder head is milled so that the original clearance is reduced to 0.4 cm, by what percentage will the engine efficiency change?

Assume in all cases that the ideal air-standard efficiency is desired.

6.59E A hot-rod mechanic has an engine from a 1966 British sports car. The engine displaces 4.2 liters, has an original compression ratio of 10.2:1, and has six cylinders. Assume that the pressure at the end of the power stroke is 40 psia just before the exhaust valve opens. The pistons in the original engine are 4.75 in diameter.

(*a*) If the mechanic bores the cylinders and replaces the pistons with new pistons that are 0.2 in larger in diameter than the originals, by what percentage will the engine efficiency change (assuming the 40-psia pressure is unchanged)?

(*b*) If the mechanic replaces the engine crankshaft and piston connecting rods, so that the original pistons have their stroke increased by 0.2 in, by what percentage will the engine efficiency change?

(*c*) If the cylinder head is milled so that the original clearance is reduced to 0.2 in, by what percentage will the engine efficiency change?

Assume in all cases that the ideal air-standard efficiency is desired.

6.60 An air-standard Otto cycle operates with inlet conditions of 1 atm and 27°C, and the compression ratio of the engine is 9 : 1. The maximum temperature reached in the cycle is 1800 K. The temperature-dependent specific heat at constant volume for air can be approximated by

$$c_v = (2.4 \times 10^{-3})T - (1.04 \times 10^{-6})T^2 \qquad \text{kJ/(kg}\cdot\text{K)}$$

where T is the absolute temperature in the range $300\text{ K} \le T \le 1800\text{ K}$. Derive an expression for the efficiency of this Otto cycle. The gas constant for air is 0.2870 kJ/(kg·K).

6.61 An engine operates on an air-standard Otto cycle. The net work of the cycle is 900 kJ/kg when the maximum cycle temperature is 3000°C. If the temperature at the end of the isentropic compression is 600°C, determine the compression ratio of the engine.

$$\left(\text{Hint: } \eta_{\text{th}} = 1 - \frac{1}{r_v^{k-1}}\right)$$

6.62 An engine operates on an air-standard Otto cycle. The pressure and temperature at the beginning of the compression stroke are 110 kPa and 40°C, respectively. The temperature at the end of the compression stroke is 500°C. If there is a heat transfer to the cycle of 1500 kJ/kg, calculate the maximum pressure that the walls must withstand.

6.63 An engine operates on an air-standard Otto cycle. The pressure and temperature at the beginning of the isentropic compression are 100 kPa and 40°C, respectively. If the pressure at the end of the compression is 2.0 MPa and the net cycle work is 300 kJ/kg, determine the maximum temperature of the cycle.

6.64 An engine operates on an air-standard Otto cycle with a compression ratio of 10 : 1. If the pressure and temperature at the beginning of the compression stroke are 0.15

MPa and 30°C, respectively, determine the work done on the air during the compression stroke (kJ/kg).

6.65 An engine operates on an air-standard Otto cycle. The pressure and temperature at the begining of the isentropic compression are 0.12 MPa and 35°C, respectively. If the peak pressure and peak temperature of the cycle are 4.8 MPa and 2500°C, determine the net work done per cycle (kJ/kg) and the cycle thermal efficiency.

6.66 Match the changes in engine parameters listed in column A with the effects in column B. Items in column A may apply to more than one item in column B.

Column A	Column B
A Decrease r_v.	______ 1. Decrease knock
B Advance spark timing.	______ 2. Increase knock
C Decrease surface volume ratio.	______ 3. Increase HC, CO
D Add tetraethyl lead to fuel.	______ 4. Increase NO_x
E Increase air-fuel ratio. toward stoichiometric	______ 5. Decrease HC, CO
F Increase air-fuel ratio beyond stoichiometric.	______ 6. Decrease NO_x

6.67 An engine operates on an air-standard Otto cycle with a compression ratio of 9 : 1. If the pressure and temperature at the end of the compression stroke are 0.8 MPa and 700°C, respectively, determine the net work of the cycle if the pressure at the end of the heat addition process is 3.0 MPa.

6.68 A supercharged spark-ignition engine operates on an air-standard Otto cycle. The inlet conditions are 210 kPa and 110°C, and the maximum temperature and pressure are limited to 2250°C and 7.5 MPa. Find the net work and thermal efficiency.

6.69 The heat transfer to an engine operating on an air-standard Otto cycle is 2100 kJ/kg. If the initial temperature and pressure are 15°C and 100 kPa, respectively, calculate the net cycle work and peak pressure. The compression ratio is 7 : 1.

6.70 An engine operating on the air-standard Otto cycle produces a net cycle work of 350 kJ/kg. Determine the cycle efficiency and peak cycle temperature if the compression ratio is 9.5 : 1 and the temperature at the beginning of the compression stroke is 25°C.

6.71 Air in an Otto cycle engine is compressed to a temperature and pressure of 450°C and 2.5 MPa. After the power stroke, the conditions are 600°C and 0.45 MPa. Find the peak cycle temperature, heat transfer addition, and efficiency.

6.72 An engine operating on the ideal (constant-specific-heat) air-standard Otto cycle has a thermal efficiency of 35 percent. At the beginning of the compression stroke (state 1), the temperature and pressure of the air in the cylinder are 300 K and 1 atm. The maximum temperature reached in the cycle is 2000 K. Determine the temperature, pressure, and specific volume of the air at the end of each process in the cycle.

6.73 A "four-stroke" internal combustion engine operates on an air-standard Otto cycle. The engine is a 454-in^3 displacement (7.44×10^{-3} m^3) V-8 engine, and it has a compression ratio of 8.5 : 1.

(*a*) What is the air-standard efficiency of the engine (percentage)?

(*b*) If the pressure at the end of the power stroke is 200 kPa just before the exhaust valve opens, how much power (kW) is produced by the engine at 3600 rpm?

6.74 A six-cylinder air-standard Otto cycle engine has an air-standard efficiency of 60 percent, and the state at the beginning of the compression stroke is $T = 27$°C, $P =$ 1 atm. The clearance volume (i.e., the volume of air in the cylinder when the piston is at the top of the stroke, top dead center) is 7.085×10^{-5} m^3. What is the engine displacement (m^3)?

6.75 An engine operating on an air-standard Otto cycle has a compression ratio of 8 : 1. Heat transferred to the working medium during the cycle is 2000 kJ/kg. If $P_1 = 0.11$ MPa and $T_1 = 22$°C, find P_2, P_3, P_4, T_2, T_3, T_4, and the thermal efficiency.

6.76 An engine operates on the air-standard diesel cycle with a compression ratio of 18. The air pressure and temperature at the beinning of compression are 0.12 MPa and 43°C. If the maximum temperature is 1992 K and the heat transfer is 1274 kJ/kg, find the maximum pressure for which the cylinder must be designed. Also find T_2.

6.77 The given data were measured for an air-standard diesel cycle (see Fig. 6.33). Find the thermal efficiency.

$P_1 = 0.1$ MPa
$T_1 = 300$ K
$P_2 = 4.0$ MPa
$T_3 = 2850$ K

6.78 The combustion process in an air-standard diesel cycle adds 800 kJ/kg to the medium. The minimum temperature and pressure are 20°C and 0.1 MPa while the maximum temperature is 1000°C. Find the thermal efficiency.

6.79 An air-standard diesel engine has a compression ratio of 20 and a cutoff ratio of 5. If the intake air pressure and temperature are 0.1 MPa and 27°C, find the heat transfer to the engine and net work per cycle.

6.80 An engine operating on the air-standard diesel cycle has a compression ratio of 16. The pressure and temperature at the beginning of compression are 0.1 MPa and 25°C. If the heat transfer to the engine is 1000 kJ/kg, find the cycle thermal efficiency.

6.81 The following data apply to an air-standard diesel cycle. Find the thermal efficiency and the pressure at the end of the expansion process.

Begin compression:	$P = 0.1$ MPa, $T = 30$°C
End compression:	$T = 500$°C
End of combustion:	$T = 4000$ K

6.82 An air-standard diesel cycle engine operates as follows: At the end of the power stroke, the pressure is 0.24 MPa and the temperature is 550°C, and at the end of the compression stroke the pressure and temperature are 4.5 MPa and 700°C, respectively. Determine the (*a*) compression ratio, (*b*) cutoff ratio, (*c*) heat transfer to the engine per cycle (kJ/kg), (*d*) net work output per cycle (kJ/kg), and (*e*) thermal efficiency.

6.83 An air-standard diesel cycle engine operates as follows: At the beginning of the compression stroke the temperature and pressure are 30°C and 0.1013 MPa, and the temperature and pressure at the end of the combustion process are 1650°C and 4.5 MPa. Determine (*a*) compression ratio, (*b*) cutoff ratio, (*c*) heat transfer input per cycle (kJ/kg), (*d*) net work per cycle, and (*e*) η_{th}.

6.84 An air-standard diesel cycle engine operates as follows: The temperatures at the beginning and end of the compression stroke are 30 and 700°C, respectively. The net work per cycle is 590.1 kJ/kg, and the heat transfer input per cycle is 925 kJ/kg. Determine the (*a*) compression ratio, (*b*) maximum temperature of the cycle, and (*c*) cutoff ratio.

6.85 An air-standard diesel cycle operates as follows: The temperature at the beginning of the compression stroke is 25°C and at the end of the expansion process is 550°C. The compression ratio is 16, and the cutoff ratio is 2.5. Determine (*a*) the maximum temperature of the cycle, (*b*) the heat transfer input per cycle (kJ/kg), (*c*) net work per cycle, and (*d*) the thermal efficiency.

6.86 An air-standard diesel cycle engine operates as follows: The compression ratio is 15 while there is a heat transfer input of 950 kJ/kg per cycle. The maximum temperature is 1800°C, and the pressure prior to the compression process is 0.1013 MPa. Determine the (*a*) cutoff ratio, (*b*) net work per cycle (kJ/kg), and (*c*) thermal efficiency.

6.87 An air-standard diesel cycle engine operates as follows: At the beginning of the compression stroke the temperature and pressure are 20°C and 0.1013 MPa, and the temperature and pressure at the end of the combustion process are 1800°C and 5 MPa. Determine the (*a*) compression ratio, (*b*) cutoff ratio, and (*c*) thermal efficiency.

6.88 An air-standard diesel cycle has a compression ratio of 12 : 1 and an efficiency of $\eta_D = 60$ percent. The state at the start of the compression process is $T = 27$°C and $P = 1$ atm. What is the temperature at the end of the constant-pressure process (start of the isentropic expansion process)? (Sketch the *P*-*v* diagram, and number the states. Note that the compression ratio must always be greater than the expansion ratio and that some iteration may be necessary.)

6.89 A diesel engine displaces 6 liters and operates at 1600 rpm. The compression ratio of the engine is 12. The pressure in the cylinder reaches 2 atm at the end of the power stroke just as the exhaust valve opens. Graph the power developed by the air-standard engine for expansion ratios in the range of 2 to 12.

6.90 A diesel engine has a compression ratio of $r_1 = 12$. Plot the diesel engine efficiency versus the expansion ratio r_2 for $1 \leq r_2 \leq 12$. Pay special attention to the case of $r_2 = 1$, and derive a relation for the cycle efficiency for that special case. Plot the cycle for $r_2 = 1$ on a *P*-*v* diagram.

6.91 Using the definition of efficiency for a control mass derived in Chap. 5, derive expressions for the cycle efficiency of air-standard Otto and diesel cycles with nonideal compression and expansion.

6.92S The COP of a reversible refrigeration cycle is 4.0. The cycle is able to remove 5 MJ/h from the refrigerator.
(*a*) What is the rate of work necessary to run the cycle?
(*b*) What temperature is maintained in the refrigerator if the cycle rejects a heat transfer to a room at 27°C?

6.92E The COP of a reversible refrigeration cycle is 4.0 The cycle is able to remove 5000 Btu/h from the refrigerator.
(*a*) What is the rate of work necessary to run the cycle?
(*b*) What temperature is maintained in the refrigerator if the cycle rejects a heat transfer to a room at 80°F?

6.93 A *reverse* air-standard Brayton cycle operates with conditions shown in Fig. P6.93. The compressor has an efficiency of 68 percent, and the turbine has an efficiency of 72 percent. What is the coefficient of performance of this refrigeration cycle?

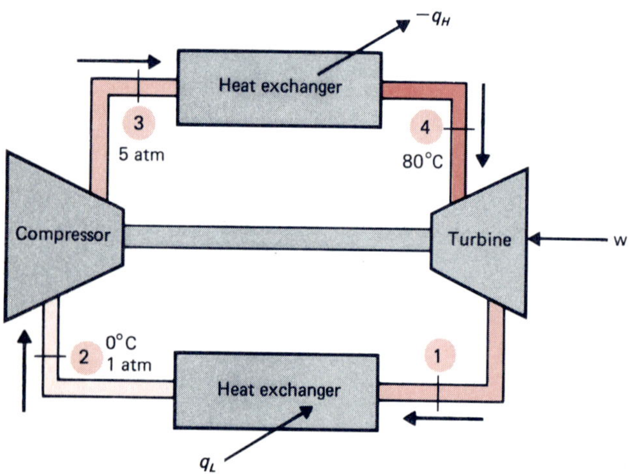

Figure P6.93

6.94S Aircraft commonly employ open reverse Brayton cooling systems for providing cabin and avionics cooling as diagrammed in Fig. 6.39.

For a particular design (Fig. P6.94S), air entering the engine compressor is at 70 kPa and 20°C, has **V** = 250 m/s, and the compressor has efficiency of 70 percent. The main compressor outlet pressure is 2.5 MPa. How much heat transfer per kg of cabin air must be provided by the heat exchanger to meet the design requirements?

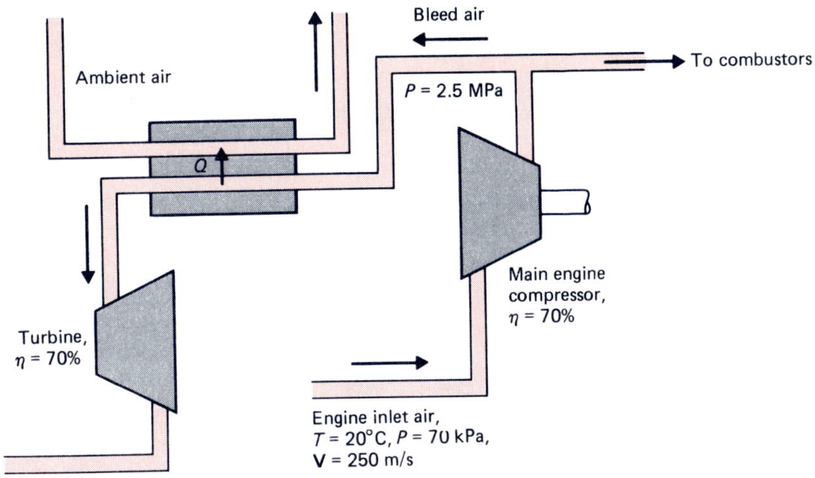

Figure P6.94S

6.94E Aircraft commonly employ open reverse Brayton cooling systems for providing cabin and avionics cooling as diagrammed in Fig. 6.39.

For a particular design (Fig. P6.94E), air entering the engine compressor is at 10 psia and 70°F, has velocity **V** = 800 ft/s, and the compressor has efficiency of 70 percent. The main compressor outlet pressure is 225

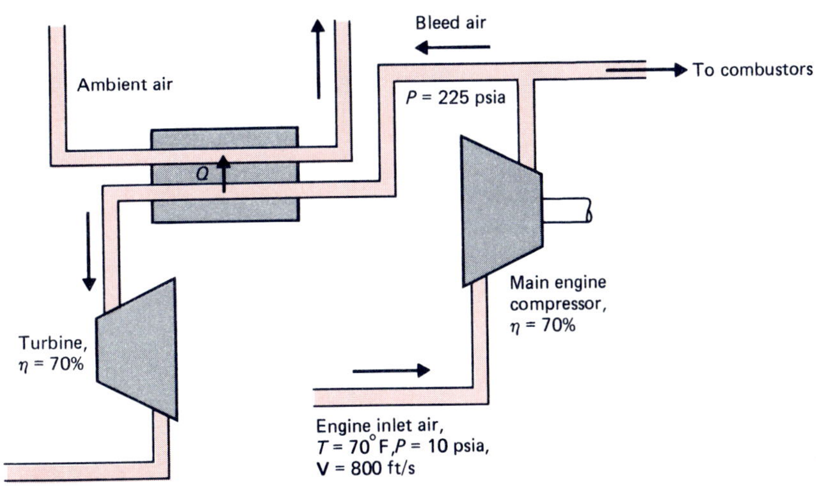

Figure P6.94E

psia. How much heat transfer per lbm of cabin air must be provided by the heat exchanger to meet the design requirements?

6.95 Derive the COP of a reverse Ericsson refrigeration cycle with a compressor isothermal efficiency of η_c and a turbine isothermal efficiency of η_T, where the isothermal efficiency is defined by $\eta_T = w_a/w_{\text{isothermal, rev}}$.

6.96 Derive an expression for the COP of a reverse Brayton cooling cycle in terms of the turbine efficiency η_T, the compressor efficiency η_c, the pressure ratio, and the ratio of turbine to compressor inlet temperatures. From the result, find expressions for the minimum allowable compressor efficiency for a given turbine efficiency and the minimum allowable turbine efficiency for a given compressor efficiency that will provide a given COP.

6.97 Find relations for the COP of cooling systems made of (*a*) a Stirling heat engine driving a reverse Stirling cooling cycle and (*b*) a Brayton heat engine driving a reverse Brayton cooling cycle. In both cases, assume that all components of the cycle are reversible and that both cycles receive energy at a maximum temperature of T_H and reject energy to the surroundings at T_E, while providing cooling at T_L. The Brayton engine and Brayton cooling cycle have no regenerators and operate at atmospheric pressure when they are at T_E. Assume temperature-independent specific heats for the working fluids in each cycle.

6.98S An inventor plans a system that operates as follows: A heat engine receives energy from a source at 540°C, gives a work output of 0.50 kJ/kg for each 1 kJ/kg of heat transferred to the working fluid at that temperature, and rejects the remaining energy to the environment at 25°C. Part of the work output from the heat engine drives a heat pump with COP_{HP} of 4. The heat pump is used to take energy from the environment at 25°C and pump it to 540°C, where it becomes the energy source for the heat engine. Since COP_{HP} is 4, only 0.25 kJ/kg of work from the heat engine is required to drive the heat pump for each 1 kJ/kg delivered to the heat engine. Thus, no outside energy source is required to drive the heat engine, and 0.25 kJ/kg is left of the engine work output to do useful work.

Is there a flaw in this device? Discuss your answer in terms of (*a*) the second law of thermodynamics and (*b*) a detailed analysis of the possible efficiency of the heat engine and the possible COP_{HP}.

6.98E An inventor plans a system that operates as follows: A heat engine receives energy from a source at 1000°F, gives a work output of 0.50 Btu/lbm for each Btu/lbm of heat transferred to the working fluid at that temperature, and rejects the remaining energy to the environment at 70°F. Part of the work output from the heat engine drives a heat pump with a COP_{HP} of 4. The heat pump is used to take energy from the environment at 70°F and pump it to 1000°F, where it becomes the energy source for the heat engine. Since COP_{HP} is 4, only 0.25 Btu/lbm of work from the heat engine is required to drive the heat pump for each 1 Btu/lbm delivered to the heat engine. Thus, no outside energy source is required to drive the heat engine, and 0.25 Btu/lbm is left of the engine work output to do useful work.

Is there a flaw in this device? Discuss your answer in terms (*a*) the second law of thermodynamics and (*b*) a detailed analysis of the possible efficiency of the heat engine and the possible COP_{HP}.

6.99 A Humphrey pump operates by using the liquid in the pump as a piston. The cycle of operation is shown in Fig. 6.36. Use assumptions similar to those used for the Otto cycle.

(*a*) Draw the *P-v* and *T-s* diagrams for the air-standard Humphrey pump cycle.

(*b*) Derive an expression for the thermodynamic efficiency of the air-standard Humphrey pump cycle.

7

Vapor Cycle Analysis

The Science of heat is not only of great curiosity, from the multitude of astonishing phenomena it offers to our contemplation, but it is likewise extremely interesting from its interesting connection with all the useful arts, and generally with all the mechanical occupations of human life.
Benjamin Thompson, Count Rumford.

The turbine shaft and bucket (blade) assembly from a utility steam turbine during maintenance. (*General Electric.*)

7.1 Vapor Cycle Analysis

As noted at the beginning of Chap. 6, heat engine cycles can be categorized in a number of ways, and a convenient division is into cycles that use a single-phase working fluid and those, such as the steam engine, in which the working fluid undergoes a change of phase during the cycle. This chapter is devoted to the study of cycles in which the working fluid undergoes a vapor-liquid phase change.

If phase change occurs, then the properties of the particular working fluid (water, ammonia, refrigerant 12, etc.) must be considered in the analysis of a cycle. The convenience of the air-standard cycle approximations is no longer useful, because the heat of vaporization and the differences in vapor and liquid properties must be considered, and these properties are quite different for different working fluids. Thus, vapor cycle analysis becomes quite dependent upon the availability of accurate thermodynamic property data for the particular working fluid.

Heat transfer during evaporation and condensation process takes place at close to isothermal conditions in a constant-pressure process. Such heat transfer also requires a much smaller temperature difference between the surroundings (for example, the boiler tube surface) and the system (working fluid) than is the case for heat transer between a hot surface and a gas. Thus, the evaporation and condensation processes approach the ideal of an isothermal reversible heat transfer process much more closely than can be obtained in a gas cycle. There is thus the promise of a practical approach to the Carnot cycle by using phase-change processes. Further, the efficient heat transfer possible in phase-change processes allows much smaller areas for a given amount of heat transfer than can be used in heat transfer processes to gases. Thus, vapor cycle devices can usually be designed to be much more compact for a given power than is the case for gas cycles.

Most vapor cycles are external heat transfer cycles, because of the difficulty of removing combustion products before condensation and reuse of the working fluid.

7.1.1 Vapor Cycle Heat Engines

Heat engines operating on vapor cycles are the standard large-scale devices for producing mechanical energy from an external heat transfer source. Commercial power plants using coal, natural gas, fuel oil, and nuclear energy generally use a vapor cycle, as do more unusual plants tied to geothermal and solar energy sources. (The exception is the fairly widespread use of Brayton cycle gas turbines for electric power production. Diesel and gasoline engines are often used for standby and small-generator applications.)

In this chapter, the simple Rankine cycle and the enhancements to the simple cycle that can be used to improve efficiency are examined. The changes necessary for the adaption of the Rankine cycle to use with nuclear and solar plants are also analyzed, and the use of waste energy from other sources is discussed.

7.1.2 Vapor Cycle Cooling Devices and Heat Pumps

The standard air conditioning cycle used for residential and commercial-scale air conditioning systems uses a refrigerant (working fluid) that is chosen to have an appropriately large saturation temperature change with pressure. The properties must be such that the saturation temperature at low pressure will allow heat transfer from the surroundings, to cause evaporation and provide effective cooling of air for comfort purposes; when the refrigerant vapor is compressed to an easily attainable higher pressure, its saturation temperature should be high enough that heat transfer to the cooler surroundings will cause condensation.

At present, the most widely used standard refrigerants with these characteristics are chloro-fluorinated hydrocarbons and ammonia. Environmental concerns are causing the gradual replacement of chloro-fluorinated hydrocarbons, and the search for replacement refrigerants with comparable thermodynamic properties is ongoing. Ammonia provides lower temperatures than the chloro-fluorinated hydrocarbons and is used in many commercial applications; however, its toxicity and the possibility of major leaks at the pressures of ammonia systems make it unsuitable for residential use.

The various phase-change cycles used for cooling are examined in this chapter.

7.2 Rankine Cycle

> I showed him round last week, o'er all—an' at the last says he:
> "Mister McAndrew, don't you think steam spoils romance at sea?"
> Damned ijjit! I'd been doon that morn to see what ailed the throws.
> Manholin', on my back—the cranks three inches off my nose.
> Romance! Those first-class passengers they like it very well.
> Printed an' bound in little books: but why don't poets tell?
> I'm sick of all their quirks an' turns—the loves an' doves they dream—
> Lord, send a man like Robby Burns to sing the Song o' Steam!
> *Kipling,* McAndrew's Hymn

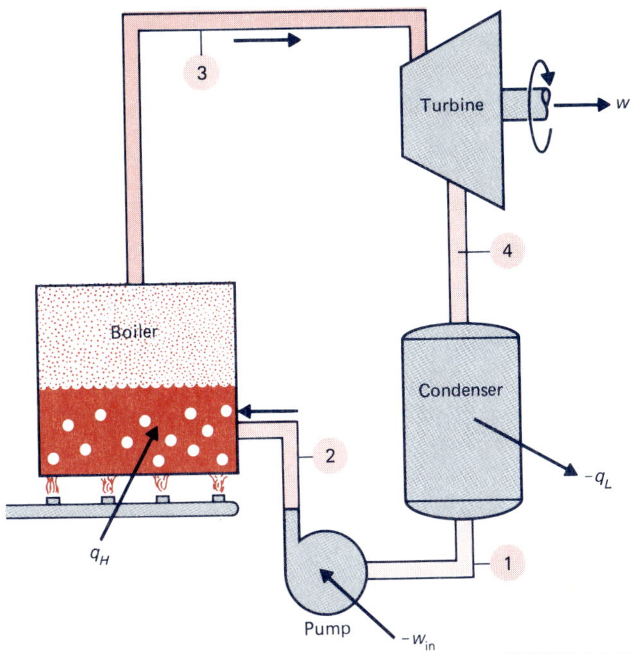

Figure 7.1 Idealized simple Rankine cycle.

The simple Rankine cycle is the common cycle used in all steam power plants. The Rankine cycle was devised to make use of the characteristics of water as a working fluid and to handle the phase change between liquid and vapor. Of course, many other materials can be chosen instead of water as working fluids. The choice depends on many factors, including the need to accommodate the temperatures of heat transfer to and from the vapor and liquid states while maintaining low vapor pressures in the system.

In an idealized simple Rankine cycle (Fig. 7.1), heat transfer takes place to the working fluid in a constant-pressure process in the boiler. The *P-v* diagram is shown in Fig. 7.2. Liquid enters the pump and is increased in pressure to state 2. It enters the boiler from the pump at a relatively low temperature (state 2) and is heated to saturation along line 2-2′. The liquid experiences a slight increase in volume owing to thermal expansion. At point 2′, the saturation temperature is reached. Since the boiler is operated at effectively constant pressure, further heat transfer takes place at constant temperature. And the energy added to the working fluid goes into the heat of vaporization, producing vapor until a quality of 100 percent is reached at point 3. Then the vapor is expanded (in an assumed isentropic process for the simple cycle) through a turbine to produce work and leaves at state 4. The expansion is limited in the practical cycle by the appearance of vapor condensation in the turbine and by the saturation pressure available at the temperature of the cooling medium used in the

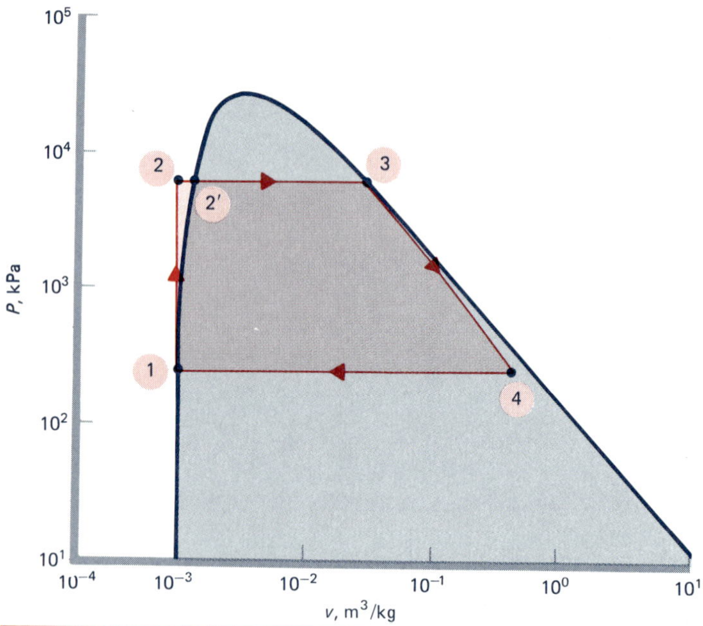

Figure 7.2 Diagram for simple ideal Rankine cycle.

condenser. If excessive condensation is allowed to occur, the condensed liquid droplets will rapidly erode the turbine blades.

At the end of the expansion process (state 4), low-pressure but fairly high-quality vapor leaves the turbine. The vapor is then condensed to a liquid as the vapor comes into contact with surfaces in the condenser that are cooled, usually by a cold-water stream. Because the condenser operates near the temperature of the cooling water, the condensation process occurs at temperatures well below the normal (atmospheric) boiling point of most working fluids. And the condenser pressure on the working fluid side is, therefore, often subatmospheric. The liquid leaves the condenser at state 1.

Following condensation, the liquid enters a pump. The working

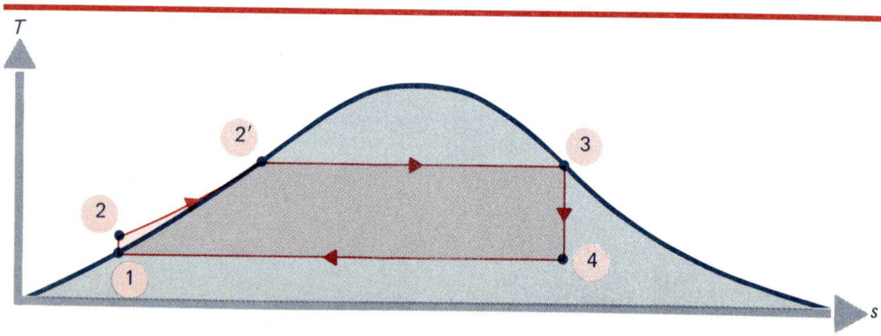

Figure 7.3 *T-s* diagram for the ideal simple Rankine cycle.

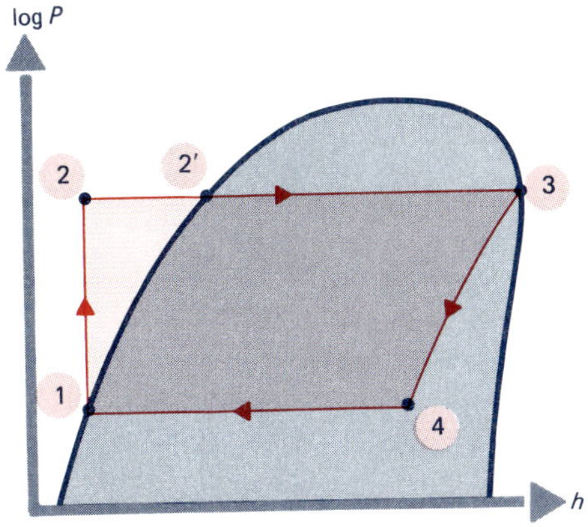

Figure 7.4 *P-h* diagram for the simple ideal Rankine cycle.

fluid is returned to the high pressure necessary for energy addition at the higher boiler temperature, and the cycle is repeated.

Note that the *P-v* diagram used to illustrate the Rankine cycle (Fig. 7.2) is a log-log plot. This is necessary to illustrate the cycle because the specific volume ranges over four decades over the pressure range of interest in steam cycles. However, such a plot greatly distorts the *P-v* behavior of water that would be presented on a linear *P-v* diagram. For example, on a linear plot, the saturated liquid line is essentially vertical, while the saturated vapor line is greatly curved.

The *T-s, P-h,* and *h-s* diagrams for the cycle are shown to scale in Figs. 7.3 through 7.5, respectively. Each plot is on a linear scale except for *P* on a *P-h* diagram (see Chap. 3). Only a portion of the *h-s* plot is shown. This allows the boiler-turbine part of the Rankine cycle to be shown in detail.

Note that this is the first cycle which cannot be treated as an air-standard cycle, since the ideal gas law is invalid for saturated vapor and liquid. Rather, the Rankine cycle efficiency is dependent not only on the states of the working fluid but also on the particular working fluid used and its properties.

Figure 7.5 The *h-s* diagram for the simple Rankine cycle.

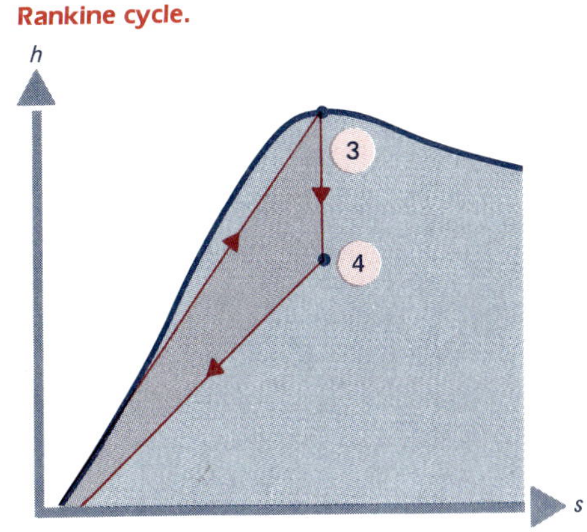

Example Problem 7.1

Find and compare the efficiency of the simple ideal Rankine cycle engine operating between 20 and 100°C for working fluids of water and refrigerant 12. Compare the results to the Carnot efficiency. See the diagrams.

Solution

The *T-s* diagrams for each working fluid are shown. The properties at each state are found from the property tables. Note that all points are for saturated vapor or liquid states except for states 2 and 4. The second law across the ideal turbine and pump is used to fill in the entropy values at states 2 and 4.

State Diagrams

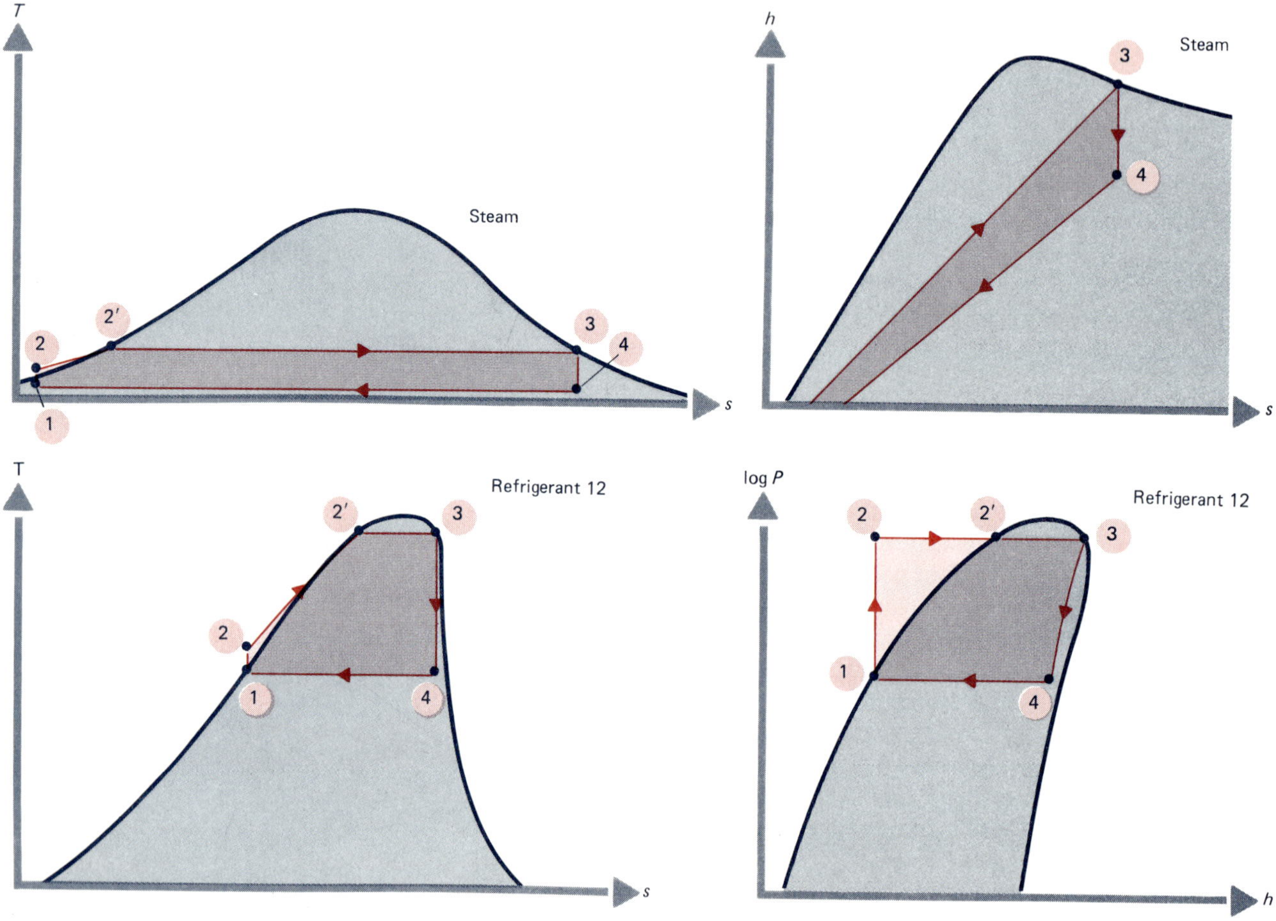

State	P, kPa	T, °C	v, m³/kg	h, kJ/kg	s, kJ/(kg·K)	x
Steam:						
1	(2.34)	20	(0.001002)	(83.835)	(0.2962)	0.0
2	101.32	—	—	—	s_1	—
2′	(101.32)	100	(0.001043)	(419.06)	(1.3069)	0.0
3	(101.32)	100	(1.6736)	(2675.7)	(7.3545)	1.0
4	(2.34)	20	—	—	s_3	—
Refrigerant 12:						
1	(567.0)	20	(0.0007524)	(218.89)	(1.0663)	0.0
2	3343.2	—	—	—	s_1	—
2′	(3343.2)	100	(0.0011129)	(314.98)	(1.3476)	0.0
3	(3343.2)	100	(0.003906)	(376.86)	(1.5134)	1.0
4	(567.0)	20	—	—	s_3	—

To determine the work output of the cycle, the first law can be written for the ideal turbine as

$$-{}_3w_{4,s} = h_{4,s} - h_3$$

The value of $h_{4,s}$ is not known at this point; however, since $s_{4,s} = s_3$, the quality $x_{4,s}$ can be found from

$$x_{4,s} = \frac{s_{4,s} - s_{l,4}}{s_{lg,4}}$$

and $h_{4,s}$ is given by

$$h_{4,s} = h_{l,4} + x_{4,s}h_{lg,4}$$

For water, this results in $x_{4,s} = 0.8434$, $h_{4,s} = 2153.0$ kJ/kg, and ${}_3w_{4,s} = 522.7$ kJ/kg.

The heat transfer to the working fluid in process 2-2′-3 may be found in two parts. That in the constant-temperature portion 2′-3 is easily found from

$$\begin{aligned} {}_{2'}q_3 &= T_{2'}(s_3 - s_{2'}) \\ &= (100 + 273.15)(7.3545 - 1.3069) \\ &= 2256.7 \text{ kJ/kg} \end{aligned}$$

In portion 2-2′, the first law gives

$${}_2q_{2'} = h_{2'} - h_2$$

To find h_2, the work done by the ideal pump is taken from Eq. (5.171):

$$-{}_1w_2 = v_1(P_2 - P_1) = 0.001002(101.32 - 2.34) = 0.099 \text{ kJ/kg}$$

Now the first law, written for the pump, gives

$$h_2 = h_1 - {}_1w_2 = 83.835 + 0.099 = 83.93 \text{ kJ/kg}$$

and the heat transfer to the water in process 2-2′ is

$${}_2q_{2'} = 419.06 - 83.93 = 335.13 \text{ kJ/kg}$$

The total energy added to the working fluid during the cycle is composed of the energy to heat the water to the boiling point, process 2-2′, and the energy needed to vaporize the water, process 2′-3. The total energy added is

$${}_2q_3 = {}_2q_{2'} + {}_{2'}q_3 = 335.13 + 2256.7 = 2591.8 \text{ kJ/kg}$$

Note that this heat transfer from state 2 to state 3 can also be evaluated by applying the first law to the boiler. This yields

$${}_2q_3 = h_3 - h_2 = 2675.7 - 83.835 = 2591.9 \text{ kJ/kg}$$

The cycle efficiency is then

$$\eta_R = \frac{{}_3w_4 + {}_1w_2}{|{}_1q_3|} = \frac{522.7 - 0.099}{2591.9} = 0.202 = 20.2\%$$

A similar analysis can be carried through for the refrigerant 12 cycle; a cycle efficiency of 15.9 percent results.

The various work, heat transfer, and quality terms found in the calculation are gathered in this table:

	Steam	**Refrigerant 12**
${}_3w_4$, kJ/kg	522.7	26.90
${}_2q_{2'}$, kJ/kg	335.1	94.00
${}_{2'}q_3$, kJ/kg	2256.7	61.84
${}_1w_2$, kJ/kg	−0.099	−2.09
x_4	0.8434	0.9301
η_R, %	20.2	15.9

A Carnot cycle operating between the same two temperature limits would have an efficiency of

$$\eta_C = 1 - \frac{293}{373} = 0.214 = 21.4\%$$

Comments

In determining the efficiency of the Rankine cycle, the pump work is included in the net work done by the cycle. The pump may not be driven by energy derived from the cycle itself, as was the case, for example, with the compressor in the Brayton cycle. The pump work for the steam cycle is very small in comparison with the energy required to heat and vaporize the working fluid. In steam plants in particular, then, pump work can usually be neglected in determining the cycle efficiency. For other working fluids, pump work can become significant; and for the refrigerant 12 cycle, it is 7.8 percent of the work done by the turbine.

A further observation is that the simple Rankine cycle efficiency for steam of 20.2 percent approaches the Carnot efficiency of 21.4 percent more closely than does the refrigerant 12 efficiency of 15.9 percent, even though both Rankine cycles work between the same two temperatures. Obviously the choice of working fluid is important. Again, referring to the table of energies, we see that 87.1 percent of the energy added in the steam cycle is added during the constant-temperature vaporization process. For refrigerant 12, only 39.7 percent of the energy is added during vaporization. Thus, for the refrigerant 12 cycle much of the energy is added at temperatures below the maximum. To most closely approach the Carnot efficiency, then, a working fluid should be chosen for which most of the heat transfer is added during vaporization and least is added during the liquid heating process. This requires a fluid with a large enthalpy of vaporization and a small liquid specific heat.

A final observation is that the quality of the working fluid leaving the turbine x_4 is higher for the refrigerant 12 cycle than for the steam cycle. This is an important turbine design consideration. Normally exit vapor quality should be greater than 90 percent to avoid excess liquid droplet formation.

7.2.1 Inefficiencies of Real Cycles

The idealized Rankine cycles examined in Example Problem 7.1 approach the Carnot efficiency rather closely. This is not true for actual engines that operate on the Rankine cycle. For real engines, many irreversibilities are present, chiefly in the turbine and pump. Also note that we are studying the *cycle* efficiency here; the efficiency of a power plant is usually defined in terms of the work (or electric energy) output of the plant compared with the energy carried by the fuel entering the *boiler*, not the energy added to the working fluid. This overall plant efficiency is called the *heat rate* of the plant, and it is often expressed in the mixed units of Btu of energy input per kilowatthour of electric output. Since boiler efficiencies may be as low as 60 percent, overall plant efficiencies will be considerably lower than cycle efficiencies.

Treating Turbine and Pump Inefficiencies

The effect on the cycle of turbine and pump irreversibilities can be included through the use of pump and turbine efficiencies, as defined in Sec. 5.11. The method of treatment is shown most easily by an example.

Example Problem 7.2

A Rankine cycle operates with steam under the same conditions as in Example Problem 7.1. The turbine has an efficiency of 70 percent, and the pump efficiency is 60 percent. Find the cycle efficiency, and sketch the *T-s* and *h-s* diagrams. See the cycle diagrams.

Solution

The entropy at states 4 and 2 must be modified from the values in Example Problem 7.1, since the pump and turbine can no longer be assumed isentropic. The table is now modified to include the properties for isentropic conditions (noted, as before, with subscript s) and for actual conditions (denoted by subscript a).

Cycle Diagrams

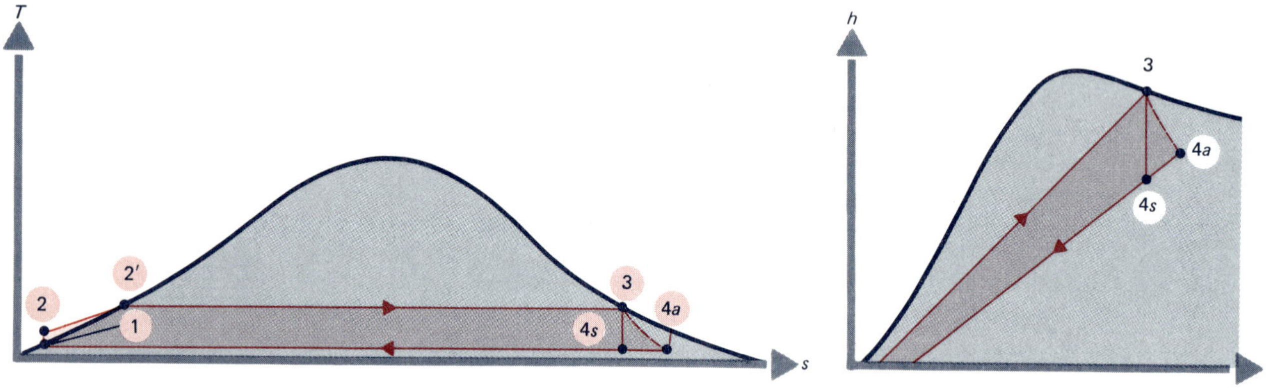

State	P, kPa	T, °C	v, m³/kg	h, kJ/kg	s, kJ/(kg·K)	x
Steam:						
1	(2.34)	20	(0.001002)	(83.835)	(0.2962)	0.0
2s	101.32	—	—	—	s_1	—
2a	101.32	—	—	84.0	—	—
2′	(101.32)	100	(0.001043)	(419.06)	(1.3069)	0.0
3	(101.32)	100	(1.6736)	(2675.7)	(7.3545)	1.0
4s	(2.34)	20	—	—	s_3	—
4a	(2.34)	20	—	2309.8	—	(0.9073)

The actual state points at 2 and 4 are now found by using the definitions of the pump and turbine efficiencies. First, let us examine the turbine. If the turbine were isentropic, the work output would be as found in Example Problem 7.1, or ${}_3w_{4,s} = 522.7$ kJ/kg. By using the turbine efficiency of 70 percent, the actual work done by the turbine is

$${}_3w_{4,a} = (0.7)(522.7) = 365.9 \text{ kJ/kg}$$

The first law can be used to find $h_{4,a}$ from

$$h_{4,a} = h_3 - {}_3w_{4,a} = 2675.7 - 365.9 = 2309.8 \text{ kJ/kg}$$

and the other properties in state 4 can now be determined. Solving for $x_{4,a}$, for example, gives

$$x_{4,a} = \frac{h_{4,a} - h_{l,4}}{h_{lg,4}} = \frac{2309.8 - 83.835}{2453.4} = 0.9073$$

For the pump, the isentropic work was found previously to be -0.099 kJ/kg, so the actual work is

$${}_1w_{2,a} = \frac{-0.099}{0.6} = -0.165 \text{ kJ/kg}$$

which results in

$$h_{2,a} = h_1 - {}_1w_{2,a} = 83.835 + 0.165 = 84.0 \text{ kJ/kg}$$

The value of ${}_2q_{2',a}$ is

$${}_2q_{2',a} = h_{2'} - h_{2,a} = 419.06 - 84.0 = 335.06 \text{ kJ/kg}$$

compared with 335.12 kJ/kg for the ideal cycle. The difference cannot be plotted for point 2 on the cycle diagrams. The cycle efficiency can be determined from

$$\eta_R = \frac{{}_3w_{4,a} + {}_1w_{2,a}}{{}_2q_{2'} + {}_{2'}q_3}$$

$$= \frac{365.7}{335.06 + 2256.7} = 0.141 = 14.1\%$$

compared with 20.2 percent for the ideal Rankine cycle and 21.4 percent for the Carnot efficiency between the same two temperatures.

Comments

The cycle efficiency has been reduced, but the change is due to only one factor; the work output of the turbine has been reduced to 70 percent of its ideal value.

Although slightly more work is required to drive the real pump, this additional work increases the enthalpy of the water leaving the pump, which in turn reduces the amount of heat transfer needed to bring the water to state 2′. The result is that the net energy *added* to the cycle is actually greater for the ideal than for the real cycle. However, if the cost of energy in the form of work to the pump is higher than that of heat transfer in the boiler, the real cycle energy input will cost more.

Line 3-4*a* on the diagrams for the real turbine expansions are really not defined, since this is not a quasi-equilibrium process. It is shown on the diagram as a dashed line to indicate the series of states within the turbine, but actually only the final state 4*a* is defined. The final state is found by application of the first and second laws to the turbine, and knowledge of the intermediate states is not required.

In problems of this type, complete specification of state 2′ is not really necessary because only ${}_2q_3 = h_3 - h_2$ is needed to find the heat transfer to the cycle. Inclusion of state 2′ does point out the relative amounts of energy needed to increase the boiler feed temperature to saturation and then to evaporate the saturated liquid.

Note finally that the exit quality for the real cycle is above 90 percent, because the less efficient real turbine did not reduce the exit enthalpy to as low a value as was predicted for the isentropic turbine. As noted earlier, the higher quality is of great practical importance because the presence of liquid droplets at lower qualities causes rapid turbine blade erosion.

7.2.2 Increasing the Rankine Cycle Efficiency

Increased Operating Pressure

From a look at the *P-v* diagram for the Rankine cycle, it might be inferred that the cycle efficiency could be increased by increasing the boiler pressure, since the area under the *P-v* diagram will be increased. This is generally the case.

A number of interrelated factors, however, determine the overall cycle efficiency change with increasing pressure. Let us try an example, building on the data from Example Problem 7.2.

Example Problem 7.3

Determine the cycle efficiency of the Rankine engine in Example Problem 7.2 when the boiler pressure is raised to 4 MPa. Assume that the pump and turbine efficiencies are unchanged. Sketch the *P-v, T-s, P-h,* and *h-s* diagrams.

Cycle Diagrams

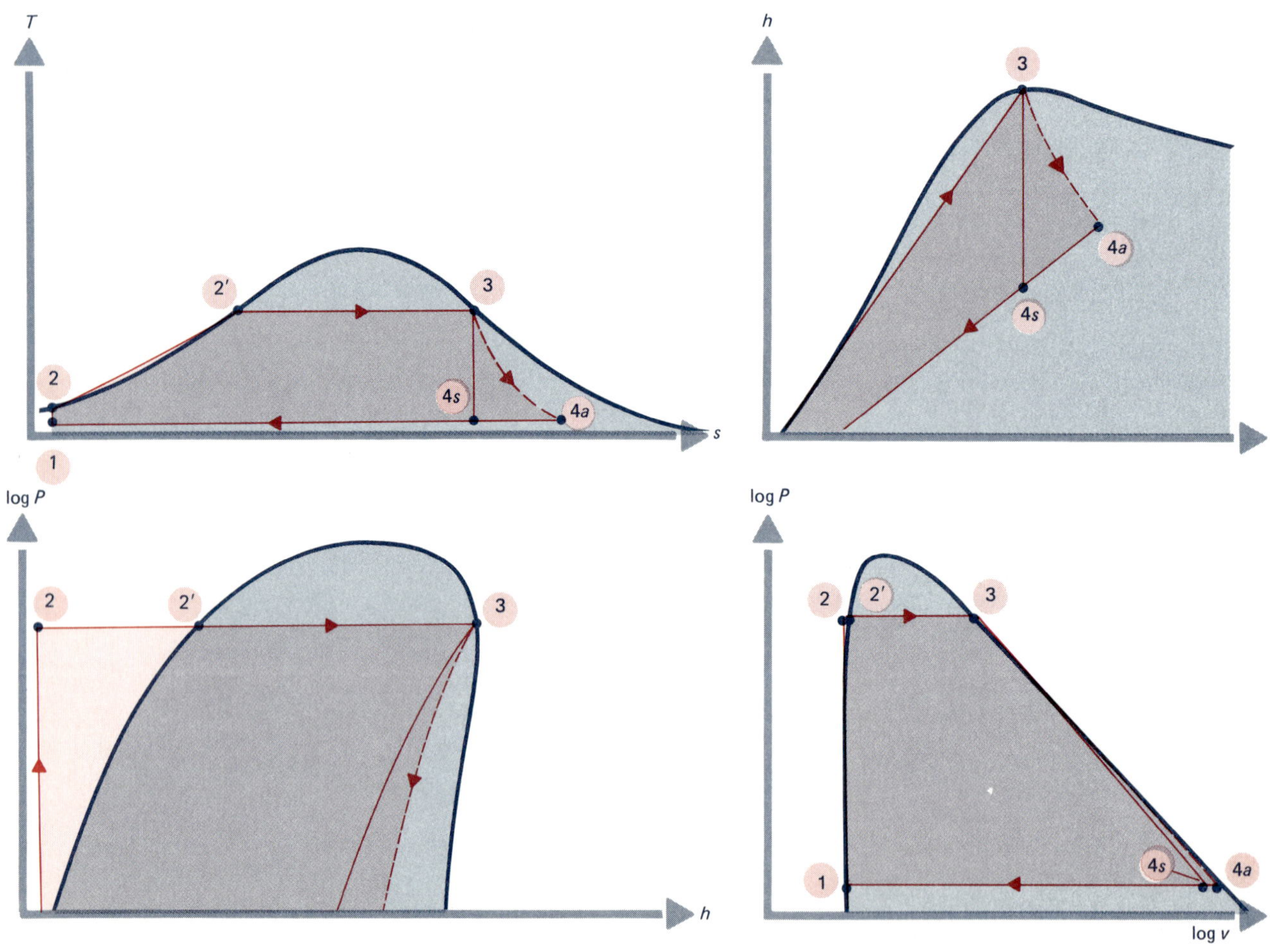

Solution

The solution follows exactly the method of Example Problem 7.2. This table of values results:

State	P, kPa	T, °C	v, m³/kg	h, kJ/kg	s, kJ/(kg·K)	x
1	(2.34)	20	(0.001002)	(83.835)	(0.2962)	0.0
$2s$	4000	—	—	(87.84)	s_1	—
$2a$	4000	—	—	90.51	—	—
$2'$	4000	(250.4)	(0.001252)	(1087.2)	(2.7962)	0.0
3	4000	(250.4)	(0.04977)	(2800.6)	(6.0689)	1.0
$4s$	(2.34)	20	—	(1776.2)	s_3	(0.6898)
$4a$	(2.34)	20	—	2083.5	—	(0.8151)

Some of the entries are not calculated for the table, since they are not required for solution of the cycle efficiency.

The actual energy values for this operating boiler pressure are

$$_3w_{4,a} = 717.1 \text{ kJ/kg} = 308.3 \text{ Btu/lbm}$$

$$_2q_{2'} = 996.7 \text{ kJ/kg} = 428.5 \text{ Btu/lbm}$$

$$_{2'}q_3 = 1713.4 \text{ kJ/kg} = 736.6 \text{ Btu/lbm}$$

$$_1w_{2,a} = -6.7 \text{ kJ/kg} = -2.9 \text{ Btu/lbm}$$

$$x_4 = 0.8151$$

The cycle efficiency is then

$$\eta_R = \frac{717.1 - 6.7}{996.7 + 1713.4} = 0.262 = 26.2\%$$

compared with the efficiency of 14.1 percent for the 1-atm operating boiler pressure.

Comments

The Carnot efficiency for operation between the boiler and condenser temperatures of this cycle would be

$$\eta_C = 1 - \frac{20 + 273.15}{250.3 + 273.15} = 0.439 = 43.9\%$$

The Rankine cycle for the 4-MPa operating pressure thus reaches 59.7 percent of the Carnot efficiency. For the 1-atm steam engine of Example Problem 7.2, the cycle reached $(0.141/0.214)100 = 65.9$ percent of the Carnot efficiency. Thus, increasing the boiler steam pressure has increased the Rankine cycle efficiency from 14.1 to 26.2 percent, but the Rankine cycle has become relatively less efficient when compared with the Carnot efficiency between the same temperatures. This decrease in relative efficiency occurs because at higher pressures the Rankine cycle has a larger fraction of its energy addition in the subcooled heating region ${}_2q_{2'}$, rather than in the constant-temperature region ${}_{2'}q_3$.

The exit quality of the steam leaving the turbine is also quite low at 81.5 percent, compared with the 90.7 percent quality leaving the turbine of the 1-atm Rankine engine. Such a low quality is not acceptable in practice. Some other method of increasing cycle efficiency must be found aside from simply raising the boiler pressure.

. . . the auld fleet engineer
That started as a boiler-whelp—when steam and he were low. I mind
the time we used to serve a broken pipe wi' tow!
Ten pound was all the pressure then—Eh! Eh!—a man wad drive;
An' here, our working' gauges give one hunder sixty-five!
Kipling, M'Andrew's Hymn

Superheating

Another way to improve Ranking cycle efficiency is to superheat the vapor leaving the boiler (Fig. 7.6). This is usually done by passing the steam through tubes exposed to hot combustion gases or some other energy source at a temperature above the boiler saturation temperature.

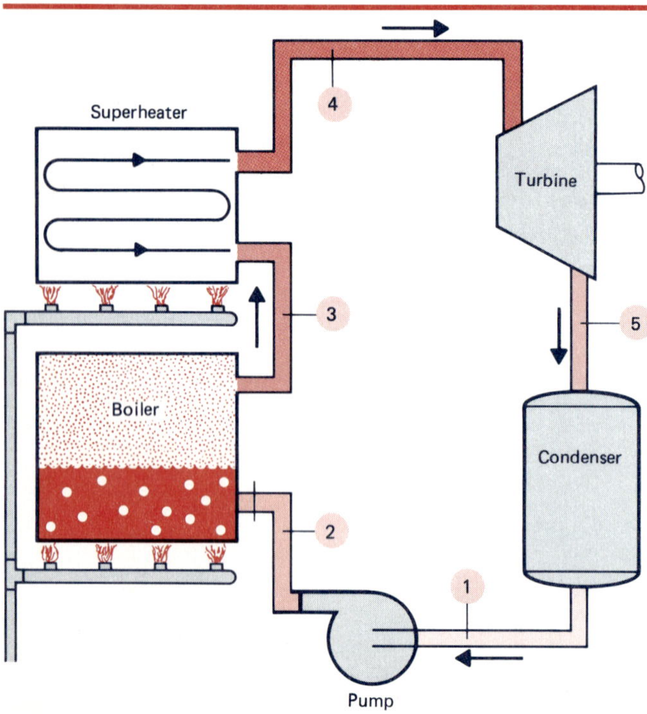

Figure 7.6 Rankine engine with superheater.

The steam entering the turbine thus is at higher enthalpy than was the case for the simple Rankine cycle of Fig. 7.5.

Superheating the steam leaving the boiler has the effect of allowing the quality leaving the turbine to be higher than for the cycle using saturated steam, as illustrated in Example Problem 7.4

Example Problem 7.4

Determine the cycle efficiency and turbine exit quality of the Rankine engine in Example Problem 7.3 when the boiler exit steam is superheated to 400°C. Sketch the *P-v*, *T-s*, *P-h*, and *h-s* diagrams.

Cycle Diagrams

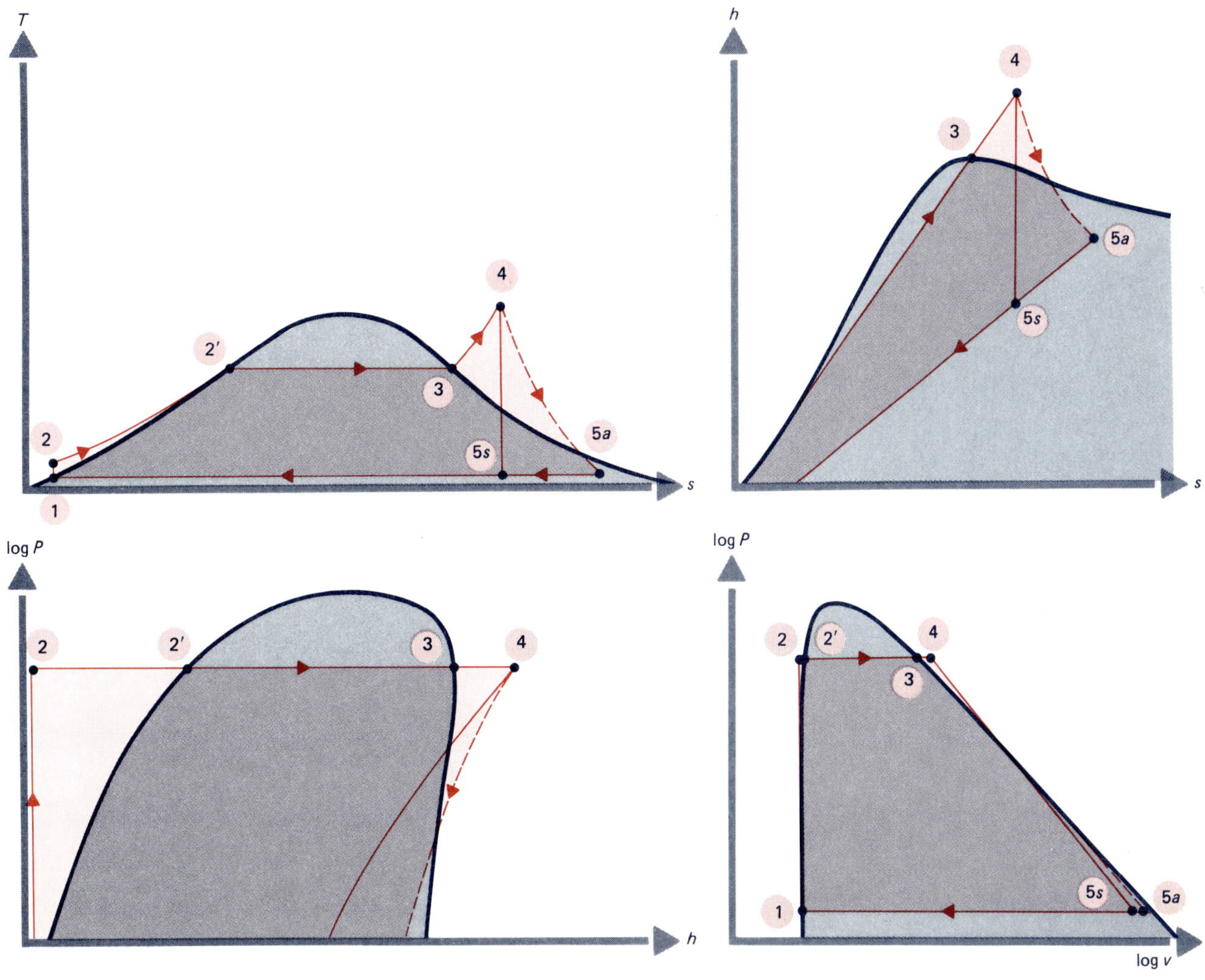

Solution

The superheater is assumed to be a constant-pressure heat transfer device (This is not true in practice, since a considerable pressure drop owing to friction occurs as the steam passes through the superheater tubes.) The cycle diagrams for the cycle are shown, and these are the state point values:

State	P, kPa	T, °C	v, m³/kg	h, kJ/kg	s, kJ/(kg·K)	x
1	(2.34)	20	(0.001002)	(83.835)	(0.2962)	0.0
2s	4000	—	—	(87.84)	s_1	—
2a	4000	--	—	90.51	—	—
2′	4000	(250.4)	(0.001252)	(1087.2)	(2.7962)	0.0
3	4000	(250.4)	(0.04977)	(2800.6)	(6.0689)	1.0
4	4000	400	(0.07340)	(3213.4)	(6.7688)	—
5s	(2.34)	20	—	(1981.3)	s_4	(0.7734)
5a	(2.34)	20	—	2350.9	—	(0.9241)

The energy values become

$${}_4w_5 = 862.5 \text{ kJ/kg} = 370.8 \text{ Btu/lbm}$$

$${}_2q_{2'} = 996.7 \text{ kJ/kg} = 428.5 \text{ Btu/lbm}$$

$${}_{2'}q_3 = 1713.4 \text{ kJ/kg} = 736.6 \text{ Btu/lbm}$$

$${}_3q_4 = 412.8 \text{ kJ/kg} = 177.5 \text{ Btu/lbm}$$

$${}_1w_2 = -6.7 \text{ kJ/kg} = -2.9 \text{ Btu/lbm}$$

$$x_{5,a} = 0.9241$$

The cycle efficiency is then

$$\eta_R = \frac{862.5 - 6.7}{996.7 + 1713.4 + 412.8} = 0.274 = 27.4\%$$

compared with the engine of Example Problem 7.3, which used saturated steam from the boiler and no superheat to attain an efficiency of 26.2 percent.

Comments

Superheating usually has a marginal effect on cycle efficiency, but it does allow an increase in turbine output per unit mass of steam flow because the enthalpy of the steam entering the turbine has been increased.

Much of the energy added to the working fluid in the cycle with superheating occurs over a range of temperatures, and it is difficult to decide how to compare with the efficiency of a Carnot engine. If the maximum temperature in the cycle, 400°C, is used to find the Carnot efficiency,

$$\eta_C = 1 - \frac{20 + 273}{400 + 273} = 0.565 = 56.5\%$$

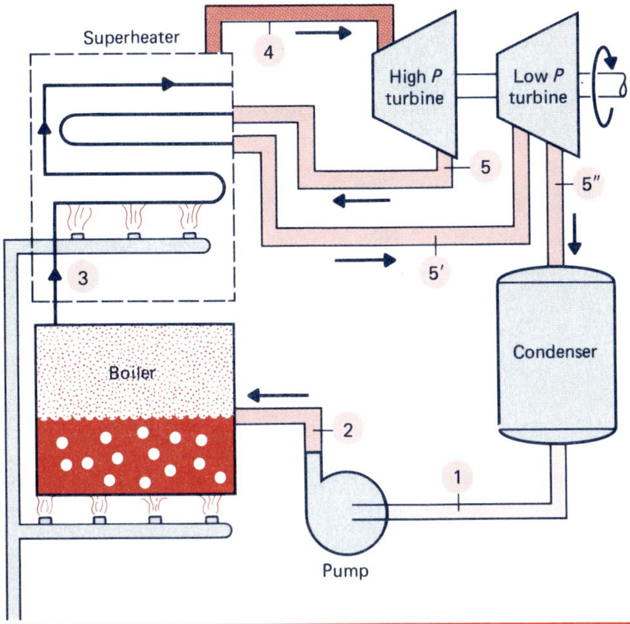

Figure 7.7 Ideal reheat in Rankine cycle.

and we have reached (27.4/56.5)100 = 48.5 percent of the Carnot efficiency. The quality of the steam leaving the turbine has reached 92.41 percent, providing better conditions for practical design than the 81.5 percent value for the system of Example Problem 7.3

Reheating

To further increase Rankine cycle efficiency, the steam passing through the turbine can be reheated. In the simplest form, this process can be carried out as indicated in the system diagram of Fig. 7.7. Here a two-stage turbine is used. As the steam leaves the high-pressure stage, it is piped back to the superheater and reheated before returning to the second (low-pressure) stage of the turbine for further expansion. A major benefit of reheating is that the quality of steam leaving each stage of the turbine can be maintained at a high value, and much higher boiler pressures can be used.

Example Problem 7.5

Using the basic system described in Example Problem 7.4, examine the effect on cycle efficiency of adding reheat to the cycle. Assume that the outlet pressure from the high-pressure turbine is 1000 kPa and that the steam is reheated to the original superheat temperature of 400°C. So that we can compare the results to the cycle without reheat, assume also that each stage of the turbine has an efficiency of 70 percent. Sketch the cycle on *P-v*, *T-s*, *P-h*, and *h-s* diagrams.

Cycle Diagrams

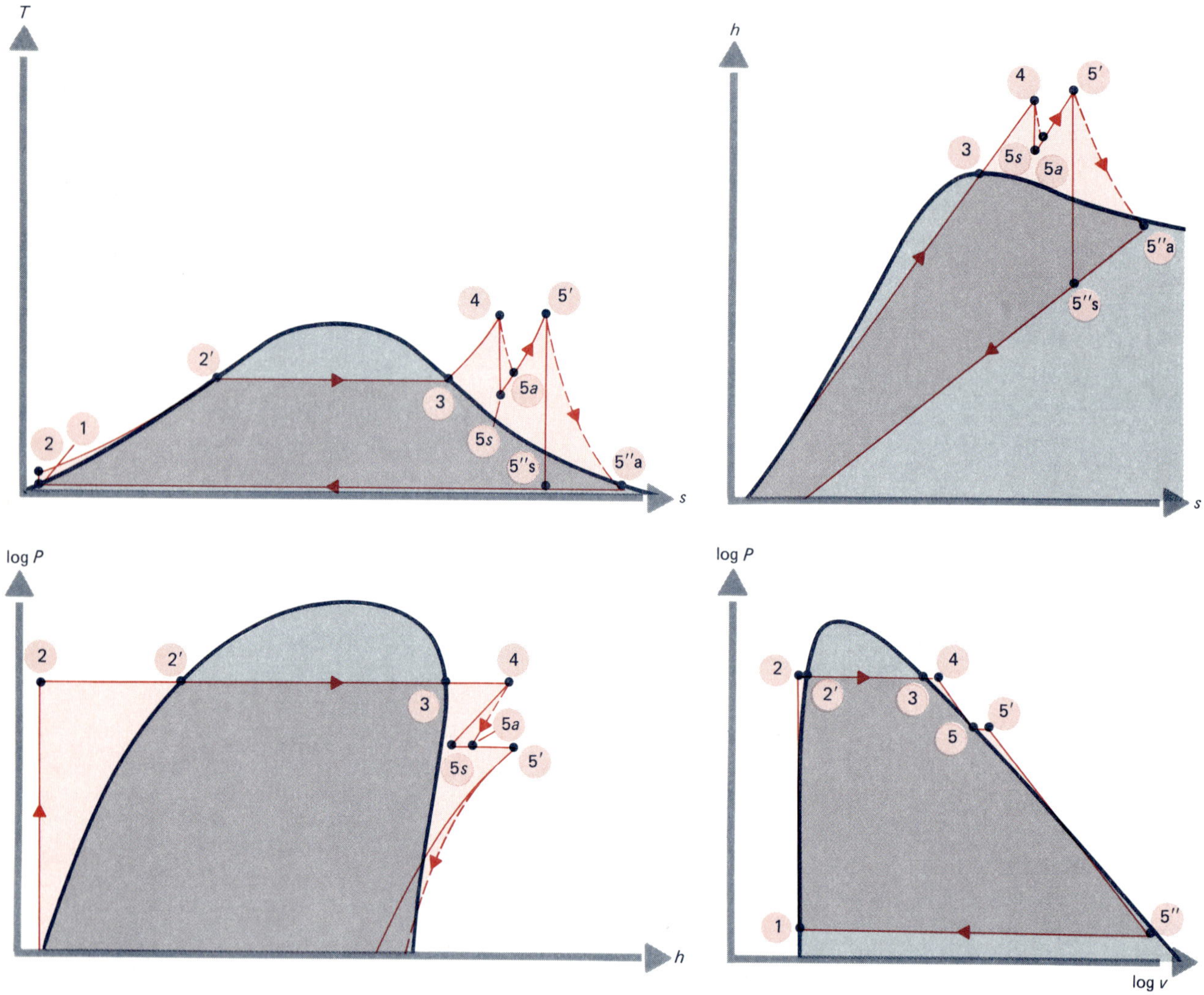

Solution

The state points for the reheat cycle in this problem are given in the table. The conditions at state 5 are known for the isentropic case, and the turbine efficiency can be used to find the actual outlet conditions of the high-pressure turbine. The result of these calculations for $P_5 = 1000$ kPa and $s_{5,s} = 6.7688$ kJ/(kg·K) is by interpolation using Table D.12,

$$h_{5,s} = \frac{6.7688 - 6.6932}{6.9235 - 6.6932} 2941.9 + 2827.4 = 2865.0 \text{ kJ/kg}$$

State	P, kPa	T, °C	v, m³/kg	h, kJ/kg	s, kJ/(kg·K)	x
1	(2.34)	20	(0.001002)	(83.835)	(0.2962)	0.0
$2s$	4000	—	—	(87.84)	s_1	—
$2a$	4000	—	—	90.51	—	—
$2'$	4000	(250.4)	(0.001252)	(1087.2)	(2.7962)	0.0
3	4000	(250.4)	(0.04977)	(2800.6)	(6.0689)	1.0
4	4000	400	(0.07340)	(3213.4)	(6.7688)	—
$5s$	1000	—	—	(2865.0)	s_4	—
$5a$	1000	—	—	2969.5	(6.9739)	—
$5'$	1000	400	—	(3263.8)	(7.4648)	—
$5''s$	(2.34)	20	—	(2185.4)	$s_{5'}$	(0.8566)
$5''a$	(2.34)	20	—	(2508.9)	—	(0.9885)

and ${}_4w_{5,s} = 348.4$ kJ/kg. Using the turbine efficiency of 70 percent then gives the actual work output of the turbine as 243.9 kJ/kg. The actual outlet conditions from the high-pressure turbine are then found to be $h_{5,a} = 2969.5$ kJ/kg and $s_{5,a} = 6.9739$ kJ/(kg·K). This is indicated in the above table.

The heat transfer required to raise the temperature to 400°C along line $5a$-$5'$ is

$$ {}_5q_{5'} = h_{5'} - h_{5,a} = 3263.8 - 2969.5 = 294.3 \text{ kJ/kg} $$

The usual calculation for the ideal turbine is now applied to the low-pressure turbine, and the actual turbine outlet conditions at state $5''$ are found by again using the turbine efficiency. The actual work output of the low-pressure turbine is found to be ${}_{5'}w_{5''} = 754.9$ kJ/kg. Thus the energy values for the reheat cycle are

$$ {}_4w_5 = 243.9 \text{ kJ/kg} = 104.9 \text{ Btu/lbm} $$

$$ {}_{5'}w_{5''} = 754.9 \text{ kJ/kg} = 324.5 \text{ Btu/lbm} $$

$$ {}_2q_{2'} = 996.7 \text{ kJ/kg} = 428.5 \text{ Btu/lbm} $$

$$ {}_2q_3 = 1713.4 \text{ kJ/kg} = 736.6 \text{ Btu/lbm} $$

$$ {}_3q_4 = 412.8 \text{ kJ/kg} = 177.5 \text{ Btu/lbm} $$

$$ {}_5q_{5'} = 294.3 \text{ kJ/kg} = 126.5 \text{ Btu/lbm} $$

$$ {}_1w_2 = -6.7 \text{ kJ/kg} = -2.9 \text{ Btu/lbm} $$

$$ x_{5'',a} = 0.9885 $$

The work output of the combined turbine is 998.8 kJ/kg. The cycle efficiency is

$$ \eta_R = \frac{{}_4w_5 + {}_{5'}w_{5''} + {}_1w_2}{{}_2q_{2'} + {}_{2'}q_3 + {}_3q_4 + {}_5q_{5'}} $$

$$ = \frac{243.9 + 754.9 - 6.7}{996.7 + 1713.4 + 412.8 + 294.3} = 0.290 = 29.0\% $$

Comments

Adding reheat to the cycle has increased the efficiency of the Rankine cycle with superheat from 27.4 to 29.0 percent, or to 51.3 percent of the efficiency of a Carnot engine operating between the maximum and minimum temperatures of the steam cycle. Such an increase does not always occur; certain extraction pressures may actually cause a decrease in cycle efficiency.

Reheating has several advantages aside from the possible increase in cycle efficiency. A major advantage is that the turbine stages in the system each work over a smaller range of pressure. It is easier to design these individual turbines to operate at high turbine efficiency than to design a single high-efficiency turbine that must handle the entire expansion process. In large power plants, multiple turbine stages with reheat between each stage are used to increase turbine and cycle efficiency and to spread the mechanical load along the turbine shaft.

The outlet quality from the turbine exceeds the exit quality of 92.4 percent found at the outlet of the turbine in Example Problem 7.4, where no reheat was included.

Regeneration

As noted earlier in this chapter, the Rankine cycle efficiency is lower than the efficiency of a corresponding Carnot engine for a number of reasons, but one of the major deviations results because energy must be added to the liquid phase in order to bring the cool working fluid leaving the pump to the saturation temperature. Further isothermal heat transfer can then occur in the boiler at saturation conditions. The heat transfer to the liquid leaving the condenser and pump occurs over a range of temperatures. Thus a portion of the cycle heat transfer input is always at nonisothermal conditions, and so the Carnot efficiency cannot be reached.

It is possible, however, to use some of the energy carried by the working fluid after it leaves the boiler to heat the liquid phase. Then a larger *fraction* of the heat transfer added to the working fluid from *external* sources will be added during vaporization in the boiler, which takes place at isothermal conditions. The various methods of using energy from one portion of the cycle to add to another portion are called *recuperation.* The term *regeneration* more generally refers to a process in which energy is stored for a time and is then transferred into the working fluid during a later process, as in the Stirling cycle, for example. Common usage in steam power plants refers to the recuperation process in these plants as regeneration, and we use this conventional term. We have seen applications of this idea in the Ericsson and Brayton cycles.

In the Rankine cycle, the only energy available at temperatures that will allow a heat transfer to the liquid in or leaving the condenser must come from vapor passing through the turbine. Let us examine the ideal case in which exactly enough energy is extracted from the turbine vapor to

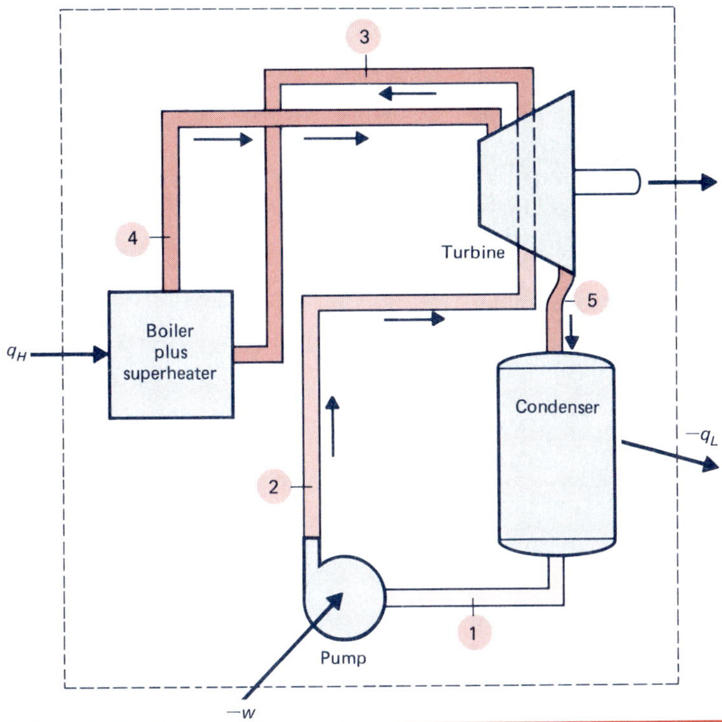

Figure 7.8 Rankine cycle with ideal regeneration.

completely preheat the working fluid from the condenser exit conditions to saturated liquid at the boiler pressure. In such a case, all heat transfer to the cycle (if there is no superheating) will take place isothermally; and if all components in the system could be made reversible, the Rankine cycle would have the same efficiency as the Carnot cycle.

One possible method is to take working fluid leaving the condenser and pump it through cooling channels in the turbine in such a way that reversible heat transfer occurs (Fig. 7.8). The *T-s* diagram for such a process is shown in Fig. 7.9.

The temperature and entropy changes for the vapor and the regenerated liquid passing through the turbine will yield the same reversible

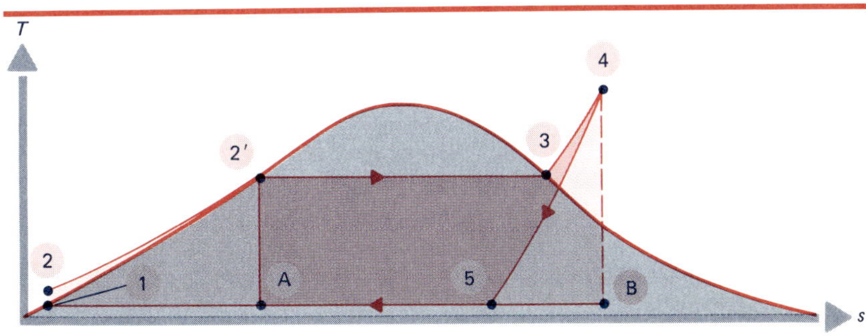

Figure 7.9 *T-s* diagram for Rankine cycle with ideal regeneration.

heat transfer process. It follows that the heat transferred from the vapor in the turbine (area under 4-B-5) is exactly equal to the energy gained by the liquid (area under 1-2-2′-A). The heat transfer from the boiler to the working fluid takes place isothermally at T_{sat} in the boiler except in the superheat region, since all preheating of the fluid has been done regeneratively. Thus, since heat rejection in the condenser is also an isothermal process, the cycle efficiency approaches the Carnot efficiency. If no superheating is present, the efficiency for this idealized cycle is equal to the Carnot efficiency.

Nature is not so obliging as to make such a system feasible in actual applications. It is not possible to design a turbine so that the heat transfer process is truly reversible; in practice, the heat transfer to the boiler feed is done externally to the turbine. Usually, only some portion of the turbine vapor is bled from one or more locations at different pressures from various turbine stages and is passed through one or more *feedwater heaters*, as shown in Fig. 7.10. The quality of the vapor-liquid mixture leaving the turbine would be considerably reduced if all the vapor were used for regeneration, an undesirable effect, as noted earlier.

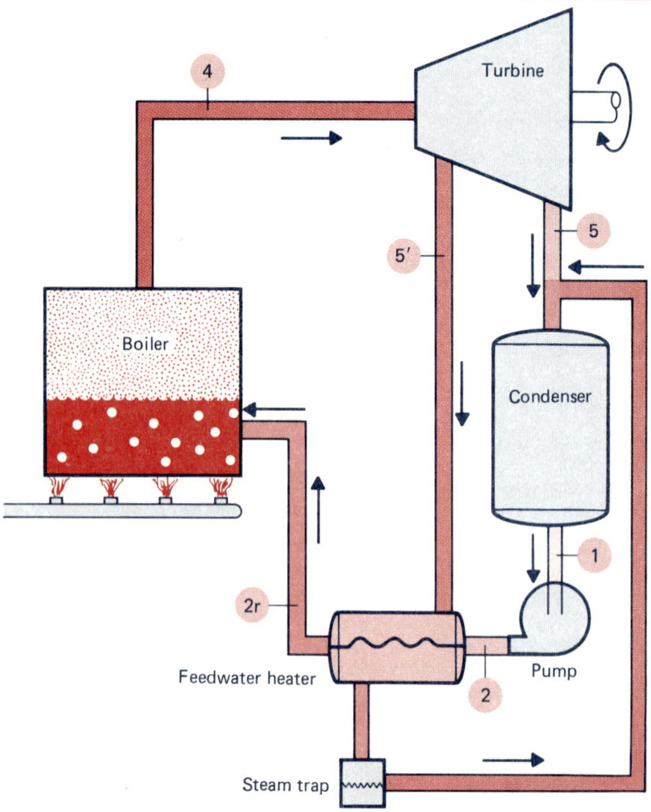

Figure 7.10 Rankine cycle with regeneration by use of a feedwater heater.

Types of Feedwater Heaters

In the type of feedwater heater shown in Fig. 7.10, called a *closed* feedwater heater, the bleed steam is not in direct contact with the fluid leaving the pump. Rather, the steam is condensed on the outer surface of tubes carrying the water leaving the pump, and the bleed condensate is simply returned to the condenser entrance and mixed with the turbine exit steam, which slightly reduces the required condenser heat transfer.

In some feedwater heaters, the bleed steam is mixed directly with the water leaving the pump and enters the boiler feed stream at that point. Such a system is called an *open feedwater heater.* These heaters offer the advantage that air or other dissolved gases are usually driven out of the feed stream by the heating process and can be removed by an appropriate pump. Removal of inert gases aids in condenser performance and reduces corrosion in the system.

Example Problem 7.6

The Rankine cycle with superheat of Example Problem 7.4 is modified by the addition of a single closed feedwater heater. For 1 kg of mass flowing into the turbine, 0.1 kg is bled from the turbine at 1000 kPa and sent to a closed feedwater heater. Assume that the turbine has a 70 percent efficiency at every pressure. Find the cycle efficiency, and compare it with the efficiency of the cycle without regeneration.

Cycle Diagram

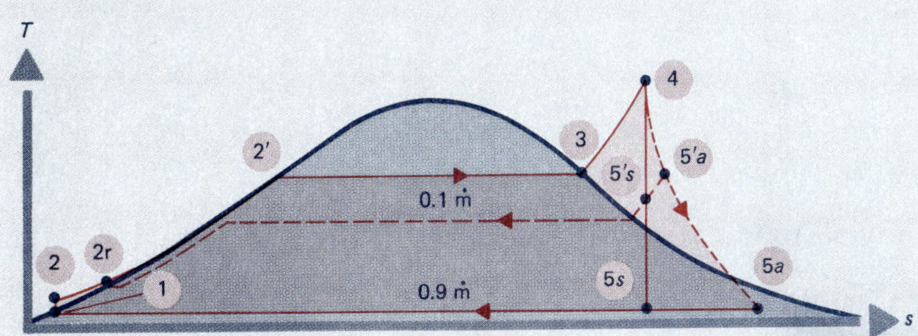

Solution

For the system shown in Fig. 7.10, the bleed flow is shown by the dashed line $5'a$-$2r$ on the T-s diagram. The line 1-2-2′-3-4-$5'a$-$5a$-1 shows the states of the main flow through the turbine and the remainder of the Rankine cycle. The flows combine along path $2r$-2′-3-4-5′. See the state table.

The work output of the turbine between states 5′ and 5 *per unit mass of fluid passing through the boiler* ($\dot{m}_5 = 0.9\dot{m}_4$) is

$$_{5'}w_5 = -0.9(h_{5a} - h_{5'a}) = -0.9(2350.9 - 2969.5)) = 556.3 \text{ kJ/kg}$$

State	P, kPa	T, °C	v, m³/kg	h, kJ/kg	s, kJ/(kg·K)	x
1	(2.34)	20	(0.001002)	(83.835)	(0.2962)	0.0
$2s$	4000	—	—	(87.84)	s_1	—
$2a$	4000	—	—	90.51	—	—
$2r$	4000	—	—	—	—	—
$2'$	4000	(250.4)	(0.001252)	(1087.2)	(2.7962)	0.0
3	4000	(250.4)	(0.04977)	(2800.6)	(6.0689)	1.0
4	4000	400	(0.07340)	(3213.4)	(6.7688)	—
$5's$	1000	—	—	(2865.0)	s_4	—
$5'a$	1000	—	—	2969.5	(6.9739)	—
$5s$	(2.34)	20	—	(1981.3)	s_4	(0.7734)
$5a$	(2.34)	20	—	(2350.9)	—	(0.9241)

It is now necessary to find the enthalpy of the feedwater at state $2r$, so that the heat transfer still necessary to bring the feedwater to saturation can be determined. This is done by writing the first law for the feedwater heater and assuming that the outlet water to the boiler and the condensed bleed steam reach the same temperature, which is an idealization. For an adiabatic feedwater heater, assuming $h_6 = h_l(P_{5'}) \cong h_{2r}$ gives

$$\dot{m}_{5'}(h_{5'a} - h_{2r}) = \dot{m}_2(h_{2r} - h_2)$$

or

$$h_{2r} = \frac{0.1h_{5'a} + h_2}{1.1} = \frac{(0.1)2969.5 + 90.5}{1.1} = 352.2 \text{ kJ/kg}$$

and

$${}_{2r}q_{2'} = h_{2'} - h_{2r} = 1087.2 - 352.2 = 735.0 \text{ kJ/kg}$$

The energy values are now (per unit of mass flow through the boiler)

$${}_4w_{5'} = 243.9 \text{ kJ/kg} = 105.3 \text{ Btu/lbm}$$

$${}_{5'}w_5 = 556.7 \text{ kJ/kg} = 239.3 \text{ Btu/lbm}$$

$${}_{2r}q_{2'} = 735.0 \text{ kJ/kg} = 316.0 \text{ Btu/lbm}$$

$${}_{2'}q_3 = 1713.4 \text{ kJ/kg} = 736.6 \text{ Btu/lbm}$$

$${}_3q_4 = 412.8 \text{ kJ/kg} = 177.5 \text{ Btu/lbm}$$

$${}_1w_2 = -6.7 \text{ kJ/kg} = -2.9 \text{ Btu/lbm}$$

$$x_{5,a} = 0.9241$$

The cycle efficiency is then

$$\eta_R = \frac{{}_4w_{5'} + {}_{5'}w_5 + {}_1w_2}{{}_{2'}q_{2'} + {}_{2'}q_3 + {}_3q_4}$$
$$= \frac{243.9 + 556.7 - 6.7}{735.0 + 1713.4 + 412.8} = 0.278 = 27.8\%$$

This compares with 27.4 percent for the cycle with superheat but no regeneration.

Comments

Is 10 percent of the flow stream a reasonable value to bleed off for the regeneration stream? The amount depends on the other conditions in the cycle such as boiler pressure and condenser temperature; the bleed flow is usually found by specifying the desired feedwater heater outlet condition. For example, sufficient bleed flow might be provided to bring the condensate stream to the saturation temperature of the bleed stream. Note that the outlet quality from the turbine is quite low; it is generally advantageous to combine reheat with regeneration to avoid low turbine exit quality.

It is worthwhile now to gather the results of the example problems that have been worked in this section and to show the comparisons of various cycle enhancements. Table 7.1 shows the overall thermodynamic

TABLE 7.1 Results of Example Problems

Cycle	P_{boiler}, kPa	η_R, %	x
Simple Rankine with ideal turbine and pump (Example Problem 7.1)	101.32	20.2	0.8434
Simple Rankine with 70 percent turbine efficiency and 60 percent pump efficiency (Example Problem 7.2)	101.32	14.1	0.9072
Simple Rankine with 70 percent turbine efficiency and 60 percent pump efficiency at higher boiler pressure (Example Problem 7.3)	4000	26.2	0.8151
Rankine with 70 percent turbine efficiency and 60 percent pump efficiency with superheat to 150°C above saturation (Example Problem 7.4)	4000	27.4	0.9241
Rankine with 70 percent turbine efficiency and 60 percent pump efficiency with superheat to 150°C above saturation and reheat to superheat temperature between stages of two-stage turbine at 1000 kPa (Example Problem 7.5)	4000	29.0	0.9885
Same as Example Problem 7.4 with single closed feedwater heater using 10 percent of total mass flow diverted from turbine at 1000 kPa (Example Problem 7.6)	4000	27.8	0.9241

efficiency of the cycle and the turbine exit quality for the various examples presented. The effect on cycle efficiency for each modification is easily seen in Table 7.1, at least for the particular conditions chosen for the examples.

Because of the great amounts of fuel used by large power plants that produce electrical outputs in the range of many hundreds to a few thousands of megawatts, it is economical to increase plant efficiency by a few tenths of a percent even at the expense of large capital investment. Thus, modern plants use all the enhancements noted above, and much research is devoted to increasing turbine efficiencies. Complete plant simulations are carried out to optimize the overall plant efficiency by choosing the best bleed flow rates to individual feedwater heaters, the best pressures for these bleed flows, the optimum reheat pressures and temperatures, the optimum boiler pressures and superheat temperatures as limited by materials of construction, etc.

Typical operating conditions in modern plants at present include boiler pressures of about 18 MPa (2600 psig) with average sizes of 300 to 400 MW of electrical output using fuels of oil, natural gas, or pulverized coal. In larger units, 800 MW and above, supercritical units are economical, and boiler pressures of 31 MPa (4500 psig) at 560°C (1050°F) with two reheat loops, also at 560°C, are typical. These design figures reflect the long experience with similar plants as well as the economics and growth rate of power requirements for utilities.

Other Rankine Cycle Inefficiencies

The Rankine cycle usually does not attain the efficiencies predicted by the example problems presented here. Although the examples include the turbine and pump efficiencies and thus account for many real losses in the Rankine cycle, other important factors have been ignored. For example, it is difficult to design a condenser that will produce saturated liquid at the outlet. Real condensers are designed to subcool the liquid; this allows the system to operate in off-design conditions without passing vapor through the condenser and avoids having nearly saturated liquid enter the pump. If the latter is allowed, the pump can be destroyed by *cavitation,* an effect created when boiling and recondensation at low-pressure points behind the pump impeller rapidly erode the impeller.

Further inefficiency is caused by the frictional losses in pumping the working fluid through the cycle, heat losses from the high-temperature components in the system to the environment, reduced condenser performance because of air leakage into the subatmospheric condenser, and energy requirements to operate auxiliary equipment such as combustion air fans and other blowers.

Other Modifications to the Rankine Cycle

Modern Rankine cycle power plants operate with both reheat and regeneration and at as high a pressure as present mechanical design and

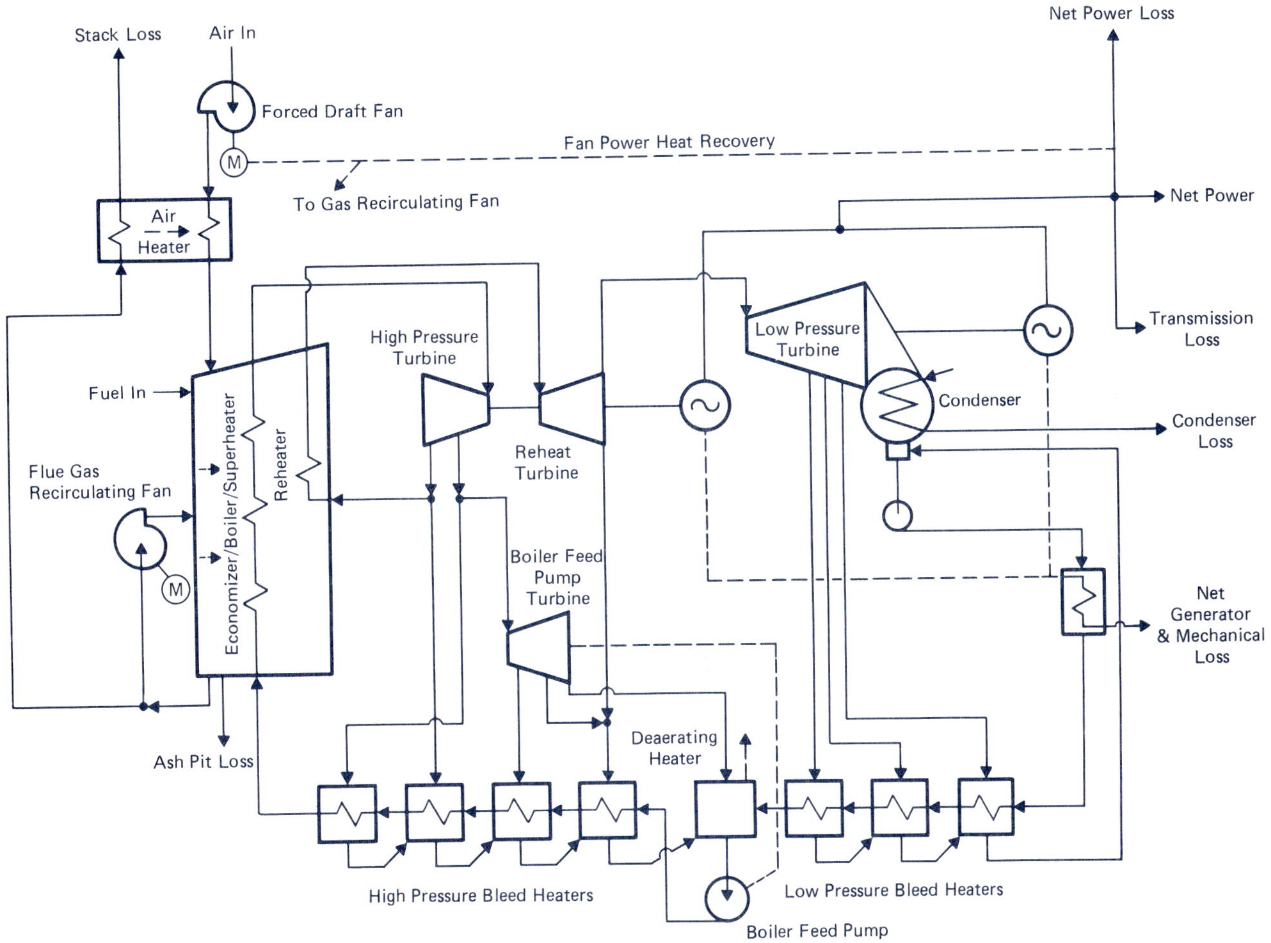

Figure 7.11 Rankine fossil-fuel power cycle with single reheat and eight-stage regeneration. $P = 24.2$ MPa (3515 psia), $T = 538°C$ (1000°F) steam. (From *Steam, Its Generation and Use,* 39th ed., Babcock and Wilcox Company, 1978, used with permission.)

materials properties allow. Some present steam cycles are designed to operate at pressures above the critical point. A diagram of a typical power plant cycle using reheat and regeneration is shown in Fig. 7.11. Efficiency also can be increased by reducing the pressure in the condenser, at the price of increased moisture content in the turbine.

Attention is also being paid to combined cycles of various types. For example, the rejected energy from a Brayton engine (gas turbine) can be used as the energy input to a Rankine engine, so that some fraction of the energy rejected by the gas turbine is recovered as useful work. The energy rejected by a well-engineered Rankine cycle using steam for power generation is usually at such a low temperature that its utility in driving another cycle is limited. However, by increasing the heat rejection temperature and lowering the power output, the waste energy can be used for space heating, plant steam, or other low-temperature application. Such combined power-heat cycles are called *cogeneration* systems. The economics of such systems depend on the particular use and value of the waste energy.

7.2.3 Applications of the Rankine Cycle

Modifications of the Rankine cycle are used in certain applications. In this section, the modifications that are necessary because of the peculiarities of the energy source for the cycle are discussed.

In Secs. 7.2.1 and 7.2.2, the Rankine cycle is assumed to operate at steady state, and all heat transfer to the working fluid is assumed to occur in the boiler. This is the case for larger power plants using the combustion of fossil fuel (oil, natural gas, or coal) as the energy source. Other sources of energy necessitate some modifications to the cycle, and these are discussed briefly here.

Nuclear Power Systems

When nuclear energy is used as the power source, modification of the cycle is desirable. If the working fluid in the Rankine cycle is passed through the nuclear reactor itself, then it is difficult to avoid radioactive contamination of the microscopic impurities in the working fluid, no matter how carefully the fluid is cleaned and purified. This radioactivity is then slowly deposited in the turbine, pumps, condenser, and other components of the cycle, and it may escape from possible leaks in these components. Maintenance is thus made increasingly difficult. Also, from an operational standpoint, if the turbogenerator in a nuclear power plant must be shut down, the time required to shut down the energy generation by the reactor will differ from the shutdown time of the turbogenerator, so it is usually preferable not to couple them directly.

For these reasons, a secondary flow loop is usually incorporated into nuclear plants cycles. Figure 7.12 shows the primary loop between the reactor and the *steam generator*. The steam generator is a large heat exchanger in which energy transfer takes place from the higher-temperature fluid in the primary loop to the somewhat lower-temperature fluid in the secondary loop. The entire primary loop plus steam generator is contained in a containment building so that any leakage of mildly radioactive primary loop fluid is contained and not vented to the environment.

The primary loop may carry water at a high enough pressure that no phase change occurs (a pressurized water system); or boiling may occur in the reactor, and then the saturated steam is condensed in the steam generator (a *boiling water* reactor system).

Nuclear power plants typically operate at much lower temperatures in the reactor core heat transfer portion than do fossil-fuel plants. This is because of the necessity to maintain the peak temperature anywhere in the reactor core well below the temperature at which material failure can

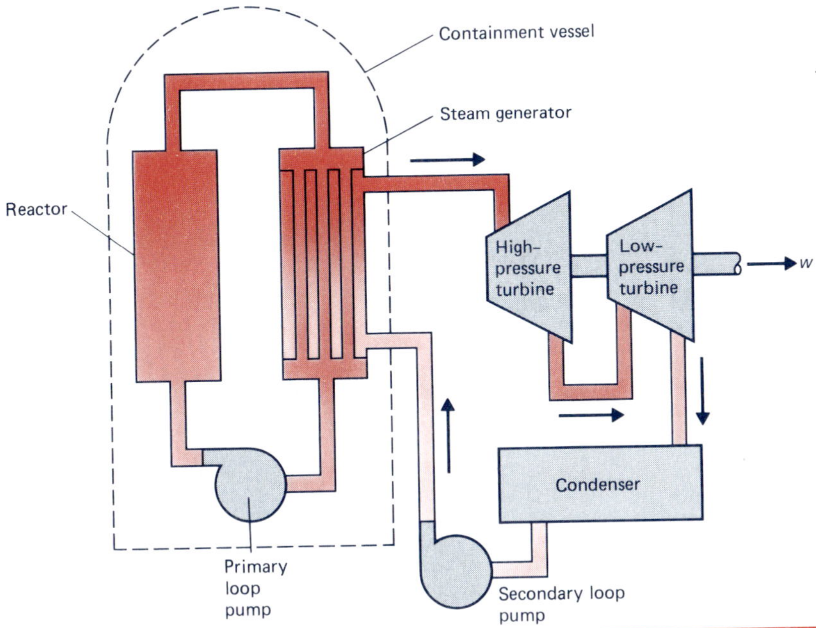

Figure 7.12 Rankine cycle as adapted to nuclear reactor energy source.

occur; thus, average core temperatures are often on the order of 350°C (about 650°F). The overall thermodynamic efficiency of these plants is thus lower than for fossil-fuel plants, causing significantly higher heat rejection and thus more serious problems in finding sites with adequate cooling capacity.

Solar and Other Renewable Energy Source Power Systems

The major design factor in solar energy driven power systems is that the source of energy is intermittent, and thus the system must be made to operate in a transient manner. To avoid this, energy storage may be incorporated, so that excess energy collected during peak periods of solar energy availability (i.e., near solar noon) is stored for use during periods of cloudiness or darkness. The Rankine cycle then can draw on storage when needed and operate at near-steady conditions for much of the day.

Because of the difficulty of providing enough storage in a solar energy system to provide for 24-h operation, such systems are generally designed to be coupled to other energy sources. For example, many utility systems in the southern United States are "summer-peaking," i.e., their greatest demand for power occurs in the summer months when residential and commercial air conditioning demand is high. They are "afternoon-peaking" for the same reason. For such a system (Fig. 7.13) a coal-fired boiler might provide continuous energy to the Rankine cycle, with auxiliary energy during the peaking periods provided by steam produced from a solar energy facility. Thus, the need for large-high-temperature energy storage capacity is eliminated.

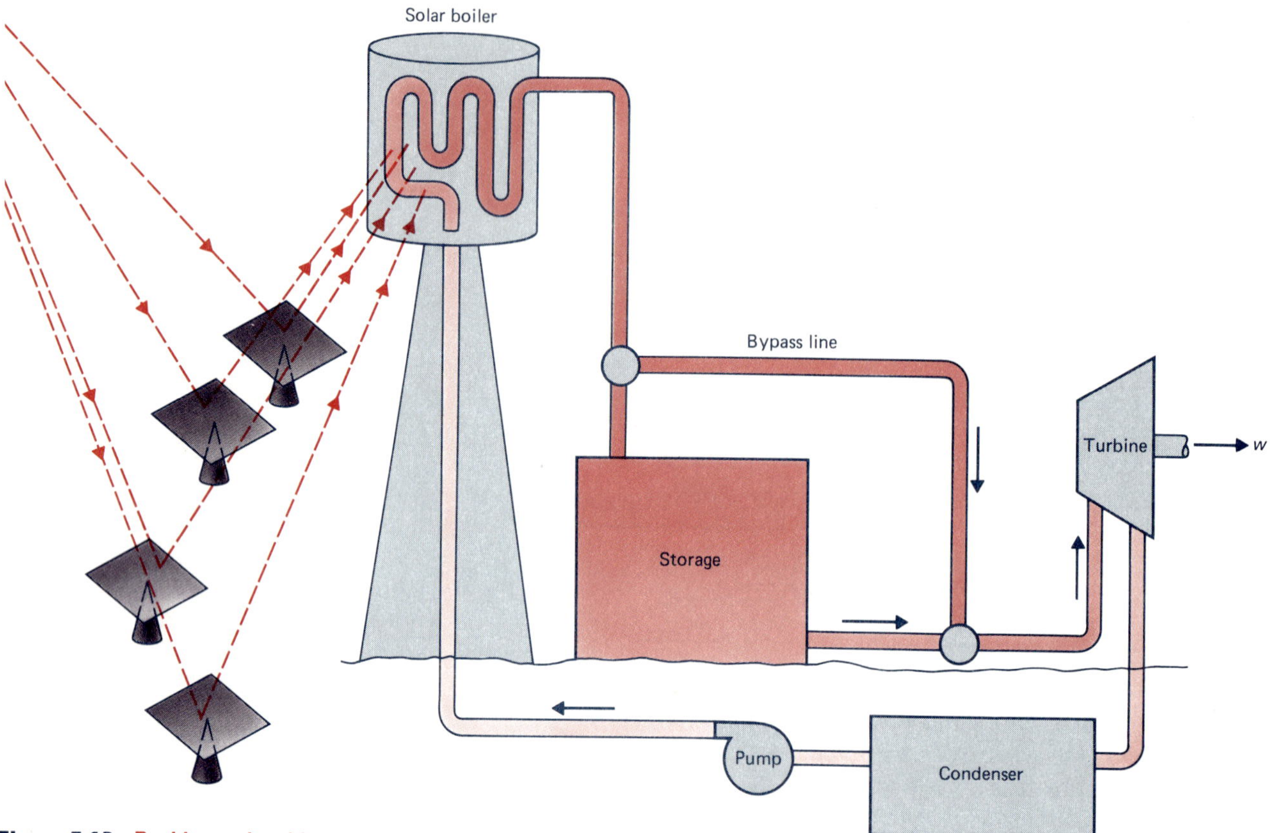

Figure 7.13 Rankine cycle with energy storage as adapted to solar energy source.

A second difficulty with solar plants is that the energy available for input to the cycle is diffusely distributed; i.e., incident sunlight on the earth's surface rarely exceeds 1 kW/m^2 [or about 300 Btu/(h$\cdot$ft^2)]. To gather a sufficient amount of energy for a large power plant obviously requires a significant land area. In addition, because the energy is distributed in this way, high temperatures for heat transfer can be obtained only by the use of solar concentrators. These are typically mirrors of various configurations that can reflect the incident solar energy to a central location where it has much greater energy density, i.e., much greater power per unit area, than is present in incident sunlight.

These two factors, the intermittence and diffuse distribution of solar energy, have made early power plants using conversion of solar energy to thermal energy for power generation uneconomical except in special applications. Direct conversion of solar energy to electricity by the use of photovoltaic cells bypasses the limitations imposed by the heat engine efficiencies and may prove to be a less expensive technology.

7.3 Other Vapor Heat Engine Cycles

In this section, some other vapor heat engine cycles are examined.

7.3.1 The Kalina Cycle

The Kalina cycle makes use of the fact that a mixture of a low-boiling-point liquid and a high-boiling-point liquid will have a boiling point that changes as the concentration of the low-boiling-point component is changed. The boiling point composition relations are discussed more fully in Chap. 10. By using such a mixture in a boiler as the working fluid, the temperature difference between the boiler energy source and the working fluid can be continually increased as the lower-boiling-point constituent in the working fluid mixture is selectively evaporated. Thus, the temperature difference between the energy source and the working fluid decreases as the working fluid passes through the boiler. The smaller temperature difference provides a closer approach to reversible heat transfer (which occurs only with a differential temperature difference between the source and the working fluid). Thus, the Kalina cycle provides a more nearly reversible heat transfer process than is possible with a single-fluid Rankine cycle.

By taking the lower-boiling-point component and selectively adding it to the higher-boiling component in the condenser, a similar change in saturation temperature can be effected across the condenser. Reheat and recuperation can be added to this cycle.

To carry out the processes necessary for the Kalina cycle, the hardware for the system is configured as shown in Fig. 7.14. In this discussion, high concentration means a high percentage of the low-boiling-point (easily vaporized) component. The two-component liquid mixture leaves the condenser at state 1 and enters the boiler. The lower-boiling component is vaporized first, and vaporization continues at increasing temperature until both components are completely vaporized, and the two-component vapor mixture leaves the boiler at state 2. The mixture does work in passing through the turbine, and it enters the heater at state 3. Here it transfers energy to a lower-temperature stream that has left pump 1 and then leaves the heater at state 4. The mixture now passes through a valve to a low pressure, where it "flashes" into vapor (state 5), and then enters a chamber called an *absorber*. Here the vapor is absorbed into a low-concentration liquid stream, and the resulting mixture is simultaneously cooled by a separate cooling stream. The cooled low-concentration liquid mixture leaves the absorber at state 6 and is pumped to a high pressure by pump 1. After leaving the pump at state 7, the low-concentration liquid mixture is split into two streams. One stream goes directly to a flash valve, where the pressure is reduced and the mixture flashes to a vapor at state 8. The other portion of the stream goes to the heater, where it is heated by the mixture leaving the turbine, and leaves the heater at state 9. The heated

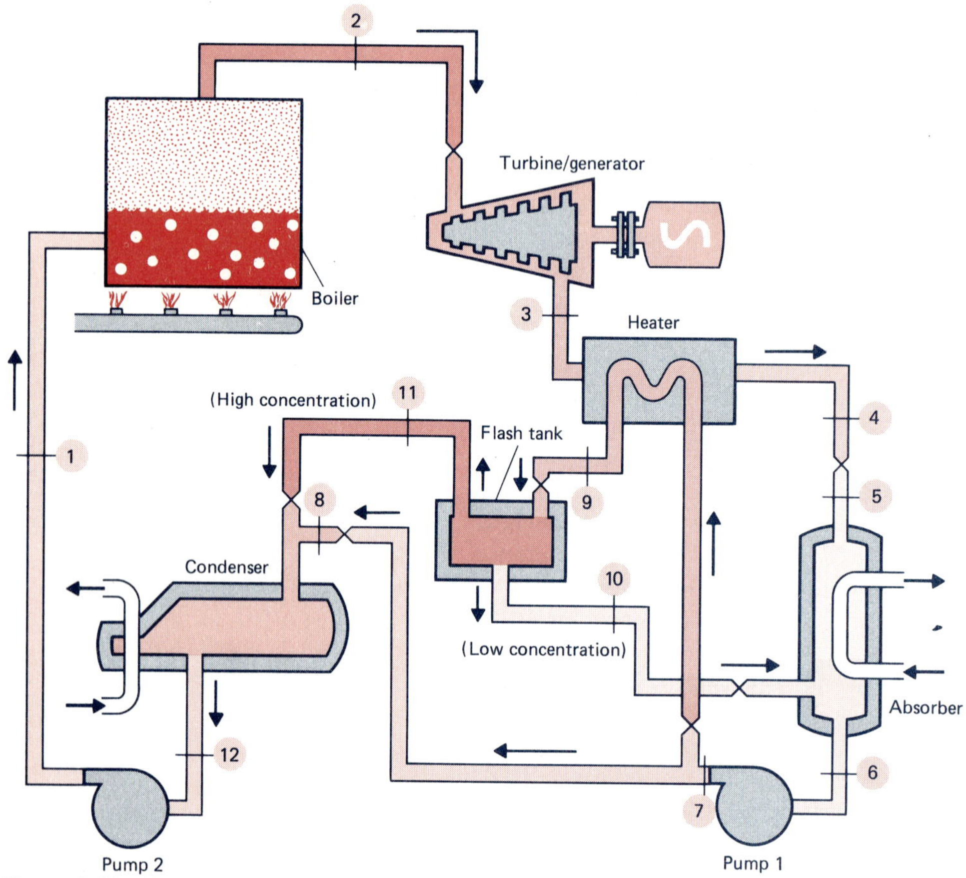

Figure 7.14 The Kalina cycle, using a two-component working fluid.

low-concentration liquid at high pressure is next passed through a flashing valve into a flash tank; the low-boiling-point component will be largely vaporized in this process, while the high-boiling-point component will remain mostly in the liquid state. The low-concentration liquid is then directed to the absorber at state 10, to act as the absorber for the mixture entering the absorber at state 5. The high-concentration vapor leaving the flash tank at state 11 is mixed with the low-concentration stream at state 8 in a controlled way, so that the temperature gradient along the condenser is established. The condensed mixture leaves the condenser at state 12 and is pumped to state 1 by the feed pump 2, to complete the cycle.

In a proposed Kalina commercial utility system, the working fluid is 70 percent ammonia and 30 percent water. Analysis of the system shows an increase of up to 10 percent in thermodynamic efficiency over a Rankine steam cycle operating over the same temperature difference.

7.3.2 Congeneration and Combined-Cycle Plants

In Chap. 6, we mentioned that the exhaust from a stationary gas-turbine cycle can be used as the energy source for a vapor cycle heat engine (called a *topping cycle* in this configuration). Such a system will generally have a higher overall thermodynamic efficiency than either cycle operating by itself.

Another method for increasing plant efficiency is to combine the need for process steam, which is common in many industries such as in the production of petrochemicals, with the production of electric power. Such a combined system is called *cogeneration.* The increased efficiency arises because most energy sources for producing steam operate at high temperature; yet process steam is usually required at much lower temperatures. By operating an electric power plant and using some of or all the exhaust steam for the process needs, a more efficient plant results. Less energy is rejected to the environment in the condenser of the power cycle because the turbine outlet stream is at least partially used for process needs. Recent federal legislation in the United States makes such plants particularly attractive.

Figure 7.15 Combined cycle with no supplementary energy addition.

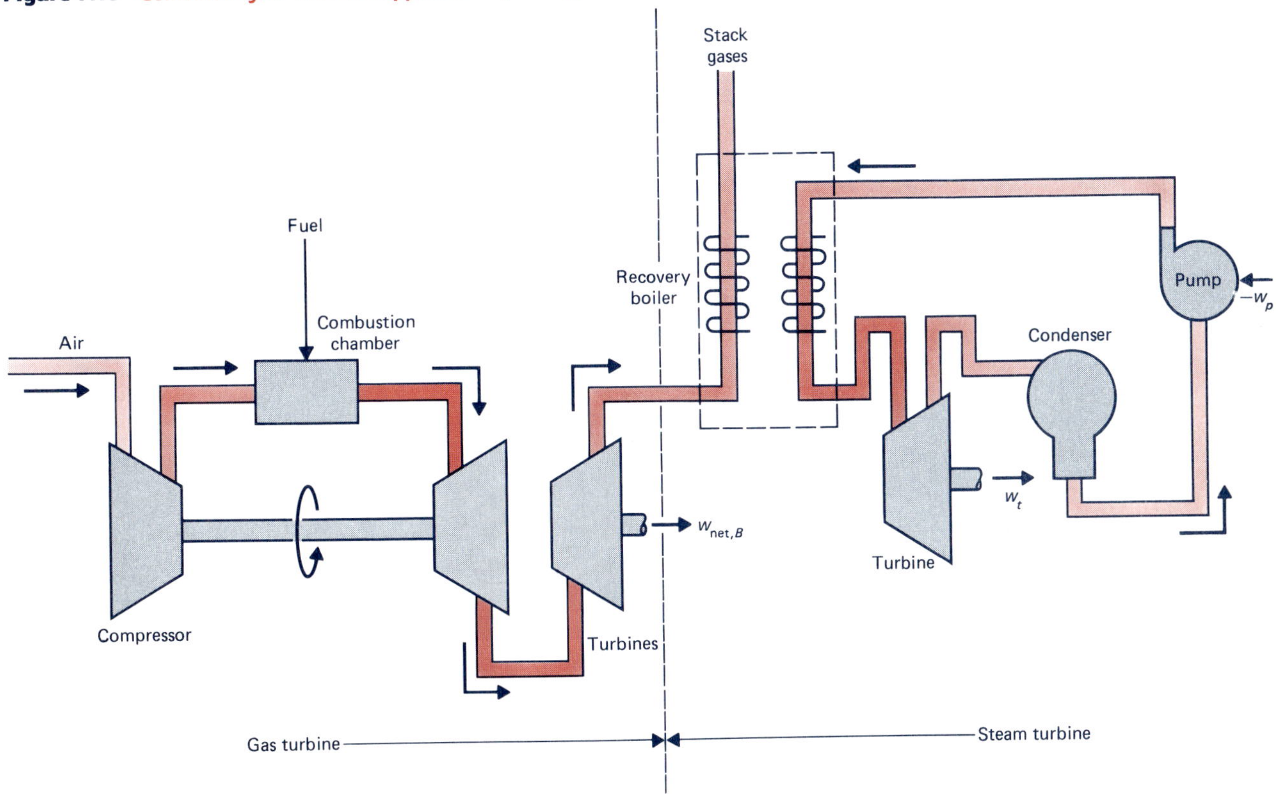

Combined-Cycle Plants

The idea of a combined cycle can be embodied in a number of ways. The most straightforward one is to route the entire exhaust stream from a gas turbine into a "heat recovery boiler," which acts as the steam generator for a Rankine cycle (Fig. 7.15). The efficiency of such a plant is then

$$\eta = \frac{w_{\text{net}, B} + w_{\text{net}, R}}{q_{H, B}} \tag{7.1}$$

where $w_{\text{net},B}$ is the net work output of the Brayton (gas turbine) portion of the plant, $w_{\text{net},R}$ is the net work output of the Rankine cycle portion of the plant (turbine less pump work), and $q_{H, B}$ is the net heat transfer into the Brayton engine (there is no additional heat transfer to the Rankine cycle from an external source in this case).

If the Rankine cycle is inefficient because of the low exhaust temperature available from the turbine, then the heat recovery boiler can operate at higher temperature by adding an auxiliary heater to the system, as shown in Fig. 7.16. In this case, the efficiency becomes

$$\eta = \frac{w_{\text{net}, B} + w_{\text{net}, R}}{q_{H, B} + q_{\text{aux}}} \tag{7.2}$$

Figure 7.16 Combined cycle with additional combustion chamber between Brayton and Rankine cycles.

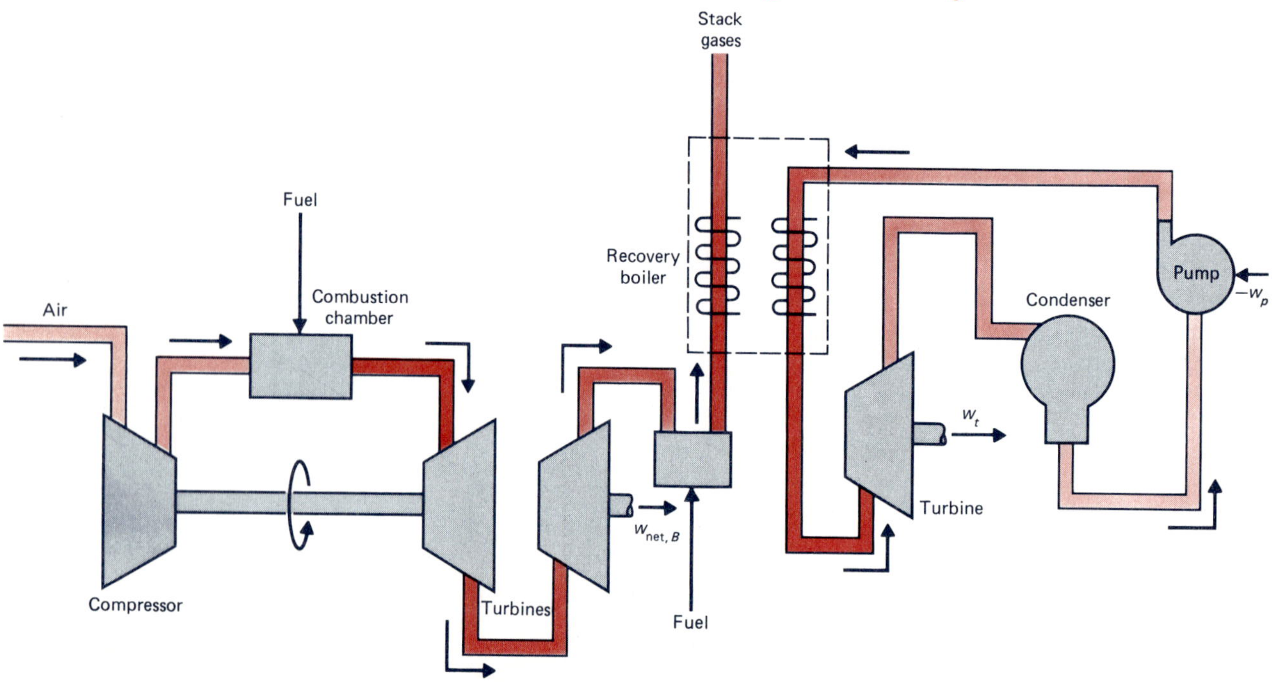

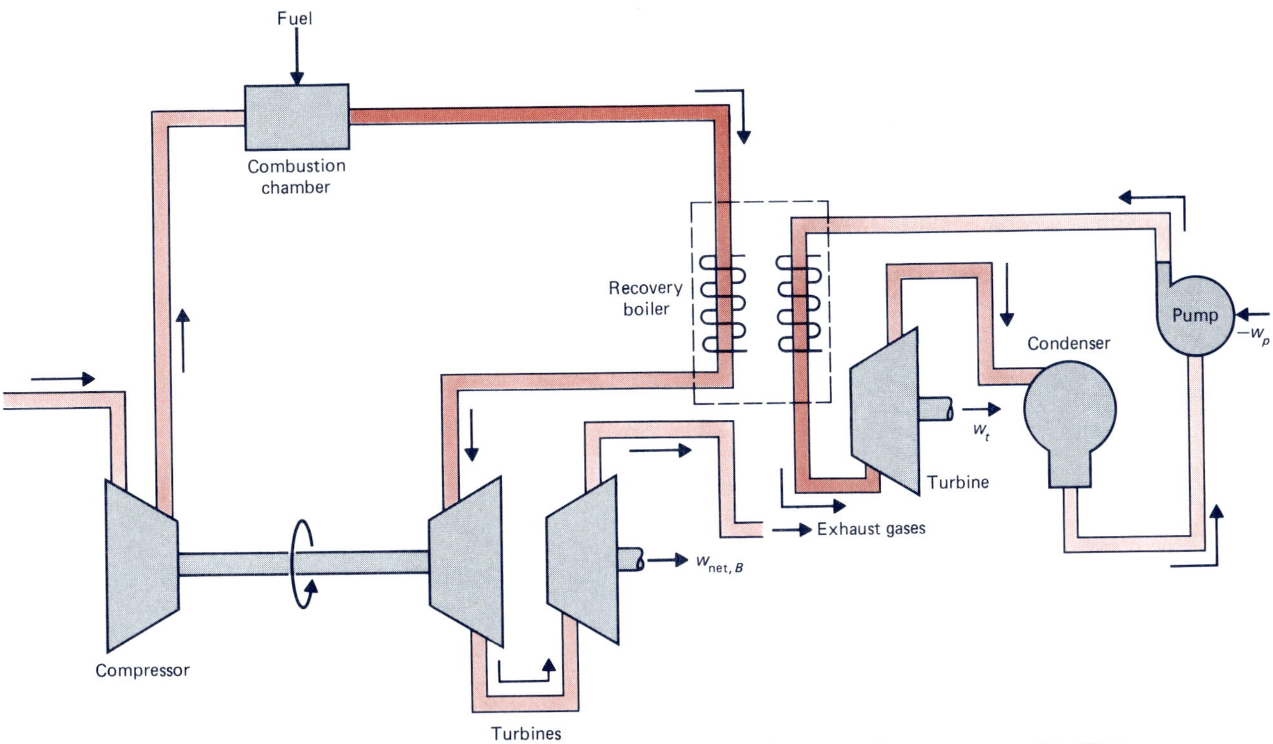

Figure 7.17 Combined cycle with heat transfer to the Rankine cycle between Brayton cycle compressor and turbines.

Another possibility is to insert the heat recovery boiler between the Brayton cycle combustion chamber and the gas turbine, as shown in Fig. 7.17. Now the boiler will operate at higher temperature, and the exhaust from the power turbine can be brought to as low an enthalpy as the turbine design will allow. The efficiency is given by

$$\eta = \frac{w_{\text{net}, B} + w_{\text{net}, R}}{q_{H, B}} \tag{7.3}$$

This is the same form as Eq. (7.1) for the simple combined cycle, but the values of the terms will be different because the relative net work outputs of the Brayton and Rankine cycles will change.

Another combined cycle may be constructed by using the rejected heat transfer from a Rankine plant using a working fluid that allows operation at high temperatures (such as a liquid metal) to drive a second Rankine plant operating with another working fluid (such as ammonia, water, or an organic fluid) at lower temperatures. Such an arrangement is sometimes called a *bottoming cycle.*

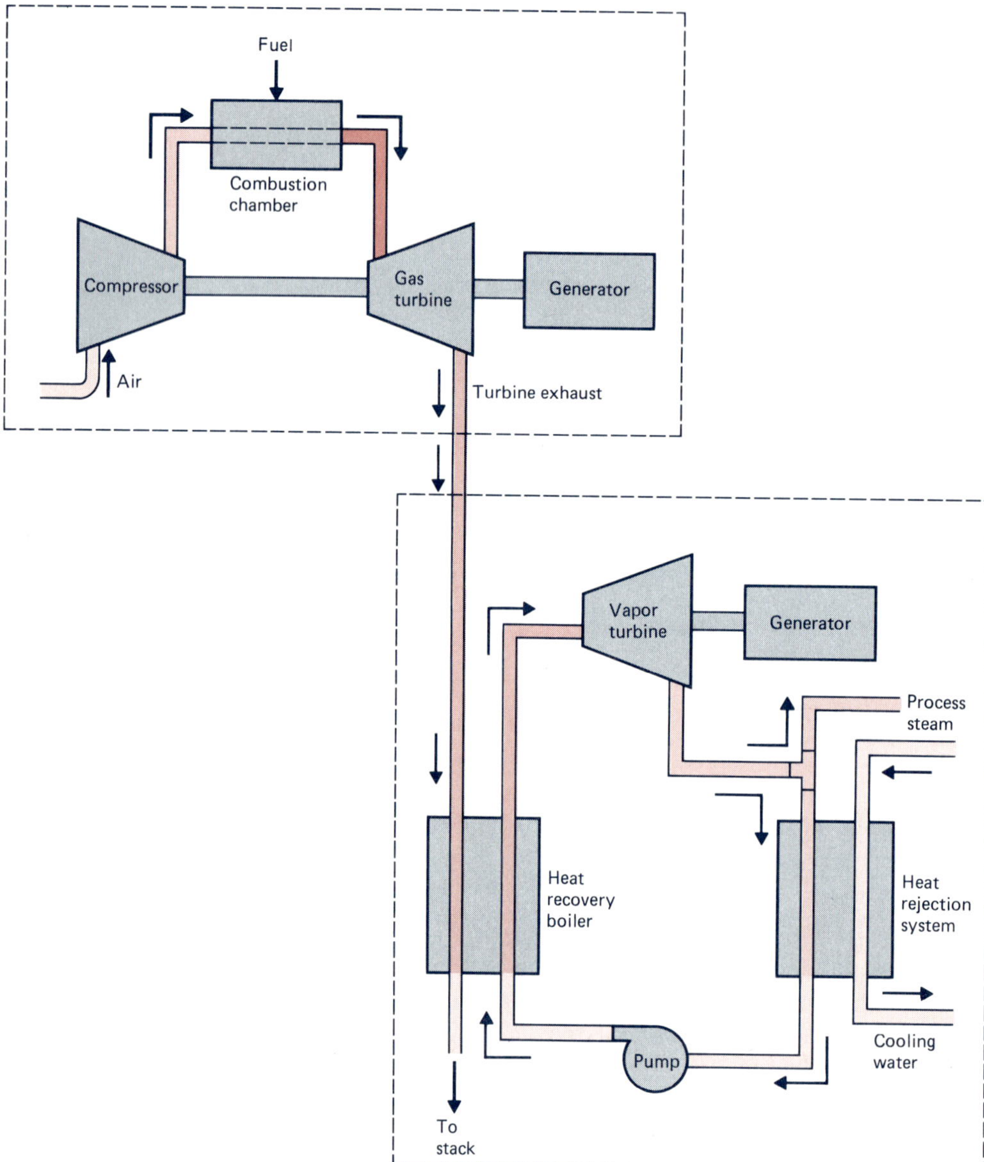

Figure 7.18 Combined cycle heat engine with cogeneration or process steam.

Cogeneration

Cogeneration plants are effectively Rankine cycle plants (or modifications such as the Kalina cycle) where the turbine output is used to drive an electric generator and steam is bled from the cycle for process use in a chemical or other plant. The steam bleed can come either directly from the turbine exhaust or, if higher pressure is desired, from a higher-pressure turbine state. The increase in the efficiency of energy use in this type of plant can be seen from the overall plant energy balance (an application of the first law) compared with the energy balance that occurs for two separate plants, one for electrical production and the other for making steam. However, analysis on the basis of the second law is also of value for this type of plant in particular, and it is addressed in Chap. 8.

There is no reason that a combined-cycle plant should not also include cogeneration for still higher efficiency. Such a combined-cycle cogeneration plant is shown schematically in Fig. 7.18.

7.4 Vapor Cooling and Heat Pump Cycles

The definitions of coefficient of performance for cooling and heat pump cycles were given in Secs. 6.5.1 and 6.5.2, and they apply equally well to gas and phase-change cooling and heat pump cycles.

7.4.1 Vapor Compression Systems

The most common method of providing air conditioning and chilling as well as heat pumping is probably by the vapor compression cycle. First, the use of the cycle in air conditioning and chilling is described.

The cycle hardware is shown in Fig. 7.19, and the *P-h* and *T-s* diagrams are seen in Figs. 7.20 and 7.21. In this cycle, the working fluid is initially a saturated or slightly superheated vapor at a relatively low pres-

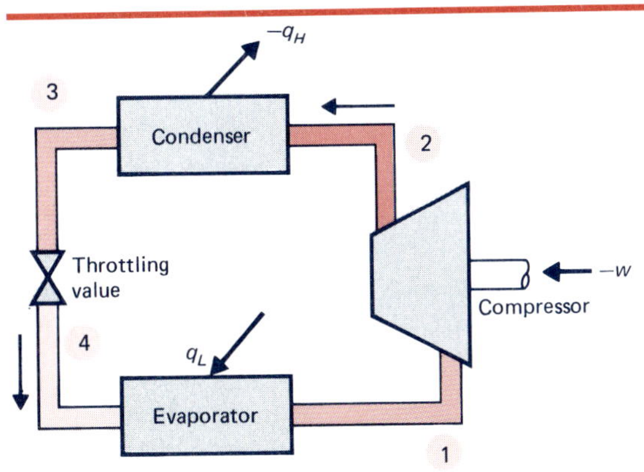

Figure 7.19 Vapor-compression air-conditioning cycle.

sure (state 1), and it is compressed to a high pressure (state 2), where it is condensed to a saturated or slightly subcooled liquid (state 3). Condensation may occur either by heat loss to the surroundings, if the saturation temperature of the condensing vapor is sufficiently greater than that of the surroundings, or by the use of a cooling tower to provide a lower-temperature heat sink. The condensed liquid is then expanded adiabatically through a throttling valve to a low pressure (state 4), where it becomes a mixture of saturated liquid and vapor of low quality. Its saturation temperature at this pressure is low, and a heat transfer from the surroundings can be used to evaporate the liquid fraction, thus providing the cooling effect and leaving the working fluid in initial state 1. The resulting saturated or slightly superheated vapor is fed to the compressor, completing the cycle.

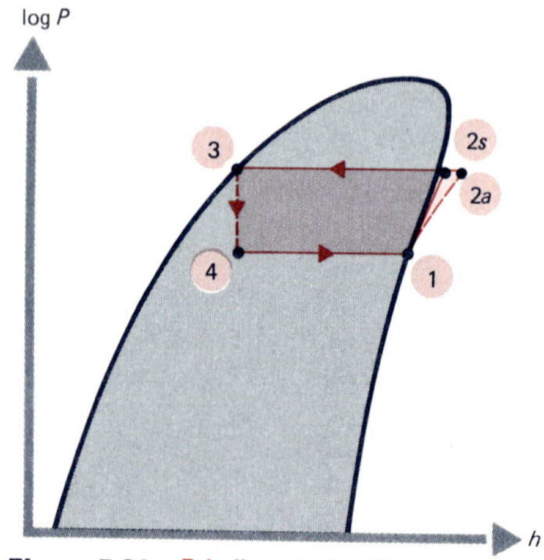

Figure 7.20 *P-h* diagram for the vapor compression cycle.

The working fluid for this cycle is chosen so that the heat transfer in the condenser can take place at pressures obtainable from economical compressors, and the evaporation temperature in the evaporator can be easily compatible with the inlet pressure to the compressor while allowing efficient heat transfer from the surroundings. For residential and commercial air conditioners, various chloro-fluorinated hydrocarbons have the necessary properties; where lower temperatures are required, such as in commercial freezing equipment, ammonia is often used.

The vapor compression heat pump cycle is exactly as shown in Figs. 7.19 through 7.21; only the objective of the cycle is changed. Now the evaporator is placed outdoors where there is a heat transfer to the working fluid at environmental temperatures (i.e., heat transfer from the environment). The evaporator may be placed in conjunction with any energy source that is at a lower temperature than desired from the cycle. In domestic heating systems, the heat transfer can be obtained from the outdoor air (an *air-source* heat pump) or from a convenient body of water such as a lake, river, or well (a *water-source* heat pump). The condenser is placed where the higher-temperature heat transfer being rejected from the cycle is to be used. This can be in the residence for heating or in a heat exchanger in an industrial process.

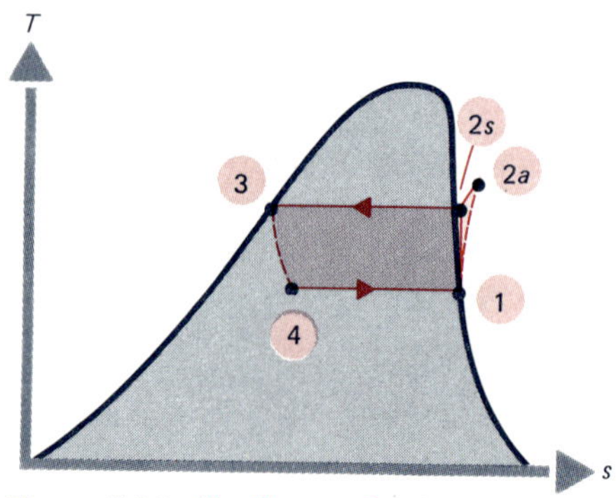

Figure 7.21 *T-s* diagram for the vapor compression cycle.

Example Problem 7.7

Find the COP for a vapor compression cycle that uses refrigerant 12 as the working fluid and is designed to operate at an evaporator temperature of 50°F and a condenser temperature of 100°F. The compressor efficiency is 65 percent.

Cycle Diagrams

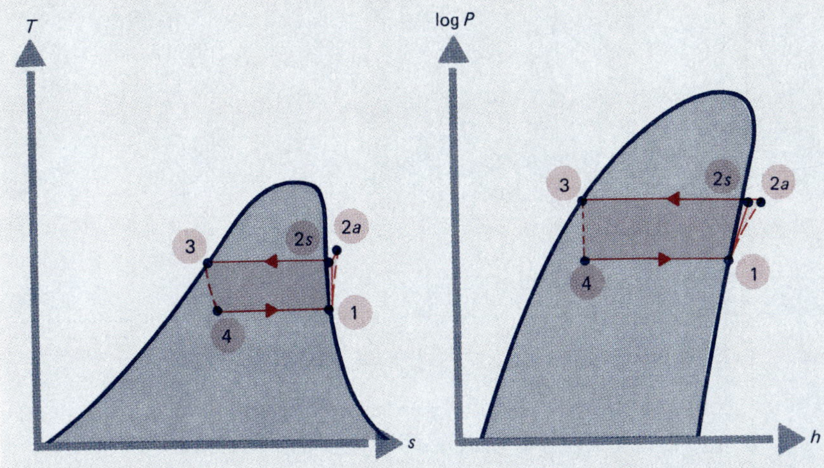

Solution

Assume that the vapor entering the compressor is saturated. Two properties are known for each state. Now, by using the compressor efficiency, all properties in state 2 can be calculated. For the adiabatic expansion between states 3 and 4, the first law indicates that the enthalpy remains constant; this fact can be used along with the known final temperature to calculate all properties in state 4. These calculations result in

State	P, psia	T, °F	v, ft³/lbm	h, Btu/lbm	s, Btu/(lbm·°R)	x
1	(61.4)	50	(0.65569)	(82.452)	(0.16548)	1.0
2s	131.8	—	—	(87.77)	s_1	—
2a	131.8	(119.4)	—	90.63	—	—
3	(131.8)	100	(0.012692)	(31.168)	(0.06362)	0.0
4	(61.4)	50	(0.13091)	31.17	(0.06486)	(0.185)

Now the COP can be computed:

$$\text{COP} = \frac{|q_L|}{|w|} = \frac{h_1 - h_4}{h_{2,a} - h_1} = \frac{82.452 - 31.168}{90.63 - 82.452} = 6.27$$

So for every unit of work done on the compressor, this cycle can remove 6.27 units of heat transfer from the evaporator. The ideal COP [Eq. (6.72)] is

$$(\text{COP})_{\text{rev}} = \frac{1}{T_H/T_L - 1} = \frac{1}{(459.67 + 119.4)/(459.67 + 50) - 1} = 7.34$$

where T_H has been taken as the maximum cycle temperature, not the condenser temperature.

Comments

The vapor compression system for cooling is efficient in the sense that each unit of work input can provide more than one unit of cooling. Because of the various heat transfers in the system, frictional losses in the working fluid, electric motor efficiencies, etc., the usual COP values of commercially marketed systems range up to about 3.

Example Problem 7.8

Find the COP of the system described in Example Problem 7.7 when it is operated as a heat pump.

Solution

The final state table of Example Problem 7.7 applies, and the COP is given by

$$\mathrm{COP_{HP}} = \frac{|q_H|}{|w|} = \frac{h_{2,a} - h_3}{h_{2,a} - h_1}$$

$$= \frac{90.63 - 31.168}{90.63 - 82.452} = 7.27$$

And now 7.27 units of heating at the condenser can be obtained at the expense of each unit of work to the compressor.

The ideal $\mathrm{COP_{HP}}$ is [Eq. (6.73*b*)]

$$\mathrm{COP_{HP,\,rev}} = \frac{1}{1 - T_L/T_H} = \frac{1}{1 - 509.67/579.07} = 8.34$$

Comments

Note that this system is also efficient as a heat pump. Because of the definitions of COP for cooling and heating modes, it is also true in all cases that [see Eq. (5.221)]

$$\mathrm{COP_{HP}} - \mathrm{COP} = 1 \tag{7.4}$$

which is seen to be true for the results of this example and Example Problem 7.7.

As for the cooling mode, the heat pump described here will have significant deviations from the ideal calculated behavior.

When a vapor-compression cycle is used both as a heat pump and for cooling, as is common in residential and commercial heating, ventilating, and air conditioning (HVAC) systems, the performance is usually somewhat below the cycle performance that could be obtained by two separate vapor-compression units. This occurs because the evaporator and condenser coils of the combined system are usually sized for best performance in the air conditioning mode (which is often the primary use) and are thus somewhat off-design for the heat pumping mode of operation.

The efficiency of the vapor compression cycle can be improved by increasing the evaporator pressure or decreasing the condenser pressure. However, the conditions for operating the system place limits on these factors.

7.4.2 Vapor Cooling Cycles Driven by Heat Transfer

Some cooling cycles can be driven directly by heat transfer from a high-temperature source such as solar energy, geothermal energy, or waste energy from a heat engine or other source. This may be a more efficient method than to use the heat transfer to drive a heat engine, which in turn is used to provide mechanical work to drive, for example, a vapor compression cooling system. We describe two such systems that involve phase change in the cooling process.

Absorption Cycles

The most common heat-driven cooling cycle is the *absorption* cycle. The hardware used in the cycle is shown in Fig. 7.22. In this cycle, two working fluids are used, the *refrigerant* and the *absorbent*. Common pairs are ammonia and water and water and lithium bromide. The choice of the refrigerant-absorbent pair depends mainly on the ability of the absorbent to absorb the refrigerant at a suitable temperature and on the absorption process being reversible, so the refrigerant can be driven from the absorbent at a higher temperature.

The process works as follows. A solution of absorbent that is rich in refrigerant enters the *generator*, where a heat transfer drives off the refrigerant. This process occurs at a relatively high pressure (which is nevertheless subatmospheric for some refrigerant-absorbent pairs). The refrigerant leaving the generator at state 1 is then condensed to liquid at state 2 and is expanded to a saturated mixture at low relative pressure through an expansion valve (state 3). The refrigerant then enters an evaporator, where heat transfer from the surroundings is used to evaporate the liquid fraction of the saturated refrigerant, producing the cooling effect and a saturated refrigerant vapor of 100 percent quality (state 4). The refrigerant vapor next enters an *absorber*, where it is absorbed in the rich absorbent which has been transferred from the generator after the refrigerant was

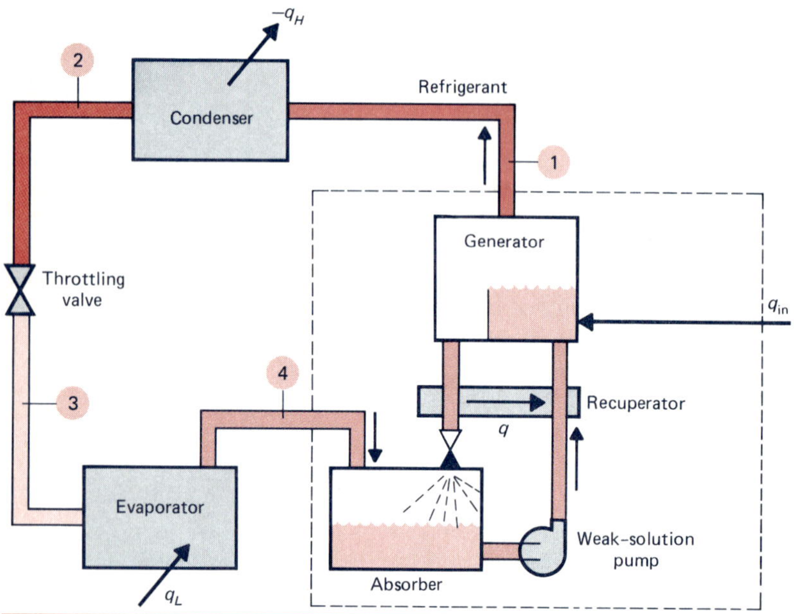

Figure 7.22 Schematic diagram of absorption cooling cycle.

driven off. The refrigerant-absorbent mixture in the absorber is then returned to the higher pressure of the generator by a pump. The heat exchanger (recuperator) between the absorber and the generator improves the system efficiency (COP) by cooling the hot absorbent before it enters the absorber (thus improving its absorption capability, which is strongly temperature-dependent). In addition, the mixture returning from the absorber to the generator is preheated by the recuperator, thus requiring less heat transfer from an external source to the generator.

The absorption system is identical in operation to the vapor compression cycle except that the generator-recuperator-absorber system of the absorption cycle replaced the compressor in the vapor compression cycle. The remaining components are the same. The advantage of the absorption cycle is that the only mechanical work needed is that necessary to drive the small liquid-return pump between the absorber and the generator (and even this can be eliminated by using an inert gas in the system, which can circulate the working fluid by using a heat transfer to expand and contract it in a "thermosyphon"). The work to drive this pump is very small compared with the work needed to drive the compressor in the vapor compression system. However, in terms of actual COP, the absorption systems are generally poor in comparison with vapor compression systems; for example, the water–lithium bromide system has a COP of about 0.6. This can be improved by various cycle modifications, but at an increase in system capital cost.

The methods of analyzing the generation and absorption processes must await the development of two-component vapor-liquid equilibrium relations in Chap. 12.

Desiccant Cycles

Another class of cooling cycles that uses heat transfer as the driving energy operates by removing almost all water vapor from the air being conditioned. Cooling is then obtained by evaporating some water to re-humidify the air while evaporatively cooling. These processes are examined further in Chap. 10 in the sections on psychrometrics. The drying is carried out by contacting the humid air with a *desiccant,* i.e., a solid material with a strong affinity for absorbing water vapor on its surface (silica gel) or a liquid with the ability to absorb water vapor into solution (lithium chloride, lithium bromide, triethylene glycol, and others). After the desiccant becomes saturated with water, it is dried by a heat transfer to increase the desiccant temperature; and the water vapor is driven off, so the desiccant can be reused. These systems can operate with regeneration temperatures as low as 330 K = 594°R, so low-grade energy sources can be used to drive them.

7.5 Concluding Remarks

Cycles and devices that can be used to harness energy transfers for useful purposes have been described in this and the preceding chapter, and the methods of analyzing the efficiency of these devices have been outlined. Note that the cycles and their analysis as presented in this chapter and Chap. 6 are the result of some hundreds of years of study, experimentation, and development by engineers, scientists, inventors, and entrepreneurs. The excitement and romance of this rich history are difficult to impart in a text meant for an introduction to the principles of the subject. But this is a good opportunity for the reader to read App. A on the historical development of classical thermodynamics and then to visit a power plant or climb over the remains of a steam locomotive and reflect on what the human mind has brought us from this history.

> It was a fine, handsome old engine, much older than Jake Holman himself. He looked at it, massive, dully gleaming brass and steel in columns and rods and links arching above drive rods from twinned eccentrics, great crossheads hung midway, and above them valve spindles and piston rods disappearing into the cylinder block. He knew them all, each part in its place in the whole, and his eye followed the pattern, three times repeated from forward to aft, each one-third of the circle out of phase, and it was all poised and balanced there like three chunks of frozen music.
>
> *Richard McKenna,* The Sand Pebbles, *Harper and Row, New York, 1962 (used with permission).*

Problems

Problems marked with a star are lengthy; thus, computerized steam tables or an accurate Mollier chart should be used if problems are to be completely worked.

7.1 A steam power plant has a mass flow rate of steam of 20,000 kg/h. Water enters the boiler (state 1) at 10 MPa and leaves the boiler (state 2) as saturated vapor. The working fluid leaves the condenser (state 4) at 10 kPa as saturated liquid. Cooling water to the condenser circulates at 1×10^6 kg/h and rises in temperature by 8.0°C in passing through the condenser. Find (*a*) the thermal efficiency of the cycle, (*b*) the turbine efficiency, and (*c*) the power output of the turbine (kW). Construct a state table to aid in your solution.

7.2S Thomas Newcomen's engines operated by spraying cold water into a cylinder filled with saturated steam near 1-atm pressure. The condensing steam then reduced the volume of the resulting liquid-vapor mixture, and atmospheric pressure pushed a piston into the cylinder (there is a diagram of the original Newcomen engine in App. A). A Newcomen engine with a piston diameter of 1 m and a stroke of 2 m reaches a minimum volume in the cylinder at the end of the stroke of 0.2 m^3.

(*a*) Find the amount of water at 20°C necessary to inject into the cylinder to just complete the stroke.

(*b*) Find the work output per stroke from the engine.

(*c*) No work was obtained from the Newcomen engine on the return stroke, which was carried out by allowing a system of weights to return the piston to its original position. If the engine could operate at 1 cycle/min and the piston is assumed to be of negligible weight, find the power output of the engine.

7.2E Thomas Newcomen's engines operated by spraying cold water into a cylinder filled with saturated steam near 1-atm pressure. The condensing steam then reduced the volume of the resulting liquid-vapor mixture, and atmospheric pressure pushed a piston into the cylinder (there is a diagram of the original Newcomen engine in App. A). A Newcomen engine with a piston diameter of 3 ft and a stroke of 6 ft reaches a minimum volume in the cylinder at the end of the stroke of 5 ft^3.

(*a*) Find the amount of water at 70°F necessary to inject into the cylinder to just complete the stroke.

(b) Find the work output per stroke from the engine.

(c) No work was obtained from the Newcomen engine on the return stroke, which was carried out by allowing a system of weights to return the piston to its original position. If the engine could operate at 1 cycle/min, and the piston is assumed to be of negligible weight, find the power output of the engine.

7.3S The P-v diagram for steam over the ranges commonly used in Rankine cycle analysis must be plotted on multicycle log-log paper for clarity. By using either the steam tables or computerized tables and plotting routines, plot the saturation curve for steam on a P-v plot for specific volume between 0.001 and 1000 m^3/kg. On this graph, show the isentropes that pass through the points (P, v) of (4000, 0.02), (4000, 0.05), and (4000, 0.5), where P is in kilopascals and v is in cubic meters per kilogram. Dicuss the linearity of the isentropes in terms of Eqs. (5.47) and (5.48).

7.3E The P-v diagram for steam over the ranges commonly used in Rankine cycle analysis must be plotted on multicycle log-log paper for clarity. By using either the steam tables or computerized tables and plotting routines, plot the saturation curve for steam on a P-v plot for specific volume between 0.01 and 1000 ft^3/lbm. On this graph, show the isentropes that pass through the points (P, v) of (600, 0.6), (600, 1), and (600, 2), where P is in psia and v is in ft^3/lbm. Discuss the linearity of the isentropes in terms of Eqs. (5.47) and (5.48).

*7.4S A simple Rankine cycle (no superheat, reheat, or regeneration) with an ideal pump and turbine is being planned with water as the working fluid. The condenser is to operate at 30°C. What boiler pressure will produce a cycle efficiency of 25 percent?

*7.4E A simple Rankine cycle (no superheat, reheat, or regeneration) with an ideal pump and turbine is being planned with water as the working fluid. The condenser is to operate at 90°F. What boiler pressure will produce a cycle efficiency of 25 percent?

*7.5S A simple Rankine cycle (no superheat, reheat, or regeneration) has a turbine with 80 percent efficiency. Water is the working fluid, and the condenser pressure is 2 kPa. What boiler pressure will provide a cycle efficiency of 25 percent?

*7.5E A simple Rankine cycle (no superheat, reheat, or regeneration) has a turbine with 80 percent efficiency.

Water is the working fluid, and the condenser pressure is 0.25 psia. What boiler pressure will provide a cycle efficiency of 25 percent?

***7.6S** Plot the simple ideal Rankine cycle efficiency versus boiler pressure for a condenser temperature of 27°C. Also plot the turbine exit quality versus boiler pressure on the same graph. Use a boiler pressure range of 0.7 to 7 MPa.

***7.6E** Plot the simple ideal Rankine cycle efficiency versus boiler pressure for a condenser temperature of 80°F. Also plot the turbine exit quality versus boiler pressure on the same graph. Use a boiler pressure range of 100 to 1000 psia.

7.7S A simple Rankine cycle operates between a condenser pressure of 8 kPa and a boiler pressure of 6 MPa. Compare the cycle efficiency of this cycle with the efficiency of a cycle that operates between the same pressures but superheats the vapor leaving the boiler to 400°C. For both systems, the turbine efficiency is 88 percent. Compare and discuss the turbine exit quality for the two cases.

7.7E A simple Rankine cycle operates between a condenser pressure of 1 psia and a boiler pressure of 800 psia. Compare the cycle efficiency of this cycle with the efficiency of a cycle that operates between the same pressures but superheats the vapor leaving the boiler to 800°F. For both systems, the turbine efficiency is 88 percent. Compare and discuss the turbine exit quality for the two cases.

***7.8S** Plot the turbine exit quality versus degrees superheat for a Rankine cycle operating at a boiler pressure of 4 MPa and a condenser pressure of 8 kPa for turbine efficiencies of 60, 70, and 80 percent. (*Degrees of superheat* are defined as the temperature *difference* between the actual superheated steam temperature and the saturation temperature, both evaluated at the boiler pressure.) Use a degrees superheat range of 0 to 150°C.

***7.8E** Plot the turbine exit quality versus degrees superheat for a Rankine cycle operating at a boiler pressure of 600 psia and a condenser pressure of 1 psia for turbine efficiencies of 60, 70, and 80 percent. (*Degrees of superheat* are defined as the temperature *difference* between the actual superheated steam temperature and the saturation temperature, both evaluated at the boiler pressure.) Use a degrees superheat range of 0 to 250°F.

7.9S Saturated steam leaves a boiler at $P = 200$ bars. Heat transfer occurs in a superheater at constant P until the temperature reaches 600°C. The steam then enters a steam turbine, where it does work on the turbine, and leaves the turbine at a pressure of $P = 7.0$ bars. The turbine has an efficiency of 75 percent.

(*a*) How much work is done on the turbine?

(*b*) How much heat transfer occurs in the superheater?

(*c*) If liquid water enters the boiler at 165°C and 200 bars, what is the Rankine cycle efficiency? (Neglect pump work.)

7.9E Saturated steam leaves a boiler at $P = 2900$ psia. Heat transfer occurs in a superheater at constant P until the temperature reaches 1100°F. The steam then enters a steam turbine, where it does work on the turbine, and leaves the turbine at a pressure of $P = 100$ psia. The turbine has an efficiency of 75 percent.

(*a*) How much work is done on the turbine?

(*b*) How much heat transfer occurs in the superheater?

(*c*) If liquid water enters the boiler at 300°F and 2900 psia, what is the Rankine cycle efficiency? (Neglect pump work.)

7.10S A Rankine cycle with superheat has a condenser pressure of 8 kPa, a boiler pressure of 6 MPa, and a turbine efficiency of 88 percent and superheats the steam leaving the boiler to 400°C. Compare the efficiency of this cycle to a reheat cycle operating with the same boiler, superheater, and condenser conditions in which steam leaves the high-pressure stage of a two-stage turbine at 300 kPa, is reheated to 400°C, and then is fed to the low-pressure turbine stage. Assume that both turbine stages have efficiencies of 88 percent.

7.10E A Rankine cycle with superheat has a condenser pressure of 1 psia, a boiler pressure of 800 psia, and a turbine efficiency of 88 percent and superheats the steam leaving the boiler to 800°F. Compare the efficiency of this cycle to a reheat cycle operating with the same boiler, superheater, and condenser conditions in which steam leaves the high-pressure stage of a two-stage turbine at 35 psia, is reheated to 800°F, and then is fed to the low-pressure turbine stage. Assume that both turbine stages have efficiencies of 88 percent.

7.11S A steam power plant operates with a boiler pressure of 1 MPa, and the steam leaving the boiler is superheated to 250°C. The turbine used in the plant is a two-stage type. The first stage operates at an efficiency of 70 percent. Steam leaving the first stage exits at 600 kPa, is reheated to 250°C, and is returned to the second (low-pressure) stage, which has an efficiency of 77 percent. The low-pressure stage exhausts to the condenser at a pressure of 8 kPa.

(*a*) Sketch the equipment layout, and label the positions of states 1 through 7 on the layout to correspond with the state table.

(*b*) Sketch the cycle on a *T-s* diagram, being as accurate as you can. Label the states to correspond to the state table given here.

(*c*) Fill in the state table (using the steam tables).

(*d*) Fill in the energy table.

State Table

Location	State	*T*, °C	*P*, kPa	*h*, kJ/kg	*s*, kJ/(kg·K)	*x*
Pump inlet	1		8			0.0
Boiler inlet	2	45	1000			—
Superheater inlet	3		1000			—
Superheater outlet	4	250	1000			—
High-pressure turbine outlet	5		600			—
Low-pressure turbine inlet	6	250	600			—
Low-pressure turbine outlet	7		8			

Energy Table

${}_2q_3$, kJ/kg =
${}_3q_4$, kJ/kg =
${}_5q_6$, kJ/kg =
${}_4w_5$, kJ/kg =
${}_6w_7$, kJ/kg =
${}_1w_2$, kJ/kg =
Cycle efficiency, % =

7.11E A steam power plant operates with a boiler pressure of 150 psia, and the steam leaving the boiler is superheated to 450°F. The turbine used in the plant is a two-stage type. The first stage operates at an efficiency of 70 percent. Steam leaving the first stage exits at 100 psia, is reheated to 450°F, and is then returned to the second (low-pressure) stage, which has an efficiency of 77 percent. The low-pressure stage exhausts to the condenser at a pressure of 1 psia.

(*a*) Sketch the equipment layout, and label the positions of states 1 through 7 on the layout to correspond with the state table.

(*b*) Sketch the cycle on a *T-s* diagram, being as accurate as you can. Label the states to correspond to the state table given here.

(*c*) Fill in the state table (using the steam tables).
(*d*) Fill in the energy table.

State Table

Location	State	T, °F	P, psia	h, Btu/lbm	s, Btu/(lbm·°F)	x
Pump inlet	1		1			0.0
Boiler inlet	2	100	150			—
Superheater inlet	3		150			—
Superheater outlet	4	450	150			—
High-pressure turbine outlet	5		100			
Low-pressure turbine inlet	6	450	100			—
Low-pressure turbine outlet	7		1			

Energy Table

${}_2q_3$, Btu/lbm =
${}_3q_4$, Btu/lbm =
${}_5q_6$, Btu/lbm =
${}_4w_5$, Btu/lbm =
${}_6w_7$, Btu/lbm =
${}_1w_2$, Btu/lbm =
Cycle efficiency, % =

7.12S A Rankine cycle with superheat has a condenser pressure of 8 kPa, a boiler pressure of 6 MPa, and a turbine efficiency of 88 percent and superheats the steam leaving the boiler to 400°C. Compare this cycle efficiency to that of a regeneration cycle using a closed feedwater heater in which 10 percent of the entering turbine mass flow is bled off at 200 kPa to the feedwater heater. Condensate is returned to the system as shown in Fig. 7.10.

7.12E A Rankine cycle with superheat has a condenser pressure of 1 psia, a boiler pressure of 800 psia, and a turbine efficiency of 88 percent and superheats the steam leaving the boiler to 800°F. Compare this cycle efficiency to that of a regeneration cycle using a closed feedwater heater in which 10 percent of the entering turbine mass flow is bled off at 25 psia to the feedwater heater. Condensate is returned to the system as shown in Fig. 7.10.

7.13 Repeat Prob. 7.12 for a regeneration cycle using an open feedwater heater.

7.14 Repeat Prob. 7.12 for bleed flows of 5, 10, 15, 20, and 25 percent of the entering turbine mass flow. Discuss why the cycle efficiency might reach a maximum at a particular bleed flow.

7.15S A Rankine cycle with superheat has a condenser pressure of 8 kPa and a boiler pressure of 6 MPa, and it superheats the steam leaving the boiler to 400°C. Steam leaves the high-pressure stage of a two-stage turbine at 300 kPa, is reheated to 400°C, and then is fed to the low-pressure turbine stage. Assume that both turbine stages have efficiencies of 88 percent. Of the entering turbine mass flow, 10 percent is bled from the turbine at 200 kPa and is sent to a closed feedwater heater. Compare the efficiency of this superheat, reheat, regeneration cycle with the results of Probs. 7.7S, 7.10S, and 7.12S.

7.15E A Rankine cycle with superheat has a condenser pressure of 1 psia and a boiler pressure of 800 psia, and it superheats the steam leaving the boiler to 800°F. Steam leaves the high-pressure stage of a two-stage turbine at 35 psia, is reheated to 800°F, and then is fed to the low-pressure turbine stage. Assume that both turbine stages have efficiencies of 88 percent. Of the entering turbine mass flow, 10 percent is bled from the turbine at 25 psia and is sent to a closed feedwater heater. Compare the efficiency of this superheat, reheat, regeneration cycle with the results of Probs. 7.7E, 7.10E, and 7.12E.

***7.16S** For a simple Rankine cycle with boiler pressure of 6 MPa and a turbine efficiency of 88 percent, plot cycle efficiency and condenser heat rejection versus condenser pressure in the range $1 \leq P_{cond} \leq 100$ kPa.

***7.16E** For a simple Rankine cycle with a boiler pressure of 800 psia and a turbine efficiency of 88 percent, plot cycle efficiency and condenser heat rejection versus condenser pressure in the range $1 \leq P_{cond} \leq 15$ psia.

***7.17S** For a simple Rankine cycle with a condenser pressure of 8 kPa and a turbine efficiency of 88 percent, plot cycle efficiency, turbine exit quality, and condenser heat rejection versus boiler pressure over the range $0.1 \leq P_{boiler} \leq 6$ MPa.

***7.17E** For a simple Rankine cycle with condenser pressure of 1 psia and a turbine efficiency of 88 percent, plot cycle efficiency, turbine exit quality, and condenser heat rejection versus boiler pressure over the range $15 \leq P_{boiler} \leq 800$ psia.

7.18S An improved Rankine cycle has been suggested that operates as diagrammed in Fig. P7.18S. Some of the turbine work is used to drive an isentropic compressor that takes a saturated mixture of quality x_5 from the condenser and compresses it to saturated liquid at state 1. The turbine is assumed isentropic, the boiler operates at 6 MPa and the condenser at 8 kPa, and the boiler steam is superheated to 400°C.

(*a*) Find the efficiency of the proposed cycle.

(*b*) Find the efficiency of the usual Rankine cycle operating under the same conditions (i.e., use a pump in place of the compressor, with $x_5 = 0$).

(*c*) What are the ratios of the proposed and usual Rankine cycle efficiencies to the appropriate Carnot efficiency?

(*d*) Calculate the efficiency of the proposed cycle with no superheat.

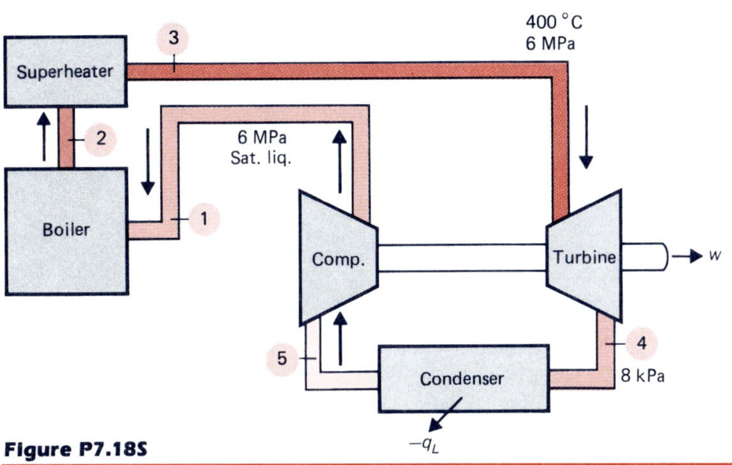

Figure P7.18S

7.18E An improved Rankine cycle has been suggested that operates as diagramed in Fig. P7.18E. Some of the turbine work is used to drive an isentropic compressor that takes a saturated mixture of quality x_5 from the condenser and compresses it to saturated liquid at state 1. The turbine is assumed isentropic, the boiler operates at 900 psia and the condenser at 1 psia, and the boiler steam is superheated to 750°F.

(*a*) Find the efficiency of the proposed cycle.

(*b*) Find the efficiency of a usual Rankine cycle operating under the same conditions (i.e., use a pump in place of the compressor, with $x_5 = 0$).

(*c*) What are the ratios of the proposed and usual Rankine cycle efficiencies to the appropriate Carnot efficiency?

(*d*) Calculate the efficiency of the proposed cycle with no superheat.

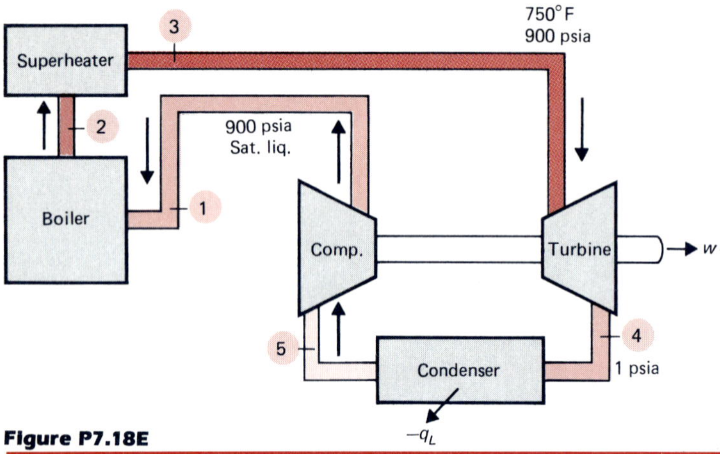

Figure P7.18E

7.19S A Rankine cycle operates with an open feedwater heater. The boiler pressure is 6 MPa, and the superheater provides a turbine inlet temperature of 400°C. Bleed steam is provided from the turbine at 500 kPa.

(*a*) What fraction of the boiler mass flow rate must be used in the bleed stream to provide a boiler feedwater inlet temperature of 150°C?

(*b*) What is the cycle efficiency?

(*c*) What is the cycle efficiency without regeneration?

7.19E A Rankine cycle operates with an open feedwater heater. The boiler pressure is 800 psia, and the superheater provides a turbine inlet temperature of 800°F. Bleed steam is provided from the turbine at 75 psia.

(*a*) What fraction of the boiler mass flow rate must be used in the bleed stream to provide a boiler feedwater inlet temperature of 300°F?

(*b*) What is the cycle efficiency?

(*c*) What is the cycle efficiency without regeneration?

7.20 An actual steam power plant operates as shown. The boiler and condenser operate at constant pressure. The turbine and pump are adiabatic. Find w_t, w_p, and η_{th}.

$T_1 = 30°C$ $\quad T_3 = 600°C$

$P_1 = 0.01$ MPa $\quad \eta_p = 0.80$

$P_2 = 3.0$ MPa $\quad \eta_t = 0.90$

Figure P7.20

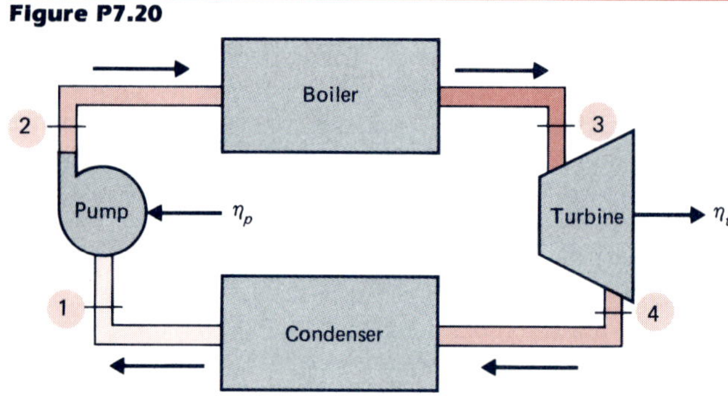

7.21 Steam enters the high-pressure turbine in an ideal reheat cycle with a mass flow rate of 1000 g/s. Find w_t, w_p, and η_{th}.

$T_3 = 600°C$ $P_4 = 0.8$ MPa
$P_3 = 15.0$ MPa $T_5 = 600°C$
$P_6 = 6$ kPa

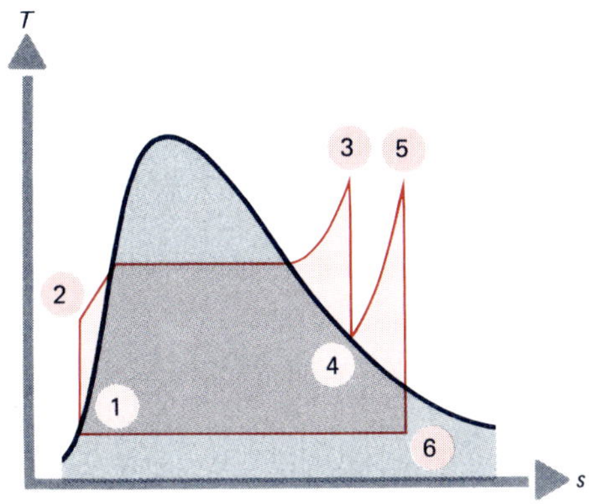

Figure P7.21

7.22 A steam power plant operates on the Rankine cycle. The turbine inlet pressure is 4.0 MPa. The steam leaves the turbine at a pressure of 0.1013 MPa and quality of 0.94. If the mass flow rate of steam is 1.2 kg/s, find the net power output and cycle thermal efficiency. Do not neglect pump work.

Figures P7.22 and P7.23

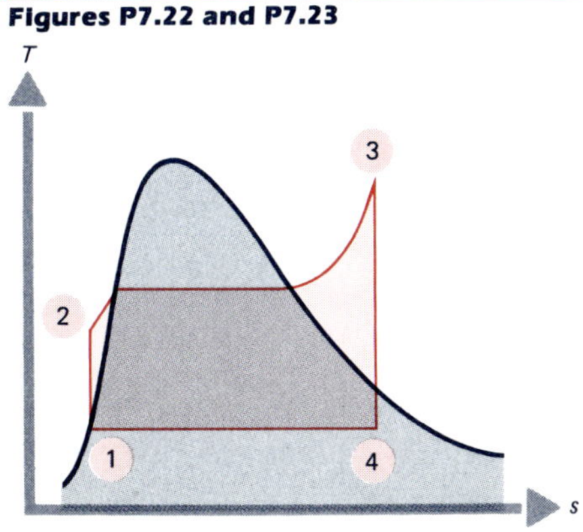

7.23 A steam power plant operates on the ideal Rankine cycle. The steam leaves the turbine at a pressure of 0.2 MPa and quality of 0.98. The turbine produces 1500 kW of power. If the mass flow rate is 1.5 kg/s, find the net power output and the cycle thermal efficiency.

7.24 Water enters the pump of an ideal Rankine cycle at 4 kPa. The H_2O exits the boiler at 0.2 MPa and 425°C. If $\dot{W}_{net} = 556.4$ hp, find w_p, q_L, q_H, w_t, $\dot{m}$, and η_{th}.

7.25 Steam enters the turbine of an ideal Rankine cycle at 10 MPa and 500°C. The steam exits the turbine as a saturated vapor. Find w_p, q_H, w_t, w_{net}, q_L, and η_{th}.

7.26 Steam enters the turbine of an ideal Rankine cycle at 500°C and exits the turbine at $x = 0.99$ and $P = 0.2$ MPa. If $\dot{m}$ of the steam is 0.2 kg/s, find w_p, q_H, q_L, w_t, $\dot{W}_{net}$ (hp), and η_{th}.

7.27 An ideal Rankine cycle operates with a high pressure of 6 MPa and a low pressure of 40 kPa. The quality of the steam leaving the turbine is 0.83. If $\dot{m} = 0.25$ kg/s, find w_p, q_H, w_t, $\dot{W}_{net}$ (hp), and q_L.

7.28 Steam leaves the boiler of an ideal Rankine cycle at 325°C. The H_2O enters the condenser at 15 kPa and $x = 0.972$. Find w_p, q_H, q_L, w_t, w_{net}, and η_{th}.

7.29 Water enters the pump in an ideal Rankine cycle at $P = 0.1$ MPa. The H_2O exits the pump at 3 MPa. The maximum temperature in the cycle is 450°C, and $\dot{m} = 0.3$ kg/s. Find w_p, q_H, w_t, $\dot{W}_{net}$ (hp), and q_L.

7.30 An ideal steam power plant with reheat operates as shown in Fig. P7.30. Determine the thermal efficiency and the mass flow rate. Do not neglect pump work ($\dot{W}_p = 2.5$ hp).

Figure P7.30

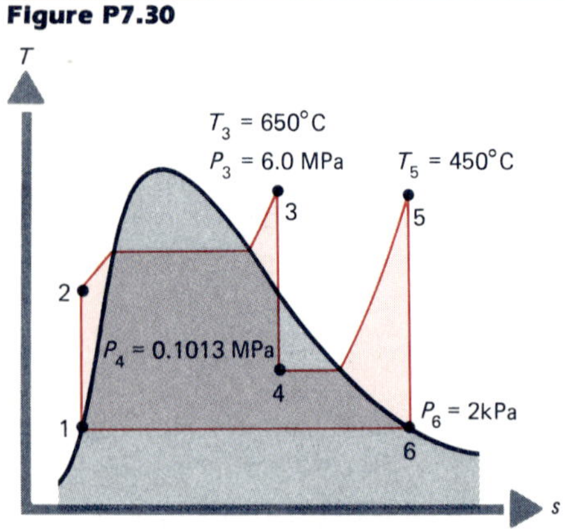

7.31 An actual steam power plant with a two-stage turbine and an interstage heater operates as shown in Fig. P7.31. Determine the mass flow rate of steam, the net power output (kW), and the total heat transfer rate (kW) to the plant.

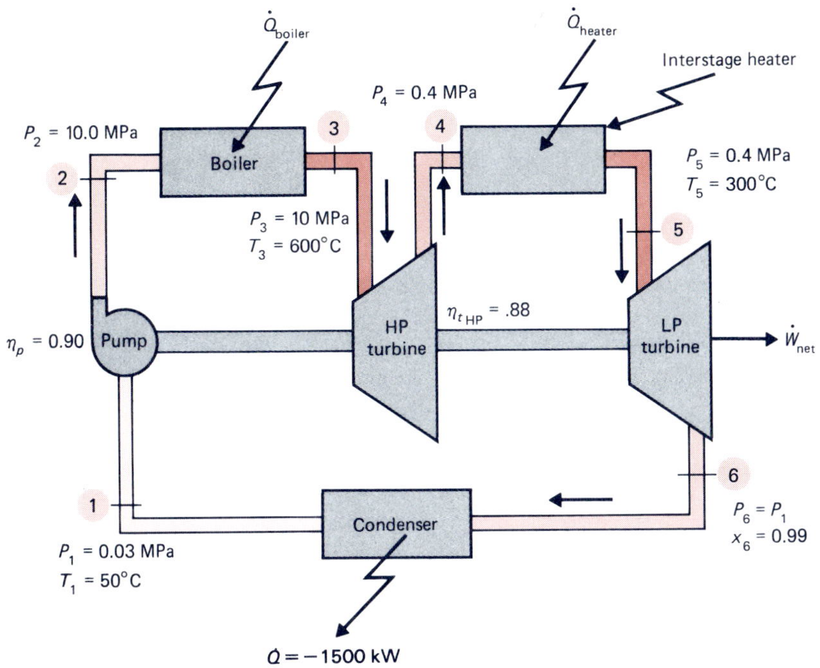

Figure P7.31

7.32 An actual steam power plant with a two-stage turbine and an interstage heater operates as shown in P7.32. Write out the equations to determine $\dot{w}_P$, h_2, $\dot{Q}_H$, $\dot{W}_t$, and η_{th}.

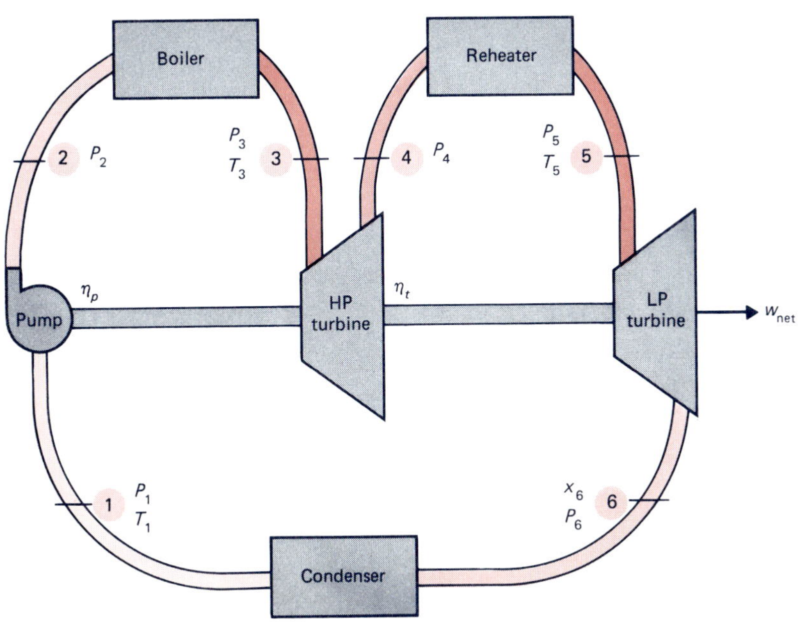

Figure P7.32

7.33 Determine the cycle thermal efficiency of a reheat cycle utilizing steam where the steam leaves the boiler and enters the turbine at 4 MPa and 400°C. After expansion in the turbine to 400 kPa, the steam is reheated to 400°C and then expanded in the low-pressure turbine to 10 kPa. The cycle is an ideal cycle.

7.34 Determine the thermal efficiency of the steam power plant shown in Fig. P7.34.

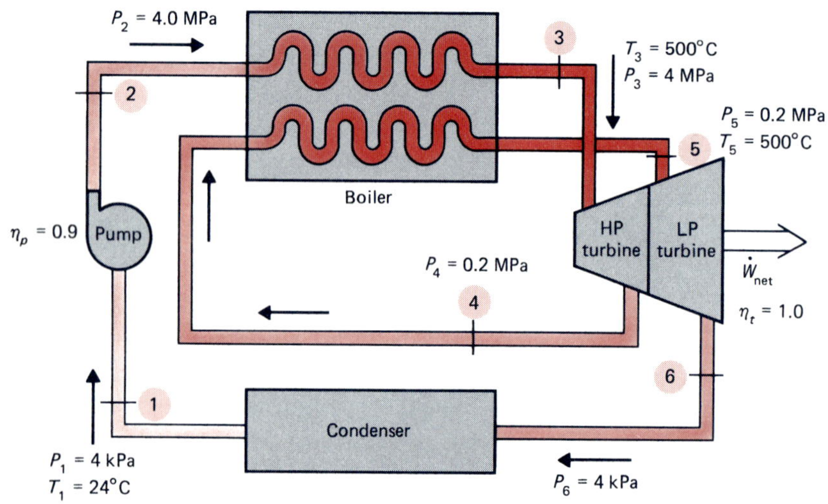

Figure P7.34

7.35 Consider the steam power plant shown in Fig. P7.35. What is η_{th} for this plant? The pump and turbine are adiabatic.

State 1
$P = 0.3$ MPa
$x = 0$

State 2
$P = 2.0$ MPa
$T = 140$°C

State 3
$P = 2.0$ MPa
$T = 350$°C

State 4
$P = 0.3$ MPa
$x = 98\%$

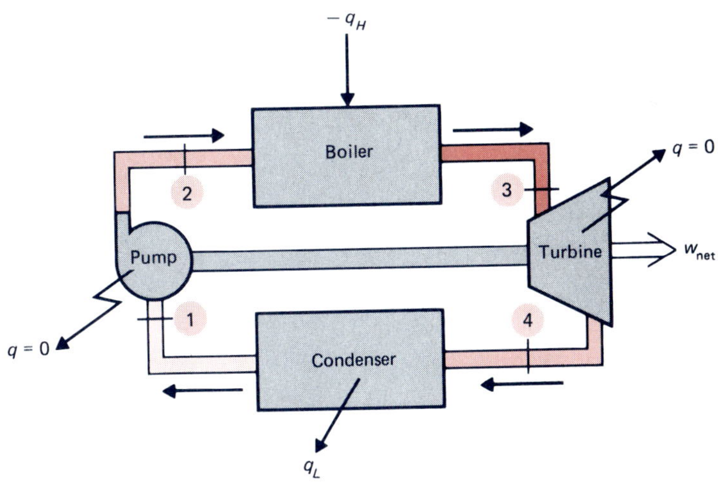

Figure P7.35

7.36 A schematic of an ideal power plant which operates on the Rankine cycle is shown in Fig. P7.36. In terms of the enthalpies at the given state points, determine q_H, q_L, w_p, w_t, w_{net}, and η_{th}.

Figure P7.36

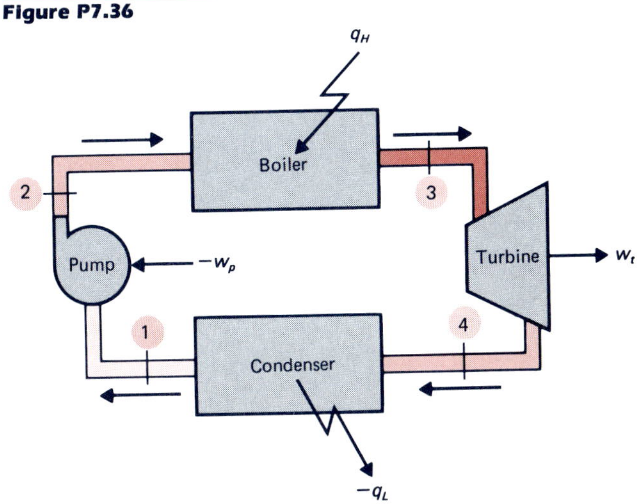

7.37 A schematic of a power plant which operates on an ideal reheat cycle is shown in Fig. P7.37. In terms of enthalpies at the given state points, determine q_H, q_L, w_p, w_t, w_{net}, and η_{th}.

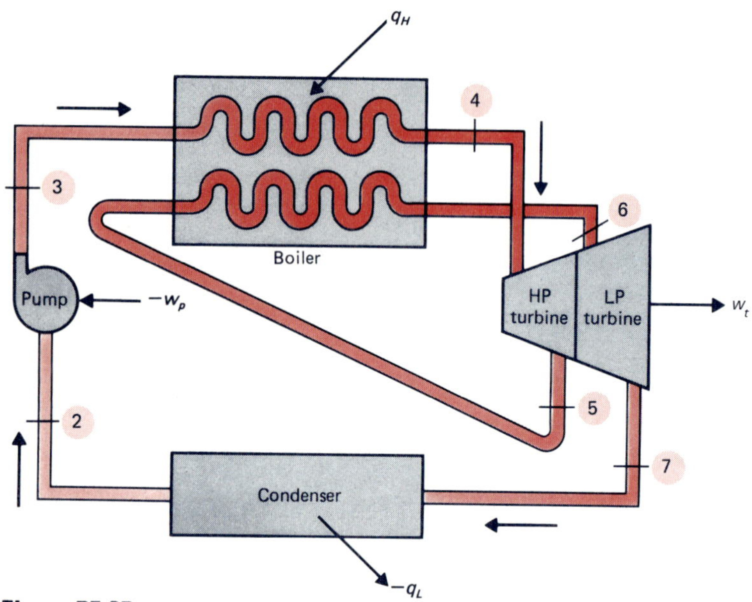

Figure P7.37

7.38 The boiler you must use in an ideal steam power cycle can only withstand a pressure of 6.0 MPa and a temperature of 600°C. To preclude damage to the turbine blades, the steam must exit the turbines with a moisture content of less than 10 percent. The condenser must operate at a pressure of 15 kPa. Draw a sketch on a Mollier chart, and show how you propose to keep the boiler from operating outside the prescribed limits.

7.39 In a Rankine cycle with superheat, steam is produced at a pressure of 2.0 MPa, and it is desired that the turbine exhaust as a saturated vapor to the condenser at 15 kPa. If the boiler cannot be operated at a steam temperature greater than 700°C, determine whether this cycle can operate without exceeding boiler temperature limitations.

7.40 What is the net work of an ideal Rankine cycle in which the working fluid enters the boiler at 2.5 MPa and exits at 450°C? The moisture content at the turbine exit is 9 percent. Do not neglect pump work.

7.41 The condenser of an ideal Rankine cycle power plant operates at 20 kPa, and steam exits the boiler at 525°C and 2.0 MPa. Compare the thermal efficiency of this power plant to that of a Carnot engine operating between the same temperature extremes. Do not neglect pump work.

7.42 A steam power plant operates on the ideal Rankine cycle. Steam leaves the boiler at $P = 4.0$ MPa and $T = 550°C$, and it leaves the turbine with a moisture content of 0.02. Find the cycle thermal efficiency. Do not neglect pump work.

7.43 An actual steam power plant operates as shown in Fig. P7.43. Find $\dot{W}_{net}$ and η_{th} if the pump and turbine are adiabatic.

$P_1 = 0.1013$ MPa	$P_5 = 1.5$ MPa
$T_1 = 36°C$	$T_5 = 300°C$
$P_2 = 2.08$ MPa	$P_6 = 0.15$ MPa
$T_2 = 38°C$	$x_6 = 0.98$
$P_3 = 2.05$ MPa	$P_7 = 0.1013$ MPa
$T_3 = 36°C$	$T_7 = 45°C$
$P_4 = 2.0$ MPa	$\dot{m} = 1$ kg/s
$T_4 = 350°C$	

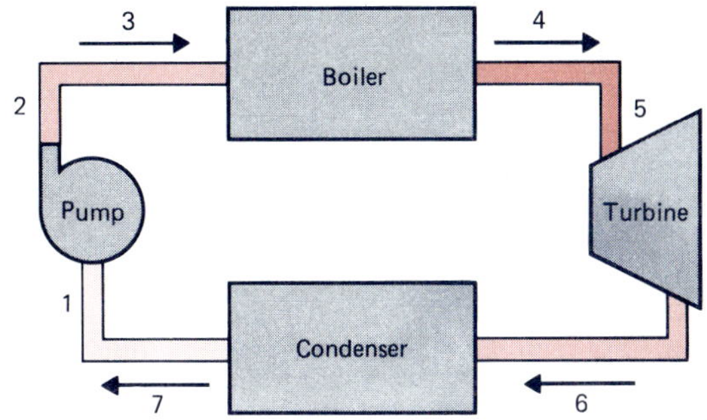

Figure P7.43

7.44 A Rankine steam power cycle produces steam from the boiler-superheater at 600 kPa and 600°C. The turbine has an efficiency of 90 percent. The pump is ideal. The condenser pressure is 4 kPa.
(*a*) Show whether the steam leaving the turbine is superheated.
(*b*) What is the work output from the turbine per kilogram of steam passing through (kJ/kg)?
(*c*) What is the cycle efficiency (percentage)?

7.45 A Rankine cycle steam power plant using geothermal energy as an energy source operates with steam leaving the steam generator at 2000 kPa and 400°C. The condenser operates at 10 kPa. The turbine efficiency is 85 percent, and the pump efficiency is 100 percent. The power output of the turbine is 50 MW.
(*a*) Fill in all entries in a state table.
(*b*) What power input is required for the boiler (kW)?

7.46 A steam Rankine cycle operates between a boiler pressure of 500 kPa and a condenser with a temperature of 20°C. The inlet steam to the turbine is saturated vapor, and the turbine is isentropic. The feedwater to the boiler has an enthalpy of 84.4 kJ/kg.
(*a*) Fill in a state table.
(*b*) What is the cycle efficiency (percentage)?

7.47 In a "total energy system," enough electricity is produced to satisfy the user's electrical demand, and as much of the waste energy as possible is used to satisfy the user's thermal demand (space heating, water heating, etc.). Considering the total energy system shown in Fig. P7.47, find the rate of heat transfer to the river.

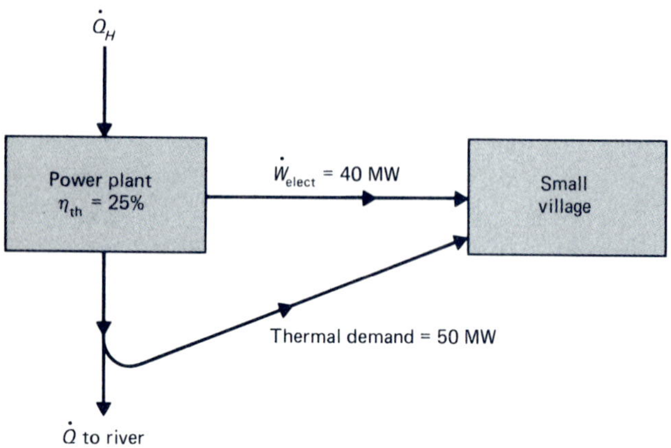

Figure P7.47

7.48 A remote Army installation employing a total energy system has an electrical demand of 10 MW and a thermal demand of 12 MW. If the installation's power facility rejects 20 MW to the environment, find the thermal efficiency $\dot{W}_{net}/\dot{Q}_H$ of the power plant.

7.49 In a "selective energy system," electricity is produced while all the waste heat is used to satisfy the user's thermal demand (space heating, water heating, etc.). Any additional electricity required is purchased from an outside utility. Considering the selective energy system shown in Fig. P7.49, find the amount of electricity purchased.

Figure P7.49

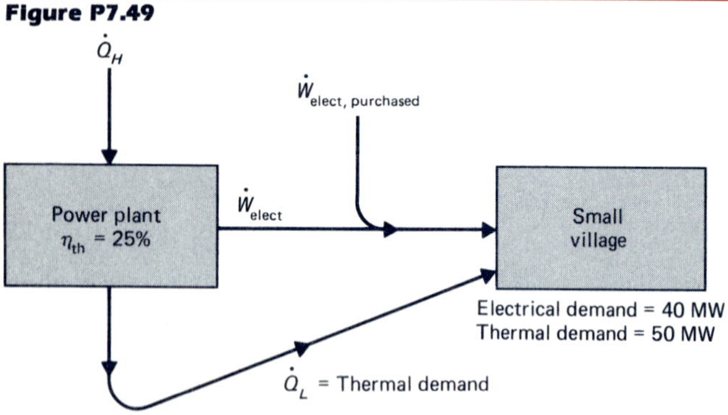

7.50 A small Army installation employing a selective energy system has an electrical demand of 10 MW and a thermal demand of 12 MW. If it is necessary to purchase 6 MW of electric power from an outside utility, find the thermal efficiency $\dot{W}_{net}/\dot{Q}_H$ of the installation's power plant. Assume 100 percent waste heat utilization.

7.51 In the ideal vapor-compression refrigerator using refrigerant 12, the evaporator temperature is $-20°C$, and the inlet temperature to the condenser is $30°C$. Sketch the cycle on a *P-h* diagram and calculate

(*a*) The work of the compressor
(*b*) The heat transfer to the evaporator
(*c*) The heat transfer from the condenser
(*d*) The coefficient of performance

7.52 A refrigeration system using refrigerant 12 is to have a capacity of 25 tons. The cycle is the ideal vapor compression cycle in which the evaporator pressure is 0.22 MPa and the condenser pressure is 1.0 MPa. Determine the horsepower required to drive the compressor.

7.53 An ideal vapor-compression refrigeration system using refrigerant 12 operates with an evaporator temperature of $-30°C$ and a condenser exit temperature of $50°C$ and requires a 100-hp motor to drive the compressor. Calculate the refrigerator's capacity in tons.

7.54 An actual vapor-compression refrigerator has a compressor whose inlet conditions are a temperature of $-20°C$ and a pressure of 0.13 MPa and whose exit temperature and pressure are $60°C$ and 0.9 MPa. The adiabatic expansion valve's inlet pressure is 0.85 MPa with a quality of zero, while its exit pressure is 0.16 MPa. Calculate the coefficient of performance if the refrigerant is refrigerant 12. The compressor is adiabatic.

7.55 An actual vapor-compression refrigerator has a 2-hp motor driving a compressor. The compressor efficiency is 70 percent, and it has an inlet pressure and temperature $-40°C$ and 0.05 MPa while the exit pressure is 0.9 MPa. The condenser exit pressure and quality are 0.75 MPa and zero. The adiabatic expansion valve exit pressure is 0.06 MPa. Calculate the rejected heat transfer, the coefficient of performance, and the refrigeration capacity in tons and in Btu per hour.

7.56 A Carnot engine operating between two temperature reservoirs of 817 and $25°C$ rejects 20 kJ/s. The engine drives the compressor of an ideal vapor compression refrigerator whose inlet and exit pressures are 0.17 and 1.1 MPa, respectively. Calculate the coefficient of performance and the refrigerator's capacity in tons if the refrigerant is refrigerant 12.

7.57 A refrigeration system using refrigerant 12 is to have a capacity of 2 tons. The cycle is the ideal vapor compression cycle. The evaporator exit pressure is 0.15

MPa, and the condenser inlet temperature is 60°C. Determine the compressor power requirement in kilowatts.

7.58 An ideal vapor-compression refrigerator uses refrigerant 12. The temperature entering the expansion valve is 50°C, and the pressure entering the evaporator is 0.2 MPa. Determine the coefficient of performance.

7.59 Refrigerant 12 enters the adiabatic compressor in an air conditioner with $P_1 = 0.33$ MPa and $x_1 = 1.0$. If $P_2 = 1.0$ MPa and the compressor efficiency is 70 percent, determine the actual work (kJ/kg).

7.60S A 10-ton vapor-compression refrigeration device uses ammonia as the working fluid. The saturated liquid entering the throttling valve has $T = 30$°C, and the saturated vapor entering the compressor is at −25°C. The compressor efficiency is 65 percent.
(*a*) What is the ratio of the COP of this cycle to the COP of a Carnot cycle operating between the same temperatures?
(*b*) How much ammonia must be circulated through the device?

7.60E A 10-ton vapor-compression refrigeration device uses ammonia as the working fluid. The saturated liquid entering the throttling valve has $T = 80$°F, and the saturated vapor entering the compressor is at −15°F. The compressor efficiency is 65 percent.
(*a*) What is the ratio of the COP of this cycle to the COP of a Carnot cycle operating between the same temperatures?
(*b*) How much ammonia must be circulated through the device?

*7.61S A vapor-compression refrigeration cycle using refrigerant 12 operates with a condenser temperature of 40°C and an evaporator temperature of 5°C. The compressor has an efficiency of 65 percent. If there is no subcooling of the liquid leaving the condenser, plot the cycle COP versus the degree of superheating of the refrigerant 12 leaving the evaporator for 0 to 20°C above the saturation temperature.

*7.61E A vapor-compression refrigeration cycle using refrigerant 12 operates with a condenser temperature of 100°F and an evaporator temperature of 40°F. The compressor has an efficiency of 65 percent. If there is no subcooling of the liquid leaving the condenser, plot the cycle COP versus the degree of superheating of the refrigerant 12 leaving the evaporator for 0 to 40°F above the saturation temperature.

***7.62S** A vapor-compression heat pump system using refrigerant 12 operates with an evaporator temperature of 10°C and a condenser temperature of 40°C. There is no superheating of the vapor leaving the evaporator, and the compressor has an efficiency of 65 percent. Plot the COP_{HP} for the system versus the degrees of subcooling of the liquid leaving the condenser for subcooling of 0 to 15°C below the saturation temperature.

***7.62E** A vapor-compression heat pump system using refrigerant 12 operates with an evaporator temperature of 50°F and a condenser temperature of 100°F. There is no superheating of the vapor leaving the evaporator, and the compressor has an efficiency of 65 percent. Plot the COP_{HP} for the system versus the degrees of subcooling of the liquid leaving the condenser for subcooling of 0 to 30°F below the saturation temperature.

7.63S A *heat pipe* operates by evaporating a liquid by heat transfer in the evaporation section of a closed conduit. The resulting vapor moves to the condenser end, where heat transfer from the heat pipe causes the vapor to condense. The liquid is then returned to the evaporator section by capillary action in a wick or by gravity. The vapor pressure is nearly constant between the evaporator and condenser, so the heat pipe is a nearly constant-temperature device for heat transfer. See Fig. P7.63. Draw the *P-v* and *T-s* diagrams for a heat pipe process using water at $P = 50$ kPa.

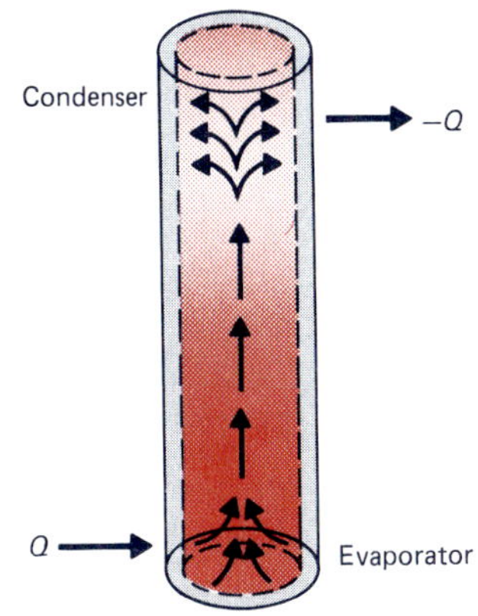

Figure P7.63

7.63E A *heat pipe* operates by evaporating a liquid by heat transfer in the evaporation section of a closed conduit. The resulting vapor moves to the condenser end, where heat transfer from the heat pipe causes the vapor to condense. The liquid is then returned to the evaporator section by capillary action in a wick or by gravity. The vapor pressure is nearly constant between the evaporator and condenser, so the heat pipe is a nearly constant-temperature device for heat transfer. See Fig. P7.63. Draw the *P-v* and *T-s* diagrams for a heat pipe process using water at $P = 8$ psia.

7.64S An irrigation pump is to be powered by a solar-driven Rankine cycle. The energy striking the solar collector during daylight hours ($0 \le t \le 10$) is given by

$$\dot{Q}_{solar} = 90A \sin \frac{\pi t}{10} \qquad \text{W}$$

where A is the collector area in square meters and t is the time after sunrise in hours. See Fig. P7.64S.

The solar collector converts 40 percent of $\dot{Q}_{solar}$ to enthalpy change of the water entering the collector. The flow rate of water $\dot{m}(t)$ is varied in such a way that the collector outlet condition is always saturated vapor at 150°C. The turbine is 70 percent efficient and exhausts to 8 kPa. The pump must lift water a distance of 8 m, and the friction in the piping system is negligible. Both pumps have an efficiency of 80 percent.

(*a*) Find the collector area required to provide 500 kg/min at solar noon ($t = 5$, maximum pumping capacity of the system).

(*b*) Plot the pumping rate $\dot{m}_p(t)$ versus time over the 10-h daylight period.

(*c*) Find the total amount of water pumped during the entire day.

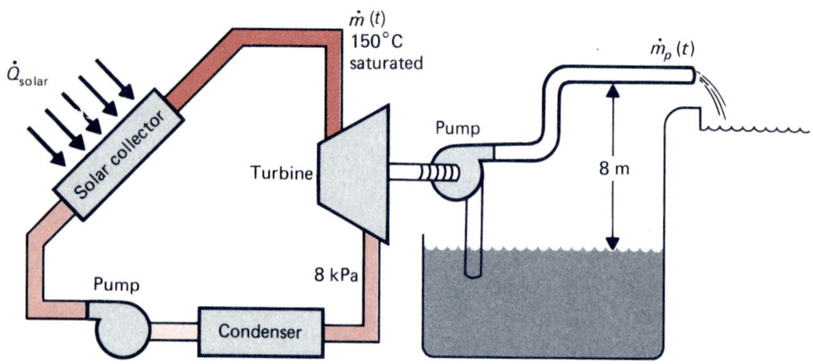

Figure P7.64S

7.64E An irrigation pump is to be powered by a solar-driven Rankine cycle. The energy striking the solar collector during daylight hours ($0 \le t \le 10$) is given by

$$\dot{Q}_{solar} = 300A \sin \frac{\pi t}{10} \qquad \text{Btu/h}$$

where A is the collector area in square feet and t is the time after sunrise in hours. See Fig. P7.64E.

The solar collector converts 40 percent of $\dot{Q}_{solar}$ to enthalpy change of the water entering the collector. The flow rate of water $\dot{m}(t)$ is varied in such a way that the collector outlet condition is always saturated vapor at 300°F. The turbine is 70 percent efficient and exhausts to 1 psia. The pump must lift water a distance of

25 ft, and the friction in the piping system is negligible. Both pumps have an efficiency of 80 percent.

(*a*) Find the collector area required to provide 120 gpm at solar noon ($t = 5$, maximum pumping capacity of the system).

(*b*) Plot the pumping rate $\dot{m}_p(t)$ versus time over the 10-h daylight period.

(*c*) Find the total amount of water pumped during the entire day.

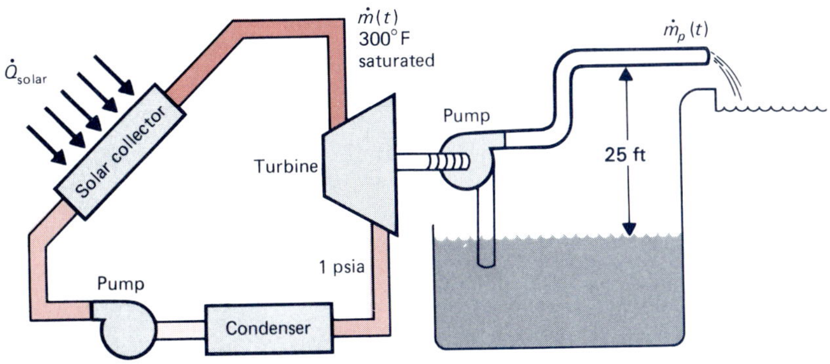

Figure P7.64E

8

Analysis Using the Second Law of Thermodynamics

A good many times I have been present at gatherings of people who, by the standards of tranditional culture, are thought highly educated and who have with considerable gusto been expressing their incredulity at the illiteracy of scientists. Once or twice I have been provoked and have asked the company how many of them could describe the Second Law of Thermodynamics. The response was cold; it was also negative.
C.P. Snow, *The Two Cultures*

Vapor-compression liquid chiller system, showing cutaway of three-stage centrifugal compressor, condenser, and evaporator. (*The Trane Company.*)

8.1 Introduction

In this chapter, a different approach from that used in Chapters 6 and 7 is applied for interpreting the performance of thermodynamic systems. The approach used here is based on the limitations placed on system behavior by the *second* law rather than the first law. This allows the comparison of the performance of a real system with that theoretically attainable within the limitations of the second law, which as we have seen may be considerably less than allowed by the first law. Such an approach gives a different insight into the inefficiencies of various components and cycles. The second law of thermodynamics is unique in its specification of the ideal limits on system processes and on system cycles.

The second law, like the first law and the conservation of mass, relates a system property to boundary transfers; unlike the first law and the conservation of mass, the relation provides only a limit on the system performance. The connection between the system property (entropy) and the boundary transfer (heat transfer divided by absolute temperature) is an equality only in the limiting case of a reversible process. Otherwise, there is an entropy production or generation which is nonzero. The second law given in Table 5.1 for a control volume is

$$\dot{S}_{\text{gen}} = \frac{dS_{\text{CV}}}{dt} + \sum_{\text{out}} s\dot{m} - \sum_{\text{in}} s\dot{m} - \sum_{i} \left(\frac{\dot{Q}_i}{T_i}\right)_{\text{CV}} \geq 0 \tag{8.1}$$

or, in differential form, from Eq. (5.160) as

$$\delta S_{\text{gen}} = dS_{\text{CV}} + \sum_{\text{out}} s\, dm - \sum_{\text{in}} s\, dm - \sum_{i} \left(\frac{\delta Q_i}{T_i}\right)_{\text{CV}} \geq 0 \tag{8.2}$$

Entropy generation is the term which quantifies the irreversibility of the process. The identification of the sources of irreversibilities and their reduction are the desired goal of second law analysis, and this is specified by the reduction of entropy generation.

Comparisons of system performance are made in Chaps. 4 through 7. The first comparison is the *thermal efficiency,* defined as the desired output divided by the required input. This is often termed a *first law efficiency,* because it is based entirely on first law quantities. The second law puts an upper limit on the first law efficiency, but does not directly enter into the definition. However, the *component efficiencies* in Sec. 5.11 are directly related to the second law. These efficiencies are ratios of the actual performance to the ideal performance. The ideal performance was usually (but not always) the performance of a device operated at the reversible adiabatic (isentropic) limit. Other reversible processes may also be used for comparison. For example, the isothermal reversible compressor is sometimes used for defining compressor efficiency [Eq. (5.190)]. Therefore, the second law specified the ideal or "upper limit" process for the component. Processes are compared in this chapter by using the entropy generation limitation given by the second law.

Second law analysis is closely tied to entropy generation and therefore to entropy transfer. Equations (8.1) and (8.2) reveal the relationship between entropy generation and boundary transfer. Boundary entropy transfer, which thermally connects the system to the heat transfer reservoir, sets the reference state for much of the subsequent analysis. The particular choice for the reference state depends on the problem, but many engineering considerations involve interactions with the earth's atmosphere. This reservoir is assumed to be an infinite heat transfer reservoir at temperature T_0. This reference state is often assumed in the analysis, and the temperature is 25°C (77°F) unless specified otherwise. The atmosphere is also referred to as the *dead state,* because it has the lowest naturally occurring temperature that can be used as a practical reservoir for a rejected heat transfer. Thus, the energy in processes ending at this state cannot be used subsequently to obtain work.

In the next section we discuss the limits on the maximum possible work obtainable for an arbitrary process. This limiting expression for maximum, or *reversible,* work is developed by simultaneously considering the constraints imposed by the second law and the relation between the system energy and energy transfer prescribed by the first law. Particular definitions of thermodynamic properties are developed from the reversible work expression to help describe the available work that can be realized. The reversible, or maximum, work possible for an arbitrary process can then be compared to the actual work obtained. The possible processes are many and depend on the particular application, so in Sec. 8.6 we discuss process comparisons. In all the analysis in this chapter, we employ the fundamental laws previously presented. We present very important definitions and analytical techniques which are based on the previous laws, particularly the concept of entropy generation.

Figure 8.1 General control volume schematic.

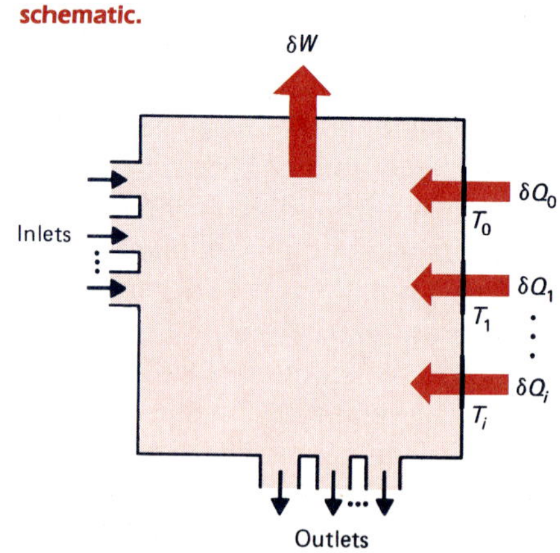

8.2 Reversible Work

The ability of the second law to predict the potential maximum work for a process is the focus of this section. The work output of a process is the desired quantity, yet the statement of the second law does not directly include the work term (since energy transfer in the form of work causes no entropy generation). A relation is developed to include the entropy generation directly in the work prediction. This is accomplished by considering the first law and second law together.

A general control volume (CV) which considers an arbitrary number of inlets and outlets is shown schematically in Fig. 8.1. Also there are various heat transfers with the surroundings. The term for index $i = 0$ represents the heat transfer with the lowest naturally occurring reservoir (usually the earth's atmosphere). The energy transfer as work is grouped into a single work term. The first law for a general process is given from Eq. (5.159) as

$$dE_{CV} + \sum_{out} (e + Pv)\, dm - \sum_{in} (e + Pv)\, dm - \sum_i \delta Q_i + \delta W = 0 \tag{8.3}$$

The second law is given by rewriting Eq. (8.2) as

$$dS_{CV} + \sum_{out} s\, dm - \sum_{in} s\, dm - \sum_i \frac{\delta Q_i}{T_i} - \delta S_{gen} = 0 \tag{8.4}$$

where the entropy generation term quantifies the irreversibilities of the process.

The first and second laws are now combined by multiplying the second law by temperature and subtracting the result from the first law. The temperature chosen is denoted by the constant value T_0, and the usual convention is to use the temperature of the surroundings or earth's atmosphere. This is the reservoir temperature corresponding to the heat transfer δQ_0. This choice is discussed further in Sec. 8.6. Multiplying Eq. (8.4) by T_0 and subtracting the result from Eq. (8.3) yield

$$d(E - T_0 S)_{CV} + \sum_{out} (e + Pv - T_0 s)\, dm - \sum_{in} (e + Pv - T_0 s)\, dm - \sum_i \delta Q_i \left(1 - \frac{T_0}{T_i}\right) + T_0\, \delta S_{gen} + \delta W = 0 \tag{8.5}$$

Solving for the work obtained for a process yields

$$\delta W = \sum_{in} (e + Pv - T_0 s)\, dm - \sum_{out} (e + Pv - T_0 s)\, dm - d(E - T_0 S)_{CV} + \sum_i \delta Q_i \left(1 - \frac{T_0}{T_i}\right) - T_0\, \delta S_{gen} \tag{8.6}$$

This is a very important relation in second law analysis.

Equation (8.6) appears to yield a somewhat arbitrary value for work since it depends on the specific choice of T_0. The mathematical manipulation of adding zero times T_0 [Eq. (8.4) times T_0] does not alter the work from the process. Alternatively, this mathematical manipulation can be viewed as an elimination of the heat transfer from the environment δQ_0 from the first law by expressing the heat transfer in terms of second law quantities. The work performed is not changed by the development and is not dependent on this mathematical manipulation.

The $(e + Pv - T_0 s)\, dm$ terms in Eq. (8.6) represent the work contributions from the outlet and inlet flows. These terms have similar counterparts in the first law equation, but the entropy terms now incorporate the eliminated heat transfer. The change in $E - T_0 S$ within the control volume [third term on the right-hand side of Eq. (8.6)] describes the work contributions resulting from change within the control volume. The last two terms in Eq. (8.6) are somewhat different. The heat transfer term does not include the contribution from δQ_0, because it was eliminated and incorporated in the entropy terms. The terms $1 - T_0/T_i$ can be viewed as the efficiency of a *reversible* engine operating between a low-temperature heat transfer reservoir at T_0 and a higher-temperature heat transfer reservoir at T_i (developed in Sec. 5.12). With a heat transfer input δQ_i into the engine, $\delta Q_i(1 - T_0/T_i)$ represents the reversible work output by this engine. This work output contributes to the work performed by the process. The $T_0\,\delta S_{\text{gen}}$ term in Eq. (8.6) represents all the irreversibility or loss of available work. This term also includes the irreversibilities that result from the heat transfers, because the engine-type heat transfer term was viewed as reversible.

Our sign convention on work makes the largest positive value the largest work output. The entropy generation is a positive quantity by definition [Eq. (8.2)], so this term reduces the work output of the process. If the process work is negative work (input), then the entropy generation term indicates that the irreversibilities require a larger work input as δS_{gen} increases.

The maximum work possible for a system undergoing a process is found in the reversible case. The reversible process has zero entropy generation, so Eq. (8.6) yields

$$\begin{aligned}
\delta W_{\text{rev}} &= \sum_{\text{in}} (e + Pv - T_0 s)\, dm - \sum_{\text{out}} (e + Pv - T_0 s)\, dm \\
&\quad - d(E - T_0 S)_{\text{CV}} + \sum_i \delta Q_i \left(1 - \frac{T_0}{T_i}\right) \\
&= \sum_{\text{in}} \left(h - T_0 s + \frac{1}{2}\frac{\mathbf{V}^2}{g_c} + \frac{gZ}{g_c}\right) dm \\
&\quad - \sum_{\text{out}} \left(h - T_0 s + \frac{1}{2}\frac{\mathbf{V}^2}{g_c} + \frac{gZ}{g_c}\right) dm \\
&\quad - d\left(U - T_0 S + \frac{1}{2}\frac{m\mathbf{V}^2_{\text{ref}}}{g_c} + \frac{mgZ}{g_c}\right)_{\text{CV}} + \sum_i \delta Q_i \left(1 - \frac{T_0}{T_i}\right)
\end{aligned} \tag{8.7}$$

where the subscript *rev* indicates reversible. This maximum work is the largest value (work output) possible and is the ideal, or maximum, work possible for the specified conditions. This expression sets limits on system processes. The heat transfer contributions, as discussed in the previous paragraph, are viewed as the work output from reversible engines operating between reservoirs at T_0 and T_i.

Example Problem 8.1

Air at 5000°F and 130 atm is contained within an expandable piston cylinder. The cylinder is cooled by a heat transfer with the surroundings at

$T_0 = 77°F$ to a final temperature of 2700°F. The polytropic process is given by $Pv^{1.5} = \text{constant}$. Evaluate the reversible work per unit mass for this process.

Solution

The properties are given by the ideal gas relations, because the compressibility chart indicates that air is ideal for the stated conditions.

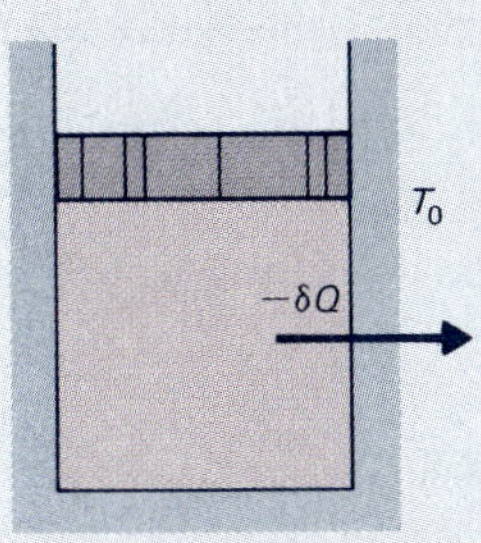

Diagram

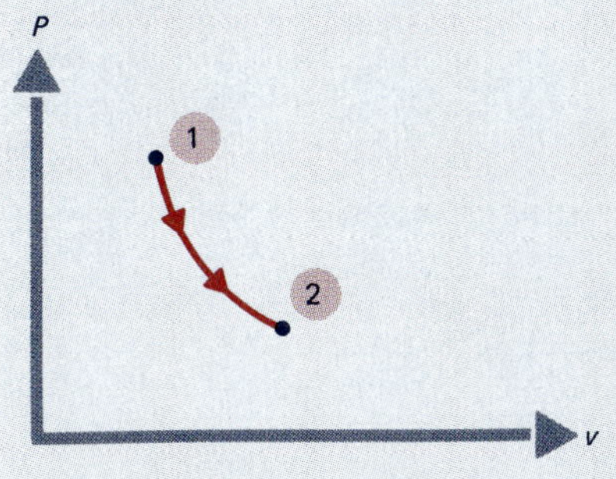

The properties are given from Table 5.4. The initial state is given, and the final state is prescribed by 2700°F and

$$P_1 v_1^{1.5} = P_2 v_2^{1.5}$$

or

$$\frac{P_1}{P_2} = \left(\frac{v_2}{v_1}\right)^{1.5} = \left(\frac{T_1}{T_2}\right)^{1.5/(1.5-1)}$$

Thus $P_2 = 25.20$ atm.

The reversible work is given by Eq. (8.7) for a control mass as

$$\delta W_{rev} = -d(E - T_0 S)$$

since there are no inlets or outlets and $1 - T_0/T_0 = 0$. Integrating this expression from the initial to the final state yields

$$_1W_{rev,2} = (E - T_0S)_1 - (E - T_0S)_2$$

or

$$_1w_{rev,2} = (e - T_0s)_1 - (e - T_0s)_2$$

Neglecting the kinetic and potential energy changes yields

$$_1w_{rev,2} = u_1 - u_2 + T_0(s_2 - s_1)$$

The change in entropy is evaluated with Eq. (5.60) to yield

$$_1w_{rev,2} = u_1 - u_2 - T_0[s_0(T_1) - s_0(T_2)] - T_0R \ln \frac{P_2}{P_1}$$

and, with values from Tables E.2 and C.2, this is

$$_1w_{rev,2} = 510.8 \text{ Btu/lbm} = 1188 \text{ kJ/kg}$$

This is the maximum work possible between the given states. This maximum possible work is independent of the stated path (polytropic) and depends on only the end states.

Comments

The work for a quasi-equilibrium polytropic process of a control mass is given in Table 5.6 as

$$_1w_2 = \frac{P_1v_1 - P_2v_2}{n-1} = \frac{R(T_1 - T_2)}{n-1} = 315.3 \text{ Btu/lbm}$$

which is a work output that is ~38 percent of the maximum reversible output. Also note that the heat transfer for this process is

$$_1q_2 = u_2 - u_1 + {}_1w_2 = -223.9 \text{ Btu/lbm}$$

which is almost three-quarters as large as the work obtained. The entropy generation for the actual process is

$$_1s_{gen,2} = s_2 - s_1 - \frac{_1q_2}{T_0} = 0.3643 \text{ Btu/(lbm}\cdot{}^\circ\text{R)} = 1.525 \text{ kJ/(kg}\cdot\text{K)}$$

which indicates, as expected, that the actual work should be less than the reversible case.

Example Problem 8.2

Determine the maximum work that can be produced by a steam turbine which has an inlet state of 3 MPa and 450°C and an outlet state as a saturated vapor at 0.1 MPa. The port properties are uniform, and there is a heat transfer with the surroundings at 25°C.

Solution

The reversible work expression for steady state conditions with a single inlet and single outlet is

$$\delta W_{rev} = (e + Pv - T_0 s)_{in}\, dm_{in} - (e + Pv - T_0 s)_{out}\, dm_{out}$$

Neglecting the changes in kinetic and potential energy and expressing the reversible work on a rate basis yield

$$\dot{W}_{rev} = (h - T_0 s)_{in}\dot{m}_{in} - (h - T_0 s)_{out}\dot{m}_{out}$$

or

$$\begin{aligned} w_{rev} &= (h - T_0 s)_{in} - (h - T_0 s)_{out} \\ &= T_0(s_{out} - s_{in}) - (h_{out} - h_{in}) \end{aligned}$$

With Tables D.11 and D.12, the properties are obtained. The reversible work is

$$\begin{aligned} w_{rev} &= T_0(7.3589 - 7.0835) - (2675.1 - 3344.1) = 751.1 \text{ kJ/kg} \\ &= 322.9 \text{ Btu/lbm} \end{aligned}$$

Comments

This is the maximum work possible between the specified states (i.e., saturated vapor out, $s_{out} \neq s_{in}$, and with the temperature of the surroundings at $T_0 = 25°C$). Remember that the inlet and outlet states are fixed. There are other processes that start at the inlet state and produce more work. However, they must terminate at a different outlet state. For example, the adiabatic reversible turbine starting at the inlet state and exhausting to the given outlet pressure produces

$$w_s = (h_{in} - h_{out})_s = 771.7 \text{ kJ/kg} = 331.8 \text{ Btu/lbm}$$

but the outlet state is at $P = 0.1$ MPa and $x = 0.9545$. This is the comparison process for the turbine efficiency, and it should not be confused with the maximum work. Many different values for work are possible if the states are changed, but the maximum work is the maximum between the *specified* states.

8.3 Availability

An important consideration in engineering analysis is the maximum work available from a particular substance in a specified state. If steam is available in a particular state, what is the maximum work that can be obtained by the use of this fluid? A compressed-air storage system contains air at a high-pressure state. What is the maximum work that can be obtained from air in this state? There are numerous questions of this type, and they are generalized as: What is the maximum available energy to perform work that is contained by a substance in a specified state?

The maximum energy available in a particular state can be used to perform work as long as the state is not at standard environmental conditions. The state in equilibrium with the environment can no longer be used to obtain work. Therefore, the environmental conditions of $T_0 = 25°\text{C} = 77°\text{F}$ and $P_0 = 1$ atm are often referred to as the *dead state.* A substance has a maximum available energy at any state that is not in thermodynamic equilibrium with this dead state.

This maximum available energy is determined by consideration of Eq. (8.7) for the reversible work. We are also interested in the available work that is useful, so the work performed as the displacement of the atmosphere must be eliminated. The expansion of the control volume or control mass against the atmosphere (which is at T_0 and P_0) is not available for other useful purposes. Thus the useful reversible work is given as

$$\delta W_{\text{rev, use}} = \delta W_{\text{rev}} - P_0\, dV \tag{8.8}$$

and substituting Eq. (8.7) into Eq. (8.8) yields

$$\begin{aligned}\delta W_{\text{rev, use}} &= \sum_{\text{in}} (e + Pv - T_0 s)\, dm - \sum_{\text{out}} (e + Pv - T_0 s)\, dm \\ &\quad - d(E + P_0 V - T_0 S)_{\text{CV}} + \sum_i \delta Q_i \left(1 - \frac{T_0}{T_i}\right) \\ &= \sum_{\text{in}} \left(h - T_0 s + \frac{\mathbf{V}^2}{2g_c} + \frac{gZ}{g_c}\right) dm \\ &\quad - \sum_{\text{out}} \left(h - T_0 s + \frac{\mathbf{V}^2}{2g_c} + \frac{gZ}{g_c}\right) dm \\ &\quad - d\left(U + P_0 V - T_0 S + \frac{m\mathbf{V}_{\text{ref}}^2}{2g_c} + \frac{mgZ}{g_c}\right)_{\text{CV}} \\ &\quad + \sum_i \delta Q_i \left(1 - \frac{T_0}{T_i}\right)\end{aligned} \tag{8.9}$$

This equation explicitly indicates the properties that prescribe the available energy in particular states and the potential for obtaining useful work through heat transfer with reservoirs other than the environment. The combination of properties at the inlet and outlet has the form $h - T_0 s$, and the combination of properties within the control volume or for a control mass is $u + P_0 v - T_0 s$. The difference lies in the pressure term.

The combinations of properties that appear in Eq. (8.9) prescribe the amount of available energy at a state to perform work. These property combinations may be used to define new state properties in relation to the environment or dead state. They are defined as the *nonflow availability per unit mass* ϕ and the *flow availability per unit mass* ψ. The nonflow availability per unit mass is

$$\begin{aligned}\phi &= (u + P_0 v - T_0 s) - (u_0 + P_0 v_0 - T_0 s_0) \\ &= (u - u_0) + P_0(v - v_0) - T_0(s - s_0)\end{aligned} \tag{8.10}$$

and the flow availability per unit mass is

$$\begin{aligned}\psi &= (h - T_0 s) - (h_0 - T_0 s_0) \\ &= (h - h_0) - T_0(s - s_0)\end{aligned} \tag{8.11}$$

These definitions do not include the kinetic and potential energies relative to a dead-state reference plane; these terms are easily included if germane to a particular problem. The properties also do not include the potential work available through communication with a thermal reservoir. The availabilities on an extensive basis are given the symbols Φ and Ψ, respectively. The quantities depend on the state of the particular substance and the environment. Therefore, they are properties relative to the defined environment. In all cases, unless specified otherwise, the environmental conditions are $P_0 = 1$ atm and $T_0 = 25°\text{C} = 77°\text{F}$.

Equation (8.10) shows that ϕ depends on three terms for available energy: $u - u_0$, the internal energy change relative to the dead state; $P_0(v - v_0)$, the work done against the environment; and $-T_0(s - s_0)$, the reversible heat transfer to the environment in coming to the dead state. Similarly, ψ in Eq. (8.11) shows that the flow availability comes from the enthalpy change between the initial and dead states less the reversible heat transfer between the control volume and the dead state.

These availability functions are quite similar. The difference lies in the pressure term. The nonflow availability includes a constant atmospheric pressure, while the flow availability includes a variable pressure within the enthalpy change term. It is important to note the physical process that each term is describing. The terms that arise from the inlet and outlet flow contributions are expressed through ψ, and the changes within the control volume are given by ϕ. Also a control mass formulation includes only the nonflow availability; so ϕ is often referred to as the control mass availability, and ψ is the control volume availability. These availability functions are special cases of the more general definitions of the Helmholtz function, or Helmoltz free energy, $a = u - Ts$, and the Gibbs function, or the Gibbs free energy, $g = h - Ts$. These general functions are discussed in greater detail in Sec. 8.5.

The maximum useful work given in Eq. (8.9) is rewritten in terms of the availability function as

$$\begin{aligned}\delta W_{\text{rev, use}} &= \sum_{\text{in}} \left(\psi + \frac{1}{2}\frac{\mathbf{V}^2}{g_c} + \frac{gZ}{g_c}\right) dm \\ &\quad - \sum_{\text{out}} \left(\psi + \frac{1}{2}\frac{\mathbf{V}^2}{g_c} + \frac{gZ}{g_c}\right) dm \\ &\quad - d\left[\left(\phi + \frac{1}{2}\frac{\mathbf{V}_{\text{ref}}^2}{g_c} + \frac{gZ}{g_c}\right) m\right]_{\text{CV}} + \sum_i \delta Q_i\left(1 - \frac{T_0}{T_i}\right)\end{aligned} \tag{8.12}$$

Conservation of mass has been used to eliminate the terms that include the state of the environment or dead state. Equation (8.12) clearly shows the importance and meaning of the availability functions. Consider steady state flow through an adiabatic control volume with a single inlet and outlet with each port having uniform properties. The useful reversible work, or maximum available work, between the specific inlet state and an outlet state that exhausts to the environment such that $P_{out} = P_0$ and $T_{out} = T_0$ is

$$\dot{W}_{rev,\,use} = \dot{m}\psi_{in} \tag{8.13}$$

where the changes in kinetic and potential energy are neglected. Thus the flow availability per unit mass at the inlet state quantifies the maximum useful work that can be obtained. If the control volume does not exit to the environment but to some other exit state, then

$$\dot{W}_{rev,\,use} = \dot{m}(\psi_{in} - \psi_{out}) \tag{8.14}$$

For a control mass, the useful reversible work between a specified initial state 1 and the final state 2 in equilibrium with the environment is

$$W_{rev,\,use} = (m\phi)_1 \tag{8.15}$$

Again the availability indicates the maximum useful work available from the specified state.

Another application of interest is the reversible work from a high-temperature reservoir at T_H. The reversible work from a high-temperature reservoir is determined from the concept of the maximum available energy to perform work. A reservoir at T_H may be used to perform work by means of some arbitrary system. The system receives a heat transfer from this reservoir. By starting from the expression for reversible, or maximum, work in Eq. (8.7), the maximum work for a single reservoir is

$$\delta W_{rev} = \delta Q_H \left(1 - \frac{T_0}{T_H}\right) \tag{8.16}$$

This is the maximum available energy to perform work for a reservoir at T_H, and it is viewed as the reversible work of the reservoir.

The above definitions indicate the close connection between the availability and the maximum useful work.

Example Problem 8.3

Determine the availability per unit mass for combustion products (treat as air) in an engine cylinder at 5000°F and 130 atm.

Solution

Availability is a property, so it requires only state information. Also consideration is directed to the combustion products contained in a cylinder, so the nonflow availability is desired. By taking the environment at

$P_0 = 1$ atm and $T_0 = 77°F$, Eq. (8.10) is

$$\phi = u - u_0 + P_0(v - v_0) - T_0(s - s_0)$$

Air is ideal in these states, but the specific heats are functions of temperature. So with Table E.2, $v = RT/P$, and Eq. (5.60),

$$\phi = u - u_0 + P_0 R\left(\frac{T}{P} - \frac{T_0}{P_0}\right) - T_0\left[s_0(T) - s_0(T_0) - R\ln\frac{P}{P_0}\right]$$

$$= 1160.1 - 91.72 + (1)\left(\frac{53.34}{778.16}\right)\left(\frac{5460}{130} - \frac{537}{1}\right)$$

$$- 537\left(2.2294 - 1.6002 - \frac{53.34}{778.16}\ln\frac{130}{1}\right)$$

$$= 1068.4 - 33.9 - 158.7 \text{ Btu/lbm}$$

$$= 876 \text{ Btu/lbm} = 2040 \text{ kJ/kg}$$

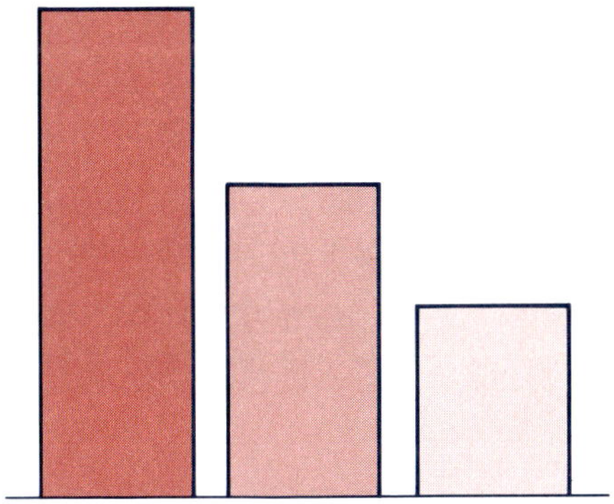

Comments

As indicated in Eq. (8.15), this value represents the maximum useful work that can be obtained for air in this state. This can be compared to the result in Example Problem 8.1, which indicates the maximum work obtained between two specified states. The results of Example Problem 8.1 and this example are indicated graphically. The available energy is a function of the specified state relative to the dead state, and the maximum work is evaluated between two process states. The actual work is less because of the irreversibilities and the unused heat transfer.

Example Problem 8.4

A geothermal energy source is the underground high-temperature water or steam which is heated by the earth's interior. Evaluate the steady flow availability per unit mass for water at 150°C and 0.6 MPa.

Solution

The availability per unit mass is given in Eq. (8.11) as

$$\psi = h - h_0 - T_0(s - s_0)$$

Table 5.5 gives the properties of an incompressible fluid as

$$\psi = c(T - T_0) + v(P - P_0) - T_0 c \ln\frac{T}{T_0}$$

For liquid water at 273 K (Table C.3), $c = 4.213$ kJ/(kg·K) and $v = 0.001$ m^3/kg, so

$$\psi = 87.3 \text{ kJ/kg} = 37.5 \text{ Btu/lbm}$$

Comments

If the temperature were 200°C, the water would be superheated at 0.6 MPa. Tables D.10 and D.12 yield

$$\psi = 2849.7 - 104.75 - (298.15)(6.9658 - 0.3670)$$
$$= 777.6 \text{ kJ/kg} = 334.3 \text{ Btu/lbm}$$

which is much larger than the liquid case. The temperature increase needed to obtain this superheated vapor was 50°C.

8.4 Irreversibility

The second law of thermodynamics puts restrictions on the *potential* of possible processes. This restriction indicates the ideal limits that are obtainable. Therefore, comparisons of actual and ideal processes are possible. Some comparisons are made in Chap. 7 for both cycles and processes. The cycle efficiency compares the desired output to the required input. The component efficiencies usually compare the actual to the isentropic processes. It is often instructive to compare an actual process and a reversible process where the inlet and initial states of the actual and reversible processes are the same and the outlet and final states of the actual and reversible processes are the same. This comparison is diagramed in Fig. 8.2.

Equation (8.6) gives the actual work obtained.

Figure 8.2 Irreversibility comparison.

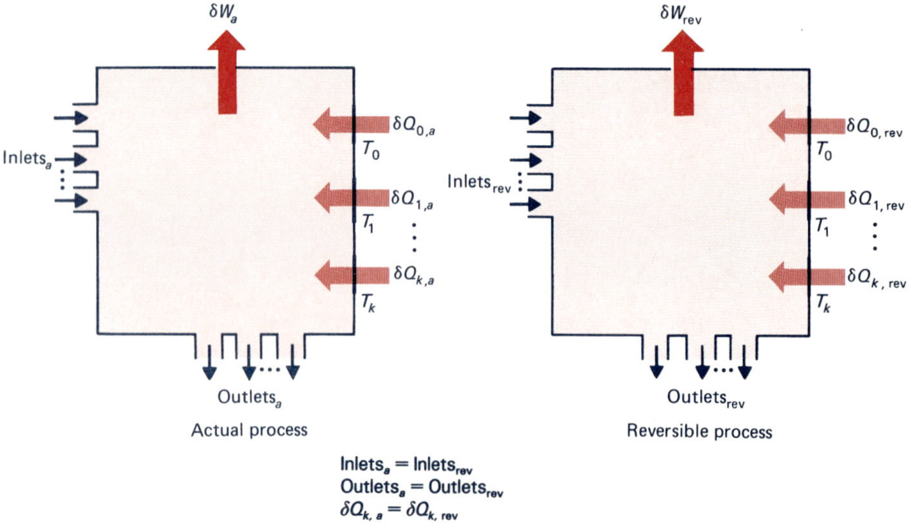

$$\delta W_a = \sum_{\text{in}} (e + Pv - T_0 s)_a \, dm - \sum_{\text{out}} (e + Pv - T_0 s)_a \, dm$$
$$- T_0 \, \delta S_{\text{gen}} - d(E - T_0 S)_{\text{CV}, a} + \sum_i \delta Q_i \left(1 - \frac{T_0}{T_i}\right) \tag{8.17}$$

where the subscript a indicates that this is the actual case. The reversible work is given from Eq. (8.7) as

$$\delta W_{\text{rev}} = \sum_{\text{in}} (e + Pv - T_0 s)_{\text{rev}} \, dm - \sum_{\text{out}} (e + Pv - T_0 s)_{\text{rev}} \, dm$$
$$- d(E - T_0 S)_{\text{CV, rev}} + \sum_i \delta Q_i \left(1 - \frac{T_0}{T_i}\right) \tag{8.18}$$

where the subscript *rev* indicates the reversible case. *Irreversibility I* is defined as the difference between the reversible work output and the actual work output with both processes originating from the same state and both terminating at the same state and with the same heat transfer with each reservoir δQ_i. Thus, subtracting Eq. (8.17) from Eq. (8.18) yields

$$\delta I = \delta W_{\text{rev}} - \delta W_a = T_0 \, \delta S_{\text{gen}} \tag{8.19}$$

On a rate basis, this is

$$\dot{I} = \dot{W}_{\text{rev}} - \dot{W}_a = T_0 \dot{S}_{\text{gen}} \tag{8.20}$$

Equations (8.1) and (8.2) can be substituted into Eqs. (8.20) and (8.19), respectively, to yield

$$\delta I = T_0 \left[dS_{\text{CV}} + \sum_{\text{out}} s \, dm - \sum_{\text{in}} s \, dm - \sum_i \left(\frac{\delta Q_i}{T_i} \right)_{\text{CV}} \right] \tag{8.21}$$

and

$$\dot{I} = T_0 \left[\frac{dS_{\text{CV}}}{dt} + \sum_{\text{out}} s\dot{m} - \sum_{\text{in}} s\dot{m} - \sum_i \left(\frac{\dot{Q}_i}{T_i} \right)_{\text{CV}} \right] \tag{8.22}$$

Irreversibility has units of work and, by its definition, measures the deviations from the ideal or reversible case. Therefore, the magnitude of δI or $\dot{I}$ is directly compared to the work performed and indicates a loss. A reduction in the entropy generation reduces the irreversibility and improves the work output of the actual process.

The irreversibility for steady state flow with uniform properties is given from Eq. (8.22) as

$$\dot{I} = T_0 \left[\sum_{\text{out}} s\dot{m} - \sum_{\text{in}} s\dot{m} - \sum_i \left(\frac{\dot{Q}_i}{T_i} \right)_{\text{CV}} \right] \tag{8.23}$$

For a single inlet and outlet, this reduces to

$$\dot{I} = T_0 (s_{\text{out}} \dot{m}_{\text{out}} - s_{\text{in}} \dot{m}_{\text{in}}) - \sum_i \dot{Q}_i \left(\frac{T_0}{T_i} \right) \tag{8.24}$$

or, on a unit-mass basis,

$$i = T_0(s_{\text{out}} - s_{\text{in}}) - \sum_i q_i \left(\frac{T_0}{T_i}\right) \tag{8.25}$$

The irreversibility for a control mass proceeding from an initial state 1 to a final state 2 is obtained from Eq. (8.21) as

$$\delta I = T_0 \left[dS_{\text{CV}} - \sum_i \delta Q_i \left(\frac{T_0}{T_i}\right) \right] \tag{8.26}$$

Integrating from the initial state to the final state yields

$$I = T_0(S_2 - S_1) - \sum_i {}_1Q_{2,i} \left(\frac{T_0}{T_i}\right) \tag{8.27}$$

On an intensive basis, this is

$$i = T_0(s_2 - s_1) - \sum_i {}_1q_{2,i} \left(\frac{T_0}{T_i}\right) \tag{8.28}$$

Example Problem 8.5

Evaluate the irreversibility per unit mass for a steady state nonadiabatic steam turbine operating between 3 MPa, 450°C and 0.1 MPa, $x = 1.0$. The work produced by the turbine is 650 kJ/kg. Assume that all heat transfer is with the environment.

Solution

The irreversibility per unit mass is given as

$$i = T_0(s_{\text{out}} - s_{\text{in}}) - q_0$$

The specified states yield the entropy difference. The first law yields

$$q_0 = h_{\text{out}} - h_{\text{in}} + w_a$$

so

$$i = T_0(s_{\text{out}} - s_{\text{in}}) - (h_{\text{out}} - h_{\text{in}}) - w_a$$

Therefore, the tables give

$$\begin{aligned} i &= T_0(7.3589 - 7.0835) - (2675.1 - 3344.1) - 650 \\ &= 101.1 \text{ kJ/kg} = 43.67 \text{ Btu/lbm} \end{aligned}$$

Comments
This could alternatively be evaluated directly from

$$i = w_{rev} - w_a$$

where w_{rev} is given in Example Problem 8.2. This yields exactly the same result.

8.5 Exergy, Helmholtz Function, and Gibbs Function

The definitions for the nonflow availability and the flow availability are given in Sec. 8.3. They represent the ability to obtain useful work from a particular state. The intensive forms for nonflow and flow availability are given in Eqs. (8.10) and (8.11) as

$$\phi = u - u_0 + P_0(v - v_0) - T_0(s - s_0)$$

and

$$\psi = h - h_0 - T_0(s - s_0)$$

These definitions are also termed *nonflow exergy* and *flow exergy*, respectively [1, 2]. There are other related definitions, and this section presents this nomenclature.

The availabilities given above express the difference between properties at the specified state and at the dead state. For the nonflow case, the property is $u + P_0v - T_0s$, and for the steady flow case, the property is $h - T_0s$. These definitions are also termed the *availability functions* [2, 3]. The difference in the availability function at the specified state and at the dead state yields the nonflow and flow availability used in this text. Essentially, the availability used here is directly referenced to the dead state, so that their values are zero at this dead state.

The *Helmholtz function*, or *Helmholtz free energy*, $a = u - Ts$ is a state function that depends on only the state of the substance (it is not tied to the dead state). The nonflow availability is related to the Helmholtz function for a special process. Consider a control mass that undergoes a constant-volume constant-temperature process. The process occurs in thermal equilibrium with the dead state (i.e., at $T = T_0$). The maximum useful work for a process from state 1 to state 2 is obtained from Eq. (8.12) as

$$_1w_{\text{rev, use, }2} = \phi_1 - \phi_2 \tag{8.29}$$

where the changes in kinetic and potential energy are neglected. With the nonflow availability, the result for a constant-volume constant-temperature ($T_1 = T_2 = T_0$) process is

$$ {}_1w_{\text{rev, use, }2} = (u_2 - T_2 s_2) - (u_1 - T_1 s_1) = a_2 - a_1 \tag{8.30} $$

Thus, the nonflow availability is the Helmholtz function for a constant-volume constant-temperature ($T_1 = T_2 = T_0$) process, and the difference gives the maximum useful work. Be sure to note that the above equation is *trivial* for a pure simple compressible substance, because the state postulate requires only two independent properties to determine the state. Therefore, with $V_1 = V_2$ and $T_1 = T_2$, the states are determined, and $a_1 = a_2$. Equation (8.30) becomes useful for chemically reacting systems, which require more properties to specify the state.

The *Gibbs function,* or *Gibbs free energy,* $g = h - Ts$ is also a state function and is related to the flow availability. The maximum useful work for a control volume undergoing a constant-pressure constant-temperature process in equilibrium with the atmosphere ($P_1 = P_2 = P_0$ and $T_1 = T_2 = T_0$) is given from Eq. (8.12) as

$$ \begin{aligned} {}_1w_{\text{rev, use, }2} &= \psi_{\text{in}} - \psi_{\text{out}} \\ &= (h - Ts)_{\text{in}} - (h - Ts)_{\text{out}} \\ &= g_{\text{in}} - g_{\text{out}} \end{aligned} \tag{8.31} $$

where the changes in kinetic and potential energy are neglected and there is a single inlet and outlet. The flow availability and Gibbs function are the same for the constant-pressure constant-temperature process indicated above. The same final result is obtained for a control mass undergoing a constant-pressure constant-temperature process. Again note that this difference is zero for a pure simple compressible substance, and Eq. (8.31) is chiefly useful for chemically reacting systems.

The Gibbs and Helmholtz functions are thus seen to be thermodynamic properties, and they are defined for a pure substance when any two independent properties are given. Therefore, they could be tabulated in additional columns of the property tables for any pure substance.

From Eqs. (8.30) and (8.31), we see that *changes* in the Helmholtz and Gibbs functions directly provide the useful reversible work for a process in thermal equilibrium with the dead state for nonflow (constant-volume) and flow (constant-pressure) processes, respectively. These functions are thus quite useful in defining the maximum work that is thermodynamically possible in these ideal processes, and they can be used for comparison with the work produced in real processes.

8.6 General Process Comparisons

This section develops a general comparison between processes and extends the previous analysis. Consider a general control volume with an arbitrary number of inlets, outlets, heat transfers, and work mechanisms. The system is depicted in Fig. 8.3. The actual process is indicated by the

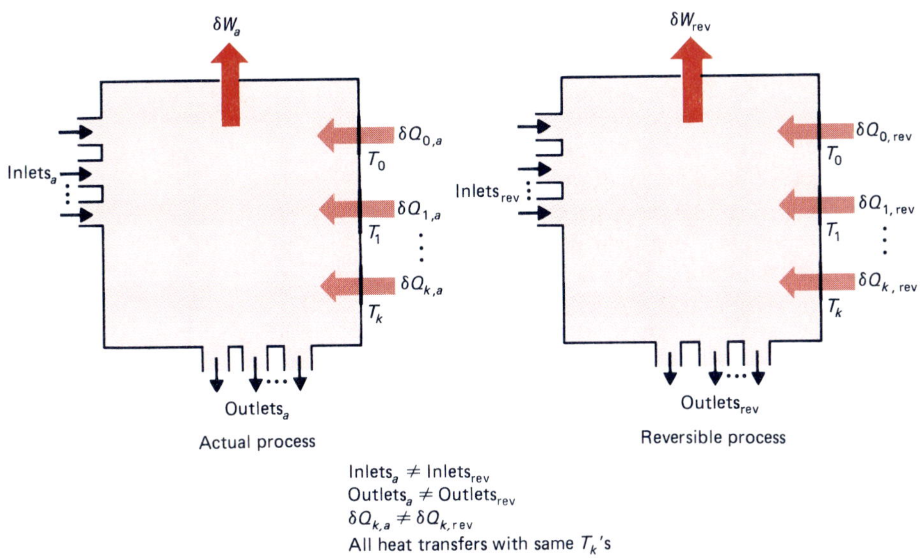

Figure 8.3 General process comparison.

subscript a. The comparison is with a totally reversible process (except for the case of multiple inlet streams, where the mixing process always introduces some irreversibility.) This process is also diagramed in Fig. 8.3, where the subscript *rev* indicates reversible. At this point in the development, the inlets, outlets, initial states, and final states may be different for the two processes. The mass flow rates are assumed to be the same at all points in each process.

The first and second laws can be written for each process and then combined. The work for the *actual* process is given in Eq. (8.6) as

$$\delta W_a = \sum_{\text{in}} (e + Pv - T_0 s)_a \, dm - \sum_{\text{out}} (e + Pv - T_0 s)_a \, dm - d(E - T_0 S)_{\text{CV}, a} + \sum_i \delta Q_{i,a}\left(1 - \frac{T_0}{T_i}\right) - T_0 \, \delta S_{\text{gen}} \tag{8.32}$$

where δS_{gen} is finite, because this is the actual process and is generally irreversible. The work for the *reversible* process is [same as Eq. (8.7)]

$$\delta W_{\text{rev}} = \sum_{\text{in}} (e + Pv - T_0 s)_{\text{rev}} \, dm - \sum_{\text{out}} (e + Pv - T_0 s)_{\text{rev}} \, dm - d(E - T_0 S)_{\text{CV, rev}} + \sum_i \delta Q_{i,\text{rev}}\left(1 - \frac{T_0}{T_i}\right) \tag{8.33}$$

There is zero entropy generation for the reversible process.

The processes are compared on the basis of the work output, δW_a and δW_{rev}. The optimum that is expected from a process is the totally reversible process, and the actual process has irreversibilities or finite entropy generation contributions. The goal of the comparison is to determine the effect of the irreversibilities on the work output of a process. Equations (8.32) and (8.33) give the expressions for the work obtained by the actual and reversible processes. If a comparison of the work of an actual process and a reversible process is to be made, not all the terms in the equations describing them can be identical. The actual process has a finite entropy generation, so one of the other terms in Eq. (8.32) must be different from the corresponding term in Eq. (8.33). The specific term that differs in the two processes accounts for the different physical processes occurring—one is actual, and the other is reversible. This physical effect or term in the governing equation accommodates the differences between the actual and reversible processes.

The term that varies between the actual and reversible processes is completely arbitrary, yet conventions have been adopted. A standard process comparison considers the states and therefore properties within the control volume and the inlet and outlet states to be the same for both the reversible and the actual process. Therefore, the heat transfers are used to accommodate the irreversibility, or entropy generation, of the actual process. The heat transfers "float" to account for the entropy generation. This comparison is presented in Sec. 8.4 and defined as irreversibility. The choice of the heat transfer as the term to account for the irreversibility is not universal. Thus, the general process comparison is left arbitrary here.

The comparison of the actual process with the reversible process is made by subtracting the actual work obtained from the reversible work. This yields [subtract Eq. (8.32) from Eq. (8.33)]

$$\begin{aligned}\delta W_{\text{rev}} - \delta W_a = {} & T_0\,\delta S_{\text{gen}} + d[(E - T_0S)_{\text{CV},a} - (E - T_0S)_{\text{CV, rev}}] \\ & + \sum_{\text{out}} [(e + Pv - T_0s)_a - (e + Pv - T_0s)_{\text{rev}}]\,dm \\ & - \sum_{\text{in}} [(e + Pv - T_0s)_a - (e + Pv - T_0s)_{\text{rev}}]\,dm \\ & - \sum_i (\delta Q_{i,a} - \delta Q_{i,\text{rev}})\left(1 - \frac{T_0}{T_i}\right) \qquad (8.34a)\end{aligned}$$

which can be written in terms of nonflow and flow availability as

$$\begin{aligned}\delta W_{\text{rev}} - \delta W_a = {} & T_0\,\delta S_{\text{gen}} \\ & + d\left\{m_{\text{CV}}\left[\phi_a - \phi_{\text{rev}} + \frac{\mathbf{V}^2_{\text{ref},a} - \mathbf{V}^2_{\text{ref, rev}}}{2g_c} + \frac{g(z_a - z_{\text{rev}})}{g_c}\right]\right\} \\ & + \sum_{\text{out}}\left[\psi_a - \psi_{\text{rev}} + \frac{\mathbf{V}_a^2 - \mathbf{V}_{\text{rev}}^2}{2g_c} + \frac{g(z_a - z_{\text{rev}})}{g_c}\right] dm \\ & - \sum_{\text{in}}\left[\psi_a - \psi_{\text{rev}} + \frac{\mathbf{V}_a^2 - \mathbf{V}_{\text{rev}}^2}{2g_c} + \frac{g(z_a - z_{\text{rev}})}{g_c}\right] dm \\ & - \sum_i (\delta Q_{i,a} - \delta Q_{i,\text{rev}})\left(1 - \frac{T_0}{T_i}\right) \qquad (8.34b)\end{aligned}$$

This equation relates the maximum work that is possible for a chosen reference process to the actual work obtained by the real process under examination. The expression is in terms of a difference and can be interpreted as *lost available work* [2]. The resulting value is a work-type term which would be zero if the actual process were the reversible process. The first term on the right-hand side represents the irreversibilities of the actual process and is the reason for the comparison. The next three terms are differences in actual and reversible quantities. They are differences in properties within the control volume and differences at the inlets and outlets. The last term represents the difference in heat transfer for the actual and reversible reference processes.

The comparison of the two processes in Fig. 8.3 and Eq. (8.34) might seem like a comparison of apples and oranges, because the processes compared appear to have no connection with each other. This is the case to this point in the development. However, the connection and the point of the comparison become clear when a few examples are considered.

Comparison 1

Consider the case of steady state conditions within a control volume with a single inlet and single outlet and only one heat transfer to the control volume from the surroundings at T_0. The sum over all the heat transfers in Eq. (8.34) includes only one term, and the temperature is T_0. Thus the heat transfer term does not appear directly in the equation, but it permits the process comparison. The reversible comparison considers the same inlet and outlet states as the actual process. Therefore, the inlet and outlet contributions in Eq. (8.34) are zero. The change within the control volume is zero, because the process is steady state. Equation (8.34) reduces to

$$\delta W_{\text{rev}} - \delta W_a = T_0\,\delta S_{\text{gen}} \tag{8.35}$$

For this steady state process comparison, Eq. (8.35) on a rate basis is

$$\dot{W}_{\text{rev}} - \dot{W}_a = T_0 \dot{S}_{\text{gen}} \tag{8.36}$$

The difference between the reversible and actual case is the entropy generation of the actual process. These are the same as Eqs. (8.19) and (8.20) for irreversibility.

Comparison 2

Consider a control mass process with a heat transfer from the surroundings at T_0 with the actual and reversible processes that originate at a common state and terminate at another common state. A control mass does not have inlet or outlet ports, so Eq. (8.34) reduces to

$$\delta W_{\text{rev}} - \delta W_a = T_0\,\delta S_{\text{gen}} \tag{8.37}$$

This is the same result obtained for the irreversibility in Sec. 8.4. The common feature in both comparisons presented is the heat transfer with the surroundings at T_0. The states of the system are the same, and the process comparison is accommodated by the heat transfer term.

Comparison 3

A final comparison considers an actual and a reversible engine operating in a cycle between two heat transfer reservoirs at T_0 and T_1. The higher-temperature source is at T_1, and the low-temperature reservoir is at T_0. The second law for the cycle is [Eq. (8.2)]

$$S_{gen} = \frac{|Q_0|}{T_0} - \frac{|Q_1|}{T_1} \geq 0 \tag{8.38}$$

Since the temperatures of the reservoirs are fixed, the heat transfers will differ for the actual and the reversible case. The integrated cyclic form of Eq. (8.34) is

$$W_{rev} - W_a = T_0 S_{gen} - (Q_{1,a} - Q_{1,rev})\left(1 - \frac{T_0}{T_1}\right) \tag{8.39}$$

The comparision is arbitrary at this point. However, we can decide for comparison to set equal the heat transfer inputs, so $Q_{1,a} = Q_{1,rev}$, and

$$W_{rev} - W_a = T_0 S_{gen} \tag{8.40}$$

Note that Q_0 floats in the comparison.

Clearly numerous comparisons can be presented. The comparisons are between the reversible work in Eq. (8.7) and the actual work. The difference is given in Eq. (8.34).

Example Problem 8.6

An idealized Rankine cycle operates with steam between a condenser temperature of 60°F and a boiler temperature of 220°F. The turbine inlet is a saturated vapor, and the pump inlet is a saturated liquid. The adiabatic turbine efficiency is 70 percent, and the adiabatic pump efficiency is 100 percent. Evaluate $w_{rev} - w_a$ for each process of the cycle as well as for the overall cycle. The ambient temperature is 50°F, and the energy source is at 300°F.

Diagram

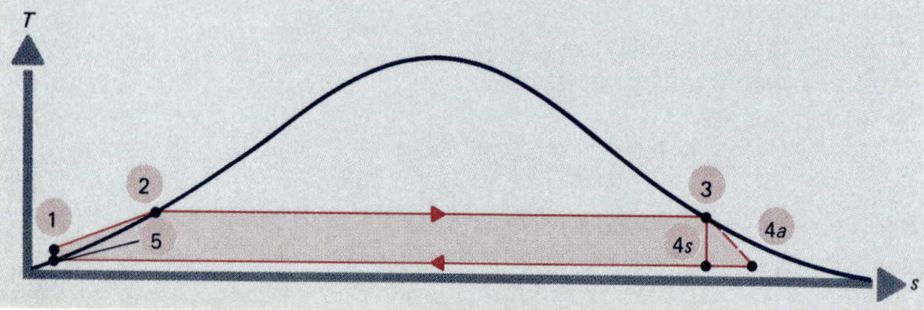

Solution

The state information is as follows:

State	P, psia	T, °F	h, Btu/lbm	s, Btu/(lbm·°R)	x
1	17.184	—	28.10	(0.05548)	
2	(17.184)	220	—	(0.3241)	0.0
3	(17.184)	220	(1153.4)	(1.7441)	1.0
4s	0.2563	(60)	(905.6)	1.7441	(0.8284)
4a	(0.2563)	60	979.9	(1.8871)	(0.8986)
5	(0.2563)	60	(28.048)	(0.05548)	0.0

Entries not in parentheses determine the state.

The actual state at the exit of the turbine is found with the component efficiency. For the turbine,

$$\eta_t = \frac{h_{4,a} - h_3}{h_{4,s} - h_3} = 0.70$$

so

$$h_{4,a} = h_3 + (0.70)(h_{4,s} - h_3) = 979.9 \text{ Btu/lbm}$$

The pump is isentropic, because the efficiency is 100 percent. Thus,

$$w_p = v_5(P_5 - P_1) = h_5 - h_1 = -0.0502 \text{ Btu/lbm}$$

and

$$h_1 = 28.10 \text{ Btu/lbm}$$

All the states are determined, and the properties are obtained from Table E.10.

The first law analysis of this cycle yields the heat transfers and the work. Since some of this information is required in the lost available work analysis, it is presented here. The work for the pump and the work for the turbine are

$$_5w_1 = w_p = v_5(P_5 - P_1) = -0.0502 \text{ Btu/lbm}$$

and

$$_3w_{4,a} = w_t = h_3 - h_{4,a} = 173.5 \text{ Btu/lbm}$$

so

$$w_{\text{net}} = w_t + w_p = 173.4 \text{ Btu/lbm}$$

The heat transfers in the boiler and condenser are

$$_1q_3 = q_b = h_3 - h_1 = 1125.3 \text{ Btu/lbm}$$
$$_{4,a}q_5 = q_c = h_5 - h_{4,a} = -951.9 \text{ Btu/lbm}$$

Equation (8.34) is the governing expression for the lost available work, and the entropy generation is given in Eq. (8.2). The steady state form eliminates the changes within the control volume. In the comparison between the actual and reversible component processes, the inlet and outlet states are the same, and the heat transfers are taken as equal. Thus, with Eq. (8.2) substituted into Eq. (8.34), the governing equation is

$$w_{rev} - w_a = T_0(s_{out} - s_{in}) - T_0 \sum_i \frac{q_i}{T_i}$$

The values of lost available work for each process, using $T_0 = 459.67 + 50 = 509.67°R$, are

$$(w_{rev} - w_a)_p = T_0(s_1 - s_5) = 0$$

$$(w_{rev} - w_a)_t = T_0(s_{4,a} - s_3) = 72.88 \text{ Btu/lbm}$$

$$(w_{rev} - w_a)_b = T_0(s_3 - s_1) - T_0\left(\frac{q_b}{459.67 + 300}\right) = 105.7 \text{ Btu/lbm}$$

$$(w_{rev} - w_a)_c = T_0(s_5 - s_{4,a}) - T_0\left(\frac{q_c}{T_0}\right) = 18.39 \text{ Btu/lbm}$$

The lost available work in the pump is zero, because the process is reversible and adiabatic or isentropic.

The lost available work for the overall cycle is found from the above expression by noting that the entropy difference for a cycle is zero. Therefore,

$$(w_{rev} - w_a)_{cycle} = -q_b\left(\frac{T_0}{759.67}\right) - q_c\left(\frac{T_0}{T_0}\right)$$

or

$$(w_{rev} - w_a)_{cycle} = |q_c| - |q_b|\left(\frac{509.67}{759.67}\right) = 196.9 \text{ Btu/lbm}$$

This value is the same as that obtained by summing the lost available work values for the turbine, boiler, and condenser.

Comments

Note that the lost available work expressions in this example are very similar to the irreversibility definition in Sec. 8.4. The difference here is that heat transfers are permitted with more than one surrounding.

This example, repeated for various turbine efficiencies, yields the following results:

η_t, %	$(w_{rev} - w_a)_t$, **Btu/lbm**	$(w_{rev} - w_a)_b$, **Btu/lbm**	$(w_{rev} - w_a)_c$, **Btu/lbm**	$(w_{rev} - w_a)_{cycle}$, **Btu/lbm**
70	72.88	105.7	18.39	196.9
80	48.57	105.7	17.90	172.1
90	24.26	105.7	17.41	147.3
100	0.0	105.7	17.27	122.9

The sum of the turbine, boiler, and condenser work values equals the cycle value, as shown. The pump value is always zero since the pump was assumed to be isentropic.

One final and important point. The lost available work for the adiabatic turbine given above is exactly the irreversibility in Sec. 8.4, as noted in comparison 1 of Sec. 8.5. It is equally important to compare the actual turbine work to that of the reversible isentropic turbine. This is a comparison which has different outlet states. The lost available work in Eq. (8.34) is

$$(w_{rev} - w_a)_t = T_0(s_{4,a} - s_3) + (h - T_0 s)_{4,a} - (h - T_0 s)_{4,s}$$
$$= h_{4,a} - h_{4,s} + T_0(s_{4,s} - s_3)$$

But $s_{4,s} = s_3$, so

$$(w_{rev} - w_a)_t = h_{4,a} - h_{4,s}$$

The values for this lost available work comparison are as follows:

η_t, %	$(w_{rev} - w_a)_t$, **Btu/lbm**
70	74.4
80	49.6
90	24.8
100	0.0

The values in this comparison are larger than those in the previous comparison. The value of w_{rev} in this last case is fixed at the isentropic value (turbine efficiency = 100 percent) and is the largest value possible. The previous comparison was between the *same* inlet and *same* outlet states for the actual and reversible work at a given turbine efficiency, so the comparison process changed with the turbine efficiency and had a smaller value than for the isentropic turbine.

8.7 Second Law Efficiencies

The efficiencies of many devices are defined in terms of "first law" efficiencies; i.e., the performance of the device is defined in terms of the ratio of the useful energy produced to the energy required to operate the device. This is the case, for example, for all the heat engine efficiencies for the various cycles discussed in Chaps. 6 and 7 as well as for the coefficient of performance used for chillers, air conditioners, and heat pumps.

Now that we have introduced and discussed availability, irreversibility, and exergy, we can use these second law concepts to consider another class of efficiencies which can be applied to heat engine and cooling cycles as well as to particular devices. This will allow us to see how closely a real device or cycle approaches the performance allowed by the restrictions of the second law.

8.7.1 Second Law Efficiency of Components that Produce Work or Require Work Input

The efficiency of turbines, compressors, pumps, and other devices requiring work input or producing work output was defined in Chap. 5 by comparing the actual work to or from the device with the work required or produced during an idealized process.

The idealized process chosen for the definition of first law efficiencies usually has some thermodynamic property that is assumed to be held constant throughout the process, allowing simple calculation of the work input or output. The ideal process used for comparison is often an *adiabatic reversible* (isentropic) one; however, for compressors in particular, the ideal process is sometimes chosen to be an *isothermal reversible* one. The choice of idealized process is arbitrary and is made on the basis of expediency; i.e., we choose for comparison the ideal process that we would like to approach with the process in the actual device.

An alternative way to evaluate a device is to compare the actual useful work output with the *maximum possible work output,* as defined by the second law. That is, for a device that gives useful work output,

$$\eta_{\mathrm{II},t} = \frac{W_{a,\,\mathrm{use}}}{W_{\mathrm{rev},\,\mathrm{use}}} \tag{8.41}$$

and for a device requiring work input

$$\eta_{\mathrm{II},c} = \frac{W_{\mathrm{rev},\,\mathrm{use}}}{W_{a,\,\mathrm{use}}} \tag{8.42}$$

Now the relations for actual and reversible useful work can be substituted for the particular conditions needed for a given device. For example, for a turbine operating between inlet state 1 and exit state 2, the first law provides the actual useful work as

$$w_{a,\text{ use},\, t} = h_1 - h_2 \tag{8.43}$$

and the useful reversible work is given by Eq. (8.14) as

$$w_{\text{rev, use},\, t} = \psi_1 - \psi_2 = (h_1 - h_2) - T_0(s_1 - s_2) \tag{8.44}$$

Thus, Eq. (8.41) can be written in the alternative forms [by noting that, from Eq. (8.35), $w_{a,\text{ use},\, t} = w_{\text{rev, use},\, t} - T_0 s_{\text{gen}}$]

$$\begin{aligned}\eta_{\text{II},\, t} &= \frac{w_{a,\text{ use},\, t}}{w_{\text{rev, use},\, t}} = \frac{h_1 - h_2}{h_1 - h_2 + T_0(s_2 - s_1)} \\ &= 1 - \frac{T_0 s_{\text{gen}}}{w_{\text{rev, use},\, t}}\end{aligned} \tag{8.45}$$

Note that $w_{\text{rev, use},\, t}$ is positive for a turbine, so the second law efficiency will always be less that one.

The maximum reversible work is defined between the given inlet and outlet states, which must be known in order to evaluate η_{II}. Thus, the standard turbine efficiency defined in Chap. 5 [Eq. (5.183)] must still be used to find the actual outlet state from the turbine, given the inlet state and outlet pressure. Equation (8.45) may then be used with the known inlet and outlet states to determine how closely the actual turbine approaches the reversible useful work output. The second law efficiency defined by Eq. (8.45) is sometimes called the *effectiveness*.

Generally, the effectiveness of a turbine will be greater than its first law efficiency. The effectiveness (second law efficiency) embodies the recognition that not all the enthalpy of the inlet fluid is available to produce useful work, and our analysis of the device should recognize this fact. The effectiveness definition recognizes that we should expect at most the useful work output from a device that is theoretically possible, and we should not compare with some process that is not attainable.

For a compressor, the second law efficiency or effectiveness is given by

$$\eta_{\text{II},\, c} = \frac{w_{\text{rev, use}}}{w_{a,\text{ use}}} = \frac{\psi_1 - \psi_2}{h_1 - h_2} \tag{8.46}$$

Example Problem 8.7

A steam turbine has a first law efficiency of 90 percent, an inlet state of 700°F and 550 psia, and an exit pressure of 14.7 psia. The turbine is adiabatic. What is the second law efficiency of this device?

Solution

The state table for the turbine is shown below (as calculated in Example Problem 5.13):

State	P, psia	T, °F	x	h, Btu/lbm	s, Btu/(lbm·°R)
1	550	700	—	(1353.5)	(1.5986)
$2s$	14.7	—	(0.8906)	(1044.3)	$s_{2s} = s_1$
$2a$	14.7	—	(0.9225)	(1075.2)	(1.6447)

The second law efficiency is given by Eq. (8.45) as

$$\eta_{\text{II},t} = \frac{h_1 - h_2}{(h_1 - h_2) + T_0(s_2 - s_1)}$$

and substituting values from the state table, we get

$$\eta_{\text{II},t} = \frac{1353.5 - 1075.2}{(1353.5 - 1075.2) + 537(1.6447 - 1.5986)}$$
$$= 0.918 = 91.8\%$$

Comments

The turbine converts 91.8 percent of the reversible useful work into actual useful work, as compared with 90 percent of the isentropic work output prescribed by the given first law efficiency.

8.7.2 Second Law Efficiency of Other Components

Components which involve no shaft work input or output must be treated differently from those that do. Each component requires a separate analysis to define the appropriate efficiency. We concentrate on the class of components used in steady-state, steady flow processes with a single flow stream.

Heat Exchangers with Heat Transfer to or from a Single Reservoir

In this specific case, we deviate somewhat from our convention in comparing the actual and ideal heat exchanger. We will assume that the heat transfer from the ideal and actual exchanger are the same; however, the outlet states are allowed to differ. We define the second law efficiency of the exchanger for the case of heat transfer to or from a single reservoir as *the change in availability of the actual flow stream divided by the reversible work to or from the exchanger.*

The reversible work from the ideal exchanger is given by Eq. (8.33) as

$$\dot{w}_{\text{rev}} = q_R \left(1 - \frac{T_0}{T_R}\right) \tag{8.47}$$

The availability change of the fluid in the real exchanger is given by Eq. (8.32), letting $dw_a = 0$ for no shaft work from the actual exchanger, and the term for changes within the control volume is zero for steady state. Neglecting kinetic and potential energy changes, Eq. (8.32) reduces to

$$(\psi_{\text{out}} - \psi_{\text{in}})_a = q_R \left(1 - \frac{T_0}{T_R}\right) - T_0 s_{\text{gen}} \tag{8.48}$$

The desired output from the exchanger is an increase or decrease in the actual availability of a fluid by heat transfer from or to a single heat transfer reservoir (as with a condenser, boiler, or steam generator). For this type of heat exchanger, then, a measure of second law efficiency is

$$\eta_{\text{II, HE}} = \frac{(\psi_{\text{out}} - \psi_{\text{in}})_a}{q_R\left(1 - \dfrac{T_0}{T_R}\right)} = \frac{q_R\left(1 - \dfrac{T_0}{T_R}\right) - T_0 s_{\text{gen}}}{q_R\left(1 - \dfrac{T_0}{T_R}\right)}$$

$$= 1 - \frac{T_0 s_{\text{gen}}}{q_R\left(1 - \dfrac{T_0}{T_R}\right)} \tag{8.49}$$

which also can be seen to be *the ratio of the actual availability change of the fluid to the change that would occur if there were no entropy generation.* This is not the only way that a second law efficiency can be defined for a heat exchanger interacting with a single reservoir, but it does give insight into the degree to which an exchanger approaches ideal behavior as predicted by the second law.

Two-Fluid Heat Exchangers

If the fluid of interest in the exchanger is receiving energy by heat transfer from a second fluid that is also changing in availability because of the heat transfer, the situation is more complex. Let us concentrate on stream 1, the stream that we wish to heat or cool in the exchanger. Stream 2 is the stream that is present to provide the necessary heat transfer to stream 1.

Various definitions of heat exchanger second law efficiency have been given for this case, including the ratio of the increase in availability of the cold fluid over the decrease in availability of the hot fluid, as well as the ratio of the outflow to the inflow of available energy.

Witte [4] has proposed a second law efficiency that follows the conventions established earlier in this chapter, i.e.: the inlet and outlet conditions of the actual and reversible exchanger are the same, and the heat transfer between the flow streams in the actual and reversible exchanger are the same. For such a case, the reversible reference exchanger must produce reversible work output. This can be seen from Eq. (8.34*b*), which reduces for the conventions imposed to

$$w_{\text{rev}} = T_0 s_{\text{gen}} \tag{8.50}$$

because all other terms drop out. The same result can be obtained from Eq. (8.36). Witte proposed that the second law efficiency now be defined as the ratio

$\eta_{\text{II, HE}}$ = actual heat transfer less reversible work that could be recovered in reversible reference exchanger/actual heat transfer

The resulting expression is

$$\eta_{\text{II ,HE}} = \frac{|q| - |w_{\text{rev}}|}{|q|} = 1 - \frac{T_0 s_{\text{gen}}}{|q|} \tag{8.51}$$

This definition can also be put in terms of the irreversibility by the use of Eq. (8.19) to give

$$\eta_{\text{II, HE}} = 1 - \frac{i}{|q|} \tag{8.52}$$

and as the irreversiblility of the real exchanger approaches zero, the second law efficiency approaches unity; as the irreversibility becomes equal to the heat transfer, the second law efficiency approaches zero.

Witte shows that if the mean temperature of the colder fluid is T_C and the mean temperature of the hotter stream is T_H, then, by evaluating the entropy generation due to heat transfer in the heat exchanger, Eq. (8.51) reduces to

$$\eta_{\text{II, HE}} = 1 + \frac{T_0}{T_H} - \frac{T_0}{T_C} \qquad T_C \leq T_H \tag{8.53}$$

Thus, the greater the temperature difference between the hotter and colder streams, the poorer will be the second law efficiency of the exchanger, as we might expect from the irreversibility introduced by heat exchange across a finite temperature difference.

8.7.3 Second Law Efficiency of Heat Engine and Refrigeration Cycles

For heat engine cycles, we can define the second law efficiency as the ratio of the actual net useful work output from the cycle $w_{a,\text{ use, net}}$ to the useful net work output $w_{\text{rev, use, net}}$ of a reversible cycle operating between the same high- and low-temperature thermal reservoirs as the actual cycle:

$$\eta_{\text{II, heat engine}} = \frac{w_{a,\text{ use, net}}}{w_{\text{rev, use, net}}} \tag{8.54}$$

Consider an engine operating between a high-temperature heat transfer reservoir at T_H and a lower-temperature reservoir at T_L. By applying the convention that the heat transfers between all reservoirs are the same for the real and the reversible process, $q_{H,\,\text{rev}} = q_{H,\,a}$, Eq. (8.54) can be written as

$$\eta_{\text{II, heat engine}} = \frac{w_{a,\,\text{use, net}}/q_H}{w_{\text{rev, use, net}}/q_H} = \frac{\eta_{\text{I},\,a}}{\eta_{\text{I, rev}}} \tag{8.55}$$

where η_I denotes the standard first law efficiencies for the engines. However, a reversible heat engine cycle operating between two heat transfer reservoirs at T_H and T_L has the Carnot efficiency, and Eq. (8.55) can be written

$$\eta_{\text{II, heat engine}} = \frac{\eta_{\text{I},\,a}}{\eta_{\text{I, Carnot}}} = \frac{\eta_{\text{I},\,a}}{1 - T_L/T_H} \tag{8.56}$$

Thus, the second law efficiency can be interpreted as a measure of how closely the actual engine first law efficiency approaches the first law efficiency of a Carnot engine operating between the same heat transfer reservoirs. This definition will apply to all heat engine cycles.

For a refrigeration cycle, similar reasoning leads to

$$\eta_{\text{II, refrig}} = \frac{w_{\text{rev, use, net}}}{w_{a,\,\text{use, net}}} = \frac{w_{\text{rev, use, net}}/q_L}{w_{a,\,\text{use, net}}/q_L} = \frac{\text{COP}_a}{\text{COP}_\text{rev}} \tag{8.57}$$

where q_L is the energy removed from the space or material being cooled and COP is the coefficient of performance of the cycle [Eq. (6.71)]. Again, a reversible cycle operating between two heat transfer reservoirs is a Carnot cycle, and Eq. (8.57) becomes [using Eq. (6.72)]

$$\eta_{\text{II, refrig}} = \frac{\text{COP}_a}{\text{COP}_\text{Carnot}} = \frac{\text{COP}_a}{1/(T_H/T_L - 1)} \tag{8.58}$$

Again, the interpretation of the second law efficiency of a refrigeration cycle is the degree to which the actual cycle approaches the ideal Carnot cycle COP. Thus, a Carnot heat engine and a Carnot refrigeration cycle both have a second law efficiency of 100 percent; all other cycles will have a lower efficiency.

Example Problem 8.8

A simple steam Rankine cycle operates with saturated steam at 100°C leaving the boiler and a condenser operating at 20°C. The turbine first law efficiency is 70 percent, and the pump first law efficiency is 60 percent. What is the second law efficiency of the cycle?

Solution

The Carnot efficiency of this cycle is $1 - 293/373 = 21.4$ percent. The Rankine cycle efficiency for this case was found in Example Problem 7.2 to be 14.1 percent. The second law efficincy is given by

$$\eta_{\text{II, heat engine}} = \frac{\eta_{\text{I}, a}}{\eta_{\text{I, Carnot}}} = \frac{\eta_{\text{I}, a}}{1 - T_L/T_H}$$

or

$$\eta_{\text{II, Rankine}} = \frac{14.1}{21.4} = 0.659 \text{ or } 65.9\%$$

Comments

The first law efficiency of this cycle is quite low, indicating that a small portion of the heat transfer to the cycle is converted to useful work. However,the second law efficiency show that almost 66 percent of the useful work that could be obtained from the ideal Carnot cycle has been recovered by the actual cycle.

8.8 Summary

The theme of this chapter is entropy generation, and the reduction of the entropy generation is the desired goal of second law analysis. Very important definitions and relations are developed in this chapter to analyze thermodynamic systems. A substance at other than the dead state contains available energy which is useful to meet our needs. Engineers try to use this available energy to meet the needs of their particular devices. Methods to calculate this available energy are presented in Sec. 8.3. Availability is directly related to the maximum useful work that can be obtained from a specified state (in relation to the dead state).

Other constraints may be imposed on thermodynamic devices that do not permit the use of all the available energy. Constraints imposed by other system components, materials, economics, etc., indicate whether a process is permitted between an initial and final state (or inlet and outlet state). Therefore, the maximum useful work between two specified states is important. In this case, the reversible work [Eq. (8.7)] and reversible useful work [Eq. (8.9)] are the relations for the maximum work that can be obtained. The measure of the inability of a real process to reach the ideal case between two specified states is the irreversibility [Eqs. (8.21) and (8.22)]. The irreversibility is directly related to entropy generation, and it can be directly compared to the difference between the actual and reversible work.

A general comparison equation for real and ideal processes is Eq. (8.34). This relation compares two arbitrary processes. The examples given in Sec. 8.6 indicate that irreversibility is a specific case of this general comparison. Specification of an actual process and a reversible comparison process enables the process comparison to be made.

Problems

8.1S The cooling water from a power plant leaves the condenser at 33°C and atmospheric pressure. The water is cooled and discharged into a reservoir at 25°C and atmospheric pressure. Determine the heat transfer required per unit mass and the entropy generation. Evaluate the reversible work that could be obtained from the specified states.

8.1E The cooling water from a power plant leaves the condenser at 90°F and atmospheric pressure. The water is cooled and discharged into a reservoir at 75°F and atmospheric pressure. Determine the heat transfer required per unit mass and the entropy generation. Evaluate the reversible work that could be obtained from the specified states.

8.2S The concept of cogeneration is that a Rankine cycle can be used with a turbine outlet condition that is above ambient conditions. The steam enthalpy at the turbine exit is then used for a second purpose such as process steam, operating a heating or cooling system, etc.

(*a*) Plot the availability of turbine exit steam versus exit pressure ($101.3 < P < 1000$ kPa) with quality ($0.5 < x < 1$) as a parameter.

(*b*) If you were assigned Prob. 7.16S, also note how the Rankine cycle efficiency in that case varies with turbine exit pressure. Discuss any limitations that you see on cogeneration.

8.2E The concept of cogeneration is that a Rankine cycle can be used with a turbine outlet condition that is above ambient conditions. The steam enthalpy at the turbine exit is then used for a second purpose such as process steam, operating a heating or cooling system, etc.

(*a*) Plot the availability of turbine exit steam versus exit pressure ($14.7 < P < 150$ psia) with quality ($0.5 < x < 1$) as a parameter.

(*b*) If you were assigned Prob. 7.16E, also note how the Rankine cycle efficiency in that case varies with turbine exit pressure. Discuss any limitations that you see on cogeneration.

8.3S Ocean thermal energy conversion (OTEC) systems seek to use the temperature gradient within the ocean as an energy source. What is the flow availability per unit mass for surface water at 25°C if the dead state is at 10°C and 100 m? Assume properties of fresh water.

8.3E Ocean thermal energy conversion systems seek to use the temperature gradient within the ocean as an energy source. What is the flow availability per unit mass for surface water at 75°F if the dead state is at 50°F and 300 ft? Assume properties of fresh water.

8.4 An air-cooled water radiator operates as shown. Find (*a*) the heat transfer $\dot{Q}$ from the radiator (kJ/h), (*b*) rate of irreversibility $\dot{I}$ of the radiator, and (*c*) availability Ψ of the air at state 4 (kJ/kg) and of the water at state 2 (kJ/kg).

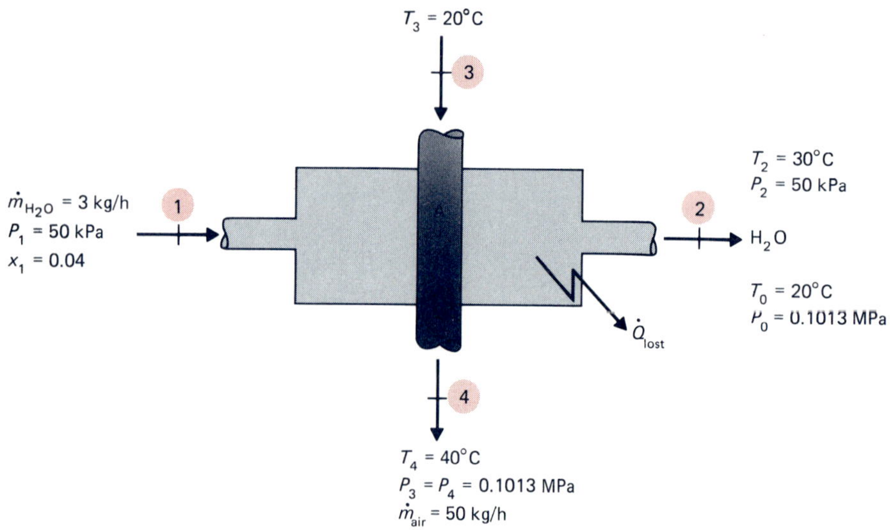

Figure P8.4

8.5 The power plant shown in Fig. P8.5 uses a low-boiling-point working medium so that the ocean may be used as a heat source and sink. Find the availability of the sea-water at the boiler inlet and the condenser exit (Ψ_1 and Ψ_4).

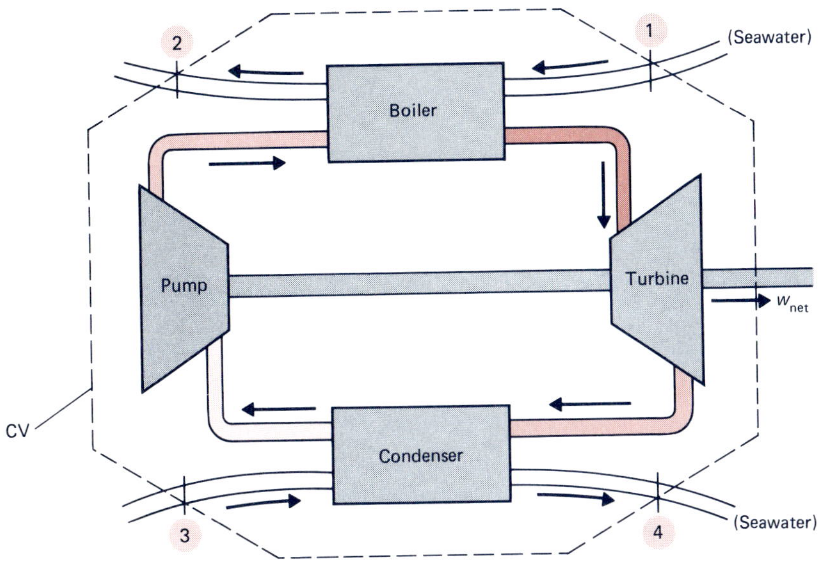

Figure P8.5

8.6 Hydrogen (2 kg/s) is cooled in the isobaric heat exchanger shown.

(*a*) What is $\dot{m}_{air}$?

(*b*) Why is $\dot{m}_{air} > \dot{m}_{12}$ for the same temperature change?

(*c*) Is this process reversible? Justify your answer. Assume that $\dot{Q}_{CV}$ occurs at $T_0 = 25°C$.

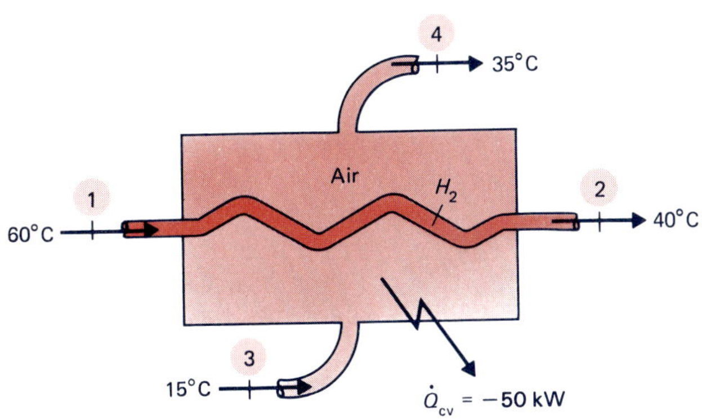

Figure P8.6

8.7 From the standpoint of the irreversibility, is the following steady state steady flow adiabatic device reversible, irreversible, or impossible? The working medium is H_2O. The direction of work is unknown (Fig. P8.7).

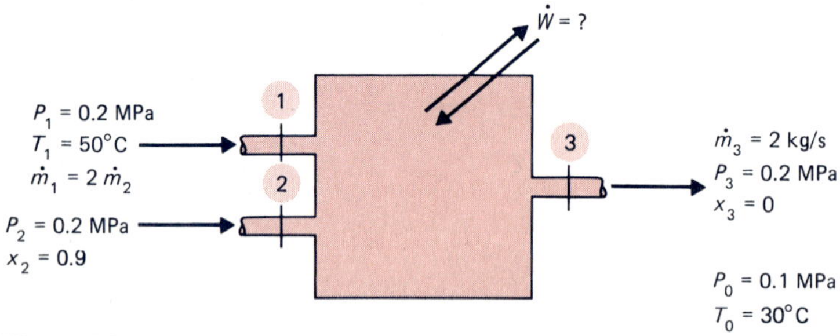

Figure P8.7

8.8 For the adiabatic turbine shown, determine the availability of the H_2O at the exit.

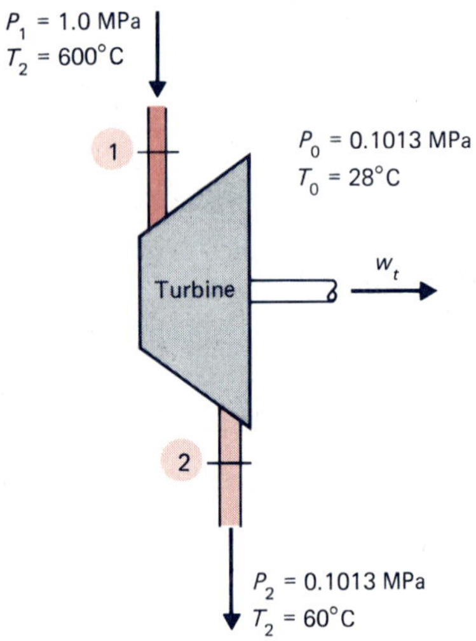

Figure P8.8

8.9S An OTEC system operates as a heat engine, taking surface water at 28°C as the high-temperature reservoir and deep water at 14°C and 50 m below the surface as the low-temperature reservoir. What is w_{rev} for this system?

8.9E An OTEC system operates as a heat engine, taking surface water at 80°F as the high-temperature reservoir and deep water at 56°F and 150 ft below the surface as the low-temperature reservoir. What is w_{rev} for this system?

8.10S Two sources of refrigerant 12 are available for use within a closed cylinder: 4 MPa, 150°C or 80 kPa, 200°C. Which source has the larger available energy? (Use P_0 = 100 kPa.)

8.10E Two sources of refrigerant 12 are available for use within a closed cylinder: 400 pisa, 300°F or 10 psia,

400°F. Which source has the larger available energy? (Use $P_0 = 14.7$ psia.)

8.11S A geothermal well produces steam at the wellhead at 220°C and 1 MPa. What are the flow availability and reversible work for this source?

8.11E A geothermal well produces steam at the wellhead at 400°F and 150 psia. What are the flow availability and reversible work for this source?

8.12S A solar pond 3 m deep produces energy at the pond bottom that can be withdrawn at up to the saturation temperature at the pond bottom. Assume, for simplicity, that the specific volume of the pond water can be taken as an average of 8.83×10^{-4} m³/kg. What are the flow availability and reversible work for this source?

8.12E A solar pond 10 ft deep produces energy at the pond bottom that can be withdrawn at up to the saturation temperature at the pond bottom. Assume, for simplicity, that the specific volume of the pond water can be taken as an average of 1.335×10^{-2} ft³/lbm. What are the flow availability and reversible work for this source?

8.13 Air is heated in a steady-state steady-flow heater as shown. Determine the reversible work w_{rev} kJ/kg as a result of the process.

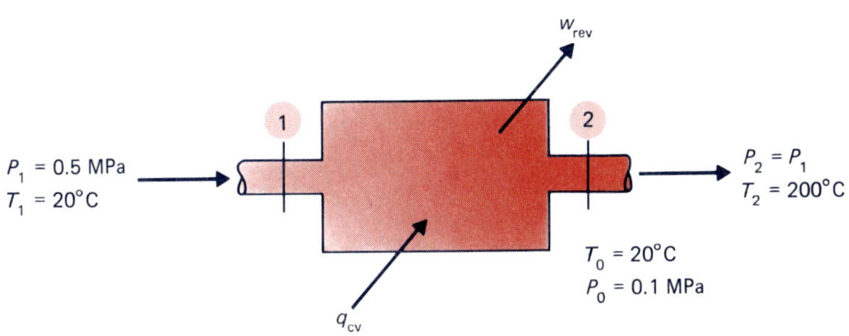

Figure P8.13

8.14 Steam enters a steady-state steady-flow device as shown. The atmospheric temperature is 22°C. Determine the reversible power $\dot{W}_{rev}$ kW.

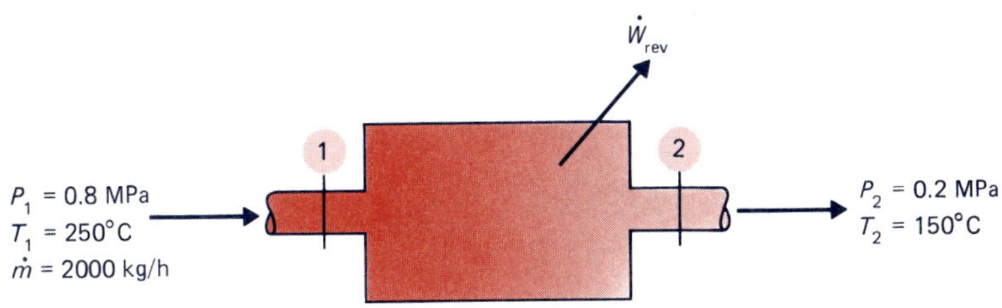

Figure P8.14

8.15S An air bottle contains air at 10 atm and 500 K. The air is cooled by heat transfer with the surroundings until it reaches 300 K.

(*a*) What is the availability ϕ of the air in the initial state?

(*b*) What is the reversible work in the cooling process?

(*c*) What is the irreversibility of the process?

8.15E An air bottle contains air at 10 atm and 900°R. The air is cooled by heat transfer with the surroundings until it reaches 540°R.

(*a*) What is the availability ϕ of the air in the initial state?

(*b*) What is the reversible work in the cooling process?

(*c*) What is the irreversibility of the process?

8.16S Evaluate the Gibbs function g and the Helmholtz function a at the inlet to a turbine receiving steam at 350°C and 2 MPa.

8.16E Evaluate the Gibbs function g and the Helmoltz function a at the inlet to a turbine receiving steam at 700°F and 300 psia.

8.17 An ideal gas is compressed isothermally at temperature T from volume V_1 to V_2. What is the ratio of the reversible to the actual work done? The surroundings are at T_0.

8.18 An ideal gas is compressed isobarically from volume V_1 to volume V_2. What is the ratio of the reversible to the actual work done? The surroundings are at T_0.

8.19 Starting with Eq. (8.34), derive a relation for the efficiency of a steam turbine $\eta_t' = w_a/w_{rev}$ operating between T and P at the inlet and P_{out} at the exhaust. Use the same assumptions made in Sec. 5.11.1 in determining the turbine efficiency. Discuss the result, and compare it with the result normally used for turbine efficiency.

8.20S Calculate the irreversibility per unit mass for an adiabatic turbine with an inlet at 4 MPa and 350°C and an outlet pressure of 0.1 MPa. The turbine efficiency is 0.80, and the working fluid is water.

8.20E Calculate the irreversibility per unit mass for an adiabatic turbine with an inlet at 600 psia and 700°F and an outlet pressure of 14.7 psia. The turbine efficiency is 0.80, and the working fluid is water.

8.21S The tank shown in Fig. P8.21 initially contains 0.3 kg of an ideal gas, N_2, at 530 K and an adiabatic weightless and frictionless piston. The 0.12-m³ tank is connected to an infinite source of air at 530 K and 0.7 MPa. Air

Figure P8.21

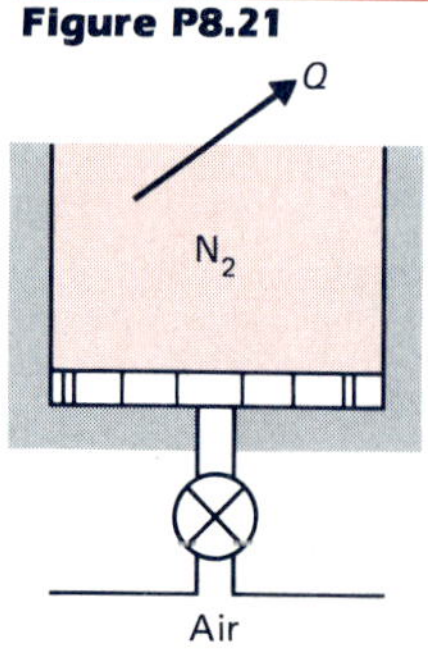

enters the tank until the temperature of the air is 560 K, and the final mass of air is 0.15 kg. Calculate the irreversibility of the process, assuming there is 8.6 kJ of heat transfer *from* the N_2 to the surroundings at 25°C.

8.21E The tank shown in Fig. P8.21 initially contains 0.7 lbm of an ideal gas, N_2, at 950°R and an adiabatic weightless and frictionless piston. The 4.2-ft³ tank is connected to an infinite source of air at 950°R and 100 psia. Air enters the tank until the temperature of the air is 1000°R, and the final mass of air is 0.35 lbm. Calculate the irreversibility of the process, assuming there is 8.2 Btu of the heat transfer *from* the N_2 to the surroundings at 75°F.

8.22S A steam turbine receives superheated steam at 350°C and 2 MPa and exhausts to a pressure of 8 kPa. The isentropic turbine efficiency is 75 percent.

(*a*) What fraction of the availability is produced as work by the turbine?

(*b*) What is the second law efficiency of the turbine?

8.22E A steam turbine receives superheated steam at 700°F and 300 psia and exhausts to a pressure of 1 psia. The isentropic turbine efficiency is 75 percent.

(*a*) What fraction of the availability is produced as work by the turbine?

(*b*) What is the second law efficiency of the turbine?

8.23 What is the irreversibility of the turbine in Prob. 8.22?

8.24S A solar tower generates steam at 515°C and 1 MPa and transfers energy to a mineral oil for storage at 304°C. Calculate the availability of the tower-generated steam and of steam produced at 300°C and 1 MPa from the oil via a steam generator.

8.24E A solar tower generates steam at 1000°F and 150 psia and transfers energy to a mineral oil for storage at 600°F. Calculate the availability of the tower-generated steam and of steam produced at 500°F and 150 psia from the oil via a steam generator.

8.25S Ammonia vapor at −10°C and 100 kPa is to enter an adiabatic compressor and be compressed to a final pressure of 250 kPa. The compressor has an isentropic efficiency of 70 percent.

(*a*) Find the actual compressor work.

(*b*) Calculate the reversible work for the prescribed process.

(*c*) What is the flow availability of the ammonia entering the compressor?

(*d*) Find the flow availability of the ammonia leaving the compressor.

8.25E Ammonia vapor at 15°F and 14.6 psia is to enter an adiabatic compressor and be compressed to a final pressure of 35 psia. The compressor has an isentropic efficiency of 70 percent.

(*a*) Find the actual compressor work.

(*b*) Calculate the reversible work for the prescribed process.

(*c*) What is the flow availability of the ammonia entering the compressor?

(*d*) Find the flow availability of the ammonia leaving the compressor.

8.26S Ammonia at −10°C and 100 kPa is to enter an isothermal compressor and be compressed to a final pressure of 250 kPa. The compressor has an isothermal efficiency of 70 percent. The heat transfer in the ideal (isothermal) compressor is −150 kJ/kg.

(*a*) Find the actual compressor work.

(*b*) What is the reversible work for the prescribed process?

(*c*) Calculate the flow availability of the ammonia entering the compressor.

(*d*) Find the flow availability of the ammonia leaving the compressor.

8.26E Ammonia at 15°F and 14.7 psia is to enter an isothermal compressor and be compressed to a final pressure of 35 psia. The compressor has an isothermal efficiency of 70 percent. The heat transfer in the ideal (isothermal) compressor is −65 Btu/lbm.

(*a*) Find the actual compressor work.

(*b*) What is the reversible work for the prescribed process?

(*c*) Calculate flow availability of the ammonia entering the compressor.

(*d*) Find flow availability of the ammonia leaving the compressor.

8.27 A process heater is to increase the temperature of a working fluid by heat transfer at a rate $\dot{Q}$ from a reservoir at temperature T_R. The working fluid receives the heat transfer in a constant-pressure process. Derive an expression for the second law efficiency for this process.

8.28 The efficiency of a given heat engine as defined in Sec. 5.12 is a first law efficiency; i.e., it describes the work output relative to the heat transfer input. Discuss how you would calculate the second law efficiency of a heat engine, and derive an expression for such an efficiency for an air-standard Stirling engine.

8.29 Derive an expression for the second law efficiency of a simple closed Brayton cycle operating between maximum and minimum temperatures of T_2 and T_4.

8.30 One form of the Clausius statement of the second law is: "It is not possible to transfer heat from a lower- to a higher-temperature reservoir without doing work on the system." Given this statement, consider an absorption cooling machine (see Sec. 7.4.2 for a description).
(*a*) Given that only heat transfer—no work—is necessary to operate the cycle, does it violate the Clausius statement of the second law?
(*b*) How would you characterize the second law efficiency of an absorption cooling system?

8.31S Find the second law efficiency of a Rankine steam cycle operating at a boiler pressure of 6 MPa with a superheat temperature of 400°C, a condenser pressure of 8 kPa, and an isentropic turbine and pump.

8.31E Find the second law efficiency of a Rankine steam cycle operating at a boiler pressure of 800 psia with a superheat temperature of 800°F, a condenser pressure of 1 psia, and an isentropic turbine and pump.

8.32S Derive an expression for the second law efficiency of a general refrigeration cycle. Using this expression, find the second law efficiency of a vapor compression refrigeration cycle using refrigerant 12 with an evaporator temperature of 5°C and a condenser temperature of 40°C. The cycle has no superheat at the evaporator exit and no subcooling at the condenser exit. The compressor has an efficiency of 65 percent.

8.32E Derive an expression for the second law efficiency of a general refrigeration cycle. Using this expression, find the second law efficiency of a vapor compression refrigeration cycle using refrigerant 12 with an evaporator temperature of 40°F and a condenser temperature of 100°F. The cycle has no superheat at the evaporator exit and no subcooling at the condenser exit. The compressor has an efficiency of 65 percent.

References

1. J. E. Ahern, *The Exergy Method of Energy Systems Analysis,* Wiley, New York, 1980.
2. A. Bejan, *Entropy Generation through Heat and Fluid Flow,* Wiley-Interscience, New York, 1982.

3. J. H. Keenan, *Thermodynamics,* Wiley, New York, 1941.
4. L. C. Witte and N. Shamsundar, "A Thermodynamic Efficiency Concept for Heat Exchanger Devices," *ASME Journal of Engineering for Power*, vol. 105, January 1983, pp. 199–203.
5. M. J. Moran and E. Sciubba (eds.), *Second Law Analysis of Energy Systems*, ASME, New York, 1987.
6. R. F. Boehm and N. Lior (eds.), *Second Law Analysis of Heat Transfer in Energy Systems*, HTD-Vol-80, ASME, New York, 1987.

9

General Property Relations and Equations of State

When a greybearded scientist says something is possible, believe him; when he says that it's impossible, he's very likely wrong.
Arthur C. Clarke

Solar collectors used for energy input to Stirling cycle engines for power generation. (*Sandia National Laboratories.*)

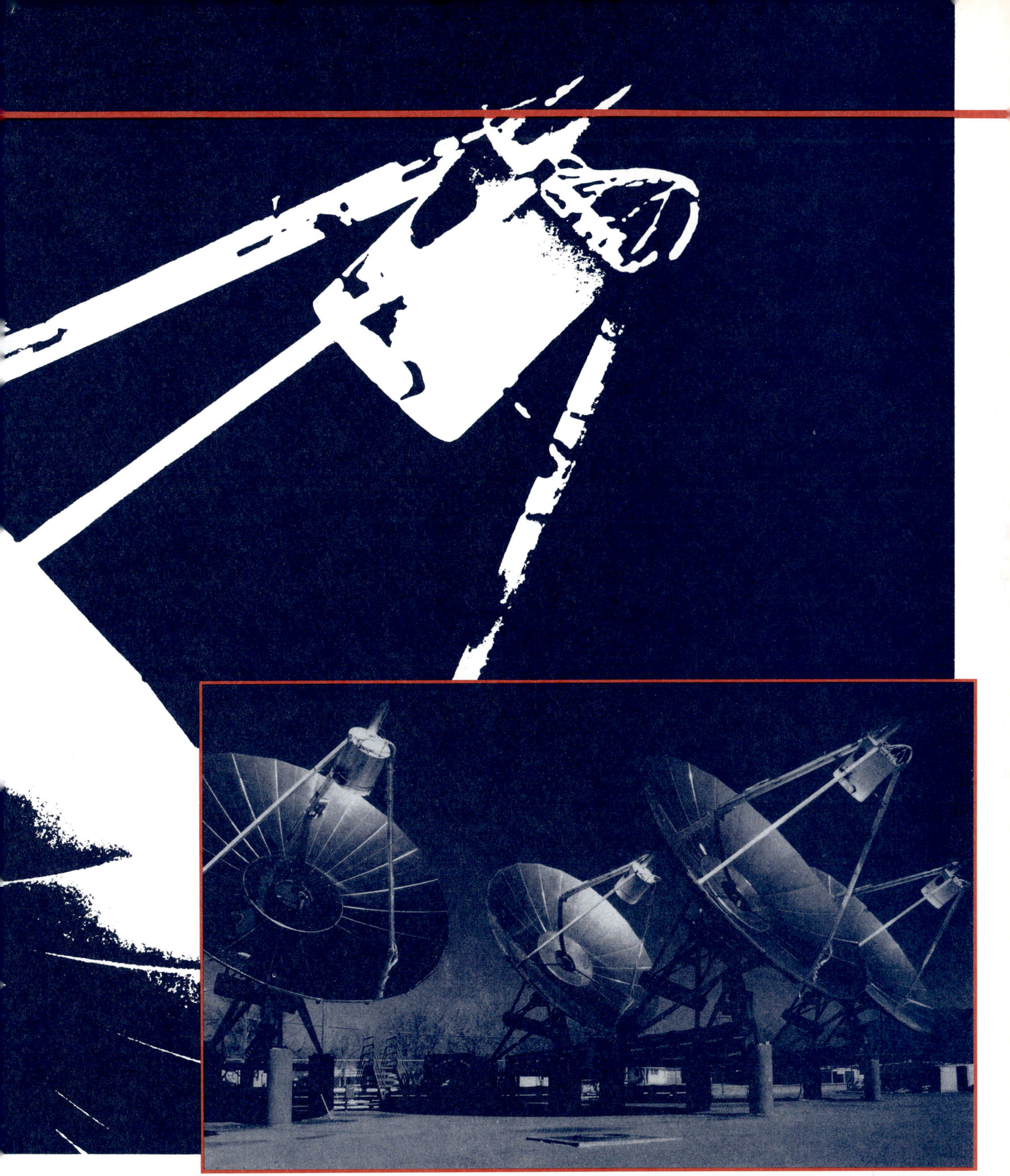

9.1 Introduction

There are many reasons for finding relationships among thermodynamic properties. First, it is, of course, desirable to minimize the amount of laboratory work, since obtaining reliable and accurate thermodynamic property data requires meticulous measurements. If a minimum set of property data is obtained and relationships are available for calculation of other properties from these measured values, much tedious work can be avoided. A second reason for developing relations among properties is that the engineer or scientist may require data on the thermodynamic properties of a particular substance for which little or no data are available in the literature. In such a case, it is certainly much easier to use the few existing data to generate other property values than to measure the required values. It is often surprising to students and practicing engineers alike that complete property data has not been measured for many common substances, and estimates must be employed.

Of course, laboratory data are usually taken in terms of the variables P, v, and T, usually as either $P(v, T)$ or $v(P, T)$. Seldom are thermodynamic property data available primarily in the form $u(s, v)$ or $h(s, P)$. We develop relations later in this chapter for finding u, h, s, g, and a in terms of data in P, v, and T.

In this chapter, we develop the general relations among thermodynamic properties, and we apply these relations to some common situations. In so doing, we see how various tables, such as those for ideal gases and for steam, have been generated from limited experimental property measurements.

9.2 Relations among Properties

9.2.1 Fundamental Equations and Maxwell's Relations

We see in Chap. 2 that for a pure simple compressible substance, the state of a system is defined by specifying any two independent intensive thermodynamic properties. It follows that all other properties are thus fixed whenever two properties are specified. In other words, any two independent properties can be chosen, and all other dependent properties are then fixed.

Let us, for example, fix the two independent properties s and v. Then the internal energy at this state is fixed at the value $u(s, v)$. Let us see what this implies. The total differential of u as a function of its partial derivatives is then [see Eq. (H.13)]

$$du = \left(\frac{\partial u}{\partial s}\right)_v ds + \left(\frac{\partial u}{\partial v}\right)_s dv \tag{9.1}$$

Now, recall Eq. (5.43):

$$du = T\,ds - P\,dv \tag{9.2}$$

Note that neither equation contains any assumptions about reversible processes, adiabatic systems, etc. Both are completely general, Eq. (9.1) being a mathematical relationship between total and partial derivatives and Eq. (9.2) being a general result of the first law providing a relation among thermodynamic properties of a simple system containing a pure substance.

Comparing the two equations, we find the identities

$$T = \left(\frac{\partial u}{\partial s}\right)_v \tag{9.3}$$

and

$$P = -\left(\frac{\partial u}{\partial v}\right)_s \tag{9.4}$$

These were the defining relations for T and P, stated in Chap. 5 in Eqs. (5.9) and (5.23). Thus, if data were available for internal energy u as a function of s and v, then *all* the thermodynamic properties at a given state could be calculated. When s and v are given, u is computed from the data; the corresponding T and P are found from Eqs. (9.3) and (9.4); and the enthalpy h, the Helmholtz function a, and the Gibbs function g are found from their definitions, $h = u + Pv$, $a = u - Ts$, and $g = h - Ts$.

It appears that the function $u(s, v)$ contains the necessary information to find all other properties at any given state. It is thus called a *fundamental equation,* or *characteristic thermodynamic function,* for the system. Is this true of *any* dependent property in terms of two independent properties? It is not. The reason is that if properties other than s and v are chosen as the independent variables with u as the dependent variable, then we find it impossible to develop a set of equations equivalent to Eqs. (9.3) and (9.4), since we cannot match the partial derivatives with the properties in the equation of state. We will have introduced some degrees of freedom into the equations; i.e., we will have fewer equations than unknown properties.

Example Problem 9.1

Develop the relations among properties starting with a known function $u(T, v)$ and Eq. (9.2).

Solution

The total derivative of the given function is

$$du = \left(\frac{\partial u}{\partial T}\right)_v dT + \left(\frac{\partial u}{\partial v}\right)_T dv$$

We also have, from Eq. (9.2),

$$du = T\,ds - P\,dv$$

Comparison of these two equations provides no way to relate property T or P to the derivatives of u which are available to us. Thus $u(T, v)$ is *not* a fundamental equation.

How do we determine what sets of independent and dependent variables *are* fundamental equations? We must choose sets having dependent and independent variables that can be matched with a set of variables in a general equation of state such as Eq. (9.2). Two such equations are developed in Sec. 5.5.1:

$$du = T\,ds - P\,dv \tag{9.5a}$$

$$dh = T\,ds + v\,dP \tag{9.5b}$$

With the definitions of the Helmholtz and Gibbs functions in Sec. 8.5 and Eqs. (9.5*a*) and (9.5*b*), two other equations are

$$da = du - T\,ds - s\,dT = -P\,dv - s\,dT \tag{9.5c}$$

$$dg = dh - T\,ds - s\,dT = v\,dP - s\,dT \tag{9.5d}$$

Thus, it appears that the fundamental equations must relate $u = f(s, v)$, $h = f(s, P)$, $a = f(v, T)$, and $g = f(P, T)$.

Let us repeat the development used in Eqs. (9.1) through (9.4), but we start with $h(s, P)$ and Eq. (9.5*b*). The total derivative dh is then

$$dh = \left(\frac{\partial h}{\partial s}\right)_P ds + \left(\frac{\partial h}{\partial P}\right)_s dP \tag{9.6}$$

However, without restriction, Eq. (8.5*b*) gives for a pure substance

$$dh = T\,ds + v\,dP \tag{9.7}$$

and by inspection we find

$$T = \left(\frac{\partial h}{\partial s}\right)_P \tag{9.8}$$

and

$$v = \left(\frac{\partial h}{\partial P}\right)_s \tag{9.9}$$

Now, if data are available for $h(s, P)$, we could, for a given s and P, find the corresponding T, v, $u = h - Pv$, $a = u - Ts$, and $g = h - Ts$.

Note that some insight into property behavior can be gleaned from the relations derived above. For example, because v is always a positive quantity, Eq. (9.9) shows that enthalpy h will always increase with pressure in an isentropic process. Similarly, Eq. (9.4) shows that internal energy u will always decrease with increasing volume in an isentropic process.

One additional piece of information has been developed. Note that Eqs. (9.3) and (9.8) both provide relations for temperature. Equating the right-hand sides of these equations results in the identity

$$\left(\frac{\partial u}{\partial s}\right)_v = \left(\frac{\partial h}{\partial s}\right)_P \tag{9.10}$$

Equations of the form of Eq. (9.10) occur because properties are exact differential quantities, and relations between the properties thus follow the relations developed for exact differentials in App. H. For example, Eq. (H.29*a*) can be invoked to develop a variety of relations among the properties. Now, applying Eq. (H.29*a*) to Eqs. (9.5*a*) to (9.5*d*) in turn results in

$$\left(\frac{\partial T}{\partial v}\right)_s = -\left(\frac{\partial P}{\partial s}\right)_v \tag{9.11a}$$

$$\left(\frac{\partial T}{\partial P}\right)_s = \left(\frac{\partial v}{\partial s}\right)_P \tag{9.11b}$$

$$\left(\frac{\partial P}{\partial T}\right)_v = \left(\frac{\partial s}{\partial v}\right)_T \tag{9.11c}$$

$$\left(\frac{\partial v}{\partial T}\right)_P = -\left(\frac{\partial s}{\partial P}\right)_T \tag{9.11d}$$

These equations are called *Maxwell's relations* after James Clerk Maxwell (1831–1879). Other contributions by Maxwell are discussed in Section 13.2.2. The Maxwell relations hold for the properties of a simple compressible substance. There are many other such relations when other work modes are considered.

Example Problem 9.2

Using Maxwell's relations, find an expression for the change in entropy with volume for an isothermal process involving an ideal gas.

Solution

Applying Eq. (9.11*c*), we find

$$\left(\frac{\partial s}{\partial v}\right)_T = \left(\frac{\partial P}{\partial T}\right)_v = \frac{R}{v}$$

so that the entropy changes inversely with the volume in such a process.

Comments

This result can also be obtained from Eq. (5.52), which is

$$ds = c_v \frac{dT}{T} + R\frac{dv}{v}$$

For a constant-temperature process of an ideal gas, the same result for $(\partial s/\partial v)_T$ is obtained as by the use of the Maxwell relation.

9.2.2 Clapeyron Equation

As a useful application of the Maxwell relations, consider Eq. (9.11*c*), repeated here:

$$\left(\frac{\partial P}{\partial T}\right)_v = \left(\frac{\partial s}{\partial v}\right)_T$$

If we consider an isothermal phase-change process, say, for the evaporation of a saturated liquid, then the right-hand side is simply

$$\frac{s_g - s_l}{v_g - v_l} = \frac{s_{lg}}{v_{lg}} \tag{9.12}$$

Also in a phase change, pressure and temperature are dependent on only each other, and the derivative of P with respect to T becomes an ordinary derivative rather than a partial derivative. Equation (9.11*c*) can now be written as

$$\left(\frac{dP}{dT}\right)_{\text{sat}} = \frac{s_{lg}}{v_{lg}} \tag{9.13}$$

For an isothermal phase change, Eq. (9.5*b*) gives $h_{lg} = T_{\text{sat}} s_{lg}$, and Eq. (9.13) finally becomes

$$\left(\frac{dP}{dT}\right)_{\text{sat}} = \frac{h_{lg}}{T_{\text{sat}} v_{lg}} \tag{9.14}$$

This equation, called the *Clapeyron equation,* provides a useful check on P-v-T data for saturated substances when h_{lg} is also available; or h_{lg} can be computed from easily measurable P-v-T data for saturated liquid and vapor. In addition, consider the P-T curve for a given substance, such as presented for water in Fig. 3.8. Equation (9.14) can be used to construct the P-T saturation curve by integrating between any two saturation states to obtain

$$(P_2 - P_1)_{\text{sat}} = \int_{T_{1,\text{sat}}}^{T_{2,\text{sat}}} \frac{h_{lg}}{Tv_{lg}}\, dT \tag{9.15}$$

Because all the properties in the integrand are temperature-dependent, the integral must be carried out numerically.

Example Problem 9.3

Discuss the shape of the solid-liquid equilibrium curve on a P-T diagram for a substance that contracts on freezing in comparison with the curve for a substance that expands on freezing.

Solution

The Clapeyron equation, Eq. (9.14), can be rewritten for a solid-liquid phase change in the form

$$\left(\frac{dP}{dT}\right)_{\text{melt}} = \frac{h_l - h_s}{T_{\text{melt}}(v_l - v_s)} = \frac{h_{sl}}{T_{\text{melt}}v_{sl}}$$

This equation predicts that the slope of the P-T curve along the solid-liquid phase-change curve for any material which contracts upon freezing ($v_l - v_s = v_{sl} > 0$) will be positive (since the heat of fusion h_{sl} is always positive). For a substance that expands on freezing ($v_{sl} < 0$), dP/dT will be negative. As noted in Chap. 3, water expands as it freezes. The slope is observed in Fig. 3.14.

Table C.13 presents some selected data for the enthalpy of fusion and sublimation.

9.2.3 Generation of Property Tables

Tables of properties and their use are discussed in Chap. 3. Here we briefly note how such tables are generated from available experimental data.

Probably the most accurate and complete tables for the properties of steam presently available are those of the U.S. National Bureau of Standards (now the National Institute for Standards and Technology, NIST)/Canadian National Research Council [1]. In these tables, the Helmholtz function a is used as the characteristic function. Values of a are readily obtained from P-v-T data, which is the most easily measured set, as will be seen.

It is assumed that a particular equation of state (an expanded form of an equation of state developed to model nearly hard ellipsoid molecules) is valid for the value of a as a function of density ρ (rather than specific volume) and temperature. This is a fundamental equation, as shown in Sec. 9.2.1. The value of the Helmholtz function given by the equation of state is called $a_{\text{basis}}(\rho, T)$. However, comparison of the chosen equation of state with experimental data shows deviations, particularly at high density or near the critical point. A correction term, called the *residual function* $a_{\text{resid}}(\rho, T)$, is thus obtained by fitting a function composed of 40 terms to the complete range of data. Finally a term is added to account for ideal gas behavior at very low densities where the other terms approach zero. This term, $a_{\text{ideal}}(T)$, is a function of temperature only. The Helmholtz function is thus given by

$$a(\rho, T) = a_{\text{basis}}(\rho, T) + a_{\text{resid}}(\rho, T) + a_{\text{ideal}}(T) \tag{9.16}$$

Once this function is constructed, all other properties can be found by taking the appropriate derivatives, and complete property tables can be constructed. The results can be validated by comparison of computed property values (say, c_v or c_P) with experimentally determined values. This is a severe test, because some of the computed properties depend on second derivatives of the Helmholtz function and can be considerably more inaccurate than the function itself. The comparisons presented in Ref. 1 show excellent agreement.

All the above argument presupposes that values of a can be found from P-v-T data so that the curve fits can be produced. This is done as follows. Values of u and s must be determined at each state, so that $a = u - Ts$ can be computed. Suppose that a graph of P versus T is constructed, with constant-volume lines superimposed from the measured P-v-T data. (Such data are obtained by placing a fixed mass into a container of known volume and measuring the pressure as the temperature of the fixed mass is varied. The experiment is then repeated for different fixed masses.) The graph will appear as in Fig. 9.1, which shows a portion of the superheat region for steam. From such graphical presentations, values of $P(T, v)$ are easily obtained, as are certain derivatives such as $(\partial P/\partial T)_v$.

Let us now develop a method for computing the internal energy from P-v-T data. This can be done by letting u be a function of T and v, as $u(T, v)$. The total derivative is then

$$du = \left(\frac{\partial u}{\partial T}\right)_v dT + \left(\frac{\partial u}{\partial v}\right)_T dv \tag{9.17}$$

If we also assume $s(T, v)$, then

$$ds = \left(\frac{\partial s}{\partial T}\right)_v dT + \left(\frac{\partial s}{\partial v}\right)_T dv \tag{9.18}$$

Substituting Eq. (9.18) into Eq. (9.5a) gives

$$du = T\left(\frac{\partial s}{\partial T}\right)_v dT + \left[T\left(\frac{\partial s}{\partial v}\right)_T - P\right] dv \tag{9.19}$$

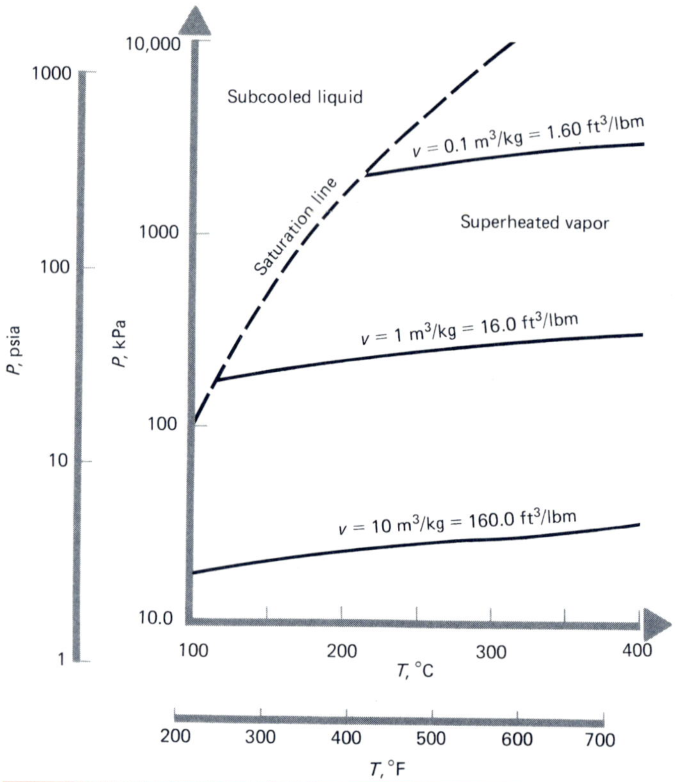

Figure 9.1 *P-T* diagram for superheated steam.

Comparing Eq. (9.19) with Eq. (9.17) shows that

$$\left(\frac{\partial u}{\partial T}\right)_v = T\left(\frac{\partial s}{\partial T}\right)_v \tag{9.20}$$

and

$$\left(\frac{\partial u}{\partial v}\right)_T = T\left(\frac{\partial s}{\partial v}\right)_T - P \tag{9.21}$$

The identity in Eq. (9.20) is not used here; nonetheless, we note that $(\partial u/\partial T)_v = c_v$. Also note that the Maxwell relation, Eq. (9.11*c*), can be used in the second identity, so that the relation in Eq. (9.21) becomes

$$\left(\frac{\partial u}{\partial v}\right)_T = T\left(\frac{\partial P}{\partial T}\right)_v - P \tag{9.22}$$

And substituting Eq. (9.22) into Eq. (9.17) results in

$$du = c_v\,dT + \left[T\left(\frac{\partial P}{\partial T}\right)_v - P\right]dv \tag{9.23}$$

To find the change in u between any two states, say, between a reference state and a final state, we can choose any path of integration between the states. The integrated form of Eq. (9.23) becomes

$$u - u_{ref} = \int_{T_{ref}}^{T} c_v\, dT + \int_{v_{ref}}^{v} \left[T\left(\frac{\partial P}{\partial T}\right)_v - P \right] dv \tag{9.24}$$

If we choose a reference state for which steam behaves as an ideal gas ($P \to 0$) and integrate along a path on Fig. 9.2 at constant v (path AB) until we intercept the desired constant-T line at point B, then the value of c_v along path AB is a function of T only. Along isotherm BC the first term is zero. The first term in Eq. (9.24) can thus be evaluated by using only ideal gas properties for water vapor. The second term has a zero value along the constant-volume line and can be evaluated along BC given only detailed $P(v, T)$ data for steam. We have reached our goal of finding u by using only P-v-T data. Numerical integration will probably be required, and the data must be quite precise because derivatives of the data appear in the relations we are using.

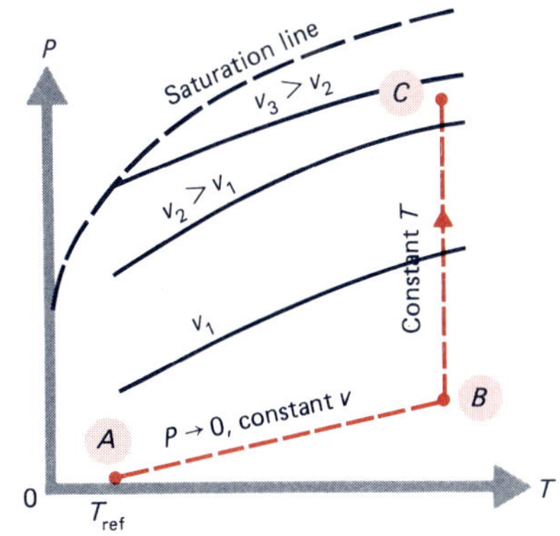

Figure 9.2 *P-T diagram.*

To determine the entropy at any state, a similar approach is used. The resulting relation for s is

$$s = s_{ref} + \left(\int_{T_{ref}}^{T} \frac{c_v}{T}\, dT \right)_{v_{ref}} + \left[\int_{v_{ref}}^{v} \left(\frac{\partial P}{\partial T}\right)_v dv \right]_T \tag{9.25}$$

Because both Eqs. (9.24) and (9.25) require a reference state as $P \to 0$, they may not be completely useful in constructing a property table, which is usually based on some other reference condition, such as the triple point. The equations are easily modified to find the difference in s or u between any two states (one of which can be a nonideal gas reference state) by simply subtracting the equation for state 2 from the equation for state 1. Then Eqs. (9.24) and (9.25) become

$$u_2 - u_1 = \left\{ \int_{v_1}^{v=\text{const}} \left[T\left(\frac{\partial P}{\partial T}\right)_v - P \right] dv \right\}_{T=T_1} + \left(\int_{T_1}^{T_2} c_v\, dT \right)_{v=\text{const}} + \left\{ \int_{v=\text{const}}^{v_2} \left[T\left(\frac{\partial P}{\partial T}\right)_v - P \right] dv \right\}_{T=T_2} \tag{9.26}$$

and

$$s_2 - s_1 = \left(\int_{T_1}^{T_2} \frac{c_v}{T}\, dT \right)_{v=\text{const}} + \left[\int_{v_1}^{v=\text{const}} \left(\frac{\partial P}{\partial T}\right)_v dv \right]_{T=T_1} + \left[\int_{v=\text{const}}^{v_2} \left(\frac{\partial P}{\partial T}\right)_v dv \right]_{T=T_2} \tag{9.27}$$

Now u and s relative to any reference point can be determined from data for $P(v, T)$, and the value of the Helmholtz function a can be found from $a = u - Ts$. The complete tables can now be constructed.

9.3 Principle of Corresponding States

In Chap. 3, the *principle of corresponding states* is expressed for a simple compressible substance as follows:

> All substances obey the same equation of state expressed in terms of the reduced properties.

This principle is an empirical statement based on some theoretical justification, but it is not a "law." That is, the results of its application are not—and should not be expected to be—exact. Indeed, we have seen by comparison of the compressibility factor at the critical point, which should have the same value for all substances if the principle is valid, that considerable deviations are exhibited by some substances (see Table C.4). Thus, although the principle provides a basis for examining the *behavior* of thermodynamic properties, we still must rely on measurements if we wish exact *values* for the properties.

We now use van der Waals' equation as a means of examining substance behavior.

9.3.1 Some Observations Based on van der Waals' Equation

As discussed in Sec. 3.5, van der Waals' equation of state [Eq. (3.26)] is

$$P = \frac{RT}{v-b} - \frac{a}{v^2} \tag{9.28}$$

To find the values of constants a and b for a particular substance, we note that the critical isotherm has a horizontal inflection point as it passes through the critical point (see, for example, Fig. 3.5). Thus, the first and second partial derivatives of P with respect to v must be zero at the critical point for any universally valid equation of state (try the ideal gas law, which fails badly around the critical point). These derivatives, evaluated at the critical point by using van der Waals' equation, are

$$\left(\frac{\partial P}{\partial v}\right)_{T=T_{cr}} = 0 = -\frac{RT_{cr}}{(v_{cr}-b)^2} + \frac{2a}{v_{cr}^3} \tag{9.29}$$

and

$$\left(\frac{\partial^2 P}{\partial v^2}\right)_{T=T_{cr}} = 0 = -\frac{2RT_{cr}}{(v_{cr}-b)^3} - \frac{6a}{v_{cr}^4} \tag{9.30}$$

Solving simultaneously and using Eq. (9.28) to eliminate v_{cr} result in

$$a = \frac{27R^2T_{cr}^2}{64P_{cr}} \quad \text{and} \quad b = \frac{RT_{cr}}{8P_{cr}} \tag{9.31}$$

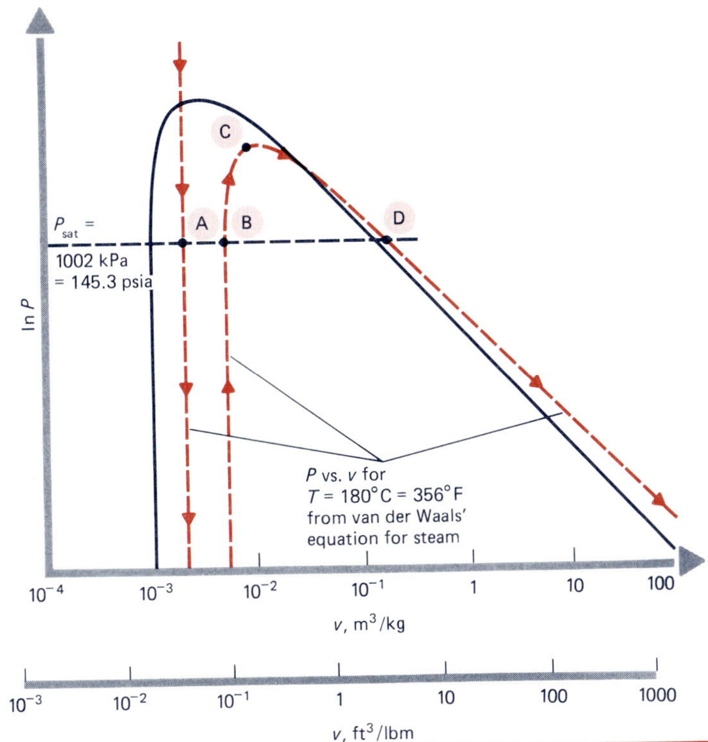

Figure 9.3 *P-v* **diagram and van der Waal's equation for water.**

Thus, van der Waals' equation follows the principle of corresponding states, because a single relation is provided for all substances which is expressed in terms of the critical properties.

If van der Waals' equation is valid, it should work throughout the entire region of properties for which a given substance is a fluid, including the saturation region. It is not *universally* valid; i.e., it is not valid, for example, for the solid phase, because it is based on a physical model for a fluid. Let us check that out. We can write Eq. (9.28) as

$$Pv^3 - (bP + RT)v^2 + av - ab = 0 \tag{9.32}$$

Note that this is a cubic equation in v, which can have three roots; that is, v can have three possible values at a given value of T and P. Let us plot the P-v curve for an isotherm that passes through the two-phase region for water ($a = 1.7034$ kPa·m^6/kg^2 and $b = 1.689 \times 10^{-3}$ m^3/kg), say, at $T = 180°C = 356°F$. The resulting curve is shown in Fig. 9.3.

The curve has three values of v at $P = 1002$ kPa $= 145.3$ psia, the saturation pressure at $T = 180°C = 356°F$. The smallest and largest values lie near the values for saturated liquid and saturated vapor, respectively. At the saturation pressure, van der Waals' equation predicts a specific volume for the saturated liquid, point A, as 0.00239 m^3/kg = 0.0383 ft^3/lbm, compared with a value from the steam tables of 0.001127 m^3/kg = 0.01806 ft^3/lbm. For the saturated vapor, the equation predicts (point D) that $v_g = 0.202$ m^3/kg = 3.24 ft^3/lbm, compared with a table value of 0.1940 m^3/kg = 3.108 ft^3/lbm. Between these values, the curve behaves much differently from the curve found experimentally. That is, between the minimum of the isotherm from van der Waals' equation (which is negative in value and so cannot be plotted on the log scale) and point C, the volume increases as the pressure increases, which is in contradiction to our observations. Keep in mind that the equation is valid only for either the vapor or the liquid, and it fails in the two-phase region. We interpret the predicted P-v curve to mean that between the minimum of the isotherm and point C, the vapor in the two-phase mixture is mechanically unstable, and slight changes in pressure cause some of the vapor to change phase. Thus, the third solution for specific volume at $P_{sat} = 1002$ kPa, $T = 180°C$, v_B, is not an equilibrium state.

Although the numerical results of this exercise show considerable error in the prediction of saturation properties when compared with experimental data, study of van der Waals' equation does give some insight into how equations of state might be constructed and the characteristics we might expect them to have.

van der Waals' equation can have greater practical use by finding values of a and b that minimize the difference from experimental data, rather than using the values given by Eq. (9.31), which provide an exact fit at the critical point only.

9.3.2 Expanded Use of the Principle of Corresponding States

In Sec. 3.5, the compressibility factor $z(T_r, P_r)$ is introduced, and its use in determining the P-v-T behavior of any substance for which critical property data are available is discussed. The principle of corresponding states implies that *any* thermodynamic property for any substance should be describable in terms of its critical properties. So we should be able to provide charts for u, h, s, etc. that are in terms of reduced temperature and pressure, for example.

Two-Parameter Reduced-Property Charts

The compressibility charts (Fig. C.1) relate P-v-T data for any substance in terms of the critical constants T_{cr} and P_{cr}. The latter two properties are used to find the *reduced* temperature and pressure $T_r = T/T_{cr}$ and $P_r = P/P_{cr}$, and the value of z is then presented as a function of those two parameters. The charts are constructed by averaging the P-v-T behavior of 30 gases, with a resulting error in the range of 4 to 6 percent except near the critical point. On the charts, the reduced specific volume v_r is cross-plotted. However, the experimental determination of v_{cr} is quite difficult to achieve, so the reduced volume shown on the charts is a *calculated* value v_r' defined by the relation

$$v_r' = \frac{v}{RT_{cr}/P_{cr}} \tag{9.33}$$

The charts are thus usable if we know only the critical pressure and temperature.

Once the compressibility chart has been generated, it can be used to find the other thermodynamic properties in reduced form. Rather than presenting the property itself in terms of reduced pressure and temperature, it is convenient to find the deviation of the property from that of an ideal gas in the same state. The ideal gas property can be found from either the gas tables (such as Tables D.2 through D.8 and Tables E.2 through E.8 for air, argon, nitrogen, oxygen, water vapor, carbon dioxide, and carbon monoxide) or by computations using ideal gas relations.

Let us see how the reduced-property chart for one of the properties, enthalpy, would be developed. Since we wish to present the data as a function of T and P, we choose as a starting point the total derivative of $h(T, P)$

$$\begin{aligned} dh &= \left(\frac{\partial h}{\partial P}\right)_T dP + \left(\frac{\partial h}{\partial T}\right)_P dT \\ &= \left(\frac{\partial h}{\partial P}\right)_T dP + c_P\, dT \end{aligned} \tag{9.34}$$

Note that $dh = T\,ds + v\,dP$ [Eq. (9.5*b*)]. If we assume that entropy is a function of temperature and pressure, then the total derivative becomes

$$ds = \left(\frac{\partial s}{\partial T}\right)_P dT + \left(\frac{\partial s}{\partial P}\right)_T dP \tag{9.35}$$

Substituting Eq. (9.35) to eliminate ds in Eq. (9.5*b*) gives

$$\begin{aligned} dh &= T\left[\left(\frac{\partial s}{\partial T}\right)_P dT + \left(\frac{\partial s}{\partial P}\right)_T dP\right] + v\,dP \\ &= T\left(\frac{\partial s}{\partial T}\right)_P dT + \left[v + T\left(\frac{\partial s}{\partial P}\right)_T\right] dP \end{aligned} \tag{9.36}$$

Comparison with Eq. (9.34) shows that

$$c_P = T\left(\frac{\partial s}{\partial T}\right)_P$$

$$\left(\frac{\partial h}{\partial P}\right)_T = v + T\left(\frac{\partial s}{\partial P}\right)_T \tag{9.37}$$

But from the Maxwell relation, Eq. (9.11d), we know that $(\partial s/\partial P)_T = -(\partial v/\partial T)_P$, so we can write Eq. (9.36) as

$$dh = c_P\,dT + \left[v - T\left(\frac{\partial v}{\partial T}\right)_P\right] dP \tag{9.38}$$

We have now generated an expression for the change in enthalpy as a function of P-v-T data and the specific heat at constant pressure c_P. Since our goal is to develop expressions for h as a function of reduced properties T_r and P_r, we can now use the equation of state in terms of the compressibility factor [Eq. (3.33)] and reduced properties as

$$v = \frac{zRT_rT_{\text{cr}}}{P_rP_{\text{cr}}} \tag{9.39}$$

and its partial derivative

$$\left(\frac{\partial v}{\partial T}\right)_P = \frac{R}{P_rP_{\text{cr}}}\left[z + T_r\left(\frac{\partial z}{\partial T_r}\right)_{P_r}\right] \tag{9.40}$$

Now, substituting Eqs. (9.40) and (9.39) into (9.38) results in

$$dh = c_PT_{\text{cr}}\,dT_r - \frac{RT_{\text{cr}}T_r^2}{P_r}\left(\frac{\partial z}{\partial T_r}\right)_{P_r} dP_r \tag{9.41}$$

or, by integrating between a reference state at $P_r = 0$ and any final state f,

$$\frac{h_f - h_{P_r=0}}{RT_{\text{cr}}} = \int_{T_{r,\text{ref}},\,P_r=0}^{T_{r,f}} \frac{c_P}{R}\,dT_r - \int_{P_r=0}^{P_{r,f}} \frac{T_r^2}{P_r}\left(\frac{\partial z}{\partial T_r}\right)_{P_r} dP_r \tag{9.42}$$

Suppose for a moment that we consider an ideal gas. For such a gas, $z = 1$ and $(\partial z/\partial T_r)_{P_r} = 0$, and Eq. (9.42) reduces to

$$\frac{h_f - h_{P_r=0}}{RT_{\text{cr}}} = \int_{T_{r,\text{ref}},P_r=0}^{T_{r,f}} \frac{c_P}{R}\,dT_r \tag{9.43}$$

At $P_r = 0$, all real gases behave as ideal gases, so $h_{P_r=0}$ is the same for the real and ideal gases. Thus, if we subtract Eq. (9.42) from (9.43), we find

$$\begin{aligned}\frac{(h_{\text{ideal}} - h)_f}{RT_{\text{cr}}} &= \int_{T_{r,\text{ref}}P_r=0}^{P_{r,f}} \frac{T_r^2}{P_r}\left(\frac{\partial z}{\partial T_r}\right)_{P_r} dP_r \\ &= \int_{P_r=0}^{P_{r,f}} T_r^2\left(\frac{\partial z}{\partial T_r}\right)_{P_r} d(\ln P_r)\end{aligned} \tag{9.44}$$

and the left-hand side of Eq. (9.44) is called the *enthalpy departure function* and is presented in Fig. C.2. It shows the deviation of the enthalpy of a real fluid (liquid or gas) from that of an ideal gas in the same state. Note, however, that choosing T and P as the independent variables in Eq. (9.34) precludes the use of this method for predicting the departure function in the saturation region, where P and T are *not* independent, although the saturation curve itself can be computed and drawn as shown on Fig. C.2.

The Clapeyron equation, Eq. (9.14), can be used to find the enthalpy change between the liquid and vapor states as a check on the saturation curve. This is done by first noting that

$$v_{lg} = v_g - v_l = \frac{RT_{\text{sat}}}{P_{\text{sat}}}(z_g - z_l) \tag{9.45}$$

and then using Eq. (9.14) to get

$$\left(\frac{dP}{dT}\right)_{\text{sat}} = \frac{h_{lg}}{T_{\text{sat}}v_{lg}} = \frac{P_{\text{sat}}h_{lg}}{RT^2_{\text{sat}}(z_g - z_l)} \tag{9.46}$$

or

$$\frac{h_{lg}}{RT_{\text{cr}}} = \left(\frac{dP_r}{dT_r}\right)_{\text{sat}} \frac{T^2_{r,\text{sat}}}{P_{r,\text{sat}}}(z_g - z_l) \tag{9.47}$$

The values of z_g and z_l are a function of $P_{r,\text{sat}}$, and the corresponding values of P_r and $(dP_r/dT_r)_{\text{sat}}$ are all available from the $z(P_r, T_r)$ charts. Note that

$$\left[\frac{h_{\text{ideal}}(T) - h_g}{RT_{\text{cr}}}\right]_g - \left[\frac{h_{\text{ideal}}(T) - h_l}{RT_{\text{cr}}}\right]_l = \frac{h_{lg}}{RT_{\text{cr}}} \tag{9.48}$$

and Eq. (9.47) can thus be used to check the h_{lg} values found on Fig. C.2 by using only $z(T_r, P_r)_{\text{sat}}$ data.

Equation (9.44) can be evaluated for any unsaturated state by using data for $z(P_r, T_r)$ as presented, for example, in Fig. C.1. The resulting variation of the enthalpy departure function with T_r and P_r is shown in Fig. C.2. Of course, similar derivations can be used to find departure functions for the other common thermodynamic properties, and charts for entropy and for fugacity (defined in Sec. 9.4.5) are also presented in Figs. C.3 and C.4, respectively. Separate charts are presented for low pressures in Fig. C.5. Note that the dimensionless ordinates used on these charts can be treated with each quantity on a molar basis if the universal gas constant is used or on a unit-mass basis if the gas constant specific to a particular gas is used.

To use the charts to find the change in a given property, for example, a change in enthalpy between two states, the change is given by

$$\frac{h_2 - h_1}{RT_{\text{cr}}} = \frac{\bar{h}_2 - \bar{h}_1}{\bar{R}T_{\text{cr}}}$$
$$= \left[\left(\frac{\bar{h}_{\text{ideal}} - \bar{h}}{\bar{R}T_{\text{cr}}}\right)_1 - \left(\frac{\bar{h}_{\text{ideal}} - \bar{h}}{\bar{R}T_{\text{cr}}}\right)_2 + \frac{\bar{h}_{2,\text{ideal}} - \bar{h}_{1,\text{ideal}}}{\bar{R}T_{\text{cr}}}\right] \tag{9.49}$$

or

$$h_2 - h_1 = RT_{\text{cr}}\left[\left(\frac{\bar{h}_{\text{ideal}} - \bar{h}}{\bar{R}T_{\text{cr}}}\right)_1 - \left(\frac{\bar{h}_{\text{ideal}} - \bar{h}}{\bar{R}T_{\text{cr}}}\right)_2 + \frac{\bar{h}_{2,\text{ideal}} - \bar{h}_{1,\text{ideal}}}{\bar{R}T_{\text{cr}}}\right]$$
$$= h_{2,\text{ideal}} - h_{1,\text{ideal}} + RT_{\text{cr}}\left[\left(\frac{\bar{h}_{\text{ideal}} - \bar{h}}{\bar{R}T_{\text{cr}}}\right)_1 - \left(\frac{\bar{h}_{\text{ideal}} - \bar{h}}{\bar{R}T_{\text{cr}}}\right)_2\right] \tag{9.50}$$

Thus, the actual change in enthalpy is given by the change in the ideal gas enthalpy modified by correction factors. The terms inside the square brackets are each taken directly from the charts (in this case, Fig. C.2).

We now can evaluate the thermodynamic properties v, h, s, and fugacity (and, of course, $u = h - Pv$) for any fluid in any state as long as the critical temperature and pressure are known for that fluid. The charts should be about as accurate as the compressibility factor chart, i.e., about 6 percent. Because of this accuracy limitation, the results are presented in graphical rather than tabular form, since the errors in reading the graphs introduce little additional uncertainty to the results.

Example Problem 9.4

Determine the change in the values of enthalpy and entropy of ethylene between an initial state at 1 atm and 460°R and a final state at 200 atm and 600°R, using the reduced-property charts.

Solution

From Table C.4, for ethylene, $T_{cr} = 508.3°\text{R}$ and $P_{cr} = 49.7$ atm. The molecular weight is 28.054. Thus, the values of reduced temperature and pressure at the initial state are

$$T_{r,1} = \frac{T_1}{T_{cr}} = \frac{460}{508.3} = 0.905$$

$$P_{r,1} = \frac{P_1}{P_{cr}} = \frac{1}{49.7} = 0.0201$$

and at the final state are

$$T_{r,2} = \frac{600}{508.3} = 1.18$$

$$P_{r,2} = \frac{200}{49.7} = 4.02$$

From Figs. C.2 and C.3, the pressure in the initial state is so low that the enthalpy and entropy departure functions are both zero, indicating that ethylene in the initial state behaves as an ideal gas. In the final state, $(h_{\text{ideal}} - h)/(RT_{cr}) = 3.20$ and $(s_{\text{ideal}} - s)/R = 1.90$. Using the particular gas constant $R = 1.98586/28.054 = 0.0708$ Btu/(lbm · °R) yields

$$\begin{aligned} h_2 &= h_{\text{ideal},2} - (3.20)(0.0708)(508.3) \\ &= h_{\text{ideal},2} - 115 \qquad \text{Btu/lbm} \end{aligned}$$

$$\begin{aligned} s_2 &= s_{\text{ideal},2} - (1.90)(0.0708) \\ &= s_{\text{ideal},2} - 0.135 \qquad \text{Btu/(lbm} \cdot \text{°R)} \end{aligned}$$

To find the change in h and s, the values of h_1 and s_1 must be subtracted. Since in this case the enthalpy and entropy departure functions have zero value in the initial state, the equations for the changes in the properties become

$$h_2 - h_1 = (h_2 - h_1)_{\text{ideal}} - 115$$

$$s_2 - s_1 = (s_2 - s_1)_{\text{ideal}} - 0.135$$

The changes in h and s for the ideal gas can be found by using the values of specific heat at constant pressure from Table E.1:

$$\begin{aligned}(h_2 - h_1)_{\text{ideal}} &= \int_{460}^{600} c_P\,dT \\ &= \frac{1}{28.054}\left[0.944T + (2.075 \times 10^{-2})\left(\frac{T^2}{2}\right) \right. \\ &\quad \left. - (0.6151 \times 10^{-5})\left(\frac{T^3}{3}\right) + (0.7326 \times 10^{-9})\left(\frac{T^4}{4}\right)\right]\Bigg|_{400}^{600} \\ &= 51.48 \text{ Btu/lbm}\end{aligned}$$

$$\begin{aligned}(s_2 - s_1)_{\text{ideal}} &= \int_{460}^{600} \frac{c_P}{T}\,dT - R\ln\frac{P_2}{P_1} \\ &= \frac{1}{28.054}\left[0.944\ln\frac{600}{460} + (2.075 \times 10^{-2})(600 - 460) \right. \\ &\quad - (0.6151 \times 10^{-5})\left(\frac{600^2 - 460^2}{2}\right) \\ &\quad \left. + (0.7326 \times 10^{-9})\left(\frac{600^3 - 460^3}{3}\right)\right] \\ &\quad - 0.0708\ln\frac{200}{1} \\ &= -0.2778 \text{ Btu/(lbm}\cdot{}^\circ\text{R)}\end{aligned}$$

Thus, the changes in the properties are finally

$$h_2 - h_1 = 51.48 - 115 = -64 \text{ Btu/lbm} = -150 \text{ kJ/kg}$$

$$\begin{aligned}s_2 - s_1 &= -0.2778 - 0.135 = -0.413 \text{ Btu/(lbm}\cdot{}^\circ\text{R)} \\ &= -1.73 \text{ kJ/(kg}\cdot\text{K)}\end{aligned}$$

Comments

Note that the entropy change is negative for this change of state even though the temperature increases because of the extremely large change in pressure. Ethylene is apparently far from an ideal gas in the final state, because the departure functions are quite large relative to the ideal gas terms in the final equations.

If tables of ideal gas properties are available for the substance under study, then the ideal gas terms can be evaluated from the tables. Care must be taken to include the pressure-dependent term in the entropy relations for the ideal gas, as is obvious in this problem. The specific entropy term in the tables does not include the pressure-dependent term, as discussed in Chap. 5.

Three-Parameter Corresponding-States Methods

At the beginning of the discussion of the principle of corresponding states, we noted that the principle is not exact but does provide guidance as to thermodynamic property behavior for many materials based on very limited data. Suppose an additional critical property were available, such as the critical volume. Could this additional piece of information be used to more accurately model the behavior of substances than is possible by using only P_{cr} and T_{cr}?

It is indeed possible. As stated earlier, the critical volume is a difficult quantity to measure accurately, so it is not an appropriate property to use in such a model; but other properties are available. Lee and Kesler [2] have presented a three-parameter corresponding-states model based on a proposal by K. S. Pitzer and coworkers that uses P_{cr}, T_{cr}, and ω as the required properties as well as a modified form of the Benedict-Webb-Rubin equation of state. The factor ω is called the *acentric factor,* and it is a measure of the deviation in the thermodynamic behavior of a real fluid from a simple fluid. A *simple fluid* is defined as a fluid with a spherical molecule, and the acentric factor indicates the degree of deviation of the molecular structure (acentricity) of the real molecule from that of the simple spherical model. For example, the compressibility factor z of a fluid with acentric factor ω is given by

$$z = z^{(0)} + \omega z^{(1)} \tag{9.51}$$

where $z^{(0)}$ is the compressibility factor for the spherical model of the molecule and $z^{(1)}$ incorporates the deviation due to the acentricity of the real molecule.

Departure functions can be evaluated by using the three-parameter method in a similar manner to that used in the two-parameter correlations. The resulting departure functions take the form

$$\frac{h_{\text{ideal}} - h}{RT_{cr}} = \left(\frac{h_{\text{ideal}} - h}{RT_{cr}}\right)^{(0)} + \omega\left(\frac{h_{\text{ideal}} - h}{RT_{cr}}\right)^{(1)} \tag{9.52}$$

and

$$\frac{s_{\text{ideal}} - s}{R} = \left(\frac{s_{\text{ideal}} - s}{R}\right)^{(0)} + \omega\left(\frac{s_{\text{ideal}} - s}{R}\right)^{(1)} \tag{9.53}$$

The terms on the right-hand sides of Eqs. (9.52) and (9.53) with superscripts (0) and (1) are termed *residual properties,* and they correspond to

$z^{(0)}$ and $z^{(1)}$ in Eq. (9.51). Values of the acentric factor are presented in Table C.4, and values of the three-parameter functions for compressibility, residual enthalpy, and residual entropy are given in Tables C.6 through C.8. Tables are used for the presentation of these data, because in most cases the accuracy of the results is greater than can be obtained from plots of the functions. The solid lines traversing the tables for $P_r < 1$ show the location of the vapor-liquid equilibrium (saturation) states. *Note that* s_{ideal} *used in Eq (9.53) must be evaluated at the actual P or v.*

Lee and Kesler also present relations for the fugacity coefficient (Sec. 9.4.5) and the constant-pressure and constant-volume specific heats. Comparison with experimental data shows that addition of the third parameter provides good accuracy. It is difficult to give a blanket number for accuracy, but Lee and Kesler show absolute average deviations in z of 1 to 2 percent over wide ranges of T_r and P_r for most hydrocarbons, with a few deviating by about 4 percent. Deviations are greatest for a given fluid near the critical point.

Example Problem 9.5

Repeat Example Problem 9.4, using the three-parameter reduced-property tables, and compare your results with those of Example Problem 9.4.

Solution

The additional information needed to use the three-parameter method is the acentric factor, which for ethylene has the value 0.089 (Table C.4). Now the enthalpy departure in state 1 ($P_r = 0.0201$ and $T_r = 0.905$) is given from Eq. (9.52) and Tables C.7 and C.8 as

$$\frac{h_{\text{ideal}} - h}{RT_{\text{cr}}} \approx 0$$

$$\frac{s_{\text{ideal}} - s}{R} \approx 0$$

so that $h_1 = h_{\text{ideal},1}$ and $s_1 = s_{\text{ideal},1}$. For state 2 ($P_r = 4.02$ and $T_r = 1.18$), by using double interpolation on Tables C.7 and C.8,

$$\frac{h_{\text{ideal}} - h}{RT_{\text{cr}}} = 3.079 + (0.089)(1.595) = 3.221$$

$$\frac{s_{\text{ideal}} - s}{R} = 1.854 + (0.089)(1.557) = 1.993$$

As in Example Problem 9.4,

$$h_2 - h_1 = (h_2 - h_1)_{\text{ideal}} - 3.221RT_{\text{cr}}$$

$$s_2 - s_1 = (s_2 - s_1)_{\text{ideal}} - 1.993R$$

or, using the ideal gas property changes from Example Problem 9.4, we find

$$h_2 - h_1 = 51.48 - (3.221)(0.0708)(508.3) = 51.48 - 115.92$$
$$= -64.44 \text{ Btu/lbm} = -149.9 \text{ kJ/kg}$$

$$s_2 - s_1 = -0.2778 - (1.993)(0.0708) = -0.2778 - 0.1411$$
$$= -0.4189 \text{ Btu/(lbm}\cdot{}^\circ\text{R)} = -1.754 \text{ kJ/(kg}\cdot\text{K)}$$

Comments

These results are in good agreement with those of Example Problem 9.4 [$h_2 - h_1 = -64$ Btu/lbm and $s_2 - s_1 = -0.413$ Btu/(lbm·°R)]. We would expect the three-parameter calculations to be more accurate. Note that it is possible to maintain accuracy to at least one more significant figure with the three-parameter results than is possible with the two-parameter graphical method.

Example Problem 9.6

Determine the value of internal energy for CO_2 at $P = 300$ atm and $T = 450$ K, using both the two-parameter and three-parameter reduced-property methods.

Solution

The critical temperature, critical pressure, and acentric factor are obtained from Table C.4 as $T_{cr} = 304.1$ K, $P_{cr} = 72.8$ atm, and $\omega = 0.239$. The reduced temperature (1.480) and reduced pressure (4.121) are used with Fig. C.1*b* to obtain

$$z = 0.80$$

which indicates that this state cannot be accurately approximated as an ideal gas. The internal energy is calculated from the enthalpy as

$$u = h - Pv = h - zRT$$

The two-parameter method uses the z value above and h determined from Fig. C.2. The three-parameter method uses the Lee and Kesler values (Tables C.6 and C.7).

Two-Parameter Method

The critical temperature and pressure given above are used with Fig. C.2 to yield $(h_{ideal} - h)/(RT_{cr}) = 1.8$. The enthalpy determined from the ideal gas approximation is given in Table D.7 as $h_{ideal} = 352.19$ kJ/kg. Therefore,

$$h = h_{ideal} - 1.8RT_{cr}$$
$$= 352.19 - (1.8)(0.18892)(304.1)$$
$$= 249 \text{ kJ/kg} = 107 \text{ Btu/lbm}$$

and the internal energy is

$$u = h - zRT$$
$$= 249 - (0.80)(0.18892)(450)$$
$$= 181 \text{ kJ/kg} = 77.8 \text{ Btu/lbm}$$

Note that no more than two significant figures should be kept.

Three-Parameter Method

The compressibility factor is given by Eq. (9.51). Double interpolation in Table C.6 yields the values for $z^{(0)}$ and $z^{(1)}$.

$$z = z^{(0)} + \omega z^{(1)}$$
$$= 0.7953 + (0.239)(0.2296) = 0.8502$$

The enthalpy is evaluated from Eq. (9.52) with Table C.7 as

$$\frac{h_{\text{ideal}} - h}{RT_{\text{cr}}} = \left(\frac{h_{\text{ideal}} - h}{RT_{\text{cr}}}\right)^{(0)} + \omega\left(\frac{h_{\text{ideal}} - h}{RT_{\text{cr}}}\right)^{(1)}$$
$$= 1.962 + (0.239)(0.097) = 1.985$$

With h_{ideal} given above as 352.19 kJ/kg, the enthalpy is

$$h = h_{\text{ideal}} - 1.985\, RT_{\text{cr}}$$
$$= 352.19 - (1.985)(0.18892)(304.1)$$
$$= 238.2 \text{ kJ/kg} = 102.4 \text{ Btu/lbm}$$

and the corresponding internal energy is

$$u = h - zRT$$
$$= 238.1 - (0.8502)(0.18892)(450)$$
$$= 165.8 \text{ kJ/kg} = 71.3 \text{ Btu/lbm}$$

Comments

The improvement in the accuracy obtained by each level of approximation is noteworthy. A very inaccurate approximation is obtained if the state is assumed to be an ideal gas since $u = h - RT = 352.19 - (0.18892)(450) = 267.2$ kJ/kg. The number of significant figures is misleading, since the equation itself is invalid. This is expected since $z = 0.80$ was obtained from Fig. C.1*b*. The two-parameter method yields $u =$ 181 kJ/kg while the more accurate, three-parameter method yields $u =$ 165.8 kJ/kg.

9.4 Some Other Properties

Although the primary thermodynamic properties P, v, T, u, h, s, a, and g and their relationships form the basis of much of engineering thermodynamics, a number of other properties are of interest in certain applications. In particular, note that the derivatives of these properties with respect to other properties are also properties. We used this fact when the specific heats were defined as $c_P = (\partial h/\partial T)_P$ and $c_v = (\partial u/\partial T)_v$.

Other derivatives of the primary properties can also be defined as separate properties where this is useful. We briefly examine some of these derived properties.

9.4.1 Isothermal Compressibility

In some applications, it is useful to know the degree to which a substance can be compressed when the pressure is changed in an isothermal process. The fractional change in volume with pressure at constant temperature is called the *isothermal compressibility* κ and is defined as

$$\kappa = -\left(\frac{1}{v}\right)\left(\frac{\partial v}{\partial P}\right)_T \tag{9.54}$$

The minus sign is used because specific volume decreases with pressure, so the partial derivative carries a negative value. Values of κ can be tabulated from P-v-T data for any substance.

9.4.2 Coefficient of Thermal Expansion

To describe the change in volume with temperature of a substance at constant pressure, we can define the *coefficient of thermal expansion* β as

$$\beta = \left(\frac{1}{v}\right)\left(\frac{\partial v}{\partial T}\right)_P \tag{9.55}$$

This definition can be combined with the isothermal compressibility to yield the total derivative of volume. The total derivative of $V(T, P)$ divided by V is

$$\frac{dV}{V} = \frac{1}{V}\left(\frac{\partial V}{\partial T}\right)_P dT + \frac{1}{V}\left(\frac{\partial V}{\partial P}\right)_T dP \tag{9.56}$$

Substituting Eqs. (9.54) and (9.55) yields

$$\frac{dV}{V} = \beta\, dT - \kappa\, dP \tag{9.57}$$

Example Problem 9.7

Derive relations for the coefficient of thermal expansion and the isothermal compressibility of an ideal gas.

Solution

The derivative $(\partial v/\partial T)_P$ for an ideal gas is R/P, so Eq. (9.55) becomes

$$\beta = \left(\frac{1}{v}\right)\left(\frac{R}{P}\right) = \frac{1}{T}$$

So the coefficient of thermal expansion of an ideal gas is simply the inverse of the absolute temperature.

The derivative $(\partial v/\partial P)_T$ for an ideal gas is $-RT/P^2$, so Eq. (9.54) becomes

$$\kappa = \left(\frac{1}{v}\right)\left(\frac{RT}{P^2}\right) = \frac{1}{P}$$

and the isothermal compressibility of an ideal gas is the inverse of the absolute pressure.

9.4.3 Joule-Thomson Coefficient

In throttling processes (Sec. 4.5.3) the enthalpy is constant across the throttling device. Of interest is whether the temperature of the throttled substance increases or decreases and the magnitude of the change. The property that tells us this is the *Joule-Thomson coefficient* μ, where

$$\mu = \left(\frac{\partial T}{\partial P}\right)_h \tag{9.58}$$

If the substance being throttled increases in temperature during passage through a throttling device, the coefficient is negative; if the temperature drops, the coefficient is positive.

Example Problem 9.8

Refrigerant 12 is being considered for use as a pressurant in a shaving cream container. The container has saturated refrigerant 12 vapor at 27°C, and the shaving cream exits the container nozzle to room conditions. What is the sign of the Joule-Thomson coefficient for this process? Will the shaving cream be warmer or colder than when it was in the can?

Solution

Assuming the velocity of the shaving cream/refrigerant 12 leaving the can is negligible, we see this is a constant-enthalpy process. From Table D.12 or Fig. D.4 or computerized properties, the initial state is $T = 27°\text{C}$, $P = 687.5$ kPa, and $h = 362.6$ kJ/kg. At the final state the enthalpy is the same, and $P = 101.3$ kPa. The final temperature is found to be 11.5°C. The coefficient is thus roughly

$$\mu = \left(\frac{11.5 - 27}{101.3 - 687.5}\right)_h = 0.026 \text{ K/kPa} = 0.10\text{°R/psia}$$

The coefficient is positive, reflecting the fact that the pressurant temperature drops during the expansion. If we want automatically warm shaving cream, we must search for a different pressurant. Only hydrogen, helium, and neon have negative coefficients near normal ambient conditions.

The Joule-Thomson coefficient can be placed in alternative form by taking the total differential of $h(T, P)$ as

$$dh = \left(\frac{\partial h}{\partial T}\right)_P dT + \left(\frac{\partial h}{\partial P}\right)_T dP = c_P\, dT + \left(\frac{\partial h}{\partial P}\right)_T dP \tag{9.59}$$

For the particular case of constant enthalpy, $dh = 0$, and the right-hand side reduces to

$$\left(\frac{\partial T}{\partial P}\right)_h = \mu = -\frac{1}{c_P}\left(\frac{\partial h}{\partial P}\right)_T \tag{9.60}$$

The enthalpy term can be modified by noting that

$$\left(\frac{\partial h}{\partial P}\right)_T = \left[\frac{\partial(u + Pv)}{\partial P}\right]_T \tag{9.61}$$

and the Joule-Thomson coefficient can be written by substituting Eq. (9.61) into Eq. (9.60), resulting in

$$\mu = \frac{1}{c_P}\left\{-\left(\frac{\partial u}{\partial P}\right)_T - \left[\frac{\partial(Pv)}{\partial P}\right]_T\right\} \tag{9.62}$$

Recall that for an ideal gas the internal energy u is independent of pressure, so the first term inside the braces will be zero for an ideal gas. Similarly, because $Pv = RT$ for an ideal gas, the second term inside the braces will be

$$\left.\frac{\partial(Pv)}{\partial P}\right|_T = \left.\frac{\partial(RT)}{\partial P}\right|_T$$
$$= 0$$

for an ideal gas. Thus, for an ideal gas the Joule-Thomson coefficient is zero.

For any real gas, neither term inside the braces will be exactly zero. In fact, at conditions near 1 atm and 25°C, the value of the terms within the braces will sum to a positive value for all gases except hydrogen, helium, and neon. That is, for all other gases with initial conditions near standard conditions, cooling will occur during a constant-enthalpy expansion to a lower pressure.

The terms inside the braces in Eq. (9.62) reflect real gas effects. When the pressure of a real gas is reduced, the molecules are separated by greater distances (because the number of molecules per volume becomes smaller). However, there are long-range attractive forces between these molecules; at greater separation distances, therefore, the potential energy per molecule *decreases.* Thus, the change in potential energy with increasing pressure is positive, and the term $(\partial u/\partial P)_T$ carries a positive value for all real gases. This term, because of its minus sign, contributes to a drop in gas temperature during an expansion. The long-range attractive forces are assumed to be nonexistent in models of gases that produce ideal gas behavior.

The second term inside the braces may be positive or negative for a real gas, with negative values occurring at low temperatures and pressures.

Because the first term inside the braces carries a positive value while the second usually carries a negative value, the difference can be zero at various states. For a given pressure, the temperature at which the value of the Joule-Thomson coefficient is zero is called the *inversion temperature.* According to Eq. (9.58), the inversion temperature could be located on the *P-h* diagram by locating the points where the isotherms are vertical. For water, such a diagram is shown in Fig. 9.4. The inversion line connects the inversion temperatures. Note that at large h and small P, in the superheat region, the isotherms all become vertical. This is the region of ideal gas behavior, where the Joule-Thomson coefficient becomes zero for a wide range of conditions. A more detailed *P-h* diagram for steam is shown in Fig. D.2.

Application to Gas Liquefaction

Liquefied gases have broad industrial use for providing cooling as well as for the ease of transport of large masses for process use. Liquid oxygen, for example, has found broad use as a rocket fuel oxidizer.

With few exceptions, the Joule-Thomson coefficient for real gases is positive near standard conditions and becomes positive for all gases at sufficiently low temperatures. Thus, isothermal compression of these gases followed by expansion back to low pressure causes the gas to be cooled. If the temperature drop is great enough, the gas can enter the saturation region and be partially liquefied. The study of gases that liquefy at conditions on the order of −100 to −150°C and the heat transfer effects at these conditions is called *cryogenics,* and such gases are *cryogenic gases.*

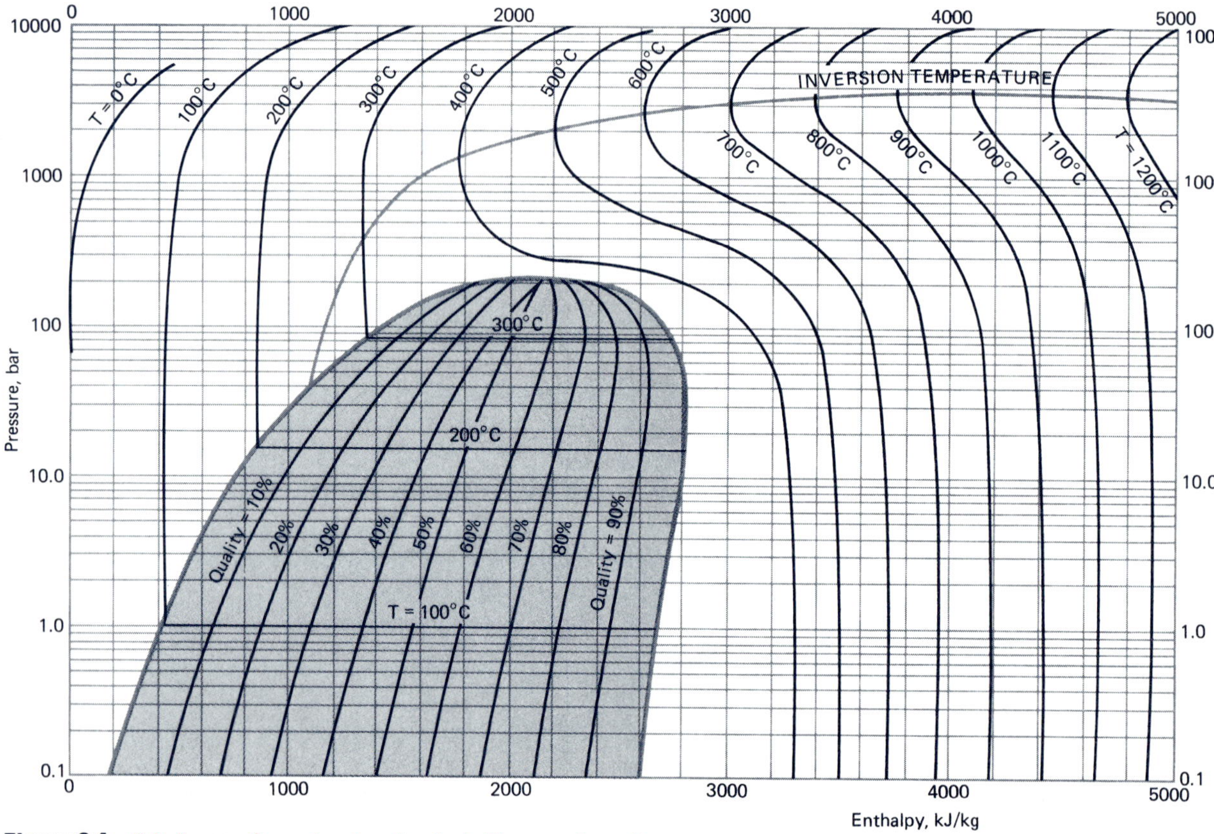

Figure 9.4 *P-h diagram for water showing Joule-Thomson inversion temperature.*

The cycle for carrying out this liquefaction process is generally called the *Linde* or *Hampson process,* and it is shown schematically in Fig. 9.5. In this process, the gas is compressed in a multistage compressor with intercooling and aftercooling to approximate the isothermal compression process. The gas is then passed through a recuperative heat exchanger to further precool it by heat transfer to the cold vapors from the holding tank for the liquefied gas. The cooled gas is next passed through a throttling value, where it partially condenses by the Joule-Thomson cooling effect. The liquefied portion is removed, often through a second throttling valve for low pressure applications, and the saturated vapor is returned through the recuperator for mixing with makeup feed gas, and then this stream is returned to the compressor.

For those gases with a negative Joule-Thomson coefficient near standard conditions (hydrogen, helium, and neon), precooling to below the inversion temperature must be done before this method can be employed.

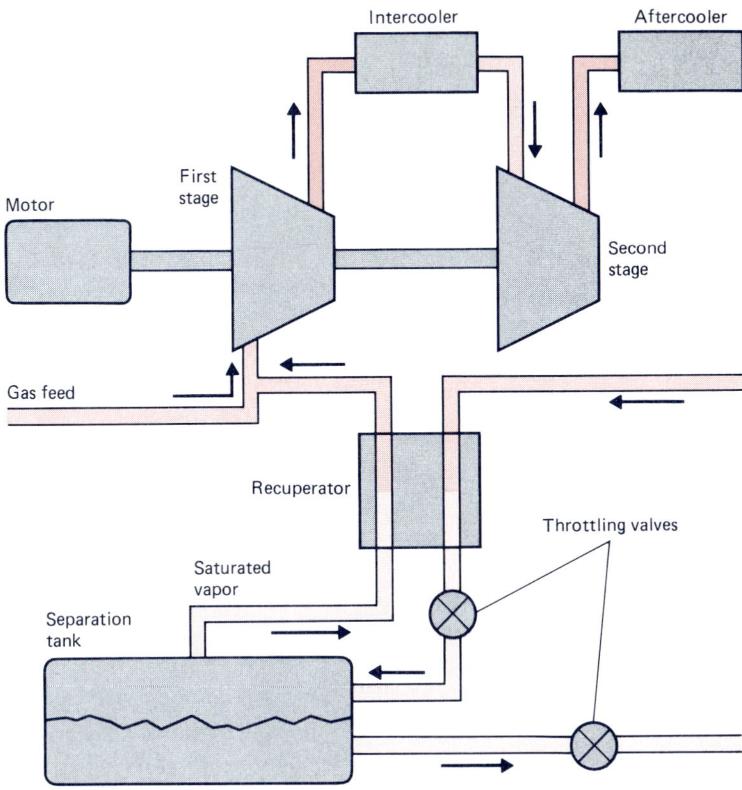

Figure 9.5 Linde or Hampson process for gas liquefaction.

9.4.4 Specific Heats

The Maxwell relations have been used to provide two new ways of relating the specific heats to other properties. In Eq. (9.20) we found

$$c_v = \left(\frac{\partial u}{\partial T}\right)_v = T\left(\frac{\partial s}{\partial T}\right)_v \tag{9.63}$$

and in Eq. (9.37)

$$c_P = \left(\frac{\partial h}{\partial T}\right)_P = T\left(\frac{\partial s}{\partial T}\right)_P \tag{9.64}$$

It follows that

$$c_P - c_v = T\left[\left(\frac{\partial s}{\partial T}\right)_P - \left(\frac{\partial s}{\partial T}\right)_v\right] \tag{9.65}$$

Let us find a relation for $c_P - c_v$ in terms of P-v-T data. Using the substitution rule, [Eq. (H.21)], in the form

$$\left(\frac{\partial s}{\partial T}\right)_P = \left(\frac{\partial s}{\partial T}\right)_v + \left(\frac{\partial s}{\partial v}\right)_T \left(\frac{\partial v}{\partial T}\right)_P$$

results in

$$c_P - c_v = T\left[\left(\frac{\partial s}{\partial v}\right)_T \left(\frac{\partial v}{\partial T}\right)_P\right] \tag{9.66}$$

Now, the Maxwell relation [Eq. (9.11*c*)] is substituted to eliminate the derivative involving *s*, and we obtain

$$c_P - c_v = T\left[\left(\frac{\partial P}{\partial T}\right)_v \left(\frac{\partial v}{\partial T}\right)_P\right] \tag{9.67}$$

We can observe the following from this equation. Since $(\partial v/\partial T)_P$ is small for liquids and solids, the values of c_P and c_v should be nearly equal for all materials in those states. This is stated for incompressible substances in Sec. 3.5. Also the values of $(\partial P/\partial T)_v$ and $(\partial v/\partial T)_P$ are easily found for an ideal gas, and for that case we find, as we have derived before, that $c_P - c_v = R$.

9.4.5 Fugacity

One other property is introduced here for use in later chapters, particularly when we deal with mixtures and equilibrium systems. The property is useful for certain single-component systems in which an isothermal process is carried out. This property is called the *fugacity*, and it is defined here by starting with Eq. (9.5*d*):

$$dg = v\,dP - s\,dT \tag{9.68}$$

For a constant-temperature process, this reduces to

$$dg = v\,dP \tag{9.69}$$

If we assume an ideal gas, then Eq. (9.69) becomes, for an isothermal process,

$$dg = RT\frac{dP}{P} = RT\,d(\ln P) \tag{9.70}$$

Suppose we now introduce the fugacity f as the quantity that will produce an exact relation for a *real* gas in this equation ($T =$ constant):

$$dg = RT\,d(\ln f) \tag{9.71}$$

From such a defining relation for fugacity, we can infer various characteristics about the behavior of f. First, f will have units of pressure; second, at low pressure, f must approach P in value, since Eq. (9.68) is correct at low pressure. Or

$$\lim_{P\to 0} \frac{f}{P} = 1 \tag{9.72}$$

The fugacity thus acts as a "pseudo-pressure" which gives the correct behavior for dg when it is used in Eq. (9.71). Values for f can be found by writing Eq. (9.69) for a real gas in terms of the compressibility factors as

$$dg = v\,dP = zRT\,d(\ln P) = RT\,d(\ln f) \tag{9.73}$$

This can be rearranged to yield

$$(z-1)\frac{dP}{P} = (z-1)\frac{dP_r}{P_r} = (z-1)\,d(\ln P_r) = \frac{d(f/P)}{f/P} = d\left(\ln \frac{f}{P}\right) \tag{9.74}$$

and it is clear that f/P (which is often referred to as the *fugacity coefficient*) can be tabulated as a function of reduced properties T_r and P_r. A graph of fugacity as a function of reduced properties is given in Figs. C.4 and C.5*c*.

Example Problem 9.9

A natural gas compressor station is to take methane at 1 atm and 77°F and compress it isothermally and reversibly to 60 atm. Find the compressor work per pound-mass of methane.

Solution

The first law gives

$$q - w = h_{out} - h_{in}$$

and for a reversible isothermal compression, $q = T(s_{out} - s_{in})$, so

$$w = (h - Ts)_{in} - (h - Ts)_{out} = g_{in} - g_{out} = RT \ln \frac{f_{in}}{f_{out}}$$

The values of the reduced properties are

$$T_r = \frac{536.7}{342.7} = 1.566$$

$$P_{r,in} = \frac{1}{45.4} = 0.0220$$

$$P_{r,out} = \frac{60}{45.4} = 1.32$$

From Fig. C.5*c*, $(f/P)_{in} = 1.0$; from Fig. C.4, $(f/P)_{out} = 0.92$; and

$$w = (0.1238)(536.7) \ln \frac{(1)(1)}{(0.92)(60)} = -270 \text{ Btu/lbm} = -630 \text{ kJ/kg}$$

Comment

Clearly, for this type of problem, the use of fugacity is much simpler than going through the complete first and second law analysis.

9.5 Summary

In this chapter, we have shown that many useful relationships exist among the thermodynamic properties; that these relationships can be used to minimize the laboratory work necessary to determine property values and to construct tables and functions which describe all properties from data; that the principle of corresponding states allows the data for one substance to provide a good estimate of the property behavior of other substances; and that many properties are available and useful for specific problems and can be derived from the set we have already studied.

We have barely touched the wealth of literature available on property measurement, relationships, equations of state, etc. This is an ongoing area of research in thermodynamics. However, the material presented here should act as a guide for the engineer in finding and applying thermodynamic data.

Problems

Problems marked with a star are lengthy; thus, computerized steam tables or an accurate Mollier chart should be used if these problems are to be completely worked.

9.1 The virial equation of state is

$$\frac{P}{T} = \frac{R}{v}\left[1 + \frac{B(T)}{v} + \frac{C(T)}{v^2} + \cdots\right]$$

Use the fact that $(\partial P/\partial v)_T$ and $(\partial^2 P/\partial v^2)_T$ must be zero at the critical point [as was used in Eqs. (9.29) and (9.30)] to find the first two constants, $B(T)$ and $C(T)$. Neglect higher-order terms, and derive an equation of state.

9.2 Show whether the Benedict-Webb-Rubin equation [Eq. (3.31)] obeys the relations that $(\partial P/\partial v)_T$ and $(\partial^2 P/\partial v^2)_T$ equal zero at the critical point for methane.

*9.3S Plot the saturation curve for water on a P-v diagram, using van der Waals' equation in the range $10\text{ kPa} < P < P_{cr}$. Do this by taking values of P_{sat} and T_{sat} from the steam tables and using van der Waals' equation to find the corresponding values of v_l and v_g. Plot the saturation curve from the steam tables on the same diagram.

***9.3E** Plot the saturation curve for water on a P-v diagram, using van der Waals' equation in the range 1 psia $< P < P_{cr}$. Do this by taking values of P_{sat} and T_{sat} from the steam tables and using van der Waals' equation to find the corresponding values of v_l and v_g. Plot the saturation curve from the steam tables on the same diagram.

9.4 Derive expressions for the isothermal compressibility $\kappa = -(1/v)(\partial v/\partial P)_T$ and the volumetric thermal expansion coefficient $\beta = (1/v)(\partial v/\partial T)_P$ for a van der Waals gas.

9.5 Show that, in general, $c_P - c_v = Tv\beta^2/\kappa$. And show that this relation reduces to $c_P - c_v = R$ for an ideal gas. Start with Eq. (9.65).

9.6 Using Eq. (9.57), show that for a constant-volume process

$$P_2 = P_1 + \frac{\beta}{\kappa}(T_2 - T_1)$$

9.7 Using the relation given in Prob. 9.5, show how c_P and c_v are related for an incompressible liquid.

9.8 Derive Eq. (9.25), following the procedure similar to that used for obtaining Eq. (9.24).

9.9 Starting with Eq. (9.38) and the fact that $h = u + Pv$, derive Eq. (9.23).

9.10 The acentric factor is defined by

$$\omega = [-\log P_{r,\text{sat}}]_{T_r=0.7} - 1$$

Find the saturation pressure of water and of refrigerant 12 at $T_r = 0.7$ from this relation. Compare the results with the values from the tables in App. D or the computerized property data.

9.11 Prove that the Joule-Thomson coefficient can be expressed by

$$\mu = \frac{1}{c_P}\left[T\left(\frac{\partial v}{\partial T}\right)_P - v\right]$$

9.12 Derive an expression for the Joule-Thomson coefficient of an ideal gas starting with the expression given in Prob. 9.11.

9.13S Using the Clapeyron equation and P-v-T data for refrigerant 12 from the tables or the computerized property relations, find h_{lg} for refrigerant 12 at $P = 50$ and $P = 200$ kPa. Compare with the values from Table D.14.

9.13E Using the Clapeyron equation and P-v-T data for refrigerant 12 from the tables or the computerized property relations, find h_{lg} for refrigerant 12 at $P = 6$ and $P = 25$ psia. Compare with the values from Table E.14.

9.14 Substitute the relations for a and b [Eq. (9.31)] into van der Waals' equation, and derive a value for the compressibility factor at the critical point for a van der Waals gas. Compare the result with the value from Fig. C.1.

9.15S n-Hexane is at $T = 507.5$ K and 17.94 atm. What is the difference between the actual enthalpy of n-hexane in this state and the enthalpy in the same state if n-hexane acts as an ideal gas? (Use the Lee-Kesler three-parameter method to obtain your result.)

9.15E n-Hexane is at $T = 913.5$°R and 17.94 atm. What is the difference between the actual enthalpy of n-hexane in this state and the enthalpy in the same state if n-hexane acts as an ideal gas? (Use the Lee-Kesler three-parameter method to obtain your result.)

9.16S Find an accurate value of g in kilojoules per kilogram for CO_2 at $P = 350$ atm and $T = 1065$ K.

9.16E Find an accurate value of g in kilojoules per kilogram for CO_2 at $P = 350$ atm and $T = 1916$°R.

9.17S Find h and s for nitrogen at 250 atm and 210 K, using (a) the departure charts and (b) the Lee-Kesler tables of departure functions.

9.17E Find h and s for nitrogen at 250 atm and 380°R, using (a) the departure charts and (b) the Lee-Kesler tables of departure functions.

9.18S You need the value of c_P for argon at $P = 30$ atm and $T = 600$ K. At this pressure the value may not be a function of T only. Use the Lee-Kesler tables to find the required value.

9.18E You need the value of c_P for argon at $P = 30$ atm and $T = 1000$°R. At this pressure the value may not be a function of T only. Use the Lee-Kesler tables to find the required value.

9.19S Find the Joule-Thompson coefficient for steam at $P = 25$ MPa and $T = 600$°C. Use the steam tables.

9.19E Find the Joule-Thompson coefficient for steam at $P = 250$ atm and $T = 1100$°R. Use the steam tables.

9.20S Ice increases in volume by about 11 percent when it is frozen from liquid water, and the enthalpy of fusion of water, h_{sl}, is given in Table C.13. How much pressure, in kilopascals, must be exerted on a block of ice to change its melting point by 2°C? Is the melting point increased or decreased when pressure is increased?

9.20E Ice increases in volume by about 11 percent when it is frozen from liquid water, and the enthalpy of fusion of water, h_{sl}, is given in Table C.13. How much pressure, in psia, must be exerted on a block of ice to change its melting point by 4°F? Is the melting point increased or decreased when pressure is increased?

9.21S An extension of the Clapeyron equation, called the *Clausius-Clapeyron equation,* is

$$\frac{d(\ln P)}{dT} = \frac{h_{lg}}{RT^2}$$

It is based on the assumptions that $v_g \gg v_l$ and that the vapor phase acts as an ideal gas.

(*a*) Derive the Clausius-Clapeyron equation from the Clapeyron equation.

(*b*) Using steam table data, make a plot of $\ln P_{sat}$ versus $1/T_{sat}$ for steam in the range $10 < P_{sat} < 100$ kPa. From the plot and using the Clausius-Clapeyron equation, find h_{lg} for steam. Check your result against the steam tables.

9.21E An extension of the Clapeyron equation, called the *Clausius-Clapeyron equation,* is

$$\frac{d(\ln P)}{dT} = \frac{h_{lg}}{RT^2}$$

It is based on the assumptions that $v_g \gg v_l$ and that the vapor phase acts as an ideal gas.

(*a*) Derive the Clausius-Clapeyron equation from the Clapeyron equation.

(*b*) Using steam table data, make a plot of $\ln P_{sat}$ versus $1/T_{sat}$ for steam in the range $1 < P_{sat} < 14.7$ psia. From the plot and using the Clausius-Clapeyron equation, find h_{lg} for steam. Check your result against the steam tables.

9.22S From the two data points given, estimate the temperature of the triple point of water. Remember that the triple point occurs where the lines for vapor-liquid and liquid-solid equilibrium meet on a P-T diagram. (The Clausius-Clapeyron equation of Prob. 9.21 will be helpful for the vapor-liquid equilibrium line.)

Solid-liquid at $T = -10$°C and $P_{sat} = 119$ MPa:

$v_{sl} = -1.1 \times 10^{-4}$ m³/kg

$h_{sl} = 333.5$ kJ/kg

Liquid-vapor at $P = 1$ kPa and $T = 7°C$:

$h_{lg} = 2393$ kJ/kg

9.22E From the two data points given, estimate the temperature of the triple point of water. Remember that the triple point occurs where the lines for vapor-liquid and liquid-solid equilibrium meet on a *P-T* diagram. (The Clausius-Clapeyron equation of Prob. 9.21 will be helpful for the vapor-liquid equilibrium line.)

Solid-liquid at $T = 14.0°F$ and $P_{sat} = 16{,}200$ psia:

$v_{sl} = -1.76 \times 10^{-3}$ ft³/lbm

$h_{sl} = 143.4$ Btu/lbm

Liquid-vapor at $P = 1$ psia and $T = 101.7°F$:

$h_{lg} = 1029$ Btu/lbm

9.23S Solar energy systems often store energy in the form of hot water in storage tanks. Because of the density difference between hot and cold water, the storage tank can often be "stratified"; i.e., hot water introduced at the top of the tank tends to remain there because of its high specific volume (low density) relative to the colder water below.

(*a*) Plot the volumetric expansion coefficient β for liquid water in the range $60 \le T \le 100°C$, using data from the steam tables.

(*b*) Plot the volumetric expansion coefficient β for liquid water in the range $0 \le T \le 15°C$.

(*c*) In light of the data generated above, discuss the feasibility of hot versus cold storage stratification for solar-powered cooling systems.

9.23E Solar energy systems often store energy in the form of hot water in storage tanks. Because of the density difference between hot and cold water, the storage tank can often be "stratified"; i.e., hot water introduced at the top of the tank tends to remain there because of its high specific volume (low density) relative to the colder water below.

(*a*) Plot the volumetric expansion coefficient β for liquid water in the range $140 \le T \le 212°F$, using data from the steam tables.

(*b*) Plot the volumetric expansion coefficient β for liquid water in the range $32 \le T \le 50°F$.

(*c*) In light of the data generated above, discuss the feasibility of hot versus cold storage stratification for solar-powered cooling systems.

9.24S Plot the volumetric expansion coefficient for water in the range $0 \le T \le 15°C$, using data from the steam tables. Using these data, discuss why lakes freeze first at the surface rather than at the bottom.

9.24E Plot the volumetric expansion coefficient for water in the range $32 \leq T \leq 50°F$, using data from the steam tables. Using these data, discuss why lakes freeze first at the surface rather than at the bottom.

9.25S In a "solar pond," injection of salt at various concentrations induces a strong gradient in specific volume with depth in the pond. Such a pond is stable when the salt specific volume gradient equals or exceeds the gradient in specific volume due to the temperature gradient caused by the absorption of solar energy in the pond, i.e.,

$$\left[\frac{1}{v}\left(\frac{\partial v}{\partial T}\right)_P dT < \frac{1}{v}\left(\frac{\partial v}{\partial X}\right)_T dX\right]_z$$

where X is the salt concentration and z is the distance beneath the pond surface.

For a pond the temperature gradient in the solar-heated layer is given by

$$T(z) = 20 + 30z \qquad °C$$

where z is in meters. Find the required minimum concentration of salt if the specific volume of a saline solution is given by

$$v = v_1(X = 0) + [v_{sat} - v_1(X = 0)]X(z)$$

where v_{sat} is the specific volume of salt-saturated brine (0.00093 m^3/kg) and $v_1(X = 0)$ is the specific volume of pure water at the local temperature. (Actually the specific volume of the brine varies with T as well as X, which makes the problem more complicated.)

9.25E In a "solar pond," injection of salt at various concentrations induces a strong gradient in specific volume with depth in the pond. Such a pond is stable when the salt specific volume gradient equals or exceeds the gradient in specific volume due to the temperature gradient caused by the absorption of solar energy in the pond, i.e.,

$$\left[\frac{1}{v}\left(\frac{\partial v}{\partial T}\right)_P dT < \frac{1}{v}\left(\frac{\partial v}{\partial X}\right)_T dX\right]_z$$

where X is the salt concentration and z is the distance beneath the pond surface.

For a pond the temperature gradient in the solar-heated layer is given by

$$T(z) = 65 + 5z \qquad °F$$

where z is in feet. Find the required minimum concentration of salt if the specific volume of a saline solution is given by

$$v = v_1(X=0) + [v_{sat} - v_1(X=0)]X(z)$$

where v_{sat} is the specific volume of salt-saturated brine (0.0149 ft^3/lbm) and $v_1(X=0)$ is the specific volume of pure water at the local temperature. (Actually the specific volume of the brine varies with T as well as X, which makes the problem more complicated.)

9.26S At what pressure does the fugacity of ammonia gas (NH_3) first deviate by more than 10 percent from the pressure when the ammonia temperature is 150°C?

9.26E At what pressure does the fugacity of ammonia gas (NH_3) first deviate by more than 10 percent from the pressure when the ammonia temperature is 300°F?

9.27 Use the data from Fig. C.8 to determine the quality of the N_2 vapor-liquid mixture leaving the throttling valve and entering the separation tank of Fig. 9.5 if the temperature and pressure at the exit to the recuperator are 165 K and 14 MPa and the separation tank is at 1 atm.

9.28 For the conditions of Prob. 9.27 draw a P-h diagram for the Linde process for nitrogen liquefaction of Fig. 9.5 if each stage of the compressor has an efficiency of 80 percent, the gas feed is at 27°C and 1 atm, the intercooler and aftercooler return the gas stream to 30°C, and the exit from the first stage of the compressor is at 1.2 MPa. Provide a state table, using data from Fig. C.8 where possible. What power is required to drive the compressor to produce 0.5 kg/s of liquid nitrogen? (Assume that all liquid leaving the throttling valve and entering the separation tank is separated and becomes liquid product, while the saturated vapor is recycled to the compressor.)

Reference

1. Lester Haar, John S. Gallagher, and George S. Kell, *NBS/NRC Steam Tables: Thermodynamic and Transport Properties and Computer Programs for Vapor and Liquid States of Water in SI Units,* Hemisphere, Washington, 1984.
2. Byung Ik Lee and Michael G. Kesler, "A Generalized Thermodynamic Correlation Based on Three-Parameter Corresponding States," *American Institute of Chemical Engineering Journal,* vol. 21, no. 3, May 1975, pp. 510–527.

10

Multicomponent Systems without Chemical Reaction

The most important thing we can experience is the mysterious. It is the source of all true art and science.
Albert Einstein

Experimental supercharged 6-cylinder 24-valve 2-stroke automotive engine. Includes direct electronic fuel injection and computerized ignition control. Produces 240 hp at 3600 rpm. (*Toyota Motor Sales, U.S.A., Inc.*)

TOYOTA S-2

10.1 Introduction

A multicomponent fluid is composed of more than one chemical species. The most familiar example is air, which is a mixture of N_2, O_2, Ar, and trace amounts of other chemical species. Another important example is moist air, which adds water vapor to the above list of chemical components. In this chapter we consider the thermodynamic properties of mixtures and chemical species as well as analysis of systems that use multicomponent fluids.

Most of the thermodynamic analysis and systems considered to this point have involved pure substances such as water or refrigerant 12. The thermodynamic properties for pure substances are considered in Chaps. 3 and 9, and the property tables are contained in the appendixes. There exists a large body of property data on pure substances. These data are available through numerical tables, equations, and programs for all phases—solid, liquid, and vapor—of numerous pure substances. We seek to use this wealth of information for multicomponent systems when possible.

One approach to obtaining thermodynamic properties for multicomponent fluids is to generate data for the specific mixture needed. This is exactly what has been done for air. As indicated, air is a mixture of various chemical species, and the propety data have been obtained for air. This approach is useful since air, with fixed proportions of chemical species, is used in many engineering systems. The same approach is also required for other nondilute mixtures; however, a separate table is needed for every imaginable combination of chemical species. In some cases this is not required, and we consider these cases in this chapter.

The first goal of this chapter is to develop methods, through idealizations, to combine the properties of pure substances to obtain the properties for multicomponent fluids. Then we analyze systems that utilize multicomponent fluids. Last, we analyze nonideal mixtures.

10.2 Multicomponent Measures

A mixture is characterized in terms of the number of moles of each component i or in terms of the mass of each component i. The number of moles of component i is denoted n_i, and the total number of moles in the mixture is

$$n = \sum_i n_i \tag{10.1}$$

The fraction of the total moles that is made up of component i is the *mole fraction,* denoted

$$y_i = \frac{n_i}{n} \tag{10.2}$$

The sum of all mole fractions is unity, so

$$\sum_i y_i = 1 \tag{10.3}$$

For a binary mixture, these relations are

$$n = n_A + n_B \tag{10.4}$$

and

$$y_A + y_B = 1 \tag{10.5}$$

A mixture is also characterized in terms of the mass of each component i, which is denoted m_i. The mass of the mixture is

$$m = \sum_i m_i \tag{10.6}$$

for nonreacting components. The *mass fraction* of component i is denoted

$$x_i = \frac{m_i}{m} \tag{10.7}$$

(The subscript denotes the component of a multicomponent fluid, while x without a subscript is the quality, which specifically means m_g/m.) Therefore, the sum of all the mass fractions of the mixture yields

$$\sum_i x_i = 1 \tag{10.8}$$

For a binary mixture this is simply

$$m = m_A + m_B \tag{10.9}$$

and

$$x_A + x_B = 1 \tag{10.10}$$

The mass of component i is related to the number of moles of component i by the molecular weight of component i, M_i, as

$$m_i = n_i M_i \tag{10.11}$$

The total mass of the mixture is given by

$$m = \sum_i m_i = \sum_i n_i M_i \tag{10.12}$$

The *mean molecular weight* of the mixture M is defined so that $nM \equiv m$.

Thus

$$M = \frac{m}{n} = \frac{1}{n} \sum_i n_i M_i = \sum_i y_i M_i \tag{10.13}$$

For a binary mixture, this is $M = y_A M_A + y_B M_B$. Also the relationship between the mass fraction and mole fraction is obtained from Eqs. (10.11) and (10.13) as

$$\frac{M_i}{M} y_i = x_i \tag{10.14}$$

The above relations indicated the notation and definitions used for the mass or mole characterization of a mixture of various components. Since there are no chemical reactions, the number of moles of each component, or the mass of each component, sums to the total amount. The next step is to prescribe properties of a mixture in terms of the properties of the individual components. *A simple sum of the individual component extensive properties does not necessarily describe the mixture property.* Consider a constant-pressure constant-temperature mixing of two real gases. Gas A occupies an initial volume V_A, and gas B occupies an initial volume V_B. They are mixed. The final volume of the mixture is V_C, and *generally*

$$V_C \neq V_A + V_B \tag{10.15}$$

This is diagrammed in Fig. 10.1. Typical results for the complete range of mole fractions are shown in Fig. 10.2. Note, from a general viewpont, that there is no reason to expect that the mixture volume is obtained from the addition of the original component volumes. Molecularly this is thought of as an interaction between the components that does not permit the volumes to add simply. The molecules of one gas can be imagined to "tuck in" between the molecules of the other, resulting in the nonlinear behavior of Fig. 10.2. The above example is also valid for mixtures of liquids. This result for a general mixture holds for all extensive properties. The internal energy, enthalpy, entropy, and specific heats of the mixture of real fluids at constant temperature and pressure are not the simple sum of pure component properties.

This complexity of mixture properties is simplified in the case of ideal gases.

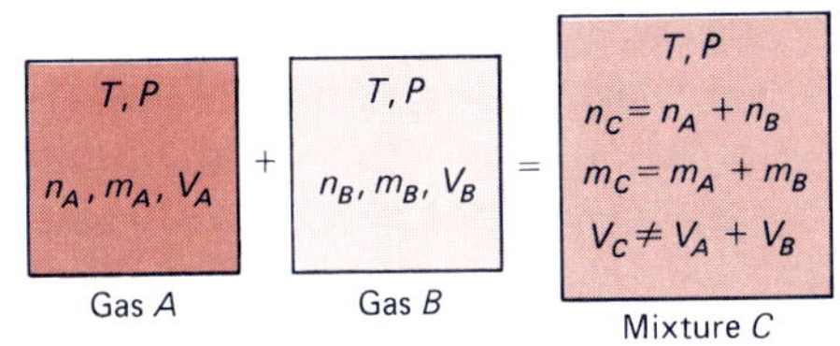

Figure 10.1 Mixture relations.

Figure 10.2 Mixture volume.

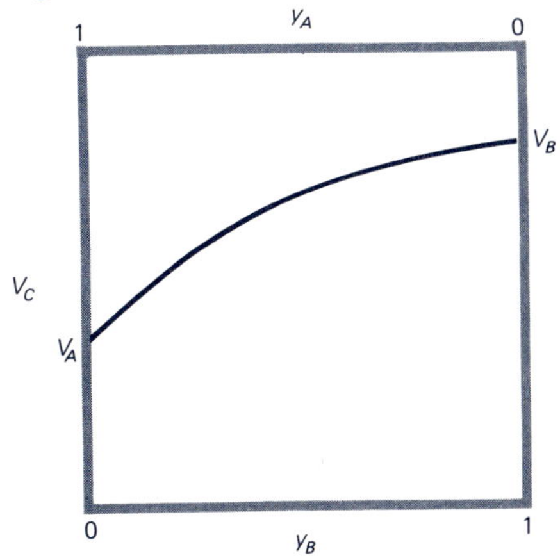

10.3 Properties of a Multicomponent Ideal Gas

An ideal gas is a special case of a general mixture but is one with significant practical utility. Physically, a gas satisfies the ideal gas relation when the gas molecules are treated as independent of one another. A mixture of ideal gases also satisfies this molecular independence idealization. As is shown later, the mixture properties are simply mole or mass weighted averages of the individual pure-substance properties.

Reconsider the mixing of two gases presented in Sec. 10.2 with each gas satisfying the ideal gas relation. Ideal gas A is mixed with ideal gas B at constant temperature and constant pressure. Since an ideal gas implies molecular independence, the final volume *is* the sum of the two volumes occupied by each gas independently. This is shown in Fig. 10.3 and is termed the *Amagat-Leduc law* of additive volumes. Repeating this experiment of various mole fractions yields a straight line on a V-y diagram (see Fig. 10.3).

This relation is demonstrated by the equation of state for a binary mixture of ideal gases. Each component is described by

$$P_A V_A = n_A \bar{R} T_A \tag{10.16a}$$

and

$$P_B V_B = n_B \bar{R} T_B \tag{10.16b}$$

and the mixture is given as

$$P_C V_C = n_C \bar{R} T_C \tag{10.16c}$$

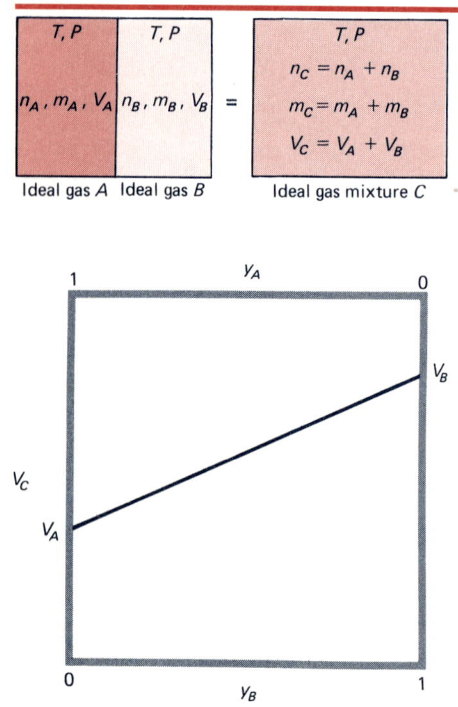

Figure 10.3 Ideal gas mixtures.

For a constant-temperature constant-pressure mixing, $T_A = T_B = T_C$ and $P_A = P_B = P_C$. Thus

$$V_A = \frac{n_A \overline{R} T_A}{P_A} = \frac{n_A \overline{R} T_C}{P_C} \tag{10.17a}$$

and

$$V_B = \frac{n_B \overline{R} T_B}{P_B} = \frac{n_B \overline{R} T_C}{P_C} \tag{10.17b}$$

The sum of the volumes yields

$$V_A + V_B = (n_A + n_B)\frac{\overline{R} T_C}{P_C} \tag{10.18}$$

but $n_A + n_B = n_C$ so that $V_A + V_B = V_C$, and the mixture relation in Eq. (10.16c) is satisfied. Thus, the mixture volume is the sum of the individual gas volumes. This development could also be viewed as proof that the mixture is an ideal gas when the individual ideal gases are combined. Also eliminating $\overline{R}T_C/P_C$ from Eqs. (10.17) and (10.16) yields the general relation

$$\frac{V_i}{V} = \frac{n_i}{n} = y_i \tag{10.19}$$

A multicomponent mixture of two or more ideal gases is given as

$$V = \sum_i V_i = \sum_i y_i V \tag{10.20}$$

where V_i is termed the *partial volume* of component i. The partial volume is the pure component volume that would result at the mixture temperature and pressure if that component were separated from the mixture.

An alternative viewpoint for a mixture of ideal gases considers the mixing of two ideal gases which are initially at different pressures but occupy equivalent volumes at the same temperature. Figure 10.4 shows this mixing. Physically, the pressure (force per unit area) exerted on a plane within an ideal gas results from the change in momentum of the independent molecules in random motion. Thus, the pressure contribution from mixing two ideal gases which have independent molecules should be the sum of the individual gas pressure contributions (with volumes and temperatures equal). This is termed *Dalton's law* of additive pressures.

Figure 10.4 Ideal gas mixture.

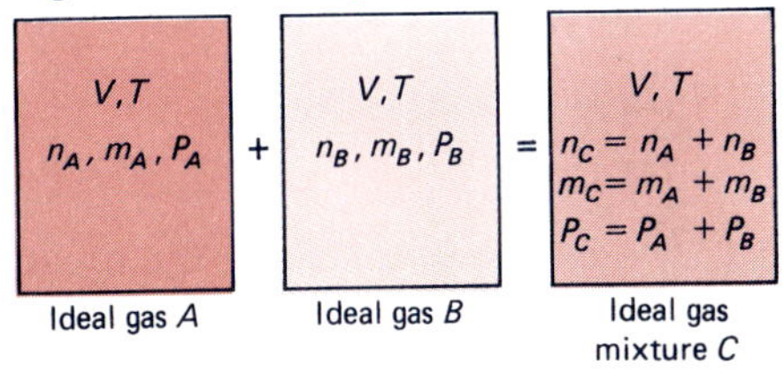

The ideal gas equation of state for the components and mixtures is given in Eq. (10.16). For a constant-temperature and equal-volume mixing, $T_A = T_B = T_C$ and $V_A = V_B = V_C$. Thus,

$$P_A = \frac{n_A \overline{R} T_A}{V_A} = \frac{n_A \overline{R} T_C}{V_C} \tag{10.21a}$$

and

$$P_B = \frac{n_B\bar{R}T_B}{V_B} = \frac{n_B\bar{R}T_C}{V_C} \qquad (10.21b)$$

Thus, the sum of the pressures is

$$P_A + P_B = (n_A + n_B)\frac{\bar{R}T_C}{V_C} \qquad (10.22)$$

But $n_A + n_B = n_C$ and $P_A + P_B = P_C$, so the mixture relation in Eq. (10.16c) is satisfied. Note that $\bar{R}T_C/V_C$ can be eliminated from Eq. (10.21) with Eq. (10.16) to yield the general relation

$$\frac{P_i}{P} = \frac{n_i}{n} = y_i \qquad (10.23)$$

where P_i is termed the *partial pressure* of component *i*. The partial pressure of component *i* is the pressure that would be exerted by the component at the mixture temperature and volume if it were separated from the mixture. A general multicomponent mixture of ideal gases yields

$$P = \sum_i P_i = \sum_i y_i P \qquad (10.24)$$

The above relations, which were stated as the Amagat-Leduc law and Dalton's law, yield identical information. Another way to state the equivalency is to say that the Amagat-Leduc law requires that Dalton's law be satisfied or Dalton's law requires the Amagat-Leduc Law be satisfied (depending on the starting point). This is seen by comparing Eqs. (10.19) and (10.23). An important point is that these statements are founded on the molecular independence concept for ideal gases and are strictly valid only for ideal gases. Modifications of these statements are available for real gas properties and are described in Sec. 10.8.

A mixture of ideal gases is an ideal gas. Therefore, it is often convenient to have an apparent ideal gas equation of state for this mixture. Equation (10.13) defines the mean molecular weight M of the ideal gas mixture, which can be viewed as the apparent molecular weight of the ideal gas mixture. On a mole basis, the mixture equation of state is [see Eq. (10.16c)]

$$PV = n\bar{R}T \qquad (10.25)$$

where the properties are those of the mixture. On a mass basis, Eq. (10.13) is used to yield

$$PV = \frac{m}{M}\bar{R}T \qquad (10.26)$$

An *apparent gas constant* for the mixture is defined as

$$R = \frac{\bar{R}}{M} \qquad (10.27a)$$

and with Eq. (10.13), $M = m/n$. So

$$R = \frac{\overline{R}}{m} n = \frac{\overline{R}}{m} \sum_i \frac{m_i}{M_i} = \sum_i x_i R_i \tag{10.27b}$$

and the mixture equation of state is

$$PV = mRT \tag{10.28}$$

with the apparent gas constant given in Eq. (10.27).

The remaining ideal gas mixture properties are obtained from an extension of Dalton's law which is referred to as the *Gibbs-Dalton law.* This states that any extensive property of a multicomponent mixture is the sum of the individual pure substance properties that would occupy the mixture volume at the mixture temperature. Therefore, the mixture properties are

$$U = \sum_i U_i(T, V) \tag{10.29a}$$

$$H = \sum_i H_i(T, V) \tag{10.29b}$$

$$S = \sum_i S_i(T, V) \tag{10.29c}$$

$$C_v = \sum_i C_{v,i}(T, V) \tag{10.29d}$$

$$C_P = \sum_i C_{P,i}(T, V) \tag{10.29e}$$

With the specific extensive property on a mole basis denoted with an overbar, Eqs. (10.29) become

$$U = n\bar{u} = n \sum_i y_i \bar{u}_i(T, V) \tag{10.30a}$$

$$H = n\bar{h} = n \sum_i y_i \bar{h}_i(T, V) \tag{10.30b}$$

$$S = n\bar{s} = n \sum_i y_i \bar{s}_i(T, V) \tag{10.30c}$$

$$C_v = n\bar{c}_v = n \sum_i y_i \bar{c}_{v,i}(T, V) \tag{10.30d}$$

$$C_P = n\bar{c}_P = n \sum_i y_i \bar{c}_{P,i}(T, V) \tag{10.30e}$$

On a mass basis, the mixture properties are

$$U = mu = m \sum_i x_i u_i(T, V) \tag{10.31a}$$

$$H = mh = m \sum_i x_i h_i(T, V) \tag{10.31b}$$

$$S = ms = m \sum_i x_i s_i(T, V) \tag{10.31c}$$

$$C_v = mc_v = m \sum_i x_i c_{v,i}(T, V) \tag{10.31d}$$

$$C_P = mc_P = m \sum_i x_i c_{P,i}(T, V) \tag{10.31e}$$

Note that all the pure-substance properties on the right-hand sides of the above equations are obtained at the volume and temperature of the mixture. This is explicitly indicated by the functional notation. From the discussion of Dalton's law of additive pressures, this is equivalent to evaluating the pure-substance properties at the mixture temperature and the component *partial* pressure. This is explicitly indicated in the following expressions:

$$U = n\bar{u} = n \sum_i y_i \bar{u}_i(T, P_i) = mu = m \sum_i x_i u_i(T, P_i) \tag{10.32a}$$

$$H = n\bar{h} = n \sum_i y_i \bar{h}_i(T, P_i) = mh = m \sum_i x_i h_i(T, P_i) \tag{10.32b}$$

$$S = n\bar{s} = n \sum_i y_i \bar{s}_i(T, P_i) = ms = m \sum_i x_i s_i(T, P_i) \tag{10.32c}$$

$$C_v = n\bar{c}_v = n \sum_i y_i \bar{c}_{v,i}(T, P_i) = mc_v = m \sum_i x_i c_{v,i}(T, P_i) \tag{10.32d}$$

$$C_P = n\bar{c}_P = n \sum_i y_i \bar{c}_{P,i}(T, P_i) = mc_P = m \sum_i x_i c_{P,i}(T, P_i) \tag{10.32e}$$

For an ideal gas, the internal energy, enthalpy, and specific heats are functions of temperature only, so entropy is the only property requiring the use of the partial pressure.

Expanding Eq. (10.32*c*) to provide the specific entropy of an ideal gas mixture [noting $s_{\text{ref}} = s\,(T = 0\text{ K} = 0°\text{R}, P = 1\text{ atm}) = 0$] results in

$$s = \sum_i x_i s_i(T, P_i) = \sum_i x_i \left[s_{0,i}(T) - R_i \ln \frac{P_i}{P_{\text{ref}}} \right] \tag{10.33}$$

This is a general relation in terms of partial pressures. (It can be alternatively expressed in terms of mole fractions.) If we define

$$s_0(T) \equiv \sum_i x_i s_{0,i}(T) \tag{10.34}$$

and use the properties of the natural logarithm to write

$$\ln \frac{P_i}{P_{\text{ref}}} = \ln \frac{P_i}{P_{\text{tot}}} + \ln \frac{P_{\text{tot}}}{P_{\text{ref}}}$$

then Eq. (10.33) becomes

$$s(T, P) = s_0(T) - \sum_i x_i R_i \ln \frac{P_{\text{tot}}}{P_{\text{ref}}} - \sum_i x_i R_i \ln \frac{P_i}{P_{\text{tot}}} \tag{10.35}$$

However, $P_i/P_{\text{tot}} = y_i$, and Eq. (10.27*b*) shows that $\sum_i x_i R_i = R$. Further, from Eq. (10.14), $x_iR_i = y_iM_iR_i/M = y_iR$. Making these substitutions results in

$$s(T, P) = s_0(T) - R \ln \frac{P_{\text{tot}}}{P_{\text{ref}}} - R \sum_i y_i \ln y_i \tag{10.36a}$$

or

$$\bar{s}(T, P) = \bar{s}_0(T) - \bar{R} \ln \frac{P_{\text{tot}}}{P_{\text{ref}}} - \bar{R} \sum_i y_i \ln y_i \tag{10.36b}$$

For a mixture with *constant composition* undergoing a process between states 1 and 2,

$$s_2 - s_1 = s_0(T_2) - s_0(T_1) - R \ln \frac{P_{\text{tot, 2}}}{P_{\text{tot, 1}}} \tag{10.36c}$$

and the final term in Eqs. (10.36*a* and 10.36*b*) drops out. In tabulations of properties for an ideal gas mixture, the $\sum_i y_i \ln y_i$ term is omitted from the tabulated s values.

Example Problem 10.1

Air is a mixture of nitrogen, oxygen, argon, and trace amounts of other components. Calculate u, h, s_0, and c_P at 1 atm and 600°F from Tables E.3 to E.5 (which all have the same reference state). Take air to be composed of 78.03 percent N_2, 20.99 percent O_2, and 0.98 percent argon on a volume basis.

Solution

Equation (10.19) indicates that the mole fractions are identically equal to the volume fractions. Thus the multicomponent measures of composition are as follows:

Component	Molecular Weight M	Volume Fraction V_i/V	Mole Fraction y_i	Mass Fraction x_i
N_2	28.013	0.7803	0.7803	0.7546
O_2	31.999	0.2099	0.2099	0.2319
Ar	39.948	0.0098	0.0098	0.0135
		1.0000	1.0000	1.0000

Here Eq. (10.13) yields

$$M = \sum_i y_iM_i = 0.7803(28.013) + 0.2099(31.999) + 0.0098(39.948)$$

$$= 28.967$$

and Eq. (10.14) gives

$$x_i = y_i \frac{M_i}{M}$$

which yields the values in the above table.

From Tables E.3, E.4, and E.5 and Eq. (10.32)

$$\begin{aligned} u &= x_{N_2}u_{N_2} + x_{O_2}u_{O_2} + x_{Ar}u_{Ar} \\ &= 0.7546(189.47) + 0.2319(170.43) + 0.0135(79.069) \\ &= 183.56 \text{ Btu/lbm} = 426.96 \text{ kJ/kg} \end{aligned}$$

$$\begin{aligned} h &= x_{N_2}h_{N2} + x_{O_2}h_{O_2} + x_{Ar}h_{Ar} \\ &= 0.7546(264.64) + 0.2319(236.24) + 0.0135(131.78) \\ &= 256.26 \text{ Btu/lbm} = 596.06 \text{ kJ/kg} \end{aligned}$$

and

$$\begin{aligned} c_P &= x_{N_2}c_{P,N_2} + x_{O_2}c_{P,O_2} + x_{Ar}c_{P,Ar} \\ &= 0.7546(0.2563) + 0.2319(0.2389) + 0.0135(0.1244) \\ &= 0.2505 \text{ Btu/(lbm} \cdot {}^\circ\text{R)} = 1.049 \text{ kJ/(kg} \cdot \text{K)} \end{aligned}$$

The computation of entropy including variable specific heat requires use of Eq. (10.34) to yield

$$\begin{aligned} s_0 &= (0.7546)(1.8044) + (0.2319)(1.6857) + (0.0135)(1.0104) \\ &= 1.7662 \text{ Btu/(lbm} \cdot {}^\circ\text{R)} = 7.3947 \text{ kJ/(kg} \cdot \text{K)} \end{aligned}$$

These values are compared to the tabulated values in Table E.2. The maximum difference is approximately 0.01 percent, which is within the above specification for the volume fractions of air.

Comments

The above calculations are possible since all the values in the tables are referenced to the same state. The usual need for properties in thermodynamics involves the differences between values as calculated in this example. The calculated mixture property values can be used directly to obtain differences in properties.

10.4 Thermodynamic Analysis of Ideal Gas Mixtures

When a mixture of nonreacting ideal gases flows through a control volume, analysis employing the first and second laws follows the same relations used for simple substances. However, the changes in the properties of the *mixture* must be determined from the property relations in Sec. 10.3 for insertion into first and second law relations.

This section applies the first and second laws to multicomponent systems. Consideration is given to a system with fixed mole and mass fractions as well as to mixing problems. Since there are no chemical reactions, the number of moles or the mass of each component can change only by the addition or deletion of a component. If a mixture of fixed composition is used in a process and there is no addition or deletion of specific components, then the mole and mass fractions are constant. This multicomponent system can be visualized as an ideal gas with fixed mixture properties. But if a process is considered in which the mole and mass fractions change, a special approach is required. This last process is referred to as a *mixing process.*

The internal energy, enthalpy, and specific heats of an ideal gas are functions of temperature only. As indicated in Table 5.4, the changes in internal energy and enthalpy are

$$du_i = c_{v,i}\,dT \tag{10.37a}$$

and

$$dh_i = c_{P,i}\,dT \tag{10.37b}$$

on a mass basis. From Eq. (10.32)

$$dU = \sum_i n_i \bar{c}_{v,i}\,dT = \sum_i m_i c_{v,i}\,dT \tag{10.38a}$$

$$dH = \sum_i n_i \bar{c}_{P,i}\,dT = \sum_i m_i c_{P,i}\,dT \tag{10.38b}$$

With the mixture definitions for the specific heats in Eqs. (10.32*d*) and (10.32*e*),

$$dU = n\bar{c}_v\,dT = mc_v\,dT \tag{10.39a}$$

$$dH = n\bar{c}_P\,dT = mc_P\,dT \tag{10.39b}$$

Again the physical viewpoint for changes in internal energy and enthalpy is one of independent nonreacting gases simply being added, and the resulting mixture is an ideal gas with new averaged properties. This is not so simple for entropy.

The entropy of a mixture is given by Eq. (10.32*c*). Entropy for an ideal gas is a function of two independent variables. Recalling that the mixture is treated as a multicomponent system with pure-substance properties evaluated at the mixture temperature and component partial pressure, we see the change in entropy for each component is given from Table 5.4 as

$$d\bar{s}_i = \bar{c}_{P,i}\frac{dT}{T} - \bar{R}\frac{dP_i}{P_i} \tag{10.40a}$$

on a mole basis and

$$ds_i = c_{P,i}\frac{dT}{T} - R_i\frac{dP_i}{P_i} \tag{10.40b}$$

on a mass basis. The change in mixture entropy is

$$dS = \sum_i n_i \bar{c}_{P,i} \frac{dT}{T} - \sum_i n_i \bar{R} \frac{dP_i}{P_i} \tag{10.41a}$$

$$= \sum_i m_i c_{P,i} \frac{dT}{T} - \sum_i m_i R_i \frac{dP_i}{P_i} \tag{10.41b}$$

The mixture specific heats are given in Eqs. (10.32*d*) and (10.32*e*), so

$$dS = n\bar{c}_P \frac{dT}{T} - \bar{R} \sum_i n_i \frac{dP_i}{P_i} \tag{10.42a}$$

$$= mc_P \frac{dT}{T} - \sum_i m_i R_i \frac{dP_i}{P_i} \tag{10.42b}$$

The change in entropy can be evaluated with constant specific heats or with variable specific heats as presented in Sec. 5.5.2. The partial pressures in Eq. (10.42) can also be expressed in terms of the mole or mass fractions [Eq. (10.23)].

First, consider a mixture of ideal gases which proceeds from one state to another but does not change composition (the mole fractions and mass fractions are constant at all points in the process). The multicomponent measures are computed, and the mole and mass fractions are determined. Because the mole and mass fractions are constants, Eq. (10.23) indicates that the partial pressures vary with the total pressure. The mixture specific heats are evaluated by using Eqs. (10.32*d*) and (10.32*e*) as

$$dU = n\bar{c}_v \, dT = mc_v \, dT \tag{10.43a}$$

$$dH = n\bar{c}_P \, dT = mc_P \, dT \tag{10.43b}$$

$$dS = n\bar{c}_P \frac{dT}{T} - n\bar{R} \frac{dP}{P} \tag{10.43c}$$

$$= mc_P \frac{dT}{T} - mR \frac{dP}{P} \tag{10.43d}$$

where $P_i = y_i P$ and Eq. (10.27*b*) are used. These equations clearly indicate that the mixture can be treated as an ideal gas with averaged properties.

10.4.1 Applications to Processes with Constant Composition

Consider the case of the flow of a mixture of constant composition through a converging adiabatic nozzle. The composition of the fluid is assumed to be known. Using the first law for this case to find the velocity at the nozzle throat, assuming negligible upstream velocity and following Eq. (4.117),

$$\mathbf{V}_{t,a} = [2g_c(h_{\text{in}} - h_{t,a})]^{1/2} \tag{10.44}$$

Now the enthalpy values must be determined for the mixture. By using Eq. (10.43b) and assuming c_P to be independent of temperature, this becomes

$$\mathbf{V}_{t,a} = [2g_c c_P(T_{\text{in}} - T_{t,a})]^{1/2} \tag{10.45}$$

where again c_P is for the mixture.

Generally the actual value of the temperature of the fluid at the throat is unknown, and the second law and nozzle efficiency must be employed in order to find $T_{t,a}$.

To find the velocity of the fluid at the throat for isentropic flow, Eq. (10.43d) is employed. For isentropic flow with temperature-independent mixture specific heats, this gives

$$\frac{T_{t,s}}{T_{\text{in}}} = \left(\frac{P_t}{P_{\text{in}}}\right)^{(k-1)/k} \tag{10.46}$$

and again k is the ratio of mixture specific heats. If the specific heats are temperature-dependent, then Eq. (10.36c) can be used with data from the gas tables to determine the value of $T_{t,s}$.

The computed value of $T_{t,s}$ may be used in Eq. (10.44) to determine the throat velocity for isentropic flow $\mathbf{V}_{t,s}$, by

$$\mathbf{V}_{t,s} = [2g_c c_P(T_{\text{in}} - T_{t,s})]^{1/2} \tag{10.47}$$

Now the actual throat velocity $\mathbf{V}_{t,a}$ is determined from the nozzle efficiency relation Eq. (5.196):

$$\eta_N = \frac{\mathbf{V}_{t,s}^2/(2g_c)}{\mathbf{V}_{t,a}^2/(2g_c)} = \frac{\mathbf{V}_{t,s}^2}{\mathbf{V}_{t,a}^2} \tag{10.48}$$

and then $\mathbf{V}_{t,a}$ is substituted into Eq. (10.45) to find $T_{t,a}$ if desired.

Note that this is the same procedure used for analysis of a nozzle when the flow is of a single-component fluid, *except* that all properties are replaced by mixture properties. This is always the case for analysis of systems employing mixtures of constant composition, so we need not further examine detailed specific applications of the laws.

10.4.2 Entropy Changes in the Mixing of Ideal Gases

Consider now the case when two or more ideal gases are mixed. The mole fractions of the gases before and after mixing are not the same. How does this affect the entropy change of the process? The general relations are given in Eqs. (10.40) through (10.42). Also note that alternative yet equivalent expressions are given in Eqs. (10.36).

Let a number of gases A, B, C, . . . be placed in a rigid container and originally separated by a partition. All the gases are intially at the same temperature T and pressure P, and thus each has entropy following Eq. (5.54) as

$$\bar{s}_{i,1} = \bar{s}_{0,i}(T) + \bar{R}\ln\frac{V_{i,1}}{V_{\mathrm{ref}}} \tag{10.49}$$

The total entropy of all the gases in the rigid container is then given by

$$S_1 = \sum_i n_i\bar{s}_{i,1} = \sum_i n_i\left[\bar{s}_{0,i}(T) + \bar{R}\ln\frac{V_{i,1}}{V_{\mathrm{ref}}}\right] \tag{10.50}$$

Note that in state 1 the partial volume V_i occupied by each gas i when all the gases are at the same temperature and pressure is related to the number of moles of each gas n_i by

$$\frac{V_i}{V} = \frac{n_i}{n} \qquad \text{or} \qquad V_i = y_iV \tag{10.51}$$

and Eq. (10.50) becomes

$$s_1 = \sum_i n_i\left[\bar{s}_{0,i}(T) + \bar{R}\ln\frac{y_iV}{V_{\mathrm{ref}}}\right] \tag{10.52}$$

Now let the partitions between the gases be removed so that after some time they are completely mixed at state 2. Each gas now fills the entire volume V of the rigid container, and because the free expansion of an ideal gas involves no change in temperature, the mixing process is isothermal. The entropy of the individual gases at state 2 is therefore given by

$$\bar{s}_{i,2} = \bar{s}_{0,i}(T) + \bar{R}\ln\frac{V}{V_{\mathrm{ref}}} \tag{10.53}$$

and the total entropy at state 2 is

$$s_2 = \sum_i n_i\left[\bar{s}_{0,i}(T) + \bar{R}\ln\frac{V}{V_{\mathrm{ref}}}\right] \tag{10.54}$$

Now, using Eqs. (10.52) and (10.54), we see that the entropy change during the isothermal and constant-pressure mixture process is

$$S_2 - S_1 = \bar{R}\sum_i n_i\left(\ln\frac{V}{V_{\mathrm{ref}}} - \ln\frac{y_iV}{V_{\mathrm{ref}}}\right) = -\bar{R}\sum_i n_i \ln y_i \tag{10.55}$$

Or, dividing through by the total number of moles of gas in the mixture n to obtain the *change in entropy per mole of mixture,* we find

$$\bar{s}_2 - \bar{s}_1 = -\bar{R}\sum_i y_i \ln y_i \tag{10.56}$$

Note that in each of these equations y_i is the mole fraction of gas i in the mixture.

From Eq. (10.56) we see that *when two or more gases are mixed at constant pressure and temperature, there is an entropy change that occurs simply because the gases change from an unmixed to a mixed state!* Because the mole fraction of each gas y_i is always less than 1 for a mixture of two or more gases, the natural log terms in the summation will each carry a negative value, and the overall change in entropy for mixing is always positive. This coincides with our expectation that such processes will not spontaneously proceed to a state of less "mixedness." Rather, we expect the gases to move to a state of being mixed when they are brought together, and the increase in entropy that we find reflects this expectation. The mixing process is an example of why entropy is linked with the concepts of "disorder" and "uncertainty." After mixing, we are less sure of the location of a given gas molecule than when the gases were confined to smaller defined volumes, and the greater entropy can be interpreted in terms of our reduced certainty of their position after mixing, or their greater disorder.

Note that the pressure of each individual gas changes in the mixing process, because the partial pressure of each gas at state 1 is P, while at state 2 it is reduced to y_iP; however, the *total* pressure before and after the mixing process is unchanged.

10.4.3 Nonisothermal Mixing of Ideal Gases at Constant Pressure

When two ideal gases initially at different temperatures but at the same pressure are mixed, which is a common industrial process and is encountered in air conditioning applications, the entropy change is determined by treating each component as having a different temperature $T_{i,1}$ before mixing but the same temperature T_2 after mixing. By following a similar derivation to Eqs. (10.49) through (10.56), the overall change in entropy is given by

$$S_2 - S_1 = \sum_i n_i[\bar{s}_{0,i}(T_2) - \bar{s}_{0,i}(T_{i,1}) - \bar{R}\ln y_i] \qquad (10.57)$$

For an ideal gas with constant specific heats, and by placing the result in terms of the specific entropy change per mole of total mixture, Eq. (10.57) can be written

$$\bar{s}_2 - \bar{s}_1 = \sum_i y_i\left(\bar{c}_{P,i}\ln\frac{T_2}{T_{i,1}} - \bar{R}\ln y_i\right) \qquad (10.58)$$

Example Problem 10.2

Nitrogen at 0.3 MPa and 100°C is to be adiabatically mixed with He at 0.15 MPa and 50°C. Determine the final temperature of the mixture and the entropy generation if the gases are mixed in the following two processes:

(*a*) Initially 0.5 kg of N_2 and 0.05 kg of He are separated by a partition, and they are mixed in a closed vessel.
(*b*) In a steady state flow device with an outlet pressure of 0.1 MPa, 0.5 kg/s of N_2 and 0.05 kg/s of He are mixed.

Diagrams

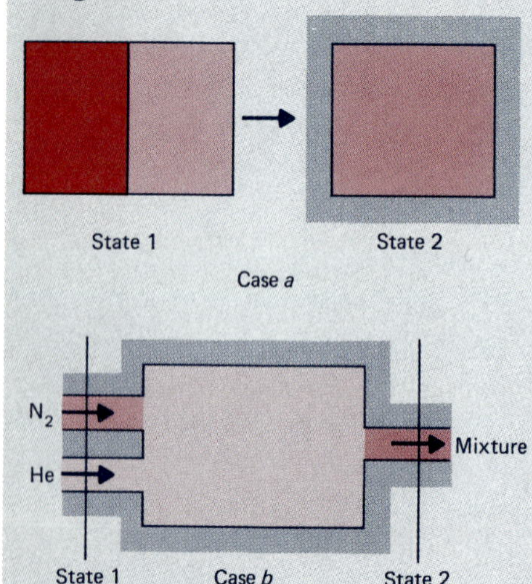

Solution

The first law is used to obtain state 2. The primary difference in the analysis of each of the mixing processes lies in the use of the internal energy for (*a*) and enthalpy for (*b*). Constant specific heats are assumed in both cases.

(*a*) The first law gives

$$U_2 - U_1 = m_2u_2 - m_1u_1 = 0$$

so

$$[m(u_2 - u_1)]_N + [m(u_2 - u_1)]_{He} = 0$$

where subscripts *N* and *He* represent nitrogen and helium, respectively. Thus

$$(mc_v)_N(T_2 - T_{N,1}) + (mc_v)_{He}(T_2 - T_{He,1}) = 0$$

since both components reach the same final equilibrium temperature. Thus

$$T_2 = \frac{(mc_v)_N T_{N,1} + (mc_v)_{He} T_{He,1}}{(mc_v)_N + (mc_v)_{He}}$$

Substituting the numerical values from Table C.2 yields

$$T_2 = 358.4 \text{ K} = 85.2°\text{C} = 185°\text{F}$$

The evaluation of the entropy change requires the partial pressures at state 2. The partial pressures are determined from $P_i = m_i R_i T/V$ for the final mixture volume. Evaluating the initial volumes and summing yield

$$V_2 = V_{N,1} + V_{He,1} = \frac{m_N R_N T_{N,1}}{P_{N,1}} + \frac{m_{He} R_{He} T_{He,1}}{P_{He,1}} = 0.408 \text{ m}^3$$

Thus

$$P_{N,2} = \frac{(0.5)(0.2968)(358.4)}{0.408} = 130 \text{ kPa}$$

$$P_{He,2} = \frac{(0.05)(2.07703)(358.4)}{0.408} = 91.2 \text{ kPa}$$

The entropy generation is evaluated from Eq. (10.41*b*) as

$$\begin{aligned} {}_1S_{gen,2} &= S_2 - S_1 \\ &= (mc_P)_N \ln \frac{T_2}{T_{N,1}} + (mc_P)_{He} \ln \frac{T_2}{T_{He,1}} \\ &\quad - (mR)_N \ln \frac{P_{N,2}}{P_{N,1}} - (mR)_{He} \ln \frac{P_{He,2}}{P_{He,1}} \\ &= 0.1813 \text{ kJ/K} = 0.0955 \text{ Btu/°R} \end{aligned}$$

(*b*) The first law for steady state adiabatic flow is

$$\dot{m}_1 h_1 = \dot{m}_2 h_2$$

$$\dot{m}_{N,1} h_{N,1} + \dot{m}_{He,1} h_{He,1} = \dot{m}_{N,2} h_{N,2} + \dot{m}_{He,2} h_{He,2}$$

Since $\dot{m}_{N,1} = \dot{m}_{N,2}$ and $\dot{m}_{He,1} = \dot{m}_{He,2}$, this yields

$$\dot{m}_N (h_2 - h_1)_N + \dot{m}_{He} (h_2 - h_1)_{He} = 0$$

$$(\dot{m}c_P)_N (T_2 - T_{N,1}) + (\dot{m}c_P)_{He} (T_2 - T_{He,1}) = 0$$

Solving for T_2 yields

$$T_2 = \frac{(\dot{m}c_P)_N T_{N,1} + (\dot{m}c_P)_{He} T_{He,1}}{(\dot{m}c_P)_N + (\dot{m}c_P)_{He}}$$

With the numerical values from Table C.2,

$$T_2 = 356.5\ \text{K} = 83.4°\text{C} = 182°\text{F}$$

The entropy generation calculation requires the outlet mole fractions. They are evaluated from $\dot{n}_N = 0.5/28.013 = 0.0178$ kmol/s and $\dot{n}_{He} = 0.05/4.003 = 0.0125$ kmol/s. Then

$$y_{N,2} = \frac{0.0178}{0.0178 + 0.0125} = 0.587$$

$$y_{He,2} = \frac{0.0125}{0.0178 + 0.0125}\ 0.413$$

The expression for the entropy generation for this steady flow process is

$$\dot{S}_{gen} = \dot{m}_2 s_2 - \dot{m}_1 s_1$$
$$= \dot{m}_N(s_{N,2} - s_{N,1}) + \dot{m}_{He}(s_{He,2} - s_{He,1})$$

With Eq. (10.40*b*), this is expressed as

$$\dot{S}_{gen} = (\dot{m}c_P)_N \ln \frac{T_2}{T_{N,1}} + (\dot{m}c_P)_{He} \ln \frac{T_2}{T_{He,1}} - (\dot{m}R)_N \ln \frac{y_{N,2}P_2}{P_{N,1}} - (\dot{m}R)_{He} \ln \frac{y_{He,2}P_2}{P_{He,1}}$$
$$= 0.378\ \text{kW/K} = 717\ \text{Btu/(h}\cdot°\text{R)}$$

Comments

The difference in the analysis for each case is slight. The specific heats are changed in the first law analysis, and the partial pressure evaluation differed in the second law analysis. Note that the properties of He differ greatly from those of N_2. Entropy generation due to mixing is positive for both gases.

10.5 Multicomponent Analysis of Ideal Gas-Vapor Mixtures

The analysis in the previous sections is now applied to a mixture which includes a component that is near its saturation state. In Chap. 3 we indicated that a vapor and a gas are both gaseous fluids, but a vapor is at a temperature below the critical temperature (see Fig. 3.6). The important point is that the vapor in the mixture is close to the saturation region for that component, so it can condense out of the mixture to form another phase.

The most common example is the mixture of air and water vapor which occurs in the atmosphere. This mixture is viewed as one component in the gas phase, air, mixed with another component in the vapor phase, water. At atmospheric conditions of 1 atm and 91°F (33°C), the meteorologist indicates that the relative humidity on that day is 90 percent. This is, indirectly, a specification of the mole and mass fractions of the mixture (this is explained subsequently). At the stated pressure and temperature, this corresponds to $x_w = 0.0269$. Thus the meteorologist has supplied the water concentration in the air-vapor mixture. Experience tells us that if this mixture were reduced in temperature while the total pressure was kept constant (atmospheric pressure), liquid droplets or dew would form, indicating the condensation process. This dew formation would occur if the ambient temperature were 87.5°F (30.8°C). This type of mixture of gas and vapor is considered in this section, and in Example Problem 10.3.

The analysis of gas-vapor mixtures is a direct application of the relations of Secs. 10.3 and 10.4. The gas and vapor are considered to obey the ideal gas relation, so the components are considered to be composed of independent molecules. The Gibbs-Dalton law states that the mixture properties are the sum of the pure-substance properties, each evaluated at the mixture temperature and the component partial pressure. Thus, the gas-vapor mixture, like any ideal gas mixture, can be treated as two independent components each at the *mixture temperature* and at the component *partial pressure.* Also the total pressure of the mixture is the sum of the partial pressures of the components. The partial pressure of the vapor component is called the *vapor pressure,* and the partial pressure of the gas component is called the *gas pressure.* The total pressure of the mixture is denoted P. The condensed phase is assumed to be pure (does not contain dissolved gas), and the saturation pressure of the condensed phase is not affected by the gas.

10.5.1 Measures and Properties

An ideal gas mixture of gas and vapor that undergoes any process is characterized by the relations in Secs. 10.3 and 10.4. The new aspect of the gas-vapor mixture is that the vapor phase can undergo a phase change. This fact requires an additional piece of information for a gas-vapor mixture which indicates the phase-change process. The mixture temperature at which the vapor will undergo a phase change to liquid while at the partial pressure within the mixture is called the *dew point temperature* T_d of the gas-vapor mixture. Consider an air–water vapor mixture with the water phase in the vapor state shown in Fig. 10.5. At the partial pressure at state 1, the dew point temperature is the saturation temperature at the water vapor partial pressure, or vapor pressure. This is indicated in Fig. 10.5 as 1*d*. Also note that any mixture state that has the vapor pressure of

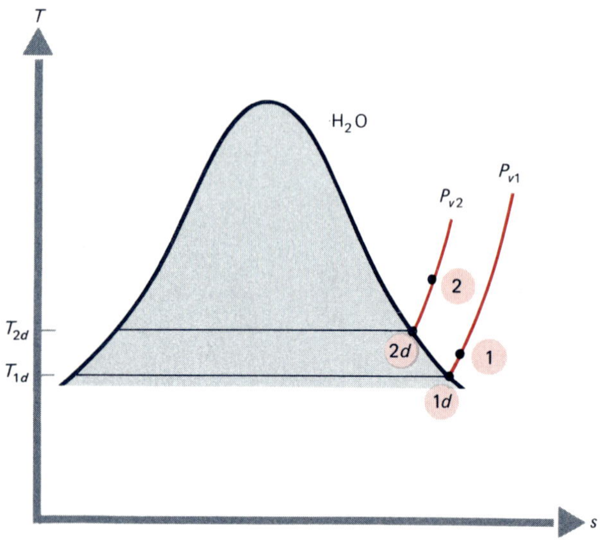

Figure 10.5 Dew point temperature.

P_{v1} has the same dew point temperature T_{1d}. For another state 2 with vapor pressure P_{v2}, the dew point temperature is T_{2d}.

A *saturated mixture* is a mixture with the vapor phase at its saturation pressure and temperature. Thus, an air–water vapor mixture with the water at state $1d$ or $2d$ in Fig. 10.5 is a saturated mixture. This mixture is also referred to as saturated air.

Relative Humidity

The concentrations of air–water mixtures are expressed in terms of two new, but related, quantities: the relative humidity ϕ and the humidity ratio. ω. The *relative humidity* ϕ is defined as

$$\phi \equiv \frac{y_v(T, P)}{y_g(T, P)} \tag{10.59}$$

where $y_v(T, P)$ is the mole fraction of the vapor component (subscript v) and $y_g(T, P)$ is the mole fraction of the vapor at the saturated state, both corresponding to the mixture temperature and total pressure. There is no v subscript in the demoninator since only the vapor component can undergo a phase change, which is indicated by the subscript g (as in Chap. 3). Since each component of the ideal gas mixture is treated as an ideal gas and the total pressures in the ratio in Eq. (10.59) are the same, the relative humidity is

$$\phi = \frac{P_v}{P_g} = \frac{P_v(T)}{P_{sat}(T)} \tag{10.60}$$

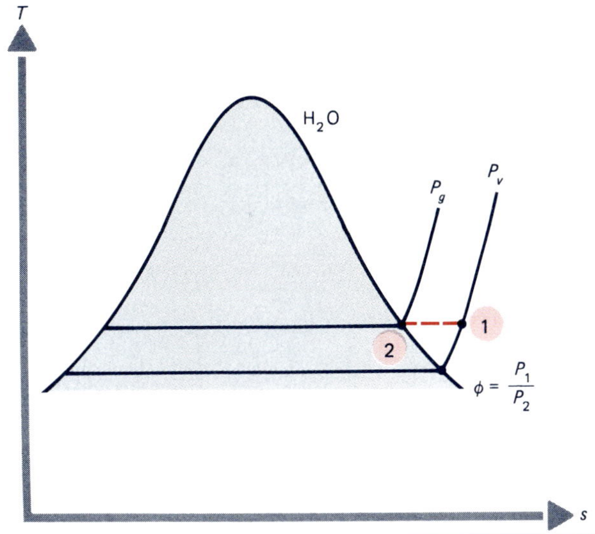

Figure 10.6 Relative humidity.

by Eq. (10.23). Figure 10.6 indicates the states that are compared in the definition of the relative humidity. This figure also shows that the relative humidity is always less than or equal to 1 since $P_v(T)$ is always less than or equal to $P_g(T)$. With the ideal gas relation for the vapor component, the relative humidity is also expressed as

$$\phi = \frac{P_v}{P_g} = \frac{RT/v_v}{RT/v_g} = \frac{v_g}{v_v} = \frac{\rho_v}{\rho_g} \tag{10.61}$$

Humidity Ratio

Another measure of the concentration of the mixture is the *humidity ratio,* or *specific humidity,* defined as

$$\omega \equiv \frac{m_v}{m_a} \tag{10.62}$$

This is the mass of water vapor divided by the mass of dry air (without water vapor). The humidity ratio is related to the mass fraction of vapor $x_v = m_v/m$ by

$$\omega = \frac{1}{1/x_v - 1} \tag{10.63}$$

The humidity ratio is also expressed in terms of partial pressures by noting that the volumes occupied by air and water vapor are the same. Thus,

$$\omega = \frac{m_v}{m_a} = \frac{M_v}{M_a}\frac{n_v}{n_a} = \frac{M_v}{M_a}\frac{P_vV/(\bar{R}T)}{P_aV/(\bar{R}T)}$$

or

$$\omega = \frac{M_v}{M_a}\frac{P_v}{P_a} \tag{10.64}$$

With the molecular weights given in Table C.2,

$$\omega = 0.622\frac{P_v}{P_a} = 0.622\frac{P_v}{P - P_v} = 0.622\frac{\phi P_g}{P - \phi P_g} \tag{10.65}$$

which is a function of the vapor pressure only. Another relation between the relative humidity and the humidity ratio is given by eliminating the vapor pressure from Eqs. (10.60) and (10.65)

$$\phi = \frac{\omega P_a}{0.622 P_g} \tag{10.66}$$

Degree of Saturation

Another measure of the mixture concentration is the *degree of saturation* μ, defined as

$$\mu \equiv \frac{\omega(T, P)}{\omega_g(T, P)} \tag{10.67}$$

With Eqs. (10.65) and (10.60), this is

$$\mu = \frac{P_v(P - P_g)}{P_g(P - P_v)} = \phi\frac{P - P_g}{P - P_v} \tag{10.68}$$

The vapor pressure and the saturation pressure are generally much less than the total pressure (except at high temperatures), so the degree of saturation is approximately equal to the relative humidity.

Mixture Enthalpy

The above measures of concentration are now used to obtain the mixture properties. The mixture properties are defined in Eq. (10.32). Many applications are based on a unit mass of dry air rather than a unit mass of mixture, so the humidity ratio is used in the property determination. In the case of a binary mixture of air and water vapor, the enthalpy *per unit mass of dry air* is given from Eq. (10.32*b*) as

$$\frac{H}{m_a} = \frac{m}{m_a}(x_a h_a + x_v h_v) = h_a + \frac{m_v}{m_a}h_v = h_a + \omega h_v \tag{10.69}$$

To find values of $(h_a + \omega h_v)$, use values of $h_v = h(T)$ from Tables D.12 or E.12 for superheated water vapor at the mixture temperature and vapor partial pressure. The value of h_a can be found from Tables D.2 or E.2. All the other properties given in Eq. (10.32) have a similar form when the enthalpy is replaced by the appropriate property.

10.5.2 Psychrometrics

The study of the particular case of air–water vapor mixtures and the changes that these mixtures undergo when heat transfer and/or multistream mixing occurs is called *psychrometrics.* Applications are in the design and analysis of heating, ventilating, and air conditioning (HVAC) equipment.

The definitions presented in the preceding section are all applicable to air–water vapor mixtures, but most work is done in terms of the dry-bulb temperature and some measure of the concentration of water vapor in the air–water vapor mixture. The *dry-bulb temperature* is what we have called to this point simply temperature. It is now given the extra adjective to differentiate it from another quantity, the *wet-bulb temperature,* which is the temperature read by a thermometer with a sensing bulb wrapped with a water-wetted wick and exposed to a moving airstream of a certain humidity.

The state of a given air–water vapor mixture must depend on any *three* independent properties, the extra property being required to specify the relative concentrations of the air and water vapor. (The need for more properties to be specified for mixtures is discussed more fully in Sec. 12.2.4.) Usually the total pressure is specified in HVAC work as 1 atm (properties at other total pressures are needed for special cases such as at high altitudes or in spacecraft or submarine environmental system design). Two additional properties then define the state of the system, and any convenient pair may be used.

The wet-bulb temperature is easily measured, and it is an approximation to the temperature of a system in which the heat and mass transfer between the wick and the surrounding air–water vapor mixture is in equilibrium; this *adiabatic saturation temperature* is discussed more fully later. The wet-bulb temperature depends on both the dry-bulb temperature and the humidity of the flowing airstream, and so it can be used with the dry-bulb temperature to determine the humidity of the air–water vapor mixture at a given total pressure.

Thus, wet-bulb and dry-bulb temperatures may be measured, and all other mixture properties can then be obtained (relative humidity, humidity ratio, degree of saturation, etc.). Any two independent properties can be used to determine all others. Clearly a plot of the properties could be prepared at any given total pressure that would allow graphical determination of all properties when the state point is determined from any two known properties. Such a plot is called a *psychrometric chart,* and it is discussed in Sec. 10.6.

Example Problem 10.3

Determine the humidity ratio, vapor mass fraction, degree of saturation, and dew point temperature for an air–water mixture at 1 atm, 91°F, and a relative humidity of 90 percent.

Solution

The relative humidity of 90 percent gives

$$\phi = \frac{P_v}{P_g} = 0.9$$

where P_g is the saturation pressure at 91°F. Table E.10 gives $P_g = 0.7238$ psia, so the vapor pressure is

$$P_v = (0.9)(0.7238) = 0.6514 \text{ psia}$$

Since the total pressure is 14.696 psia, the partial pressure of the air is

$$P_a = P - P_v = 14.05 \text{ psia}$$

Equation (10.65) gives the humidity ratio as

$$\omega = 0.622 \frac{0.6525}{14.696} = 0.0276 \frac{\text{lbm H}_2\text{O}}{\text{lbm dry air}}$$

The mass fraction is given from Eq. (10.63) as

$$x_v = \frac{m_v}{m} = \frac{m_v}{m_v + m_a}$$
$$= \frac{1}{1 + m_a/m_v} = \frac{1}{1 + 1/\omega} = 0.0269 \frac{\text{lbm H}_2\text{O}}{\text{lbm mixture}}$$

The degree of saturation is given from Eq. (10.68) as

$$\mu = 0.9 \frac{14.696 - 0.7238}{14.696 - 0.6514} = 0.895$$

Finally, the dew point temperature is obtained from Table E.10 at a pressure of 0.6514 psia:

$$T_d = 87.5°\text{F} = 30.8°\text{C}$$

Comments

The vapor pressure for water is generally small compared to the total pressure, as indicated. Also the relative humidity and the degree of saturation are almost equal.

10.5.3 Thermodynamic Analysis

The first and second laws are now applied to air–water vapor mixtures. Many specific applications are considered in Sec. 10.7, but a few processes that exhibit many important features of gas-vapor mixtures are considered here.

Consider a control mass with a movable piston, so the air–water mixture is initially at state 1 and is cooled at constant pressure through the saturation state 2 to a temperature T_3 below the dew point temperature at state 1. Figure 10.7 diagrams the process. The mixture is initially in state 1, and as the mixture is cooled at constant pressure, the partial pressures remain constant until state 2. At this point, condensate forms at the bottom of the container, which is at state 2*l*. Subsequent lowering of the temperature forms liquid and saturated vapor at state 3. The vapor pressure from state 2 to 3 decreases, so the partial pressure of the gas phase increases to keep the total pressure constant. The mass of air is constant throughout the process, and so is the mass of the water. The first law for this process is

$$U_3 - U_1 = {}_1Q_3 - {}_1W_3 \tag{10.70}$$

Figure 10.7 Vapor condensation.

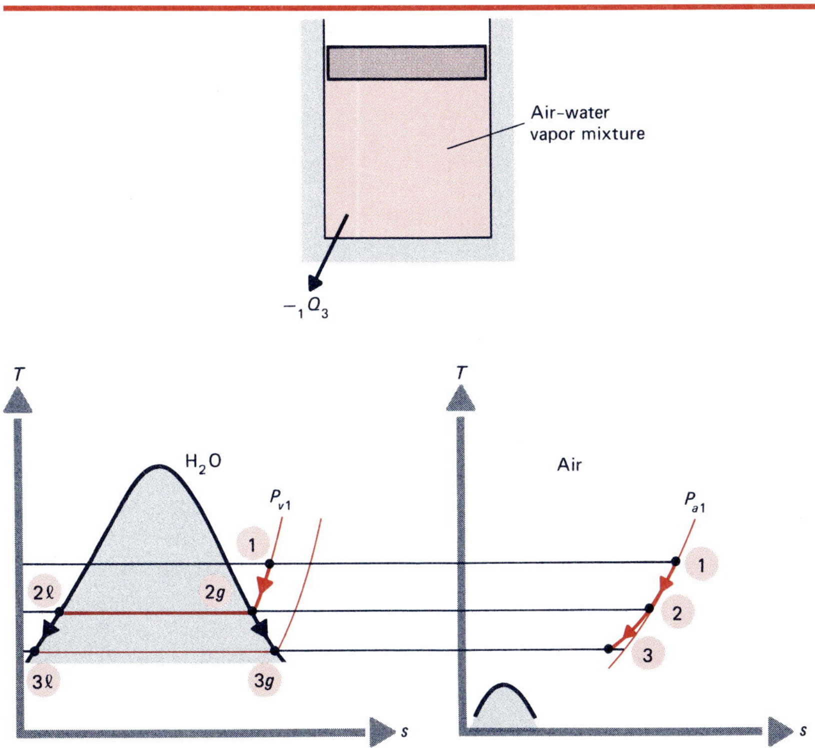

which, for constant-pressure work, is

$$U_3 - U_1 = {}_1Q_3 - P(V_3 - V_1) \tag{10.71}$$

The heat transfer is given as

$${}_1Q_3 = H_3 - H_1 \tag{10.72}$$

Explicitly denoting the components and phases yields

$${}_1Q_3 = m_a(h_{a,3} - h_{a,1}) + m_{l,3}h_{l,3} + m_{g,3}h_{g,3} - m_{v,1}h_{v,1} \tag{10.73}$$

The mass of water is constant ($m_{v,1} = m_{l,3} + m_{g,3}$), so dividing this mass relation by the mass of dry air yields

$$\omega_1 = \frac{m_{l,3}}{m_a} + \omega_3 \tag{10.74}$$

Thus the first law is

$$\begin{aligned}\frac{{}_1Q_3}{m_a} &= h_{a,3} - h_{a,1} + (\omega_1 - \omega_3)h_{l,3} + \omega_3 h_{g,3} - \omega_1 h_{v,1} \\ &= h_{a,3} - h_{a,1} + \omega_1(h_{l,3} - h_{v,1}) + \omega_3 h_{lg,3}\end{aligned} \tag{10.75}$$

This can be rearranged into the form given in Eq. (10.69) to yield

$$\frac{{}_1Q_3}{m_a} = h_{a,3} + \omega_3 h_{g,3} - (h_{a,1} + \omega_1 h_{v,1}) + (\omega_1 - \omega_3)h_{l,3} \tag{10.76}$$

Example Problem 10.4

An air–water vapor mixture is initially at 101.32 kPa, 30°C, and $\phi = 90$ percent. The mixture is cooled at constant pressure to 10°C. Plot the process for both air and water on separate *T-s* diagrams. Also plot the humidity ratio and relative humidity as a function of process temperature. Evaluate the heat transfer per unit mass of dry air for the process.

Diagrams

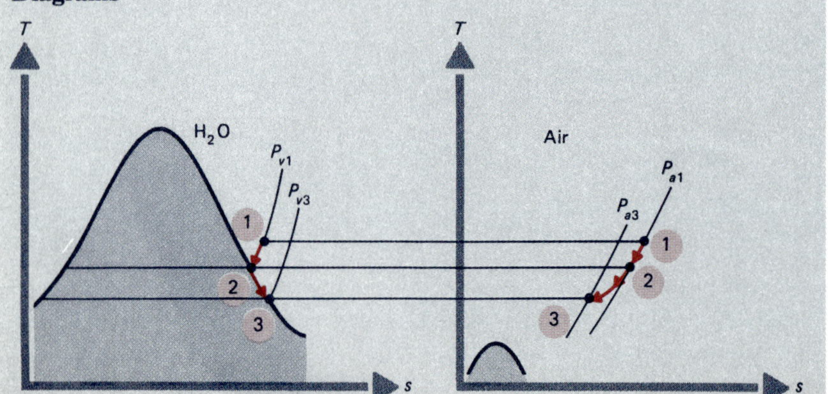

Solution

The total pressure in this process is 101.32 kPa. The humidity ratio, vapor pressure, and partial pressure of air are evaluated from Tables D.10 and D.11, as in Example Problem 10.3. At 30°C, the saturation pressure is $P_g = 4.2455$ kPa, so the vapor pressure is $P_v = \phi P_g = 3.82$ kPa. The dew point temperature at this vapor pressure is $T_d = 28.0$°C. The following table is obtained:

State	T, °C	ϕ	ω, $\frac{\text{kg } H_2O}{\text{kg dry air}}$	P_v, kPa	P_a, kPa
1	30.0	0.9	0.0244	3.82	97.50
2	28.0	1.0	0.0244	3.82	97.50
	25	1.0	0.0201	3.17	98.15
	20	1.0	0.0147	2.34	98.98
	15	1.0	0.0107	1.71	99.61
3	10	1.0	0.0076	1.23	100.09

The process plots are given, and the plot of (ϕ, ω) versus T is given here:

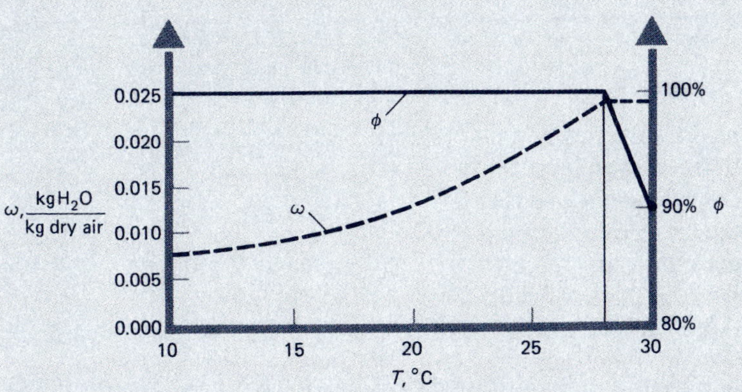

The heat transfer is evaluated from Eq. (10.75). Using $dh = c_P\, dT$ for air yields

$$\frac{{}_1Q_3}{m_a} = c_{P,a}(T_3 - T_1) + \omega_1(h_{l,3} - h_{v,1}) + \omega_3 h_{lg,3}$$

With numerical values given in Tables C.2 and D.10, this yields

$$\frac{{}_1Q_3}{m_a} = 1.0052(-20) + 0.0244(42.0 - 2555.3) + 0.0076(2476.9)$$

$$= -62.6 \text{ kJ/kg air} = -26.9 \text{ Btu/lbm air}$$

The value for the enthalpy of water vapor at $P_v = 3.82$ kPa and $T = 30$°C is taken as $h_{v,1} = h_{g,1}$ (30°C) = 2555.3 kJ/kg, since water vapor in this state acts as an ideal gas so $h = h(T \text{ only})$ (see Fig. D.1 or the compressibility chart, Fig. C.1).

Comments

The humidity ratio does not change until condensation occurs. Condensation removes water vapor from the gas-vapor mixture, so the humidity ratio decreases. The relative humidity does not change after the mixture is saturated, because ϕ depends on only the water vapor state and not that of the air.

The heat transfer per unit mass of air to reduce the temperature from 30 to 20°C at state 1 is given as above:

$$\frac{{}_{30}Q_{20}}{m_a} = 1.0052(-10) + 0.0244(83.8 - 2555.3) + 0.0147(2453.4)$$

$$= -34.3 \text{ kJ/kg air} = -14.7 \text{ Btu/lbm air}$$

The required heat transfer per unit mass of air to go from 20 to 10°C is

$$\frac{{}_{20}Q_{10}}{m_a} = \frac{{}_1Q_3}{m_a} - \frac{{}_{30}Q_{20}}{m_a} = -28.3 \text{ kJ/kg air} = -12.2 \text{ Btu/lbm air}$$

This shows the required heat transfer for various portions of the process.

Another important application is a constant-pressure steady flow process through the adiabatic system shown in Fig. 10.8. Examples of this process are "swamp coolers" which pass hot desert air through wetted mats to provide cool air in hot dry climates, and other related cooling and dehumidification processes. The air–water mixture enters at T_1 and P_1 with a humidity ratio of ω_1 and leaves the system at T_2, $P_2 = P_1$, and a humidity ratio ω_2. The air flows over the water within the system, and the water evaporates and is carried in the air as the air–water vapor mixture. Liquid water is supplied at T_2 to maintain the water level in the system.

The mass of air entering is equal to that which leaves, and the same is true for the water component. The first law, if we neglect changes in kinetic and potential energy, is

$$\dot{m}_a h_{a,1} + \dot{m}_{v,1} h_{v,1} + \dot{m}_l h_{l,2} = \dot{m}_a h_{a,2} + \dot{m}_{v,2} h_{v,2} \tag{10.77}$$

Dividing by $\dot{m}_a$ yields

$$h_{a,1} + \omega_1 h_{v,1} + \frac{\dot{m}_l}{\dot{m}_a} h_{l,2} = h_{a,2} + \omega_2 h_{v,2} \tag{10.78}$$

Conservation of mass for water is

$$\dot{m}_{v,1} + \dot{m}_l = \dot{m}_{v,2} \tag{10.79}$$

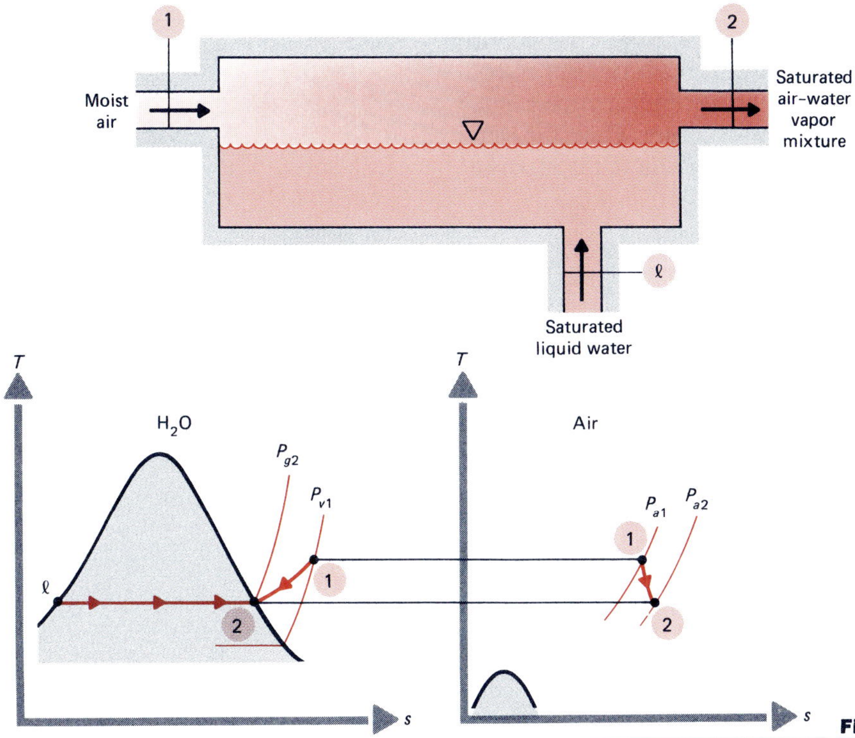

Figure 10.8 Adiabatic saturation process.

or

$$\omega_1 + \frac{\dot{m}_l}{\dot{m}_a} = \omega_2 \tag{10.80}$$

Therefore, the first law is

$$h_{a,1} + \omega_1 h_{v,1} + (\omega_2 - \omega_1)h_{l,2} = h_{a,2} + \omega_2 h_{v,2} \tag{10.81}$$

If the system is sufficiently long, then the mixture leaving is saturated and the exit temperature is referred to as the *adiabatic saturation temperature.* Equation (10.81) is rearranged to give the entering humidity ratio as

$$\omega_1 = \frac{h_{a,1} - h_{a,2} - \omega_2(h_{g,2} - h_{l,2})}{h_{l,2} - h_{v,1}} \tag{10.82}$$

Therefore, with the inlet temperature and adiabatic saturation temperature known for this constant-pressure saturation process, the inlet humidity ratio is obtained. This is one method to experimentally determine the humidity ratio of a mixture.

The adiabatic saturation temperature obtained at the outlet of the above system requires the outlet state to be saturated. This is difficult to obtain since the channel must be very long. The adiabatic saturation system is approximated by a thermometer that is covered with a wetted wick material. The mixture of unknown humidity passes over the wetted wick, and an equilibrium of the heat transfer and mass transfer is obtained. This measured temperature is called the wet-bulb temperature, and it approximates the adiabatic saturation temperature. Thus, the wet-bulb temperature is a measure of the water concentration of the mixture as given in Eq. (10.82).

10.6 Psychrometric Chart

The previous equations give the measures of concentration and properties for air–water vapor mixtures. The expressions can be used with the tables in the appendixes for air and water to yield the necessary solutions for air–water vapor systems. Since these mixtures occur often, the quantities are presented in a chart form, which is termed the *psychrometric chart.* A portion of the psychrometric charts in Fig. D.5 and Fig. E.5 is presented in Fig. 10.9.

These charts are very useful in practice for the design and analysis of components and systems used in heating, ventilating, and air conditioning. The state of the flow streams at each component can be shown on the chart, and the processes that describe the thermodynamic behavior of each component are easily diagramed.

The state of the air–water vapor mixture is determined by pressure, temperature, and concentration. The total pressure for the chart is 1 atm or 101.325 kPa or 14.69 psia (corrections are available for other pressures, but they are not discussed in this text). Therefore, if the pressure is at a single value, the state is determined by the temperature and concentra-

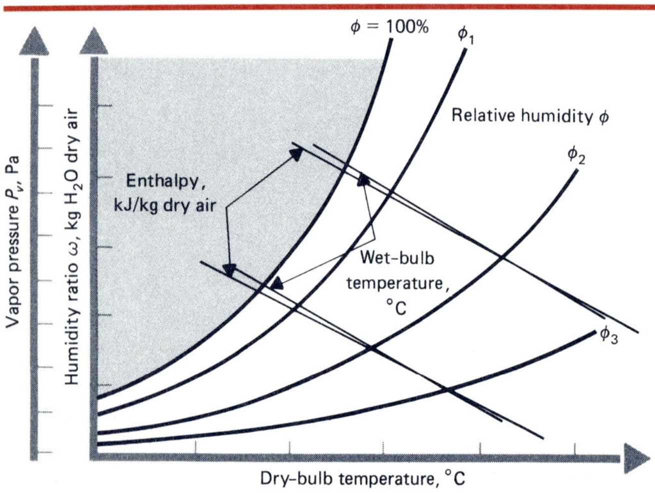

Figure 10.9 Psychrometric chart.

tion. The measure of concentration is taken as the humidity ratio. Figure 10.9 shows the two axes as the dry-bulb temperature and the humidity ratio. Since Eq. (10.60) indicates that the humidity ratio is a function of the vapor pressure only, an equivalent concentration measure is the vapor pressure.

The air–water vapor mixture is bounded by the saturated mixture condition. This is given by $\phi = 1$ ($\phi = 100$ percent) and has the basic trends of a *P-T* diagram for pure water (see Chap. 3). The gas–water vapor mixture states are below this line. Above this line, the water must be in the liquid phase. Since $\phi = 1$ is the saturated state, the dry-bulb temperature at the saturated state is also the wet-bulb temperature. This temperature at $\phi = 1$ also corresponds to the dew point temperature for the vapor pressure. Enthalpy and the wet-bulb temperature for the mixture states are also indicated. Note that the enthalpy is based on the total (air + water) enthalpy per unit mass of dry air, that is, $(h_a + \omega h_v)_i$. Thus, the first law for a process with no work, for example Eq. (10.75), is written as

$$\frac{{}_1Q_3}{\dot{m}_a} = (h_{a,3} + \omega_3 h_{g,3}) - (h_{a,1} + \omega_1 h_{v,1}) + (\omega_1 - \omega_3)h_{l,3}$$

$$= h_3 - h_1 + (\omega_1 - \omega_3)h_{lg}$$

and h_1 and h_3 are found directly from the chart.

The psychrometric chart combines the ideal gas charts with the steam tables. The reference states are different for each component, and this practice of combining properties with different reference states is not recommended in general. In this case, this is acceptable since only differences in properties are required. This is indicated in Eqs. (10.72) and (10.75).

Example Problem 10.4 can now be solved by using the psychrometric chart.

10.7 Applications

This section considers some of the basic elements of systems which operate with multicomponent mixtures of gas and vapor. Combinations of these basic elements are used to construct various kinds of complex systems.

10.7.1 Heat Transfer at Constant ω

An air–water vapor mixture that undergoes a heat transfer while the component concentrations are kept constant is the same process as considered in Sec. 10.4. For a steady state flow at atmospheric pressure, this process is diagrammed in Fig. 10.10. Since the process is at atmospheric pressure, the psychrometric chart is used. The process is a horizontal line

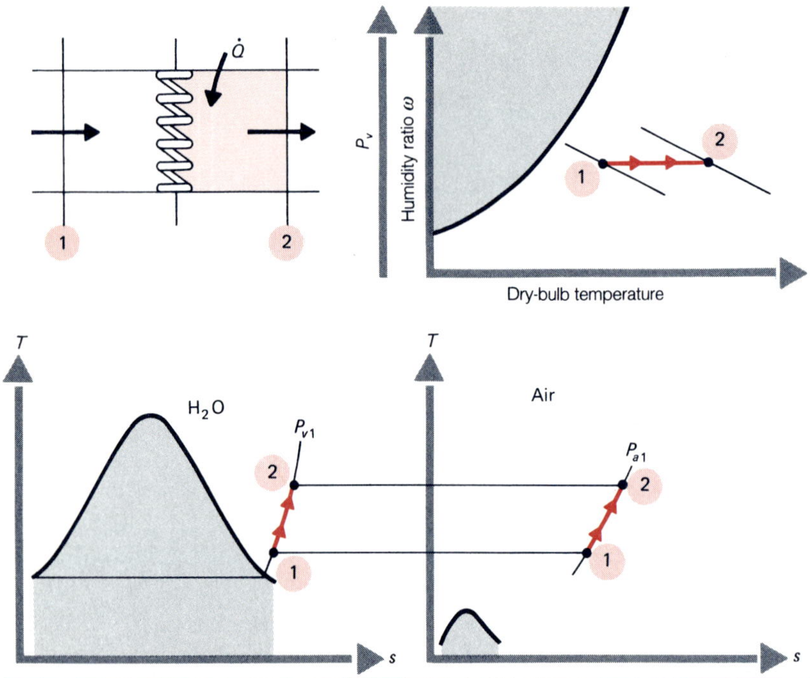

Figure 10.10 Heat transfer with constant ω.

on the psychrometric chart, and the direction of the process depends on the direction of the heat transfer. The process shown is a heating process, so the enthalpy increases and the relative humidity decreases. The vapor pressure and air partial pressure are also constants since ω is a constant, and these processes are also shown in Fig. 10.10. The *T-s* diagram for water shows a process where superheated water vapor at state 1 is heated to state 2. A similar process for air is also shown. The water vapor states for the process are near the saturation dome, while the air states are highly superheated. A cooling process goes in the opposite direction, so the enthalpy decreases and the relative humidity increases. The first law for the diagramed process is

$$h_{a,1} + \omega_1 h_{v,1} + \frac{\dot{Q}}{\dot{m}_a} = h_{a,2} + \omega_2 h_{v,2} \tag{10.83}$$

10.7.2 Humidification

Many applications require an addition of water to air. This process is termed *humidification.* Figure 10.11 diagrams a steady humidificaton process where water is injected into a mixture of air and water vapor. The addition of water increases the humidity ratio. The psychrometric chart indicates the larger value of ω at the outlet state of the process, but the

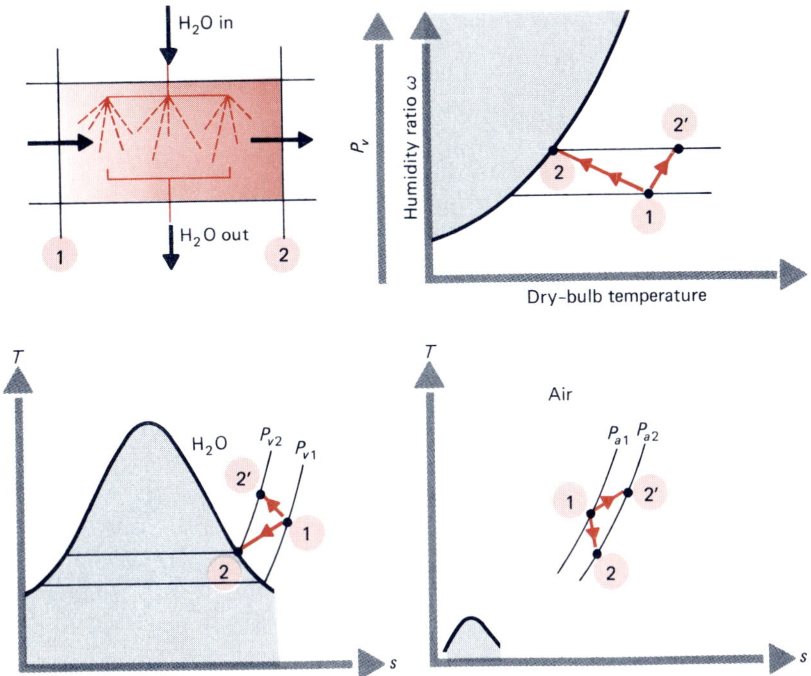

Figure 10.11 Humidification.

process direction (with respect to the dry-bulb temperature) depends on the enthalpy of the entering water. If liquid water enters at the outlet air temperature and the process ends at $\phi = 1$, then the process is the adiabatic saturation process described in Sec. 10.5.3. Equation (10.81) describes this process as

$$(h_a + \omega h_v)_1 + (\omega_2 - \omega_1)h_l = (h_a + \omega h_v)_2 \tag{10.84}$$

Since $(\omega_2 - \omega_1)h_l$ is small relative to $h_a + \omega h_v$, this process is essentially a process with constant $h_a + \omega h_v$. This enthalpy is a function of the wet-bulb temperature only. Thus, the process follows approximately a constant wet-bulb temperature line. The lowest dry-bulb temperature possible is the saturation temperature. If water vapor is added rather than liquid water, the process can reach a higher dry-bulb temperature at the outlet, since h_l in Eq. (10.84) is now $h_{v,3}$ and its value can be large. This process is also shown in Fig. 10.11, ending at state 2′.

10.7.3 Dehumidification

Dehumidification is the process of taking water out of the mixture. A process used to achieve this goal is the cooling process at constant ω to the saturation state and subsequent condensation of the liquid from the mixture. The process is discussed in Sec. 10.5 for a control mass and is

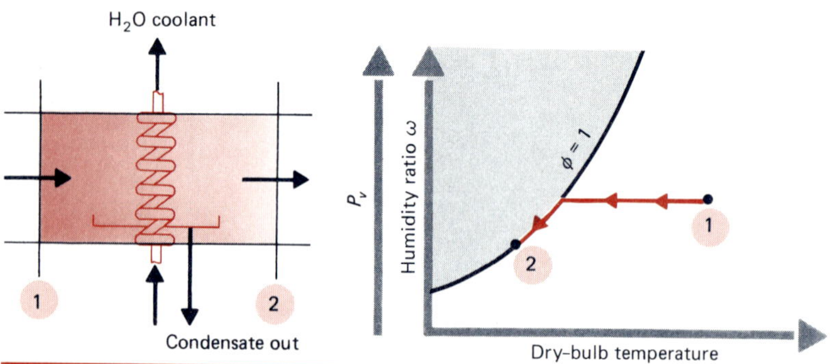

Figure 10.12 Dehumidification.

presented in Fig. 10.7. On a psychrometric chart, the mixture undergoes a process shown in Fig. 10.12.

10.7.4 Mixing of Air–Water Vapor Streams

A specific air–water vapor mixture can be obtained from separate streams by mixing the streams to obtain the desired outlet state. Figure 10.13 diagrams an adiabatic process with two entering streams and a single outlet stream. Conservation of mass yields

$$\dot{m}_{a,1} + \dot{m}_{a,2} = \dot{m}_{a,3} \tag{10.85}$$

and

$$\dot{m}_{v,1} + \dot{m}_{v,2} = \dot{m}_{v,3} \tag{10.86}$$

or

$$\dot{m}_{a,1}\omega_1 + \dot{m}_{a,2}\omega_2 = \dot{m}_{a,3}\omega_3 \tag{10.87}$$

These conservation of mass statements yield

$$\frac{\dot{m}_{a,1}}{\dot{m}_{a,2}} = \frac{\omega_2 - \omega_3}{\omega_3 - \omega_1} \tag{10.88}$$

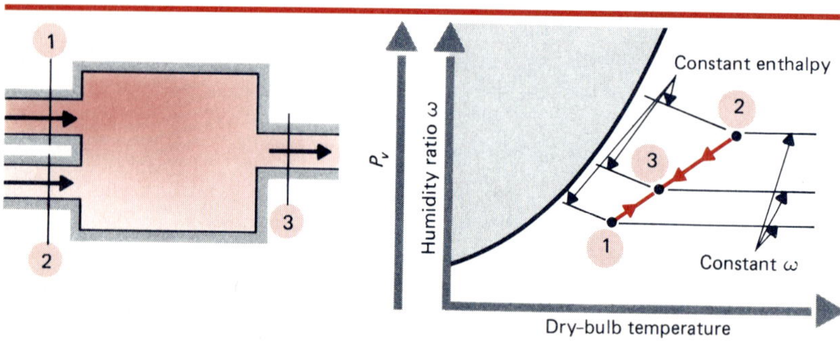

Figure 10.13 Mixing of streams.

The first law is written as

$$(\dot{m}_a h_a + \dot{m}_v h_v)_1 + (\dot{m}_a h_a + \dot{m}_v h_v)_2 = (\dot{m}_a h_a + \dot{m}_v h_v)_3 \quad (10.89)$$

Rewriting the first law for a unit mass of dry air yields

$$\dot{m}_{a,1}(h_a + \omega h_v)_1 + \dot{m}_{a,2}(h_a + \omega h_v)_2 = \dot{m}_{a,3}(h_a + \omega h_v)_3 \quad (10.90)$$

Substituting Eq. (10.85) into Eq. (10.90) yields

$$\frac{\dot{m}_{a,1}}{\dot{m}_{a,2}} = \frac{(h_a + \omega h_v)_2 - (h_a + \omega h_v)_3}{(h_a + \omega h_v)_3 - (h_a + \omega h_v)_1} \quad (10.91)$$

Combining Eqs. (10.88) and (10.91) gives the following relation:

$$\frac{\dot{m}_{a,1}}{\dot{m}_{a,2}} = \frac{\omega_2 - \omega_3}{\omega_3 - \omega_1} = \frac{(h_a + \omega h_v)_2 - (h_a + \omega h_v)_3}{(h_a + \omega h_v)_3 - (h_a + \omega h_v)_1} \quad (10.92)$$

Thus, fixing the inlet states and mass flow rates of the inlets specifies the outlet state. The graphical interpretation is shown in Fig. 10.13.

Example Problem 10.5

An important application requiring the analysis of a multicomponent mixture of gas and vapor is a cooling tower. The cooling-water outlet of a power plant is often at too high a temperature to be rejected into a lake or river, so the energy is rejected into the atmosphere by evaporation of the water. Some makeup water is added to the cooling water before the water is returned to the power plant.

The water leaving the power plant condenser is at 95°F and is to be cooled to 77°F. The water mass flow rate leaving the plant into the cooling tower is 1500 lbm/s. The air–water vapor mixture enters the cooling tower at 60°F and a relative humidity of 50 percent, and it leaves at 85°F as saturated air. Determine the mass flow rates of the dry air and the makeup water.

Diagram

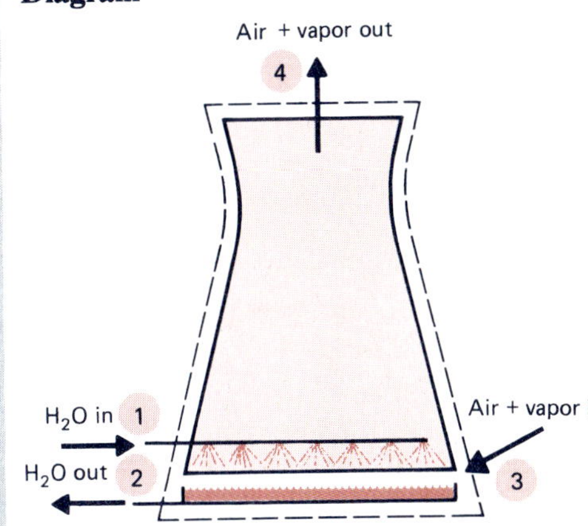

Solution

This process is similar to the humidification process discussed earlier except that the desired result is a reduction of temperature of the power plant cooling water rather than humidification of the air. Also this particular case has $\phi = 1$ at the air outlet, so this is similar to the adiabatic saturation process discussed in Sec. 10.5.2. The first law is

$$\dot{m}_{a,3}h_{a,3} + \dot{m}_{v,3}h_{v,3} + \dot{m}_{l,1}h_{l,1} = \dot{m}_{a,4}h_{a,4} + \dot{m}_{v,4}h_{v,4} + \dot{m}_{l,2}h_{l,2}$$

Dividing by $\dot{m}_{a,3} = \dot{m}_{a,4} = \dot{m}_a$ gives

$$(h_a + \omega h_v)_3 + \frac{\dot{m}_{l,1}}{\dot{m}_a} h_{l,1} = (h_a + \omega h_v)_4 + \frac{\dot{m}_{l,2}}{m_a} h_{l,2}$$

Conservation of mass for water yields

$$\frac{\dot{m}_{l,2}}{\dot{m}_a} + \omega_4 = \frac{\dot{m}_{l,1}}{\dot{m}_a} + \omega_3$$

This is substituted into the first law to yield

$$(h_a + \omega h_v)_3 + \frac{\dot{m}_{l,1}}{\dot{m}_a} h_{l,1} = (h_a + \omega h_v)_4 + \left(\frac{\dot{m}_{l,1}}{\dot{m}_a} + \omega_3 - \omega_4\right) h_{l,2}$$

or

$$\frac{\dot{m}_{l,1}}{m_a} = \frac{(h_a + \omega h_v)_4 - (h_a + \omega h_v)_3 + (\omega_3 - \omega_4) h_{l,2}}{h_{l,1} - h_{l,2}}$$

The values are obtained from the psychrometric chart in Fig. E.5 and Table E.10. This yields

$$\frac{\dot{m}_{l,1}}{\dot{m}_a} = \frac{49.4 - 20.4 + (0.0055 - 0.0266)(45.0)}{63.0 - 45.0} = 1.56$$

Therefore, $\dot{m}_a = \dot{m}_{l,1}/1.56 = 1500/1.56 = 962$ lbm/s = 437 kg/s.

The makeup water is given by $\dot{m}_{l,1} - \dot{m}_{l,2}$. Conservation of mass for water yields

$$\frac{\dot{m}_{l,1} - \dot{m}_{l,2}}{\dot{m}_a} = \omega_4 - \omega_3$$

or

$$\dot{m}_{l,1} - \dot{m}_{l,2} = \dot{m}_a(\omega_4 - \omega_3)$$

Thus, the required makeup water is

$$\dot{m}_{l,1} - \dot{m}_{l,2} = 962(0.0266 - 0.0055) = 20.2 \text{ lbm/s} = 9.2 \text{ kg/s}$$

Comments

The makeup water mass flow rate is 20.2/1500 = 0.0135, or 1.35 percent, of the inlet mass flow rate. This must be supplied to the power plant since it is lost to the surrounding atmosphere.

10.8 Nonideal Mixtures

When the components of a mixture cannot be treated as ideal gases, then the relations governing the properties of mixtures become more complex.

10.8.1 Mixtures of Nonideal Gases

As a relatively simple way to treat gases that are at pressures and/or temperatures that make their behavior nonideal, we can proceed as follows. Take Eq. (5.44) on a molar basis,

$$d\bar{h} = T\,d\bar{s} + \bar{v}\,dP \tag{10.93}$$

where we understand each of the properties to apply to the mixture. Now substitute the relations for the specific extensive properties of the mixture $\bar{h}$, $\bar{s}$, and $\bar{v}$ from Eqs. (10.30*b*), (10.30*c*), and (10.20), to give

$$d\left(\sum_i y_i\bar{h}_i\right) = T\,d\left(\sum_i y_i\bar{s}_i\right) + \left(\sum_i y_i\bar{v}_i\right)dP \tag{10.94}$$

This substitution makes the implicit assumption that the Gibbs-Dalton and Amagat-Leduc laws apply to a mixture of nonideal gases, which is, of course, an approximation.

Equation (10.94) can now be rearranged to give

$$\sum_i y_i(d\bar{h}_i - T\,d\bar{s}_i - \bar{v}_i\,dP) = 0 \tag{10.95}$$

Because y_i is independent of $\bar{h}_i$, $\bar{s}_i$, and $\bar{v}_i$ (that is, the mixture can have any concentration of the components regardless of the specific extensive properties of the components), the only way that Eq. (10.95) can be satisfied is for the term in parentheses to be zero for every term in the series; i.e.,

$$d\bar{h}_i = T\,d\bar{s}_i + \bar{v}_i\,dP \tag{10.96}$$

Now reduced-property charts for each component in the mixture could be developed by using Eq. (10.96) and the method described in Sec. 9.3.2. Note, however, that temperature and pressure in Eq. (10.96) are the *total* pressure and the temperature of the *mixture*, not the component (partial) pressure. The result of these observations is that the existing reduced-property charts can be used for determining the properties of the components of the mixture *provided* that the mixture temperature and mixture total pressure are used in finding the reduced properties, that is, $P_{r,i} = P_{tot}/P_{cr,i}$ and $T_{r,i} = T/T_{cr,i}$ for each component of the mixture. The mixture properties are computed by using the Gibbs-Dalton relations [Eqs. (10.29) to (10.31)] or their extensions, Eqs. (10.32). In using Eqs. (10.32), however, we are implicitly evaluating the component properties, using P_{tot} instead of P_i as implied by the notation.

This is not an exact method of determining the properties of a nonideal mixture, because it involves using properties of nonideal gases and then using ideal gas mixture relations to define the mixture properties. The results are, of course, more accurate than the simple ideal gas relations.

10.8.2 Other Mixture Rules

Alternative methods for finding mixture properties exist. One approach is to assume that the principle of corresponding states will apply to a mixture as well as to a single-component fluid. If this is the case, and if a suitable corresponding critical pressure and temperature for the *mixture* can be found, then any of the equations of state based on the principle (such as the compressibility charts, van der Waals' equation, etc.) can be applied to predict properties of the mixture. These corresponding critical properties of the mixture are called *pseudocritical* properties.

W. B. Kays suggests that the *pseudocritical temperature* of the mixture be defined as

$$T_{cr} = \sum_i y_i T_{cr,i} \tag{10.97}$$

This relation is accurate to within less than 2 percent if for each pair of components of the mixture i and j, $0.5 < T_{cr,i}/T_{cr,j} < 2$ and $0.5 < P_{cr,i}/P_{cr,j} < 2$. The mixture pseudocritical pressure is given by

$$P_{cr} = \frac{R\left(\sum_i y_i z_{cr,i}\right) T_{cr}}{\sum_i y_i v_{cr,i}} \tag{10.98}$$

where T_{cr} is the mixture pseudocritical temperature from Eq. (10.97).

The mixture acentric factor can be found from

$$\omega = \sum_i y_i \omega_i \tag{10.99}$$

None of these relations is exact, but they can be used along with corresponding-states relations to predict mixture properties with reasonable accuracy, at least for mixtures of nonpolar fluids. Reid, Prausnitz, and Poling [1] give additional mixture rules that are applicable when using the van der Waals, Redlich-Kwong, and Lee-Kesler equations of state, and Kyle [2] discusses the application of the Benedict-Webb-Rubin equation to mixtures.

The prediction of the properties of mixtures is an active field of research in thermodynamics. Accurate methods for mixtures involving one or more polar molecules in particular do not exist, and the properties of such mixtures must often be directly measured if they are required.

10.9 General Mixture Relations

To this point, consideration has been limited to gases and to simplified relations for computing the properties of gas mixtures. The methods presented so far apply to a wide variety of engineering problems, including air conditioning design and combustion-product analysis. However, in some cases, the relations developed so far are not applicable. In particular, the approximate relations for nonideal gases may not be accurate enough for the required calculations or, more generally, when one or more of the mixture components are a liquid or a solid. In this section, we develop relations for general mixtures and show their application to a few problems of engineering interest.

10.9.1 Partial Molal Properties

In general, any mixture extensive property Z can be expressed as a function of the two intensive properties T and P and the number of moles of the individual components $n_A, n_B, n_C, \ldots$ as

$$Z = Z(T, P, n_A, n_B, n_C, \ldots) \tag{10.100}$$

Extending this relation [Eq. (H.13)] to more than two independent variables gives

$$dZ = \left(\frac{\partial Z}{\partial T}\right)_{P, n_A, n_B, \ldots} dT + \left(\frac{\partial Z}{\partial P}\right)_{T, n_A, n_B, \ldots} dP + \left(\frac{\partial Z}{\partial n_A}\right)_{T, P, n_B, n_C, \ldots} dn_A + \left(\frac{\partial Z}{\partial n_B}\right)_{T, P, n_A, n_C, \ldots} dn_B + \ldots \tag{10.101}$$

Now divide through by the total number of moles present in the mixture n_{tot}, and define the mole fraction of each component in the mixture as $y_i = n_i/n_{tot}$. The value of $d\bar{z} = d(Z/n_{tot})$, the value of dZ per mole of mixture, is then

$$d\bar{z} = d\left(\frac{Z}{n_{tot}}\right) = \frac{1}{n_{tot}}\left(\frac{\partial Z}{\partial T}\right)_{P, n_A, n_B, \ldots} dT + \frac{1}{n_{tot}}\left(\frac{\partial Z}{\partial P}\right)_{T, n_A, n_B, \ldots} dP + \left(\frac{\partial Z}{\partial n_A}\right)_{P, T, n_B, n_C, \ldots} dy_A + \left(\frac{\partial Z}{\partial n_B}\right)_{P, T, n_A, n_C, \ldots} dy_B + \ldots \tag{10.102}$$

Let us define the properties

$$\bar{Z}_A = \bar{Z}_{T, P, n_B, n_C, \ldots} = \left(\frac{\partial Z}{\partial n_A}\right)_{T, P, n_B, n_C, \ldots}$$

$$\bar{Z}_B = \bar{Z}_{T, P, n_A, n_C, \ldots} = \left(\frac{\partial Z}{\partial n_B}\right)_{T, P, n_A, n_C, \ldots} \tag{10.103}$$

$$\bar{Z}_C = \text{etc.}$$

as *partial molal properties*. Note the use of a capital letter with an overbar to denote a partial molal property. This is the only use of this combination of notation, and it is generally different from a lowercase letter with an overbar, which is a molar intensive quantity.

Inserting these definitions into Eq. (10.102) gives

$$d\bar{z} = d\left(\frac{Z}{n_{tot}}\right) = \frac{1}{n_{tot}}\left(\frac{\partial Z}{\partial T}\right)_{P, n_A, n_B, \ldots} dT + \frac{1}{n_{tot}}\left(\frac{\partial Z}{\partial P}\right)_{T, n_A, n_B, \ldots} dP + \bar{Z}_A\, dy_A + \bar{Z}_B\, dy_B + \ldots \tag{10.104}$$

or, for the case of constant temperature and pressure,

$$(d\bar{z})_{T, P} = d\left(\frac{Z}{n_{tot}}\right)_{T, P} = \bar{Z}_A\, dy_A + \bar{Z}_B\, dy_B + \ldots \tag{10.105}$$

Note that the partial molal property $\overline{Z}$ is an intensive property that depends only on T, P, and the number of moles of the components of the mixture. *For a mixture of constant composition at constant temperature and pressure,* then, the values of $\overline{Z}$ are constant, and Eq. (10.105) can be integrated to give

$$\bar{z}_{T,P} = \left(\frac{Z}{n_{\text{tot}}}\right)_{T,P} = \overline{Z}_A y_A + \overline{Z}_B y_B + \ldots \tag{10.106}$$

or

$$\begin{aligned} Z_{T,P} &= \overline{Z}_{T,P,n_B,n_C,\ldots}\, n_A + \overline{Z}_{T,P,n_A,n_C,\ldots}\, n_B + \ldots \\ &= \overline{Z}_A n_A + \overline{Z}_B n_B + \ldots \end{aligned} \tag{10.107}$$

Let us relate Eq. (10.107) to what we know about ideal mixtures. For the case of the volume of a gas mixture composed of two gases, Eq. (10.107) becomes

$$V_{T,P} = \overline{V}_{T,P,n_B} n_A + \overline{V}_{T,P,n_A} n_B \tag{10.108}$$

For an ideal gas, $\overline{V}_{T,P} = (\partial V/\partial n)_{T,P} = \bar{v}$, and Eq. (10.108) becomes

$$V_{T,P} = \bar{v}_A n_A + \bar{v}_B n_B \tag{10.109}$$

as given by the Amagat-Leduc law.

Now the general mixture curve for a general multicomponent mixture (Fig. 10.2) can be used to show how partial molal properties can be computed from experimental data. For a two-component mixture, Fig. 10.14 shows the volume per mole of mixture versus mole fraction relation that can be obtained experimentally. For such a curve, T and P are held constant.

In this figure, the partial molal volumes of each component can be found graphically at any composition as follows: At the given mixture

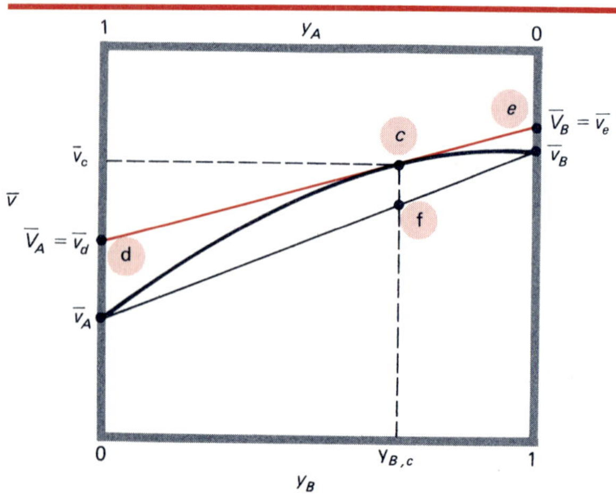

Figure 10.14 Nonideal mixture volumes.

composition (point c), draw a tangent to the composition curve and extend it to the $y_B = 0$ and $y_B = 1$ axes. At the $y_B = 0$ intercept (point d), we read the value of $\overline{V}_A$, the partial molal volume of component A; and at the $y_B = 1$ axis (point e), we read the partial molal volume of component B, or $\overline{V}_B$. Clearly these endpoints to the tangent line are the partial molal volumes given the definitions of the partial molal properties.

Note that

$$V_c = (n_A + n_B)\bar{v}_c \tag{10.110}$$

and from Eqs. (10.103) and (10.110), at the mixture composition at point c,

$$\overline{V}_A = \left(\frac{\partial V}{\partial n_A}\right)_{T,P,n_B} = \bar{v}_c + (n_A + n_B)\left(\frac{\partial \bar{v}}{\partial n_A}\right)_{T,P,n_B} \tag{10.111}$$

Since $y_A = n_A/(n_A + n_B)$, it follows that $dy_A = n_B\, dn_A/(n_A + n_B)^2$, or

$$\frac{dy_A}{1 - y_A} = \frac{dn_A}{n_A + n_B} \tag{10.112}$$

Equation (10.111) can then be written as

$$\begin{aligned}\overline{V}_A &= \bar{v}_c + (1 - y_A)\left(\frac{\partial \bar{v}}{\partial y_A}\right)_{T,P,n_B} \\ &= \bar{v}_c + (1 - y_A)\frac{\bar{v}_c - \bar{v}_d}{y_A - 1} = \bar{v}_d\end{aligned} \tag{10.113}$$

and we see that the partial molal volume of gas A at point c is indeed equal to $\bar{v}_d$, the molar specific volume at the tangent intercept where $y_B = 0$.

All mixture relations can be expressed in terms of partial molal properties. The definition of mixture internal energy, for example, is

$$U = H - PV \tag{10.114}$$

Holding constant T, P, and all molar amounts n_i except for n_A, we see the total derivative of Eq. (10.114) is simply

$$\left(\frac{\partial U}{\partial n_A}\right)_{T,P,n_i} = \left(\frac{\partial H}{\partial n_A}\right)_{T,P,n_i} - P\left(\frac{\partial V}{\partial n_A}\right)_{T,P,n_i} \tag{10.115}$$

or

$$\overline{U}_A = \overline{H}_A - P\overline{V}_A \tag{10.116}$$

Thus, all partial molal mixture properties can be developed from a minimal set of P-v-T data at a given concentration by using the methods in Chap. 9, which provide all information from a given fundamental equation.

10.9.2 Property Changes during Mixing

For a mixture of real materials, in general, the volume of the mixture is not equal to the sum of the volumes of the components. Let us restrict our attention to a two-component mixture that is mixed at constant temperature and pressure. The difference between the mixture volume and the volume of the components is

$$\Delta V = (n_A + n_B)\bar{v} - n_A\bar{v}_A - n_B\bar{v}_B = V - n_A\bar{v}_A - n_B\bar{v}_B \tag{10.117}$$

By using Eq. (10.108), this becomes

$$\begin{aligned}\Delta V &= \overline{V}_A n_A + \overline{V}_B n_B - n_A\bar{v}_A - n_B\bar{v}_B \\ &= n_A(\overline{V}_A - \bar{v}_A) + n_B(\overline{V}_B - \bar{v}_B)\end{aligned} \tag{10.118}$$

The change in volume on mixing is seen to be due to the difference between the partial molal volume of the components when mixed as compared with the molar volumes when unmixed. This difference is given by $\Delta\bar{v} = \bar{v}_c - \bar{v}_f$ in Fig. 10.14.

Changes due to mixing for other properties are found in a similar way. Note that, for an ideal gas, the partial molal volumes are given by Eq. (10.109), and substitution into Eq. (10.118) reveals that $\Delta V = 0$, as predicted by Amagat's law. We can define an *ideal solution* as one with the molar volume equal to the sum of the partial molal volumes for each component, so that the change in volume during a constant-temperature constant-pressure mixing process is zero. An ideal gas is obviously one example of an ideal solution. Ideal solutions offer considerable simplification in treating the properties of multicomponent systems with vapor-liquid equilibrium, which are examined in Chap. 12.

Changes in properties with concentration are used in engineering practice. The Kalina cycle (Chap. 7) makes use of the change in mixture saturation temperature with concentration to improve the Rankine cycle efficiency. Absorption air conditioning cycles (also Chap. 7) depend for their operation on the same phenomenon.

10.10 Summary

This chapter presents the thermodynamic analysis of multicomponent systems. General definitions and relations for systems composed of more than one component are given. Note that most of the analysis and properties are presented for ideal gases where the components are viewed as being composed of independent molecules. This independence characteristic of an ideal gas permits the combination of individual component properties to obtain mixture properties. The individual components are viewed as being independent at the mixture temperature and component partial pressure.

When one component exists in a state that is close to the saturation state for that component, a phase change is possible. Section 10.5 addresses mixtures of gas and vapor, and Sec. 10.6 presents the psychrometric chart which is applicable to mixtures of air and water vapor. The applications given in Sec. 10.7 involve air–water vapor mixtures and demonstrate the use of the psychrometric chart. Throughout Sec. 10.7, the processes for both individual components (air and water) are diagrammed. The general relations in Sec. 10.5 for mixtures of ideal gas and vapor are also valid for mixtures other than air and water vapor. The gases leaving a power plant contain the products of combustion (hydrocarbons, CO_2, H_2O, N_2, etc.) and can be analyzed with the relations presented. The water vapor in this mixture can condense and present corrosion difficulties which must be avoided. The process for each component can be diagrammed as in Sec. 10.7. (The pyschrometric chart cannot be used since it is applicable to air–water vapor mixtures only.)

Nonideal mixtures can be approximated by ideal mixture relations with modified properties. Although the result is not accurate in all cases, the relations are quite useful. General mixture relations are presented that require the partial molal properties.

Problems

10.1S Methane and oxygen are mixed in a volume ratio of 1 : 5. The total pressure of the mixture is 100 kPa. Find the following:

(*a*) The mole fraction of methane in the mixture
(*b*) The mass fraction of methane in the mixture
(*c*) The mean molecular weight of the mixture
(*d*) The partial pressure of the methane
(*e*) The apparent gas constant for the mixture
(*f*) The volume of 1 kmol of mixture at 300 K

10.1E Methane and oxygen are mixed in a volume ratio of 1 : 5. The total pressure of the mixture is 14.7 psia. Find the following:

(*a*) The mole fraction of methane in the mixture
(*b*) The mass fraction of methane in the mixture
(*c*) The mean molecular weight of the mixture
(*d*) The partial pressure of the methane
(*e*) The apparent gas constant for the mixture
(*f*) The volume of 1 lbmol of the mixture at 500°R

10.2 Given an air–water vapor mixture at:

$$T_{mix} = 60°C$$
$$P_{mix} = 0.3 \text{ MPa}$$
$$\phi = 50.15 \text{ percent}$$

Find: ω, $T_{dew\ point}$

10.3 Given an air–water vapor mixture at:

$$T_{mix} = 70°C$$
$$P_{mix} = 200 \text{ kPa}$$
$$P_a = 180 \text{ kPa}$$

Find: $T_{\text{dew point}}$, ω, ϕ

10.4 A room 4 m by 7 m by 3 m contains an air–water vapor mixture at a total pressure of 100 kPa and a temperature of 20°C. The partial pressure of the water vapor is 2 kPa. Calculate (*a*) the humidity ratio, (*b*) the dew point temperature, and (*c*) the total mass of vapor in the room.

10.5 How many grams of N_2 must be mixed with 50 g of CO_2 to produce a mixture that is 50 percent by volume of each constituent?

10.6 A mixture of O_2 and N_2 occupies a rigid container at 0.3 MPa and 150°C. The partial pressure of the O_2 is twice that of the N_2. (*a*) Determine the volumetric analysis. (*b*) Determine the specific gas constant of the mixture. (*c*) Determine the specific volume of the mixture.

10.7S For the mixture of Prob. 10.1S at 400 K, find (*a*) the specific internal energy u of the mixture in kJ/kg *and* kJ/kmol and (*b*) the specific entropy of the mixture in kJ/(kg·K) *and* kJ/(kmol·K). Assume $T_{ref} = 298$ K.

10.7E For the mixture of Prob. 10.1E at 600°R, find (*a*) the specific internal energy u of the mixture in Btu/lbm *and* Btu/lbmol and (*b*) the specific entropy of the mixture in Btu/(lbm·°R) *and* Btu/(lbmol·°R). Assume $T_{ref} = 537$°R.

10.8S A 1.0 kg/h stream of argon at 27°C and 1 atm is mixed adiabatically with 0.5 kg/h CO_2 at 1 atm and 300°C.

(*a*) What is the final temperature of the mixture?
(*b*) Find the rate of entropy generation during the process.
(*c*) Calculate the entropy of the final mixture.

10.8E A 1.0 lbm/h stream of argon at 80°F and 1 atm is mixed adiabatically with 0.5 lbm/h CO_2 at 1 atm and 550°F.

(*a*) What is the final temperature of the mixture?
(*b*) Find the rate of entropy generation during the process.
(*c*) Calculate the entropy of the final mixture.

10.9S It is desired to isothermally separate air (21 percent O_2 and 79 percent N_2 by volume) into pure oxygen and nitrogen, by carrying out the membrane separation process at 1 atm and 27°C in a continuous reactor.

(a) What are the heat transfer and entropy change for the process?
(b) What is the reversible work required if the surroundings are at 27°C? Is the process possible?

10.9E It is desired to isothermally separate air (21 percent O_2 and 79 percent N_2 by volume) into pure oxygen and nitrogen, by carrying out the membrane separation process at 1 atm and 80°F in a continuous reactor.
(a) What are the heat transfer and entropy change for the process?
(b) What is the reversible work required if the surroundings are at 80°F? Is the process possible?

10.10 The gases leaving a natural-gas-fired boiler have the following composition:

Component	Volume Percentage
CO_2	8.7
H_2O	16.0
N_2	72.1
O_2	3.2
	100.0

At what temperature will condensation of the water vapor occur if the stack gases are at 1 atm total pressure?

10.11 What is the value of h for the gas mixture of Prob. 10.10 when the given mixture is saturated with water vapor?

10.12S If the stack gases in Prob. 10.10 are initially at 600°C, how much heat transfer can occur before condensation of the water vapor occurs?

10.12E If the stack gases in Prob. 10.10 are initially at 1100°F, how much heat transfer can occur before condensation of the water vapor occurs?

10.13S A gas turbine provides combustion products at 540°C with the following composition (by weight): O_2, 17 percent; H_2O, 4 percent; N_2, 79 percent. These products are to be mixed with air (25°C, 1 atm), and the resulting gas mixture will be used as the oxidizer in a natural-gas-fired furnace. The combustion air must be at $T = 375$°C upon entering the furnace. See Fig. P10.13.
(a) How much air must be added (kilograms of air per kilogram of turbine exhaust)?
(b) What will be the value of h of the mixture entering the furnace?

(*c*) How much energy is saved by this system over an alternative system that uses fuel to preheat 100 percent ambient air to the furnace to 375°C? (Put the answer in terms of energy saved per kilogram of O_2 entering the furnace.)

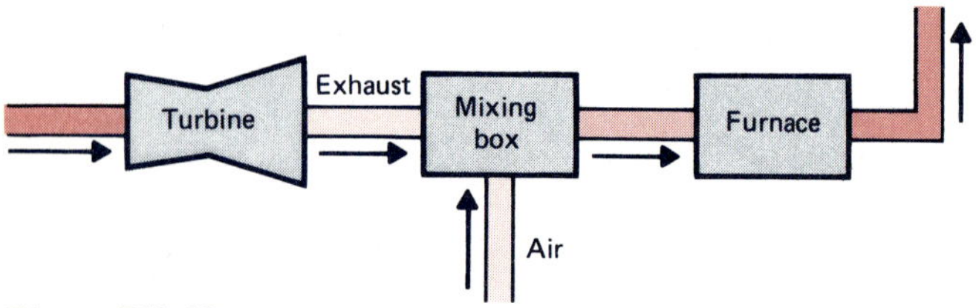

Figure P10.13

10.13E A gas turbine provides combustion products at 1000°F with the following composition (by weight): O_2, 17 percent; H_2O, 4 percent; N_2, 79 percent. These products are to be mixed with air (77°F, 1 atm), and the resulting gas mixture will be used as the oxidizer in a natural-gas-fired furnace. The combustion air must be at $T = 700$°F upon entering the furnace. See Fig. P10.13.

(*a*) How much air must be added (pound-mass of air per pound-mass of turbine exhaust)?

(*b*) What will be the value of h of the mixture entering the furnace?

(*c*) How much energy is saved by this system over an alternative system that uses fuel to preheat 100 percent ambient air to the furnace to 700°F? (Put the answer in terms of energy saved per-pound mass of O_2 entering the furnace.)

10.14S Air at 30°C and 1 atm has a relative humidity of 60 percent. Find (*a*) the dew point (*b*), the humidity ratio, and (*c*) the enthalpy of the wet air.

10.14E Air at 90°F and 1 atm has a relative humidity of 60 percent. Find (*a*) the dew point, (*b*) the humidity ratio, and (*c*) the enthalpy of the wet air.

10.15S A stream of air at 1 atm, 20°C, and relative humidity of 80 percent has a mass flow rate of 0.05 kg/s of dry air. A second stream of air at 1 atm, 35°C, and relative humidity of 40 percent is mixed adiabatically with the first stream to give a mixed stream temperature of 30°C.

(*a*) What mass flow rate of the second stream is required?

(*b*) What is the relative humidity of the mixed stream?

10.15E A stream of air at 1 atm, 60°F, and relative humidity of 80 percent has a mass flow rate of 0.1 lbm/s of dry air. A second stream of air at 1 atm, 95°F, and relative humidity of 40 percent is mixed adiabatically with the first stream to give a mixed stream temperature of 70°F.

(*a*) What mass flow rate of the second stream is required?

(*b*) What is the relative humidity of the mixed stream?

10.16 An air–water vapor mixture enters an air conditioning unit at a pressure of 100 kPa, a temperature of 30°C, and a relative humidity of 70 percent. The mass of dry air entering is 1000 g/min. The mixture leaves the air conditioner at 0.1 MPa, 20°C, and 40 percent relative humidity. Condensate leaves at 26°C. Find the heat transfer rate per minute for the process. Do not use a psychrometric chart.

10.17 An air–water vapor mixture enters a device with a pressure of 0.15 MPa, a temperature of 40°C, and an unknown relative humidity. The mass of dry air entering per minute is 200 g. The mixture leaves the device at 30°C, 15 MPa, and 80 percent relative humidity. Water is sprayed into the air and evaporated to do this cooling. If the water temperature is 30°C and the device is adiabatic, how much water is required to run for 1 h?

10.18 Given an air–water vapor mixture at:

$T_{mix} = 30°C$

$P_{mix} = 0.1$ MPa

$\phi = 60$ percent

Find: ω, $T_{dew\ point}$

10.19 Given an air–water vapor mixture at:

$T_{mix} = 90°C$

$P_{mix} = 400$ kPa

$P_a = 380$ kPa

Find: $T_{dew\ point}$, ϕ, ω

10.20 Given an air–water vapor mixture at:

$T_{dry\ bulb} = 35°C$

$T_{wet\ bulb} = 28°C$

$P = 1$ atm

Find: ϕ, ω, $T_{dew\ point}$

10.21 Given an air–water vapor mixture at:

$T_{dry\ bulb} = 25°C$

$\phi = 70$ percent

Find: $T_{wet\ bulb}$, ω, $T_{dew\ point}$

10.22 Given an air–water vapor mixture at:

$T_{\text{wet bulb}} = 20°\text{C}$
$T_{\text{dry bulb}} = 30°\text{C}$
$P = 1\ \text{atm}$

Find: ϕ, ω, P_v, $T_{\text{dew point}}$

10.23 Given an air–water vapor mixture at:

$T_{\text{dry bulb}} = 35°\text{C}$
$\phi = 80$ percent
$P = 1\ \text{atm}$

Find: $T_{\text{wet bulb}}$, ω, P_v, $T_{\text{dew point}}$

10.24 Given an air–water vapor mixture at:

$\omega = 0.016$
$\phi = 60$ percent
$P = 1\ \text{atm}$

Find: $T_{\text{dry bulb}}$, $T_{\text{wet bulb}}$, $T_{\text{dew point}}$

10.25 Given the air conditioner in Fig. P10.25, determine $\dot{Q}_{CV}$.

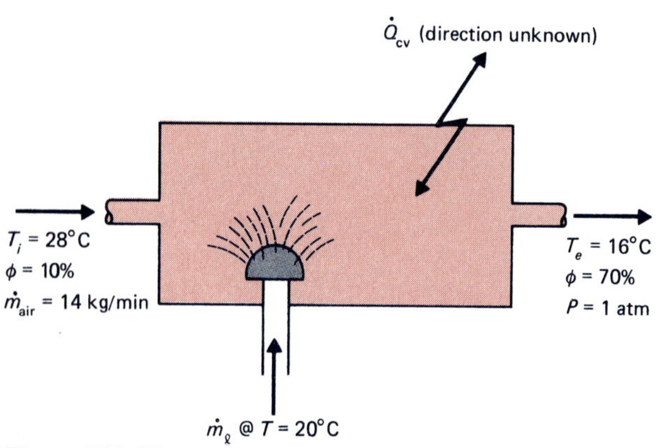

Figure P10.25

10.26 Given the device with conditions shown in Fig. P10.26, determine $\dot{Q}_{CV}$ and $\dot{m}_l$ if it operates at a pressure of 1 atm.

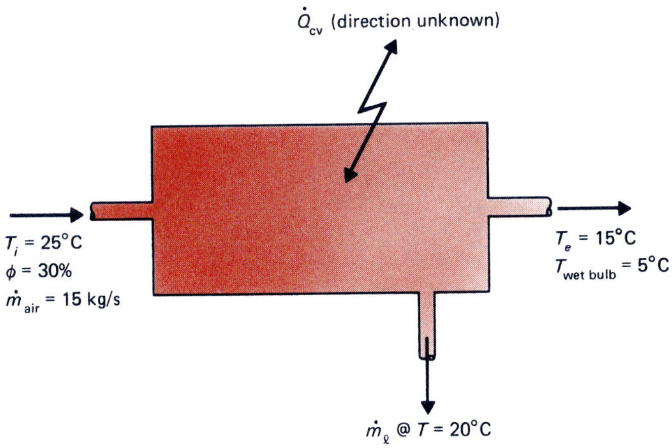

Figure P10.26

10.27 Given the device shown in Fig. P10.27, determine $\dot{Q}_{CV}$. The pressure is 1 atm.

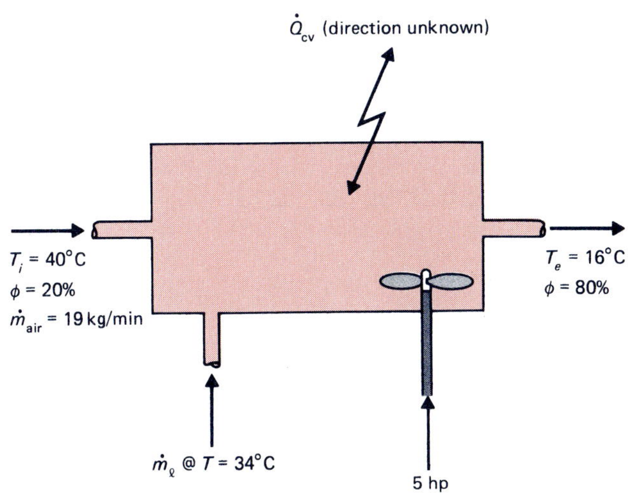

Figure P10.27

10.28 The nozzle in Fig. P10.28 expels an air–water vapor mixture and $\dot{m}_l = 0.3975$ kg/s of condensate at the exit of the nozzle at the exit temperature shown. Determine the exit velocity of the air–vapor mixture.

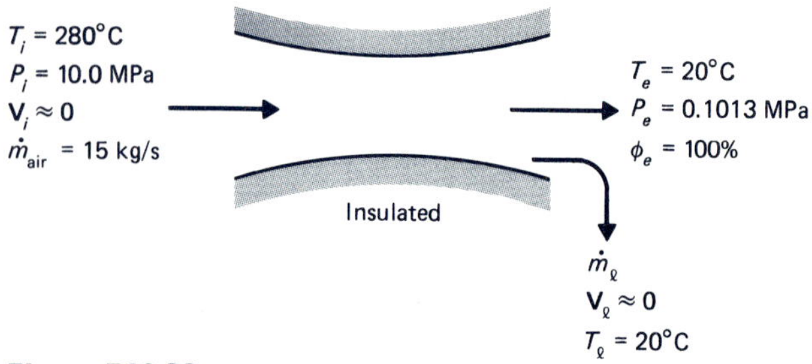

Figure P10.28

10.29 The wet-bulb temperature is 25°C, and the dry-bulb temperature is 32°C. Using the psychrometric chart (Fig. D.8), determine ϕ, ω, P_v, and $T_{\text{dew point}}$.

10.30 The wet-bulb temperature is 20°C, and the dew point temperature is 15°C. Use the psychrometric chart (Fig. D.8) to determine the dry-bulb temperature, the relative humidity, the humidity ratio, and the partial pressure of the vapor.

10.31 If the dew point temperature is 24°C and the dry-bulb temperature is 30°C, determine the wet-bulb temperature, $T_{\text{wet bulb}}$, the humidity ratio ω, the partial pressure of the vapor P_v, and the relative humidity ϕ.

10.32 The partial pressure of the vapor P_v is 0.002 MPa, and the dry-bulb temperature is 35°C. Use the psychrometric chart (Fig. D.8) to determine $T_{\text{wet bulb}}$, ω, and ϕ.

10.33 An air-vapor mixture in a steady state, steady flow process is cooled and dehumidified as shown in Fig. P10.33. The process occurs at 0.1013 MPa. Do not use a psychrometric chart.

(*a*) Determine the rate of heat transfer (kW).

(*b*) Determine the mass flow rate of the water condensed (kg/s).

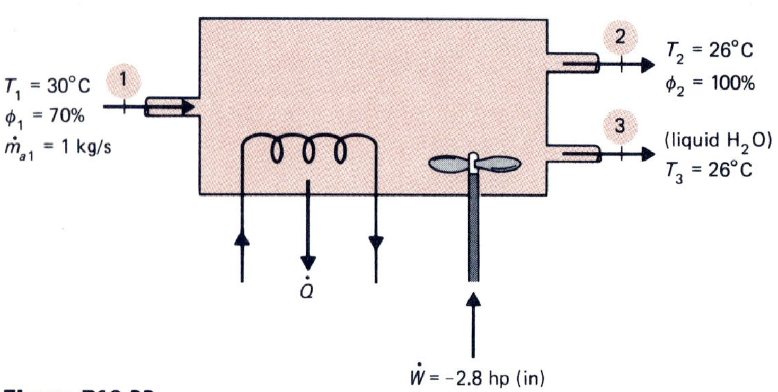

Figure P10.33

10.34 At an Army research facility in the Southwest, air is cooled from $T_{\text{dry bulb}} = 45°C$ and a relative humidity of 10 percent by evaporative cooling and a cooling coil to 24°C and a relative humidity of 60 percent. This mass flow rate of air is 2×10^3 kg/min and $T_l = 21°C$.

(*a*) Draw the entire process on a psychrometric chart.

(*b*) Determine the mass flow rate of water required for the water sprays.

(*c*) Determine the rate of heat transferred in the cooling coils.

10.35 Given the winter air conditioning system shown in Fig. P10.35. The pressure of this steady state, steady flow process is 101.3 kPa.

(*a*) Draw the entire process on a psychrometric chart.

(*b*) Determine the mass flow rate of the water in the water sprays.

(*c*) Determine the total rate of heat transfer to the system.

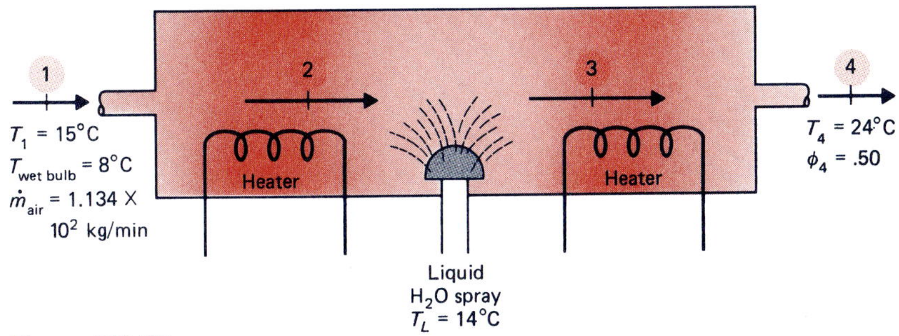

Figure P10.35

10.36S Desert air at 33°C with 10 percent relative humidity is passed through a *swamp cooler*, a device that contacts the dry air with a wet surface. Water in the swamp cooler is evaporated until the desert air is saturated with water vapor. The swamp cooler is assumed to operate adiabatically, and the water in the swamp cooler is at the temperature of the air leaving the swamp cooler. What will be the temperature of the saturated air leaving the swamp cooler?

10.36E Desert air at 90°F with 10 percent relative humidity is passed through a *swamp cooler*, a device that contacts the dry air with a wet surface. Water in the swamp cooler is evaporated until the desert air is saturated with water vapor. The swamp cooler is assumed to operate adiabatically, and the water in the swamp cooler is at the temperature of the air leaving the swamp cooler. What will be the temperature of the saturated air leaving the swamp cooler?

10.37S If 0.01 kg/s of desert air enters the swamp cooler of Prob. 10.36S, how much makeup water must be added to the swamp cooler?

10.37E If 0.02 lbm/s of desert air enters the swamp cooler of Prob. 10.36E, how much makeup water must be added to the swamp cooler?

10.38S Air at 1 atm, 30°C, and 80 percent relative humidity is passed over the cold evaporator coil of an air conditioner. The air leaves the coil at 18°C. The wet air has a mass flow rate of 0.02 kg/s. How much water is condensed from the wet air?

10.38E Air at 1 atm, 85°F, and 80 percent relative humidity is passed over the cold evaporator coil of an air conditioner. The air leaves the coil at 65°F. The wet air has a mass flow rate of 0.04 lbm/s. How much water is condensed from the wet air?

10.39 What is the rate of heat transfer from the air in Prob. 10.38?

10.40S An air conditioning system works as shown in Fig. P10.40S. Find (*a*) T_B, (*b*) ϕ_B, (*c*) T_C, (*d*) ϕ_C, and (*e*) $T_{H_2O, out}$.

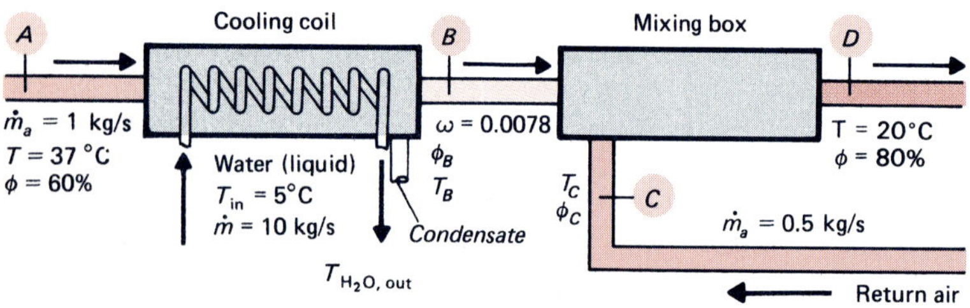

Figure P10.40S

10.40E An air conditioning system works as shown in Fig. P10.40E. Find (*a*) T_B, (*b*) ϕ_B, (*c*) T_C, (*d*) ϕ_C, and (*e*) $T_{H_2O, out}$.

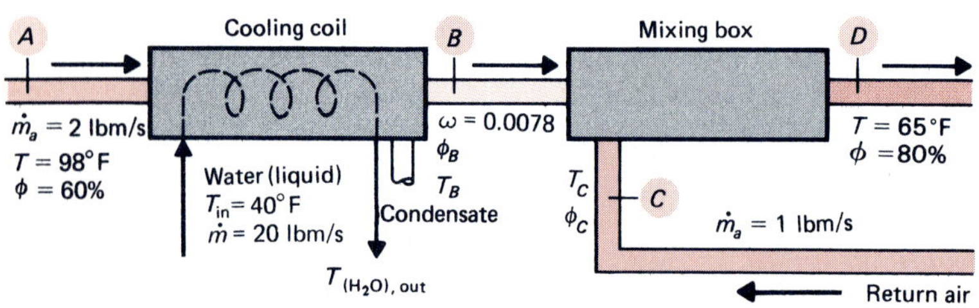

Figure P10.40E

10.41S Outside air at 35°C and $\phi = 65$ percent is cooled to 20°C and then mixed with recirculated air at 25°C and $\phi = 50$ percent. Recirculated air is mixed with outside air in the ratio of 1 : 3 on a dry air mass basis. What is the humidity ratio of the resulting mixed air? Sketch the process on a psychrometric chart.

10.41E Outside air at 95°F and $\phi = 65$ percent is cooled to 65°F and then mixed with recirculated air at 70°F and $\phi = 50$ percent. Recirculated air is mixed with outside air in the ratio 1 : 3 on a dry air mass basis. What is the humidity ratio of the resulting mixed air? Sketch the process on a psychrometric chart.

10.42S An old and inefficient air conditioning system used in commercial buildings when energy was very cheap worked as follows: Outside air (warm or cold, dry or humid) was passed over a cooling coil to condense water vapor and thus control humidity. The air was then heated to raise the temperature back to comfort conditions. The conditions are shown in Fig. P10.42S.

(*a*) Evaluate the heat transfer in the heater.

(*b*) How much heat transfer occurred in the cooling coil?

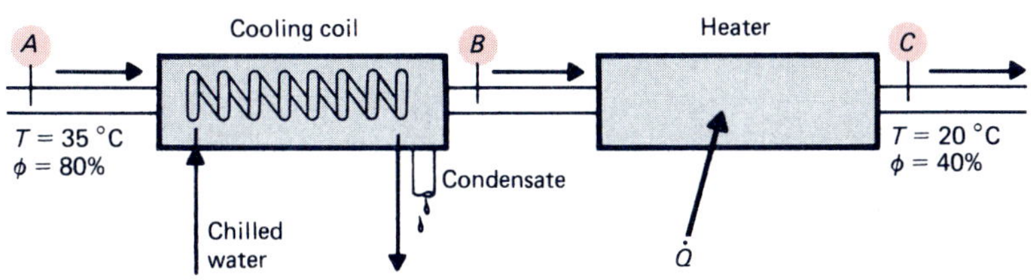

Figure P10.42S

10.42E An old and inefficient air conditioning system used in commercial buildings when energy was very cheap worked as follows: Outside air (warm or cold, dry or humid) was passed over a cooling coil to condense water vapor and thus control humidity. The air was then heated to raise the temperature back to comfort

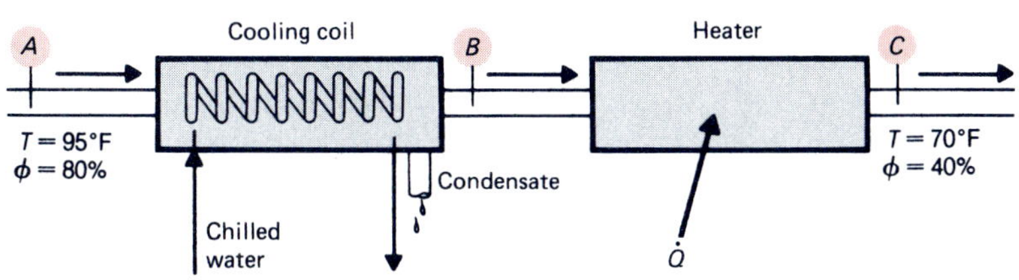

Figure P10.42E

conditions. The conditions are shown in Fig. P.10.42E.

(*a*) Evaluate the heat transfer in the heater.

(*b*) How much heat transfer occurred in the cooling coil?

10.43S Plot the density ρ ($\rho = 1/v$) for air versus relative humidity at $T = 30°C$. If aircraft lift is proportional to density at a given speed, what effect on required runway length would you expect humidity to have?

10.43E Plot the density ρ ($\rho = 1/v$) for air versus relative humidity at $T = 90°F$. If aircraft lift is proportional to density at a given speed, what effect on required runway length would you expect humidity to have?

10.44S The temperature of the atmosphere on a given day follows roughly the relation

$$T(y) = T_0 - \frac{2y}{300}$$

where T_0 (°C) is the temperature at sea level and y is the height above sea level in meters. On this day, the temperature at sea level is 30°C, and the relative humidity as sea level is 70 percent. At what altitude would you predict clouds to form if sea-level air were being carried aloft by convective currents and cooled to the temperature of the altitude it reaches?

10.44E The temperature of the atmosphere on a given day follows roughly the relation

$$T(y) = T_0 - \frac{5y}{1000}$$

where T_0 (°F) is the temperature at sea level and y is the height above sea level in feet. On this day, the temperature at sea level is 85°F, and the relative humidity at sea level is 70 percent. At what altitude would you predict clouds to form if sea-level air were being carried aloft by convective currents and cooled to the temperature of the altitude it reaches?

10.45S A passenger is on an aircraft that takes off from sea level on a day when the dry-bulb temperature is reported as 27°C and the relative humidity is 80 percent. She notes that fog appears to form in a sheet on the top of the wing as the plane takes off, and she assumes that this is due to condensation of water vapor.

(*a*) Explain this phenomenon in terms of what you know about air–water vapor mixtures and Bernoulli's equation relating velocity and pressure.

(*b*) What is the *maximum* pressure on the wing surface at which this phenomenon can occur at the given conditions?

10.45E A passenger is on an aircraft that takes off from sea level on a day when the dry-bulb temperature is reported as 80°F and the relative humidity is 80 percent. He notes that fog appears to form in a sheet on the top of the wing as the plane takes off, and he assumes that this is due to condensation of water vapor.

(*a*) Explain this phenomenon in terms of what you know about air–water vapor mixtures and Bernoulli's equation relating velocity and pressure.

(*b*) What is the *maximum* pressure on the wing surface at which this phenomenon can occur at the given conditions?

10.46 Desiccant systems control air humidity by absorbing or adsorbing water vapor from the air into liquids or onto solids with a high affinity for water molecules. During the process, enthalpy of absorption is released, and the air-desiccant system increases in temperature.

Draw a schematic of the desiccation process described on a psychrometric chart, and show how to find the enthalpy change during the desiccant absorption process.

10.47 In a certain northern city, a heating system is designed to heat outside air by injecting steam and by heating as shown in Fig. P10.47.

(*a*) Determine the mass flow rate of steam required.

(*b*) Determine the rate of heat transfer.

Figure P10.47

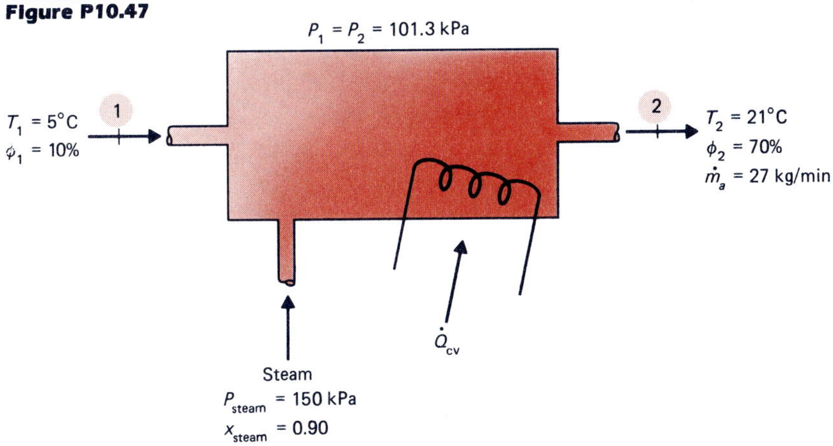

10.48 An air-vapor mixture is cooled and dehumidified in the steady-state, steady-flow device shown in Fig. P10.48. The device operates at atmospheric pressure.

(*a*) Determine the mass flow rate of the liquid.

(*b*) Determine the rate of heat transfer.

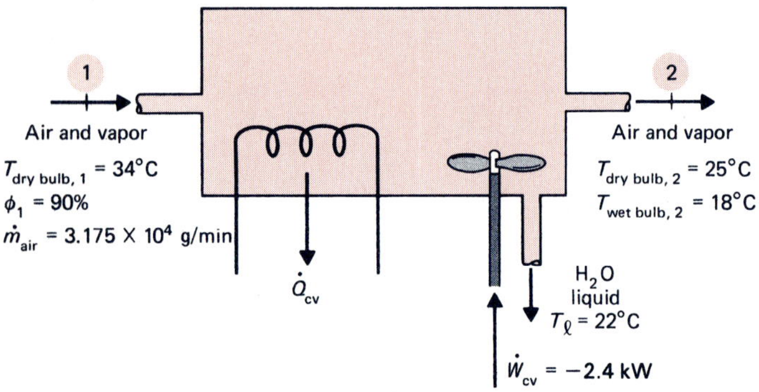

Figure P10.48

10.49 The air conditioner shown in Fig. P10.49 operates at a pressure of 101.3 kPa.

(*a*) Draw the entire process on a psychrometric chart.
(*b*) Determine the mass flow rate of water required by the spray.
(*c*) Determine the rate of heat transfer.

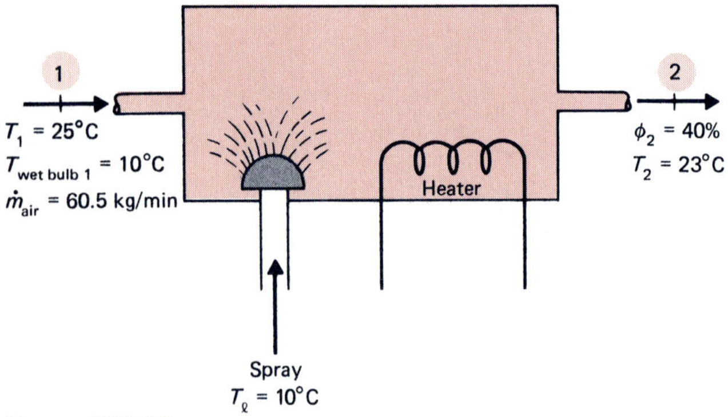

Figure P10.49

10.50 An air–water vapor mixture at 101.3 kPa has a dry-bulb temperature of 35°C and a relative humidity of 80 percent. The mixture is cooled to a dry-bulb temperature of 20°C. If any condensate has a temperature of 22°C, find the heat transfer (kJ/kg dry air) required.

10.51 A room air conditioner treats the air as shown in Fig. P10.51. Determine the mass of water removed per hour by the unit.

10.52 A desert cooler in Saudi Arabia cools the hot dry air by a cool water spray evaporation process (Fig. P10.52). The desert air is at 40°C and 10 percent relative humidity. If the desired exit conditions from the cooler

Figure P10.51

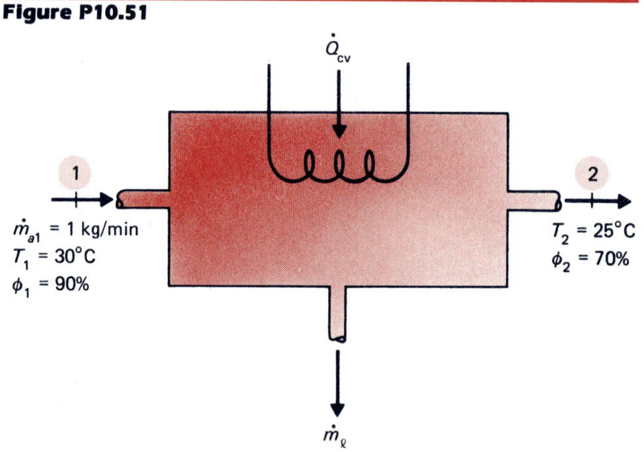

are 25 °C and 50 percent relative humidity, how much water (kg/h) must be sprayed into a dry air flow of 10 kg/min?

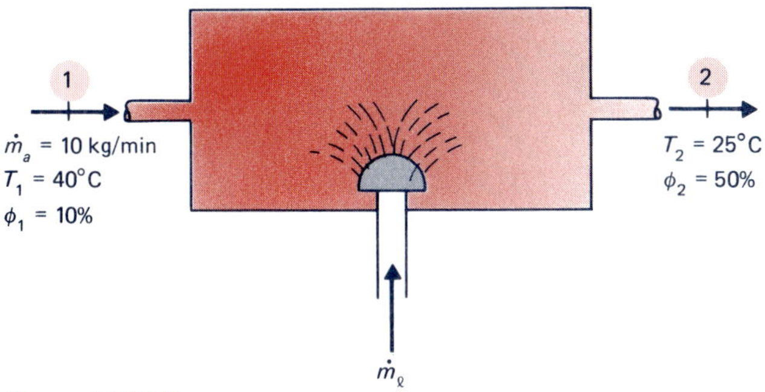

Figure P10.52

10.53 A residential air conditioner heats and humidifies outside air as shown in Fig. P10.53.
(*a*) Draw the process on a psychrometric chart.
(*b*) Calculate the rate of heat transfer (kW).
(*c*) Calculate $\dot{m}_l$.

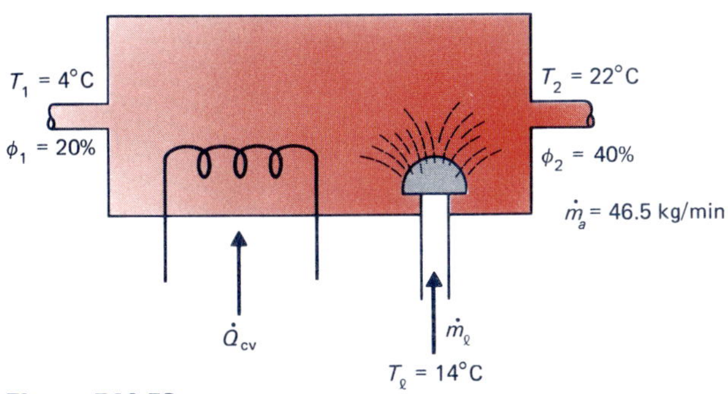

Figure P10.53

10.54 A heater-humidifier operates as shown in Fig. P10.54. The rate of heat transfer at the coils is 15 kW.
(*a*) Find the mass flow rate of dry air.
(*b*) Find the mass flow rate of spray water.

Figure P10.54

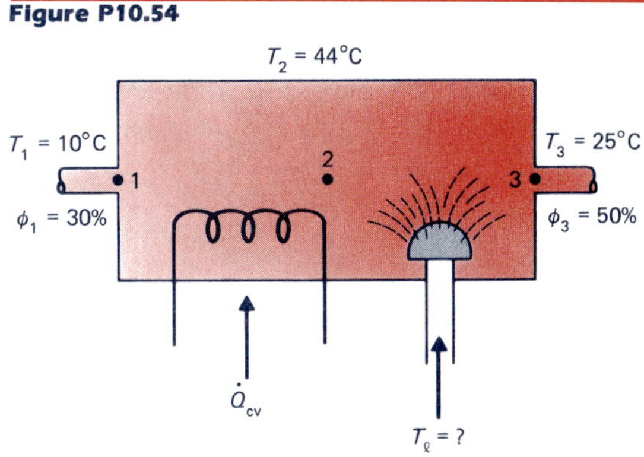

10.55 A central air conditioning system operates as shown in Fig. P10.55.
(*a*) Draw the process on a psychrometric chart.
(*b*) Calculate the required cooling capacity of the cooling coils (tons).
(*c*) Calculate $\dot{Q}_{CV}$ for the heating coils (kW).
(*d*) Determine the amount of condensate produced in 1-h operation (kg).

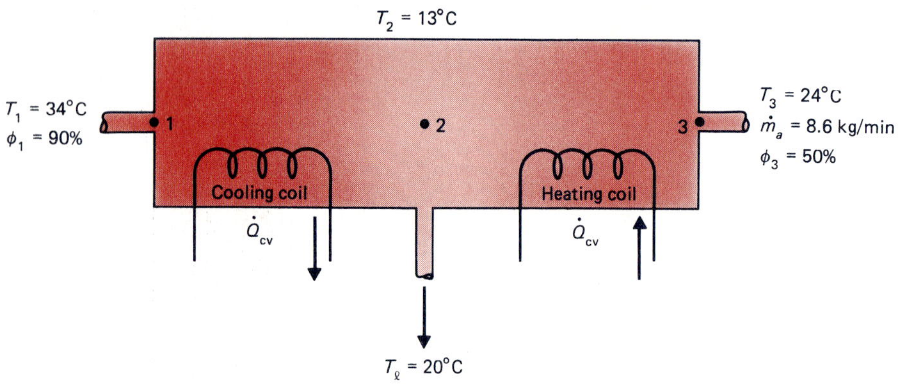

Figure P10.55

10.56 The industrial air conditioner shown (Fig. P10.56) provides warm dry air at atmospheric pressure for an industrial process.

(*a*) Find $\dot{m}_a$ if the cooling capacity is 12 tons.

(*b*) Find $\dot{Q}_{CV}$ (kW) for the heating coil.

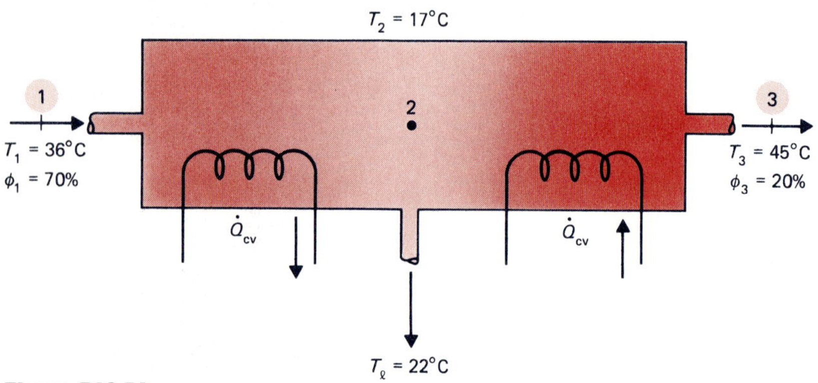

Figure P10.56

10.57 A large municipal power plant (Fig. P10.57) is cooled with an evaporative cooling tower as shown. Find the required mass flow rate of dry air through the tower and the required flow rate of basin makeup water.

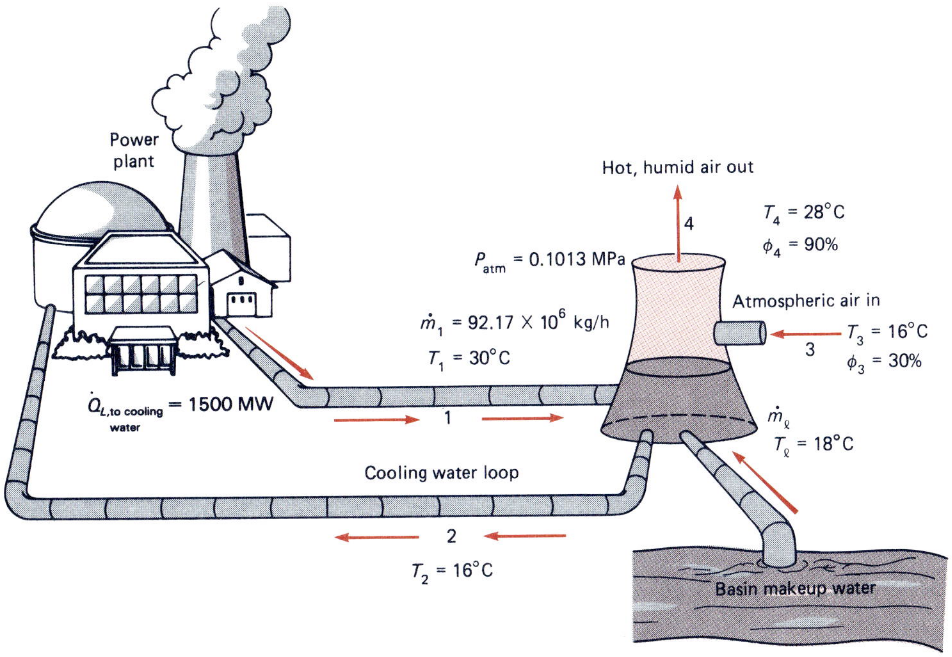

Figure P10.57

10.58 Some 100×10^6 kg/h of cooling water is cooled from 50 to 30°C in an evaporative cooling tower. The atmospheric air–water vapor mixture enters the tower at 20°C with relative humidity 40 percent and leaves at 34°C and relative humidity of 100 percent. Basin makeup water has a temperature of 20°C. Find the required mass flow rate of dry air through the tower and the required flow rate of basin makeup water.

10.59 Some 100×10^6 kg/h of cooling water is cooled from 50 to 30°C in a dry cooling tower. Atmospheric air enters the tower at 20°C and leaves at 30°C. Find the required flow rate of the air.

10.60 Some 150×10^6 kg/h of cooling water is cooled from 50 to 30°C in a mechanical draft evaporative cooling tower (Fig. P10.60). The atmospheric air–water vapor mixture enters the tower at 20°C, relative humidity of 30 percent and leaves at 40°C, relative humidity of 100 percent. The basin makeup water enters at 15°C. The fan is driven by a 10,000-hp motor. Determine the required mass flow rate of dry air through the tower and the required flow rate of basin makeup water.

Figure P10.60

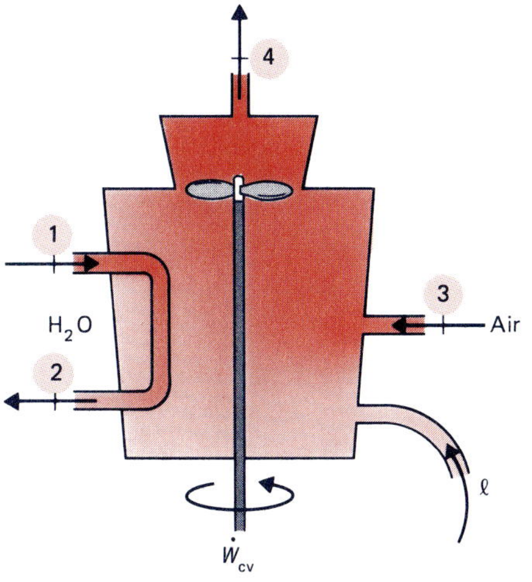

10.61 Some 150×10^6 kg/h of cooling water is cooled from 60 to 26°C in a mechanical draft dry cooling tower (Fig. P10.61). Atmospheric air enters the tower at 20°C and leaves at 35°C. The fan is driven by a 10,000-hp motor. Find the required flow rate of the air.

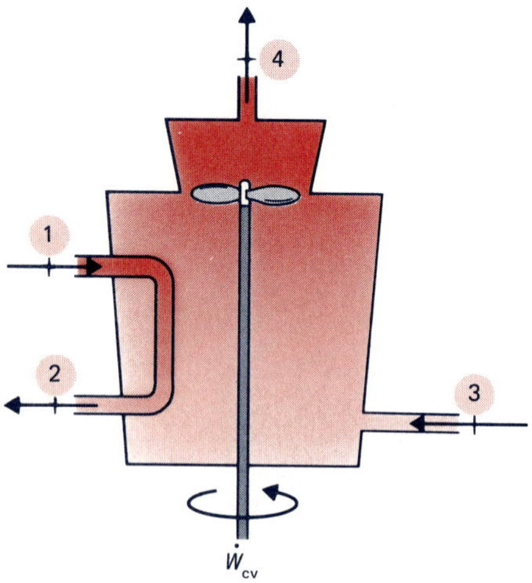

Figure P10.61

10.62 A power plant with a cooling lake operates as shown in Fig. P10.62. The lake temperature seasonally varies between 2 and 18°C. If the maximum temperature ecologically allowed at the exit is 30°C, determine the seasonal range in $\dot{m}$ of the cooling water.

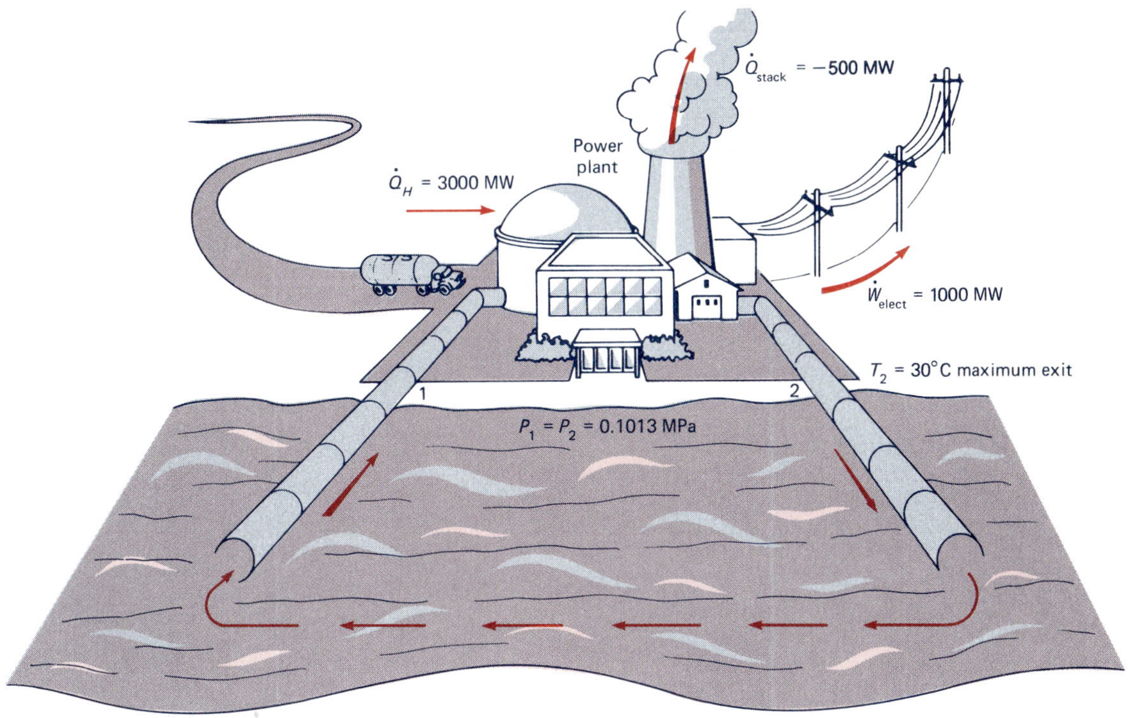

Figure P10.62

10.63 A mixture of ideal gases at 30°C and 0.2 MPa is composed of 30 g CO_2, 75 g of N_2, and 5 g of He. What is the volume of the mixture?

10.64 Equal masses of H_2 and O_2 are mixed. Determine the percentage by volume and the partial pressures of each constituent if the mixture is at 0.15 MPa.

10.65 The volumetric analysis of a mixture of ideal gases is given below.

(*a*) Determine the gravimetric analysis (mass percent).

(*b*) Determine c_v for the mixture.

(*c*) Determine the heat transfer required (kJ/kg) to increase the temperature of the mixture 100°C if it is contained in a rigid vessel.

Constituent	v_i
CO_2	30 percent
N_2	40 percent
O_2	30 percent

10.66S Using Kays' rule, find the fugacity of a mixture of methane and *n*-butane in the molar ratio of 2:1, respectively, when the mixture is at $T = 500°C$ and $P = 60$ atm.

10.66E Using Kays' rule, find the fugacity of a mixture of methane and *n*-butane in the molar ratio of 2:1, respectively, when the mixture is at $T = 950°F$ and $P = 60$ atm.

10.67S Find *h*, *s*, and *v* for a mixture of 20 mol percent methane, 40 mol percent ethane, and 40 mol percent propane at 500°C and 100 atm by using (*a*) Kays' rule and (*b*) the nonideal mixture relations and reduced-property data (Sec. 10.8).

10.67E Find *h*, *s*, and *v* for a mixture of 20 mol percent methane, 40 mol percent ethane, and 40 mol percent propane at 930°F and 100 atm by using (*a*) Kays' rule and (*b*) the nonideal mixture relations and reduced-property data (Sec. 10.8).

10.68S Using Kays' rule and the Lee-Kesler tables, plot the specific volume of a mixture of SO_2 and methane at $P = 300$ atm and $T = 400$ K versus the mole fraction of SO_2 present. Using this plot, determine the partial molal volume of SO_2 in the mixture at $y_{SO_2} = 0.5$.

10.68E Using Kays' rule and the Lee-Kesler tables, plot the specific volume of a mixture of SO_2 and methane at $P = 300$ atm and $T = 700°R$ versus the mole fraction of SO_2 present. Using this plot, determine the partial molal volume of SO_2 in the mixture at $y_{SO_2} = 0.5$.

10.69 The molar volume of a mixture of components *A* and *B* at *T* and *P* is given by

$$\bar{v}_{T,P} = y_A \bar{v}_A + (1 - y_A)^2 \bar{v}_B$$

What are the partial molal volumes of components *A* and *B* at $y_A = 0.75$?

References

1. Robert C. Reid, John M. Prausnitz, and Bruce E. Poling, *The Properties of Liquids and Gases*, 4th ed., McGraw-Hill, New York, 1987.
2. B. G. Kyle, *Chemical and Process Themodynamics,* Prentice-Hall, Englewood Cliffs, N.J., 1984.

11

Chemical Reactions and Combustion

When primitive man learned how to make fire, he had discovered controllable energy, which then became a "servant" destined to perform an endless series of "miracles." . . . This discovery may have been the single most vital factor which allowed mankind to develop modern civilization.
Herman Kahn, *The Next 200 Years*

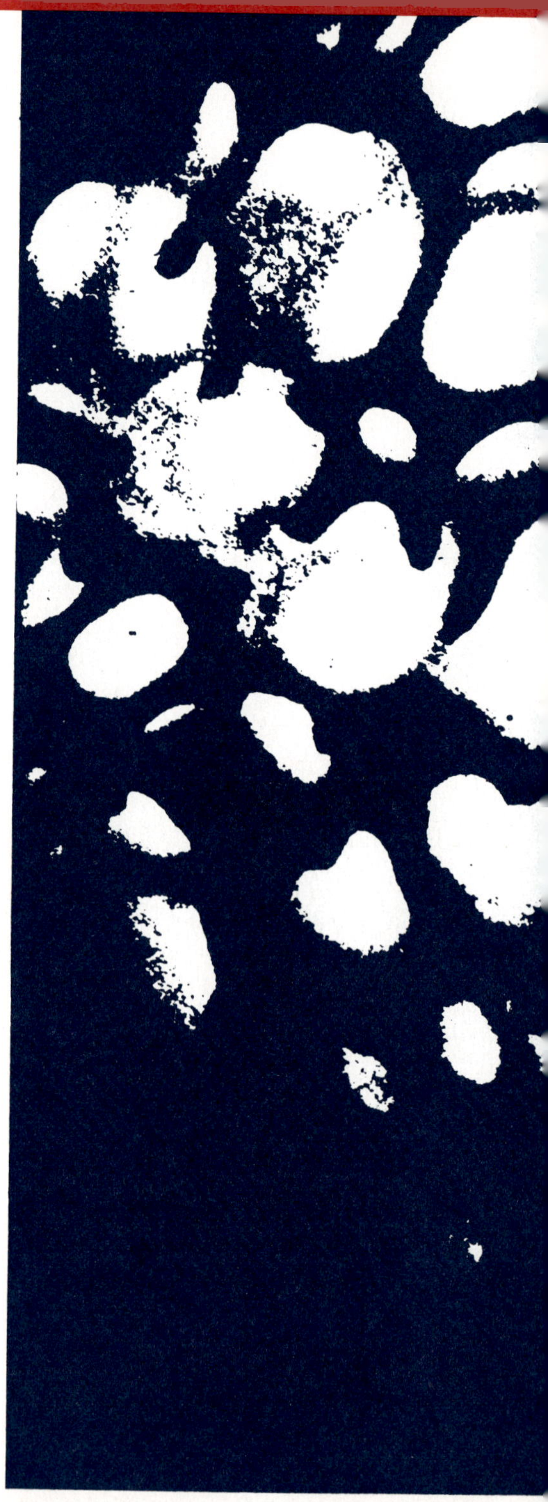

Porous ceramic insert in natural gas-fired burner used to produce high thermal radiation output. (*Center for Energy Studies, The University of Texas at Austin.*)

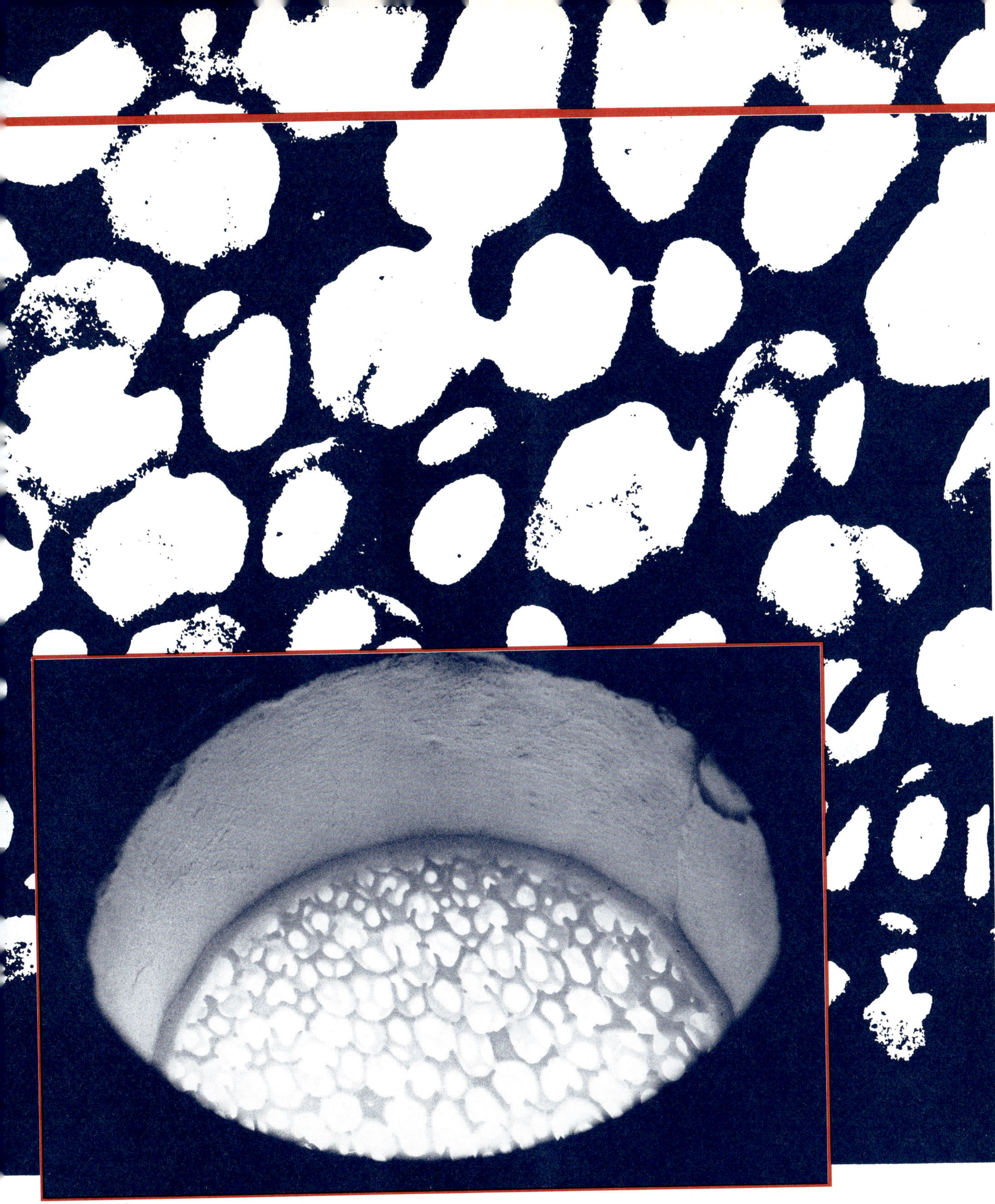

11.1 Introduction

Chemically reacting systems, and particularly systems undergoing combustion, are of interest to the engineer dealing with design in the chemical process industry, internal and external combustion engines of all types, fossil-fuel power plants, solid waste conversion systems, and studies of the initiation and spread of fires, among many other related subjects.

In this chapter, we develop the concepts and methods necessary to treat a broad variety of combustion problems of engineering interest, and we apply the methods to some problems.

11.2 Chemical Reactions in Combustion Systems

11.2.1 The General Case

Most systems that we consider in this text are systems in which a hydrocarbon or mixture of hydrocarbons with the general formula $C_lH_mO_n$ is the fuel, and this fuel is burned in air or with pure oxygen. Such a system will produce combustion products that may contain O_2, N_2, CO, CO_2, and H_2O. The general hydrocarbon combustion reaction equation is thus, for 1 mol of fuel,

$$C_lH_mO_n + A_1O_2 + A_2N_2 \rightarrow A_3CO_2 + A_4CO + A_5H_2O + A_6O_2 + A_7N_2 + A_8C_lH_mO_n \quad (11.1)$$

where the A_i are the *stoichiometric coefficients* for the reaction. In the idealized case called *complete combustion,* it is assumed that if there is excess air or oxygen in the original reactant mixture, then all the fuel will be burned, all the CO will be oxidized to CO_2, and no CO will appear in the products of combustion. If excess fuel is present, then no O_2 will appear in the combustion products because it will all be consumed, and some mixture of CO and CO_2 will be present in the products along with some unburned fuel. (In real combustion processes, CO_2 and H_2O will dissociate and CO and O_2 will be found in the combustion products, even in the presence of excess air and good mixing.)

Because air can be assumed to be a mixture of 21 percent oxygen and 79 percent nitrogen by volume (and thus, by the ideal gas relation, to be in the same molar ratio), the coefficients A_1 and A_2 must be related by $A_2 = (79/21)A_1 = 3.76\ A_1$. (It is common practice to lump the other inert atmospheric gases such as argon and CO_2 with the nitrogen, because they do not take part in the combustion process and have properties close to those of nitrogen.) Because the nitrogen is assumed to be inert and not take part in the chemical reaction, $A_7 = A_2$.

In real combustion systems, *regardless* of whether excess fuel or oxidant is present, it is impossible to achieve complete mixing of fuel and oxidant. All the combustion products shown on the right side of Eq. (11.1) will be present, at least in small amounts.

11.2.2 Combustion with Stoichiometric or Excess Air

If just enough air is present to completely oxidize the fuel (a *stoichiometric* ratio of oxidant to fuel) or if excess air is present over that required for complete combustion of the fuel, as is the case for most processes we will study, then it can be taken as an idealization that $A_4 = A_8 = 0$, and Eq. (11.1) becomes

$$C_lH_mO_n + A_1O_2 + 3.76A_1N_2 \rightarrow A_3CO_2 + A_5H_2O + A_6O_2 + 3.76A_1N_2 \quad (11.2)$$

For the stoichiometric case, A_6 is also equal to zero.

The *total number* of carbon, hydrogen, oxygen, and nitrogen atoms is conserved during combustion; however, the *structure* of the atoms within the reactant molecules is changed by the combustion process into the form of the product molecules. Because the number of atoms of each element is the same before and after combustion, detailed balances can be written that equate the number of atoms of each element present in all reactants to the number of atoms of that element present in the products. The number of atoms of a given element represented in the chemical balance equation is given by the product of the coefficient (A_1, A_3, A_5, etc.), which represents the relative number of molecules present, and the subscript on the element, which represents the number of atoms of the element that are present in the molecule. That is,

$$\begin{aligned}\text{Number of atoms of type } i \text{ present in reactant or product}\\ &= (\text{number of molecules of type } i)\\ &\quad \times (\text{atoms of type } i/\text{molecule})\\ &= A_i \times (\text{subscript on atom } i \text{ in molecule}) \quad (11.3)\end{aligned}$$

The balances on the atoms present on each side of the general hydrocarbon combustion equation (11.2) are shown in the table below.

Element	Reactant	Product
C	$1 \times l$	$A_3 \times 1$
H	$1 \times m$	$A_5 \times 2$
O	$1 \times n + A_1 \times 2$	$A_3 \times 2 + A_5 \times 1 + A_6 \times 2$
N	$3.76A_1 \times 2$	$3.76A_1 \times 2$

Thus, equating the numbers of atoms of each element in the reactants to the number in the products gives $A_3 = l$, $A_5 = m/2$, and $A_6 = (n + 2A_1 - 2A_3 - A_5)/2 = (n + 2A_1 - 2l - m/2)/2$. Equation (11.2) then can be written as

$$C_lH_mO_n + A_1O_2 + 3.76A_1N_2 \rightarrow lCO_2 + \frac{m}{2}H_2O + \frac{n + 2A_1 - 2l - m/2}{2}O_2 + 3.76A_1N_2 \quad (11.4)$$

If a particular hydrocarbon is specified as the fuel and the amount of oxygen (or air) is specified in terms of A_1 = (atoms oxygen)/(mole fuel) = 2(molecules oxygen)/(mole fuel) = 2(volume of oxygen)/(volume of fuel), then the equation can be balanced.

For the stoichiometric case, no oxygen will appear among the combustion products, so the coefficient $(n + 2A_1 - 2l - m/2)/2$ on the right-hand side of Eq. (11.4) must equal to zero. For this case, then, the value of $A_{1,\text{complete}}$, the required atoms of oxygen per mole of fuel for stoichiometric complete combustion, can be found directly, and for any given fuel $C_lH_mO_n$,

$$A_{1,\text{complete}} = \frac{2l + m/2 - n}{2} \quad (11.5)$$

In practice, it is difficult to achieve complete combustion with a stoichiometric ratio of fuel and oxidant. This is because it is nearly impossible to achieve such good mixing in an engineering system that every molecule of fuel is brought into contact with a molecule of oxidant and undergoes combustion. Invariably some unburned fuel and unused oxidant will be carried with the combustion products from such a system. To avoid this, excess air is almost always introduced so that practically all the fuel is burned. This causes a considerable penalty in other ways. Note that for every α mol of excess oxygen used from air to promote complete combustion, 3.76α molecules of nitrogen are carried through the combustion process. The unburned oxygen and nitrogen are mixed with the products of complete combustion and provide a considerable mass of material that must be heated along with the combustion products. This can considerably lower the final temperature of the mixed products. The lower temperature is an advantage in some cases, and excess air can be used to control the inlet temperature to a gas turbine to below material limits.

Example Problem 11.1

Propane gas (C_3H_8) is burned in air. Balance the combustion reaction equation for the cases when just enough air is present for complete combustion of the propane and 1.5 times the necessary air for complete combustion is present.

Solution

For this case, $l = 3$, $m = 8$, and $n = 0$. The general equation, Eq. (11.3), then becomes

$$C_3H_8 + A_1O_2 + 3.76A_1N_2 \rightarrow 3CO_2 + 4H_2O + \frac{2A_1 - 10}{2}O_2 + 3.76A_1N_2$$

If just sufficient oxygen is present for combustion (the stoichiometric case), then no oxygen will appear among the products, so it follows that $(2A_1 - 10)/2 = 0$, or $A_{1,\text{complete}} = 5$. [This can also be found from the equivalent expression Eq. (11.5).] For complete combustion, then, the equation becomes

$$C_3H_8 + 5O_2 + (3.76 \times 5)N_2 \rightarrow 3CO_2 + 4H_2O + (3.76 \times 5)N_2$$

Thus, 25.8 mol of gas appears on the products side of the equation for each mole of propane that is burned in the stoichiometric reaction.

If 50 percent excess air is present above that needed for complete combustion, then $A_1 = 1.5A_{1,\text{complete}} = 7.5$, and the equation becomes

$$C_3H_8 + 7.5O_2 + (3.76 \times 7.5)N_2 \rightarrow 3CO_2 + 4H_2O + 2.5O_2 + (3.76 \times 7.5)N_2$$

Now 37.7 mol of products appears.

Comments

It is good to avoid errors by rechecking the numbers of atoms of each element to make sure that they balance on the reactant and product sides of the chemical balance equations.

In engineering practice, it is common to measure the concentration of the products in the exhaust or stack gases from power plants. From this information, the efficiency of the combustion process can be calculated. Usually, the water vapor in the products is condensed, and the remaining dry gases are then analyzed for concentration. This was originally commonly done by chemical analysis, but now the concentrations are usually measured with electronic sensors of various types.

Example Problem 11.2

Propane gas is burned in a residential heater. The dry flue gases are analyzed, and the analysis by volume is found to be

Product (gas)	Volume percent
N_2	83
CO_2	9
CO	4
O_2	4
	100

Determine the applicable stoichiometric coefficients for the combustion reaction.

Solution

Remembering that volume percent and mole percent are equivalent, we can write the combustion equation for propane, Eq. (11.1) as

$$C_lH_mO_n + A_1O_2 + A_2N_2 \rightarrow A_3CO_2 + A_4CO + A_5H_2O + A_6O_2 + A_7N_2 + A_8C_lH_mO_n$$

For this example, $A_8 = 0$, because no unburned fuel was found, and $n = 0$ because the fuel (propane, C_3H_8) contains no oxygen. Thus

$$C_3H_8 + A_1O_2 + A_2N_2 \rightarrow A_3CO_2 + A_4CO + A_5H_2O + A_6O_2 + A_7N_2$$

To balance the reaction equation, it is necessary to know the moles of water vapor present in the flue gases. This can be estimated in either of two ways. First, the ratio of hydrogen atoms to carbon atoms must be conserved between the reactants and the products. These elements only appear in the propane on the reactant side of the equation, and it follows that

$$\frac{8}{3} = \frac{2A_5}{A_3 + A_4}$$

If $A_3 + A_4$ are taken to be the moles of CO_2 and CO found in the flue gas analysis, then the moles of water vapor expected to be present are

$$A_5 = (8/3) \times (1/2) \times (9 + 4) = 17.3$$

The second approach is to note that the ratio of nitrogen to oxygen should also balance on the two sides of the reaction equation. Thus

$$\frac{N_2}{O_2} = \frac{79}{21} = \frac{A_7}{[A_3 + (A_4/2) + (A_5/2) + A_6]}$$

or

$$A_5 = 2 \times (21/79)A_7 - 2A_3 - A_4 - 2A_6$$
$$= 2 \times (21/79) \times 83 - 2 \times 9 - 4 - 2 \times 4 = 14.1$$

The two values of A_5 obviously do not agree well. Generally, the value based on the C/H ratio in the fuel is chosen, because the air-fuel reaction will then produce the required amount of water vapor in the products. The reaction equation then becomes

$$A_0C_3H_8 + A_1O_2 + A_2N_2 \rightarrow 9CO_2 + 4CO + 17.3H_2O + 4O_2 + 84N_2$$

and balancing the atoms on each side gives

$$4.33C_3H_8 + 21.65O_2 + 84N_2 \rightarrow 9CO_2 + 4CO + 17.3H_2O + 4O_2 + 84N_2$$

Dividing by the $A_0 = 4.33$ coefficient, the equation becomes

$$C_3H_8 + 5O_2 + 19.4N_2 \rightarrow 2.08CO_2 + 0.923CO + 4.00H_2O + 0.923O_2 + 19.4N_2$$

Comments

Now, the complete reaction equation is available from the flue gas concentration measurements. This is necessary to carry through a complete energy analysis, for example, of combustion efficiency for the furnace, as will be shown in the remainder of the chapter.

11.2.3 Air-Fuel and Equivalence Ratios

It is common engineering practice to specify the *air-fuel (mass) ratio,* designated AFR, when discussing hydrocarbon combustion in flow (constant-pressure) systems. The number of moles of oxygen per mole of fuel is given by $A_1/1$, and because there is 3.76 mol of nitrogen per mole of oxygen in air, the air-fuel *mole* ratio is $4.76A_1/1$. To convert this to a *mass* ratio, the denominator is multiplied by the molecular weight of the fuel M_{fuel} and the numerator by M_{air}. Thus

$$\text{AFR} = \frac{4.76M_{\text{air}}A_1}{M_{\text{fuel}}} \tag{11.6}$$

For the general hydrocarbon $C_lH_mO_n$, M_{fuel} is given by

$$M_{\text{fuel}} = 12l + m + 16n \tag{11.7}$$

and $M_{\text{air}} = 28.97$. Thus, a general expression for air-fuel ratio for $C_lH_mO_n$ is

$$\text{AFR} = \frac{137.9A_1}{12l + m + 16n} \tag{11.8}$$

To determine the dependence of AFR on the amount of excess air, note that we can express A_1 in terms of the stoichiometrically necessary amount of oxygen required per mole of fuel $A_{1,\text{complete}}$ and the fraction of excess air used X by

$$A_1 = (1+X)A_{1,\text{complete}} = (1+X)\,\frac{2l + m/2 - n}{2}$$

where Eq. (11.5) has been used to evaluate $A_{1,\text{complete}}$. Now A_1 is substituted into Eq. (11.8) to obtain a final relation for AFR for a given hydrocarbon in terms of the fraction of excess air as

$$\text{AFR} = \frac{68.9(1+X)(2l + m/2 - n)}{12l + m + 16n} \tag{11.9}$$

The *equivalence ratio, φ*, is defined as the actual fuel air (*not* air fuel) ratio over the stoichiometric fuel air ratio, or

$$\varphi = \frac{(\text{FAR})\text{ actual}}{(\text{FAR})\text{ stoichiometric}} = \frac{(\text{AFR})\text{ stoichiometric}}{(\text{AFR})\text{ actual}} \tag{11.10}$$

A value of φ less than 1 indicates less fuel (or more air) than is necessary for stoichiometric combustion of the fuel, or a *lean* mixture; conversely, φ greater than one indicates a *rich* mixture.

Example Problem 11.3

Propane gas (C_3H_8) is burned in air. Find the air-fuel ratio and equivalence ratio for the cases when just enough air is present for complete combustion of the propane and 1.5 times the necessary air for complete combustion is present.

Solution

For this case, $l = 3$, $m = 8$, and $n = 0$. From Eq. (11.7), for propane

$$M_{\text{fuel}} = (12l + m + 16n)$$
$$= 12 \times 3 + 8 + 16 \times 0 = 44$$

which agrees with the value from Table C.4 of 44.094. From Eq. (11.6),

$$\text{AFR} = \frac{4.76 M_{\text{air}} A_1}{M_{\text{fuel}}} = \frac{4.76(28.97)(1+X)A_{1,\text{complete}}}{M_{\text{fuel}}}$$
$$= \frac{4.76(28.97)(1+X)(2l + m/2 - n)/2}{M_{\text{fuel}}}$$

For the stoichiometric case, $X = 0$ and

$$(\text{AFR})_{\text{stoich}} = \frac{4.76(28.97)(2l + m/2 - n)/2}{M_{\text{fuel}}}$$
$$= \frac{4.76(28.97)(2 \times 3 + 8/2 - 0)/2}{44}$$
$$= 15.7$$

and for 50 percent excess air (150 percent theoretical air) $X = 0.5$, so

$$\text{AFR} = \frac{4.76(28.97)(1.5)(2l + m/2 - n)/2}{M_{\text{fuel}}}$$

$$= \frac{4.76(28.97)(1.5)(2 \times 3 + 8/2 - 0)/2}{44} = 23.5$$

For case a), by definition $\varphi = 1$. For case b),

$$\varphi = \frac{(\text{AFR})_{\text{stoich}}}{(\text{AFR})_{\text{actual}}} = \frac{15.7}{23.5} = 0.668,$$

indicating a lean mixture.

In the reaction equations written for hydrocarbon combustion, Eqs. (11.1) through (11.5), we have neglected other reactions that also occur. These include the production of oxides of nitrogen and of various intermediate hydrocarbons. These products occur in minute amounts in the combustion products of even carefully controlled burners, and they are significant causes of air pollution when large amounts of hydrocarbons are burned.

We have simplified the chemistry of the hydrocarbon combustion process greatly. In reality, the combustion of a hydrocarbon proceeds through a number of intermediate parallel and series reactions which occur at differing rates and culminate in the production of water vapor and carbon dioxide only if allowed to proceed to completion. In a poorly designed burner where mixing is not complete or the residence time of the reactants is too short to allow the reactions to proceed to completion, intermediates (such as carbon in the form of soot, carbon monoxide, formaldehyde, etc.) may appear in the combustion products.

In a properly designed burner, the reaction equations presented in this section allow us to do a good job of predicting the thermodynamic behavior of combustion systems; because of the complexity of the combustion process in real systems, the equations in this section are not sufficient for a detailed look at combustion chemistry.

11.3 Establishing a Common Basis for Combustion Processes

In treating pure simple substances as in Chap. 9, we noted that any state could be chosen as the reference state for property tabulations, since only *differences* between the properties in two states were of interest. When two or more substances undergo a chemical reaction, however, the establishment of a reference state that is common to all substances participating in the reaction becomes necessary. This is because the amount of a given compound is not constant throughout a process. More than this, we must account for changes in the internal energy of a system that occur because of the chemical energy released or absorbed during the reaction of the chemical species. Thus, a reference value must be established that can be used to find the internal energy of the reactants *before* the reaction and of the products *after* the reaction. We establish the necessary reference values in this section.

11.3.1 Zero-Enthalpy Basis

A simple choice for a reference state would appear to be to set h^o to zero at $T = 0\text{ K} = 0°\text{R}$, so that all enthalpies take on positive values. (We denote property values at the chosen standard reference state by the superscript o, as in h^o.) However, as noted in Chap. 13 on examination of the microscopic behavior of materials, we find that some energy is possessed by most materials, even at absolute zero temperature. This energy is in the form of nuclear binding energy, molecular binding forces, and other forces that are not the same for all molecules. The situation is simpler for the *elements,* as opposed to more complex molecules. The elements can be assigned enthalpy values relative to one another at an arbitrary state. The effects of molecular energies can then be computed relative to the elements that make up the molecules. For convenience in comparison with laboratory data, the *standard reference state with $h^o = 0$ for all elements is assigned at $T = 25°C$ (77°F) and $P = 1$ atm.*

Thus for all elements $h^o = 0$ at the standard reference state. When the element exists in the chemically stable form of a diatomic gas at the reference state (N_2, O_2, H_2, etc.), the diatomic gas is assigned a reference value of zero at the standard reference state. The value of h for elements in any other state is found by applying the relation

$$h(T, P) - h_{\text{ref}} = h(T, P) - h^o = h(T, P) = \int_{\substack{T=25°\text{C}=77°\text{F},\\ P=1\text{ atm}}}^{T,P} dh \tag{11.11}$$

The value of u at the standard reference state u^o is easily found from

$$u^o = h^o - P^o v^o = -P^o v^o \tag{11.12}$$

and the value of v^o is found from P-v-T data at the reference state for the particular element. Not all tables in the literature use the same reference state, and care must be used when obtaining data from various sources.

11.3.2 Enthalpy of Formation

In Sec. 11.3.1 we established a common basis for the enthalpy of all elements. When compounds (molecules) are formed from the elements, chemical energy may be required to complete the reaction, or energy may be released. In either case, the energy, and therefore the enthalpy, of the molecule is different from the sum of the enthalpies of the atoms of which the molecule is composed. We can denote the enthalpy of the resulting molecule as

$$h = \Delta h_f + \sum_i N_i h_i \tag{11.13}$$

where Δh_f is the enthalpy required to form the molecule from its constituent atoms at a given state, or the *enthalpy of formation.* The summation term accounts for the enthalpy added to the compound by the N_i atoms of each element i that make up the molecule.

If we use Eq. (11.13) to find the enthalpy of the compound at the standard reference state, then

$$h^o = \Delta h_f^o + \sum_i N_i h_i^o \tag{11.14}$$

However $h_i^o = 0$ for all elements at the standard reference state, so *the enthalpy of a compound at the standard reference state is equal to the enthalpy of formation at the standard reference state.* Thus we can provide a reference value for all compounds by setting

$$h^o = \Delta h^o{}_f \tag{11.15}$$

at the standard reference state of $T = 25°\text{C} = 77°\text{F}$ and $P = 1$ atm. The values of the enthalpy of formation at the standard reference state are usually determined experimentally, although in some cases the methods of statistical thermodynamics can be used to predict values. A tabulation of Δh_f^o for various compounds is found in Table C.9. Note that the values presented in the table are actually $\Delta \overline{h}_f^o = M\, \Delta h_f^o$. That is, the values in the table are on a molar rather than a mass basis, and the molecular weight of the compound M is used to convert to the mass basis.

The enthalpy of a compound can now be computed in any state by using the enthalpy of the compound in the standard reference state as the reference value. The enthalpy in the standard reference state, given by Eq. (11.15), is called the *standardized reference enthalpy* of the compound. The enthalpy in another state is then

$$h(T, P) = h^o + \int_{\substack{T=25°\text{C}=77°\text{F},\\ P=1\,\text{atm}}}^{T,P} dh = \Delta h_f^o + \int_{\substack{T=25°\text{C}=77°\text{F},\\ P=1\,\text{atm}}}^{T,P} dh \tag{11.16}$$

where $h(T, P)$ is the *standardized enthalpy* of the compound.

The standardized reference internal energy then follows as

$$u^o = h^o - P_{\text{ref}} v(T = 25°\text{C}, P = 1 \text{ atm}) \tag{11.17}$$

and the standardized internal energy is

$$u(T,P) = u^o + \int_{\substack{T=25°C=77°F,\\ P=1\,atm}}^{T,P} du$$

$$= \Delta h_f^o - P_{ref} v(T = 25°C = 77°F, P = 1 \text{ atm}) + \int_{\substack{T=25°C=77°F,\\ P=1\,atm}}^{T,P} du \tag{11.18}$$

Example Problem 11.4

Find the standardized enthalpy of the following:
(*a*) Argon gas at 100°C (assume it acts as an ideal gas)
(*b*) Liquid water at 100°C
(*c*) Water vapor at 100°C and 1 atm

Solution

(*a*) The change in h for argon between 25 and 100°C can be found from Table D.3 as

$$\int_{25°C}^{100°C} dh = h(T = 100°C) - h(T = 25°C)$$
$$= 194.29 - 155.24 = 39.05 \text{ kJ/kg}$$

The ideal gas assumption allows us to consider h as independent of pressure. Because argon is an element, the value of the enthalpy at the standard reference state of 25°C and 1 atm is taken as zero. Thus, the value of the enthalpy at $T = 100°C$ is given by

$$h(T = 100°C) = h^o + \int_{25°C}^{100°C} dh = 0 + 39.05 = 39.05 \text{ kJ/kg}$$
$$= 16.79 \text{ Btu/lbm}$$

(*b*) For liquid water, assuming incompressibility, we use

$$h(T = 100°C) = h^o + \int_{25°C}^{100°C} dh = \Delta h_f^o + \int_{25°C}^{100°C} dh$$
$$= \Delta h_f^o + c_P(100 - 25)$$

The value of Δh_f^o is found from Table C.9 as $(-286.0 \times 1000)/18.02 = -15{,}870$ kJ/kg, and c_P for water is found in Table C.3 as 4.178 kJ/(kg·K) (by using the value at 323 K = 50°C). Substituting results in

$$h(T = 100°C) = -15{,}870 + 4.178(75) = -15{,}560 \text{ kJ/kg}$$
$$= -6690 \text{ Btu/lbm}$$

(*c*) For water vapor, we can take the value found in (*b*) for the liquid water standardized enthalpy and simply add h_{lg} from the steam tables. From Table D.10, $h_{lg} = 2257$ kJ/kg, so

$$h(T = 100°\text{C}) = -15{,}560 + 2257 = -13{,}300 \text{ kJ/kg}$$
$$= -5718 \text{ Btu/lbm}$$

Comments

Note the negative enthalpy value for liquid water as well as water vapor at the final state. This results from the negative enthalpy of formation at the standard reference state. Tabular data or equations of state that allow calculations of changes in enthalpy between states are found to be convenient aids in converting to standardized enthalpies.

11.3.3 Zero-Entropy Basis

In Sec. 11.3.2 we found a standard reference state for enthalpy and inferred reference values for internal energy at the same state. When we discuss second law considerations in combustion systems, we also need a standard set of entropy values. Here a key observation from statistical thermodynamics is brought into play. As noted in Chap. 5, the property of entropy is related to our knowledge about the state of a particular system. As discussed more fully in Chap. 13, as the temperature of a pure substance is reduced, the possible range of energies that can be possessed by the atoms and molecules making up the system becomes smaller. Finally, as absolute zero is approached, the energy of all the constituent particles becomes small and indeed also approaches zero. Our knowledge about the state of the system becomes complete (the system is ordered), and our uncertainty about the state of the system (the entropy) becomes zero! This observation, originally formulated by Nernst and expanded by Max Planck, is embodied in the *third law of thermodynamics:*

> The entropy of each pure substance in thermodynamic equilibrium approaches zero as the temperature approaches absolute zero.

Thus, to find standard values of entropy in the standard reference state of 25°C = 77°F and 1 atm, we integrate from absolute zero to the standardized state and tabulate the resulting values. We must account for all phase transitions between absolute zero and the standardized reference state. Entropies computed in this way are referred to as *absolute entropies.* Since the absolute entropies for all substances are referred to a common reference state, the entropies on this scale are appropriate for use in analyzing chemical reactions and combustion problems in particular. Some absolute entropies in the standard reference state are presented in Table C.10. To find absolute entropies in other states, the same methods are used as for the standardized enthalpy values carried out in Example Problem 11.3.

All the material presented above for standardized properties is general. For the particular case of combustion, it is usually valid to consider the products of the combustion reaction, the combustion air, and quite often the fuel to be ideal gases. When that is the case, many of the property data necessary for the treatment of combustion reactions are at hand in Tables D.2 through D.8 or Tables E.2 through E.8, which list the temperature-dependent portion of the absolute entropy s_0 for various ideal gases.

11.4 Combustion Processes

In flow processes such as in the combustion chambers of turbine engines, in industrial furnaces, and in rocket engines, combustion is assumed to occur at constant pressure. This is also the case for the ideal air-standard diesel internal combustion engine, which is treated as a control mass system. In calorimetric equipment used for measurement of the energy release during reaction, and during very fast combustion reactions such as at top dead center in an internal combustion engine or during the combustion period of an explosion (followed by the expansion of the combustion gases), combustion is assumed to occur at constant volume. We can restrict our discussion, then, to the two special but useful cases of constant-pressure and constant-volume combustion.

For each of these two special cases, there are two quantities of interest to the engineer: the energy released and the temperature reached in the combustion process.

First we examine combustion in constant-pressure systems.

11.4.1 Combustion at Constant Pressure

Let us examine the case of an isothermal constant-flow reactor to which are fed the fuel at unit molar flow rate and air at a rate that gives the molar ratios to the fuel specified by Eq. (11.4). The reactor is shown in Fig. 11.1. In general, there is a heat transfer from the reactor in order for the control volume to remain isothermal. The first law for this control volume is

$$\dot{Q} = \sum_p \dot{m}_p h_p - \sum_r \dot{m}_r h_r \tag{11.19}$$

where the subscript r denotes reactants and the subscript p denotes products of the reaction. Let us carry out the reaction *under the particular conditions of the standard reference state,* that is, 1-atm pressure and with the temperature maintained at 25°C = 77°F by heat transfer from the reactor. Then Eq. (11.19) on a molar basis becomes

$$\dot{Q}^o = \sum_p \dot{n}_p \bar{h}_p^o - \sum_r \dot{n}_r \bar{h}_r^o \tag{11.20}$$

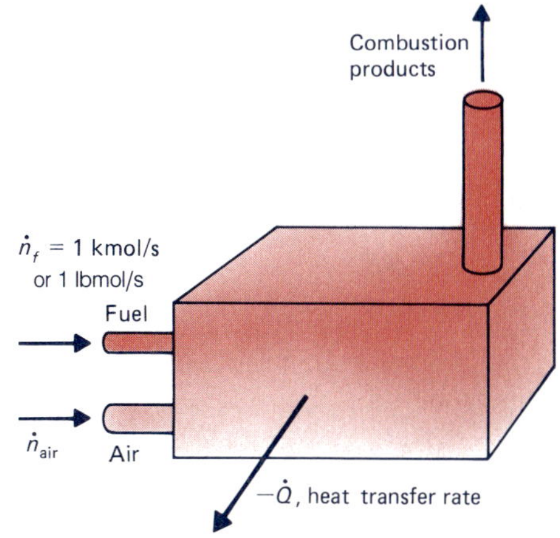

Figure 11.1 Isothermal constant-flow reactor.

where, as before, the overbar on a quantity indicates that it is on a molar basis. Now divide through Eq. (11.20) by the molar flow rate of the fuel $\dot{n}_f$, and the equation becomes

$$\frac{\dot{Q}^o}{\dot{n}_f} = \bar{q}^o = \frac{1}{\dot{n}_f}\left(\sum_p \dot{n}_p \bar{h}_p^o - \sum_r \dot{n}_r \bar{h}_r^o\right) \tag{11.21}$$

Equation (11.21) gives the amount of heat transfer required per mole of fuel from a reactor operating at the standard reference condition, which is equal to the net enthalpy difference between the products leaving the reactor and the enthalpy of the reactants entering. The quantity $\bar{q}^o$ is called the *enthalpy of combustion* or sometimes the *heat of combustion,* and it is also denoted by h_{rp}, because it is an enthalpy. Because we interpret the first law sign convention for Q as heat transfer to the control volume being positive, we expect $\bar{q}^o$ to generally carry a negative sign. The value of $\bar{q}^o$ as defined by Eq. (11.21) is fixed for a given fuel, and it is a measure of the chemical energy content per mole of the fuel when burned stoichiometrically with oxygen (or air) at constant pressure.

Example Problem 11.5

Determine the enthalpy of combustion of methane.

Solution

For methane (CH_4) the combustion equation, Eq. (11.4), becomes

$$CH_4 + (1+1)O_2 + 3.76(1+1)N_2 \rightarrow CO_2 + 2H_2O + 3.76(1+1)N_2$$

We must determine the values of $\bar{h}^o$ for each species in the reaction for use in Eq. (11.21). These are as follows:

Species	M	$\bar{h}^o$, Btu/lbmol	
CH_4	16.042	-32.20×10^3	(Table C.9)
O_2	32.00	0	(element)
N_2	28.016	0	(element)
CO_2	44.01	-169.3×10^3	(Table C.9)
H_2O: liquid	18.016	-123.0×10^3	(Table C.9)
vapor		-104.0×10^3	(Table C.9)

Substituting into Eq. (11.21) and assuming that the water is produced in the form of vapor, we find

$$\bar{q}^o = \frac{1}{n_{CH_4}}[(n\bar{h}^o)_{CO_2} + (n\bar{h}^o)_{H_2O} + (n\bar{h}^o)_{N_2} - (n\bar{h}^o)_{CH_4} - (n\bar{h}^o)_{O_2} - (n\bar{h}^o)_{N_2}]$$

$$= \frac{1}{1}[-169.3 \times 10^3 + (2)(-104.0 \times 10^3) + 0 - (1)(-32.20 \times 10^3) - 0 - 0]$$

$$= -345{,}100 \text{ Btu/lbmol} = -802{,}700 \text{ kJ/kmol}$$

Note that $\dot{n}/\dot{n}_f = n/n_f$ for a steady flow process. The enthalpy of combustion per mole of methane is thus $\bar{q}^o = -345{,}100$ Btu/lbmol or $q^o = -345{,}100/16.042 = -21{,}510$ Btu/lbm.

Comments

The nitrogen in the air did not affect the result, so the enthalpy of combustion in the standard reference state is the same whether the methane is burned in stoichiometric air, pure oxygen, or excess air.

Note that if we had assumed that the water produced in the combustion reaction was in the form of liquid rather than vapor, then the calculated enthalpy of combustion would have been larger (−383,100 Btu/lbmol, or −23,880 Btu/lbm of fuel). This occurs because more of the combustion energy is required to vaporize the water, and then less is available to remove from the combustion reactor in the form of heat transfer. The enthalpy of combustion when liquid water is produced is often called the *higher heating value* (HHV) of the fuel (usually the minus sign is dropped when this name is used). And the enthalpy of combustion is called the *lower heating value* (LHV) when water vapor is produced (again we drop the minus sign). In practical devices, the LHV is the only choice that we have, since combustion reactions in practice usually produce water vapor as a combustion product.

Values of the enthalpy of combustion (lower heating values) are tabulated for many substances, and some are given in Table C.11. A fuel with a larger enthalpy of combustion per unit mass is obviously more desirable than one with a smaller enthalpy of combustion per unit mass, if their costs are the same.

11.4.2 Combustion at Constant Volume

For constant-volume combustion of a stoichiometric mixture of fuel and oxidant, the first law for a control mass gives

$$Q = \sum_p m_p u_p - \sum_r m_r u_r \tag{11.22}$$

where the subscript r denotes reactants and the subscript p denotes the products of the reaction. Let us now assume that we will carry out the combustion at the constant temperature of the reference state, so that Q is the heat transfer necessary to maintain isothermal conditions, and the products and reactants are both at the reference state temperature. Equation (11.22) then becomes, after writing it on a molar basis,

$$\left(\frac{\bar{Q}^o}{n_f}\right)_v = (\bar{q}^o)_v = \frac{1}{n_f}\left(\sum_p n_p\bar{u}_p^o - \sum_r n_r\bar{u}_r^o\right) \tag{11.23}$$

where we have divided through by the number of moles of fuel n_f. Thus, Eq. (11.23) predicts the heat transfer that occurs per mole of fuel from a constant-temperature, constant-volume combustion process at standard reference conditions.

Using Eq. (11.17) and assuming all products and reactants act as ideal gases, we see that Eq. (11.23) reduces to

$$(\bar{q}^o)_v = \bar{q}^o - \frac{\bar{R}T^o}{n_f}\left(\sum_p n_p - \sum_f n_r\right) \tag{11.24}$$

Thus, the standard internal energy of combustion can be found directly from the standard enthalpy of combustion, which is more often found in tabulated form than are the values of standard internal energy.

11.4.3 Adiabatic Flame Temperature

One measure of the usefulness of a particular fuel is its enthalpy of combustion. However, in some instances, the engineer is more interested in producing a high temperature than a large energy release. This is the case in the design of process equipment that requires a given minimum temperature of operation, such as dryers and chemical process furnaces. In this section, we determine a method for comparing the temperatures available from the combustion of a given fuel.

Suppose that the combustion reactor of Fig. 11.1 is operated in a different mode from that used to determine the enthalpy of combustion. Rather than removing sufficient energy from the reactor by heat transfer, so that the reactor is maintained at isothermal conditions, let us heavily insulate the reactor so that *no* heat transfer occurs and the reactor operates adiabatically. The reactor then appears as in Fig. 11.2. In such a case, the first law for the reactor as given by Eq. (11.19) reduces to

$$\sum_p \dot{m}_p h_p = \sum_r \dot{m}_r h_r \tag{11.25}$$

Suppose that the *reactants enter the reactor at the standard reference state.* (Cases with the reactants at temperatures other than the reference state are

Figure 11.2 Adiabatic reactor.

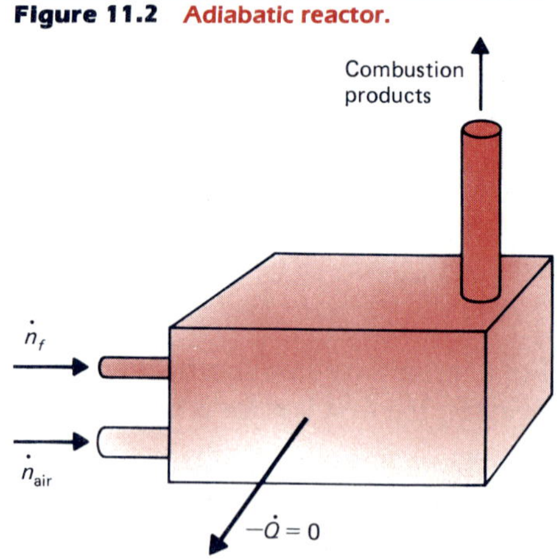

considered in Sec. 11.5.2.) Equation (11.25) on a per-mole-of-fuel basis then becomes

$$\frac{1}{n_f}\sum_p n_p\bar{h}_p = \frac{1}{n_f}\sum_r n_r\bar{h}_r^o \tag{11.26}$$

If enthalpy-of-combustion data are available for the fuel being used, then it is convenient to substitute Eq. (11.21) into Eq. (11.26) to eliminate the reactant term:

$$\frac{1}{n_f}\sum_p n_p\bar{h}_p = \frac{1}{n_f}\sum_p n_p\bar{h}_p^o - \bar{q}^o \tag{11.27}$$

The enthalpy carried from the reactor by the products of combustion equals the enthalpy of the products at the standard reference state plus (since $\bar{q}^o$ carries a negative value) the energy released by the combustion reaction.

Solution of Eq. (11.26) for the temperature of the combustion products (assumed to be the same for all products because of mixing within the reactor) is iterative. A temperature must be assumed, and the values of $\bar{h}_p$ are found from tables or are computed. The $\bar{h}_p$ values are substituted into the product side of Eq. (11.26), and the result is checked against the reactant side of Eq. (11.26). The process is repeated until a temperature is found for which the equality is satisfied.

For the case of no excess air, the temperature of the products is called the *theoretical adiabatic flame temperature* for the fuel. The theoretical adiabatic flame temperature is the highest temperature that can be obtained from the fuel used in the combustion reaction (unless we preheat some of the reactants to temperatures above the standard reference state), because no heat transfer is allowed to occur to the surroundings and no excess air is present to absorb any of the enthalpy of combustion.

Figure 11.3 shows schematically the lines of enthalpy versus temperature for the reactants and products of a typical combustion reaction. The lines are fairly straight because of the weak variation of specific heat with temperature. At a given temperature, the enthalpy of the products is always *smaller* than that of the reactants because our sign convention is taken to mean that the enthalpy of combustion is *removed* from the reactants to maintain the temperature of the products at the given constant temperature. This is easily seen from Eq. (11.19), noting that $\dot{Q}$ carries a negative numerical value.

The difference between the enthalpy lines on Fig. 11.3 at any temperature is the enthalpy of combustion at that temperature (which is negative in value). If all the energy released by the combustion reaction goes to raise the temperature of the products in an adiabatic combustor, then the enthalpy of the products must rise by the negative of the enthalpy of combustion. As shown in Fig. 11.3, this rise in enthalpy causes the products to reach the adiabatic flame temperature.

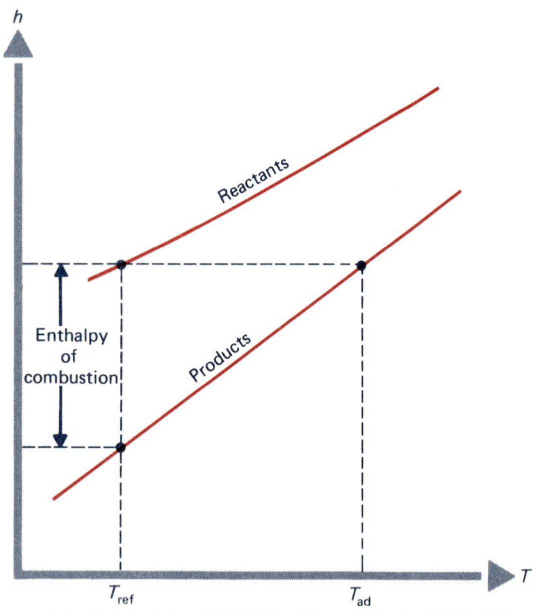

Figure 11.3 Relation of enthalpy of combustion to adiabatic flame temperature.

Values of theoretical adiabatic flame temperature for some common fuels are shown in Table C.11. Note that the value of the theoretical adiabatic flame temperature is based on the assumed chemical reaction equation. The actual temperature is often lower because of dissociation or ionization of the combustion products or the presence of reactions which we have assumed to not occur. (In some cases, operation at stoichiometric conditions may actually produce a lower flame temperature than at slightly off-stoichiometric conditions. For example, additional reactions may occur at the temperatures in the combustor, the CO_2 may dissociate to form CO, or significant amounts of oxides of nitrogen may form. In that case, higher adiabatic flame temperatures may actually occur by running at conditions away from stoichiometric, where these reactions are suppressed.)

If tables of computer programs are available for $\bar{h}(T_p, P)$ for each product, then these can be used to determine the adiabatic flame temperature by iteratively choosing values of T_p until Eq. (11.27) is satisfied. Care must be taken that the sources of all the data use the standard reference state, or the user must convert the source data to the standard reference state. For such data, based on ideal gas properties, Refs. 1 to 3 are useful.

Another method of finding T_p is to rewrite Eq. (11.27) as

$$\sum_p n_p(\bar{h}_p - \bar{h}_p^o) = -n_f\bar{q}^o \tag{11.28}$$

and then to replace the enthalpy changes in terms of the specific heat for a constant-pressure combustor, or by assuming the enthalpy to be independent of pressure (the ideal gas assumption), so that Eq. (11.27) becomes

$$\sum_p n_p \int_{T^o}^{T_p} \bar{c}_{P,p}\, dT = -n_f \bar{q}^o \tag{11.29}$$

Now the polynomial expressions of Table D.1 can be used in the left-hand side and iteration used until convergence. Note that for a fixed T^o, the individual integrals for each product species have the form $\int_{T^o}^{T_p} \bar{c}_P(T)\, dT$. As shown by Eq. (6.26), we could define a temperature-averaged value of $\bar{c}_P$ by the relation

$$\bar{c}_P^*(T) = \frac{1}{T - T^o} \int_{T^o}^{T} \bar{c}_{P,p}(T)\, dT = \frac{h(T_p) - h(T^o)}{T - T^o} \tag{11.30}$$

and Eq. (11.29) reduces to

$$(T_p - T^o) \sum_p n_p \bar{c}_{P,p}^*(T_p) = -n_f \bar{q}^o \tag{11.31}$$

Now the iteration procedure is somewhat simplified. A T_p value is assumed, values of $\bar{c}_P^*(T)$ are found at that T_p for each species, and convergence is checked by testing the equality in Eq. (11.31). A graph of $\bar{c}_P^*(T)$ is presented in Fig. 11.4.

If mean specific heats are computed from the polynomial expressions of Table D.1 and E.1, care must be taken that the temperature limits of Eq. (11.30) are within the range of applicability of the polynomials. A table of c_p^* vaues is given in Table C.14 for various ideal gases, computed from the enthalpy values.

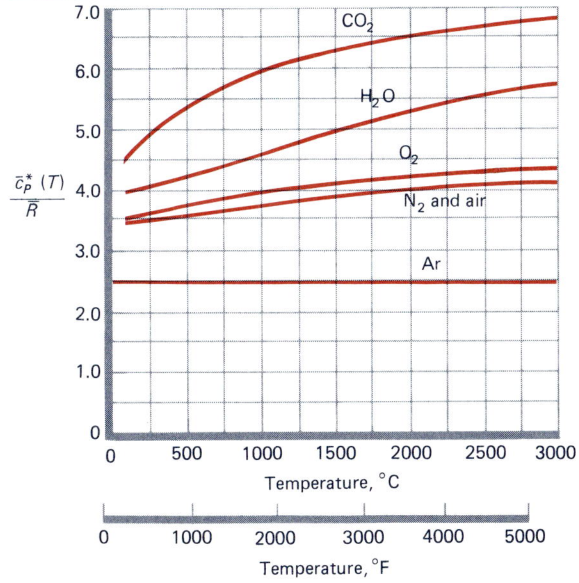

Figure 11.4 Mean specific heats of gases.

Example Problem 11.6

Calculate the theoretical adiabatic flame temperature for methane.

Solution

The enthalpy of combustion for methane (for the production of water vapor) is found in Example Problem 11.5 to be $-802{,}700$ kJ/kmol. The enthalpies at the standard reference state for the products are also given in that example. The chemical reaction is

$$CH_4 + 2O_2 + 7.52N_2 \rightarrow CO_2 + 2H_2O + 7.52N_2$$

Substituting the known values into Eq. (11.27) for 1 kmol of fuel results in

$$\begin{aligned}\sum_p n_p\bar{h}_p &= \bar{h}^o_{CO_2} + 2\bar{h}^o_{H_2O} - \bar{q}^o \\ &= -393.8 \times 10^3 + 2(-242 \times 10^3) - (-802{,}700) \\ &= -75{,}100 \text{ kJ/kmol}\end{aligned}$$

The right-hand side of this equation does not include N_2 since its standard reference state enthalpy is zero. The term $\sum_p n_p\bar{h}_p$ on the left-hand side (LHS) is

$$\sum_p n_p\bar{h}_p = \bar{h}_{CO_2} + 2\bar{h}_{H_2O} + 7.52\bar{h}_{N_2}$$

The enthalpy of the N_2 becomes important and must be included since this last expression is at the temperature of the products leaving the reactor.

Method 1

Let us first use the polynomial expressions for specific heats to determine the adiabatic flame temperature. The expressions for the LHS by this method are as follows: For CO_2,

$$\begin{aligned}\int_{T^o}^{T_p} \bar{c}_P(T)\,dT &= 22.26(T_p - 298) + (5.981 \times 10^{-2})\left(\frac{T_p^2 - 298^2}{2}\right) \\ &\quad - (3.501 \times 10^{-5})\left(\frac{T_p^3 - 298^3}{3}\right) + (7.469 \times 10^{-9})\left(\frac{T_p^4 - 298^4}{4}\right)\end{aligned}$$

For H_2O,

$$\begin{aligned}2\int_{T^o}^{T_p} \bar{c}_P(T)\,dT = 2\Bigg[&32.24(T_p - 298) \\ &+ (0.1923 \times 10^{-2})\left(\frac{T_p^2 - 298^2}{2}\right) \\ &+ (1.055 \times 10^{-5})\left(\frac{T_p^3 - 298^3}{3}\right) - (3.595 \times 10^{-9})\left(\frac{T_p^4 - 298^4}{4}\right)\Bigg]\end{aligned}$$

and for N_2,

$$7.52 \int_{T^o}^{T_p} \bar{c}_P(T)\, dT = 7.52 \Bigg[28.90(T_p - 298) - (0.1571 \times 10^{-2}) \left(\frac{T_p^2 - 298^2}{2} \right) + (0.8081 \times 10^{-5}) \left(\frac{T_p^3 - 298^3}{3} \right) - (2.873 \times 10^{-9}) \left(\frac{T_p^4 - 298^4}{4} \right) \Bigg]$$

We now choose values of T_p, evaluate the LHS of the first law relation by summing the three terms above, and compare to the right-hand side (RHS). Graphical presentation of the iterations is usually helpful in interpreting the results.

	$n[\bar{h}(T_p) - \bar{h}^o]$, kJ				
T_p (guessed), K	**CO_2**	**H_2O**	**N_2**	**Sum, kJ**	**$-q^o$, kJ**
2300	110,700	174,200	499,200	784,100	802,700
2400	117,300	183,900	523,700	824,900	802,700
2346	113,700	178,700	510,500	802,900	802,700

The graph of the sum $= \sum_p n_p[\bar{h}(T_p) - \bar{h}_p^o]$ versus T_p(guessed) is shown here.

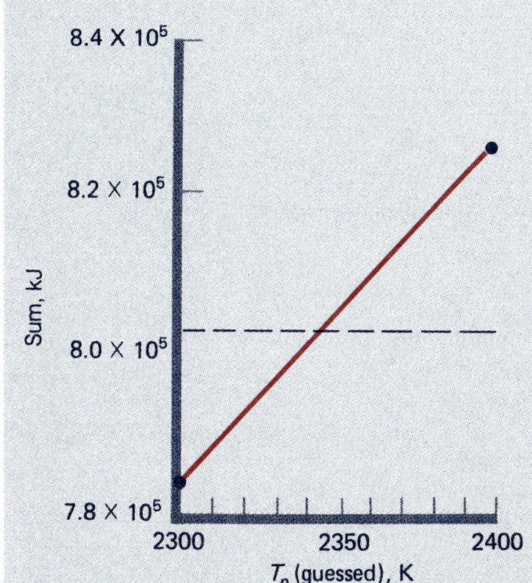

Note that the third guess was chosen based on linear interpolation between the guesses at 2300 and 2400 K, and this yielded the correct result. This is often the case for combustion problems because $\bar{h}$ is very nearly linear with T over fairly large temperature ranges, and the iterative procedure converges very quickly.

Note also that the largest portion of the product enthalpy is carried by the nitrogen rather than the reaction products CO_2 and H_2O.

The computed value does not agree well with the tabulated value of 2285 K for the adiabatic flame temperature of methane shown in Table C.11. This discrepancy is partly due to the use of the $\bar{c}_P(T)$ relations outside the temperature range for which they were intended (the range is usually 273 to 1800 K).

Method 2

Using the definition of mean specific heat, we can write Eq. (11.31) for this case as

$$(T_p - T^o)\{[n\bar{c}_P^*(T_p)]_{CO_2} + [n\bar{c}_P^*(T_p)]_{H_2O} + [n\bar{c}_P^*(T_p)]_{N_2}\} = -\bar{q}^o$$

A simple procedure which converges quickly is to rewrite this equation as

$$T_p = T^o - \frac{\bar{q}^o}{[n\bar{c}_P^*(T_p)]_{CO_2} + [n\bar{c}_P^*(T_p)]_{H_2O} + [n\bar{c}_P^*(T_p)]_{N_2}}$$

Now values of T_p are assumed and of $\bar{c}_P^*(T_p)$ are taken from Fig. 11.4 or Table C.14 and substituted into the RHS of the last equation. A new value of T_p is then computed from the equation, and the procedure is repeated until convergence. For example,

	$\bar{c}_P^*(T)$			
T_p (guessed), K	**CO_2**	**H_2O**	**N_2**	**RHS, K**
2300	$6.56\bar{R}$	$5.31\bar{R}$	$4.05\bar{R}$	2325

The first guess in this case produced a new value for T_p that is within the accuracy of the graphs.

Method 3

For the final method, tabulated values of $\bar{h}(T_p) - \bar{h}^o$ from the ideal gas tables are used directly in the equation

$$(\bar{h} - \bar{h}^o)_{CO_2} + 2(\bar{h} - \bar{h}^o)_{H_2O} + 7.52(\bar{h} - \bar{h}^o)_{N_2} = 802{,}700$$

Again values of T_p are assumed, and $h - h^o$ values are found directly from the tables. The process is continued until the equation is satisfied.

	$\bar{h} - \bar{h}^o = M(h - h^o)$, kJ/kmol			
T_p (guessed), K	**CO_2**	**H_2O**	**N_2**	**LHS, kJ/kmol**
2300	109,100	88,420	67,250	791,700
2350	112,300	91,070	68,860	812,300
2329	110,800	89,960	68,310	804,400

Linear interpolation between the first and third table entries gives $T =$ 2325 K.

Comments

The three methods have produced theoretical flame temperature predictions for methane of 2346, 2325, and 2325 K, respectively, while Table C.11 gives 2285 K. Which should we believe? As noted, method 1 required the use of polynomial expressions that were extended to temperatures outside their range of validity, so the results of that method are less accurate. In method 2, graphical values good to no more than three significant figures were used, so the results cannot be precise to more than ± 20 K at best. The final method makes use of quite precise tables with five significant figures; however, the tables are for ideal gases and do not account for nonideal effects or for dissociation of the gases. Dissociation does occur to a small degree at the temperatures in the flame, and chemical energy is needed to sever the bonds holding the molecules together. This lowers the flame temperature slightly, and the values in Table C.11 account for this effect.

11.4.4 Explosion Temperature (Constant-Volume Combustion)

When the products of combustion are contained at constant volume, the energy released by combustion causes an increase in temperature and usually pressure. For chemical reactions that occur so quickly that the products of reaction do not expand significantly during the combustion process, this process is effectively an explosion, and the final temperature reached in the constant-volume system is referred to as the *explosion temperature.* Of course, if the explosion is not contained in a rigid vessel, the rapid constant-volume combustion is immediately followed by a destructive expansion.

For slower reactions or for small quantities of explosive materials, a rigid container can be used to carry out the reaction; the energy released by the reaction can then be determined by measuring the heat transfer necessary to return the combustion products to the original temperature.

Analysis of constant-volume combustion is carried out as for the constant-pressure case, except that the enthalpies or specific heats at constant pressure are replaced by internal energies or specific heats at constant volume in the energy equations derived for the constant-pressure case of adiabatic flame temperature.

Example Problem 11.7

Calculate the stoichiometric explosion temperature and final pressure for methane combustion with air. Assume standard initial conditions.

Solution

The energy equation for the explosion temperature is, by analogy with Eq. (11.23),

$$\sum_p n_p(\bar{u}_p - \bar{u}_p^o) = -n_f(\bar{q}^o)_v$$

From Eq. (11.24),

$$(\bar{q}^o)_v = \bar{q}^o - \frac{\bar{R}T^o}{n_f}\left(\sum_p n_p - \sum_r n_r\right)$$

For the stoichiometric combustion of methane,

$$CH_4 + 2O_2 + 7.52N_2 \rightarrow CO_2 + 2H_2O + 7.52N_2$$

Thus, for 1 mol of fuel, Eq. (11.24) gives

$$\begin{aligned}(q^o)_v &= -802{,}700 \text{ kJ/kmol} \\ &\quad - [8.31441 \text{ kJ/(mol}\cdot\text{K)}](298 \text{ K})(10.52 - 10.52) \\ &= -802{,}700 \text{ kJ/kmol}\end{aligned}$$

We note that for this particular chemical reaction, because the number of moles of products is equal to the number of moles of reactants, the internal energy of combustion is equal to the enthalpy of combustion.

Now, using values of internal energy from the gas tables and paralleling the solution of method 3 of Example Problem 11.6, we guess temperatures, compute the change in internal energy of the products, and compare with internal energy of combustion. The equation being used is

$$(\bar{u} - \bar{u}^o)_{CO_2} + 2(\bar{u} - \bar{u}^o)_{H_2O} + 7.52(\bar{u} - \bar{u}^o)_{N_2} = 802{,}700$$

Constructing a table of values gives

	$\bar{u} - \bar{u}^o = M(u - u^o)$, kJ/kmol			
T_p (guessed), °C	CO_2	H_2O	N_2	LHS, kJ/kmol
2600	123,620	195,960	500,850	820,430
2500	118,270	186,750	479,650	784,670

Interpolating between these two entries gives $T_p = 2550°C = 2823$ K. This temperature should be compared with the adiabatic flame temperature for the same case of 2325 K. The higher explosion temperature results because the combustion products did not do work against the atmosphere in the constant-volume process; the combustion products do work while they increase their volume during the constant-pressure combustion process assumed for the calculation of the adiabatic flame temperature.

The final volume is equal to the initial volume, V_1. For initial conditions of one mole of CH_4, and noting that *in this case* the total moles of reactants and products are equal, the ideal gas law results in

$$P_2 = P_1\left(\frac{T_2}{T_1}\right) = 1 \text{ atm}\left(\frac{2823}{297}\right) = 9.51 \text{ atm}$$

11.5 Applications to Combustion Systems

Now that we have found methods for comparing fuels on the basis of their energy release (enthalpy of combustion) and relative combustion temperatures (adiabatic flame temperature), we can examine the application of thermodynamics to some engineering combustion problems. In particular, we examine systems in which combustion occurs, but in which we deviate from the idealized conditions used in Sec. 11.4, i.e., reactants all entering at the standard reference state, adiabatic or isothermal combustion, stoichiometric amounts of oxygen (air) provided for combustion, etc. We assume that the combustor operates at constant pressure.

11.5.1 Accounting for Excess Air

In practice, more air must be provided for the combustion process than is indicated by the balanced chemical reaction equation. Excess air is necessary because it is difficult, it not impossible, to achieve perfect mixing of the fuel with the oxygen carried by the air during the period of residence of these components in the combustion reactor.

If air in an amount above that necessary for stoichiometric combustion is provided, then we refer to the extra as *excess air.* And, for example, if 20 percent more air by volume (or mole) is provided than the stoichiometric amount, then the system is receiving 20 percent excess air. (Sometimes the term *theoretical air* is used. Thus 100 percent theoretical air means the stoichiometric amount of air, and 120 percent theoretical air is the same as 20 percent excess air.)

For a combustor receiving $X \times 100$ percent excess air, the chemical reaction Eq. (11.4) for combustion of 1 mol of a general hydrocarbon C_lH_m now becomes

$$\begin{aligned} C_lH_m + (1+X)\left(l+\frac{m}{4}\right)O_2 + 3.76(1+X)\left(l+\frac{m}{4}\right)N_2 \rightarrow \\ lCO_2 + \frac{m}{2}H_2O + 3.76(1+X)\left(l+\frac{m}{4}\right)N_2 \\ + X\left(l+\frac{m}{4}\right)O_2 \end{aligned} \tag{11.32}$$

The general first law for the combustor is given by Eq. (11.19). If we assume that the reactants enter at the standard reference state, then changing to a molar basis and subtracting Eq. (11.20) result in

$$\bar{q} - \frac{1}{n_f}\sum_p n_p[\bar{h}_p(T) - \bar{h}_p^o] = \bar{q}^o \tag{11.33}$$

For many large combustors, the energy loss to the combustor surroundings by heat transfer is small; however, heat transferred to a reacting chemical stream (in a chemical furnace) or to steam (as in a power plant boiler) must be included in $\bar{q}$. Let us, for a moment, restrict our attention to an adiabatic furnace, so that the effect of excess air becomes obvious. In that case, Eq. (11.33) reduces to

$$\frac{1}{n_f}\sum_p n_p[\bar{h}_p(T) - \bar{h}_p^o] = -\bar{q}^o \tag{11.34}$$

which is the same form as was used in determining the adiabatic flame temperature, Eq. (11.27). Now, however, the left-hand side of the first law equation contains more terms in the series. Now the excess unused oxygen must be included, and the number of moles of nitrogen is also larger because of the additional nitrogen carried into (and out of) the combustor as a result of the excess air. Both factors decrease the temperature reached by the combustion products in comparison with the theoretical adiabatic flame temperature.

Example Problem 11.8

Determine the AFR (Sec. 11.2.3) and the temperature of the combustion products of methane burning with 20 percent excess air. Assume the reactants enter at the standard reference state.

Solution

The chemical balance equation is

$$CH_4 + 2.4O_2 + 9.024N_2 \rightarrow CO_2 + 2H_2O + 9.024N_2 + 0.4O_2$$

Then, with Eq. (11.9),

$$AFR = \frac{68.9(1+0.2)[2 \times (1) + 4/2 - 0]}{12+4} = 20.7$$

The first law, Eq. (11.34), becomes

$$[\bar{h}(T_p) - \bar{h}^o]_{CO_2} + 2[\bar{h}(T_p) - \bar{h}^o]_{H_2O} + 9.024[\bar{h}(T_p) - \bar{h}^o]_{N_2} + 0.4[\bar{h}(T_p) - \bar{h}^o]_{O_2} = -(-345{,}000)\ \text{Btu/lbmol}$$

An iterative solution using method 1 of Example Problem 11.6 yields $T_p = 3730°R = 2070$ K.

Comments

The presence of the excess air, while necessary for such practical reasons as preventing unburned fuel and the products of incomplete combustion from leaving the combustor, has also lowered the flame temperature by 492°R from that predicted for stoichiometric conditions with complete mixing.

11.5.2 Accounting for Air or Fuel Preheating

We have assumed so far that the reactants enter the combustor at the standard reference state. This is rarely the case, so in this section we include the effect of preheating the fuel and/or air entering the combustor. As we shall see, the enthalpy of the combustion products can be raised significantly by such preheating, especially preheating of the combustion air. Industrial furnaces in particular use such preheating whenever a convenient source of waste energy is available.

If the reactants are preheated, then the first law is given by Eq. (11.19) which, given an adiabatic combustor and a molar basis, becomes

$$\sum_p n_p \bar{h}_p = \sum_r n_r \bar{h}_r \tag{11.35}$$

Subtracting Eq. (11.21) from Eq. (11.35) and assuming 1 mol of fuel result in

$$\sum_p n_p[\bar{h}_p(T_p) - \bar{h}_p^o] = \sum_r n_r[\bar{h}_r(T_r) - \bar{h}_r^o] - \bar{q}^o \tag{11.36}$$

Now the effect of each of the reactants (fuel and air) entering the combustor at their respective temperatures T_r can be considered. Examination of Eq. (11.36) shows that the higher the temperature T_r of each reactant, and therefore the higher the $\bar{h}_r(T_r)$ of each reactant, the larger will be the right-hand side of Eq. (11.36). Thus, the larger must be the value of T_p on the left-hand side of the equation.

Equation (11.36) illustrates why evaluation of enthalpy for all substances based on a standard reference state is so convenient. If this were not done, the summations in that term could not be conveniently evaluated, since each substance would have enthalpy changes based on different reference states.

Example Problem 11.9

Determine the temperature of the combustion products leaving an adiabatic furnace in which methane is burned with 20 percent excess air. The combustion air is preheated to 200°C, and the methane is preheated to 80°C.

Solution

This problem is similar to Example Problem 11.8, except that the fuel and air are preheated. The chemical balance equation remains the same. The first law, Eq. (11.36), for this case is

$$[\bar{h}(T_p) - \bar{h}^o]_{CO_2} + 2[\bar{h}(T_p) - \bar{h}^o]_{H_2O} + 9.024[\bar{h}(T_p) - \bar{h}^o]_{N_2}$$
$$+ 0.4[\bar{h}(T_p) - \bar{h}^o]_{O_2} = [\bar{h}(T = 80°C) - \bar{h}^o]_{CH_4}$$
$$+ 2.4[\bar{h}(T = 200°C) - \bar{h}^o]_{O_2} + 9.024[\bar{h}(T = 200°C) - \bar{h}^o]_{N_2}$$
$$-(-802{,}900) \qquad \text{kJ/kmol}$$

Using method 3 as in Example Problem 11.6 and iterating to find T_p yields a temperature of 2241 K = 4034°R. The values of terms on the right-hand side for the enthalpies of the reactants at their respective temperatures are found to be dominated by the nitrogen (46,200 kJ), with a lesser contribution by the oxygen (12,600 kJ) and a small contribution by the methane (2000 kJ).

Comments

Preheating the reactants increases the resulting enthalpy and thus the temperature of the products. When a preheater can be used to withdraw the available energy from stack gases or other waste heat source and increase the temperature of the reactants, particularly the air, it is often a good investment.

11.5.3 Applications

Now that we are able to handle combustion calculations under most conditions, there remains only the case of combustors in which significant heat transfer occurs from the combustor. For such a case, the heat transfer and shaft work terms are retained in the first law, and Eq. (11.36) becomes

$$\bar{q} - \bar{w} = \sum_p n_p[\bar{h}_p(T_p) - \bar{h}_p^o] - \sum_r n_r[\bar{h}_r(T_r) - \bar{h}_r^o] + \bar{q}^o \qquad (11.37)$$

Again, each term is based on 1 mol of fuel. The shaft work term $\bar{w}$ is usually zero for combustion systems.

Example Problem 11.10

A power plant boiler is being designed to transfer 1700 MBtu/h of thermal power to the working fluid. The boiler is to be fired by natural gas (assume methane properties) with 20 percent excess air. The stack gases from the plant are to be maintained at or above a temperature that exceeds the dew point of the combustion products by at least 300°F (so that water vapor will not condense in the stack and cause corrosion problems). The methane fuel is preheated to 175°F, and the air is preheated to 400°F. What mass flow rates of methane and air will be required?

Solution

First, the dew point of the combustion gases is determined. When that is known, the first law can be used to determine the heat transfer per mole of fuel, and that value plus the required power output sets the fuel flow rate.

The chemical reaction equation is the same as that given in Example Problem 11.8. The total moles of product gas per mole of fuel equal $n_{CO_2} + n_{H_2O} + n_{N_2} + n_{O_2} = 1 + 2 + 9.024 + 0.4 = 12.42$. The mole fraction of H_2O vapor is then $2/12.42 = 0.161$. Assuming that water vapor acts as an ideal gas mixture, we see the partial pressure of water vapor is then $P_{H_2O} = 0.161 \times 14.7$ psia $= 2.37$ psia. The saturation temperature corresponding to this partial pressure is (Table E.11) $T_{sat} = 129$°F. At 300°F above the dew point of 129°F or 429°F = 889°R, the first law [Eq. (11.37)] is

$$\begin{aligned}
\bar{q} &= \sum_p n_p[\bar{h}_p(T_p) - \bar{h}^o_p] - \sum_r n_r[\bar{h}_r(T_r) - \bar{h}^o_r] + \bar{q}^o \\
&= [\bar{h}(T_p = 889°\text{R}) - \bar{h}^o]_{CO_2} + 2[\bar{h}(T_p = 889°\text{R}) - \bar{h}^o]_{H_2O} \\
&\quad + 9.024[\bar{h}(T_p = 889°\text{R}) - \bar{h}^o]_{N_2} + 0.4[\bar{h}(T_p = 889°\text{R}) - \bar{h}^o]_{O_2} \\
&\quad - [\bar{h}(T = 635°\text{R}) - \bar{h}^o]_{CH_4} - 2.4[\bar{h}(T = 860°\text{R}) - \bar{h}^o]_{O_2} \\
&\quad - 9.024[\bar{h}(T = 860°\text{R}) - \bar{h}^o]_{N_2} + (-345{,}100) \qquad \text{Btu/lbmol}
\end{aligned}$$

Note that iteration is not required here, since the temperatures are known for evaluation of every property. Using the polynomial expressions for $\bar{c}_P$ to find the enthalpy changes results in $\bar{q} = -339{,}500$ Btu/lbmol of fuel. (Note that the polynomials are within their correct temperature ranges here; the ideal gas tables could also be used.) The required molar flow rate of methane is then

$$\dot{n}_{CH_4} = \frac{\dot{Q}_{plant}}{\bar{q}} = \frac{1.7 \times 10^9 \text{ Btu/h}}{339{,}500 \text{ Btu/lbmol}} = 5.01 \times 10^3 \text{ lbmol/h}$$

The required mass flow rate of methane is then

$$\dot{m} = M\dot{n}_{CH_4} = 16 \times 5.01 \times 10^3 = 8.02 \times 10^4 \text{ lbm/h}$$

From Example Problem 11.8, the AFR was found to be 20.7, so the mass flow rate of air required is

$$\dot{m}_{air} = \text{AFR} \times \dot{m}_{CH_4} = 20.7 \times 80.2 \times 10^3 = 1.66 \times 10^6 \text{ lbm/h}$$

Comments

As the stack gas (products) temperature is allowed to rise, less and less energy is available from the combustion process to provide heat transfer to the steam in the boiler. The figure shows the variation in heat transfer with increasing stack gas temperature for this fuel, excess air, and preheating. Note that the available energy for heat transfer approaches zero as the temperature approaches that found for the adiabatic combustor of Example Problem 11.9.

Diagram

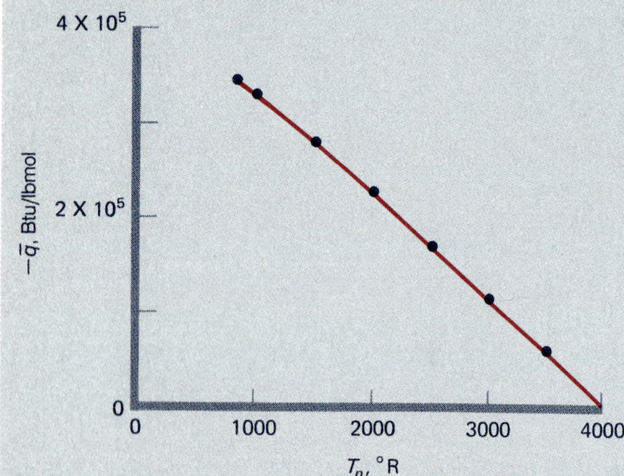

11.6 Applying the Second Law to Combustion Processes

All the calculations of fuel properties (enthalpy of combustion, adiabatic flame temperature) and applications of these properties (fuel and air preheating, excess-air effects, furnace and boiler calculations) have depended on the application of only the first law for reacting systems. The second law also has important applications in combustion and chemical reactions. We discuss applications of the second law of two general types: the determination of whether a particular chemical reaction can occur under a given set of conditions and the analysis of the overall thermodynamic efficiency of a system in which a chemical reaction plays a part.

11.6.1 Determining the Possibility of Reaction: Adiabatic Combustion

One consequence of the second law is that the entropy generation in any adiabatic process must be greater than or equal to zero. We have tacitly assumed so far that the chemical reactions used in this chapter in fact occur; that is, methane when mixed with air can undergo a combustion reaction that results in the formation of combustion products in the form of CO_2 and H_2O. Experience tells us that this can occur; however, unless special conditions are met (the presence of an ignition source, for example), the reaction may not occur. The second law can be used to tell us whether a given chemical reaction is possible, as we shall see. Whether the reaction *will,* in fact, occur and the *rate* at which it can occur depend on other factors (the presence of an ignition source or a catalyst, for example). These other factors are studied under the heading of chemical kinetics and are outside the scope of this text.

Application of the second law to determine whether a given reaction can occur is straightforward. We check to see whether, in an adiabatic steady flow reactor, the entropy generation is positive. If so, the reaction can possibly occur. If, however, the calculated entropy generation is negative, then the proposed reaction is not possible. If the entropy generation is zero, then the chemical reaction is reversible. (This idea is further developed in Chap. 12 for the case of reactions that do not proceed to completion, where chemical equilibrium is reached.)

The second law (Table 5.3) requires, for an adiabatic steady flow reactor, that

$$\dot{S}_{\text{gen}} = \sum_p \dot{n}_p \bar{s}_p - \sum_r \dot{n}_r \bar{s}_r \geq 0 \tag{11.38}$$

Example Problem 11.11

Methane is preheated to 80°C and is burned in 20 percent excess air that is preheated to 200°C. Is this reaction possible according to the second law?

Solution

The chemical reaction equation is the same as that used in Example Problems 11.8 and 11.9. The product temperature was found in Example Problem 11.9 to be 2241 K. Required data are taken from Table C.10:

Reactant	Moles	$\bar{s}^o$, kJ/(kmol·K)
CH_4	1	186.3
O_2	2.4	205.03
N_2	9.024	191.5
	12.424	

Product	Moles	$\bar{s}^o$, kJ/(kmol·K)
CO_2	1	213.7
O_2	0.4	205.03
N_2	9.024	191.5
H_2O (vapor)	2	188.7
	12.424	

The values of entropy at the conditions present for the reactants and products must now be computed. Given that the combustion occurs at a total pressure of 1 atm and that the materials act as a mixture of ideal gases, the partial pressure of each constituent i of the reactants and the products is given by [Eq. (10.23)] $P_i = (n_i/n)(101.3\ \text{kPa})$. The entropy of each constituent at the known conditions of temperature and partial pressure is given by modifying Eq. (10.40*a*) to give the standard entropy:

$$\begin{aligned}\bar{s} &= \bar{s}^o + \int_{298}^{T} \frac{\bar{c}_P(T)}{T}\,dT - \bar{R}\ln\frac{P_i}{P^o} \\ &= \bar{s}^o + \int_{298}^{T} \frac{\bar{c}_P(T)}{T}\,dT - \bar{R}\ln\frac{n_i}{n} \\ &= \bar{s}^o + \int_{298}^{T} \frac{\bar{c}_P(T)}{T}\,dT - \bar{R}\ln P_i\end{aligned}$$

Note that the final two forms of the equation only hold for a total pressure in the combustor of 1 atm. For all substances in this problem except methane, Tables D.2 through D.8 or the computerized tables are convenient and give values of

$$\bar{s}^o + \int_{298}^{T} \frac{\bar{c}_P(T)}{T}\,dT$$

directly. (*Note:* The table values are per unit mass rather than per mole.) For methane the polynomial expressions for $\bar{c}_P(T)$ in Table D.1 may be used.

These calculations using the tables result in the following:

Reactant	T, K	$\frac{n_i}{n}$	$\bar{s}(T, P)$, kJ/(kmol·K)
CH_4	353	0.0805	213.5
O_2	473	0.1932	232.6
N_2	473	0.7263	207.8

Product	T, K	$\frac{n_i}{n}$	$\bar{s}(T, P)$, kJ/(kmol·K)
CO_2	2241	0.0805	316.2
O_2	2241	0.0322	273.1
N_2	2241	0.7263	256.1
H_2O	2241	0.1610	270.3

Substituting into Eq. (11.38) for 1 kmol of fuel results in

$$\begin{aligned}\bar{s}_{\text{gen}} &= (1 \times 316.2 + 0.4 \times 273.1 + 9.024 \times 256.1 + 2 \times 270.3) \\ &\quad - (1 \times 213.5 + 2.4 \times 232.6 + 9.024 \times 207.8) \\ &= 3277.1 - 2646.9 = 630.2\ \text{kJ/(kmol·K)} > 0\end{aligned}$$

so the reaction is possible, as we expected.

Comment

References 1 to 3 contain excellent data sources to aid in the solution of this type of problem, which greatly simplifies the computations necessary for accurate results.

11.6.2 Determining the Possibility of Reaction: General Combustion Problems

In a reactor that is not adiabatic, it is possible to remove or add energy by means of heat transfer with the surroundings. The second law must be modified to include the entropy changes that occur because of the heat transfer. The entropy generation must remain positive for the reaction being considered to have the possibility of going to completion. In such a case, the entropy generation of a system composed of the combustor and the surroundings is

$$S_{gen} = \sum_p n_p \bar{s}_p - \sum_r n_r \bar{s}_r - \frac{Q}{T_0} \tag{11.39}$$

where T_0 is the absolute temperature of the surroundings to which there is a heat transfer. In this equation, the entropy of mixing (Chap. 10) has not been explicitly stated but is included in the $\Sigma n\bar{s}$ terms. If the reactants are not mixed before entering the combustor, then the entropy-of-mixing term should not be included; quite often, however, they are mixed. The products are invariably mixed, and the mixing term should be included. In most engineering problems, the effect of the entropy-of-mixing term on the results for S_{gen} is minor. Let us divide Eq. (11.39) by the number of moles of fuel and find the entropy generation per mole of fuel, or

$$\bar{s}_{gen} = \frac{1}{n_f}\left(\sum_p n_p \bar{s}_p - \sum_r n_r \bar{s}_r\right) - \frac{\bar{q}}{T_0} \tag{11.40}$$

Now the general first law from Eq. (11.37) with $\bar{w} = 0$ can be used to eliminate $\bar{q}$, resulting in (for 1 mol of fuel)

$$\bar{s}_{gen} = \sum_p n_p \bar{s}_p - \sum_r n_r \bar{s}_r - \frac{\sum_p n_p[\bar{h}_p(T_p) - \bar{h}_p^o] - \sum_r n_r[\bar{h}_r(T_r) - \bar{h}_r^o] + q^o}{T_0} \tag{11.41}$$

This last equation can also be obtained from Eq. (8.6) with modifications for chemical reactions and zero work. Rearranging Eq. (11.41) gives

$$\bar{s}_{gen} = \frac{1}{T_0}\left\{\sum_p n_p[T_0\bar{s}(T) - \bar{h}(T)]_p - \sum_r n_r[T_0\bar{s}(T) - \bar{h}(T)]_r\right\} + \frac{1}{T_0}\left(\sum_p n_p\bar{h}_p^o - \sum_r n_r\bar{h}_r^o - \bar{q}^o\right) \tag{11.42}$$

Noting that the expression in parentheses of the final term is zero [Eq. (11.21)] results in the form

$$\bar{s}_{gen} = \frac{1}{T_0}\left\{\sum_p n_p[T_0\bar{s}(T) - \bar{h}(T)]_p - \sum_r n_r[T_0\bar{s}(T) - \bar{h}(T)]_r\right\} \tag{11.43}$$

Some special cases can be examined. For an adiabatic combustor, the final term in Eq. (11.40) or (11.41) goes to zero, and the equation reduces to Eq. (11.38).

Standard Gibbs Function of Reaction

For an isothermal combustor operated at the temperature $T_0 = T_r = T_p$, Eq. (11.43) becomes, after the Gibbs function ($\bar{g} = \bar{h} - T\bar{s}$) is used, for 1 mol of fuel,

$$\bar{s}_{\text{gen}} = \frac{1}{T_0}\left[\sum_r n_r \bar{g}_r(T_0) - \sum_p n_p \bar{g}_p(T_0)\right] \tag{11.44}$$

Since $\bar{s}_{\text{gen}}$ must be greater than or equal to zero for the reaction to proceed, to make the reaction proceed isothermally and reversibly at the temperature of the surroundings requires that

$$\sum_p n_p \bar{g}_p \leq \sum_r n_r \bar{g}_r \tag{11.45}$$

The Gibbs function is obviously a very important property for determining whether a particular reaction will proceed.

Values of the Gibbs function at the standard reference state $\bar{g}^o$ are easily calculated from

$$\bar{g}^o = \bar{h}^o - 298\bar{s}^o \qquad \text{kJ/kmol} \tag{11.46a}$$

or

$$\bar{g}^o = \bar{h}^o - 537\bar{s}^o \qquad \text{Btu/lbmol} \tag{11.46b}$$

If the reaction in question is carried out at the standard reference state, then

$$\Delta\bar{g}_f^o = \sum_p n_p \bar{g}_p^o - \sum_r n_r \bar{g}_r^o \leq 0 \tag{11.47}$$

The value of $\Delta\bar{g}_f^o$ is called the *standard Gibbs function of reaction,* or the *standard free energy of reaction,* and it must be negative in order for the reaction to have the possibility of proceeding spontaneously, as shown by Eq. (11.45). Values of $\Delta\bar{g}_f^o$ are given in Table C.9. Applications and extensions of the second law relations developed here to other types of chemical reactions are discussed in Chap. 12.

Useful Work Output from a Combustor

To determine the useful reversible work output available from a combustor operating in steady flow, we can refer to Eq. (8.9) and modify that relation for the presence of the combusting mixture, to give

$$\bar{w}_{\text{rev}} = \sum_r n_r(\bar{h} - T_0\bar{s})_r - \sum_p n_p(\bar{h} - T_0\bar{s})_p \tag{11.48}$$

for 1 mol of fuel, where changes in kinetic and potential energy have been neglected. If the combustion is carried out isothermally at the temperature of the surroundings, then

$$\bar{w}_{\text{rev}} = \sum_r n_r \bar{g}(T_{\text{surr}})_r - \sum_p n_p \bar{g}(T_{\text{surr}})_p \tag{11.49}$$

and again we see the importance of the Gibbs function. If the temperature of the surroundings is taken to be the standard reference temperature of 298 K (537°R), then

$$\bar{w}_{\text{rev}} = \sum_r n_r \bar{g}_r^o - \sum_p n_p \bar{g}_p^o = -\Delta \bar{g}_f^o \tag{11.50}$$

and the reversible work from the combustion reaction carried out adiabatically at the standard reference temperature and pressure is seen to be simply equal to the negative of the standard Gibbs free energy of reaction. Noting that

$$\bar{w}_{\text{rev}} = -\Delta \bar{g}_f^o = -\Delta \bar{h}_f^o + 298\ \Delta \bar{s}^o \qquad \text{kJ/kmol} \tag{11.51}$$

we can interpret the reversible work on the surroundings as coming from the standard enthalpy of reaction less the "unavailable work" 298 $\Delta \bar{s}^o$ kJ/kmol.

Example Problem 11.12

The combustion system described in Example Problem 11.10 is used to drive a generator. Determine the maximum mechanical power that can be obtained from the combustor when the surroundings are at 25°C.

Solution

The values of the entropy of the products and reactants must be found for use in Eq. (11.48) to determine the reversible work done per mole of fuel. These values are available for the reactants from Example Problem 11.11. For the products, $\bar{s}$ must be calculated at the outlet temperature of 494 K found in Example Problem 11.10. The values are computed as in Example Problem 11.11.

Reactant	T, K	$\frac{n_i}{n}$	$\bar{s}(T, P)$, kJ/(kmol·K)
CH_4	353	0.0805	213.5
O_2	473	0.1932	232.6
N_2	473	0.7263	207.8

Product	T, K	$\frac{n_i}{n}$	$\bar{s}(T, P)$, kJ/(kmol·K)
CO_2	494	0.0805	255.4
O_2	494	0.0322	248.9
N_2	494	0.7263	209.1
H_2O	494	0.1610	221.2

Substituting into Eq. (11.48) gives

$$\bar{w}_{rev} = \sum_r n_r(\bar{h} - T_0\bar{s})_r - \sum_p n_p(\bar{h} - T_0\bar{s})_p$$
$$= \sum_r (n\bar{h})_r - \sum_p (n\bar{h})_p - 298\left[\sum_r (n\bar{s})_r - \sum_p (n\bar{s})_p\right]$$

Note that the value of $\sum_r (n\bar{h})_r - \sum_p (n\bar{h})_p$ is equal to 789,400 kJ/kmol of fuel, since it is equal to $\bar{q}$ by the first law. Making this substitution and entering the entropies of the products and reactants results in

$$\bar{w}_{rev} = 789{,}400 + 298(2684.1 - 2646.9) = 800{,}500 \text{ kJ/kmol}$$
$$= 342{,}000 \text{ Btu/lbmol}$$

Using the molar flow rate of $\dot{n} = 0.630$ kmol/s found in Example Problem 11.10 results in a reversible power output for the combustor of

$$\dot{W}_{rev} = \dot{n}\bar{w}_{rev} = (0.630 \text{ kmol/s})(800{,}500 \text{ kJ/kmol}) = 504 \text{ MW}$$
$$= 1720 \text{ MBtu/h}$$

Comments

Note that the reversible work (800,500 kJ/kmol) is close to the value of the energy removed by heat transfer from the combustor 789,400 kJ/kmol), because the value of the $T_0\bar{s}$ terms is relatively small in comparison with the value of the enthalpy terms. This is usually the case for the combustion of hydrocarbons. Also the reversible power is greater than the power provided by the furnace in Example Problem 11.10 (494 MW), because some power is carried from that furnace by the stack gases.

11.7 Applications to Real Devices: Efficiency of Combustion Equipment

For an actual device in which work output is the objective, such as an internal combustion engine, the amount of energy used is less than that calculated as being available from the combustion reaction. We have seen that the reversible work available from the combustion reaction is less than the energy release calculated from the first law, so that the upper limit of the energy that can be obtained from any real device where combustion occurs is set by Eq. (11.49). It would seem reasonable, then, to define the efficiency of a real device in terms of the fraction of the reversible work attained, as discussed in Chap. 8. However, historically a "first law" efficiency has been used for combustion processes in which work is produced, as is the case for power cycles described in Chap. 6. Thus, an alternative efficiency for a heat engine is defined as the actual work output over the negative of the enthalpy of combustion of the fuel, or

$$\eta_{\text{th}} = \frac{w}{-q^o} \tag{11.52}$$

Remember that the value of q^o depends on whether the water in the combustion products is assumed to be liquid or vapor. The choice obviously changes the value of the thermal efficiency computed for a given engine.

For devices in which maximum heat transfer is the desired result rather than work, such as furnaces, water heaters, steam generators, etc., the *heat transfer efficiency* η_{HT} is defined as the actual heat transferred in the device divided by the higher heating value of the fuel, or

$$\eta_{\text{HT}} = \frac{-q}{\text{HHV}} \tag{11.53}$$

Finally, as mentioned earlier, in some devices the desired result of combustion is to attain a high temperature. This is true in certain combustion processes in internal combustion engines with efficiencies that depend on the combustion temperature and in some process furnaces where a particular temperature must be reached. In such cases, attainment of the desired temperature often requires that additional fuel be injected over that necessary for stoichiometric combustion; in other words, a slightly fuel-rich mixture is needed. Thus, the actual AFR is reduced below the stoichiometric value. The *combustion efficiency* is then defined as

$$\eta_{\text{comb}} = \frac{(\text{AFR})_{\text{actual}}}{(\text{AFR})_{\text{ideal}}} = \frac{(\text{FAR})_{\text{ideal}}}{(\text{FAR})_{\text{actual}}} \tag{11.54}$$

where FAR is the fuel-air ratio (the inverse of AFR.)

Example Problem 11.13

For the combustor described in Example Problem 11.10, find the heat transfer efficiency.

Solution

The amount of heat transfer per mole of fuel was found to be −339,400 Btu/lbmol of fuel. The HHV of methane was found in Example Problem 11.5 to be 383,100 Btu/lbmol, so the heat transfer efficiency [Eq. (11.53)] is

$$\eta_{\text{HT}} = \frac{-339{,}400}{383{,}100} = 0.89 = 89\%$$

Example Problem 11.14

An adiabatic rocket motor burns kerosene (approximate chemical composition assumed to be $C_{10}H_{22}$) with pure oxygen in the combustion chamber. Twenty percent excess oxygen is present. The nozzle has the shape shown in Fig. 5.24. Take the standard enthalpy of formation of kerosene to be −22,500 kJ/kg. The kerosene enters the combustion chamber as a gas at 25°C, as does the oxygen, and the combustion chamber is maintained at 30 atm. Assuming one-dimensional flow, find out whether choked flow exists at the throat as well as the temperature and total pressure of the combustion products at the throat.

Solution

For a kerosene-oxygen combustion system (note that no nitrogen is present in this case), the reaction relation Eq. (11.32) is

$$C_{10}H_{22} + (1 + 0.20)(\tfrac{31}{2})O_2 \rightarrow 10CO_2 + 11H_2O + 0.20(\tfrac{31}{2})O_2$$

To solve the problem, the conditions in the combustion chamber first must be determined. The adiabatic combustion temperature of the combustion products is found from Eq. (11.34), which requires knowledge of the enthalpy of combustion $\bar{q}^o$. Using Eq. (11.21) for the stoichiometric reaction (i.e., the excess oxygen is not considered in finding the enthalpy of combustion) and data from Table C.9, we get

$$\begin{aligned}\bar{q}^o &= \sum_1^3 n_p \bar{h}_p^o - \sum_1^2 n_r \bar{h}_r^o \\ &= [11(-242.0) + 10(-393.8)]\ \text{MJ/kmol fuel} \\ &\quad - [1(-22.5)\ \text{MJ/kg} \times 142\ \text{kg/kmol fuel} + 15.5 \times 0] \\ &= -6600 + 3195 = -3405\ \text{MJ/kmol fuel}\end{aligned}$$

Now by using Eq. (11.34) the adiabatic flame temperature T_p can be found:

$$10[\bar{h}(T_p) - \bar{h}^o]_{CO_2} + 11[\bar{h}(T_p) - \bar{h}^o]_{H_2O} + 3.1[\bar{h}(T_p) - \bar{h}^o]_{O_2} = 3405\ \text{MJ/kmol fuel}$$

We use method 3 from Example Problem 11.6 for finding the adiabatic flame temperature. Noting that the molar enthalpy difference terms can be replaced by $M[h(T_p) - h(T = 25°\text{C})]$, we see that the equation can be written as

$$\begin{aligned}&(10 \times 44)\,[h(T_p) - h(T = 25°\text{C})]_{CO_2} + (11 \times 18) \\ &\quad \times [h(T_p) - h(T = 25°\text{C})]_{H_2O} + (3.1 \times 32)\,[h(T_p) - h(T = 25°\text{C})]_{O_2} \\ &\qquad = 3405\ \text{MJ/kmol fuel}\end{aligned}$$

Substituting the ideal gas table values of $h(T = 25°\text{C})$ for each product results in

$$h(T_p)_{CO_2} + 0.45h(T_p)_{H_2O} + 0.2255h(T_p)_{O_2} = 8260.2$$

where all $h(T_p)$ values are in kilojoules per kilogram. Various values of T_p are tried from the gas tables until convergence is achieved, giving a final value of $T_p = 2859°C = 3132$ K.

It is now assumed that the mixture of combustion products passes through the converging section of the nozzle isentropically. We follow the procedure used in Example Problem 5.17, except that mixture properties of the combustion products are used at each step.

We must *check to see whether choked flow* (Sec. 5.11.3) *exists at the throat.* Because we have a gas mixture rather than a single component, the problem again becomes iterative. The values of H for the gas mixture in the combustion chamber and for the change in S as the mixture passes through the nozzle are found from Eqs. (10.32*b*) and (10.36*c*).

The mass fraction of each component in the combustion products is, from Eq. (10.14), $x_i = y_i M_i/(\Sigma y_i M_i)$. Thus

$$x_{CO_2} = 0.415(44)/[0.415(44) + 0.456(18) + 0.129(32)] = 0.597$$

$$x_{H_2O} = 0.456(18)/30.596 = 0.268; \; x_{O_2} = 0.129(32)/30.596 = 0.135$$

Using the ideal gas assumptions, we see that the inlet enthalpy of the combustion products entering the nozzle is

$$h_{in} = \Sigma x_i h_i = 0.597 \times 3875.7 + 0.268 \times 7988.1 + 0.135 \times 3497.8$$
$$= 4925.5 \text{ kJ/kg}$$

and the first law for the entrance section of the nozzle can be expressed as

$$\mathbf{V}_t = [2g_c(h_{in} - h_t)]^{1/2}$$

Isentropic Flow

The temperature at the throat can be found for an isentropic process from Eq. (10.36*c*) as

$$s_t - s_{in} = 0 = s_0(T_t) - s_0(T_{in}) - R \ln \frac{P_{tot,\,t}}{P_{tot,\,in}}$$

The value of R for the mixture is, from Eq. (10.27*b*),

$$R = \Sigma x_i R_i = 0.597(0.18892) + 0.268(0.46152) + 0.135(0.25983)$$
$$= 0.2715 \text{ kJ/(kg·K)}$$

and, using the data from Tables D.5 through D.7,

$$s_0(T_{in} = 2859°C) = 0.597(7.6564) + 0.268(16.029) + 0.135(8.9431)$$
$$= 10.0740 \text{ kJ/(kg·K)}$$

Thus, by the entropy relation,

$$s_0(T_{t,\,s}) = s_0(T_{in} = 2859°C) + R \ln \frac{P_{tot,\,t}}{P_{tot,\,in}} = 10.074 + 0.2715 \ln \frac{0.1}{30}$$
$$= 8.525 \text{ kJ/(kg·K)}$$

Now we must find the corresponding value of $T_{t,s}$ that produces this value of $s_0(T_{t,s})$ for the mixture. This can be done by assuming various values of $T_{t,s}$ and calculating the corresponding values of $s_0(T_{t,s})$ until agreement is reached. A table of computed values is shown:

Assumed $T_{t,s}$, °C	Calculated $s_0(T_{t,s})$, kJ/(kg·K)
2859	10.0740
2000	9.4929
1000	8.5139
1100	8.6360

By interpolation between the final two values, $T_{t,s} = 1009°C$. Now the enthalpy of the combustion products for isentropic flow at the throat is

$$h(T_{t,s}) = 0.597(322.55) + 0.268(680.76) + 0.135(336.97) \\ = 420.6 \text{ kJ/kg}$$

and the *throat velocity for isentropic flow* is

$$\mathbf{V}_{t,s} = [2g_c(h_{in} - h_{t,s})]^{1/2} \\ = [2(4925.5 - 420.6) \text{ kJ/kg} \times 1000 \text{ kg}\cdot\text{m}^2/(\text{kJ}\cdot\text{s}^2)]^{1/2} \\ = 3001.6 \text{ m/s}$$

Actual Flow (One-Dimensional)

The flow velocity for isentropic flow in the nozzle can be checked against the acoustic velocity at the throat given in Eq. (5.197) (based on the calculated $T_{t,s} = 1009°C = 1281$ K for the unchoked case) by first finding the value of k at the throat for the mixture as

$$k = \frac{c_{P,\text{mix}}}{c_{v,\text{mix}}} \\ = \frac{0.597(1.3117) + 0.268(2.4833) + 0.135(1.2184)}{0.597(1.1226) + 0.268(2.0215) + 0.135(0.8635)} \\ = \frac{1.6131}{1.3285} = 1.2142$$

and then

$$\mathbf{V}_{ac} = (g_c kRT_t)^{1/2} \\ = [1.2142 \times 0.2715 \text{ kJ/(kg}\cdot\text{K)} \times 1281 \text{ K} \times 1000 \text{ J/kJ}]^{1/2} \\ = 650 \text{ m/s}$$

Because the predicted acoustic velocity at the nozzle throat is lower than the computed value for an unchoked nozzle, and because the actual velocity at the throat cannot exceed the acoustic velocity at the throat temperature, this nozzle must be operating in choked flow.

Actual Temperature and Velocity at the Throat

To find the actual temperature and velocity at the throat is again an iterative problem. Because the flow is choked, the acoustic velocity relation can be substituted into the first law to obtain

$$\mathbf{V}_{ac} = (g_c k R T_t)^{1/2} = [2g_c(h_{in} - h_t)]^{1/2}$$

which reduces to

$$T_t = \frac{2}{kR}(h_{in} - h_t)$$

To solve for T_t, we must guess values of T_t, compute corresponding values of k and h_t for the mixture, and see whether this equation is satisfied. To compute a first guess, assume that $h_{in} - h_t = c_P(T_{in} - T_t)$, which results in

$$T_t = \frac{2}{k+1} T_{in} = \frac{2}{2.2412}(3232) = 2884 \text{ K} = 2611°\text{C}$$

Note that this relation contains the assumption that c_P is the mean value for the mixture over the temperature range T_{in} to T_t, and that this c_P is the same as the value used in computing the ratio of specific heats at the throat.

Using the approximate value of 2611°C as a first guess, we calculate values of k and h_t for the mixture, and we find a new throat temperature from the first law expression. The process is repeated until convergence is achieved.

For the first guess, a table can be constructed to aid in calculation:

		T_t(guessed) = 2611°C		
Gas	**Mass fraction, x**	c_P, **kJ/(kg·K)**	c_v, **kJ/(kg·K)**	h, **kJ/kg**
CO_2	0.597	1.4113	1.2222	3524.6
H_2O	0.268	3.0742	2.6124	7216.1
O_2	0.135	1.2352	0.9752	3189.7
Mixture		1.8332	1.5613	4468.7

$$k = \frac{1.8332}{1.5613} = 1.1741$$

This information is entered into the first law relation, and a new throat temperature is calculated. The process is repeated, and the information for this and succeeding iterations is shown in another table:

T_t**(guessed), °C**	k	h_t, **kJ/kg**	T_t**(calculated), °C**
2611	1.1741	4468.7	2593
2593	1.1742	4435.7	2800
2600	1.1742	4448.5	2720
2609.5	1.1741	4466	2610.0

After computing the first two values in the table above, we noted that the guess of 2611 produced a calculated temperature of 2593, lower than the guess. A guess of 2593 produced a calculated value of 2800, much higher than the guess. Thus, the correct answer must lie between the first two guessed values. The third guess was chosen as 2600, and it still produced too high a calculated value. The three initial guesses were then plotted in the figure against the calculated temperatures, resulting in the dashed line. The solid line is the locus of where equal guessed and calculated values lie. Where these lines cross should be where the equation will give equal guessed and calculated temperatures, and this point appears to be near 2609.5°C, which was chosen as the final guess. The calculated value is within 0.5°C for this guess, so the final answer is 2610°C.

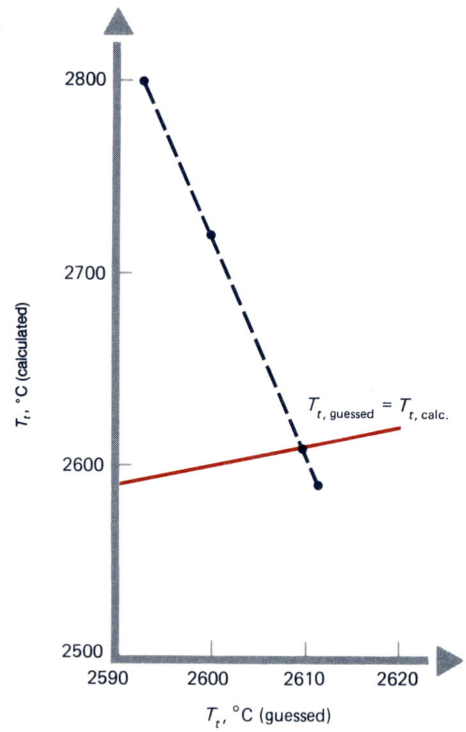

The actual velocity at the throat is now calculated as

$$\mathbf{V}_a = (g_c kRT_t)^{1/2} = [(1.1741)(0.2715)(2610 + 273)(1000)]^{1/2} = 959 \text{ m/s}$$

where the actual acoustic velocity at the throat is based on the actual temperature at the throat rather than the isentropic temperature, which was used earlier to check for the presence of choked flow.

The pressure at the throat is given for choked flow by Eq. (5.200*b*) as

$$P_t = P_i\left(\frac{2}{k+1}\right)^{k/(k-1)} = 100\left(\frac{2}{2.1741}\right)^{1.1741/0.1741} = 63.2 \text{ atm}$$

Comments

In this solution, we checked for the presence of choked flow by finding the isentropic conditions at the throat, and for the isentropic conditions we calculated the velocity from the first law and compared it with the acoustic velocity *at the isentropic temperature.* Choked flow was determined to exist, so we proceeded to find the throat conditions for choked flow by substituting the acoustic velocity into the first law relation.

Example Problem 11.15

Find the specific impulse, exhaust velocity relative to the nozzle, and exit diameter for the rocket engine shown. The rocket motor is adiabatic and burns kerosene (approximate chemical composition assumed to be $C_{10}H_{22}$) with pure oxygen in the combustion chamber. Twenty percent excess oxygen is present. Take the standard enthalpy of formation of kerosene to be −22,500 kJ/kg. Assume that the kerosene enters the combustion chamber as a gas at 25°C, as does the oxygen, and that the combustion chamber is maintained at 30 atm.

Solution

The throat conditions for the temperature and pressure will be the same as those found for Example Problem 11.14, because the fuel, inlet conditions, and throat diameter for this problem are the same. Thus, $T_t = 2883$ K and $P_t = 63.2$ atm. We now calculate the mass flow rate through the nozzle for the choked flow case from Eq. (5.201) as

$$\Gamma = k^{1/2}\left(\frac{2}{k+1}\right)^{(k+1)/[2(k-1)]}$$

$$= 1.1741^{1/2}\left(\frac{2}{1.1741+1}\right)^{(1.1741+1)/[2(1.1741-1)]} = 0.9801$$

$$\dot{m}_{\text{choked}} = \Gamma A_t P_i \left(\frac{1}{RT_i}\right)^{1/2}$$

$$= 0.9801\left(\frac{\pi D_t^2}{4}\right)\text{ m}^2 \times 10\text{ atm} \times 101.3 \times 10^3\text{ Pa/atm}$$

$$\times\left[\frac{1}{0.2715\text{ kJ/(kg}\cdot\text{K)} \times 1000\text{ K} \times 1000\text{ J/kJ}}\right]^{1/2}$$

$$= 0.9801\left(\frac{\pi}{16}\right) \times 10 \times 101.3 \times 10^3 \times \left(\frac{1}{2.715 \times 10^5}\right)^{1/2}$$

$$= 37.4\text{ kg/s}$$

The actual outlet velocity may be found directly from the first law

$$\mathbf{V}_{e,a} = [2g_c(h_i - h_{e,a})]^{1/2}$$

The exit enthalpy is easily found as $h_e = 283.2$ kJ/kg, s

$$\mathbf{V}_{e,a} = \{[2(4925.5 - 283.2)(\text{kJ/kg})](1000\text{ J/kJ})\}^{1/2}$$
$$= 3047\text{ m/s}$$

The specific impulse for this engine is given by Eq. (6.66) as

$$I = \frac{\mathbf{V}_{e,a}}{g_c} = \frac{3047\text{m/s}}{1\text{ N}\cdot\text{s}^2/(\text{kg}\cdot\text{m})} = 3047\text{ N/(kg/s)}$$

The outlet diameter for this nozzle can be found from the continuity equation and the known mass flow rate to be

$$D_e = \left(\frac{4\dot{m}v_e}{\mathbf{V}_{e,a}}\right)^{1/2} = \left(\frac{4\dot{m}RT_e}{P_e\mathbf{V}_{e,a}}\right)^{1/2}$$

$$= \left[\frac{4 \times 37.4\text{ kg/s} \times 0.2715\text{ kJ/(kg}\cdot\text{K)} \times 273\text{ K}}{101.3\text{ kPa} \times 3047\text{ m/s}}\right]^{1/2}$$

$$= 0.190\text{ m}$$

Another application of combustion chemistry is the *fuel cell*, which is a device used for direct conversion of chemical energy to work in the form of electricity. It can be pictured as a battery in which the chemicals producing the potential difference and current are continuously replaced and the reaction products are continually removed.

Many reactions are possible, but for illustration let us look at the hydrogen-oxygen fuel cell. In one type of such cell, hydrogen is fed to the cathode and oxygen to the anode, as shown in Fig. 11.5. In this type of cell, gaseous hydrogen is fed to the cathode, where a catalytic reaction causes ionization of the hydrogen and release of electrons, which travel through the external electric circuit. Oxygen is simultaneously fed to the anode, where it combines with the ionized hydrogen and the electrons returned through the external circuit to form liquid water, which is continuously removed from the cell.

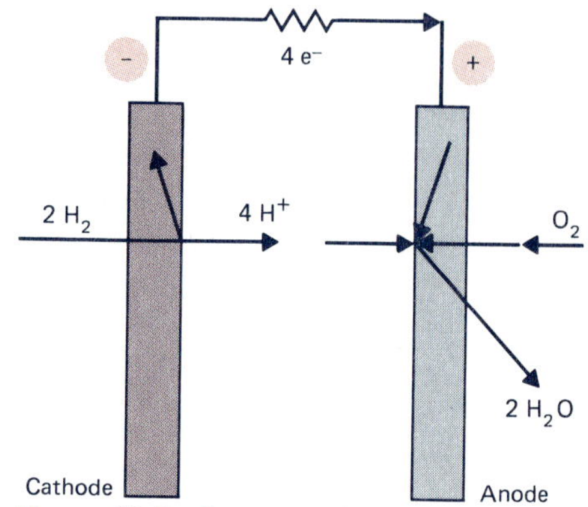

Figure 11.5 One type of hydrogen-oxygen fuel cell.

Because no heat engine is involved in this conversion of chemical energy to work, the fuel cell is not limited by the Carnot efficiency; there is no set of thermal reservoirs involved in converting heat transfer to work.

The behavior of the fuel cell can be analyzed by the methods of second law analysis discussed in Sec. 8.7. The fuel cell can be treated as a control volume in steady state, so that the second law fuel cell efficiency $\eta_{II,\,fc}$ can be defined as

$$\eta_{II,\,fc} = \frac{w_{a,\,use}}{w_{rev,\,use}} = 1 - \frac{T_0 s_{gen}}{w_{rev,\,use}} \tag{11.55}$$

and the value of $w_{rev,\,use}$ is equal to the difference in flow availability of the inlet streams (gaseous hydrogen and oxygen) and the outlet stream (liquid water). The flow availability for either the inlet or the outlet streams is given by

$$\psi = \frac{1}{n_f} \sum_i n_i[(h - h^0) - T_0(s - s^0)]_i \tag{11.56}$$

where the sum is over the product or reactant stream as required. If the fuel cell operates isothermally, the availability becomes equal to the Gibbs function of the mixture stream, so that

$$w_{rev,\,use} = \psi_{in} - \psi_{out} = g_{in} - g_{out} \tag{11.57}$$

Example Problem 11.16

A hydrogen-oxygen fuel cell has a measured second law efficiency of 85 percent when operated isothermally at 25°C and with a hydrogen flow rate of 5×10^{-4} kg/s. Find the electrical power output of the cell.

Solution

The chemical reaction is

$$2H_2 + O_2 \rightarrow 2H_2O$$

For an isothermal cell, Eq. (11.57) gives

$$\dot{W}_{rev,\,use} = \Sigma(\dot{m}\psi)_{in} - (\dot{m}\psi)_{out} = \Sigma(\dot{m}g)_{in} - (\dot{m}g)_{out}$$

For the inlet stream,

$$\Sigma(\dot{m}g)_{in} = \dot{m}_{H_2}g_{H_2} + \dot{m}_{O_2}g_{O_2} = \dot{m}_{H_2}g_{H_2} + \dot{m}_{H_2}\left[\frac{(nM)_{O_2}}{(nM)_{H_2}}\right]g_{O_2}$$
$$= 5\times10^{-4}(h - Ts)_{H_2} + 5\times10^{-4}\left[\frac{(1\times32)}{(2\times2)}\right](h - Ts)_{O_2}$$

The Gibbs function for the reactants hydrogen and oxygen at 25°C are, because $h^o(25°C) = 0$ for H_2 and N_2,

$$(h - Ts)_{H_2} = -(273 + 25)\times s_{H_2}(25°C)$$
$$= -298(K)\times 130.58\ (kJ/kmol\cdot K)\times(1/2)\ (kmol/kg)$$
$$= -19{,}456\ kJ/kg$$

$$(h - Ts)_{O_2} = -(273 + 25)\times s_{O_2}(25°C)$$
$$= -298(K)\times 205.03\ (kJ/kmol\cdot K)\times(1/32)\ (kmol/kg)$$
$$= -1{,}909\ kJ/kg$$

and for the product water (taken as liquid),

$$\dot{m}_{H_2O}(h - Ts)_{H_2O} = 5\times10^{-4}\left[\frac{(2\times18)}{(2\times2)}\right]\times(h - Ts)_{H_2O}$$

$$(h - Ts)_{H_2O} = \Delta h_f - (273 + 25)\times s_{H_2O}(25°C)$$
$$= [-286{,}000 - 298(K)\times 69.96]\ (kJ/kmol\cdot K)$$
$$\times(1/18)\ (kmol/kg)$$
$$= -17{,}050\ kJ/kg$$

Equation (11.57) then gives

$$\dot{W}_{rev,\,use} = 5\times10^{-4}\times(-19{,}456) + 5\times10^{-4}\left[\frac{(1\times32)}{(2\times2)}\right]\times(-1{,}909)$$
$$- 5\times10^{-4}\left[\frac{(2\times18)}{(2\times2)}\right]\times(-17{,}050) = 59.4\ kW$$

and, from Eq. (11.55)

$$\dot{W}_{a,\,use} = \eta_{II,\,fc}\times\dot{W}_{rev,\,use} = 0.85\times59.4 = 50.5\ kW$$

This problem illustrates the use of the Gibbs function for determining the work output in thermodynamic devices.

Problems

11.1 One mole of ethene is burned with 70 percent excess air. What is the molecular weight of the mixture of the product gases (including H_2O vapor)?

11.2 A sample of the combustion products from a reaction yields the following volumetric analysis: $H_2 = 3$ percent, $CO = 27$ percent, $CO_2 = 10$ percent, and $N_2 = 60$ percent. Determine the composition of this mixture on a *mass* basis, and find R and c_v for this mixture.

11.3 An engine burns ethene with 200 percent theoretical air.
(*a*) How much excess air is used?
(*b*) Write the balanced combustion equation for 200 percent theoretical air.
(*c*) Find the air-fuel ratio on a mass basis.

11.4 An experimental combustion chamber uses ethane as the fuel. The ethane is combusted with 125 percent theoretical air.
(*a*) How much excess air is used?
(*b*) Write the balanced combustion equation for 125 percent theoretical air.
(*c*) Find the air-fuel ratio on a mass basis.

11.5 Combustion of a hydrocarbon of unkown composition resulted in the following equation (based on 100 mol of products):

$$C_aH_b + d(O_2 + 3.76N_2) \rightarrow 8.0CO_2 + 0.9CO + 8.8O_2 + eH_2O + 82.3N_2$$

(*a*) Determine the composition of the fuel.
(*b*) Determine the air-fuel ratio on a mass basis.
(*c*) Find the 100 percent theoretical air (you may neglect the CO contribution here).

11.6 A stove burns propane with 100 percent excess air.
(*a*) Balance the actual combustion equation.
(*b*) Determine the air-fuel ratio on a mass basis.

11.7 Determine the standardized molal enthalpy $\bar{h}$ for 1 mol of H_2O at 1 atm and 800 K.

11.8 Calculate the theoretical air-fuel ratio for the combustion of methane.

11.9 In a rocket engine complete combustion occurs when 2 mol of hydrogen combines with 1 mol of oxygen. The rocket carries 21 m^3 of O_2 at 100°C and 4 MPa. Determine the number of kilomoles of O_2 and the mass of H_2 required for complete combustion.

11.10 Liquid octane burns with 100 percent excess air. The air and octane are at 25°C and 100 kPa. The products of combustion are at 627°C and 100 kPa.
(*a*) How much theoretical air is used?
(*b*) Write the balanced combustion equation.
(*c*) Find the air-fuel ratio on a mass basis.
(*d*) Find H_r.
(*e*) Find H_p.
(*f*) Find the heat transfer released during the combustion.
(*g*) Find the dew point temperature of the H_2O.

11.11 A new engine receives its energy by burning ethene with 150 percent theoretical air. The fuel and air are at 25°C. The products of combustion are at 627°C and $P = 100$ kPa.
(*a*) How much excess air is used?
(*b*) Write the balanced combustion equation.
(*c*) Find the air-fuel ratio on a mass basis.
(*d*) Find H_p.
(*e*) Find H_r.
(*f*) Find the heat transfer released during the combustion.
(*g*) Find the dew point temperature of the H_2O.

11.12 The burning of garbage can be approximated by the combustion of methane (CH_4). Assume that the garbage burns with 50 percent excess air and that the garbage and air are at 100 kPa and 25°C. The products of combustion are at 100 kPa and 227°C.
(*a*) How much theoretical air is used?
(*b*) Write the balanced combustion equation.
(*c*) Find the air-fuel ratio on a mass basis.
(*d*) Find H_r.
(*e*) Find H_p.
(*f*) Find the heat transfer released during the combustion.
(*g*) Find the dew point temperature of the H_2O.

11.13 A cigarette lighter burns gaseous butane with 200 percent theoretical air. The fuel and air are at 25°C, and the products of combustion are at 127°C and $P = 100$ kPa.
(*a*) How much excess air is used?
(*b*) Write the balanced combustion equation.
(*c*) Find the air-fuel ratio on a mass basis.
(*d*) Find H_r.
(*e*) Find H_p.
(*f*) How much heat transfer is released during the combustion?
(*g*) Find the dew point temperature of the H_2O.

11.14 Using the enthalpy-of-formation data of Table C.9, calculate the enthalpy of combustion for methane (CH_4), ethane (C_2H_6), and propane (C_3H_8). Based on the differences between succeeding members of this hydrocarbon family, predict the value of the enthalpy of combustion of butane (C_4H_{10}). Check your prediction against the result in Table C.11.

11.15 Determine the standard enthalpy of combustion of (*a*) H_2, (*b*) CO, and (*c*) C_2H_5OH.

11.16S Plot the standard enthalpy of methane versus temperature in the range $25 < T < 200°C$. From these data and data for $\Delta\bar{g}_f^o$ from Table C.9, plus the Gibbs-Helmholtz relation, estimate the Gibbs function of combustion at 200°C.

11.16E Plot the standard enthalpy for methane versus temperature in the range $80 < T < 400°F$. From these data and data for $\Delta\bar{g}_f^o$ from Table C.9, plus the Gibbs-Helmoltz relation, estimate the Gibbs function of combustion at 400°F.

11.17 Derive a relationship between the lower and higher heating values of a general hydrocarbon fuel C_nH_m in terms of the mass fraction of hydrogen in the fuel and the heat of vaporization of water.

11.18S One mole of ethane is reacted stoichiometrically with air in an adiabatic constant-volume vessel. If the reactants are initially at 25°C and 1 atm, what final temperature is reached in the vessel? Compare your result with the result for the adiabatic flame temperature of ethane (Table C.11).

11.18E One mole of ethane is reacted stoichiometrically with air in an adiabatic constant-volume vessel. If the reactants are initially at 77°F and 1 atm, what final temperature is reached in the vessel? Compare your result with the result for the adiabatic flame temperature of ethane (Table C.11).

11.19 The gasohol sold by one service station is effectively 10 percent ethyl alcohol (C_2H_5OH) and 90 percent octane (C_8H_{18}) by volume. Calculate the enthalpy of combustion of this mixture, assuming that it is initially in gaseous form and that all combustion products are gaseous.

11.20S The materials often used for biomass conversion to methane have the effective chemical formulas and heating values listed here. For each, compute the amount of energy released by combustion of the resulting methane per kilogram of original biomass.

Compare the result with the energy released per kilogram of each material by direct combustion. Assume all combustion processes occur at standard conditions of 1 atm and 25°C and with no excess air. (Note that H_2 must often be added to obtain complete conversion to methane.)

Substance	Formula	HHV, kJ/kg
Wood chips (oak)	$C_{10}H_{24}O_{11}$	14,600
Cattle manure	$CH_{2.4} \cdot 4H_2O$	13,900
Bagasse (sugar cane processing by-product)	$C_4H_6O_5 \cdot 6H_2O$	9,300
Typical municipal waste	$CH_3 \cdot 3H_2O$	20,000
Chemical plant waste	$CH_{1.6}$	32,600

11.20E The materials often used for biomass conversion to methane have the effective chemical formulas and heating values listed here. For each, compute the amount of energy released by combustion of the resulting methane per pound-mass of original biomass. Compare the result with the energy released per pound-mass of each material by direct combustion. Assume all combustion processes occur at standard conditions of 1 atm and 77°F and with no excess air. (Note that H_2 must often be added to obtain complete combustion to methane.)

Substance	Formula	HHV, Btu/lbm
Wood chips (oak)	$C_{10}H_{24}O_{11}$	6,300
Cattle manure	$CH_{2.4} \cdot 4H_2O$	5,970
Bagasse (sugar cane processing by-product)	$C_4H_6O_5 \cdot 6H_2O$	4,000
Typical municipal waste	$CH_3 \cdot 3H_2O$	8,580
Chemical plant waste	$CH_{1.6}$	14,000

11.21 Set up the stoichiometric coefficient relations for the combustion reaction with X mol of excess air per mole of fuel in the form

$$C_nH_m + A_1\text{air} \rightarrow A_2CO_2 + A_3H_2O + A_4\text{air} + A_5N_2$$

and determine $A_1, A_2, \ldots, A_5$ in terms of m, n, and X.

11.22 Find the theoretical adiabatic flame temperature for the constant-pressure combustion of butane.

11.23S Find the temperature of the combustion products of butane when it is burned in an adiabatic steady flow combustor at 1 atm with 40 percent excess air preheated to 200°C.

11.23E Find the temperature of the combustion products of butane when it is burned in an adiabatic steady flow combustor at 1 atm with 40 percent excess air preheated to 400°F.

11.24S Plot the stack gas temperature leaving an adiabatic steady flow combustor versus excess air in the range 0 to 150 percent. The fuel is propane, and the fuel and air both enter the combustor at 1 atm and 25°C.

11.24E Plot the stack gas temperature leaving an adiabatic steady flow combustor versus excess air in the range 0 to 150 percent. The fuel is propane, and the fuel and air both enter the combustor at 1 atm and 77°F.

11.25S Plot the stack gas temperature versus air preheat temperature for an adiabatic steady flow combustor in which propane is burned with 20 percent excess air. The fuel and air both enter the combustor at 1 atm, and the fuel enters at 25°C. Air preheating temperature ranges from 25 to 400°C.

11.25E Plot the stack gas temperature versus air preheat temperature for an adiabatic steady flow combustor in which propane is burned with 20 percent excess air. The fuel and air both enter the combustor at 1 atm, and the fuel enters at 77°F. Air preheating temperatures range from 77 to 750°F.

11.26S A natural-gas-fired furnace is used to evaporate saturated water at 300 kPa. The natural gas has a volume flow rate of 10 m^3/s and is burned with 20 percent excess air. The air and gas enter the furnace at 25°C and 1 atm. If the combustion gases must leave the furnace at a temperature no lower than 300°C, what flow rate, in kilogams per second, of water can be evaporated in the furnace? (Assume that the natural gas has the same properties as methane.)

11.26E A natural-gas-fired furnace is used to evaporate saturated water at 50 psia. The natural gas has a volume flow rate of 300 ft^3/s and is burned with 20 percent excess air. The air and gas enter the furnace at 77°F and 1 atm. If the combustion gases must leave the furnace at a temperature no lower than 575°F, what flow rate, in pound-mass per second, of water can be evaporated in the furnace? (Assume that the natural gas has the same properties as methane).

11.27S A water heater in a farmhouse uses butane to heat the water. The butane burns with 50 percent excess air, and the air and butane enter the burner of the heater at 25°C and 1 atm. The flue gases leave the heater at 130°C. If the maximum butane flow rate to the burner when the burner runs continuously is 0.005 m^3/s, how much water, in kilograms per second, can be heated from 20 to 75°C?

11.27E A water heater in a farmhouse uses butane to heat the water. The butane burns with 50 percent excess air, and the air and butane enter the butane of the heater at 77°F and 1 atm. The flue gases leave the heater at 250°F. If the maximum butane flow rate to the burner when the burner runs continuously is 0.15 ft^3/s, how much water, in pound-mass per second, can be heated from 65 to 160°F?

11.28S The farmer who owns the farm in Prob. 11.27S decides to change bottle-gas dealers. The new supplier provides propane instead of butane. All conditions remain the same.

(*a*) How much water can be heated, in kilograms per second, by using propane?

(*b*) What ratio of costs per kilogram should the farmer expect to pay for the two fuels in relation to their energy content?

11.28E The farmer who owns the farm in Prob. 11.27E decides to change bottle-gas dealers. The new supplier provides propane instead of butane. All conditions remain the same.

(*a*) How much water can be heated, in pound-mass per second, by using propane?

(*b*) What ratio of costs per pound-mass should the farmer expect to pay for the two fuels in relation to their energy content?

11.29S Plot the entropy generation per mole of fuel versus temperature when carbon monoxide is burned adiabatically with the stoichiometric amount of air. Use temperatures from 600 to 2000°C. The reactants enter at 25°C and 1 atm. What do you conclude from this plot?

11.29E Plot the entropy generation per mole of fuel versus temperature when carbon monoxide is burned adiabatically with the stoichiometric amount of air. Use temperatures from 1000 to 3600°F. The reactants enter at 77°F and 1 atm. What do you conclude from this plot?

11.30S Calculate the reversible work per mole of fuel that is given by the combuston of carbon monoxide with stoichiometric air when the combustion products are mixed at 400°C and the reactants enter at 25°C and 1 atm. The surrounding temperature is 25°C.

11.30E Calculate the reversible work per mole of fuel that is given by the combustion of carbon monoxide with stoichiometric air when the combustion products are mixed at 750°F and the reactants enter at 77°F and 1 atm. The surrounding temperature is 77°F.

11.31S Jet *A* fuel has a standard enthalpy of formation of −22,570 kJ/kmol and an equivalent chemical formula of $CH_{1.9}$. In the combustor of an aircraft turbine engine, jet *A* is burned with 10 percent excess air. The air enters the combustion chamber at 1 atm and −50°C. The fuel enters at the standard reference state and is gaseous. What is the temperature of the gas leaving the combustor?

11.31E Jet *A* fuel has a standard enthalpy of formation of −21,390 Btu/lbmol and an equivalent chemical formula $CH_{1.9}$. In the combustor of an aircraft turbine engine, jet *A* is burned with 10 percent excess air. The air enters the combustion chamber at 1 atm and −60°F. The fuel enters at the standard reference state and is gaseous. What is the temperature of the gas leaving the combustor?

11.32 Coal can be gasified by heating and mixing with steam by the *water-shift* reaction

$$C + H_2O \Leftrightarrow CO + H_2$$

The resulting water gas ($CO + H_2$) was sold in many cities as illuminating gas. Assume that the water gas has equal volumes of CO and H_2.

(*a*) Find the higher heating value of water gas.
(*b*) Find the adiabatic flame temperature of water gas when it is burned with stoichiometric air.

11.33S An advanced supersonic transport (super SST, or SSST) has been proposed with a goal of engine efficiency = 0.55. The turbine pressure ratio will be 20, and the thrust-to-weight ratio is 8.0. The turbine inlet temperature of this design will be 1500°C. Assuming that jet *A* fuel ($CH_{1.9}$, heat of formation = −22,570 kJ/kmol) is used, how much excess air must be used to obtain this temperature? Assume that the inlet fuel is gaseous at the reference state and that inlet air is at −30°C and 30 kPa.

11.33E An advanced supersonic transport (super SST or SSST) has been proposed with a goal of engine efficiency = 0.55. The turbine pressure ratio will be 20, and the thrust-to-weight ratio is 8.0. The turbine inlet temperature of this design will be 2800°F. Assuming that jet *A* fuel ($CH_{1.9}$, heat of formation = −21,3890 Btu/lbmol) is used, how much excess air must be used to obtain this temperature? Assume that the inlet fuel is gaseous at the reference state and that inlet air is at −20°F and 5 psia.

11.34 Hydrogen gas is to be burned with 20 percent excess air in an adiabatic combustor. Both enter at standard reference conditions. What is the dew point of the combustion products?

11.35S Propane is to be burned with 10 percent excess air in an adiabatic furnace. The combustion air is to be preheated in a heat exchanger (air preheater) by heat transfer from the combustion products. The temperature of the combustion products is reduced to T_{out} in the preheater. The air enters the preheater at 25°C. See Fig. P11.35S.

What will be the temperature of the combustion products leaving the preheater? (The propane enters the furnace at the standard reference condition.) Discuss the effect of the presence of the preheater on your result. How would you find the temperature T of the air leaving the preheater?

11.35E Propane is to be burned with 10 percent excess air in an adiabatic furnace. The combustion air is to be preheated in a heat exchanger (air preheater) by heat transfer from the combustion products. The temperature of the combustion products is reduced to T_{out} in the preheater. The air enters the preheater at 77°F. See Fig. P11.35E.

What will be the temperature of the combustion products leaving the preheater? (The propane enters the furnace at the standard reference condition.) Discuss the effect of the presence of the preheater on your result. How would you find the temperature T of the air leaving the preheater?

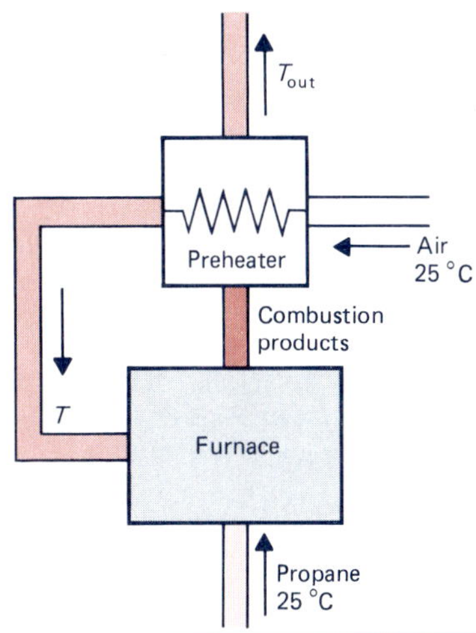

Figure P11.35S

Figure P11.35E

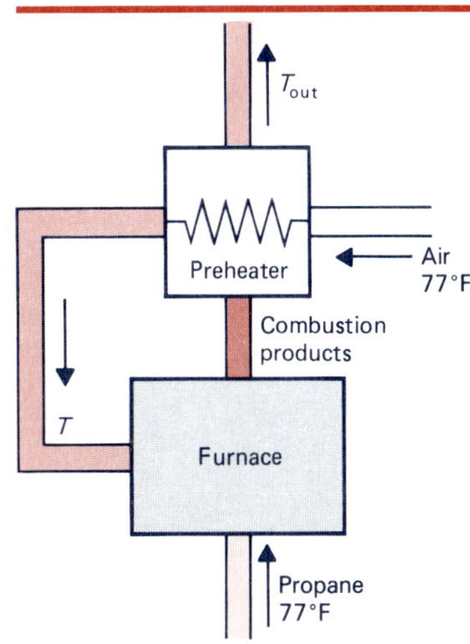

11.36S Butane and 30 percent excess air enter the combustor of a simple Brayton engine at 5 atm. The air enters at the temperature at which it left the compressor (efficiency 70 percent). The butane enters at standard reference temperature. After the combustion process, the hot combustion products enter the turbine (75 percent efficiency) and leave at 1 atm. Air enters the compressor at 25°C and 1 atm. See Fig. P11.36S.

What is the cycle efficiency? (Do not forget to do mass balances!)

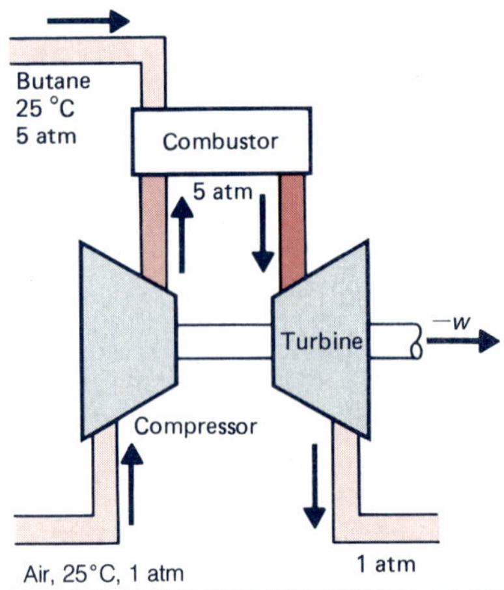

Figure P11.36S

11.36E Butane and 30 percent excess air enter the combustor of a simple Brayton engine at 5 atm. The air enters at the temperature at which it left the compressor (efficiency 70 percent). The butane enters at standard reference temperature. After the combustion process, the hot combustion products enter the turbine (75 percent efficiency) and leave at 1 atm. Air enters the compressor at 77°F and 1 atm. See Fig. P11.36E.

What is the cycle efficiency? (Do not forget to do mass balances!)

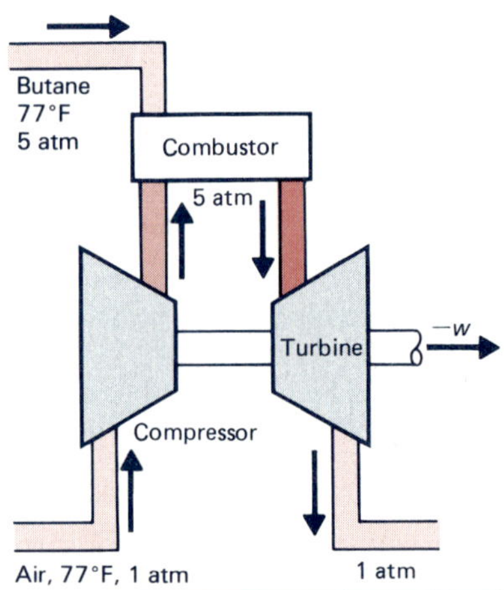

Figure P11.36E

11.37 For an adiabatic combustor operating with a given fuel and given amount of excess air, derive an expression for the second law efficiency of the furnace.

11.38S Methane is burned with 15 percent excess air, both of which enter at standard reference conditions. The combustion products are then mixed with 10 percent by volume of cold air at 25°C. See Fig. P11.38. What are the temperature and dew point of the resulting mixture?

11.38E Methane is burned with 15 percent excess air, both of which enter at standard reference conditions. The combustion products are then mixed with 10 percent by volume of cold air at 77°F. See Fig. P11.38. What are the temperature and dew point of the resulting mixture?

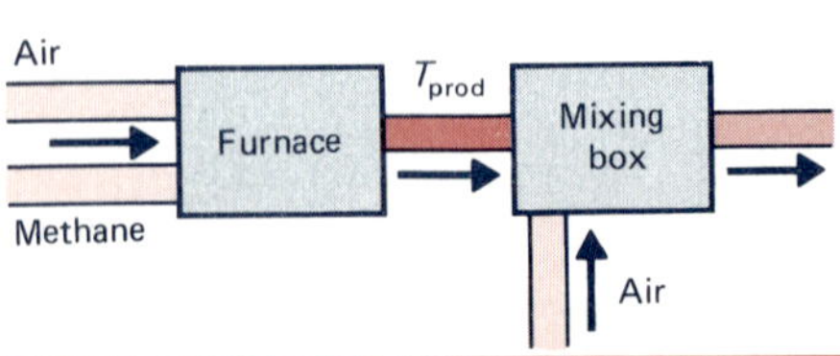

Figure P11.38

References

1. Joseph H. Keenan, Jing Chao, and Joseph Kaye, *Gas Tables: International Version,* 2d ed., Wiley, New York, 1983.
2. Joseph H. Keenan, J. Chao, and Joseph Kaye, *Gas Tables* (*English Units*), 2d ed., Wiley, New York, 1980.
3. GASPROPS, "A Computer Program for the Calculation of the Thermodynamic Properties of Gases and Combustion Products," Wiley, New York, 1984.
4. Henry C. Barnett and Robert R. Hibbard, "Basic Considerations in the Combustion of Hydrocarbon Fuels with Air," NACA Rept. 1300, 1959.

12

Phase and Chemical Equilibrium

Human subtlety . . . will never devise an invention more beautiful, more simple, or more direct than does nature, because in her inventions nothing is lacking and nothing is superfluous.
Leonardo da Vinci

Four-stroke turbocharged V-16 diesel engine for marine propulsion service. (*General Electric.*)

12.1 Introduction

In this chapter, we investigate the behavior of systems in which either mass is transferred between two or more phases during a change of state or an equilibrium chemical reaction occurs during a change of state. An equilibrium chemical reaction is a reaction that is *reversible,* unlike the combustion reactions studied in Chap. 11, where the reactions were assumed to proceed to completion and either all the fuel or all the oxidant was consumed. In an equilibrium reaction, some of the products and some of the reactants are present at all states, and their relative concentrations change as the state of the system is changed.

In systems involving phase change or chemical reaction, additional variables are necessary to specify the system state; simply specifying two independent properties is no longer sufficient.

12.1.1 Chemical Equilibrium

In simple systems (i.e., systems with a single chemical species), we have found that the state of the system is defined in terms of any two independent properties. When a system has more than one chemical species that do not undergo a chemical reaction with one another, then it is necessary to specify one or more additional properties, the concentrations of the components. If the chemical species can react chemically with one another, the system state cannot be specified until the chemical reaction reaches a steady condition in which the concentrations of the species present are not changing. We can define *chemical equilibrium* as the condition when *all chemical reactions have reached equilibrium so that the concentration of all chemical components is unchanging.*

Specification of the state of a system that contains chemically reacting species as being in thermodynamic equilibrium requires that three types of equilibrium be present: *thermal equilibrium* and *mechanical equilibrium,* which we have discussed in previous chapters, as well as *chemical equilibrium.*

In Chap. 5, entropy is discussed as an extensive equilibrium property, and it is shown that for an isolated system, the entropy could only remain constant or increase. The focus was limited to a simple compressible substance. In this chapter, we expand the treatment of equilibrium systems to include the discussion of single-component and multicomponent systems in which more than one phase is present and to systems in which chemical reactions can occur. Much of the groundwork for this chapter has already been laid—the ideas of thermal and mechanical equilibrium in Chap. 5, the introduction of partial molal properties in Chap. 10, and the use of standard reference properties in Chap. 11.

The Gibbs function for simple systems $g = h - Ts$ has some limited usefulness in problems involving single components and for some problems involving mixtures of nonideal gases. However, the Gibbs function enjoys a central place in the discussion of systems involving phase and chemical equilibrium, as does the Helmholtz function $a = u - Ts$. The Gibbs function appears in formulations involving constant-pressure processes, while the Helmholtz function usually is found in descriptions of constant-volume processes. Because many practical engineering problems occur at constant pressure, we begin with a review of the properties of the Gibbs function.

12.1.2 The Gibbs-Duhem Relation

Let us assume that the Gibbs function is dependent on temperature, pressure, and the concentrations of the various chemical species present in the system. Thus, the independent variables are T, P, n_A, n_B, n_C, ... Remember that $g(T, P)$ is a fundamental equation, so that if it is known, then all other thermodynamic properties may be derived at any state. We can then write the total differential [following Eq. (10.101)] as

$$dG = \left(\frac{\partial G}{\partial T}\right)_{P,\, n_A,\, n_B,\, n_C,\, \ldots} dT + \left(\frac{\partial G}{\partial P}\right)_{T,\, n_A,\, n_B,\, n_C,\, \ldots} dP + \left(\frac{\partial G}{\partial n_A}\right)_{T,\, P,\, n_B,\, n_C,\, \ldots} dn_A + \left(\frac{\partial G}{\partial n_B}\right)_{T,\, P,\, n_A,\, n_C,\, \ldots} dn_B + \cdots \tag{12.1}$$

Note that $\underline{(\partial G/\partial n_i)_{T,\, P,\, n_{j(j \neq i)}}}$ is defined in Sec. 10.9 as the *partial molal Gibbs function* $\overline{G}_A$. For a process at constant pressure and temperature, Eq. (12.1) can be expressed as

$$dG_{T,\, P} = \overline{G}_A\, dn_A + \overline{G}_B\, dn_B + \overline{G}_C\, dn_C + \cdots \tag{12.2}$$

Commonly the partial molal Gibbs function is called the *electrochemical potential,* or simply the *chemical potential,* and is given the symbol $\bar{\mu}$ and defined as

$$\left(\frac{\partial G}{\partial n_i}\right)_{T,\, P,\, n_A,\, \ldots,\, n_{i-1},\, n_{i+1},\, \ldots} \equiv \bar{\mu}_i \tag{12.3}$$

so that Eq. (12.2) becomes

$$dG_{T,\, P} = \bar{\mu}_A\, dn_A + \bar{\mu}_B\, dn_B + \bar{\mu}_C\, dn_C + \cdots \tag{12.4a}$$

or

$$d\bar{g}_{T,\, P} = \bar{\mu}_A\, dy_A + \bar{\mu}_B\, dy_B + \cdots \tag{12.4b}$$

As we discuss in the next section, the chemical potential is the driving potential for the molar transfer of components. Equation (12.4) can be integrated (if we remember that T and P are held constant) by using a particular approach. First, observe that $\bar{\mu}_i$ is an *intensive* property that depends on the *relative* amount of each species present. If the dn_i are changed in such a way that the mole fractions y_i are unchanged (by adding or withdrawing mass from the phase under consideration), then $\bar{\mu}_i$ is held constant and Eq. (12.4*a*) can be integrated to give

$$G_{T,P} = \bar{\mu}_A n_A + \bar{\mu}_B n_B + \bar{\mu}_C n_C + \cdots \tag{12.5}$$

This is an application of the general argument presented in Eqs. (10.101) through (10.107). This results for a single phase with multiple components of fixed composition at any constant temperature and pressure, and it relates the Gibbs function of a solution or mixture to the chemical potentials and amounts of each component in the phase. Now the total differential of Eq. (12.5) for a system with varying composition and varying chemical potential at constant temperature and pressure is

$$dG_{T,P} = \bar{\mu}_A\, dn_A + \bar{\mu}_B\, dn_B + \bar{\mu}_C\, dn_C + \cdots + n_A\, d\bar{\mu}_A + n_B\, d\bar{\mu}_B + n_C\, d\bar{\mu}_C + \cdots \tag{12.6}$$

Comparing Eq. (12.6) with Eq. (12.4), both of which are general relations for a constant-T and constant-P process, indicates that

$$n_A\, d\bar{\mu}_A + n_B\, d\bar{\mu}_B + n_C\, d\bar{\mu}_C + \cdots = 0 \tag{12.7}$$

This is an important relation, as we shall see. It is called the *Gibbs-Duhem equation,* and it applies to any given phase in a multicomponent, multiphase system.

12.2 Equilibrium in Nonreacting Systems

Let us examine a system in which two or more components exist in two or more phases. If we limit discussion to a control mass maintained at constant temperature and pressure with no changes in kinetic or potential energy, then the first law for the system is

$$dU = \delta Q - \delta W \tag{12.8}$$

If the system exchanges energy with the surroundings at T_0 only, then Eq. (5.74) can be substituted to give the inequality

$$T_0\, dS - \delta W - dU \geq \delta Q - \delta W - dU = 0 \tag{12.9}$$

where the inequality becomes an equality only when the control mass undergoes reversible heat transfer. Further, if the only work mode present is by boundary movement, Eq. (12.9) becomes

$$T_0\, dS - P\, dV - dU \geq 0 \tag{12.10}$$

For constant temperature and pressure, the differential of the Gibbs function $G = H - TS$ is

$$dG_{T,P} = dH - T\,dS = d(U + PV)_{T,P} - T\,dS = dU + P\,dV - T\,dS \tag{12.11}$$

and substituting Eq. (12.10), we find

$$dG_{T,P} + (T - T_0)\,dS \leq 0 \tag{12.12}$$

12.2.1 Isolated Systems

For an isolated system in thermodynamic equilibrium, $dS = 0$, so that Eq. (12.12) becomes, by using Eq. (12.4),

$$\begin{aligned} dG_{T,P} &= \sum_j \left[\sum_i (\bar{\mu}_i\,dn_i) \right]_j \\ &= (\bar{\mu}_A\,dn_A + \bar{\mu}_B\,dn_B + \cdots)_a + (\bar{\mu}_A\,dn_A + \bar{\mu}_B\,dn_B + \cdots)_b \\ &\quad + (\cdots)_c + \cdots \leq 0 \end{aligned} \tag{12.13}$$

where the sum on i indicates the sum over each component in phase j and the sum on j is the sum over the phases. Again, in thermodynamic equilibrium the equality holds.

An important identity for the chemical potential is also obtained from Eq. (12.5). A single component existing in a single phase eliminates all the terms in Eq. (12.5) except one, so

$$\bar{\mu} = \bar{g} \tag{12.14}$$

Thus, the chemical potential for a single component in one phase is equal to the molar Gibbs function.

12.2.2 Single-Component Phase Equilibrium

Let us now examine a two-phase system with a single component (the subscript i is therefore eliminated). For such a system, the total number of moles of material is constant. So if mass is transferred from phase a to phase b, it follows that

$$d(n_a + n_b) = 0 \tag{12.15}$$

or

$$dn_a = -dn_b \tag{12.16}$$

Equation (12.13) then gives, for thermodynamic equilibrium,

$$dG_{T,P} = \bar{\mu}_a\,dn_a - \bar{\mu}_b\,dn_a = 0 \tag{12.17}$$

or

$$\bar{G}_a = \bar{\mu}_a = \bar{\mu}_b = \bar{G}_b \tag{12.18}$$

Thus, the chemical potential (partial molal Gibbs function) in both phases of a single-pure-component system in thermodynamic equilibrium must be the same. This argument can be extended to multiple components in multiple phases, with the conclusion that *the chemical potential of a given component is the same in every phase.*

An important physical interpretation of the chemical potential is obtained from Eq. (12.13). For a single-component two-phase system, Eq. (12.13) yields

$$(\bar{\mu}_a - \bar{\mu}_b)\, dn_a \le 0 \tag{12.19}$$

The inequality implies that if $\bar{\mu}_a > \bar{\mu}_b$, then $dn_a < 0$, or the mass in phase a tries to move away from a higher chemical potential. Alternatively, if $\bar{\mu}_b > \bar{\mu}_a$, then $dn_a > 0$, which again indicates the motion from a higher to lower chemical potential. These arguments presenting the chemical potential as the driving force for mass transfer should be compared to the presentation of temperature and pressure as the driving forces for heat transfer and boundary movement work, respectively, given in Sec. 5.4.

Now rewrite Eq. (12.17) with Eq. (12.14) as

$$\begin{aligned} dG_{T,P} &= \sum_j dG_j = \bar{g}_a\, dn_a + \bar{g}_b\, dn_b \\ &= (\bar{h} - T\bar{s})_a\, dn_a + (\bar{h} - T\bar{s})_b\, dn_b = 0 \end{aligned} \tag{12.20}$$

which, after substituting Eq. (12.16), results in

$$\bar{s}_a - \bar{s}_b = \frac{\bar{h}_a - \bar{h}_b}{T} \tag{12.21}$$

If Eq. (9.11c) is substituted into Eq. (12.21) and we remember that all the development in this section is for an isothermal and isobaric process, the result is

$$\left(\frac{dP}{dT}\right)_{\text{phase change}} = \frac{\bar{h}_{ab}}{T\bar{v}_{ab}} = \frac{h_{ab}}{Tv_{ab}} \tag{12.22}$$

which is the Clapeyron equation [Eq. (9.14)], derived in a different way in Chap. 9.

Example Problem 12.1

Ammonia exists in the saturated state at a pressure of 150 psia. Find the values of the chemical potentials of the liquid and vapor phase by the use of Table E.16 and compare them. Also, using Eq. (12.22), derive a value for h_{fg} from P-v-T data for ammonia from the same table, and compare the result to the tabulated value.

Solution

The chemical potential is given by

$$\mu = \bar{g} = \bar{h} - T\bar{s}$$

Using data for the saturated liquid and the saturated vapor from Table E.16 results in

$$\bar{\mu}_l = M(h - T_{sat}s)_l = 17.031[132.01 - (459.67 + 78.8)(0.2755)]$$
$$= -278.3 \text{ Btu/lbmol}$$

and

$$\bar{\mu}_g = M(h - T_{sat}s)_g = 17.031[631.08 - (459.67 + 78.8)(1.2023)]$$
$$= -278.0 \text{ Btu/lbmol}$$

The difference in the chemical potentials of the saturated liquid and saturated vapor, which is about 0.1 percent, is attributable to inaccuracies in the tabulated data.

Rewriting Eq. (12.22) for the case of a saturated liquid–vapor mixture gives

$$h_{lg} = T_{sat}v_{lg}\left(\frac{\Delta P}{\Delta T}\right)_{sat}$$

where the derivative $(dP/dT)_{sat}$, the slope of the saturation curve, has been replaced by a finite difference. Taking the two table entries on either side of $P = 150$ psia to evaluate the derivative and the values at $P = 150$ psia to obtain the other values gives

$$h_{lg} = [(459.67 + 78.8)°\text{R}](1.9657 \text{ ft}^3/\text{lbm})\left(\frac{160 - 140}{82.6 - 74.8}\right)\frac{\text{psia}}{°\text{R}}$$
$$= (2714.0 \text{ psia ft}^3/\text{lbm})(144 \text{ in}^2/\text{ft}^2)$$
$$= 390{,}820 \text{ ft}\cdot\text{lbf/lbm} = 502.2 \text{ Btu/lbm}$$

This value compares with the tabulated value of 499.07 Btu/lbm. Note that the calculated value exhibits much less accuracy than is implied by the number of significant figures shown because of the small temperature difference in the denominator, which has only two significant figures.

Comments

As expected, the chemical potentials of the two phases are equal, as predicted by Eq. (12.18), and the Clapeyron equation, Eq. (12.22), predicts the change in enthalpy between the two phases within the accuracy of the *P-v-T* data.

12.2.3 Ideal Solutions

Ideal solutions are introduced in Chap. 10, where they are described as mixtures of two or more components in which the total volume of the solution equals the sum of the volumes of the components of the solution; i.e., the partial molal volumes equal the specific molar volumes. We now extend the idea of an ideal solution to other properties, and we reach some useful conclusions about the equilibrium properties of ideal solutions.

Suppose that we consider a general liquid solution of two components that is in equilibrium with the vapors of the components, with the entire control mass held at constant temperature and pressure. Remember that the chemical potential of each component has the same value in the liquid and vapor phases. For this case, if there are $n_{A,\,l}$ and $n_{B,\,l}$ moles of the two components in the liquid phase, then the Gibbs-Duhem equation, Eq. (12.7), can be divided by the total number of moles in the liquid $(n_A + n_B)_l$ and then can be written for the liquid phase (denoted by a subscript l) as

$$(y_A\, d\bar{\mu}_A + y_B\, d\bar{\mu}_B)_l = 0 \tag{12.23}$$

The change in chemical potential of component i depends on only the amount of component i present (if T and P are constant), so that

$$d\bar{\mu}_i = \left(\frac{\partial \bar{\mu}_i}{\partial y_i}\right)_{T,\,P} dy_i \tag{12.24}$$

Noting that $y_A + y_B = 1$ and substituting Eq. (12.24), we see that Eq. (12.23) becomes

$$\left\{\left[\frac{\partial \bar{\mu}_A}{\partial(\ln y_A)}\right]_{T,\,P} = \left[\frac{\partial \bar{\mu}_B}{\partial(\ln y_B)}\right]_{T,\,P}\right\}_l \tag{12.25}$$

Let us now consider the question of how to provide numerical values for the chemical potential, since we have derived in Eq. (12.25) a means of relating the chemical potentials of two components in a liquid solution. In Chap. 9, we define the fugacity f of a gas or vapor by the relation [Eq. (9.71)]

$$d\bar{g}_T = \bar{R}T\, d(\ln f) \tag{12.26}$$

However, from the definition of the chemical potential as the partial molal Gibbs function, it also follows that the value of $\bar{\mu}_i$ for any component of a vapor or gas mixture at constant temperature and pressure is

$$d\bar{\mu}_i = (d\bar{g}_i)_{T,\,P} = \bar{R}T\, d(\ln f_i) \tag{12.27}$$

Since the chemical potential of component i is the same in both phases, it also follows that Eq. (12.27) can be used to determine the chemical potential of each component of the liquid. In addition, it follows that the fugacity of component i must be the same in every phase!

Substituting Eq. (12.27) into Eq. (12.25) results in

$$\left\{\left[\frac{\partial(\ln f_A)}{\partial(\ln y_A)}\right]_{T,\,P} = \left[\frac{\partial(\ln f_B)}{\partial(\ln y_B)}\right]_{T,\,P}\right\}_l \tag{12.28}$$

which is called *Duhem-Margules equation.* We have made no assumptions as to ideality in its derivation, except that it applies for changes in composition that occur at constant temperature and pressure. If, however, the vapor phase is in a state where the vapor mixture can be treated as an ideal gas mixture, then, from Chap. 9, $f \rightarrow P$, and Eq. (12.28) can be revised to

$$\left\{\left[\frac{\partial(\ln P_A)}{\partial(\ln y_A)}\right]_{T,P} = \left[\frac{\partial(\ln P_B)}{\partial(\ln y_B)}\right]_{T,P}\right\}_l \tag{12.29}$$

where P_A and P_B are the vapor pressures of the two constituents that are in equilibrium with the liquid, which has mole fractions y_A and y_B.

We can now define and *ideal liquid solution* as one which obeys the relation

$$f_i^* \equiv y_{i,l} f_i \tag{12.30a}$$

or, at low pressures,

$$P_i = y_{i,l} P_{\text{sat},i}(T) \tag{12.30b}$$

where P_i and f_i^* are the vapor pressure and vapor-phase fugacity of component i in the solution, respectively; $y_{i,l}$ is the mole fraction of component i in the solution; and $P_{\text{sat},i}(T)$ is the saturation pressure of component i at the control mass temperature. Equation (12.30*a*) is called the *Lewis-Randall rule.* If the vapor acts as an ideal gas mixture, then Dalton's law can be used to give

$$P_i = y_{i,g} P_{\text{tot}} = y_{i,l} P_{\text{sat},i}(T) \tag{12.31}$$

This relation is called *Raoult's law.*

If Raoult's law holds for either component of a two-component ideal solution, then Eq. (12.29) shows that it must also hold for the second component. Although we have limited the discussion to two-component mixtures, the results are easily extended to multiple components. We can now prescribe all the thermodynamic properties of an ideal liquid solution and the vapor in equilibrium with the solution.

Example Problem 12.2

An ideal liquid solution is composed of 80 mol % ammonia and 20 mol % water. What are the mole percentages of the ammonia and water in the vapor that is in equilibrium with the liquid solution if the temperature of the control mass is 20°C?

Solution

The mole fractions in the liquid are given. According to Raoult's law, the partial pressures of the vapor are given by

$$P_{H_2O} = 0.2P_{sat,\,H_2O}(T = 20°C) = 0.2(2.34) = 0.468 \text{ kPa}$$

$$P_{NH_3} = 0.8P_{sat,\,NH_3}(T = 20°C) = 0.8(857.6) = 686.1 \text{ kPa}$$

and the total pressure of the vapor is 686.6 kPa. The mole fractions in the vapor are then

$$y_{g,\,H_2O} = \frac{P_{H_2O}}{P_{tot}} = \frac{0.468}{686.6} = 6.82 \times 10^{-4}$$

$$y_{g,\,NH_3} = 1 - y_{g,\,H_2O} = 0.9993$$

Given the mole fractions in each phase, all the thermodynamic properties of the ideal mixtures can be computed by using the relations in Chap. 10.

Comments

Note how much larger the mole fraction of the volatile component (ammonia) is in the vapor phase than in the liquid phase. If the mole fraction of component A in the vapor is plotted against the mole fraction of A in the liquid, generally the relation is not linear even for an ideal solution. Note that the nonideality of a mixture can be treated more fully by assuming Raoult's law to be of the general form

$$f_i^* = y_{i,\,l} f_{sat,\,i}(T)$$

which allows use of real gas properties in the vapor phase. However, for this problem, examination of the reduced-property charts shows that $f/P = 1$, so the use of the ideal gas form of Raoult's law is justified.

12.2.4 Phase Rule

We have limited discussion so far to one- and two-component two-phase mixtures. In the single-component two-phase case, we found it sufficient to specify only one intensive property (usually T or P) plus the amount of each phase to completely describe the state of the system. For a two-component system with two phases, it was necessary to know the values of two intensive variables, normally T and P, plus the concentration (usually mole fraction) of *one* of the components in each phase. The concentration of the second component is then known, since the mole fractions in each phase must sum to unity. How many independent intensive properties N_{pr} must be specified to determine the state of a system of N_{ph} phases made up of N_c components in thermodynamic equilibrium? In general, the number that must be specified is given by

$$N_{pr} = N_c - N_{ph} + 2 \tag{12.32}$$

which is called the *Gibbs phase rule.* The number 2 results from the need to specify two independent properties for a single-phase simple compressible substance. There is a need to specify $N_c - 1$ mole fractions (there is 1 less than N_c since the sum of all mole fractions must equal 1) for each of the N_{ph} phases or $N_{ph}(N_c - 1)$ properties. However, there are $N_{ph} - 1$ independent equations of the form of Eq. (12.18) that equate the chemical potential equalities for each component, or $N_c(N_{ph} - 1)$ equations. Thus, this gives $N_{ph}(N_c - 1) - N_c(N_{ph} - 1) = N_c - N_{ph}$ other variables required, as indicated in Eq. (12.32).

For a single-component, single-phase control mass ($N_c = 1$ and $N_{ph} = 1$), two independent intensive properties must be known to specify the state of an equilibrium system. For a single-component, two-phase system, only one intensive variable needs to be specified. For a single-component, three-phase system, *no* independent intensive properties can be specified. These requirements are in accord with our observations on simple compressible substances. The three-phase system for water, for example (Fig. 12.1), corresponds to the behavior at the triple point, where only one pressure and one temperature correspond to the equilibrium three-phase mixture, and these cannot be independently specified. Along

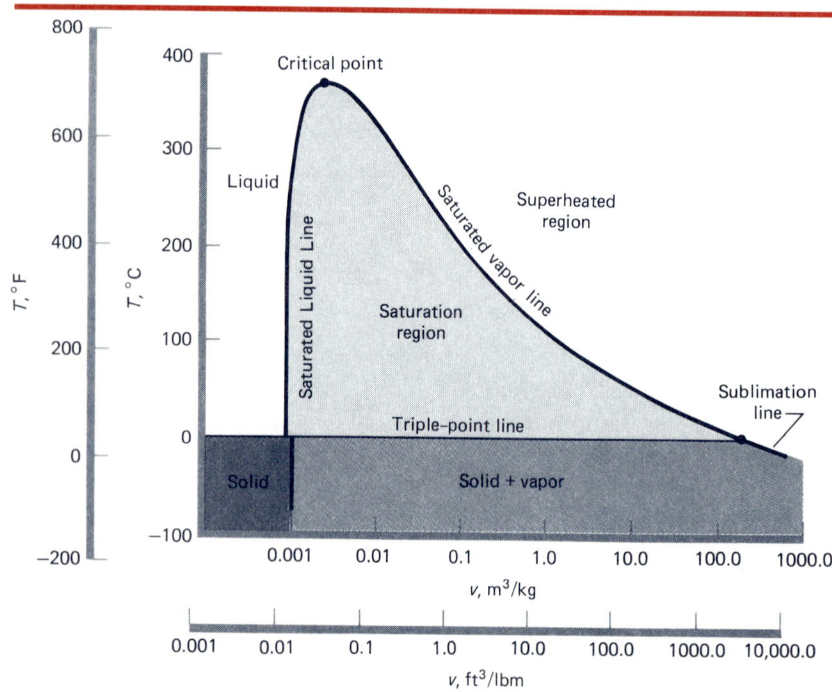

Figure 12.1 *P-T* **diagram for water.**

the saturation line, either P or T may be independently specified to define a saturated state. In the liquid or vapor regions, *both* P and T must be specified.

For a two-component, two-phase system, we apparently must set two intensive properties to define the system. We have done this in Example Problem 12.2, for example, by fixing the system temperature and the mole fraction of one component of the system.

The phase rule tells us nothing about the relative amounts of each phase present; it tells us only that if we specify a certain number of intensive properties, the state of the system will be defined. For steam in a saturated state, then, if either P or T is specified and two phases are present, the state will be an equilibrium state for *any* proportions of liquid and vapor; but those proportions are not fixed by the phase rule.

12.3 Equilibrium in Systems with Chemical Reaction

Let us now extend our considerations of equilibrium states to systems in which chemical reactions occur. Many chemical reactions proceed to completion; i.e., the reactants undergo complete reaction and are consumed in the production of reaction products. This was the case for all the combustion reactions studied in Chap 11. In other cases, however, the reaction may not proceed to completion because the products themselves react to re-form into the original reactants. An equilibrium mixture of products and reactants will result. In this section, we examine such situations, limiting our discussion to *homogeneous* reactions in which all the reaction takes place in the gas phase.

Examine a general chemical reaction in which elements or compounds A and B react to form elements or compounds C and D. (The discussion is easily extended to more or fewer reactants and products.) The general reaction equation is

$$\nu_A A + \nu_B B \leftrightarrow \nu_C C + \nu_D D \tag{12.33}$$

where the ν's are the stoichiometric coefficients.

The rate at which the number of moles of A, n_A, is being depleted because of reaction with B at any time depends on the concentrations of A and B; that is, if either A or B is missing, then no reaction can occur. If we denote the concentration of A in arbitrary units by $[A]$, then the rate of depletion of A mol/s can be written as

$$\left(\frac{d[A]}{dt}\right)^+ = -k_1[A]^{\nu_A}[B]^{\nu_B} \tag{12.34}$$

where k_1 is a proportionality constant, called the *rate constant*, that describes the rate of the particular reaction. The stoichiometric coefficients appear as exponents because the reaction rate must depend on how many molecules (moles) of B are required to react with ν_A molecules (moles) of A. If, for example, $\nu_A = 1$ and $\nu_B = 2$, then the reaction rate must depend on $[B]^2$, because 2 molecules of B must be present for each A molecule. The reaction rate for this case is

$$\left(\frac{d[A]}{dt}\right)^+ = -k_1[A][B][B] = -k_1[A][B]^2 \tag{12.35}$$

The value of k_1 depends not only on the particular reaction but also on the units chosen for expressing concentration (for example, partial pressure, moles per liter, etc). The minus sign appears in Eq. (12.34) because the concentration of A must decrease as the reaction proceeds, so $(d[A]/dt)^+$ must be negative.

For many of the reactions described by Eq. (12.33), products C and D will themselves react to form A and B. For such a case, the rate at which A is formed by this *reverse reaction* is given by reasoning similar to that used for Eq. (12.34) as

$$\left(\frac{d[A]}{dt}\right)^- = k_2[C]^{\nu_C}[D]^{\nu_D} \tag{12.36}$$

When equilibrium is reached, the rates of depletion $(d[A]/dt)^+$ and of formation $(d[A]/dt)^-$ of A must sum to zero. Summing Eqs. (12.34) and (12.36) results in

$$K = \frac{k_1}{k_2} = \frac{[C]^{\nu_C}[D]^{\nu_D}}{[A]^{\nu_A}[B]^{\nu_B}} \tag{12.37}$$

where K is a new constant, called the *equilibrium constant* for the reaction. We see that K is related to the equilibrium concentrations of the products, $[C]$ and $[D]$, and of the reactants, $[A]$ and $[B]$. The value of K also depends on the units chosen for concentration (unless $\nu_C + \nu_D$ happens to be equal to $\nu_A + \nu_B$, in which case K is dimensionless for any concentration units).

Because the individual rate constants k_1 and k_2 are often not known for a given reaction, K must usually be found by other means than Eq. (12.37). We now use the information previously developed about equilibrium to find values for K for a given reaction.

12.3.1 Finding the Equilibrium Constant

If the reaction of Eq. (12.33) is not complete and equilibrium final concentrations are reached, then we can see what happens if the reaction proceeds to an extent α, where $\alpha = 0$ indicates only reactants are present and $\alpha = 1$ indicates a reaction that proceeds to completion. For an arbitrary α, we are left with a mixture of reactants and products that has the following composition:

Component	Final Number of Moles
A	$\nu_A(1-\alpha)$
B	$\nu_B(1-\alpha)$
C	$\alpha\nu_C$
D	$\alpha\nu_D$

The *change* in the number of moles of each component is then $dn_A = -\nu_A\, d\alpha$, $dn_B = -\nu_B\, d\alpha$, $dn_C = \nu_C\, d\alpha$, and $dn_D = \nu_D\, d\alpha$, which shows that

$$d\alpha = \frac{-dn_A}{\nu_A} = \frac{-dn_B}{\nu_B} = \frac{dn_C}{\nu_C} = \frac{dn_D}{\nu_D} \tag{12.38}$$

Given the value of α, then, the final number of moles of each component in the equilibrium chemical reaction is known from the table.

For stoichiometric initial concentrations of reactants, the number of moles of reactant remaining at any degree of reaction α is given by integrating Eq. (12.38) to obtain

$$\frac{n_i}{\nu_i} = 1 - \alpha \tag{12.39a}$$

and for the products

$$\frac{n_i}{\nu_i} = \alpha \tag{12.39b}$$

The fraction of the initial reactants and the resulting fraction of possible products are then linear with the degree of reaction α, and a plot of Eq. (12.39) is shown in Fig. 12.2. Our immediate problem is to find the value of α for a given chemical reaction at a given temperature and pressure, so that the concentrations of all components may be determined.

Figure 12.2 Concentrations of products and reactants as a function of degree of reaction.

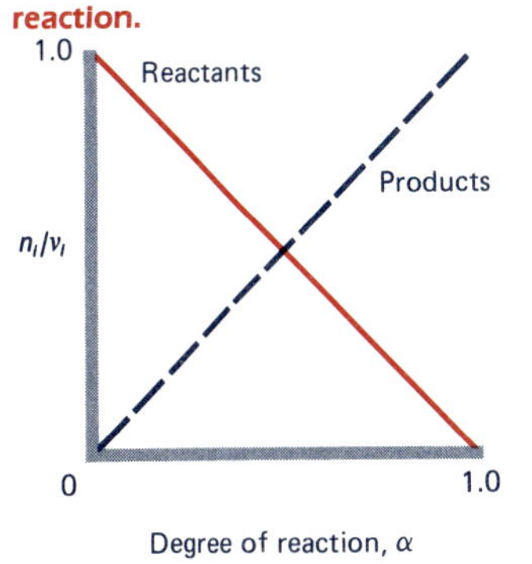

At the condition of chemical equilibrium, we can substitute Eq. (12.38) into the relation between the Gibbs function and the chemical potential of each component, as given by Eq. (12.4), to find

$$dG_{T,P} = (-\bar{\mu}_A\nu_A - \bar{\mu}_B\nu_B + \bar{\mu}_C\nu_C + \bar{\mu}_D\nu_D)\, d\alpha \tag{12.40}$$

We now must find a relation for the chemical potential of a component of the system after the reaction comes to equilibrium at a given temperature and pressure. Equation (12.27) gives the differential of the chemical potential of a component of a mixture. To define a *standard* chemical potential, we must define a standard reference state. *For the equilibrium mixture, the standard reference state is taken to be an ideal solution with the same mole fraction of component i as in the equilibrium mixture.* The ideal reference solution is assumed to be *at 1-atm total pressure and at the equilibrium mixture temperature.* Let us integrate Eq. (12.27) at constant temperature from the ideal solution reference state to the equilibrium mixture state:

$$\int_{\bar{\mu}_i^o}^{\bar{\mu}_i} (d\bar{\mu}_i)_T = \bar{\mu}_i(T) - \bar{\mu}_i^o(T) = \int_{f_i^{*o}}^{f_i^*} \bar{R}T\, d(\ln f_i^*)_T$$

$$= \bar{R}T \ln \frac{f_i^*}{f_i^{*o}} \tag{12.41}$$

Or, since $f_i^{*o} = y_i f_i^o$ in an ideal solution, where f_i^o is the fugacity of the pure component at 1 atm and T, this becomes

$$\bar{\mu}_i(T) = \bar{\mu}_i^o(T) + \bar{R}T \ln \frac{f_i^*}{y_i f_i^o}$$

$$= \bar{\mu}_i^o(T) - \bar{R}T \ln y_i + \bar{R}T \ln \frac{f_i^*}{f_i^o} \tag{12.42}$$

Now, remembering from Chap. 10 [for example, Eqs. (10.106) and (10.107)] that the general thermodynamic relations among properties can also be expressed for partial molal properties, we can write

$$\bar{\mu}_i^o(T) - \bar{R}T \ln y_i = (\bar{G}_{T,P}^o)_i - \bar{R}T \ln y_i$$

$$= (\bar{H}^o - T\bar{S}^o)_{T,P,i} - \bar{R}T \ln y_i \tag{12.43}$$

For an ideal solution, which was taken as the reference state, the partial molal properties are equal to the actual properties of the component, so $\bar{H}_i^o = \bar{h}_i^o$ and $\bar{S}_i^o + \bar{R} \ln y_i = \bar{s}_i^o$. Equation (12.42) then finally becomes

$$\bar{\mu}_i(T) = \bar{h}_i^o(T) - T\bar{s}_i^o(T) + \bar{R}T \ln \frac{f_i^*}{f_i^o}$$

$$= \bar{g}_i^o(T) + \bar{R}T \ln \frac{f_i^*}{f_i^o} \tag{12.44}$$

which is the equation for the standard chemical potential, with a reference state taken as an ideal solution at 1 atm and T.

In Chap. 11, Eq. (11.47) is given for the standard Gibbs function per mole of fuel for a combustion reaction that goes to completion while at standard conditions of 25°C = 77°F and 1 atm. Here it is convenient to define a slightly different although closely related function which we denote $\Delta\bar{G}^o(T)$. This is the change in Gibbs' function that occurs for a given reaction *that goes to completion* at a given temperature (not necessarily the reference temperature), and from this definition

$$\Delta\bar{G}^o(T) = (\nu_C \bar{g}_C^o + \nu_D \bar{g}_D^o - \nu_A \bar{g}_A^o - \nu_B \bar{g}_B^o)_T \tag{12.45}$$

where the $\bar{g}_i^o$ are the molar Gibbs functions for the components in their pure state at standard pressure and at the temperature of the reaction. This implies that $\Delta\bar{G}^o(T)$ applies to a reaction in which the reactants were initally separated but were brought together, the reaction between them went to completion, and the products were then separated.

Now we can substitute Eqs. (12.44) and (12.45) into Eq. (12.40), which results in

$$dG_{T,P} = \left[\Delta \overline{G}^o(T) + \overline{R}T \ln \frac{a_C^{\nu_C} a_D^{\nu_D}}{a_A^{\nu_A} a_B^{\nu_B}} \right] d\alpha \tag{12.46}$$

where we have used the definition $a_i \equiv f_i^*/f_i^o$ and a_i is called the *activity* of component i. The activity should not be confused with the Helmoltz function, which uses the same symbol.

Because $dG_{T,P} = 0$ at equilibrium, this equation becomes

$$\ln \frac{a_C^{\nu_C} a_D^{\nu_D}}{a_A^{\nu_A} a_B^{\nu_B}} = \frac{-\Delta \overline{G}^o(T)}{\overline{R}T} = \ln K \tag{12.47}$$

The K is the equilibrium constant found in Eq. (12.37) by using the concentrations in terms of activities, which are dimensionless. Now K is defined by

$$K = \frac{a_C^{\nu_C} a_D^{\nu_D}}{a_A^{\nu_A} a_B^{\nu_B}} \tag{12.48}$$

Equation (12.47) is a complete definition of the equilibrium constant; however, it is difficult to find the activities (or fugacities) of each component in many cases. If we treat the equilibrium gas phase as an ideal gas mixture, then

$$a_i = \frac{f_i^*}{f_i^o} = \frac{f_i^*}{P^o} = y_i \frac{P}{P^o} \tag{12.49}$$

and the equilibrium constant becomes

$$K = \frac{y_C^{\nu_C} y_D^{\nu_D}}{y_A^{\nu_A} y_B^{\nu_B}} \left(\frac{P}{P^o} \right)^{\nu_C + \nu_D - \nu_A - \nu_B} \tag{12.50}$$

and the equilibrium constant is determined solely by the mole fractions of the components, the stoichiometric coefficients, and the pressure of the equilibrium mixture (although K itself is *not* a function of pressure, as we will see). The mole fractions themselves can be put in the form of the degree of reaction α; and if K is known for a given reaction, the degree of reaction is then fixed. The equilibrium constant can be obtained through Eq. (12.47) as

$$K = \exp \left[\frac{-\Delta \overline{G}^o(T)}{\overline{R}T} \right] \tag{12.51}$$

This evaluation of α was the goal of our derivation, and Eq. (12.51) is the useful form for finding numerical values of K; or if K is known, for finding equilibrium species concentrations though Eq. (12.50).

Example Problem 12.3

The reaction $O_2 + (1/2)N_2 \leftrightarrow NO_2$ takes place at 3500 K and a total pressure of 1 atm. What are the mole fractions of each component in the equilibrium mixture?

Solution

Table C.12 gives the value of the equilibrium constant for this reaction as $\log_{10}K = -3.783$, or $K = 1.648 \times 10^{-4}$. The *relative* concentrations of the final constituents are $[NO_2] = \alpha$; $[O_2] = 1 - \alpha$; $[N_2] = (1/2)(1 - \alpha)$. The mole fractions of each constituent in the final equilibrium mixture are then $y_{NO_2} = \alpha/[1.5 - (\alpha/2)]$; $y_{O_2} = (1 - \alpha)/[1.5 - (\alpha/2)]$; $y_{N_2} = (1/2)(1 - \alpha)/[1.5 - (\alpha/2)]$. Thus, Equation (12.50) gives

$$K = 1.648 \times 10^{-4} = \frac{\dfrac{\alpha}{[1.5 - (\alpha/2)]}}{\left[\dfrac{1 - \alpha}{1.5 - (\alpha/2)}\right]\left[\dfrac{(0.5)(1 - \alpha)}{1.5 - (\alpha/2)}\right]^{1/2}}$$

$$= \frac{\alpha[1.5 - (\alpha/2)]^{1/2}}{(1 - \alpha)[(0.5)(1 - \alpha)]^{1/2}}$$

Solving for α results in $\alpha = 9.515 \times 10^{-5}$, and the mole fractions in the equilibrium mixture are $y_{NO_2} = 6.433 \times 10^{-4}$; $y_{O_2} = 0.6666$; $y_{N_2} = 0.3333$.

Comment

This is the standard use of the equilibrium constant. When K values are available directly from tabulated information, the concentrations present after a given reaction reaches completion are easily found.

12.3.2 Values of the Equilibrium Constant

Table C.12 gives values of the logarithm of the equilibrium constant for many common reactions. When the equilibrium constant for a particular reaction is not available, often it can be found by combining the constants for an appropriate set of reactions. This is done as follows. Consider the two reactions and their corresponding equilibrium coefficients

$$\nu_{A,1}A_1 \leftrightarrow \nu_{B,1}B_1 + \nu_{C,1}C$$

$$K_1 = \frac{y_{B,1}^{\nu_{B,1}} y_{C,1}^{\nu_{C,1}}}{y_{A,1}^{\nu_{A,1}}} \left(\frac{P}{P^o}\right)^{\nu_{B,1} + \nu_{C,1} - \nu_{A,1}} \qquad (12.52a)$$

$$\nu_{A,2}A_2 \leftrightarrow \nu_{B,2}B_2 + \nu_{C,2}C$$

$$K_2 = \frac{y_{B,2}^{\nu_{B,2}} y_{C,2}^{\nu_{C,2}}}{y_{A,2}^{\nu_{A,2}}} \left(\frac{P}{P^o}\right)^{\nu_{B,2} + \nu_{C,2} - \nu_{A,2}} \qquad (12.52b)$$

where the same product C appears in both reactions. Now multiply through Eq. (12.52b) by the raio $\nu_{C,1}/\nu_{C,2}$ and subtract the resulting form of Eq. (12.52b) from Eq. (12.52a) to eliminate the C terms. The result is

$$\nu_{A,1}A_1 + \frac{\nu_{C,1}}{\nu_{C,2}}\nu_{B,2}B_2 \leftrightarrow \nu_{B,1}B_1 + \frac{\nu_{C,1}}{\nu_{C,2}}\nu_{A,2}A_2 \tag{12.53}$$

and the equilibrium constant for the reaction described by Eq. (12.53) is

$$K = \frac{y_{B,1}^{\nu_{B,1}} y_{A,2}^{\nu_{A,2}(\nu_{C,1}/\nu_{C,2})}}{y_{A,1}^{\nu_{A,1}} y_{B,2}^{\nu_{B,2}(\nu_{C,1}/\nu_{C,2})}} \left(\frac{P}{P^o}\right)^{\nu_{B,1}+\nu_{A,2}(\nu_{C,1}/\nu_{C,2})-\nu_{A,1}-\nu_{B,2}(\nu_{C,1}/\nu_{C,2})} \tag{12.54}$$

Note that

$$K = \frac{K_1}{(K_2)^{\nu_{C,1}/\nu_{C,2}}} \tag{12.55}$$

Therefore, if the reaction in Eq. (12.53) can be manufactured by adding or subtracting two other reactions, Eqs. (12.51) and (12.52), for which equilibrium constants are known, then the equilibrium constant for the resulting reaction is easily found.

The rule of thumb is that *when reactions are added, the equilibrium constant of the resulting reaction will be the product of the equilibrium constants of the reactions being added;* however, *when reaction 2 is subtracted from reaction 1 to obtain another reaction, the equilibrium constant of the resulting reaction will be the equilibrium constant of reaction 1 divided by the equilibrium constant of reaction 2.* If any of the reactions are multiplied by a constant so that terms could be canceled, then the equilibrium constant for that reaction must be raised to the power of the constant.

Note that Eq. (12.55) can be rewritten by taking the logarithm (base 10) of both sides to obtain

$$\begin{aligned}\log K &= \log \frac{K_1}{(K_2)^{\nu_{C,1}/\nu_{C,2}}}\\ &= \log K_1 - \log K_2{}^{\nu_{C,1}/\nu_{C,2}}\\ &= \log K_1 - \frac{\nu_{C,1}}{\nu_{C,2}}\log K_2\end{aligned} \tag{12.56}$$

This form is convenient because the tabulated values of the equilibrium constants in Table C.12 are given in terms of log K.

Example Problem 12.4

Find the equilibrium constant at 2000 K for the reaction

$$CH_4 + H_2O \leftrightarrow 3H_2 + CO$$

Solution

Searching Table C.12, we find the reactions

$$C + \tfrac{1}{2}O_2 \leftrightarrow CO \qquad \log K_1 = 7.469 \qquad K_1 = 2.944 \times 10^7$$

$$C + 2H_2 \leftrightarrow CH_4 \qquad \log K_2 = -3.408 \qquad K_2 = 3.908 \times 10^{-4}$$

$$H_2 + \tfrac{1}{2}O_2 \leftrightarrow H_2O \qquad \log K_3 = 3.540 \qquad K_3 = 3.467 \times 10^3$$

By the methodology of Sec. 12.3.2, the second equation can be subtracted from the first to give

$$\begin{array}{l} C + \tfrac{1}{2}O_2 \quad \leftrightarrow CO \\ -(C \quad + 2H_2 \leftrightarrow \quad + CH_4) \\ \hline \tfrac{1}{2}O_2 - 2H_2 \leftrightarrow CO - CH_4 \end{array}$$

or

$$\tfrac{1}{2}O_2 + CH_4 \leftrightarrow CO + 2H_2$$

and for this reaction, according to Eq. (12.55),

$$K_4 = \frac{K_1}{K_2} = \frac{2.944 \times 10^7}{3.908 \times 10^{-4}} = 7.533 \times 10^{10}$$

The third equation can now be subtracted to eliminate the O_2, to obtain

$$\begin{array}{l} \tfrac{1}{2}O_2 + CH_4 \leftrightarrow CO + 2H_2 \\ -(H_2 + \tfrac{1}{2}O_2 \quad \leftrightarrow \quad H_2O) \\ \hline -H_2 \quad + CH_4 \leftrightarrow CO + 2H_2 - H_2O \end{array}$$

or

$$CH_4 + H_2O \leftrightarrow 3H_2 + CO$$

which is the desired equation. The equilibrium constant for this reaction is, again following Eq. (12.55),

$$K = \frac{K_4}{K_3} = \frac{7.533 \times 10^{10}}{3.467 \times 10^3} = 2.173 \times 10^7$$

In this particular example, it was not necessary to multiply through any of the equations when eliminating terms.

Example Problem 12.5

The reaction

$$CO_2 + H_2 \leftrightarrow CO + H_2O$$

takes place at 3600°R and 1 atm. What are the mole fractions of each component in the resulting equilibrium mixture?

Solution

Table C.12 gives the value of the equilibrium constant for various reactions but does not give this specific reaction. The equilibrium constant for this reaction is obtained by noting that the reaction equation is the sum of three basic reactions:

Reaction 1: $C + O_2 \leftrightarrow CO_2 \qquad K_1 = \dfrac{y_{CO_2}}{y_C y_{O_2}}$

Reaction 2: $C + \frac{1}{2}O_2 \leftrightarrow CO \qquad K_2 = \dfrac{y_{CO}}{y_C y_{O_2}^{1/2}}$

Reaction 3: $H_2 + \frac{1}{2}O_2 \leftrightarrow H_2O \qquad K_3 = \dfrac{y_{H_2O}}{y_{H_2} y_{O_2}^{1/2}}$

With Eq. (12.33) and the definition of K in Eq. (12.50), the sum of reactions 2 and 3 minus reaction 1 yields $H_2 \leftrightarrow -CO_2 + CO + H_2O$, or

$$H_2 + CO_2 \leftrightarrow CO + H_2O \qquad K = \frac{y_C y_{O_2} y_{CO} y_{H_2O}}{y_{CO_2} y_C y_{O_2}^{1/2} y_{H_2} y_{O_2}^{1/2}} = \frac{K_2 K_3}{K_1}$$

Table C.12 gives the values $K_1 = 12.25 \times 10^{10}$, $K_2 = 2.94 \times 10^7$, and $K_3 = 3470$ at 3600°R. Thus

$$K = \frac{(2.94 \times 10^7)(3470)}{2.25 \times 10^{10}} = 4.53$$

The final relative concentrations are as follows:

Component	**Relative Equilibrium Concentration**
CO_2	$1-\alpha$
H_2	$1-\alpha$
CO	α
H_2O	α

Equation (12.50) then gives

$$K = 4.53 = \frac{y_{CO} y_{H_2O}}{y_{CO} y_{H_2}} = \frac{\alpha^2}{(1-\alpha)^2}$$

Solving the resulting quadratic equation for the value of α that lies in the range $0 < \alpha < 1$ gives $\alpha = 0.681$. The equilibrium mole fractions are then as follows:

Component	Relative Equilibrium Concentration	Relative Mole Fraction
CO_2	0.319	0.1595
H_2	0.319	0.1595
CO	0.681	0.3405
H_2O	0.681	0.3405
	2.000	1.0000

Comment

If not all the stoichiometric coefficients are unity, a more complex equation is obtained to determine the value of α.

12.3.3 Temperature Dependence of K

The definition of the equilibrium constant is given through Eq. (12.47). For an ideal gas mixture, combining Eqs. (12.50) and (12.51) yields

$$K = \frac{y_C^{\nu_C} y_D^{\nu_D}}{y_A^{\nu_A} y_B^{\nu_B}} \left(\frac{P}{P^o}\right)^{\nu_C+\nu_D-\nu_A-\nu_B} = \exp\left[-\frac{\Delta \overline{G}^o(T)}{\overline{R}T}\right] \tag{12.57}$$

The right-hand side is a function of temperature only, so the equilibrium constant is also a function of temperature only. Differentiating Eq. (12.57) with respect to temperature yields

$$\frac{dK}{dT} = \frac{K}{\overline{R}}\left[\frac{\Delta \overline{G}^o(T)}{T^2} - \frac{1}{T}\frac{d\,\Delta \overline{G}^o(T)}{dT}\right] \tag{12.58}$$

or

$$\frac{1}{K}\frac{dK}{dT} = \frac{d(\ln K)}{dT} = \frac{1}{\overline{R}T^2}\left[\Delta \overline{G}^o(T) - T\frac{d\,\Delta \overline{G}^o(T)}{dT}\right] \tag{12.59}$$

The total differential of the Gibbs function given in Eq. (9.5d) shows that $(\partial g/\partial T)_{P,\,n_i} = -s$. This relation is used to replace the temperature derivative of the Gibbs function in Eq. (12.59) by

$$\frac{d\,\Delta \overline{G}^o(T)}{dT} = -\Delta \overline{S}^o \tag{12.60}$$

The definition of the Gibbs function gives

$$\Delta \overline{G}^o(T) = \Delta \overline{H}^o - T\,\Delta \overline{S}^o \tag{12.61}$$

Therefore, Eq. (12.59) yields

$$\frac{d(\ln K)}{dT} = \frac{\Delta \overline{H}^o}{\overline{R}T^2} \tag{12.62}$$

This equation is called the *van't Hoff equation.* This expression can be used to evaluate $\Delta \overline{H}^o$ from equilibrium mole fractions or, alternatively, to evaluate the equilibrium constant from $\Delta \overline{H}^o$ information. Specifically, if $\Delta \overline{H}^o$ is independent of temperature, then $\ln K$ is linear in $1/T$. The slope on a plot of $\ln K$ versus $1/T$ then yields $\Delta \overline{H}^o/\overline{R}$.

Frozen Equilibrium

When a chemical reaction is carried out at a high temperature, the equilibrium concentrations at that temperature will, of course, be produced. However, in some cases the equilibrium mixture is then cooled very rapidly to a lower temperature by some process. The equilibrium constant for the governing reaction may be much different at the lower temperature. If the cooling occurs so rapidly that the rate of chemical reaction is much slower than the cooling rate, then the concentrations of the species at the new low temperature may not change significantly during the cooling process. Also because the equilibrium constant at the low temperature may not favor further reaction, no reaction may occur at the lower temperature. Thus, the cooled species may reflect the concentrations reached at the higher temperature, and not the equilibrium concentrations expected if the reaction had occurred at the lower temperature to begin with.

Such a situation is called *frozen equilibrium,* because the final concentrations are frozen at values typical of the higher temperature at which the reaction actually occurred, but may differ considerably from the concentrations predicted for equilibrium at the lower temperature. This situation occurs in automotive engines, where oxides of nitrogen are produced within the engine combustion chamber and are then rapidly cooled during the expansion and exhaust processes. These oxides would not be present in the same concentrations if the combustion reaction were carried out at the final exhaust temperature. This is a significant problem in producing air pollution by nitrogen oxides.

Determination of whether frozen equilibrium occurs in a given case requires knowledge of the individual rate constants for the forward and reverse reactions [Eqs. (12.34) and (12.36)]. This information is studied in the field of *chemical kinetics* and is outside the scope of this text.

Example Problem 12.6

In an automotive engine, NO is formed by the reaction of oxygen and nitrogen at the peak combustion chamber temperature of 2000 K. Determine the relative equilibrium concentrations of NO at the peak combustion temperature and at an exhaust temperature of 600 K.

Solution

The reaction is

$$\tfrac{1}{2}N_2 + \tfrac{1}{2}O_2 \leftrightarrow NO$$

and if the reaction is not affected by the other reactions present in the combustion chamber, the final equilibrium concentrations will be α for the NO and $\frac{1}{2}(1-\alpha)$ for both N_2 and O_2. Thus, the equilibrium constant is

$$K = \frac{\alpha}{[0.5(1-\alpha)]^{1/2}[0.5(1-\alpha)]^{1/2}} = \frac{2\alpha}{1-\alpha}$$

From Table C.12, $\log K(T = 2000\text{ K}) = -1.699$ and $\log K(T = 600\text{ K}) = -7.210$. Solving for α gives $\alpha(T = 2000\text{ K}) = 0.0200$ and $\alpha(T = 600\text{ K}) = 3.083 \times 10^{-8}$, or a concentration at 2000 K that is $0.0200/3.083 \times 10^{-8} = 6.49 \times 10^{5}$ times as large at the higher temperature! If the equilibrium is frozen during the rapid cooling ("quenching") of the combustion products, then the much higher concentrations of NO produced during combustion will occur in the exhaust products, and the lower concentrations predicted at the exhaust temperature will not have time to come to equilibrium.

12.3.4 Pressure Dependence of Equilibrium Concentrations

In the last section we found that K itself is independent of pressure. Yet pressure appears explicitly in Eq. (12.50). It follows that the equilibrium concentrations in Eq. (12.50) must change in such a way that K remains constant when P is varied. At constant T, Eq. (12.50) can be rewritten in the form

$$\frac{y_C^{\nu_C} y_D^{\nu_D}}{y_A^{\nu_A} y_B^{\nu_B}} = \frac{K}{(P/P^o)^{\nu_C+\nu_D-\nu_A-\nu_B}} \tag{12.63}$$

Examining the exponent of P/P^o shows that if $\nu_C + \nu_D = \nu_A + \nu_B$, then pressure has no effect on the equilibrium concentrations. However, if $\nu_C + \nu_D > \nu_A + \nu_B$, then the concentrations of the products will be decreased relative to the reactants as the pressure is increased; and if $\nu_C + \nu_D < \nu_A + \nu_B$, the concentrations of the reactants will be decreased relative to the products as the pressure is increased. Thus, *increasing pressure tends to increase the concentrations of the constituents that are on the side of the reaction with the smallest sum of stoichiometric coefficients.* This observation is known as *LeChatelier's principle.*

Example Problem 12.7

Methane and water vapor react to form CO and H_2. To attain a high concentration of the products, should the reaction be carried out at high or low pressure?

Solution

The reaction is balanced to give

$$CH_4 + H_2O \leftrightarrow CO + 3H_2$$

The sum of the stoichiometric coefficients is greater on the right, indicating that higher pressure will hinder product formation. Reaction at low pressure is thus preferred.

12.3.5 Phase Rule

The Gibbs phase rule for nonreacting equilibrium properties is given in Eq. (12.32). For reacting mixtures N_{ch} additional chemical reaction equations must be satisfied. The number of independent intensive properties is therefore the same as for the previous case [Eq. (12.32)] minus N_{ch}. Thus,

$$N_{pr} = N_c - N_{ph} - N_{ch} + 2 \tag{12.64}$$

Example Problem 12.8

The gas-phase reaction

$$CH_4 + H_2O \leftrightarrow CO + 3H_2$$

is to be carried out. The initial concentrations of the reactants are unknown. How many independent intensive properties must be known to specify the thermodynamic equilibrium state when the reaction reaches equilibrium?

Solution

There is one phase with four components, and a single chemical reaction is present. According to the phase rule for a chemically reacting system, Eq. (12.64), the number of independent intensive properties to be specified is

$$N_{pr} = 4 - 1 - 1 + 2 = 4$$

Thus, we might specify T, P, and the mole fractions of any two of the components. In this case, the concentrations of the remaining components can be determined from the chemical balance equation and the fact that at equilibrium the mole fractions must sum to unity.

Comments

In a nonreacting system with the same components, we would be required to specify five independent intensive properties because the additional information on relative concentrations supplied by the balanced chemical reaction equation would not be available.

12.4 General Equilibrium

To generalize the concepts of phase and chemical equilibrium, take the general definition of the Gibbs function $G = H - TS$ and then take the total derivative to find

$$dG = dH - T\,dS - S\,dT \tag{12.65}$$

We can now substitute Eq. (5.42), which is a general relation among properties, to give

$$dG = V\,dP - S\,dT \tag{12.66}$$

Now examine Eq. (12.1), and substitute Eq. (12.3) into that equation. The result is

$$dG = \left(\frac{\partial G}{\partial T}\right)_{P,\,n_A,\,n_B\ldots} dT + \left(\frac{\partial G}{\partial P}\right)_{T,\,n_A,\,n_B,\ldots} dP + \bar{\mu}_A\,dn_A + \bar{\mu}_B\,dn_B + \cdots \tag{12.67}$$

For a system in chemical and phase equilibrium, we can substitute Eq. (12.13), which gives

$$dG|_{n_A,\,n_B,\ldots} = \left(\frac{\partial G}{\partial T}\right)_{P,\,n_A,\,n_B,\ldots} dT + \left(\frac{\partial G}{\partial P}\right)_{T,\,n_A,\,n_B,\ldots} dP \tag{12.68}$$

Comparing Eqs. (12.68) and (12.66) gives the identities

$$\left(\frac{\partial G}{\partial T}\right)_{P,\,n_A,\,n_B,\ldots} = -S \tag{12.69}$$

and

$$\left(\frac{\partial G}{\partial P}\right)_{T,\,n_A,\,n_B,\ldots} = V \tag{12.70}$$

Now, Eq. (12.1) can be written in the completely general form

$$dG = -S\,dT + V\,dP + \bar{\mu}_A\,dn_A + \bar{\mu}_B\,dn_B + \cdots \tag{12.71}$$

We observed previously that thermal equilibrium requires $dT = 0$, mechanical equilibrium requires $dP = 0$, and chemical and phase equilibrium require that $\sum_j \sum_i (\bar{\mu}_i\,dn_i)_j = 0$. If all these requirements are met simultaneously, then Eq. (12.71) predicts that $dG = 0$. Thus, the Gibbs function incorporates a general description of the thermodynamic properties of any system [since $G(T, P, n_A, n_B, \ldots)$ is a fundamental equation (Chap. 9)], but in addition we note that any thermodynamic system must have $dG = 0$ if the system is in thermodynamic equilibrium.

The Gibbs function is extremely useful in describing equilibrium in control volumes, where T and P and the component concentrations are the useful properties. For a control mass, the Helmholtz function $A = U - TS$ plays a similar role. For a general isolated system (Fig. 12.3) containing a control mass and its surroundings, we know from Eq. (5.73) that

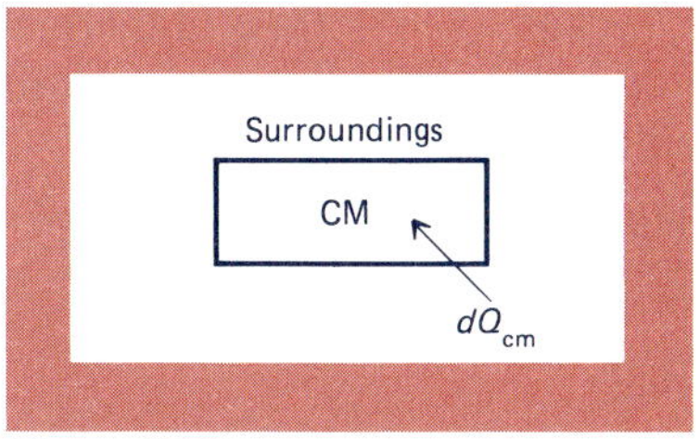

Figure 12.3 General isolated system composed of control mass and surroundings.

$$(dS)_{CM} + (dS)_{surr} \geq 0 \tag{12.72}$$

If the surroundings are a thermal reservoir at T_0, then $(dS)_{surr}$ can be expressed in terms of the heat transfer from the reservoir to the control mass as

$$(dS)_{surr} = \frac{-(\delta Q)_{CM}}{T_0} \tag{12.73}$$

and from the first law for a control mass, $(\delta Q)_{CM} = (dU + \delta W)_{CM} = (dU + P\,dV)_{CM}$,

$$(dS)_{surr} = -\frac{(dU + P\,dV)_{CM}}{T_0} \tag{12.74}$$

and Eq. (12.72) becomes

$$(dU + P\,dV - T_0\,dS)_{CM} \leq 0 \tag{12.75}$$

which is a general expression for a control mass exchanging heat transfer with the surroundings. The equality applies for a control mass in thermodynamic equilibrium with its surroundings.

We can examine three specific cases of interest in chemically reacting systems.

12.4.1 Isolated System in Chemical Equilibrium

In this case the volume is constant, and because there is no work or heat transfer for an isolated system (by definition), the first law gives $dU = 0$. Then Eq. (12.75) reduces to $dS \geq 0$ which we have previously derived for an isolated system as the basis for the existence of the property entropy (5.70). For a mixture,

$$dS = d(n\bar{s}) = \sum_i n_i\,d\bar{s}_i + \sum_i \bar{s}_i\,dn_i \geq 0 \tag{12.76}$$

We can eliminate $d\bar{s}_i$ by using Gibbs' equation

$$d\bar{s}_i = \frac{d\bar{u}_i + P_i\,d\bar{v}_i}{T} \tag{12.77}$$

where, because the system is in thermal equilibrium, $T_i = T_0 = T_{CM}$ and the subscripts on T are dropped. Substituting into Eq. (12.76) for the case of equilibrium ($dS = 0$),

$$\sum_i n_i \frac{d\bar{u}_i + P_i\, d\bar{v}_i}{T} + \sum_i \bar{s}_i\, dn_i = 0 \tag{12.78}$$

Note that

$$U = \sum_i n_i \bar{u}_i \tag{12.79}$$

or, because $dU = 0$ for the isolated system,

$$\sum_i n_i\, d\bar{u}_i = \sum_i \bar{u}_i\, dn_i \tag{12.80}$$

Similarly, because $V = \sum_i n_i \bar{v}_i$ for an ideal mixture and V is constant,

$$\sum_i n_i\, d\bar{v}_i = -\sum_i \bar{v}_i\, dn_i \tag{12.81}$$

Substituting Eqs. (12.80) and (12.81) into Eq. (12.78) to eliminate $d\bar{v}_i$ and $d\bar{u}_i$ results in

$$-\sum_i \frac{\bar{u}_i + P_i \bar{v}_i}{T}\, dn_i + \sum_i \bar{s}_i\, dn_i = 0 \tag{12.82}$$

Using the definition of enthalpy $\bar{h} = \bar{u} + P\bar{v}$ and gathering terms under the summation give

$$\sum_i (\bar{h}_i - T\bar{s}_i)\, dn_i = 0 \tag{12.83}$$

We recognize the term in parentheses as the Gibbs function $\bar{g}_i$.

Note that in a mixture the change in the number of moles of any component dn_i is not independent of the changes for the other components because of the relation imposed by the chemical balance equation, Eq. (12.38). Substituting Eq. (12.38) to eliminate dn_i in Eq. (12.83) results in

$$\sum_{\text{products}} \bar{g}_i \nu_i = \sum_{\text{reactants}} \bar{g}_i \nu_i \tag{12.84}$$

Thus, as long as Eq. (12.84) is valid, the entropy change of the isolated system is zero, and the isolated system is in chemical and therefore complete thermodynamic equilibrium.

12.4.2 Control Mass at Constant Volume

This is the case for a closed container in which a reaction occurs and continues until equilibrium is reached at some final temperature and pressure. For a reversible reaction, the temperatures of the surroundings and of the control mass must be equal ($T_0 = T_{CM}$) throughout the reaction, so Eq. (12.75) becomes

$$(dU - T\,dS)_{CM} = d(U - TS)_{CM,\,T} \le 0 \tag{12.85}$$

The group of properties $U - TS$ is called the *Helmholtz function A*. For the case of a chemical reaction in a constant-volume system, the Helmholtz function must move to a minimum at chemical equilibrium according to Eq. (12.85), or

$$(dA)_{\mathrm{CM},T,V} = 0 \tag{12.86}$$

The Helmholtz function can be evaluated from

$$A = \sum_i n_i\bar{a}_i = \sum_i n_i(\bar{u}_i - T\bar{s}_i) \tag{12.87}$$

For the particular case of a reaction in which all reactants and products are ideal gases, we can use Eq. (10.38*a*) for $\bar{u}$ and Eq. (10.36*b*) for $\bar{s}$, that is,

$$\bar{u}_i = \bar{u}_{i,\mathrm{ref}} + \int_{T_\mathrm{ref}}^{T} \bar{c}_{v,i}\,dT \tag{12.88}$$

and

$$\bar{s}_i = \bar{s}_{i,\mathrm{ref}} + \int_{T_\mathrm{ref}}^{T} \bar{c}_{P,i}\frac{dT}{T} - \bar{R}\ln\frac{P}{P_\mathrm{ref}} - \bar{R}\ln n_i \tag{12.89}$$

Using the ideal gas relation to replace P and letting $P_\mathrm{ref} = 1$ atm yield

$$\bar{s}_i = \bar{s}_{i,\mathrm{ref}} + \int_{T_\mathrm{ref}}^{T} \bar{c}_{P,i}\frac{dT}{T} - \bar{R}\ln\frac{\bar{R}T}{\bar{v}} - \bar{R}\ln n_i \tag{12.90}$$

Substituting Eqs. (12.88) and (12.90) into $\bar{a}_i = \bar{u}_i - T\bar{s}_i$ gives

$$\bar{a}_i = \bar{a}_i^*(T, \bar{v}) - \bar{R}\ln n_i \tag{12.91}$$

where

$$\bar{a}_i^*(T,\bar{v}) = \bar{u}_{i,\mathrm{ref}} + \int_{T_\mathrm{ref}}^{T} \bar{c}_{v,i}\,dT - T\bar{s}_{i,\mathrm{ref}} - T\int_{T_\mathrm{ref}}^{T} \bar{c}_{P,i}\frac{dT}{T} + \bar{R}T\ln\frac{\bar{R}T}{\bar{v}}$$

and we note that $\bar{a}^*$ is the Helmholtz function of pure component i at $(T, \bar{v})$.

For a control mass held at constant temperature and volume, then, from Eq. (12.86) and substituting Eq. (12.91),

$$dA_{\mathrm{CM},T,v} = d\sum_i n_i\bar{a}_{i,T,v} = \sum_i (dn_i\,\bar{a}_{i,T,v} + n_i\,d\bar{a}_{i,T,v}) = 0 \tag{12.92}$$

Now using Eq. (12.38) in the form $dn_i = \nu_i\,d\alpha$ in Eq. (12.92), we see that the variables in Eq. (12.92) become T, $\bar{v}$, and α. Thus, for constant T and $\bar{v}$,

$$\begin{aligned}(dA)_{\mathrm{CM},T,v} &= \frac{\partial}{\partial\alpha}\left(\sum_i n_i\bar{a}_i\right)_{\mathrm{CM},T,v} = \left[\sum_i\left(\frac{\partial n_i}{\partial\alpha}\bar{a}_i + n_i\frac{\partial\bar{a}_i}{\partial\alpha}\right)\right]_{\mathrm{CM},T,v}\\ &= \left(\sum_i\left\{\frac{\partial n_i}{\partial\alpha}\bar{a}_i + n_i\frac{\partial[\bar{a}_i^*(T,\bar{v}) + \bar{R}\ln n_i]}{\partial\alpha}\right\}\right)_{\mathrm{CM},T,v} = 0\end{aligned} \tag{12.93}$$

The final derivative in the last term of Eq. (12.93) is zero, because $\bar{a}_i^*(T, \bar{v})$ is independent of α, and

$$\sum_i \left[n_i \frac{\partial (\ln n_i)}{\partial \alpha}\right]_{\text{CM}, T, v} = \sum_i \left[n_i \frac{\partial (\ln y_i n)}{\partial \alpha}\right]_{\text{CM}, T, v}$$
$$= \sum_i \left[n_i \frac{\partial (\ln y_i)}{\partial \alpha}\right]_{\text{CM}, T, v} = \sum_i \left(\frac{n_i}{y_i}\frac{\partial y_i}{\partial \alpha}\right)_{\text{CM}, T, v}$$
$$= \sum_i \left(n \frac{\partial y_i}{\partial \alpha}\right)_{\text{CM}, T, v} = n \frac{\partial}{\partial \alpha}\left(\sum_i y_i\right)_{\text{CM}, T, v} = 0 \tag{12.94}$$

Thus, Eq. (12.93) becomes simply

$$\left(\sum_i \frac{\partial n_i}{\partial \alpha} \bar{a}_i\right)_{\text{CM}, T, v} = 0 \tag{12.95}$$

Using Eq. (12.38) to replace $\partial n_i/\partial \alpha$ in Eq. (12.95) then results in

$$\sum_{\text{products}} \bar{a}_{i, T, v} \nu_i = \sum_{\text{reactants}} \bar{a}_{i, T, v} \nu_i \tag{12.96}$$

Thus, *in a chemical reaction in which the temperature and volume are held constant, the total Helmholtz function of the reactants will be equal to the total Helmholtz function of the products when chemical equilibrium is reached.*

Chemical Potential in a Constant-Volume System

If the Helmholtz function A is taken as a function of T, V, n_A, n_B, . . . , then

$$dA = \left(\frac{\partial A}{\partial T}\right)_{V, n_A, n_B, \ldots} dT + \left(\frac{\partial A}{\partial V}\right)_{T, n_A, n_B, \ldots} dV$$
$$+ \left(\frac{\partial A}{\partial n_A}\right)_{T, V, n_B, n_C, \ldots} dn_A + \left(\frac{\partial A}{\partial n_B}\right)_{T, V, n_A, n_C, \ldots} dn_B + \cdots \tag{12.97}$$

Referring to Eq. (9.11), the Maxwell relations for the Helmholtz function may be used to place Eq. (12.97) in the form

$$dA = -S\,dT - P\,dV + \left(\frac{\partial A}{\partial n_A}\right)_{T, V, n_B, n_C, \ldots} dn_A$$
$$+ \left(\frac{\partial A}{\partial n_B}\right)_{T, V, n_A, n_C, \ldots} dn_B + \cdots \tag{12.98}$$

Now note that

$$G = H - TS = (U + PV) - TS = A + PV \tag{12.99}$$

and therefore

$$dG = dA + P\,dV + V\,dP \tag{12.100}$$

Substituting Eq. (12.98) to eliminate the Helmholtz function from Eq. (12.100) results in

$$dG = -S\,dT + V\,dP + \left(\frac{\partial A}{\partial n_A}\right)_{T,\,V,\,n_B,\,n_C,\,\ldots} dn_A + \left(\frac{\partial A}{\partial n_B}\right)_{T,\,V,\,n_A,\,n_C,\,\ldots} dn_B + \cdots \tag{12.101}$$

For the particular case of a system at constant temperature and pressure,

$$dG_{T,\,P} = \left(\frac{\partial A}{\partial n_A}\right)_{T,\,V,\,n_B,\,n_C,\,\ldots} dn_A + \left(\frac{\partial A}{\partial n_B}\right)_{T,\,V,\,n_A,\,n_C,\,\ldots} dn_B + \cdots \tag{12.102}$$

Comparing with Eq. (12.6), we find that

$$\mu_i = \left(\frac{\partial A}{\partial n_i}\right)_{T,\,V,\,\text{all } n_j \text{ with } j \neq i} \tag{12.103}$$

and Eq. (12.98) becomes

$$dA = -S\,dT - P\,dV + \sum_i \mu_i\,dn_i \tag{12.104}$$

This should be compared with Eq. (12.71) which gives the total differential of the Gibbs function with respect to changes in T, P, and n_i.

From Eq. (12.103) we observe that the partial molal Helmholtz function for a constant-temperature constant-volume system is equal to the partial molal Gibbs function, which was in turn defined as the chemical potential $\bar{\mu}_i$. Thus, the chemical potential describes the tendency to chemical equilibrium in *both* constant-volume and constant-pressure systems in which an isothermal chemical reaction occurs.

12.4.3 Control Mass at Constant Temperature and Pressure

For this case, Eq. (12.75) becomes

$$d(H - TS)_{T,\,P} = dG_{T,\,P} \leq 0 \tag{12.105}$$

The Gibbs function can be expressed in terms of the Gibbs functions of the individual components

$$G_{T,\,P} = \sum_i n_i \bar{g}_i \tag{12.106}$$

Proceeding with a parallel derivation to that used for the constant-volume constant-temperature case [Eq. (12.76) through (12.96)] results in

$$\sum_{\text{products}} \bar{g}_{i,\,T,\,P}\nu_i = \sum_{\text{reactants}} \bar{g}_{i,\,T,\,P}\nu_i \tag{12.107}$$

and we find that *for a chemical reaction carried out in a control mass at constant temperature and constant pressure, the total Gibbs function for the reactants must equal the total Gibbs function of the products at chemical (and therefore thermodynamic) equilibrium.*

This is the same result reached for the isolated system, and it reduces to Eq. (12.18) for a system with a single phase and a single component.

12.4.4 General Criterion for Chemical Equilibrium

At equilibrium, we have found that for an isolated system Eqs. (12.71) and (12.83) give

$$dG_{T,P} = \sum_i \bar{g}_{i,T,P}\, dn_i = \sum_i \nu_i \bar{g}_{i,T,P}\, d\alpha \le 0 \tag{12.108}$$

and for a system at constant temperature and volume, Eqs. (12.85) and (12.86) give

$$dA_{T,V} = \sum_i \bar{a}_{i,T,P}\, dn_i = \sum_i \nu_i \bar{a}_{i,T,P}\, d\alpha \le 0 \tag{12.109}$$

Finally, for a system at constant temperature and pressure, Eq. (12.105) gives

$$dG_{T,P} = \sum_i \bar{g}_{i,T,P}\, dn_i = \sum_i \nu_i \bar{g}_{i,T,P}\, d\alpha \le 0 \tag{12.110}$$

These three equations express the fact that the Gibbs function (or Helmholtz function for a constant-temperature constant-volume system) proceeds to a minimum value, reaching zero when the degree of reaction α reaches an equilibrium condition where the reactants and products have balanced rate equations.

If this is the case, then taking the partial derivative of each of Eqs. (12.108) through (12.110) and setting the result equal to zero will be the condition for a minimum, and thus a condition for chemical equilibrium in the system described by the equation.

Let us examine first Eqs. (12.108) and (12.110), since these are identical. Taking the derivative with respect to the degree of reaction α gives

$$\frac{\partial}{\partial\alpha}\left(\sum_i \nu_i \bar{g}_{i,T,P}\, d\alpha\right) = \sum_i \nu_i \left(\bar{g}_{i,T,P} + \alpha \frac{\partial \bar{g}_{i,T,P}}{\partial\alpha}\right) = 0 \tag{12.111}$$

At equilibrium, the equality in Eq. (12.83) or (12.105) can be used to note that the first term in the second summation is zero, and the result is that at chemical equilibrium,

$$\sum_i \nu_i \frac{\partial \bar{g}_{T,P}}{\partial\alpha}\, d\alpha = \sum_i \frac{\partial \bar{g}_{T,P}}{\partial n_i}\, dn_i = \sum_i \mu_i\, dn_i = 0 \tag{12.112}$$

Following a similar derivation starting with Eq. (12.109) for a constant-temperature constant-volume control mass and using Eq. (12.103) give exactly the same result!

Therefore, for the various cases we have examined, the general criterion for chemical equilibrium is simply that the sum of the chemical potentials of each component in the system multiplied by the change in the number of moles of that component must be zero. We can apply Eq. (12.38) to get another form as

$$\sum_{\text{products}} \nu_i \mu_i = \sum_{\text{reactants}} \nu_i \mu_i \qquad (12.113)$$

which is a general statement of the criterion for chemical equilibrium.

In this section, we used many of the most basic ideas that are introduced in various chapters. The use of a fundamental equation is introduced in Chap. 9, along with the identification of various properties with the corresponding partial derivatives. The concept of equilibrium requiring that the changes in certain properties be zero is used in Chap. 5 to relate entropy changes to other properties and in this chapter in discussing chemical and phase equilibrium. We have gathered this information to provide a comprehensive prescription for describing a system, its properties, and its deviation from thermodynamic equilibrium.

12.5 Concluding Remarks

In this chapter, some general concepts of equilibrium are applied to problems of phase and chemical equilibrium. Much wider applications are found in the study of chemical thermodynamics and physical chemistry, but the fundamentals given here (mostly formulated originally by J. Willard Gibbs) form the basis for all these studies.

Note that this chapter deals entirely with thermodynamic systems in thermodynamic equilibrium. We have not discussed the question of how long it will take an initial mixture of reactants to reach the predicted equilibrium concentration of products plus reactants. Some reactions proceed very quickly (explosions) while others take years (rust, decay). Such questions are the province of the field of chemical kinetics.

It should be obvious from Sec. 12.4 that the ideas of mechanical, thermal, and chemical equilibrium form a coherent whole. Indeed, one approach to classical thermodynamics is to consider the requirements for general thermodynamic equilibrium and then to generate the necessary properties and relationships that describe changes in the general equilibrium state from that viewpoint.

Problems

12.1 For the following nonreacting mixtures, how many independent intensive properties must be specified for an equilibrium mixture?

(a) A three-phase two-component mixture
(b) A solid that consists of two crystalline forms and has three components
(c) A LiBr-H_2O mixture that has a liquid and a vapor phase

12.2S For saturated steam at 200°C, calculate the value of the Gibbs function for saturated liquid, saturated vapor, and a saturated mixture with quality $x = 40$ percent.

12.2E For saturated steam at 400°F, calculate the value of the Gibbs function for saturated liquid, saturated vapor, and a saturated mixture with quality $x = 40$ percent.

12.3S Without using any other data, find the value of h_l for saturated refrigerant 12 at $T = 20°C$, given $T_{sat} = 20°C$, $h_g = 359.81$ kJ/kg, and $s_{lg} = 0.48071$ kJ/(kg·K).

12.3E Without using any other data, find the value of h_l for saturated refrigerant 12 at $T = 80°F$, given $T_{sat} = 80°F$, $h_g = 1096.7$ Btu/lbm, $s_{lg} = 1.9426$ Btu/(lbm·°R).

12.4S For an ammonia-water mixture at $T = 20°C$, construct a graph of ammonia vapor pressure versus mole fraction of ammonia in the liquid phase. On the same graph, plot ammonia vapor pressure versus mole fraction in the vapor phase. Assume ideal behavior.

12.4E For an ammonia-water mixture at $T = 70°F$, construct a graph of ammonia vapor pressure versus mole fraction of ammonia in the liquid phase. On the same graph, plot ammonia vapor pressure versus mole fraction in the vapor phase. Assume ideal behavior.

12.5S An ammonia-water absorption chiller operates with an evaporator temperature 2°C and a condenser temperature of 50°C with no superheating of vapor in the evaporator and no subcooling in the condenser. The NH_3 constitutes 95 mol percent of the vapor in both the absorber and the generator. What is the pressure in (a) the absorber and (b) the generator? What is the mole fraction of ammonia in the liquid solution in (c) the absorber and (d) the generator? Assume ideal solution behavior.

12.5E An ammonia-water absorption chiller operates with an evaporator temperature of 35°F and a condenser temperature of 100°F with no superheating of vapor in the evaporator and no subcooling in the condenser. The NH_3 constitutes 95 mol percent of the vapor in both the absorber and the generator. What is the pressure in (*a*) the absorber and (*b*) the generator? What is the mole fraction ammonia in the liquid solution in (*c*) the absorber and (*d*) the generator? Assume ideal solution behavior.

12.6S Plot the vapor mole fraction versus liquid mole fraction for ammonia-H_2O ideal mixtures at $T = 2°C$ and at $T = 50°C$ on the same graph. Comment on the meaning of such a graph for an absorption chilling device (Sec. 7.4.2) using ammonia and H_2O as the refrigerant-absorbent pair.

12.6E Plot the vapor mole fraction versus liquid mole fraction for ammonia-H_2O ideal mixtures at $T = 35°F$ and at $T = 100°F$ on the same graph. Comment on the meaning of such a graph for an absorption chilling device (Sec. 7.4.2) using ammonia and H_2O as the refrigerant-absorbent pair.

12.7S A LiCl-H_2O absorption air conditioning system is to be designed for no vapor superheating in the evaporator and no liquid subcooling in the condenser. The LiCl liquid concentration in the generator is 45 weight percent and in the absorber is 20 weight percent. Using the data given and assuming ideal solution behavior where necessary, find for an absorber temperature of 5°C and a generator temperature of 30°C:
(*a*) Absorber pressure
(*b*) Generator pressure
(*c*) Mole fraction of LiCl in the generator

The vapor pressure of H_2O in the LiCl solution is

$$\log P = A - \frac{B}{T + 230}$$

$$A = 7.5713 + (3.7964 \times 10^{-2})(X) - (9.0133 \times 10^{-4})(X^2)$$

$$B = 1.5821 \times 10^3 + 10.364X - (1.4462 \times 10^{-1})(X^2)$$

where P = vapor pressure, mmHg
T = temperature, °C
X = concentration, wt % LiCl

12.7E A LiCl-H_2O absorption air conditioning system is to be designed for no vapor superheating in the evaporator and no liquid subcooling in the condenser. The LiCl liquid concentration in the generator is 45 weight percent and in the absorber is 20 weight percent. Using the data given and assuming ideal solution behavior where necessary, find for an absorber temperature of 40°F and a generator temperature of 85°F:
(*a*) Absorber pressure
(*b*) Generator pressure

(*c*) Mole fraction of LiCl in the generator

The vapor pressure of H_2O in the LiCl solution is

$$\log P = A - \frac{B}{0.555T + 198}$$

$$A = 7.2615 + (3.7964 \times 10^{-2})(X) - (9.0133 \times 10^{-4})(X^2)$$

$$B = 1.5821 \times 10^3 + 10.364\,X - (1.4462 \times 10^{-1})(X^2)$$

where P = vapor pressure, mmHg
T = temperature, °F
X = concentration, wt % LiCl

12.8S One mole each of nonreacting refrigerant 12 and water is in a closed vessel at 1 atm and 25°C. Find the concentration (mole fraction) of refrigerant 12 in each phase, liquid and vapor, assuming ideal behavior in both phases.

12.8E One mole each of nonreacting refrigerant 12 and water is in a closed vessel at 1 atm and 77°F. Find the concentration (mole fraction) of refrigerant 12 in each phase, liquid and vapor, assuming ideal behavior in both phases.

12.9 For a system of I nonreacting components in single phase, derive the Gibbs-Duhem equation, Eq. (12.7), starting with Eq. (12.13).

12.10 Show that the Gibbs-Duhem equation [Eq. (12.7)] applies for the cases of any component i among J phases and for the I components in any phase j. [Start with Eq. (12.13)].

12.11 For the case of one component in J phases, show that the chemical potential $\bar{\mu}_j$ is the same in all phases. [Start with Eq. (12.13)].

12.12 For the case of I components in J phases, prove that the chemical potential of any component i, $[\bar{\mu}_i]_j$, is the same in all phases. [Start with Eq. (12.13)].

12.13 For the reaction

$$2CO + O_2 \leftrightarrow 2CO_2$$

show the relationship of K for this reaction to the value of K: (*a*) when N_2 is present during the reaction and (*b*) for the reaction $2CO_2 \leftrightarrow 2CO + O_2$.

12.14S An inventor claims that she has a method for producing hydrogen and oxygen from water vapor with a new catalyst by the reaction

$$2H_2O \leftrightarrow 2H_2 + O_2$$

The reaction is carried out at 4000 K and 0.1 atm.

(a) What are the equilibrium mole fractions of H_2 and O_2 at this condition?
(b) In a continuous flow reactor with inlet at 25°C, how much heat transfer is necessary per kilogram of H_2 produced?
(c) How much of the heat transfer could be recovered by cooling the products to 40°C?

12.14E An inventor claims that he has a method for producing hydrogen and oxygen from water vapor with a new catalyst by the reaction

$$2H_2O \leftrightarrow 2H_2 + O_2$$

The reaction is carried out at 7200°R and 0.1 atm.
(a) What are the equilibrium mole fractions of H_2 and O_2 at the condition?
(b) In a continuous flow reactor with inlet at 77°F, how much heat transfer is necessary per pound-mass of H_2 produced?
(c) How much of the heat transfer could be recovered by cooling the products to 100°F?

12.15S At 3000 K and $P = 1$ atm, the reversible reaction

$$CO + \tfrac{1}{2}O_2 \leftrightarrow CO_2$$

takes place.
(a) Find the value of y_{CO_2} in the mixture.
(b) Find the value of y_{CO_2} if the pressure is 10 atm.

12.15E At 5400°F and $P = 1$ atm, the revesible reaction

$$CO + \tfrac{1}{2}O_2 \leftrightarrow CO_2$$

takes place.
(a) Find the value of y_{CO_2} in the mixture.
(b) Find the value of y_{CO_2} if the pressure is 10 atm.

12.16S The reaction

$$N_2 + O_2 \leftrightarrow 2NO$$

occurs in automobile engines. Calculate the equilibrium concentration of NO at $T = 327$°C and $P = 1$ atm (assumed conditions at the tailpipe exit). Discuss how the NO concentration in the real case, which is much higher than found by this calculation, could occur.

12.16E The reaction

$$N_2 + O_2 \leftrightarrow 2NO$$

occurs in automobile engines. Calculate the equilibrium concentration of NO at $T = 620$°F and $P = 1$ atm (assumed conditions at the tailpipe exit). Discuss how the NO concentration in the real case, which is much higher than found by this calculation, could occur.

12.17S Ammonia is synthesized for fertilizer production by reacting nitrogen and hydrogen at high pressure and temperature through the reaction

$N_2 + 3H_2 \leftrightarrow 2NH_3$

(*a*) At $T = 1000$ K and $P = 40$ atm, what will be the mole fraction of NH_3 in the mixture of gases? (Assume ideal behavior.)

(*b*) Would increasing the pressure of the synthesis aid or hurt ammonia production?

12.17E Ammonia is synthesized for fertilizer production by reacting nitrogen and hydrogen at high pressure and temperature through the reaction

$N_2 + 3H_2 \leftrightarrow 2NH_3$

(*a*) At $T = 2000°$R and $P = 40$ atm, what will be the mole fraction of NH_3 in the mixture of gases? (Assume ideal behavior.)

(*b*) Would increasing the pressure of the synthesis aid or hurt ammonia production?

12.18 After nitrogen and hydrogen react to form ammonia at any given P and T, the nitrogen and hydrogen in the equilibrium mixture are in the mole ratio $1/r$ when the mole fraction of ammonia formed is x. Show that $r = 3$ for the maximum value of x that is possible

12.19S Carbon dioxide dissociates to form carbon monoxide and oxygen under certain conditions. For pure CO_2 at 3000 K, plot the mole fraction of CO in the equilibrium mixture versus total pressure in the range $0.01 < P_{tot} < 5$ atm. Use semilog paper for the plot, or plot mole fraction versus log P.

12.19E Carbon dioxide dissociates to form carbon monoxide and oxygen under certain conditions. For pure CO_2 at 5000°R, plot the mole fraction of CO in the equilibrium mixture versus total pressure in the range $0.01 < P_{tot} < 5$ atm. Use semilog paper for the plot, or plot mole fraction versus log P.

12.20S Carbon dioxide at 1 atm may dissociate to form CO and O_2. Plot the mole fraction of CO versus T in the range $600 < T < 3000$ K on semilog paper, or plot mole fraction versus log T.

12.20E Carbon dioxide at 1 atm may dissociate to form CO and O_2. Plot the mole fraction of CO versus T in the range $900 < T < 5500°$R on semilog paper, or plot mole fraction versus log T.

12.21 Following the development of Eqs. (12.8) through (12.12), derive an expression for chemical equilibrium in a constant-volume constant-temperature control mass, using the Helmholtz function.

12.22S Find $\Delta \overline{G}^o(T)$ for the reaction

$$CO + \tfrac{1}{2}O_2 \leftrightarrow CO_2$$

at $T = 1000$ K, and use the result to calculate $K(T)$ for this reaction. Compare the result with $K(T)$ from Table C.12.

12.22E Find $\Delta \overline{G}^o(T)$ for the reaction

$$CO + \tfrac{1}{2}O_2 \leftrightarrow CO_2$$

at $T = 2000°$R, and use the result to calculate $K(T)$ for this reaction. Compare the result with $K(T)$ from Table C.12.

12.23S A certain reaction has an equilibrium constant of $K(T = 500 \text{ K})$ with $\Delta \overline{h}^o = 8812$ kJ/kmol. Estimate the value of $K(T = 700 \text{ K})$ in terms of $K(T = 500 \text{ K})$.

12.23E A certain reaction has an equilibrium constant of $K(T = 1000°\text{R})$ with $\Delta \overline{h}^o = 3788$ Btu/lbmol. Estimate the value of $K(T = 1250°\text{R})$ in terms of $K(T = 1000°\text{R})$.

12.24S Near $T = 1000$ K, a plot of the equilibrium constant for a particular reaction is curve-fitted and yields the relation

$$\ln K = -\frac{4200}{T}$$

Estimate the value of $\Delta \overline{H}^o(T = 1000 \text{ K})$ for this reaction.

12.24E Near $T = 2000°$R, a plot of the equilibrium constant for a particular reaction is curve-fitted and yields the relation

$$\ln K = -\frac{7560}{T}$$

Estimate the value of $\Delta \overline{H}^o(T = 2000°\text{R})$ for this reaction.

12.25S For the reaction described in Prob. 12.24S, estimate the value of K at $T = 1200$ K.

12.25E For the reaction described in Prob. 12.24E, estimate the value of K at $T = 2200°$R.

12.26 For a control mass, it is often convenient to work with the Helmholtz function $A = U - TS$ rather than the Gibbs function. Following the development of Eqs. (12.65) through (12.71), derive a general expression for dA in terms of the appropriate control mass properties. Discuss the implied conditions for the various types of equilibrium (thermal, mechanical, and chemical) that result.

13

Introduction to Microscopic Thermodynamics

We have no right to assume that any physical laws exist, or if they have existed up to now, that they will continue to exist in a similar manner.
Max Planck

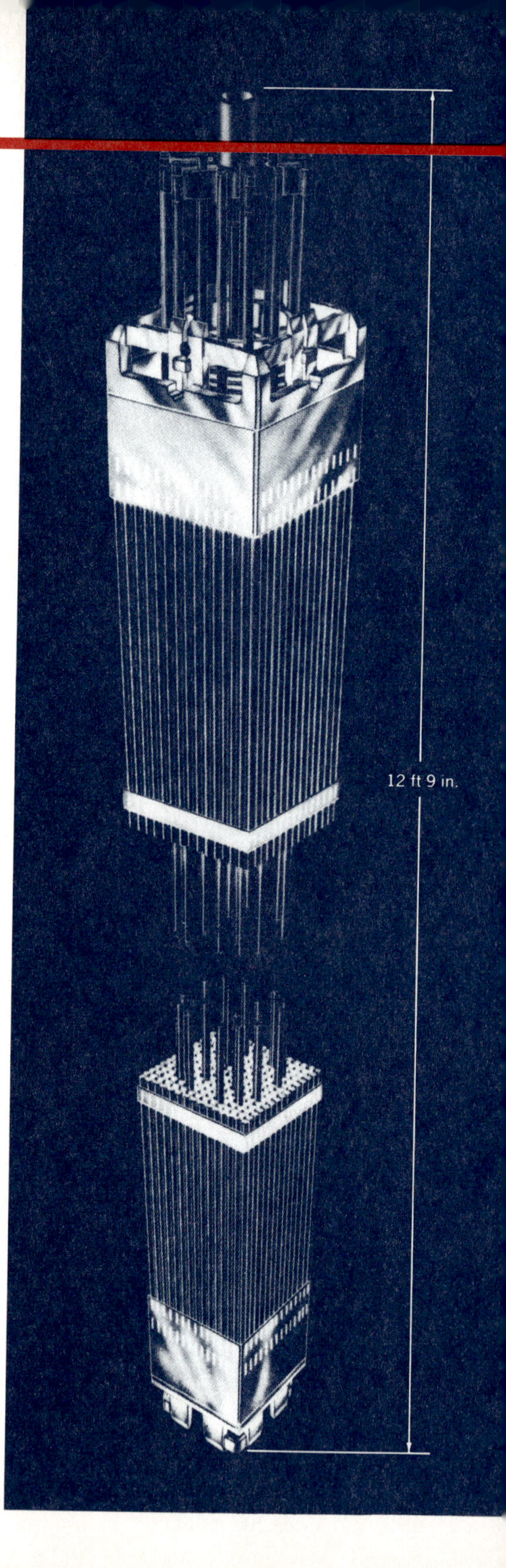

Fuel assembly for pressurized water nuclear reactor used for commercial electric power generation. Fuel is uranium dioxide with a zirconium alloy cladding to contain fission products and avoid corrosion, with an assembly of aluminum alloy plates. (*Babcock and Wilcox.*)

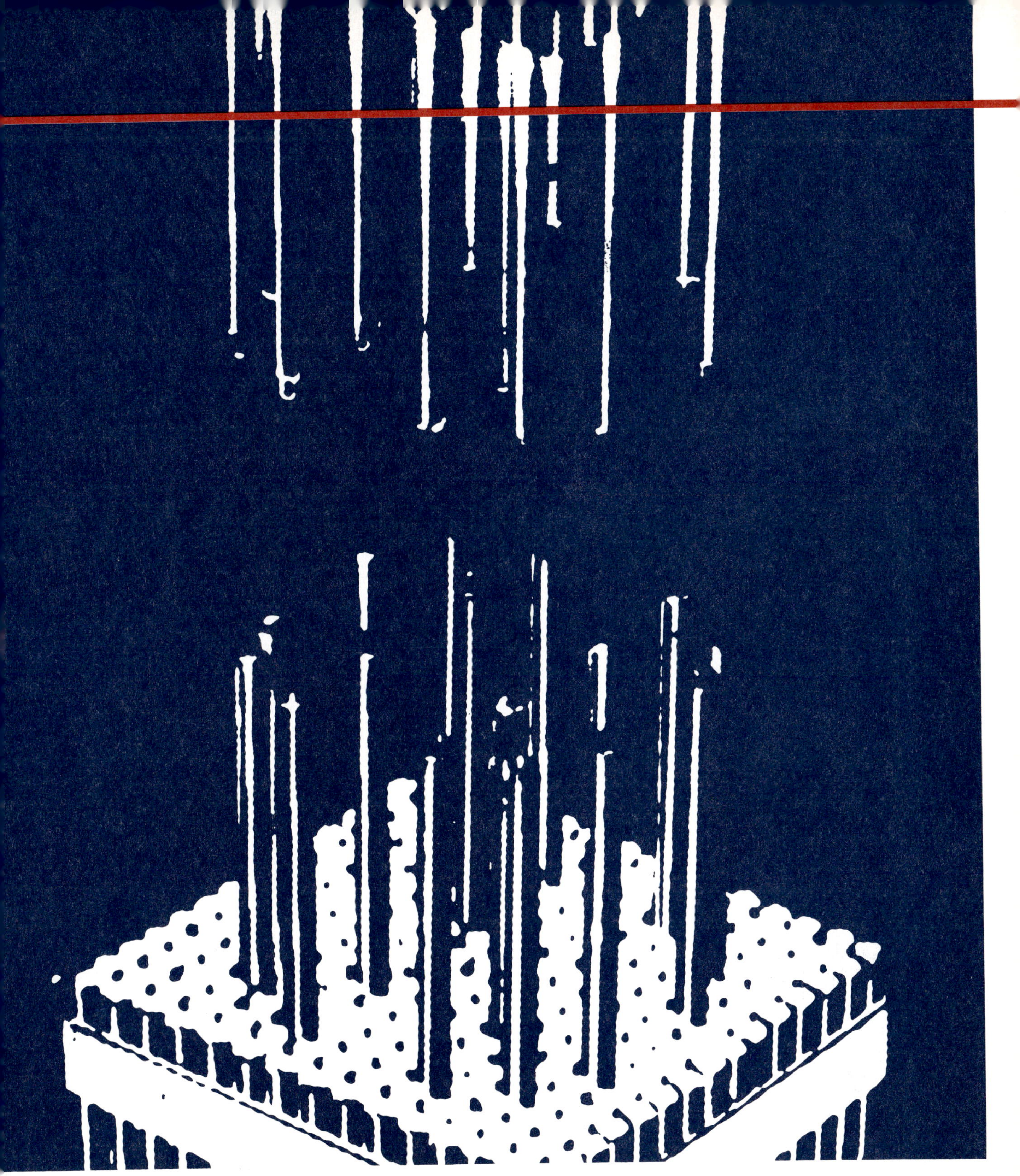

13.1 Introduction

Reference has been made to this chapter at earlier points, generally with respect to the ability to use a microscopic approach to thermodynamics that provides insight into the behavior of thermodynamic systems which is unavailable from the classical approach. To do so, we present here an overview of the microscopic approach to thermodynamics, with emphasis on providing that insight rather than on a rigorous introduction to statistical thermodynamics. Nevertheless, on completion of this chapter, the reader should have the tools to do some things that are not available from the classical approach. In particular, we see how properties can be computed for materials that can be modeled in a simple way on a microscopic level; how entropy is related to our knowledge about the state of a system, and what this implies about irreversibility; and how we might develop equations of state that are based on more than empirical observation of macroscopic behavior.

The system of units used in this chapter is exclusively SI. The use of USCS units for statistical thermodynamics is uncommon, so the dual presentation is abandoned here. The reader should note that some texts use the centimeter-gram-second system (cgs system) of units for statistical thermodynamics which is slightly different from SI.

13.2 Defining a Microscopic System

What do we need to know to completely describe a microscopic system in thermodynamic equilibrium? This question has led to detailed and interesting investigations, and the answer depends to some extent on how we intend to apply the information. The description of a plasma composed of free electrons and ions may require us to provide different sorts of information than is necessary for describing an ideal gas. For our purposes, we might agree on the following: We must specify as a minimum the number of individual particles that compose the system under study, the position and velocity of each particle, and the internal energy of each particle. With all this information, it is possible to calculate the quantities that we have found to be of interest in our study of classical thermodynamics.

How do we obtain such information for a given system? Generally we can find the number of particles by use of the standard methods of chemistry and physics that allow us to relate the number of particles N_p to the total mass of the system m_{sys}, the molecular weight M, and Avogadro's number $N_a = 6.023 \times 10^{26}$ particles per mole by

$$N_p = \frac{m_{sys}}{M} N_a \tag{13.1}$$

How many of these particles have a given microscopic property value, and what are their locations? Generally, we cannot answer these questions, because the particle energies and positions change with time. This is obvious for the case of a gas, where the individual gas molecules are in constant motion and the molecules exchange energy with one another through collisions. Even if we somehow were able to know the positions and energies of all particles at a given time, a short time later the values would have changed and our information would no longer be valid. (At one period in the history of physics, it was believed that given the required information at one particular time, Newton's laws would allow us to trace the history of all particles through time so that our required information would be available at any later time. As we will see, this hope of a "deterministic" solution is no longer a possibility, even if we had a computer with enough speed and memory to carry out such a computation.)

If we cannot expect to obtain the required information, how can we proceed with microscopic calculations? A different approach is obviously required, and we must work with the data that we can obtain. Such information is usually in the form of statistical distributions and their averages rather than the details of the properties of each individual particle that make up the averages. Much information is lost to us in this way. Consider knowing the average age of a group of 150 people at a cocktail party. Such information tells us little about the actual age of any individual. We might have some other information, such as that the minimum age of any individual is, say, 16, while the upper bound on ages might be on the order of 85. We might further speculate on the statistical distribution of ages within this range, perhaps based on the distribution in the national population. But we can never predict with certainty the exact age of individual A or B at the party from the given information. This is the sort of dilemma that we face in the microscopic approach.

In this section, we address two questions: First, if we have statistical information about how many particles have a given microscopic property value (kinetic energy, for example) in the sense of a known statistical distribution of the property per particle, how are these distributions related to properties that we can observe? Second, how can the required distributions be developed?

13.2.1 General Properties

We are familiar with properties that describe the macroscopic state of a system in thermodynamic equilibrium, such as P, T, v, u, h, s, and others. For our discussion of microscopic thermodynamics to be useful, we should be able to relate the microscopic system properties to the macroscopic properties.

Suppose we examine particle i from the N_p particles that make up a control mass or that are instantaneously in a control volume under study. Further, let us make the same assumptions on a microscopic level that we do on a macroscopic level, i.e., that the system is homogeneous, so that if we look at any subvolume dV within the system, the average properties within that subvolume are the same as in any other subvolume. On a microscopic level, this means that particles may move in and out of dV, but the net number of particles within any dV at any time is, on average, the same. We can now define some properties in terms of averages. The simplest example is the system mass, which must be made up of the masses of the individual particles, or

$$m_{\text{sys}} = \sum_{i=0}^{N_p} m_i \tag{13.2}$$

An alternative way of writing Eq. (13.2) is to realize that there may be J groups of particles, and each group will have N_j particles of the same mass m_j, so that the total mass is

$$m_{\text{sys}} = \sum_{j=0}^{J} N_j m_j \tag{13.3}$$

or, if all particles have identical mass,

$$m_{\text{sys}} = mN_p = N_p \frac{M}{N_a} \tag{13.4}$$

The specific volume of a system of N_p identical particles is then

$$v = \left(\frac{V}{m}\right)_{\text{sys}} = \frac{VN_a}{N_p M} \tag{13.5}$$

For particle i, there is some associated energy ε_i. This energy can be in various forms; if the particle is a gas molecule, the energy may be made up of translational kinetic energy, vibrational or rotational energy, etc. At the moment, the form of the energy is immaterial. The total energy of the N_p particles in the system is the sum of the energies of all the particles. However, more than one particle may have the same energy, so we can take the number of particles N_j with energy ε_j and sum over the resulting J groups, to give

$$U_{\text{sys}} = \sum_{j=0}^{J} N_j \varepsilon_j \tag{13.6}$$

The specific energy of the material in the system is then

$$u_{\text{sys}} = \left(\frac{U}{m}\right)_{\text{sys}} = \frac{1}{m_{\text{sys}}} \sum_{j=0}^{J} N_j \varepsilon_j \tag{13.7}$$

Any specific extensive property ϕ, such as system energy, can be written in a similar form as

$$\phi_{sys} = \left(\frac{\Phi}{m}\right)_{sys} = \frac{1}{m_{sys}} \sum_{j=0}^{J} N_j \phi_j \tag{13.8}$$

Thus, if we know how many particles have individual property ϕ_j, then the macroscopic specific extensive property ϕ_{sys} can be computed. We are left, then, with the problem of determining how many particles N_j have various values of property ϕ_j.

13.2.2 Allowable Microstates

Answering the question of how many particles N_j have a certain value of property ϕ_j at a given state would lead us through much of the development of modern physics, which is not our purpose here. However, we can observe the results of the reasoning that went into answering the question and some of the important discoveries that ensued.

Maxwell-Boltzmann Energy Distribution

Classical physics allows us to proceed a certain distance toward answering the question of the distribution of a given property ϕ_j among the N_p particles being studied. James Clerk Maxwell derived the distribution of velocities that exists among the molecules of a monatomic gas, and Ludwig Boltzmann extended Maxwell's result to general forms of energy, assuming that a particle could have any value of energy. This assumption will prove to be only an approximation and leads to significant errors in certain cases.

The derivation by Maxwell and Boltzmann shows that N_p particles in a system of fixed mass have energy distributed among the particles so that $dN_p(\varepsilon)$ particles have energy in the range $d\varepsilon$ around ε according to the relation

$$\frac{dN_p(\varepsilon)}{N_p} = 2(kT)^{-3/2}\left(\frac{\varepsilon}{\pi}\right)^{1/2} \exp\left(\frac{-\varepsilon}{kT}\right) d\varepsilon \tag{13.9}$$

In this equation, $k = \overline{R}/N_a$ is the *Boltzmann constant,* which has magnitude in SI units of 1.3806×10^{-26} kJ/K and can be interpreted as the gas constant per particle. Note that if we plot $(1/N_p)\, dN_p(\varepsilon)/d\varepsilon$ versus ε, the only parameter is the absolute temperature, or more commonly the product kT. Such a plot is shown in Fig. 13.1.

The peak of the curve, or most probable (MP) energy, occurs at $\varepsilon_{MP} = kT/2$.

From Eq. (13.9), the average energy per particle can be found:

$$\begin{aligned}\bar{\varepsilon} &= \frac{1}{N_p} \sum_{j=0}^{J} N_j \varepsilon_j \\ &= 2(kT)^{-3/2}\left(\frac{1}{\pi}\right)^{1/2} \int_0^{\infty} \varepsilon^{3/2} \exp\left(\frac{-\varepsilon}{kT}\right) d\varepsilon \end{aligned} \tag{13.10}$$

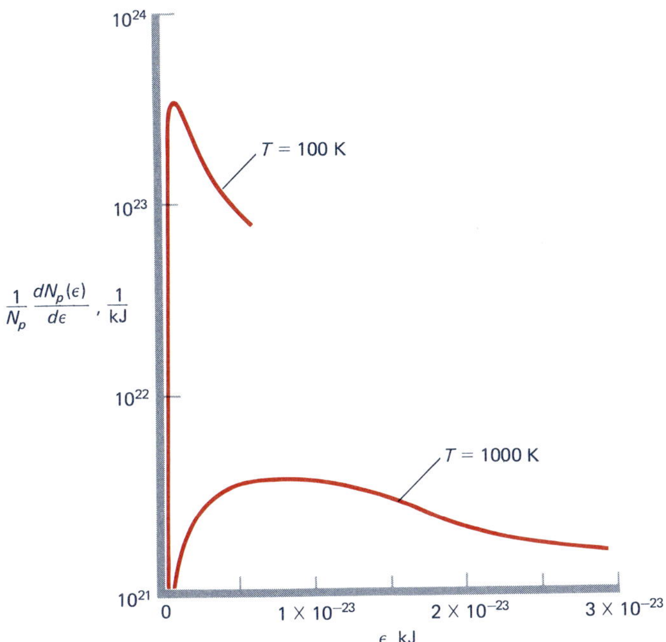

Figure 13.1 Maxwell-Boltzmann energy distribution.

In Eq. (13.10), the summation has been replaced by an integration because of the assumption that *any* energy can be possessed by the particles, and so the energy levels form a continuous distribution, as plotted in Fig. 13.1. The integration in Eq. (13.10) can be done by a change of variables, by letting $X^2 = \varepsilon/(kT)$, which results in

$$\bar{\varepsilon} = \frac{4kT}{\pi^{1/2}} \int_0^{\infty} X^4 \exp(-X^2)\, dX = \frac{4kT}{\pi^{1/2}} \frac{3(\pi^{1/2})}{8}$$

$$= \tfrac{3}{2}kT \tag{13.11}$$

Some observations are in order here. As temperature increases, both $\bar{\varepsilon}$ and ε_{MP} increase in proportion. Also the distributions of particles with respect to energy in Fig. 13.1 become more spread out as temperature increases. This is in accordance with the arguments made in Chap. 5 concerning the relation of entropy change and heat transfer, where we observed that the entropy, or "uncertainty," decreased with a heat transfer from a control mass. We see from Fig. 13.1 that a heat transfer from a control mass reduces the temperature of the control mass, but the energy distribution among the particles becomes compacted; i.e., we are more sure of the energy of any given particle, since the range of values is not so spread out at lower temperatures.

Some Predictions Using the Maxwell-Boltzmann Distribution

Let us see whether the Maxwell-Boltzmann distribution can lead us even further. Suppose the particles under study are assumed to be point masses within an adiabatic container, that the particles undergo perfectly elastic collisions with one another and the surface of their container, and that the particles have no other energy but their kinetic energy. This model has most of the characteristics of a monatomic gas.

For such an idealized gas, the pressure exerted on the container surface can be found from the rate of change of momentum, and thus force, exerted on the wall as the particles strike and rebound in the perfectly elastic collisions from the surface. For each particle, the change in momentum $\Delta \mathbf{p}$ on striking the wall is (Fig. 13.2)

$$\Delta \mathbf{p} = m\mathbf{V}_z - m(-\mathbf{V}_z) = 2m\mathbf{V}_z \tag{13.12}$$

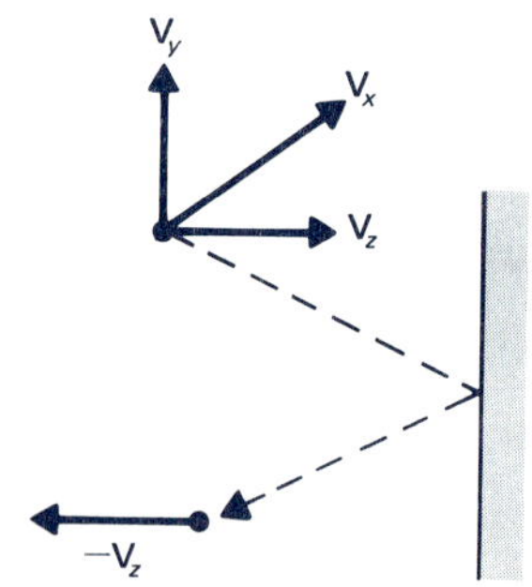

Figure 13.2 Particle-wall interaction.

The number of particles at speed $\mathbf{V}_z$ striking the wall per unit time and area is found by examining Fig. 13.3. Any particle with velocity component $\mathbf{V}_z$ that is within a distance $\Delta L = \mathbf{V}_z\,\Delta t$ of the wall will strike the wall within time interval Δt. If there are $dN(\mathbf{V}_z)/V$ particles per unit volume with speed $\mathbf{V}_z$, then $dN(\mathbf{V}_z)/\Delta t = A\,\Delta L\, dN(\mathbf{V}_z)/(V\,\Delta t) = A\mathbf{V}_z\, dN(\mathbf{V}_z)/V$ particles with speed $\mathbf{V}_z$ will strike the wall per unit time.

The total rate of change of momentum at the wall is due to the number of particles per unit time striking the wall times the momentum change per particle, integrated over all particle speeds, or

$$F = \frac{dp}{dt} = \frac{m}{dt}\int_0^\infty \mathbf{V}_z\, dN(\mathbf{V}_z) = \frac{Am}{V}\int_0^\infty \mathbf{V}_z^2\, dN(\mathbf{V}_z) \tag{13.13}$$

However, the integral on the right is simply the mean square speed times the number of particles per unit volume, i.e.,

$$\int_{\mathbf{V}_z=0}^\infty \mathbf{V}_z^2\, dN(\mathbf{V}_z) = N\overline{\mathbf{V}_z^2} \tag{13.14}$$

so that Eq. (13.13) can be written as

$$V\left(\frac{F}{A}\right) = PV = mN\overline{\mathbf{V}_z^2} \tag{13.15}$$

The average kinetic energy of a gas particle is given by summing the three components of velocity, to give

$$\bar{\varepsilon} = \left(\frac{m}{2}\right)(\overline{\mathbf{V}_x^2} + \overline{\mathbf{V}_y^2} + \overline{\mathbf{V}_z^2}) \tag{13.16}$$

If we assume that the three components of mean square speed are equal (otherwise, all the particles would end up in one portion of the container), then the average energy per particle is

$$\bar{\varepsilon} = \left(\frac{3m}{2}\right)\overline{\mathbf{V}_z^2} \tag{13.17}$$

Figure 13.3 Rate of particles striking wall.

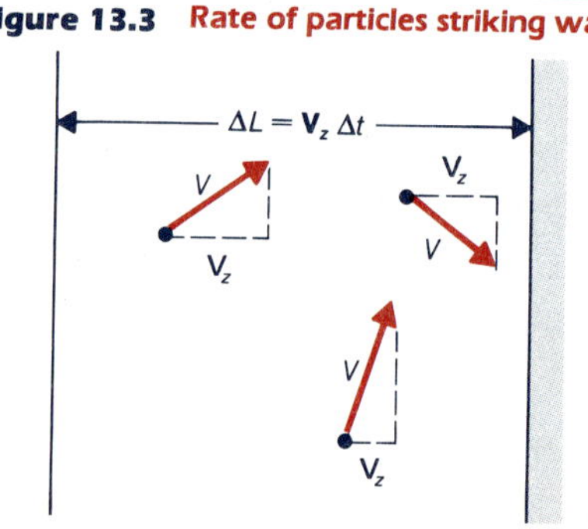

According to the Maxwell-Boltzmann distribution, the average energy per particle [Eq. (13.11)] is simply $\frac{3}{2}kT$, so

$$\overline{\mathbf{V}_z^2} = \frac{2}{3m}\bar{\varepsilon} = \frac{2}{3m}\frac{3kT}{2} = \frac{kT}{m} \tag{13.18}$$

Substituting into Eq. (13.15) gives

$$PV = NkT = \left(\frac{N}{N_a}\right)(kN_a)T \tag{13.19}$$

or since $kN_a = \bar{R}$ and $N/N_a = n$, the number of moles of gas, Eq. (13.19), becomes

$$PV = n\bar{R}T \tag{13.20}$$

Our model of a gas follows the ideal gas law! In itself, this is perhaps not so surprising; however, we have developed some powerful insight into just what an ideal gas is on the microscopic level. In addition, because we know something about the energy contained in such a gas and how it is related to temperature, we may be able to go further with our analysis.

Let us examine the energy contained by the gas. Since we have assumed that only translational kinetic energy is present (in order to invoke the Maxwell-Boltzmann distribution), the energy contained by 1 mol of gas is

$$\bar{u} = N_a\bar{\varepsilon} = N_a(\tfrac{3}{2}kT) = \tfrac{3}{2}\bar{R}T \tag{13.21}$$

and the internal energy of the gas depends on the temperature only, which we know to be correct for an ideal gas. Taking the derivative of $\bar{u}$ with respect to temperature gives

$$\frac{d\bar{u}}{dT} = \bar{c}_v = \frac{3}{2}\bar{R} \tag{13.22}$$

We have thus predicted the value of a property $\bar{c}_v$ for an ideal gas. Indeed, we predict $\bar{c}_v$ for all gases to be the same! This is obviously not true; however, if we examine molar values of $\bar{c}_v$ for *monatomic* gases, we find the prediction to be quite accurate over broad ranges of temperature. For more complex molecules, the assumption that only translational energy makes up the internal energy $\bar{u}$ is not valid, and we must extend the model.

In the classical approach, we were unable to predict property values; we could develop relations among properties, but we were always forced to make a certain minimal set of experimental measurements to develop the remaining properties.

Extensions to Diatomic Molecules

Is it possible to extend the results of the simple model to diatomic or even more complex molecules? Many attempts at such an extension have been made. One simple and useful one is to note that the translational kinetic energy is made up of the three components of velocity according to Eq. (13.16). Further, this translational energy was set equal to the average energy found from the Maxwell-Boltzmann distribution, $\frac{3}{2}kT$. Thus it might be possible to assign to each component of kinetic energy the value $\frac{1}{2}kT$; in other words, we can *partition* the total kinetic energy of a particle into its components, each with value $\frac{1}{2}kT$.

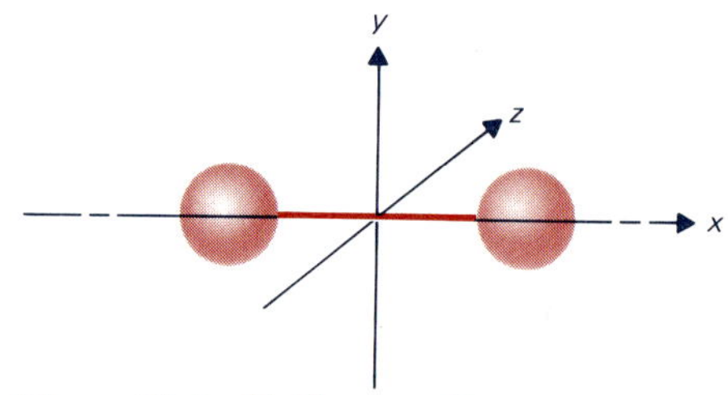

Figure 13.4 Rigid rotator diatomic molecule.

Suppose we now examine a diatomic molecule (Fig. 13.4) and assume that it acts as two masses attached by a rigid rod. Such a model is called a *rigid rotator.* Let an axis pass through the two masses, and two other mutually perpendicular axes originate at the center of mass of the molecule. If we assume that the moment of inertia of the molecule is negligible for rotation around the axis through the masses, there are two remaining rotational axes for storing kinetic energy of rotation in the rigid rotator. If each of these axes can store rotational kinetic energy in the amount $\frac{1}{2}kT$, then the total internal energy of a diatomic molecule is $\frac{5}{2}kT$, that is, $\frac{3}{2}kT$ in translational kinetic energy and $\frac{2}{2}kT$ in rotational energy. The internal energy of 1 mol of a diatomic gas is then

$$\bar{u} = \frac{5}{2}N_a kT \tag{13.23}$$

resulting in a prediction of

$$\bar{c}_v = \frac{5}{2}\bar{R} \tag{13.24}$$

This value for the molar specific heat of a diatomic ideal gas is also quite accurate at low temperatures; however, as the absolute temperature of the gas is increased, the measured specific heat increases. For nitrogen, for example, the $\bar{c}_v$ versus T data from Table D.4 are plotted in Fig. 13.5. At higher temperatures, the value of $\bar{c}_v$ appears to be approaching $\frac{7}{2}\bar{R}$. It appears, then, that if our partitioning method is correct, two more ways of storing internal energy in the diatomic molecule are occurring at higher temperature. The usual model invoked is that the two atoms making up the molecule are no longer rigidly attached; rather, the rigid rod should instead be modeled as a spring, which allows both potential and kinetic energy to be stored as the spring vibrates more and more intensely at increasing temperature. The value of $\bar{c}_v$ for a vibrating diatomic molecule should then approach $\frac{7}{2}\bar{R}$ at high temperature.

Figure 13.5 Specific heat versus temperature for nitrogen.

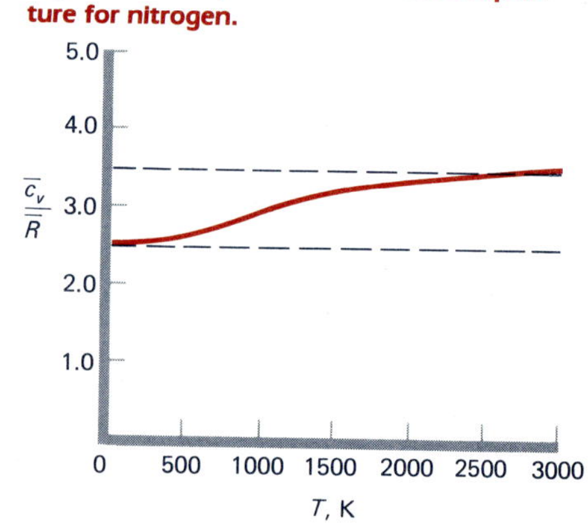

These results should raise a flag as to the correctness of the model. Based on our model, the vibrating diatomic molecule should have 7 energy storage modes, usually called *degrees of freedom,* while the rigid rotator should have 5. However, either the degrees of freedom exist for the molecule, or they do not. They should not slowly become active as a function of temperature. Why does the observed slow transition from 5 to 7 degrees of freedom occur as the temperature is raised? Additionally, it seems artificial to assign a separate degree of freedom to both the potential and the kinetic energy of vibration. After all, the kinetic and potential energies are simply traded back and forth during the vibration process such that their sum remains constant. Is this really two *modes* of energy storage? Yet we must find two such modes to make the model agree with experimental data. Our model must be still incomplete or at least not completely understood.

As more complex (polyatomic) molecules must be modeled, the kinetic theory approach we have been using becomes less fruitful. However, it does provide a simple model for the behavior of a gas and, along with the principle of the equipartition of energy (which assigns $\frac{1}{2}kT$ to each possible energy mode), is historically important in that together they give some useful insight into the macroscopic observations available in classical thermodynamics. We have also found some useful equations for the internal energy and specific heat of ideal gases that agree with observation and indeed allow accurate predictions for these properties when no experimental data are available. It is not possible to predict properties only from knowledge of classical thermodynamics. Complete property data is based on some minimal experimental set of property data. Kinetic theory has allowed successful property prediction without resort to experiment.

With this success, how do we proceed to the prediction of other properties and to a better understanding of macroscopic thermodynamics? When we try to take further steps in kinetic theory, we find that some earlier assumptions prove to be invalid in important cases. It is worthwhile to examine the consequences of our earlier assumptions: In particular, what happens if all possible energies are not available to the particles which we are studying, as we assumed for Eq. (13.9)? And do all particles then follow the Maxwell-Boltzmann distribution, or must some other distribution of energies be used?

It proves quite difficult to build models of the detailed behavior of atoms and molecules based on their assumed microscopic structure. A more fruitful approach is to look at the statistical average behavior of a group of molecules whose exact structure is unknown but whose average behavior can be inferred. Such an approach is called *statistical mechanics,* as opposed to the kinetic theory approach we have used so far, which is closely tied to the assumed structural properties of atoms and molecules.

The mathematical foundations of statistical mechanics, indeed almost all the detail that is required, were laid out in the early twentieth century by U.S. thermodynamicist Josiah Willard Gibbs. At that time, however, the statistical description of molecular behavior was limited to the Maxwell-Boltzmann distribution, which, as we have implied, has some shortcomings. Thus, although the *method* proposed by Gibbs is complete and correct, the information available at the time he proposed the method was incomplete and led to predictions which were in error. To avoid that pitfall, we take a detour through a most interesting area, quantum mechanics, and then return to our goal of surveying statistical approaches to thermodynamics with more complete information than was available to Gibbs.

13.3 Influence of Quantum Effects

So far, all our modeling (including the Maxwell-Boltzmann distribution) has been based on the assumption that newtonian mechanics provides a valid description of particle interactions at the microscopic level. There is much evidence to indicate that this is not the case, particularly in the observations of the behavior of the properties of solids at low temperatures, the absorption of radiant energy by gases, and even in our observations of the predicted versus observed specific heats of gases.

The inability of classical mechanics to provide satisfactory explanations of observed phenomena led physicists in the early twentieth century to try a different approach. Max Planck, in trying to predict the observed distribution of radiant energy with respect to wavelength from an ideal radiating surface, was forced to model the possible values possessed by the radiated energy as being discrete; i.e., not all energies were possible, only discrete multiples of a basic value. Only by making such an assumption could he reconcile predictions with experiment, where predictions based on classical physics had all failed. Neils Bohr used this idea of discrete energy states in constructing a model of the hydrogen atom, which was successful in exactly predicting the radiation absorption characteristics of hydrogen gas, again after all attempts based on classical physics had failed. These successes, along with others by Albert Einstein, led to further development and final acceptance of Planck's original idea.

A further major advancement in the development of quantum theory was triggered by Louis de Broglie's suggestion that energy and matter were closely related and that both should therefore possess wavelike characteristics. His suggestion was almost immediately given striking confirmation by the results of experimental work by Davisson and Weeks on the patterns resulting from the scattering of electrons from solid surfaces, which could be explained only by ascribing a wavelength to the particles used in their experiments. This work led to a mathematical description of the wavelike properties of matter by Erwin Schrödinger (1887–1961). His equation describes the probability that matter will exist in a given location at a given time, and it gave impetus to the idea that our knowledge about the behavior of microscopic systems could, at best, be on a statistical or probabilistic basis, rather than on the deterministic basis prescribed by newtonian mechanics. We need not go into detailed solutions of the Schrödinger equation, except to note its time-independent form in cartesian coordinates:

$$\frac{\partial^2\psi}{\partial x^2}+\frac{\partial^2\psi}{\partial y^2}+\frac{\partial^2\psi}{\partial z^2}+\frac{8\pi^2 m}{h^2}(\varepsilon-\mathrm{PE})\psi=0 \tag{13.25}$$

where ψ is the *wave function,* which describes the amplitude or "presence" of a "matter wave" in space, PE is the potential energy, and h is Planck's constant, 6.6262×10^{-34} J · s.

13.3.1 An Example of Quantization

Let us examine a single particle which moves only along the x coordinate between two infinite parallel planes separated by a distance L. There is no potential energy for such a case, and Schrödinger's equation reduces to

$$\frac{d^2\psi_x}{dx^2} + \left(\frac{8\pi^2 m}{h^2}\right)\varepsilon\psi_x = 0 \tag{13.26}$$

This is a second-order linear ordinary differential equation, and we must prescribe two boundary conditions to obtain a complete solution. If the particle cannot escape from the enclosure and ψ_x is related to the probability that the particle exists at some position, then we can require $\psi_x = 0$ at $x \le 0$ and at $x \ge L$. The solution of Eq. (13.26) is

$$\psi_x = A \sin\left(\frac{8\pi^2 m\varepsilon}{h^2}\right)^{1/2} x + B \cos\left(\frac{8\pi^2 m\varepsilon}{h^2}\right)^{1/2} x \tag{13.27}$$

Applying the boundary conditions $\psi_x = 0$ at $x = 0$ gives $B = 0$, and applying the condition $\psi_x = 0$ at $x = L$ results in

$$0 = A \sin\left(\frac{8\pi^2 m\varepsilon}{h^2}\right)^{1/2} L \tag{13.28}$$

For Eq. (13.28) to hold without the trivial solution $A = 0$ requires that the argument of the sine term be a multiple of π, or

$$n'\pi = \left(\frac{8\pi^2 m\varepsilon}{h^2}\right)^{1/2} L \qquad n' = 0, 1, 2, 3, \ldots \tag{13.29}$$

Thus, because all of the quantities within the parentheses are constants except for the particle energy ε, the energy of the particle must be

$$\varepsilon = \frac{n'^2 h^2}{8mL^2} \qquad n' = 0, 1, 2, 3, \ldots \tag{13.30}$$

If Schrödinger's equation is indeed correct (and there is ample evidence from other predictions that it is), then this is a profound result. The energy of the particle as given by Eq. (13.30) can have only the values $h^2/(8mL^2)$, $h^2/(2mL^2)$, $9h^2/(8mL^2)$,. . . . This *quantization* of the energy into a certain discrete progression of values is seen to be forced on the system by the form of the equation and the boundary conditions; yet the prediction by Planck that such discrete energy values must exist for massless radiative energy is borne out here for a single particle of mass m.

If we bat a perfectly elastic tennis ball and allow it to bounce continuously between two walls, Eq. (13.30) should apply for the energy values that can be possessed by the ball. We have, after all, placed no restriction on the mass of the particle or the size of the enclosure in our analysis. Yet we do not observe tennis balls to have only discrete energies (velocities). Is our analysis flawed? Let us see how large the difference is between allowed energy states j and k. Equation (13.30) predicts

$$\varepsilon_j - \varepsilon_k = (j^2 - k^2)\frac{h^2}{8mL^2} \tag{13.31}$$

For a 0.1-kg ball and walls 1 m apart, the energy difference between the two states is

$$\begin{aligned}\varepsilon_j - \varepsilon_k &= \frac{(j^2 - k^2)(6.6262 \times 10^{-34}\ \mathrm{J \cdot s})^2}{8 \times 0.1\ \mathrm{kg} \times 1^2\ \mathrm{m}^2} \\ &= (j^2 - k^2)(5.49 \times 10^{-67}) \qquad \mathrm{J}\end{aligned} \tag{13.32}$$

For any pair of integers that differ by unity, the difference between the quantized energy values is so small that we could not observe it. Thus our intuition is not violated. For very low-mass particles in small "containers," however, we can see that the difference in the energies between the quantized states could become appreciable.

The values of the integer n' that appear in the solution are called the *quantum numbers* for this system. If more coordinates than simply x are present, then additional quantum numbers appear in the mathematical solution of the resulting partial differential equation.

We could continue the solution to express the final form and properties of the wave function and further examine the probability of the particle being at a given point (which is proportional to ψ^2). However, our purpose here is simply to show that quantized energy states are a *result* of the probabilistic view of the behavior of matter. They are not imposed on the system by the analyst; they result from a model that has been proved capable of explaining observed behavior to a very high degree of accuracy.

If you are uncomfortable with the thought that you are composed of matter waves that interact in such a way that mass appears with the probability required in the locations necessary to compose your particular shape, then you are in good company. Einstein is widely quoted as saying, "God does not play dice," and Schrödinger said, "I don't like it, and I'm sorry I ever had anything to do with it." Nevertheless, the quantum theory has proved remarkably successful, and the understanding of the behavior of matter that it provides has led, among other things, to the laser, the transistor and integrated circuits, nuclear energy, the development of molecular biology, and our understanding of DNA, and seems on the brink of major breakthroughs in understanding the behavior of subatomic particles.

13.3.2 Uncertainty Principle

One consequence of the Schrödinger equation (and other approaches to quantum theory) is that the product of our uncertainty about the momentum (and thus velocity) of a particle $m\,\Delta\mathbf{V}$ and the uncertainty in the particle position Δx is greater than or equal to a constant. More exactly,

$$m\,\Delta\mathbf{V}\,\Delta x \geq \frac{h}{4\pi} \tag{13.33}$$

This relation was first observed and commented on by Werner Heisenberg in 1926. Equation (13.33) indicates that it is impossible to measure both the momentum and the position of a particle to an arbitrary degree of precision, since the product of the uncertainties in the two quantities must always exceed $h/(4\pi)$. The relation comes out of the interpretation of the wave equation, which implies that a particle has only a probability, not a certainty, of being at a given place at a given time. *Heisenberg's uncertainty principle* gives the limits on our possible knowledge about the location and momentum of a particle. This inherent uncertainty is what finally ends the dream of classical physics of a deterministic universe in which we could track the paths of all particles into the future if we could only know their present states. The more accurately we determine the position of a particle, the less accurately we can know its momentum, and vice versa.

The uncertainty principle applies not only to the product of the uncertainties in momentum and position, but also to any pair of *conjugate variables,* i.e., variables that multiply to give the units possessed by h, J · s, such as the obvious pair, energy × time.

The size of h (6.6262×10^{-34} J · s) is assurance that the uncertainty principle will not be a factor in our everyday laboratory experiments; it does become important when we deal with particles such as electrons.

13.3.3 Bose-Einstein Statistics

Satyendra Bose developed and Albert Einstein extended the study of the statistics that govern the possible energy states in a collection of particles that are indistinguishable. In this context, *indistinguishable* means that if we exchange two particles in the set under study, we cannot distinguish any difference in the set. Photons and single-component ideal gases fall in this category. Particles of this type are called *bosons.*

Bose and Einstein showed that the number of ways N_p bosons could be arranged among all the possible energy levels, when a given energy level can have g_j different quantum states (i.e., sets of quantum numbers) with the same energy, is given by

$$w_{\mathbf{BE}} = \prod_{j=0}^{\infty} \frac{(g_j + N_j - 1)!}{(g_j - 1)!N_j!} \tag{13.34}$$

where w represents the number of possible particle arrangements among the possible energy states. Here, the symbol Π means the product of all the terms from $j = 0$ to $j = \infty$. The quantity g_j is called the *degeneracy* of level

j, and it reflects the fact that various sets of quantum numbers can predict a state with the same energy. It is convenient to group all states with the same energy, rather than to sum them separately.

Using Eq. (13.34) for the number of possible ways of arranging the particles among the energy levels plus general Eq. (13.3), which requires the number of particles in all levels to add to the total number of particles, and Eq. (13.6), which requires the energy of the particles in all levels to sum to the total energy in the system, we find the Bose-Einstein (BE) energy distribution for bosons to be

$$\frac{N_j}{N_p} = \frac{g_j}{Q_{BE}\{\exp[\varepsilon_j/(kT)] - 1\}} \tag{13.35}$$

where

$$Q_{BE} = \sum_{j=0}^{\infty} \frac{g_j}{\exp[\varepsilon_j/(kT)] - 1}$$

Equation (13.35) describes the distribution of particles among the possible energy levels for a system in thermodynamic equilibrium.

13.3.4 Fermi-Dirac Statistics

Wolfgang Pauli in 1925 observed that no more than one particle of certain types could have the same set of quantum numbers. In such a case, more than one indistinguishable particle could not occupy the same quantum state. Since g_j degenerate quantum states might have the same energy, it is still possible for indistinguishable particles to be arranged in many ways among the degenerate states, as long as no more than one particle is in any given one of the degenerate states in level j.

For such a case, the statistics are slightly different from the Bose-Einstein case. The problem was worked out independently by Enrico Fermi and Paul Dirac (hence the FD subscript) in 1925 and 1926. The results are

$$w_{FD} = \prod_{j=0}^{\infty} \frac{g_j!}{N_j!(g_j - N_j)!} \tag{13.36}$$

and

$$\frac{N_j}{N_p} = \frac{g_j}{Q_{FD}\{\exp[\varepsilon_j/(kT)] + 1\}} \tag{13.37}$$

where

$$Q_{FD} = \sum_{j=0}^{\infty} \frac{g_j}{\exp[\varepsilon_j/(kT)] + 1}$$

As before, the distribution of particles among the possible energy levels [Eq. (13.37)] is for a system in thermodynamic equilibrium. Particles that obey these statistics include electrons and protons and are usually called *fermions.*

Let us make an important observation applicable to particles that make up the substances studied in this text. Examination of Eqs. (13.35) and (13.37) indicates that they are much different in form from the Maxwell-Boltzmann distribution, Eq. (13.9).

For dilute gases, the number of degenerate states g_j at any energy level is very large compared with the number of particles at that level N_j. Note that in the limit $g_j \gg N_j$, for Bose-Einstein statistics,

$$\frac{(g_j + N_j - 1)!}{(g_j - 1)!} = (g_j + N_j - 1)(g_j + N_j - 2) \cdots (g_j + 1)(g_j)\frac{(g_j - 1)!}{(g_j - 1)!}$$
$$= (g_j + N_j - 1) \cdots (g_j + 1)g_j \approx g_j^{N_j} \tag{13.38}$$

Similarly, for Fermi-Dirac statistics

$$\frac{g_j!}{(g_j - N_j)!} = \frac{g_j(g_j - 1)(g_j - 2) \cdots (g_j - N_j + 1)(g_j - N_j)!}{(g_j - N_j)!}$$
$$= g_j(g_j - 1) \cdots (g_j - N_j + 1) \approx g_j^{N_j} \tag{13.39}$$

Substituting Eqs. (13.38) and (13.39) into (13.34) and (13.36), respectively, gives

$$w_{\mathrm{BE}} = w_{\mathrm{FD}} = \prod_{j=1}^{\infty} \frac{g_j^{N_j}}{N_j!} \tag{13.40a}$$

in the limit of $g_j \gg N_j$. Thus, in the limit of a *dilute gas* with many more available energy states than particles, the bosons and fermions both obey the same statistics.

13.3.5 Maxwell-Boltzmann Statistics

If the particles are *distinguishable,* i.e., interchanging any two particles gives a definable new state, then the possible number of arrangements is given by modifying Eq. (13.40*a*) to

$$w_{\mathrm{MB}} = N_p! \prod_{j=0}^{\infty} \frac{g_j^{N_j}}{N_j!} \tag{13.40b}$$

which is the result for Maxwell-Boltzmann (MB) statistics. Using Eq. (13.40*a*) along with Eqs. (13.3) and (13.6) to develop an energy distribution for a dilute gas results in

$$\frac{N_j}{N_p} = \frac{g_j}{Q \exp\,[\epsilon_j/(kT)]} \tag{13.41}$$

where $Q = \sum_j g_j/\exp\,[\varepsilon_j/(kT)]$. This particular form of Q is usually given the symbol Z (from the German word *Zustandssumme,* or sum of states) and is called the *partition function.*

Equation (13.40*b*) can also be derived directly from the statistics governing *distinguishable* particles, i.e., particles in systems where we can tell the effect of interchanging any two particles. Note from Eq. (13.40*b*) that the number of possible states w for such a system is much greater than that for systems described by Bose-Einstein (BE) or Fermi-Dirac (FD) statistics, since every interchange of any two particles in such a system results in another state. These additional states do not occur if the particles are indistinguishable.

Particles belonging to systems described by Eqs. (13.40*a*) and (13.41) are called *boltzons,* because these equations result for a quantized system that otherwise is described by the classical or continuous Maxwell-Boltzmann distribution. So Eq. (13.41) is also called the Maxwell-Boltzmann distribution. Note that a plot of Eq. (13.41) is not possible until the system and the particles in the system are fully described through quantum mechanics. When that is done, the values of the energy levels ε_j and the degeneracy of each level g_j can be found, and Z for that system can be evaluated.

13.4 Application of Microsystem Information: Entropy and Other Properties

Properties

Entropy is described in Chap. 5 as a measure of randomness, disorder, or uncertainty. We can now examine the meaning of such concepts. Note that the quantity w is the number of ways that a system of particles can be arranged among the possible energy states available. As w increases, we are less and less certain of the particular distribution of particle energies that does exist, since more possibilities exist.

For boltzons, which are representative of the particles in an ideal gas, the number of possible arrangements of particles is given by Eq. (13.40*a*). If entropy is a measure of uncertainty about the actual state of the system, we expect S to be directly related to w. However, w has a very large value for any system composed of N_p particles, so we might try an equation of the form

$$S = C \ln w \tag{13.42}$$

where C is a constant of proportionality. There are more fundamental reasons for the choice of this form, but here we simply observe that it works well. It also has the advantage that for a system which has only one possible state ($w = 1$), the value of S is zero, reflecting the fact that we are certain of the state of such a system!

In Sec. 11.3.3, "Third Law of Thermodynamics" was stated that "the entropy of each pure substance in thermodynamic equilibrium approaches zero as the temperature approaches absolute zero." For this law to be valid, the number of possible states w must approach one as absolute temperature approaches zero. There is no experimental way to determine whether this in fact occurs, and some strong evidence suggests that it does not. Because changes in entropy are all that are important in engineering calculations, the third law provides a simple reference state for property values.

We can now proceed with Eq. (13.42) for entropy, apply it to a dilute gas, and see what results are obtained. Substituting Eq. (13.40*a*) to eliminate w results in

$$S = C \ln \prod_{j=0}^{\infty} \frac{g_j^{N_j}}{N_j!} = C \sum_{j=0}^{\infty} \ln \frac{g_j^{N_j}}{N_j!} = C \sum_{j=0}^{\infty} (\ln g_j^{N_j} - \ln N_j!)$$

$$= C \sum_{j=0}^{\infty} (N_j \ln g_j - \ln N_j!) \tag{13.43}$$

The final term can be evaluated by the use of *Stirling's approximation,* which states that, for large N_j

$$\ln N_j! \approx N_j \ln N_j - N_j \tag{13.44}$$

Substituting Eq. (13.44) into (13.43) results in

$$S = C \sum_{j=0}^{\infty} N_j \left(\ln \frac{g_j}{N_j} + 1 \right) \tag{13.45}$$

Now, Eq. (13.41) for the equilibrium distribution of particles among the energy levels can be solved for g_j/N_j and substituted into Eq. (13.45) to give

$$S = C \sum_{j=0}^{\infty} N_j \left\{ \ln \frac{Z \exp\left[\varepsilon_j/(kT)\right]}{N_p} + 1 \right\}$$

$$= C \sum_{j=0}^{\infty} N_j \left(\ln \frac{Z}{N_p} + \frac{\varepsilon_j}{kT} + 1 \right) \tag{13.46}$$

Finally, we can substitute Eqs. (13.2) and (13.6) to obtain

$$S = C \left(N_p \ln \frac{Z}{N_p} + \frac{U}{kT} + N_p \right) \tag{13.47}$$

Can we identify the constant of proportionality C? We can take the partial derivative of S with respect to U at constant V (in which case Z is constant) and obtain

$$\left(\frac{\partial S}{\partial U} \right)_V = \frac{C}{kT} \tag{13.48}$$

However, Eq. (5.9) gives $(\partial U/\partial S)_V = T$, so the constant must be $C = k$, the Boltzmann constant, if our microscopic and macroscopic viewpoints are to agree. Thus, for a dilute gas of boltzons, the entropy is given by

$$S = k \ln w = k\left(N_p \ln \frac{Z}{N_p} + \frac{U}{kT} + N_p\right) \tag{13.49}$$

Note that this is the absolute entropy of the system. We have succeeded in relating the entropy directly to the quantum behavior of the system through the partition function Z.

We can use Eq. (13.6) to find U by substituting Eq. (13.41) to eliminate N_j:

$$U = \sum_{j=0}^{\infty} N_j \varepsilon_j = \frac{N_p}{Z} \sum_{j=0}^{\infty} g_j \varepsilon_j \exp\left(\frac{-\varepsilon_j}{kT}\right) \tag{13.50}$$

Note that

$$\frac{dZ}{dT} = \frac{1}{kT^2} \sum_{j=0}^{\infty} g_j \varepsilon_j \exp\left(\frac{-\varepsilon_j}{kT}\right) \tag{13.51}$$

so that

$$U = \frac{N_p kT^2}{Z}\frac{dZ}{dT} = N_p kT^2 \frac{d(\ln Z)}{dT} \tag{13.52}$$

This result can be substituted into the final relation for S [Eq. (13.49)] so that S depends on only N_p, T, and Z.

Finally, the Helmholtz function A is given by

$$A = U - TS = -N_p kT\left(\ln \frac{Z}{N_p} + 1\right) \tag{13.53}$$

Recall that the Helmholtz function as used here is a fundamental equation; now that we have a relation for it, given in Eq. (13.53), we can find all other thermodynamic properties for a system of boltzons at thermodynamic equilibrium. The only missing piece of information is the partition function; once it is known, all the properties of the gas can be derived.

Partition Function for a Monatomic Gas

Our purpose is to show only the methodology and power of statistical thermodynamics, not the details, but we can take one case in some detail. Suppose we examine an ideal monatomic gas which has only translational energy. (Actually, at higher temperatures, the energy stored in electronic energy levels can become important, but we ignore that effect here.)

For such a case, the quantized translational kinetic energy is $\varepsilon_j = m\mathbf{V}_j^2/2$. The partition function is then

$$Z = \sum_{j=0}^{\infty} g_j \exp\left(\frac{-\varepsilon_j}{kT}\right) = \sum_{j=0}^{\infty} g_j \exp\left(\frac{-m\mathbf{V}_j^2}{2kT}\right) \tag{13.54}$$

The degeneracy g_j can be estimated from the uncertainty principle. For a particle with energy in the range $\varepsilon_j \pm \Delta\varepsilon_j/2$, the mean square speed must be in a range given by

$$\frac{m\overline{\mathbf{V}_j^2}}{2} \pm \frac{m}{2}\,\Delta\overline{\mathbf{V}_j^2} = \varepsilon_j \pm \frac{\Delta\varepsilon_j}{2} \tag{13.55}$$

If we can determine the number of indistinguishable particles that have speeds in the range $\Delta\mathbf{V}_j = (\Delta\overline{\mathbf{V}_j^2})^{1/2}$, then that is the number of particles with energies in the corresponding range given by Eq. (13.55). The number of particles within $\Delta\varepsilon_j$ around ε_j is just the degeneracy g_j.

The product of momentum and position in space for all particles with energy $\varepsilon_j \pm \Delta\varepsilon_j$ is $m\,\Delta\mathbf{V}_x \cdot m\,\Delta\mathbf{V}_y \cdot m\,\Delta\mathbf{V}_z \cdot \Delta x\,\Delta y\,\Delta z$. This six-dimensional quantity is a part of what is called *momentum space.* The momentum space occupied by a single particle is given by Eq. (13.33) as $m\,\Delta\mathbf{V}_x \cdot m\,\Delta\mathbf{V}_y \cdot m\,\Delta\mathbf{V}_z \cdot \Delta x\,\Delta y\,\Delta z \approx h^3$. The degeneracy is then the momentum space occupied by all particles with energies within $\Delta\varepsilon_j$ divided by the momentum space per particle, or

$$\begin{aligned} g_j &= \frac{m\,\Delta\mathbf{V}_x \cdot m\,\Delta\mathbf{V}_y \cdot m\,\Delta\mathbf{V}_z \cdot \Delta x\,\Delta y\,\Delta z}{h^3} \\ &\approx \left(\frac{m}{h}\right)^3 d\mathbf{V}_x\,d\mathbf{V}_y\,d\mathbf{V}_z\,dx\,dy\,dz \end{aligned} \tag{13.56}$$

The partition function now becomes

$$Z = \left(\frac{m}{h}\right)^3 \sum_{j=0}^{\infty} \left\{dx\,dy\,dz\,\exp\left[\frac{-m(\mathbf{V}_x^2+\mathbf{V}_y^2+\mathbf{V}_z^2)}{kT}\right] d\mathbf{V}_x\,d\mathbf{V}_y\,d\mathbf{V}_z\right\}_j \tag{13.57}$$

As we did for the tennis ball in Sec. 13.3.1, we could show that the difference between energy levels for monatomic gases at most conditions is very small. This allows us to replace the summation in Eq. (13.57) with an integration over speed and direction, so that Z becomes

$$\begin{aligned} Z = \left(\frac{m}{h}\right)^3 &\int_{z=0}^{\infty}\int_{y=0}^{\infty}\int_{x=0}^{\infty} dx\,dy\,dz \\ &\times\left[\int_{\mathbf{V}_x=-\infty}^{\infty} \exp\left(\frac{-m\mathbf{V}_x^2}{2kT}\right) d\mathbf{V}_x \int_{\mathbf{V}_y=-\infty}^{\infty} \exp\left(\frac{-m\mathbf{V}_y^2}{2kT}\right) d\mathbf{V}_y \right. \\ &\left. \times \int_{\mathbf{V}_z=-\infty}^{\infty} \exp\left(\frac{-m\mathbf{V}_z^2}{2kT}\right) d\mathbf{V}_z\right] \end{aligned} \tag{13.58}$$

The triple integral over $dx\,dy\,dz$ is just the system volume V. The remaining three integrals can be evaluated with the aid of a table of integrals, and each gives $(2\pi kT/m)^{1/2}$. The partition function for a monatomic gas with no electronic energy states finally becomes

$$Z = \left(\frac{V}{h^3}\right)(2\pi mkT)^{3/2} \tag{13.59}$$

Example Problem 13.1

Derive the expression for the internal energy u, specific heat at constant volume c_v, and absolute entropy of argon. Compare the prediction from the derived expressions with the results of Table D.3 at 300°C and 1 atm.

Solution

Starting with Eq. (13.52), we have

$$U = N_p k T^2 \frac{d(\ln Z)}{dT}$$

$$= N_p k T^2 \frac{d}{dT}\left(\ln \frac{V}{h^3} + \frac{3}{2}\ln 2\pi m k + \frac{3}{2}\ln T\right)$$

$$= \frac{3N_p kT}{2}$$

Letting $N_p = N_a$ for 1 mol of gas, dividing through by the molecular weight, and noting that $N_a k/M = R$, we get

$$u = \frac{U}{M} = \frac{3RT}{2}$$

Then

$$c_v = \left(\frac{\partial u}{\partial T}\right)_v = \frac{3R}{2}$$

These are the same results that we obtained for an ideal monatomic gas using the classical approach.

Substituting the result for U into Eq. (13.49) gives

$$S = kN_p\left(\ln \frac{Z}{N_p} + \frac{5}{2}\right)$$

Substituting Eq. (13.59) for Z gives

$$S = kN_p\left\{\ln\left[\frac{V}{N_p h^3}(2\pi m k T)^{3/2}\right] + \frac{5}{2}\right\}$$

If we set $N_p = N_a$, then S will be the entropy per mole. Dividing by the molecular weight of the gas M results in the specific entropy, kJ/(kg · K), as

$$s = R\left\{\ln\left[\frac{V}{N_a h^3}(2\pi m k T)^{3/2}\right] + \frac{5}{2}\right\}$$

Note that the Helmholtz function [Eq. (13.53)] can be used to find the equation of state for this gas. The total differential of the Helmholtz function in Eq. (9.5c) gives P as

$$\begin{aligned} P &= -\left(\frac{\partial A}{\partial V}\right)_T \\ &= -N_p kT \frac{\partial}{\partial V}\left[\ln V + \ln \frac{(2\pi mkT)^{3/2}}{N_p h^3} + 1\right]_T \\ &= \frac{N_p kT}{V} \end{aligned}$$

or, with $N_p = N_a$,

$$PV = N_a kT = \bar{R}T$$

which should not be surprising by now. Substituting into the last relation for s to eliminate V results in

$$\begin{aligned} s &= R\left(\frac{3}{2}\ln T + \ln \frac{V}{N_a} + \frac{3}{2}\ln 2\pi mk - 3\ln h + \frac{5}{2}\right) \\ &= R(\tfrac{5}{2}\ln T - \ln P + \ln N_a k + \tfrac{3}{2}\ln 2\pi mk - 3\ln h + \tfrac{5}{2}) \\ &= c_P \ln T - R\ln P + \text{const} \end{aligned}$$

where we have used $c_P = c_v + R = 5R/2$, and

$$\begin{aligned} \text{Const} &= R(\tfrac{3}{2}\ln 2\pi m - 3\ln h + \tfrac{5}{2}\ln k + \tfrac{5}{2}) \\ &= R\left[\ln \frac{(2\pi m)^{3/2}k^{5/2}}{h^3} + \frac{5}{2}\right] \end{aligned}$$

The final form for the entropy is called the *Sackur-Tetrode equation,* and it predicts the absolute entropy of an ideal gas. The dependence on T and P is seen to be the same as indicated in Table 5.4 for an ideal gas; now, however, no arbitrary constant has been introduced into the relation for entropy. All constants in the Sackur-Tetrode equation are known for a given monatomic gas.

Now we can compare the predictions with the table values for argon. Using $k = 1.3806 \times 10^{-26}$ kJ/K, $h = 6.6262 \times 10^{-37}$ kJ · s (Table B.1), $R = 0.20813$ kJ/(kg · K), $m = M/N_a = 39.948/(6.023 \times 10^{26}) = 6.6326 \times 10^{-26}$ kg, we find

$$u = \frac{3RT}{2} = 3(0.20813)\,\frac{273.15 + 300}{2} = 178.93 \text{ kJ/kg}$$

which compares with 179.05 from Table D.3. For c_v,

$$c_v = \frac{3R}{2} = \frac{3(0.20813)}{2} = 0.3122 \text{ kJ/(kg} \cdot \text{K)}$$

which compares with 0.3124 kJ/(kg · K) from Table D.3.

Using the Sackur-Tetrode equation, after combining the log terms, we find

$$s = R\left[\ln \frac{(2\pi m)^{3/2}(kT)^{5/2}}{Ph^3} + \frac{5}{2}\right]$$

$$= 0.20815$$

$$\times \left\{\ln \frac{[2\pi(6.6326 \times 10^{-26})/1000]^{3/2}[(1.3806 \times 10^{-26})(573.15)]^{5/2}}{(101.32)(6.6262 \times 10^{-37})^3} + \frac{5}{2}\right\}$$

$$= 0.20815[\ln (5.0833 \times 10^7) + \tfrac{5}{2}]$$

$$= 4.2138 \text{ kJ/(kg} \cdot \text{K)}$$

which compares with $s = s_0 = 4.2163$ kJ/(kg · K) from Table D.3.

Comments

The differences between the calculated values and the table values arise for two reasons. Small differences are introduced by the use of slightly different values for the fundamental constants. More importantly, the tables are developed by including the effect of electronic energy states, which increases the values of u and s above those calculated by ignoring the electronic contributions as we did here. At low temperatures, the effect of the electronic states is small for most gases. The tables include the effect of the first 285 electronic states for argon.

In this section, we use the quantum statistics for general classes of particles that obey various types of restrictions (bosons, fermions) and show that the statistics reduce to the Maxwell-Boltzmann statistics for a dilute gas composed of indistinguishable particles. In describing the behavior of matter that is not in the form of a dilute gas, the original Bose-Einstein or Fermi-Dirac statistics must be used, because the assumption that $g_j \gg N_j$ is not valid. In such cases the Maxwell-Boltzmann statistics and distribution give erroneous predictions.

Given the case of a dilute gas, we predicted the values of the thermodynamic properties for the case of a monatomic gas. Note that we made no assumptions about the structure of the gas—only that it obeyed the applicable statistics for the number of particles that exist in a given energy level. This procedure can be continued for a diatomic or polyatomic gas, with allowance for the greater number of energy states that can exist because of the presence of quantized rotational, vibrational, and electronic energies. The gas tables in App. D were constructed from the resulting relations.

The fact that all dilute gases follow similar P-v-T behavior was observed and exploited in the property relations of Chaps. 3 and 9, and is embodied in the classical principal of corresponding states.

13.5 First Law

Equation (13.6) gives a general relation for the internal energy of a system of particles in terms of the energy levels of each individual particle. The internal energy can evidently be changed in two ways: Either the energy level distribution ε_j can be altered, or the distribution of particles among the energy levels N_j can be changed. Thus

$$dU = \sum_{j=0}^{\infty} \varepsilon_j \, dN_j + \sum_{j=0}^{\infty} N_j \, d\varepsilon_j \tag{13.60}$$

Suppose we consider a control mass; i.e., the total number of particles N_p in the system is fixed, so the total mass remains constant. First, consider a change in U that occurs without a change in the distribution of N_j, that is, $dN_j = 0$. For such a case, the thermodynamic probability w of the system is unchanged by the process [Eq. (13.34), (13.36), or (13.40*a*), depending on the applicable statistics]. Since $S = k \ln w$, such a process is *isentropic.* We noted in Chap. 5 that a process that changes the internal energy isentropically is classified as work, so we can identify

$$\delta W = -\sum_{j=0}^{\infty} N_j \, d\varepsilon_j \tag{13.61}$$

Consider now a case in which U is changed by changing the distribution of particles among the energy states while holding the energy state distribution constant, that is, $d\varepsilon_j = 0$. In that case, w is changed, so there is a change in the value of S. If the process is reversible, then this can occur only by the mechanism of heat transfer, so we can identify

$$\delta Q = \sum_{j=0}^{\infty} \varepsilon_j \, dN_j \tag{13.62}$$

and we find that Eq. (13.60) is indeed a statement of the first law for a control mass: $\delta Q - \delta W = dU$.

13.6 Concluding Remarks

After careful reading of this chapter, some new insights should be found. First, entropy is directly related to our uncertainty about the microscopic state of a system of particles, since it is directly proportional to $\ln w$, the natural log of the possible number of ways of arranging the particles in the system among the available energy levels and the available geometric positions. Thus, we can see why entropy is called a measure of uncertainty, chaos, etc. These terms simply reflect the fact that the more ways there are to arrange the particles in the system, the less we know about the particular state of the particles and the greater the entropy of the system. We can see why the entropy of a system of N_p particles confined to volume

V increases when the particles are allowed to expand to a volume $2V$; there are now more positions for the particles to occupy, and we become less certain about the state of the set of particles.

The relation $S = k \ln w$ is fundamental to much of science and describes not only the behavior of the quantity we know as entropy but also information about many other types of systems, including biological systems and information transfer in communication and computer systems. Ludwig Boltzmann, who first derived the relation, saw it as important enough that he requested to have it engraved on his headstone, and it was. We have observed that different formulations for w may result, depending on the quantum mechanical rules obeyed by the particles under study. We studied briefly only the particular case of a dilute monatomic gas; however, the methods apply to many other types of particles, including electrons, photons, atoms, and molecules constrained in the form of solids, liquids, superfluids, etc. We need only use the correct formulation for w and the partition function for these particles to predict their thermodynamic behavior, although the correct formulation may become quite complex.

We leave this introductory study of thermodynamics at this point, having examined many of the practical applications of the subject to engineering problems as based on the fundamental principles. The broad and powerful applications of the subject to engineering problems have been stressed. The application to diverse fields of science, which, without exception, make use of some of the tools outlined here, also extends to broad fields of the liberal arts. It is particularly interesting to see the impact of such concepts as entropy, the wave equation, and the uncertainty principle on fields such as theology and philosophy, which allow and apply these ideas much as we do in science and engineering.

Problems

13.1 Derive the result $\varepsilon_{MP} = kT/2$ for the most probable energy in the Maxwell-Boltzmann distribution, Eq. (13.9).

13.2 Compare the values of $\bar{c}_v$ for the monatomic and diatomic gases in Table C.2 to the predictions of Eqs. (13.22) and (13.24), respectively.

13.3S Find the uncertainty in position Δx (m) relative to the accuracy of the speed measurement of the tennis ball of Sec. 13.3.1 when it moves with a speed of 10 m/s.

13.3E Find the uncertainty in position Δx (ft) relative to the accuracy of the speed measurement of the tennis ball of Sec. 13.3.1 when it moves with a speed of 30 ft/s.

13.4 A laboratory experiment is able to measure the speed of an electron within 0.5 percent. The mass of an electron is given in Table B.1. What is the minimum uncertainty in the position of the electron that can be determined in the experiment if the measured electron speed is 3×10^5 m/s?

13.5S If the position of a 1-g pellet can be measured within 10^{-3} cm as it moves through the air, what is the minimum uncertainty possible in a measurement of the pellet speed?

13.5E If the position of a 0.002-lbm pellet can be measured within 10^{-3} in as it moves through the air, what is the minimum uncertainty possible in a measurement of the pellet speed?

13.6 Using Eq. (13.59) for the partition function and the complete form of the Maxwell-Boltzmann distribution for boltzons, Eq. (13.41), see whether the classical Maxwell-Boltzmann distribution of Eq. (13.9) is obtained for a monatomic gas.

13.7 Develop a relation for the enthalpy of an ideal gas in terms of the partition function Z, T, and V.

13.8 A gas has three particles. There are four energy states, each with degeneracy of 2, available to each particle. The populations of the states are $N_0 = 2$, $N_1 = 1$, $N_2 = N_3 = 0$. Find the thermodynamic probability and the entropy of the gas if the gas particles obey (*a*) Bose-Einstein statistics, (*b*) Fermi-Dirac statistics, and (*c*) Maxwell-Boltzmann statistics.

13.9 (*a*) What is the value of the partition function for a gas with four energy states, each with degeneracy of 10, available to each particle? The energy states have values of $\varepsilon_0 = 0$, $\varepsilon_1 = kT$, $\varepsilon_2 = 2kT$, $\varepsilon_3 = 3kT$. (*b*) What are the relative populations of the four energy states?

13.10S An ideal monatomic gas has an atomic weight of 40 and is in thermodynamic equilibrium at $T = 300$ K. For this gas what is the value of (*a*) the mean speed of the atoms, (*b*) c_P, (*c*) u, (*d*) h, and (*e*) s?

13.10E An ideal monatomic gas has an atomic weight of 40 and is in thermodynamic equilibrium at $T = 500°$R. For this gas what is the value of (*a*) the mean speed of the atoms, (*b*) c_P, (*c*) u, (*d*) h, and (*e*) s?

13.11S Plot the absolute entropy of krypton at $P = 2$ atm in the range $0 < T \leq 2000$ K. Ignore electronic energy states.

13.11E Plot the absolute entropy of krypton at $P = 2$ atm in the range $0 < T \leq 3600°R$. Ignore electronic energy states.

13.12 Starting with $S = k \ln w$, derive a relation for the change in entropy of an ideal monatomic gas during a constant-temperature expansion from V_1 to V_2. Compare the result with the classical result for the same process.

Contents

APPENDIX A

A Short History of the Development of Themodynamics

Confound these ancients. They've stolen all our best ideas.
Ben Jonson

The development of classical thermodynamics provides a look into the thought processes of scientists and engineers, and it allows us a glimpse of human failings as well as the triumphs of intellect that were produced.

The history of this subject is usually presented as an orderly progression of ideas, each building on the foundation laid by earlier investigators. This view has some validity. But, as with present-day research, there were often long periods when invalid ideas were tenaciously held in the face of decisive evidence of their falsity. In other cases, a number of investigators almost simultaneously adopted a whole new block of theory and built upon it. It is simply not accurate to describe this history in a linear fashion. In 1889, Samuel P. Langley, no mean researcher in his own right, examined another branch of research in his address as the retiring president of the American Association for the Advancement of Science:

> We often hear (the progress of science) likened to the march of an army towards some definite end; but this, it has seemed to me, is not the way science usually does move, but only the way it seems to move in the retrospective view of the compiler, who probably knows almost nothing of the real confusion, diversity and retrograde motion of the individuals comprising the body, and only shows us such parts of it as he, looking backward from his present standpoint, now sees to have been the right direction.
>
> I believe this comparison of the progress of science to that of an army, which obeys an impulse from one head, has more error than truth in it;

> and, though all similes are more or less misleading, I would prefer to ask you to think rather of a moving crowd, where the direction of the whole comes somehow from the independent impulses of the individual members; not wholly unlike a pack of hounds, which, in the long run perhaps catches its game, but where, nevertheless, when at fault, each individual goes his own way, by scent not by sight, some running back and some forward; where the louder-voiced bring many to follow them nearly as often in a wrong path as in a right one; where the entire pack even has been known to move off bodily on a false scent; for this, if a less dignified illustration, would be one which had the merit of having a truth left in it, left out of sight by the writers of textbooks.

With this warning in mind, we can embark on a brief historical voyage through the development of classical thermodynamics.

The concept of energy was introduced in the field of mechanics by Galileo Galelei (1564–1642). As early as the middle of the seventeenth century, Sir Isaac Newton and Christian Huygens (1629–1695) had used it as a convenient method of solving problems involving the calculation of the height reached by a swinging pendulum. In the more general sense in which energy is used in thermodynamics, however, early attempts to quantify energy were hampered by the lack of understanding that heat transfer and work were simply different forms of energy transfer. Even more basically, researchers were plagued by confusion between the concepts of heat transfer and temperature. Temperature measurement was necessary before ideas of heat transfer could be developed. The early development of the air thermometer by Galileo (1592), the sealed alcohol thermometer by Ferdinand II, Grand Duke of Tuscany in 1641, the suggestion of an oil-filled thermometer by Newton (1701) (who proposed a temperature scale that began at 0° at the melting point of ice and had a second fixed point of 12° as "the maximum heat the thermometer can attain by contact with the human body"), and finally the development of the mercury-in-glass thermometer by Gabriel D. Fahrenheit in 1715—all laid the foundation for later investigations.

A.1 Heat Engines and the Theory of Thermodynamics

Engineers were designing various types of heat engines even before the development of careful temperature measurement and thermodynamic theory. Huygens and Daniel Bernoulli (1700–1782), among others both earlier and later, had considered the idea of using gunpowder as a fuel, but for rather obvious reasons such engines were not developed into practical devices. Thomas Savery (1650?–1715) patented a steam-operated pump in England in 1698 based on a principle suggested by Edward Somerset, the Marquis of Worcester, around 1663. Savery's pump used a valve system that was controlled by hand, making it cumbersome, slow, and inefficient. Thomas Newcomen (1663–1729), an ironmonger from Dartmouth, put his first coal-burning engine into service in England in 1712. The engine used atmospheric pressure on one side of a piston to drive the piston against the lower pressure of condensing steam on the other side, and in later versions the valves were operated by the pump linkage so that the engine could operate with scant attention (Fig. A.1).

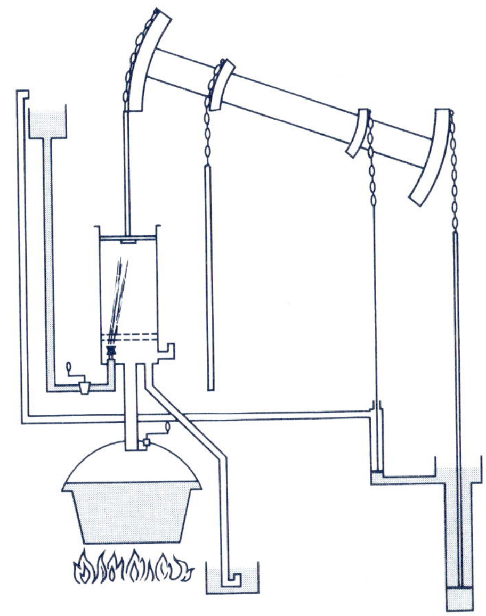

Figure A.1 Newcomen's engine.

Not until 1760 did Joseph Black (1728–1799), a professor of medicine and chemistry at the University of Glasgow, finally lay the foundation of the quantitative science of heat transfer. In that year, he measured the heat capacity (specific heat) of various materials and noted the difference between temperature and heat transfer. From 1761 to 1764 he demonstrated the concepts of latent heat of fusion and evaporation. These concepts became the basis for what was to be called the *caloric theory,* which envisioned heat transfer as a colorless fluid that migrated from a body at higher temperature to one at lower temperature. This caloric fluid was envisioned at the time to be indestructible.

James Watt (1736–1819), a Scottish instrument maker who worked for some time in Black's laboratory, had in the meantime realized some of the limitations of the Newcomen engine, and he was building engines that used steam pressure on one side of the piston to drive against the lower pressure of condensing steam on the other. He used some of Black's results to design an external condenser (1765), rather than relying on condensation in the cylinder as Newcomen had done. This allowed the cylinder to remain at higher temperatures during the entire cycle and greatly improved the engine efficiency. By 1778, Watt and his partner Matthew Boulton (1728–1809) were in competition with other manufacturers. To compare his engines with others, Watt introduced the concept of *duty,* which is a form of engine efficiency, defined as "the number of pounds raised through a height of 1 foot per bushel of coal used." By 1781–1782, Watt had introduced the double-acting engine that turned a flywheel, allowing continuous rotary motion. This had not been possible with earlier engines, because power was produced only on the downstroke. He defined the horsepower as the rate of work from a mill horse, calculating that value as 33,000 pounds of force per minute exerted through a distance of 1 foot (1782–1783). This value was conservative by a factor of 2, to avoid dissatisfied customers.

Benjamin Thompson (1753–1814) was an American who made the unfortunate choice of backing the British and spying for them at the time of the revolution. So he found it discreet to emigrate to England, where he became a lieutenant colonel in the British army. He invented, among other things, the drip coffee pot, improved fireplace designs, and efficient roasters and pots. He took a commission to aid in providing armaments for a Bavarian prince, who gave him the title of Count Rumford, and while so engaged (1798), he observed that there was a continuous heat release during the boring of a cannon. If this were so, how could the "caloric fluid" be conserved when it was apparently being continuously produced? Rumford deduced that heat transfer was "a kind of motion." This observation was, of course, available to anyone who observed the frictional production of thermal energy, but Rumford is generally given credit for bringing it to the attention of the scientific community. The caloric theory continued to be widely accepted as correct for more than 50 years, and indeed much of the mathematical interpretation of the caloric theory was taken completely into the modern view of thermodynamics. Rumford had earlier (1787 – 1799) carried out careful experiments which showed that the weight of a given amount of water was unchanged during the freezing process, to an accuracy of 1 part in 1 million. Thus, if there was a caloric fluid, it was essentially weightless. Rumford ended his days in Paris after marrying (1805) the widow of Lavoisier, who had been one of the strongest advocates of the caloric theory.

In the early years of the nineteenth century, Richard Trevithick (1771 – 1819) in England and Oliver Evans (1755 – 1819) in the United States were producing steam engines that worked at boiler pressures above atmospheric pressure (Figs. A.2 and A.3), thus considerably increasing cycle efficiency. Trevithick used pressures of about 15 psig, while Evans built boilers that operated at up to 220 psig.

The Rev. Robert Stirling (1790 – 1878), an English parish minister, patented a practical heat engine in 1816 that used air as the working fluid. In theory, the cycle used in the Stirling engine approaches the ideal cycle soon to be proposed by Carnot. The engine had practical importance because it operated at low pressure, eliminating the safety problems arising from the increasing use of higher pressures in steam boilers.

In 1824, the brilliant French military engineer Nicolas Leonard Sadi Carnot (1796 – 1832), son of Napoleon's minister of war, presented his only published paper, "Reflexions on the Motive Power of Fire, and on Machines Fitted to Develop that Power," which presented one form of what we now call the second law of thermodynamics as applied to the amount of work that can be obtained from an engine using a heat transfer as its driving energy. He also stated a reasoned form of the first law. Carnot's work was presented in terms of the caloric theory, and this fact led many later researchers to cling to that theory, since Carnot's predictions were evidently valid and were confirmed by experiment. However, Carnot himself was apparently beginning to question the caloric theory basis of his own work, as evidenced by changes he made in proofs of his manuscript and by some of his unpublished notes. He died at age 36 of cholera while recovering from scarlet fever, having provided probably the single most important individual contribution to classical thermodynamics.

In the early 1840s, James Prescott Joule (1818 – 1889) and Julius Robert Mayer (1814 – 1878) almost simultaneously set forth the idea that heat transfer and mechanical work were simply different forms of the same quantity, which we now recognize as energy transfer. Neither of these representations of the "mechanical theory of heat" was accepted at the time.

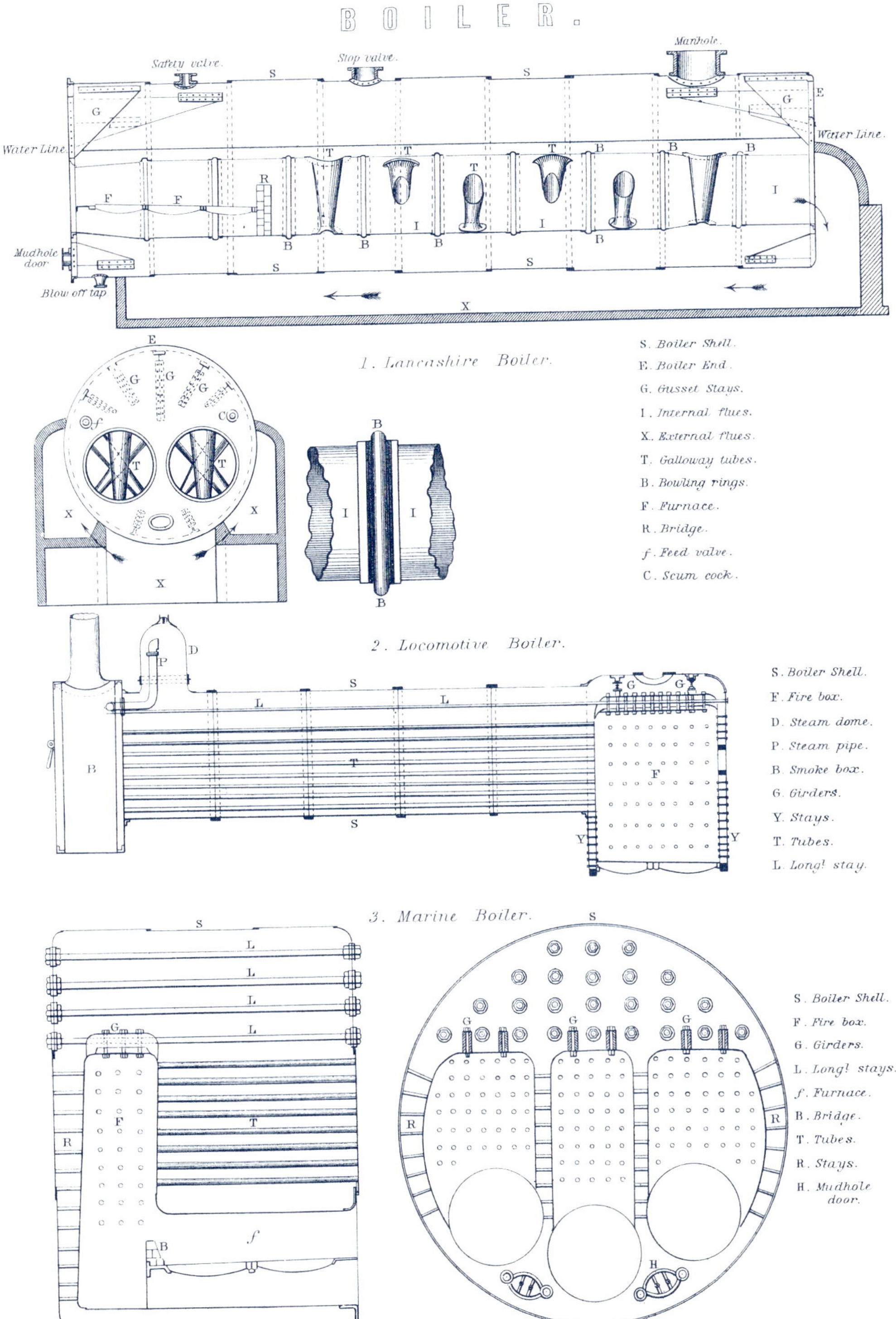

Figure A.2 Early steam boiler designs.

22 **WILLIAMS' PERRAN FOUNDRY CO.**

IMPROVED PORTABLE ENGINES.

SINGLE-CYLINDER ENGINES, from 6 to 10 Horse-Power.
DOUBLE-CYLINDER ENGINES, from 10 to 20 Horse-Power.

These Engines are especially constructed to suit the wants of the Australian Colonies, the Foreign British Possessions, and all those Countries where such power is wanted, and where the greatest results must be attained by the smallest outlay both of fuel and manual labor. It is useless to give a long written description of these Engines, now employed in almost every country, either for Agricultural and Mining operations, or as auxiliaries in the construction of Public Works, but a specification of each size can always be had on application. It is sufficient to say these Engines are built on the latest and most improved principles, and of the very best material. They have plenty of boiler power, and will make steam (with a like amount of fuel) faster than any other Engine of this class. They are at once the most durable and economical of portable powers.

PRICES AND PARTICULARS OF ENGINES.

HORSE-POWER.	S. 5	S. 6	S. 8	S. 10	D. 10	D. 12	D. 14	S. 16	D. 20
Diameter of Cylinder	$7\frac{1}{4}''$	$8\frac{1}{2}''$	$9\frac{1}{2}''$	11″	$7\frac{1}{4}''$	$7\frac{3}{4}''$	$8\frac{1}{2}''$	$13\frac{1}{4}''$	11″
Length of Stroke	10″	13″	13″	13″	13″	13″	13″	16″	15″
Diameter and face of Fly-Wheel	4′ 6″ × 5″	5′ 0″ × $5\frac{1}{2}''$	5′ 0″ × 7″	5′ 0″ × 7″	5′ 0″ × 7″	5′ 0″ × 7″	5′ 0″ × 7″	6′ 0″ × 7″	5′ 0″ × 9″
Number of Revolutions	160	120	120	120	120	120	120	96	100
Coal per hour	35 lbs.	42 lbs.	56 lbs.	70 lbs.	70 lbs.	84 lbs.	98 lbs.	112 lbs.	140 lbs.
Water per hour	$6\frac{1}{2}$ gals.	$7\frac{3}{4}$ gals.	$10\frac{1}{4}$ gals.	12 gals.	12 gals.	15 gals.	18 gals.	$20\frac{1}{2}$ gals.	25 gals.
Tested Pressure	200	200	200	200	200	200	200	200	200
Working Pressure	80	80	80	80	80	80	80	80	80
Net Weight of Engine	2 9 0 0	3 4 0 0	3 13 0 0	4 0 0 0	4 5 0 0	4 15 0 0	5 5 0 0	6 0 0 0	7 15 0 0
Gross Weight with Case	3 0 0 0	3 15 0 0	4 5 0 0	4 15 0 0	5 0 0 0	5 5 0 0	6 0 0 0	7 0 0 0	8 10 0 0
Cubic Contents Packed	220	356	413	430	430	460	490	520	540
Price in London	£165	£180	£210	£240	£260	£300	£335	£335	£445
Extra for Large Fire Box (for burning wood)	£2 10	£3	£4	£5	£5	£6	£7	£8	£10
Charge for Packing for Export	£5	£6	£8	£10	£10	£12	£14	£16	£20

PERRANARWORTHAL, CORNWALL,

AND

1 & 2 GREAT WINCHESTER STREET BUILDINGS, LONDON, E.C.

Figure A.3 Catalog page for steam engines. Note proposed boiler pressure of 80 psig.

Joule's ideas were based on a remarkable series of experiments. He immersed the armature of a dynamo in a rotating vessel of water and measured the heating of the water when a current flowed through the armature with (1) the armature stationary, (2) the armature rotating in the forward direction, and (3) the armature rotating in the reverse direction. He found that the heating could be increased or decreased depending on the direction of rotation. He measured a work input of 4.60 N·m/°C per kilogram of water. He then designed a series of experiments, including forcing water through holes in a piston, 4.25 N·m/(kg·°C); friction between two surfaces submerged in water or mercury, 4.25 N·m/(kg·°C); air pumped into a closed cylinder, 4.60 N·m/(kg·°C); air in the cylinder allowed to escape slowly, 4.38 N·m/(kg·°C); and air from one cylinder allowed to escape slowly into a second cylinder, with both cylinders submerged in the same tank (no net work done, and no change in temperature, since no energy crosses the system boundary). Joule published these results, and the mechanical theory of heat began to attract more attention. Joule later made even more careful measurements of the temperature change of an insulated canister of water when stirred by the measured work input from a paddle wheel stirrer. While on his honeymoon in Switzerland, he hoped to find a high enough waterfall to provide sufficient energy in the form of work that he could make still more careful measurements.

In the meantime, Mayer, who had conceived the mechanical theory of heat while a ship's physician in the East Indies, had tried to publish similar results based on data obtained by Black. Mayer had great difficulty getting his work into print in the scientific journals; his first submitted manuscript was not even acknowledged. Some of his later work (1842) was ridiculed. His despair was so great that he attempted suicide by jumping from a window, but he only broke both legs. Because attempted suicide was considered proof of insanity, he was placed in an asylum for some time. He was finally given some measure of recognition in later years as having been equal with Joule in establishing the mechanical theory of heat.

Another physician who had difficulty in publishing his work was Herman Ludwig von Helmholtz (1821–1894), self-taught in mathematics and physics. He gave the first clear analytical exposition of energy on a generalized basis in 1847 in a privately published paper that he also could not get accepted by professional journals.

In the meantime, Carnot's work had been expanded and clarified by Emile Clapeyron (1799–1864). During the later 1840s, many thermodynamicists, including the physicists William Thomson (1824–1907) (later Lord Kelvin), Rudolf Julius Emanuel Clausius (1822–1888), and the Scottish engineer William John Macquorn Rankine (1820–1872), struggled with the problem of how to reconcile Carnot's work, which was based on the caloric theory, with the experimental confirmation of the mechanical theory of heat being provided by Joule and Mayer. Because Joule's results depended on temperature difference measurements of the order of 0.01°F, there was room for considerable skepticism over his results.

In 1848, Kelvin, a professor of natural philosophy at the University of Glasgow and then 24 years of age, had suggested an absolute temperature scale, based in part on Carnot's results. Kelvin was engaged in work on telegraphy in support of the laying of the Atlantic cable, but he found time to publish work in 1849 in which appeared the first use of the term *thermodynamic,* and the term *mechanical energy* was also introduced. In 1850, he finally abandoned the caloric theory entirely, and from 1852 to 1862 he worked with Joule in a series of experiments to measure the temperature change of a gas during a controlled expansion. This was done to attempt to disprove a supposed assumption of Mayer's that such a temperature change was always zero, which Joule and Thomson (Kelvin) believed had introduced an error into Mayer's value for the mechanical equivalent of heat. (Mayer had actually only assumed the zero value was valid for air in the range of conditions where air acts as an ideal gas, which it is.) The experimental work led to very important results, including determination of the so-called Joule-Thomson coefficient for real gases.

Clausius in the meantime realized that there were two distinct laws at work, the second law, as expounded by Carnot, and the first law, which Clausius first constructed with a logical theory in 1850. In this work, he defined the internal energy U and showed clearly the difference between the specific heats measured at constant volume and constant pressure. He also showed that this formulation agreed with experiment. Although both Clausius and Kelvin had been using the function Q_{rev}/T for some years, Clausius recognized the value of the function as a property and coined the word *entropy* to describe it. He also assigned it the symbol S. His statement of the first law as *"Die Energie der Welt ist konstant"* remains a concise and generally valid one.

Rankine, in applying the theory of thermodynamics to heat engines, in 1853 defined the thermodynamic efficiency of a heat engine and in 1854 showed the usefulness of P-v diagrams as related to work. He wrote the first thermodynamics textbook in 1859.

In 1862, the cycle used in modern gasoline-powered internal combustion engines was proposed in a French patent issued to Alphonse Beau de Rochas (1815–1893), although no engine is known to have been built as a result. The cycle was introduced in a practical engine by a young merchant in Cologne, Nikolaus August Otto (1832–1891), and his partner Eugen Langen (1833–1895) in 1876 and was demonstrated at the Paris Exposition of 1878. Otto's work was first applied to large single-cylinder engines using illuminating gas. The successful engine followed a series of attempts which were ridiculed at the time, but which led to continual improvement and final success. Otto was unaware of Beau de Rochas' work and fought many legal battles to maintain rights of production of his engines, finally losing.

Captain John Ericsson (1803–1889) was a Swedish engineer who spent a long and productive career in the United States. He perfected the screw propeller for ships, replacing the paddle wheel common at that time, and he built the ship *Monitor* for the Union forces in the Civil War, which set the standard for iron-clad ships for the next 50 years. He had earlier invented the shell-and-tube heat exchanger for use as a condenser for marine engines. He, with others, became interested in air engines (Fig. A.4), again because of the number of catastrophic boiler explosions, particularly on ships where they resulted from a combination of higher-pressure boilers and careless operation. In 1850, he built and demonstrated a large hot-air engine for the 2200-ton ship *Ericsson.* The engine was a technical success, but took so much space from the cargo hold with its four 14-ft-diameter cylinders with 6-ft strokes that the ship probably could not have competed economically with steam-powered machinery. The *Ericsson* sank in a storm before conclusive tests could be performed. Ericsson later marketed smaller solar-powered and coal-fired hot-air engines, the latter with some success.

In 1873, George Bailey Brayton (1830–1892), born in East Greenwich, Vermont, introduced and in 1876 marketed an internal combustion engine that operated by injecting compressed air through a hot grating into the combustion chamber. Heavy oil or other liquid fuel was injected directly into the air before it entered the combustion chamber. Because there was no sudden explosion, the combustion occurred at nearly constant pressure. An earlier model had operated with gas as the fuel, but the flame had passed back through the grating into the fuel supply, making the engine dangerous to operate. The heavy-oil engine did not enjoy continued commercial success, but the thermodynamics of the engine have become the basis for modern gas-turbine engines.

Josiah Willard Gibbs (1839–1903) is often noted as the most brilliant but least recognized thermodynamicist produced in the United States. He received the first doctoral degree in engineering awarded in the United States (usually assumed to be in mechanical engineering since the subject was gearing). He developed the T-s diagram as a tool for analyzing heat transfer in thermodynamic systems and provided methods for analyzing thermodynamic equilibrium in the general sense. In 1878 he published work defining the phase rule, which made thermodynamics the basis for the field of physical chemistry. His later contributions in establishing the fundamentals of statistical thermodynamics fall outside the realm of classical thermodynamics, but are equally important.

Gottlieb Daimler (1834–1900) was superintendent of Otto's gas engine works in Deutz, Germany, and realized that the Otto engine must be made to operate on volatile liquid fuel if it was to become practical for use in transportation. In 1879, he obtained a patent for a multicylinder engine operating on a common crank, and in 1883–1884 he, along with his brilliant design engineer Wilhelm Maybach (1846–1929), produced

HARPER'S WEEKLY

THE DIFFERENT USES FOR COMPRESSED AIR.—DRAWN BY W. LOUIS SONNTAG, JUN.—[SEE PAGE 1202.]

1. Car using the Hardie Compressed-Air Motor on 125th Street, New York City. 2. The Mekarski Compressed-Air System used in Paris—recharging a Car *en route*. 3. Motor and Train in Paris (Mekarski System). 4. Compressed-Air Locomotive to be operated on the Elevated Railroad in New York City. 5. Switches moved by Compressed Air and controlled by Electricity in the Pennsylvania Railroad Yards. 6. The Use of Compressed Air for drilling, excavating, etc., on the Chicago Drainage Canal.

Figure A.4 Some proposed and actual uses of compressed air for transportation systems.

the first commercially successful automotive engine. Their engine used a constant-burning flame that was exposed to the fuel-air charge for ignition. Improvements were made by Karl Benz, who introduced a controlled-timing electric ignition system with a vibrating make-break switch to provide multiple sparks during the spark dwell time in each cylinder, and Robert Bosch, who introduced a magneto-driven ignition system. Daimler switched from flame-tube ignition in 1899.

The gasoline engine did not immediately capture the automotive market. As late as 1900, of the total automobile production in the United States, 1575 were electric, 1681 were steam-powered, and 936 used spark ignition engines.

Dr. Rudolph Christian Karl Diesel (1858 – 1913) was born in France but was sent to Germany by his parents as a child, where he later attended the *Technische Hochschule* in Munich as a student in machine design. He designed large steam engines and boilers, but he kept studying to find a replacement for the steam engine that would be more efficient than the 6 to 10 percent then available from the steam cycles. He developed his cycle of operation based on the use of the compression stroke to reach high temperatures, with the combustion process then occurring at constant temperature by metering of the rate of fuel injection, believing this to be the nearest practical approach to the Carnot cycle for internal combustion engines. He demonstrated a modification of his cycle in 1893, but his first attempt resulted in an explosion of the engine which came close to ending his life as well as his experiments. He continued to develop the concept, and finally he demonstrated a practical operating engine in 1897. Fuel development was crucial to the acceptance of the diesel engine. Early engines were run on almost any liquid hydrocarbon. Usually these came from crude stocks with a range of sulfur and asphalt content, but sometimes from oil shale, coal tar, or lignite. Resulting operational problems with the engines, including the carbonizing of fuel injection nozzles and catastrophic failure of air injection compressors, created dissatisfied users and investors. A truck engine of 40 bhp (30 kW) using solid fuel injection was tested in 1910, but it smoked so badly that the project was canceled. The use of diesel engines for marine service was more successful. Fuel could be obtained wherever it was cheap and of consistent quality, and the engines had advantages over their steam competitors in reduced size and increased range. Failing health, continuing criticism, and serious financial setbacks beset Diesel, and in 1913 he disappeared from a boat crossing the English Channel on a calm moonlit night.

Steam was still the major motive force for industrial use, and steam engines grew to immense size (Fig. A.5). All the engines developed to this point were reciprocating, using a piston in a cylinder to drive a flywheel to provide circular motion. As early as 1791, the Englishman John Barber had patented an engine with all the elements of a contemporary gas turbine, and further paper designs were made by others until practical turbines finally were developed for steam cycles almost simultaneously by

OUR CENTENNIAL—PRESIDENT GRANT AND DOM PEDRO STARTING THE CORLISS ENGINE.—FROM A SKETCH BY THEO. R. DAVIS.—[SEE PAGE 422.]

Figure A.5 Corliss Engine at U.S. Centennial Celebration, 1876.

THE ICE INDUSTRY OF NEW YORK.—Drawn by F. Ray.—[See Page 565.]

Figure A.6 Use of natural ice for refrigeration.

Sir Charles A. Parsons (1854–1931) in England and Carl G. P. DeLaval (1845–1913) in Sweden during the 1880s.

A.2 Refrigeration and Thermodynamics

To follow the development of refrigeration, we must back up in time to the middle of the seventeenth century, when the Englishman Robert Boyle (1627–1691) observed the reduction of the boiling temperature of water when the pressure was reduced. Dr. William Cullen, a professor of medicine at the University of Glasgow, observed in 1755 that an isolated container of water underwent a drop in temperature during evaporation. In 1844, Thomas Masters of London patented an ice cream maker that used a mixture of ice and salt to lower the brine temperature. Charles E. Monroe of Cambridge, Massachusetts, in 1871 patented a food cooler based on the idea of evaporation of water from the porous lining of a refrigerator. However, the most notable of the entrepreneurs who used the methods of "natural" refrigeration was Frederic Tudor (1783–1864), who cut and stored natural ice (Fig. A.6). By 1804, he was making regular shipments of ice to the South from the North, and by 1834 he had extended his trade to the West Indies, South Africa, and finally to Europe.

Sir John Leslie (1766–1832), professor of mathematics at Edinburgh University, building on the observations by Boyle and Cullen, used sulfuric acid to absorb water vapor from a vessel containing water, and thus produce a vacuum in the closed container. The vacuum in turn caused the saturation temperature of the water producing the vapor to be low enough that ice could be formed. In 1810, 1-lb blocks were made by this process. By 1881, Franz Windhausen was producing six 672-lb blocks per cycle in a commercial-scale machine in Germany. The sulfuric acid was regenerated for further use after each cycle by heating it with steam to drive off the absorbed water.

In 1858, a system based on the observation that ammonia could reach lower temperatures than water when boiled at the same pressure was marketed by Ferdinand P. E. Carre (1824–?) (Fig. A.7). In the early 1930s, the Crosley "Icyball" system based on exactly Carre's cycle was sold widely in the rural United States. The gas refrigerator operates on the same cycle in a continuous rather than batch fashion.

In 1755, M. Hoell observed that compressed air leaving a pressurized air line cooled when it escaped from the line. In 1828, Trevithick proposed a refrigeration machine based on Hoell's observation, and in 1851 Dr. John Gorrie (1803–1855) was issued a U. S. patent for the first machine to operate successfully on the air compression–expansion cycle.

Jacob Perkins (1766–1849), an American living in London, observed that working fluids other than air could operate more efficiently, especially if they could be condensed easily after compression. He built the first practical vapor compression machine, which was patented in 1834. The use of ammonia in such a machine was demonstrated success-

The New York Ice Machine Company

Room 54, No. 21 Cortlandt Street, New York.

Patents of C. M. Tessie du Motay and Leonard F. Beckwith, and of C. M. Tessie du Motay and A. J. Rossi.

LOW PRESSURE BINARY ABSORPTION SYSTEM. NON-INFLAMMABLE.

Ice and Refrigerating Machines ¼ ton to 50 tons per day. Cost of Ice, everything included, less than $1.00 per ton.

ADVANTAGES OVER ALL OTHER MACHINES.

SELF LUBRICATING. NO LEAKS. BINARY LIQUID MIXED WITH WATER HAS NO ACTION ON METALS EVEN IF MIXED WITH WATER. Makes 25 to 30 per cent. more ice or cold air. Uses only ¼ of water of condensation. Pressure at rest 0 instead of 20 to 100 lbs. Pressure when running, 1½ lbs. instead of 45 to 300 lbs. Machines as easy of attendance as Low-Pressure Steam Engines. Send for descriptive Circulars and Pamphlet.

AUTOMATIC FAMILY MACHINES making 5 lbs. and 10 lbs. per hour. These Machines can be run by any boy or servant. No danger whatever of any kind in running them. A 6-tons Ice Machine can be seen at work at C. H. Delamater & Co., Foot of West 14th St., New York.

All the Machines are Guaranteed. E. GILLET, Sec'y. Reference, C. H. DELAMATER & CO., New York City

Figure A.7 Commercial icemaking machine in 1880 advertisement.

fully by David Boyle of Chicago, who developed the machine between 1869 and 1873 and operated a 1 ton/day ice-making plant in Jefferson, Texas, in 1873. Carl P. G. Linde (1842–1934) of Munich used an advanced cycle with much better mechanical detail which was developed to an experimental stage in 1873 and to commercial use in 1875 (Fig. A.8).

A.3 Summary

The work of this diverse group of scientists and engineers drawn from many nations follows Langley's picture referred to at the beginning of this

EXAMINING SLAB OF ICE IN PROCESS OF FREEZING.

COMPLETELY FROZEN—WEIGHT, FOUR AND A HALF TONS.

READY FOR THE CIRCULAR SAWS.

SAWING INTO SMALL PIECES.

ARTIFICIAL ICE MANUFACTURE.—From Sketches by Horace Bradley.—[See Page 67.]

Figure A.8 Early systems for large-scale ice production.

section—a pack of hunting hounds questing after an elusive quarry. Classical thermodynamics presents a beautifully coherent theory that finally joins the many threads that were followed by the early investigators. However, the history of that theory is not a continuous line of advancement, as perhaps best indicated by the fact that formulation of what we now call the second law preceded the formulation of the first law by more than 25 years. The history is a tangled string filled with loops and knots.

For further information on the early history of thermodynamics, see Refs. 1 to 15.

References

1. R. Bruce Lindsay, *Julius Robert Mayer—Prophet of Energy,* Pergamon, Elmsford, N.Y., 1973.
2. R. Bruce Lindsay, *Energy, Historical Development of the Concept,* vol. 1, *Benchmark Papers on Energy,* Hutchinson & Ross, Stroudsburg, Pa., 1975.
3. R. Bruce Lindsay, *Applications of Energy: Nineteenth Century,* vol 2, *Benchmark Papers on Energy,* Dowden, Hutchinson and Ross, Stroudsburg, Pa., 1976.
4. William Edgar Knowles Middleton, *A History of the Thermometer and Its Use in Meteorology,* Johns Hopkins, Baltimore, Md., 1966.
5. E. Mendoza, "A Sketch for a History of Early Thermodynamics," *Physics Today,* vol. 14, no. 2, February 1961, pp. 32–42.
6. Charles Singer, *A Short History of Science to the Nineteenth Century,* Oxford at the Clarendon Press, London, 1941.
7. Abbot Payson Usher, *A History of Mechanical Inventions,* rev. ed., Harvard University Press, Cambridge, Mass., 1954.
8. Carl J. Echkardt, *Men and Power,* privately published.
9. C. Lyle Cummins, Jr., *Internal Fire,* Carnot Press, Lake Osweago, Ore., 1976.
10. Arthur W. J. G. Orde-Hume, *Perpetual Motion, The History of an Obsession,* St. Martin's Press, New York, 1977.
11. P. W. Atkins, *The Second Law,* Scientific American Library, New York, 1984.
12. H. O. Hardenberg, "An Historical Overview of Gunpowder Engine Development—1508–1868," in E. F. C. Somerscales and A. A. Zagotta (eds.), *History of the Internal Combustion Engine,* vol. 8, pp. 1–9, ASME, October 1989.
13. C. A. Amann, "Power to Burn—A History of the Spark Ignition Engine," *Mechanical Engineering,* pp. 46–54, April 1990.
14. C. L. Cummins, Jr., "Diesel Engines of 2,000 BHP per Cylinder—Pre-1914 Marine Engine Developments," in E. F. C. Somerscales and A. A. Zagotta (eds.), *History of the Internal Combustion Engine,* vol. 8, pp. 19–32, ASME, October 1989.
15. Richard J. Goldstein, "Count Rumford on the Art and Science of Heat Transfer," in G. Hetsroni (ed.), *Heat Transfer 1990: Proc. of Ninth Int. Heat Transfer Conf.,* vol. 1, pp. 3–13, Hemisphere, New York, 1990.

APPENDIX B

Conversion Factors

TABLE B.1 Fundamental Numerical Values

First Bohr electron radius	$a_0 = 0.5292 \times 10^{-10}$ m
Speed of light in a vacuum	$c_0 = 2.9979 \times 10^{8}$ m/s
Electronic charge	$e = 1.6022 \times 10^{-19}$ C
Gravitational acceleration	$g = 9.81$ m/s^2
Planck's constant	$h = 6.6262 \times 10^{-34}$ J·s
Boltzmann constant	$k = 1.3806 \times 10^{-23}$ J/K
Electron rest mass	$m_e = 9.1096 \times 10^{-31}$ kg
Classical electron radius	$r_0 = 2.8179 \times 10^{-15}$ m
Electron volt	1 eV $= 1.6022 \times 10^{-19}$ J
Temperature associated with 1 eV	1 eV/k = 11,605 K
Ionization potential of hydrogen atom	13.606 eV
Avogadro's number	$N_a = 6.023 \times 10^{26}$ particles/kmol

TABLE B.2 Values for the Universal Gas Constant

$\overline{R}$ = 8.31441 kPa·m^3/(kmol·K)
= 8.31441 kJ/(kmol·K)
= 0.0820568 liter·atm/(gmol·K)
= 0.0820568 m^3·atm/(kmol·K)
= 1.98586 Btu/(lbmol·°R)
= 0.730235 ft^3·atm/(lbmol·°R)
= 1545.3 ft·lbf/(lbmol·°R)
= 10.73 psia·ft^3/(lbmol·°R)

TABLE B.3 Conversion Factors for Length

	Mile, mi	Kilometer, km	Meter, m	Foot, ft	Inch, in
1 mile =	1	1.609	1609	5280	6.336×10^4
1 kilometer =	0.6214	1	10^3	3.281×10^3	3.937×10^4
1 meter =	6.214×10^{-4}	10^{-3}	1	3.281	39.37
1 foot =	1.894×10^{-4}	3.048×10^{-4}	0.3048	1	12
1 inch =	1.578×10^{-5}	2.540×10^{-5}	2.540×10^{-2}	8.333×10^{-2}	1
1 centimeter =	6.214×10^{-6}	10^{-5}	10^{-2}	3.281×10^{-2}	0.3937
1 millimeter =	6.214×10^{-7}	10^{-6}	10^{-3}	3.281×10^{-3}	0.03937
1 micrometer =	6.214×10^{-10}	10^{-9}	10^{-6}	3.281×10^{-6}	3.937×10^{-5}
1 nanometer =	6.214×10^{-13}	10^{-12}	10^{-9}	3.281×10^{-9}	3.937×10^{-8}
1 angstrom =	6.214×10^{-14}	10^{-13}	10^{-10}	3.281×10^{-10}	3.937×10^{-9}
	Centimeter, cm	**Millimeter, mm**	**Micrometer, μm**	**Nanometer, nm**	**Angstrom, Å**
1 mile =	1.609×10^5	1.609×10^6	1.609×10^9	1.609×10^{12}	1.609×10^{13}
1 kilometer =	10^5	10^6	10^9	10^{12}	10^{13}
1 meter =	10^2	10^3	10^6	10^9	10^{10}
1 foot =	30.48	3.048×10^2	3.048×10^5	3.048×10^8	3.048×10^9
1 inch =	2.540	25.40	2.540×10^4	2.540×10^7	2.540×10^8
1 centimeter =	1	10	10^4	10^7	10^8
1 millimeter =	10^{-1}	1	10^3	10^6	10^7
1 micrometer =	10^{-4}	10^{-3}	1	10^3	10^4
1 nanometer =	10^{-7}	10^{-6}	10^{-3}	1	10
1 angstrom =	10^{-8}	10^{-7}	10^{-4}	10^{-1}	1

TABLE B.4 Useful Conversion Factors†

Area
- $1\ m^2 = 10.764\ ft^2$
- $1\ ft^2 = 0.092903\ m^2$
- $1\ in^2 = 6.4516 \times 10^{-4}\ m^2$

Mass
- 1 kg = 2.2046 lbm
- 1 lbm = 0.45359 kg

Volume
- $1\ m^3 = 35.315\ ft^3$
- $1\ ft^3 = 0.028317\ m^3$
- $1\ gal = 0.13368\ ft^3$
- $1\ liter = 1000.0\ cm^3$

Specific Volume
- $1\ m^3/kg = 16.019\ ft^3/lbm$
- $1\ ft^3/lbm = 0.062428\ m^3/kg$

Density
- $1\ lbm/ft^3 = 16.019\ kg/m^3$
- $1\ kg/m^3 = 0.062428\ lbm/ft^3$

Force
- $1\ N = 1\ kg \cdot m/s^2 = 0.22481\ lbf$
- 1 lbf = 4.4482 N

TABLE B.4 (*Continued*)

Pressure

1 Pa = 1 N/m^2 = 1 × 10^{-5} bar = 1.4504 × 10^{-4} psia = 9.8692 × 10^{-6} atm = 0.020886 lbf/ft^2

1 kPa = 0.14504 psia

1 bar = 10^5 Pa = 0.98692 atm = 14.504 psia = 2088.6 lbf/ft^2

1 lbf/in^2 (psia) = 144 lbf/ft^2 = 6894.8 Pa = 6.8948 × 10^{-2} bar = 0.068046 atm

1 atm = 101.325 kPa = 14.696 psia = 1.0133 bar = 2116.2 lbf/ft^2

Energy

1 J = 1 N·m = 1 kg·m^2/s^2

1 kJ = 1 kW·s = 0.94783 Btu = 0.23885 kcal = 737.56 ft·lbf

1 Btu = 1.0550 kJ = 0.25200 kcal = 778.16 ft·lbf

1 kcal = 4.1868 kJ = 3.9684 Btu = 3088.0 ft·lbf

1 kWh = 3.60 × 10^3 kJ = 2655.2 × 10^3 ft·lbf = 3412.2 Btu = 859.86 kcal/h

1 ft·lbf = 1.2851 × 10^{-3} Btu = 1.3558 × 10^{-3} kJ

Energy Rate (Power)

1 W = 1 J/s

1 W = 3.4122 Btu/h = 0.85987 kcal/h = 1.34102 × 10^{-3} hp = 0.73756 ft·lbf/s

1 Btu/h = 0.29307 W = 0.25200 kcal/h = 3.9300 × 10^{-4} hp = 0.21616 ft·lbf/s

1 kcal/h = 1.1630 W = 3.9683 Btu/h = 1.5595 × 10^{-3} hp = 0.85778 ft·lbf/s

1 horsepower (hp) = 550 ft·lbf/s = 2544.5 Btu/h = 745.70 W

1 ft·lbf/s = 4.6262 Btu/h = 1.3558 W = 1.8182 × 10^{-3} hp

1 ton (cooling capacity) = 12,000 Btu/h = 3.5168 kW

Specific Energy, Specific Enthalpy

1 kJ/kg = 0.42992 Btu/lbm = 0.23885 kcal/kg = 334.55 ft·lbf/lbm

1 Btu/lbm = 2.3260 kJ/kg = 0.55556 kcal/kg = 778.16 ft·lbf/lbm

1 kcal/kg = 4.1868 kJ/kg = 1.8000 Btu/lbm = 1400.7 ft·lbf/lbm

1 ft·lbf/lbm = 2.9891 × 10^{-3} kJ/kg = 1.2851 × 10^{-3} Btu/lbm = 7.1394 × 10^{-4} kcal/kg

Energy Rate per Unit Area

1 W/m^2 = 0.31700 Btu/(h·ft^2) = 0.85986 kcal/(h·m^2)

1 Btu/(h·ft^2) = 3.1546 W/m^2 = 2.7125 kcal/(h·m^2)

1 kcal/(h·m^2) = 1.1630 W/m^2 = 0.36867 Btu/(h·ft^2)

Entropy

1 kJ/K = 0.52657 Btu/°R = 0.23885 kcal/K

1 Btu/°R = 1.8991 kJ/K = 0.45359 kcal/K

1 kcal/K = 4.1868 kJ/K = 2.2047 Btu/°R

Specific Entropy, Specific Heat, Gas Constant

1 kJ/(kg·K) = 0.23885 Btu/(lbm·°R) = 0.23885 kcal/(kg·K)

1 Btu/(lbm·°R) = 4.1868 kJ/(kg·K) = 1.0000 kcal/(kg·K)

1 kcal/(kg·K) = 4.1868 kJ/(kg·K) = 1.0000 Btu/(lbm·°R)

Temperature

T,K = $\frac{5}{9}T$,°R = $\frac{5}{9}$(T,°F + 459.67) = T,°C + 273.15

T,°R = $\frac{9}{5}T$, K = $\frac{9}{5}$(T,°C + 273.15) = T,°F + 459.67

T,°F = $\frac{9}{5}T$, °C + 32

T,°C = $\frac{5}{9}$(T,°F − 32)

Velocity

1 m/s = 3.281 ft/s

1 ft/s = 0.3048 m/s

† All energy conversions based on International Steam Table values.

APPENDIX C

Thermodynamic Properties in Dimensionless Form or for Both SI and USCS Units

TABLE C.1 Triple-Point Properties for Common Substances

Substance	T, K	T, °R	P, kPa	P, psia
Ammonia (NH_3)	195.4	351.7	6.18	0.896
Carbon dioxide (CO_2)	216.6	389.9	516.6	74.93
Helium 4 (λ point)	2.17	3.91	5.07	0.735
Hydrogen (H_2)	13.84	24.91	7.09	1.03
Nitrogen (N_2)	63.18	113.7	12.56	1.822
Oxygen (O_2)	54.36	97.85	0.152	0.0220
Water (H_2O)	273.16	491.69	0.6113	0.08866

SI values from William Z. Black and James G. Hartley, *Thermodynamics,* Harper & Row, New York, 1985 (used with permission).

TABLE C.2 Gas Constants and Zero-Pressure Specific Heats for Various Ideal Gases at 300 K (540°R)

		Gas Constant R		c_p^o		c_v^o		
Gas	Molecular Weight	kJ/(kg·K)	ft·lbf/(lbm·°R)	kJ/(kg·K)	Btu/(lbm·°R)	kJ/(kg·K)	Btu/(lbm·°R)	$k = \frac{c_p^o}{c_v^o}$
Air	28.97	0.28700	53.34	1.0052	0.2401	0.7180	0.1716	1.400
Argon (Ar)	39.948	0.20813	38.68	0.5207	0.1244	0.3124	0.0746	1.667
Butane (C_4H_{10})	58.124	0.14304	26.58	1.7164	0.415	1.5734	0.381	1.09
Carbon dioxide (CO_2)	44.010	0.18892	35.10	0.8464	0.2021	0.6573	0.1569	1.288
Carbon monoxide (CO)	28.010	0.29683	55.16	1.0411	0.2487	0.7441	0.1777	1.399
Ethane (C_2H_6)	30.070	0.27650	51.38	1.7662	0.427	1.4897	0.361	1.183
Ethylene (C_2H_4)	28.054	0.29637	55.07	1.5482	0.411	1.2518	0.340	1.208
Helium (He)	4.003	2.07703	386.0	5.1926	1.25	3.1156	0.753	1.667
Hydrogen (H_2)	2.016	4.12418	766.4	14.3193	3.4198	10.1919	2.4340	1.405
Methane (CH_4)	16.043	0.51835	96.35	2.2537	0.532	1.7354	0.403	1.32
Neon (Ne)	20.183	0.41195	76.55	1.0299	0.246	0.6179	0.1477	1.667
Nitrogen (N_2)	28.013	0.29680	55.15	1.0404	0.2485	0.7434	0.1776	1.400
Octane (C_8H_{18})	114.232	0.07279	13.53	1.7113	0.409	1.6385	0.392	1.044
Oxygen (O_2)	31.999	0.25983	48.28	0.9190	0.2195	0.6590	0.1574	1.395
Propane (C_3H_8)	44.094	0.18855	35.04	1.6794	0.407	1.4909	0.362	1.124
Water (H_2O)	18.015	0.46152	85.76	1.8649	0.4454	1.4031	0.3351	1.329

From Gordon Van Wylen and Richard Sonntag, *Fundamentals of Classical Thermodynamics,* 2d ed., Wiley, New York, 1976 (with permission) and *GASPROPS,* Wiley Professional Software, Wiley, New York, 1984.

TABLE C.3 Specific Heats and Specific Volumes of Various Liquids and Solids

Substance	T		c_P		v	
	K	°R	kJ/(kg·K)	Btu/(lbm·°R)	m³/kg	ft³/lbm
Liquid						
Ammonia						
(sat)	253	455	4.52	1.08	15.04×10^{-4}	24.06×10^{-3}
(sat)	323	581	5.10	1.22	17.76×10^{-4}	28.41×10^{-3}
Benzene						
(1 atm)	288	518	1.80	0.430	11.4×10^{-4}	18.3×10^{-3}
(1 atm)	338	608	1.92	0.459	—	—
Bismuth						
(1 atm)	698	1256	0.144	0.0344	1.02×10^{-4}	1.63×10^{-3}
(1 atm)	1033	1859	0.164	0.0392	—	—
Ethyl alcohol						
(1 atm)	298	536	2.43	0.580	12.7×10^{-4}	20.3×10^{-3}
Glycerin						
(1 atm)	283	509	2.32	0.554	7.94×10^{-4}	12.7×10^{-3}
(1 atm)	323	581	2.58	0.616	—	—
Mercury						
(1 atm)	283	509	0.138	0.0330	0.738×10^{-4}	1.18×10^{-3}
(1 atm)	588	1058	0.134	0.0320	—	—
Refrigerant 12						
(sat)	233	419	0.883	0.211	6.59×10^{-4}	10.66×10^{-3}
(sat)	253	455	0.908	0.217	6.85×10^{-4}	10.97×10^{-3}
(sat)	323	581	1.02	0.244	8.26×10^{-4}	13.20×10^{-3}
Sodium						
(1 atm)	368	662	1.38	0.330	—	—
(1 atm)	813	1463	1.26	0.301	—	—
Water						
(1 atm)	273	492	4.213	1.006	10.00×10^{-4}	16.02×10^{-3}
(1 atm)	298	536	4.177	0.998	10.03×10^{-4}	16.06×10^{-3}
(1 atm)	323	581	4.178	0.998	10.12×10^{-4}	16.21×10^{-3}
(1 atm)	373	671	4.213	1.006	10.43×10^{-4}	16.70×10^{-3}
Solid						
Aluminum	23	41	0.0163	0.00389	—	—
	73	131	0.318	0.0760	—	—
	173	311	0.699	0.167	—	—
	273	492	0.870	0.208	3.7×10^{-4}	5.9×10^{-3}
	373	671	0.941	0.225	—	—
	573	1031	1.04	0.248	—	—
Carbon						
(diamond)	298	536	0.519	0.124	2.86×10^{-4}	4.58×10^{-3}
(graphite)	298	536	0.711	0.170	5.12×10^{-4}	8.20×10^{-3}
Chromium	298	536	0.448	0.107	1.40×10^{-4}	2.24×10^{-3}
Copper	50	90	0.0967	0.0231	—	—
	100	180	0.252	0.0602	—	—
	173	311	0.328	0.0783	—	—
	223	401	0.361	0.0862	—	—
	273	492	0.381	0.0910	—	—
	300	540	0.385	0.0920	1.12×10^{-4}	1.79×10^{-3}
	373	671	0.393	0.0939	—	—
	473	851	0.403	0.0963	—	—
Gold	298	536	0.129	0.0308	0.518×10^{-4}	0.830×10^{-3}

TABLE C.3 Specific Heats and Specific Volumes of Various Liquids and Solids (*Continued*)

Substance	T K	T °R	c_P kJ/(kg·K)	c_P Btu/(lbm·°R)	v m³/kg	v ft³/lbm
Solid						
Ice	73	131	0.678	0.162	—	—
	133	239	1.096	0.2618	—	—
	213	383	1.640	0.3917	—	—
	262	472	2.033	0.4856	—	—
	270.8	487.4	1.682	0.4017	11.1×10^{-4}	17.8×10^{-3}
Iron	293	527	0.448	0.107	0.127×10^{-4}	0.203×10^{-3}
Lead	3	5	0.0033	0.00079	—	—
	14	25	0.0305	0.00728	—	—
	173	311	0.118	0.0282	—	—
	273	492	0.124	0.0296	0.882×10^{-4}	1.41×10^{-3}
	373	671	0.134	0.0320	—	—
	573	1031	0.149	0.0356	—	—
Nickel	298	536	0.444	0.1060	1.12×10^{-4}	1.79×10^{-3}
Silver	293	527	0.233	0.0557	0.952×10^{-4}	1.52×10^{-3}
	773	1391	0.243	0.0580	—	—
Sodium	298	536	1.226	0.2928	10.3×10^{-4}	16.5×10^{-3}
Tungsten	298	536	0.134	0.0320	0.518×10^{-4}	0.830×10^{-3}
Zinc	298	536	0.385	0.0920	1.40×10^{-4}	2.24×10^{-3}

Values of c_P from Kenneth G. Wark, *Thermodynamics*, 4th ed., McGraw-Hill, New York, 1983 (used with permission). Values of specific volume from various sources.

TABLE C.4 Critical Constants

Gas	Formula	Molecular Weight	Critical Temperature T_{cr} K	Critical Temperature T_{cr} °R	Critical Pressure P_{cr} MPa	Critical Pressure P_{cr} atm	Critical Volume v_{cr} m³/kmol	Critical Volume v_{cr} ft³/lbmol	Acentric Factor ω	Critical Compressibility Factor Z_{cr}
Acetic acid	$C_2H_4O_2$	60.052	592.7	1067.	5.79	57.1	0.171	2.74	0.447	0.201
Acetone	C_3H_6O	58.080	508.1	914.6	4.70	46.4	0.209	3.35	0.304	0.232
Acetonitrile	CH_3CN	41.053	545.5	981.9	4.83	47.7	0.173	2.77	0.327	0.184
Acetylene	C_2H_2	26.038	308.3	554.9	6.14	60.6	0.1127	1.805	0.190	0.270
Air	—	28.97	132.5	238.5	3.77	37.2	—	—	—	—

TABLE C.4 Critical Constants (*Continued*)

Gas	Formula	Molecular Weight	Critical Temperature T_{cr} K	Critical Temperature T_{cr} °R	Critical Pressure P_{cr} MPa	Critical Pressure P_{cr} atm	Critical Volume v_{cr} m³/kmol	Critical Volume v_{cr} ft³/lbmol	Acentric Factor ω	Critical Compressibility Factor z_{cr}
Ammonia	NH_3	17.031	405.5	729.9	11.35	112.0	0.0725	1.16	0.250	0.244
Argon	Ar	39.948	150.8	271.4	4.87	48.1	0.0749	1.20	0.001	0.291
Benzene	C_6H_6	78.114	562.2	1012.	4.89	48.3	0.259	4.15	0.212	0.271
Bromine	Br_2	159.808	588.	1058.	10.3	102.	0.1272	2.038	0.108	0.268
n-Butane	C_4H_{10}	58.124	425.	765.4	3.80	37.5	0.255	4.08	0.199	0.274
Carbon dioxide	CO_2	44.010	304.1	547.4	7.38	72.8	0.0939	1.50	0.239	0.274
Carbon disulfide	CS_2	76.131	552.	994.	7.90	78.0	0.160	2.56	0.109	0.276
Carbon monoxide	CO	28.010	132.9	239.2	3.50	34.5	0.0932	1.49	0.066	0.295
Carbon tetrachloride	CCl_4	153.823	556.4	1001.5	4.56	45.0	0.2759	4.420	0.193	0.272
Chlorine	Cl_2	70.906	416.9	750.4	7.98	78.8	0.1238	1.983	0.090	0.285
Chloroform	$CHCl_3$	119.378	536.4	965.5	5.37	53.0	0.2389	3.827	0.218	0.293
Cyclohexane	C_6H_{12}	84.162	553.5	996.3	4.07	40.2	0.308	4.93	0.212	0.273
Decane	$C_{10}H_{22}$	142.286	617.7	1112.	2.12	20.9	0.603	9.66	0.489	0.249
Deuterium (normal)	D_2	4.032	38.4	69.1	1.66	16.4	—	—	−0.160	0.249
Dichlorodifluoromethane (refrigerant 12)	CCl_2F_2	120.914	385.0	693.0	4.14	40.9	0.2167	3.471	0.204	0.280
Dichlorofluoromethane (refrigerant 21)	$CHCl_2F$	102.923	451.6	812.9	5.18	51.1	0.1964	3.15	0.210	0.271
Diethyl ether	$C_4H_{10}O$	74.123	466.7	840.1	3.64	35.9	0.280	4.49	0.281	0.262
Dioxane	$C_4H_8O_2$	88.107	587.	1057.	5.21	51.4	0.238	3.81	0.281	0.254
Ethane	C_2H_6	30.070	305.4	549.7	4.88	48.2	0.1483	2.376	0.099	0.285
Ethyl acetate	$C_4H_8O_2$	88.107	523.2	941.8	3.83	37.8	0.286	4.58	0.362	0.252
Ethyl alcohol	C_2H_5OH	46.069	513.9	925.0	6.14	60.6	0.1671	2.677	0.644	0.240
Ethyl chloride	C_2H_5Cl	64.515	460.4	828.7	5.27	52.0	0.199	3.19	0.191	0.274
Ethylene	C_2H_4	28.054	282.4	508.3	5.04	49.7	0.1304	2.089	0.089	0.280
Ethylene oxide	C_2H_4O	44.054	469.	844.	7.19	71.0	0.140	2.24	0.202	0.259
Ethyl methyl ketone	C_4H_8O	72.107	536.8	966.2	4.21	41.6	0.267	4.28	0.320	0.252
Helium	^4He	4.003	5.19	9.34	0.227	2.24	0.0574	0.9195	−0.365	0.302
Helium 3	^3He	3.017	3.31	5.96	0.114	1.13	0.0729	1.168	−0.473	0.302
n-Heptane	C_7H_{16}	100.205	540.3	972.5	2.74	27.0	0.432	6.92	0.349	0.263
n-Hexane	C_6H_{14}	86.178	507.5	913.5	3.01	29.7	0.370	5.93	0.299	0.264
Hydrazine	N_2H_4	32.045	653.	1175.	14.7	145.	—	—	0.316	—
Hydrogen (normal)	H_2	2.016	33.2	59.8	1.30	12.8	0.0651	1.043	−0.218	0.306
Hydrogen chloride	HCl	36.461	324.7	584.5	8.31	82.0	0.0809	1.296	0.133	0.249
Hydrogen cyanide	HCN	27.026	456.7	822.1	5.39	53.2	0.1388	2.223	0.388	0.197
Hydrogen sulfide	H_2S	34.080	373.2	671.8	8.94	88.2	0.986	15.8	0.081	0.284
Isobutane	C_4H_{10}	58.124	408.2	734.8	3.65	36.0	0.263	4.21	0.183	0.283
Isopropyl alcohol	C_3H_7OH	60.096	508.3	914.9	4.76	47.0	0.220	3.52	0.665	0.248
Krypton	Kr	83.800	209.4	376.9	5.50	54.3	0.0912	1.46	0.005	0.288
Methane	CH_4	16.043	190.4	342.7	4.60	45.4	0.0992	1.59	0.011	0.288
Methyl alcohol	CH_3OH	32.042	512.6	922.7	8.09	79.8	0.1180	1.890	0.556	0.224
Methyl chloride	CH_3Cl	50.488	416.3	749.3	6.70	66.1	0.1389	2.225	0.153	0.269
Neon	Ne	20.183	44.4	79.9	2.76	27.2	0.0416	0.666	−0.029	0.311
Nitric oxide	NO	30.006	180.	324.	6.48	64.0	0.0577	0.924	0.588	0.250
Nitrogen	N_2	28.013	126.2	227.2	3.39	33.5	0.0898	1.44	0.039	0.290
Nitrogen peroxide	NO_2	46.006	431.	776.	10.1	99.7	0.1678	2.688	0.834	0.473
Nitrous oxide	N_2O	44.013	309.6	557.3	7.24	71.5	0.0974	1.56	0.165	0.274
Nonane	C_9H_{20}	128.259	594.6	1070.	2.29	22.6	0.548	8.78	0.445	0.26
Octane	C_8H_{18}	114.232	568.8	1024.	2.49	24.6	0.492	7.88	0.398	0.259
Oxygen	O_2	31.999	154.6	278.3	5.04	49.7	0.0734	1.18	0.025	0.288
Pentane	C_5H_{12}	72.151	469.7	845.5	3.37	33.3	0.304	4.87	0.251	0.263
Propane	C_3H_8	44.094	369.8	665.6	4.25	41.9	0.203	3.25	0.153	0.281
Propene	C_3H_6	42.081	364.9	656.8	4.60	45.4	0.181	2.90	0.144	0.274

TABLE C.4 Critical Constants (*Continued*)

Gas	Formula	Molecular Weight	Critical Temperature T_{cr} K	°R	Critical Pressure P_{cr} MPa	atm	Critical Volume v_{cr} m³/kmol	ft³/lbmol	Acentric Factor ω	Critical Compressibility Factor Z_{cr}
n-Propyl alcohol	C_3H_7OH	60.096	536.8	966.2	5.17	51.0	0.219	3.51	0.623	0.253
Propyne	C_3H_4	40.065	393.	707.	5.47	54.0	0.162	2.60	0.313	0.271
Sulfur dioxide	SO_2	64.063	430.8	775.4	7.88	77.8	0.1222	1.957	0.256	0.269
Sulfur trioxide	SO_3	80.058	491.0	883.8	8.21	81.0	0.1273	2.039	0.481	0.256
Toluene	C_7H_8	92.141	591.8	1065.	4.10	40.5	0.316	5.06	0.263	0.263
Trichlorofluorormethane (refrigerant 11)	CCl_3F	137.378	471.2	848.2	4.41	43.5	0.2478	3.969	0.189	0.279
Water	H_2O	18.015	647.1	1165.	22.06	217.7	0.0560	0.8971	0.344	0.2439
Xenon	Xe	131.300	289.7	521.5	5.84	57.6	0.1184	1.897	0.008	0.287

Data excerpted from Robert C. Reid, John M. Prausnitz, and Bruce E. Poling, *The Properties of Liquids and Gases,* 4th ed., McGraw-Hill, New York, 1987.

TABLE C.5 Empirical Constants for the Benedict-Webb-Rubin Equation

Gas	Formula	A_0 N·m⁴/kg²	A_0 lbf·ft⁴/lbm²	B_0 m³/kg	B_0 ft³/lbm	C_0 N·m⁴·K²/kg²	C_0 lbf·ft⁴·°R²/lbm²
Methane	CH_4	731.195	3918.49	2.65735×10^{-3}	4.25667×10^{-2}	0.889635×10^{7}	1.54469×10^{8}
Ethylene	C_2H_4	430.550	2307.33	1.98649×10^{-3}	3.18205×10^{-2}	1.69071×10^{7}	2.93562×10^{8}
Ethane	C_2H_6	466.269	2498.75	2.08914×10^{-3}	3.34648×10^{-2}	2.01509×10^{7}	3.49885×10^{8}
Propylene	C_3H_6	350.217	1876.82	2.02308×10^{-3}	3.24066×10^{-2}	2.51642×10^{7}	4.36932×10^{8}
Propane	C_3H_8	358.575	1921.61	2.20855×10^{-3}	3.53776×10^{-2}	2.65194×10^{7}	4.60462×10^{8}
i-Butane	C_4H_{10}	307.308	1646.87	2.36826×10^{-3}	3.79359×10^{-2}	2.55256×10^{7}	4.43207×10^{8}
i-Butylene	C_4H_8	288.571	1546.46	2.06958×10^{-3}	3.31515×10^{-2}	2.98871×10^{7}	5.18937×10^{8}
n-Butane	C_4H_{10}	302.865	1623.06	2.14127×10^{-3}	3.43000×10^{-2}	2.98168×10^{7}	5.17716×10^{8}
i-Pentane	C_5H_{12}	249.391	1336.49	2.22006×10^{-3}	3.55620×10^{-2}	3.40357×10^{7}	5.90970×10^{8}
n-Pentane	C_5H_{12}	237.376	1272.10	2.17426×10^{-3}	3.48283×10^{-2}	4.13424×10^{7}	7.17838×10^{8}
n-Hexane	C_6H_{14}	197.242	1057.02	2.06498×10^{-3}	3.30778×10^{-2}	4.53487×10^{7}	7.87400×10^{8}
n-Heptane	C_7H_{16}	177.041	948.77	1.98756×10^{-3}	3.18377×10^{-2}	4.79543×10^{7}	8.32642×10^{8}

Gas	Formula	a N·m⁷/kg³	a lbf·ft⁷/lbm³	b m⁶/kg²	b ft⁶/lbm²	c N·m⁷·K²/kg³	c lbf·ft⁷·°R²/lbm³
Methane	CH_4	1.21466	104.270	1.31523×10^{-5}	3.37476×10^{-3}	0.62577×10^{5}	1.74047×10^{7}
Ethylene	C_2H_4	1.19119	102.256	1.09451×10^{-5}	2.80842×10^{-3}	0.97139×10^{5}	2.70175×10^{7}
Ethane	C_2H_6	1.28892	110.645	1.23191×10^{-5}	3.16097×10^{-3}	1.22361×10^{5}	3.40325×10^{7}
Propylene	C_3H_6	1.05482	90.5491	1.05806×10^{-5}	2.71489×10^{-3}	1.39829×10^{5}	3.88910×10^{7}
Propane	C_3H_8	1.12224	96.3367	1.15892×10^{-5}	2.97369×10^{-3}	1.52759×10^{5}	4.24872×10^{7}
i-Butane	C_4H_{10}	1.00195	86.0106	1.25806×10^{-5}	3.22807×10^{-3}	1.47891×10^{5}	4.11333×10^{7}
i-Butylene	C_4H_8	0.97316	83.539	1.10774×10^{-5}	2.84236×10^{-3}	1.58056×10^{5}	4.39605×10^{7}
n-Butane	C_4H_{10}	0.97334	83.555	1.18582×10^{-5}	3.04271×10^{-3}	1.63610×10^{5}	4.55052×10^{7}
i-Pentane	C_5H_{12}	1.01546	87.1704	1.28545×10^{-5}	3.29835×10^{-3}	1.87887×10^{5}	5.22574×10^{7}
n-Pentane	C_5H_{12}	1.10159	94.5640	1.28545×10^{-5}	3.29835×10^{-3}	2.22807×10^{5}	6.19698×10^{7}
n-Hexane	C_6H_{14}	1.12913	96.9282	1.47181×10^{-5}	3.77654×10^{-3}	2.40013×10^{5}	6.67554×10^{7}
n-Heptane	C_7H_{16}	1.04602	89.7937	1.51575×10^{-5}	3.88928×10^{-3}	2.49275×10^{5}	6.93314×10^{7}

TABLE C.5 Empirical Constants for the Benedict-Webb-Rubin Equation (*Continued*)

Gas	Formula	α m^9/kg^3	α ft^9/lbm^3	γ m^6/kg^2	γ ft^6/lbm^2
Methane	CH_4	30.1853×10^{-9}	12.4068×10^{-4}	23.3469×10^{-6}	5.99061×10^{-3}
Ethylene	C_2H_4	8.08173×10^{-9}	3.32175×10^{-4}	11.7469×10^{-6}	3.01415×10^{-3}
Ethane	C_2H_6	8.97220×10^{-9}	3.68775×10^{-4}	13.0701×10^{-6}	3.35367×10^{-3}
Propylene	C_3H_6	6.13014×10^{-9}	2.51961×10^{-4}	10.3453×10^{-6}	2.65451×10^{-3}
Propane	C_3H_8	7.09776×10^{-9}	2.91732×10^{-4}	11.3317×10^{-6}	2.90761×10^{-3}
i-Butane	C_4H_{10}	5.48279×10^{-9}	2.25353×10^{-4}	10.0799×10^{-6}	2.58641×10^{-3}
i-Butylene	C_4H_8	5.16963×10^{-9}	2.12482×10^{-4}	9.41616×10^{-6}	2.41610×10^{-3}
n-Butane	C_4H_{10}	5.62184×10^{-9}	2.31069×10^{-4}	10.0799×10^{-6}	2.58641×10^{-3}
i-Pentane	C_5H_{12}	4.53682×10^{-9}	1.86472×10^{-4}	8.90805×10^{-6}	2.28573×10^{-3}
n-Pentane	C_5H_{12}	4.83038×10^{-9}	1.98538×10^{-4}	9.13893×10^{-6}	2.34497×10^{-3}
n-Hexane	C_6H_{14}	4.40244×10^{-9}	1.80949×10^{-4}	8.99353×10^{-6}	2.30766×10^{-3}
n-Heptane	C_7H_{16}	4.33982×10^{-9}	1.78375×10^{-4}	8.97754×10^{-6}	2.30356×10^{-3}

For SI units, R is $Pa \cdot m^3/(kg \cdot K)$ or $J/(kg \cdot K)$, P in Pa, v in m^3/kg, and T in K.

For USCS units, R is in $ft \cdot lbf/(lbm \cdot {}^\circ R)$, P in lbf/ft^2, v in ft^3/lbm, and T in °R.

SI data, as corrected, from Ernest Cravalho and Joseph L. Smith, *Engineering Thermodynamics,* Pitman, Marshfield, Mass., 1981 (used with permission).

TABLE C.6 Lee-Kesler Values for the COMPRESSIBILITY FACTOR

(a) Values for $z^{(0)}$

	P_r						
T_r	0.010	0.050	0.100	0.200	0.400	0.600	0.800
0.30	0.0029	0.0145	0.0290	0.0579	0.1158	0.1737	0.2315
0.35	0.0026	0.0130	0.0261	0.0522	0.1043	0.1564	0.2084
0.40	0.0024	0.0119	0.0239	0.0477	0.0953	0.1429	0.1904
0.45	0.0022	0.0110	0.0221	0.0442	0.0882	0.1322	0.1762
0.50	0.0021	0.0103	0.0207	0.0413	0.0825	0.1236	0.1647
0.55	0.9804	0.0098	0.0195	0.0390	0.0778	0.1166	0.1553
0.60	0.9849	0.0093	0.0186	0.0371	0.0741	0.1109	0.1476
0.65	0.9881	0.9377	0.0178	0.0356	0.0710	0.1063	0.1415
0.70	0.9904	0.9504	0.8958	0.0344	0.0687	0.1027	0.1366
0.75	0.9922	0.9598	0.9165	0.0336	0.0670	0.1001	0.1330
0.80	0.9935	0.9669	0.9319	0.8539	0.0661	0.0985	0.1307
0.85	0.9946	0.9725	0.9436	0.8810	0.0661	0.0983	0.1301
0.90	0.9954	0.9768	0.9528	0.9015	0.7800	0.1006	0.1321
0.93	0.9959	0.9790	0.9573	0.9115	0.8059	0.6635	0.1359
0.95	0.9961	0.9803	0.9600	0.9174	0.8206	0.6967	0.1410
0.97	0.9963	0.9815	0.9625	0.9227	0.8338	0.7240	0.5580
0.98	0.9965	0.9821	0.9637	0.9253	0.8398	0.7360	0.5887
0.99	0.9966	0.9826	0.9648	0.9277	0.8455	0.7471	0.6138
1.00	0.9967	0.9832	0.9659	0.9300	0.8509	0.7574	0.6353
1.01	0.9968	0.9837	0.9669	0.9322	0.8561	0.7671	0.6542
1.02	0.9969	0.9842	0.9679	0.9343	0.8610	0.7761	0.6710
1.05	0.9971	0.9855	0.9707	0.9401	0.8743	0.8002	0.7130
1.10	0.9975	0.9874	0.9747	0.9485	0.8930	0.8323	0.7649
1.15	0.9978	0.9891	0.9780	0.9554	0.9081	0.8576	0.8032
1.20	0.9981	0.9904	0.9808	0.9611	0.9205	0.8779	0.8330
1.30	0.9985	0.9926	0.9852	0.9702	0.9396	0.9083	0.8764
1.40	0.9988	0.9942	0.9884	0.9768	0.9534	0.9298	0.9062
1.50	0.9991	0.9954	0.9909	0.9818	0.9636	0.9456	0.9278
1.60	0.9993	0.9964	0.9928	0.9856	0.9714	0.9575	0.9439
1.70	0.9994	0.9971	0.9943	0.9886	0.9775	0.9667	0.9563
1.80	0.9995	0.9977	0.9955	0.9910	0.9823	0.9739	0.9659
1.90	0.9996	0.9982	0.9964	0.9929	0.9861	0.9796	0.9735
2.00	0.9997	0.9986	0.9972	0.9944	0.9892	0.9842	0.9796
2.20	0.9998	0.9992	0.9983	0.9967	0.9937	0.9910	0.9886
2.40	0.9999	0.9996	0.9991	0.9983	0.9969	0.9957	0.9948
2.60	1.0000	0.9998	0.9997	0.9994	0.9991	0.9990	0.9990
2.80	1.0000	1.0000	1.0001	1.0002	1.0007	1.0013	1.0021
3.00	1.0000	1.0002	1.0004	1.0008	1.0018	1.0030	1.0043
3.50	1.0001	1.0004	1.0008	1.0017	1.0035	1.0055	1.0075
4.00	1.0001	1.0005	1.0010	1.0021	1.0043	1.0066	1.0090

TABLE C.6 Lee-Kesler Values for the COMPRESSIBILITY FACTOR (*Continued*)

(*a*) Values for $z^{(0)}$

P_r							
1.000	1.200	1.500	2.000	3.000	5.000	7.000	10.000
0.2892	0.3470	0.4335	0.5775	0.8648	1.4366	2.0048	2.8507
0.2604	0.3123	0.3901	0.5195	0.7775	1.2902	1.7987	2.5539
0.2379	0.2853	0.3563	0.4744	0.7095	1.1758	1.6373	2.3211
0.2200	0.2638	0.3294	0.4384	0.6551	1.0841	1.5077	2.1338
0.2056	0.2465	0.3077	0.4092	0.6110	1.0094	1.4017	1.9801
0.1939	0.2323	0.2899	0.3853	0.5747	0.9475	1.3137	1.8520
0.1842	0.2207	0.2753	0.3657	0.5446	0.8959	1.2398	1.7440
0.1765	0.2113	0.2634	0.3495	0.5197	0.8526	1.1773	1.6519
0.1703	0.2038	0.2538	0.3364	0.4991	0.8161	1.1241	1.5729
0.1656	0.1981	0.2464	0.3260	0.4823	0.7854	1.0787	1.5047
0.1626	0.1942	0.2411	0.3182	0.4690	0.7598	1.0400	1.4456
0.1614	0.1924	0.2382	0.3132	0.4591	0.7388	1.0071	1.3943
0.1630	0.1935	0.2383	0.3114	0.4527	0.7220	0.9793	1.3496
0.1664	0.1963	0.2405	0.3122	0.4507	0.7138	0.9648	1.3257
0.1705	0.1998	0.2432	0.3138	0.4501	0.7092	0.9561	1.3108
0.1779	0.2055	0.2474	0.3164	0.4504	0.7052	0.9480	1.2968
0.1844	0.2097	0.2503	0.3182	0.4508	0.7035	0.9442	1.2901
0.1959	0.2154	0.2538	0.3204	0.4514	0.7018	0.9406	1.2835
0.2901	0.2237	0.2583	0.3229	0.4522	0.7004	0.9372	1.2772
0.4648	0.2370	0.2640	0.3260	0.4533	0.6991	0.9339	1.2710
0.5146	0.2629	0.2715	0.3297	0.4547	0.6980	0.9307	1.2650
0.6026	0.4437	0.3131	0.3452	0.4604	0.6956	0.9222	1.2481
0.6880	0.5984	0.4580	0.3953	0.4770	0.6950	0.9110	1.2232
0.7443	0.6803	0.5798	0.4760	0.5042	0.6987	0.9033	1.2021
0.7858	0.7363	0.6605	0.5605	0.5425	0.7069	0.8990	1.1844
0.8438	0.8111	0.7624	0.6908	0.6344	0.7358	0.8998	1.1580
0.8827	0.8595	0.8256	0.7753	0.7202	0.7761	0.9112	1.1419
0.9103	0.8933	0.8689	0.8328	0.7887	0.8200	0.9297	1.1339
0.9308	0.9180	0.9000	0.8738	0.8410	0.8617	0.9518	1.1320
0.9463	0.9367	0.9234	0.9043	0.8809	0.8984	0.9745	1.1343
0.9583	0.9511	0.9413	0.9275	0.9118	0.9297	0.9961	1.1391
0.9678	0.9624	0.9552	0.9456	0.9359	0.9557	1.0157	1.1452
0.9754	0.9715	0.9664	0.9599	0.9550	0.9772	1.0328	1.1516
0.9865	0.9847	0.9826	0.9806	0.9827	1.0094	1.0600	1.1635
0.9941	0.9936	0.9935	0.9945	1.0011	1.0313	1.0793	1.1728
0.9993	0.9998	1.0010	1.0040	1.0137	1.0463	1.0926	1.1792
1.0031	1.0042	1.0063	1.0106	1.0223	1.0565	1.1016	1.1830
1.0057	1.0074	1.0101	1.0153	1.0284	1.0635	1.1075	1.1848
1.0097	1.0120	1.0156	1.0221	1.0368	1.0723	1.1138	1.1834
1.0115	1.0140	1.0179	1.0249	1.0401	1.0747	1.1136	1.1773

TABLE C.6 Lee-Kesler Values for the COMPRESSIBILITY FACTOR (*Continued*)

(*b*) Values for $z^{(1)}$

T_r	P_r = 0.010	0.050	0.100	0.200	0.400	0.600	0.800
0.30	−0.0008	−0.0040	−0.0081	−0.0161	−0.0323	−0.0484	−0.0645
0.35	−0.0009	−0.0046	−0.0093	−0.0185	−0.0370	−0.0554	−0.0738
0.40	−0.0010	−0.0048	−0.0095	−0.0190	−0.0380	−0.0570	−0.0758
0.45	−0.0009	−0.0047	−0.0094	−0.0187	−0.0374	−0.0560	−0.0745
0.50	−0.0009	−0.0045	−0.0090	−0.0181	−0.0360	−0.0539	−0.0716
0.55	−0.0314	−0.0043	−0.0086	−0.0172	−0.0343	−0.0513	−0.0682
0.60	−0.0205	−0.0041	−0.0082	−0.0164	−0.0326	−0.0487	−0.0646
0.65	−0.0137	−0.0772	−0.0078	−0.0156	−0.0309	−0.0461	−0.0611
0.70	−0.0093	−0.0507	−0.1161	−0.0148	−0.0294	−0.0438	−0.0579
0.75	−0.0064	−0.0339	−0.0744	−0.0143	−0.0282	−0.0417	−0.0550
0.80	−0.0044	−0.0228	−0.0487	−0.1160	−0.0272	−0.0401	−0.0526
0.85	−0.0029	−0.0152	−0.0319	−0.0715	−0.0268	−0.0391	−0.0509
0.90	−0.0019	−0.0099	−0.0205	−0.0442	−0.1118	−0.0396	−0.0503
0.93	−0.0015	−0.0075	−0.0154	−0.0326	−0.0763	−0.1662	−0.0514
0.95	−0.0012	−0.0062	−0.0126	−0.0262	−0.0589	−0.1110	−0.0540
0.97	−0.0010	−0.0050	−0.0101	−0.0208	−0.0450	−0.0770	−0.1647
0.98	−0.0009	−0.0044	−0.0090	−0.0184	−0.0390	−0.0641	−0.1100
0.99	−0.0008	−0.0039	−0.0079	−0.0161	−0.0335	−0.0531	−0.0796
1.00	−0.0007	−0.0034	−0.0069	−0.0140	−0.0285	−0.0435	−0.0588
1.01	−0.0006	−0.0030	−0.0060	−0.0120	−0.0240	−0.0351	−0.0429
1.02	−0.0005	−0.0026	−0.0051	−0.0102	−0.0198	−0.0277	−0.0303
1.05	−0.0003	−0.0015	−0.0029	−0.0054	−0.0092	−0.0097	−0.0032
1.10	−0.0000	0.0000	0.0001	0.0007	0.0038	0.0106	0.0236
1.15	0.0002	0.0011	0.0023	0.0052	0.0127	0.0237	0.0396
1.20	0.0004	0.0019	0.0039	0.0084	0.0190	0.0326	0.0499
1.30	0.0006	0.0030	0.0061	0.0125	0.0267	0.0429	0.0612
1.40	0.0007	0.0036	0.0072	0.0147	0.0306	0.0477	0.0661
1.50	0.0008	0.0039	0.0078	0.0158	0.0323	0.0497	0.0677
1.60	0.0008	0.0040	0.0080	0.0162	0.0330	0.0501	0.0677
1.70	0.0008	0.0040	0.0081	0.0163	0.0329	0.0497	0.0667
1.80	0.0008	0.0040	0.0081	0.0162	0.0325	0.0488	0.0652
1.90	0.0008	0.0040	0.0079	0.0159	0.0318	0.0477	0.0635
2.00	0.0008	0.0039	0.0078	0.0155	0.0310	0.0464	0.0617
2.20	0.0007	0.0037	0.0074	0.0147	0.0293	0.0437	0.0579
2.40	0.0007	0.0035	0.0070	0.0139	0.0276	0.0411	0.0544
2.60	0.0007	0.0033	0.0066	0.0131	0.0260	0.0387	0.0512
2.80	0.0006	0.0031	0.0062	0.0124	0.0245	0.0365	0.0483
3.00	0.0006	0.0029	0.0059	0.0117	0.0232	0.0345	0.0456
3.50	0.0005	0.0026	0.0052	0.0103	0.0204	0.0303	0.0401
4.00	0.0005	0.0023	0.0046	0.0091	0.0182	0.0270	0.0357

TABLE C.6 Lee-Kesler Values for the COMPRESSIBILITY FACTOR (*Continued*)

(*b*) Values for $z^{(1)}$

P_r							
1.000	1.200	1.500	2.000	3.000	5.000	7.000	10.000
−0.0806	−0.0966	−0.1207	−0.1608	−0.2407	−0.3996	−0.5572	−0.7915
−0.0921	−0.1105	−0.1379	−0.1834	−0.2738	−0.4523	−0.6279	−0.8863
−0.0946	−0.1134	−0.1414	−0.1879	−0.2799	−0.4603	−0.6365	−0.8936
−0.0929	−0.1113	−0.1387	−0.1840	−0.2734	−0.4475	−0.6162	−0.8606
−0.0893	−0.1069	−0.1330	−0.1762	−0.2611	−0.4253	−0.5831	−0.8099
−0.0849	−0.1015	−0.1263	−0.1669	−0.2465	−0.3991	−0.5446	−0.7521
−0.0803	−0.0960	−0.1192	−0.1572	−0.2312	−0.3718	−0.5047	−0.6928
−0.0759	−0.0906	−0.1122	−0.1476	−0.2160	−0.3447	−0.4653	−0.6346
−0.0718	−0.0855	−0.1057	−0.1385	−0.2013	−0.3184	−0.4270	−0.5785
−0.0681	−0.0808	−0.0996	−0.1298	−0.1872	−0.2929	−0.3901	−0.5250
−0.0648	−0.0767	−0.0940	−0.1217	−0.1736	−0.2682	−0.3545	−0.4740
−0.0622	−0.0731	−0.0888	−0.1138	−0.1602	−0.2439	−0.3201	−0.4254
−0.0604	−0.0701	−0.0840	−0.1059	−0.1463	−0.2195	−0.2862	−0.3788
−0.0602	−0.0687	−0.0810	−0.1007	−0.1374	−0.2045	−0.2661	−0.3516
−0.0607	−0.0678	−0.0788	−0.0967	−0.1310	−0.1943	−0.2526	−0.3339
−0.0623	−0.0669	−0.0759	−0.0921	−0.1240	−0.1837	−0.2391	−0.3163
−0.0641	−0.0661	−0.0740	−0.0893	−0.1202	−0.1783	−0.2322	−0.3075
−0.0680	−0.0646	−0.0715	−0.0861	−0.1162	−0.1728	−0.2254	−0.2989
−0.0879	−0.0609	−0.0678	−0.0824	−0.1118	−0.1672	−0.2185	−0.2902
−0.0223	−0.0473	−0.0621	−0.0778	−0.1072	−0.1615	−0.2116	−0.2816
−0.0062	0.0227	−0.0524	−0.0722	−0.1021	−0.1556	−0.2047	−0.2731
0.0220	0.1059	0.0451	−0.0432	−0.0838	−0.1370	−0.1835	−0.2476
0.0476	0.0897	0.1630	0.0698	−0.0373	−0.1021	−0.1469	−0.2056
0.0625	0.0943	0.1548	0.1667	0.0332	−0.0611	−0.1084	−0.1642
0.0719	0.0991	0.1477	0.1990	0.1095	−0.0141	−0.0678	−0.1231
0.0819	0.1048	0.1420	0.1991	0.2079	0.0875	0.0176	−0.0423
0.0857	0.1063	0.1383	0.1894	0.2397	0.1737	0.1008	0.0350
0.0864	0.1055	0.1345	0.1806	0.2433	0.2309	0.1717	0.1058
0.0855	0.1035	0.1303	0.1729	0.2381	0.2631	0.2255	0.1673
0.0838	0.1008	0.1259	0.1658	0.2305	0.2788	0.2628	0.2179
0.0816	0.0978	0.1216	0.1593	0.2224	0.2846	0.2871	0.2576
0.0792	0.0947	0.1173	0.1532	0.2144	0.2848	0.3017	0.2876
0.0767	0.0916	0.1133	0.1476	0.2069	0.2819	0.3097	0.3096
0.0719	0.0857	0.1057	0.1374	0.1932	0.2720	0.3135	0.3355
0.0675	0.0803	0.0989	0.1285	0.1812	0.2602	0.3089	0.3459
0.0634	0.0754	0.0929	0.1207	0.1706	0.2484	0.3009	0.3475
0.0598	0.0711	0.0876	0.1138	0.1613	0.2372	0.2915	0.3443
0.0565	0.0672	0.0828	0.1076	0.1529	0.2268	0.2817	0.3385
0.0497	0.0591	0.0728	0.0949	0.1356	0.2042	0.2584	0.3194
0.0443	0.0527	0.0651	0.0849	0.1219	0.1857	0.2378	0.2994

From Ernest G. Cravalho and Joseph L. Smith, Jr., *Engineering Thermodynamics*, Pitman, Marshfield, Mass., 1981 (used with permission) as calculated from data in Byung Ik Lee and Michael G. Kesler, "A Generalized Thermodynamic Correlation Based on Three-Parameter Corresponding States," *American Institute of Chemical Engineering Journal*, vol. 21, no. 3, May 1975, pp. 510–527.

TABLE C.7 Lee-Kesler Values for RESIDUAL ENTHALPY

(a) Values of $[(h_{ideal} - h_{actual})/(RT_{cr})]^{(0)}$

T_r	P_r = 0.010	0.050	0.100	0.200	0.400	0.600	0.800
0.30	6.045	6.043	6.040	6.034	6.022	6.011	5.999
0.35	5.906	5.904	5.901	5.895	5.882	5.870	5.858
0.40	5.763	5.761	5.757	5.751	5.738	5.726	5.713
0.45	5.615	5.612	5.609	5.603	5.590	5.577	5.564
0.50	5.465	5.463	4.459	5.453	5.440	5.427	5.414
0.55	0.032	5.312	5.309	5.303	5.290	5.278	5.265
0.60	0.027	5.162	5.159	5.153	5.141	5.129	5.116
0.65	0.023	0.118	5.008	5.002	4.991	4.980	4.968
0.70	0.020	0.101	0.213	4.848	4.838	4.828	4.818
0.75	0.017	0.088	0.183	4.687	4.679	4.672	4.664
0.80	0.015	0.078	0.160	0.345	4.507	4.504	4.499
0.85	0.014	0.069	0.141	0.300	4.309	4.313	4.316
0.90	0.012	0.062	0.126	0.264	0.596	4.074	4.094
0.93	0.011	0.058	0.118	0.246	0.545	0.960	3.920
0.95	0.011	0.056	0.113	0.235	0.516	0.885	3.763
0.97	0.011	0.054	0.109	0.225	0.490	0.824	1.356
0.98	0.010	0.053	0.107	0.221	0.478	0.797	1.273
0.99	0.010	0.052	0.105	0.216	0.466	0.773	1.206
1.00	0.010	0.051	0.103	0.212	0.455	0.750	1.151
1.01	0.010	0.050	0.101	0.208	0.445	0.728	1.102
1.02	0.010	0.049	0.099	0.203	0.434	0.708	1.060
1.05	0.009	0.046	0.094	0.192	0.407	0.654	0.955
1.10	0.008	0.042	0.086	0.175	0.367	0.581	0.827
1.15	0.008	0.039	0.079	0.160	0.334	0.523	0.732
1.20	0.007	0.036	0.073	0.148	0.305	0.474	0.657
1.30	0.006	0.031	0.063	0.127	0.259	0.399	0.545
1.40	0.005	0.027	0.055	0.110	0.224	0.341	0.463
1.50	0.005	0.024	0.048	0.097	0.196	0.297	0.400
1.60	0.004	0.021	0.043	0.086	0.173	0.261	0.350
1.70	0.004	0.019	0.038	0.076	0.153	0.231	0.309
1.80	0.003	0.017	0.034	0.068	0.137	0.206	0.275
1.90	0.003	0.015	0.031	0.062	0.123	0.185	0.246
2.00	0.003	0.014	0.028	0.056	0.111	0.167	0.222
2.20	0.002	0.012	0.023	0.046	0.092	0.137	0.182
2.40	0.002	0.010	0.019	0.038	0.076	0.114	0.150
2.60	0.002	0.008	0.016	0.032	0.064	0.095	0.125
2.80	0.001	0.007	0.014	0.027	0.054	0.080	0.105
3.00	0.001	0.006	0.011	0.023	0.045	0.067	0.088
3.50	0.001	0.004	0.007	0.015	0.029	0.043	0.056
4.00	0.000	0.002	0.005	0.009	0.017	0.026	0.033

TABLE C.7 Lee-Kesler Values for RESIDUAL ENTHALPY (*Continued*)

(*a*) Values of $[(h_{ideal} - h_{actual})/(RT_{cr})]^{(0)}$

P_r							
1.000	1.200	1.500	2.000	3.000	5.000	7.000	10.000
5.987	5.975	5.957	5.927	5.868	5.748	5.628	5.446
5.845	5.833	5.814	5.783	5.721	5.595	5.469	5.278
5.700	5.687	5.668	5.636	5.572	5.442	5.311	5.113
5.551	5.538	5.519	5.486	5.421	5.288	5.154	4.950
5.401	5.388	5.369	5.336	5.270	5.135	4.999	4.791
5.252	5.239	5.220	5.187	5.121	4.986	4.849	4.638
5.104	5.091	5.073	5.041	4.976	4.842	4.704	4.492
4.956	4.945	4.927	4.896	4.833	4.702	4.565	4.353
4.808	4.797	4.781	4.752	4.693	4.566	4.432	4.221
4.655	4.646	4.632	4.607	4.554	4.434	4.303	4.095
4.494	4.488	4.478	4.459	4.413	4.303	4.178	3.974
4.316	4.316	4.312	4.302	4.269	4.173	4.056	3.857
4.108	4.118	4.127	4.132	4.119	4.043	3.935	3.744
3.953	3.976	4.000	4.020	4.024	3.963	3.863	3.678
3.825	3.865	3.904	3.940	3.958	3.910	3.815	3.634
3.658	3.732	3.796	3.853	3.890	3.856	3.767	3.591
3.544	3.652	3.736	3.806	3.854	3.829	3.743	3.569
3.376	3.558	3.670	3.758	3.818	3.801	3.719	3.548
2.584	3.441	3.598	3.706	3.782	3.774	3.695	3.526
1.796	3.283	3.516	3.652	3.744	3.746	3.671	3.505
1.627	3.039	3.442	3.595	3.705	3.718	3.647	3.484
1.359	2.034	3.030	3.398	3.583	3.632	3.575	3.420
1.120	1.487	2.203	2.965	3.353	3.484	3.453	3.315
0.968	1.239	1.719	2.479	3.091	3.329	3.329	3.211
0.857	1.076	1.443	2.079	2.807	3.166	3.202	3.107
0.698	0.860	1.116	1.560	2.274	2.825	2.942	2.899
0.588	0.716	0.915	1.253	1.857	2.486	2.679	2.692
0.505	0.611	0.774	1.046	1.549	2.175	2.421	2.486
0.440	0.531	0.667	0.894	1.318	1.904	2.177	2.285
0.387	0.466	0.583	0.777	1.139	1.672	1.953	2.091
0.344	0.413	0.515	0.683	0.996	1.476	1.751	1.908
0.307	0.368	0.458	0.606	0.880	1.309	1.571	1.736
0.276	0.330	0.411	0.541	0.782	1.167	1.411	1.577
0.226	0.269	0.334	0.437	0.629	0.937	1.143	1.295
0.187	0.222	0.275	0.359	0.513	0.761	0.929	1.058
0.155	0.185	0.228	0.297	0.422	0.621	0.756	0.858
0.130	0.154	0.190	0.246	0.348	0.508	0.614	0.689
0.109	0.129	0.159	0.205	0.288	0.415	0.495	0.545
0.069	0.081	0.099	0.127	0.174	0.239	0.270	0.264
0.041	0.048	0.058	0.072	0.095	0.116	0.110	0.061

TABLE C.7 Lee-Kesler Values for RESIDUAL ENTHALPY (*Continued*)

(*b*) Values of $[(h_{ideal} - h_{actual})/(RT_{cr})]^{(1)}$

T_r	P_r = 0.010	0.050	0.100	0.200	0.400	0.600	0.800
0.30	11.098	11.096	11.095	11.091	11.083	11.076	11.069
0.35	10.656	10.655	10.654	10.653	10.650	10.646	10.643
0.40	10.121	10.121	10.121	10.120	10.121	10.121	10.121
0.45	9.515	9.515	9.515	9.517	9.519	9.521	9.523
0.50	8.868	8.869	8.870	8.872	8.876	8.880	8.884
0.55	0.080	8.211	8.212	8.215	8.221	8.226	8.232
0.60	0.059	7.568	7.570	7.573	7.579	7.585	7.591
0.65	0.045	0.247	6.949	6.952	6.959	6.966	6.973
0.70	0.034	0.185	0.415	6.360	6.367	6.373	6.381
0.75	0.027	0.142	0.306	5.796	5.802	5.809	5.816
0.80	0.021	0.110	0.234	0.542	5.266	5.271	5.278
0.85	0.017	0.087	0.182	0.401	4.753	4.754	4.758
0.90	0.014	0.070	0.144	0.308	0.751	4.254	4.248
0.93	0.012	0.061	0.126	0.265	0.612	1.236	3.942
0.95	0.011	0.056	0.115	0.241	0.542	0.994	3.737
0.97	0.010	0.052	0.105	0.219	0.483	0.837	1.616
0.98	0.010	0.050	0.101	0.209	0.457	0.776	1.324
0.99	0.009	0.048	0.097	0.200	0.433	0.722	1.154
1.00	0.009	0.046	0.093	0.191	0.410	0.675	1.034
1.01	0.009	0.044	0.089	0.183	0.389	0.632	0.940
1.02	0.008	0.042	0.085	0.175	0.370	0.594	0.863
1.05	0.007	0.037	0.075	0.153	0.318	0.498	0.691
1.10	0.006	0.030	0.061	0.123	0.251	0.381	0.507
1.15	0.005	0.025	0.050	0.099	0.199	0.296	0.385
1.20	0.004	0.020	0.040	0.080	0.158	0.232	0.297
1.30	0.003	0.013	0.026	0.052	0.100	0.142	0.177
1.40	0.002	0.008	0.016	0.032	0.060	0.083	0.100
1.50	0.001	0.005	0.009	0.018	0.032	0.042	0.048
1.60	0.000	0.002	0.004	0.007	0.012	0.013	0.011
1.70	0.000	0.000	0.000	−0.000	−0.003	−0.009	−0.017
1.80	−0.000	−0.001	−0.003	−0.006	−0.015	−0.025	−0.037
1.90	−0.001	−0.003	−0.005	−0.011	−0.023	−0.037	−0.053
2.00	−0.001	−0.003	−0.007	−0.015	−0.030	−0.047	−0.065
2.20	−0.001	−0.005	−0.010	−0.020	−0.040	−0.062	−0.083
2.40	−0.001	−0.006	−0.012	−0.023	−0.047	−0.071	−0.095
2.60	−0.001	−0.006	−0.013	−0.026	−0.052	−0.078	−0.104
2.80	−0.001	−0.007	−0.014	−0.028	−0.055	−0.082	−0.110
3.00	−0.001	−0.007	−0.014	−0.029	−0.058	−0.086	−0.114
3.50	−0.002	−0.008	−0.016	−0.031	−0.062	−0.092	−0.122
4.00	−0.002	−0.008	−0.016	−0.032	−0.064	−0.096	−0.127

TABLE C.7 Lee-Kesler Values for RESIDUAL ENTHALPY (*Continued*)

(*b*) Values of $[(h_{\text{ideal}} - h_{\text{actual}})/(RT_{\text{cr}})]^{(1)}$

P_r							
1.000	**1.200**	**1.500**	**2.000**	**3.000**	**5.000**	**7.000**	**10.000**
11.062	11.055	11.044	11.027	10.992	10.935	10.872	10.781
10.640	10.637	10.632	10.624	10.609	10.581	10.554	10.529
10.121	10.121	10.121	10.122	10.123	10.128	10.135	10.150
9.525	9.527	9.531	9.537	9.549	9.576	9.611	9.663
8.888	8.892	8.899	8.909	8.932	8.978	9.030	9.111
8.238	8.243	8.252	8.267	8.298	8.360	8.425	8.531
7.596	7.603	7.614	7.632	7.669	7.745	7.824	7.950
6.980	6.987	6.997	7.017	7.059	7.147	7.239	7.381
6.388	6.395	6.407	6.429	6.475	6.574	6.677	6.837
5.824	5.832	5.845	5.868	5.918	6.027	6.142	6.318
5.285	5.293	5.306	5.330	5.385	5.506	5.632	5.824
4.763	4.771	4.784	4.810	4.872	5.008	5.149	5.358
4.249	4.255	4.268	4.298	4.371	4.530	4.688	4.916
3.934	3.937	3.951	3.987	4.073	4.251	4.422	4.662
3.712	3.713	3.730	3.773	3.873	4.068	4.248	4.497
3.470	3.467	3.492	3.551	3.670	3.885	4.077	4.336
3.332	3.327	3.363	3.434	3.568	3.795	3.992	4.257
3.164	3.164	3.223	3.313	3.464	3.705	3.909	4.178
2.471	2.952	3.065	3.186	3.358	3.615	3.825	4.100
1.375	2.595	2.880	3.051	3.251	3.525	3.742	4.023
1.180	1.723	2.650	2.906	3.142	3.435	3.661	3.947
0.877	0.878	1.496	2.381	2.800	3.167	3.418	3.722
0.617	0.673	0.617	1.261	2.167	2.720	3.023	3.362
0.459	0.503	0.487	0.604	1.497	2.275	2.641	3.019
0.349	0.381	0.381	0.361	0.934	1.840	2.273	2.692
0.203	0.218	0.218	0.178	0.300	1.066	1.592	2.086
0.111	0.115	0.108	0.070	0.044	0.504	1.012	1.547
0.049	0.046	0.032	−0.008	−0.078	0.142	0.556	1.080
0.005	−0.004	−0.023	−0.065	−0.151	−0.082	0.217	0.689
−0.027	−0.040	−0.063	−0.109	−0.202	−0.223	−0.028	0.369
−0.051	−0.067	−0.094	−0.143	−0.241	−0.317	−0.203	0.112
−0.070	−0.088	−0.117	−0.169	−0.271	−0.381	−0.330	−0.092
−0.085	−0.105	−0.136	−0.190	−0.295	−0.428	−0.424	−0.255
−0.106	−0.128	−0.163	−0.221	−0.331	−0.493	−0.551	−0.489
−0.120	−0.144	−0.181	−0.242	−0.356	−0.535	−0.631	−0.645
−0.130	−0.156	−0.194	−0.257	−0.376	−0.567	−0.687	−0.754
−0.137	−0.164	−0.204	−0.269	−0.391	−0.591	−0.729	−0.836
−0.142	−0.170	−0.211	−0.278	−0.403	−0.611	−0.763	−0.899
−0.152	−0.181	−0.224	−0.294	−0.425	−0.650	−0.827	−1.015
−0.158	−0.188	−0.233	−0.306	−0.442	−0.680	−0.874	−1.097

From Ernest G. Cravalho and Joseph L. Smith, Jr., *Engineering Thermodynamics,* Pitman, Marshfield, Mass., 1981 (used with premission), as calculated from data in Byung Ik Lee and Michael G. Kesler, "A Generalized Thermodynamic Correlation Based on Three-Parameter Corresponding States," *American Institute of Chemical Engineering Journal,* vol. 21, no. 3, May 1975, pp. 510–527.

TABLE C.8 Lee-Kesler Values for RESIDUAL ENTROPY

(a) Values of $[(s_{ideal} - s_{actual})/R]^{(0)}$

T_r	P_r 0.010	0.050	0.100	0.200	0.400	0.600	0.800
0.30	11.614	10.008	9.319	8.635	7.961	7.574	7.304
0.35	11.185	9.579	8.890	8.205	7.529	7.140	6.869
0.40	10.802	9.196	8.506	7.821	7.144	6.755	6.483
0.45	10.453	8.847	8.157	7.472	6.794	6.404	6.132
0.50	10.137	8.531	7.841	7.156	6.479	6.089	5.816
0.55	0.038	8.245	7.555	6.870	6.193	5.803	5.531
0.60	0.029	7.983	7.294	6.610	5.933	5.544	5.273
0.65	0.023	0.122	7.052	6.368	5.694	5.306	5.036
0.70	0.018	0.096	0.206	6.140	5.467	5.082	4.814
0.75	0.015	0.078	0.164	5.917	5.248	4.866	4.600
0.80	0.013	0.064	0.134	0.294	5.026	4.649	4.388
0.85	0.011	0.054	0.111	0.239	4.785	4.418	4.166
0.90	0.009	0.046	0.094	0.199	0.463	4.145	3.912
0.93	0.008	0.042	0.085	0.179	0.408	0.750	3.723
0.95	0.008	0.039	0.080	0.168	0.377	0.671	3.556
0.97	0.007	0.037	0.075	0.157	0.350	0.607	1.056
0.98	0.007	0.036	0.073	0.153	0.337	0.580	0.971
0.99	0.007	0.035	0.071	0.148	0.326	0.555	0.903
1.00	0.007	0.034	0.069	0.144	0.315	0.532	0.847
1.01	0.007	0.033	0.067	0.139	0.304	0.510	0.799
1.02	0.006	0.032	0.065	0.135	0.294	0.491	0.757
1.05	0.006	0.030	0.060	0.124	0.267	0.439	0.656
1.10	0.005	0.026	0.053	0.108	0.230	0.371	0.537
1.15	0.005	0.023	0.047	0.096	0.201	0.319	0.452
1.20	0.004	0.021	0.042	0.085	0.177	0.277	0.389
1.30	0.003	0.017	0.033	0.068	0.140	0.217	0.298
1.40	0.003	0.014	0.027	0.056	0.114	0.174	0.237
1.50	0.002	0.011	0.023	0.046	0.094	0.143	0.194
1.60	0.002	0.010	0.019	0.039	0.079	0.120	0.162
1.70	0.002	0.008	0.017	0.033	0.067	0.102	0.137
1.80	0.001	0.007	0.014	0.029	0.058	0.088	0.117
1.90	0.001	0.006	0.013	0.025	0.051	0.076	0.102
2.00	0.001	0.006	0.011	0.022	0.044	0.067	0.089
2.20	0.001	0.004	0.009	0.018	0.035	0.053	0.070
2.40	0.001	0.004	0.007	0.014	0.028	0.042	0.056
2.60	0.001	0.003	0.006	0.012	0.023	0.035	0.046
2.80	0.000	0.002	0.005	0.010	0.020	0.029	0.039
3.00	0.000	0.002	0.004	0.008	0.017	0.025	0.033
3.50	0.000	0.001	0.003	0.006	0.012	0.017	0.023
4.00	0.000	0.001	0.002	0.004	0.009	0.013	0.017

TABLE C.8 Lee-Kesler Values for RESIDUAL ENTROPY (*Continued*)

(*a*) Values of $[(s_{ideal} - s_{actual})/R]^{(0)}$

P_r							
1.000	1.200	1.500	2.000	3.000	5.000	7.000	10.000
7.099	6.935	6.740	6.497	6.182	5.847	5.683	5.578
6.663	6.497	6.299	6.052	5.728	5.376	5.194	5.060
6.275	6.109	5.909	5.660	5.330	4.967	4.772	4.619
5.924	5.757	5.557	5.306	4.974	4.603	4.401	4.234
5.608	5.441	5.240	4.989	4.656	4.282	4.074	3.899
5.324	5.157	4.956	4.706	4.373	3.998	3.788	3.607
5.066	4.900	4.700	4.451	4.120	3.747	3.537	3.353
4.830	4.665	4.467	4.220	3.892	3.523	3.315	3.131
4.610	4.446	4.250	4.007	3.684	3.322	3.117	2.935
4.399	4.238	4.045	3.807	3.491	3.138	2.939	2.761
4.191	4.034	3.846	3.615	3.310	2.970	2.777	2.605
3.976	3.825	3.646	3.425	3.135	2.812	2.629	2.463
3.738	3.599	3.434	3.231	2.964	2.663	2.491	2.334
3.569	3.444	3.295	3.108	2.860	2.577	2.412	2.262
3.433	3.326	3.193	3.023	2.790	2.520	2.362	2.215
3.259	3.188	3.081	2.932	2.719	2.463	2.312	2.170
3.142	3.106	3.019	2.884	2.682	2.436	2.287	2.148
2.972	3.010	2.953	2.835	2.646	2.408	2.263	2.126
2.178	2.893	2.879	2.784	2.609	2.380	2.239	2.105
1.391	2.736	2.798	2.730	2.571	2.352	2.215	2.083
1.225	2.495	2.706	2.673	2.533	2.325	2.191	2.062
0.965	1.523	2.328	2.483	2.415	2.242	2.121	2.001
0.742	1.012	1.557	2.081	2.202	2.104	2.007	1.903
0.607	0.790	1.126	1.649	1.968	1.966	1.897	1.810
0.512	0.651	0.890	1.308	1.727	1.827	1.789	1.722
0.385	0.478	0.628	0.891	1.299	1.554	1.581	1.556
0.303	0.372	0.478	0.663	0.990	1.303	1.386	1.402
0.246	0.299	0.381	0.520	0.777	1.088	1.208	1.260
0.204	0.247	0.312	0.421	0.628	0.913	1.050	1.130
0.172	0.208	0.261	0.350	0.519	0.773	0.915	1.013
0.147	0.177	0.222	0.296	0.438	0.661	0.799	0.908
0.127	0.153	0.191	0.255	0.375	0.570	0.702	0.815
0.111	0.134	0.167	0.221	0.325	0.497	0.620	0.733
0.087	0.105	0.130	0.172	0.251	0.388	0.492	0.599
0.070	0.084	0.104	0.138	0.201	0.311	0.399	0.496
0.058	0.069	0.086	0.113	0.164	0.255	0.329	0.416
0.048	0.058	0.072	0.094	0.137	0.213	0.277	0.353
0.041	0.049	0.061	0.080	0.116	0.181	0.236	0.303
0.029	0.034	0.042	0.056	0.081	0.126	0.166	0.216
0.021	0.025	0.031	0.041	0.059	0.093	0.123	0.162

TABLE C.8 Lee-Kesler Values for RESIDUAL ENTROPY (*Continued*)

(*b*) Values of $[(s_{ideal} - s_{actual})/R]^{(1)}$

T_r	P_r = 0.010	0.050	0.100	0.200	0.400	0.600	0.800
0.30	16.782	16.774	16.764	16.744	16.705	16.665	16.626
0.35	15.413	15.408	15.401	15.387	15.359	15.333	15.305
0.40	13.990	13.986	13.981	13.972	13.953	13.934	13.915
0.45	12.564	12.561	12.558	12.551	12.537	12.523	12.509
0.50	11.202	11.200	11.197	11.192	11.182	11.172	11.162
0.55	0.115	9.948	9.946	9.942	9.935	9.928	9.921
0.60	0.078	8.828	8.826	8.823	8.817	8.811	8.806
0.65	0.055	0.309	7.832	7.829	7.824	7.819	7.815
0.70	0.040	0.216	0.491	6.951	6.945	6.941	6.937
0.75	0.029	0.156	0.340	6.173	6.167	6.162	6.158
0.80	0.022	0.116	0.246	0.578	5.475	5.468	5.462
0.85	0.017	0.088	0.183	0.408	4.853	4.841	4.832
0.90	0.013	0.068	0.140	0.301	0.744	4.269	4.249
0.93	0.011	0.058	0.120	0.254	0.593	1.219	3.914
0.95	0.010	0.053	0.109	0.228	0.517	0.961	3.697
0.97	0.010	0.048	0.099	0.206	0.456	0.797	1.570
0.98	0.009	0.046	0.094	0.196	0.429	0.734	1.270
0.99	0.009	0.044	0.090	0.186	0.405	0.680	1.098
1.00	0.008	0.042	0.086	0.177	0.382	0.632	0.977
1.01	0.008	0.040	0.082	0.169	0.361	0.590	0.883
1.02	0.008	0.039	0.078	0.161	0.342	0.552	0.807
1.05	0.007	0.034	0.069	0.140	0.292	0.460	0.642
1.10	0.005	0.028	0.055	0.112	0.229	0.350	0.470
1.15	0.005	0.023	0.045	0.091	0.183	0.275	0.361
1.20	0.004	0.019	0.037	0.075	0.149	0.220	0.286
1.30	0.003	0.013	0.026	0.052	0.102	0.148	0.190
1.40	0.002	0.010	0.019	0.037	0.072	0.104	0.133
1.50	0.001	0.007	0.014	0.027	0.053	0.076	0.097
1.60	0.001	0.005	0.011	0.021	0.040	0.057	0.073
1.70	0.001	0.004	0.008	0.016	0.031	0.044	0.056
1.80	0.001	0.003	0.006	0.013	0.024	0.035	0.044
1.90	0.001	0.003	0.005	0.010	0.019	0.028	0.036
2.00	0.000	0.002	0.004	0.008	0.016	0.023	0.029
2.20	0.000	0.001	0.003	0.006	0.011	0.016	0.021
2.40	0.000	0.001	0.002	0.004	0.008	0.012	0.015
2.60	0.000	0.001	0.002	0.003	0.006	0.009	0.012
2.80	0.000	0.001	0.001	0.003	0.005	0.008	0.010
300	0.000	0.001	0.001	0.002	0.004	0.006	0.008
3.50	0.000	0.000	0.001	0.001	0.003	0.004	0.006
4.00	0.000	0.000	0.001	0.001	0.002	0.003	0.005

TABLE C.8 Lee-Kesler Values for RESIDUAL ENTROPY (*Continued*)

(*b*) Values of $[(s_{ideal} - s_{actual})/R]^{(1)}$

P_r							
1.000	1.200	1.500	2.000	3.000	5.000	7.000	10.000
16.586	16.547	16.488	16.390	16.195	15.837	15.468	14.925
15.278	15.251	15.211	15.144	15.011	14.751	14.496	14.153
13.896	13.877	13.849	13.803	13.714	13.541	13.376	13.144
12.496	12.482	12.462	12.430	12.367	12.248	12.145	11.999
11.153	11.143	11.129	11.107	11.063	10.985	10.920	10.836
9.914	9.907	9.897	9.882	9.853	9.806	9.769	9.732
8.799	8.794	8.787	8.777	8.760	8.736	8.723	8.720
7.810	7.807	7.801	7.794	7.784	7.779	7.785	7.811
6.933	6.930	6.926	6.922	6.919	6.929	6.952	7.002
6.155	6.152	6.149	6.147	6.149	6.174	6.213	6.285
5.458	5.455	5.453	5.452	5.461	5.501	5.555	5.648
4.826	4.822	4.820	4.822	4.839	4.898	4.969	5.082
4.238	4.232	4.230	4.236	4.267	4.351	4.442	4.578
3.894	3.885	3.884	3.896	3.941	4.046	4.151	4.300
3.658	3.647	3.648	3.669	3.728	3.851	3.966	4.125
3.406	3.391	3.401	3.437	3.517	3.661	3.788	3.957
3.264	3.247	3.268	3.318	3.412	3.569	3.701	3.875
3.093	3.082	3.126	3.195	3.306	3.477	3.616	3.796
2.399	2.868	2.967	3.067	3.200	3.387	3.532	3.717
1.306	2.513	2.784	2.933	3.094	3.297	3.450	3.640
1.113	1.655	2.557	2.790	2.986	3.209	3.369	3.565
0.820	0.831	1.443	2.283	2.655	2.949	3.134	3.348
0.577	0.640	0.618	1.241	2.067	2.534	2.767	3.013
0.437	0.489	0.502	0.654	1.471	2.138	2.428	2.708
0.343	0.385	0.412	0.447	0.991	1.767	2.115	2.430
0.226	0.254	0.282	0.300	0.481	1.147	1.569	1.944
0.158	0.178	0.200	0.220	0.290	0.730	1.138	1.544
0.115	0.130	0.147	0.166	0.206	0.479	0.823	1.222
0.086	0.098	0.112	0.129	0.159	0.334	0.604	0.969
0.067	0.076	0.087	0.102	0.127	0.248	0.456	0.775
0.053	0.060	0.070	0.083	0.105	0.195	0.355	0.628
0.043	0.049	0.057	0.069	0.089	0.160	0.286	0.518
0.035	0.040	0.048	0.058	0.077	0.136	0.238	0.434
0.025	0.029	0.035	0.043	0.060	0.105	0.178	0.322
0.019	0.022	0.027	0.034	0.048	0.086	0.143	0.254
0.015	0.018	0.021	0.028	0.041	0.074	0.120	0.210
0.012	0.014	0.018	0.023	0.035	0.065	0.104	0.188
0.010	0.012	0.015	0.020	0.031	0.058	0.093	0.158
0.007	0.009	0.011	0.015	0.024	0.046	0.073	0.122
0.006	0.007	0.009	0.012	0.020	0.038	0.060	0.100

From Ernest G. Cravalho and Joseph L. Smith, Jr., *Engineering Thermodynamics,* Pitman, Marshfield, Mass., 1981 (used with premission), as calculated from data in Byung Ik Lee and Michael G. Kesler, "A Generalized Thermodynamic Correlation Based on Three-Parameter Corresponding States," *American Institute of Chemical Engineering Journal,* vol. 21, no. 3, May 1975, pp. 510–527.

TABLE C.9 Enthalpies and Gibbs' Function of Formation of Selected Organic Compounds at 298.15 K (77°F)

Substance	Formula	Molecular Weight	State	$\Delta \bar{h}_f^\circ$ MJ/kmol	$\Delta \bar{h}_f^\circ$ Btu/lbmol	$\Delta \bar{g}_f^\circ$ MJ/kmol	$\Delta \bar{g}_f^\circ$ Btu/lbmol
Methane	CH_4	16.043	*g*	−74.90	−32,200	−50.83	−21,850
Ethane	C_2H_6	30.070	*g*	−84.72	−36,420	−32.9	−14,100
Propane	C_3H_8	44.094	*g*	−103.9	−44,670	−23.5	−10,100
n-Butane	C_4H_{10}	58.124	*g*	−126.2	−54,260	−17.2	−7,390
i-Butane	C_4H_{10}	58.124	*g*	−134.6	−57,870	−20.9	−8,990
n-Pentane	C_5H_{12}	72.151	*g*	−146.5	−62,980	−8.37	−3,600
n-Hexane	C_6H_{14}	86.178	*g*	−167.3	−71,930	−0.29	−120
n-Heptane	C_7H_{16}	100.205	*g*	−187.9	−80,780	8.12	3,490
n-Octane	C_8H_{18}	114.232	*g*	−208.6	−89,680	16.5	7,090
Ethylene	C_2H_4	28.054	*g*	52.32	22,490	68.17	29,310
Propylene	C_3H_6	42.081	*g*	20.4	8,770	62.76	26,980
Acetylene	C_2H_2	26.038	*g*	226.9	97,550	209.3	89,980
Benzene	C_6H_6	78.114	*g*	82.98	35,670	129.7	55,760
			l	49.07	21,100	124.4	53,480
Methanol	CH_4O	32.042	*g*	−201.3	−86,540	−162.6	−69,910
			l	−238.7	−102,600	−166.3	−71,500
Ethanol	C_2H_6O	46.069	*g*	−235.0	−101,000	−168.4	−72,400
			l	−277.2	−119,200	−174.3	−74,940
Ammonia	NH_3	17.031	*g*	−46.22	−19,870	−16.6	−7,140
			aq	−80.89	−34,780	−26.7	−11,500
Carbon monoxide	CO	28.010	*g*	−110.6	−47,550	−137.4	−59,070
Carbon dioxide	CO_2	44.010	*g*	−393.8	−169,300	−394.6	−169,600
Hydrogen chloride	HCl	36.461	*g*	−92.37	−39,710	−95.33	−40,980
			aq	−167.6	−72,060	−131.3	−56,450
Nitric oxide	NO	30.006	*g*	90.44	38,880	86.75	37,300
Nitrogen dioxide	NO_2	46.006	*g*	33.9	14,600	51.87	22,300
Nitrous oxide	N_2O	44.013	*g*	81.60	35,100	103.7	44,580
Sulfur dioxide	SO_2	64.063	*g*	−297.1	−127,700	−300.6	−129,200
Sulfur trioxide	SO_3	80.058	*g*	−395.4	−170,000	−370.6	−159,300
Sulfuric acid	H_2SO_4	98.08	*aq*	−908.12	−390,420	−742.49	−319,210
			l	−814.12	−350,010	−690.11	−296,690
Water	H_2O	18.015	*g*	−242.0	−104,000	−228.7	−90,320
			l	−286.0	−123,000	237.4	−102,100

SI values from B. G. Kyle, *Chemical and Process Thermodynamics,* Prentice-Hall, Englewood Cliffs, N.J., 1984 (with permission) from data in:

Selected Values of Chemical Thermodynamic Properties, National Bureau of Standards Circular 500, 1952.

Selected Values of Properties of Hydrocarbons, National Bureau of Standards Circular 461, 1947.

D. R. Stull, E. F. Westrum, Jr., and G. C. Sinke, *The Chemical Thermodynamics of Organic Compounds,* Wiley, New York, 1969.

TABLE C.10 Absolute Entropy, Standardized Enthalpy, and Ideal Gas Enthalpy at 25°C (77°F) and 1 atm

Gas	Formula	State	Molecular Weight	$\bar{s}^o$ kJ/kmol·K	$\bar{s}^o$ Btu/lbmol·°	$\bar{h}^o$ kJ/kmol	$\bar{h}^o$ Btu/lbmol	$\bar{h}_{\text{ideal gas}}(T_{\text{ref}})$ kJ/kmol	$\bar{h}_{\text{ideal gas}}(T_{\text{ref}})$ Btu/lbm
Acetylene	C_2H_2	*g*	26.038	200.8	48.00	226,900	97,550	10,020	4,307.7
Ammonia	NH_3	*g*	17.031	192.8	46.05	− 46,220	− 19,870	10,042	4,317.4
Argon	Ar	*g*	39.948	154.84	36.98	0	0	6,201.5	2,666.3
Benzene	C_6H_6	*l*	78.114	173.3	41.41	49,070	21,100	—	—
		g		269.2	64.34	82,980	35,670	—	—
n-Butane	C_4H_{10}	*l*	58.124	231.0	55.2	−147,240	− 63,300	—	—
		g		310.1	74.12	−126,200	− 54,260	—	—
Carbon (graphite)	C	*s*	14.01	5.694	1.361	—	—	—	—
Carbon dioxide	CO_2	*g*	44.010	213.7	51.07	−393,800	−169,300	9,371.0	4,028.8
Carbon monoxide	CO	*g*	28.010	197.9	47.30	−110,600	− 47,550	8,684.5	5,414.3
Ethane	C_2H_6	*g*	30.070	229.5	54.85	− 84,720	− 36,420	—	—
Ethanol	C_2H_5OH	*l*	46.069	—	—	−277,200	−119,200	—	—
		g		282.0	67.35	−234,800	−101,000	—	—
Ethylene	C_2H_4	*g*	28.054	219.5	52.45	52,320	22,490	10,575	4,546.2
Hydrogen (normal)	H_2	*g*	2.016	130.58	31.211	0	0	8,473.6	3,643.0
Hydrogen chloride	HCl	*g*	36.461	186.8	44.64	− 92,370	− 39,710	8,645.6	3,716.8
Methane	CH_4	*g*	16.043	186.3	44.52	− 74,900	− 32,200	10,016	4,305.8
Methanol	CH_3OH	*l*	32.042	—	—	−238,700	−102,600	—	—
		g		239.70	57.25	−201,300	− 86,540	—	—
Nitric oxide	NO	*g*	30.006	210.79	50.35	90,440	38,880	9,198.0	3,954.5
Nitrogen	N_2	*g*	28.013	191.5	45.767	0	0	7,611.7	3,272.5
Nitrogen peroxide	NO_2	*g*	46.006	240.12	57.35	33,900	14,600	10,204	4,386.7
Nitrous oxide	N_2O	*g*	44.013	220.04	52.56	81,600	35,100	9,586.0	4,121.4
Oxygen	O_2	*g*	31.999	205.03	49.003	0	0	9,919.4	4,264.5
Propane	C_3H_8	*g*	44.094	270.09	64.51	−103,900	− 44,670	—	—
Sulfur dioxide	SO_2	*g*	64.063	248.26	59.30	−297,100	−127,700	10,137	4,358.2
Sulfur trioxide	SO_3	*g*	80.058	256.84	61.35	−395,400	−170,000	11,705	5,032.4
Water	H_2O	*l*	18.015	69.96	16.72	−286,000	−123,000	—	—
		v		188.7	45.11	−242,000	−104,000	9,903.7	4,257.8

Data from Daniel R. Stull, Edgar F. Westrum, Jr., and Gerard C. Sinke, *The Chemical Thermodynamics of Organic Compounds,* Wiley, New York, 1969 (as gathered from various sources) and *GASPROPS,* Wiley Professional Software, Wiley, New York, 1984, and *HYCARB,* Software Systems, Austin, TX, 1988.

TABLE C.11 Lower Values of Enthalpy of Combustion and Adiabatic Flame Temperatures *(Stoichiometric combustion with dry air, fuel, and air at 298 K)*

Substance	Formula	Enthalpy of Combustion kJ/kmol	Enthalpy of Combustion Btu/lbmol	Adiabatic Flame Temperature K	Adiabatic Flame Temperature °R
Acetylene	C_2H_2	-12.5×10^5	-5.37×10^5	2859	5146
Benzene	C_6H_6	-31.7×10^5	-13.6×10^5	2484	4471
n-Butane	C_4H_{10}	-26.5×10^5	-11.4×10^5	2357	4243
Carbon monoxide	CO	-2.83×10^5	-1.22×10^5	2615	4707
Ethane	C_2H_6	-14.3×10^5	-6.15×10^5	2338	4208
Ethylene	C_2H_4	-13.2×10^5	-5.67×10^5	2523	4541
Hydrogen	H_2	-2.42×10^5	-1.04×10^5	2490	4482
Methane	CH_4	-8.03×10^5	-3.45×10^5	2285	4113
Propane	C_3H_8	-20.5×10^5	-8.81×10^5	2629	4732

Selected data from A. G. Gaydon and H. G. Wolfhard, *Flames, Their Structure, Radiation and Temperature,* 2d ed., Macmillan, New York, 1960, and from Henry C. Barnett and Robert R. Hibbard (eds.), "Basic Considerations in the Combustion of Hydrocarbon Fuels," NACA Report 1300, 1957.

TABLE C.12 Logarithms to Base 10 of the EQUILIBRIUM CONSTANT K

Temperature, K	Temperature, °R	$\frac{1}{2}O_2 \leftrightarrow O$	$\frac{1}{2}H_2 \leftrightarrow H$	$\frac{1}{2}N_2 \leftrightarrow N$	$H_2 + \frac{1}{2}O_2 \leftrightarrow H_2O$	$C + O_2 \leftrightarrow CO_2$	$C + \frac{1}{2}O_2 \leftrightarrow CO$
600	1080	−18.574	−16.336	−38.081	18.633	34.405	14.318
700	1260	−15.449	−13.599	−32.177	15.583	29.506	12.946
800	1440	−13.101	−11.539	−27.744	13.289	25.830	11.914
900	1620	−11.272	−9.934	−24.292	11.498	22.970	11.108
1000	1800	−9.807	−8.646	−21.528	10.062	20.680	10.459
1100	1980	−8.606	−7.589	−19.265	8.883	18.806	9.926
1200	2160	−7.604	−6.707	−17.377	7.899	17.243	9.479
1300	2340	−6.755	−5.958	−15.778	7.064	15.920	9.099
1400	2520	−6.027	−5.315	−14.406	6.347	14.785	8.771
1500	2700	−5.395	−4.756	−13.217	5.725	13.801	8.485
1600	2880	−4.842	−4.266	−12.175	5.180	12.940	8.234
1700	3060	−4.353	−3.833	−11.256	4.699	12.180	8.011
1800	3240	−3.918	−3.448	−10.437	4.270	11.504	7.811
1900	3420	−3.529	−3.102	−9.705	3.886	10.898	7.631
2000	3600	−3.178	−2.790	−9.046	3.540	10.353	7.469
2100	3780	−2.860	−2.508	−8.449	3.227	9.860	7.321
2200	3960	−2.571	−2.251	−7.905	2.942	9.411	7.185
2300	4140	−2.307	−2.016	−7.409	2.682	9.001	7.061
2400	4320	−2.065	−1.800	−6.954	2.443	8.625	6.946
2500	4500	−1.842	−1.601	−6.535	2.224	8.280	6.840
2600	4680	−1.636	−1.417	−6.149	2.021	7.960	6.741
2700	4860	−1.446	−1.247	−5.790	1.833	7.664	6.649
2800	5040	−1.268	−1.089	−5.457	1.658	7.388	6.563
2900	5220	−1.103	−0.941	−5.147	1.495	7.132	6.483
3000	5400	−0.949	−0.803	−4.858	1.343	6.892	6.407
3100	5580	−0.805	−0.674	−4.587	1.201	6.668	6.336
3200	5760	−0.670	−0.553	−4.332	1.067	6.458	6.269
3300	5940	−0.543	−0.439	−4.093	0.942	6.260	6.206
3400	6120	−0.423	−0.332	−3.868	0.824	6.074	6.145
3500	6300	−0.310	−0.231	−3.656	0.712	5.898	6.088
3600	6480	−0.204	−0.135	−3.455	0.607	5.732	6.034
3700	6660	−0.103	−0.044	−3.265	0.507	5.574	5.982
3800	6840	−0.007	0.042	−3.086	0.413	5.425	5.933
3900	7020	0.084	0.123	−2.915	0.323	5.283	5.886
4000	7200	0.170	0.201	−2.752	0.238	5.149	5.841

TABLE C.12 Logarithms to Base 10 of the EQUILIBRIUM CONSTANT K (*Continued*)

Temperature, K	Temperature, °R	$\frac{1}{2}O_2 + \frac{1}{2}N_2 \leftrightarrow NO$	$O_2 + \frac{1}{2}N_2 \leftrightarrow NO_2$	$\frac{1}{2}O_2 + \frac{1}{2}H_2 \leftrightarrow OH$	$\frac{1}{2}O_2 + N_2 \leftrightarrow N_2O$	$C + 2H_2 \leftrightarrow CH_4$	$C + \frac{3}{2}H_2 \rightarrow CH_3$
600	1080	−7.210	−6.111	−2.568	−11.040	2.001	−13.212
700	1260	−6.086	−5.714	−2.085	−10.021	0.951	−11.458
800	1440	−5.243	−5.417	−1.724	−9.253	0.146	−10.152
900	1620	−4.587	−5.185	−1.444	−8.654	−0.493	−9.145
1000	1800	−4.062	−5.000	−1.222	−8.171	−1.011	−8.344
1100	1980	−3.633	−4.848	−1.041	−7.774	−1.440	−7.693
1200	2160	−3.275	−4.721	−0.890	−7.442	−1.801	−7.153
1300	2340	−2.972	−4.612	−0.764	−7.158	−2.107	−6.698
1400	2520	−2.712	−4.519	−0.656	−6.914	−2.372	−6.309
1500	2700	−2.487	−4.438	−0.563	−6.701	−2.602	−5.974
1600	2880	−2.290	−4.367	−0.482	−6.514	−2.803	−5.681
1700	3060	−2.116	−4.304	−0.410	−6.347	−2.981	−5.423
1800	3240	−1.962	−4.248	−0.347	−6.198	−3.139	−5.195
1900	3420	−1.823	−4.198	−0.291	−6.065	−3.281	−4.991
2000	3600	−1.699	−4.152	−0.240	−5.943	−3.408	−4.808
2100	3780	−1.586	−4.111	−0.195	−5.833	−3.523	−4.642
2200	3960	−1.484	−4.074	−0.153	−5.732	−3.627	−4.492
2300	4140	−1.391	−4.040	−0.116	−5.639	−3.722	−4.355
2400	4320	−1.305	−4.008	−0.082	−5.554	−3.809	−4.230
2500	4500	−1.227	−3.979	−0.050	−5.475	−3.889	−4.115
2600	4680	−1.154	−3.953	−0.021	−5.401	−3.962	−4.009
2700	4860	−1.087	−3.928	0.005	−5.333	−4.030	−3.911
2800	5040	−1.025	−3.905	0.030	−5.270	−4.093	−3.820
2900	5220	−0.967	−3.884	0.053	−5.210	−4.152	−3.736
3000	5400	−0.913	−3.864	0.074	−5.154	−4.206	−3.659
3100	5580	−0.863	−3.846	0.094	−5.102	−4.257	−3.584
3200	5760	−0.815	−3.828	0.112	−5.052	−4.304	−3.515
3300	5940	−0.771	−3.812	0.129	−5.006	−4.349	−3.451
3400	6120	−0.729	−3.797	0.145	−4.962	−4.391	−3.391
3500	6300	−0.690	−3.783	0.160	−4.920	−4.430	−3.334
3600	6480	−0.653	−3.770	0.174	−4.881	−4.467	−3.280
3700	6660	−0.618	−3.757	0.188	−4.843	−4.503	−3.230
3800	6840	−0.585	−3.746	0.200	−4.807	−4.536	−3.182
3900	7020	−0.554	−3.734	0.212	−4.773	−4.568	−3.137
4000	7200	−0.524	−3.724	0.223	−4.741	−4.598	−3.095

TABLE C.12 Logarithms to Base 10 of the EQUILIBRIUM CONSTANT K (*Continued*)

Temperature, K	Temperature, °R	$C + H_2 \leftrightarrow CH_2$	$C + \frac{1}{2}H_2 \leftrightarrow CH$	$2C + H_2 \leftrightarrow C_2H_2$	$C_2 + 2H_2 \leftrightarrow C_2H_4$	$\frac{1}{2}N_2 + \frac{3}{2}H_2 \leftrightarrow NH_3$	$\frac{1}{2}N_2 + H_2 \leftrightarrow NH_2$	$\frac{1}{2}N_2 + \frac{1}{2}H_2 \leftrightarrow NH$
600	1080	−30.678	−45.842	−16.687	−7.652	−1.377	−18.326	−31.732
700	1260	−25.898	−38.448	−13.882	−7.114	−2.023	−15.996	−27.049
800	1440	−22.319	−32.905	−11.784	−6.728	−2.518	−14.255	−23.537
900	1620	−19.540	−28.597	−10.155	−6.438	−2.910	−12.905	−20.806
1000	1800	−17.321	−25.152	−8.856	−6.213	−3.228	−11.827	−18.621
1100	1980	−15.508	−22.336	−7.795	−6.034	−3.490	−10.948	−16.834
1200	2160	−14.000	−19.991	−6.913	−5.889	−3.710	−10.216	−15.345
1300	2340	−12.726	−18.008	−6.168	−5.766	−3.897	−9.598	−14.084
1400	2520	−11.635	−16.310	−5.531	−5.664	−4.058	−9.069	−13.004
1500	2700	−10.691	−14.838	−4.979	−5.575	−4.197	−8.610	−12.068
1600	2880	−9.866	−13.551	−4.497	−5.497	−4.319	−8.210	−11.249
1700	3060	−9.139	−12.417	−4.072	−5.430	−4.426	−7.856	−10.526
1800	3240	−8.493	−11.409	−3.695	−5.369	−4.521	−7.542	−9.883
1900	3420	−7.916	−10.507	−3.358	−5.316	−4.605	−7.261	−9.308
2000	3600	−7.397	−9.696	−3.055	−5.267	−4.681	−7.009	−8.790
2100	3780	−6.929	−8.963	−2.782	−5.223	−4.749	−6.780	−8.322
2200	3960	−6.503	−8.296	−2.532	−5.183	−4.810	−6.572	−7.896
2300	4140	−6.115	−7.687	−2.306	−5.146	−4.866	−6.382	−7.507
2400	4320	−5.760	−7.130	−2.098	−5.113	−4.916	−6.208	−7.151
2500	4500	−5.433	−6.617	−1.906	−5.081	−4.963	−6.048	−6.823
2600	4680	−5.133	−6.144	−1.730	−5.052	−5.005	−5.899	−6.520
2700	4860	−4.854	−5.706	−1.566	−5.025	−5.044	−5.762	−6.240
2800	5040	−4.596	−5.300	−1.415	−5.000	−5.079	−5.635	−5.979
2900	5220	−4.356	−4.922	−1.274	−4.977	−5.112	−5.516	−5.737
3000	5400	−4.132	−4.569	−1.142	−4.955	−5.143	−5.405	−5.511
3100	5580	−3.923	−4.239	−1.019	−4.934	−5.171	−5.300	−5.299
3200	5760	−3.728	−3.930	−0.903	−4.915	−5.197	−5.203	−5.100
3300	5940	−3.544	−3.639	−0.795	−4.897	−5.221	−5.111	−4.914
3400	6120	−3.372	−3.366	−0.693	−4.880	−5.244	−5.024	−4.738
3500	6300	−3.210	−3.108	−0.597	−4.864	−5.265	−4.942	−4.572
3600	6480	−3.056	−2.865	−0.506	−4.848	−5.285	−4.865	−4.416
3700	6660	−2.912	−2.636	−0.420	−4.834	−5.304	−4.791	−4.267
3800	6840	−2.775	−2.418	−0.339	−4.821	−5.321	−4.721	−4.127
3900	7020	−2.646	−2.212	−0.262	−4.808	−5.338	−4.655	−3.994
4000	7200	−2.523	−2.016	−0.189	−4.796	−5.353	−4.592	−3.867

Values taken from Roger A. Strehlow, *Combustion Fundamentals,* McGraw-Hill, New York, 1984 (with permission). The original source is the JANAF thermodynamic tables including supplements.

TABLE C.13 Enthalpy of Fusion (or Sublimation) for Selected Substances

Substance	Phase Change Temperature		Enthalpy of Fusion or Sublimation	
	K	°R	kJ/kg	Btu/lbm
Carbon dioxide, CO_2 (S)	216.6	389.9	18.4	7.91
Copper, Cu (F)	1,357	2,443	206.8	88.91
Ethane, C_2H_6 (F)	89.9	162	95.1	40.9
Gold, Au (F)	1,337	2,407	62.8	27.0
Iron, Fe (F)	1,811	3,260	247.3	106.3
Methane, CH_4 (F)	90.7	163	58.7	25.2
Nitrogen, N_2 (F)	63.1	114	25.7	11.0
Oxygen, O_2 (F)	54.4	97.9	13.9	5.98
Propane, C_3H_8 (F)	86	155	79.91	34.35
Refrigerant 12, CF_2Cl_2 (F)	115.4	207.7	34.3	14.7
Silicon, Si (F)	1,684	3,031	1,802	774.7
Tungsten, W (F)	3,660	6,588	192.5	82.76
Water, H_2O (F)	273.2	491.7	333.5	143.4

(F) fusion, (S), sublimation

TABLE C.14(a) Mean Specific Heat at Constant Pressure between 25°C and *T* for Various Gases at Low Pressure (SI Units)

		Mean Specific Heat, kJ/(kg·K)			
T, °C	*T*, K	Air	Nitrogen	Carbon Dioxide	Water Vapor
25	298	1.0051	1.0403	0.8443	1.8644
50	323	1.0060	1.0408	0.8572	1.8668
100	373	1.0076	1.0415	0.8815	1.8755
200	473	1.0131	1.0447	0.9251	1.8974
300	573	1.0209	1.0505	0.9630	1.9237
400	673	1.0305	1.0586	0.9964	1.9523
500	773	1.0411	1.0682	1.0260	1.9828
600	873	1.0521	1.0787	1.0524	2.0145
700	973	1.0631	1.0895	1.0761	2.0473
800	1073	1.0736	1.1001	1.0974	2.0809
900	1173	1.0839	1.1106	1.1165	2.1146
1000	1273	1.0935	1.1206	1.1338	2.1485
1100	1373	1.1014	1.1301	1.1495	2.1821
1200	1473	1.1111	1.1391	1.1638	2.2151
1300	1573	1.1192	1.1475	1.1768	2.2474
1400	1673	1.1267	1.1555	1.1889	2.2788
1500	1773	1.1337	1.1629	1.1998	2.3091
1600	1873	1.1404	1.1699	1.2101	2.3385
1700	1973	1.1466	1.1765	1.2194	2.3667
1800	2073	1.1525	1.1827	1.2274	2.3939
1900	2173	1.1581	1.1884	1.2374	2.4181
2000	2273	1.1633	1.1939	1.2355	2.4451
2100	2373	1.1683	1.1990	1.2480	2.4691
2200	2473	1.1731	1.2039	1.2579	2.4922
2300	2573	1.1776	1.2084	1.2633	2.5143
2400	2673	1.1819	1.2128	1.2694	2.5356
2500	2773	1.1860	1.2169	1.2749	2.5560
2600	2873	1.1899	1.2207	1.2801	2.5757

TABLE C.14(b) Mean Specific Heat at Constant Pressure between 77°F and T for Various Gases at Low Pressure (USCS Units)

T, °F	T, °R	Mean Specific Heat, Btu/(lbm · °R)			
		Air	Nitrogen	Carbon Dioxide	Water Vapor
77	537	0.2401	0.2485	0.2017	0.4453
100	560	0.2400	0.2487	0.2034	0.4457
200	660	0.2406	0.2487	0.2098	0.4478
300	760	0.2412	0.2491	0.2158	0.4504
400	860	0.2420	0.2496	0.2213	0.4535
500	960	0.2430	0.2503	0.2265	0.4569
600	1060	0.2442	0.2512	0.2313	0.4605
700	1160	0.2454	0.2522	0.2358	0.4643
800	1260	0.2468	0.2534	0.2400	0.4682
900	1360	0.2482	0.2547	0.2439	0.4723
1000	1460	0.2497	0.2561	0.2475	0.4764
1100	1560	0.2511	0.2575	0.2510	0.4807
1200	1660	0.2526	0.2589	0.2542	0.4850
1300	1760	0.2540	0.2603	0.2573	0.4894
1400	1860	0.2554	0.2618	0.2601	0.4938
1500	1960	0.2568	0.2632	0.2628	0.4983
1600	2060	0.2582	0.2646	0.2654	0.5027
1700	2160	0.2595	0.2659	0.2678	0.5072
1800	2260	0.2608	0.2672	0.2701	0.5118
1900	2360	0.2620	0.2685	0.2723	0.5162
2000	2460	0.2632	0.2698	0.2743	0.5207
2200	2660	0.2655	0.2722	0.2781	0.5294
2400	2860	0.2676	0.2744	0.2815	0.5379
2600	3060	0.2696	0.2765	0.2847	0.5462
2800	3260	0.2710	0.2784	0.2875	0.5542
3000	3460	0.2731	0.2802	0.2902	0.5619
3200	3660	0.2747	0.2819	0.2923	0.5692
3400	3860	0.2762	0.2835	0.2950	0.5762
3600	4060	0.2776	0.2849	0.2952	0.5830
3800	4260	0.2790	0.2863	0.2978	0.5894
4000	4460	0.2802	0.2876	0.3005	0.5955

(*a*) $0 < P_r < 1.0$

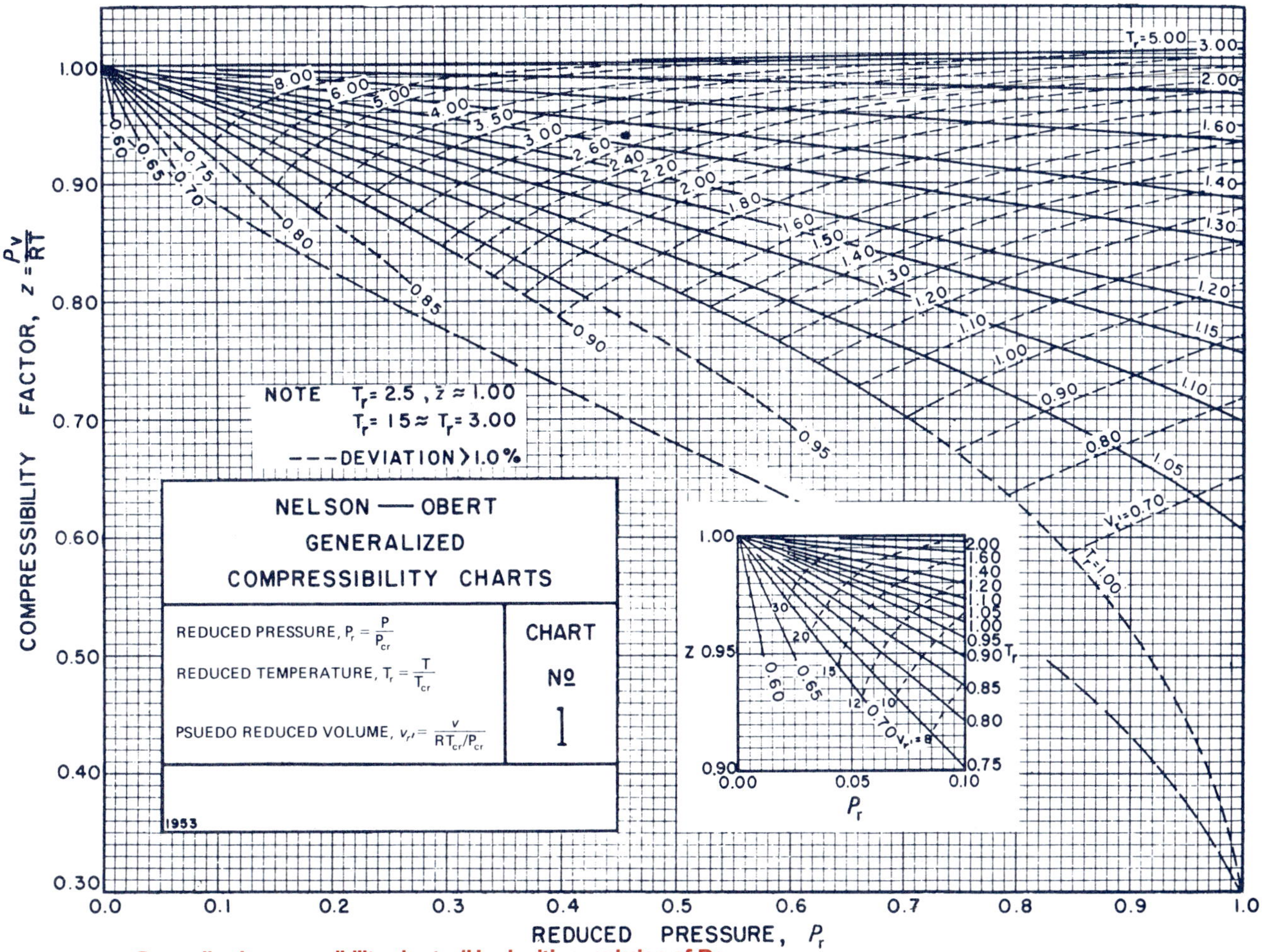

Figure C.1 Generalized compressibility charts. (Used with permission of Dr. Edward E. Obert, University of Wisconsin.)

$(b)\ 0 < P_r < 7$

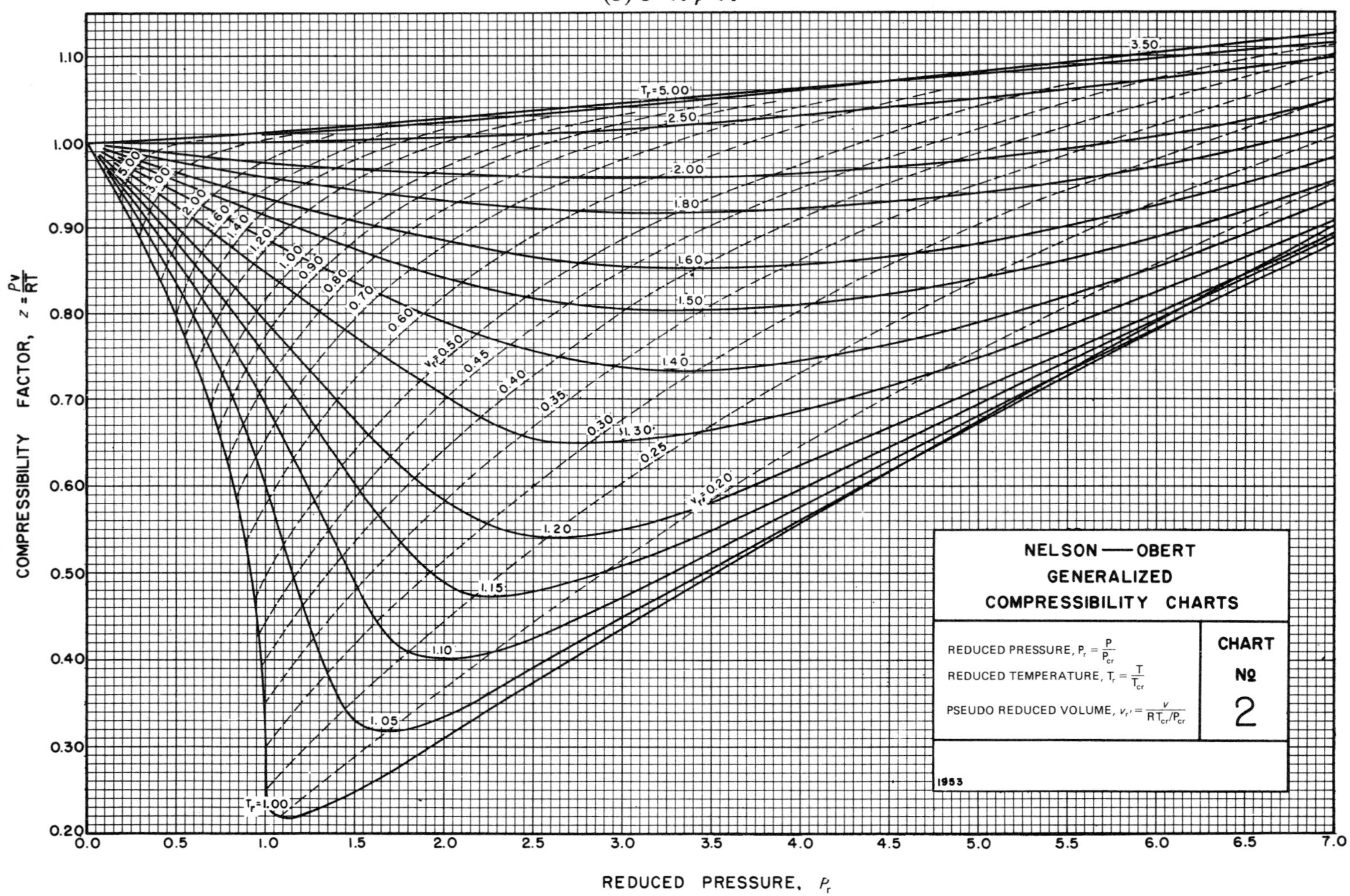

Figure C.1 Generalized compressibility charts. (Continued)

(c) $0 < P_r < 40$

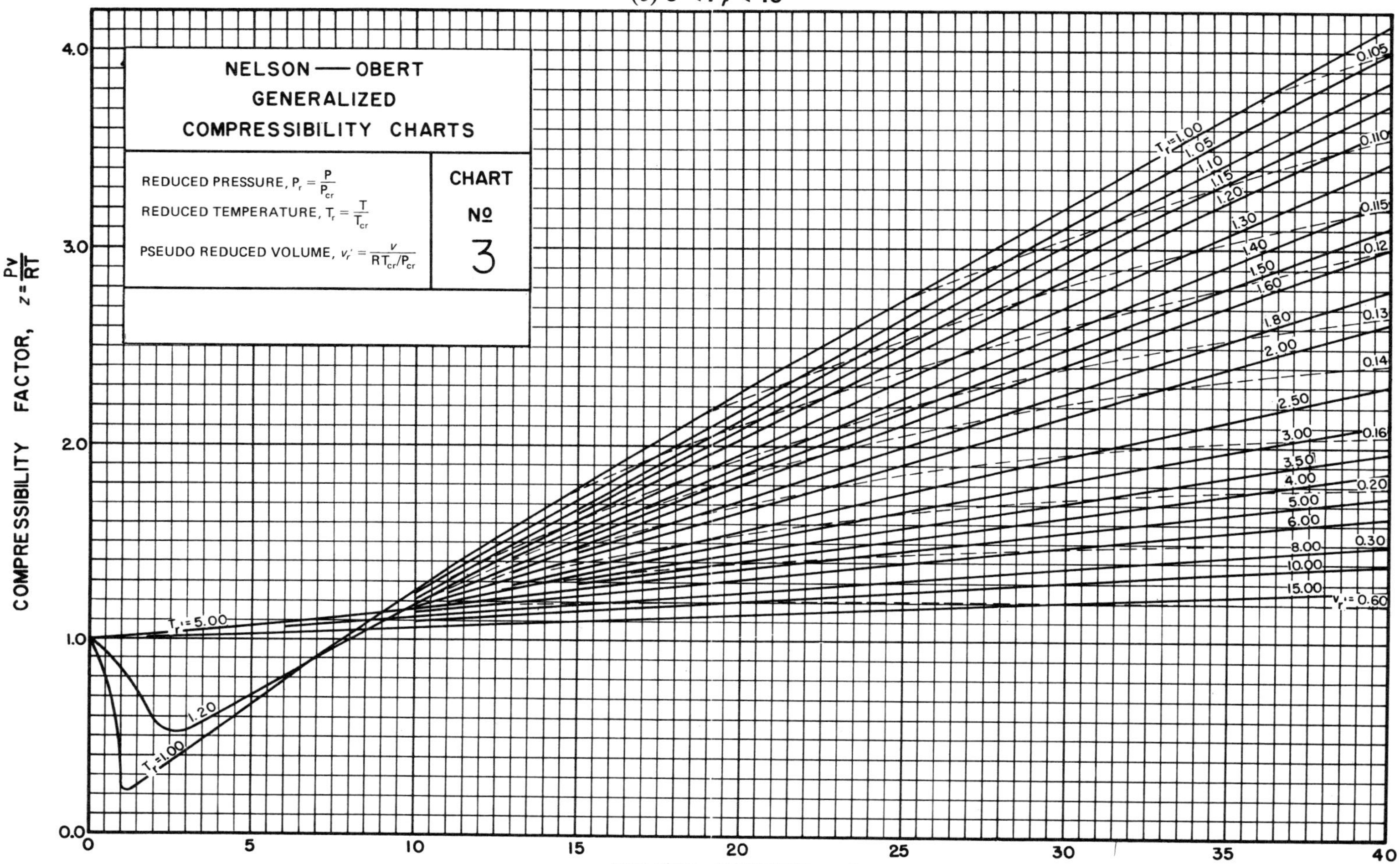

Figure C.1 Generalized compressibility charts. (Continued)

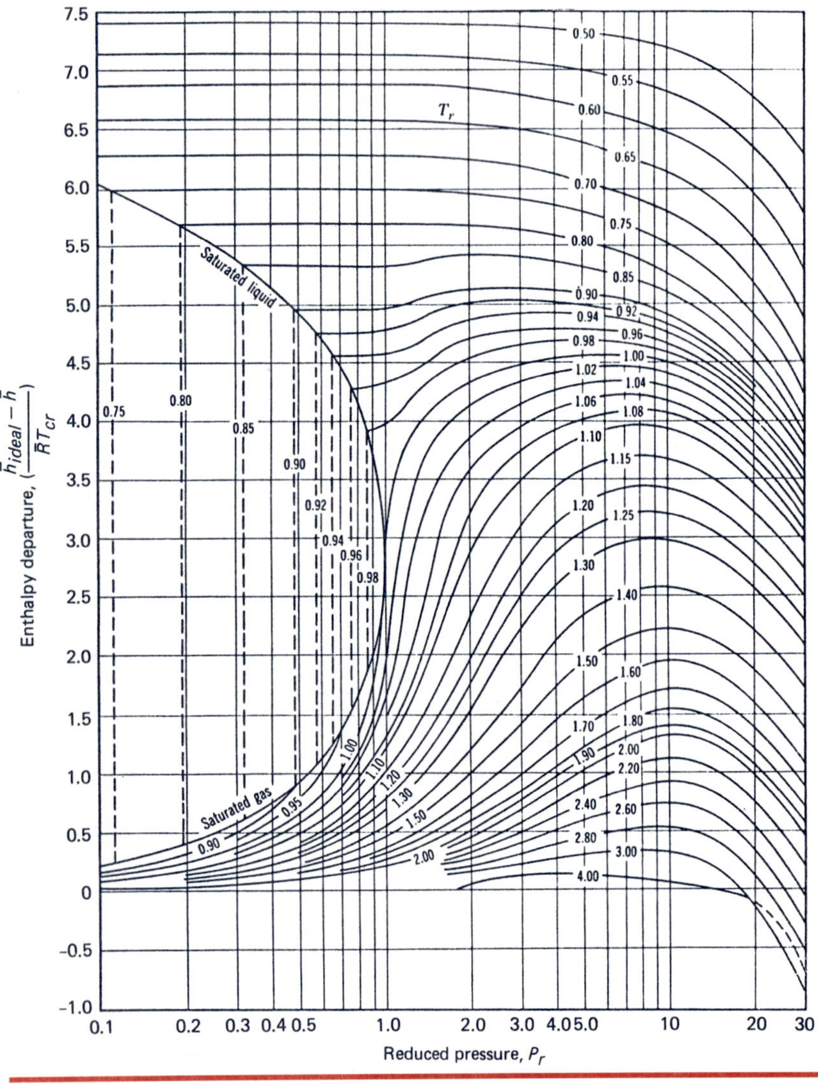

Figure C.2 Generalized ENTHALPY departure chart. [Redrawn from Gordon van Wylen and Richard Sonntag, *Fundamentals of Classical Thermodynamics* (SI version), 2d ed., Wiley, New York, 1976. Reprinted by permission of John Wiley & Sons, Inc.]

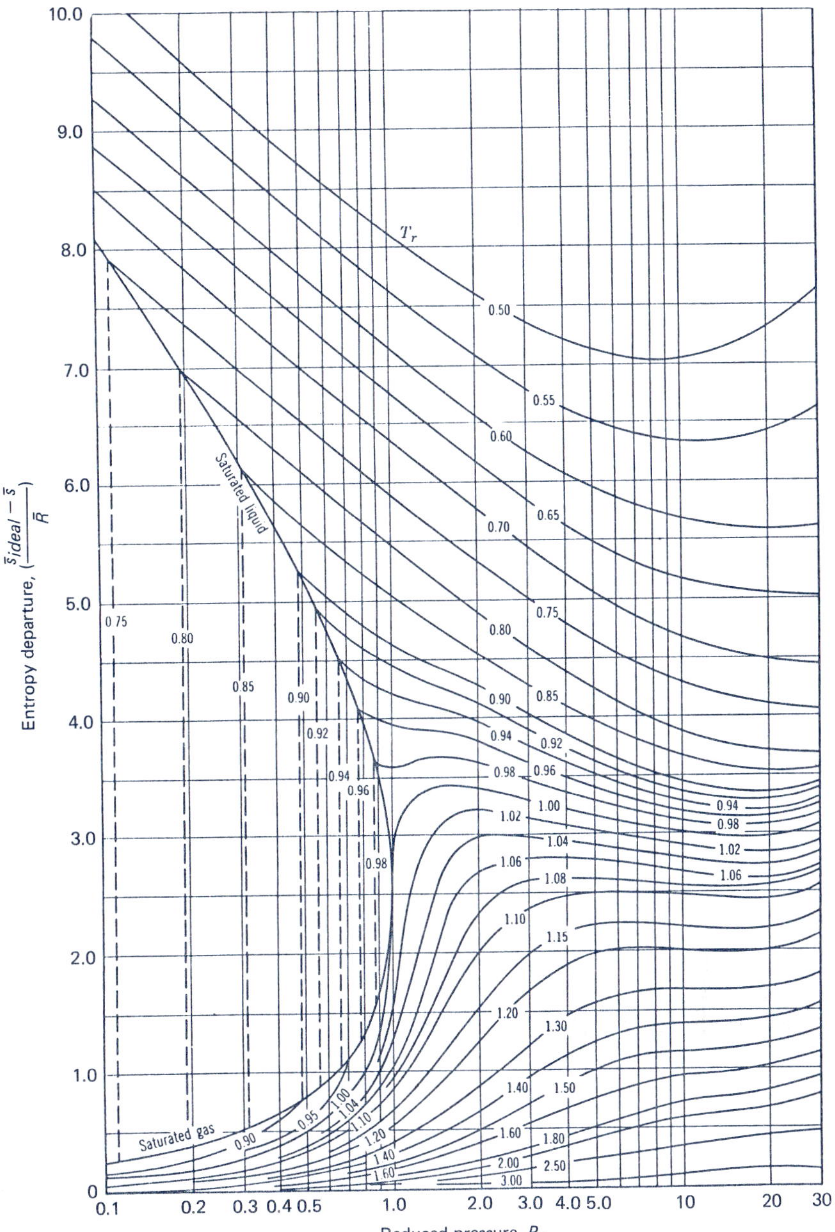

Figure C.3 Generalized ENTROPY departure chart. [Redrawn from Gordon van Wylen and Richard Sonntag, *Fundamentals of Classical Thermodynamics* (SI version), 2d ed., Wiley, New York, 1976. Reprinted by permission of John Wiley & Sons, Inc.]

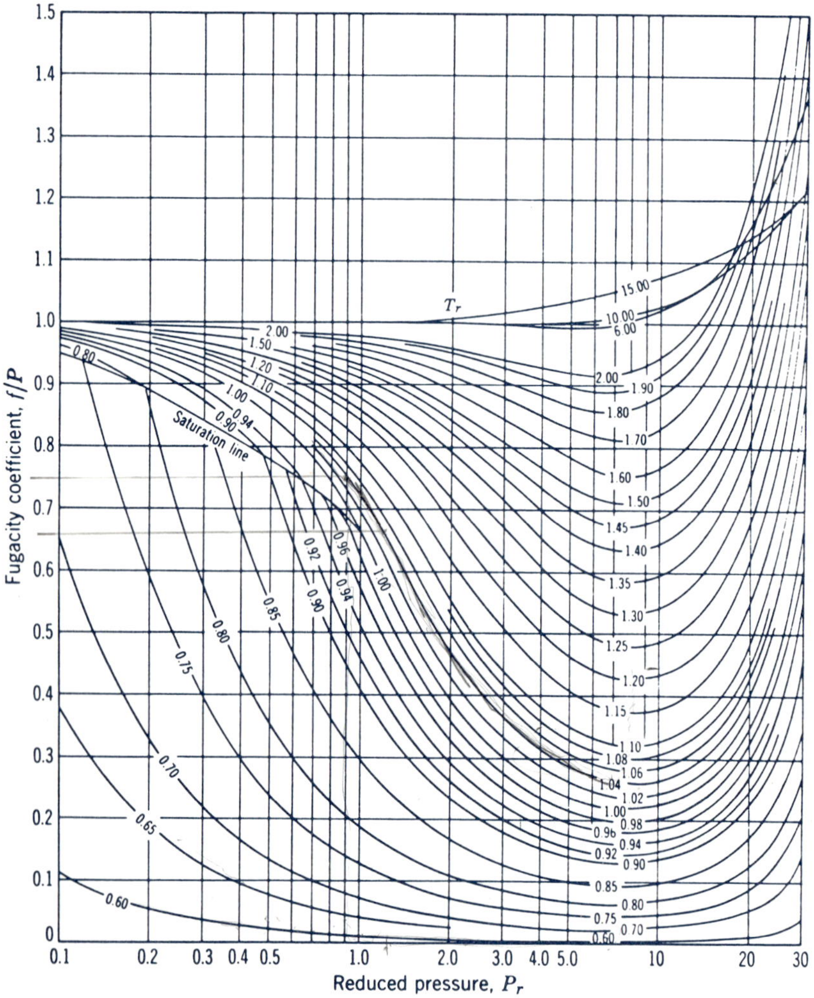

Figure C.4 Generalized FUGACITY chart. [Redrawn from Gordon van Wylen and Richard Sonntag, *Fundamentals of Classical Thermodynamics* (SI version), 2d ed., Wiley, New York, 1976. Reprinted by permission of John Wiley & Sons, Inc.]

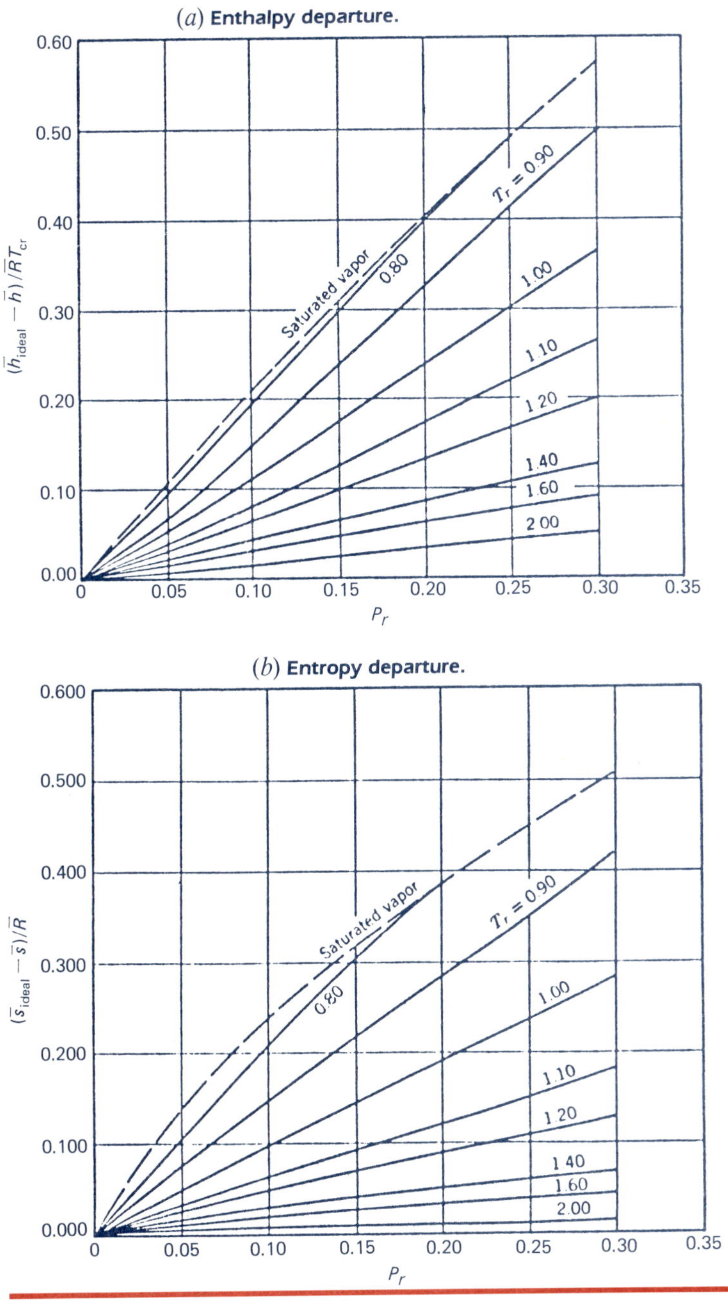

Figure C.5 Generalized departure charts at low pressures. [Redrawn from Gordon van Wylen and Richard Sonntag, *Fundamentals of Classical Thermodynamics* (SI version), 2d ed., Wiley, New York, 1976. Reprinted by permission of John Wiley & Sons, Inc.]

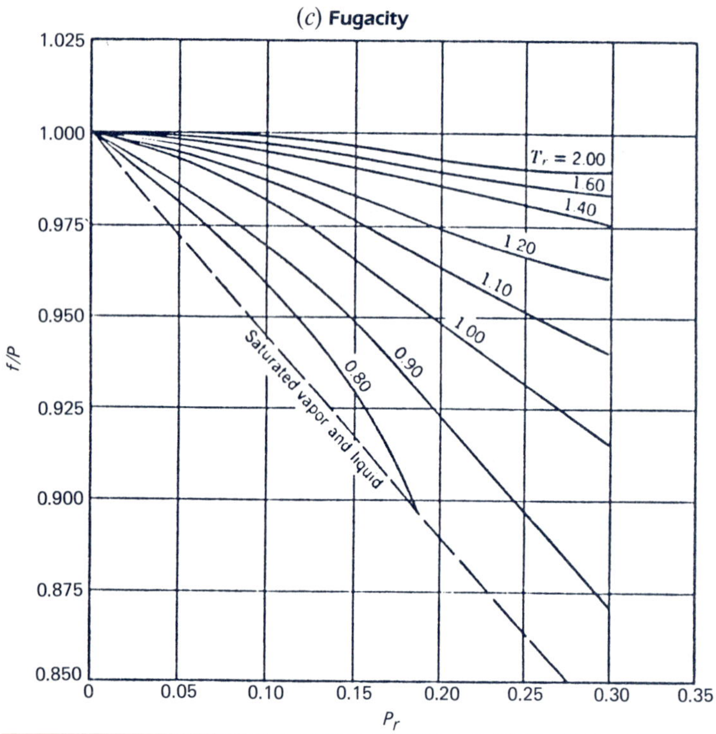

Figure C.5 Generalized departure charts at low pressure (Continued)

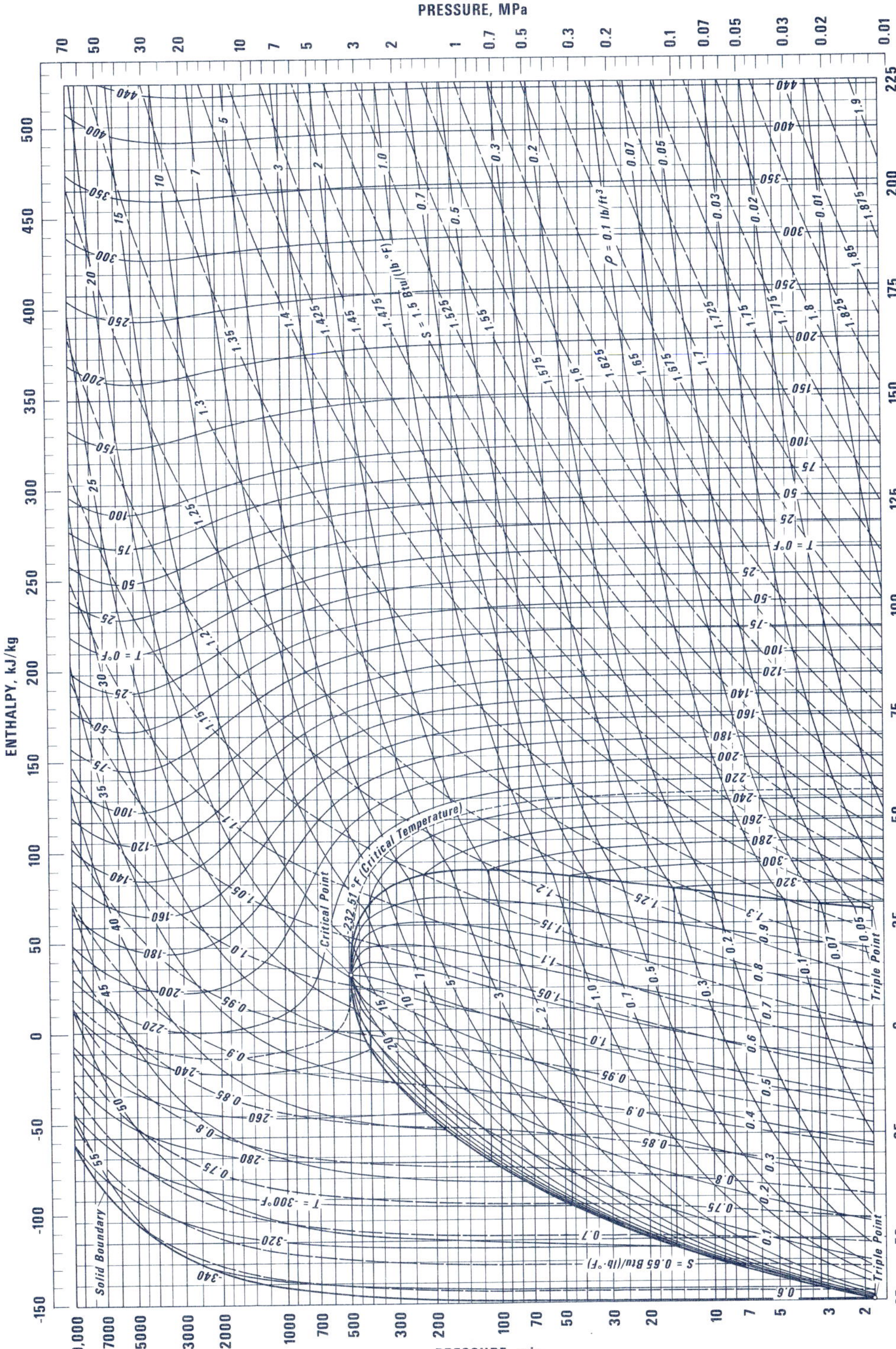

Figure C.6 Pressure-enthalpy diagram for NITROGEN, SI and USCS units.
(Courtesy, Dr. James F. Ely, U.S. National Institute of Standards and Technology.)

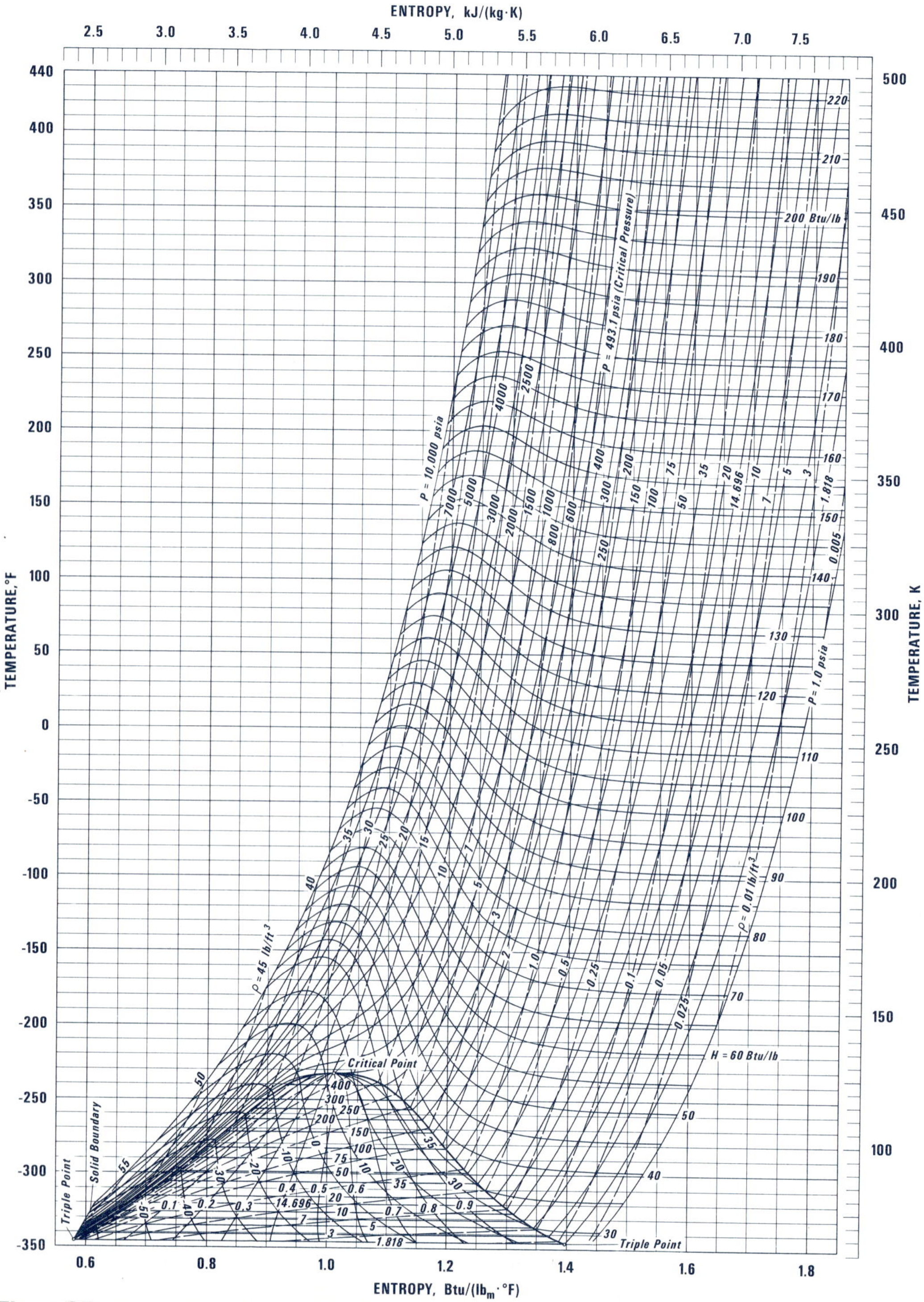

Figure C.7 Temperature-entropy diagram for NITROGEN, SI and USCS units. (Courtesy, Dr. James F. Ely, U.S. National Institute of Standards and Technology.)

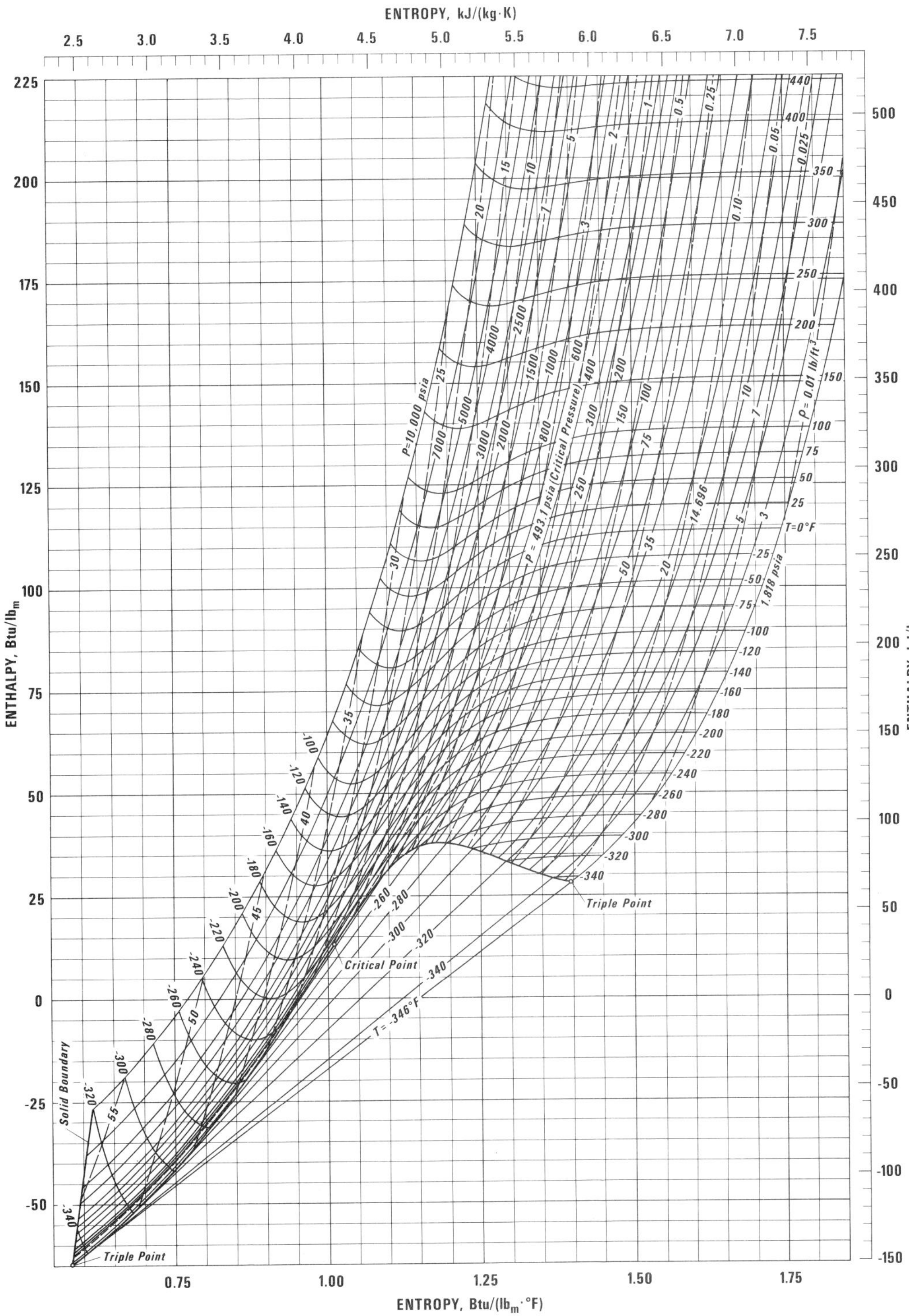

Figure C.8 Enthalpy-entropy (Mollier) diagram for NITROGEN, SI and USCS units. (Courtesy, Dr. James F. Ely, U.S. National Institute of Standards and Technology.)

METHANE
Pressure-Enthalpy (P-H)

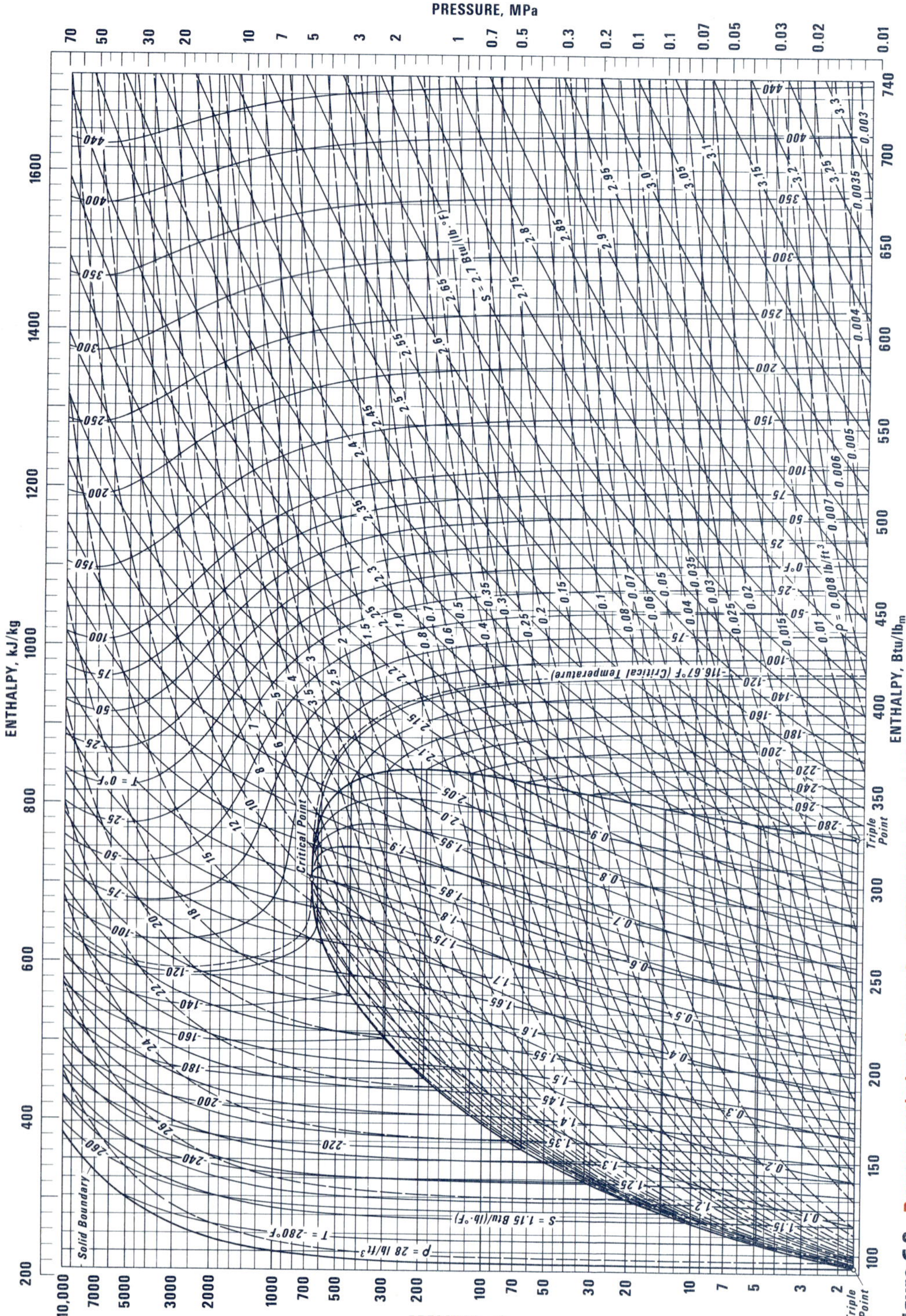

Cryogenics Division, NBS-IBS, Boulder, Colorado
Chart 2029W (4-1-77)

Figure C.9 Pressure-enthalpy diagram for METHANE, SI and USCS units. (Courtesy, Dr. James F. Ely, U.S. National Institute of Standards and Technology.)

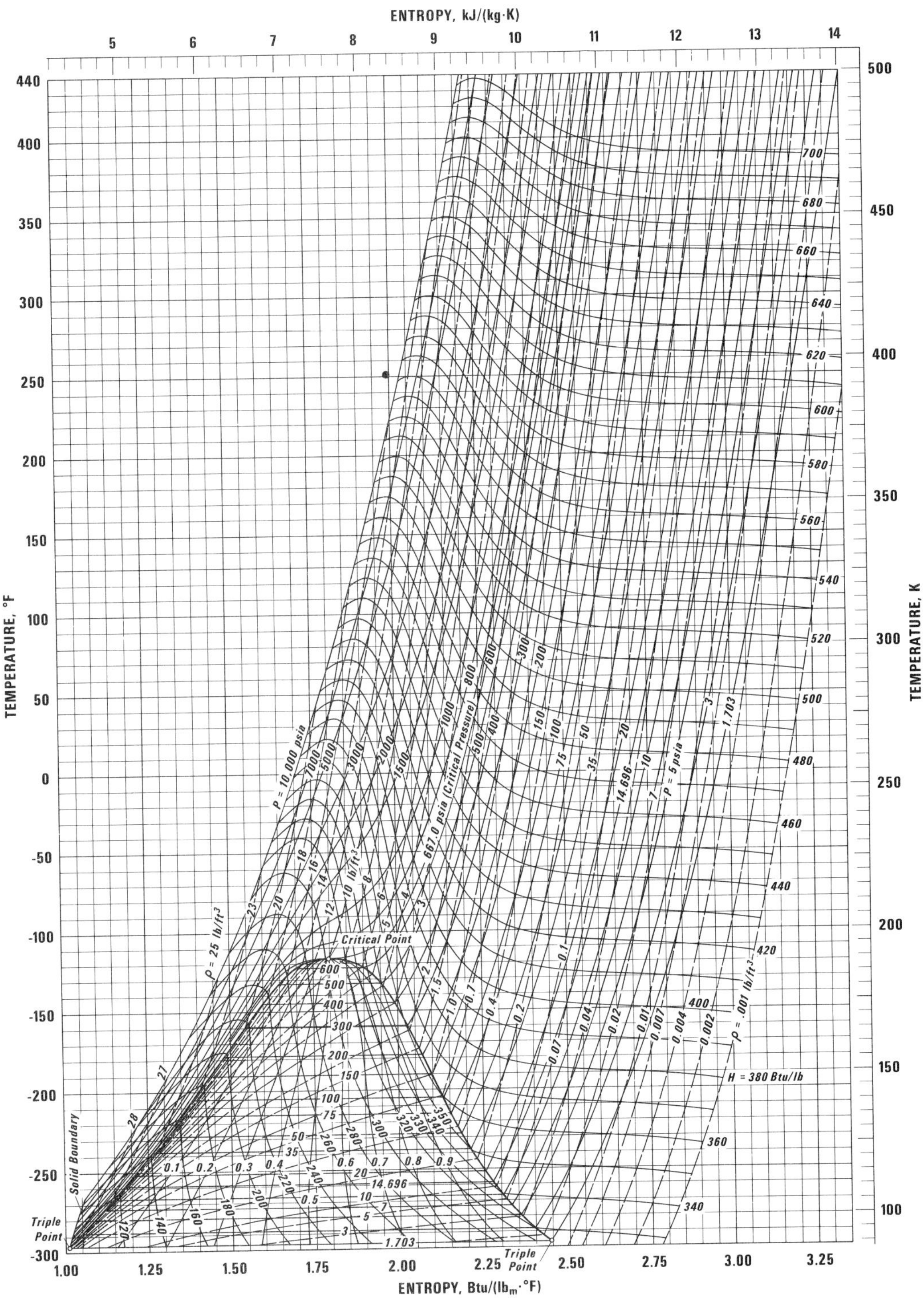

Figure C.10 Temperature-entropy diagram for METHANE, SI and USCS units. (Courtesy, Dr. James F. Ely, U.S. National Institute of Standards and Technology.)

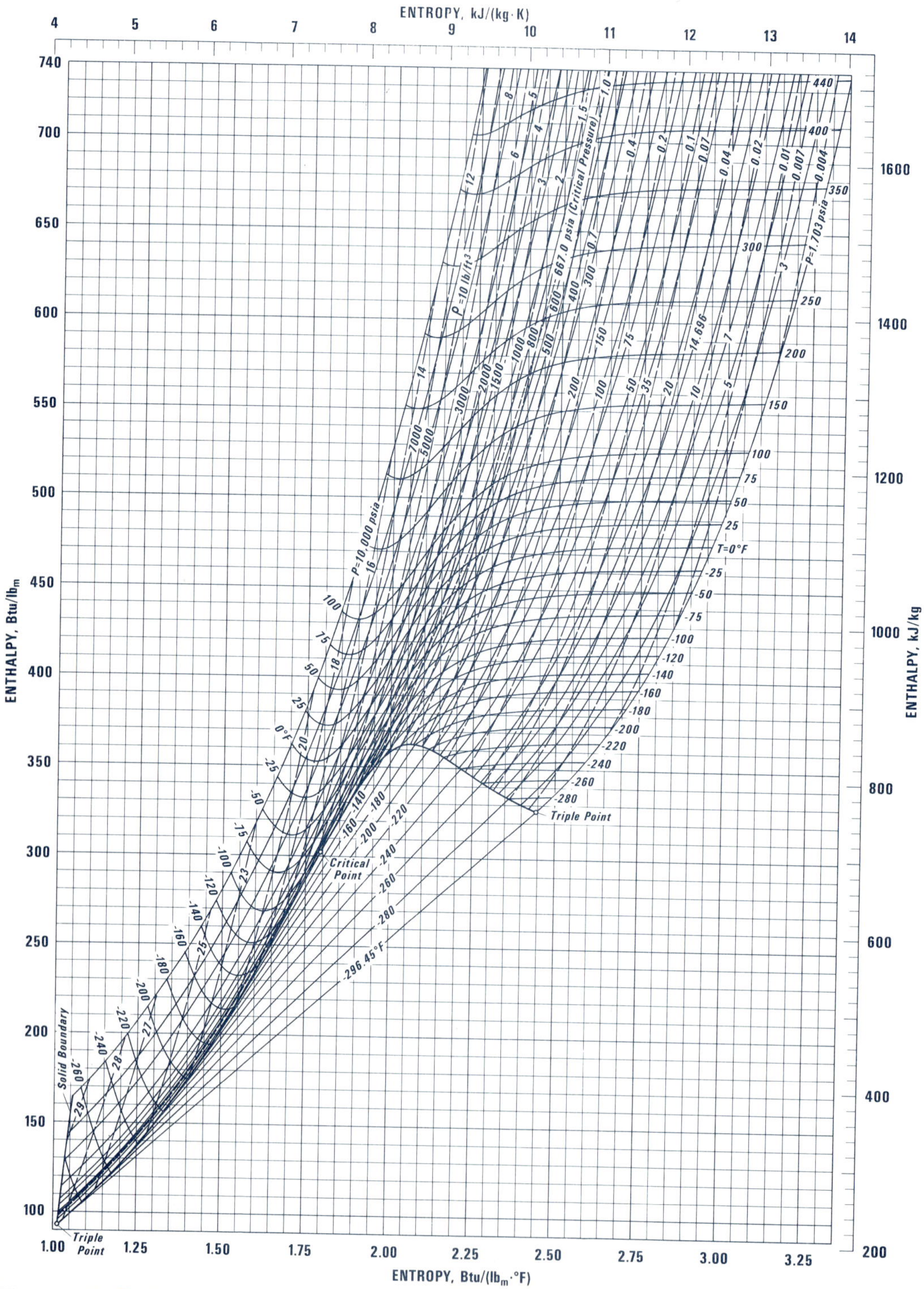

Figure C.11 Enthalpy-entropy (Mollier) diagram for METHANE, SI and USCS units. (Courtesy, Dr. James F. Ely, U.S. National Institute of Standards and Technology.)

APPENDIX D

Tables And Diagrams of Thermodynamic Data for Various Substances—SI Units

TABLE D.1 Temperature-Dependent Molar Specific Heats of Gases at Zero Pressure—SI Units

$$\bar{c}_P = a + bT + cT^2 + dT^3$$

[T in K, $\bar{c}_P$ in kJ/(kmol · K)]

Substance	Formula	a	b	c	d	Temperature Range, K	Error, % Max.	Error, % Avg.
Nitrogen	N_2	28.90	-0.1571×10^{-2}	0.8081×10^{-5}	-2.873×10^{-9}	273–1800	0.59	0.34
Oxygen	O_2	25.48	1.520×10^{-2}	-0.7155×10^{-5}	1.312×10^{-9}	273–1800	1.19	0.28
Air		28.11	0.1967×10^{-2}	0.4802×10^{-5}	-1.966×10^{-9}	273–1800	0.72	0.33
Hydrogen	H_2	29.11	-0.1916×10^{-2}	0.4003×10^{-5}	-0.8704×10^{-9}	273–1800	1.01	0.26
Carbon monoxide	CO	28.16	0.1675×10^{-2}	0.5372×10^{-5}	-2.222×10^{-9}	273–1800	0.89	0.37
Carbon dioxide	CO_2	22.26	5.981×10^{-2}	-3.501×10^{-5}	7.469×10^{-9}	273–1800	0.67	0.22
Water vapor	H_2O	32.24	0.1923×10^{-2}	1.055×10^{-5}	-3.595×10^{-9}	273–1800	0.53	0.24
Nitric oxide	NO	29.34	-0.09395×10^{-2}	0.9747×10^{-5}	-4.187×10^{-9}	273–1500	0.97	0.36
Nitrous oxide	N_2O	24.11	5.8632×10^{-2}	-3.562×10^{-5}	10.58×10^{-9}	273–1500	0.59	0.26
Nitrogen dioxide	NO_2	22.9	5.715×10^{-2}	-3.52×10^{-5}	7.87×10^{-9}	273–1500	0.46	0.18
Ammonia	NH_3	27.568	2.5630×10^{-2}	0.99072×10^{-5}	-6.6909×10^{-9}	273–1500	0.91	0.36
Sulfur	S_2	27.21	2.218×10^{-2}	-1.628×10^{-5}	3.986×10^{-9}	273–1800	0.99	0.38
Sulfur dioxide	SO_2	25.78	5.795×10^{-2}	-3.812×10^{-5}	8.612×10^{-9}	273–1800	0.45	0.24
Sulfur trioxide	SO_3	16.40	14.58×10^{-2}	-11.20×10^{-5}	32.42×10^{-9}	273–1300	0.29	0.13
Acetylene	C_2H_2	21.8	9.2143×10^{-2}	-6.527×10^{-5}	18.21×10^{-9}	273–1500	1.46	0.59
Benzene	C_6H_6	−36.22	48.475×10^{-2}	-31.57×10^{-5}	77.62×10^{-9}	273–1500	0.34	0.20
Methanol	CH_4O	19.0	9.152×10^{-2}	-1.22×10^{-5}	-8.039×10^{-9}	273–1000	0.18	0.08
Ethanol	C_2H_6O	19.9	20.96×10^{-2}	-10.38×10^{-5}	20.05×10^{-9}	273–1500	0.40	0.22
Hydrogen chloride	HCl	30.33	-0.7620×10^{-2}	1.327×10^{-5}	-4.338×10^{-9}	273–1500	0.22	0.08
Methane	CH_4	19.89	5.024×10^{-2}	1.269×10^{-5}	-11.01×10^{-9}	273–1500	1.33	0.57
Ethane	C_2H_6	6.900	17.27×10^{-2}	-6.406×10^{-5}	7.285×10^{-9}	273–1500	0.83	0.28
Propane	C_3H_8	−4.04	30.48×10^{-2}	-15.72×10^{-5}	31.74×10^{-9}	273–1500	0.40	0.12
n-Butane	C_4H_{10}	3.96	37.15×10^{-2}	-18.34×10^{-5}	35.00×10^{-9}	273–1500	0.54	0.24
i-Butane	C_4H_{10}	−7.913	41.60×10^{-2}	-23.01×10^{-5}	49.91×10^{-9}	273–1500	0.25	0.13
n-Pentane	C_5H_{12}	6.774	45.43×10^{-2}	-22.46×10^{-5}	42.29×10^{-9}	273–1500	0.56	0.21
n-Hexane	C_6H_{14}	6.938	55.22×10^{-2}	-28.65×10^{-5}	57.69×10^{-9}	273–1500	0.72	0.20
Ethylene	C_2H_4	3.95	15.64×10^{-2}	-8.344×10^{-5}	17.67×10^{-9}	273–1500	0.54	0.13
Propylene	C_3H_6	3.15	23.83×10^{-2}	-12.18×10^{-5}	24.62×10^{-9}	273–1500	0.73	0.17

Data from B. G. Kyle, *Chemical and Process Thermodynamics,* Prentice-Hall, Englewood Cliffs, N.J., 1984 (used with permission).

TABLE D.2 Ideal Gas Properties of AIR at Low Pressure—SI Units

T, °C	u, kJ/kg	h, kJ/kg	s_0, kJ/(kg·K)	P_0	v_0	c_P, kJ/(kg·K)	c_v, kJ/(kg·K)	k
−200	52.32	73.33	5.2906	0.0101	20,488.	1.0025	0.7152	1.4016
−150	88.09	123.46	5.8128	0.0624	5,592.1	1.0025	0.7153	1.4015
−100	123.85	173.58	6.1544	0.2051	2,391.5	1.0026	0.7153	1.4015
−50	159.65	223.74	6.4090	0.4979	1,269.6	1.0030	0.7158	1.4013
0	195.46	273.92	6.6119	1.0095	766.52	1.0041	0.7169	1.4007
25	213.40	299.03	6.6999	1.3715	615.80	1.0051	0.7179	1.4001
50	231.36	324.18	6.7808	1.8186	503.36	1.0065	0.7193	1.3993
100	267.42	374.60	6.9259	3.0147	350.63	1.0107	0.7235	1.3970
150	303.74	425.28	7.0533	4.6998	255.05	1.0169	0.7296	1.3937
200	340.42	476.32	7.1673	6.9913	191.71	1.0249	0.7376	1.3894
250	377.54	527.80	7.2707	10.024	147.85	1.0345	0.7472	1.3844
300	415.17	579.79	7.3656	13.952	116.37	1.0452	0.7580	1.3789
350	453.36	632.34	7.4535	18.950	93.152	1.0568	0.7695	1.3732
400	492.13	685.48	7.5356	25.218	75.617	1.0687	0.7815	1.3675
450	531.51	739.22	7.6126	32.977	62.120	1.0808	0.7935	1.3620
500	571.49	793.56	7.6852	42.475	51.563	1.0927	0.8054	1.3566
550	612.06	848.49	7.7540	53.987	43.192	1.1042	0.8170	1.3516
600	653.19	903.98	7.8195	67.812	36.475	1.1154	0.8281	1.3468
650	694.87	960.02	7.8819	84.280	31.028	1.1260	0.8388	1.3424
700	737.07	1,016.6	7.9415	103.75	26.571	1.1360	0.8488	1.3384
750	779.75	1,073.6	7.9987	126.61	22.892	1.1455	0.8583	1.3346
800	822.89	1,131.1	8.0536	153.28	19.883	1.1544	0.8672	1.3312
850	866.47	1,189.1	8.1063	184.21	17.271	1.1628	0.8756	1.3280
900	910.45	1,247.4	8.1571	219.89	15.113	1.1706	0.8834	1.3251
950	954.81	1,306.1	8.2062	260.84	13.284	1.1779	0.8907	1.3225
1,000	999.52	1,365.2	8.2535	307.60	11.725	1.1848	0.8976	1.3200
1,100	1,088.6	1,483.0	8.3436	420.98	9.2400	1.1973	0.9101	1.3156
1,200	1,181.5	1,604.6	8.4281	565.19	7.3836	1.2083	0.9210	1.3118
1,300	1,274.1	1,726.0	8.5078	746.01	5.9736	1.2180	0.9308	1.3086
1,400	1,367.7	1,848.2	8.5831	969.91	4.8867	1.2267	0.9395	1.3057
1,500	1,462.0	1,971.3	8.6546	1,244.0	4.0377	1.2345	0.9473	1.3032
1,600	1,557.1	2,095.1	8.7225	1,576.1	3.3666	1.2416	0.9544	1.3010
1,700	1,652.9	2,219.6	8.7872	1,974.9	2.8302	1.2480	0.9608	1.2989
1,800	1,749.3	2,344.7	8.8491	2,449.8	2.3972	1.2539	0.9667	1.2971
1,900	1,846.2	2,470.4	8.9083	3,011.0	2.0446	1.2593	0.9721	1.2955
2,000	1,943.7	2,596.6	8.9650	3,669.4	1.7549	1.2644	0.9771	1.2939
2,100	2,041.7	2,723.3	9.0196	4,437.2	1.5150	1.2690	0.9818	1.2925
2,200	2,140.1	2,850.5	9.0720	5,327.2	1.3151	1.2734	0.9862	1.2913
2,300	2,239.0	2,978.0	9.1226	6,353.2	1.1473	1.2775	0.9902	1.2901
2,400	2,338.2	3,106.0	9.1714	7,530.1	1.0056	1.2813	0.9940	1.2889
2,500	2,437.8	3,234.3	9.2185	8,873.5	0.8853	1.2848	0.9976	1.2879
2,600	2,537.7	3,363.0	9.2641	10,401.	0.7826	1.2882	1.0010	1.2869
2,700	2,638.0	3,492.0	9.3082	12,129.	0.6944	1.2914	1.0042	1.2860
2,800	2,738.6	3,621.3	9.3510	14,078.	0.6184	1.2944	1.0071	1.2852
2,900	2,839.5	3,750.9	9.3925	16,267.	0.5526	1.2972	1.0100	1.2844
3,000	2,940.6	3,880.8	9.4328	18,719.	0.4953	1.2999	1.0126	1.2836

Properties generated from the program *GASPROPS*, Wiley Professional Software, Wiley, New York, 1984.

TABLE D.3 Ideal Gas Properties of ARGON at Low Pressure — SI Units

T, °C	u, kJ/kg	h, kJ/kg	s_0, kJ/(kg·K)	P_0	v_0	c_P, kJ/(kg·K)	c_v, kJ/(kg·K)	k
−200	22.85	38.09	3.1444	0.3641	412.66	0.5207	0.3124	1.667
−150	38.47	64.12	3.4156	1.3402	188.75	0.5207	0.3124	1.667
−100	54.09	90.15	3.5930	3.1433	113.15	0.5207	0.3124	1.667
−50	69.71	116.19	3.7251	5.9294	77.305	0.5207	0.3124	1.667
0	85.33	142.22	3.8304	9.8325	57.063	0.5207	0.3124	1.667
25	93.14	155.24	3.8760	12.241	50.031	0.5207	0.3124	1.667
50	100.95	168.26	3.9179	14.973	44.333	0.5207	0.3124	1.667
100	116.57	194.29	3.9928	21.458	35.719	0.5207	0.3124	1.667
150	132.19	220.32	4.0583	29.391	29.573	0.5207	0.3124	1.667
200	147.81	246.36	4.1165	38.865	25.007	0.5207	0.3124	1.667
250	163.43	272.39	4.1688	49.969	21.505	0.5207	0.3124	1.667
300	179.05	298.42	4.2163	62.787	18.751	0.5207	0.3124	1.667
350	194.68	324.46	4.2598	77.400	16.538	0.5207	0.3124	1.667
400	210.30	350.49	4.3000	93.885	14.728	0.5207	0.3124	1.667
450	225.92	376.53	4.3373	112.32	13.225	0.5207	0.3124	1.667
500	241.54	402.56	4.3721	132.76	11.962	0.5207	0.3124	1.667
550	257.16	428.59	4.4048	155.20	10.888	0.5207	0.3124	1.667
600	272.78	454.63	4.4355	179.98	9.9651	0.5207	0.3124	1.667
650	288.40	480.66	4.4645	206.88	9.1657	0.5207	0.3124	1.667
700	304.02	506.69	4.4919	236.07	8.4677	0.5207	0.3124	1.667
750	319.64	532.73	4.5180	267.59	7.8540	0.5207	0.3124	1.667
800	335.26	558.76	4.5429	301.51	7.3109	0.5207	0.3124	1.667
850	350.88	584.80	4.5666	337.90	6.8277	0.5207	0.3124	1.667
900	366.50	610.83	4.5892	376.79	6.3954	0.5207	0.3124	1.667
950	382.12	636.86	4.6110	418.26	6.0069	0.5207	0.3124	1.667
1,000	397.74	662.90	4.6318	462.36	5.6562	0.5207	0.3124	1.667
1,100	428.98	714.96	4.6712	558.64	5.0490	0.5207	0.3124	1.667
1,200	460.22	767.03	4.7078	666.04	4.5432	0.5207	0.3124	1.667
1,300	491.46	819.10	4.7420	784.98	4.1165	0.5207	0.3124	1.667
1,400	522.70	871.17	4.7741	915.83	3.7526	0.5207	0.3124	1.667
1,500	553.94	923.23	4.8043	1,059.0	3.4394	0.5207	0.3124	1.667
1,600	585.18	975.30	4.8329	1,214.8	3.1674	0.5207	0.3124	1.667
1,700	616.42	1,027.4	4.8600	1,383.6	2.9294	0.5207	0.3124	1.667
1,800	647.66	1,079.4	4.8857	1,565.7	2.7198	0.5207	0.3124	1.667
1,900	678.90	1,131.5	4.9102	1,761.5	2.5341	0.5207	0.3124	1.667
2,000	710.14	1,183.6	4.9337	1,971.4	2.3685	0.5207	0.3124	1.667
2,100	741.38	1,235.6	4.9561	2,195.6	2.2202	0.5207	0.3124	1.667
2,200	772.62	1,287.7	4.9776	2,434.4	2.0868	0.5207	0.3124	1.667
2,300	803.86	1,339.8	4.9982	2,688.1	1.9662	0.5207	0.3124	1.667
2,400	835.11	1,391.8	5.0181	2,957.2	1.8568	0.5207	0.3124	1.667
2,500	866.35	1,443.9	5.0372	3,241.7	1.7572	0.5207	0.3124	1.667
2,600	897.59	1,496.0	5.0556	3,542.2	1.6661	0.5207	0.3124	1.667
2,700	928.83	1,548.0	5.0734	3,858.7	1.5827	0.5207	0.3124	1.667
2,800	960.07	1,600.1	5.0907	4,191.6	1.5060	0.5207	0.3124	1.667
2,900	991.31	1,652.2	5.1073	4,541.2	1.4353	0.5207	0.3124	1.667
3,000	1,022.6	1,704.3	5.1235	4,907.7	1.3700	0.5207	0.3124	1.667

Properties generated from the program *GASPROPS*, Wiley Professional Software, Wiley, New York, 1984.

TABLE D.4 Ideal Gas Properties of NITROGEN at Low Pressure — SI Units

T, °C	u, kJ/kg	h, kJ/kg	s_0, kJ/(kg·K)	P_0	v_0	c_P, kJ/(kg·K)	c_v, kJ/(kg·K)	k
−200	54.31	76.04	5.3795	0.0074	28,801.	1.0395	0.7425	1.4000
−150	91.44	128.01	5.9210	0.0461	7,822.2	1.0395	0.7425	1.4000
−100	128.56	179.99	6.2752	0.1521	3,334.4	1.0395	0.7425	1.4000
−50	165.72	231.99	6.5392	0.3702	1,765.7	1.0398	0.7428	1.3998
0	202.86	283.99	6.7494	0.7517	1,064.4	1.0400	0.7430	1.3997
25	221.44	309.99	6.8405	1.0218	854.74	1.0403	0.7433	1.3996
50	240.03	336.01	6.9243	1.3550	698.56	1.0409	0.7439	1.3993
100	277.27	388.10	7.0742	2.2452	486.83	1.0430	0.7460	1.3981
150	314.66	440.34	7.2055	3.4953	354.62	1.0468	0.7498	1.3961
200	352.29	492.82	7.3227	5.1880	267.15	1.0527	0.7557	1.3930
250	390.26	545.64	7.4289	7.4179	206.58	1.0605	0.7635	1.3890
300	428.67	598.89	7.5261	10.293	163.11	1.0699	0.7729	1.3842
350	467.58	652.65	7.6160	13.935	130.99	1.0806	0.7836	1.3790
400	507.05	706.97	7.6998	18.484	106.68	1.0922	0.7952	1.3735
450	547.10	761.88	7.7785	24.094	87.916	1.1042	0.8072	1.3679
500	587.77	817.39	7.8527	30.940	73.197	1.1163	0.8193	1.3625
550	629.03	873.51	7.9231	39.213	61.489	1.1284	0.8314	1.3572
600	670.90	930.22	7.9900	49.124	52.064	1.1401	0.8431	1.3523
650	713.34	987.52	8.0538	60.906	44.398	1.1515	0.8545	1.3476
700	756.34	1,045.4	8.1148	74.808	38.105	1.1623	0.8653	1.3432
750	799.86	1,103.7	8.1733	91.104	32.897	1.1726	0.8756	1.3392
800	843.89	1,162.6	8.2294	110.09	28.554	1.1824	0.8854	1.3355
850	888.39	1,222.0	8.2835	132.08	24.909	1.1915	0.8945	1.3320
900	933.34	1,281.8	8.3356	157.42	21.830	1.2002	0.9032	1.3288
950	978.70	1,342.0	8.3859	186.47	19.215	1.2082	0.9112	1.3259
1,000	1,024.5	1,402.6	8.4344	219.61	16.982	1.2158	0.9188	1.3232
1,100	1,117.0	1,524.9	8.5269	299.87	13.413	1.2295	0.9325	1.3185
1,200	1,210.9	1,648.4	8.6137	401.81	10.739	1.2415	0.9445	1.3145
1,300	1,305.9	1,773.1	8.6956	529.47	8.7032	1.2520	0.9550	1.3110
1,400	1,401.8	1,898.8	8.7730	687.32	7.1306	1.2612	0.9642	1.3080
1,500	1,498.7	2,025.3	8.8465	880.31	5.9001	1.2693	0.9723	1.3055
1,600	1,596.3	2,152.6	8.9163	1,113.9	4.9260	1.2765	0.9795	1.3032
1,700	1,694.6	2,280.6	8.9827	1,393.9	4.1466	1.2829	0.9859	1.3013
1,800	1,793.4	2,409.2	9.0465	1,726.8	3.5168	1.2886	0.9916	1.2995
1,900	1,892.9	2,538.3	9.1073	2,119.5	3.0033	1.2937	0.9967	1.2980
2,000	1,992.8	2,667.9	9.1656	2,579.6	2.5812	1.2983	1.0013	1.2966
2,100	2,093.1	2,797.9	9.2216	3,115.0	2.2315	1.3025	1.0055	1.2954
2,200	2,193.8	2,928.4	9.2754	3,734.6	1.9398	1.3062	1.0092	1.2943
2,300	2,294.9	3,059.2	9.3273	4,447.4	1.6947	1.3097	1.0127	1.2933
2,400	2,396.4	3,190.3	9.3773	5,263.4	1.4877	1.3129	1.0159	1.2924
2,500	2,498.1	3,321.7	9.4255	6,193.0	1.3117	1.3158	1.0188	1.2915
2,600	2,600.1	3,453.4	9.4722	7,247.3	1.1613	1.3184	1.0214	1.2908
2,700	2,702.4	3,585.4	9.5173	8,438.0	1.0321	1.3209	1.0239	1.2901
2,800	2,804.9	3,717.6	9.5611	9,777.7	0.9207	1.3232	1.0262	1.2894
2,900	2,907.6	3,850.0	9.6035	11,279.	0.8241	1.3254	1.0284	1.2888
3,000	3,010.6	3,982.7	9.6446	12,957.	0.7400	1.3274	1.0304	1.2882

Properties generated from the program *GASPROPS*, Wiley Professional Software, Wiley, New York, 1984.

TABLE D.5 Ideal Gas Properties of OXYGEN at Low Pressure—SI Units

T, °C	u, kJ/kg	h, kJ/kg	s_0, kJ/(kg·K)	P_0	v_0	c_P, kJ/(kg·K)	c_v, kJ/(kg·K)	k
−200	47.55	66.57	5.1302	0.0376	4,993.6	0.9100	0.6500	1.4000
−150	80.06	112.08	5.6045	0.2330	1,355.1	0.9102	0.6502	1.3999
−100	112.58	157.60	5.9147	0.7689	577.44	0.9104	0.6504	1.3998
−50	145.12	203.14	6.1457	1.8711	305.83	0.9116	0.6516	1.3991
0	177.78	248.80	6.3303	3.8077	183.96	0.9154	0.6554	1.3967
25	194.20	271.72	6.4107	5.1865	147.41	0.9187	0.6587	1.3947
50	210.72	294.74	6.4848	6.8990	120.12	0.9230	0.6630	1.3922
100	244.14	341.16	6.6183	11.534	82.963	0.9342	0.6742	1.3856
150	278.19	388.21	6.7366	18.186	59.668	0.9482	0.6882	1.3778
200	312.98	436.01	6.8434	27.425	44.241	0.9637	0.7037	1.3695
250	348.57	484.59	6.9410	39.928	33.599	0.9797	0.7197	1.3613
300	384.95	533.97	7.0311	56.485	26.021	0.9954	0.7354	1.3536
350	422.09	584.12	7.1150	78.005	20.486	1.0103	0.7503	1.3465
400	459.96	634.99	7.1935	105.53	16.358	1.0243	0.7643	1.3402
450	498.50	686.53	7.2673	140.22	13.226	1.0372	0.7772	1.3345
500	537.66	738.69	7.3371	183.39	10.811	1.0490	0.7890	1.3295
550	577.39	791.41	7.4032	236.48	8.9260	1.0597	0.7997	1.3251
600	617.62	844.65	7.4659	301.12	7.4359	1.0695	0.8095	1.3212
650	658.32	898.34	7.5257	379.04	6.2455	1.0783	0.8183	1.3177
700	699.44	952.47	7.5828	472.19	5.2850	1.0864	0.8264	1.3146
750	740.94	1,007.0	7.6374	582.64	4.5031	1.0937	0.8337	1.3119
800	782.80	1,061.8	7.6898	712.68	3.8614	1.1005	0.8404	1.3094
850	824.98	1,117.0	7.7400	864.76	3.3306	1.1066	0.8466	1.3071
900	867.46	1,172.5	7.7884	1,041.5	2.8885	1.1124	0.8524	1.3050
950	910.21	1,228.2	7.8349	1,245.8	2.5177	1.1177	0.8577	1.3032
1,000	953.22	1,284.3	7.8798	1,480.7	2.2049	1.1227	0.8627	1.3014
1,100	1,040.0	1,397.0	7.9650	2,055.6	1.7130	1.1319	0.8719	1.2982
1,200	1,127.6	1,510.6	8.0449	2,795.2	1.3515	1.1403	0.8803	1.2954
1,300	1,216.0	1,625.0	8.1200	3,732.7	1.0808	1.1481	0.8881	1.2928
1,400	1,305.2	1,740.2	8.1910	4,905.3	0.8747	1.1556	0.8956	1.2903
1,500	1,395.1	1,856.1	8.2583	6,355.3	0.7155	1.1629	0.9028	1.2880
1,600	1,485.7	1,972.8	8.3223	8,130.1	0.5908	1.1699	0.9099	1.2857
1,700	1,577.1	2,090.1	8.3833	10,283.	0.4921	1.1769	0.9169	1.2836
1,800	1,669.1	2,208.2	8.4417	12,872.	0.4130	1.1838	0.9238	1.2815
1,900	1,761.8	2,326.9	8.4976	15,963.	0.3491	1.1906	0.9306	1.2794
2,000	1,855.2	2,446.3	8.5513	19,629.	0.2970	1.1973	0.9373	1.2774
2,100	1,949.3	2,566.3	8.6030	23,949.	0.2541	1.2039	0.9438	1.2755
2,200	2,044.0	2,687.0	8.6528	29,010.	0.2186	1.2103	0.9503	1.2736
2,300	2,139.4	2,808.4	8.7009	34,910.	0.1890	1.2166	0.9566	1.2718
2,400	2,235.3	2,930.4	8.7474	41,752.	0.1642	1.2228	0.9628	1.2701
2,500	2,331.9	3,052.9	8.7925	49,650.	0.1432	1.2288	0.9688	1.2684
2,600	2,429.1	3,176.1	8.8361	58,728.	0.1255	1.2346	0.9746	1.2668
2,700	2,526.8	3,299.9	8.8784	69,121.	0.1103	1.2402	0.9802	1.2653
2,800	2,625.1	3,424.1	8.9195	80,971.	0.0973	1.2456	0.9856	1.2638
2,900	2,723.9	3,549.0	8.9595	94,435.	0.0862	1.2508	0.9908	1.2624
3,000	2,823.2	3,674.3	8.9984	109,680.	0.0765	1.2557	0.9957	1.2611

Properties generated from the program *GASPROPS*, Wiley Professional Software, Wiley, New York, 1984.

TABLE D.6 Ideal Gas Properties of WATER VAPOR at Low Pressure—SI Units

T, °C	u, kJ/kg	h, kJ/kg	s_0, kJ/(kg·K)	P_0	v_0	c_P, kJ/(kg·K)	c_v, kJ/(kg·K)	k
0	377.07	503.22	10.310	0.0503	24,725.	1.8590	1.3972	1.3305
25	412.05	549.75	10.423	0.0643	21,130.	1.8644	1.4026	1.3292
50	477.22	596.46	10.470	0.0712	20,660.	1.8713	1.4095	1.3277
100	518.11	690.45	10.751	0.1309	12,988.	1.8897	1.4279	1.3234
150	590.08	785.51	11.135	0.3009	6,405.3	1.9132	1.4514	1.3182
200	663.32	881.84	11.350	0.4795	4,494.2	1.9401	1.4782	1.3124
250	737.94	979.55	11.547	0.7338	3,247.2	1.9691	1.5073	1.3064
300	814.07	1,078.8	11.728	1.0864	2,403.1	1.9995	1.5377	1.3003
350	891.69	1,179.5	11.896	1.5650	1,813.6	2.0311	1.5693	1.2943
400	971.00	1,281.9	12.054	2.2036	1,391.4	2.0636	1.6018	1.2883
450	1,051.9	1,385.9	12.203	3.0441	1,082.0	2.0969	1.6351	1.2825
500	1,134.5	1,491.6	12.345	4.1338	851.91	2.1308	1.6689	1.2767
550	1,218.8	1,599.0	12.479	5.5338	677.53	2.1654	1.7035	1.2711
600	1,304.9	1,708.1	12.608	7.3140	543.76	2.2006	1.7388	1.2656
650	1,392.7	1,819.0	12.731	9.5579	439.93	2.2361	1.7743	1.2603
700	1,482.3	1,931.7	12.850	12.367	358.43	2.2717	1.8099	1.2552
750	1,573.7	2,046.2	12.965	15.860	293.84	2.3072	1.8454	1.2503
800	1,666.8	2,162.5	13.076	20.161	242.46	2.3424	1.8806	1.2456
850	1,761.7	2,280.4	13.183	25.446	201.04	2.3772	1.9154	1.2411
900	1,858.3	2,400.1	13.288	31.904	167.49	2.4114	1.9496	1.2369
950	1,956.7	2,521.6	13.389	39.729	140.23	2.4448	1.9830	1.2329
1,000	2,056.7	2,644.6	13.488	49.192	117.89	2.4777	2.0159	1.2291
1,100	2,261.4	2,895.6	13.677	74.225	84.264	2.5402	2.0784	1.2222
1,200	2,472.2	3,152.5	13.858	109.78	61.122	2.5988	2.1370	1.2161
1,300	2,688.6	3,415.2	14.030	159.48	44.931	2.6533	2.1915	1.2107
1,400	2,910.3	3,683.1	14.195	228.04	33.419	2.7034	2.2416	1.2060
1,500	3,136.8	3,955.7	14.354	321.26	25.140	2.7498	2.2880	1.2018
1,600	3,367.8	4,232.9	14.503	444.36	19.201	2.7923	2.3304	1.1982
1,700	3,602.8	4,514.0	14.619	570.41	15.756	2.8310	2.3692	1.1949
1,800	3,841.5	4,799.0	14.793	832.17	11.347	2.8667	2.4048	1.1920
1,900	4,083.7	5,087.3	14.929	1,116.9	8.8623	2.8996	2.4378	1.1894
2,000	4,329.0	5,378.8	15.060	1,483.8	6.9780	2.9299	2.4681	1.1871
2,100	4,577.2	5,673.2	15.187	1,952.8	5.5353	2.9577	2.4959	1.1850
2,200	4,828.2	5,970.3	15.309	2,547.2	4.4225	2.9837	2.5218	1.1831
2,300	5,081.6	6,269.9	15.428	3,294.7	3.5574	3.0080	2.5461	1.1814
2,400	5,337.3	6,571.9	15.543	4,227.8	2.8799	3.0306	2.5687	1.1798
2,500	5,595.3	6,876.0	15.655	5,384.9	2.3457	3.0519	2.5901	1.1783
2,600	5,855.3	7,182.2	15.763	6,814.4	1.9205	3.0721	2.6103	1.1769
2,700	6,117.3	7,490.4	15.869	8,560.8	1.5819	3.0911	2.6293	1.1756
2,800	6,383.9	7,803.1	15.971	10,674.	1.3114	3.1095	2.6477	1.1744
2,900	6,651.2	8,116.7	16.069	13,213.	1.0939	3.1271	2.6653	1.1733
3,000	6,918.7	8,430.2	16.164	16,247.	0.9176	—	—	—

Properties generated from the program *GASPROPS*, Wiley Professional Software, Wiley, New York, 1984.

TABLE D.7 Ideal Gas Properties of CARBON DIOXIDE at Low Pressure—SI Units

T, °C	u, kJ/kg	h, kJ/kg	s_0, kJ/(kg·K)	P_0	v_0	c_P, kJ/(kg·K)	c_v, kJ/(kg·K)	k
−200	34.56	48.39	3.8607	0.0075	18,191.	0.6618	0.4728	1.3999
−150	58.63	81.91	4.2065	0.0468	4,911.	0.6712	0.4821	1.3921
−100	73.96	106.69	4.4403	0.1612	2,002.4	0.7078	0.5188	1.3644
−50	110.29	152.47	4.6262	0.4313	964.6	0.7616	0.5725	1.3302
0	140.51	192.15	4.7857	1.0033	507.6	0.8176	0.6286	1.3008
25	156.57	212.93	4.8585	1.4747	376.97	0.8443	0.6553	1.2884
50	173.27	234.36	4.9275	2.1249	283.55	0.8697	0.6807	1.2777
100	208.50	279.04	5.0560	4.1943	165.8	0.9166	0.7275	1.2599
150	245.94	325.94	5.1738	7.8275	100.79	0.9584	0.7694	1.2457
200	285.37	374.82	5.2830	13.948	63.25	0.9960	0.8070	1.2343
250	326.58	425.48	5.3847	23.903	40.80	1.0299	0.8409	1.2248
300	369.40	477.75	5.4801	39.606	26.98	1.0606	0.8716	1.2169
350	413.69	531.50	5.5700	63.737	18.23	1.0886	0.8996	1.2102
400	459.32	586.57	5.6550	99.962	12.55	1.1141	0.9251	1.2044
450	506.16	642.87	5.7357	153.20	8.8011	1.1375	0.9484	1.1993
500	554.12	700.29	5.8125	230.00	6.2676	1.1588	0.9697	1.1949
550	603.10	758.72	5.8857	338.87	4.5292	1.1782	0.9892	1.1911
600	653.02	818.08	5.9557	490.89	3.3164	1.1961	1.0071	1.1877
650	703.78	878.30	6.0228	700.12	2.4585	1.2124	1.0234	1.1847
700	755.33	939.30	6.0871	984.37	1.8433	1.2273	1.0383	1.1821
750	807.58	1,001.0	6.1472	1,352.6	1.4103	1.2410	1.0520	1.1797
800	860.48	1,063.4	6.2153	1,939.5	1.0317	1.2535	1.0645	1.1776
850	913.99	1,126.3	6.2345	2,147.4	0.9752	1.2650	1.0760	1.1757
900	968.07	1,189.9	6.2842	2,794.0	0.7829	1.2756	1.0866	1.1740
950	1,022.7	1,253.9	6.3821	4,689.7	0.4863	1.2855	1.0964	1.1724
1,000	1,077.7	1,318.4	6.4243	5,865.1	0.4047	1.2946	1.1055	1.1710
1,100	1,189.1	1,448.6	6.5248	9,983.1	0.2565	1.3105	1.1214	1.1686
1,200	1,301.9	1,580.4	6.6174	16,299.	0.1685	1.3240	1.1349	1.1666
1,300	1,416.0	1,713.4	6.7047	25,872.	0.1134	1.3359	1.1469	1.1648
1,400	1,531.3	1,847.6	6.7873	40,053.	0.0779	1.3462	1.1571	1.1634
1,500	1,647.4	1,982.6	6.8657	60,656.	0.0545	1.3551	1.1661	1.1621
1,600	1,764.8	2,118.9	6.9403	90,050.	0.0388	1.3633	1.1742	1.1610
1,700	1,882.5	2,255.5	7.0114	131,170.	0.0280	1.3704	1.1813	1.1600
1,800	1,999.7	2,391.6	7.0793	187,920.	0.0206	1.3767	1.1877	1.1592
1,900	2,122.2	2,533.0	7.1443	265,110.	0.0153	1.3824	1.1934	1.1584
2,000	2,223.2	2,653.0	7.2067	368,730.	0.0115	1.3876	1.1986	1.1577
2,100	2,354.0	2,802.6	7.2665	506,000.	0.0087	1.3923	1.2033	1.1571
2,200	2,481.3	2,948.9	7.3241	686,580.	0.0067	1.3965	1.2075	1.1566
2,300	2,600.5	3,087.0	7.3795	920,640.	0.0052	1.4006	1.2115	1.1560
2,400	2,722.4	3,227.7	7.4329	1,221,400.	0.0041	1.4043	1.2152	1.1556
2,500	2,843.9	3,368.2	7.4846	1,605,700.	0.0032	1.4077	1.2187	1.1551
2,600	2,966.0	3,509.1	7.5346	2,092,100.	0.0026	1.4110	1.2219	1.1547
2,700	3,088.3	3,650.4	7.5829	2,701,300.	0.0021	1.4140	1.2250	1.1543
2,800	3,211.0	3,792.0	7.6296	3,458,500.	0.0017	1.4169	1.2279	1.1540
2,900	3,333.9	3,933.8	7.6751	4,399,900.	0.0013	1.4196	1.2306	1.1536
3,000	3,457.1	4,075.9	7.7193	5,561,300.	0.0011	1.4222	1.2331	1.1533

Properties generated from the program *GASPROPS*, Wiley Professional Software, Wiley, New York, 1984.

TABLE D.8 Ideal Gas Properties of CARBON MONOXIDE at Low Pressure—SI Units

T, °C	u, kJ/kg	h, kJ/kg	s_0, kJ/(kg·K)	P_0	v_0	c_P, kJ/(kg·K)	c_v, kJ/(kg·K)	k
−200	54.32	76.05	5.5948	0.0153	13,973.	1.0396	0.7426	1.4000
−150	91.45	128.03	6.1363	0.0951	3,794.9	1.0396	0.7426	1.4000
−100	128.60	180.03	6.4908	0.3138	1,616.4	1.0398	0.7428	1.3999
−50	165.74	232.02	6.7546	0.7632	856.52	1.0400	0.7429	1.3998
−0	202.90	284.03	6.9649	1.5500	516.25	1.0405	0.7434	1.3995
25	221.49	310.05	7.0561	2.1071	414.52	1.0411	0.7440	1.3992
50	240.10	336.09	7.1399	2.7950	338.70	1.0421	0.7450	1.3987
100	277.43	388.27	7.2901	4.6350	235.85	1.0455	0.7485	1.3968
150	314.99	440.68	7.4219	7.2258	171.55	1.0514	0.7543	1.3938
200	352.91	493.45	7.5397	10.748	128.97	1.0596	0.7626	1.3895
250	391.28	546.68	7.6467	15.409	99.460	1.0699	0.7729	1.3843
300	430.22	600.46	7.7448	21.450	78.278	1.0818	0.7847	1.3785
350	469.78	654.87	7.8359	29.146	62.634	1.0947	0.7976	1.3724
400	509.99	709.94	7.9208	38.809	50.813	1.1081	0.8111	1.3662
450	550.88	765.68	8.0007	50.792	41.709	1.1216	0.8246	1.3602
500	592.45	822.10	8.0762	65.488	34.586	1.1350	0.8380	1.3545
550	634.68	879.18	8.1477	83.333	28.937	1.1480	0.8509	1.3491
600	677.54	936.89	8.2157	104.81	24.405	1.1604	0.8634	1.3440
650	721.00	995.21	8.2807	130.44	20.732	1.1722	0.8752	1.3394
700	765.04	1,054.1	8.3428	160.81	17.728	1.1834	0.8863	1.3351
750	809.62	1,113.5	8.4024	196.54	15.251	1.1938	0.8968	1.3312
800	854.71	1,173.5	8.4596	238.30	13.193	1.2036	0.9066	1.3276
850	900.27	1,233.9	8.5146	286.83	11.471	1.2128	0.9157	1.3244
900	946.28	1,294.7	8.5676	342.91	10.022	1.2213	0.9243	1.3214
950	992.69	1,356.0	8.6187	407.39	8.7957	1.2293	0.9322	1.3186
1,000	1,039.5	1,417.7	8.6681	481.15	7.7516	1.2367	0.9396	1.3161
1,100	1,134.1	1,542.0	8.7621	660.45	6.0908	1.2500	0.9530	1.3117
1,200	1,230.0	1,667.6	8.8504	889.22	4.8533	1.2616	0.9646	1.3079
1,300	1,327.0	1,794.3	8.9336	1,176.9	3.9159	1.2718	0.9748	1.3047
1,400	1,424.9	1,921.9	9.0123	1,534.0	3.1953	1.2807	0.9837	1.3019
1,500	1,523.7	2,050.4	9.0869	1,972.1	2.6340	1.2886	0.9916	1.2995
1,600	1,623.2	2,179.6	9.1578	2,504.1	2.1913	1.2957	0.9986	1.2974
1,700	1,723.4	2,309.5	9.2253	3,144.1	1.8385	1.3019	1.0049	1.2956
1,800	1,824.2	2,440.0	9.2898	3,907.3	1.5544	1.3076	1.0106	1.2939
1,900	1,925.5	2,571.0	9.3515	4,810.3	1.3235	1.3127	1.0157	1.2924
2,000	2,027.3	2,702.5	9.4107	5,871.3	1.1342	1.3174	1.0203	1.2911
2,100	2,129.6	2,834.5	9.4675	7,109.7	0.9778	1.3216	1.0246	1.2899
2,200	2,232.2	2,966.8	9.5221	8,546.3	0.8477	1.3255	1.0285	1.2888
2,300	2,335.3	3,099.6	9.5748	10,204.	0.7388	1.3291	1.0321	1.2878
2,400	2,438.6	3,232.6	9.6255	12,106.	0.6469	1.3325	1.0354	1.2869
2,500	2,542.3	3,366.1	9.6745	14,278.	0.5690	1.3356	1.0386	1.2860
2,600	2,646.3	3,499.8	9.7219	16,749.	0.5025	1.3385	1.0415	1.2852
2,700	2,750.6	3,633.8	9.7677	19,546.	0.4456	1.3413	1.0442	1.2844
2,800	2,855.2	3,768.0	9.8121	22,700.	0.3966	1.3439	1.0468	1.2837
2,900	2,960.0	3,902.5	9.8552	26,245.	0.3542	1.3463	1.0493	1.2831
3,000	3,065.0	4,037.3	9.8970	30,215.	0.3174	1.3487	1.0516	1.2824

TABLE D.9 Properties of WATER in Compressed Liquid State—SI Units

P, kPa	T, °C	v, m³/kg	u, kJ/kg	h, kJ/kg	s, kJ/(kg·K)
5,000.0					
	10	0.000998	41.860	46.850	0.1505
	30	0.001002	125.21	130.22	0.4350
	50	0.001010	208.58	213.63	0.7014
	70	0.001020	291.97	297.07	0.9520
	90	0.001034	375.59	380.75	1.1890
	110	0.001049	459.63	464.88	1.4145
	130	0.001067	544.28	549.61	1.6301
	150	0.001087	629.68	635.12	1.8371
	170	0.001111	716.04	721.59	2.0367
	190	0.001138	803.64	809.33	2.2304
	210	0.001169	892.89	898.74	2.4193
	230	0.001206	984.40	990.43	2.6053
	250	0.001250	1079.0	1085.3	2.7901
10,000					
	10	0.000996	41.729	51.685	0.1500
	30	0.001000	124.75	134.75	0.4334
	50	0.001008	207.85	217.93	0.6991
	70	0.001018	290.98	301.16	0.9491
	90	0.001031	374.32	384.63	1.1855
	110	0.001046	458.07	468.53	1.4104
	130	0.001064	542.38	553.02	1.6253
	150	0.001084	627.41	638.25	1.8316
	170	0.001107	713.31	724.39	2.0305
	190	0.001133	800.37	811.70	2.2232
	210	0.001164	888.94	900.57	2.4111
	230	0.001199	979.55	991.54	2.5955
	250	0.001241	1073.0	1085.4	2.7784
	270	0.001292	1170.3	1183.2	2.9619
	290	0.001357	1273.5	1287.1	3.1497
15,000					
	10	0.000993	41.593	56.493	0.1494
	30	0.000998	124.30	139.27	0.4319
	50	0.001006	207.13	222.22	0.6968
	70	0.001016	290.01	305.25	0.9461
	90	0.001029	373.08	388.51	1.1820
	110	0.001044	456.54	472.20	1.4063
	130	0.001061	540.53	556.45	1.6206
	150	0.001081	625.19	641.41	1.8263
	170	0.001104	710.67	727.22	2.0244
	190	0.001129	797.20	814.14	2.2163
	210	0.001159	885.13	902.51	2.4030
	230	0.001193	974.91	992.80	2.5862
	250	0.001233	1067.2	1085.7	2.7672
	270	0.001281	1163.0	1182.2	2.9481
	290	0.001341	1263.7	1283.8	3.1319
	310	0.001421	1372.1	1393.5	3.3231
	330	0.001539	1494.7	1517.8	3.5327

TABLE D.9 Properties of WATER in Compressed Liquid State — SI Units (*Continued*)

P, kPa	*T*, °C	*v*, m³/kg	*u*, kJ/kg	*h*, kJ/kg	*s*, kJ/(kg·K)
20,000					
	10	0.000991	41.453	61.274	0.1488
	30	0.000996	123.86	143.77	0.4303
	50	0.001003	206.43	226.50	0.6946
	70	0.001014	289.05	309.33	0.9432
	90	0.001026	371.86	392.39	1.1785
	110	0.001041	455.04	475.87	1.4023
	130	0.001058	538.73	559.90	1.6160
	150	0.001078	623.03	644.59	1.8211
	170	0.001100	708.10	730.10	2.0185
	190	0.001125	794.14	816.64	2.2095
	210	0.001154	881.46	904.53	2.3953
	230	0.001187	970.48	994.21	2.5771
	250	0.001225	1061.8	1086.3	2.7565
	270	0.001271	1156.1	1181.5	2.9352
	290	0.001327	1254.7	1281.3	3.1155
	310	0.001399	1359.7	1387.7	3.3011
	330	0.001499	1475.0	1505.0	3.4988
	350	0.001665	1612.1	1645.4	3.7277
30,000					
	10	0.000987	41.158	70.757	0.1473
	30	0.000992	122.99	152.73	0.4271
	50	0.000999	205.07	235.05	0.6900
	70	0.001010	287.20	317.49	0.9375
	90	0.001022	369.51	400.16	1.1717
	110	0.001036	452.15	483.24	1.3944
	130	0.001053	535.24	566.83	1.6070
	150	0.001072	618.87	651.03	1.8109
	170	0.001093	703.17	735.96	2.0070
	190	0.001117	788.29	821.81	2.1965
	210	0.001144	874.50	908.83	2.3804
	230	0.001175	962.13	997.39	2.5600
	250	0.001211	1051.6	1088.0	2.7365
	270	0.001253	1143.6	1181.2	2.9113
	290	0.001303	1238.8	1277.9	3.0862
	310	0.001364	1338.6	1379.5	3.2635
	330	0.001443	1444.9	1488.1	3.4466
	350	0.001552	1561.5	1608.1	3.6421
	370	0.001726	1697.8	1749.6	3.8656
50,000					
	10	0.000978	40.519	89.428	0.1439
	30	0.000984	121.31	170.48	0.4205
	50	0.000991	202.46	252.03	0.6810
	70	0.001001	283.69	333.76	0.9264
	90	0.001013	365.06	415.73	1.1585
	110	0.001027	446.70	498.06	1.3792
	130	0.001043	528.70	580.85	1.5898

TABLE D.9 Properties of WATER in Compressed Liquid State — SI Units (*Continued*)

P, kPa	*T*, °C	*v*, m³/kg	*u*, kJ/kg	*h*, kJ/kg	*s*, kJ/(kg·K)
	150	0.001061	611.12	664.16	1.7915
	170	0.001081	694.04	748.07	1.9853
	190	0.001103	777.56	832.70	2.1720
	210	0.001128	861.87	918.25	2.3529
	230	0.001155	947.19	1005.0	2.5287
	250	0.001187	1033.8	1093.2	2.7006
	270	0.001223	1122.1	1183.3	2.8696
	290	0.001264	1212.5	1275.8	3.0368
	310	0.001313	1305.6	1371.3	3.2035
	330	0.001371	1402.2	1470.7	3.3712
	350	0.001442	1503.2	1575.3	3.5417
	370	0.001533	1610.2	1686.9	3.7179

TABLE D.10 Properties of SATURATED WATER–Temperature Table, SI Units

T °C	P kPa	v_l m³/kg	v_{lg} m³/kg	v_g m³/kg	u_l kJ/kg	u_{lg} kJ/kg	u_g kJ/kg	h_l kJ/kg	h_{lg} kJ/kg	h_g kJ/kg	s_l kJ/(kg·K)	s_{lg} kJ/(kg·K)	s_g kJ/(kg·K)
5	0.8726	0.001000	147.02	147.02	21.020	2360.4	2381.4	21.021	2488.7	2509.7	0.07626	8.9473	9.0236
10	1.2281	0.001000	106.32	106.32	41.986	2346.3	2388.3	41.988	2476.9	2518.9	0.1510	8.7476	8.8986
15	1.7056	0.001001	77.896	77.897	62.915	2332.3	2395.2	62.917	2465.1	2528.0	0.2242	8.5550	8.7792
20	2.3388	0.001002	57.777	57.778	83.833	2318.2	2402.0	83.835	2453.4	2537.2	0.2962	8.3689	8.6651
25	3.1690	0.001003	43.356	43.357	104.75	2304.1	2408.9	104.75	2441.6	2546.3	0.3670	8.1888	8.5558
30	4.2455	0.001004	32.895	32.896	125.67	2290.0	2415.7	125.67	2429.6	2555.3	0.4365	8.0148	8.4513
35	5.6267	0.001006	25.219	25.220	146.58	2275.9	2422.5	146.59	2417.8	2564.4	0.5050	7.8461	8.3511
40	7.3814	0.001008	19.527	19.528	167.50	2261.7	2429.2	167.50	2405.9	2573.4	0.5723	7.6827	8.2550
45	9.5898	0.001010	15.262	15.263	188.41	2247.5	2435.9	188.42	2393.9	2582.3	0.6385	7.5244	8.1629
50	12.344	0.001012	12.036	12.037	209.31	2233.3	2442.6	209.33	2381.9	2591.2	0.7037	7.3708	8.0745
55	15.752	0.001015	9.5716	9.5726	230.22	2219.0	2449.2	230.24	2369.8	2600.0	0.7679	7.2217	7.9896
60	19.932	0.001017	7.6733	7.6743	251.13	2204.7	2455.8	251.15	2357.7	2608.8	0.8312	7.0768	7.9080
65	25.022	0.001020	6.1986	6.1996	272.05	2190.3	2462.4	272.08	2345.4	2617.5	0.8935	6.9360	7.8295
70	31.176	0.001023	5.0437	5.0447	292.98	2175.8	2468.8	293.01	2333.1	2626.1	0.9549	6.7991	7.7540
75	38.563	0.001026	4.1323	4.1333	313.92	2161.3	2475.2	313.96	2320.6	2634.6	1.0155	6.6658	7.6813
80	47.373	0.001029	3.4078	3.4088	334.88	2146.7	2481.6	334.93	2308.2	2643.1	1.0753	6.5359	7.6112
85	57.815	0.001032	2.8279	2.8289	355.86	2132.0	2487.9	355.92	2295.5	2651.4	1.1343	6.4093	7.5436
90	70.117	0.001036	2.3607	2.3617	376.86	2117.1	2494.0	376.93	2282.7	2659.6	1.1925	6.2859	7.4784
95	84.529	0.001040	1.9818	1.9828	397.89	2102.2	2500.1	397.98	2269.7	2667.7	1.2501	6.1653	7.4154
100	101.32	0.001043	1.6726	1.6736	418.96	2087.1	2506.1	419.06	2256.6	2675.7	1.3069	6.0476	7.3545
105	120.79	0.001047	1.4190	1.4200	440.05	2072.1	2512.1	440.18	2243.4	2683.6	1.3630	5.9326	7.2956
110	143.24	0.001052	1.2095	1.2106	461.19	2056.7	2517.9	461.34	2230.0	2691.3	1.4186	5.8200	7.2386
115	169.02	0.001056	1.0359	1.0370	482.36	2041.1	2523.5	482.54	2216.3	2698.8	1.4735	5.7098	7.1833
120	198.48	0.001060	0.8911	0.8922	503.57	2025.5	2529.1	503.78	2202.4	2706.2	1.5278	5.6019	7.1297
125	232.01	0.001065	0.7698	0.7709	524.82	2009.7	2534.5	525.07	2188.3	2713.4	1.5815	5.4962	7.0777
130	270.02	0.001070	0.6676	0.6687	546.12	1993.7	2539.8	546.41	2174.0	2720.4	1.6346	5.3926	7.0272
135	312.93	0.001075	0.5813	0.5824	567.46	1977.5	2545.0	567.80	2159.4	2727.2	1.6873	5.2907	6.9780
140	361.19	0.001080	0.5079	0.5090	588.85	1961.2	2550.0	589.24	2144.6	2733.8	1.7394	5.1908	6.9302
145	415.29	0.001085	0.4453	0.4464	610.30	1944.5	2554.8	610.75	2129.4	2740.2	1.7910	5.0926	6.8836
150	475.72	0.001090	0.3918	0.3929	631.80	1927.7	2559.5	632.32	2114.1	2746.4	1.8421	4.9960	6.8381
155	542.99	0.001096	0.3457	0.3468	653.35	1910.7	2564.0	653.95	2098.4	2752.3	1.8927	4.9010	6.7937
160	617.66	0.001102	0.3060	0.3071	674.97	1893.3	2568.3	675.65	2082.3	2758.0	1.9429	4.8074	6.7503
165	700.29	0.001108	0.2716	0.2727	696.65	1875.7	2572.4	697.43	2065.9	2763.3	1.9927	4.7151	6.7078
170	791.47	0.001114	0.2417	0.2428	718.40	1857.9	2576.3	719.28	2049.2	2768.5	2.0421	4.6241	6.6662
175	891.80	0.001121	0.2157	0.2168	740.22	1839.7	2579.9	741.22	2032.1	2773.3	2.0910	4.5344	6.6254
180	1001.9	0.001127	0.1929	0.1940	762.12	1821.3	2583.4	763.25	2014.6	2777.8	2.1397	4.4456	6.5853
185	1122.5	0.001134	0.1730	0.1741	784.10	1802.5	2586.6	785.37	1996.6	2782.0	2.1879	4.3580	6.5459
190	1254.2	0.001141	0.1554	0.1565	806.17	1783.4	2589.6	807.60	1978.2	2785.8	2.2358	4.2713	6.5071
195	1397.6	0.001149	0.1399	0.1410	828.33	1764.0	2592.3	829.93	1959.5	2789.4	2.2834	4.1855	6.4689
200	1553.6	0.001156	0.1261	0.1273	850.58	1744.1	2594.7	852.38	1940.1	2792.5	2.3308	4.1004	6.4312
205	1722.9	0.001164	0.1140	0.1152	872.95	1723.9	2596.9	874.95	1920.4	2795.3	2.3778	4.0162	6.3940
210	1906.2	0.001173	0.1032	0.1044	895.43	1703.3	2598.7	897.66	1900.0	2797.7	2.4246	3.9326	6.3572
215	2104.2	0.001181	0.09357	0.09475	918.02	1682.3	2600.3	920.51	1879.2	2799.7	2.4712	3.8496	6.3208
220	2317.8	0.001190	0.08497	0.08616	940.75	1660.9	2601.6	943.51	1857.8	2801.3	2.5175	3.7672	6.2847
225	2547.9	0.001199	0.07726	0.07846	963.61	1638.9	2602.5	966.67	1835.7	2802.4	2.5637	3.6851	6.2488
230	2795.1	0.001209	0.07034	0.07155	986.62	1616.5	2603.1	990.00	1813.1	2803.1	2.6097	3.6034	6.2131
235	3060.4	0.001219	0.06412	0.06534	1009.8	1593.5	2603.3	1013.5	1789.8	2803.3	2.6556	3.5221	6.1777
240	3344.7	0.001229	0.05851	0.05974	1033.1	1570.0	2603.1	1037.2	1765.8	2803.0	2.7013	3.4410	6.1423
245	3648.8	0.001240	0.05345	0.05469	1056.6	1546.0	2602.6	1061.2	1740.9	2802.1	2.7470	3.3600	6.1070
250	3973.6	0.001251	0.04886	0.05011	1080.3	1521.3	2601.6	1085.3	1715.4	2800.7	2.7926	3.2791	6.0717
255	4320.2	0.001263	0.04470	0.04596	1104.3	1495.9	2600.2	1109.7	1689.1	2798.8	2.8382	3.1981	6.0363
260	4689.4	0.001276	0.04091	0.04219	1128.4	1470.0	2598.4	1134.4	1661.8	2796.2	2.8838	3.1171	6.0009
265	5082.3	0.001289	0.03747	0.03876	1152.8	1443.2	2596.0	1159.3	1633.7	2793.0	2.9294	3.0358	5.9652
270	5499.9	0.001303	0.03434	0.03564	1177.4	1415.8	2593.2	1184.6	1604.5	2789.1	2.9751	2.9542	5.9293
275	5943.1	0.001317	0.03146	0.03278	1202.3	1387.4	2589.7	1210.1	1574.4	2784.5	3.0209	2.8722	5.8931
280	6413.2	0.001332	0.02883	0.03016	1227.5	1358.2	2585.7	1236.1	1543.1	2779.2	3.0669	2.7896	5.8565
285	6911.1	0.001349	0.02642	0.02777	1253.1	1328.0	2581.1	1262.4	1510.6	2773.0	3.1131	2.7064	5.8195

TABLE D.10 Properties of SATURATED WATER–Temperature Table, SI Units (*Continued*)

T °C	P kPa	v_l m³/kg	v_{lg} m³/kg	v_g m³/kg	u_l kJ/kg	u_{lg} kJ/kg	u_g kJ/kg	h_l kJ/kg	h_{lg} kJ/kg	h_g kJ/kg	s_l kJ/(kg·K)	s_{lg} kJ/(kg·K)	s_g kJ/(kg·K)
290	7438.0	0.001366	0.02419	0.02556	1279.0	1296.7	2575.7	1289.1	1476.8	2765.9	3.1595	2.6223	5.7818
295	7995.2	0.001384	0.02216	0.02354	1305.3	1264.4	2569.7	1316.3	1441.5	2757.8	3.2062	2.5372	5.7434
300	8583.8	0.001404	0.02027	0.02167	1332.0	1230.8	2562.8	1344.1	1404.6	2748.7	3.2534	2.4508	5.7042
305	9205.1	0.001425	0.01852	0.01994	1359.2	1195.8	2555.0	1372.3	1366.2	2738.5	3.3010	2.3630	5.6640
310	9860.5	0.001447	0.01689	0.01834	1387.0	1159.2	2546.2	1401.2	1325.8	2727.0	3.3491	2.2735	5.6226
315	10,550	0.001472	0.01539	0.01686	1415.3	1121.0	2536.3	1430.8	1283.4	2714.2	3.3979	2.1820	5.5799
320	11,280	0.001498	0.01398	0.01548	1444.4	1080.8	2525.2	1461.3	1238.4	2699.7	3.4476	2.0880	5.5356
325	12,050	0.001528	0.01266	0.01419	1474.2	1038.4	2512.6	1492.6	1190.9	2683.5	3.4983	1.9910	5.4893
330	12,850	0.001560	0.01142	0.01298	1504.9	993.50	2498.4	1525.0	1140.3	2665.3	3.5501	1.8906	5.4407
335	13,700	0.001596	0.01025	0.01185	1536.8	945.50	2482.3	1558.6	1086.1	2644.7	3.6035	1.7859	5.3894
340	14,590	0.001637	0.009153	0.01079	1569.9	894.00	2463.9	1593.8	1027.5	2621.3	3.6587	1.6758	5.3345
345	15,530	0.001684	0.008094	0.009778	1604.7	838.00	2442.7	1630.9	963.60	2594.5	3.7164	1.5589	5.2753
350	16,520	0.001740	0.007072	0.008812	1641.7	776.20	2417.9	1670.4	893.10	2563.5	3.7774	1.4331	5.2105
355	17,560	0.001808	0.006071	0.007879	1681.5	706.90	2388.4	1713.3	813.40	2526.7	3.8429	1.2950	5.1379
360	18,660	0.001894	0.005068	0.006962	1725.6	626.60	2352.2	1761.0	721.00	2482.0	3.9153	1.1389	5.0542
365	19,810	0.002012	0.004017	0.006029	1776.8	528.40	2305.2	1816.7	607.90	2424.6	3.9994	0.9526	4.9520
370	21,030	0.002207	0.002786	0.004993	1843.3	391.90	2235.2	1889.7	450.50	2340.2	4.1094	0.7004	4.8098
373	22,055	0.00311	—	0.00311	2017	—	2017	2086	—	2086	4.409	—	4.409

TABLE D.11 Properties of SATURATED WATER–Pressure Table, SI Units

P kPa	T °C	v_l m³/kg	v_{lg} m³/kg	v_g m³/kg	u_l kJ/kg	u_{lg} kJ/kg	u_g kJ/kg	h_l kJ/kg	h_{lg} kJ/kg	h_g kJ/kg	s_l kJ/(kg·K)	s_{lg} kJ/(kg·K)	s_g kJ/(kg·K)
1.0	6.9696	0.001000	129.19	129.19	29.287	2354.8	2384.1	29.288	2484.0	2513.3	0.1059	8.8678	8.9737
1.5	13.021	0.001001	87.970	87.971	54.634	2337.9	2392.5	54.635	2469.8	2524.4	0.1954	8.6304	8.8258
2.0	17.497	0.001001	66.997	66.998	73.364	2325.2	2398.6	73.366	2459.2	2532.6	0.2603	8.4613	8.7216
2.5	21.080	0.001002	54.248	54.249	88.353	2315.1	2403.5	88.356	2450.7	2539.1	0.3116	8.3295	8.6411
3.0	24.083	0.001003	45.660	45.661	100.92	2306.7	2407.6	100.92	2443.7	2544.6	0.3541	8.2214	8.5755
3.5	26.677	0.001003	39.473	39.474	111.77	2299.4	2411.2	111.77	2437.5	2549.3	0.3904	8.1299	8.5203
4.0	28.966	0.001004	34.797	34.798	121.34	2293.0	2414.3	121.35	2432.2	2553.5	0.4222	8.0503	8.4725
4.5	31.018	0.001005	31.136	31.137	129.93	2287.2	2417.1	129.93	2427.3	2557.2	0.4506	7.9799	8.4305
5.0	32.881	0.001005	28.190	28.191	137.72	2281.9	2419.6	137.72	2422.8	2560.5	0.4761	7.9169	8.3930
5.5	34.589	0.001006	25.767	25.768	144.86	2277.0	2421.9	144.87	2418.7	2563.6	0.4994	7.8598	8.3592
6.0	36.167	0.001006	23.737	23.738	151.46	2272.5	2424.0	151.47	2415.0	2566.5	0.5208	7.8075	8.3283
6.5	37.635	0.001007	22.013	22.014	157.60	2268.4	2426.0	157.61	2411.5	2569.1	0.5406	7.7594	8.3000
7.0	39.008	0.001008	20.528	20.529	163.35	2264.5	2427.9	163.35	2408.3	2571.6	0.5590	7.7148	8.2738
7.5	40.299	0.001008	19.236	19.237	168.75	2260.9	2429.6	168.76	2405.1	2573.9	0.5763	7.6731	8.2494
8.0	41.518	0.001008	18.102	18.103	173.85	2257.5	2431.3	173.85	2402.3	2576.1	0.5925	7.6342	8.2267
8.5	42.673	0.001009	17.098	17.099	178.68	2254.1	2432.8	178.68	2399.4	2578.1	0.6078	7.5975	8.2053
9.0	43.771	0.001009	16.202	16.203	183.27	2251.0	2434.3	183.27	2396.8	2580.1	0.6223	7.5629	8.1852
9.5	44.817	0.001010	15.398	15.399	187.64	2248.1	2435.7	187.65	2394.4	2582.0	0.6361	7.5301	8.1662
10	45.817	0.001010	14.673	14.674	191.82	2245.2	2437.0	191.83	2392.0	2583.8	0.6493	7.4989	8.1482
15	53.983	0.001014	10.022	10.023	225.97	2221.9	2447.9	225.98	2372.2	2598.2	0.7550	7.2516	8.0066
20	60.073	0.001017	7.6489	7.6499	251.44	2204.5	2455.9	251.46	2357.4	2608.9	0.8321	7.0747	7.9068
25	64.980	0.001020	6.2038	6.2048	271.97	2190.3	2462.3	271.99	2345.4	2617.4	0.8933	6.9365	7.8298
30	69.114	0.001022	5.2288	5.2298	289.27	2178.4	2467.7	289.30	2335.3	2624.6	0.9441	6.8231	7.7672
35	72.700	0.001024	4.5252	4.5262	304.28	2168.0	2472.3	304.32	2326.4	2630.7	0.9878	6.7266	7.7144
40	75.877	0.001026	3.9930	3.9940	317.59	2158.8	2476.4	317.64	2318.5	2636.1	1.0261	6.6427	7.6688
45	78.736	0.001028	3.5759	3.5769	329.58	2150.4	2480.0	329.62	2311.3	2640.9	1.0603	6.5684	7.6287
50	81.339	0.001030	3.2398	3.2408	340.49	2142.8	2483.3	340.54	2304.8	2645.3	1.0912	6.5016	7.5928
60	85.949	0.001033	2.7314	2.7324	359.84	2129.2	2489.0	359.90	2293.1	2653.0	1.1454	6.3856	7.5310
70	89.956	0.001036	2.3644	2.3654	376.68	2117.3	2494.0	376.75	2282.9	2659.6	1.1920	6.2869	7.4789
80	93.511	0.001038	2.0866	2.0876	391.63	2106.7	2498.3	391.71	2273.6	2665.3	1.2330	6.2009	7.4339
90	96.713	0.001041	1.8688	1.8698	405.11	2097.1	2502.2	405.20	2265.3	2670.5	1.2696	6.1247	7.3943
100	99.632	0.001043	1.6933	1.6943	417.41	2088.3	2505.7	417.51	2257.6	2675.1	1.3027	6.0562	7.3589
101.32	100.00	0.001043	1.6727	1.6737	418.96	2087.1	2506.1	419.06	2256.6	2675.7	1.3069	6.0476	7.3545
125	105.99	0.001048	1.3742	1.3752	444.25	2068.9	2513.2	444.38	2240.7	2685.1	1.3741	5.9100	7.2841
150	111.38	0.001053	1.1584	1.1595	467.02	2052.4	2519.4	467.18	2226.2	2693.4	1.4338	5.7894	7.2232
175	116.07	0.001057	1.0027	1.0038	486.89	2037.8	2524.7	487.08	2213.3	2700.4	1.4851	5.6866	7.1717
200	120.24	0.001060	0.8848	0.8859	504.59	2024.8	2529.4	504.80	2201.7	2706.5	1.5304	5.5968	7.1272
225	124.01	0.001064	0.7923	0.7934	520.59	2012.9	2533.5	520.83	2191.2	2712.0	1.5708	5.5172	7.0880
250	127.44	0.001067	0.7177	0.7188	535.22	2001.9	2537.1	535.49	2181.3	2716.8	1.6075	5.4454	7.0529
275	130.61	0.001070	0.6563	0.6574	548.73	1991.8	2540.5	549.02	2172.3	2721.3	1.6411	5.3800	7.0211
300	133.56	0.001073	0.6048	0.6059	561.29	1982.2	2543.5	561.61	2163.7	2725.3	1.6721	5.3200	6.9921
325	136.31	0.001076	0.5609	0.5620	573.04	1973.3	2546.3	573.39	2155.6	2729.0	1.7009	5.2645	6.9654
350	138.89	0.001079	0.5232	0.5243	584.10	1964.8	2548.9	584.48	2147.9	2732.4	1.7278	5.2129	6.9407
375	141.33	0.001081	0.4903	0.4914	594.56	1956.7	2551.3	594.96	2140.6	2735.6	1.7531	5.1646	6.9177
400	143.64	0.001084	0.4614	0.4625	604.47	1949.0	2553.5	604.91	2133.6	2738.5	1.7770	5.1191	6.8961
425	145.84	0.001086	0.4357	0.4368	613.91	1941.7	2555.6	614.37	2126.9	2741.3	1.7996	5.0762	6.8758
450	147.94	0.001088	0.4129	0.4140	622.93	1934.7	2557.6	623.42	2120.5	2743.9	1.8211	5.0356	6.8567
475	149.94	0.001090	0.3923	0.3934	631.56	1927.8	2559.4	632.07	2114.2	2746.3	1.8415	4.9971	6.8386
500	151.87	0.001093	0.3738	0.3749	639.84	1921.4	2561.2	640.38	2108.2	2748.6	1.8610	4.9604	6.8214
550	155.49	0.001097	0.3415	0.3426	655.48	1908.9	2564.4	656.08	2096.8	2752.9	1.8977	4.8917	6.7894
600	158.86	0.001101	0.3145	0.3156	670.05	1897.3	2567.3	670.71	2086.0	2756.7	1.9315	4.8286	6.7601
650	162.02	0.001104	0.2915	0.2926	683.71	1886.2	2569.9	684.42	2075.8	2760.2	1.9631	4.7699	6.7330
700	164.98	0.001108	0.2717	0.2728	696.58	1875.8	2572.4	697.35	2066.0	2763.3	1.9925	4.7154	6.7079
750	167.79	0.001111	0.2544	0.2555	708.76	1865.8	2574.6	709.59	2056.6	2766.2	2.0203	4.6642	6.6845
800	170.44	0.001115	0.2393	0.2404	720.33	1856.3	2576.6	721.23	2047.7	2768.9	2.0464	4.6161	6.6625
850	172.97	0.001118	0.2258	0.2269	731.37	1847.1	2578.5	732.32	2039.1	2771.4	2.0712	4.5706	6.6418
900	175.39	0.001121	0.2138	0.2149	741.92	1838.3	2580.2	742.93	2030.7	2773.6	2.0948	4.5274	6.6222

TABLE D.11 Properties of SATURATED WATER – Pressure Table, SI Units (*Continued*)

P kPa	T °C	v_l m³/kg	v_{lg} m³/kg	v_g m³/kg	u_l kJ/kg	u_{lg} kJ/kg	u_g kJ/kg	h_l kJ/kg	h_{lg} kJ/kg	h_g kJ/kg	s_l kJ/(kg·K)	s_{lg} kJ/(kg·K)	s_g kJ/(kg·K)
950	177.70	0.001124	0.2030	0.2041	752.03	1829.8	2581.8	753.10	2022.6	2775.7	2.1173	4.4863	6.6036
1000	179.92	0.001127	0.1933	0.1944	761.75	1821.6	2583.3	762.88	2014.8	2777.7	2.1388	4.4471	6.5859
1100	184.10	0.001133	0.1764	0.1775	780.14	1805.9	2586.0	781.38	1999.8	2781.2	2.1793	4.3736	6.5529
1200	188.00	0.001138	0.1622	0.1633	797.31	1791.1	2588.4	798.68	1985.6	2784.3	2.2167	4.3059	6.5226
1300	191.64	0.001144	0.1501	0.1512	813.44	1777.1	2590.5	814.93	1972.1	2787.0	2.2515	4.2430	6.4945
1400	195.08	0.001149	0.1397	0.1408	828.67	1763.6	2592.3	830.28	1959.1	2789.4	2.2842	4.1841	6.4683
1500	198.33	0.001154	0.1305	0.1317	843.12	1750.8	2593.9	844.85	1946.7	2791.5	2.3150	4.1288	6.4438
1600	201.41	0.001159	0.1225	0.1237	856.88	1738.4	2595.3	858.73	1934.6	2793.3	2.3441	4.0766	6.4207
1700	204.35	0.001163	0.1155	0.1167	870.02	1726.6	2596.6	872.00	1923.0	2795.0	2.3717	4.0272	6.3989
1800	207.15	0.001168	0.1092	0.1104	882.61	1715.1	2597.7	884.71	1911.7	2796.4	2.3980	3.9801	6.3781
1900	209.84	0.001172	0.1035	0.1047	894.70	1704.0	2598.7	896.92	1900.7	2797.6	2.4231	3.9353	6.3584
2000	212.42	0.001177	0.09841	0.09959	906.33	1693.2	2599.5	908.69	1890.0	2798.7	2.4471	3.8925	6.3396
2250	218.45	0.001187	0.08753	0.08872	933.70	1667.5	2601.2	936.37	1864.4	2800.8	2.5032	3.7926	6.2958
2500	223.99	0.001197	0.07875	0.07995	958.98	1643.3	2602.3	961.97	1840.2	2802.2	2.5544	3.7016	6.2560
2750	229.11	0.001207	0.07151	0.07272	982.53	1620.5	2603.0	985.85	1817.2	2803.0	2.6016	3.6178	6.2194
3000	233.89	0.001217	0.06544	0.06666	1004.6	1598.7	2603.3	1008.3	1795.0	2803.3	2.6454	3.5401	6.1855
3250	238.37	0.001226	0.06027	0.06150	1025.5	1577.7	2603.2	1029.5	1773.6	2803.1	2.6865	3.4673	6.1538
3500	242.60	0.001235	0.05582	0.05705	1045.3	1557.6	2602.9	1049.6	1753.0	2802.6	2.7251	3.3989	6.1240
3750	246.59	0.001244	0.05194	0.05318	1064.2	1538.1	2602.3	1068.8	1732.9	2801.7	2.7616	3.3341	6.0957
4000	250.39	0.001252	0.04852	0.04977	1082.2	1519.3	2601.5	1087.2	1713.4	2800.6	2.7962	3.2727	6.0689
5000	263.98	0.001286	0.03815	0.03944	1147.8	1448.7	2596.5	1154.2	1639.5	2793.7	2.9201	3.0524	5.9725
6000	275.62	0.001319	0.03112	0.03244	1205.4	1383.9	2589.3	1213.3	1570.6	2783.9	3.0266	2.8620	5.8886
7000	285.86	0.001352	0.02602	0.02737	1257.5	1322.7	2580.2	1267.0	1504.8	2771.8	3.1211	2.6919	5.8130
8000	295.04	0.001384	0.02214	0.02352	1305.5	1264.1	2569.6	1316.6	1441.2	2757.8	3.2066	2.5365	5.7431
9000	303.38	0.001418	0.01906	0.02048	1350.3	1207.3	2557.6	1363.1	1378.9	2742.0	3.2855	2.3916	5.6771
10,000	311.03	0.001452	0.01658	0.01803	1392.8	1151.4	2544.2	1407.3	1317.2	2724.5	3.3591	2.2548	5.6139
11,000	318.11	0.001488	0.01450	0.01599	1433.3	1096.2	2529.5	1449.7	1255.7	2705.4	3.4287	2.1238	5.5525
12,000	324.71	0.001526	0.01273	0.01426	1472.4	1041.0	2513.4	1490.7	1193.8	2684.5	3.4953	1.9968	5.4921
13,000	330.89	0.001566	0.01121	0.01278	1510.5	985.20	2495.7	1530.9	1130.9	2661.8	3.5595	1.8723	5.4318
14,000	336.70	0.001610	0.009870	0.01148	1547.9	928.40	2476.3	1570.4	1066.7	2637.1	3.6220	1.7491	5.3711
15,000	342.19	0.001657	0.008683	0.01034	1585.0	870.00	2455.0	1609.8	1000.3	2610.1	3.6837	1.6255	5.3092
16,000	347.39	0.001710	0.007600	0.009310	1622.1	809.20	2431.3	1649.5	930.80	2580.3	3.7452	1.4999	5.2451
17,000	352.34	0.001770	0.006603	0.008373	1659.9	744.90	2404.8	1690.0	857.10	2547.1	3.8073	1.3704	5.1777
18,000	357.04	0.001840	0.005665	0.007505	1698.9	675.70	2374.6	1732.0	777.70	2509.7	3.8714	1.2340	5.1054
19,000	361.52	0.001925	0.004756	0.006681	1740.3	599.00	2339.3	1776.8	689.40	2466.2	3.9393	1.0862	5.0255
20,000	365.80	0.002036	0.003838	0.005874	1786.0	510.10	2296.1	1826.7	586.90	2413.6	4.0146	0.9184	4.9330
21,000	369.88	0.002200	0.002820	0.005020	1841.4	396.00	2237.4	1887.6	455.20	2342.8	4.1062	0.7079	4.8141
22,000	373.77	0.002702	0.000952	0.003654	1953.4	142.80	2096.2	2012.8	163.60	2176.5	4.2866	0.2530	4.5486
22,055	373.98	0.00311	—	.00311	2017	—	2017	2086	—	2086	4.409	—	4.409

TABLE D.12 Properties of SUPERHEATED STEAM—SI Units

P, kPa	T, °C	v, m³/kg	u, kJ/kg	h, kJ/kg	s, kJ/(kg·K)
10	(45.82)	(14.674)	(2437.0)	(2583.8)	(8.1482)
	50	14.869	2443.1	2591.8	8.1731
	100	17.196	2515.0	2687.0	8.4471
	150	19.513	2587.4	2782.5	8.6873
	200	21.826	2660.8	2879.0	8.9030
	250	24.136	2735.5	2976.9	9.0995
	300	26.446	2811.7	3076.2	9.2808
	350	28.755	2889.5	3177.0	9.4494
	400	31.063	2968.8	3279.4	9.6075
	450	33.372	3049.7	3383.4	9.7565
	500	35.680	3132.4	3489.2	9.8979
	550	37.988	3216.7	3596.6	10.032
	600	40.296	3302.8	3705.7	10.161
	650	42.604	3390.6	3816.7	10.285
	700	44.912	3480.2	3929.4	10.404
	750	47.220	3571.6	4043.8	10.518
	800	49.527	3664.8	4160.1	10.629
	850	51.835	3759.7	4278.1	10.737
50	(81.33)	(3.2408)	(2483.3)	(2645.3)	(7.5928)
	100	3.4188	2511.2	2682.1	7.6941
	150	3.8895	2585.2	2779.7	7.9394
	200	4.3560	2659.4	2877.2	8.1572
	250	4.8205	2734.5	2975.6	8.3548
	300	5.2840	2811.0	3075.2	8.5367
	350	5.7469	2888.9	3176.2	8.7057
	400	6.2094	2968.3	3278.8	8.8640
	450	6.6717	3049.3	3382.9	9.0132
	500	7.1338	3132.0	3488.7	9.1547
	550	7.5958	3216.4	3596.2	9.2894
	600	8.0577	3302.5	3705.4	9.4182
	650	8.5195	3390.4	3816.4	9.5417
	700	8.9813	3480.1	3929.1	9.6606
	750	9.4430	3571.5	4043.6	9.7754
	800	9.9047	3664.7	4159.9	9.8863
	850	10.366	3759.6	4277.9	9.9938
100	(99.63)	(1.6943)	(2505.7)	(2675.1)	(7.3589)
	100	1.6961	2506.3	2675.9	7.3609
	150	1.9364	2582.4	2776.1	7.6129
	200	2.1723	2657.6	2874.8	7.8335
	250	2.4061	2733.3	2973.9	8.0325
	300	2.6388	2810.1	3073.9	8.2152
	350	2.8709	2888.2	3175.3	8.3846
	400	3.1027	2967.7	3278.0	8.5432
	450	3.3342	3048.9	3382.3	8.6927
	500	3.5655	3131.6	3488.2	8.8342
	550	3.7968	3216.1	3595.8	8.9690
	600	4.0279	3302.3	3705.0	9.0979
	650	4.2590	3390.2	3816.1	9.2216
	700	4.4900	3479.8	3928.8	9.3405
	750	4.7210	3571.3	4043.4	9.4553
	800	4.9519	3664.5	4159.7	9.5662
	850	5.1828	3759.4	4277.7	9.6738

TABLE D.12 Properties of SUPERHEATED STEAM — SI Units *(Continued)*

P, kPa	T, °C	v, m³/kg	u, kJ/kg	h, kJ/kg	s, kJ/(kg·K)
101.33	(100.00)	(1.6737)	(2506.1)	(2675.7)	(7.3545)
	150	1.9108	2582.3	2776.0	7.6066
	200	2.1436	2657.5	2874.8	7.8273
	250	2.3744	2733.3	2973.8	8.0264
	300	2.6041	2810.0	3073.9	8.2090
	350	2.8332	2888.1	3175.2	8.3785
	400	3.0619	2967.7	3278.0	8.5371
	450	3.2904	3048.9	3382.3	8.6865
	500	3.5187	3131.6	3488.2	8.8281
	550	3.7469	3216.1	3595.8	8.9629
	600	3.9750	3302.2	3705.0	9.0918
	650	4.2030	3390.2	3816.1	9.2154
	700	4.4310	3479.8	3928.8	9.3344
	750	4.6590	3571.3	4043.4	9.4492
	800	4.8869	3664.5	4159.7	9.5601
	850	5.1148	3759.4	4277.7	9.6677
200	(120.24)	(0.8859)	(2529.4)	(2706.5)	(7.1272)
	150	0.9597	2576.7	2768.6	7.2793
	200	1.0803	2653.9	2870.0	7.5059
	250	1.1988	2730.8	2970.5	7.7078
	300	1.3162	2808.2	3071.4	7.8920
	350	1.4329	2886.7	3173.3	8.0624
	400	1.5493	2966.6	3276.4	8.2216
	450	1.6655	3047.9	3381.0	8.3714
	500	1.7814	3130.8	3487.1	8.5133
	550	1.8973	3215.4	3594.9	8.6483
	600	2.0130	3301.7	3704.3	8.7773
	650	2.1287	3389.7	3815.4	8.9011
	700	2.2443	3479.4	3928.3	9.0201
	750	2.3599	3570.9	4042.9	9.1350
	800	2.4755	3664.1	4159.2	9.2460
	850	2.5910	3759.1	4277.3	9.3536
300	(133.56)	(0.6059)	(2543.5)	(2725.3)	(6.9921)
	150	0.6339	2570.7	2760.9	7.0779
	200	0.7163	2650.2	2865.1	7.3108
	250	0.7963	2728.2	2967.1	7.5157
	300	0.8753	2806.3	3068.9	7.7015
	350	0.9536	2885.3	3171.3	7.8729
	400	1.0315	2965.4	3274.9	8.0327
	450	1.1092	3047.0	3379.7	8.1830
	500	1.1867	3130.1	3486.1	8.3252
	550	1.2641	3214.7	3594.0	8.4604
	600	1.3414	3301.1	3703.5	8.5895
	650	1.4186	3389.1	3814.7	8.7134
	700	1.4958	3478.9	3927.7	8.8325
	750	1.5729	3570.5	4042.3	8.9475
	800	1.6500	3663.8	4158.8	9.0585
	850	1.7271	3758.8	4276.9	9.1661

TABLE D.12 Properties of SUPERHEATED STEAM — SI Units (*Continued*)

P, kPa	*T*, °C	*v*, m³/kg	*u*, kJ/kg	*h*, kJ/kg	*s*, kJ/(kg·K)
400	(143.64)	(0.4625)	(2553.5)	(2738.5)	(6.8961)
	150	0.4708	2564.4	2752.8	6.9300
	200	0.5342	2646.4	2860.1	7.1699
	250	0.5951	2725.6	2963.6	7.3779
	300	0.6548	2804.4	3066.3	7.5654
	350	0.7139	2883.8	3169.4	7.7378
	400	0.7726	2964.3	3273.3	7.8982
	450	0.8311	3046.0	3378.5	8.0489
	500	0.8894	3129.3	3485.0	8.1914
	550	0.9475	3214.1	3593.1	8.3268
	600	1.0056	3300.5	3702.7	8.4561
	650	1.0636	3388.6	3814.1	8.5801
	700	1.1215	3478.5	3927.1	8.6993
	750	1.1794	3570.1	4041.8	8.8143
	800	1.2373	3663.4	4158.3	8.9254
	850	1.2951	3758.4	4276.5	9.0331
600	(158.86)	(0.3156)	(2567.3)	(2756.7)	(6.7601)
	200	0.3520	2638.5	2849.7	6.9658
	250	0.3938	2720.3	2956.6	7.1806
	300	0.4344	2800.5	3061.2	7.3716
	350	0.4742	2880.9	3165.4	7.5459
	400	0.5137	2961.9	3270.2	7.7076
	450	0.5529	3044.1	3375.9	7.8591
	500	0.5920	3127.7	3482.9	8.0022
	550	0.6309	3212.7	3591.2	8.1380
	600	0.6697	3299.3	3701.2	8.2676
	650	0.7085	3387.6	3812.7	8.3918
	700	0.7472	3477.6	3925.9	8.5112
	750	0.7859	3569.2	4040.8	8.6264
	800	0.8246	3662.7	4157.4	8.7376
	850	0.8632	3757.8	4275.7	8.8453
800	(170.44)	(0.2404)	(2576.6)	(2768.9)	(6.6625)
	200	0.2607	2630.2	2838.8	6.8151
	250	0.2931	2714.8	2949.3	7.0373
	300	0.3241	2796.6	3055.9	7.2319
	350	0.3544	2877.9	3161.4	7.4084
	400	0.3843	2959.6	3267.0	7.5713
	450	0.4139	3042.2	3373.3	7.7237
	500	0.4433	3126.1	3480.7	7.8673
	550	0.4726	3211.3	3589.4	8.0036
	600	0.5018	3298.1	3699.6	8.1335
	650	0.5310	3386.6	3811.4	8.2579
	700	0.5601	3476.7	3924.7	8.3775
	750	0.5892	3568.4	4039.8	8.4928
	800	0.6182	3661.9	4156.5	8.6041
	850	0.6472	3757.1	4274.9	8.7120

TABLE D.12 Properties of SUPERHEATED STEAM—SI Units (*Continued*)

P, kPa	*T*, °C	*v*, m³/kg	*u*, kJ/kg	*h*, kJ/kg	*s*, kJ/(kg·K)
1000	(179.92)	(0.1944)	(2583.3)	(2781.2)	(6.5529)
	200	0.2059	2621.5	2827.4	6.6932
	250	0.2326	2709.2	2941.9	6.9235
	300	0.2579	2792.7	3050.6	7.1219
	350	0.2825	2874.9	3157.3	7.3005
	400	0.3066	2957.2	3263.8	7.4648
	450	0.3304	3040.3	3370.7	7.6180
	500	0.3541	3124.5	3478.6	7.7622
	550	0.3776	3210.0	3587.6	7.8989
	600	0.4011	3297.0	3698.1	8.0292
	650	0.4245	3385.5	3810.0	8.1538
	700	0.4478	3475.7	3923.6	8.2736
	750	0.4711	3567.6	4038.7	8.3890
	800	0.4944	3661.2	4155.5	8.5005
	850	0.5176	3756.4	4274.0	8.6084
1500	(198.33)	(0.1317)	(2593.9)	(2791.5)	(6.4438)
	200	0.1324	2597.5	2796.1	6.4536
	250	0.1519	2694.6	2922.4	6.7077
	300	0.1696	2782.5	3036.9	6.9168
	350	0.1866	2867.2	3147.1	7.1011
	400	0.2030	2951.2	3255.7	7.2687
	450	0.2192	3035.4	3364.2	7.4242
	500	0.2351	3120.4	3473.1	7.5699
	550	0.2510	3206.5	3583.0	7.7076
	600	0.2668	3294.0	3694.2	7.8386
	650	0.2825	3382.9	3806.6	7.9639
	700	0.2981	3473.4	3920.6	8.0841
	750	0.3137	3565.6	4036.1	8.1999
	800	0.3293	3659.3	4153.2	8.3116
	850	0.3448	3754.8	4272.0	8.4198
2000	(212.42)	(0.09959)	(2599.5)	(2798.7)	(6.3396)
	250	0.1114	2678.8	2901.6	6.5438
	300	0.1254	2771.8	3022.7	6.7651
	350	0.1386	2859.4	3136.6	6.9556
	400	0.1512	2945.1	3247.5	7.1269
	450	0.1635	3030.5	3357.5	7.2845
	500	0.1757	3116.3	3467.7	7.4318
	550	0.1877	3203.1	3578.4	7.5706
	600	0.1996	3291.0	3690.2	7.7024
	650	0.2114	3380.3	3803.2	7.8283
	700	0.2232	3471.1	3917.6	7.9490
	750	0.2350	3563.5	4033.5	8.0651
	800	0.2467	3657.5	4150.9	8.1771
	850	0.2584	3753.1	4269.9	8.2855

TABLE D.12 Properties of SUPERHEATED STEAM—SI Units (*Continued*)

P, kPa	*T*, °C	*v*, m³/kg	*u*, kJ/kg	*h*, kJ/kg	*s*, kJ/(kg·K)
2500	(223.99)	(0.07995)	(2602.3)	(2802.2)	(6.2560)
	250	0.08698	2661.7	2879.1	6.4069
	300	0.09888	2760.8	3008.0	6.6424
	350	0.1097	2851.4	3125.8	6.8395
	400	0.1201	2938.9	3239.2	7.0146
	450	0.1301	3025.5	3350.9	7.1746
	500	0.1400	3112.2	3462.2	7.3235
	550	0.1497	3199.6	3573.8	7.4634
	600	0.1593	3288.0	3686.3	7.5960
	650	0.1688	3377.7	3799.8	7.7225
	700	0.1783	3468.8	3914.7	7.8436
	750	0.1878	3561.4	4030.9	7.9601
	800	0.1972	3655.6	4148.6	8.0724
	850	0.2066	3751.4	4267.9	8.1810
3000	(233.89)	(0.06666)	(2603.3)	(2803.3)	(6.1855)
	250	0.07056	2643.1	2854.8	6.2857
	300	0.08113	2749.2	2992.6	6.5375
	350	0.09052	2843.2	3114.8	6.7420
	400	0.09935	2932.7	3230.7	6.9210
	450	0.1079	3020.5	3344.1	7.0835
	500	0.1162	3108.1	3456.6	7.2339
	550	0.1244	3196.1	3569.1	7.3750
	600	0.1324	3285.0	3682.3	7.5084
	650	0.1404	3375.1	3796.4	7.6355
	700	0.1484	3466.5	3911.7	7.7571
	750	0.1563	3559.4	4028.3	7.8739
	800	0.1642	3653.8	4146.3	7.9865
	850	0.1720	3749.7	4265.8	8.0954
4000	(250.39)	(0.04977)	(2601.5)	(2800.6)	(6.0689)
	300	0.05882	2724.4	2959.7	6.3598
	350	0.06644	2826.1	3091.8	6.5811
	400	0.07340	2919.8	3213.4	6.7688
	450	0.08002	3010.3	3330.4	6.9364
	500	0.08642	3099.7	3445.4	7.0902
	550	0.09268	3189.0	3559.7	7.2335
	600	0.09884	3278.9	3674.3	7.3687
	650	0.1049	3369.8	3789.5	7.4970
	700	0.1110	3461.8	3905.7	7.6195
	750	0.1170	3555.2	4023.0	7.7371
	800	0.1229	3650.0	4141.7	7.8503
	850	0.1288	3746.3	4261.7	7.9596

TABLE D.12 Properties of SUPERHEATED STEAM—SI Units (*Continued*)

P, kPa	*T*, °C	*v*, m³/kg	*u*, kJ/kg	*h*, kJ/kg	*s*, kJ/(kg·K)
5000	(263.98)	(0.03944)	(2596.5)	(2793.7)	(5.9725)
	300	0.04530	2697.0	2923.5	6.2067
	350	0.05193	2808.0	3067.7	6.4482
	400	0.05781	2906.5	3195.5	6.6456
	450	0.06330	2999.8	3316.3	6.8187
	500	0.06856	3091.1	3433.9	6.9760
	550	0.07367	3181.8	3550.2	7.1218
	600	0.07869	3272.8	3666.2	7.2586
	650	0.08362	3364.5	3782.6	7.3882
	700	0.08850	3457.1	3899.7	7.5117
	750	0.09334	3551.0	4017.7	7.6300
	800	0.09815	3646.3	4137.0	7.7438
	850	0.1029	3742.9	4257.5	7.8536
6000	(275.62)	(0.03244)	(2589.3)	(2783.9)	(5.8886)
	300	0.03615	2666.3	2883.2	6.0659
	350	0.04222	2788.9	3042.2	6.3322
	400	0.04739	2892.7	3177.0	6.5404
	450	0.05214	2989.1	3301.9	6.7195
	500	0.05665	3082.4	3422.3	6.8805
	550	0.06100	3174.6	3540.6	7.0287
	600	0.06525	3266.6	3658.1	7.1673
	650	0.06942	3359.1	3775.6	7.2982
	700	0.07353	3452.4	3893.6	7.4227
	750	0.07760	3546.8	4012.4	7.5418
	800	0.08164	3642.5	4132.3	7.6561
	850	0.08565	3739.5	4253.4	7.7664
7000	(285.86)	(0.02737)	(2580.2)	(2771.8)	(5.8130)
	300	0.02946	2631.4	2837.6	5.9293
	350	0.03523	2768.5	3015.1	6.2269
	400	0.03993	2878.4	3157.9	6.4474
	450	0.04416	2978.1	3287.3	6.6329
	500	0.04813	3073.6	3410.5	6.7978
	550	0.05194	3167.2	3530.8	6.9486
	600	0.05565	3260.3	3649.8	7.0889
	650	0.05927	3353.6	3768.5	7.2211
	700	0.06284	3447.6	3887.5	7.3466
	750	0.06636	3542.6	4007.1	7.4665
	800	0.06985	3638.7	4127.6	7.5815
	850	0.07331	3736.1	4249.2	7.6922

TABLE D.12 Properties of SUPERHEATED STEAM — SI Units (*Continued*)

P, kPa	T, °C	v, m³/kg	u, kJ/kg	h, kJ/kg	s, kJ/(kg·K)
8000	(295.04)	(0.02352)	(2569.6)	(2757.8)	(5.7431)
	300	0.02426	2590.5	2784.6	5.7901
	350	0.02995	2746.7	2986.3	6.1286
	400	0.03431	2863.5	3138.0	6.3630
	450	0.03816	2966.9	3272.2	6.5554
	500	0.04174	3064.6	3398.5	6.7243
	550	0.04515	3159.8	3521.0	6.8778
	600	0.04845	3254.0	3641.5	7.0200
	650	0.05166	3348.1	3761.4	7.1535
	700	0.05482	3442.8	3881.4	7.2800
	750	0.05793	3538.3	4001.7	7.4007
	800	0.06101	3634.9	4122.9	7.5163
	850	0.06406	3732.6	4245.1	7.6275
10,000	(311.03)	(0.01803)	(2544.2)	(2724.5)	(5.6139)
	350	0.02242	2698.1	2922.2	5.9425
	400	0.02641	2832.0	3096.1	6.2114
	450	0.02975	2943.6	3241.1	6.4194
	500	0.03278	3046.2	3374.0	6.5971
	550	0.03563	3144.6	3500.9	6.7561
	600	0.03836	3241.1	3624.7	6.9022
	650	0.04101	3337.1	3747.1	7.0385
	700	0.04359	3433.1	3869.0	7.1671
	750	0.04613	3529.7	3991.0	7.2893
	800	0.04863	3627.2	4113.5	7.4062
	850	0.05110	3725.7	4236.7	7.5184
15,000	(342.19)	(0.01034)	(2455.0)	(2610.1)	(5.3092)
	350	0.01147	2519.3	2691.3	5.4404
	400	0.01565	2739.9	2974.7	5.8799
	450	0.01845	2879.9	3156.6	6.1410
	500	0.02080	2997.3	3309.3	6.3452
	550	0.02292	3104.9	3448.8	6.5201
	600	0.02490	3207.9	3581.5	6.6767
	650	0.02679	3308.6	3710.5	6.8204
	700	0.02862	3408.3	3837.6	6.9544
	750	0.03039	3507.8	3963.7	7.0808
	800	0.03213	3607.6	4089.6	7.2009
	850	0.03384	3708.1	4215.6	7.3158

TABLE D.12 Properties of SUPERHEATED STEAM — SI Units (*Continued*)

P, kPa	*T*, °C	*v*, m³/kg	*u*, kJ/kg	*h*, kJ/kg	*s*, kJ/(kg·K)
20,000	(365.80)	(0.005874)	(2296.1)	(2413.6)	(4.9330)
	400	0.009946	2617.9	2816.9	5.5521
	450	0.01270	2806.8	3060.8	5.9026
	500	0.01477	2944.1	3239.4	6.1417
	550	0.01655	3063.0	3393.9	6.3355
	600	0.01817	3173.3	3536.7	6.5039
	650	0.01969	3279.2	3672.9	6.6557
	700	0.02113	3382.8	3805.5	6.7955
	750	0.02253	3485.4	3936.0	6.9263
	800	0.02388	3587.8	4065.4	7.0498
	850	0.02521	3690.3	4194.4	7.1673
25,000	500	0.01112	2886.1	3164.2	5.9616
	550	0.01272	3018.6	3336.5	6.1778
	600	0.01413	3137.3	3490.4	6.3593
	650	0.01542	3249.0	3634.5	6.5198
	700	0.01664	3356.8	3773.0	6.6659
	750	0.01781	3462.7	3908.0	6.8012
	800	0.01894	3567.6	4041.1	6.9282
	850	0.02003	3672.3	4173.1	7.0485
30,000	500	0.008676	2823.2	3083.5	5.7936
	550	0.01016	2972.0	3276.8	6.0362
	600	0.01143	3100.1	3443.1	6.2324
	650	0.01258	3218.0	3595.5	6.4022
	700	0.01365	3330.4	3740.1	6.5547
	750	0.01467	3439.7	3879.8	6.6948
	800	0.01564	3547.3	4016.7	6.8254
	850	0.01659	3654.2	4151.8	6.9484
35,000	600	0.009511	3061.9	3394.7	6.1174
	650	0.01056	3186.5	3556.0	6.2971
	700	0.01152	3303.6	3706.9	6.4563
	750	0.01243	3416.5	3851.6	6.6012
	800	0.01330	3526.9	3992.2	6.7355
	850	0.01413	3636.0	4130.4	6.8614
40,000	650	0.009046	3154.5	3516.3	6.2012
	700	0.009930	3276.6	3673.8	6.3673
	750	0.01075	3393.1	3823.3	6.5171
	800	0.01154	3506.4	3967.8	6.6551
	850	0.01229	3617.7	4109.2	6.7838

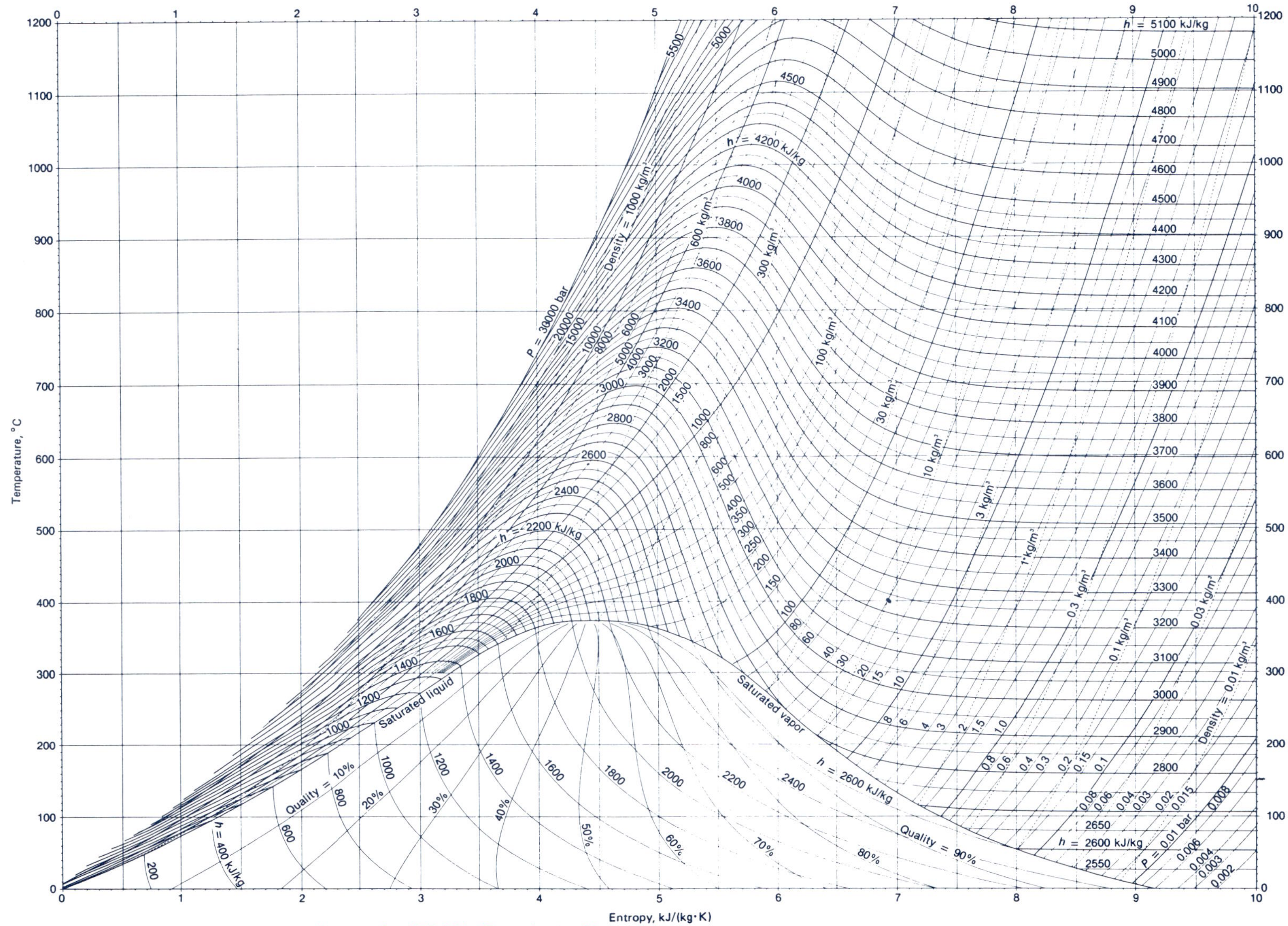

Figure D.1 Temperature-entropy diagram for STEAM. (From Lester Haar, John S. Gallagher, and George S. Kell, NBS/NRC Steam Tables, 1984. With permission from Hemisphere Publishing Corporation, New York.)

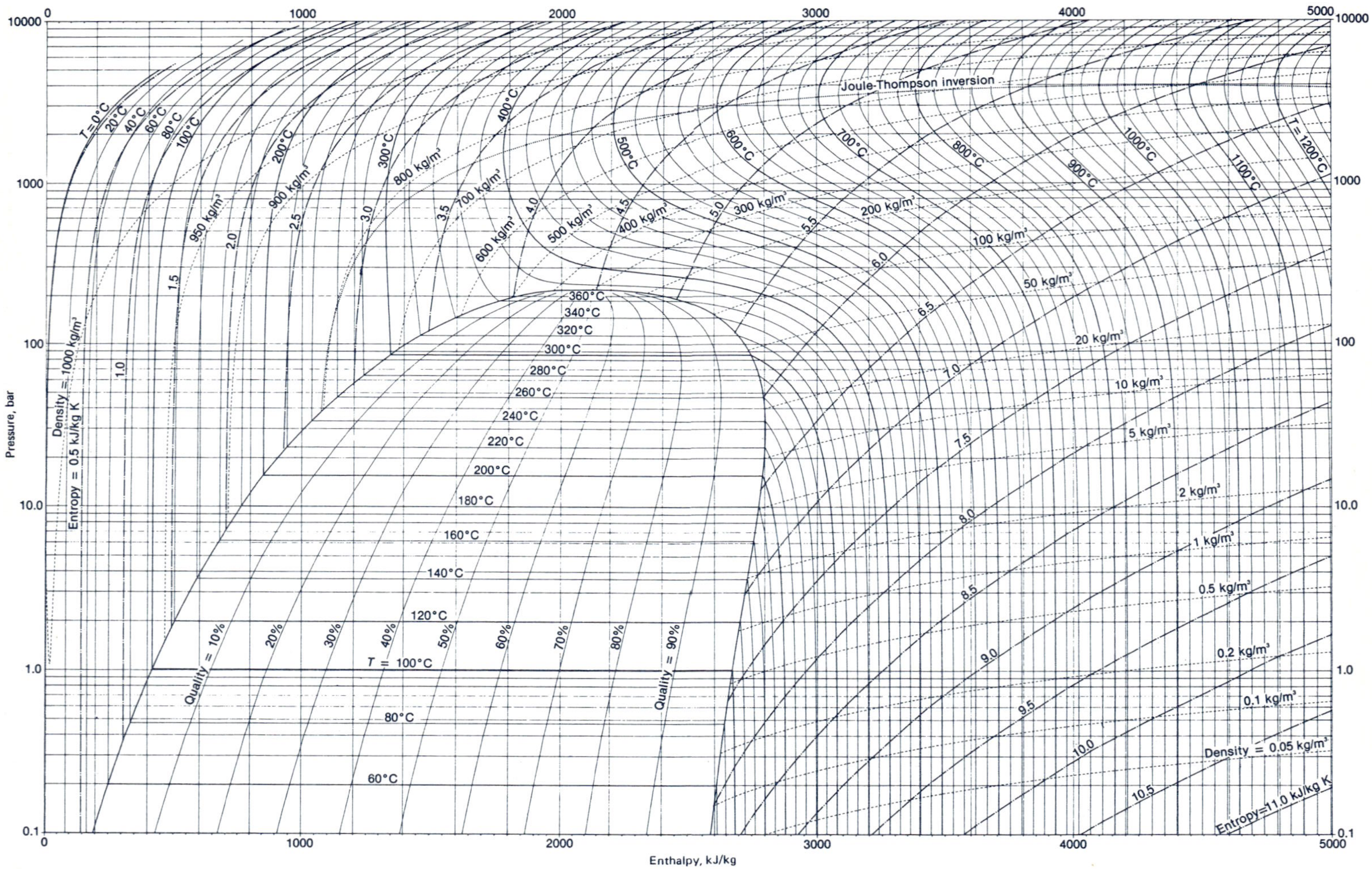

Figure D.2 Pressure-enthalpy diagram for STEAM. (From Lester Haar, John S. Gallagher, and George S. Kell, NBS/NRC Steam Tables, 1984. With permission from Hemisphere Publishing Corporation, New York.)

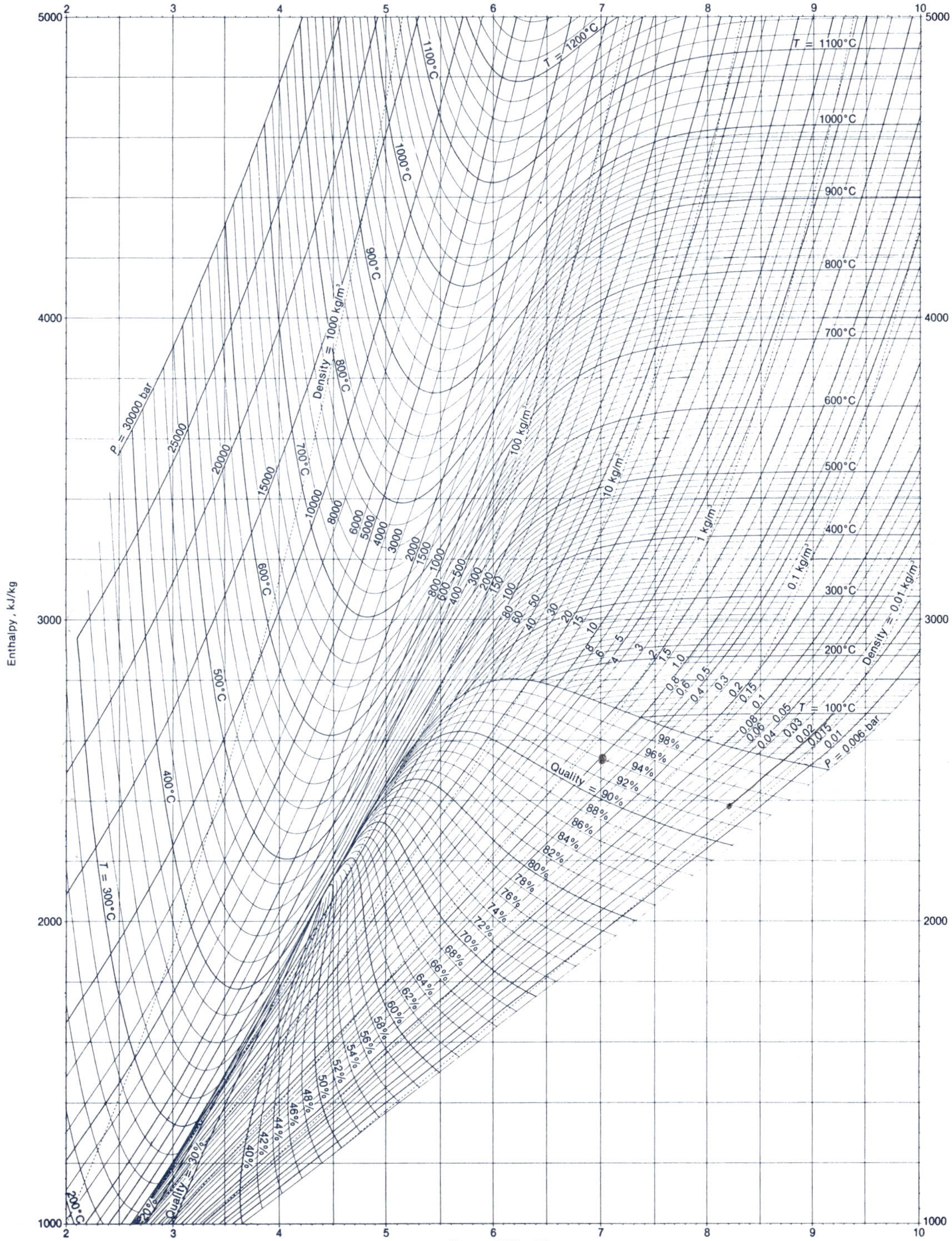

Figure D.3 Mollier diagram for STEAM. (From Lester Haar, John S. Gallagher, and Geroge S. Kell, NBS/NRC Steam Tables, 1984. With permission from Hemisphere Publishing Corporation, New York.)

TABLE D.13 Properties of Saturated REFRIGERANT 12—Temperature Table, SI Units

Reference State: $h_l = 200$ kJ/kg and $s_l = 1$ kJ/(kg·K) at 0°C

		Specific Volume, m³/kg			Internal Energy, kJ/kg			Enthalpy, kJ/kg			Entropy, kJ/(kg·K)		
T_{sat}, °C	P_{sat}, kPa	v_l	v_{lg}	v_g	u_l	u_{lg}	u_g	h_l	h_{lg}	h_g	s_l	s_{lg}	s_g
−100	1.2	0.0005991	10.114	10.115	112.09	181.96	294.05	112.09	193.85	305.94	0.60086	1.1195	1.7204
−90	2.8	0.0006078	4.4206	4.4211	120.64	177.21	297.85	120.64	189.76	310.40	0.64889	1.0360	1.6849
−80	6.2	0.0006169	2.1402	2.1408	129.20	172.56	301.76	129.20	185.75	314.96	0.69439	0.96171	1.6561
−70	12.3	0.0006266	1.1279	1.1285	137.78	167.97	305.75	137.79	181.79	319.58	0.73770	0.89485	1.6326
−60	22.6	0.0006369	0.63783	0.63847	146.43	163.37	309.80	146.45	177.78	324.23	0.77930	0.83407	1.6134
−50	39.1	0.0006478	0.38275	0.38340	155.13	158.77	313.89	155.15	173.74	328.89	0.81920	0.77857	1.5978
−40	64.1	0.0006594	0.24142	0.24208	163.89	154.12	318.01	163.93	169.60	333.53	0.85765	0.72744	1.5851
−30	100.3	0.0006720	0.15881	0.15948	172.73	149.41	322.14	172.80	165.34	338.14	0.89483	0.68001	1.5748
−20	150.8	0.0006854	0.10823	0.10891	181.65	144.61	326.25	181.75	160.93	342.68	0.93085	0.63571	1.5666
−10	219.0	0.0007000	0.075992	0.076692	190.66	139.69	330.35	190.81	156.33	347.14	0.96586	0.59408	1.5599
0	308.4	0.0007159	0.054710	0.055426	199.77	134.64	334.41	200.00	151.51	351.50	1.0000	0.55469	1.5547
10	423.1	0.0007332	0.040200	0.040934	209.05	129.37	338.41	209.36	146.37	355.73	1.0334	0.51695	1.5504
20	567.0	0.0007524	0.030042	0.030794	218.46	123.89	342.34	218.89	140.92	359.81	1.0663	0.48071	1.5470
30	744.6	0.0007738	0.022745	0.023519	228.07	118.11	346.18	228.65	135.05	363.70	1.0988	0.44548	1.5443
40	960.3	0.0007980	0.017383	0.018181	237.94	111.95	349.90	238.71	128.65	367.35	1.1311	0.41082	1.5419
50	1218.9	0.0008257	0.013350	0.014176	248.18	105.26	353.43	249.18	121.53	370.71	1.1636	0.37608	1.5397
60	1525.4	0.0008581	0.010259	0.011117	258.83	97.90	356.74	260.14	113.55	373.69	1.1965	0.34084	1.5374
70	1885.2	0.0008971	0.007832	0.008729	270.11	89.58	359.70	271.81	104.35	376.15	1.2304	0.30409	1.5344
80	2303.9	0.0009460	0.005879	0.006825	282.24	79.92	362.16	284.42	93.46	377.88	1.2658	0.26466	1.5304
90	2787.7	0.0010117	0.004249	0.005261	295.63	68.16	363.79	298.45	80.01	378.46	1.3040	0.22031	1.5243
100	3343.2	0.0011129	0.002793	0.003906	311.26	52.54	363.80	314.98	61.87	376.86	1.3476	0.16582	1.5134
111.6	4010	0.001792	0	0.001792	340.0	0	340.0	347.37	0	347.37	1.4267	0	1.4267

Properties generated from the program *REFRIG,* Wiley Professional Software, New York, 1984.

TABLE D.14 Properties of Saturated REFRIGERANT 12—Pressure Table, SI Units

Reference State: $h_l = 200$ kJ/kg and $s_l = 1$ kJ/(kg·K) at 0°C

		Specific Volume, m³/kg			Internal Energy, kJ/kg			Enthalpy, kJ/kg			Entropy, kJ/(kg·K)		
P_{sat}, kPa	T_{sat}, °C	v_l	v_{lg}	v_g	u_l	u_{lg}	u_g	h_l	h_{lg}	h_g	s_l	s_{lg}	s_g
2	−94.1	0.0006041	6.1375	6.1381	117.11	179.16	296.27	117.11	191.43	308.55	0.62939	1.0693	1.6987
4	−85.7	0.0006116	3.2067	3.2073	124.29	175.21	299.51	124.30	188.04	312.34	0.66860	1.0033	1.6719
6	−80.4	0.0006166	2.1954	2.1960	128.88	172.73	301.61	128.88	185.90	314.79	0.69273	0.96434	1.6571
8	−76.3	0.0006204	1.6785	1.6791	132.33	170.88	303.20	132.33	184.30	316.64	0.71044	0.93649	1.6469
10	−73.1	0.0006235	1.3632	1.3638	135.13	169.38	304.51	135.13	183.01	318.15	0.72455	0.91476	1.6393
20	−62.1	0.0006347	0.71472	0.71535	144.62	164.32	308.95	144.64	178.62	323.25	0.77078	0.84628	1.6171
30	−55.0	0.0006423	0.49002	0.49067	150.80	161.06	311.85	150.82	175.76	326.57	0.79956	0.80555	1.6051
40	−49.6	0.0006483	0.37487	0.37552	155.50	158.57	314.07	155.53	173.56	329.09	0.82089	0.77627	1.5972
50	−45.2	0.0006533	0.30449	0.30514	159.36	156.52	315.88	159.39	171.75	331.14	0.83798	0.75332	1.5913
60	−41.4	0.0006578	0.25688	0.25754	162.66	154.78	317.43	162.69	170.19	332.88	0.85233	0.73438	1.5867
70	−38.1	0.0006617	0.22245	0.22311	165.55	153.24	318.79	165.60	168.81	334.41	0.86475	0.71824	1.5830
80	−35.2	0.0006654	0.19634	0.19701	168.14	151.86	320.00	168.19	167.57	335.76	0.87573	0.70415	1.5799
90	−32.5	0.0006687	0.17585	0.17652	170.50	150.60	321.10	170.56	166.43	336.99	0.88558	0.69164	1.5772
100	−30.1	0.0006719	0.15932	0.15999	172.66	149.45	322.10	172.73	165.38	338.10	0.89454	0.68037	1.5749
200	−12.5	0.0006962	0.082849	0.083545	188.38	140.94	329.32	188.52	157.51	346.03	0.95715	0.60432	1.5615
300	−0.8	0.0007145	0.056197	0.056912	199.00	135.07	334.07	199.21	151.93	351.14	0.99712	0.55793	1.5551
400	8.2	0.0007299	0.042478	0.043208	207.34	130.34	337.68	207.63	147.34	354.97	1.0274	0.52373	1.5511
500	15.6	0.0007437	0.034075	0.034819	214.31	126.32	340.63	214.68	143.36	358.04	1.0520	0.49647	1.5484
600	22.0	0.0007565	0.028376	0.029133	220.38	122.75	343.13	220.84	139.77	360.61	1.0729	0.47354	1.5465
700	27.7	0.0007686	0.024244	0.025013	225.81	119.49	345.30	226.35	136.46	362.81	1.0913	0.45362	1.5449
800	32.8	0.0007802	0.021105	0.021886	230.76	116.46	347.22	231.39	133.34	364.73	1.1077	0.43589	1.5436
900	37.4	0.0007913	0.018635	0.019426	235.33	113.60	348.94	236.05	130.38	366.42	1.1227	0.41984	1.5425
1000	41.7	0.0008023	0.016637	0.017439	239.60	110.89	350.49	240.40	127.53	367.93	1.1365	0.40510	1.5416
1500	59.2	0.0008554	0.010470	0.011326	258.00	98.50	356.49	259.28	114.20	373.48	1.1940	0.34359	1.5375
2000	72.9	0.0009100	0.007226	0.008136	273.52	86.96	360.47	275.34	101.41	376.75	1.2404	0.29305	1.5335
2500	84.2	0.0009711	0.005159	0.006130	287.70	75.28	362.98	290.13	88.18	378.31	1.2815	0.24674	1.5282
3000	94.0	0.0010459	0.003659	0.004704	301.50	62.57	364.07	304.64	73.55	378.19	1.3205	0.20032	1.5208
3500	102.6	0.0011515	0.002416	0.003568	315.96	47.32	363.29	319.99	55.78	375.77	1.3606	0.14844	1.5090
4000	110.3	0.0013844	0.001023	0.002408	334.75	22.56	357.31	340.29	26.65	366.94	1.4126	0.06949	1.4821
4010	111.6	0.001792	0	0.001792	340.0	0	340.0	347.37	0	347.37	1.4267	0	1.4267

Properties generated from the program *REFRIG*, Wiley Professional Software, Wiley, New York, 1984.

TABLE D.15 Properties of SUPERHEATED REFRIGERANT 12—SI Units

Reference State: h_l = 200 kJ/kg and s_l = 1 kJ/(kg·K) at 0°C

P, kPa	T, °C	v, m³/kg	u, kJ/kg	h, kJ/kg	s, kJ/(kg·K)
20					
(T_{sat} = −62.1°C)		(0.7154)	(308.95)	(323.25)	(1.6171)
	−50.0	0.7585	314.36	329.53	1.6460
	−25.0	0.8466	326.09	343.02	1.7032
	0.0	0.9339	338.52	357.20	1.7576
	25.0	1.0209	351.60	372.02	1.8096
	50.0	1.1075	365.29	387.44	1.8592
	75.0	1.1940	379.53	403.41	1.9068
	100.0	1.2804	394.28	419.89	1.9525
	125.0	1.3666	409.49	436.83	1.9964
	150.0	1.4528	425.13	454.18	2.0387
	175.0	1.5390	441.14	471.92	2.0794
	200.0	1.6251	457.51	490.01	2.1187
40					
(T_{sat} = −49.6°C)		(0.3755)	(314.07)	(329.09)	(1.5972)
	−25.0	0.4200	325.74	342.53	1.6542
	0.0	0.4645	338.26	356.83	1.7091
	25.0	0.5084	351.40	371.74	1.7612
	50.0	0.5521	365.13	387.22	1.8111
	75.0	0.5956	379.40	403.23	1.8588
	100.0	0.6390	394.17	419.73	1.9046
	125.0	0.6823	409.40	436.69	1.9485
	150.0	0.7255	425.05	454.07	1.9909
	175.0	0.7687	441.07	471.82	2.0316
	200.0	0.8119	457.44	489.92	2.0709
60					
(T_{sat} = −41.4°C)		(0.25754)	(317.43)	(332.88)	(1.5867)
	−25.0	0.2777	325.38	342.04	1.6249
	0.0	0.3079	337.99	356.46	1.6802
	25.0	0.3376	351.20	371.45	1.7327
	50.0	0.3670	364.97	386.99	1.7827
	75.0	0.3962	379.27	403.04	1.8306
	100.0	0.4252	394.06	419.58	1.8764
	125.0	0.4542	409.31	436.56	1.9205
	150.0	0.4831	424.96	453.95	1.9628
	175.0	0.5120	441.00	471.72	2.0036
	200.0	0.5408	457.38	489.83	2.0429
80					
(T_{sat} = −35.2°C)		(0.19701)	(320.00)	(335.76)	(1.5799)
	−25.0	0.2066	325.01	341.54	1.6037
	0.0	0.2296	337.72	356.09	1.6595
	25.0	0.2522	350.99	371.16	1.7122
	50.0	0.2744	364.81	386.76	1.7625
	75.0	0.2965	379.14	402.86	1.8104
	100.0	0.3183	393.95	419.42	1.8564
	125.0	0.3402	409.21	436.43	1.9005
	150.0	0.3619	424.88	453.84	1.9429
	175.0	0.3836	440.93	471.62	1.9837
	200.0	0.4053	457.31	489.74	2.0230
100					
(T_{sat} = −30.1°C)		(0.1600)	(332.10)	(338.10)	(1.5749)
	−25.0	0.1639	324.65	341.04	1.5868
	0.0	0.1826	337.45	355.71	1.6432
	25.0	0.2009	350.78	370.87	1.6962
	50.0	0.2188	364.64	386.53	1.7466
	75.0	0.2366	379.01	402.67	1.7947
	100.0	0.2542	393.84	419.27	1.8408
	125.0	0.2717	409.12	436.29	1.8849
	150.0	0.2892	424.80	453.72	1.9273
	175.0	0.3066	440.85	471.51	1.9682
	200.0	0.3240	457.25	489.64	2.0075
200					
(T_{sat} = −12.5°C)		(0.08355)	(329.32)	(346.03)	(1.5615)
	0.0	0.08861	336.04	353.76	1.5904
	25.0	0.09833	349.72	369.38	1.6451
	50.0	0.10771	363.81	385.35	1.6965
	75.0	0.11687	378.34	401.71	1.7452
	100.0	0.12590	393.29	418.47	1.7917
	125.0	0.13483	408.65	435.62	1.8362
	150.0	0.14370	424.39	453.13	1.8788
	175.0	0.15254	440.49	471.00	1.9198
	200.0	0.16131	456.92	489.18	1.9593
400					
(T_{sat} = 8.2°C)		(0.04321)	(337.68)	(354.97)	(1.5511)
	25.0	0.04692	347.46	366.22	1.5900
	50.0	0.05207	362.08	382.91	1.6436
	75.0	0.05698	376.96	399.75	1.6938
	100.0	0.06173	392.15	416.85	1.7412
	125.0	0.06638	407.69	434.24	1.7863
	150.0	0.07095	423.56	451.94	1.8294
	175.0	0.07548	439.76	469.95	1.8707
	200.0	0.07997	456.26	488.25	1.9104
600					
(T_{sat} = 22.0°C)		(0.02913)	(343.13)	(360.61)	(1.5465)
	25.0	0.02962	344.97	362.74	1.5536
	50.0	0.03345	360.23	380.30	1.6101
	75.0	0.03698	375.52	397.70	1.6619
	100.0	0.04032	390.98	415.17	1.7103
	125.0	0.04355	406.71	432.84	1.7561
	150.0	0.04670	422.72	450.74	1.7997
	175.0	0.04980	439.01	468.89	1.8413
	200.0	0.05286	455.59	487.31	1.8813

TABLE D.15 Properties of SUPERHEATED REFRIGERANT 12—SI Units (*Continued*)

Reference State: $h_l = 200$ kJ/kg and $s_l = 1$ kJ/(kg·K) at 0°C

P, kPa	T, °C	v, m³/kg	u, kJ/kg	h, kJ/kg	s, kJ/(kg·K)
800					
($T_{sat} = 32.8$°C)		(0.02189)	(347.22)	(364.73)	(1.5436)
	50.0	0.02407	358.25	377.50	1.5842
	75.0	0.02693	374.00	395.55	1.6379
	100.0	0.02959	389.77	413.45	1.6874
	125.0	0.03212	405.70	431.40	1.7339
	150.0	0.03456	421.86	449.51	1.7780
	175.0	0.03695	438.26	467.82	1.8200
	200.0	0.03930	454.92	486.36	1.8602
1000					
($T_{sat} = 41.7$°C)		(0.01744)	(350.49)	(367.93)	(1.5416)
	50.0	0.01837	356.09	374.45	1.5620
	75.0	0.02087	372.41	393.28	1.6180
	100.0	0.02313	388.52	411.65	1.6688
	125.0	0.02525	404.67	429.92	1.7162
	150.0	0.02728	420.98	448.26	1.7607
	175.0	0.02924	437.50	466.74	1.8031
	200.0	0.03116	454.24	485.40	1.8436
2000					
($T_{sat} = 72.9$°C)		(0.00814)	(360.47)	(376.75)	(1.5335)
	75.0	0.00831	362.23	378.86	1.5395
	100.0	0.01003	381.30	401.37	1.6017
	125.0	0.01142	399.02	421.86	1.6547
	150.0	0.01265	416.31	441.61	1.7027
	175.0	0.01379	433.50	461.08	1.7472
	200.0	0.01488	450.73	480.48	1.7893
4000					
($T_{sat} = 110.3$°C)		(0.002408)	(357.31)	(366.94)	(1.4821)
	125.0	0.004056	382.49	398.71	1.5633
	150.0	0.005172	404.82	425.51	1.6281
	175.0	0.005999	424.38	448.38	1.6803
	200.0	0.006708	443.05	469.89	1.7267

Properties generated from the program *REFRIG*, Wiley Professional Software, Wiley, New York, 1984.

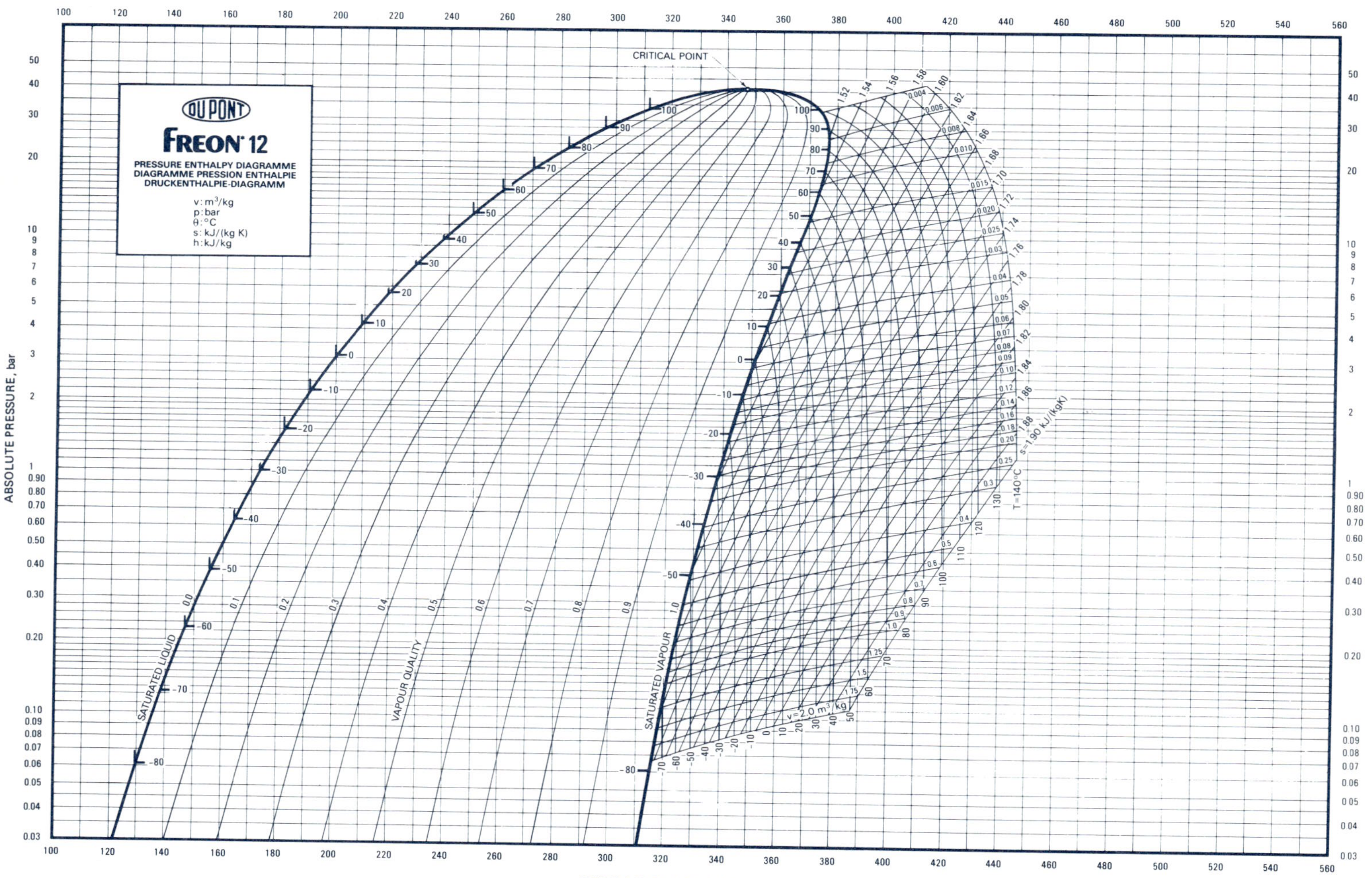

E-52455 * FREON is Du Pont's registered trademark for its fluorocarbon refrigerants

Figure D.4 Pressure-enthalpy diagram for REFRIGERANT 12. (Freon 12 is the DuPont trademark for refrigerant 12.)

TABLE D.16 Properties of Saturated AMMONIA—Temperature Table, SI Units

Reference State: $h_l = 200$ kJ/kg and $s_l = 1$ kJ/(kg·K) at 0°C and 1 atm

T_{sat}, °C	P_{sat}, kPa	Specific Volume, m³/kg			Internal Energy, kJ/kg			Enthalpy, kJ/kg			Entropy, kJ/(kg·K)		
		v_l	v_{lg}	v_g	u_l	u_{lg}	u_g	h_l	h_{lg}	h_g	s_l	s_{lg}	s_g
−50	40.7	0.001424	2.6357	2.6371	−33.62	1318.5	1284.9	−33.6	1425.8	1392.3	0.0566	6.3895	6.4461
−48	45.8	0.001429	2.3606	2.3621	−23.45	1310.9	1287.4	−23.4	1419.0	1395.6	0.1020	6.3023	6.4043
−46	51.4	0.001434	2.1192	2.1206	−13.47	1303.3	1289.9	−13.4	1412.3	1398.9	0.1461	6.2173	6.3635
−44	57.5	0.001439	1.9067	1.9081	−3.65	1296.0	1292.3	−3.6	1405.7	1402.1	0.1892	6.1344	6.3236
−42	64.3	0.001444	1.7192	1.7207	6.03	1288.7	1294.7	6.1	1399.2	1405.3	0.2312	6.0533	6.2845
−40	71.6	0.001449	1.5530	1.5545	15.86	1281.3	1297.1	16.0	1392.5	1408.5	0.2736	5.9727	6.2463
−38	79.7	0.001454	1.4060	1.4075	25.33	1274.1	1299.5	25.4	1386.2	1411.6	0.3140	5.8948	6.2089
−36	88.4	0.001460	1.2753	1.2768	34.71	1267.1	1301.8	34.8	1379.9	1414.7	0.3538	5.8185	6.1723
−34	97.9	0.001465	1.1590	1.1604	44.01	1260.1	1304.1	44.2	1373.6	1417.7	0.3928	5.7436	6.1365
−32	108.3	0.001470	1.0551	1.0566	53.25	1253.1	1306.3	53.4	1367.3	1420.7	0.4313	5.6701	6.1014
−30	119.5	0.001476	0.9623	0.9637	62.44	1246.1	1308.6	62.6	1361.1	1423.7	0.4692	5.5977	6.0670
−28	131.6	0.001481	0.8790	0.8805	71.58	1239.2	1310.8	71.8	1354.8	1426.6	0.5067	5.5265	6.0332
−26	144.6	0.001487	0.8043	0.8058	80.70	1232.2	1312.9	80.9	1348.5	1429.5	0.5437	5.4564	6.0001
−24	158.7	0.001492	0.7371	0.7386	89.79	1225.3	1315.0	90.0	1342.2	1432.3	0.5804	5.3873	5.9677
−22	173.9	0.001498	0.6765	0.6780	98.87	1218.3	1317.1	99.1	1335.9	1435.0	0.6167	5.3192	5.9358
−20	190.2	0.001504	0.6219	0.6234	107.94	1211.3	1319.2	108.2	1329.5	1437.7	0.6527	5.2519	5.9046
−18	207.7	0.001509	0.5724	0.5739	117.02	1204.2	1321.2	117.3	1323.1	1440.4	0.6884	5.1855	5.8739
−16	226.4	0.001515	0.5277	0.5292	126.09	1197.1	1323.2	126.4	1316.6	1443.0	0.7238	5.1199	5.8437
−14	246.5	0.001521	0.4870	0.4885	135.18	1190.0	1325.1	135.6	1310.0	1445.6	0.7590	5.0551	5.8141
−12	268.0	0.001527	0.4501	0.4516	144.28	1182.8	1327.1	144.7	1303.4	1448.1	0.7940	4.9910	5.7850
−10	290.9	0.001534	0.4165	0.4180	153.39	1175.5	1328.9	153.8	1296.7	1450.5	0.8288	4.9276	5.7564
−8	315.3	0.001540	0.3859	0.3874	162.53	1168.2	1330.8	163.0	1289.9	1452.9	0.8634	4.8649	5.7282
−6	341.3	0.001546	0.3579	0.3595	171.68	1160.9	1332.6	172.2	1283.1	1455.3	0.8978	4.8028	5.7005
−4	369.0	0.001553	0.3324	0.3340	180.86	1153.5	1334.3	181.4	1276.1	1457.6	0.9320	4.7413	5.6733
−2	398.4	0.001559	0.3090	0.3106	190.06	1146.0	1336.0	190.7	1269.1	1459.8	0.9661	4.6804	5.6465
0	429.6	0.001566	0.2876	0.2892	199.29	1138.4	1337.7	200.0	1262.0	1461.9	1.0000	4.6201	5.6201
2	462.7	0.001573	0.2679	0.2695	208.54	1130.8	1339.3	209.3	1254.8	1464.0	1.0337	4.5603	5.5941
4	497.7	0.001579	0.2498	0.2514	217.81	1123.1	1340.9	218.6	1247.5	1466.1	1.0673	4.5011	5.5684
6	534.8	0.001586	0.2332	0.2348	227.11	1115.4	1342.5	228.0	1240.1	1468.1	1.1008	4.4424	5.5432
8	573.9	0.001594	0.2179	0.2195	236.44	1107.5	1344.0	237.4	1232.6	1470.0	1.1341	4.3842	5.5183
10	615.3	0.001601	0.2038	0.2054	245.79	1099.6	1345.4	246.8	1225.0	1471.8	1.1672	4.3264	5.4937
12	658.9	0.001608	0.1907	0.1923	255.16	1091.7	1346.8	256.2	1217.4	1473.6	1.2002	4.2692	5.4694
14	704.9	0.001615	0.1787	0.1803	264.55	1083.7	1348.2	265.7	1209.6	1475.3	1.2331	4.2124	5.4455
16	753.2	0.001623	0.1675	0.1691	273.97	1075.5	1349.5	275.2	1201.7	1476.9	1.2658	4.1560	5.4218
18	804.1	0.001631	0.1572	0.1588	283.41	1067.4	1350.8	284.7	1193.7	1478.5	1.2984	4.1001	5.3984
20	857.6	0.001638	0.1476	0.1492	292.87	1059.1	1352.0	294.3	1185.7	1479.9	1.3308	4.0446	5.3753
22	913.8	0.001646	0.1387	0.1403	302.35	1050.8	1353.1	303.9	1177.5	1481.3	1.3630	3.9894	5.3524
24	972.7	0.001655	0.1304	0.1320	311.86	1042.4	1354.2	313.5	1169.2	1482.7	1.3951	3.9347	5.3298
26	1034.5	0.001663	0.1227	0.1243	321.38	1033.9	1355.3	323.1	1160.8	1483.9	1.4271	3.8803	5.3074
28	1099.3	0.001671	0.1155	0.1172	330.92	1025.3	1356.2	332.8	1152.3	1485.0	1.4589	3.8263	5.2852
30	1167.1	0.001680	0.1088	0.1105	340.48	1016.7	1357.2	342.4	1143.7	1486.1	1.4906	3.7726	5.2632
32	1238.0	0.001689	0.1026	0.1043	350.07	1007.9	1358.0	352.2	1134.9	1487.1	1.5222	3.7193	5.2414
34	1312.2	0.001698	0.0967	0.0984	359.67	999.1	1358.8	361.9	1126.1	1488.0	1.5536	3.6662	5.2198
36	1389.6	0.001707	0.0913	0.0930	369.30	990.2	1359.5	371.7	1117.1	1488.8	1.5849	3.6134	5.1983
38	1470.5	0.001716	0.0862	0.0879	378.96	981.2	1360.2	381.5	1108.0	1489.4	1.6160	3.5609	5.1769
40	1554.9	0.001725	0.0814	0.0831	388.64	972.1	1360.8	391.3	1098.7	1490.0	1.6471	3.5086	5.1557
42	1642.9	0.001735	0.0769	0.0787	398.34	962.9	1361.3	401.2	1089.3	1490.5	1.6780	3.4565	5.1346
44	1734.7	0.001745	0.0727	0.0745	408.08	953.6	1361.7	411.1	1079.8	1490.9	1.7089	3.4047	5.1135
46	1830.2	0.001755	0.0688	0.0706	417.84	944.2	1362.0	421.1	1070.1	1491.2	1.7396	3.3530	5.0926
48	1929.6	0.001766	0.0651	0.0669	427.64	934.7	1362.3	431.0	1060.3	1491.3	1.7703	3.3015	5.0717
50	2033.1	0.001776	0.0616	0.0634	437.47	925.0	1362.5	441.1	1050.3	1491.3	1.8009	3.2501	5.0509

Properties generated from the program *REFRIG*, Wiley Professional Software, Wiley, New York, 1984.

TABLE D.17 Properties of Saturated AMMONIA—Pressure Table, SI Units

Reference State: $h_l = 200$ kJ/kg and $s_l = 1$ kJ/(kg·K) at 0°C

		Specific Volume, m³/kg			Internal Energy, kJ/kg			Enthalpy, kJ/kg			Entropy, kJ/(kg·K)		
P_{sat}, kPa	T_{sat}, °C	v_l	v_{lg}	v_g	u_l	u_{lg}	u_g	h_l	h_{lg}	h_g	s_l	s_{lg}	s_g
40.0	−50.3	0.001424	2.6792	2.6806	−35.13	1319.7	1284.5	−35.1	1426.8	1391.8	0.0498	6.4025	6.4523
45.0	−48.3	0.001429	2.3997	2.4012	−24.97	1312.0	1287.0	−24.9	1420.0	1395.1	0.0952	6.3153	6.4105
50.0	−46.5	0.001433	2.1745	2.1759	−15.85	1305.1	1289.3	−15.8	1413.9	1398.1	0.1356	6.2376	6.3732
55.0	−44.8	0.001437	1.9892	1.9906	−7.59	1298.9	1291.3	−7.5	1408.3	1400.8	0.1720	6.1676	6.3396
60.0	−43.2	0.001441	1.8336	1.8351	−0.00	1293.2	1293.2	0.1	1403.3	1403.3	0.2051	6.1037	6.3088
65.0	−41.8	0.001445	1.7014	1.7028	7.01	1288.0	1295.0	7.1	1398.6	1405.7	0.2355	6.0451	6.2806
70.0	−40.4	0.001448	1.5871	1.5886	13.80	1282.8	1296.6	13.9	1393.9	1407.8	0.2648	5.9897	6.2544
75.0	−39.1	0.001451	1.4879	1.4893	19.93	1278.2	1298.1	20.0	1389.8	1409.8	0.2910	5.9391	6.2301
80.0	−37.9	0.001455	1.4007	1.4021	25.69	1273.9	1299.6	25.8	1385.9	1411.7	0.3156	5.8919	6.2075
85.0	−36.8	0.001458	1.3235	1.3249	31.14	1269.8	1300.9	31.3	1382.3	1413.5	0.3387	5.8475	6.1862
90.0	−35.7	0.001460	1.2545	1.2560	36.30	1265.9	1302.2	36.4	1378.8	1415.2	0.3605	5.8057	6.1661
95.0	−34.6	0.001463	1.1925	1.1940	41.22	1262.2	1303.4	41.4	1375.5	1416.8	0.3812	5.7660	6.1471
100.0	−33.6	0.001466	1.1367	1.1381	45.92	1258.6	1304.6	46.1	1372.3	1418.4	0.4008	5.7284	6.1292
120.0	−29.9	0.001476	0.9582	0.9596	62.87	1245.8	1308.7	63.0	1360.8	1423.8	0.4710	5.5943	6.0654
140.0	−26.7	0.001485	0.8292	0.8307	77.55	1234.6	1312.2	77.8	1350.7	1428.5	0.5310	5.4805	6.0115
160.0	−23.8	0.001493	0.7315	0.7330	90.59	1224.6	1315.2	90.8	1341.7	1432.5	0.5836	5.3813	5.9649
180.0	−21.2	0.001500	0.6549	0.6564	102.36	1215.6	1317.9	102.6	1333.5	1436.1	0.6306	5.2932	5.9237
200.0	−18.9	0.001507	0.5931	0.5946	113.12	1207.2	1320.4	113.4	1325.9	1439.3	0.6731	5.2139	5.8870
220.0	−16.7	0.001513	0.5422	0.5437	123.05	1199.5	1322.5	123.4	1318.8	1442.2	0.7120	5.1418	5.8538
240.0	−14.6	0.001519	0.4995	0.5010	132.30	1192.2	1324.5	132.7	1312.1	1444.8	0.7479	5.0755	5.8234
260.0	−12.7	0.001525	0.4631	0.4646	140.97	1185.4	1326.4	141.4	1305.8	1447.2	0.7813	5.0142	5.7955
280.0	−10.9	0.001531	0.4318	0.4333	149.14	1178.9	1328.1	149.6	1299.8	1449.4	0.8126	4.9571	5.7697
300.0	−9.2	0.001536	0.4045	0.4061	156.87	1172.8	1329.6	157.3	1294.1	1451.5	0.8420	4.9036	5.7456
320.0	−7.6	0.001541	0.3805	0.3821	164.21	1166.9	1331.1	164.7	1288.7	1453.4	0.8697	4.8534	5.7231
340.0	−6.1	0.001546	0.3593	0.3608	171.23	1161.2	1332.5	171.8	1283.4	1455.2	0.8961	4.8058	5.7019
360.0	−4.6	0.001551	0.3403	0.3419	177.93	1155.8	1333.8	178.5	1278.3	1456.8	0.9211	4.7608	5.6819
380.0	−3.2	0.001555	0.3233	0.3248	184.36	1150.6	1335.0	184.9	1273.5	1458.4	0.9450	4.7181	5.6631
400.0	−1.9	0.001560	0.3078	0.3094	190.55	1145.6	1336.1	191.2	1268.7	1459.9	0.9679	4.6772	5.6451
450.0	1.2	0.001570	0.2751	0.2767	205.05	1133.7	1338.7	205.8	1257.5	1463.3	1.0210	4.5828	5.6038
500.0	4.1	0.001580	0.2488	0.2503	218.40	1122.6	1341.0	219.2	1247.0	1466.2	1.0694	4.4974	5.5668
550.0	6.8	0.001589	0.2270	0.2286	230.80	1112.3	1343.1	231.7	1237.1	1468.8	1.1140	4.4193	5.5333
600.0	9.3	0.001598	0.2088	0.2104	242.39	1102.5	1344.9	243.3	1227.8	1471.1	1.1552	4.3474	5.5026
650.0	11.6	0.001607	0.1933	0.1949	253.28	1093.3	1346.6	254.3	1218.9	1473.2	1.1937	4.2806	5.4742
700.0	13.8	0.001615	0.1799	0.1815	263.59	1084.5	1348.1	264.7	1210.4	1475.1	1.2297	4.2182	5.4479
750.0	15.9	0.001622	0.1682	0.1698	273.36	1076.1	1349.4	274.6	1202.2	1476.8	1.2637	4.1597	5.4233
800.0	17.8	0.001630	0.1579	0.1596	282.67	1068.0	1350.7	284.0	1194.4	1478.3	1.2958	4.1045	5.4003
850.0	19.7	0.001637	0.1489	0.1505	291.55	1060.3	1351.8	292.9	1186.8	1479.7	1.3263	4.0523	5.3785
900.0	21.5	0.001645	0.1407	0.1424	300.07	1052.8	1352.9	301.5	1179.5	1481.0	1.3553	4.0027	5.3579
950.0	23.2	0.001651	0.1335	0.1351	308.24	1045.6	1353.8	309.8	1172.4	1482.2	1.3829	3.9555	5.3384
1000.0	24.9	0.001658	0.1269	0.1285	316.10	1038.6	1354.7	317.8	1165.5	1483.2	1.4094	3.9104	5.3198
1050.0	26.5	0.001665	0.1209	0.1225	323.69	1031.8	1355.5	325.4	1158.7	1484.2	1.4348	3.8672	5.3020
1100.0	28.0	0.001671	0.1154	0.1171	331.02	1025.2	1356.3	332.9	1152.2	1485.1	1.4592	3.8257	5.2850
1150.0	29.5	0.001678	0.1104	0.1121	338.12	1018.8	1356.9	340.0	1145.8	1485.9	1.4828	3.7858	5.2686
1200.0	30.9	0.001684	0.1058	0.1075	344.98	1012.6	1357.6	347.0	1139.6	1486.6	1.5054	3.7475	5.2530
1250.0	32.3	0.001690	0.1016	0.1033	351.65	1006.5	1358.2	353.8	1133.5	1487.2	1.5273	3.7105	5.2378
1300.0	33.7	0.001696	0.0977	0.0993	358.13	1000.6	1358.7	360.3	1127.5	1487.8	1.5485	3.6747	5.2232
1350.0	35.0	0.001702	0.0940	0.0957	364.42	994.7	1359.2	366.7	1121.7	1488.4	1.5690	3.6401	5.2091
1400.0	36.3	0.001708	0.0906	0.0923	370.57	989.0	1359.6	373.0	1115.9	1488.9	1.5889	3.6065	5.1955
1450.0	37.5	0.001714	0.0874	0.0891	376.55	983.5	1360.0	379.0	1110.2	1489.3	1.6083	3.5739	5.1822
1500.0	38.7	0.001719	0.0845	0.0862	382.39	978.0	1360.4	385.0	1104.7	1489.7	1.6271	3.5423	5.1694
1600.0	41.0	0.001730	0.0791	0.0808	393.65	967.4	1361.0	396.4	1093.9	1490.3	1.6631	3.4817	5.1447
1700.0	43.3	0.001741	0.0743	0.0760	404.44	957.1	1361.5	407.4	1083.4	1490.8	1.6974	3.4240	5.1214
1800.0	45.4	0.001752	0.0700	0.0718	414.79	947.2	1361.9	417.9	1073.1	1491.1	1.7300	3.3691	5.0991
1900.0	47.4	0.001763	0.0662	0.0679	424.75	937.5	1362.2	428.1	1063.2	1491.3	1.7613	3.3166	5.0779
2000.0	49.4	0.001773	0.0627	0.0645	434.37	928.1	1362.4	437.9	1053.4	1491.3	1.7913	3.2662	5.0575

Properties generated from the program *REFRIG*, Wiley Professional Software, Wiley, New York, 1984.

TABLE D.18 Properties of SUPERHEATED AMMONIA—SI Units

Reference State: $h_l = 200$ kJ/kg and $s_l = 1$ kJ/(kg·K) at 0°C

P, kPa	*T*, °C	*v*, m³/kg	*u*, kJ/kg	*h*, kJ/kg	*s*, kJ/(kg·K)
50 ($T_{sat} = -46.5$°C)		(2.1759)	(1289.3)	(1398.1)	(6.3732)
	−40	2.2427	1299.7	1411.9	6.4332
	−30	2.3450	1315.9	1433.1	6.5223
	−20	2.4464	1332.0	1454.3	6.6077
	−10	2.5471	1348.1	1475.5	6.6898
	0	2.6474	1364.3	1496.6	6.7687
	10	2.7472	1380.5	1517.8	6.8449
	20	2.8466	1396.7	1539.1	6.9187
	30	2.9458	1413.1	1560.4	6.9902
	40	3.0447	1429.6	1581.8	7.0596
	50	3.1435	1446.1	1603.3	7.1273
	60	3.2417	1462.8	1624.9	7.1931
	70	3.3406	1479.7	1646.7	7.2576
	80	3.4389	1496.7	1668.6	7.3205
	90	3.5373	1513.8	1690.7	7.3821
	100	3.6355	1531.1	1712.9	7.4425
75 ($T_{sat} = -39.1$°C)		(1.4893)	(1298.1)	(1409.8)	(6.2301)
	−30	1.5534	1313.3	1429.8	6.3137
	−20	1.6223	1329.8	1451.4	6.4011
	−10	1.6906	1346.2	1473.0	6.4846
	0	1.7583	1362.6	1494.5	6.5648
	10	1.8255	1379.1	1516.0	6.6420
	20	1.8924	1395.5	1537.5	6.7166
	30	1.9591	1412.0	1559.0	6.7887
	40	2.0255	1428.6	1580.5	6.8587
	50	2.0917	1445.3	1602.2	6.9267
	60	2.1574	1462.1	1623.9	6.9929
	70	2.2237	1479.0	1645.8	7.0576
	80	2.2895	1496.0	1667.8	7.1208
	90	2.3553	1513.3	1689.9	7.1826
	100	2.4210	1530.6	1712.2	7.2431
100 ($T_{sat} = -33.6$°C)		(1.1381)	(1304.6)	(1418.4)	(6.1292)
	−30	1.1573	1310.7	1426.4	6.1424
	−20	1.2102	1327.5	1448.6	6.2518
	−10	1.2622	1344.3	1470.5	6.3369
	0	1.3137	1361.0	1492.4	6.4184
	10	1.3647	1377.7	1514.1	6.4966
	20	1.4153	1394.3	1535.8	6.5719
	30	1.4657	1411.0	1557.5	6.6447
	40	1.5158	1427.7	1579.3	6.7152
	50	1.5658	1444.5	1601.0	6.7837
	60	1.6153	1461.3	1622.9	6.8502
	70	1.6652	1478.3	1644.8	6.9152
	80	1.7148	1495.4	1666.9	6.9786
	90	1.7643	1512.7	1689.1	7.0406
	100	1.8137	1530.1	1711.5	7.1013
125 ($T_{sat} = -29.1$°C)		(0.9237)	(1309.6)	(1425.1)	(6.0511)
	−20	0.96270	1325.3	1445.6	6.1337
	−10	1.0052	1342.4	1468.0	6.2206
	0	1.0469	1359.4	1490.2	6.3034
	10	1.0881	1376.2	1512.3	6.3826
	20	1.1290	1393.1	1534.2	6.4588
	30	1.1696	1409.9	1556.1	6.5322
	40	1.2100	1426.7	1578.0	6.6032
	50	1.2502	1443.6	1599.9	6.6721
	60	1.2903	1460.6	1621.9	6.7390
	70	1.3302	1477.6	1643.9	6.8042
	80	1.3700	1494.8	1666.0	6.8679
	90	1.4097	1512.1	1688.3	6.9300
	100	1.4494	1529.6	1710.7	6.9909
150 ($T_{sat} = -25.2$°F)		(0.7787)	(1313.8)	(1430.6)	(5.9874)
	−20	0.79774	1323.0	1442.6	6.0355
	−10	0.83380	1340.4	1465.5	6.1241
	0	0.86901	1357.7	1488.0	6.2082
	10	0.90377	1374.8	1510.4	6.2885
	20	0.93817	1391.8	1532.6	6.3655
	30	0.97227	1408.8	1554.6	6.4396
	40	1.0062	1425.8	1576.7	6.5111
	50	1.0398	1442.8	1598.7	6.5804
	60	1.0734	1459.8	1620.8	6.6477
	70	1.1068	1476.9	1643.0	6.7132
	80	1.1401	1494.2	1665.2	6.7771
	90	1.1733	1511.5	1687.5	6.8394
	100	1.2065	1529.0	1710.0	6.9005
200 ($T_{sat} = -18.9$°C)		(0.5946)	(1320.4)	(1439.3)	(5.8870)
	−10	0.6192	1336.4	1460.3	5.9683
	0	0.6466	1354.3	1483.6	6.0553
	10	0.6733	1371.9	1506.5	6.1377
	20	0.6995	1389.3	1529.2	6.2164
	30	0.7255	1406.6	1551.7	6.2919
	40	0.7513	1423.8	1574.1	6.3645
	50	0.7768	1441.0	1596.4	6.4347
	60	0.8023	1458.3	1618.7	6.5027
	70	0.8275	1475.6	1641.1	6.5687
	80	0.8527	1492.9	1663.5	6.6330
	90	0.8778	1510.4	1685.9	6.6958
	100	0.9028	1528.0	1708.5	6.7572

TABLE D.18 Properties of SUPERHEATED AMMONIA—SI Units (*Continued*)

Reference State: $h_l = 200$ kJ/kg and $s_l = 1$ kJ/(kg·K) at 0°C

P, kPa	*T*, °C	*v*, m³/kg	*u*, kJ/kg	*h*, kJ/kg	*s*, kJ/(kg·K)
250 ($T_{sat} = -13.7$°C)		(0.4821)	(1325.5)	(1446.0)	(5.8092)
	−10	0.4905	1332.4	1455.0	5.8436
	0	0.5129	1350.8	1479.1	5.9334
	10	0.5349	1368.9	1502.7	6.0182
	20	0.5563	1386.8	1525.8	6.0987
	30	0.5775	1404.4	1548.8	6.1755
	40.	0.5983	1421.9	1571.5	6.2493
	50	0.6190	1439.3	1594.1	6.3203
	60	0.6396	1456.7	1616.6	6.3891
	70	0.6600	1474.2	1639.2	6.4557
	80	0.6803	1491.6	1661.7	6.5205
	90	0.7005	1509.2	1684.3	6.5837
	100	0.7206	1526.9	1707.0	6.6453

P, kPa	*T*, °C	*v*, m³/kg	*u*, kJ/kg	*h*, kJ/kg	*s*, kJ/(kg·K)
300 ($T_{sat} = -9.2$°C)		(0.4061)	(1329.6)	(1451.5)	(5.7456)
	0	0.4238	1347.3	1474.4	5.8312
	10	0.4425	1365.9	1498.7	5.9183
	20	0.4608	1384.2	1522.4	6.0008
	30	0.4787	1402.1	1545.7	6.0790
	40	0.4964	1419.9	1568.8	6.1539
	50	0.5138	1437.6	1591.7	6.2259
	60	0.5311	1455.2	1614.5	6.2953
	70	0.5483	1472.8	1637.2	6.3626
	80	0.5653	1490.4	1660.0	6.4279
	90	0.5823	1508.0	1682.7	6.4914
	100	0.5992	1525.8	1705.6	6.5534

P, kPa	*T*, °C	*v*, m³/kg	*u*, kJ/kg	*h*, kJ/kg	*s*, kJ/(kg·K)
350 ($T_{sat} = -5.4$°C)		(0.3511)	(1333.1)	(1456.0)	(5.6918)
	0	0.3601	1343.7	1469.7	5.7424
	10	0.3765	1362.8	1494.6	5.8320
	20	0.3926	1381.5	1518.9	5.9164
	30	0.4082	1399.9	1542.7	5.9962
	40	0.4235	1417.9	1566.1	6.0722
	50	0.4386	1435.8	1589.3	6.1451
	60	0.4536	1453.6	1612.4	6.2153
	70	0.4685	1471.3	1635.3	6.2832
	80	0.4832	1489.1	1658.2	6.3490
	90	0.4978	1506.9	1681.1	6.4129
	100	0.5124	1524.7	1704.1	6.4753

P, kPa	*T*, °C	*v*, m³/kg	*u*, kJ/kg	*h*, kJ/kg	*s*, kJ/(kg·K)
400 ($T_{sat} = -1.9$°C)		(0.3094)	(1336.1)	(1459.9)	(5.6451)
	0	0.3123	1340.0	1464.9	5.6634
	10	0.3270	1359.7	1490.5	5.7556
	20	0.3413	1378.8	1515.4	5.8418
	30	0.3552	1397.5	1539.6	5.9233
	40	0.3688	1415.9	1563.4	6.0005
	50	0.3823	1434.0	1586.9	6.0744
	60	0.3954	1452.0	1610.2	6.1452
	70	0.4086	1469.9	1633.4	6.2138
	80	0.4216	1487.8	1656.4	6.2801
	90	0.4345	1505.7	1679.5	6.3445
	100	0.4473	1523.6	1702.6	6.4072

P, kPa	*T*, °C	*v*, m³/kg	*u*, kJ/kg	*h*, kJ/kg	*s*, kJ/(kg·K)
450 ($T_{sat} = 1.3$°C)		(0.2767)	(1338.7)	(1463.3)	(5.6038)
	10	0.2885	1356.5	1486.3	5.6865
	20	0.3014	1376.1	1511.8	5.7749
	30	0.3141	1395.2	1536.5	5.8579
	40	0.3263	1413.9	1560.7	5.9364
	50	0.3384	1432.2	1584.5	6.0113
	60	0.3502	1450.4	1608.0	6.0829
	70	0.3620	1468.5	1631.4	6.1521
	80	0.3737	1486.5	1654.7	6.2189
	90	0.3852	1504.5	1677.9	6.2837
	100	0.3967	1522.5	1701.1	6.3467

P, kPa	*T*, °C	*v*, m³/kg	*u*, kJ/kg	*h*, kJ/kg	*s*, kJ/(kg·K)
500 ($T_{sat} = 4.1$°C)		(0.2503)	(1341.0)	(1466.2)	(5.6668)
	20	0.2695	1373.3	1508.1	5.7138
	40	0.2923	1411.8	1557.9	5.8783
	60	0.3141	1448.8	1605.9	6.0267
	80	0.3354	1485.2	1652.9	6.1637
	100	0.3562	1521.5	1699.6	6.2923
	120	0.3768	1557.9	1746.3	6.4144
	140	0.3973	1594.8	1793.4	6.5313
	160	0.4176	1632.2	1841.0	6.6437
	180	0.4377	1670.3	1889.2	6.7524

TABLE D.18 Properties of SUPERHEATED AMMONIA—SI Units (*Continued*)

Reference State: $h_l = 200$ kJ/kg and $s_l = 1$ kJ/(kg·K) at 0°C

P, kPa	T, °C	v, m³/kg	u, kJ/kg	h, kJ/kg	s, kJ/(kg·K)
600 ($T_{sat} = 9.3$°C)		(0.2104)	(1344.9)	(1471.1)	(5.5026)
	10	0.2112	1346.5	1473.2	5.5098
	20	0.2215	1367.7	1500.6	5.6049
	40	0.2412	1407.6	1552.3	5.7756
	60	0.2598	1445.5	1601.4	5.9277
	80	0.2778	1482.6	1649.3	6.0672
	100	0.2955	1519.2	1696.5	6.1974
	120	0.3129	1556.0	1743.7	6.3206
	140	0.3300	1593.1	1791.1	6.4382
	160	0.3470	1630.8	1839.0	6.5513
	180	0.3639	1669.0	1887.4	6.6605
700 ($T_{sat} = 13.8$°C)		(0.1815)	(1348.1)	(1475.1)	(5.4479)
	20	0.1872	1361.8	1492.8	5.5090
	40	0.2047	1403.3	1546.5	5.6863
	60	0.2210	1442.2	1596.9	5.8423
	80	0.2367	1479.9	1645.6	5.9842
	100	0.2521	1517.0	1693.5	6.1161
	120	0.2671	1554.1	1741.1	6.2405
	140	0.2820	1591.5	1788.9	6.3589
	160	0.2967	1629.3	1837.0	6.4726
	180	0.3113	1667.7	1885.6	6.5824
800 ($T_{sat} = 17.8$°C)		(0.1596)	(1350.7)	(1478.3)	(5.4003)
	20	0.1614	1355.6	1484.7	5.4222
	40	0.1772	1398.8	1540.6	5.6065
	60	0.1919	1438.8	1592.3	5.7668
	80	0.2059	1477.2	1641.9	5.9113
	100	0.2196	1514.8	1690.4	6.0449
	120	0.2328	1552.2	1738.5	6.1704
	140	0.2459	1589.8	1786.5	6.2897
	160	0.2589	1627.8	1834.9	6.4040
	180	0.2717	1666.4	1883.8	6.5142
1000 ($T_{sat} = 24.9$°C)		(0.1285)	(1354.7)	(1483.2)	(5.3198)
	40	0.1387	1389.6	1528.3	5.4672
	60	0.1511	1431.8	1582.9	5.6365
	80	0.1627	1471.6	1634.3	5.7864
	100	0.1739	1510.2	1684.1	5.9236
	120	0.1848	1548.3	1733.1	6.0515
	140	0.1955	1586.4	1781.9	6.1725
	160	0.2060	1624.8	1830.8	6.2881
	180	0.2164	1663.7	1880.1	6.3994
1200 ($T_{sat} = 30.9$°C)		(0.1075)	(1357.6)	(1486.6)	(5.2530)
	40	0.1129	1379.8	1515.2	5.3458
	60	0.1238	1424.6	1573.1	5.5251
	80	0.1339	1465.9	1626.6	5.6810
	100	0.1435	1505.5	1677.7	5.8219
	120	0.1528	1544.4	1727.7	5.9523
	140	0.1618	1583.0	1777.2	6.0751
	160	0.1707	1621.8	1826.7	6.1921
	180	0.1795	1661.0	1876.5	6.3044
1400 ($T_{sat} = 36.3$°C)		(0.0923)	(1359.6)	(1488.9)	(5.1955)
	40	0.0943	1369.3	1501.4	5.2357
	60	0.1042	1417.0	1563.0	5.4265
	80	0.1132	1460.1	1618.6	5.5888
	100	0.1217	1500.8	1671.2	5.7337
	120	0.1299	1540.4	1722.2	5.8667
	140	0.1378	1579.5	1772.4	5.9914
	160	0.1455	1618.8	1822.5	6.1098
	180	0.1532	1658.3	1872.8	6.2232
1600 ($T_{sat} = 41.0$°C)		(0.0808)	(1361.0)	(1490.3)	(5.1447)
	60	0.0895	1409.1	1552.3	5.3366
	80	0.0977	1454.0	1610.4	5.5060
	100	0.1054	1495.9	1664.6	5.6552
	120	0.1127	1536.3	1716.6	5.7911
	140	0.1198	1576.0	1767.7	5.9177
	160	0.1266	1615.7	1818.3	6.0375
	180	0.1334	1655.6	1869.1	6.1520
1800 ($T_{sat} = 45.4$°C)		(0.0718)	(1361.9)	(1491.1)	(5.0991)
	60	0.0780	1400.8	1541.2	5.2531
	80	0.0857	1447.8	1602.0	5.4303
	100	0.0927	1491.0	1657.8	5.5841
	120	0.0993	1532.2	1711.0	5.7229
	140	0.1057	1572.5	1762.8	5.8515
	160	0.1119	1612.6	1814.1	5.9728
	180	0.1180	1652.9	1865.3	6.0884
2000 ($T_{sat} = 49.4$°C)		(0.0645)	(1362.4)	(1491.3)	(5.0575)
	60	0.0687	1392.1	1529.6	5.1742
	80	0.0760	1441.4	1593.3	5.3600
	100	0.0825	1485.9	1650.9	5.5187
	120	0.0886	1528.0	1705.2	5.6606
	140	0.0945	1568.9	1757.9	5.7913
	160	0.1002	1609.5	1809.9	5.9141
	180	0.1057	1650.1	1861.6	6.0308

Properties generated from the program *REFRIG*, Wiley Professional Software, Wiley, New York, 1984.

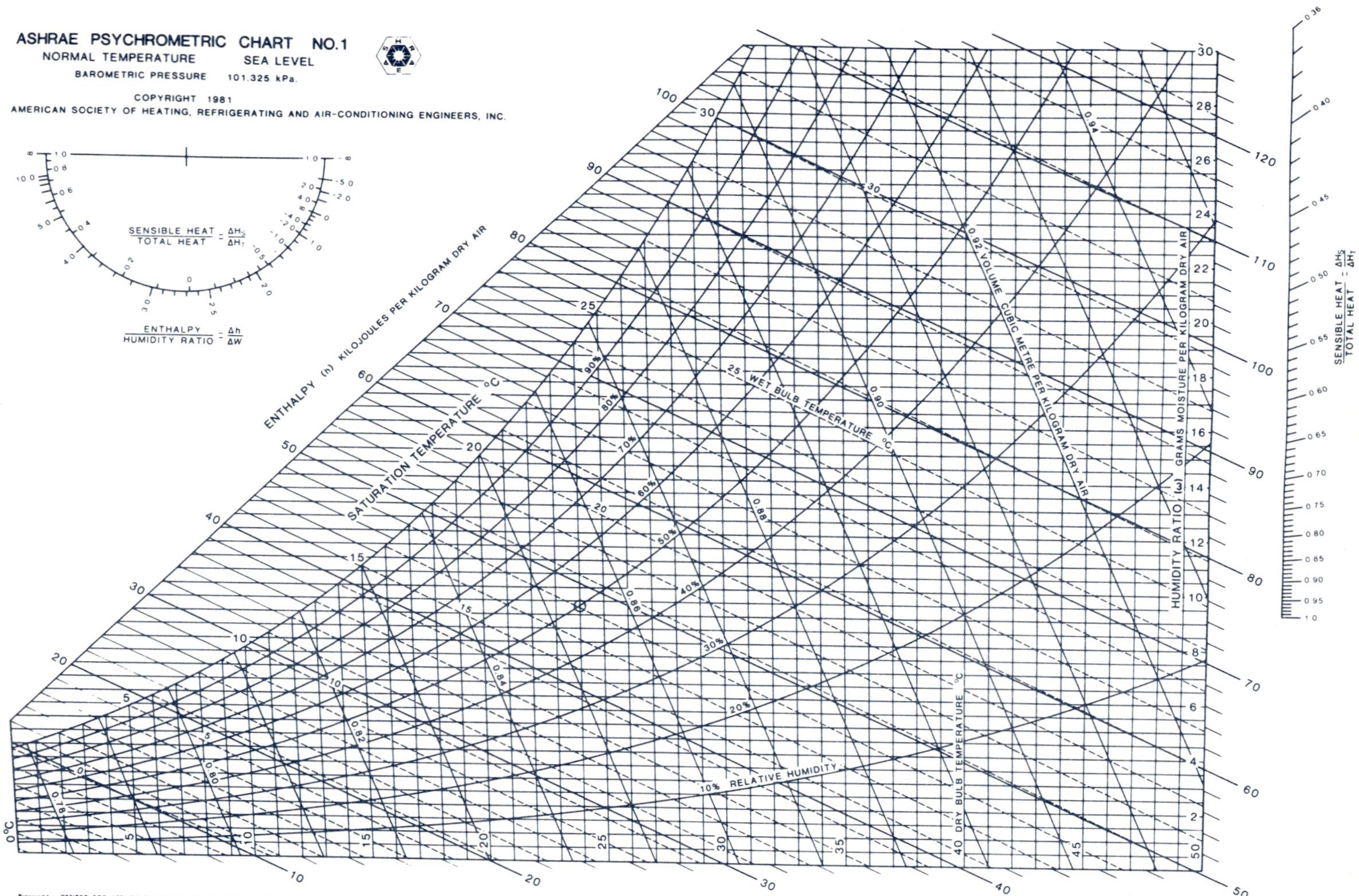

Figure D.5 Psychrometric chart. *(From American Society of Heating, Refrigerating and Air-Conditioning Engineers, used with permission.)*

APPENDIX E

Thermodynamic Data for Various Substances—USCS Units

TABLE E.1 Temperature-Dependent Molar Specific Heats of Gases at Zero Pressure—USCS Units

$$\bar{c}_P = a + bT + cT^2 + dT^3$$

[T in °R, $\bar{c}_P$ in Btu/(lbmol · °R)]

Substance	Formula	a	b	c	d	Temperature Range, °R	Error, % Max.	Error, % Avg.
Nitrogen	N_2	6.903	-0.02085×10^{-2}	0.05957×10^{-5}	-0.1176×10^{-9}	492–3240	0.59	0.34
Oxygen	O_2	6.085	0.2017×10^{-2}	-0.05275×10^{-5}	0.05372×10^{-9}	492–3240	1.19	0.28
Air		6.713	0.02609×10^{-2}	0.03540×10^{-5}	-0.08052×10^{-9}	492–3240	0.72	0.33
Hydrogen	H_2	6.952	-0.02542×10^{-2}	0.02952×10^{-5}	-0.03565×10^{-9}	492–3240	1.01	0.26
Carbon monoxide	CO	6.726	0.02222×10^{-2}	0.03960×10^{-5}	-0.09100×10^{-9}	492–3240	0.89	0.37
Carbon dioxide	CO_2	5.316	0.79361×10^{-2}	-0.2581×10^{-5}	0.3059×10^{-9}	492–3240	0.67	0.22
Water vapor	H_2O	7.700	0.02552×10^{-2}	0.07781×10^{-5}	-0.1472×10^{-9}	492–3240	0.53	0.24
Nitric oxide	NO	7.008	-0.01247×10^{-2}	0.07185×10^{-5}	-0.1715×10^{-9}	492–2700	0.97	0.36
Nitrous oxide	N_2O	5.758	0.7780×10^{-2}	-0.2596×10^{-5}	0.4331×10^{-9}	492–2700	0.59	0.26
Nitrogen dioxide	NO_2	5.48	0.7583×10^{-2}	-0.260×10^{-5}	0.322×10^{-9}	492–2700	0.46	0.18
Ammonia	NH_3	6.5846	0.34028×10^{-2}	0.073034×10^{-5}	-0.27402×10^{-9}	492–2700	0.91	0.36
Sulfur	S_2	6.499	0.2943×10^{-2}	-0.1200×10^{-5}	0.1632×10^{-9}	492–3240	0.99	0.38
Sulfur dioxide	SO_2	6.157	0.7689×10^{-2}	-0.2810×10^{-5}	0.3527×10^{-9}	492–3240	0.45	0.24
Sulfur trioxide	SO_3	3.918	1.935×10^{-2}	-0.8256×10^{-5}	1.328×10^{-9}	492–2340	0.29	0.13
Acetylene	C_2H_2	5.21	1.2227×10^{-2}	-0.4812×10^{-5}	0.7457×10^{-9}	492–2700	1.46	0.59
Benzene	C_6H_6	−8.650	6.4322×10^{-2}	-2.327×10^{-5}	3.179×10^{-9}	492–2700	0.34	0.20
Methanol	CH_4O	4.55	1.214×10^{-2}	-0.0898×10^{-5}	-0.329×10^{-9}	492–1800	0.18	0.08
Ethanol	C_2H_6O	4.75	2.781×10^{-2}	-0.7651×10^{-5}	0.821×10^{-9}	492–2700	0.40	0.22
Hydrogen chloride	HCl	7.244	-0.1011×10^{-2}	0.09783×10^{-5}	-0.1776×10^{-9}	492–2740	0.22	0.08
Methane	CH_4	4.750	0.6666×10^{-2}	0.09352×10^{-5}	-0.4510×10^{-9}	492–2740	1.33	0.57
Ethane	C_2H_6	1.648	2.291×10^{-2}	-0.4722×10^{-5}	0.2984×10^{-9}	492–2740	0.83	0.28
Propane	C_3H_8	−0.966	4.044×10^{-2}	-1.159×10^{-5}	1.300×10^{-9}	492–2740	0.40	0.12
n-Butane	C_4H_{10}	0.945	4.929×10^{-2}	-1.352×10^{-5}	1.433×10^{-9}	492–2740	0.54	0.24
i-Butane	C_4H_{10}	−1.890	5.520×10^{-2}	-1.696×10^{-5}	2.044×10^{-9}	492–2740	0.25	0.13
n-Pentane	C_5H_{12}	1.618	6.028×10^{-2}	-1.656×10^{-5}	1.732×10^{-9}	492–2740	0.56	0.21
n-Hexane	C_6H_{14}	1.657	7.328×10^{-2}	-2.112×10^{-5}	2.363×10^{-9}	492–2740	0.72	0.20
Ethylene	C_2H_4	0.944	2.075×10^{-2}	-0.6151×10^{-5}	0.7326×10^{-9}	492–2740	0.54	0.13
Propylene	C_3H_6	0.753	3.162×10^{-2}	-0.8981×10^{-5}	1.008×10^{-9}	492–2740	0.73	0.17

Data from B. G. Kyle, *Chemical and Process Thermodynamics,* Prentice-Hall, Englewood Cliffs, N.J., 1984 (used with permission).

TABLE E.2 Ideal Gas Properties of AIR at Low Pressure—USCS Units

T, °F	u, Btu/lbm	h, Btu/lbm	s_0, Btu/(lbm·°R)	P_0	v_0	c_p, Btu/(lbm·°R)	c_v, Btu/(lbm·°R)	k
−300	27.28	38.231	1.3098	0.1958	17,248.	0.2394	0.1708	1.402
−200	44.36	62.176	1.4262	1.0692	5,137.2	0.2394	0.1708	1.402
−100	61.46	86.130	1.5043	3.3357	2,280.8	0.2395	0.1709	1.401
0	78.56	110.09	1.5631	7.8574	1,237.5	0.2397	0.1711	1.401
77	91.72	128.56	1.6002	13.504	840.65	0.2401	0.1715	1.400
100	95.69	134.08	1.6103	15.641	756.91	0.2402	0.1716	1.400
200	112.89	158.15	1.6499	27.843	501.15	0.2412	0.1726	1.397
300	130.23	182.35	1.6840	45.807	350.80	0.2428	0.1742	1.394
400	147.77	206.74	1.7142	71.093	255.78	0.2450	0.1764	1.389
500	165.53	231.37	1.7413	105.52	192.37	0.2476	0.1790	1.383
600	183.57	256.27	1.7660	151.21	148.23	0.2505	0.1819	1.377
700	201.92	281.47	1.7887	210.60	116.48	0.2536	0.1850	1.371
800	220.58	306.99	1.8098	286.47	93.013	0.2568	0.1882	1.365
900	239.56	332.83	1.8295	381.97	75.295	0.2600	0.1914	1.358
1,000	258.85	358.99	1.8481	500.64	61.673	0.2631	0.1945	1.353
1,100	278.45	385.45	1.8656	646.40	51.038	0.2661	0.1975	1.347
1,200	298.34	412.20	1.8822	823.62	42.624	0.2689	0.2003	1.343
1,300	318.50	439.22	1.8980	1,037.1	35.891	0.2715	0.2029	1.338
1,400	338.93	466.50	1.9131	1,292.0	30.446	0.2740	0.2054	1.334
1,500	359.59	494.03	1.9275	1,594.1	26.003	0.2764	0.2078	1.330
1,600	380.48	521.78	1.9413	1,949.6	22.347	0.2785	0.2099	1.327
1,700	401.57	549.73	1.9546	2,365.0	19.316	0.2805	0.2119	1.324
1,800	422.87	577.88	1.9673	2,847.7	16.785	0.2824	0.2138	1.321
1,900	444.34	606.22	1.9796	3,405.3	14.657	0.2842	0.2156	1.318
2,000	465.98	634.72	1.9914	4,046.1	12.859	0.2858	0.2172	1.316
2,100	487.77	663.37	2.0029	4,778.9	11.330	0.2873	0.2187	1.314
2,200	509.71	692.18	2.0139	5,613.0	10.023	0.2887	0.2201	1.312
2,300	531.79	721.11	2.0246	6,558.6	8.9005	0.2900	0.2214	1.310
2,400	554.00	750.18	2.0349	7,626.1	7.9320	0.2913	0.2227	1.308
2,500	576.33	779.37	2.0449	8,826.8	7.0926	0.2924	0.2238	1.307
2,600	598.77	808.67	2.0547	10,173.	6.3622	0.2935	0.2249	1.305
2,800	642.99	866.61	2.0733	13,350.	5.1648	0.2955	0.2269	1.302
3,000	689.53	926.87	2.0910	17,268.	4.2381	0.2973	0.2287	1.300
3,200	735.45	986.51	2.1077	22,045.	3.5116	0.2989	0.2303	1.298
3,400	781.67	1,046.5	2.1237	27,813.	2.9354	0.3004	0.2318	1.296
3,600	828.18	1,106.7	2.1389	34,718.	2.4735	0.3018	0.2332	1.294
3,800	874.95	1,167.1	2.1534	42,916.	2.0995	0.3030	0.2344	1.293
4,000	921.97	1,227.9	2.1674	52,578.	1.7942	0.3042	0.2356	1.291
4,200	969.20	1,288.9	2.1807	63,891.	1.5427	0.3053	0.2367	1.290
4,400	1,016.6	1,350.0	2.1936	77,052.	1.3341	0.3063	0.2377	1.289
4,600	1,064.3	1,411.4	2.2060	92,278.	1.1598	0.3072	0.2386	1.288
4,800	1,112.1	1,472.9	2.2179	109,800.	1.0133	0.3081	0.2395	1.286
5,000	1,160.1	1,534.6	2.2294	129,860.	0.8893	0.3089	0.2403	1.286

Properties generated from the program *GASPROPS,* Wiley Professional Software, Wiley, New York, 1984.

TABLE E.3 Ideal Gas Properties of ARGON at Low Pressure — USCS Units

T, °F	u, Btu/lbm	h, Btu/lbm	s_0, Btu/(lbm · °R)	P_0	v_0	c_P, Btu/(lbm · °R)	c_v, Btu/(lbm · °R)	k
−300	11.914	19.857	0.7750	0.5837	7,979.2	0.1244	0.0746	1.667
−200	19.376	32.293	0.8355	1.9689	3,847.4	0.1244	0.0746	1.667
−100	26.837	44.729	0.8760	4.4455	2,360.1	0.1244	0.0746	1.667
0	34.299	57.165	0.9065	8.2088	1,633.5	0.1244	0.0746	1.667
77	40.046	66.744	0.9258	12.090	1,294.9	0.1244	0.0746	1.667
100	41.761	69.601	0.9310	13.427	1,215.9	0.1244	0.0746	1.667
200	49.222	82.037	0.9514	20.253	950.16	0.1244	0.0746	1.667
300	56.684	94.473	0.9690	28.822	768.88	0.1244	0.0746	1.667
400	64.146	106.91	0.9844	39.264	638.70	0.1244	0.0746	1.667
500	71.607	119.35	0.9980	51.697	541.52	0.1244	0.0746	1.667
600	79.069	131.78	1.0104	66.235	466.70	0.1244	0.0746	1.667
700	86.531	144.22	1.0216	82.985	407.66	0.1244	0.0746	1.667
800	93.992	156.65	1.0319	102.05	360.09	0.1244	0.0746	1.667
900	101.45	169.09	1.0414	123.52	321.10	0.1244	0.0746	1.667
1,000	108.92	181.53	1.0502	147.50	288.68	0.1244	0.0746	1.667
1,100	116.38	193.96	1.0584	174.08	261.36	0.1244	0.0746	1.667
1,200	123.84	206.40	1.0662	203.34	238.10	0.1244	0.0746	1.667
1,300	131.30	218.83	1.0734	235.36	218.10	0.1244	0.0746	1.667
1,400	138.76	231.27	1.0803	270.24	200.74	0.1244	0.0746	1.667
1,500	146.22	243.71	1.0868	308.05	185.58	0.1244	0.0746	1.667
1,600	153.68	256.14	1.0930	348.86	172.23	0.1244	0.0746	1.667
1,700	161.15	268.58	1.0989	392.76	160.40	0.1244	0.0746	1.667
1,800	168.61	281.01	1.1045	439.82	149.87	0.1244	0.0746	1.667
1,900	176.07	293.45	1.1099	490.11	140.45	0.1244	0.0746	1.667
2,000	183.53	305.89	1.1151	543.69	131.97	0.1244	0.0746	1.667
2,100	190.99	318.32	1.1201	600.65	124.31	0.1244	0.0746	1.667
2,200	198.46	330.76	1.1248	661.05	117.37	0.1244	0.0746	1.667
2,300	205.92	343.20	1.1294	724.95	111.05	0.1244	0.0746	1.667
2,400	213.38	355.63	1.1338	792.41	105.27	0.1244	0.0746	1.667
2,500	220.84	368.07	1.1381	863.52	99.984	0.1244	0.0746	1.667
2,600	228.30	380.50	1.1422	938.32	95.123	0.1244	0.0746	1.667
2,800	243.23	405.38	1.1501	1,099.2	86.504	0.1244	0.0746	1.667
3,000	258.15	430.25	1.1575	1,275.7	79.112	0.1244	0.0746	1.667
3,200	273.07	455.12	1.1645	1,468.1	72.716	0.1244	0.0746	1.667
3,400	288.00	479.99	1.1711	1,677.0	67.138	0.1244	0.0746	1.667
3,600	302.92	504.87	1.1774	1,902.8	62.239	0.1244	0.0746	1.667
3,800	317.84	529.74	1.1834	2,145.9	57.907	0.1244	0.0746	1.667
4,000	332.77	554.61	1.1891	2,406.7	54.056	0.1244	0.0746	1.667
4,200	347.69	579.48	1.1946	2,685.7	50.613	0.1244	0.0746	1.667
4,400	362.61	604.35	1.1998	2,983.2	47.521	0.1244	0.0746	1.667
4,600	377.54	629.23	1.2048	3,299.6	44.731	0.1244	0.0746	1.667
4,800	392.46	654.10	1.2096	3,635.5	42.204	0.1244	0.0746	1.667
5,000	407.33	678.97	1.2143	3,991.0	39.907	0.1244	0.0746	1.667

Properties generated from the program *GASPROPS*, Wiley Professional Software, Wiley, New York, 1984.

TABLE E.4 Ideal Gas Properties of NITROGEN at Low Pressure—USCS Units

T, °F	u, Btu/lbm	h, Btu/lbm	s_0, Btu/(lbm · °R)	P_0	v_0	c_P, Btu/(lbm · °R)	c_v, Btu/(lbm · °R)	k
−300	28.32	39.643	1.3327	0.1443	22,629.	0.2483	0.1773	1.400
−200	46.05	64.471	1.4535	0.7917	6,709.1	0.2483	0.1773	1.400
−100	63.80	89.309	1.5344	2.4781	2,969.0	0.2483	0.1774	1.400
0	81.54	114.15	1.5954	5.8497	1,607.5	0.2484	0.1774	1.400
77	95.20	133.27	1.6338	10.062	1,091.1	0.2485	0.1775	1.400
100	99.29	138.99	1.6443	11.655	982.28	0.2485	0.1776	1.399
200	117.07	163.86	1.6851	20.743	650.56	0.2490	0.1781	1.398
300	134.92	188.81	1.7204	34.075	456.06	0.2500	0.1791	1.396
400	152.90	213.89	1.7514	52.754	333.35	0.2516	0.1806	1.393
500	171.07	239.15	1.7792	78.059	251.49	0.2537	0.1828	1.388
600	189.47	264.64	1.8044	111.47	194.47	0.2563	0.1854	1.383
700	208.16	290.42	1.8277	154.68	153.36	0.2593	0.1883	1.377
800	227.14	316.50	1.8492	209.66	122.91	0.2624	0.1914	1.371
900	246.45	342.90	1.8694	278.59	99.838	0.2656	0.1947	1.364
1,000	266.08	369.62	1.8884	363.96	82.041	0.2688	0.1979	1.359
1,100	286.02	396.66	1.9063	468.53	68.096	0.2719	0.2010	1.353
1,200	306.27	424.01	1.9233	595.35	57.026	0.2750	0.2040	1.348
1,300	326.82	451.65	1.9394	747.79	48.137	0.2778	0.2069	1.343
1,400	347.65	479.57	1.9549	929.52	40.926	0.2806	0.2096	1.338
1,500	368.74	507.75	1.9696	1,144.5	35.025	0.2831	0.2122	1.334
1,600	390.08	536.18	1.9838	1,397.2	30.156	0.2855	0.2145	1.331
1,700	411.64	564.84	1.9974	1,692.2	26.108	0.2877	0.2168	1.327
1,800	433.42	593.72	2.0104	2,034.5	22.721	0.2898	0.2188	1.324
1,900	455.40	622.79	2.0230	2,429.6	19.868	0.2917	0.2207	1.321
2,000	477.57	652.05	2.0352	2,883.2	17.452	0.2935	0.2225	1.319
2,100	499.90	681.48	2.0469	3,401.5	15.394	0.2951	0.2242	1.316
2,200	522.40	711.06	2.0582	3,991.1	13.632	0.2966	0.2257	1.314
2,300	545.04	740.80	2.0692	4,658.9	12.117	0.2981	0.2271	1.312
2,400	567.82	770.67	2.0799	5,412.3	10.808	0.2994	0.2285	1.311
2,500	590.73	800.68	2.0902	6,259.2	9.6729	0.3006	0.2297	1.309
2,600	613.75	830.80	2.1002	7,207.6	8.6838	0.3018	0.2308	1.307
2,800	660.13	891.36	2.1194	9,444.5	7.0603	0.3038	0.2329	1.305
3,000	706.90	952.32	2.1375	12,198.	5.8020	0.3057	0.2347	1.302
3,200	754.00	1,013.6	2.1547	15,550.	4.8144	0.3072	0.2363	1.300
3,400	801.41	1,075.2	2.1711	19,591.	4.0302	0.3087	0.2377	1.298
3,600	849.08	1,137.1	2.1867	24,418.	3.4010	0.3099	0.2390	1.297
3,800	896.99	1,199.2	2.2017	30,138.	2.8912	0.3110	0.2401	1.295
4,000	945.11	1,261.5	2.2160	36,866.	2.4746	0.3120	0.2411	1.294
4,200	993.42	1,324.0	2.2297	44,726.	2.1312	0.3129	0.2420	1.293
4,400	1,041.9	1,386.6	2.2428	53,850.	1.8461	0.3138	0.2428	1.292
4,600	1,090.5	1,449.5	2.2555	64,380.	1.6077	0.3145	0.2436	1.291
4,800	1,139.3	1,512.4	2.2677	76,468.	1.4070	0.3152	0.2443	1.290
5,000	1,188.2	1,575.5	2.2795	90,275.	1.2372	0.3158	0.2449	1.290

Properties generated from the program *GASPROPS*, Wiley Professional Software, Wiley, New York, 1984.

TABLE E.5 Ideal Gas Properties of OXYGEN at Low Pressure—USCS Units

T, °F	u, Btu/lbm	h, Btu/lbm	s_0, Btu/(lbm·°R)	P_0	v_0	c_P, Btu/(lbm·°R)	c_v, Btu/(lbm·°R)	k
−300	24.79	34.705	1.2672	0.7280	5,124.7	0.2174	0.1553	1.400
−200	40.32	56.449	1.3730	3.9972	1,518.0	0.2174	0.1553	1.400
−100	55.86	78.195	1.4438	12.508	671.94	0.2175	0.1554	1.400
0	71.43	99.976	1.4973	29.568	363.26	0.2182	0.1561	1.398
77	83.49	116.82	1.5312	51.020	245.79	0.2194	0.1573	1.395
100	87.12	121.87	1.5404	59.183	220.97	0.2199	0.1578	1.394
200	103.03	144.00	1.5767	106.29	145.02	0.2227	0.1606	1.387
300	119.27	166.45	1.6084	177.04	100.27	0.2264	0.1643	1.378
400	135.91	189.29	1.6367	278.99	72.001	0.2305	0.1684	1.369
500	152.96	212.56	1.6623	421.29	53.227	0.2347	0.1726	1.360
600	170.43	236.24	1.6857	614.79	40.275	0.2389	0.1768	1.351
700	188.30	260.32	1.7074	872.14	31.070	0.2428	0.1807	1.344
800	206.55	284.78	1.7277	1,207.9	24.368	0.2463	0.1842	1.337
900	225.14	309.58	1.7466	1,638.7	19.387	0.2496	0.1875	1.331
1,000	244.04	334.68	1.7644	2,183.3	15.622	0.2525	0.1904	1.326
1,100	263.21	360.07	1.7813	2,862.4	12.732	0.2551	0.1930	1.322
1,200	282.64	385.70	1.7972	3,699.5	10.483	0.2575	0.1954	1.318
1,300	302.28	411.56	1.8123	4,720.0	8.7114	0.2596	0.1975	1.314
1,400	322.14	437.63	1.8267	5,952.2	7.3006	0.2616	0.1995	1.311
1,500	342.17	463.87	1.8405	7,426.9	6.1656	0.2633	0.2012	1.309
1,600	362.37	490.28	1.8536	9,177.6	5.2440	0.2649	0.2028	1.306
1,700	382.73	516.85	1.8662	11,241.	4.4893	0.2664	0.2043	1.304
1,800	403.22	543.55	1.8783	13,656.	3.8664	0.2677	0.2056	1.302
1,900	423.85	570.39	1.8899	16,467.	3.3484	0.2690	0.2069	1.300
2,000	444.60	597.35	1.9011	19,718.	2.9148	0.2702	0.2081	1.298
2,100	465.47	624.43	1.9119	23,460.	2.5495	0.2713	0.2092	1.297
2,200	486.45	651.62	1.9223	27,745.	2.2399	0.2724	0.2103	1.295
2,300	507.53	678.92	1.9324	32,632.	1.9761	0.2735	0.2114	1.294
2,400	528.72	706.32	1.9421	38,181.	1.7501	0.2745	0.2124	1.292
2,500	550.01	733.82	1.9516	44,458.	1.5556	0.2755	0.2134	1.291
2,600	571.40	761.42	1.9608	51,533.	1.3873	0.2765	0.2144	1.290
2,800	614.47	816.90	1.9783	68,381.	1.1139	0.2784	0.2163	1.287
3,000	657.92	872.77	1.9950	89,382.	0.9044	0.2803	0.2182	1.285
3,200	701.73	929.00	2.0108	115,280.	0.7418	0.2821	0.2200	1.282
3,400	745.91	985.60	2.0258	146,910.	0.6139	0.2839	0.2218	1.280
3,600	790.45	1,042.6	2.0402	185,210.	0.5122	0.2857	0.2236	1.278
3,800	835.34	1,099.9	2.0540	231,230.	0.4305	0.2874	0.2253	1.276
4,000	880.58	1,157.5	2.0672	286,110.	0.3642	0.2891	0.2270	1.274
4,200	926.16	1,215.5	2.0799	351,160.	0.3101	0.2908	0.2287	1.272
4,400	972.06	1,273.9	2.0922	427,770.	0.2655	0.2924	0.2303	1.270
4,600	1,018.3	1,332.5	2.1040	517,510.	0.2285	0.2940	0.2319	1.268
4,800	1,064.8	1,391.5	2.1154	622,060.	0.1976	0.2955	0.2334	1.266
5,000	1,111.7	1,450.7	2.1265	743,290.	0.1716	0.2970	0.2349	1.264

Properties generated from the program *GASPROPS,* Wiley Professional Software, Wiley, New York, 1984.

TABLE E.6 Ideal Gas Properties of WATER VAPOR at Low Pressure—USCS Units

T, °F	u, Btu/lbm	h, Btu/lbm	s_0, Btu/(lbm·°R)	P_0	v_0	c_p, Btu/(lbm·°R)	c_v, Btu/(lbm·°R)	k
50	168.12	224.34	2.4802	0.0582	115,150.	0.4445	0.3342	1.330
77	177.15	236.35	2.4895	0.0610	115,720.	0.4453	0.3350	1.329
100	184.86	246.60	2.4896	0.0634	116,210.	0.4461	0.3358	1.328
200	218.66	291.43	2.5595	0.1194	72,654.	0.4507	0.3404	1.324
300	253.00	336.79	2.6584	0.2929	34,120.	0.4568	0.3465	1.318
400	288.00	382.83	2.7153	0.4907	23,049.	0.4640	0.3537	1.312
500	323.75	429.61	2.7668	0.7825	16,134.	0.4717	0.3614	1.305
600	360.30	477.19	2.8140	1.1997	11,620.	0.4799	0.3696	1.298
700	397.68	525.60	2.8576	1.7820	8,561.0	0.4884	0.3781	1.292
800	435.92	574.87	2.8984	2.5790	6,425.6	0.4971	0.3868	1.285
900	475.04	625.02	2.9367	3.6491	4,901.7	0.5060	0.3957	1.279
1,000	515.07	676.08	2.9729	5.0681	3,788.9	0.5152	0.4048	1.272
1,100	556.02	728.06	3.0073	6.9246	2,963.1	0.5245	0.4142	1.266
1,200	597.91	780.98	3.0402	9.3293	2,340.3	0.5339	0.4236	1.260
1,300	640.74	834.84	3.0717	12.415	1,864.6	0.5433	0.4330	1.255
1,400	684.51	889.64	3.1020	16.341	1,497.1	0.5528	0.4424	1.249
1,500	729.23	945.40	3.1312	21.284	1,211.3	0.5621	0.4518	1.244
1,600	774.84	1,002.0	3.1594	27.490	985.67	0.5713	0.4610	1.239
1,700	821.39	1,059.6	3.1867	35.027	806.98	0.5802	0.4699	1.235
1,800	868.85	1,118.1	3.2132	44.744	664.38	0.5890	0.4787	1.230
1,900	917.14	1,177.4	3.2389	56.491	549.51	0.5975	0.4872	1.226
2,000	966.27	1,237.6	3.2639	70.853	456.69	0.6057	0.4954	1.223
2,100	1,016.2	1,298.6	3.2882	88.311	381.30	0.6137	0.5034	1.219
2,200	1,066.9	1,360.3	3.3118	109.45	319.69	0.6213	0.5110	1.216
2,300	1,118.4	1,422.8	3.3349	134.89	269.14	0.6286	0.5183	1.213
2,400	1,170.0	1,486.0	3.3574	165.37	227.48	0.6357	0.5254	1.210
2,500	1,223.1	1,549.9	3.3793	201.79	192.95	0.6424	0.5320	1.207
2,600	1,277.0	1,614.5	3.4008	245.17	164.18	0.6488	0.5384	1.205
2,700	1,331.1	1,679.7	3.4217	296.32	140.28	0.6549	0.5446	1.203
2,800	1,385.9	1,745.4	3.4424	357.48	119.96	0.6607	0.5504	1.200
2,900	1,441.2	1,811.8	3.4617	425.76	103.81	0.6663	0.5560	1.198
3,000	1,497.1	1,878.7	3.4842	522.10	87.172	0.6716	0.5612	1.197
3,200	1,610.3	2,014.0	3.5199	722.06	66.676	0.6814	0.5711	1.193
3,400	1,725.5	2,151.2	3.5564	1,005.0	50.522	0.6904	0.5800	1.190
3,600	1,842.3	2,290.1	3.5915	1,381.5	38.658	0.6986	0.5882	1.188
3,800	1,960.7	2,430.6	3.6253	1,876.4	29.865	0.7060	0.5957	1.185
4,000	2,080.6	2,572.5	3.6578	2,520.5	23.277	0.7129	0.6026	1.183
4,200	2,201.7	2,715.7	3.6892	3,350.9	18.294	0.7193	0.6090	1.181
4,400	2,324.1	2,860.2	3.7196	4,413.5	14.485	0.7252	0.6149	1.179
4,600	2,447.7	3,005.8	3.7490	5,757.7	11.560	0.7308	0.6205	1.178
4,800	2,572.3	3,152.5	3.7774	7,452.8	9.2841	0.7360	0.6257	1.176
5,000	2,698.6	3,300.8	3.8049	9,559.1	7.5136	0.7409	0.6306	1.175

Properties generated from the program *GASPROPS*, Wiley Professional Software, Wiley, New York, 1984.

TABLE E.7 Ideal Gas Properties of CARBON DIOXIDE at Low Pressure—USCS Units

T, °F	u, Btu/lbm	h, Btu/lbm	s_0, Btu/(lbm·°R)	P_0	v_0	c_p, Btu/(lbm·°R)	c_v, Btu/(lbm·°R)	k
−300	18.07	25.278	0.9525	0.0145	353,760.	0.1583	0.1131	1.399
−200	30.10	41.823	1.0303	0.0812	102,820.	0.1632	0.1181	1.382
−100	41.87	58.109	1.0852	0.2741	42,172.	0.1757	0.1305	1.346
0	55.70	76.456	1.1301	0.7400	19,964.	0.1906	0.1454	1.310
77	67.31	91.543	1.1604	1.4496	11,898.	0.2017	0.1565	1.289
100	70.95	96.221	1.1690	1.7510	10,272.	0.2048	0.1596	1.283
200	87.56	117.35	1.2037	3.7764	5,613.9	0.2175	0.1723	1.262
300	105.37	139.67	1.2351	7.5842	3,219.1	0.2287	0.1836	1.246
400	124.23	163.03	1.2640	14.383	1,920.9	0.2386	0.1935	1.233
500	144.03	187.36	1.2908	26.010	1,185.8	0.2475	0.2024	1.223
600	164.67	212.52	1.3157	45.174	753.87	0.2555	0.2103	1.215
700	186.07	238.43	1.3391	75.792	491.72	0.2627	0.2175	1.208
800	208.15	265.03	1.3611	123.36	328.17	0.2691	0.2240	1.202
900	230.85	292.24	1.3819	195.47	223.55	0.2750	0.2299	1.196
1,000	254.10	320.01	1.4016	302.42	155.12	0.2803	0.2352	1.192
1,100	277.86	348.29	1.4203	457.95	109.45	0.2851	0.2400	1.188
1,200	302.08	377.02	1.4382	680.12	78.424	0.2895	0.2443	1.185
1,300	326.72	406.17	1.4552	992.79	56.962	0.2934	0.2483	1.182
1,400	351.72	435.69	1.4711	1,409.0	42.417	0.2970	0.2519	1.179
1,500	377.07	465.56	1.4888	1,752.9	36.992	0.3003	0.2551	1.177
1,600	402.73	495.73	1.4890	2,096.9	31.568	0.3032	0.2581	1.175
1,700	428.69	526.20	1.5166	3,860.8	17.977	0.3060	0.2608	1.173
1,800	454.90	556.93	1.5303	5,235.1	13.872	0.3084	0.2633	1.171
1,900	481.35	587.89	1.5441	7,098.7	10.683	0.3107	0.2656	1.170
2,000	507.99	619.05	1.5569	9,432.3	8.3805	0.3128	0.2676	1.169
2,100	534.85	650.43	1.5694	12,446.	6.6096	0.3146	0.2695	1.168
2,200	561.88	681.98	1.5815	16,264.	5.2556	0.3164	0.2712	1.166
2,300	589.08	713.69	1.5932	21,082.	4.2068	0.3180	0.2728	1.166
2,400	616.45	745.57	1.6045	27,080.	3.3937	0.3195	0.2743	1.165
2,500	643.97	777.60	1.6155	34,544.	2.7535	0.3209	0.2757	1.164
2,600	671.60	809.76	1.6262	43,774.	2.2463	0.3221	0.2770	1.163
2,700	699.33	842.00	1.6366	55,083.	1.8435	0.3233	0.2781	1.162
2,800	727.24	874.43	1.6467	68,876.	1.5210	0.3244	0.2793	1.162
2,900	755.34	907.04	1.6565	85,636.	1.2608	0.3255	0.2803	1.161
3,000	783.48	939.70	1.6661	105,844.	1.0505	0.3265	0.2813	1.161
3,200	839.25	1,004.5	1.6845	159,051.	0.7395	0.3282	0.2831	1.160
3,400	897.53	1,071.8	1.7020	234,410.	0.5292	0.3298	0.2847	1.159
3,600	948.22	1,131.5	1.7187	339,340.	0.3845	0.3312	0.2861	1.158
3,800	1,007.9	1,200.2	1.7346	483,080.	0.2834	0.3325	0.2873	1.157
4,000	1,069.0	1,270.4	1.7499	678,010.	0.2114	0.3336	0.2884	1.157
4,200	1,126.2	1,336.6	1.7646	937,850.	0.1597	0.3347	0.2895	1.156
4,400	1,184.3	1,403.7	1.7787	1,280,900.	0.1219	0.3356	0.2905	1.155
4,600	1,242.5	1,470.9	1.7922	1,730,000.	0.0940	0.3365	0.2914	1.155
4,800	1,300.8	1,538.3	1.8053	2,310,700.	0.0732	0.3374	0.2922	1.155
5,000	1,359.4	1,605.9	1.8179	3,052,700.	0.0575	0.3382	0.2930	1.154

Properties generated from the program *GASPROPS*, Wiley Professional Software, Wiley, New York, 1984.

TABLE E.8 Ideal Gas Properties of CARBON MONOXIDE at Low Pressure—USCS Units

T, °F	u, Btu/lbm	h, Btu/lbm	s_0, Btu/(lbm·°R)	P_0	v_0	c_P, Btu/(lbm·°R)	c_v, Btu/(lbm·°R)	k
−300	28.319	39.647	1.3842	0.0297	109,810.	0.2483	0.1774	1.400
−200	46.056	64.478	1.5049	0.1631	32,557.	0.2483	0.1774	1.400
−100	63.804	89.320	1.5859	0.5107	14,406.	0.2484	0.1774	1.400
−0	81.549	114.16	1.6468	1.2055	7,799.1	0.2485	0.1775	1.400
77	95.224	133.30	1.6853	2.0740	5,292.7	0.2487	0.1777	1.399
100	99.313	139.02	1.6958	2.4027	4,764.4	0.2488	0.1778	1.399
200	117.13	163.93	1.7367	4.2795	3,153.0	0.2496	0.1786	1.397
300	135.06	188.96	1.7720	7.0406	2,207.0	0.2511	0.1801	1.394
400	153.18	214.17	1.8032	10.926	1,609.4	0.2533	0.1823	1.389
500	171.55	239.63	1.8312	16.217	1,210.4	0.2561	0.1851	1.383
600	190.22	265.40	1.8568	23.244	932.48	0.2593	0.1884	1.377
700	209.23	291.51	1.8803	32.389	732.34	0.2628	0.1919	1.370
800	228.60	317.96	1.9022	44.092	584.35	0.2664	0.1954	1.363
900	248.32	344.78	1.9227	58.853	472.55	0.2700	0.1990	1.356
1,000	268.40	371.95	1.9419	77.233	386.58	0.2734	0.2025	1.350
1,100	288.82	399.47	1.9602	99.861	319.46	0.2768	0.2058	1.345
1,200	309.56	427.30	1.9775	127.43	266.39	0.2799	0.2090	1.339
1,300	330.60	455.44	1.9939	160.72	223.95	0.2829	0.2119	1.335
1,400	351.94	483.87	2.0096	200.56	189.66	0.2856	0.2147	1.330
1,500	373.53	512.56	2.0247	247.88	161.71	0.2882	0.2172	1.327
1,600	395.38	541.50	2.0391	303.67	138.73	0.2905	0.2196	1.323
1,700	417.45	570.66	2.0529	369.01	119.71	0.2927	0.2218	1.320
1,800	439.73	600.04	2.0662	445.07	103.85	0.2948	0.2238	1.317
1,900	462.20	629.61	2.0790	533.11	90.535	0.2966	0.2257	1.314
2,000	484.86	659.36	2.0914	634.46	79.296	0.2984	0.2274	1.312
2,100	507.68	689.28	2.1033	750.57	69.755	0.3000	0.2290	1.310
2,200	530.66	719.35	2.1148	882.95	61.613	0.3015	0.2305	1.308
2,300	553.78	749.57	2.1260	1,033.2	54.631	0.3028	0.2319	1.306
2,400	577.04	779.91	2.1368	1,203.2	48.615	0.3041	0.2332	1.304
2,500	600.41	810.39	2.1472	1,394.6	43.409	0.3053	0.2344	1.303
2,600	623.91	840.97	2.1574	1,609.4	38.886	0.3064	0.2355	1.301
2,800	671.21	902.47	2.1769	2,117.6	31.486	0.3084	0.2375	1.299
3,000	718.89	964.34	2.1953	2,745.3	25.776	0.3102	0.2393	1.296
3,200	766.91	1,026.5	2.2128	3,512.3	21.312	0.3118	0.2408	1.295
3,400	815.22	1,089.0	2.2294	4,440.0	17.781	0.3132	0.2422	1.293
3,600	863.80	1,151.8	2.2452	5,551.9	14.957	0.3145	0.2435	1.291
3,800	912.61	1,214.8	2.2604	6,873.6	12.676	0.3156	0.2446	1.290
4,000	961.65	1,278.0	2.2749	8,432.8	10.817	0.3166	0.2457	1.289
4,200	1,010.9	1,341.5	2.2888	10,260.	9.2898	0.3176	0.2466	1.288
4,400	1,060.3	1,405.1	2.3022	12,387.	8.0248	0.3185	0.2475	1.287
4,600	1,109.9	1,468.8	2.3150	14,848.	6.9699	0.3193	0.2483	1.286
4,800	1,159.6	1,532.8	2.3274	17,682.	6.0843	0.3200	0.2491	1.285
5,000	1,209.5	1,596.8	2.3394	20,928.	5.3361	0.3207	0.2498	1.284

TABLE E.9 Properties of WATER in the Compressed Liquid State — USCS Units

P, psia	T, °F	v, ft³/lbm	u, Btu/lbm	h, Btu/lbm	s, Btu/(lbm·°R)
750					
	40	0.01598	8.0303	10.248	0.01620
	80	0.01604	47.849	50.075	0.09287
	120	0.01617	87.670	89.914	0.1641
	160	0.01636	127.50	129.77	0.2306
	200	0.01659	167.46	169.76	0.2931
	240	0.01688	207.66	210	0.3523
	280	0.01721	248.20	250.59	0.4087
	320	0.01760	289.18	291.63	0.4628
	360	0.01805	330.74	333.24	0.5148
	400	0.01858	373.08	375.65	0.5653
	440	0.01921	416.51	419.17	0.6148
	480	0.01997	461.51	464.28	0.6639
2000					
	40	0.01591	8.0144	13.904	0.01613
	80	0.01598	47.549	53.462	0.09229
	120	0.01611	87.145	93.108	0.1632
	160	0.01629	126.76	132.79	0.2294
	200	0.01653	166.49	172.61	0.2916
	240	0.01680	206.45	212.67	0.3506
	280	0.01713	246.71	253.05	0.4067
	320	0.01750	287.36	293.84	0.4604
	360	0.01794	328.51	335.15	0.5121
	400	0.01844	370.34	377.16	0.5621
	440	0.01903	413.10	420.14	0.6110
	480	0.01974	457.16	464.46	0.6592
	520	0.02061	503.08	510.71	0.7074
3250					
	40	0.01585	7.9866	17.519	0.01601
	80	0.01592	47.255	56.828	0.09169
	120	0.01605	86.637	96.290	0.1622
	160	0.01623	126.05	135.81	0.2282
	200	0.01646	165.56	175.46	0.2902
	240	0.01673	205.28	215.35	0.3489
	280	0.01705	245.28	255.53	0.4047
	320	0.01741	285.62	296.09	0.4581
	360	0.01783	326.40	337.12	0.5094
	400	0.01831	367.76	378.78	0.5590
	440	0.01887	409.93	421.28	0.6074
	480	0.01953	453.19	464.93	0.6548
	520	0.02033	497.96	510.19	0.7020
	560	0.02132	544.91	557.74	0.7496
	600	0.02263	595.16	608.77	0.7986
4500					
	40	0.01579	7.9476	21.094	0.01584
	80	0.01586	46.966	60.173	0.09108
	120	0.01599	86.143	99.462	0.1613
	160	0.01617	125.36	138.83	0.2270
	200	0.01640	164.66	178.32	0.2887

TABLE E.9 Properties of WATER in the Compressed Liquid State — USCS Units (*Continued*)

P, psia	*T*, °F	*v*, ft³/lbm	*u*, Btu/lbm	*h*, Btu/lbm	*s*, Btu/(lbm·°R)
	200	0.01640	164.66	178.32	0.2887
	240	0.01667	204.16	218.04	0.3472
	280	0.01697	243.90	258.04	0.4028
	320	0.01733	283.95	298.38	0.4559
	360	0.01773	324.39	339.15	0.5069
	400	0.01819	365.34	380.48	0.5561
	440	0.01872	406.97	422.56	0.6040
	480	0.01934	449.53	465.64	0.6508
	520	0.02008	493.36	510.08	0.6971
	560	0.02098	538.92	556.39	0.7434
	600	0.02211	586.94	605.35	0.7905
	640	0.02363	638.70	658.38	0.8396
5750					
	40	0.01573	7.8981	24.631	0.01563
	80	0.01580	46.681	63.497	0.09045
	120	0.01594	85.663	102.62	0.1604
	160	0.01612	124.68	141.84	0.2258
	200	0.01634	163.79	181.18	0.2873
	240	0.01660	203.07	220.74	0.3455
	280	0.01690	242.58	260.56	0.4009
	320	0.01724	282.36	300.71	0.4537
	360	0.01763	322.48	341.24	0.5044
	400	0.01808	363.04	382.27	0.5533
	440	0.01858	404.19	423.97	0.6007
	480	0.01917	446.15	466.55	0.6470
	520	0.01986	489.16	510.30	0.6926
	560	0.02068	533.60	555.61	0.7379
	600	0.02169	579.95	603.04	0.7836
	640	0.02298	629.01	653.46	0.8303
	680	0.02472	682.11	708.41	0.8793
7000					
	40	0.01567	7.8389	28.133	0.01537
	80	0.01575	46.402	66.802	0.08981
	120	0.01588	85.196	105.77	0.1595
	160	0.01606	124.03	144.84	0.2246
	200	0.01628	162.95	184.04	0.2859
	240	0.01654	202.03	223.45	0.3439
	280	0.01683	241.31	263.11	0.3991
	320	0.01716	280.83	303.06	0.4517
	360	0.01754	320.65	343.37	0.5021
	400	0.01797	360.86	384.13	0.5506
	440	0.01845	401.58	425.48	0.5976
	480	0.01901	442.99	467.62	0.6435
	520	0.01966	485.30	510.77	0.6884
	560	0.02042	528.80	555.25	0.7329
	600	0.02134	573.84	601.48	0.7774
	640	0.02247	620.94	650.05	0.8224
	680	0.02392	670.88	701.86	0.8686

TABLE E.10 Properties of SATURATED WATER—Temperature Table, USCS Units

T °F	P psia	v_l ft³/lbm	v_{lg} ft³/lbm	v_g ft³/lbm	u_l Btu/lbm	u_{lg} Btu/lbm	u_g Btu/lbm	h_l Btu/lbm	h_{lg} Btu/lbm	h_g Btu/lbm	s_l Btu/(lbm·°R)	s_{lg} Btu/(lbm·°R)	s_g Btu/(lbm·°R)
40	0.1217	0.01602	2443.5	2443.5	8.0337	1015.5	1023.5	8.0341	1070.5	1078.5	0.01621	2.1424	2.1586
50	0.1781	0.01602	1703.1	1703.1	18.051	1008.7	1026.8	18.051	1064.8	1082.9	0.03606	2.0893	2.1254
60	0.2563	0.01603	1206.3	1206.3	28.048	1002.1	1030.1	28.049	1059.3	1087.3	0.05548	2.0383	2.0938
70	0.3633	0.01605	867.42	867.44	38.040	995.26	1033.3	38.041	1053.7	1091.7	0.07453	1.9892	2.0637
80	0.5073	0.01607	632.66	632.68	48.032	988.57	1036.6	48.033	1048.0	1096.0	0.09322	1.9419	2.0351
90	0.6987	0.01610	467.64	467.66	58.023	981.88	1039.9	58.026	1042.3	1100.3	0.1116	1.8962	2.0078
100	0.9500	0.01613	350.04	350.06	68.014	975.09	1043.1	68.017	1036.6	1104.6	0.1296	1.8522	1.9818
110	1.2759	0.01617	265.14	265.16	78.004	968.30	1046.3	78.007	1030.9	1108.9	0.1473	1.8096	1.9569
120	1.6939	0.01620	203.09	203.11	87.992	961.51	1049.5	87.997	1025.2	1113.2	0.1646	1.7686	1.9332
130	2.2244	0.01625	157.21	157.23	97.979	954.72	1052.7	97.986	1019.4	1117.4	0.1817	1.7288	1.9105
140	2.8908	0.01629	122.91	122.93	107.97	947.83	1055.8	107.98	1013.6	1121.6	0.1985	1.6903	1.8888
150	3.7201	0.01634	97.009	97.025	117.96	940.94	1058.9	117.97	1007.7	1125.7	0.2151	1.6529	1.8680
160	4.7431	0.01639	77.246	77.262	127.96	934.04	1062.0	127.97	1001.8	1129.8	0.2313	1.6168	1.8481
170	5.9943	0.01645	62.029	62.045	137.96	927.14	1065.1	137.98	995.92	1133.9	0.2473	1.5817	1.8290
180	7.5125	0.01651	50.206	50.223	147.98	920.12	1068.1	148.00	989.90	1137.9	0.2631	1.5476	1.8107
190	9.3406	0.01657	40.944	40.961	158.01	913.09	1071.1	158.04	983.86	1141.9	0.2787	1.5143	1.7930
200	11.526	0.01663	33.630	33.647	168.05	905.95	1074.0	168.08	977.72	1145.8	0.2940	1.4821	1.7761
210	14.122	0.01670	27.809	27.826	178.11	898.79	1076.9	178.15	971.45	1149.6	0.3091	1.4507	1.7598
212.02	14.700	0.01671	26.783	26.800	180.14	897.36	1077.5	180.18	970.22	1150.4	0.3122	1.4444	1.7566
220	17.184	0.01677	23.142	23.159	188.18	891.52	1079.7	188.23	965.17	1153.4	0.3241	1.4200	1.7441
230	20.774	0.01684	19.376	19.393	198.27	884.23	1082.5	198.34	958.66	1157.0	0.3388	1.3901	1.7289
240	24.960	0.01692	16.315	16.332	208.39	876.81	1085.2	208.47	952.13	1160.6	0.3534	1.3609	1.7143
250	29.814	0.01700	13.813	13.830	218.52	869.28	1087.8	218.62	945.48	1164.1	0.3678	1.3323	1.7001
260	35.411	0.01708	11.755	11.772	228.68	861.72	1090.4	228.79	938.81	1167.6	0.3820	1.3044	1.6864
270	41.835	0.01717	10.052	10.069	238.86	854.04	1092.9	239.00	931.90	1170.9	0.3960	1.2772	1.6732
280	49.173	0.01726	8.6350	8.6523	249.07	846.33	1095.4	249.23	924.87	1174.1	0.4099	1.2504	1.6603
290	57.516	0.01735	7.4505	7.4679	259.31	838.39	1097.7	259.49	917.71	1177.2	0.4237	1.2241	1.6478
300	66.963	0.01745	6.4550	6.4725	269.57	830.33	1099.9	269.78	910.42	1180.2	0.4373	1.1983	1.6356
310	77.615	0.01755	5.6146	5.6321	279.86	822.24	1102.1	280.11	902.89	1183.0	0.4507	1.1731	1.6238
320	89.580	0.01765	4.9014	4.9191	290.18	814.02	1104.2	290.48	895.22	1185.7	0.4641	1.1482	1.6123
330	102.97	0.01776	4.2939	4.3117	300.54	805.56	1106.1	300.88	887.42	1188.3	0.4773	1.1237	1.6010
340	117.90	0.01787	3.7742	3.7921	310.94	797.06	1108.0	311.33	879.37	1190.7	0.4903	1.0997	1.5900
350	134.50	0.01799	3.3277	3.3457	321.37	788.33	1109.7	321.82	871.18	1193.0	0.5033	1.0759	1.5792
360	152.89	0.01811	2.9426	2.9607	331.85	779.45	1111.3	332.36	862.74	1195.1	0.5162	1.0525	1.5687
370	173.20	0.01823	2.6094	2.6276	342.37	770.43	1112.8	342.95	854.05	1197.0	0.5289	1.0294	1.5583
380	195.57	0.01836	2.3198	2.3382	352.94	761.16	1114.1	353.60	845.10	1198.7	0.5416	1.0065	1.5481
390	220.14	0.01850	2.0675	2.0860	363.56	751.74	1115.3	364.31	835.99	1200.3	0.5542	0.9839	1.5381
400	247.05	0.01864	1.8468	1.8654	374.23	742.17	1116.4	375.08	826.52	1201.6	0.5667	0.9615	1.5282
410	276.45	0.01878	1.6532	1.6720	384.96	732.34	1117.3	385.92	816.88	1202.8	0.5791	0.9393	1.5184
420	308.50	0.01894	1.4828	1.5017	395.76	722.24	1118.0	396.84	806.86	1203.7	0.5915	0.9172	1.5087
430	343.36	0.01909	1.3324	1.3515	406.63	711.97	1118.6	407.84	796.66	1204.5	0.6038	0.8954	1.4992

TABLE E.10 Properties of SATURATED WATER—Temperature Table, USCS Units *(Continued)*

T °F	P psia	v_l ft³/lbm	v_{lg} ft³/lbm	v_g ft³/lbm	u_l Btu/lbm	u_{lg} Btu/lbm	u_g Btu/lbm	h_l Btu/lbm	h_{lg} Btu/lbm	h_g Btu/lbm	s_l Btu/(lbm·°R)	s_{lg} Btu/(lbm·°R)	s_g Btu/(lbm·°R)
440	381.18	0.01926	1.1993	1.2186	417.57	701.43	1119.0	418.93	785.97	1204.9	0.6160	0.8737	1.4897
450	422.14	0.01943	1.0812	1.1006	428.59	690.61	1119.2	430.11	775.09	1205.2	0.6282	0.8520	1.4802
460	466.41	0.01962	0.9761	0.9957	439.70	679.50	1119.2	441.39	763.71	1205.1	0.6404	0.8304	1.4708
470	514.15	0.01981	0.8823	0.9021	450.89	668.11	1119.0	452.78	752.02	1204.8	0.6525	0.8089	1.4614
480	565.56	0.02001	0.7984	0.8184	462.19	656.41	1118.6	464.29	740.01	1204.3	0.6646	0.7875	1.4521
490	620.82	0.02022	0.7231	0.7433	473.60	644.40	1118.0	475.92	727.48	1203.4	0.6767	0.7660	1.4427
500	680.11	0.02044	0.6555	0.6759	485.12	631.98	1117.1	487.70	714.50	1202.2	0.6888	0.7445	1.4333
510	743.64	0.02067	0.5944	0.6151	496.77	619.23	1116.0	499.62	700.98	1200.6	0.7009	0.7229	1.4238
520	811.60	0.02092	0.5394	0.5603	508.56	605.94	1114.5	511.70	687.00	1198.7	0.7130	0.7013	1.4143
530	884.21	0.02118	0.4895	0.5107	520.50	592.30	1112.8	523.97	672.43	1196.4	0.7252	0.6795	1.4047
540	961.69	0.02146	0.4442	0.4657	532.61	578.19	1110.8	536.42	657.28	1193.7	0.7374	0.6575	1.3949
550	1044.2	0.02175	0.4031	0.4248	544.89	563.51	1108.4	549.10	641.40	1190.5	0.7497	0.6353	1.3850
560	1132.1	0.02207	0.3654	0.3875	557.38	548.32	1105.7	562.00	624.90	1186.9	0.7621	0.6128	1.3749
570	1225.5	0.02241	0.3311	0.3535	570.09	532.41	1102.5	575.17	607.53	1182.7	0.7745	0.5900	1.3645
580	1324.8	0.02278	0.2996	0.3224	583.05	515.75	1098.8	588.63	589.27	1177.9	0.7871	0.5668	1.3539
590	1430.1	0.02318	0.2706	0.2938	596.29	498.41	1094.7	602.42	569.98	1172.4	0.7999	0.5430	1.3429
600	1541.7	0.02362	0.2439	0.2675	609.85	480.05	1089.9	616.59	549.61	1166.2	0.8129	0.5187	1.3316
610	1660.0	0.02410	0.2191	0.2432	623.78	460.72	1084.5	631.18	528.02	1159.2	0.8261	0.4936	1.3197
620	1785.3	0.02464	0.1961	0.2207	638.14	440.16	1078.3	646.28	504.92	1151.2	0.8396	0.4677	1.3073
630	1917.9	0.02524	0.1746	0.1998	653.02	418.18	1071.2	661.98	480.12	1142.1	0.8535	0.4406	1.2941
640	2058.2	0.02592	0.1544	0.1803	668.53	394.37	1062.9	678.41	453.19	1131.6	0.8679	0.4122	1.2801
650	2206.6	0.02672	0.1352	0.1619	684.84	368.56	1053.4	695.75	423.75	1119.5	0.8830	0.3818	1.2648
660	2363.6	0.02766	0.1168	0.1445	702.18	339.82	1042.0	714.28	390.92	1105.2	0.8989	0.3492	1.2481
670	2529.7	0.02882	0.09908	0.1279	720.95	307.45	1028.4	734.45	353.75	1088.2	0.9161	0.3131	1.2292
680	2705.6	0.03033	0.08117	0.1115	741.89	269.31	1011.2	757.08	310.02	1067.1	0.9352	0.2720	1.2072
690	2892.1	0.03249	0.06236	0.09485	766.63	221.72	988.35	784.02	255.08	1039.1	0.9577	0.2219	1.1796
700	3090.8	0.03647	0.03912	0.07559	801.02	150.07	951.09	821.88	172.45	994.33	0.9894	0.1487	1.1381
705.18	3198.9	0.0498	—	0.0498	867.1	—	867.1	896.8	—	896.8	1.053	—	1.053

TABLE E.11 Properties of SATURATED WATER—Pressure Table, USCS Units

P psia	T °F	v_l ft³/lbm	v_{lg} ft³/lbm	v_g ft³/lbm	u_l Btu/lbm	u_{lg} Btu/lbm	u_g Btu/lbm	h_l Btu/lbm	h_{lg} Btu/lbm	h_g Btu/lbm	s_l Btu/(lbm·°R)	s_{lg} Btu/(lbm·°R)	s_g Btu/(lbm·°R)
1.0	101.71	0.01614	333.53	333.55	69.722	973.88	1043.6	69.725	1035.7	1105.4	0.1326	1.8448	1.9774
3.0	141.45	0.01630	118.70	118.72	109.41	946.89	1056.3	109.42	1012.8	1122.2	0.2009	1.6848	1.8857
5.0	162.22	0.01641	73.520	73.536	130.18	932.52	1062.7	130.19	1000.5	1130.7	0.2349	1.6089	1.8438
7.0	176.83	0.01649	53.642	53.658	144.80	922.30	1067.1	144.83	991.77	1136.6	0.2581	1.5583	1.8164
9.0	188.27	0.01656	42.393	42.410	156.27	914.33	1070.6	156.30	984.90	1141.2	0.2760	1.5200	1.7960
11.0	197.75	0.01662	35.133	35.150	165.78	907.52	1073.3	165.82	979.08	1144.9	0.2906	1.4892	1.7798
13.0	205.88	0.01667	30.048	30.065	173.96	901.74	1075.7	174.00	974.00	1148.0	0.3029	1.4635	1.7664
14.7	212.02	0.01671	26.783	26.800	180.14	897.36	1077.5	180.18	970.22	1150.4	0.3122	1.4444	1.7566
15.0	213.04	0.01672	26.281	26.298	181.17	896.53	1077.7	181.21	969.49	1150.7	0.3137	1.4412	1.7549
20.0	227.97	0.01683	20.077	20.094	196.23	885.67	1081.9	196.29	960.01	1156.3	0.3358	1.3961	1.7319
25.0	240.09	0.01692	16.291	16.308	208.48	876.72	1085.2	208.56	952.14	1160.7	0.3535	1.3606	1.7141
30.0	250.36	0.01700	13.732	13.749	218.89	869.01	1087.9	218.98	945.32	1164.3	0.3683	1.3313	1.6996
35.0	259.31	0.01708	11.884	11.901	227.98	862.22	1090.2	228.09	939.21	1167.3	0.3810	1.3064	1.6874
40.0	267.28	0.01715	10.484	10.501	236.09	856.21	1092.3	236.22	933.78	1170.0	0.3922	1.2845	1.6767
45.0	274.47	0.01721	9.3859	9.4031	243.43	850.57	1094.0	243.57	928.73	1172.3	0.4023	1.2651	1.6674

TABLE E.11 Properties of SATURATED WATER—Pressure Table, USCS Units *(Continued)*

P psia	T °F	v_l ft³/lbm	v_{lg} ft³/lbm	v_g ft³/lbm	u_l Btu/lbm	u_{lg} Btu/lbm	u_g Btu/lbm	h_l Btu/lbm	h_{lg} Btu/lbm	h_g Btu/lbm	s_l Btu/(lbm·°R)	s_{lg} Btu/(lbm·°R)	s_g Btu/(lbm·°R)
50.0	281.05	0.01727	8.5005	8.5178	250.14	845.46	1095.6	250.30	924.10	1174.4	0.4114	1.2476	1.6590
55.0	287.11	0.01732	7.7712	7.7885	256.35	840.65	1097.0	256.53	919.77	1176.3	0.4197	1.2317	1.6514
60.0	292.75	0.01738	7.1594	7.1768	262.13	836.17	1098.3	262.32	915.68	1178.0	0.4274	1.2170	1.6444
65.0	298.02	0.01743	6.6389	6.6563	267.53	831.97	1099.5	267.74	911.86	1179.6	0.4346	1.2034	1.6380
70.0	302.97	0.01748	6.1902	6.2077	272.63	827.97	1100.6	272.85	908.15	1181.0	0.4413	1.1908	1.6321
75.0	307.65	0.01752	5.7995	5.8170	277.44	824.16	1101.6	277.68	904.62	1182.3	0.4476	1.1790	1.6266
80.0	312.09	0.01757	5.4559	5.4735	282.01	820.49	1102.5	282.27	901.33	1183.6	0.4535	1.1679	1.6214
85.0	316.31	0.01761	5.1515	5.1691	286.37	817.03	1103.4	286.64	898.06	1184.7	0.4592	1.1573	1.6165
90.0	320.33	0.01765	4.8797	4.8974	290.53	813.67	1104.2	290.82	894.98	1185.8	0.4645	1.1474	1.6119
95.0	324.18	0.01770	4.6357	4.6534	294.51	810.49	1105.0	294.82	891.98	1186.8	0.4696	1.1379	1.6075
100	327.88	0.01774	4.4152	4.4329	298.34	807.36	1105.7	298.67	889.03	1187.7	0.4745	1.1289	1.6034
110	334.84	0.01781	4.0323	4.0501	305.57	801.43	1107.0	305.93	883.57	1189.5	0.4836	1.1121	1.5957
120	341.32	0.01789	3.7112	3.7291	312.32	795.88	1108.2	312.71	878.29	1191.0	0.4921	1.0965	1.5886
130	347.39	0.01796	3.4378	3.4558	318.64	790.56	1109.2	319.08	873.32	1192.4	0.4999	1.0821	1.5820
140	353.10	0.01802	3.2023	3.2203	324.61	785.59	1110.2	325.08	868.52	1193.6	0.5073	1.0686	1.5759
150	358.49	0.01809	2.9971	3.0152	330.27	780.73	1111.0	330.77	863.93	1194.7	0.5143	1.0559	1.5702
160	363.61	0.01815	2.8167	2.8348	335.64	776.16	1111.8	336.18	859.62	1195.8	0.5208	1.0441	1.5649
170	368.49	0.01821	2.6567	2.6749	340.77	771.73	1112.5	341.35	855.35	1196.7	0.5270	1.0329	1.5599
180	373.14	0.01827	2.5139	2.5322	345.68	767.52	1113.2	346.29	851.21	1197.5	0.5329	1.0222	1.5551
190	377.60	0.01833	2.3858	2.4041	350.39	763.41	1113.8	351.04	847.26	1198.3	0.5386	1.0119	1.5505
200	381.87	0.01839	2.2699	2.2883	354.92	759.38	1114.3	355.60	843.40	1199.0	0.5440	1.0022	1.5462
250	401.04	0.01865	1.8253	1.8440	375.35	741.15	1116.5	376.21	825.59	1201.8	0.5680	0.9591	1.5271
300	417.43	0.01890	1.5246	1.5435	392.98	724.82	1117.8	394.03	809.47	1203.5	0.5883	0.9229	1.5112
350	431.81	0.01912	1.3071	1.3262	408.61	710.09	1118.7	409.85	794.75	1204.6	0.6060	0.8914	1.4974
400	444.69	0.01934	1.1422	1.1615	422.73	696.37	1119.1	424.16	780.94	1205.1	0.6217	0.8635	1.4852
450	456.38	0.01955	1.0127	1.0323	435.67	683.53	1119.2	437.30	767.90	1205.2	0.6360	0.8382	1.4742
500	467.11	0.01975	0.9082	0.9280	447.65	671.45	1119.1	449.48	755.52	1205.0	0.6490	0.8151	1.4641
550	477.05	0.01995	0.8222	0.8421	458.85	659.85	1118.7	460.88	743.62	1204.5	0.6610	0.7938	1.4548
600	486.32	0.02014	0.7499	0.7700	469.38	648.82	1118.2	471.62	732.08	1203.7	0.6722	0.7739	1.4461
650	495.01	0.02032	0.6883	0.7086	479.36	638.24	1117.6	481.80	721.00	1202.8	0.6827	0.7553	1.4380
700	503.20	0.02051	0.6352	0.6557	488.84	627.96	1116.8	491.50	710.20	1201.7	0.6927	0.7376	1.4303
750	510.96	0.02069	0.5889	0.6096	497.90	617.90	1115.8	500.78	699.62	1200.4	0.7021	0.7208	1.4229
800	518.34	0.02087	0.5481	0.5690	506.59	608.21	1114.8	509.68	689.32	1199.0	0.7110	0.7049	1.4159
850	525.37	0.02106	0.5120	0.5331	514.95	598.75	1113.7	518.27	679.23	1197.5	0.7196	0.6895	1.4091
900	532.09	0.02124	0.4797	0.5009	523.02	589.38	1112.4	526.55	669.35	1195.9	0.7277	0.6749	1.4026
950	538.53	0.02141	0.4506	0.4720	530.82	580.28	1111.1	534.58	659.52	1194.1	0.7356	0.6607	1.3963
1000	544.72	0.02159	0.4243	0.4459	538.38	571.32	1109.7	542.38	649.82	1192.2	0.7432	0.6470	1.3902
1100	556.42	0.02196	0.3785	0.4005	552.88	553.82	1106.7	557.35	630.85	1188.2	0.7576	0.6209	1.3785
1200	567.33	0.02232	0.3400	0.3623	566.67	536.73	1103.4	571.62	612.18	1183.8	0.7712	0.5961	1.3673
1300	577.56	0.02269	0.3070	0.3297	579.86	519.94	1099.8	585.32	593.78	1179.1	0.7841	0.5724	1.3565
1400	587.20	0.02307	0.2784	0.3015	592.55	503.35	1095.9	598.53	575.47	1174.0	0.7963	0.5497	1.3460
1500	596.33	0.02346	0.2534	0.2769	604.83	486.87	1091.7	611.34	557.26	1168.6	0.8081	0.5277	1.3358
1600	605.00	0.02386	0.2312	0.2551	616.76	470.54	1087.3	623.82	538.98	1162.8	0.8195	0.5062	1.3257
1700	613.25	0.02427	0.2114	0.2357	628.40	454.20	1082.6	636.03	520.67	1156.7	0.8305	0.4853	1.3158
1800	621.13	0.02470	0.1936	0.2183	639.80	437.70	1077.5	648.03	502.17	1150.2	0.8412	0.4646	1.3058
1900	628.68	0.02516	0.1773	0.2025	651.03	421.17	1072.2	659.87	483.53	1143.4	0.8517	0.4442	1.2959
2000	635.92	0.02563	0.1625	0.1881	662.12	404.28	1066.4	671.60	464.50	1136.1	0.8620	0.4239	1.2859
2100	642.87	0.02614	0.1488	0.1749	673.13	387.17	1060.3	683.29	445.01	1128.3	0.8722	0.4036	1.2758
2200	649.57	0.02668	0.1360	0.1627	684.11	369.69	1053.8	694.97	425.03	1120.0	0.8823	0.3832	1.2655
2300	656.02	0.02726	0.1241	0.1514	695.13	351.67	1046.8	706.73	404.47	1111.2	0.8924	0.3626	1.2550
2400	662.24	0.02790	0.1129	0.1408	706.24	332.96	1039.2	718.63	383.07	1101.7	0.9026	0.3415	1.2441
2500	668.25	0.02860	0.1021	0.1307	717.54	313.36	1030.9	730.78	360.62	1091.4	0.9129	0.3198	1.2327
2600	674.07	0.02938	0.09182	0.1212	729.15	292.75	1021.9	743.28	336.92	1080.2	0.9235	0.2972	1.2207
2700	679.69	0.03028	0.08172	0.1120	741.20	270.60	1011.8	756.33	311.47	1067.8	0.9345	0.2734	1.2079
2800	685.14	0.03133	0.07177	0.1031	753.95	246.55	1000.5	770.18	283.72	1053.9	0.9462	0.2478	1.1940
2900	690.41	0.03260	0.06154	0.09414	767.77	219.45	987.22	785.26	252.44	1037.7	0.9588	0.2195	1.1783
3000	695.51	0.03427	0.05065	0.08492	783.43	187.54	970.97	802.45	215.65	1018.1	0.9732	0.1867	1.1599
3198.9	705.18	0.0498	—	0.0498	867.1	—	867.1	896.8	—	896.8	1.053	—	1.053

TABLE E.12 Properties of SUPERHEATED STEAM—USCS Units

P, psia	*T*, °F	*v*, ft³/lbm	*u*, Btu/lbm	*h*, Btu/lbm	*s*, Btu/(lbm·°R)
1.0					
	(101.71)	(333.55)	(1043.6)	(1105.4)	(1.9774)
	200.0	392.51	1077.3	1149.9	2.0505
	300.0	452.24	1111.7	1195.4	2.1148
	400.0	511.89	1146.8	1241.5	2.1718
	500.0	571.51	1182.6	1288.4	2.2233
	600.0	631.11	1219.2	1336.0	2.2705
	700.0	690.70	1256.6	1384.4	2.3142
	800.0	750.28	1294.8	1433.7	2.3549
	900.0	809.85	1334.0	1483.8	2.3932
	1000	869.43	1374.0	1534.9	2.4295
	1100	929.00	1415.0	1586.9	2.4639
	1200	988.57	1456.9	1639.8	2.4968
	1300	1048.1	1499.7	1693.7	2.5283
	1400	1107.7	1543.5	1748.5	2.5586
	1500	1167.3	1588.2	1804.2	2.5878
	1600	1226.9	1633.9	1860.9	2.6160
5.0					
	(162.22)	(73.536)	(1062.7)	(1130.7)	(1.8438)
	200.0	78.152	1076.0	1148.4	1.8713
	300.0	90.238	1111.1	1194.6	1.9365
	400.0	102.24	1146.4	1241.0	1.9939
	500.0	114.20	1182.3	1288.0	2.0456
	600.0	126.15	1219.0	1335.7	2.0929
	700.0	138.08	1256.4	1384.2	2.1366
	800.0	150.01	1294.7	1433.5	2.1774
	900.0	161.93	1333.9	1483.7	2.2158
	1000	173.86	1373.9	1534.8	2.2520
	1100	185.78	1414.9	1586.8	2.2865
	1200	197.69	1456.8	1639.7	2.3194
	1300	209.61	1499.7	1693.6	2.3509
	1400	221.53	1543.4	1748.4	2.3812
	1500	233.45	1588.2	1804.2	2.4104
	1600	245.36	1633.8	1860.8	2.4386
10.0					
	(193.21)	(38.412)	(1072.0)	(1143.1)	(1.7875)
	200.0	38.852	1074.5	1146.4	1.7925
	300.0	44.985	1110.3	1193.5	1.8590
	400.0	51.030	1145.9	1240.3	1.9169
	500.0	57.037	1182.0	1287.5	1.9689
	600.0	63.025	1218.7	1335.4	2.0163
	700.0	69.004	1256.2	1383.9	2.0600
	800.0	74.976	1294.6	1433.3	2.1009
	900.0	80.944	1333.7	1483.5	2.1393
	1000	86.909	1373.8	1534.7	2.1755
	1100	92.873	1414.8	1586.7	2.2100
	1200	98.835	1456.7	1639.6	2.2429
	1300	104.80	1499.6	1693.5	2.2744
	1400	110.76	1543.4	1748.3	2.3047
	1500	116.72	1588.1	1804.1	2.3339
	1600	122.68	1633.8	1860.8	2.3621

TABLE E.12 Properties of SUPERHEATED STEAM—USCS Units (Continued)

P, psia	T, °F	v, ft³/lbm	u, Btu/lbm	h, Btu/lbm	s, Btu/(lbm·°R)
14.70					
	(212.02)	(26.800)	(1077.5)	(1150.4)	(1.7566)
	300.0	30.516	1109.5	1192.5	1.8155
	400.0	34.657	1145.4	1239.7	1.8739
	500.0	38.760	1181.6	1287.1	1.9260
	600.0	42.844	1218.5	1335.0	1.9736
	700.0	46.918	1256.0	1383.7	2.0174
	800.0	50.985	1294.4	1433.1	2.0583
	900.0	55.049	1333.6	1483.4	2.0967
	1000	59.110	1373.7	1534.5	2.1330
	1100	63.169	1414.7	1586.6	2.1675
	1200	67.226	1456.7	1639.5	2.2004
	1300	71.283	1499.5	1693.4	2.2319
	1400	75.339	1543.3	1748.3	2.2622
	1500	79.394	1588.1	1804.0	2.2914
	1600	83.449	1633.7	1860.7	2.3197
25.0					
	(240.09)	(16.308)	(1085.2)	(1160.7)	(1.7141)
	300.0	17.830	1107.7	1190.2	1.7547
	400.0	20.304	1144.3	1238.3	1.8141
	500.0	22.738	1180.9	1286.1	1.8668
	600.0	25.153	1218.0	1334.3	1.9145
	700.0	27.557	1255.6	1383.1	1.9585
	800.0	29.955	1294.1	1432.7	1.9995
	900.0	32.350	1333.4	1483.0	2.0380
	1000	34.741	1373.5	1534.2	2.0743
	1100	37.131	1414.5	1586.3	2.1088
	1200	39.519	1456.5	1639.3	2.1418
	1300	41.906	1499.4	1693.3	2.1733
	1400	44.292	1543.2	1748.1	2.2036
	1500	46.678	1588.0	1803.9	2.2329
	1600	49.063	1633.6	1860.6	2.2611
50.0					
	(281.05)	(8.5178)	(1095.6)	(1174.4)	(1.6590)
	300.0	8.7713	1103.2	1184.4	1.6722
	400.0	10.060	1141.7	1234.8	1.7346
	500.0	11.304	1179.1	1283.7	1.7885
	600.0	12.528	1216.7	1332.6	1.8369
	700.0	13.742	1254.7	1381.8	1.8813
	800.0	14.949	1293.3	1431.6	1.9225
	900.0	16.152	1332.7	1482.2	1.9611
	1000	17.352	1373.0	1533.5	1.9975
	1100	18.550	1414.1	1585.7	2.0321
	1200	19.747	1456.1	1638.8	2.0651
	1300	20.943	1499.1	1692.8	2.0967
	1400	22.138	1542.9	1747.7	2.1271
	1500	23.332	1587.7	1803.6	2.1563
	1600	24.526	1633.4	1860.3	2.1846

TABLE E.12 Properties of SUPERHEATED STEAM — USCS Units *(Continued)*

P, psia	*T*, °F	*v*, ft³/lbm	*u*, Btu/lbm	*h*, Btu/lbm	*s*, Btu/(lbm · °R)
75.0					
	(307.65)	(5.8170)	(1101.6)	(1182.3)	(1.6266)
	400.0	6.6431	1138.9	1231.1	1.6866
	500.0	7.4923	1177.3	1281.3	1.7419
	600.0	8.3199	1215.4	1330.8	1.7910
	700.0	9.1364	1253.7	1380.5	1.8357
	800.0	9.9463	1292.5	1430.6	1.8772
	900.0	10.752	1332.1	1481.3	1.9160
	1000	11.555	1372.5	1532.8	1.9525
	1100	12.356	1413.7	1585.2	1.9872
	1200	13.156	1455.7	1638.3	2.0202
	1300	13.955	1498.7	1692.4	2.0518
	1400	14.753	1542.6	1747.4	2.0822
	1500	15.550	1587.4	1803.3	2.1115
	1600	16.347	1633.1	1860.0	2.1397
100					
	(327.88)	(4.4329)	(1105.7)	(1187.7)	(1.6034)
	400.0	4.9333	1136.0	1227.3	1.6515
	500.0	5.5858	1175.5	1278.8	1.7082
	600.0	6.2155	1214.1	1329.1	1.7580
	700.0	6.8336	1252.7	1379.2	1.8032
	800.0	7.4451	1291.8	1429.5	1.8449
	900.0	8.0524	1331.5	1480.5	1.8838
	1000	8.6569	1371.9	1532.1	1.9204
	1100	9.2595	1413.2	1584.6	1.9552
	1200	9.8607	1455.4	1637.8	1.9883
	1300	10.461	1498.4	1692.0	2.0199
	1400	11.060	1542.3	1747.0	2.0503
	1500	11.659	1587.2	1802.9	2.0796
	1600	12.257	1632.9	1859.7	2.1079
125					
	(344.40)	(3.5872)	(1108.7)	(1191.7)	(1.5852)
	400.0	3.9062	1133.0	1223.4	1.6233
	500.0	4.4414	1173.6	1276.3	1.6816
	600.0	4.9526	1212.7	1327.3	1.7322
	700.0	5.4519	1251.7	1377.8	1.7777
	800.0	5.9443	1291.0	1428.5	1.8196
	900.0	6.4325	1330.9	1479.6	1.8587
	1000	6.9179	1371.4	1531.4	1.8955
	1100	7.4014	1412.8	1584.0	1.9303
	1200	7.8835	1455.0	1637.3	1.9634
	1300	8.3645	1498.0	1691.5	1.9951
	1400	8.8448	1542.0	1746.6	2.0256
	1500	9.3245	1586.9	1802.6	2.0549
	1600	9.8037	1632.7	1859.5	2.0832

TABLE E.12 Properties of SUPERHEATED STEAM—USCS Units (*Continued*)

P, psia	*T*, °F	*v*, ft³/lbm	*u*, Btu/lbm	*h*, Btu/lbm	*s*, Btu/(lbm·°R)
150					
	(358.49)	(3.0152)	(1111.0)	(1194.7)	(1.5702)
	400.0	3.2202	1129.9	1219.3	1.5995
	500.0	3.6781	1171.6	1273.7	1.6595
	600.0	4.1106	1211.4	1325.5	1.7108
	700.0	4.5306	1250.7	1376.5	1.7567
	800.0	4.9437	1290.2	1427.4	1.7989
	900.0	5.3526	1330.2	1478.8	1.8381
	1000	5.7586	1370.9	1530.7	1.8750
	1100	6.1626	1412.3	1583.4	1.9099
	1200	6.5653	1454.6	1636.8	1.9431
	1300	6.9670	1497.7	1691.1	1.9749
	1400	7.3678	1541.7	1746.3	2.0053
	1500	7.7681	1586.6	1802.3	2.0347
	1600	8.1679	1632.4	1859.2	2.0630
175					
	(370.84)	(2.6016)	(1112.9)	(1197.1)	(1.5574)
	400.0	2.7292	1126.6	1215.0	1.5786
	500.0	3.1326	1169.7	1271.1	1.6404
	600.0	3.5090	1210.0	1323.7	1.6925
	700.0	3.8725	1249.7	1375.1	1.7389
	800.0	4.2290	1289.4	1426.4	1.7813
	900.0	4.5812	1329.6	1477.9	1.8207
	1000	4.9305	1370.4	1530.0	1.8577
	1100	5.2778	1411.9	1582.8	1.8926
	1200	5.6238	1454.2	1636.3	1.9259
	1300	5.9687	1497.4	1690.7	1.9577
	1400	6.3129	1541.4	1745.9	1.9882
	1500	6.6564	1586.4	1801.9	2.0175
	1600	6.9995	1632.2	1858.9	2.0459
200					
	(381.87)	(2.2883)	(1114.3)	(1199.0)	(1.5462)
	400.0	2.3600	1123.2	1210.6	1.5598
	500.0	2.7232	1167.6	1268.4	1.6235
	600.0	3.0576	1208.6	1321.8	1.6765
	700.0	3.3789	1248.7	1373.7	1.7233
	800.0	3.6929	1288.6	1425.3	1.7659
	900.0	4.0026	1328.9	1477.1	1.8055
	1000	4.3094	1369.8	1529.3	1.8426
	1100	4.6142	1411.4	1582.2	1.8776
	1200	4.9176	1453.8	1635.8	1.9109
	1300	5.2200	1497.0	1690.2	1.9428
	1400	5.5216	1541.1	1745.5	1.9733
	1500	5.8226	1586.1	1801.6	2.0027
	1600	6.1232	1632.0	1858.6	2.0310

TABLE E.12 Properties of SUPERHEATED STEAM—USCS Units (Continued)

P, psia	T, °F	v, ft³/lbm	u, Btu/lbm	h, Btu/lbm	s, Btu/(lbm·°R)
225					
	(391.88)	(2.0423)	(1115.5)	(1200.6)	(1.5362)
	400.0	2.0719	1119.6	1205.9	1.5425
	500.0	2.4044	1165.6	1265.7	1.6083
	600.0	2.7065	1207.2	1319.9	1.6621
	700.0	2.9949	1247.6	1372.3	1.7094
	800.0	3.2760	1287.8	1424.2	1.7523
	900.0	3.5526	1328.3	1476.2	1.7920
	1000	3.8263	1369.3	1528.6	1.8292
	1100	4.0981	1411.0	1581.6	1.8643
	1200	4.3684	1453.4	1635.3	1.8977
	1300	4.6377	1496.7	1689.8	1.9296
	1400	4.9062	1540.8	1745.1	1.9602
	1500	5.1742	1585.8	1801.3	1.9896
	1600	5.4416	1631.7	1858.3	2.0179
250					
	(401.04)	(1.8440)	(1116.5)	(1201.8)	(1.5271)
	500.0	2.1492	1163.5	1262.9	1.5945
	600.0	2.4255	1205.8	1318.1	1.6492
	700.0	2.6877	1246.6	1371.0	1.6969
	800.0	2.9424	1287.0	1423.2	1.7401
	900.0	3.1926	1327.7	1475.4	1.7799
	1000	3.4398	1368.8	1527.9	1.8172
	1100	3.6851	1410.5	1581.0	1.8524
	1200	3.9290	1453.0	1634.8	1.8859
	1300	4.1719	1496.4	1689.4	1.9178
	1400	4.4139	1540.5	1744.7	1.9484
	1500	4.6554	1585.6	1801.0	1.9778
	1600	4.8963	1631.5	1858.0	2.0062
275					
	(409.53)	(1.6806)	(1117.2)	(1202.7)	(1.5189)
	500.0	1.9401	1161.3	1260.0	1.5817
	600.0	2.1955	1204.4	1316.1	1.6373
	700.0	2.4363	1245.6	1369.6	1.6855
	800.0	2.6694	1286.2	1422.1	1.7289
	900.0	2.8980	1327.0	1474.5	1.7690
	1000	3.1236	1368.2	1527.2	1.8064
	1100	3.3473	1410.1	1580.4	1.8416
	1200	3.5695	1452.7	1634.3	1.8751
	1300	3.7907	1496.0	1688.9	1.9071
	1400	4.0111	1540.2	1744.4	1.9377
	1500	4.2309	1585.3	1800.6	1.9672
	1600	4.4502	1631.2	1857.7	1.9956

TABLE E.12 Properties of SUPERHEATED STEAM—USCS Units (*Continued*)

P, psia	*T*, °F	*v*, ft³/lbm	*u*, Btu/lbm	*h*, Btu/lbm	*s*, Btu/(lbm·°R)
300					
	(417.43)	(1.5435)	(1117.8)	(1203.5)	(1.5112)
	500.0	1.7656	1159.1	1257.1	1.5697
	600.0	2.0037	1203.0	1314.2	1.6263
	700.0	2.2267	1244.5	1368.1	1.6750
	800.0	2.4419	1285.4	1421.0	1.7187
	900.0	2.6525	1326.4	1473.6	1.7589
	1000	2.8601	1367.7	1526.5	1.7964
	1100	3.0657	1409.6	1579.8	1.8318
	1200	3.2699	1452.3	1633.8	1.8653
	1300	3.4731	1495.7	1688.5	1.8973
	1400	3.6754	1539.9	1744.0	1.9280
	1500	3.8772	1585.0	1800.3	1.9574
	1600	4.0784	1631.0	1857.4	1.9859
350					
	(431.81)	(1.3262)	(1118.7)	(1204.6)	(1.4974)
	500.0	1.4908	1154.5	1251.1	1.5478
	600.0	1.7021	1200.0	1310.3	1.6065
	700.0	1.8973	1242.4	1365.3	1.6561
	800.0	2.0844	1283.8	1418.8	1.7004
	900.0	2.2667	1325.1	1471.9	1.7409
	1000	2.4460	1366.6	1525.1	1.7787
	1100	2.6233	1408.7	1578.6	1.8142
	1200	2.7992	1451.5	1632.8	1.8478
	1300	2.9740	1495.0	1687.6	1.8799
	1400	3.1479	1539.3	1743.2	1.9106
	1500	3.3213	1584.5	1799.6	1.9402
	1600	3.4942	1630.5	1856.8	1.9687
400					
	(444.69)	(1.1615)	(1119.1)	(1205.1)	(1.4852)
	500.0	1.2839	1149.7	1244.8	1.5278
	600.0	1.4757	1197.0	1306.2	1.5889
	700.0	1.6501	1240.3	1362.4	1.6395
	800.0	1.8161	1282.2	1416.6	1.6844
	900.0	1.9773	1323.8	1470.1	1.7253
	1000	2.1354	1365.6	1523.6	1.7632
	1100	2.2915	1407.8	1577.4	1.7989
	1200	2.4461	1450.7	1631.8	1.8326
	1300	2.5996	1494.3	1686.8	1.8648
	1400	2.7523	1538.7	1742.5	1.8956
	1500	2.9045	1584.0	1799.0	1.9252
	1600	3.0561	1630.0	1856.3	1.9537

TABLE E.12 Properties of SUPERHEATED STEAM—USCS Units (Continued)

P, psia	T, °F	v, ft³/lbm	u, Btu/lbm	h, Btu/lbm	s, Btu/(lbm·°R)
450					
	(456.38)	(1.0323)	(1119.2)	(1205.2)	(1.4742)
	500.0	1.1222	1144.7	1238.1	1.5093
	600.0	1.2993	1193.9	1302.1	1.5729
	700.0	1.4578	1238.1	1359.5	1.6246
	800.0	1.6075	1280.5	1414.4	1.6700
	900.0	1.7522	1322.4	1468.4	1.7113
	1000	1.8938	1364.5	1522.2	1.7495
	1100	2.0334	1406.9	1576.2	1.7853
	1200	2.1714	1449.9	1630.8	1.8192
	1300	2.3084	1493.7	1685.9	1.8514
	1400	2.4446	1538.1	1741.7	1.8823
	1500	2.5802	1583.4	1798.3	1.9119
	1600	2.7153	1629.6	1855.7	1.9405
500					
	(467.11)	(0.9280)	(1119.1)	(1205.0)	(1.4641)
	500.0	0.9920	1139.3	1231.1	1.4919
	600.0	1.1580	1190.7	1297.9	1.5582
	700.0	1.3038	1235.9	1356.5	1.6111
	800.0	1.4405	1278.8	1412.1	1.6571
	900.0	1.5721	1321.1	1466.6	1.6987
	1000	1.7005	1363.4	1520.7	1.7371
	1100	1.8269	1406.0	1575.0	1.7731
	1200	1.9517	1449.1	1629.7	1.8071
	1300	2.0755	1493.0	1685.0	1.8394
	1400	2.1985	1537.5	1741.0	1.8704
	1500	2.3208	1582.9	1797.6	1.9000
	1600	2.4427	1629.1	1855.1	1.9286
550					
	(477.05)	(0.8421)	(1118.7)	(1204.5)	(1.4548)
	500.0	0.8848	1133.7	1223.7	1.4751
	600.0	1.0421	1187.5	1293.5	1.5445
	700.0	1.1777	1233.6	1353.5	1.5986
	800.0	1.3038	1277.2	1409.9	1.6452
	900.0	1.4247	1319.8	1464.8	1.6872
	1000	1.5424	1362.3	1519.3	1.7259
	1100	1.6579	1405.1	1573.8	1.7620
	1200	1.7720	1448.4	1628.7	1.7961
	1300	1.8849	1492.3	1684.1	1.8285
	1400	1.9971	1536.9	1740.2	1.8595
	1500	2.1086	1582.4	1797.0	1.8893
	1600	2.2197	1628.6	1854.5	1.9179

TABLE E.12 Properties of SUPERHEATED STEAM—USCS Units (*Continued*)

P, psia	*T*, °F	*v*, ft³/lbm	*u*, Btu/lbm	*h*, Btu/lbm	*s*, Btu/(lbm·°R)
600					
	(486.32)	(0.7700)	(1118.2)	(1203.7)	(1.4461)
	500.0	0.7945	1127.6	1215.9	1.4589
	550.0	0.8746	1157.8	1254.9	1.4986
	600.0	0.9453	1184.1	1289.1	1.5316
	650.0	1.0107	1208.4	1320.6	1.5607
	700.0	1.0726	1231.4	1350.5	1.5870
	750.0	1.1321	1253.6	1379.3	1.6114
	800.0	1.1899	1275.5	1407.6	1.6343
	850.0	1.2464	1297.0	1435.4	1.6559
	900.0	1.3019	1318.4	1463.0	1.6766
	950.0	1.3565	1339.8	1490.4	1.6964
	1000	1.4106	1361.2	1517.8	1.7155
	1050	1.4641	1382.6	1545.2	1.7340
	1100	1.5171	1404.1	1572.6	1.7518
	1150	1.5698	1425.8	1600.1	1.7692
	1200	1.6222	1447.6	1627.7	1.7860
	1250	1.6743	1469.5	1655.4	1.8025
	1300	1.7261	1491.6	1683.3	1.8186
	1350	1.7778	1513.9	1711.3	1.8343
	1400	1.8293	1536.3	1739.4	1.8496
	1450	1.8806	1559.0	1767.8	1.8647
	1500	1.9318	1581.8	1796.3	1.8794
	1550	1.9828	1604.9	1825.0	1.8939
	1600	2.0338	1628.1	1853.9	1.9081
800					
	(518.34)	(0.5690)	(1114.8)	(1199.0)	(1.4159)
	550.0	0.6152	1138.5	1229.5	1.4466
	600.0	0.6773	1169.7	1270.0	1.4857
	650.0	0.7322	1196.9	1305.3	1.5183
	700.0	0.7827	1221.9	1337.8	1.5469
	750.0	0.8305	1245.6	1368.5	1.5729
	800.0	0.8763	1268.5	1398.2	1.5969
	850.0	0.9206	1290.9	1427.2	1.6195
	900.0	0.9639	1313.0	1455.7	1.6408
	950.0	1.0063	1334.9	1483.9	1.6612
	1000	1.0480	1356.7	1511.9	1.6807
	1050	1.0892	1378.6	1539.8	1.6995
	1100	1.1299	1400.4	1567.7	1.7177
	1150	1.1702	1422.3	1595.6	1.7353
	1200	1.2102	1444.4	1623.5	1.7524
	1250	1.2499	1466.5	1651.6	1.7691
	1300	1.2894	1488.8	1679.1	1.7853
	1350	1.3286	1511.3	1708.0	1.8011
	1400	1.3677	1533.9	1736.4	1.8166
	1450	1.4067	1556.7	1764.9	1.8317
	1500	1.4454	1579.7	1793.7	1.8466
	1550	1.4841	1602.8	1822.5	1.8611
	1600	1.5227	1626.2	1851.6	1.8754

TABLE E.12 Properties of SUPERHEATED STEAM—USCS Units (Continued)

P, psia	T, °F	v, ft³/lbm	u, Btu/lbm	h, Btu/lbm	s, Btu/(lbm·°R)
1000					
	(544.72)	(0.4459)	(1109.7)	(1192.2)	(1.3902)
	550.0	0.4534	1114.6	1198.5	1.3964
	600.0	0.5137	1153.2	1248.3	1.4446
	650.0	0.5636	1184.4	1288.6	1.4819
	700.0	0.6079	1211.8	1324.3	1.5133
	750.0	0.6489	1237.1	1357.2	1.5411
	800.0	0.6877	1261.2	1388.5	1.5664
	850.0	0.7249	1284.6	1418.7	1.5900
	900.0	0.7609	1307.4	1448.2	1.6121
	950.0	0.7960	1329.9	1477.2	1.6330
	1000	0.8304	1352.2	1505.9	1.6530
	1050	0.8642	1374.4	1534.4	1.6722
	1100	0.8975	1396.6	1562.7	1.6907
	1150	0.9304	1418.9	1591.0	1.7085
	1200	0.9630	1441.2	1619.4	1.7259
	1250	0.9953	1463.5	1647.7	1.7427
	1300	1.0273	1486.0	1676.2	1.7591
	1350	1.0592	1508.7	1704.7	1.7751
	1400	1.0908	1531.4	1733.3	1.7907
	1450	1.1223	1554.4	1762.1	1.8059
	1500	1.1537	1577.5	1791.0	1.8209
	1550	1.1849	1600.8	1820.0	1.8355
	1600	1.2160	1624.2	1849.2	1.8499
1200					
	(567.33)	(0.3623)	(1103.4)	(1183.8)	(1.3673)
	600.0	0.4015	1134.0	1223.2	1.4051
	650.0	0.4496	1170.5	1270.3	1.4486
	700.0	0.4905	1200.9	1309.9	1.4834
	750.0	0.5273	1228.2	1345.4	1.5134
	800.0	0.5616	1253.7	1378.4	1.5402
	850.0	0.5942	1278.0	1410.0	1.5648
	900.0	0.6254	1301.6	1440.5	1.5876
	950.0	0.6557	1324.8	1470.4	1.6092
	1000	0.6852	1347.6	1499.7	1.6297
	1050	0.7141	1370.2	1528.8	1.6493
	1100	0.7425	1392.8	1557.7	1.6681
	1150	0.7705	1415.3	1586.4	1.6862
	1200	0.7982	1437.9	1615.1	1.7038
	1250	0.8255	1460.5	1643.8	1.7208
	1300	0.8526	1483.2	1672.6	1.7374
	1350	0.8795	1506.0	1701.3	1.7535
	1400	0.9062	1529.0	1730.2	1.7693
	1450	0.9328	1552.0	1759.2	1.7846
	1500	0.9592	1575.3	1788.3	1.7997
	1550	0.9854	1598.7	1817.5	1.8144
	1600	1.0116	1622.2	1846.9	1.8288

TABLE E.12 Properties of SUPERHEATED STEAM — USCS Units (*Continued*)

P, psia	*T*, °F	*v*, ft³/lbm	*u*, Btu/lbm	*h*, Btu/lbm	*s*, Btu/(lbm·°R)
1400					
	(587.20)	(0.3015)	(1095.9)	(1174.0)	(1.3460)
	600.0	0.3174	1110.6	1192.9	1.3640
	650.0	0.3666	1155.0	1250.0	1.4167
	700.0	0.4058	1189.3	1294.4	1.4559
	750.0	0.4400	1218.9	1332.8	1.4883
	800.0	0.4713	1245.9	1368.0	1.5168
	850.0	0.5006	1271.3	1401.0	1.5425
	900.0	0.5285	1295.7	1432.6	1.5662
	950.0	0.5554	1319.5	1463.4	1.5884
	1000	0.5814	1342.9	1493.5	1.6094
	1050	0.6069	1366.0	1523.2	1.6294
	1100	0.6318	1388.9	1552.6	1.6486
	1150	0.6563	1411.8	1581.8	1.6670
	1200	0.6804	1434.6	1610.9	1.6848
	1250	0.7043	1457.4	1639.9	1.7020
	1300	0.7278	1480.4	1668.9	1.7188
	1350	0.7512	1503.4	1698.0	1.7351
	1400	0.7744	1526.5	1727.1	1.7509
	1450	0.7974	1549.7	1756.3	1.7664
	1500	0.8203	1573.1	1785.6	1.7816
	1550	0.8430	1596.6	1815.0	1.7964
	1600	0.8656	1620.3	1844.5	1.8109
1600					
	(605.00)	(0.2551)	(1087.3)	(1162.8)	(1.3257)
	650.0	0.3025	1137.3	1226.9	1.3847
	700.0	0.3414	1176.6	1277.6	1.4295
	750.0	0.3740	1208.9	1319.6	1.4650
	800.0	0.4032	1237.7	1357.0	1.4953
	850.0	0.4302	1264.3	1391.7	1.5223
	900.0	0.4557	1289.7	1424.6	1.5469
	950.0	0.4800	1314.2	1456.3	1.5698
	1000	0.5035	1338.1	1487.2	1.5914
	1050	0.5264	1361.6	1517.5	1.6118
	1100	0.5487	1385.0	1547.4	1.6313
	1150	0.5706	1408.1	1577.1	1.6500
	1200	0.5921	1431.3	1606.6	1.6681
	1250	0.6133	1454.4	1636.0	1.6855
	1300	0.6343	1477.5	1665.3	1.7024
	1350	0.6550	1500.7	1694.6	1.7189
	1400	0.6755	1524.0	1724.0	1.7349
	1450	0.6959	1547.3	1753.4	1.7505
	1500	0.7161	1570.9	1782.9	1.7657
	1550	0.7362	1594.5	1812.5	1.7806
	1600	0.7561	1618.3	1842.2	1.7952

TABLE E.12 Properties of SUPERHEATED STEAM—USCS Units (*Continued*)

P, psia	*T*, °F	*v*, ft³/lbm	*u*, Btu/lbm	*h*, Btu/lbm	*s*, Btu/(lbm·°R)
1800					
	(621.13)	(0.2183)	(1077.5)	(1150.2)	(1.3058)
	650.0	0.2504	1116.4	1199.9	1.3512
	700.0	0.2904	1162.6	1259.4	1.4037
	750.0	0.3222	1198.3	1305.7	1.4428
	800.0	0.3500	1229.1	1345.7	1.4753
	850.0	0.3753	1257.1	1382.1	1.5036
	900.0	0.3989	1283.4	1416.3	1.5293
	950.0	0.4214	1308.7	1449.0	1.5529
	1000	0.4429	1333.2	1480.7	1.5750
	1050	0.4638	1357.2	1511.7	1.5959
	1100	0.4841	1381.0	1542.2	1.6157
	1150	0.5039	1404.5	1572.3	1.6348
	1200	0.5234	1427.9	1602.2	1.6531
	1250	0.5426	1451.2	1632.0	1.6707
	1300	0.5615	1474.6	1661.6	1.6878
	1350	0.5801	1498.0	1691.2	1.7044
	1400	0.5986	1521.4	1720.8	1.7205
	1450	0.6169	1545.0	1750.5	1.7362
	1500	0.6351	1568.6	1780.2	1.7516
	1550	0.6531	1592.4	1809.9	1.7666
	1600	0.6710	1616.3	1839.8	1.7813
2000					
	(635.92)	(0.1881)	(1066.4)	(1136.1)	(1.2859)
	650.0	0.2056	1090.6	1166.7	1.3138
	700.0	0.2487	1147.2	1239.2	1.3777
	750.0	0.2804	1187.0	1290.8	1.4213
	800.0	0.3072	1220.1	1333.8	1.4562
	850.0	0.3312	1249.7	1372.2	1.4861
	900.0	0.3534	1277.1	1407.9	1.5128
	950.0	0.3744	1303.1	1441.7	1.5372
	1000	0.3944	1328.2	1474.2	1.5599
	1050	0.4136	1352.8	1505.9	1.5813
	1100	0.4323	1376.9	1536.9	1.6015
	1150	0.4506	1400.8	1567.6	1.6208
	1200	0.4684	1424.5	1597.9	1.6394
	1250	0.4860	1448.1	1628.0	1.6572
	1300	0.5032	1471.7	1657.9	1.6745
	1350	0.5203	1495.3	1687.8	1.6913
	1400	0.5371	1518.9	1717.7	1.7075
	1450	0.5537	1542.6	1747.5	1.7234
	1500	0.5702	1566.4	1777.4	1.7388
	1550	0.5866	1590.3	1807.4	1.7539
	1600	0.6029	1614.3	1837.4	1.7687

TABLE E.12 Properties of SUPERHEATED STEAM—USCS Units (*Continued*)

P, psia	*T*, °F	*v*, ft³/lbm	*u*, Btu/lbm	*h*, Btu/lbm	*s*, Btu/(lbm·°R)
2500					
	(668.25)	(0.1307)	(1030.9)	(1091.4)	(1.2327)
	700.0	0.1684	1098.0	1175.9	1.3067
	750.0	0.2031	1154.8	1248.7	1.3683
	800.0	0.2291	1195.6	1301.7	1.4112
	850.0	0.2513	1229.8	1346.1	1.4458
	900.0	0.2712	1260.3	1385.8	1.4755
	950.0	0.2896	1288.6	1422.5	1.5021
	1000	0.3069	1315.4	1457.4	1.5264
	1050	0.3233	1341.3	1490.9	1.5490
	1100	0.3392	1366.6	1523.5	1.5702
	1150	0.3545	1391.4	1555.4	1.5903
	1200	0.3695	1415.9	1586.8	1.6096
	1250	0.3841	1440.1	1617.8	1.6280
	1300	0.3984	1464.3	1648.6	1.6457
	1350	0.4125	1488.4	1679.2	1.6629
	1400	0.4264	1512.5	1709.7	1.6795
	1450	0.4401	1536.6	1740.2	1.6957
	1500	0.4536	1560.7	1770.6	1.7114
	1550	0.4670	1585.0	1801.0	1.7267
	1600	0.4803	1609.3	1831.5	1.7417
3000					
	(695.51)	(0.08492)	(970.97)	(1018.1)	(1.1599)
	700.0	0.09792	1003.9	1058.3	1.1945
	750.0	0.1483	1114.1	1196.5	1.3117
	800.0	0.1758	1167.6	1265.2	1.3674
	850.0	0.1974	1208.0	1317.6	1.4083
	900.0	0.2160	1242.3	1362.2	1.4418
	950.0	0.2328	1273.3	1402.5	1.4709
	1000	0.2484	1302.1	1440.0	1.4970
	1050	0.2630	1329.5	1475.5	1.5209
	1100	0.2770	1355.9	1509.7	1.5432
	1150	0.2905	1381.7	1543.0	1.5642
	1200	0.3035	1407.0	1575.5	1.5841
	1250	0.3162	1432.0	1607.6	1.6031
	1300	0.3286	1456.8	1639.2	1.6214
	1350	0.3407	1481.4	1670.6	1.6390
	1400	0.3526	1506.0	1701.7	1.6559
	1450	0.3643	1530.5	1732.8	1.6724
	1500	0.3759	1555.0	1763.7	1.6884
	1550	0.3873	1579.6	1794.6	1.7040
	1600	0.3986	1604.2	1825.5	1.7192

TABLE E.12 Properties of SUPERHEATED STEAM — USCS Units (Continued)

P, psia	T, °F	v, ft³/lbm	u, Btu/lbm	h, Btu/lbm	s, Btu/(lbm·°R)
3500					
	800.0	0.1363	1134.6	1222.9	1.3225
	850.0	0.1583	1183.8	1286.3	1.3719
	900.0	0.1763	1223.0	1337.2	1.4101
	950.0	0.1921	1257.1	1381.6	1.4421
	1000	0.2065	1288.2	1421.9	1.4703
	1050	0.2199	1317.2	1459.7	1.4957
	1100	0.2326	1345.0	1495.7	1.5192
	1150	0.2447	1371.8	1530.4	1.5410
	1200	0.2564	1398.0	1564.1	1.5617
	1250	0.2677	1423.8	1597.2	1.5813
	1300	0.2787	1449.2	1629.7	1.6001
	1350	0.2894	1474.4	1661.8	1.6181
	1400	0.2999	1499.4	1693.7	1.6354
	1450	0.3103	1524.4	1725.3	1.6522
	1500	0.3204	1549.3	1756.8	1.6685
	1550	0.3304	1574.2	1788.2	1.6843
	1600	0.3403	1599.1	1819.5	1.6997
4000					
	800.0	0.1052	1094.8	1172.7	1.2738
	850.0	0.1284	1156.9	1252.0	1.3356
	900.0	0.1462	1202.3	1310.6	1.3795
	950.0	0.1615	1240.1	1359.6	1.4150
	1000	0.1751	1273.7	1403.3	1.4454
	1050	0.1876	1304.6	1443.5	1.4725
	1100	0.1993	1333.8	1481.3	1.4972
	1150	0.2104	1361.8	1517.6	1.5200
	1200	0.2211	1388.9	1552.6	1.5415
	1250	0.2314	1415.4	1586.7	1.5617
	1300	0.2413	1441.5	1620.1	1.5810
	1350	0.2510	1467.2	1653.0	1.5994
	1400	0.2604	1492.8	1685.6	1.6172
	1450	0.2697	1518.2	1717.8	1.6343
	1500	0.2788	1543.5	1749.9	1.6509
	1550	0.2878	1568.7	1781.8	1.6669
	1600	0.2966	1594.0	1813.5	1.6825

TABLE E.12 Properties of SUPERHEATED STEAM—USCS Units (*Continued*)

P, psia	*T*, °F	*v*, ft³/lbm	*u*, Btu/lbm	*h*, Btu/lbm	*s*, Btu/(lbm·°R)
4500					
	800.0	0.07965	1045.3	1111.6	1.2186
	850.0	0.1047	1126.8	1214.0	1.2984
	900.0	0.1227	1180.0	1282.2	1.3496
	950.0	0.1376	1222.2	1336.8	1.3890
	1000	0.1506	1258.7	1384.1	1.4220
	1050	0.1624	1291.6	1426.9	1.4508
	1100	0.1734	1322.3	1466.8	1.4768
	1150	0.1838	1351.5	1504.6	1.5007
	1200	0.1936	1379.6	1540.9	1.5229
	1250	0.2031	1407.0	1576.1	1.5438
	1300	0.2122	1433.7	1610.5	1.5636
	1350	0.2211	1460.1	1644.2	1.5825
	1400	0.2298	1486.1	1677.5	1.6007
	1450	0.2382	1512.0	1710.3	1.6181
	1500	0.2465	1537.7	1742.9	1.6349
	1550	0.2546	1563.3	1775.3	1.6513
	1600	0.2626	1588.8	1807.5	1.6671
5000					
	800.0	0.05926	986.62	1041.5	1.1578
	850.0	0.08553	1093.1	1172.3	1.2599
	900.0	0.1038	1156.2	1252.2	1.3198
	950.0	0.1184	1203.5	1313.1	1.3638
	1000	0.1310	1243.1	1364.4	1.3996
	1050	0.1423	1278.3	1410.0	1.4303
	1100	0.1528	1310.7	1452.0	1.4577
	1150	0.1625	1341.1	1491.5	1.4826
	1200	0.1717	1370.2	1529.1	1.5057
	1250	0.1805	1398.4	1565.5	1.5272
	1300	0.1890	1425.9	1600.8	1.5476
	1350	0.1973	1452.8	1635.3	1.5670
	1400	0.2052	1479.4	1669.3	1.5855
	1450	0.2130	1505.7	1702.8	1.6033
	1500	0.2207	1531.8	1736.0	1.6204
	1550	0.2281	1557.8	1768.9	1.6370
	1600	0.2355	1583.7	1801.6	1.6530

Figure E.1 Temperature-entropy diagram for STEAM. (From J. H. Keenan, F. G. Keyes, P. G. Hill, and J. G. Moore, *Steam Tables*, Wiley, New York, 1969, used with permission.)

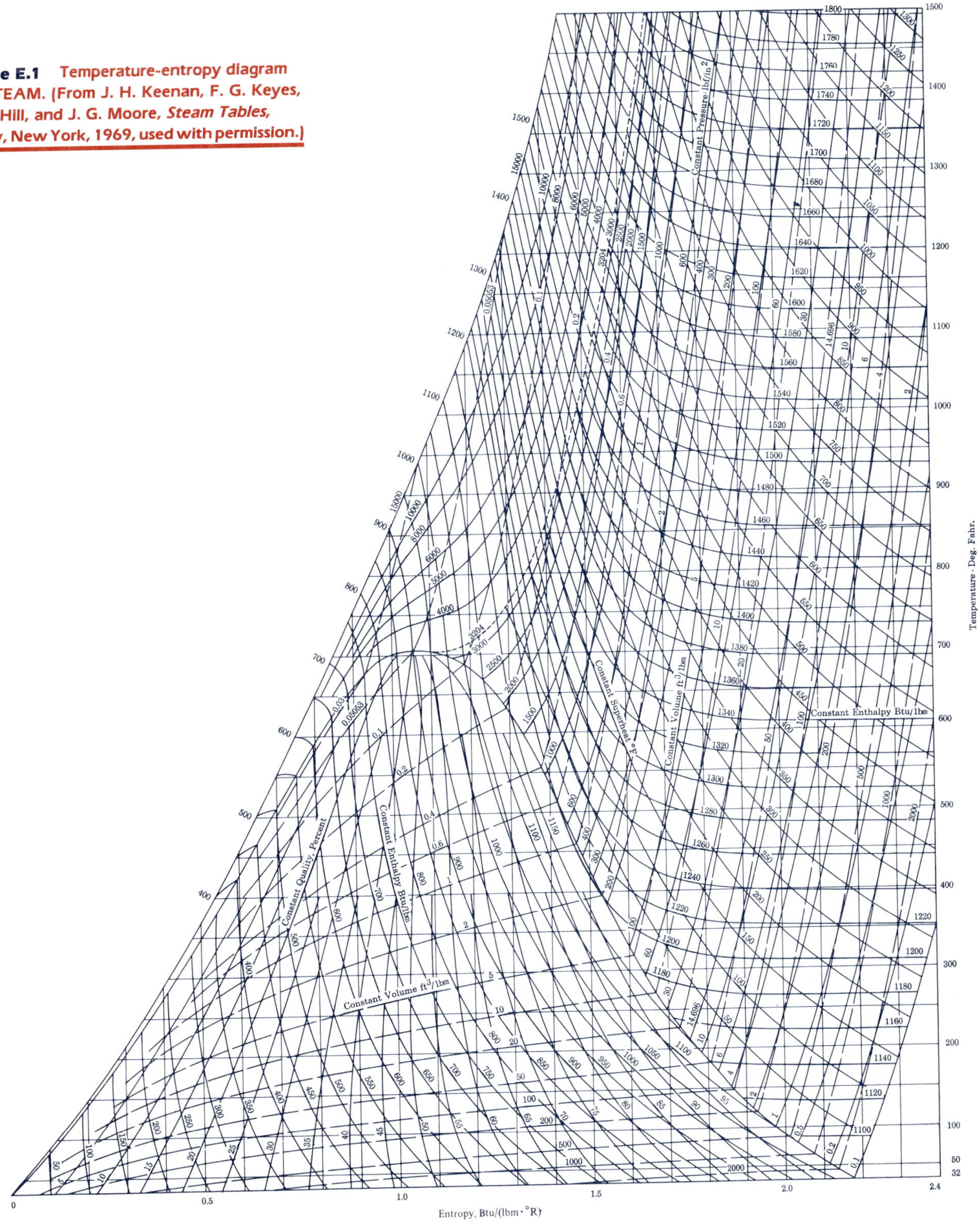

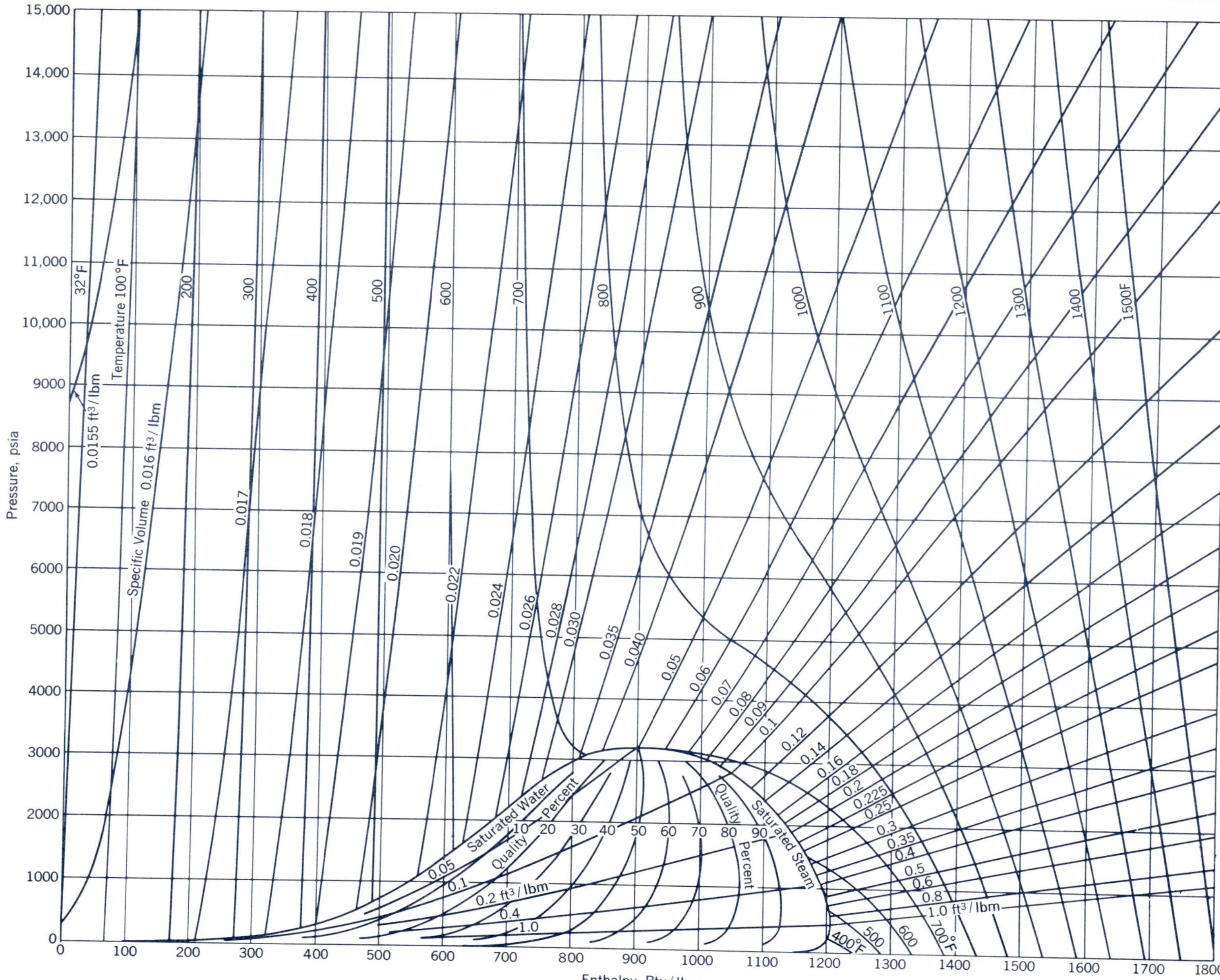

Figure E.2 Pressure-enthalpy diagram for STEAM. (From *Steam, Its Generation and Use*, 39th ed., Babcock and Wilcox Co., 1978.)

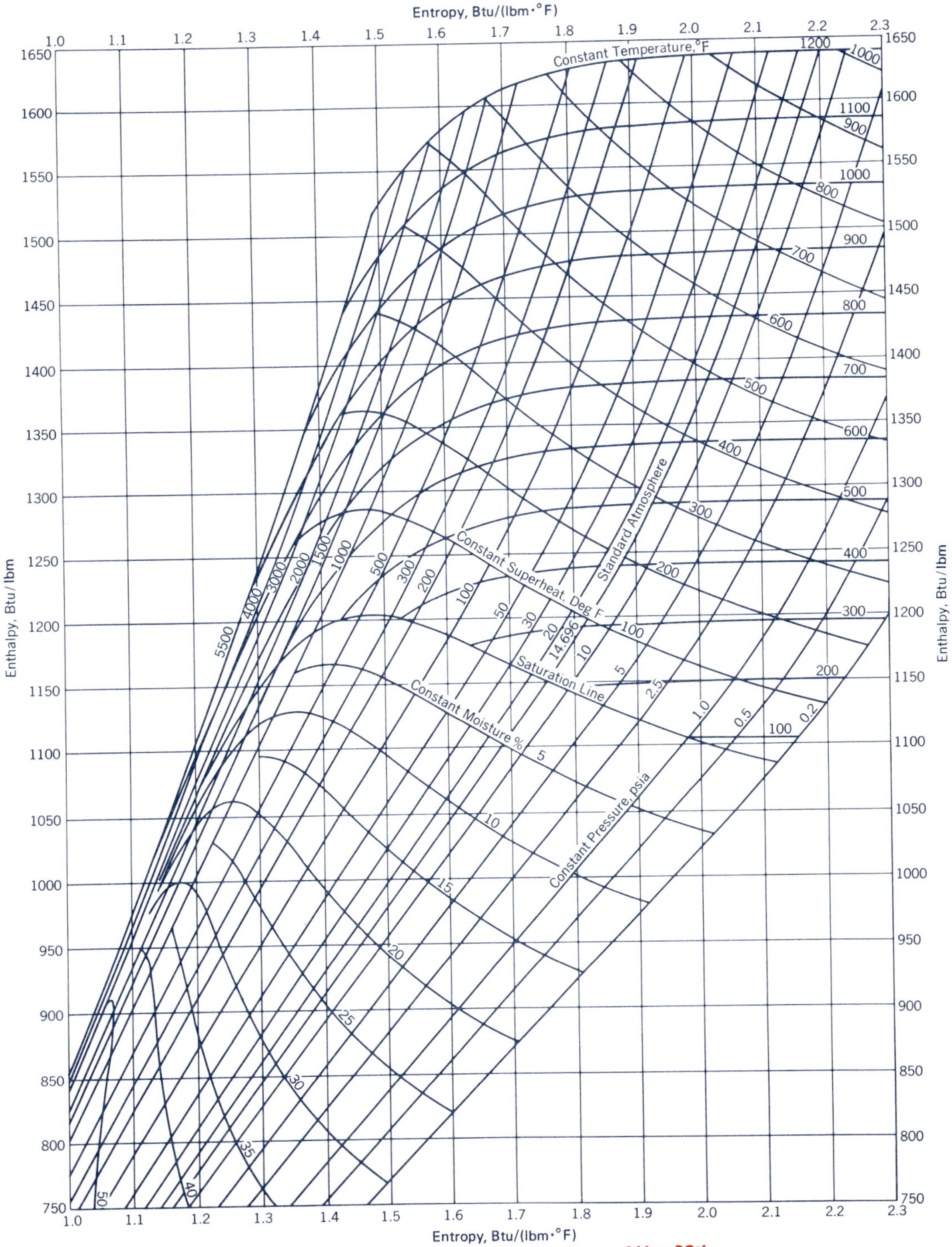

Figure E.3 Mollier diagram for STEAM. (From *Steam, Its Generation and Use*, 39th ed., Babcock and Wilcox Co., 1978.)

TABLE E.13 Properties of SATURATED REFRIGERANT 12—Temperature Table, USCS Units

Reference State: $h_l = s_l = 0$ at -40°F

		Specific Volume, ft³/lbm			Internal Energy, Btu/lbm			Enthalpy, Btu/lbm			Entropy, Btu/(lbm·°R)		
T_{sat}, °F	P_{sat}, psia	v_l	v_{lg}	v_g	u_l	u_{lg}	u_g	h_l	h_{lg}	h_g	s_l	s_{lg}	s_g
−150	0.2	0.009582	178.90	178.91	−22.703	78.461	55.758	−22.703	83.538	60.835	−0.06264	0.26976	0.20712
−140	0.3	0.009658	110.60	110.61	−20.659	77.315	56.656	−20.659	82.552	61.893	−0.05614	0.25824	0.20210
−130	0.4	0.009736	70.811	70.820	−18.617	76.186	57.570	−18.616	81.581	62.965	−0.04985	0.24746	0.19761
−120	0.6	0.009816	46.787	46.797	−16.574	75.071	58.497	−16.573	80.621	64.049	−0.04374	0.23735	0.19361
−110	1.0	0.009899	31.803	31.813	−14.529	73.966	59.437	−14.527	79.670	65.143	−0.03781	0.22784	0.19003
−100	1.4	0.009984	22.178	22.188	−12.481	72.869	60.388	−12.478	78.723	66.245	−0.03203	0.21887	0.18684
−90	2.0	0.010073	15.828	15.838	−10.428	71.777	61.348	−10.425	77.777	67.353	−0.02640	0.21040	0.18399
−80	2.9	0.010164	11.533	11.543	−8.357	70.673	62.316	−8.351	76.815	68.464	−0.02087	0.20232	0.18145
−70	4.0	0.010259	8.5656	8.5759	−6.286	69.577	63.290	−6.279	75.856	69.577	−0.01548	0.19467	0.17918
−60	5.4	0.010357	6.4722	6.4826	−4.208	68.477	64.269	−4.198	74.888	70.691	−0.01021	0.18738	0.17716
−50	7.1	0.010458	4.9675	4.9779	−2.121	67.372	65.252	−2.107	73.909	71.802	−0.00505	0.18041	0.17536
−40	9.3	0.010564	3.8672	3.8778	−0.024	66.260	66.236	0.000	72.910	72.910	0.00000	0.17376	0.17376
−30	12.0	0.010673	3.0500	3.0605	2.083	65.139	67.222	2.107	71.906	74.013	0.00498	0.16735	0.17233
−20	15.3	0.010787	2.4337	2.4445	4.200	64.007	68.207	4.231	70.878	75.109	0.00986	0.16121	0.17107
−10	19.2	0.010906	1.9630	1.9739	6.329	62.862	69.191	6.367	69.828	76.196	0.01466	0.15529	0.16994
0	23.8	0.011030	1.5988	1.6098	8.468	61.704	70.173	8.517	68.756	77.273	0.01937	0.14958	0.16895
10	29.3	0.011160	1.3137	1.3249	10.620	60.531	71.151	10.681	67.658	78.339	0.02402	0.14405	0.16807
20	35.7	0.011295	1.0882	1.0995	12.785	59.339	72.124	12.860	66.532	79.391	0.02859	0.13870	0.16730
30	43.1	0.011438	0.90797	0.91941	14.964	58.129	73.093	15.055	65.375	80.430	0.03311	0.13351	0.16661
40	51.6	0.011588	0.76235	0.77394	17.172	56.882	74.054	17.282	64.167	81.449	0.03759	0.12842	0.16601
50	61.4	0.011745	0.64395	0.65569	19.389	55.618	75.007	19.523	62.930	82.452	0.04201	0.12347	0.16548
60	72.4	0.011912	0.54673	0.55864	21.628	54.322	75.951	21.788	61.647	83.435	0.04638	0.11863	0.16501
70	84.9	0.012089	0.46631	0.47840	23.891	52.992	76.883	24.081	60.314	84.395	0.05072	0.11387	0.16459
80	98.8	0.012277	0.39926	0.41154	26.181	51.622	77.803	26.406	58.924	85.330	0.05503	0.10918	0.16422
90	114.4	0.012477	0.34299	0.35546	28.502	50.206	78.708	28.766	57.469	86.236	0.05933	0.10455	0.16388
100	131.8	0.012692	0.29541	0.30810	30.859	48.736	79.595	31.168	55.942	87.110	0.06362	0.09996	0.16357
110	151.1	0.012923	0.25494	0.26786	33.255	47.206	80.461	33.616	54.332	87.948	0.06790	0.09537	0.16328
120	172.3	0.013173	0.22019	0.23336	35.716	45.584	81.300	36.136	52.604	88.740	0.07223	0.09075	0.16298
130	195.6	0.013446	0.19028	0.20373	38.222	43.888	82.111	38.709	50.777	89.486	0.07657	0.08611	0.16268
140	221.2	0.013745	0.16433	0.17807	40.795	42.090	82.885	41.358	48.818	90.175	0.08095	0.08141	0.16236
150	249.2	0.014077	0.14166	0.15574	43.444	40.171	83.616	44.093	46.705	90.798	0.08539	0.07661	0.16200
160	279.7	0.014448	0.12165	0.13610	46.203	38.086	84.289	46.951	44.383	91.334	0.08995	0.07162	0.16157
170	312.9	0.014870	0.10393	0.11880	49.065	35.831	84.896	49.926	41.849	91.775	0.09461	0.06646	0.16107
180	348.9	0.015359	0.08800	0.10336	52.079	33.332	85.411	53.070	39.013	92.084	0.09945	0.06099	0.16044
190	387.9	0.015940	0.07352	0.08946	55.275	30.530	85.805	56.419	35.807	92.226	0.10451	0.05512	0.15963
200	430.0	0.016657	0.06012	0.07678	58.716	27.311	86.027	60.042	32.094	92.136	0.10989	0.04865	0.15855
210	475.4	0.017597	0.04731	0.06491	62.511	23.471	85.982	64.059	27.633	91.692	0.11577	0.04126	0.15703
220	524.3	0.018981	0.03423	0.05321	66.921	18.512	85.433	68.762	21.833	90.595	0.12253	0.03212	0.15466
232.7	582.0	0.02870	0	0.02870	75.69	0	75.69	78.86	0	78.86	0.1359	0	0.1359

Properties generated from the program *REFRIG*, Wiley Professional Software, Wiley, New York, 1985.

TABLE E.14 Properties of SATURATED REFRIGERANT 12—Pressure Table, USCS Units

Reference State: $h_l = s_l = 0$ at -40°F

		Specific Volume, ft³/lbm			Internal Energy, Btu/lbm			Enthalpy, Btu/lbm			Entropy, Btu/(lbm·°R)		
P_{sat}, psia	T_{sat}, °F	v_l	v_{lg}	v_g	u_l	u_{lg}	u_g	h_l	h_{lg}	h_g	s_l	s_{lg}	s_g
0.2	−144.9	0.009620	139.37	139.38	−21.662	77.875	56.213	−21.661	83.033	61.372	−0.05930	0.26380	0.20449
0.4	−130.6	0.009731	72.747	72.756	−18.745	76.257	57.512	−18.744	81.641	62.897	−0.05024	0.24812	0.19788
0.6	−121.5	0.009803	49.776	49.786	−16.889	75.242	58.353	−16.888	80.769	63.881	−0.04467	0.23887	0.19420
0.8	−114.7	0.009859	38.042	38.052	−15.496	74.487	58.991	−15.494	80.119	64.624	−0.04060	0.23227	0.19167
1.0	−109.2	0.009906	30.888	30.898	−14.368	73.879	59.511	−14.366	79.595	65.229	−0.03735	0.22712	0.18977
2.0	−90.7	0.010067	16.187	16.197	−10.569	71.851	61.282	−10.565	77.842	67.277	−0.02678	0.21096	0.18418
4.0	−69.7	0.010262	8.4890	8.4992	−6.222	69.542	63.321	−6.214	75.826	69.612	−0.01532	0.19443	0.17912
6.0	−56.0	0.010396	5.8196	5.8300	−3.385	68.042	64.657	−3.373	74.503	71.130	−0.00816	0.18459	0.17643
8.0	−45.7	0.010503	4.4508	4.4613	−1.216	66.893	65.677	−1.200	73.482	72.282	−0.00285	0.17750	0.17465
10.0	−37.2	0.010594	3.6140	3.6246	0.565	65.947	66.152	0.585	72.635	73.220	0.00141	0.17193	0.17334
20.0	−8.1	0.010929	1.8868	1.8977	6.733	62.644	69.377	6.774	69.627	76.401	0.01556	0.15419	0.16975
40.0	26.0	0.011379	0.97617	0.98755	14.081	58.621	72.702	14.165	65.847	80.011	0.03129	0.13559	0.16688
60.0	48.7	0.011724	0.65834	0.67006	19.094	55.787	74.881	19.224	63.097	82.321	0.04142	0.12412	0.16555
80.0	66.2	0.012021	0.49478	0.50680	23.038	53.496	76.534	23.216	60.821	84.037	0.04909	0.11565	0.16474
100.0	80.8	0.012292	0.39446	0.40675	26.364	51.512	77.876	26.591	58.812	85.403	0.05538	0.10882	0.16419
200.0	131.8	0.013497	0.18543	0.19892	38.673	43.578	82.251	39.172	50.441	89.613	0.07734	0.08528	0.16262
300.0	166.2	0.014703	0.11041	0.12512	47.964	36.711	84.675	48.781	42.840	91.621	0.09283	0.06845	0.16127
400.0	193.0	0.016135	0.06947	0.08560	56.263	29.629	85.893	57.458	34.771	92.229	0.10606	0.05328	0.15934
500.0	215.1	0.018225	0.04072	0.05895	64.670	21.120	85.789	66.356	24.888	91.243	0.11908	0.03688	0.15596
582.0	232.7	0.02870	0	0.02870	75.69	0	75.69	78.86	0	78.86	0.1359	0	0.1359

Properties generated from the program *REFRIG*, Wiley Professional Software, Wiley, New York, 1985.

TABLE E.15 Properties of SUPERHEATED REFRIGERANT 12—USCS Units

Reference State: $h_l = s_l = 0$ at -40°F

P, psia	T, °F	v, ft³/lbm	u, Btu/lbm	h, Btu/lbm	s, Btu/(lbm·°R)
0.4 (T_{sat} = −130.6°F)		(72.756)	(57.512)	(62.897)	(0.19788)
	−100	79.601	60.485	66.377	0.20799
	−50	90.748	65.703	72.420	0.22371
	0	101.875	71.339	78.880	0.23857
	50	112.989	77.356	85.720	0.25269
	100	124.096	83.720	92.905	0.26614
	150	135.199	90.397	100.404	0.27897
	200	146.299	97.356	108.185	0.29123
	250	157.397	104.570	116.221	0.30297
	300	168.493	112.013	124.485	0.31422
	350	179.620	119.662	132.957	0.32502
	400	190.711	127.496	141.612	0.33539

P, psia	T, °F	v, ft³/lbm	u, Btu/lbm	h, Btu/lbm	s, Btu/(lbm·°R)
0.6 (T_{sat} = −121.5°F)		(49.786)	(58.353)	(63.881)	(0.19420)
	−100	53.005	60.466	66.351	0.20128
	−50	60.455	65.690	72.402	0.21702
	0	67.885	71.330	78.867	0.23190
	50	75.302	77.349	85.710	0.24602
	100	82.712	83.715	92.898	0.25947
	150	90.117	90.392	100.398	0.27230
	200	97.520	97.353	108.180	0.28457
	250	104.920	104.567	116.217	0.29631
	300	112.319	112.011	124.482	0.30756
	350	119.718	119.660	132.952	0.31836
	400	127.116	127.494	141.608	0.32873

TABLE E.15 Properties of SUPERHEATED REFRIGERANT 12—USCS Units (*Continued*)

P, psia	*T*, °F	*v*, ft³/lbm	*u*, Btu/lbm	*h*, Btu/lbm	*s*, Btu/(lbm·°R)
0.8					
(T_{sat} = −114.7°F)		(38.052)	(58.991)	(64.624)	(0.19167)
	−100	39.708	60.447	66.326	0.19650
	−50	45.309	65.677	72.384	0.21226
	0	50.889	71.320	78.854	0.22715
	50	56.458	77.342	85.701	0.24128
	100	62.019	83.709	92.891	0.25474
	150	67.576	90.388	100.392	0.26757
	200	73.130	97.349	108.175	0.27984
	250	78.682	104.564	116.213	0.29158
	300	84.233	112.008	124.478	0.30283
	350	89.782	119.658	132.949	0.31363
	400	95.331	127.492	141.605	0.32400
1					
(T_{sat} = −109.2°F)		(30.898)	(59.511)	(65.229)	(0.18977)
	−100	31.729	60.428	66.300	0.19278
	−50	36.221	65.663	72.366	0.20857
	0	40.692	71.311	78.841	0.22347
	50	45.152	77.336	85.691	0.23761
	100	49.604	83.704	92.883	0.25106
	150	54.051	90.384	100.386	0.26390
	200	58.496	97.346	108.171	0.27617
	250	62.939	104.562	116.208	0.28791
	300	67.380	112.006	124.475	0.29917
	350	71.821	119.656	132.946	0.30997
	400	76.261	127.490	141.602	0.32034
2					
(T_{sat} = −90.7°F)		(16.197)	(61.282)	(67.277)	(0.18418)
	−50	18.045	65.597	72.276	0.19702
	0	20.298	71.263	78.776	0.21199
	50	22.539	77.301	85.642	0.22616
	100	24.773	83.677	92.846	0.23964
	150	27.002	90.363	100.357	0.25249
	200	29.229	97.329	108.146	0.26476
	250	31.453	104.548	116.188	0.27651
	300	33.676	111.994	124.457	0.28777
	350	35.898	119.645	132.931	0.29857
	400	38.120	127.481	141.589	0.30895
4					
(T_{sat} = −69.7°F)		(8.4992)	(63.321)	(69.612)	(0.17912)
	−50	8.955	65.464	72.092	0.18532
	0	10.101	71.168	78.645	0.20040
	50	11.233	77.230	85.545	0.21464
	100	12.358	83.624	92.771	0.22816
	150	13.478	90.321	100.297	0.24104
	200	14.595	97.295	108.098	0.25333
	250	15.710	104.519	116.148	0.26509
	300	16.824	111.970	124.423	0.27636
	350	17.937	119.624	132.901	0.28717
	400	19.049	127.462	141.562	0.29755

P, psia	*T*, °F	*v*, ft³/lbm	*u*, Btu/lbm	*h*, Btu/lbm	*s*, Btu/(lbm·°R)
6					
(T_{sat} = −56.0°F)		(5.8300)	(64.657)	(71.130)	(0.17643)
	−50	5.925	65.328	71.907	0.17833
	0	6.701	71.072	78.512	0.19354
	50	7.464	77.160	85.447	0.20785
	100	8.219	83.570	92.696	0.22141
	150	8.969	90.279	100.238	0.23432
	200	9.717	97.261	108.050	0.24663
	250	10.462	104.491	116.107	0.25840
	300	11.207	111.945	124.388	0.26967
	350	11.950	119.603	132.871	0.28049
	400	12.693	127.443	141.535	0.29087
8					
(T_{sat} = −45.7°F)		(4.4613)	(65.677)	(72.282)	(0.17465)
	0	5.0008	70.975	78.378	0.18861
	50	5.5789	77.089	85.348	0.20299
	100	6.1500	83.517	92.621	0.21660
	150	6.7154	90.237	100.179	0.22953
	200	7.2779	97.227	108.001	0.24186
	250	7.8386	104.463	116.067	0.25364
	300	8.3980	111.921	124.354	0.26492
	350	8.9564	119.581	132.841	0.27574
	400	9.5142	127.424	141.509	0.28612
10					
(T_{sat} = −37.2°F)		(3.6246)	(66.152)	(73.220)	(0.17334)
	0	3.9807	70.877	78.243	0.18473
	50	4.4481	77.017	85.248	0.19919
	100	4.9085	83.463	92.546	0.21284
	150	5.3630	90.195	100.119	0.22580
	200	5.8146	97.193	107.953	0.23815
	250	6.2643	104.434	116.027	0.24994
	300	6.7128	111.897	124.319	0.26123
	350	7.1603	119.560	132.810	0.27205
	400	7.6072	127.405	141.482	0.28244
20					
(T_{sat} = −8.1°F)		(1.8977)	(69.377)	(76.401)	(0.16975)
	0	1.9389	70.373	77.549	0.17226
	50	2.1857	76.654	84.743	0.18711
	100	2.4240	83.190	92.161	0.20099
	150	2.6572	89.983	99.817	0.21408
	200	2.8873	97.022	107.708	0.22652
	250	3.1159	104.292	115.824	0.23837
	300	3.3424	111.775	124.146	0.24970
	350	3.5681	119.454	132.659	0.26055
	400	3.7932	127.310	141.348	0.27096

TABLE E.15 Properties of SUPERHEATED REFRIGERANT 12—USCS Units (*Continued*)

P, psia	T, °F	v, ft³/lbm	u, Btu/lbm	h, Btu/lbm	s, Btu/(lbm·°R)
40					
(T_{sat} = 26.0°F)		(0.98755)	(72.702)	(80.011)	(0.16688)
	50	1.0526	75.889	83.680	0.17425
	100	1.1811	82.625	91.368	0.18863
	150	1.3040	89.548	99.200	0.20202
	200	1.4235	96.674	107.210	0.21464
	250	1.5408	104.004	115.409	0.22661
	300	1.6567	111.530	123.793	0.23802
	350	1.7717	119.239	132.353	0.24893
	400	1.8859	127.118	141.078	0.25938
60					
(T_{sat} = 48.7°F)		(0.67006)	(74.881)	(82.321)	(0.16555)
	50	0.6727	75.066	82.534	0.16596
	100	0.7659	82.034	90.538	0.18093
	150	0.8524	89.099	98.563	0.19465
	200	0.9353	96.318	106.702	0.20747
	250	1.0158	103.711	114.990	0.21957
	300	1.0948	111.281	123.437	0.23107
	350	1.1729	119.023	132.045	0.24203
	400	1.2502	126.925	140.806	0.25253
80					
(T_{sat} = 66.2°F)		(0.50680)	(76.543)	(84.037)	(0.16474)
	100	0.55730	81.410	89.661	0.17509
	150	0.62624	88.634	97.905	0.18919
	200	0.69093	95.953	106.182	0.20222
	250	0.75317	103.413	114.563	0.21446
	300	0.81383	111.030	123.078	0.22604
	350	0.87343	118.804	131.734	0.23707
	400	0.93230	126.732	140.533	0.24760
100					
(T_{sat} = 80.8°F)		(0.40673)	(77.876)	(85.403)	(0.16419)
	100	0.43135	80.749	88.731	0.17023
	150	0.49005	88.152	97.220	0.18474
	200	0.54414	95.580	105.649	0.19801
	250	0.59548	103.110	114.129	0.21039
	300	0.64516	110.775	122.714	0.22207
	350	0.69374	118.584	131.421	0.23316
	400	0.74155	126.536	140.259	0.24374
200					
(T_{sat} = 131.8°F)		(0.19892)	(82.251)	(89.613)	(0.16262)
	150	0.21368	85.370	93.278	0.16871
	200	0.24859	93.540	102.740	0.18359
	250	0.27909	101.504	111.833	0.19684
	300	0.30734	109.449	120.824	0.20906
	350	0.33409	117.449	129.814	0.22050
	400	0.35994	125.540	138.861	0.23132
400					
(T_{sat} = 193.0°F)		(0.08560)	(85.893)	(92.229)	(0.15934)
	200	0.09100	87.605	94.341	0.16254
	250	0.11743	97.606	106.298	0.17994
	300	0.13680	106.457	116.583	0.19389
	350	0.15352	114.992	126.356	0.20629
	400	0.16884	123.435	135.932	0.21773

Properties generated from the program *REFRIG*, Wiley Professional Software, Wiley, New York, 1985.

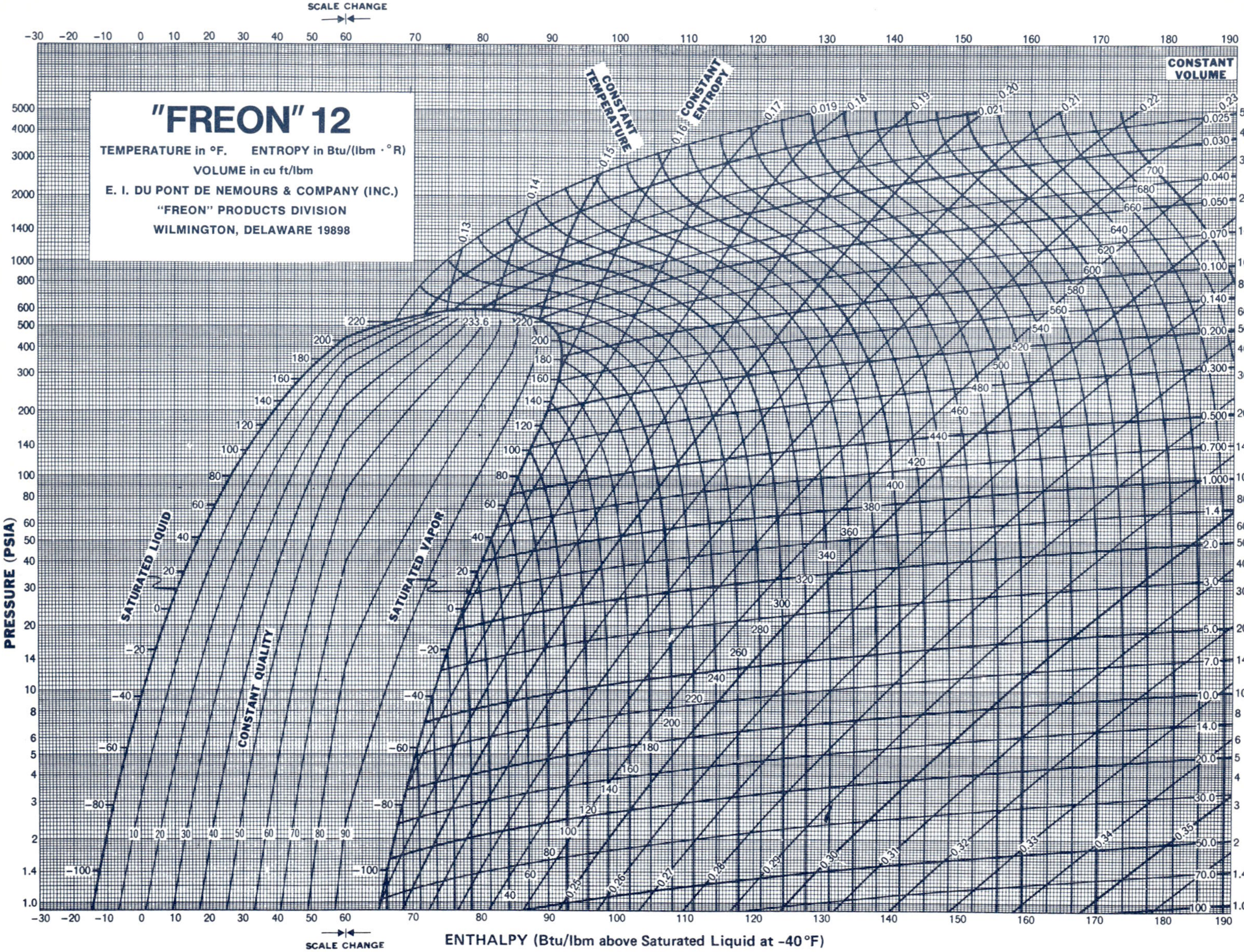

Figure E.4 Pressure-enthalpy diagram for REFRIGERANT 12. (Freon® 12 is the Dupont trademark for refrigerant 12.) (Copyright, E. I. Dupont de Nemours & Company, used with permission.)

TABLE E.16 Properties of SATURATED AMMONIA—Temperature Table, USCS Units

Reference State: $h_l = s_l = 0$ at -40°F and 1 atm

		Specific Volume, ft³/lbm			Internal Energy, Btu/lbm			Enthalpy, Btu/lbm			Entropy, Btu/(lbm·°R)		
T_{sat}, °F	P_{sat}, psia	v_l	v_{lg}	v_g	u_l	u_{lg}	u_g	h_l	h_{lg}	h_g	s_l	s_{lg}	s_g
−60	5.5	0.02277	44.933	44.955	−23.80	568.73	544.93	−23.78	614.66	590.88	−0.0580	1.5379	1.4799
−55	6.5	0.02288	38.509	38.532	−17.68	564.11	546.43	−17.66	610.53	592.88	−0.0428	1.5087	1.4659
−50	7.6	0.02299	33.152	33.175	−11.72	559.63	547.90	−11.69	606.53	594.84	−0.0281	1.4805	1.4524
−45	8.9	0.02310	28.663	28.686	−5.90	555.25	549.36	−5.86	602.63	596.77	−0.0140	1.4533	1.4393
−40	10.4	0.02321	24.877	24.901	−0.06	550.84	550.78	0.00	598.66	598.66	0.0000	1.4265	1.4265
−35	12.0	0.02333	21.679	21.702	5.58	546.60	552.18	5.64	594.89	600.53	0.0134	1.4008	1.4142
−30	13.9	0.02345	18.961	18.985	11.16	542.40	553.56	11.22	591.14	602.35	0.0264	1.3758	1.4022
−25	16.0	0.02356	16.642	16.666	16.67	538.24	554.91	16.74	587.40	604.14	0.0392	1.3514	1.3906
−20	18.3	0.02369	14.656	14.679	22.15	534.08	556.23	22.23	583.67	605.90	0.0517	1.3275	1.3792
−15	20.9	0.02381	12.947	12.971	27.60	529.93	557.53	27.69	579.92	607.61	0.0640	1.3042	1.3682
−10	23.7	0.02393	11.473	11.497	33.03	525.77	558.79	33.13	576.15	609.29	0.0762	1.2813	1.3575
−5	26.9	0.02406	10.196	10.220	38.44	521.59	560.03	38.56	572.36	610.92	0.0882	1.2588	1.3470
0	30.4	0.02419	9.0863	9.1105	43.86	517.38	561.24	44.00	568.51	612.51	0.1000	1.2368	1.3368
5	34.3	0.02432	8.1189	8.1433	49.28	513.13	562.42	49.44	564.62	614.06	0.1117	1.2151	1.3269
10	38.5	0.02446	7.2730	7.2975	54.71	508.85	563.56	54.89	560.68	615.57	0.1234	1.1938	1.3171
15	43.2	0.02459	6.5311	6.5557	60.16	504.52	564.68	60.35	556.67	617.03	0.1349	1.1728	1.3077
20	48.2	0.02473	5.8785	5.9032	65.62	500.15	565.77	65.84	552.60	618.44	0.1463	1.1521	1.2984
25	53.8	0.02488	5.3029	5.3277	71.10	495.72	566.82	71.34	548.47	619.81	0.1577	1.1316	1.2893
30	59.8	0.02502	4.7938	4.8188	76.59	491.24	567.84	76.87	544.26	621.13	0.1690	1.1115	1.2805
35	66.3	0.02517	4.3423	4.3675	82.11	486.71	568.82	82.42	539.97	622.40	0.1802	1.0916	1.2718
40	73.4	0.02533	3.9411	3.9664	87.65	482.12	569.77	88.00	535.62	623.61	0.1914	1.0719	1.2633
45	81.0	0.02548	3.5835	3.6090	93.21	477.47	570.68	93.59	531.18	624.77	0.2024	1.0525	1.2550
50	89.2	0.02564	3.2641	3.2898	98.79	472.76	571.55	99.21	526.67	625.88	0.2134	1.0334	1.2468
55	98.1	0.02580	2.9782	3.0040	104.39	468.00	572.39	104.86	522.07	626.93	0.2244	1.0144	1.2388
60	107.7	0.02597	2.7217	2.7476	110.01	463.18	573.18	110.52	517.40	627.92	0.2352	0.9956	1.2309
65	117.9	0.02614	2.4910	2.5172	115.64	458.29	573.94	116.21	512.64	628.85	0.2460	0.9771	1.2231
70	128.9	0.02632	2.2832	2.3095	121.30	453.35	574.65	121.92	507.79	629.72	0.2568	0.9587	1.2155
75	140.6	0.02650	2.0956	2.1221	126.97	448.34	575.31	127.66	502.86	630.52	0.2674	0.9405	1.2080
80	153.1	0.02668	1.9258	1.9525	132.66	443.27	575.92	133.41	497.84	631.25	0.2780	0.9225	1.2005
85	166.5	0.02687	1.7720	1.7989	138.36	438.13	576.49	139.19	492.72	631.91	0.2886	0.9046	1.1932
90	180.7	0.02706	1.6323	1.6594	144.08	432.92	577.00	144.99	487.51	632.50	0.2990	0.8869	1.1860
95	195.9	0.02726	1.5052	1.5325	149.82	427.64	577.46	150.81	482.20	633.01	0.3095	0.8693	1.1788
100	212.0	0.02747	1.3895	1.4169	155.58	422.28	577.86	156.66	476.78	633.44	0.3198	0.8519	1.1717
105	229.0	0.02768	1.2838	1.3114	161.36	416.84	578.21	162.54	471.25	633.78	0.3301	0.8346	1.1647
110	247.1	0.02790	1.1871	1.2150	167.17	411.32	578.49	168.44	465.60	634.04	0.3404	0.8173	1.1577
115	266.2	0.02813	1.0986	1.1267	172.99	405.71	578.70	174.38	459.83	634.21	0.3506	0.8002	1.1507
120	286.5	0.02836	1.0174	1.0458	178.85	399.99	578.84	180.35	453.93	634.28	0.3607	0.7831	1.1438

Properties generated from the program *REFRIG*, Wiley Professional Software, Wiley, New York, 1985.

TABLE E.17 Properties of SATURATED AMMONIA—Pressure Table, USCS Units

Reference State: $h_l = s_l = 0$ at -40°F and 1 atm

		Specific Volume, ft³/lbm			Internal Energy, Btu/lbm			Enthalpy, Btu/lbm			Entropy, Btu/(lbm·°R)		
P_{sat}, psia	T_{sat}, °F	v_l	v_{lg}	v_g	u_l	u_{lg}	u_g	h_l	h_{lg}	h_g	s_l	s_{lg}	s_g
6.0	−57.5	0.02283	41.588	41.612	−20.74	566.41	545.68	−20.71	612.59	591.88	−0.0503	1.5233	1.4729
7.0	−52.8	0.02293	36.005	36.027	−15.01	562.10	547.09	−14.98	608.74	593.75	−0.0362	1.4960	1.4599
8.0	−48.6	0.02302	31.776	31.799	−10.03	558.36	548.33	−10.00	605.40	595.40	−0.0240	1.4726	1.4486
9.0	−44.8	0.02311	28.463	28.486	−5.61	555.04	549.43	−5.58	602.44	596.87	−0.0133	1.4520	1.4387
10.0	−41.3	0.02319	25.786	25.809	−1.52	551.94	550.42	−1.48	599.66	598.18	−0.0035	1.4333	1.4298
11.0	−38.1	0.02326	23.585	23.608	2.12	549.21	551.32	2.17	597.22	599.38	0.0052	1.4166	1.4218
12.0	−35.1	0.02333	21.741	21.765	5.47	546.69	552.16	5.52	594.97	600.49	0.0131	1.4014	1.4145
13.0	−32.3	0.02339	20.172	20.196	8.57	544.35	552.92	8.63	592.88	601.51	0.0204	1.3874	1.4077
14.0	−29.7	0.02345	18.820	18.843	11.47	542.17	553.64	11.53	590.92	602.46	0.0271	1.3744	1.4015
15.0	−27.3	0.02351	17.642	17.666	14.19	540.11	554.31	14.26	589.08	603.34	0.0335	1.3623	1.3958
16.0	−24.9	0.02357	16.607	16.631	16.76	538.17	554.93	16.83	587.34	604.17	0.0394	1.3510	1.3904
17.0	−22.7	0.02362	15.690	15.713	19.20	536.32	555.52	19.27	585.68	604.96	0.0450	1.3403	1.3853
18.0	−20.6	0.02367	14.872	14.895	21.51	534.57	556.08	21.59	584.10	605.69	0.0503	1.3303	1.3805
19.0	−18.6	0.02372	14.137	14.161	23.72	532.89	556.61	23.81	582.59	606.39	0.0553	1.3207	1.3760
20.0	−16.6	0.02377	13.472	13.496	25.84	531.27	557.11	25.93	581.14	607.06	0.0601	1.3117	1.3717
25.0	−7.9	0.02398	10.926	10.950	35.25	524.05	559.30	35.36	574.60	609.96	0.0811	1.2720	1.3531
30.0	−0.6	0.02418	9.2042	9.2284	43.25	517.85	561.10	43.38	568.95	612.33	0.0987	1.2393	1.3379
35.0	5.9	0.02435	7.9598	7.9842	50.25	512.37	562.62	50.41	563.93	614.33	0.1138	1.2113	1.3251
40.0	11.7	0.02450	7.0166	7.0411	56.52	507.42	563.94	56.70	559.36	616.06	0.1272	1.1868	1.3140
45.0	16.9	0.02465	6.2767	6.3013	62.20	502.89	565.09	62.41	555.16	617.56	0.1392	1.1650	1.3042
50.0	21.7	0.02478	5.6799	5.7047	67.43	498.69	566.12	67.66	551.24	618.90	0.1501	1.1453	1.2954
55.0	26.1	0.02491	5.1882	5.2131	72.28	494.76	567.04	72.53	547.57	620.10	0.1601	1.1273	1.2874
60.0	30.2	0.02503	4.7758	4.8009	76.80	491.07	567.87	77.08	544.10	621.18	0.1694	1.1107	1.2801
65.0	34.0	0.02514	4.4245	4.4496	81.06	487.58	568.63	81.36	540.80	622.16	0.1781	1.0954	1.2734
70.0	37.7	0.02525	4.1218	4.1471	85.07	484.26	569.33	85.40	537.65	623.05	0.1862	1.0810	1.2672
75.0	41.1	0.02536	3.8582	3.8836	88.88	481.09	569.97	89.23	534.64	623.87	0.1938	1.0676	1.2614
80.0	44.4	0.02546	3.6265	3.6520	92.51	478.06	570.57	92.88	531.75	624.63	0.2010	1.0550	1.2560
85.0	47.5	0.02556	3.4210	3.4466	95.97	475.15	571.12	96.37	528.96	625.33	0.2079	1.0430	1.2509
90.0	50.4	0.02565	3.2375	3.2632	99.28	472.35	571.63	99.71	526.26	625.98	0.2144	1.0317	1.2461
95.0	53.3	0.02575	3.0728	3.0985	102.47	469.64	572.11	102.92	523.66	626.58	0.2206	1.0209	1.2415
100.0	56.0	0.02584	2.9238	2.9497	105.53	467.03	572.55	106.01	521.13	627.14	0.2266	1.0106	1.2372
110.0	61.2	0.02601	2.6651	2.6912	111.33	462.03	573.37	111.86	516.29	628.15	0.2378	0.9912	1.2290
120.0	66.0	0.02618	2.4484	2.4746	116.75	457.33	574.08	117.33	511.69	629.03	0.2482	0.9734	1.2216
130.0	70.5	0.02633	2.2636	2.2900	121.86	452.85	574.71	122.50	507.31	629.80	0.2578	0.9569	1.2147
140.0	74.8	0.02649	2.1043	2.1308	126.69	448.59	575.28	127.38	503.11	630.48	0.2669	0.9414	1.2083
150.0	78.8	0.02664	1.9657	1.9923	131.27	444.51	575.78	132.01	499.07	631.08	0.2755	0.9269	1.2023
160.0	82.6	0.02678	1.8437	1.8705	135.63	440.60	576.22	136.42	495.18	631.61	0.2835	0.9132	1.1967
170.0	86.3	0.02692	1.7355	1.7624	139.80	436.82	576.62	140.65	491.42	632.07	0.2912	0.9001	1.1914
180.0	89.8	0.02705	1.6389	1.6660	143.80	433.18	576.98	144.70	487.77	632.47	0.2985	0.8878	1.1863
190.0	93.1	0.02719	1.5522	1.5793	147.64	429.66	577.29	148.59	484.23	632.82	0.3055	0.8760	1.1815
200.0	96.3	0.02732	1.4739	1.5012	151.33	426.24	577.57	152.34	480.79	633.13	0.3122	0.8648	1.1769
220.0	102.4	0.02757	1.3377	1.3653	158.34	419.69	578.03	159.47	474.15	633.61	0.3247	0.8436	1.1683
240.0	108.1	0.02782	1.2234	1.2512	164.93	413.46	578.39	166.16	467.79	633.95	0.3364	0.8240	1.1604
260.0	113.4	0.02805	1.1261	1.1541	171.13	407.51	578.64	172.48	461.69	634.17	0.3473	0.8056	1.1529
280.0	118.4	0.02829	1.0421	1.0704	177.01	401.80	578.81	178.47	455.79	634.27	0.3575	0.7884	1.1460

Properties generated from the program *REFRIG*, Wiley Professional Software, Wiley, New York, 1985.

TABLE E.18 Properties of SUPERHEATED AMMONIA—USCS Units

Reference State: $h_l = s_l = 0$ at -40°F

P, psia	T, °F	v, ft³/lbm	u, Btu/lbm	h, Btu/lbm	s, Btu/(lbm·°R)
6					
($T_{sat} = -57.5$°F)		(41.612)	(545.68)	(591.88)	(1.4729)
	−40	43.535	552.35	600.68	1.4943
	−20	45.716	559.98	610.73	1.5177
	0	47.881	567.62	620.78	1.5401
	20	50.034	575.28	630.83	1.5615
	40	52.177	582.97	640.90	1.5820
	60	54.313	590.69	650.99	1.6019
	80	56.442	598.46	661.13	1.6210
	100	58.566	606.29	671.31	1.6395
	120	60.700	614.18	681.57	1.6575
	140	62.800	622.14	691.87	1.6750
	160	64.920	630.18	702.26	1.6920
	180	67.034	638.30	712.72	1.7087
	200	69.146	646.50	723.28	1.7249
	220	71.256	654.80	733.92	1.7408
8					
($T_{sat} = -48.6$°F)		(31.799)	(548.33)	(595.40)	(1.4486)
	−40	32.514	551.64	599.77	1.4591
	−20	34.171	559.38	609.97	1.4828
	0	35.811	567.12	620.13	1.5054
	20	37.439	574.86	630.28	1.5271
	40	39.057	582.61	640.43	1.5478
	60	40.668	590.39	650.59	1.5677
	80	42.272	598.20	660.78	1.5870
	100	43.872	606.06	671.01	1.6056
	120	45.467	613.98	681.29	1.6236
	140	47.055	621.96	691.62	1.6411
	160	48.650	630.02	702.04	1.6582
	180	50.239	638.15	712.53	1.6749
	200	51.826	646.37	723.09	1.6912
	220	53.411	654.68	733.75	1.7071
10					
($T_{sat} = -41.3$°F)		(25.809)	(550.42)	(598.18)	(1.4298)
	−40	25.896	550.92	598.84	1.4314
	−20	27.243	558.78	609.19	1.4554
	0	28.569	566.61	619.48	1.4783
	20	29.882	574.43	629.73	1.5002
	40	31.185	582.25	639.96	1.5210
	60	32.481	590.08	650.18	1.5411
	80	33.770	597.93	660.43	1.5604
	100	35.055	605.83	670.70	1.5791
	120	36.335	613.78	681.01	1.5972
	140	37.608	621.78	691.38	1.6148
	160	38.888	629.86	701.82	1.6320
	180	40.162	638.01	712.33	1.6486
	200	41.433	646.24	722.91	1.6649
	220	42.704	654.56	733.58	1.6809
12					
($T_{sat} = -35.1$°F)		(21.765)	(552.16)	(600.49)	(1.4145)
	−30	22.055	554.19	603.17	1.4207
	−10	23.184	562.14	613.63	1.4445
	0	23.740	566.11	618.82	1.4558
	10	24.293	570.06	624.00	1.4671
	30	25.392	577.95	634.33	1.4886
	50	26.481	585.83	644.63	1.5093
	70	27.563	593.72	654.92	1.5291
	90	28.640	601.63	665.23	1.5482
	110	29.712	609.58	675.56	1.5666
	130	30.785	617.58	685.94	1.5845
	150	31.848	625.64	696.36	1.6019
	170	32.912	633.77	706.86	1.6188
	190	33.975	641.98	717.42	1.6354
	210	35.036	650.26	728.06	1.6515
14					
($T_{sat} = -29.7$°F)		(18.843)	(553.64)	(602.46)	(1.4015)
	−20	19.320	557.56	607.61	1.4134
	0	20.291	565.59	618.16	1.4369
	20	21.245	573.58	628.61	1.4591
	40	22.188	581.52	639.01	1.4804
	60	23.124	589.46	649.37	1.5007
	80	24.053	597.40	659.72	1.5202
	100	24.978	605.37	670.08	1.5391
	120	25.899	613.37	680.47	1.5573
	140	26.812	621.42	690.89	1.5750
	160	27.732	629.54	701.38	1.5922
	180	28.645	637.72	711.93	1.6090
	200	29.557	645.98	722.55	1.6253
	220	30.467	654.31	733.25	1.6413
16					
($T_{sat} = -24.9$°F)		(16.631)	(554.93)	(604.17)	(1.3904)
	−20	16.844	556.94	606.82	1.3964
	0	17.704	565.08	617.50	1.4202
	20	18.545	573.14	628.05	1.4426
	40	19.376	581.16	638.53	1.4640
	60	20.200	589.15	648.96	1.4845
	80	21.017	597.14	659.36	1.5042
	100	21.829	605.14	669.77	1.5231
	120	22.637	613.17	680.19	1.5414
	140	23.438	621.25	690.64	1.5591
	160	24.245	629.38	701.16	1.5764
	180	25.046	637.57	711.73	1.5932
	200	25.845	645.84	722.37	1.6095
	220	26.643	654.19	733.08	1.6255

TABLE E.18 Properties of SUPERHEATED AMMONIA—USCS Units (*Continued*)

P, psia	T, °F	v, ft³/lbm	u, Btu/lbm	h, Btu/lbm	s, Btu/(lbm·°R)
18					
($T_{sat} = -20.6$°F)		(14.895)	(556.08)	(605.69)	(1.3805)
	−20	14.918	556.32	606.01	1.3812
	0	15.688	564.56	616.81	1.4053
	20	16.445	572.71	627.49	1.4280
	40	17.189	580.79	638.05	1.4496
	60	17.925	588.84	648.55	1.4702
	80	18.655	596.87	659.01	1.4899
	100	19.379	604.91	669.46	1.5089
	120	20.100	612.97	679.92	1.5273
	140	20.818	621.07	690.41	1.5451
	160	21.533	629.22	700.94	1.5624
	180	22.247	637.43	711.53	1.5792
	200	22.958	645.71	722.18	1.5956
	220	23.669	654.07	732.91	1.6116
20					
($T_{sat} = -16.6$°F)		(13.496)	(557.11)	(607.06)	(1.3717)
	0	14.077	564.03	616.13	1.3918
	20	14.765	572.27	626.92	1.4148
	40	15.439	580.42	637.57	1.4365
	60	16.105	588.53	648.13	1.4573
	80	16.765	596.60	658.65	1.4771
	100	17.420	604.67	669.14	1.4962
	120	18.071	612.76	679.64	1.5147
	140	18.719	620.88	690.16	1.5325
	160	19.364	629.05	700.72	1.5498
	180	20.007	637.28	711.33	1.5667
	200	20.649	645.58	722.00	1.5831
	220	21.289	653.95	732.74	1.5991
30					
($T_{sat} = -0.6$°F)		(9.2284)	(561.10)	(612.33)	(1.3379)
	0	9.2422	561.35	612.66	1.3386
	20	9.7206	570.04	624.00	1.3628
	40	10.189	578.55	635.11	1.3855
	60	10.646	586.94	646.04	1.4069
	80	11.096	595.24	656.84	1.4273
	100	11.541	603.50	667.57	1.4469
	120	11.982	611.74	678.25	1.4656
	140	12.420	619.98	688.93	1.4837
	160	12.855	628.24	699.61	1.5012
	180	13.289	636.55	710.32	1.5183
	200	13.721	644.91	721.08	1.5348
	220	14.151	653.34	731.90	1.5510

P, psia	T, °F	v, ft³/lbm	u, Btu/lbm	h, Btu/lbm	s, Btu/(lbm·°R)
40					
($T_{sat} = 11.7$°F)		(7.0411)	(563.94)	(616.06)	(1.3140)
	20	7.1963	567.73	620.99	1.3244
	40	7.5595	576.62	632.58	1.3480
	60	7.9143	585.31	643.90	1.3702
	80	8.2600	593.86	655.00	1.3912
	100	8.6005	602.30	665.97	1.4112
	120	8.9369	610.69	676.85	1.4303
	140	9.2701	619.06	687.68	1.4486
	160	9.6008	627.42	698.49	1.4664
	180	9.9294	635.81	709.31	1.4835
	200	10.256	644.24	720.16	1.5002
	220	10.582	652.72	731.05	1.5165
50					
($T_{sat} = 21.7$°F)		(5.7047)	(566.12)	(618.90)	(1.2954)
	40	5.9813	574.64	629.98	1.3180
	60	6.2731	583.65	641.69	1.3410
	80	6.5579	592.45	653.12	1.3625
	100	6.8358	601.09	664.34	1.3829
	120	7.1095	609.64	675.42	1.4024
	140	7.3800	618.13	686.41	1.4210
	160	7.6478	626.60	697.36	1.4390
	180	7.9135	635.07	708.29	1.4564
	200	8.1776	643.57	719.23	1.4732
	220	8.4403	652.11	730.20	1.4896
75					
($T_{sat} = 41.1$°F)		(3.8836)	(569.97)	(623.87)	(1.2614)
	60	4.0827	579.31	635.97	1.2852
	80	4.2850	588.79	648.26	1.3084
	100	4.4812	597.97	660.16	1.3300
	120	4.6717	606.93	671.77	1.3504
	140	4.8589	615.76	683.19	1.3698
	160	5.0431	624.49	694.48	1.3883
	180	5.2252	633.18	705.70	1.4061
	200	5.4056	641.86	716.88	1.4233
	220	5.5845	650.54	728.05	1.4400

TABLE E.18 Properties of SUPERHEATED AMMONIA—USCS Units (*Continued*)

P, psia	*T*, °F	*v*, ft³/lbm	*u*, Btu/lbm	*h*, Btu/lbm	*s*, Btu/(lbm·°R)
100 (T_{sat} = 56.0°F)		(2.9497)	(572.55)	(627.14)	(1.2372)
	60	2.9832	574.67	629.88	1.2424
	80	3.1459	584.93	643.14	1.2675
	100	3.3012	594.70	655.79	1.2905
	120	3.4514	604.13	667.99	1.3119
	140	3.5971	613.31	679.88	1.3321
	160	3.7400	622.33	691.54	1.3512
	180	3.8805	631.25	703.06	1.3695
	200	4.0192	640.12	714.49	1.3871
	220	4.1563	648.96	725.87	1.4041
125 (T_{sat} = 68.3°F)		(2.379)	(574.41)	(629.43)	(1.2181)
	80	2.4597	580.84	637.74	1.2337
	100	2.5917	591.28	651.23	1.2582
	120	2.7180	601.21	664.08	1.2808
	140	2.8392	610.79	676.46	1.3018
	160	2.9574	620.12	688.52	1.3216
	180	3.0731	629.28	700.37	1.3404
	200	3.1869	638.34	712.06	1.3584
	220	3.2989	647.35	723.66	1.3757
150 (T_{sat} = 78.8°F)		(1.9923)	(575.78)	(631.08)	(1.2023)
	80	1.9997	576.49	632.00	1.2040
	100	2.1170	587.68	646.44	1.2303
	120	2.2275	598.17	660.00	1.2541
	140	2.3331	608.18	672.94	1.2761
	160	2.4351	617.84	685.43	1.2966
	180	2.5344	627.27	697.62	1.3159
	200	2.6318	636.54	709.59	1.3344
	220	2.7272	645.71	721.42	1.3520
175 (T_{sat} = 88.02°F)		(1.713)	(576.81)	(632.28)	(1.1888)
	100	1.7762	583.89	641.41	1.2053
	125	1.9003	597.68	659.21	1.2364
	150	2.0165	610.53	675.84	1.2643
	175	2.1275	622.80	691.70	1.2898
	200	2.2348	634.70	707.07	1.3135
	225	2.3392	646.38	722.13	1.3360
	250	2.4413	657.94	737.00	1.3573
	275	2.5418	669.48	751.79	1.3778
	300	2.6411	681.04	766.57	1.3975
200 (T_{sat} = 96.3°F)		(1.5012)	(577.57)	(633.13)	(1.1769)
	100	1.5191	579.87	636.09	1.1822
	125	1.6338	594.51	654.97	1.2153
	150	1.7400	607.94	672.34	1.2443
	175	1.8404	620.62	688.73	1.2707
	200	1.9369	632.82	704.50	1.2951
	225	2.0302	644.73	719.86	1.3179
	250	2.1213	656.48	734.98	1.3396
	275	2.2106	668.16	749.97	1.3604
	300	2.2987	679.83	764.91	1.3804
225 (T_{sat} = 103.9°F)		(1.335)	(578.13)	(633.71)	(1.1663)
	125	1.4257	591.19	650.55	1.1956
	150	1.5243	605.26	668.73	1.2261
	175	1.6166	618.38	685.69	1.2533
	200	1.7049	630.90	701.89	1.2784
	225	1.7898	643.05	717.57	1.3017
	250	1.8723	654.99	732.95	1.3238
	275	1.9531	666.82	748.14	1.3448
	300	2.0324	678.62	763.24	1.3650
250 (T_{sat} = 110.8°F)		(1.201)	(578.52)	(634.07)	(1.1566)
	125	1.2582	587.71	645.92	1.1771
	150	1.3513	602.48	664.99	1.2091
	175	1.4373	616.08	682.57	1.2373
	200	1.5190	628.94	699.22	1.2630
	225	1.5972	641.35	715.24	1.2869
	250	1.6729	653.48	730.88	1.3093
	275	1.7468	665.46	746.28	1.3307
	300	1.8194	677.39	761.56	1.3511
275 (T_{sat} = 117.2°F)		(1.090)	(578.77)	(634.25)	(1.1477)
	125	1.1201	584.05	641.05	1.1594
	150	1.2092	599.60	661.13	1.1930
	175	1.2902	613.71	679.37	1.2223
	200	1.3667	626.94	696.49	1.2488
	225	1.4395	639.62	712.87	1.2732
	250	1.5098	651.95	728.79	1.2960
	275	1.5781	664.10	744.41	1.3177
	300	1.6450	676.15	759.86	1.3383

Properties generated from the program *REFRIG*, Wiley Professional Software, Wiley, New York, 1985.

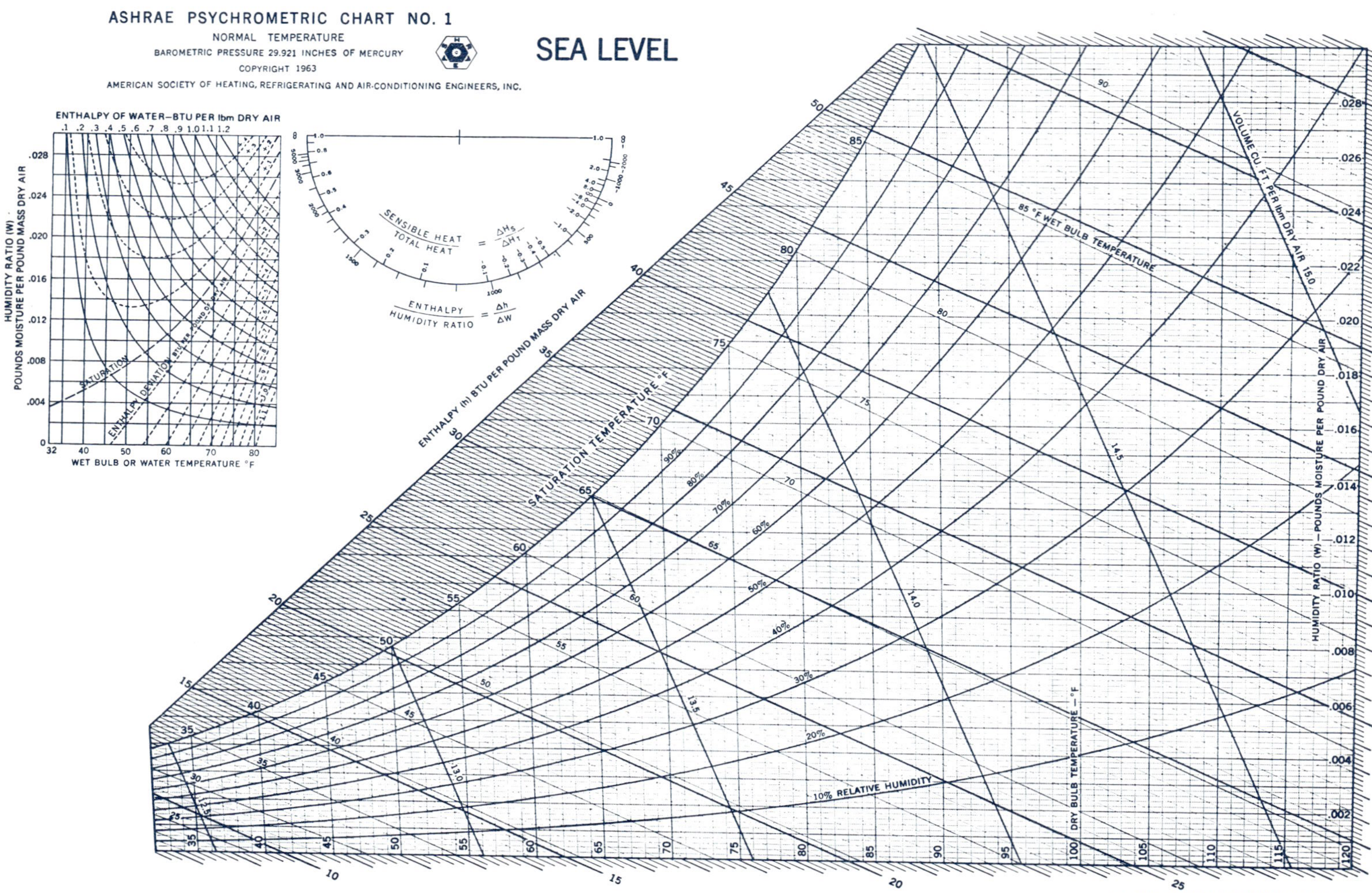

Figure E.5 Psychrometric chart. (From American Society of Heating, Refrigerating, and Air-Conditioning Engineers, used with permission.)

APPENDIX F

Reynolds' Transport Theorem

The movement of mass through the system requires a transformation of fundamental statements presented in Sec. 4.2 to a form useful to control volume systems. The laws in that section do not change, but incorporation of the movement of fluid must be addressed. This appendix provides the rigorous transformation of the control mass expression to control volume form. This transformation should not be considered as a new physical law but rather as a manipulation of the conservation principle.

The expressions for the conservation of mass and the conservation of energy have been presented for a control mass. These equations are

$$\frac{d}{dt}\left(\int_V \rho \, dV\right) = 0 \tag{F.1}$$

$$\frac{d}{dt}\left(\int_V e\rho \, dV\right) = \dot{Q}_{\mathrm{CV}} - \dot{W} \tag{F.2}$$

The left-hand sides of these equations represent the rates of change of mass and of energy of the control mass. These rates of change of the control mass properties are the quantities of interest in this transformation.

Figure F.1 presents a control mass (CM) which is passing through a particular component. The control volume (CV) is fixed to the component and coincides with the control mass at the initial time. The control volume also expands to raise the piston and performs work on the surroundings during the time interval. Thus, the boundary B of the control volume is fixed at the inlets and outlets of the system and expands to do work on the surroundings.

The control mass moves from its initial position at time t to its later position at $t + \Delta t$ as a result of the velocity of the fluid elements. The

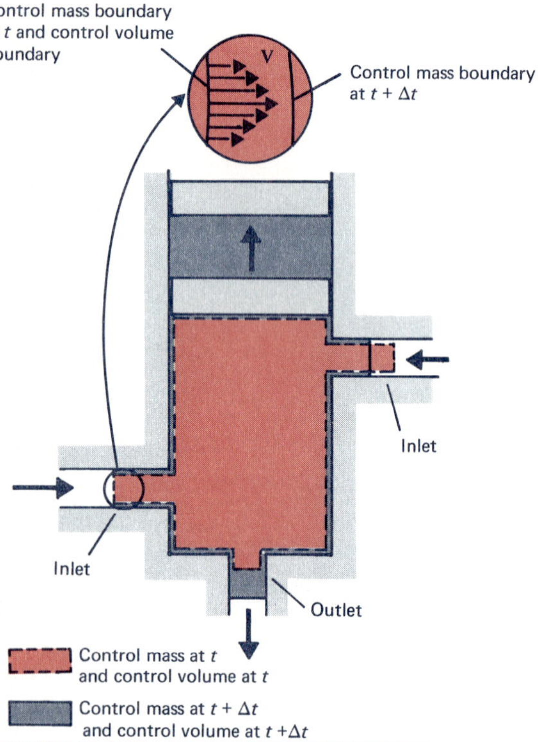

Figure F.1 The Reynolds' transport theorem transformation.

specific system diagramed indicates two inlets and a single outlet, although the specific number is arbitrary. During this motion, the fluid elements transport mass and energy into and out of the control volume. We consider the movement of the control mass through the control volume in a limiting process as the time interval Δt tends to zero.

The total derivatives in the conservation of mass and conservation of energy principles for a control mass are of interest. The property in these equations is represented in terms of the extensive property Λ and the corresponding specific intensive property λ as

$$\Lambda = \int d\Lambda = \int_V \rho\lambda \, dV \tag{F.3}$$

where $\lambda = 1$ and $\Lambda = m$ for mass and $\lambda = e$ and $\Lambda = E$ for total energy. The total derivatives are then given by

$$\frac{d\Lambda_{\mathrm{CM}}}{dt} = \lim_{\Delta t \to 0} \frac{\Lambda_{\mathrm{CM},t+\Delta t} - \Lambda_{\mathrm{CM},t}}{\Delta t} \tag{F.4}$$

The control mass property is an integral over the volume occupied. The volume occupied by the control mass is expressed in terms of the control volume. The fluid has swept the control mass into a new volume at the outlets, and the control mass has left behind an old volume at the inlets. Therefore, since the control mass properties are volume integrals, the integral is reformulated in terms of the control volume. The terms in Eq. (F.4) are

$$\Lambda_{\text{CM},t} = \Lambda_{\text{CV},t}$$
$$\Lambda_{\text{CM},t+\Delta t} = \Lambda_{\text{CV},t+\Delta t} + \Lambda_{\text{CM,out},t+\Delta t} - \Lambda_{\text{CM,in},t+\Delta t} \tag{F.5}$$

Substituting these expressions into Eq. (F.4) yields

$$\begin{aligned}\frac{d\Lambda_{\text{CM}}}{dt} &= \lim_{\Delta t\to 0} \frac{\Lambda_{\text{CV},t+\Delta t} - \Lambda_{\text{CV},t} + \Lambda_{\text{CM,out},t+\Delta t} - \Lambda_{\text{CM,in},t+\Delta t}}{\Delta t}\\ &= \lim_{\Delta t\to 0} \frac{\Lambda_{\text{CV},t+\Delta t} - \Lambda_{\text{CV},t}}{\Delta t} + \lim_{\Delta t\to 0} \frac{\Lambda_{\text{CM,out},t+\Delta t}}{\Delta t}\\ &\quad - \lim_{\Delta t\to 0} \frac{\Lambda_{\text{CM,in},t+\Delta t}}{\Delta t}\end{aligned} \tag{F.6}$$

The derivative of the control mass property is represented by three limits —the first is a change within the control volume while the second and third limits result from the flows in and out.

The first limit is the definition of the partial derivative of the control volume property, i.e.,

$$\lim_{\Delta t\to 0} \frac{\Lambda_{\text{CV},t+\Delta t} - \Lambda_{\text{CV},t}}{\Delta t} = \frac{\partial \Lambda_{\text{CV}}}{\partial t} \tag{F.7}$$

This is the change in the property within the control volume.

The second and third limits are similar and require a detailed consideration of the inlets and outlets. An expanded view of an outlet is shown in Fig. F.2. The velocity **V** is a vector denoted at an area element dA. The area orientation is represented by a normal vector **n** which is the

Figure F.2 Flow at the outlet.

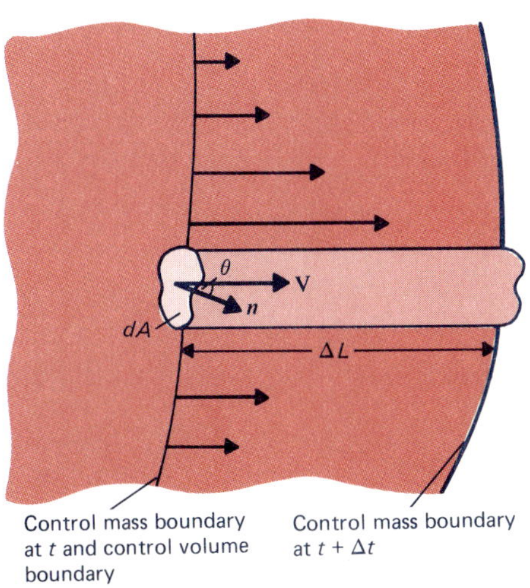

outward normal to the area element. The angle between the velocity vector and normal vector is θ. This area sweeps out a volume of

$$dV = \Delta L \, dA \cos\theta \tag{F.8}$$

The volume integral of the control mass property within the second limit of Eq. (F.6) is

$$\frac{\Lambda_{\mathrm{CM,out},t+\Delta t}}{\Delta t} = \frac{1}{\Delta t}\int_{V_{\mathrm{out}}} \rho\lambda \, dV = \frac{1}{\Delta t}\int_{A_{\mathrm{out}}} \rho\lambda \, \Delta L \cos\theta \, dA \tag{F.9}$$

The area of the integration is the boundary of the outlet. With the magnitude of the fluid velocity given as

$$\lim_{\Delta t \to 0} \frac{\Delta L \cos\theta}{\Delta t} = |\mathbf{V} \cdot \mathbf{n}| \tag{F.10}$$

The second limit is

$$\lim_{\Delta t \to 0} \frac{\Lambda_{\mathrm{CM,out},t+\Delta t}}{\Delta t} = \int_{A_{\mathrm{out}}} \rho\lambda |\mathbf{V} \cdot \mathbf{n}| \, dA \tag{F.11}$$

Note that

$$|\mathbf{V} \cdot \mathbf{n}| = \mathbf{V} \cos\theta \tag{F.12}$$

The absolute value sign is used since the volume swept out in Eq. (F.8) represents a positive quantity. The $\rho|\mathbf{V} \cdot \mathbf{n}| \, dA$ term in the integrand represents the mass flow rate for an area element

$$d\dot{m} = \rho|\mathbf{V} \cdot \mathbf{n}| \, dA \tag{F.13}$$

Thus Eq. (F.11) is alternatively represented as

$$\lim_{\Delta t \to 0} \frac{\Lambda_{\mathrm{CM,out},t+\Delta t}}{\Delta t} = \int_{A_{\mathrm{out}}} \lambda \, d\dot{m} \tag{F.14}$$

The third limit is considered in exactly the same manner as the second. The only difference is that the angle between the outward normal of the area elements and the velocity vector is larger than 90°. Thus the absolute value sign ensures a positive value for the volume. Thus,

$$\begin{aligned} \lim_{\Delta t \to 0} \frac{\Lambda_{\mathrm{CM,in},t+\Delta t}}{\Delta t} &= \lim_{\Delta t \to 0} \frac{1}{\Delta t}\int_{A_{\mathrm{in}}} \rho\lambda \, \Delta L \cos\theta \, dA \\ &= \int_{A_{\mathrm{in}}} \rho\lambda |\mathbf{V} \cdot \mathbf{n}| \, dA \\ &= \int_{A_{\mathrm{in}}} \lambda \, d\dot{m} \end{aligned} \tag{F.15}$$

Combining the three limits in Eqs. (F.7), (F.11), (F.14), and (F.15) into the control mass derivatives in Eq. (F.6) yields

$$\frac{d\Lambda_{CM}}{dt} = \frac{\partial \Lambda_{CV}}{\partial t} + \int_{A_{out}} \rho\lambda |\mathbf{V}\cdot\mathbf{n}|\, dA - \int_{A_{in}} \rho\lambda |\mathbf{V}\cdot\mathbf{n}|\, dA$$

$$= \frac{\partial \Lambda_{CV}}{\partial t} + \int_{A_{out}} \lambda\, d\dot{m} - \int_{A_{in}} \lambda\, d\dot{m} \tag{F.16}$$

The last two integrals are area integrals over the crossing boundary. The velocity of the fluid elements crossing the inside boundary is zero (no fluid is permitted to cross—see Sec. 2.2.1), so these two integrals are represented by a single integral over the boundary B. With the definition of Λ given in Eq. (F.3), Eq. (F.16) becomes

$$\frac{d\Lambda_{CM}}{dt} = \frac{\partial}{\partial t}\left(\int_{CV} \rho\lambda\, dV\right) + \int_B \rho\lambda \mathbf{V}\cdot\mathbf{n}\, dA$$

$$= \frac{\partial}{\partial t}\left(\int_{CV} \rho\lambda\, dV\right) + \int_B \lambda\, d\dot{m} \tag{F.17}$$

The absolute value sign is removed in this expression, so the negative contribution at the inlet is incorporated. This desired relation between the derivative of a control mass property and the control volume property itself is termed *Reynolds' transport theorem.* The change in the control mass property is equal to the change of the property within the control volume plus the net transport out of the control volume.

Conservation of mass and conservation of energy for a control volume are obtained from the original conservation statements with Eq. (F.17). These expressions are

$$\frac{\partial}{\partial t}\left(\int_{CV} \rho\, dV\right) + \int_B d\dot{m} = 0 \tag{F.18}$$

$$\frac{\partial}{\partial t}\left(\int_{CV} \rho e\, dV\right) + \int_B e\, d\dot{m} = \dot{Q}_{CV} - \dot{W} \tag{F.19}$$

If there are several separate inlet streams and several separate outlet streams, then these are expressed in terms of sums over the streams. Equations (F.18) and (F.19) become

$$\frac{\partial}{\partial t}\left(\int_{CV} \rho\, dV\right) + \sum_{out} \int_{A_{out}} d\dot{m} - \sum_{in} \int_{A_{in}} d\dot{m} = 0 \tag{F.20}$$

$$\frac{\partial}{\partial t}\left(\int_{CV} \rho e\, dV\right) + \sum_{out} \int_{A_{out}} e\, d\dot{m} - \sum_{in} \int_{A_{in}} e\, d\dot{m} = \dot{Q}_{CV} - \dot{W} \tag{F.21}$$

These are recognized as Eqs. (4.68) and (4.73), respectively. By noting that $\int_A d\dot{m} = \dot{m}$, the mass flow rate across the boundary at a stream, and denoting $\int_A e\, d\dot{m} = \dot{E}$, the energy transport across the boundary at a stream, Eqs. (F.20) and (F.21) become

$$\frac{\partial}{\partial t}\left(\int_{\mathrm{CV}} \rho \, dV\right) + \sum_{\mathrm{out}} \dot{m} - \sum_{\mathrm{in}} \dot{m} = 0 \tag{F.22}$$

$$\frac{\partial}{\partial t}\left(\int_{\mathrm{CV}} \rho e \, dV\right) + \sum_{\mathrm{out}} \dot{E} - \sum_{\mathrm{in}} \dot{E} = \dot{Q}_{\mathrm{CV}} - \dot{W} \tag{F.23}$$

These are directly related to Eqs. (4.27) and (4.37).

APPENDIX G

Computerized Tables of Thermodynamic Properties

In the pocket inside the rear cover of this text is a floppy disk that contains computerized tables of the thermodynamic properties of steam, argon, nitrogen, air, and refrigerant 12. The program, called PROPERTIES, is in compiled code in a format that can be executed on the IBM PC, IBM/XT, and many compatible computers that operate on PC-DOS or MS-DOS, versions 1.1 and above. These programs are also available in Macintosh-compatible format. Before you try to use the programs, please read this appendix carefully.

The programs are completely menu-driven and are essentially self-explanatory. The tables in Apps. D and E for steam, gases, and refrigerant 12 were generated from expanded versions of these programs, so solutions generated from either the tables or the programs are identical except for minor truncation or roundoff errors when the program has an iterative solution.

G.1 What the Programs Can Do

The programs are to be used as an aid in solving problems in this text. They can help avoid the single and double interpolations necessary if the tables of Apps. D and E are used for property data. Many of the problems are extremely tedious if the programs are not used as an aid in their solution. The computerized tables cover the range of temperatures and pressures necessary for solution of the problems in this text and allow consideration of more interesting and useful problems than is possible by using tabular data alone.

Each program provides results in the user's choice of either SI or USCS units.

Steam Tables

The computerized tables provide the properties of steam in the subcooled, saturated, and superheated regions. The program is a modified version of the NIST (National Institute of Standards and Testing) program used to generate the standard NIST/NRC Tables [1]. It covers all regions of common use for steam properties. Any combination of independent properties may be entered to define a state, and the program will return the remaining properties at that state. The properties are computed from the fundamental thermodynamic equations and thus do not contain any interpolation errors at any state.

Gas Tables

Ideal gas properties for argon, nitrogen, and air are provided over the full range given in the tables in Apps. D and E. The complete statistical thermodynamic equations for ideal gases are used in the program. Rather than the function $s_0(T)$ as presented in the tables, the actual value of $s(T, P)$ is given by the program. At a given state, any two independent properties from the set of P, T, v, u, h, and s can be specified, and the rest will be computed and displayed. Note that T, h, and u are not independent for an ideal gas. The values of c_P, c_v, and $k = c_P/c_v$ are also displayed at the chosen state.

Refrigerant 12 Tables

The program for refrigerant 12 calculates properties in the saturated, superheated, and subcooled regions for states that do not have a pressure in excess of 1379 kPa (200 psia). At some temperature levels, states are outside the stated limits of accuracy of the program, and a message is displayed. This program also uses the fundamental thermodynamic relations to compute the properties. In the subcooled region, the program provides v, u, h, and s as a function of T and P. In the saturation region, an input of T or P and x, v, u, h, or s provides the remaining properties. In the superheated region, an input of any two properties from the set of P, v, and T provides all others, including c_P and c_v. If the region is unknown, this particular program finds it from an input of P and T and displays the other properties at that state.

Hydrocarbon Tables

The program for hydrocarbons uses the Benedict-Webb-Rubin equation along with the Lee-Kesler reduced property correlations to predict vapor phase specific volume, enthalpy, and entropy. Other standard correlations are used for the liquid phase. The pressure-temperature relation for the saturation curve is based on a three-parameter expansion in terms of reduced pressure and temperature and the acentric factor.

Properties are available for 78 hydrocarbons, separated into eight groups (alkanes, alkenes, chlorinated hydrocarbons, alcohols, aldehydes and acids, ketones-ethers-acetates-esters, benzene compounds and alkynes, and miscellaneous).

Once the particular group and hydrocarbon are selected, a choice of SI or USCS units on a per unit mass or per mole basis is made. The program requires a choice of the saturation or superheat region, and will provide all properties (v, u, h, s, and g) at the chosen state. In the saturation region, either P or T must be specified along with one specific extensive property or quality. In the superheat region, T and P must be specified. The reference values at 25°C and 1 atm are displayed, and may be changed if a different reference state is desired.

G.2 Hardware Requirements

The specific equipment required for the IBM disk included with this text includes

1. IBM Personal Computer (PC, XT or higher, and most compatibles, each with at least 128K of memory)
2. At least one double-sided disk drive
3. A color or monochrome monitor that is compatible with the computer used
4. An 80-column or wider printer (optional)

Macintosh-compatible equipment is required for the Macintosh disc.

G.3 Running the Programs

The disk contains five programs plus a data file. Four of the programs provide the properties outlined in the preceding paragraphs, and are called STEAM, GASPROPS, REFRIG, and HYCARB. The fifth program, START, is a master program that allows the user to call each of the others, and to switch among the programs when a succession of problems is to be solved. The data file, HYDAT, contains the property data used in HYCARB correlations.

It is convenient to begin by loading START, although any of the programs may be loaded individually. For example, if a steam Rankine cycle problem is to be worked, only STEAM might be loaded. If this is done, START will automatically be called when you are finished with STEAM, and you must exit the program from START.

The programs are all self-explanatory and easy to use.

Reference

1. L. Haar, J. S. Gallagher, and G. S. Kell, *NBS/NRC Steam Tables*, Hemisphere, New York, 1984.

APPENDIX H

Fundamentals of Mathematics for Thermodynamics

In this appendix we review the mathematics needed for an introductory study of thermodynamics. The mathematics required for thermodynamics is minimal, yet it is important to have a clear understanding of the necessary elements. This makes it possible to distinguish between the physical laws of thermodynamics and the mathematical manipulations. So some of the difficulties that arise in the understanding of thermodynamics can be attacked at their source—by a clear division between mathematical manipulation and physical principle. This appendix should be referred to throughout the text as needed.

In this section we review the necessary mathematics rather than offer a rigorous development of the subject. Excellent references are available for further details [1, 2, 3].

H.1 Function Representation

Consider the function

$$f(x, y, z) = c = \text{constant} \tag{H.1}$$

which represents a surface in the three-dimensional space of coordinates x, y, z. An example of such a function is shown in Fig. H.1. Only the octant with positive x, y, and z values is shown because thermodynamic properties usually have positive values. Any two of the three variables uniquely specify the value of the third. Thus, an alternative representation is

$$z = z(x, y) \tag{H.2}$$

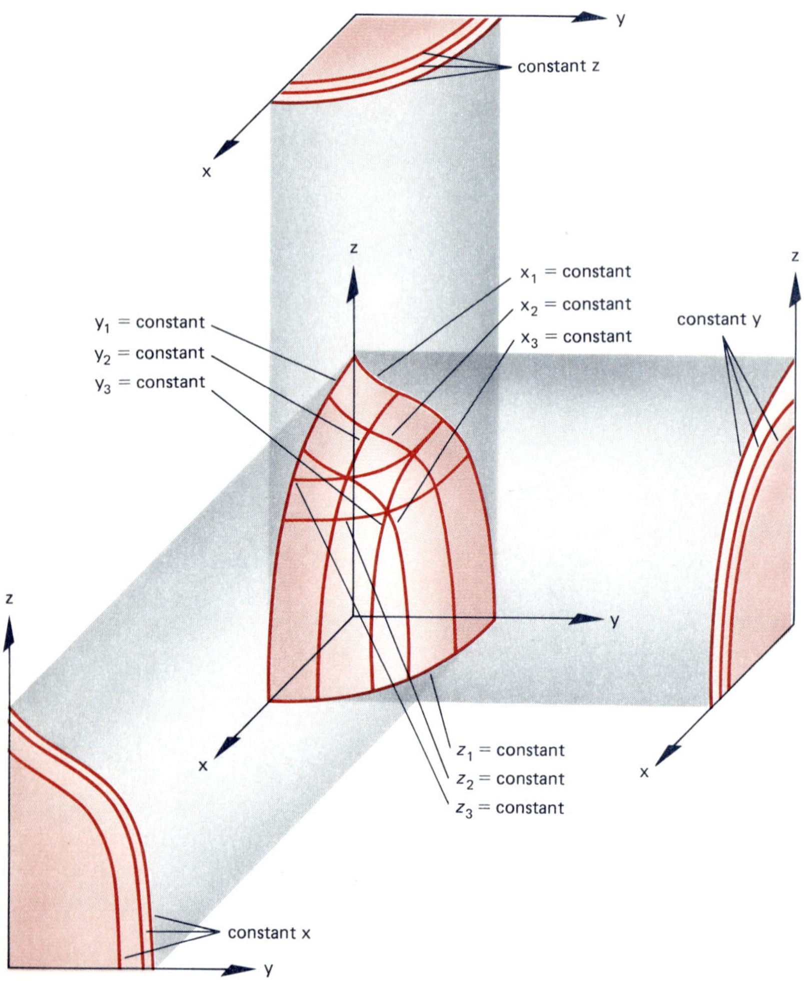

Figure H.1 Surface representing $f(x, y, z) = c$.

where the value of z is uniquely specified by the values of x and y. A general surface is mathematically specified in terms of n variables. The physical representation of the general surface with n variables is difficult to show pictorially if n is greater than 3.

The intersection of the surface with a plane that is parallel to two of the coordinates forms a line. For example, in Fig. H.1, an xz plane can intersect the surface at y coordinates of y_1, y_2, y_3, etc. The lines of intersection are shown as $y =$ constant lines. These lines can be projected onto the plane of the xz coordinates to form a two-dimensional representation of the surface. Figure H.1 shows the xz and xy representations and a similar representation on the yz surface.

A general coordinate system is composed of *independent* variables since each can be varied independently without varying the others. The coordinates x, y, and z are independent variables for a general coordinate system. The concept of a *dependent* variable arises when a function like Eq. (H.1) is introduced. As indicated above, two variables uniquely specify the third for this surface, so one is dependent on the two other independent variables. Standard convention indicates the independent variables in parenthetical functional notation; so Eq. (H.2), for example, represents the dependent variable z as a function of the independent variables x and y. The denotation of independent and dependent variables is arbitrary. Thus, alternative forms are

$$x = x(y, z) \tag{H.3}$$

and

$$y = y(x, z) \tag{H.4}$$

Equation (H.1) is called an *implicit* representation of the function or surface. The dependent and independent variables are not directly indicated and can be varied from consideration to consideration. Equations (H.2) through (H.4) are termed *explicit* representations.

H.2 Partial Derivatives

A *partial derivative* represents the rate of change of the dependent variable with respect to a single independent variable, with all other independent variables held constant. This is mathematically denoted as

$$\left(\frac{\partial z}{\partial x}\right)_y \equiv \lim_{\Delta x \to 0} \frac{z(x + \Delta x, y) - z(x, y)}{\Delta x} \tag{H.5}$$

for the surface prescribed in Eq. (H.2). The derivative is taken with respect to the independent variable x, with y held constant. This is geometrically represented by the slope of a curve obtained by passing a plane parallel to the x and z coordinates (at constant y) through the surface. This is shown in Fig. H.2. For the three-dimensional example in Eq. (H.2), there are two independent variables, so that another possible partial derivative is

$$\left(\frac{\partial z}{\partial y}\right)_x = \lim_{\Delta y \to 0} \frac{z(x, y + \Delta y) - z(x, y)}{\Delta y} \tag{H.6}$$

The partial derivatives generally depend on the particular values of the independent variables; i.e., the slope of the line on the surface depends on the specific y plane considered and the specific x point on the line.

Figure H.2 Geometric representation of a partial derivative $(\partial z/\partial x)_y$.

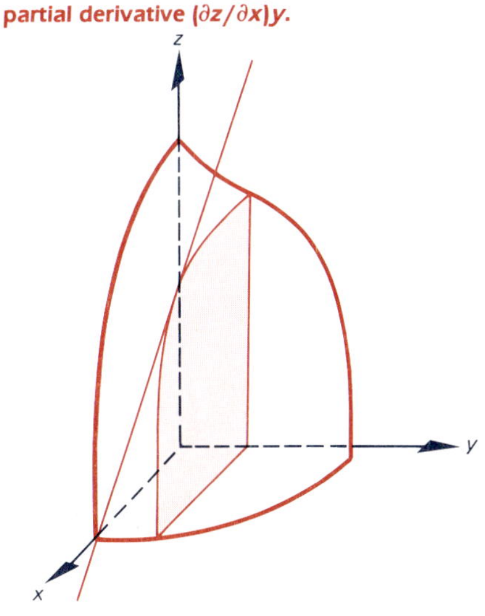

Thus, partial derivatives of the partial derivatives are possible. The second derivatives are denoted as

$$\left[\frac{\partial}{\partial x}\left(\frac{\partial z}{\partial x}\right)_y\right]_y = \frac{\partial^2 z}{\partial x^2} \tag{H.7}$$

$$\left[\frac{\partial}{\partial y}\left(\frac{\partial z}{\partial x}\right)_y\right]_x = \frac{\partial^2 z}{\partial y\,\partial x} \tag{H.8}$$

For single-valued, continuous functions (the usual case in thermodynamics), the result is independent of the order of differentiation, so

$$\frac{\partial^2 z}{\partial y\,\partial x} = \frac{\partial^2 z}{\partial x\,\partial y} \tag{H.9}$$

Example Problem H.1

The ideal gas relation is $P = mRT/V$, where P is pressure, V is volume, m is mass, R is the gas constant, and T is temperature. Verify Eq. (H.9) with m and R constant.

Solution

Let $P = P(V, T)$. The first derivatives for the two independent variables V and T are

$$\left(\frac{\partial P}{\partial V}\right)_T = -\frac{mRT}{V^2}$$

$$\left(\frac{\partial P}{\partial T}\right)_V = \frac{mR}{V}$$

The second derivatives are

$$\left[\frac{\partial}{\partial T}\left(\frac{\partial P}{\partial V}\right)_T\right]_V = -\frac{mR}{V^2}$$

$$\left[\frac{\partial}{\partial V}\left(\frac{\partial P}{\partial T}\right)_V\right]_T = -\frac{mR}{V^2}$$

This satisfies the equality in Eq. (H.9).

Total Differential

The partial derivatives represent the slopes of lines tangent to the surface for a plane parallel to the coordinate axes. The change in z corresponding to an infinitesimal change in x for constant y is

$$dz = \left(\frac{\partial z}{\partial x}\right)_y dx \tag{H.10}$$

If the change in z is sought for simultaneous changes in x and y, then

$$\Delta z = z(x + \Delta x, y + \Delta y) - z(x, y) \tag{H.11}$$

Adding and subtracting $z(x, y + \Delta y)$ yield

$$\Delta z = \frac{z(x + \Delta x, y + \Delta y) - z(x, y + \Delta y)}{\Delta x} \Delta x + \frac{z(x, y + \Delta y) - z(x, y)}{\Delta y} \Delta y \tag{H.12}$$

and considering infinitesimal changes (taking the limits as $\Delta x \to 0$ and $\Delta y \to 0$) yields

$$dz = \left(\frac{\partial z}{\partial x}\right)_y dx + \left(\frac{\partial z}{\partial y}\right)_x dy \tag{H.13}$$

This is termed the *total differential* of z. The partial derivative and total differential expressions can be extended to any number of independent variables.

Relations between Partial Derivatives

As in the case of ordinary derivatives, the following product identities are valid:

$$\left(\frac{\partial z}{\partial x}\right)_y \left(\frac{\partial x}{\partial y}\right)_z \left(\frac{\partial y}{\partial z}\right)_x = -1 \tag{H.14}$$

and

$$\left(\frac{\partial y}{\partial x}\right)_z \left(\frac{\partial x}{\partial y}\right)_z = 1 \quad \text{so} \quad \left(\frac{\partial y}{\partial x}\right)_z = \frac{1}{(\partial x/\partial y)_z} \tag{H.15}$$

Another important relation among partial derivatives is needed to change from one set of independent variables to another. If $z(x, y)$ is desired in the form $z(x, w)$, then y must be expressed as $y = y(x, w)$. The total differential of $z(x, y)$ is

$$dz = \left(\frac{\partial z}{\partial x}\right)_y dx + \left(\frac{\partial z}{\partial y}\right)_x dy \tag{H.16}$$

With $y = y(x, w)$,

$$dy = \left(\frac{\partial y}{\partial x}\right)_w dx + \left(\frac{\partial y}{\partial w}\right)_x dw \tag{H.17}$$

Substituting Eq. (H.17) into (H.16) yields

$$dz = \left[\left(\frac{\partial z}{\partial x}\right)_y + \left(\frac{\partial z}{\partial y}\right)_x \left(\frac{\partial y}{\partial x}\right)_w\right] dx + \left(\frac{\partial z}{\partial y}\right)_x \left(\frac{\partial y}{\partial w}\right)_x dw \tag{H.18}$$

This last equation is exactly the form desired, $z(x, w)$. The total differential of $z(x, w)$ is

$$dz = \left(\frac{\partial z}{\partial x}\right)_w dx + \left(\frac{\partial z}{\partial w}\right)_x dw \tag{H.19}$$

Comparison of Eqs. (H.18) and (H.19) shows that

$$\left(\frac{\partial z}{\partial w}\right)_x = \left(\frac{\partial z}{\partial y}\right)_x \left(\frac{\partial y}{\partial w}\right)_x \tag{H.20a}$$

or

$$\left(\frac{\partial z}{\partial y}\right)_x \left(\frac{\partial y}{\partial w}\right)_x \left(\frac{\partial w}{\partial z}\right)_x = 1 \tag{H.20b}$$

and

$$\left(\frac{\partial z}{\partial x}\right)_w = \left(\frac{\partial z}{\partial x}\right)_y + \left(\frac{\partial z}{\partial y}\right)_x \left(\frac{\partial y}{\partial x}\right)_w \tag{H.21}$$

This last expression is called the *substitution rule,* and it is used to change independent variables.

H.3 Integration

Two types of integration commonly arise in thermodynamics. The first type is ordinary integration, which parallels the ordinary derivative in differentiation. The second type is the line integral, which parallels the partial derivative in differentiation. This parallelism is not complete but does serve as a distinction for the two types of integration.

The ordinary integral of the continuous function $f(x)$ between two limits x_1 and x_2 is denoted as

$$\int_{x_1}^{x_2} f(x)\, dx \tag{H.22}$$

This is geometrically displayed as the area under the curve $f(x)$, as indicated in Fig. H.3. The ordinary integral is thought of as an infinite sum of infinitesimal strips forming the area indicated. The fundamental theorem of calculus connects differentiation and integration as

$$\int_{x_1}^{x_2} f(x)\, dx = g(x_2) - g(x_1) \tag{H.23}$$

where

$$\frac{dg(x)}{dx} = f(x) \tag{H.24}$$

and $g(x)$ is often termed an *antiderivative.* An alternate representation for ordinary integration is given by combining Eqs. (H.23) and (H.24):

Figure H.3 Ordinary integral.

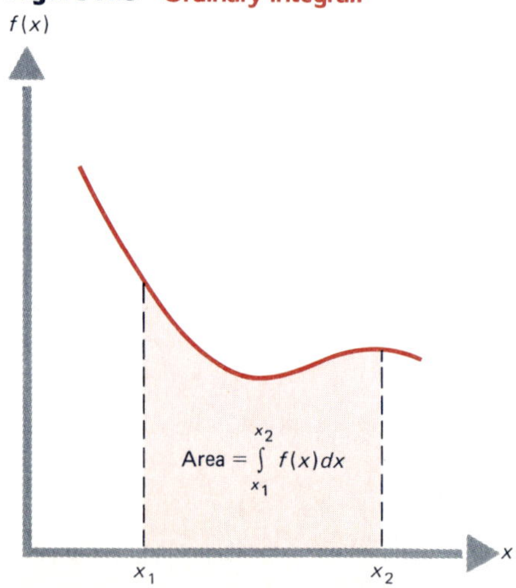

$$\int_{x_1}^{x_2} dg(x) = g(x_2) - g(x_1) \tag{H.25}$$

The extension of ordinary integration to more than one variable represents the summation of a differential on a specified curve. This is written as

$$\int_{x_1,y_1}^{x_2,y_2} dg(x, y) \tag{H.26}$$

The line integral does not have the area representation, as is the case for ordinary integration. Line integration is most easily presented by an example.

Example Problem H.2

The work required to move a particle through a path from position 1 to 2 in a force field **F** is

$${}_1W_2 = -\int_1^2 \mathbf{F} \cdot d\mathbf{s}$$

where $d\mathbf{s}$ is the differential length and the boldface center dot denotes the dot product. For the general path and force indicated below, the components are given as $\mathbf{F} = F_x\hat{\mathbf{i}} + F_y\hat{\mathbf{j}}$ and $d\mathbf{s} = dx\,\hat{\mathbf{i}} + dy\,\hat{\mathbf{j}}$. Thus, the expression for the work is

$${}_1W_2 = -\int_1^2 (F_x\,dx + F_y\,dy)$$

Evaluate the work done for the two paths from 1 to 2 indicated below if $F_x = -y\ N$ and $F_y = -c = -1$ N.

Diagrams

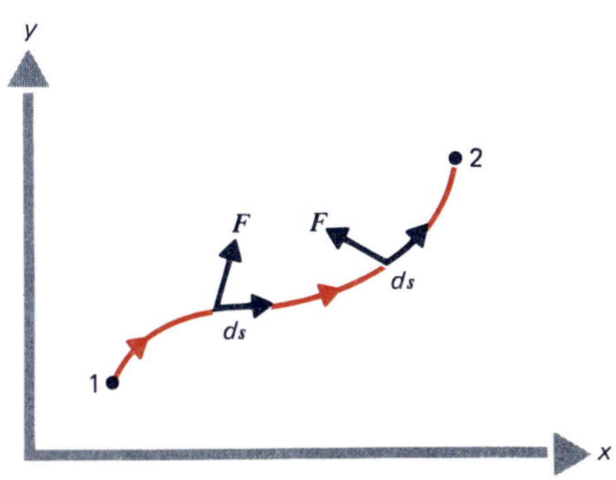

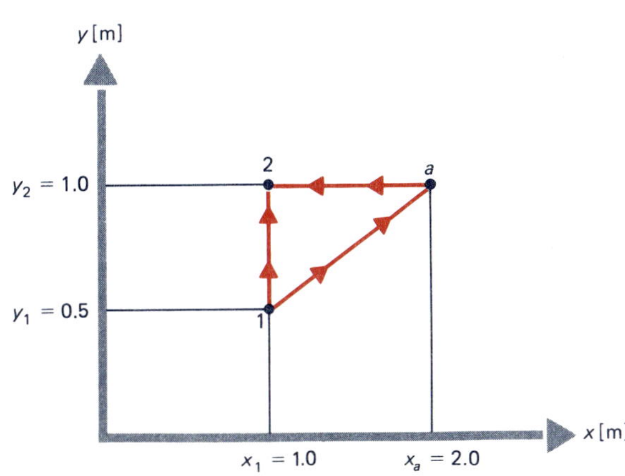

Solution

The required line integral for the *direct* path from 1 to 2 is

$$ {}_1W_2 = \int_1^2 (y\,dx + c\,dy) $$

Along this path $x = \text{constant} = x_1$, so $dx = 0$ and

$$ {}_1W_2 = \int_1^2 y\,\cancelto{0}{dx} + c\int_1^2 dy = c(y_2 - y_1) $$

$$ = (1)\,(1 - 0.5)\ \text{N} \cdot \text{m} = 0.5\ \text{N} \cdot \text{m} = 0.5\ \text{J} = 0.37\ \text{ft} \cdot \text{lbf} $$

The required line integral for the *indirect* path 1-a-2 through position a is

$$ {}_1W_2 = \int_1^a (y\,dx + c\,dy) + \int_a^2 (y\,dx + c\,dy) $$

Each integral is considered separately, and each has particular characteristics. The first integral on the right-hand side is along a line given by

$$ \frac{y - y_1}{x - x_1} = \frac{y_2 - y_1}{x_a - x_1} = \text{slope} \equiv m $$

or

$$ y = y_1 + \left(\frac{y_2 - y_1}{x_a - x_1}\right)(x - x_1) $$
$$ = y_1 + m(x - x_1) $$

Thus, this first integral is

$$ \int_1^a (y\,dx + c\,dy) = \int_1^a [y_1 + m(x - x_1)]\,dx + \int_1^2 c\,dy $$
$$ = y_1(x_a - x_1) + \frac{m}{2}(x_a - x_1)^2 + c(y_2 - y_1) $$

Substituting m yields

$$ \int_1^a (y\,dx + c\,dy) = y_1(x_a - x_1) + \tfrac{1}{2}(y_2 - y_1)(x_a - x_1) + c(y_2 - y_1) $$

The second integral is along a path parallel to the x axis, so $y =$ constant $= y_2$. Thus,

$$ \int_a^2 (y\,dx + c\,dy) = \int_a^2 y\,dx + c\int_a^2 \cancelto{0}{dy} $$
$$ = y_2(x_2 - x_a) = y_2(x_1 - x_a) $$

Therefore, the work done for the indirect path is

$$
\begin{aligned}
{}_1W_2 &= y_1(x_a - x_1) + \tfrac{1}{2}(y_2 - y_1)(x_a - x_1) + c(y_2 - y_1) + y_2(x_1 - x_a) \\
&= (y_1 - y_2)(x_a - x_1) + \tfrac{1}{2}(y_2 - y_1)(x_a - x_1) + c(y_2 - y_1) \\
&= (-0.5)(1) + (0.5)(0.5)(1) + (1)(0.5) = 0.25\ \text{J} = 0.18\ \text{ft}\cdot\text{lbf}
\end{aligned}
$$

Comments

The line integration is written for each variable, and the rules of ordinary integration are used. The work done is different along each path. Work generally depends on the path considered.

Line integration depends on the direction of integration, so that for a given path

$$\int_{x_1,y_1}^{x_2,y_2} dg(x, y) = -\int_{x_2,y_2}^{x_1,y_1} dg(x, y) \tag{H.27}$$

H.4 Exact and Inexact Differentials

Exact and inexact differentials play important roles in thermodynamics. Knowledge of whether a differential is exact indicates certain facts about the quantity. Consider a general differential expression given by

$$dz(x, y) = g(x, y)\, dx + h(x, y)\, dy \tag{H.28}$$

This is an *exact differential* if there exists a function $z(x, y)$ such that $dz(x, y)$ is a total differential [see Eq. (H.13)]. This is an *inexact differential* if there exists no function of x and y yielding Eq. (H.28). Listed here are three conditions which are valid for exact differentials:

$$\left[\frac{\partial g(x, y)}{\partial y}\right]_x = \left[\frac{\partial h(x, y)}{\partial x}\right]_y \tag{H.29a}$$

$$\int_{x_1,y_1}^{x_2,y_2} dz(x, y) = \text{function of endpoints only and independent of path} \tag{H.29b}$$

$$\oint_c dz(x, y) = 0 \quad \text{i.e., the integral around every closed path } C \text{ is zero} \tag{H.29c}$$

Equation (H.29*a*) is useful to test for exactness. Complete proofs of these conditions are not presented, but illustrative examples that demonstrate these conditions follow.

Inexact differentials are denoted by δz rather than dz to clearly distinguish these path-dependent functions from exact differentials.

Example Problem H.3

Determine whether the following differentials are exact or inexact, using the condition given in Eq. (H.29*a*):

(*a*) $y\,dx + x\,dy$
(*b*) $y\,dx - x\,dy$

Solution

Comparing these differentials with Eq. (H.28) and performing the derivatives indicated in Eq. (H.29*a*) yield

$$(a)\quad g(x, y) = y \quad \text{and} \quad \left[\frac{\partial g(x, y)}{\partial y}\right]_x = 1$$

$$h(x, y) = x \quad \text{and} \quad \left[\frac{\partial h(x, y)}{\partial x}\right]_y = 1$$

Then Eq. (H.29*a*) indicates that this differential is *exact.*

$$(b)\quad g(x, y) = y \quad \text{and} \quad \left[\frac{\partial g(x, y)}{\partial y}\right]_x = 1$$

$$h(x, y) = -x \quad \text{and} \quad \left[\frac{\partial h(x, y)}{\partial x}\right]_y = -1$$

so that this differential is *inexact.*

Comment

Is the differential in Example Problem H.2 exact or inexact? (It is inexact.)

Example Problem H.4

For the two paths between 1 and 2 shown, determine which of the differentials in Example Problem H.3 satisfy the condition given in Eq. (H.29*b*).

Solution

(*a*) The integral along path 1-*a*-2 is

$${}_1I_2 = \int_1^2 (y\,dx + x\,dy)$$

For path 1-*a*-2,

$${}_1I_2 = \int_1^a (y\,dx + x\,dy) + \int_a^2 (y\,dx + x\,dy)$$

where y is constant at y_1 from 1 to a and x is constant at x_2 from a to 2.

Thus,

$$ {}_1I_2 = y_1 \int_1^a dx + x_2 \int_a^2 dy $$
$$ = y_1(x_2 - x_1) + x_2(y_2 - y_1) = x_2y_2 - x_1y_1 $$

For path 1-*b*-2,

$$ {}_1I_2 = \int_1^b (y\,dx + x\,dy) + \int_b^2 (y\,dx + x\,dy) $$

where $x = x_1$ from 1 to b and $y = y_2$ from b to 2. Thus

$$ {}_1I_2 = x_1 \int_1^b dy + y_2 \int_b^2 dx $$
$$ = x_1(y_2 - y_1) + y_2(x_2 - x_1) = x_2y_2 - x_1y_1 $$

The values of the integrals along path 1-*a*-2 and path 1-*b*-2 are identical and independent of the path since this differential is *exact.*

(*b*) The same process is repeated for this differential, so some of the steps can be eliminated. For path 1-*a*-2,

$$ {}_1I_2 = y_1 \int_1^a dx - x_2 \int_a^2 dy $$
$$ = y_1(x_2 - x_1) - x_2(y_2 - y_1) = 2x_2y_1 - x_1y_1 - x_2y_2 $$

For path 1-*b*-2,

$$ {}_1I_2 = -x_1 \int_1^b dy + y_2 \int_b^2 dx $$
$$ = -x_1(y_2 - y_1) + y_2(x_2 - x_1) = -2x_1y_2 + x_1y_1 + x_2y_2 $$

This integral depends on the path since the differential is *inexact.*

Diagram

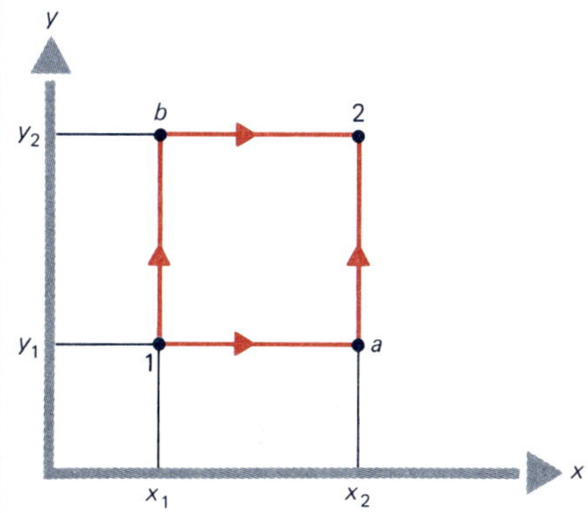

Example Problem H.5

Consider the differentials in Example Problem H.3 for the closed path 1-*a*-2-*b*-1 shown in Example Problem H.4. Determine which differentials satisfy the condition given in Eq. (H.29*c*).

Solution

The cyclic integral can be decomposed into integrals along the four separate sides, and much of the detail in Example Problem H.4 can be used. A key point is the negative sign arising from the direction of integration, as indicated in Eq. (H.27).

(*a*) The integral is

$$\oint (x\,dy + y\,dx) = \left[\int_1^a (x\,dy + y\,dx) + \int_a^2 (x\,dy + y\,dx)\right] + \left[\int_2^b (x\,dy + y\,dx) + \int_b^1 (x\,dy + y\,dx)\right]$$

Using the details in Example Problem H.4 yields

$$\oint (x\,dy + y\,dx) = (x_2y_2 - x_1y_1) - (x_2y_2 - x_1y_1) = 0$$

This verifies the third condition given in Eq. (H.29*c*) for an *exact* differential.

(*b*) The integral is

$$\oint (y\,dx - x\,dy) = \left[\int_1^a (y\,dx - x\,dy) + \int_a^2 (y\,dx - x\,dy)\right] + \left[\int_2^b (y\,dx - x\,dy) + \int_b^1 (y\,dx - x\,dy)\right]$$

Using the details in Example Problem H.4 yields

$$\begin{aligned}\oint (y\,dx - x\,dy) &= (2x_2y_1 - x_1y_1 - x_2y_2) - (-2x_1y_2 + x_1y_1 + x_2y_2)\\ &= 2(x_2y_1 + x_1y_2) - 2x_1y_1 - 2x_2y_2\\ &\neq 0\end{aligned}$$

This is indicated by the third condition in (H.29*c*) for an *inexact* differential.

Comments

These three example problems are applications of the conditions for exact and inexact differentials. Reconsider the differential in Example Problem H.2. This differential is inexact, as can be found from Eq. (H.29).

Problems

H.1 A function is represented as

$$P = C\left(\frac{T}{V}\right)$$

where P is pressure, T is temperature, V is volume, and C is a constant.

(*a*) Is this an implicit or explicit representation?

(*b*) What are the dependent and independent variables?

(*c*) Graph the function. Note that the dependent variable is normally on the y-axis, one independent variable on the x-axis, and a second independent variable can be shown as a parameter.

H.2 What are the mathematical representations in terms of partial derivatives that represent the following infinitesimal changes? Consider a gas enclosed in a tank. All processes take place at constant pressure.

(*a*) The temperature of the gas is altered by changing the gas mass. The tank is at a constant volume during this process.

(*b*) The temperature of the gas is altered by changing the volume of the gas. The gas mass is maintained at a constant value during this process.

Write the total differential of the dependent variable in terms of the partial derivatives.

H.3 The implicit representation is given for a function as

$$f(U, S, V) = 0$$

where U is internal energy, S is entropy, and V is volume.

(*a*) Write the functional form indicating that U is the dependent variable and S and V are the independent variables.

(*b*) Write the expression for the total differential of U, dU, in terms of partial derivatives and differentials.

(*c*) Sketch the figure that represents the two partial derivatives in (*b*).

H.4 Sketch the following partial derivatives for an arbitrary surface: $(\partial z/\partial x)_y$, $(\partial T/\partial P)_V$, and $(\partial U/\partial S)_V$.

H.5 What is the total differential of the function $z = z(y_1, y_2, y_3, \ldots y_n)$, where there are n independent variables? Write the total differential dz in terms of the appropriate partial derivatives.

H.6 Develop the relationship given in Eq. (H.14) of App. H as

$$\left(\frac{\partial z}{\partial x}\right)_y \left(\frac{\partial x}{\partial y}\right)_z \left(\frac{\partial y}{\partial z}\right)_x = -1$$

Start by considering the total differential of $z = z(x, y)$ for the special case of a constant z ($dz = 0$).

H.7 Evaluate the following ordinary integrals and represent the values as areas on the appropriate diagrams:

(a) $\int_{x=1}^{x=2} x\, dx$

(b) $5\int_{x=1}^{x=2} dx$

(c) $\int_{x=0}^{x=3} (3 - x)\, dx$

H.8 Evaluate the line integrals from A to C for the paths indicated in the figure. (Note that the ideal gas relation $PV = mRT$ is used in this problem with m and R as constants.)

(a) $\int_{A\text{-}B\text{-}C} P\, dV$

(b) $\int_{A\text{-}C} P\, dV \qquad$ where $P = \frac{mRT}{V} = \frac{\text{const}}{V}$

(c) $\int_{A\text{-}B\text{-}C} \left(\frac{P}{MR}\, dV + \frac{V}{mR}\, dP\right)$

(d) $\int_{A\text{-}C} \left(\frac{P}{mR}\, dV + \frac{V}{mR}\, dP\right)$

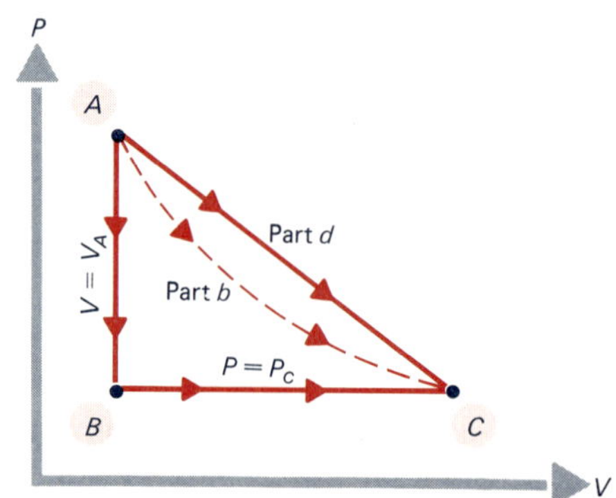

Figure PH.8

H.9 Evaluate the line integrals from A to C for the specific paths indicated in Fig. PH.9. (Take m and R as constants.)

(a) $\int_{A\text{-}B\text{-}C} \left(\frac{mR}{P}\, dT - \frac{mRT}{P^2}\, dP\right)$

(b) $\int_{A\text{-}B'\text{-}C} \left(\frac{mR}{P}\, dT - \frac{mRT}{P^2}\, dP\right)$

Figure PH.9

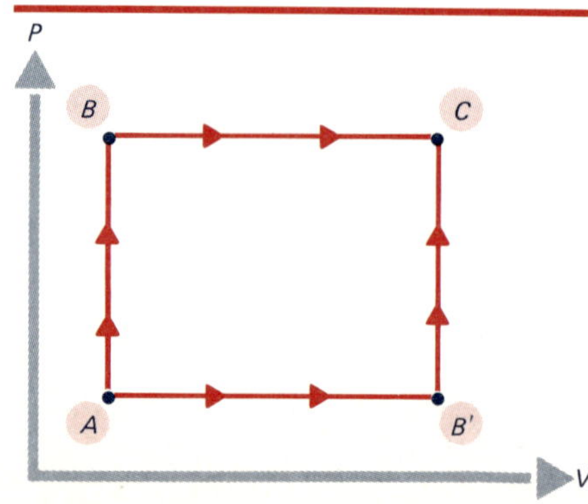

H.10 Evaluate the line integrals from A to B for the specific paths indicated in Fig. PH.10.

(a) $\int_{A\text{-}B} P\, dV$ along the path given by $P = \frac{\text{const}}{V} = \frac{C}{V}$

(b) $\int_{A\text{-}B} P\, dV$ along a straight path from A to B

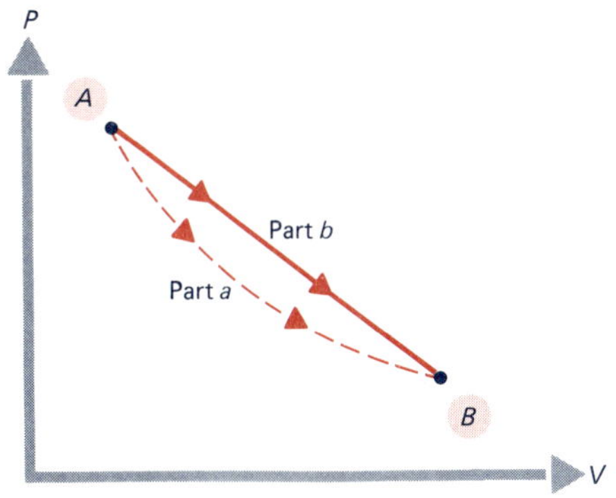

Figure PH.10

H.11 Which of the following functions $z = z(x, y)$ are exact and which are inexact? Prove your answers.
(a) $dz = x\,dx + x\,dy$
(b) $dz = y\,dx$
(c) $dz = \sin y\,dx - \cos x\,dy$
(d) $dz = xy^2\,dx + x^2y\,dy$

H.12 The differential of the function $z(x, y)$ is given by

$$dz = xy^3\,dx + \tfrac{3}{2}(xy)^2\,dy$$

(a) Is dz an exact differential? (Prove your answer.)
(b) What is the value of $z_2 - z_1$ between the points $(x_1 = 1, y_1 = 0)$ and $(x_2 = 3, y_2 = 5)$ along the path shown?

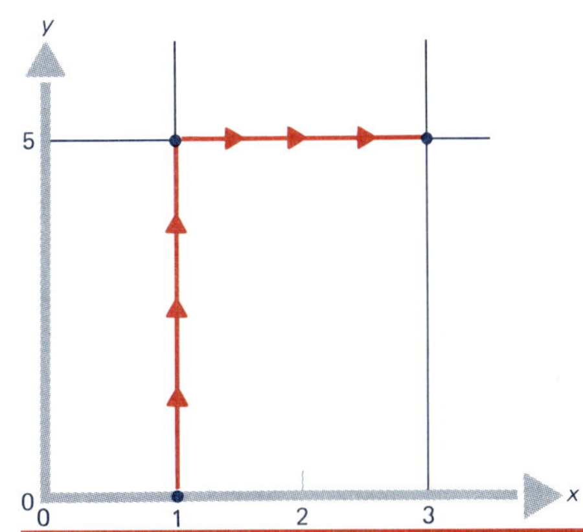

Figure PH.12

H.13 What are the required relations between partial derivatives for the following differentials to be exact?
(a) $-P\,dV - S\,dT$
(b) $V\,dP - S\,dT$

H.14 The total differential of a property called *enthalpy* H is an exact differential given by

$$dH = T\,dS + V\,dP$$

where T is temperature, S is entropy, V is volume, and P is pressure. Represent temperature and volume as partial derivatives, and determine a relation between the partial derivatives of temperature and volume.

H.15 You are given that dU is an exact differential and that

$$dU = T\,dS - P\,dV$$

where U is internal energy, S is entropy, and T, P, and V are temperature, pressure, and volume, respectively.
(a) What are the dependent and independent variables?

(*b*) Express T and P as partial derivatives.
(*c*) What is the relation between partial derivatives of T and P?

H.16 Given that

$$dz = -2dx + \tfrac{3}{2}x^2y^2\,dy$$

and

$$dw = xy^3\,dx - y\,dy$$

(*a*) Are dz and dw exact differentials?
(*b*) Is $dz + dw$ an exact differential?

H.17 Determine whether any values of $A(y)$ exist such that $d\phi$ and $d\mu$ given here are exact differentials. Then find any values of $A(y)$ that will make $dz = d\mu + d\phi$ an exact differential.

$$d\mu = A(y)y\,dx + dy$$
$$d\phi = A(y)\,dx + x\,dy$$

H.18 A particular total differential has the form

$$dz = 2xy\,dx + (x^2 + \cos y)\,dy$$

(*a*) Prove the dz is exact.
(*b*) Find an expression for $z = z(x, y)$, and find the value of $z_2 - z_1$ between the points $(x = 1, y = 0)$ and $(z = 3, y = \pi)$.
(*c*) Find the value of $z_2 - z_1$ between the points $(x = 1, y = 0)$ and $(x = 3, y = \pi)$ by integrating the expression given for dz along the particular path shown below.

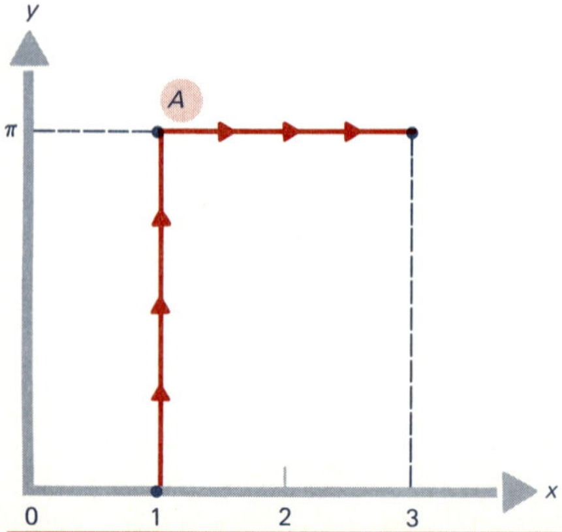

Figure PH.18

H.19 Given that du and dv are exact differentials and are given by

$$du = a(y)\,dx + y^2x\,dy = g_1(y)\,dx + h_1(x, y)\,dy$$

and

$$dv = 3y^2x\,dx + b(x, y)\,a(y)\,dy = g_2(x, y)\,dx + h_2(x, y)\,dy$$

Find expressions for $a(y)$ and $b(x, y)$, where these functions are the same in each equation.

H.20

(*a*) Show whether the differential du is exact or inexact:

$$du = x \ln y\,dx + \frac{x^2}{2y}\,dy$$

(*b*) Find the value of the constant A that is necessary to make dw an exact differential:

$$dw = \frac{Ax^2}{2y}\,dx + \frac{x^3}{y^2}\,dy$$

H.21 Given that

$$dz = a(3x^2y^2 + x^2y)\,dx + a(2x^3y - x^2y^2)\,dy$$

and

$$dw = ax^2y\,dx - ax^2y^2\,dy$$

(*a*) Prove whether dz and dw are exact differentials.
(*b*) Prove whether $(dz - dw)$ is an exact differential.

References

1. E. Kreyszig, *Advanced Engineering Mathematics,* 4th ed., Wiley, New York, 1979.
2. P. H. Badger, *Equilibrium Thermodynamics,* Allyn and Bacon, Boston, 1967.
3. S. M. Blinder, "Mathematical Methods in Elementary Thermodynamics," *Journal of Chemical Education,* vol. 43, no. 2, 1966, pp. 85–92.

APPENDIX I

Answers to Selected Homework Problems

Chapter 1

1.1	1 slug = 1 lbf·s²/ft.
1.3	3.1525 W/m²; 30,238 kJ/kg; 14,654 W.
1.5S	$m_{\text{Earth}} = m_{\text{moon}} = 76.48$ kg.
1.7E	181 lbf.
1.9S	682 N.
1.11	36.29 N, or 8.16 lbf.
1.13E	102 ft/s² (upward).

Chapter 2

2.1S	8.66×10^6 Pa.
2.3E	71.3 kPa.
2.5	2.5 MPa.
2.7S	0.1836 MPa.
2.9E	29,400 lbf.
2.11S	150 kPa.
2.13E	19.8 psia.
2.15S	0.2094 m³/kg, 4.775 kg/m³.
2.17E	2030 lbm.
2.19S	$v_A = 0.9873$ m³/kg.
2.21E	0.0174 ft³/lbm.
2.23S	$v_A = 50$ m³/kg; $v_B = 75$ m³/kg.
2.25E	8.88 psi.
2.27S	0.178 m³.
2.29E	12.52 ft.
2.31	480.5 kPa.
2.33S	$L_1 = 20.4$ m; $P_2 = 187$ kPa; $L_3 = 3.21$ m.
2.35	215.7 kPa.
2.37E	304.4 ft.
2.39	1.954 kPa.
2.41	$P_A - P_B = \rho_1 h_1 g + \rho_2 h_2 g - \rho_3 h_3 g$.
2.43S	2.883 kPa.
2.45	$x = (L\rho_{H20} - R\rho_{Hg})/\rho_{H20}$.
2.47E	−10.11 psi.
2.49S	2.94 kPa.
2.51	$h_{hot} = h_{cold}(\rho_{H2O,cold}/(\rho_{H2O,hot})$.
2.53E	5540°R.
2.55S	KE = 9 J; PE = 196 J; F = 19.6 N.
2.57E	$_1W_2 = -10.8 \times 10^3$ lbf·ft.
2.59S	(*a*) $_1w_2 = 1750$ kJ/kg; (*b*) no change.
2.61	(*b*) −12.28 kJ.
2.63	(*a*) 572 kN/m.
2.65	25×10^7 kN/m².
2.67E	(*b*) 46.7 lbf·ft; (*c*) 625°F.
2.69S	(*a*) 279.0 J; (*b*) 64 percent; (*c*) 41 percent.
2.71E	−159 lbf·ft.
2.73S	22.6 kN/m.
2.75E	(*b*) 192 ft; (*c*) 17,600 lbf·ft = 22.6 Btu.
2.77S	(*a*) 817 kJ; (*b*) KE, 23.1 percent; PE, 24 percent; U, 52.9 percent; (*c*) −432 kJ.
2.79E	(*a*) −60 lbf·ft; (*b*) −60 lbf·ft.

Chapter 3

3.1S (*a*) 0.105; (*b*) 0.105; (*c*) 0.1050; (*d*) same as *a*.

3.3E (*a*) 0.0; (*b*) nonideal; (*c*) $\Delta u = 9.77$ Btu/lbm, $\Delta h = 13.67$ Btu/lbm.

3.5S 119 atm.

3.9E 1, liquid; 2, liquid; 3, saturation; 4, superheat; 5, saturation.

3.11S $v = 6.086$ m³/kg; $m = 9.859$ kg.

3.13E 17.3 lbm/lbmol.

3.15S 53,770 kg.

3.17E Hydrogen.

3.19S $P = 1.5536$ MPa; $v = 1.156 \times 10^{-3}$ m³/kg.

3.21E $v = 0.01705$ Btu/lbm.

3.23S $P = 791.47$ kPa; $v = 0.1099$ m³/kg.

3.25E $u = 371.44$ Btu/lbm.

3.27S $h = 503.78$ kJ/kg.

3.29E $P = 680.11$ psia; $v = 0.3070$ ft³/lbm.

3.31S $u = 895.43$ kJ/kg; $h = 898.94$ kJ/kg.

3.33E $T = 385.7$°F; $h = 365.8$ Btu/lbm.

3.35S $x =$ undefined; $P = 600$ kPa; $h = 3270.2$ kJ/kg.

3.37E 1, subcooled; 2-6, superheat.

3.39S 897.0 kJ/kg.

3.41E From polynomials, $c_P(200) = 0.2422$ Btu/lbm·°R, $c_P(2800) = 0.2946$ Btu/lbm·°R.

3.43 (*a*) 0.554 Btu/(lbm·°F); (*b*) 0.769; (*c*) 671.01 Btu/lbm.

3.45 (*a*) 0.02685 m³/kg; (*b*) 0.221; (*c*) 1511.3 kJ/kg.

3.47S Tables, 1720.5 kJ/kg; polynomial, 1721.3 kJ/kg; at avg *T*, 1738.3 kJ/kg.

3.49E 180.3 Btu/lbm.

3.51S (*a*) saturated vapor, 304.8°C; (*b*) saturated liquid, 364.5°C.

3.55E (*b*) 81.1 Btu; (*c*) 0.0162 ft³.

3.57S 15 kJ.

3.59E About 95 Btu/lbm.

Chapter 4

4.1S 0.721 kJ/kg·K.

4.3E (*b*) 98.1 Btu/lbm; (*c*) 447.3 Btu/lbm; (*d*) 1704.4 Btu/lbm.

4.5S 355 kJ.

4.7 ${}_1Q_2 = c_P P_1 (V_2 - V_1)/R$.

4.9 -588.9 kJ.

4.11E 30.68 Btu/lbm.

4.13S ${}_2w_3 = 221.6$ kJ/kg; $T_3 = 431.4$°C, $v_3 = 0.3216$ m³/kg.

4.15 (*a*) 190.3 kPa; (*b*) -243.7 kJ.

4.17 (*b*) 64.7°C, 1694.5 kPa; (*c*) 301.6 kJ/kg.

4.19E 495.1°F.

4.21S 2736 kJ.

4.23E $Q = 1602$ Btu; $W = 199$ kJ.

4.25S -185 kJ.

4.27E -54.3 Btu.

4.29S ${}_1w_2 = 0$; $\Delta P = -3.899$ MPa.

4.31S ${}_1w_2 = -196.1$ kJ/kg; $m = 0.153$ kg.

4.33 1.022×10^8 J.

4.35 $v_2 = 0.9475$ m³/kg; ${}_1q_2 = 3550.7$ kJ/kg.

4.37 ${}_1W_2 = -473.0$ kJ; ${}_1Q_2 = -2658.4$ kJ.

4.39 (*a*) 0.475 kg; (*b*) -98.53 kJ.

4.41 (*a*) $m_i = m_f = 0.014$ kg; (*b*) -1.322 kJ.

4.43 $m = 0.006379$ kg; ${}_1W_3 = 0.1613$ kJ.

4.45 ${}_1w_3 = 263.7$ kJ/kg; ${}_1q_3 = 2697$ kJ/kg.

4.47 ${}_1w_3 = -123.6$ kJ/kg; ${}_1q_3 = -3451$ kJ/kg.

4.49E -190.5 Btu.

4.51 -1065 kJ/kg.

4.53 -1571 kJ.

4.55S 70°C.

4.57E ${}_1w_2 = 0$; ${}_1q_2 = -26.17$ Btu/lbm.

4.59S ${}_1w_2 = 119.7$ kJ/kg; ${}_1Q_2 = 4662$ kJ.

4.61E ${}_1Q_3 = 4068$ Btu.

4.63 117.3°F.

4.65 ${}_1w_2 = 111.93$ kJ/kg; ${}_1q_2 = 401.4$ kJ/kg.

4.67 351.5 K; 204.9 kPa.

4.69 $\Delta U = 0$; $V_2 = 0.0152$m³.

4.71 No work on system.

4.73 175 kJ/kg, work by system.

4.75 (*b*) 4.113 kJ; (*c*) by the system.

4.77 314.5 kJ/kg.

4.79 (*a*) 0; (*b*) 1.709 MPa.

4.81 -30.0 kJ/kg; 1.0 kg.

4.83 47,288 kJ.

4.85 1201 kJ; 200 kJ.

4.87 (*b*) 2.29 MJ; (*c*) 14.55 MJ.

4.89 (*a*) 7 kg; (*b*) no work; (*c*) 1563 kJ.

4.91 (*a*) 0 kJ; (*b*) 7.6 kPa.

4.93 Heat transfer; -5004 kJ; lost.

4.95 4400 kJ; 1140 kJ.

4.97	284.6°C.
4.99	0 kJ/kg; −304.4 kJ/kg.
4.101	30.08 kJ/kg; 263.5 kJ.
4.103	−449.8 kJ/kg.
4.105	−421 kJ.
4.107	935 kJ.
4.109	21.71 MW.
4.111S	(*a*) 0.9798 kg/s and 82.1 kW; (*b*) 1.960 kg/s and 164.2 kW.
4.113S	16.4 m; 1640 m^3.
4.115E	(*a*) 812.6 Btu/s; (*b*) 822.8 Btu/s; (*c*) 2.73 in, 5.66 in.
4.117	1.103×10^4 kW.
4.119	441.7 kJ/kg.
4.121	−257.8 kW.
4.123	917.7 kJ/kg.
4.125	44.26 kW.
4.127	−2077 kJ/kg (In).
4.129S	(*a*) −605.0 kJ/kg; (*b*) −1210.0 kW.
4.131	−181.9 kJ/kg.
4.133	2.0 MPa.
4.135	448.3°C.
4.137	(*a*) 0.258 kg/s; (*b*) same.
4.139	21.8 W.
4.141	998 kg/s.
4.143	355.6 kJ/kg.
4.145	718.8°C.
4.147	180.9 kW.
4.149	−25.39 kW.
4.151	12 kg/s; 197.9°C.
4.153	−28.39 kW.
4.155	−8523 kW.
4.157	9716 kW.
4.159	339.4 m/s.
4.161	1082 m/s.
4.163	239.1 m/s.
4.165	451.6 m/s.
4.167	0.26.
4.169S	(*a*) $x_4 = 0.08473$; (*b*) 0.97 cm.
4.171E	(*a*) 791.6°F; (*b*) 800°F.
4.175	1211 kg/s.
4.177	96 kg/s.
4.179S	2.39 kg/s.
4.181E	(*a*) 0.977.
4.183S	1600 MW.
4.185E	(*a*) 22,975 Btu; (*b*) −216°F.
4.187S	−496 kJ.
4.189E	-1.72×10^6 Btu.
4.191	−163.6 kJ/kg.
4.193S	(*a*) 1.26; (*b*) 0.135 kg/s.
4.195	106.15 kPa; 225 m/s; −62.1°C.
4.197	0.0174 m^3/s; $P_A = 78.0$ kPa; $P_B = -20.1$ kPa.
4.199	0.126 m^3/s.

Chapter 5

5.5	650°C; 3775 kJ/kg; undefined.
5.7	2515 kJ/kg; 7.6 kJ/kg·K; 60°C.
5.9	500°C; 7.8 kJ/kg·K.
5.11	382 kJ/kg: 1.585 kJ/kg·K.
5.13	372 kJ/kg; 0.65 MPa.
5.17S	$\Delta s = 0.0522$ kJ/kg·K.
5.21	(*b*) Final *s* smaller for isothermal process.
5.23E	(*a*) 55.06 psia; (*b*) 4.7808 Btu/°R; (*c*) −4.4630 Btu/°R; (*d*) 0.3178 Btu/°R.
5.25S	109.3 kJ/kg.
5.27	Impossible.
5.29	−237.0; −216.9 kJ/kg.
5.31	−291.5 kJ.
5.33	−1096.3 kJ/kg.
5.35	311.6 kJ.
5.37	$\Delta S = -0.4673$ kJ/K.
5.39	${}_1w_2 = -1474$ kJ/kg; ${}_2w_3 = 180.7$ kJ/kg.
5.41	${}_1W_2 = 0.200$ kJ; $\Delta S = 2.43 \times 10^{-5}$ kJ/K.
5.43	(*a*) $\Delta s = 5.7284$ kJ/(kg·K); (*b*) $(\Delta s)_{max} =$ 5.732 kJ/(kg·K).
5.45	−0.2449 m^3/kg.
5.47	(*a*) $\Delta s = 0.2645$ kJ/(kg·K); (*b*) ${}_1w_2 =$ −45.36 kJ/kg.
5.49	${}_1w_2 = {}_1q_2 = 693.1$ kJ/kg.
5.51	Impossible
5.53	−474.3 kJ.
5.55	171.0 kJ/kg.
5.57	1562 kJ.
5.59	${}_1W_2 = {}_1Q_2 = -4134$ kJ.
5.61E	35,880 Btu.
5.63S	(*b*) 30 kPa; (*b*) −3.2 kJ/K.
5.65S	0.510 kJ/kg.
5.67E	(*a*) 0.2545 Btu/(lbm·°R); (*b*) 0.0056 Btu/(lbm·°R).
5.69	−353.4 kJ/kg.
5.71	179.6 kJ; 179.6 kJ.

5.73	90.2 percent.
5.75	−87.3 kW.
5.77	Yes.
5.79	2860 kW or 3830 hp.
5.81	422.8 kJ/kg.
5.83	Will produce 30.24 hp.
5.85	78.34 kPa; 0.358 m^3/s.
5.87	Yes.
5.89	83.9 percent, 12,600 hp.
5.91	Efficiencies are 87.2 percent and 92.7 percent.
5.93	3.98 kg/s.
5.95	1141 kW.
5.97	6706 hp.
5.99	(*a*) 0.969; (*b*) 955.3 kJ/kg.
5.101S	73.7 percent.
5.103E	$h_{2,a}$ = 1100.5 Btu/lbm; w_a = −344 Btu/lbm.
5.105S	(*a*) 2345.7 kJ/kg; (*b*) 907.9 kJ/kg; (*c*) 79.2 percent.
5.107E	(*a*) −5380 hp; (*b*) −4573 hp; (*c*) 1205.9 Btu/lbm.
5.109S	(*a*) −401 kJ/kg out; (*b*) −0.8157 kJ/(kg·K); (*c*) 0.5210 kJ/(kg·K).
5.111E	−143 hp.
5.113S	174°C.
5.115	Impossible.
5.117	−105.1 kW; −105.1 kW.
5.119	−256.1 kJ/kg.
5.121	250°F.
5.123	86.0%.
5.125	0 kJ/kg; −2295 kW.
5.127	-1.46×10^4 kJ/kg.
5.129	75%.
5.131	(*a*) 45,900 kW; −46,200 kW; (*b*) 0 kW; −58,000 kW.
5.133	53.7%.
5.135	4871.1 kPa.
5.137	(*a*) 58.6 percent; (*b*) 22°C; (*c*) 5.86 hp.
5.139	−0.337 hp.
5.141E	89.4°F.
5.143	1.990 kJ/kg.
5.145S	36.4°C.
5.147E	0.636 lbm/s.
5.149S	$w_1 = -159$ kJ/kg, $w_2 = -243$ kJ/kg; s_{gen} = 0.39 kJ/(kg·K).
5.151E	(*a*) 11.5 percent.
5.153	570 K.
5.155	From 2 to 1.
5.157	Impossible.
5.159	−2130 hp; Possible.
5.161	(*a*) 220°C; (*b*) No.
5.163	Yes.
5.165	−100 kJ/kg, work is done on CV.
5.167S	158 kW.
5.171	767.7 m/s.
5.173	0.5065 (choked); 105.7 kPa; 308.3°C; 351.9 m/s.
5.175	0.5652; 130 kPa; 347.9 m/s; 13.9 kg/s.
5.177	0.996; 220 kPa; 281.1 m/s; 210.7 K; 16.4 kg/s.
5.179	576 m/s.
5.181	(*a*) 5.10 kN; (*b*) Yes.
5.183	0.9491; 119.1 kPa; 370.1 K; 0.059 m^2.
5.185	0.462 kJ/(kg·K).
5.187	(*a*) 67.7 kg/min; (*b*) Yes.
5.189E	(*a*) 333.4°F; (*b*) 0.619 Btu/°R.
5.191	62.43 kJ.
5.193	−119 Btu/lbm.
5.195	1.245 atm.
5.197S	427 kPa.
5.199E	0.0425 Btu/lbm.
5.201S	(*b*) 0.193 kJ/K.
5.203E	(*a*) 632°R; (*b*) 0.402 lbm; (*c*) −0.674 Btu/°R; (*d*) 1.676 Btu/°R.
5.205	3.25 kg.
5.207S	(*a*) Valid; (*b*) Valid; (*c*) Invalid.
5.209	2.150
5.211	1.4×10^3 kJ; −75.3°C; 1.27.
5.213	61.7%; 501.8°C.
5.215	Possible.
5.217	10.6 hp.
5.219	1190 K.
5.221	(*a*) 50 kJ/kg; (*b*) 477°C.
5.223	(*a*) 332°C; (*b*) 1.82 kJ; (*c*) Smaller.
5.225	Independent of T_m.
5.227	(*a*) 0.0180; (*b*) 101.2 kW.
5.229	$38,200.
5.231	3.25×10^3 kJ; 9.229.
5.233	(*a*) 266 kW; (*b*) 28.2%.
5.235	No.
5.237	92.2×10^6 kg/hr.

5.239	37%.
5.241	(*a*) 12.4 kW; (*b*) 49.1°C.
5.243	13.2.
5.247E	40,000 Btu·°R.
5.249S	1 students/s.
5.251E	792 ft².

Chapter 6

6.1	40 kPa; 1.5 m³/kg; 208 K.
6.3	9.9 kW
6.5	(*b*) 0.606
6.7	1-2, 0 kJ/kg; 2-3, 6013 kJ/kg; 3-1, −6013 kJ/kg.
6.9	−202.5 kJ/kg.
6.11	−232.2 kJ/kg.
6.13S	(*a*) 372 kJ/kg; (*b*) 372 kJ/kg.
6.15E	(*a*) 99.5 atm; (*b*) 0.731; (*c*) 1476°F.
6.19	(*a*) 7.60; (*b*) 239.2 kJ/kg; (*c*) 0.4693 kg/s; (*d*) 44.0 percent.
6.21S	(*b*) 31.0 percent; (*c*) 36.3 percent.
6.23E	57.3 percent.
6.27	(*a*) −176.8 kJ/kg; (*b*) 362.9 kJ/kg; (*c*) 26.7 percent; (*d*) 96.5 percent.
6.29	(*a*) 88.4 percent; (*b*) 191 kJ/kg; (*c*) 29.5 percent.
6.31	(*a*) 258.3 kJ/kg; (*b*) 750.1 K; (*c*) 26.7 percent.
6.33	(*a*) 0.2682 kg/s; (*b*) 38.8 percent.
6.35	(*a*) 48.24 percent; (*b*) 48.23 percent; Yes, 48.24 percent.
6.37	(*a*) 68.9 percent; (*b*) −254.3 kJ/kg; (*c*) 705.6 m/s.
6.41	63.6 percent.
6.43	195 kJ/kg.
6.45	89.7 percent (or 80.4 percent if use h values.).
6.47	(*a*) 629.6 m/s; (*b*) 10.9 MJ/s; (*c*) 918.9 m/s.
6.49	6.375
6.51	(*a*) 159.7 kPa; (*b*) 52.35 percent; (*c*) 241 percent. (*d*) No.
6.53	(*a*) −230.2 kJ/kg; (*b*) 522.0 K; (*c*) 526.3 m/s.
6.55	700 kg/s.
6.57	220.9 kJ/kg.
6.59S	(*a*) 60.5 percent; (*b*) 11.7 percent; (*c*) 11.4 percent.
6.61	6.374.
6.63	1463 K.
6.65	713.8 kJ/kg; 44.9 percent.
6.67	1123 kJ/kg.
6.69	1136 kJ/kg; 10.42 kPa.
6.71	1869 K; 823 kJ/kg; 53.3 percent.
6.73	(*a*) 0.575; (*b*) 85.3 kW.
6.75	$P_2 = 2.022$ MPa; $P_3 = 10.33$ MPa; $P_4 = 0.5620$ MPa; $T_2 = 677.7$ K; $T_3 = 3463$ K; $T_4 = 1507$ K; 56.5%.
6.77	53.2 percent.
6.79	3998 kJ/kg; 2163 kJ/kg.
6.81	39.7 percent; 0.9887 kPa.
6.83	(*a*) 15.0; (*b*) 2.15; (*c*) 1032.6 kJ/kg; (*d*) 59.6%; (*e*) 615.6 kJ/kg.
6.85	(*a*) 1729 K; (*b*) 829.9 kJ/kg; (*c*) 453.0 kJ/kg; (*d*) 54.6 percent.
6.87	(*a*) 16.2; (*b*) 2.322; (*c*) 60.1 percent.
6.93	0.098.
6.97	(*a*) $COP_S = [1 - (T_E/T_H)]/[(T_C/T_E) - 1]$; (*b*) $COP_B = (1/r_B^{0.286}) - 1$.
6.99	$\eta_H = (T_3 - T_1)/(T_3 - T_2)$.

Chapter 7

7.1	(*a*) 33.2 percent; (*b*) 89.3 percent; (*c*) 4.73 MW.
7.5S	850 kPa.
7.7E	$\eta_R = 31.9$ percent; $\eta_{SH} = 33.6$ percent.
7.9S	(*a*) −647.1 kJ/kg; (*b*) 1123.9 kJ/kg; (*c*) 22.9 percent.
7.13	34.7 percent.
7.15E	35.5 percent.
7.19S	(*a*) 0.187; (*b*) 0.40; (*c*) 37.5 percent.
7.21	(*a*) 2000 kW; (*b*) 15.1 kW; (*c*) 45.7 percent.
7.23	1492 MW; 33.4 percent.
7.25	10.2 kJ/kg; 2607.2 kJ/kg; 597.5 kJ/kg; 587.3 kJ/kg; 2020 kJ/kg; 22.5 percent.
7.27	6.12 kJ/kg; 2852.7 kJ/kg; 934.4 kJ/kg; 311.2 hp; 1922.4 kJ/kg.
7.29	3.03 kJ/kg; 2923.7 kJ/kg; 768.6 kJ/kg; 309.2 hp; 2155.1 kJ/kg.

7.31	0.6271 kg/s; 772.3 kW; 2273.3 kW.
7.33	35.9 percent.
7.35	16.7 percent.
7.39	It cannot.
7.41	$\eta_R = 32.0$ percent; 58.3 percent; 82.2 percent more efficient.
7.43	378 kW; 12.7 percent.
7.45	(*b*) 182 MW.
7.47	70 MW.
7.49	23.3 MW.
7.51	(*a*) −23.8 kJ/kg; (*b*) −146.2 kJ/kg; (*c*) 122.4 kJ/kg; (*d*) 5.14.
7.53	57.1 tons.
7.55	181.7 kJ/kg; 1.41; 7169 Btu/h or 0.597 tons.
7.57	2.73 kW.
7.59	27.6 kJ/kg.

Chapter 8

8.1S	−33.4 kJ/kg; 0.0014 kJ/(kg.·K); 0.43 kJ/kg.
8.3E	0.207 Btu/lbm.
8.5	0.7629 kJ/kg; −0.3585 kJ/kg.
8.7	Possible.
8.9S	−0.9 kJ/kg.
8.11E	362.3 Btu/lbm; 362.3 Btu/lbm.
8.13	−39.84 kJ/kg.
8.15S	(*a*) 116.3 kJ/kg; (*b*) 31.3 kJ/kg; (*c*) 31.3 kJ/kg.
8.21E	8.0 Btu
8.23	227.4 kJ/kg or 103.4 Btu.
8.25S	(*a*) −184.3 kJ/kg; (*b*) −135.5 kJ/kg; (*c*) 4.78 kJ/kg; (*d*) 140.3 kJ/kg.
8.31E	68.9 percent.

Chapter 9

9.13S	169.5 kJ/kg; 168.0 kJ/kg.
9.15E	−20.04 kJ/kg.
9.17S	(*a*) 154.8 kJ/kg, 4.5944 kJ/(kg·K); (*b*) 141.1 kJ/kg, 4.5392 kJ/(kg·K).
9.19E	0.0296 °F/psia.
9.25S	0.1206z kg salt/kg water (z = depth in m.)
9.27	0.64.

Chapter 10

10.1S	(*a*) 0.1667; (*b*) 0.0911; (*c*) 29.340; (*d*) 16.67 kPa; (*e*) 0.283 kJ/(kg·K); (*f*) 24.9 m³.
10.3	60.1°C; 0.06911; 64.1 percent.
10.5	31.83 g.
10.7E	(*a*) 27.03 Btu/lbm; (*b*) 0.06566 Btu/(lbm·R).
10.9S	(*a*) 0 kJ/(kg·K); (*b*) −0.0620 kJ/kg; (*c*) −18.6 kJ/kg; (*d*) not possible.
10.11	558.8 kJ/kg (using steam tables for h_{H2O}.)
10.13E	(*a*) 0.35 lbm air/lbm exhaust; (*b*) 310.3 Btu/lbm; (*c*) 432.3 Btu/lbm O_2.
10.15S	(*a*) 0.075 kg/s; (*b*) 50.0 percent.
10.17	50.5 g/h.
10.19	60.1°C; 28.5 percent; 0.0327.
10.21	21.0°C; 0.0138; 19°C.
10.23	32°C; 0.0289; 4.50 kPa; 31°C.
10.25	−1.305 kW.
10.27	−11.66 kW.
10.29	57%; 0.0171; 2.7 kPa; 22.5°C.
10.31	25.5°C; 0.0189; 2.92 kPa; 70 percent.
10.33	(*a*) −13.98 kW; (*b*) 0.00302 kg/s.
10.35	(*b*) 0.6237 kg/s; (*b*) 43.9 kW.
10.37E	12.9 lbm/day.
10.39	−0.67 kW.
10.41S	0.0135
10.45E	0.656 psia.
10.47	(*a*) 281 g/min; (*b*) 7.56 kW.
10.49	(*b*) 0.3328 kg/min; (*c*) 11.84 kW.
10.51	0.636 kg/h.
10.53	(*b*) 25.0 kW; (*c*) 0.265 kg/min.
10.55	(*b*) −3.15 tons; (*c*) 1.612 kW; (*d*) 10.84 kg/h.
10.57	477.2 kg/s.
10.59	2.31×10^5 kg/s.
10.61	3.94×10^5 kg/s.
10.63	0.479 m³.
10.65	(*c*) 69.0 kJ/kg.
10.67S	Differences from ideal gas are (*a*) −23.4 kJ/kg; −0.0304 kJ/(kg·K); 1.01 v_{ideal} m³/kg; (*b*) −26.1 kJ/kg; −0.0317 kJ/(kg·K); 0.998 v_{ideal} m³/kg.

Chapter 11

11.1	28.82 kg/kmol.
11.3	(*a*) 100 percent; (*c*) 29.49 kg air/kg fuel.
11.5	(*b*) 24.08 kg air/kg fuel; (*c*) 161.6 percent.
11.7	−224,000 kJ/kmol.
11.9	27.087 kmol; 109.2 kg.
11.11	(*a*) 50 percent; (*c*) 22.12; (*d*) −834,200 kJ/kg; (*e*) 52,320 kJ/kg; (*f*) −886,500 kJ/kg; (*g*) 43.6°C.
11.13	(*a*) 100 percent; (*c*) 30.84; (*d*) −126,200 kJ/kg; (*e*) −2,587,000 kJ/kg; (*f*) −2,460,000 kJ/kg; (*g*) 40.9°C.
11.15	(*a*) -1.2×10^5 kJ/kg; (*b*) -1.01×10^4 kJ/kg; (*c*) -2.8×10^4 kJ/kmol.
11.19	−4.736 MJ/kg.
11.21	$A_1 = \mathrm{n} + (\mathrm{m}/4) + \mathrm{X}$; $A_2 = \mathrm{n}$; $A_3 = \mathrm{m}/2$; $A_4 = \mathrm{X}$; $A_5 = 0.79A_1$.
11.23S	1764°C.
11.27E	4.34 lbm/s.
11.31S	1925°C.
11.33E	41.2 percent.
11.35S	1982°C.

Chapter 12

12.1	(*a*) 1; (*b*) 4; (*c*) 2.
12.3S	218.9 kJ/kg.
12.5E	(*a*) 1.95 psia; (*b*) 17.5 psia; (*c*) 0.0279; (*d*) 0.0785.
12.7S	(*a*) 0.917 kPa; (*b*) 4.08 kPa; (*c*) 0.145.
12.15E	(*a*) 0.370; (*b*) 0.679.
12.17S	(*a*) 0.004; (*b*) Favored at higher pressure.
12.23E	K(T = 1250°R) = 1.83 K(T = 1000°R)
12.25S	0.030

Chapter 13

13.3S	$5.273 \times 10^{-5} / (\Delta V/V)$ m.
13.5E	7.51×10^{-27} ft/s.
13.9	15.53; 0.6439, 0.2369, 0.0871, 0.0321.

Appendix H

H.3	(*a*) $U = f(S,V)$ or $U = U(V,S)$; (*b*) $dU = (\partial U/\partial S)_V\, dS + (\partial U/\partial V)_S\, dV$.
H.5	$dz = (\partial z/\partial y_1)_{y2,y3,\ldots yn}\, dy_1 + (\partial z/\partial y_2)_{y1,y3,\ldots yn}\, dy_2 \ldots + (\partial z/\partial y_{n-1})_{y1,y2,\ldots yn}\, dy_{n-1} + (\partial z/\partial y_n)_{y1,y2,\ldots yn-1}\, dy_n$.
H.7	(*a*) 3/2; (*b*) 5; (*c*) 9/2.
H.9	(*a*) $V_C - V_B$; (*b*) $V_C - V_B$.
H.11	(*a*) Inexact; (*b*) inexact; (*c*) inexact; (*d*) exact.
H.13	(*a*) $(\partial P/\partial T)_V = (\partial S/\partial V)_T$; (*b*) $(\partial V/\partial T)_P = -(\partial S/\partial P)_T$.
H.15	(*a*) U is dependent, S, V are independent; (*b*) $T = (\partial U/\partial S)_V$, $P = -(\partial U/\partial V)_S$; (*c*) $(\partial T/\partial V)_S = -(\partial P/\partial S)_V$.
H.17	$A = 1 + 1/[C(1 + y)]$.
H.19	$a = y^3/3 + C_1$; $b = 9x^2/y^2 + C_2$.
H.21	(*a*) Inexact; (*b*) exact.

Index